D1231120

Multiples and Prefixes for Metric Units*

Multiple	Prefix (and Abbreviation)	Pronunciation
10^{24}	yotta- (Y)	yot′ta (*a* as in *a*bout)
10^{21}	zetta- (Z)	zet′ta (*a* as in *a*bout)
10^{18}	exa- (E)	ex′a (*a* as in *a*bout)
10^{15}	peta- (P)	pet′a (as in *petal*)
10^{12}	tera- (T)	ter′a (as in *terra*ce)
10^{9}	giga- (G)	ji′ga (*ji* as in *ji*ggle, *a* as in *a*bout)
10^{6}	mega- (M)	meg′a (as in *mega*phone)
10^{3}	kilo- (k)	kil′o (as in *kilo*watt)
10^{2}	hecto- (h)	hek′to (*heck-toe*)
10	deka- (da)	dek′a (*deck* plus *a* as in *a*bout)
10^{-1}	deci- (d)	des′i (as in *deci*mal)
10^{-2}	centi- (c)	sen′ti (as in *senti*mental)
10^{-3}	milli- (m)	mil′li (as in *mili*tary)
10^{-6}	micro- (μ)	mi′kro (as in *micro*phone)
10^{-9}	nano- (n)	nan′oh (*an* as in *an*nual)
10^{-12}	pico- (p)	pe′ko (*peek-oh*)
10^{-15}	femto- (f)	fem′toe (*fem* as in *fem*inine)
10^{-18}	atto- (a)	at′toe (as in an*atom*y)
10^{-21}	zepto- (z)	zep′toe (as in *zep*pelin)
10^{-24}	yocto- (y)	yock′toe (as in *sock*)

*For example, 1 gram (g) multiplied by 1000 (10^3) is 1 kilogram (kg); 1 gram multiplied by 1/1000 (10^{-3}) is 1 milligram (mg).

SI Base Units

Physical Quantity	Name of Unit	Symbol
Length	meter	m
Mass	kilogram	kg
Time	second	s
Electric current	ampere	A
Temperature	kelvin	K
Amount of substance	mole	mol
Luminous intensity	candela	cd

Some SI Derived Units

Physical Quantity	Name of Unit	Symbol	SI Unit
Frequency	hertz	Hz	s^{-1}
Energy	joule	J	$kg{\cdot}m^2/s^2$
Force	newton	N	$kg{\cdot}m/s^2$
Pressure	pascal	Pa	$kg/(m{\cdot}s^2)$
Power	watt	W	$kg{\cdot}m^2/s^3$
Electric charge	coulomb	C	$A{\cdot}s$
Electric potential	volt	V	$kg{\cdot}m^2/(A{\cdot}s^3)$
Electric resistance	ohm	Ω	$kg{\cdot}m^2/(A^2{\cdot}s^3)$
Capacitance	farad	F	$A^2{\cdot}s^4/(kg{\cdot}m^2)$
Inductance	henry	H	$kg{\cdot}m^2/(A^2{\cdot}s^2)$
Magnetic field	tesla	T	$kg/(A{\cdot}s^2)$

Quadratic Formula

If $ax^2 + bx + c = 0$, then

$$x = \frac{-b \pm \sqrt{b^2 - 4ac}}{2a}$$

Pythagorean Theorem (right triangle)

$$r = \sqrt{x^2 + y^2}$$

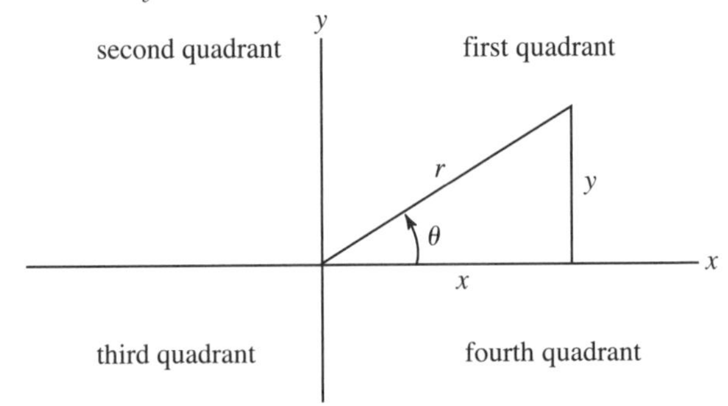

Trigonometric Relationships

Definitions of Trigonometric Functions

$$\sin\theta = \frac{y}{r} \qquad \cos\theta = \frac{x}{r} \qquad \tan\theta = \frac{\sin\theta}{\cos\theta} = \frac{y}{x}$$

θ° (rad)	sin θ	cos θ	tan θ
0° (0)	0	1	0
30° ($\pi/6$)	0.500	$\sqrt{3}/2 \approx 0.866$	$\sqrt{3}/3 \approx 0.577$
45° ($\pi/4$)	$\sqrt{2}/2 \approx 0.707$	$\sqrt{2}/2 \approx 0.707$	1.00
60° ($\pi/3$)	$\sqrt{3}/2 \approx 0.866$	0.500	$\sqrt{3} \approx 1.73$
90° ($\pi/2$)	1	0	∞

Law of Cosines

$$a^2 = b^2 + c^2 - 2bc\cos A$$

Law of Sines

$$\frac{a}{\sin A} = \frac{b}{\sin B} = \frac{c}{\sin C}$$

(a, b, c sides, A, B, C angles opposite sides for any plane triangle)

Conversion Factors

Mass

$1\ \text{g} = 10^{-3}\ \text{kg}$
$1\ \text{kg} = 10^{3}\ \text{g}$
$1\ \text{u} = 1.66 \times 10^{-24}\ \text{g} = 1.66 \times 10^{-27}\ \text{kg}$
$1\ \text{metric ton} = 1000\ \text{kg}$

Length

$1\ \text{nm} = 10^{-9}\ \text{m}$
$1\ \text{cm} = 10^{-2}\ \text{m} = 0.394\ \text{in.}$
$1\ \text{m} = 10^{-3}\ \text{km} = 3.28\ \text{ft} = 39.4\ \text{in.}$
$1\ \text{km} = 10^{3}\ \text{m} = 0.621\ \text{mi}$
$1\ \text{in.} = 2.54\ \text{cm} = 2.54 \times 10^{-2}\ \text{m}$
$1\ \text{ft} = 0.305\ \text{m} = 30.5\ \text{cm}$
$1\ \text{mi} = 5280\ \text{ft} = 1609\ \text{m} = 1.609\ \text{km}$

Area

$1\ \text{cm}^2 = 10^{-4}\ \text{m}^2 = 0.1550\ \text{in}^2$
$= 1.08 \times 10^{-3}\ \text{ft}^2$
$1\ \text{m}^2 = 10^{4}\ \text{cm}^2 = 10.76\ \text{ft}^2 = 1550\ \text{in}^2$
$1\ \text{in}^2 = 6.94 \times 10^{-3}\ \text{ft}^2 = 6.45\ \text{cm}^2$
$= 6.45 \times 10^{-4}\ \text{m}^2$
$1\ \text{ft}^2 = 144\ \text{in}^2 = 9.29 \times 10^{-2}\ \text{m}^2 = 929\ \text{cm}^2$

Volume

$1\ \text{cm}^3 = 10^{-6}\ \text{m}^3 = 3.53 \times 10^{-5}\ \text{ft}^3$
$= 6.10 \times 10^{-2}\ \text{in}^3$
$1\ \text{m}^3 = 10^{6}\ \text{cm}^3 = 10^{3}\ \text{L} = 35.3\ \text{ft}^3$
$= 6.10 \times 10^{4}\ \text{in}^3 = 264\ \text{gal}$
$1\ \text{liter} = 10^{3}\ \text{cm}^3 = 10^{-3}\ \text{m}^3 = 1.056\ \text{qt}$
$= 0.264\ \text{gal} = 0.0353\ \text{ft}^3$
$1\ \text{in}^3 = 5.79 \times 10^{-4}\ \text{ft}^3 = 16.4\ \text{cm}^3$
$= 1.64 \times 10^{-5}\ \text{m}^3$
$1\ \text{ft}^3 = 1728\ \text{in}^3 = 7.48\ \text{gal} = 0.0283\ \text{m}^3$
$= 28.3\ \text{L}$
$1\ \text{qt} = 2\ \text{pt} = 946\ \text{cm}^3 = 0.946\ \text{L}$
$1\ \text{gal} = 4\ \text{qt} = 231\ \text{in}^3 = 0.134\ \text{ft}^3 = 3.785\ \text{L}$

Time

$1\ \text{h} = 60\ \text{min} = 3600\ \text{s}$
$1\ \text{day} = 24\ \text{h} = 1440\ \text{min} = 8.64 \times 10^{4}\ \text{s}$
$1\ \text{y} = 365\ \text{days} = 8.76 \times 10^{3}\ \text{h}$
$= 5.26 \times 10^{5}\ \text{min} = 3.16 \times 10^{7}\ \text{s}$

Angle

$1\ \text{rad} = 57.3^\circ$

$1^\circ = 0.0175\ \text{rad}$	$60^\circ = \pi/3\ \text{rad}$
$15^\circ = \pi/12\ \text{rad}$	$90^\circ = \pi/2\ \text{rad}$
$30^\circ = \pi/6\ \text{rad}$	$180^\circ = \pi\ \text{rad}$
$45^\circ = \pi/4\ \text{rad}$	$360^\circ = 2\pi\ \text{rad}$

$1\ \text{rev/min} = (\pi/30)\ \text{rad/s} = 0.1047\ \text{rad/s}$

Speed

$1\ \text{m/s} = 3.60\ \text{km/h} = 3.28\ \text{ft/s}$
$= 2.24\ \text{mi/h}$
$1\ \text{km/h} = 0.278\ \text{m/s} = 0.621\ \text{mi/h}$
$= 0.911\ \text{ft/s}$
$1\ \text{ft/s} = 0.682\ \text{mi/h} = 0.305\ \text{m/s}$
$= 1.10\ \text{km/h}$
$1\ \text{mi/h} = 1.467\ \text{ft/s} = 1.609\ \text{km/h}$
$= 0.447\ \text{m/s}$
$60\ \text{mi/h} = 88\ \text{ft/s}$

Force

$1\ \text{N} = 0.225\ \text{lb}$
$1\ \text{lb} = 4.45\ \text{N}$
Equivalent weight of a mass of 1 kg on Earth's surface = 2.2 lb = 9.8 N

Pressure

$1\ \text{Pa}\ (\text{N/m}^2) = 1.45 \times 10^{-4}\ \text{lb/in}^2$
$= 7.5 \times 10^{-3}\ \text{torr (mm Hg)}$
$1\ \text{torr (mm Hg)} = 133\ \text{Pa}\ (\text{N/m}^2)$
$= 0.02\ \text{lb/in}^2$
$1\ \text{atm} = 14.7\ \text{lb/in}^2 = 1.013 \times 10^{5}\ \text{N/m}^2$
$= 30\ \text{in. Hg} = 76\ \text{cm Hg}$
$1\ \text{lb/in}^2 = 6.90 \times 10^{3}\ \text{Pa}\ (\text{N/m}^2)$
$1\ \text{bar} = 10^{5}\ \text{Pa}$
$1\ \text{millibar} = 10^{2}\ \text{Pa}$

Energy

$1\ \text{J} = 0.738\ \text{ft·lb} = 0.239\ \text{cal}$
$= 9.48 \times 10^{-4}\ \text{Btu} = 6.24 \times 10^{18}\ \text{eV}$
$1\ \text{kcal} = 4186\ \text{J} = 3.968\ \text{Btu}$
$1\ \text{Btu} = 1055\ \text{J} = 778\ \text{ft·lb} = 0.252\ \text{kcal}$
$1\ \text{cal} = 4.186\ \text{J} = 3.97 \times 10^{-3}\ \text{Btu}$
$= 3.09\ \text{ft·lb}$
$1\ \text{ft·lb} = 1.36\ \text{J} = 1.29 \times 10^{-3}\ \text{Btu}$
$1\ \text{eV} = 1.60 \times 10^{-19}\ \text{J}$
$1\ \text{kWh} = 3.6 \times 10^{6}\ \text{J}$

Power

$1\ \text{W} = 0.738\ \text{ft·lb/s} = 1.34 \times 10^{-3}\ \text{hp}$
$= 3.41\ \text{Btu/h}$
$1\ \text{ft·lb/s} = 1.36\ \text{W} = 1.82 \times 10^{-3}\ \text{hp}$
$1\ \text{hp} = 550\ \text{ft·lb/s} = 745.7\ \text{W}$
$= 2545\ \text{Btu/h}$

Mass–Energy Equivalents

$1\ \text{u} = 1.66 \times 10^{-27}\ \text{kg} \leftrightarrow 931.5\ \text{MeV}$
$1\ \text{electron mass} = 9.11 \times 10^{-31}\ \text{kg}$
$= 5.49 \times 10^{-4}\ \text{u} \leftrightarrow 0.511\ \text{MeV}$
$1\ \text{proton mass} = 1.67262 \times 10^{-27}\ \text{kg}$
$= 1.007276\ \text{u} \leftrightarrow 938.27\ \text{MeV}$
$1\ \text{neutron mass} = 1.67493 \times 10^{-27}\ \text{kg}$
$= 1.008665\ \text{u} \leftrightarrow 939.57\ \text{MeV}$

Temperature

$T_F = \frac{9}{5} T_C + 32$
$T_C = \frac{5}{9} (T_F - 32)$
$T_K = T_C + 273$

cgs Force

$1\ \text{dyne} = 10^{-5}\ \text{N} = 2.25 \times 10^{-6}\ \text{lb}$

cgs Energy

$1\ \text{erg} = 10^{-7}\ \text{J} = 7.38 \times 10^{-6}\ \text{ft·lb}$

YOUR ACCESS TO SUCCESS

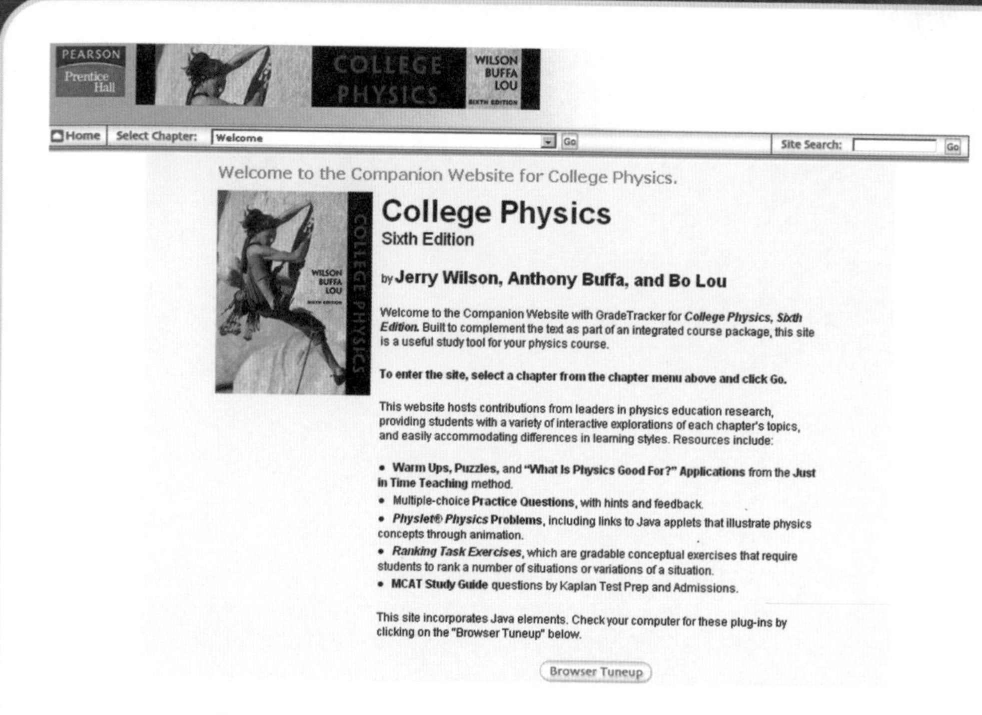

Students with new copies of Wilson/Buffa/Lou *College Physics, 6th Edition* have full access to the book's Companion Website with GradeTracker—a 24/7 study tool with quizzes, self study tools, and other features designed to help you make the most of your limited study time.

Just follow the easy website registration steps listed below...

Registration Instructions for the Wilson/Buffa/Lou, *College Physics, Sixth Edition Companion Website with GradeTracker*

1. Go to *www.prenhall.com/wilson.*
2. Click the cover for Wilson/Buffa/Lou, *College Physics, Sixth Edition.*
3. Click on the link to the *Companion Website with GradeTracker.*
4. Click "*Register*".
5. Using a coin, scratch off the metallic coating below to reveal your Access Code.
6. Complete the online registration form, choosing your own personal Login Name and Password.
7. Enter your pre-assigned Access Code exactly as it appears below.
8. Complete the online registration form by entering your School Location information.
9. After your personal Login Name and Password are confirmed by e-mail, repeat steps 1-3 above, and enter your new Login Name and Password to log in.

Your Access Code is:

If there is no metallic coating covering the access code above, the code may no longer be valid and you will need to purchase online access using a major credit card to use the website. To do so, go to www.prenhall.com/wilson, click the cover for Wilson/Buffa/Lou, *College Physics, Sixth Edition*, and follow the instructions to purchase access to the Companion Website with Grade Tracker.

Important: Please read the Subscription and End-User License Agreement located on the "Log In" screen before using the Wilson/Buffa/Lou, *College Physics, Sixth Edition Companion Website with Grade Tracker*. By using the website, you indicate that you have read, understood and accepted the terms of the agreement.

Minimum system requirements

PC Operating Systems:

Windows 2000/XP

Pentium II 233 MHz processor. 64 MB RAM
In addition to the minimum memory required by your OS

Internet Explorer (TM) 5.5 or 6, Netscape(TM) 7, Firefox 1.0

Macintosh Operating Systems:

Macintosh Power PC with OS X (10.2 and 10.3)

In addition to the RAM required by your OS, this application requires 64 MB RAM, with 40 MB Free RAM, with Virtual Memory enabled

Netscape(TM) 7, Safari 1.3, Firefox 1.0

Acrobat Reader 6.0.1

4x CD-ROM drive

800 x 600 pixel screen resolution

Technical Support Call 1-800-677-6337. Phone support is available Monday-Friday, 8am to 8pm and Sunday 5pm to 12am, Eastern time. Visit our support site at *http://247.pearsoned.com*
E-mail support is available 24/7.

COLLEGE PHYSICS

SIXTH EDITION

COLLEGE PHYSICS

Volume 2

Jerry D. Wilson
Lander University
Greenwood, SC

Anthony J. Buffa
California Polytechnic State University
San Luis Obispo, CA

Bo Lou
Ferris State University
Big Rapids, MI

Upper Saddle River, NJ 07458

Library of Congress Cataloging-in-Publication Data

Wilson, Jerry D.
College physics.— 6th ed. / Jerry D. Wilson, Anthony J. Buffa, Bo Lou.
p. cm.
Includes index.
ISBN 0-13-195111-4
1. Physics—Textbooks. I. Buffa, Anthony J. II. Lou, Bo. III. Title.

QC21.3.W35 2007
530—dc22 2005051527

Senior Editor: Erik Fahlgren
Associate Editor: Christian Botting
Editor in Chief, Science: Dan Kaveney
Executive Managing Editor: Kathleen Schiaparelli
Assistant Managing Editor: Beth Sweeten
Manufacturing Buyer: Alan Fischer
Manufacturing Manager: Alexis Heydt-Long
Director of Creative Services: Paul Belfanti
Creative Director: Juan López
Art Director: Heather Scott
Director of Marketing, Science: Patrick Lynch
Media Editor: Michael J. Richards
Senior Managing Editor, Art Production and Management: Patricia Burns
Manager, Production Technologies: Matthew Haas
Managing Editor, Art Management: Abigail Bass
Art Editor: Eric Day
Art Studio: ArtWorks
Image Coordinator: Cathy Mazzucca
Mgr Rights & Permissions: Zina Arabia
Photo Researchers: Alexandra Truitt & Jerry Marshall
Research Manager: Beth Brenzel
Interior and Cover Design: Tamara Newnam
Cover Image: Greg Epperson/Index Stock Imagery
Managing Editor, Science Media: Nicole Jackson
Media Production Editors: William Wells, Dana Dunn
Editorial Assistant: Jessica Berta
Production Assistant: Nancy Bauer
Production Supervision/Composition: Prepare, Inc.

Pearson Prentice Hall
Pearson Education, Inc.
Upper Saddle River, New Jersey 07458

Printed in the United States of America
10 9 8 7 6 5 4 3 2 1

ISBN 0-13-195111-4

Pearson Education LTD., *London*
Pearson Education Australia PTY, Limited, *Sydney*
Pearson Education Singapore, Pte. Ltd
Pearson Education North Asia Ltd. *Hong Kong*
Pearson Education Canada, Ltd., *Toronto*
Pearson Educación de Mexico, S.A. de C.V.
Pearson Education — Japan, *Tokyo*
Pearson Education Malaysia, Pte. Ltd

ABOUT THE AUTHORS

Jerry D. Wilson, a native of Ohio, is Emeritus Professor of Physics and former chair of the Division of Biological and Physical Sciences at Lander University in Greenwood, South Carolina. He received a B.S. degree from Ohio University, an M.S. degree from Union College, and, in 1970, a Ph.D. from Ohio University. He earned his M.S. degree while employed as a Materials Behavior physicist.

As a doctoral graduate student, Professor Wilson held the faculty rank of Instructor and began teaching physical science courses. During this time, he coauthored a physical science text that is now in its eleventh edition. In conjunction with his teaching career, Professor Wilson continued his writing and has authored or coauthored six titles. Having retired from full-time teaching, he continues to write, producing, among other works, *The Curiosity Corner*, a weekly column for local newspapers that can also be found on the Internet.

Anthony J. Buffa received his B.S. degree in physics from Rensselaer Polytechnic Institute in Troy, New York, and M.S. and Ph.D. degrees in physics from the University of Illinois, Urbana–Champaign. In 1970, Professor Buffa joined the faculty at California Polytechnic State University, San Luis Obispo. Recently retired, he now teaches halftime at Cal Poly as an Emeritus Professor of Physics. During his career he was involved in nuclear physics research at several accelerators, including LAMPF at Los Alamos National Laboratory. On campus, he was a research associate with the physics department's radioanalytical facility for sixteen years.

Professor Buffa's main interest remains teaching. During his tenure at Cal Poly, he taught courses ranging from introductory physical science to quantum mechanics, developed and revised many laboratory experiments, and taught physics to local K-12 teachers at an NSF-sponsored workshop. Combining physics with his interests in art and architecture, Dr. Buffa develops his own artwork and sketches, which he uses to increase his effectiveness in teaching physics. In addition to continued teaching, during his (partial) retirement, he and his wife intend to travel more and hopefully enjoy their future grandkids for a long time to come.

Bo Lou is currently Professor of Physics at Ferris State University in Michigan. His primary teaching responsibilities are undergraduate introductory physics lectures and laboratories. Professor Lou emphasizes the importance of conceptual understanding of the basic laws and principles of physics and their practical applications to the real world. He is also an enthusiastic advocate of using technology in teaching and learning.

Professor Lou received B.S. and M.S. degrees in optical engineering from Zhejiang University (China) in 1982 and 1985, respectively, and a Ph.D. in condensed matter physics from Emory University in 1989.

Dr. Lou, his wife Lingfei, and their daughter Alina reside in Big Rapids, Michigan. The family enjoys travel, nature, and tennis.

BRIEF CONTENTS

Preface XV

Part One: Mechanics

1 Measurement and Problem Solving 1
2 Kinematics: Description of Motion 32
3 Motion in Two Dimensions 67
4 Force and Motion 103
5 Work and Energy 140
6 Linear Momentum and Collisions 177
7 Circular Motion and Gravitation 216
8 Rotational Motion and Equilibrium 256
9 Solids and Fluids 297

Part Two: Thermodynamics

10 Temperature and Kinetic Theory 338
11 Heat 367
12 Thermodynamics 397

Part Three: Oscillations and Wave Motion

13 Vibrations and Waves 433
14 Sound 467

Part Four: Electricity and Magnetism

15 Electric Charge, Forces, and Fields 505
16 Electric Potential, Energy, and Capacitance 536
17 Electric Current and Resistance 568
18 Basic Electric Circuits 591
19 Magnetism 623
20 Electromagnetic Induction and Waves 656
21 AC Circuits 686

Part Five: Optics

22 Reflection and Refraction of Light 705
23 Mirrors and Lenses 729
24 Physical Optics: The Wave Nature of Light 760
25 Vision and Optical Instruments 792

Part Six: Modern Physics

26 Relativity 819
27 Quantum Physics 851
28 Quantum Mechanics and Atomic Physics 877
29 The Nucleus 902
30 Nuclear Reactions and Elementary Particles 935

Appendices

I Mathematical Review (with Examples) for College Physics A-1
II Kinetic Theory of Gases A-5
III Planetary Data A-6
IV Alphabetical Listing of the Chemical Elements A-7
V Properties of Selected Isotopes A-7

Answers to Follow-Up Exercises A-10
Answers to Odd-Numbered Exercises A-18
Photo Credits P-1
Index I-1

CONTENTS

Preface XV

15 ELECTRIC CHARGE, FORCES, AND FIELDS 505

15.1 Electric Charge 506
15.2 Electrostatic Charging 508
15.3 Electric Force 512
15.4 Electric Field 517
LEARN BY DRAWING: Using the Superposition Principle to Determine the Electric Field Direction 518
LEARN BY DRAWING: Sketching Electric Lines of Force 521
INSIGHT: 15.1 Lightning and Lightning Rods 523
INSIGHT: 15.2 Electric Fields in Law Enforcement and Nature: Stun Guns and Electric Fish 524
15.5 Conductors and Electric Fields 526
*15.6 Gauss's Law for Electric Fields: A Qualitative Approach 528
Chapter Review 529 Exercises 530

16 ELECTRIC POTENTIAL, ENERGY, AND CAPACITANCE 536

16.1 Electric Potential Energy and Electric Potential Difference 537
LEARN BY DRAWING: ΔV Is Independent of Reference Point 538
16.2 Equipotential Surfaces and the Electric Field 543
LEARN BY DRAWING: Graphical Relationship between Electric Field Lines and Equipotentials 547
16.3 Capacitance 549
INSIGHT: 16.1 Electric Potential and Nerve Signal Transmission 552
16.4 Dielectrics 552
16.5 Capacitors in Series and in Parallel 557
Chapter Review 561 Exercises 562

17 ELECTRIC CURRENT AND RESISTANCE 568

17.1 Batteries and Direct Current 569
LEARN BY DRAWING: Sketching Circuits 571
17.2 Current and Drift Velocity 571
17.3 Resistance and Ohm's Law 573
INSIGHT: 17.1 The "Bio-Generation" of High Voltage 575
INSIGHT: 17.2 Bioelectrical Impedance Analysis (BIA) 578
17.4 Electric Power 580
Chapter Review 585 Exercises 586

18 BASIC ELECTRIC CIRCUITS 591

18.1 Resistances in Series, Parallel, and Series–Parallel Combinations 592
18.2 Multiloop Circuits and Kirchhoff's Rules 599
LEARN BY DRAWING: Kirchhoff Plots: A Graphical Interpretation of Kirchhoff's Loop Theorem 602
18.3 RC Circuits 604
18.4 Ammeters and Voltmeters 607
INSIGHT: 18.1 Applications of RC Circuits to Cardiac Medicine 608
18.5 Household Circuits and Electrical Safety 611
INSIGHT: 18.2 Electricity and Personal Safety 614
Chapter Review 615 Exercises 616

19 MAGNETISM 623

19.1 Magnets, Magnetic Poles, and Magnetic Field Direction 624
19.2 Magnetic Field Strength and Magnetic Force 626
19.3 Applications: Charged Particles in Magnetic Fields 629
19.4 Magnetic Forces on Current-Carrying Wires 632
19.5 Applications: Current-Carrying Wires in Magnetic Fields 635
19.6 Electromagnetism: The Source of Magnetic Fields 637
19.7 Magnetic Materials 641
INSIGHT: 19.1 The Magnetic Force in Future Medicine 642
*19.8 Geomagnetism: The Earth's Magnetic Field 644
INSIGHT: 19.2 Magnetism in Nature 645
Chapter Review 647 Exercises 648

20 ELECTROMAGNETIC INDUCTION AND WAVES 656

20.1 Induced emf: Faraday's Law and Lenz's Law 657
20.2 Electric Generators and Back emf 663
INSIGHT: 20.1 Electromagnetic Induction at Work: Flashlights and Antiterrorism 664
INSIGHT: 20.2 Electromagnetic Induction at Play: Hobbies and Transportation 666
20.3 Transformers and Power Transmission 668
20.4 Electromagnetic Waves 672
Chapter Review 679 Exercises 679

21 AC CIRCUITS 686

21.1 Resistance in an AC Circuit 687
21.2 Capacitive Reactance 689
21.3 Inductive Reactance 691
21.4 Impedance: RLC Circuits 693
21.5 Circuit Resonance 697
INSIGHT: 21.1 Oscillator Circuits: Broadcasters of Electromagnetic Radiation 699
Chapter Review 700 Exercises 701

22 REFLECTION AND REFRACTION OF LIGHT 705

22.1 Wave Fronts and Rays 706
22.2 Reflection 707
22.3 Refraction 708
LEARN BY DRAWING: Tracing the Reflected Rays 708
INSIGHT: 22.1 A Dark, Rainy Night 709
INSIGHT: 22.2 Negative Index of Refraction and the "Perfect" Lens 715
22.4 Total Internal Reflection and Fiber Optics 717
INSIGHT: 22.3 Fiber Optics: Medical Applications 720
22.5 Dispersion 721
INSIGHT: 22.4 The Rainbow 722
Chapter Review 723 Exercises 724

23 MIRRORS AND LENSES 729

23.1 Plane Mirrors 730
23.2 Spherical Mirrors 732
INSIGHT: 23.1 It's All Done with Mirrors 733
LEARN BY DRAWING: A Mirror Ray Diagram (see Example 23.2) 734
23.3 Lenses 740
LEARN BY DRAWING: A Lens Ray Diagram (see Example 23.5) 743
INSIGHT: 23.2 Fresnel Lenses 748
23.4 The Lens Maker's Equation 750
*23.5 Lens Aberrations 752
Chapter Review 753 Exercises 754

24 PHYSICAL OPTICS: THE WAVE NATURE OF LIGHT 760

24.1 Young's Double-Slit Experiment 761
24.2 Thin-Film Interference 764
INSIGHT: 24.1 Nonreflecting Lenses 768
24.3 Diffraction 768
24.4 Polarization 775
LEARN BY DRAWING: Three Polarizers (see Integrated Example 24.6.) 778
*24.5 Atmospheric Scattering of Light 782
INSIGHT: 24.2 LCDs and Polarized Light 783
INSIGHT: 24.3 Optical Biopsy 785
Chapter Review 785 Exercises 786

25 VISION AND OPTICAL INSTRUMENTS 792

25.1 The Human Eye 793
INSIGHT: 25.1 Cornea "Orthodontics" and Surgery 797
25.2 Microscopes 799
25.3 Telescopes 803
25.4 Diffraction and Resolution 807
INSIGHT: 25.2 Telescopes Using Nonvisible Radiation 808
*25.5 Color 810
Chapter Review 813 Exercises 814

26 RELATIVITY 819

26.1 Classical Relativity and the Michelson–Morley Experiment 820
26.2 The Postulates of Special Relativity and the Relativity of Simultaneity 822
26.3 The Relativity of Length and Time: Time Dilation and Length Contraction 825
26.4 Relativistic Kinetic Energy, Momentum, Total Energy, and Mass–Energy Equivalence 833
26.5 The General Theory of Relativity 837
INSIGHT: 26.1 Relativity in Everyday Living 838
*26.6 Relativistic Velocity Addition 841
INSIGHT: 26.2 Black Holes, Gravitational Waves, and LIGO 842
Chapter Review 844 **Exercises** 845

27 QUANTUM PHYSICS 851

27.1 Quantization: Planck's Hypothesis 852
27.2 Quanta of Light: Photons and the Photoelectric Effect 854
LEARN BY DRAWING: The Photoelectric Effect and Energy Conservation 856
27.3 Quantum "Particles": The Compton Effect 858
27.4 The Bohr Theory of the Hydrogen Atom 860
27.5 A Quantum Success: The Laser 866
INSIGHT: 27.1 CD and DVD Systems 869
INSIGHT: 27.2 Lasers in Modern Medicine 870
Chapter Review 871 **Exercises** 873

28 QUANTUM MECHANICS AND ATOMIC PHYSICS 877

28.1 Matter Waves: The de Broglie Hypothesis 878
28.2 The Schrödinger Wave Equation 881
INSIGHT: 28.1 The Electron Microscope 883
INSIGHT: 28.2 The Scanning Tunneling Microscope (STM) 884
28.3 Atomic Quantum Numbers and the Periodic Table 885
INSIGHT: 28.3 Magnetic Resonance Imaging (MRI) 888
28.4 The Heisenberg Uncertainty Principle 894
28.5 Particles and Antiparticles 896
Chapter Review 897 **Exercises** 898

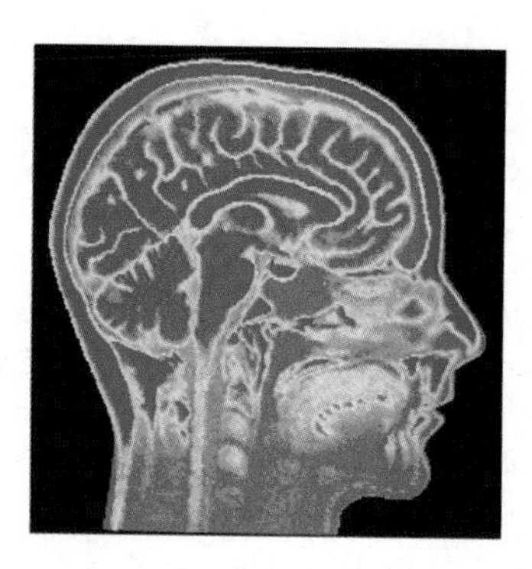

29 THE NUCLEUS 902

29.1 Nuclear Structure and the Nuclear Force 903
29.2 Radioactivity 906
29.3 Decay Rate and Half-Life 911
29.4 Nuclear Stability and Binding Energy 917
29.5 Radiation Detection, Dosage, and Applications 922
INSIGHT: 29.1 Biological and Medical Applications of Radiation 927
Chapter Review 929 **Exercises** 930

30 NUCLEAR REACTIONS AND ELEMENTARY PARTICLES 935

30.1 Nuclear Reactions 936
30.2 Nuclear Fission 939
30.3 Nuclear Fusion 944
30.4 Beta Decay and the Neutrino 946
30.5 Fundamental Forces and Exchange Particles 948
30.6 Elementary Particles 951
30.7 The Quark Model 953
30.8 Force Unification Theories, the Standard Model, and the Early Universe 954
Chapter Review 956 **Exercises** 957

APPENDIX I Mathematical Review (with Examples) for College Physics A-1
APPENDIX II Kinetic Theory of Gases A-5
APPENDIX III Planetary Data A-6
APPENDIX IV Alphabetical Listing of the Chemical Elements A-7
APPENDIX V Properties of Selected Isotopes A-7
Answers to Follow-Up Exercises A-10
Answers to Odd-Numbered Exercises A-18
Photo Credits P-1
Index I-1

LEARN BY DRAWING

Cartesian Coordinates and One-Dimensional Displacement **35**
Signs of Velocity and Acceleration **42**
Make a Sketch and Add Them Up **80**
Forces on an Object on an Inclined Plane and Free-body Diagram **116**
Work: Area under the *F*-versus-Curve **142**
Determining the Sign of Work **143**
Energy Exchanges: A Falling Ball **161**
The Small-Angle Approximation **219**
Thermal Area Expansion **351**
From Cold Ice To Hot Steam **377**
Leaning on Isotherms **409**
Representing Work in Thermal Cycles **415**
Oscillating in a Parabolic Potential Well **437**
Using the Superposition Principle to Determine the Electric Field Direction **518**
Sketching Electric Lines of Force **521**
ΔV Is Independent of Reference Point **538**
Graphical Relationship between Electric Field Lines and Equipotentials **547**
Sketching Circuits **571**
Kirchhoff Plots: A Graphical Interpretation of Kirchhoff's Loop Theorem **602**
Tracing the Reflected Rays **708**
A Mirror Ray Diagram (see Example 23.2) **734**
A Lens Ray Diagram (see Example 23.5) **743**
Three Polarizers (see Integrated Example 24.6) **778**
The Photoelectric Effect and Energy Conservation **856**

Applications (Insights appear in **boldface**, and "**(bio)**" indicates a biomedical application)

Chapter 1
Why Study Physics? 2
Capillary system (bio) **11**
Is Unit Conversion Important? 16
Drawing blood (bio) **23**
How many red cells in blood (bio) **24**
Capillary length (bio) **14, 27**
Circulatory system (bio) **27**
Heart beat (bio) **28**
Red blood cells (bio) **28**
Nutrition labels (bio) **30**
White cells and platelets (bio) **30**
Human hair (bio) **30**

Chapter 2
Free-fall on the Moon **50**
Galileo Galilei and the Leaning Tower of Pisa 51
Reaction time (bio) **53**
Free-fall on Mars **56**
Taipei 101 Tower **65**

Chapter 3
Air resistance and range **88**
The longest jump (bio) **88**
Air-to-air refueling **91**

Chapter 4
***g*'s of Force and Effects on the Human Body (bio) 108**
Sailing into the Wind—Tacking 115
Leg traction (bio) **119**
On your toes (bio) **120**
Burning in (racer tires) **122**
Air foils **128**
Sky diving and terminal velocity **129**
Aerobraking **129**
Down force and race cars **137**
Racing tires versus passenger-car tires **137**
Down-slope run **138**

Chapter 5
People Power: Using Body Energy (bio) 156
Hybrid Energy Conversion 164
Power ratings of motors **166**
Weightlifting (bio) **169**

Chapter 6
How to catch a fastball **184**
Impulse force and body injury (bio) **184**
Following through in sports **185**
The Automobile Airbag and Martian Airbags 186
Center of mass of a high jumper **204**
Squid jet propulsion (bio) **204**
Recoil of a rifle **205**
Rocket thrust **205**
Reverse thrust of jet aircraft **206**
Karate chop **209**
Propulsion of fan boats **210**
Bird catching fish (bio) **212**
Flamingo on one leg (bio) **214**

Chapter 7
Measuring angular distance **218**
Merry-go-round and rotational speeds **221**
Centrifuge speed **224**
The Centrifuge: Separating Blood Components (bio) 225
Driving on a curved road **227**
Compact discs (CDs) and angular acceleration **229**
Cooking evenly in a microwave oven **230**
Geosynchronous satellite orbit **234**
Space Exploration: Gravity Assists 238
Satellite orbits **242**
"Weightlessness" (zero gravity) and apparent weightlessness **244**
"Weightlessness": Effects on the Human Body (bio) 245
Space colonies and artificial gravity **246**
Thrown outward when driving on a curved road **250**
Banked roads **250**
Space walks **253**

Chapter 8
Muscle torque (bio) **260**
My aching back (bio) **261**
No net torque: the Iron Cross (bio) **266**
Low bases and center of gravity of race cars **268**

The center-of-gravity challenge (bio) 268
Stabilizing the Leaning Tower of Pisa 269
Stability in Action 271
Yo-yo torque 276
Slide or Roll to a Stop? Antilock Brakes 281
Angular momentum in diving and skating (bio) 283
Tornadoes and Hurricanes 283
Throwing a spiraling football 285
Gyrocompass 285
Precession of the Earth's axis 285
Helicopter rotors 286
Back pain (bio) 288
Gymnast and balance (bio) 288
Muscle force (bio) 289
Russell traction (bio) 290
Knee physical therapy (bio) 290
Tightrope walkers 291
Roller coaster loop-the-loop 294
Falling cat (bio) 295

Chapter 9
Bone (femur) extension (bio) 300
Osteoporosis and Bone Mineral Density (BMD) (bio) 304
Hydraulic brakes, shock absorbers, lifts, and jacks 307
Manometers, tire gauges, and barometers 309
An Atmospheric Effect: Possible Earaches (bio) 311
Blood Pressure and Its Measurement (bio) 312
An IV: gravity assist (bio) 312
Fish swim bladders or gas bladders (bio) 317
Tip of the iceberg 318
Blood flow: cholesterol and plaque (bio) 321
Speed of blood in the aorta (bio) 321
Chimneys, smokestacks, and the Bernoulli effect 322
Airplane lift 322
Water strider (bio) 324
The Lungs and Baby's First Breath (bio) 325
Motor oils and viscosity 327
Poiseuille's law: a blood transfusion (bio) 328
A bed of nails (bio) 331
Shape of water towers 331
Pet water dispenser 332
Pimsoll mark for depth loading 334
Perpetual motion machine 334
Zepplins 334
Indy race cars and Venturi tunnel 335
Speed of blood flow (bio) 335
Blood flow in the pulmonary artery (bio) 336
Blood transfusion (bio) 337
Drawing blood (bio) 337

Chapter 10
Thermometers and thermostats 340
Human Body Temperature (bio) 343
Warm-Blooded versus Cold-Blooded (bio) 344
Expansion gaps 352
Why lakes freeze at top first 353
Osmosis and kidneys (bio) 357
Physiologal Diffusion in Life Processes (bio) 357
Highest and lowest recorded temperatures 361
Cooling in open-heart surgery (bio) 361
Lung capacity (bio) 362
Gaseous diffusion and the atomic bomb 365

Chapter 11
Working off that birthday cake (bio) 369
Specific heat and burning your mouth (bio) 371
Cooking at Pike's Peak 378
Keeping organs ready for transplant (bio) 378
Physiological Regulation of Body Temperature (bio) 380
Copper-bottomed pots 381
Thermal insulation: Helping prevent heat loss 382
Physics, the Construction Industry, and Energy Conservation 384
Day–night atmospheric convection cycles 384
R-values 384
Forced convection in refrigerators, heating and cooling systems, and the body (bio) 385
Polymer-foam insulation 385
Thermography (bio) 387
The Greenhouse Effect (bio) 388
Solar panels 389
Saving fruit trees from frost (bio) 389
Dressing for the desert 389
Thermal bottle 389
Passive solar design 390
Solar collectors for heating 395

Chapter 12
Energy balancing: Exercising using physics (bio) 401
How not to recycle a spray can 406
Exhaling: Blowing hot and cold (bio) 407
Perpetual-motion machines 410
Life, Order, and the Second Law (bio) 414
Thermal efficiency of engines 416
Internal Combustion Engines and the Otto Cycle 417
Thermodynamics and the Human Body (bio) 420
Refrigerators as thermal pumps 421
Air conditioner/heat pump: Thermal switch hitting 422

Chapter 13
Damping: bathroom scales, shock absorbers, and earthquake protection 445
Surf 449
Earthquakes, Seismic Waves, and Seismology 450
Destructive interference: pilot's headphones 452
Stringed musical instruments 455
Tuning a guitar 457
Desirable and Undesirable Resonances 458
Pushing a swing in resonance 459
Radio frequencies 464

Chapter 14
Infrasonic and ultrasonic hearing in animals (bio) 468
Sonar 469
Ultrasound in Medicine (bio) 470
Low-frequency fog horns 474
The Physiology and Physics of the Ear and Hearing (bio) 475
Protect your hearing (bio) 480
Beats and stringed instruments 484
Traffic radar 488
Sonic booms 489
Crack of a whip 489
Doppler Applications: Blood Cells and Raindrops (bio) 490
Pipe organs 492
Wind and brass instruments 493
Ultrasound in medical diagnosis (bio) 499
Ultrasound and dolphins (bio) 499
Speed of sound in human tissue (bio) 499
Size of eardrum (bio) 499
Fundamental frequency of ear canal (bio) 503
Helium and "Donald Duck" sound (bio) 503

Chapter 15
Uses of semiconductors 508
Application of electrostatic charging 512
Lightning and Lightning Rods 523
Electric Fields in Law Enforcement and Nature (bio) 524
Safety in lightning storms (Ex. 15.70) (bio) 533
Electric fields in a computer monitor 535

Chapter 16
Creation of X-rays 540
The water molecule: the molecule of life (bio) 542
Common voltages (Table 16.1) 546
Cardiac defibrillators (bio) 551
Electric Potential and Nerve Signal Transmission (bio) 552
Computer keyboard design 556
Computer monitor operation (Ex. 16.29) 563
Nerve signal transmission (Exs. 16.106 and 16.107) (bio) 567

Chapter 17
Battery operation 569
Automotive batteries in operation 570
Electrical hazards in the house 574
The "Bio-generation" of High Voltage (bio) 575
Bioelectrical Impedance Analysis (BIA) (bio) 578
An electrical thermometer 579
Applications of superconductivity 579
Power requirements of appliances 581
Electrical repair of appliances 582
Cost of electric energy 583
Energy efficiency and natural resources 583
Various appliance applications in exercises 590

Chapter 18
Strings of Christmas tree lights 596
Flash photography 606
Blinker (flashing) circuit operation 606
Ammeter design 607
Applications of RC Circuits to Cardiac Medicine (bio) 608
Voltmeter design 610
Multimeter design 610
Household circuit wiring 611
Fuses and circuit breakers 612
Electrical safety and grounding (bio) 613
Electricity and Personal Safety (bio) 614
Polarized plugs 614
Various circuit applications to medicine and safety in exercises (bio) 622

Chapter 19
Magnetically levitated trains 623
Cathode-ray tubes, oscilloscopes, TV screens and monitors 629
Mass spectrometer operation 629
Submarine propulsion using magnetohydrodynamics 631
dc motor operation 636
Electronic balance 636
Electromagnets and magnetic materials 642
The Magnetic Force in Future Medicine (bio) 642
The Earth's magnetic field and geomagnetism 644
Magnetism in Nature (bio) 645
Navigating with compasses 646
The aurorae 647
The Hall effect in solid state engineering (Ex. 19.16) 649
Charged pions and cancer treatment (Ex. 19.31) (bio) 650
Doorbell and chime operation (Ex. 19.41) 651

Chapter 20
Induced currents and equipment hazards 661
Electric generators 663
Electromagnetic Induction at Work: Flashlights and Antiterrorism 664
ac generators from waterfalls 665
Electromagnetic Induction at Play: Hobbies and Transportation 666
dc motors 667
Transformers 669
Eddy currents in braking rapid-transit railcars 671
Electric energy transmission 672
Radiation pressure and space exploration 675
Power waves and electrical noise 675
Radio and TV waves 676
Microwaves 677
IR radiation: heat lamps and the greenhouse effect (bio) 677
Visible light and eyesight (bio) 677
UV light, ozone layer, sunburn and skin cancer (bio) 677
UV light and photogray sunglasses (bio) 677
X-rays, TV tube, medical applications and CT scans (bio) 678
Old-fashioned telephone operation (Ex. 20.11) 680
Microwave Ovens (Ex. 20.84) 684

Chapter 21
British versus U.S. electrical systems 689
Oscillator Circuits: Broadcasters of Electromagnetic Radiation 699
Resonance circuits and radio tuning 699
AM versus FM radio broadcasting 699

Chapter 22
How we see things 706
Diffuse reflection and seeing illuminated objects 707
A Dark, Rainy Night 709
The human eye: Refraction and wavelength (bio) 713
Mirages 714
Negative Index of Refraction and the "Perfect" Lens 715
Refraction and depth perception 716
Atmospheric effects 716
Diamond brilliance and fire 718
Fiber Optics: Medical Applications (bio) 720
Optical networking and information 720
Endoscopy and cardioscopes (bio) 720
Glass prisms 721
The Rainbow 722

Chapter 23
Coating of mirrors 730
Plane mirrors 730
Spherical mirrors 732
Store-monitoring diverging mirror 732
It's All Done with Mirrors 733
Spherical aberration of mirrors 740
Converging lenses 740
Diverging lenses 740
Fresnel Lenses 748
Combination of lenses 748
Lens power and optometry (bio) 751
Lens aberrations 752
Day–night rearview mirror 754
Backward lettering on emergency vehicles 754
Dual mirrors for driving 755
Compound-microscope geometry 758
Autocollimation 759

Chapter 24
Measuring the wavelength of light 762
Interference of oil and soap films 765
Peacock feathers (bio) 766
Optical flats 767
Newton's rings 767
Diffraction of water around natural barriers 768
Nonreflecting Lenses 768
Diffraction around a razor blade 769
Diffraction and radio reception 770
Diffraction gratings 772
Compact-disc and DVD diffraction 773
Spectrometers 773
X-ray diffraction 774
Polaroid™ and dichroism 776
Polarizing sunglasses and glare reduction (bio) 779
Glare reduction 780
Birefringent crystals 781
Optical activity and stress 781
LCDs and Polarized Light 783
The blue sky 783
Red sunrises and sunsets 784
Mars, the red planet 784
Optical Biopsy (bio) 785
TV interference 786

Chapter 25
The human eye (bio) 793
Simple cameras 793
Nearsightedness and corrective lenses (bio) 795
Farsightedness and corrective lenses (bio) 795
Cornea "Orthodontics" and Surgery (bio) 797
Bifocals (bio) 796
Astigmatism and corrective lenses (bio) 798
The magnifying glass (bio) 799
The compound microscope (bio) 801
Refracting telescopes 803
Prism binoculars 804
Reflecting telescopes 805
Hubble space telescope 807
Telescopes Using Nonvisual Radiation 808
Resolution of eye and telescope (bio) 809
Viewing the Great Wall of China—from space? 810
Oil-immersion lenses 810
Color vision (bio) 811
Paint and mixing of pigments (bio) 812
Photographic filters 812
"Red eye" in flash photos (bio) 814
A camera's f-stops 818

Chapter 26
Relativity and space travel 832
Relativity in Everyday Living 838
Gravitational lensing 839
Black holes 840
Black Holes, Gravitational Waves, and LIGO 842
Conventional and nuclear power plants (Ex. 26.79 and 26.81) 849

Chapter 27
Blackbody radiation, star color and temperature 853
Photocell applications, electric eyes and garage door safety 858
Photographic light meters 858
Solar-energy conversion 858
Fluorescence in nature and mineral detection 866
Lasers 866
Phosphorescent materials 867
Industrial lasers 868
CD and DVD Systems 869
Holography 869
Lasers in Modern Medicine (bio) 870

Chapter 28
Crystallography using electron diffraction 881
The Electron Microscope (bio) 883
The Scanning Tunneling Microscope (STM) 884
Magnetic Resonance Imaging (MRI) (bio) 888
Structure of the Elements, Chemistry, and the Periodic Chart 891
Molecular binding 893

Chapter 29
Bone scans (bio) 902
Damage by radiation (bio) 910
Radioactive half-lives and medicine (bio) 913
Thyroid treatment with I-131 (bio) 914
Radioactive dating (bio) 915
Carbon-14 dating of bones (bio) 917
Radiation detectors 923
Biological radiation hazards (bio) 924
Radiation dosage (bio) 924
Biological effects and medical applications of radiation exposure (bio) 925
Radiation Dosage for Thyroid Cancer Treatment (bio) 926
Biological and Medical Applications of Radiation (bio) 927
Radioactive tracers in medicine (bio) 927
SPET and PET (bio) 928
Domestic and industrial applications of radiation 928
Smoke detectors 928
Radioactive tracers in industry 928
Neutron activation in screening for bombs 929
Gamma radiation for sterilizing food (bio) 930

Chapter 30
Energy from fission: the power reactor 941
Energy from fission: the breeder reactor 942
Nuclear electrical generation 943
Nuclear-reactor safety 943
Fusion as an energy source 944
Energy from fusion: magnetic confinement 946
Energy from fusion: inertial confinement 946

PREFACE

We believe there are two basic goals in any introductory physics course: (1) to impart an understanding of the basic concepts of physics and (2) to enable students to use these concepts to solve a variety of problems.

These goals are linked. We want students to apply their conceptual understanding as they solve problems. Unfortunately, students often begin the problem-solving process by searching for an equation. There is the temptation to try to plug numbers into equations before visualizing the situation or considering the physical concepts that could be used to solve the problem. In addition, students often do not check their numerical answer to see if it matches their understanding of the relevant physical concept.

We feel, and users agree, that the strengths of this textbook are as follows:

Conceptual Basis. Giving students a secure grasp of physical principles will almost invariably enhance their problem-solving abilities. We have organized discussions and incorporated pedagogical tools to ensure that conceptual insight drives the development of practical skills.

Concise Coverage. To maintain a sharp focus on the essentials, we have avoided topics of marginal interest. We do not derive relationships when they shed no additional light on the principle involved. It is usually more important for students in this course to understand what a relationship means and how it can be used than to understand the mathematical or analytical techniques employed to derive it.

Applications. *College Physics* is known for the strong mix of applications related to medicine, science, technology, and everyday life in its text narrative and Insight boxes. While the Sixth Edition continues to have a wide range of applications, we have also increased the number of biological and biomedical applications, in recognition of the high percentage of premed and allied health majors who take this course. A complete list of applications, with page references, is found on pages X-XIII.

The Sixth Edition

While we worked to reduce the total number of pages in this edition, we have added material to further student understanding and make physics more relevant, interesting, and memorable for students.

Physics Facts. Each chapter begins with four to six interesting facts about discoveries or everyday phenomena applicable to the chapter.

Visual Summary. Each end-of-chapter Summary includes visual representations of the key concepts from the chapter to serve as a reminder for students as they review.

130 CHAPTER 4 Force and Motion

Chapter Review

- A **force** is something that is capable of changing an object's state of motion. To produce a change in motion, there must be a nonzero net, or unbalanced, force:

$$\vec{\mathbf{F}}_{net} = \Sigma\vec{\mathbf{F}}_i$$

- **Newton's first law of motion** is also called the *law of inertia,* where inertia is the natural tendency of an object to maintain its state of motion. It states that in the absence of a net applied force, a body at rest remains at rest, and a body in motion remains in motion with constant velocity.
- **Newton's second law** relates the net force acting on an object or system to the (total) mass and the resulting acceleration. It defines the cause-and-effect relationship between force and acceleration:

$$\Sigma\vec{\mathbf{F}}_i = \vec{\mathbf{F}}_{net} = m\vec{\mathbf{a}} \quad (4.1)$$

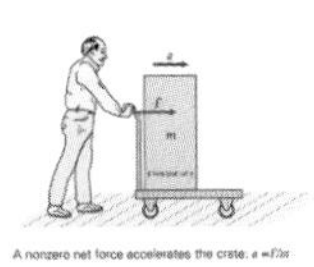

A nonzero net force accelerates the crate: $a = F/m$

The equation for **weight** in terms of mass is a form of Newton's second law:

$$w = mg \quad (4.2)$$

- **Newton's third law** states that for every force, there is an equal and opposite reaction force. The opposing forces of a third-law force pair always act on different objects.

- An object is said to be in **translational equilibrium** when it either is at rest or moves with a constant velocity. When remaining at rest, an object is said to be in *static translational equilibrium.* The condition for translational equilibrium is represented as

$$\Sigma\vec{\mathbf{F}}_i = 0 \quad (4.4)$$

or

$$\Sigma F_x = 0 \quad \text{and} \quad \Sigma F_y = 0 \quad (4.5)$$

- **Friction** is the resistance to motion that occurs between contacting surfaces. (In general, friction occurs for all types of media—solids, liquids, and gases.)

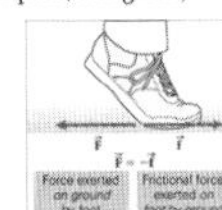

- The frictional force between surfaces is characterized by coefficients of friction (μ), one for the static case and one for the kinetic (moving) case. In many cases, $f = \mu N$, where N is the normal force—the force perpendicular to the surface (that is, the force exerted *by* the surface

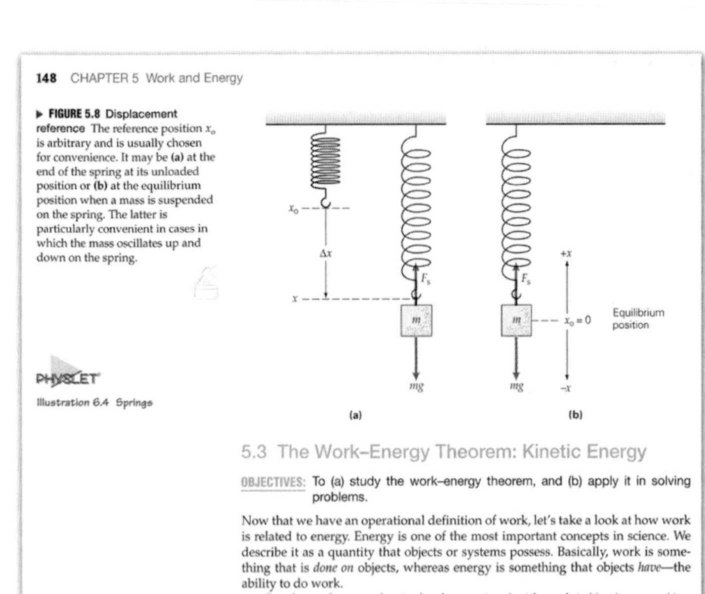
148 CHAPTER 5 Work and Energy

▶ **FIGURE 5.8 Displacement reference** The reference position x_o is arbitrary and is usually chosen for convenience. It may be **(a)** at the end of the spring at its unloaded position or **(b)** at the equilibrium position when a mass is suspended on the spring. The latter is particularly convenient in cases in which the mass oscillates up and down on the spring.

PHYSLET Illustration 6.4 Springs

5.3 The Work–Energy Theorem: Kinetic Energy

OBJECTIVES: To (a) study the work–energy theorem, and (b) apply it in solving problems.

Now that we have an operational definition of work, let's take a look at how work is related to energy. Energy is one of the most important concepts in science. We describe it as a quantity that objects or systems possess. Basically, work is something that is *done on* objects, whereas energy is something that objects *have*—the ability to do work.

One form of energy that is closely associated with work is *kinetic energy*. (Another form of energy, *potential energy*, will be described in Section 5.4.) Consider an object at rest on a frictionless surface. Let a horizontal force act on the object and set it in motion. Work is done *on* the object, but where does the work "go," so to speak? It goes into setting the object into motion, or changing its *kinetic* conditions. Because of its motion, we say the object has gained energy—kinetic energy, which gives it the capability to do work.

PHYSLET Exploration 6.1 An Operational Definition of Work

Integration of Physlet® Physics. Physlets are Java-based applets that illustrate physics concepts through animation. *Physlet Physics* is a best-selling book and CD-ROM containing more than 800 Physlets in three different formats: Physlet Illustrations, Physlet Explorations, and Physlet Problems. In the Sixth Edition of *College Physics*, the Physlets from *Physlet Physics* are denoted by an icon so students know when an alternate explanation and animation are available to further their understanding. The *Physlet Physics* CD-ROM is included with the purchase of a new textbook.

Biological Applications. We have greatly increased the number and scope of biological and biomedical applications. Examples of new biological applications include "People Power: Using Body Energy," "Osteoporosis and Bone Mineral Density (BMD)," and the Magnetic Force in Future Medicine.

We have enhanced the following pedagogical features in the Sixth Edition:

116 CHAPTER 4 Force and Motion

PHYSLET Illustration 4.2 Free-Body Diagrams

LEARN BY DRAWING
Forces on an Object on an Inclined Plane and Free-body Diagrams

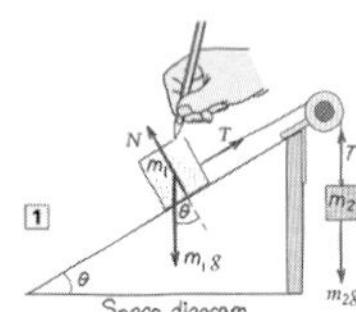

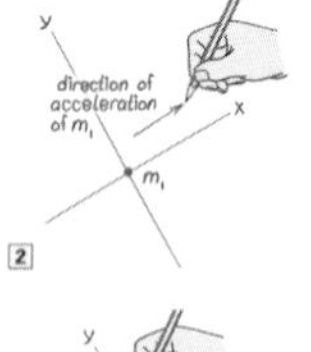

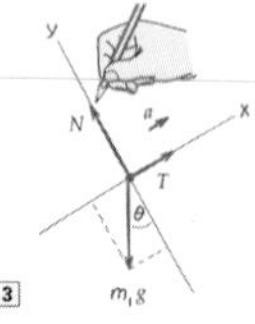

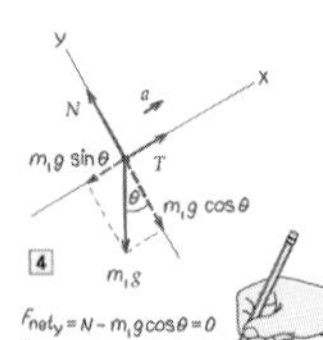

4.5 More on Newton's Laws: Free-Body Diagrams and Translational Equilibrium

OBJECTIVES: To (a) apply Newton's laws in analyzing various situations using free-body diagrams, and (b) understand the concept of translational equilibrium.

Now that you have been introduced to Newton's laws and some applications in analyzing motion, the importance of these laws should be evident. They are so simply stated, yet so far-reaching. The second law is probably the most often applied, because of its mathematical relationship. However, the first and third laws are often used in qualitative analysis, as our continuing study of the different areas of physics will reveal.

In general, we will be concerned with applications that involve constant forces. Constant forces result in constant accelerations and allow the use of the kinematic equations from Chapter 2 in analyzing the motion. When there is a variable force, Newton's second law holds for the *instantaneous* force and acceleration, but the acceleration will vary with time, requiring advanced mathematics to analyze. We will generally limit ourselves to constant accelerations and forces. Several examples of applications of Newton's second law are presented in this section so that you can become familiar with its use. This small but powerful equation will be used again and again throughout this text.

There is still one more item in the problem-solving arsenal that is a great help with force applications—free-body diagrams. These are explained in the following Problem-Solving Strategy.

Problem-Solving Strategy: Free-Body Diagrams

In illustrations of physical situations, sometimes called *space diagrams*, force vectors may be drawn at different locations to indicate their points of application. However, because we are presently concerned only with linear motions, vectors in *free-body diagrams* (FBD) may be shown as emanating from a common point, which is usually chosen as the origin of the x–y axes. One of the axes is generally chosen along the direction of the net force acting on a body, since that is the direction in which the body will accelerate. Also, it is often important to resolve force vectors into components, and properly chosen x–y axes simplify this task.

In a free-body diagram, the vector arrows do not have to be drawn exactly to scale. However, the diagram should clearly show whether there is a net force and whether forces balance each other in a particular direction. When the forces aren't balanced, we know from Newton's second law that there must be an acceleration.

In summary, the general steps in constructing and using free-body diagrams are as follows (refer to the accompanying Learn by Drawing as you read):

1. Make a sketch, or space diagram, of the situation (if one is not already available) and identify the forces acting on each body of the system. A space diagram is an illustration of the physical situation that identifies the force vectors.
2. Isolate the body for which the free-body diagram is to be constructed. Draw a set of Cartesian axes, with the origin at a point through which the forces act and with one of the axes along the direction of the body's acceleration. (The acceleration will be in the direction of the net force, if there is one.)
3. Draw properly oriented force vectors (including angles) on the diagram, emanating from the origin of the axes. If there is an unbalanced force, assume a direction of acceleration and indicate it with an acceleration vector. Be sure to include only those forces that act on the isolated body of interest.
4. Resolve any forces that are not directed along the x- or y-axis into x- or y-components (use plus and minus signs to indicate direction). Use the free-body diagram and force components to analyze the situation in terms of Newton's second law of motion. (*Note*: If you assume that the acceleration is in one direction, and in the solution it comes out with the opposite sign, then the acceleration is actually in the opposite direction from that assumed. For example, if you assume that $\vec{a}$ is in the $+x$-direction, but you get a negative answer, then $\vec{a}$ is in the $-x$-direction.)

Learn by Drawing Boxes. Visualization is one of the most important steps in problem solving. In many cases, if students can make a sketch of a problem, they can solve it. "Learn by Drawing" features offer students specific help on making certain types of sketches and graphs that will provide key insights into a variety of physical situations.

Suggested Problem-Solving Procedure. Section 1.7 provides a framework for thinking about problem solving. This section includes the following:

- An overview of problem-solving strategies
- A six-step procedure that is general enough to apply to most problems in physics, but is easily used in specific situations
- Examples that illustrate the detailed problem-solving process, showing how the general procedure is applied in practice

Problem-Solving Strategies and Hints. The initial treatment of problem solving is followed throughout with an abundance of suggestions, tips, cautions, shortcuts, and useful techniques for solving specific kinds of problems. These strategies and hints help students apply general principles to specific contexts as well as avoid common pitfalls and misunderstandings.

Conceptual Examples. These Examples ask students to think about a physical situation and conceptually solve a question or choose the correct prediction out of a set of possible outcomes, on the basis of an understanding of relevant principles. The discussion that follows ("Reasoning and Answer") explains clearly how the correct answer can be identified, as well as why the other answers are wrong.

Worked Examples. We have tried to make in-text Examples as clear and detailed as possible. The aim is not merely to show students which equations to use, but to explain the strategy being employed and the role of each step in the overall plan. Students are encouraged to learn the "why" of each step along with the "how." Our goal is to provide a model for students to use as they solve problems. Each worked Example includes the following:

- ***Thinking It Through*** focuses students on the critical thinking and analysis they should undertake before beginning to use equations.
- ***Given*** and ***Find*** are provided as the first part of every ***Solution*** to remind students the importance of identifying what is known and what needs to be solved.
- ***Follow-Up Exercises*** at the end of each Conceptual Example and each worked Example further reinforce the importance of conceptual understanding and offer additional practice. (Answers to Follow-Up Exercises are given at the back of the text.)

Integrated Examples. In order to further emphasize the connection between conceptual understanding and quantitative problem solving, we have developed Integrated Examples for each chapter. These Examples work through a physical situation both qualitatively and quantitatively. The qualitative portion is solved by conceptually choosing the correct answer from a set of possible answers. The quantitative portion involves a mathematical solution related to the conceptual part, demonstrating how conceptual understanding and numerical calculations go hand in hand.

End-of-Chapter Exercises. Each section of the end-of-chapter material begins with multiple-choice questions (**MC**) to allow students a quick self-test for that section. These are followed by short-answer conceptual questions (CQ) that test students' conceptual understanding and ask students to reason from principles. Quantitative problems round out the Exercises in each section. *College Physics* provides short answers to all odd-numbered Exercises (quantitative *and* conceptual) at the back of the text, so students can check their understanding.

Paired Exercises. To encourage students to work problems on their own, most sections include at least one set of Paired Exercises that deal with similar situations. The first problem in a pair is solved in the *Student Study Guide and Solutions Manuals*; the second problem, which explores a similar situation to that presented in the first problem, has only an answer at the back of the book.

Integrated Exercises. Like the Integrated Examples in the chapter, Integrated Exercises (IE) ask students to solve a problem quantitatively as well as answer a conceptual question dealing with the exercise. By answering both parts, students can see if their numerical answer matches their conceptual understanding.

Comprehensive Exercises. To ensure that students can synthesize concepts, each chapter concludes with a section of comprehensive exercises drawn from all sections of the chapter and perhaps basic principles from previous chapters.

Absolutely Zero Tolerance for Errors Club (AZTEC)

We have continued to ensure accuracy through the Absolutely Zero Tolerance for Errors Club (AZTEC). Bo Lou of Ferris State University, new co-author of the text and author of our *Instructor's Solutions Manual* and *Student Study Guide and Solutions Manuals,* headed the AZTEC team and was supported by the text's co-authors as well as Billy Younger, College of the Albemarle; Mike LoPresto, Henry Ford Community College; David Curott, University of North Alabama; and Daniel Lottis, Panconsult Ciências Físicas e Naturais, Curitiba, Brazil. Every end-of-chapter Exercise was worked by three of the five team members individually and independently. The results were collected and discrepancies were resolved by a team discussion. While there has never been a text that was absolutely free of errors, that was our goal; we worked very hard to make the book error-free.

The Sixth Edition is supplemented by a media and print ancillary package developed to address the needs of both students and instructors.

For the Instructor

***Annotated Instructor's Edition* (0-13-149706-5).** The margins of the *Annotated Instructor's Edition* (*AIE*) contain an abundance of suggestions for classroom demonstrations and activities, along with teaching tips (points to emphasize, discussion suggestions, and common misunderstandings to avoid). In addition, the *AIE* contains the following:

- Icons that identify each illustration reproduced as a transparency in the *Transparency Pack*
- Answers to end-of-chapter exercises (following each exercise)

***Instructor's Resource Manual and Instructor Notes on ConcepTest Questions* (0-13-149711-1).** This manual has two parts. The first part, prepared by Katherine Whatley and Judith Beck (both of the University of The North Carolina–Asheville), contains sample syllabi, lecture outlines, notes, demonstration suggestions, readings, and additional references and resources. The second part, prepared by Cornelius Bennhold and Gerald Feldman (both of The George Washington University) contains an overview of the development and implementation of ConcepTests, as well as instructor notes for each ConcepTest found on the *Instructor Resource Center on CD-ROM.*

***Instructor's Solutions Manual* (0-13-149710-3).** Prepared by Bo Lou of Ferris State University, the *Instructor's Solutions Manual* supplies answers with complete, worked-out solutions to all end-of-chapter exercises. This manual is also available electronically on the *Instructor Resource Center on CD-ROM.*

***Test Item File* (0-13-149713-8).** Fully revised by Delena Gatch (University of North Alabama), the *Test Item File* offers more than 2800 multiple-choice, true/false, and short-answer/essay questions, approximately 50% conceptual. The questions are organized and referenced by chapter section and by question type. This manual is also available both within TestGenerator and as chapter-by-chapter Microsoft Word files on the *Instructor Resource Center on CD-ROM* (see below).

***Transparency Pack* (0-13-149708-1).** The *Transparency Pack* contains more than 400 full-color acetates of text illustrations.

***"Physics You Can See" Video Demonstrations* (0-205-12393-7).** Each segment, two to five minutes long, demonstrates a classical physics experiment. Eleven segments are included, such as "Coin & Feather" (acceleration due to gravity), "Monkey & Gun" (rate of vertical free fall), "Swivel Hips" (force pairs), and "Collapse a Can" (atmospheric pressure). Digital versions of the videos are available on the *Instructor Resource Center on CD-ROM.*

***Instructor Resource Center on CD-ROM* (0-13-149712-X).** This multiple CD-ROM set, new to the Sixth Edition, provides virtually every electronic asset you'll need in and out of the classroom. Though you can navigate freely through the CDs to find the resources you want, the software allows you to browse and search through the catalog of assets. The CD-ROMs are organized by chapter and include all text illustrations and tables from the Sixth Edition in JPEG and PowerPoint formats. The IRC/CDs also contain *TestGenerator*, a powerful dual-platform, fully networkable software program for creating tests ranging from short quizzes to long exams. Questions from the Sixth Edition *Test Item File*, including randomized versions, are supplied, and professors can use the Question Editor to modify existing questions or create new questions. The IRC/CDs also contain the instructor's version of *Physlet Physics*, chapter-by-chapter lecture outlines in PowerPoint, *ConcepTest*

"Clicker" Questions in PowerPoint, chapter-by-chapter Microsoft Word files of all numbered equations, the eleven *Physics You Can See* demonstration videos, and Microsoft Word and PDF versions of the *Test Item File*, the *Instructor's Solutions Manual*, the *Instructor's Resource Manual*, and the end-of-chapter exercises from *College Physics*, Sixth Edition.

***Peer Instruction* (0-13-656441-6)**. Authored by Eric Mazur (Harvard University), this manual explains an interactive teaching style that actively involves students in the learning process by focusing their attention on underlying concepts through interactive "ConcepTests," reading quizzes, and conceptual exam questions.

***Just-in-Time Teaching: Blending Active Learning with Web Technology* (0-13-085034-9).** Gregor Novak (Indiana University–Purdue University, Indianapolis), Evelyn Patterson (United States Air Force Academy), Andrew Gavrin (Indiana University–Purdue University, Indianapolis), and Wolfgang Christian (Davidson College) authored this resource book for educators. Just-in-time teaching (JITT) is a teaching and learning methodology designed to engage students. Using feedback from preclass Web assignments, instructors can adjust classroom lessons so that students receive rapid response to the specific questions and problems they are having.

***Physlets: Teaching Physics with Interactive Curricular Material* (0-13-029341-5).** Authored by Wolfgang Christian and Mario Belloni (both of Davidson College), this text is a teacher's resource book with an accompanying CD for instructors who are interested in incorporating Physlets into their physics courses. The book and CD discuss the pedagogy behind the use of Physlets.

For the Student

***Student Study Guide and Selected Solutions Manual* (Volume 1: 0-13-149716-2, Volume 2: 0-13-174405-4).** Written by Bo Lou of Ferris State University, the *Student Study Guide and Selected Solutions Manual* presents chapter-by-chapter reviews, chapter summaries, additional worked examples, practice quizzes, and solutions to paired and selected exercises.

***Student Pocket Guide* (0-13-149718-9).** Written by Biman Das (State University of New York–Potsdam), this easy-to-carry 5" x 7" paperback contains a summary of the entire text, including all key concepts and equations, as well as tips and hints.

***E&M TIPERs: Electricity & Magnetism Tasks Inspired by Physics Education Research* (0-13-185499-2).** Curtis J. Hieggelke (Joliet Junior College), David P. Maloney (Indiana University Purdue University Fort Wayne), Stephen E. Kanim (New Mexico State University), and Thomas L. O'Kuma (Lee College) created this comprehensive set of conceptual exercises for electricity and magnetism based on the results of education research into how students learn physics. This workbook contains more than 300 tasks in eleven different task formats.

***Tutorials in Introductory Physics* (0-13-097069-7).** Written by Lillian C. McDermott, Peter S. Schaffer, and the Physics Education Group at the University of Washington, this landmark book presents a series of physics tutorials designed by a leading physics education research group. Emphasizing the development of concepts and scientific reasoning skills, the tutorials focus on the specific conceptual and reasoning difficulties that students tend to encounter. The tutorials cover a range of topics in Mechanics, E & M, and Waves and Optics.

***Ranking Task Exercises in Physics: Student Edition* (0-13-144851-X).** Developed by Thomas L. O'Kuma (Lee College), David P. Maloney (Indiana University–Purdue University, Fort Wayne), and Curtis J. Hieggelke (Joliet Junior College), ranking tasks are an innovative type of conceptual exercises that ask students to make comparative judgments about a set of variations on a particular physical situation. This text is a unique resource for physics instructors who are looking for tools to incorporate more conceptual analysis in their courses. This supplement contains 218 ranking task exercises that cover all classical physics topics.

***Interactive Physics Player Workbook*, Second Edition (0-13-067108-8).** Created by Cindy Schwarz (Vassar College), John P. Ertel (U.S. Naval Academy), and MSC Software, this interactive workbook, tutorial-oriented worksheets, and CD-ROM package are designed to help students visualize and work with specific physics problems through simulations created with Interactive Physics files. Forty problems of varying degrees of difficulty require students to make predictions, change variables, run, and visualize motion on the computer. The accompanying workbook/study guide provides instructions, physics review, hints, and questions. The CD-ROM contains everything students need to run the simulations.

***Mathematics for College Physics* (0-13-141427-5).** Written by Biman Das (SUNY–Potsdam), this text, for students who need help with the necessary mathematical tools, shows how mathematics is directly applied to physics and discusses how to overcome math anxiety.

***MCAT Physics Study Guide* (0–13–627951–1).** Because most MCAT questions require more thought and reasoning than simply plugging numbers into an equation, this study guide, by Joseph Boone (California Polytechnic State University–San Luis Obispo), is designed to refresh students' memory about the topics they've covered in class. Additional review, practice problems, and review questions are included.

Companion Website with GradeTracker (http://www.prenhall.com/wilson)

This text-specific Web site provides students and instructors a wealth of innovative online materials for use with *College Physics*, Sixth Edition. The Companion Website with GradeTracker includes the following:

- Integration of Just-in-Time Teaching (JiTT) Warm-Ups, Puzzles, & Applications by Gregor Novak and Andrew Gavrin (Indiana University–Purdue University, Indianapolis): Warm-Up questions are short-answer questions based on important concepts in the text chapters. Puzzles are more complex questions and often require the integration of more than one concept. Thus professors can assign Warm-Up questions as a reading quiz before the class lecture on that topic, and Puzzle questions as follow-up assignments submitted after class. The Applications modules answer the question, "What is physics good for?" by connecting physics concepts to real-world phenomena and new developments in science and technology. Each Application module contains short-answer/essay questions.
- Practice Questions: A module of twenty to thirty gradable multiple-choice practice questions are available for review with each chapter.
- Ranking Task Exercises, edited by Thomas O'Kuma (Lee College), David Maloney (Indiana University–Purdue University, Fort Wayne), and Curtis Hieggelke (Joliet Junior College): Ranking tasks are gradable conceptual exercises that require students to rank a number of situations or variations of a situation.
- *Physlet Physics* Problems: Physlets are Java-based applets that illustrate physics concepts through animation. *Physlet Physics* is a best-selling book and CD-ROM containing more than 800 Physlets. Physlet problems are interactive versions of the kind of exercises typically assigned for homework. Gradable problems from *Physlet Physics* are available for students to self-test. To access the Physlets, students use their copy of the *Physlet Physics* CD-ROM, included with the purchase of a new copy of this text.
- *MCAT Study Guide*, by Kaplan Test Prep and Admissions: The MCAT Study Guide module provides students with ten quizzes on topics and concepts covered on the MCAT exam.

Blackboard. Blackboard is a comprehensive and flexible software platform that delivers a course management system, customizable institution-wide portals, online communities, and an advanced architecture that allows for Web-based integration of multiple administrative systems. Its features include the following:

- Progress tracking, class and student management, grade book, communication, assignments, and reporting tools.
- Testing programs that enable instructors to create online quizzes and tests from the *College Physics*, Sixth Edition course content and automatically grade and track results. All tests can be entered into the grade book for easy course management.
- Communications tools such as the Virtual Classroom (chat rooms, whiteboard, slides), document sharing, and bulletin boards.

CourseCompass, Powered by Blackboard. With the highest level of service, support, and training available today, CourseCompass combines tested, quality online course resources for *College Physics*, Sixth Edition with easy-to-use online course management tools. CourseCompass is designed to address the individual needs of instructors, who will be able to create an online course without any special technical skills or training. Its features include the following:

- Higher flexibility—Professors can adapt Prentice Hall content to match their own teaching goals, with little or no outside assistance needed.
- Assessment, customization, class administration, and communication tools.
- Point-and-click access—*College Physics*, Sixth Edition course resources are available to instructors at a click of the mouse.
- A nationally hosted and fully supported system that relieves individuals and institutions of the burdens of troubleshooting and maintenance.

WebCT. WebCT offers a powerful set of tools that enables instructors to create practical Web-based educational programs—ideal resources to enhance a campus course or to construct one entirely online. The WebCT shell and tools, integrated with the *College Physics*, Sixth Edition content, results in a versatile, course-enhancing teaching and learning system. Its features include the following:

- Page tracking, progress tracking, class and student management, grade book, communication, calendar, and reporting tools.
- Communication tools including chat rooms, bulletin boards, private e-mail, and whiteboard.
- Testing tools that help create and administer timed online quizzes and tests and automatically grade and track all results.

***WebAssign* (http://www.webassign.com).** WebAssign's homework delivery service gives you the freedom to get back to create assignments from a database of exercises from *College Physics*, Sixth Edition, or write and customize your own exercises. You have complete control over the homework your students receive, including due date, content, feedback, and question formats. Its features include the following:

- Create, post, and review assignments twenty-four hours a day, seven days a week.
- Deliver, collect, grade, and record assignments instantly.
- Offer more practice exercises, quizzes, homework, labs, and tests.
- Randomize numerical values or phrases to create unique questions.
- Assess student performance to keep abreast of individual progress.
- Grade algebraic formulas with enhanced math-type display.
- Capture the attention of your online and distance-learning students.

Acknowledgments

The members of AZTEC—Billy Younger, Michael LoPresto, David Curott, and Daniel Lottis—as well as accuracy reviewers Michael Ottinger, and Mark Sprague deserve more than a special thanks for their tireless, timely, and extremely thorough review of this book.

Dozens of other colleagues, listed in the upcoming section, helped us identify ways to make the sixth edition a better learning tool for students. We are indebted to them, as their thoughtful and constructive suggestions benefited the book greatly.

We owe many thanks to the editorial and production team at Prentice Hall including Erik Fahlgren, Senior Acquisitions Editor, Heather Scott, Art Director, Christian Botting, Associate Editor, and Jessica Berta, Editorial Assistant. In particular, the authors wish to acknowledge the outstanding performance of Simone Lukashov, Production Editor. His courteous, coscientious, and cheerful manner made for an efficient and enjoyable production process. Also, we thank Karen Karlin, Prentice Hall Development Editor, for her helpful assistance and editing.

In addition, I (Tony Buffa) once again extend many thanks to my co-authors, Jerry Wilson and Bo Lou, for their cheerful helpfulness and professional approach to the work on this edition. As always, several colleagues of mine at Cal Poly gave of their time for fruitful discussions. Among them are Professors Joseph Boone, Ronald Brown, and Theodore Foster. My family—my wife, Connie, and daughters, Jeanne and Julie—was, as always, a continuous and welcomed source of support. I also acknowledge the support from my father, Anthony Buffa, Sr., and my aunt, Dorothy Abbott. Last, I thank the students in my classes who contributed excellent ideas over the past few years.

Finally, we would like to urge anyone using the book—student or instructor—to pass on to us any suggestions that you have for its improvement. We look forward to hearing from you.

—Jerry D. Wilson
jwilson@greenwood.net
—Anthony J. Buffa
abuffa@calpoly.edu
—Bo Lou
loub@ferris.edu

Reviewers of the Sixth Edition:

David Aaron
South Dakota State University

E. Daniel Akpanumoh
Houston Community College, Southwest

Ifran Azeem
Embry-Riddle Aeronautical University

Raymond D. Benge
Tarrant County College

Frederick Bingham
University of North Carolina, Wilmington

Timothy C. Black
University of North Carolina, Wilmington

Mary Boleware
Jones County Junior College

Art Braundmeier
Southern Illinois University, Edwardsville

Michael L. Broyles
Collin County Community College

Debra L. Burris
Oklahoma City Community College

Jason Donav
University of Puget Sound

Robert M. Drosd
Portland Community College

Bruce Emerson
Central Oregon Community College

Milton W. Ferguson
Norfolk State University

Phillip Gilmour
Tri-County Technical College

Allen Grommet
East Arkansas Community College

Brian Hinderliter
North Dakota State University

Ben Yu-Kuang Hu
University of Akron

Porter Johnson
Illinois Institute of Technology

Andrew W. Kerr
University of Findlay

Jim Ketter
Linn-Benton Community College

Terrence Maher
Alamance Community College

Kevin McKone
Copiah Lincoln Community College

Kenneth L. Menningen
University of Wisconsin, Stevens Point

Michael Mikhaiel
Passaic County Community College

Ramesh C. Misra
Minnesota State University, Mankato

Sandra Moffet
Linn Benton Community College

Michael Ottinger
Missouri Western State College

James Palmer
University of Toledo

Kent J. Price
Morehead State University

Salvatore J. Rodano
Harford Community College

John B. Ross
Indiana University-Purdue University, Indianapolis

Terry Scott
University of Northern Colorado

Rahim Setoodeh
Milwaukee Area Technical College

Martin Shingler
Lakeland Community College

Mark Sprague
East Carolina State University

Steven M. Stinnett
McNeese State University

John Underwood
Austin Community College

Tristan T. Utschig
Lewis-Clark State College

Steven P. Wells
Louisiana Technical University

Christopher White
Illinois Institute of Technology

Anthony Zable
Portland Community College

John Zelinsky
Community College of Baltimore County, Essex

Reviewers of Previous Editions

William Achor
Western Maryland College

Alice Hawthorne Allen
Virginia Tech

Arthur Alt
College of Great Falls

Zaven Altounian
McGill University

Frederick Anderson
University of Vermont

Charles Bacon
Ferris State College

Ali Badakhshan
University of Northern Iowa

Anand Batra
Howard University

Michael Berger
Indiana University

William Berres
Wayne State University

James Borgardt
Juniata College

Hugo Borja
Macomb Community College

Bennet Brabson
Indiana University

Jeffrey Braun
University of Evansville

Michael Browne
University of Idaho

David Bushnell
Northern Illinois University

Lyle Campbell
Oklahoma Christian University

James Carroll
Eastern Michigan State University

Aaron Chesir
Lucent Technologies

Lowell Christensen
American River College

Philip A. Chute
University of Wisconsin–Eau Claire

Robert Coakley
University of Southern Maine

Lawrence Coleman
University of California–Davis

Lattie F. Collins
East Tennessee State University

Sergio Conetti
University of Virginia, Charlottesville

James Cook
Middle Tennessee State University

David M. Cordes
Belleville Area Community College

James R. Crawford
Southwest Texas State University

William Dabby
Edison Community College

Purna Das
Purdue University

J. P. Davidson
University of Kansas

Donald Day
Montgomery College

Richard Delaney
College of Aeronautics

James Ellingson
College of DuPage

Donald Elliott
Carroll College

Arnold Feldman
University of Hawaii

John Flaherty
Yuba College

Rober J. Foley
University of Wisconsin–Stout

Lewis Ford
Texas A&M University

Donald Foster
Wichita State University

Donald R. Franceschetti
Memphis State University

Frank Gaev
ITT Technical Institute–Ft. Lauderdale

Rex Gandy
Auburn University

Simon George
California State–Long Beach

Barry Gilbert
Rhode Island College

Richard Grahm
Ricks College

Tom J. Gray
University of Nebraska

Douglas Al Harrington
Northeastern State University

Gary Hastings
Georgia State University

Xiaochun He
Georgia State University

J. Erik Hendrickson
University of Wisconsin–Eau Claire

Al Hilgendorf
University of Wisconsin–Stout

Joseph M. Hoffman
Frostburg State University

Andy Hollerman
University of Louisiana, Layfayette

Jacob W. Huang
Towson University

Randall Jones
Loyola University

Omar Ahmad Karim
University of North Carolina–Wilmington

S. D. Kaviani
El Camino College

Victor Keh
ITT Technical Institute–Norwalk, California

John Kenny
Bradley University

James Kettler
Ohio University, Eastern Campus

Dana Klinck
Hillsborough Community College

Chantana Lane
University of Tennessee–Chattanooga

Phillip Laroe
Carroll College

Rubin Laudan
Oregon State University

Bruce A. Layton
Mississippi Gulf Coast Community College

R. Gary Layton
Northern Arizona University

Kevin Lee
University of Nebraska

Paul Lee
California State University, Northridge

Federic Liebrand
Walla Walla College

Mark Lindsay
University of Louisville

Bryan Long
Columbia State Community College

Michael LoPresto
Henry Ford Community College

Dan MacIsaac
Northern Arizona University

Robert March
University of Wisconsin

Trecia Markes
University of Nebraska–Kearney

Aaron McAlexander
Central Piedmont Community College

William McCorkle
West Liberty State University

John D. McCullen
University of Arizona

Michael McGie
California State University–Chico

Paul Morris
Abilene Christian University

Gary Motta
Lassen College

J. Ronald Mowrey
Harrisburg Area Community College

Gerhard Muller
University of Rhode Island

K. W. Nicholson
Central Alabama Community College

Erin O'Connor
Allan Hancock College

Anthony Pitucco
Glendale Community College

William Pollard
Valdosta State University

R. Daryl Pedigo
Austin Community College

T. A. K. Pillai
University of Wisconsin–La Crosse

Darden Powers
Baylor University

Donald S. Presel
University of Massachusetts–Dartmouth

E. W. Prohofsky
Purdue University

Dan R. Quisenberry
Mercer University

W. Steve Quon
Ventura College

David Rafaelle
Glendale Community College

George Rainey
California State Polytechnic University

Michael Ram
SUNY–Buffalo

William Riley
Ohio State University

William Rolnick
Wayne State University

Robert Ross
University of Detroit–Mercy

Craig Rottman
North Dakota State University

Gerald Royce
Mary Washington College

Roy Rubins
University of Texas, Arlington

Sid Rudolph
University of Utah

Om Rustgi
Buffalo State College

Anne Schmiedekamp
Pennsylvania State University–Ogontz

Cindy Schwarz
Vassar College

Ray Sears
University of North Texas

Mark Semon
Bates College

Bartlett Sheinberg
Houston Community College

Jerry Shi
Pasadena City College

Peter Shull
Oklahoma State University

Thomas Sills
Wilbur Wright College

Larry Silva
Appalachian State University

Michael Simon
Housatonic Community Technical College

Christopher Sirola
Tri-County Technical College

Gene Skluzacek
St. Petersburg College

Soren P. Sorensen
University of Tennessee–Knoxville

Ross Spencer
Brigham Young University

Dennis W. Suchecki
San Diego Mesa College

Frederick J. Thomas
Sinclair Community College

Jacqueline Thornton
St. Petersburg Junior College

Anthony Trippe
ITT Technical Institute–San Diego

Gabriel Umerah
Florida Community College–Jacksonville

Lorin Vant-Hull
University of Houston

Pieter B. Visscher
University of Alabama

Karl Vogler
Northern Kentucky University

John Walkup
California Polytechnic State University

Arthur J. Ward
Nashville State Technical Institute

Larry Weinstein
Old Dominion University

John C. Wells
Tennessee Technical University

Arthur Wiggins
Oakland Community College

Kevin Williams
ITT Technical Institute–Earth City

Linda Winkler
Appalachian State University

Jeffery Wragg
College of Charleston

Rob Wylie
Carl Albert State University

John Zelinsky
Southern Illinois University

Dean Zollman
Kansas State University

CHAPTER 15

ELECTRIC CHARGE, FORCES, AND FIELDS

15.1 Electric Charge 506

15.2 Electrostatic Charging 508

15.3 Electric Force 512

15.4 Electric Field 517

15.5 Conductors and Electric Fields 526

*15.6 Gauss's Law for Electric Fields: A Qualitative Approach 528

PHYSICS FACTS

- Charles Augustin de Coulomb (1736–1806), a French scientist and the discoverer of the force law between charged objects, had a diverse career. In addition, he made significant contributions in hospital reform, the cleanup of the Parisian water supply, Earth magnetism, soils engineering, and the construction of forts, the latter two while he served in the military.
- The Taser stun gun, as used by law enforcement agencies, works by generating a large electric charge separation and applying it to parts of the body, disrupting the normal electrical signals and causing temporary incapacity. The stun gun needs to physically contact the body with its two electrodes, and the shock can be delivered even through thick clothing. A long-distance version of the Taser works by firing barbed electrodes with trailing wires.
- The electric eel (which can grow up to 6 feet in length and is actually a fish) acts electrically in a similar way to a Taser. More than 80% of the eel's body is tail, with its vital organs located behind its small head. It uses the electric field it creates for both locating prey and stunning them before eating.
- Home air purifiers use the electric force to reduce dust, bacteria, and other particulates in the air. The electric force removes electrons from the pollutants, making them positively charged. These particles are attracted to negatively charged plates, where they stay until manually removed. When working properly, these purifiers can reduce the particulate level by more than 99%.

Few natural processes deliver such an enormous amount of energy in a fraction of a second as a lightning bolt. Yet most people have never experienced its power at close range; luckily, only a few hundred people per year are struck by lightning in the United States.

It might surprise you to realize that you have almost certainly had a similar experience, at least in a physics context. Have you ever walked across a carpeted room and gotten a shock when you reached for a metallic doorknob? Although the scale is dramatically different, the physical process involved (static electricity discharge) is much the same as being struck by lightning—mini-lightning, so to speak.

Electricity sometimes gives rise to dramatic effects such as sparking electrical outlets or lightning strikes. We know that electricity can sometimes be dangerous, but we also know that electricity can be "domesticated." In the home or office, its usefulness is taken for granted. Indeed, our dependence on electric energy becomes evident only when the power goes off unexpectedly, providing a dramatic reminder of the role that it plays in our daily lives. Yet less than a century ago there were no power lines crossing the country, no electric lights or appliances—none of the electrical applications that are all around us today.

Physicists now know that the electric force is related to the magnetic force (see Chapter 20). Together they are called the "electromagnetic force," which is one of the four fundamental forces in nature. (Gravity [Chapter 7] and two types of short-range nuclear forces discussed in Chapters 29 and 30 are the other three.) We begin here by studying the electric force and its properties. Eventually (Chapter 20) the magnetic and electric forces will be interconnected.

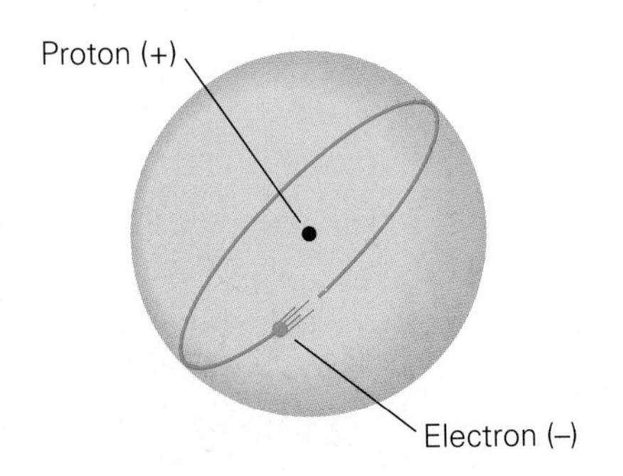

(a) Hydrogen atom

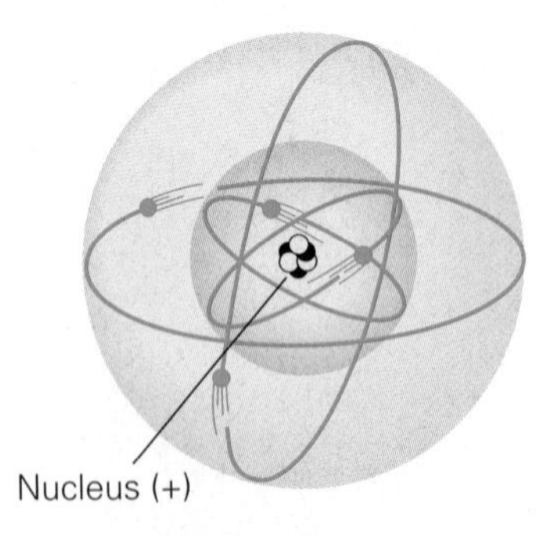

(b) Beryllium atom

▲ **FIGURE 15.1 Simplistic model of atoms** The so-called solar system model of **(a)** a hydrogen atom and **(b)** a beryllium atom views the electrons (negatively charged) as orbiting the nucleus (positively charged), analogously to the planets orbiting the Sun. The electronic structure of atoms is actually much more complicated than this.

15.1 Electric Charge

OBJECTIVES: To (a) distinguish between the two types of electric charge, (b) state the charge–force law that operates between charged objects, and (c) understand and use the law of charge conservation.

What is *electricity*? One simple answer is that it is a term describing phenomena associated with the electricity we have in our homes. But fundamentally it really involves the study of the interaction between *electrically charged* objects. To demonstrate this, our study will start with the simplest situation, electro*statics*, when electrically charged objects are *at rest*.

Like mass, **electric charge** is a fundamental property of matter (Chapter 1). Electric charge is associated with particles that make up the atom: the electron and the proton. The simplistic solar system model of the atom, as illustrated in ◂Fig. 15.1, likens its structure to that of the planets orbiting the Sun. The *electrons* are viewed as orbiting a nucleus, a core containing most of the atom's mass in the form of *protons* and electrically neutral particles called *neutrons*. As seen in Section 7.5, the centripetal force that keeps the planets in orbit about the Sun is supplied by gravity. Similarly, the force that keeps the electrons in orbit around the nucleus is the electrical force. However, there are important distinctions between gravitational and electrical forces.

One difference is that there is only one type of mass and gravitational forces are only attractive. Electric charge, however, comes in two types, distinguished by the labels positive (+) and negative (−). Protons carry a positive charge, and electrons carry a negative charge. Different combinations of the two types of charge can produce *either* attractive *or* repulsive electrical forces.

The directions of the electric forces when charges interact with one another are given by the following principle, called the **law of charges** or the **charge–force law**:

Like charges repel, and unlike charges attract.

That is, two negatively charged particles or two positively charged particles repel each other, whereas particles with opposite charges attract each other (▾Fig. 15.2). The repulsive and attractive forces are equal and opposite, and act on different objects, in keeping with Newton's third law (action–reaction).

Note: Recall the discussion of Newton's third law in Section 4.4.

The charge on an electron and that on a proton are equal in magnitude, but opposite in sign. The magnitude of the charge on an electron is abbreviated as e and is the fundamental unit of charge, because it is the smallest charge observed in nature.* The SI unit of charge is the **coulomb (C)**, named for the French physicist/engineer Charles A. de Coulomb (1736–1806), who discovered a relationship between electric force and charge (Section 15.3). The charges and masses of the electron, proton, and

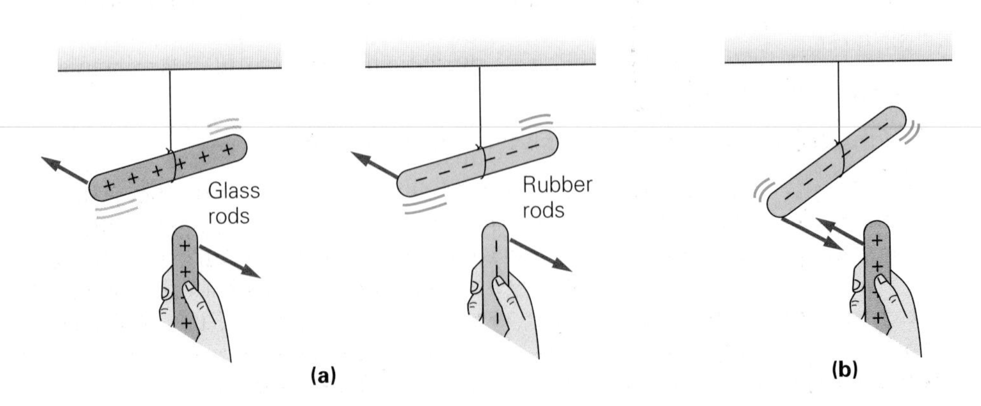

▸ **FIGURE 15.2 The charge–force law, or law of charges (a)** Like charges repel. **(b)** Unlike charges attract.

*Protons, as well as neutrons and other particles, are now known to be made up of more fundamental particles called *quarks*, which carry charges of $\pm\frac{1}{3}$ and $\pm\frac{2}{3}$ of the electronic charge. There is experimental evidence of the existence of quarks within the nucleus, but free quarks have not been detected. Current theory implies that direct detection of quarks may, in principle, be impossible (Chapter 30).

TABLE 15.1 Sub-atomic Particles and Their Electric Charge

Particle	*Electric Charge**	*Mass**
Electron	-1.602×10^{-19} C	$m_e = 9.109 \times 10^{-31}$ kg
Proton	$+1.602 \times 10^{-19}$ C	$m_p = 1.673 \times 10^{-27}$ kg
Neutron	0	$m_n = 1.675 \times 10^{-27}$ kg

*Even though the values are displayed to four significant figures, we will usually use only two or three in our calculations.

neutron are given in Table 15.1, where we see that $e = 1.602 \times 10^{-19}$ C. Our general symbol for charge will be q or Q. The charge on the electron is written as $q_e = -e = -1.602 \times 10^{-19}$ C, and that on the proton as $q_p = +e = +1.602 \times 10^{-19}$ C.

Other terms are frequently used when discussing charged objects. Saying that an object has a **net charge** means that the object has an excess of either positive or negative charges. (It is common, however, to ask about the "charge" of an object when we really mean the net charge.) As you will see in Section 15.2, excess charge is most commonly produced by a transfer of electrons, *not* protons. (Protons are bound in the nucleus and, under most common situations, do not leave.) For example, if an object has a (net) charge of $+1.6 \times 10^{-18}$ C, then it has had electrons removed from it. Specifically it has a deficiency of *ten* electrons, because $10 \times 1.6 \times 10^{-19}\ \text{C} = 1.6 \times 10^{-18}$ C. That is, the total number of electrons on the object no longer completely cancels the positive charge of all the protons—resulting in a net positive charge. On an atomic level, some of the atoms that compose the object are deficient in electrons. Such positively charged atoms are termed *positive ions*. Atoms with an excess of electrons are *negative ions*.

Since the charge of the electron is such a tiny fraction of a coulomb, an object having a net charge on the order of one coulomb is rarely seen in everyday situations. Therefore, it is common to express amounts of charge using *microcoulombs* (μC, or 10^{-6} C), *nanocoulombs* (nC, or 10^{-9} C), and *picocoulombs* (pC, or 10^{-12} C).

Because the (net) electric charge on an object is caused by either a deficiency or an excess of electrons, it must always be an integer multiple of the charge on an electron. A plus sign or a minus sign will indicate whether the object has a deficiency or an excess of electrons respectively. Thus, for the (net) charge of an object, we may write

$$q = \pm ne \tag{15.1}$$

SI unit of charge: coulomb (C)

where $n = 1, 2, 3, \ldots$. It is sometimes said that charge is "quantized," which means that it occurs only in integral multiples of the fundamental electronic charge.

In dealing with any electrical phenomena, another important principle is **conservation of charge**:

> The net charge of an isolated system remains constant.

That is, the net charge remains constant even though it may not be zero. Suppose, for example, that a system consists initially of two electrically neutral objects, and one million electrons are transferred from one to the other. The object with the added electrons will then have a net negative charge, and the object with the reduced number of electrons will have a net positive charge of equal magnitude. (See Example 15.1.) But the net charge of the *system* remains zero. If the universe is considered as a whole, conservation of charge means that the net charge *of the universe* is constant.

Note that this principle doesn't prohibit the creation or destruction of charged particles. In fact, physicists have known for a long time that charged particles can be created and destroyed on the atomic and nuclear levels. However, because of charge conservation, charged particles are created or destroyed only in pairs with equal and opposite charges.

Integrated Example 15.1 ■ On the Carpet: Conservation of Quantized Charge

You shuffle across a carpeted floor on a dry day and the carpet acquires a net positive charge (for details on this mechanism, see Section 15.2). (a) Will you have a (1) deficiency or (2) an excess of electrons? (b) If the charge the carpet acquired has a magnitude of 2.15 nC, how many electrons were transferred?

(a) Conceptual Reasoning. (a) Since the carpet has a net positive charge, it must have lost electrons and you must have gained them. Thus, your charge is negative, indicating an excess of electrons, and the correct answer is (2).

(b) Quantitative Reasoning and Solution. Because the charge of one electron is known, we can quantify the excess of electrons. Express the charge in coulombs, and state what is to be found.

Given: $q_c = +(2.15\ \text{nC})\left(\frac{10^{-9}\ \text{C}}{1\ \text{nC}}\right) = +2.15 \times 10^{-9}\ \text{C}$
$q_e = -1.60 \times 10^{-19}\ \text{C}$ (from Table 15.1)

Find: n, number of transferred electrons

The net charge on you is

$$q = -q_c = -2.15 \times 10^{-9}\ \text{C}$$

Thus

$$n = \frac{q}{q_e} = \frac{-2.15 \times 10^{-9}\ \text{C}}{-1.60 \times 10^{-19}\ \text{C/electron}} = 1.34 \times 10^{10}\ \text{electrons}$$

As can be seen, net charges, even in everyday situations, can involve huge numbers of electrons (here, more than 13 billion), because the charge of any one electron is very small.

Follow-Up Exercise. In this Example, if your mass is 80 kg, by what percentage has your mass increased due to the excess electrons? *(Answers to all Follow-Up Exercises are at the back of the text.)*

15.2 Electrostatic Charging

OBJECTIVES: To (a) distinguish between conductors and insulators, (b) explain the operation of the electroscope, and (c) distinguish among charging by friction, conduction, induction, and polarization.

The existence of two types of electric charge along with the attractive and repulsive electrical forces can be easily demonstrated. Before learning how this is done, let's distinguish between electrical conductors and insulators. What distinguishes these broad groups of substances is their ability to conduct, or transmit, electric charge. Some materials, particularly metals, are good **conductors** of electric charge. Others, such as glass, rubber, and most plastics, are **insulators**, or poor electrical conductors. A comparison of the relative magnitudes of the conductivities of some materials is given in ▸Fig. 15.3.

In conductors, the *valence* electrons of the atoms—that is, the electrons in the outermost orbits—are loosely bound. As a result, they can be easily removed from the atom and moved about in the conductor, or they can leave the conductor altogether. That is, the valence electrons are not permanently bound to a particular atom. In insulators, however, even the loosest bound electrons are too tightly bound to be easily removed from their atoms. Thus, charge does not readily move through, nor is it readily removed from, an insulator.

As Fig. 15.3 shows, there is also a "middle" class of materials called **semiconductors**. Their ability to conduct charge is intermediate between that of insulators and conductors. The movement of electrons in semiconductors is much more difficult to describe than the simple valence electron approach used for insulators and conductors. In fact, the details of semiconductor properties can be understood only with the aid of quantum mechanics, which is beyond the scope of this book.

However, it is interesting to note that the conductivity of semiconductors can be adjusted by adding atomic impurities in varying concentrations. Beginning in the 1940s, scientists undertook research into the properties of semiconductors to create applications for such materials. Scientists used semiconductors to create transistors, then solid-

Relative magnitude of conductivity | Material

CONDUCTORS
10^8 Silver
Copper
10^7 Aluminum
Iron
Mercury
Carbon
10^3
SEMICONDUCTORS
Germanium (Transistors)
Silicon (Computer chips)
10^{-9}
INSULATORS
10^{-10} Wood
10^{-12} Glass
10^{-15} Rubber

◀ **FIGURE 15.3 Conductors, semiconductors, and insulators** A comparison of the relative magnitudes of the electrical conductivities of various materials (not drawn to scale).

state circuits, and, eventually, modern computer microchips. The microchip is one of the major developments responsible for the high-speed computer technology of today.

Now that we know a bit about conductors and insulators, let's investigate a way of determining the sign of the charge on an object. The *electroscope* is one of the simplest devices used to determine electric charge (▼Fig. 15.4). In its simplest form, it consists of a metal rod with a metallic bulb at one end. The rod is attached to a solid, rectangular piece of metal that has an attached foil "leaf," usually made of gold or aluminum. This arrangement is insulated from its protective glass container by a nonconducting frame. When charged objects are brought close to the bulb, electrons in the bulb are either attracted to or repelled by the charged objects. For example, if a negatively charged rod is brought near the bulb, electrons in the bulb are repelled, and the bulb is left with a positive charge. The electrons are conducted down to the metal rectangle and its attached foil leaf, which then will swing away, because they have like charges (Fig. 15.4b). Similarly, if a positively charged rod is brought near the bulb, the leaf also swings away. (Can you explain why?)

Notice that the net charge on the electroscope remains zero in these instances. Because the device is isolated, only the *distribution* of charge is altered. However, it

Note: An uncharged electroscope can detect only whether an object is electrically charged. If the electroscope is previously charged with a known sign, it can then also determine the sign of the charge on the object.

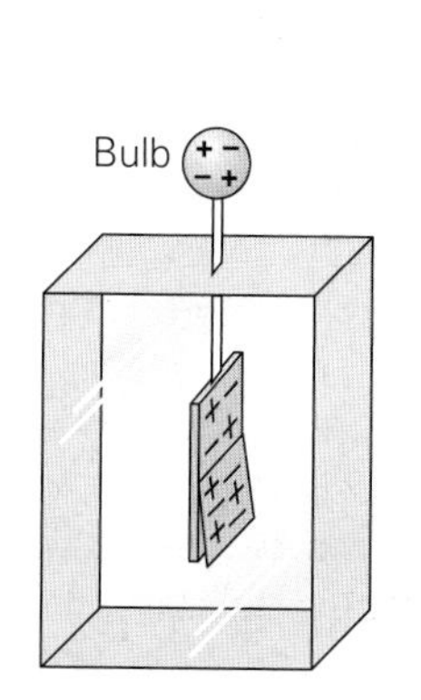

(a) Neutral electroscope has charges evenly distributed; leaf is vertical.

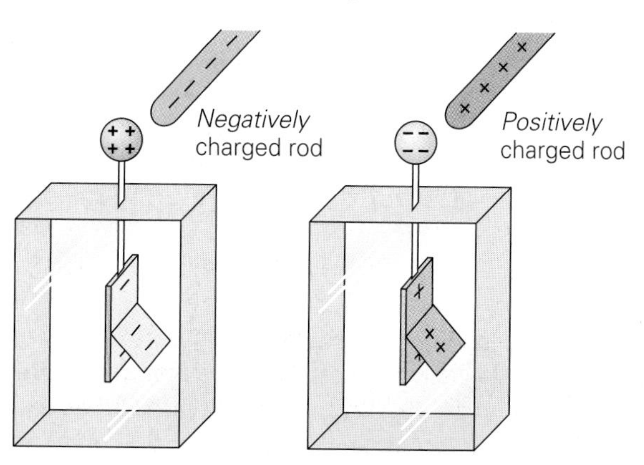

(b) Electrostatic forces cause leaf to separate away. (only excess or net charge is shown)

◀ **FIGURE 15.4 The electroscope** An electroscope can be used to determine whether an object is electrically charged. When a charged object is brought near the bulb, the leaf moves away from the metal piece.

is possible to give an electroscope (and other objects) a net charge by different methods, all of which are said to involve **electrostatic charging**. Consider the following types of processes that produce electrostatic charging.

Charging by Friction

In the frictional charging process, certain insulator materials are rubbed with cloth or fur, and they become electrically charged by a transfer of charge. For example, if a hard rubber rod is rubbed with fur, the rod will acquire a net negative charge; rubbing a glass rod with silk will give the rod a net positive charge. This process is called **charging by friction**. The transfer of charge is due to the contact between the materials, and the amount of charge transferred depends, as you might expect, on the nature of those materials.

Note: From an external viewpoint, you cannot tell whether the rubber rod gained negative charges or the fur gained positive charges. In other words, moving electrons to the rubber rod results in the same physical situation as moving positive charges to the fur. However, because the rubber is an insulator and its electrons are therefore tightly bound, we might suspect that the fur lost electrons and the rubber gained them. In solids, the protons, being in the nuclei of the atoms, do not move; only electrons move. It is just a question of which material most easily loses electrons.

Example 15.1 was an example of frictional charging, in which a net charge was picked up from the carpet. If you had reached for a metal object such as a doorknob, you might have been "zapped" by a spark. As your hand approaches, the knob becomes positively changed, thus attracting the electrons from your hand. As they travel, they collide with, and excite, the atoms of the air, which give off light as they de-excite (lose energy). This light is seen as the spark of "mini-lightning" between your hand and the knob.

Charging by Conduction (Contact)

Bringing a charged rod close to an electroscope will reveal that the rod is charged, but it does not tell you what type of charge the rod has (positive or negative). The sign of the charge can be determined, however, if the electroscope is first given a known type of (net) charge. For example, electrons can be transferred to the electroscope from a negatively charged object, as illustrated in ▾Fig. 15.5a. The electrons in

▸ **FIGURE 15.5 Charging by conduction** **(a)** The electroscope is initially neutral (but the charges are separated), as a charged rod touches the bulb. **(b)** Charge is transferred to the electroscope. **(c)** When a rod of the same charge is brought near the bulb, the leaf moves farther apart. **(d)** When an oppositely charged rod is brought nearby, the leaf collapses.

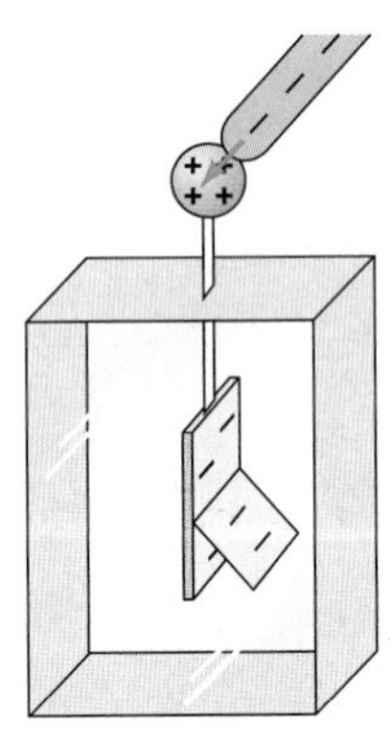

(a) Neutral electroscope is touched with negatively charged rod.

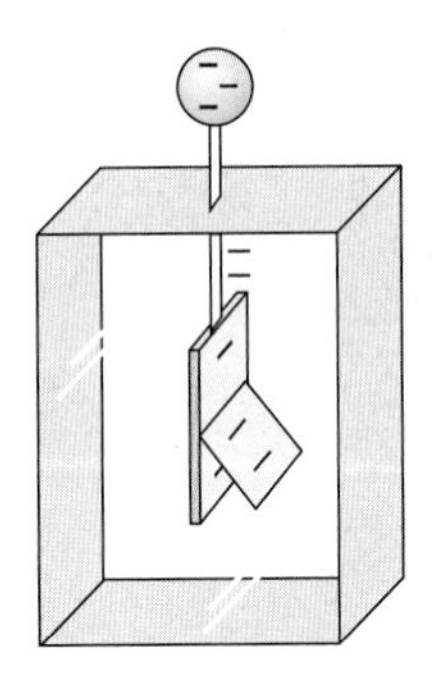

(b) Charges are transferred to bulb; electroscope has net negative charge.

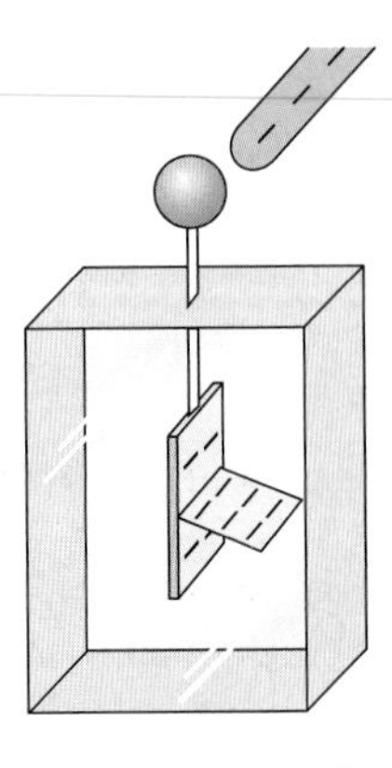

(c) Negatively charged rod repels electrons; leaf moves further.

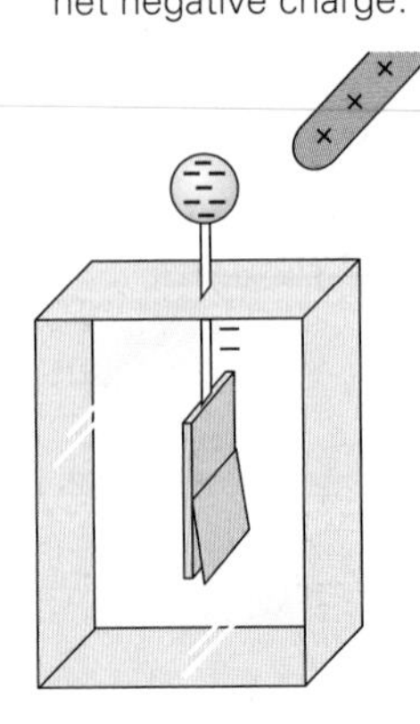

(d) Positively charged rod attracts electrons; leaf collapses.

the rod repel one another, and some will transfer onto the electroscope. Notice that the leaf is now permanently diverged from the metal. In this case, we say that the electroscope has been **charged by contact** or by **conduction** (Fig. 15.5b). "Conduction" in this case refers to the flow of charge during the short period of time the electrons are transferred.

If a negatively charged rod is brought close to the now negatively charged electroscope, the leaf will diverge even further as more electrons are repelled down from the bulb (Fig. 15.5c). A positively charged rod will cause the leaf to collapse by attracting electrons up to the bulb and away from the leaf area (Fig. 15.5d).

Charging by Induction

Using a negatively charged rubber rod (already charged by friction), you might ask whether it is possible to create an electroscope that is positively charged. The answer is yes. This can be accomplished by **charging by induction**. Starting with an uncharged electroscope, you touch the bulb with a finger, which *grounds* the electroscope—that is, provides a path by which electrons can escape from the bulb (▾Fig. 15.6). Then, when a negatively charged rod is brought close to (but not touching) the bulb, the rod repels electrons from the bulb through your finger and body and down into the Earth (hence the term *ground*). Removing your finger *while the charged rod is kept nearby* leaves the electroscope with a net positive charge. This is because when the rod is removed, the electrons that moved to the Earth have no way back because the return path is gone.

Charge Separation by Polarization

Charging by contact and charging by induction create a net charge through the removal of charge from an object. However, charge can be moved *within the object* while keeping its net charge zero. For example, the induction process described previously initially causes **polarization**, or separation of positive and negative charge. If the object is not grounded, it will remain electrically neutral, but have equal and opposite amounts of charge at its ends. In this situation, we say that it has become an *electric dipole* (Section 15.4). On the molecular level, electric dipoles can be permanent; that is, they don't need a nearby charged object to retain their charge separation. A good example of this is the water molecule. Examples of both permanent and nonpermanent electric dipoles and forces

▾ **FIGURE 15.6 Charging by induction** **(a)** Touching the bulb with a finger provides a path to ground for charge transfer. The symbol e^- stands for "electron." **(b)** When the finger is removed, the electroscope has a net positive charge, opposite that of the rod.

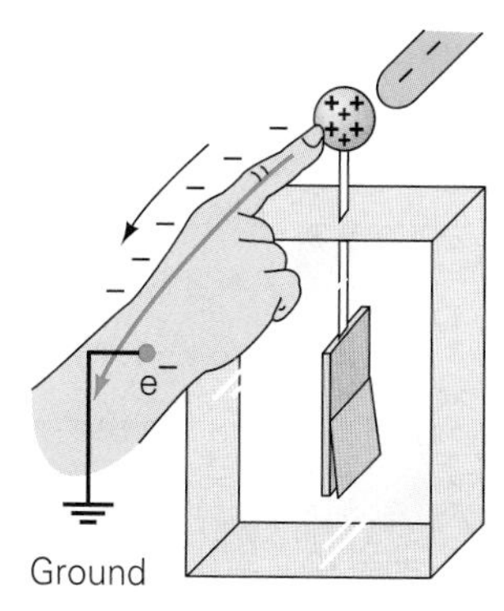

(a) Repelled by the nearby negatively charged rod, electrons are transferred to ground through hand.

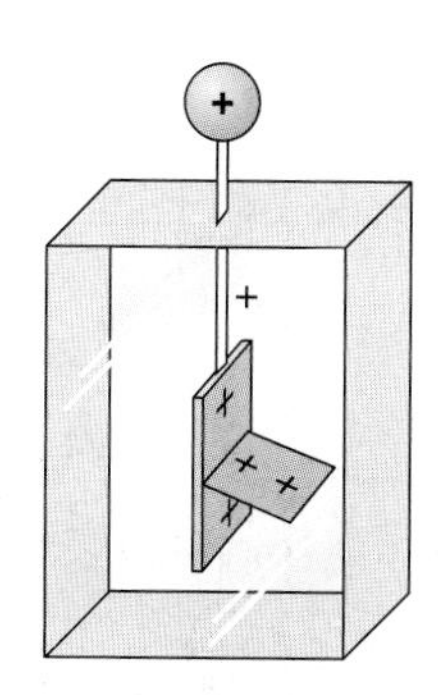

(b) After removing the finger first, then later the rod, the electroscope is left positively charged.

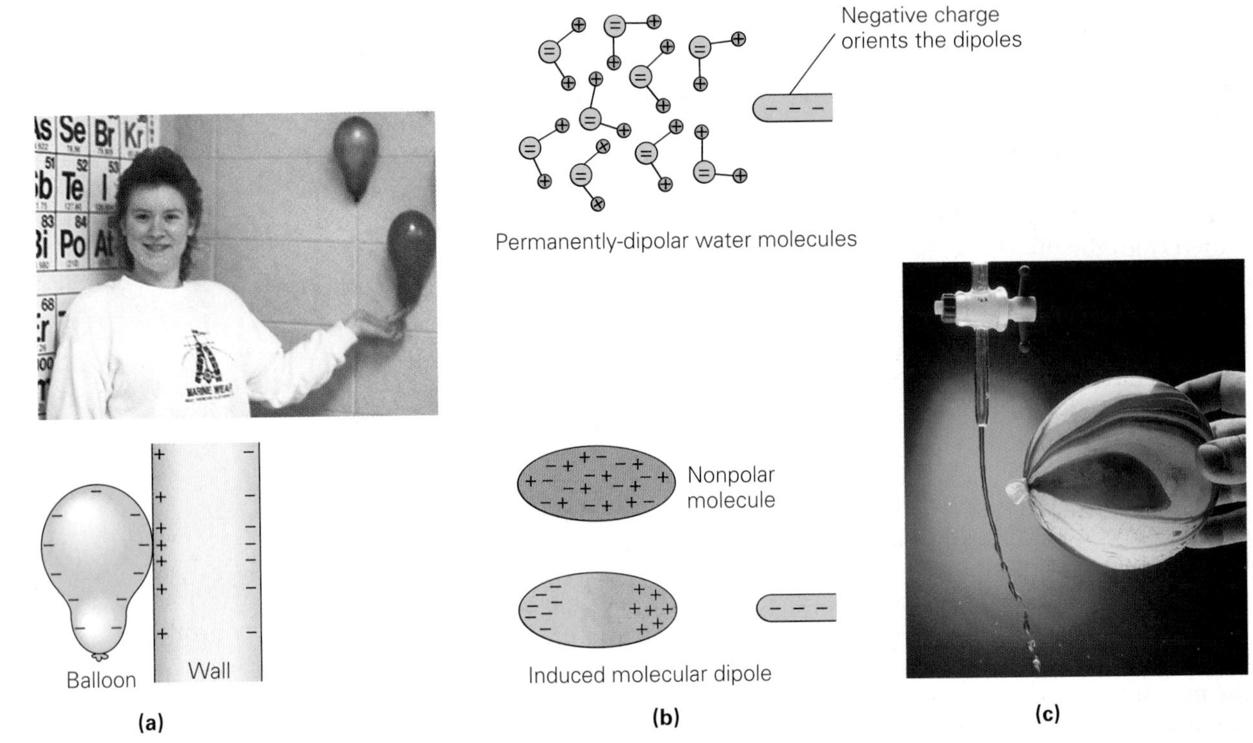

▲ **FIGURE 15.7 Polarization** **(a)** When the balloons are charged by friction and placed in contact with the wall, the wall is polarized. That is, an opposite charge is induced on the wall's surface, to which the balloons then stick by the force of electrostatic attraction. The electrons on the balloon do not leave the balloon because its material (rubber) is a poor conductor. **(b)** Some molecules, such as those of water, are polar by nature; that is, they have permanently separated regions of positive and negative charge. But even some molecules that are not normally dipolar can be polarized temporarily by the presence of a nearby charged object. The electric force induces a separation of charge and, consequently, temporary molecular dipoles. **(c)** A stream of water bends toward a charged balloon. The negatively charged balloon attracts the positive ends of the water molecules, thus causing the stream to bend.

that can act on them are shown in ▲Fig. 15.7. Now you can understand why, when you rub a balloon on your sweater, it can stick to the wall. The balloon is charged by friction, and bringing it near to the wall polarizes the wall. The opposite sign charge on the wall's nearest surface creates a net attractive force.

PHYSLET®
Illustration 22.4 *Charging Objects and Static Cling*

Electrostatic charge can be annoying, as when static cling causes clothes and papers to stick together, or even dangerous, such as when electrostatic spark discharges start a fire or cause an explosion in the presence of a flammable gas. To discharge electric charge, many large trucks have dangling metal chains in contact with the ground. At gas stations, there are warnings to fill your gas cans while they are on the ground, not on the truck bed or car trunk surface (why?).

However, electrostatic forces can also be beneficial. For example, the air we breathe is cleaner because of electrostatic precipitators used in smokestacks. In these devices, electrical discharges cause the particles (by-products of fuel combustion) to acquire a net charge. The charged particles can then be removed from the flue gases by attracting them to electrically charged surfaces. On a smaller scale, electrostatic air cleaners are available for the home (see opening Physics Fact).

15.3 Electric Force

OBJECTIVES: To (a) understand Coulomb's law, and (b) use it to calculate the electric force between charged particles.

We know that the *directions* of electric forces on interacting charges are given by the charge–force law. However, what about their *magnitudes*? This was investigated by Coulomb, who found that the magnitude of the electric force between two "point" (very small) charges q_1 and q_2 depended directly on the product of the magnitude of the charges and inversely on the square of the distance between

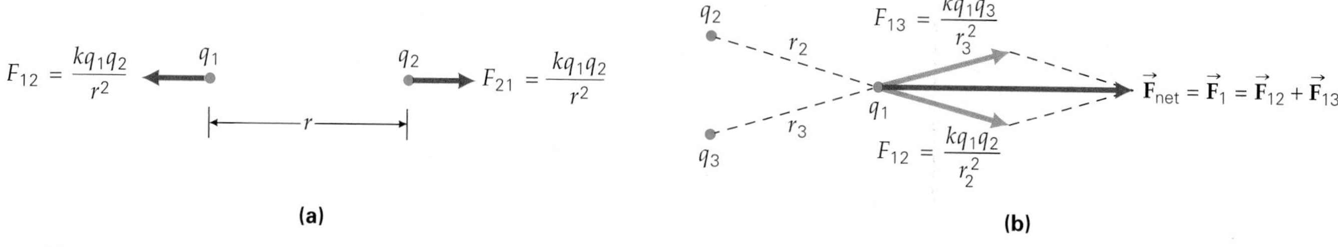

▲ **FIGURE 15.8 Coulomb's law (a)** The mutual electrostatic forces on two point charges are equal and opposite. **(b)** For a configuration of two or more point charges, the force on a particular charge is the vector sum of the forces on it due to all the other charges. (*Note:* In each of these situations, all of the charges are of the same sign. How can we tell that this is true? Can you tell their *sign*? What is the direction of the force on q_2 due to q_3?)

them. That is, $F_e \propto q_1q_2/r^2$. (q is the charge *magnitude*; thus, q_1 means the magnitude of q_1. This relationship is mathematically similar to that for the force of gravity between two point masses ($F_g \propto m_1m_2/r^2$); see Chapter 7.

Like Cavendish's measurements to determine the universal gravitational constant G (Section 7.5), Coulomb's measurements provided a constant of proportionality, k, so that the electric force could be written in equation form. Thus, the magnitude of the electric force between two point charges is described by an equation called **Coulomb's law**:

$$F_e = \frac{kq_1q_2}{r^2} \quad \textit{(point charges only, q means charge magnitude)} \qquad (15.2)$$

Here, r is the distance between the charges (▲Fig. 15.8a) and k a constant whose experimental value is

$$k = 8.988 \times 10^9\ \mathrm{N \cdot m^2/C^2} \approx 9.00 \times 10^9\ \mathrm{N \cdot m^2/C^2}$$

Equation 15.2 gives the force between any two charged particles, but in many instances, we are concerned with the forces between more than two charges. In this situation, the net electric force on any particular charge is the vector sum of the forces on that charge due to all the other charges (Fig. 15.8b). For a review of vector addition, using electric forces, see the next two Examples.

PHYSLET® Illustration 22.1 Charge and Coulomb's Law

PHYSLET® Exploration 22.6 Run Coulomb's Gauntlet

Note: Coulomb's law gives the electric force only between point charges, not objects with extended charged areas.

Note: In calculations, we will take k to be exact at $9.00 \times 10^9\ \mathrm{N \cdot m^2/C^2}$ for significant-figure purposes.

Conceptual Example 15.2 ■ Free of Charge: Electric Forces

You may have done this. A rubber comb combed through dry hair can acquire a net negative charge. That comb will then attract small pieces of *uncharged* paper. This would seem to violate Coulomb's force law. Since the paper has no net charge, you might expect there to be no electric force on it. Which charging mechanism explains this phenomenon, and how does it explain it: (a) conduction, (b) friction, or (c) polarization?

Reasoning and Answer. Because the comb doesn't touch the paper, the paper cannot be charged by either conduction or friction, because both of these require contact. Thus, the answer must be (c). When the charged comb is near the paper, the paper becomes polarized (▸Fig. 15.9). The key to understanding the attraction is to observe that the charged ends of the paper are *not* the same distance from the comb. The positive end of the paper is closer to the comb than the negative end. Since the electric force decreases with distance, the attraction ($\vec{F}_1$) between the comb and the positive end of the paper is greater than the repulsion ($\vec{F}_2$) between the comb and the paper's negative end. Therefore, after adding these two forces vectorially, we find that the net force on the paper points toward the comb, and if it is light enough, the paper will accelerate in that direction.

Follow-Up Exercise. Does the phenomenon described in this Example tell you the sign of the charge on the comb? Why or why not?

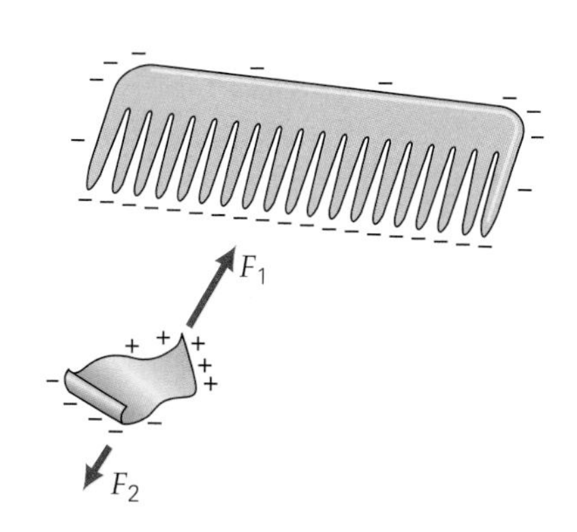

▲ **FIGURE 15.9 Comb and paper** See Conceptual Example 15.2.

$q_1 = -1.0$ nC $q_2 = +2.0$ nC
F_{12} F_{21}
0.30 m
(a)

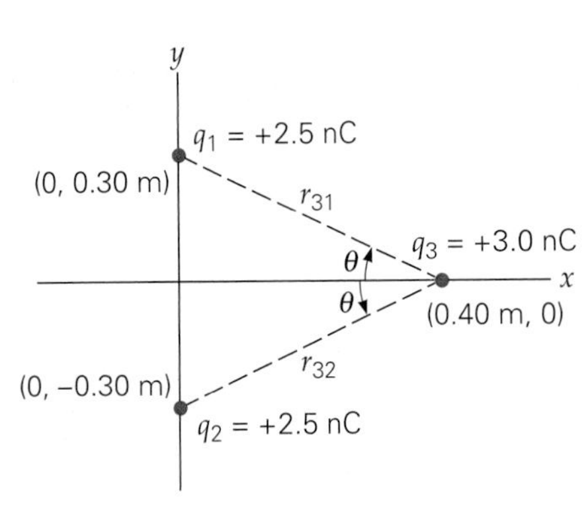

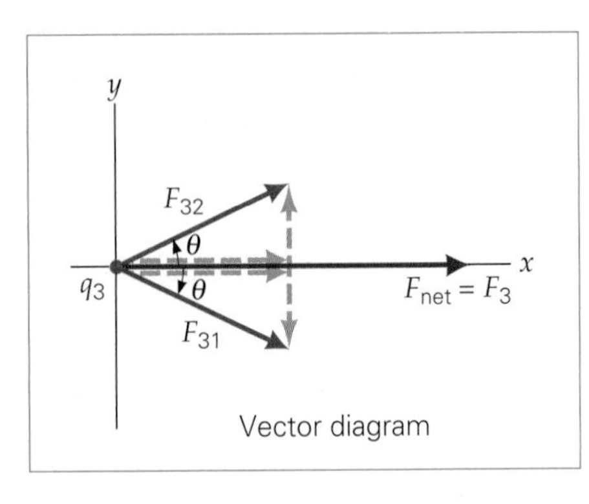

(b)

▲ **FIGURE 15.10 Coulomb's law and electrostatic forces** See Example 15.3.

PHYSLET®

Exploration 22.4 Dipole Symmetry

PHYSLET®

Exploration 22.1 Equilibrium

Example 15.3 ■ Coulomb's Law: Vector Addition Involving Trigonometry

(a) Two point charges of -1.0 nC and $+2.0$ nC are separated by a distance of 0.30 m (▲Fig. 15.10a). What is the electric force on each particle? (b) A configuration of three charges is shown in Fig. 15.10b. What is the net electric force on q_3?

Thinking It Through. Adding electric forces is no different from adding any other type of force. The only difference here is that we first must use Coulomb's law to calculate the force *magnitudes*. Then it is just a matter of computing components. (a) For the two point charges, we use Coulomb's law (Eq. 15.2), noting that the forces are attractive. (Why?) (b) Here we must use components to vectorially add the two forces acting on q_3 due to q_1 and q_2. We can find θ from the distances between charges. This angle is necessary to calculate the x and y force components. (See the Problem-Solving Hint on p. 515.)

Solution. Listing the data and converting nanocoulombs to coulombs, we have

Given: (a) $q_1 = -(1.0\text{ nC})\left(\dfrac{10^{-9}\text{ C}}{1\text{ nC}}\right) = -1.0 \times 10^{-9}\text{ C}$

$q_2 = +(2.0\text{ nC})\left(\dfrac{10^{-9}\text{ C}}{1\text{ nC}}\right) = +2.0 \times 10^{-9}\text{ C}$

$r = 0.30$ m

(b) Data given in Figure 15.10b. Convert charges to coulombs as in part (a).

Find: (a) $\vec{\mathbf{F}}_{12}$ and $\vec{\mathbf{F}}_{21}$
(b) $\vec{\mathbf{F}}_3$

(a) Equation 15.2 gives the magnitude of the force acting on each charge using the charge magnitudes and distance between them:

$$F_{12} = F_{21} = \frac{kq_1q_2}{r^2} = \frac{(9.00 \times 10^9\text{ N}\cdot\text{m}^2/\text{C}^2)(1.0 \times 10^{-9}\text{ C})(2.0 \times 10^{-9}\text{ C})}{(0.30\text{ m})^2}$$
$$= 0.20 \times 10^{-6}\text{ N} = 0.20\ \mu\text{N}$$

Note that Coulomb's law gives only the force's magnitude. However, because the charges are of opposite sign, the forces must be mutually attractive as shown in Fig. 15.10a.

(b) The forces $\vec{\mathbf{F}}_{31}$ and $\vec{\mathbf{F}}_{32}$ must be added vectorially, using trigonometry and components, to find the net force. Since all the charges are positive, the forces are repulsive, as shown in the vector diagram in Fig. 15.10b. Since $q_1 = q_2$ and the charges are equidistant from q_3, it follows that $\vec{\mathbf{F}}_{31}$ and $\vec{\mathbf{F}}_{32}$ have the same magnitude.

Also from the figure, we can see that $r_{31} = r_{32} = 0.50$ m. (Why?) With data from the figure, using Eq. 15.2:

$$F_{32} = \frac{kq_2q_3}{r_{32}^2} = \frac{(9.00 \times 10^9\text{ N}\cdot\text{m}^2/\text{C}^2)(2.5 \times 10^{-9}\text{ C})(3.0 \times 10^{-9}\text{ C})}{(0.50\text{ m})^2}$$
$$= 0.27 \times 10^{-6}\text{ N} = 0.27\ \mu\text{N}$$

Taking into account the directions of $\vec{\mathbf{F}}_{31}$ and $\vec{\mathbf{F}}_{32}$ we see by symmetry that the y-components cancel. Thus, $\vec{\mathbf{F}}_3$ (the net force on q_3) acts along the positive x-axis and has a magnitude of

$$F_3 = F_{31_x} + F_{32_x} = 2\,F_{31_x}$$

because $F_{31} = F_{32}$.

The angle θ can be determined from the triangles; that is, $\theta = \tan^{-1}\left(\frac{0.30\text{ m}}{0.40\text{ m}}\right) = 37°$. Thus $\vec{\mathbf{F}}_3$ has a magnitude of

$$F_3 = 2\,F_{31_x} = 2\,F_{32}\cos\theta$$
$$= 2(0.27\ \mu\text{N})\cos 37° = 0.43\ \mu\text{N}$$

and acts in the positive x-direction (to the right).

Follow-Up Exercise. In part (b) of this Example, calculate the net force $\vec{\mathbf{F}}_1$ on q_1.

The magnitudes of the charges in Example 15.3 are typical of static charges produced by frictional rubbing; that is, they are tiny. Thus, the forces involved are very small by everyday standards, much smaller than any force we have studied so far. However, on the atomic scale, even tiny forces can produce huge accelerations, because the particles (such as electrons and protons) have extremely small mass. Consider the answers in Example 15.4 compared with the answers in Example 15.3.

Problem-Solving Hint

The signs of the charges can be used explicitly in Eq. 15.2 with a positive value for F meaning a repulsive force and a negative value an attractive force. *However, such an approach is not recommended*, because this sign convention is useful only for one-dimensional forces, that is, those that have only one component, as in Example 15.3a. When forces are two-dimensional, thus requiring components, Eq. 15.2 should instead be used to calculate the *magnitude* of the force, using only the *magnitude* of the charges (as in Example 15.3b). Then the charge–force law determines the direction of the force between each pair of charges. (Draw a sketch and put in the angles.) Lastly, use trigonometry to calculate each force's components and then combine them appropriately. This latter approach is recommended and the one that will be used in this text.

Example 15.4 ■ Inside the Nucleus: Repulsive Electrostatic Forces

(a) What is the magnitude of the repulsive electrostatic force between two protons in a nucleus? Take the distance from center to center of these protons to be 3.00×10^{-15} m. (b) If the protons were released from rest, how would the magnitude of their initial acceleration compare with that of the acceleration due to gravity on the Earth's surface, g?

Thinking It Through. (a) Coulomb's law must be applied to find the repulsive force. (b) To find the initial acceleration, we use Newton's second law ($F_{\text{net}} = ma$).

Solution. Listing the known quantities, we have the following:

Given: $r = 3.00 \times 10^{-15}$ m
$q_1 = q_2 = +1.60 \times 10^{-19}$ C (from Table 15.1)
$m_p = 1.67 \times 10^{-27}$ kg (from Table 15.1)

Find: (a) F_e (magnitude of force)
(b) $\frac{a}{g}$ (magnitude of acceleration compared with g)

(a) Using Coulomb's law (Eq. 15.2), we have

$$F_e = \frac{kq_1q_2}{r^2} = \frac{(9.00 \times 10^9\ \text{N}\cdot\text{m}^2/\text{C}^2)(1.60 \times 10^{-19}\ \text{C})(1.60 \times 10^{-19}\ \text{C})}{(3.00 \times 10^{-15}\ \text{m})^2} = 25.6\ \text{N}$$

This force is much larger than that in the previous Example and is equivalent to the weight of an object with a mass of about 2.5 kg. Thus, with its small mass, we expect the proton to experience a huge acceleration.

(b) If it acted alone on a proton, this force would produce an acceleration of

$$a = \frac{F_e}{m_p} = \frac{25.6\ \text{N}}{1.67 \times 10^{-27}\ \text{kg}} = 1.53 \times 10^{28}\ \text{m/s}^2$$

Then

$$\frac{a}{g} = \frac{1.53 \times 10^{28}\ \text{m/s}^2}{9.8\ \text{m/s}^2} = 1.56 \times 10^{27}$$

(continues on next page)

That is, $a \approx 10^{27}\,g$. The factor of 10^{27} is enormous. To help see how large it is, if a uranium atom were subject to this acceleration, the net force required would be about the same as the weight of a polar bear (a thousand pounds or so)!

Most atoms contain more than two protons in their nuclei. With these enormous repulsive forces, you would expect nuclei to fly apart. Because this doesn't generally happen, there must be a stronger attractive force holding the nucleus together. This is called the nuclear (or strong) force, and will be discussed in Chapters 29 and 30.

Follow-Up Exercise. Suppose you could anchor a proton to the ground and you wished to place a second one directly above the first so that the second proton was in equilibrium (that is, so the electrical repulsion force acting on the second proton balanced its weight force). How far apart would the protons be?

Although there is a striking similarity between the mathematical form of the expressions for the electric and gravitational forces, there is a huge difference in the relative strengths of the two forces, as is shown in the next Example.

Example 15.5 ■ Inside the Atom: Electric Force versus Gravitational Force

Determine the ratio of the electric to gravitational force between a proton and an electron. In other words, how many times larger is the electric force than the gravitational force?

Thinking It Through. The distance between the proton and electron is not given. However, both the electrical force and the gravitational force vary as the inverse square of the distance, so the distance will cancel out in a ratio. By using Coulomb's law and Newton's law of gravitation (Chapter 7), the ratio can be determined if one knows the charges, masses, and appropriate electric and gravitational constants.

Solution. The charges and masses of the particles are known (Table 15.1), as are the electrical constant k and the universal gravitational constant G.

Given:
$q_e = -1.60 \times 10^{-19}\ \mathrm{C}$
$q_p = +1.60 \times 10^{-19}\ \mathrm{C}$
$m_e = 9.11 \times 10^{-31}\ \mathrm{kg}$
$m_p = 1.67 \times 10^{-27}\ \mathrm{kg}$

Find: $\dfrac{F_e}{F_g}$ (ratio of forces)

The expressions for the forces are

$$F_e = \frac{kq_eq_p}{r^2} \quad \text{and} \quad F_g = \frac{Gm_em_p}{r^2}$$

Forming a ratio of magnitudes for comparison purposes (and to cancel r) gives

$$\frac{F_e}{F_g} = \frac{kq_eq_p}{Gm_em_p}$$

$$= \frac{(9.00 \times 10^{9}\ \mathrm{N\cdot m^2/C^2})(1.60 \times 10^{-19}\ \mathrm{C})^2}{(6.67 \times 10^{-11}\ \mathrm{N\cdot m^2/kg^2})(9.11 \times 10^{-31}\ \mathrm{kg})(1.67 \times 10^{-27}\ \mathrm{kg})} = 2.27 \times 10^{39}$$

or

$$F_e = (2.27 \times 10^{39})F_g$$

The magnitude of the electrostatic force between a proton and an electron is more than 10^{39} times the magnitude of the gravitational force. While a factor of 10^{39} is incomprehensible to most, it should be perfectly clear that because of this large value, the gravitational force between charged particles can generally be neglected in our study of electrostatics.

Follow-Up Exercise. With respect to this Example, show that gravity is even more negligible compared with the electric force between two electrons. Explain why this is so.

15.4 Electric Field

OBJECTIVES: To (a) understand the definition of the electric field, and (b) plot electric field lines and calculate electric fields for simple charge distributions.

The electric force, like the gravitational force, is an "action-at-a-distance" force. Since the range of the electric force is infinite ($F_e \propto 1/r^2$ and approaches zero only as r approaches infinity), a particular configuration of charges can have an effect on an additional charge placed anywhere nearby.

The idea of a force acting across space was difficult for early investigators to accept, and the modern concept of a *force field*, or simply a field, was introduced. An *electric field* is envisioned as surrounding every arrangement of charges. Thus, the electric field represents the *physical effect* of a particular configuration of charges on the nearby space. The field represents what is different about the nearby space because those charges are there. The concept treats charges as interacting with the electric field created by other charges, not directly with the charges "at a distance." The main idea of the electric field concept is as follows: A configuration of charges creates an electric field in the space nearby. When another charge is placed in this field, *the field* will exert an electric force on that charge. Thus:

Note: Charges set up an electric field, and this field then acts on other charges placed in it.

> Charges create electric fields, and these fields in turn exert electric forces on other charges.

An electric field is actually a *vector field* (it has direction as well as magnitude). It enables us to determine the force (including direction) exerted on a charge at a location. *However, the electric field is not that force.* Instead, the magnitude (or strength) of the field is defined as the electric force exerted per unit charge. Determining an electric field's strength may be theoretically imagined using the following procedure. Place a very small charge (called a *test charge*) at the location of interest. Measure the force acting on that test charge and divide by the amount of its charge, thus determining the force that would be exerted *per coulomb*. Next imagine removing the test charge. The force disappears (why?), but the field remains, because it is created by the nearby charges, which remain. When the electric field is determined in many locations, we have a "map" of the electric field strength but no direction. Thus the "mapping" is incomplete.

Exploration 23.1 Fields and Test Charges

Note: A *test charge* (q_+) is small and positive.

Because the field direction is specified by the direction of the force on the test charge, it depends on whether the test charge is chosen as positive or negative. The convention is that a *positive test charge* (q_+) is used for measuring electric field direction (see ▸Fig. 15.11). That is:

> The electric field direction is in the direction of the force experienced by a positive test charge.

Once the electric field's magnitude and direction due to a charge configuration is known, you can ignore the "source" charge configuration and talk in terms of the field they produce. This way of visualizing electric interactions between charges often facilitates calculations.

The **electric field $\vec{\mathbf{E}}$** at any location is defined as follows

$$\vec{\mathbf{E}} = \frac{\vec{\mathbf{F}}_{\text{on } q_+}}{q_+} \tag{15.3}$$

SI unit of electric field: newton/coulomb (N/C)

The direction of $\vec{\mathbf{E}}$ is in the direction of the force on a small *positive* test charge at that location.

For the special case of a point charge, we can use Coulomb's force law. To determine the magnitude of the electric field due to a point charge at a distance r from that point charge, Eq. 15.3 can be used:

$$E = \frac{F_{\text{on } q_+}}{q_+} = \frac{(kqq_+/r^2)}{q_+} = \frac{kq}{r^2}$$

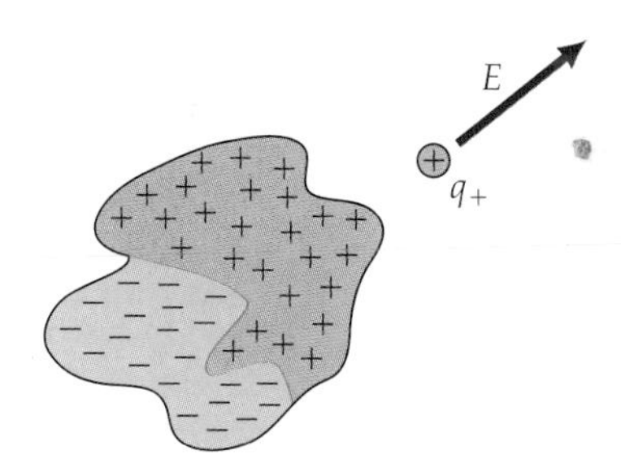

▲ **FIGURE 15.11 Electric field direction** By convention, the direction of the electric field $\vec{\mathbf{E}}$ is in the direction of the force experienced by an imaginary (positive) test charge. To see the direction, ask which way the test charge would accelerate if released. Here the "system of charges" produces a (net) electric field upward and to the right at the location of the test charge. In this particular arrangement, can you explain this direction by observing the signs and locations of charges in the system?

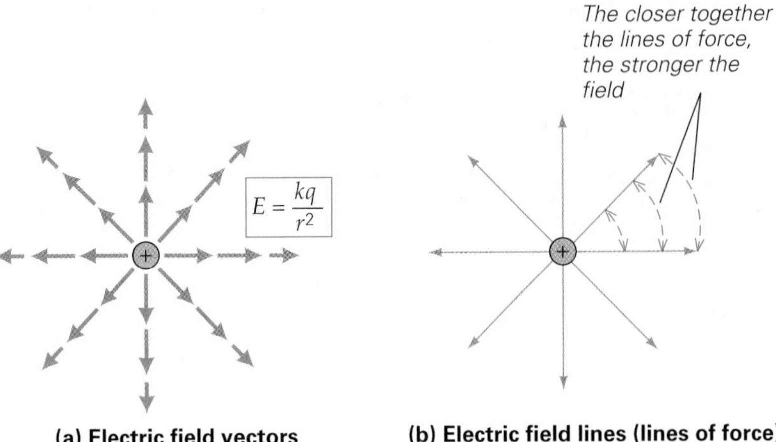

▲ **FIGURE 15.12 Electric field** **(a)** The electric field points away from a positive point charge, in the direction a force would be exerted on a small positive test charge. The field's magnitude (the lengths of the vectors) decreases as the distance from the charge increases, reflecting the inverse-square distance relationship characteristic of the field produced by a point charge. **(b)** In this simple case, the vectors are easily connected to give the electric field line pattern due to a positive point charge.

Illustration 23.2 Electric Fields from Point Charges

That is,

$$E = \frac{kq}{r^2} \quad \textit{(magnitude of electric field due to point charge } q\textit{)} \tag{15.4}$$

Notice that in deriving Eq. 15.4, q_+ canceled out. *This must always happen*, because the field is produced by the other charges, *not* the test charge q_+.

Some electric field vectors in the vicinity of a positive charge are illustrated in ▲Fig. 15.12a. Note that their directions are *away from the positive charge* because a positive test charge would feel a force in this direction. Notice also that the magnitude of the field (the arrow length) decreases with increasing distance r.

Note: Think of the electric field definition as being useful in the same way that the price per pound is for food items. Knowing how much you want of an item, you can compute how much it will cost if you know the price per pound. Similarly, given the magnitude of a charge placed in an electric field, you can compute the force on it if you know the field strength in newtons per coulomb.

If there is more than one charge creating an electric field, then the total, or net, electric field at any point is found using the **superposition principle for electric fields**, which can be stated as follows.

> For a configuration of charges, the total, or net, electric field at any point is the vector sum of the electric fields due to the individual charges.

This principle is demonstrated in the next two Examples, and a way to qualitatively determine the direction of the electric field when more than one charge is involved is shown in the accompanying Learn by Drawing on Using the Superposition Principle to Determine the Electric Field Direction.

LEARN BY DRAWING USING THE SUPERPOSITION PRINCIPLE TO DETERMINE THE ELECTRIC FIELD DIRECTION

To estimate the direction of the electric field at any point P, draw the individual electric fields vectors and add them, taking into account the relative field magnitudes if you can. In this situation, $\vec{\mathbf{E}}_1$ is much smaller than $\vec{\mathbf{E}}_2$ because of both distance and charge factors. Can you explain why the vector representing $\vec{\mathbf{E}}_2$, if drawn accurately, would be about eight times as long as that of $\vec{\mathbf{E}}_1$? The final step is to complete the vector addition.

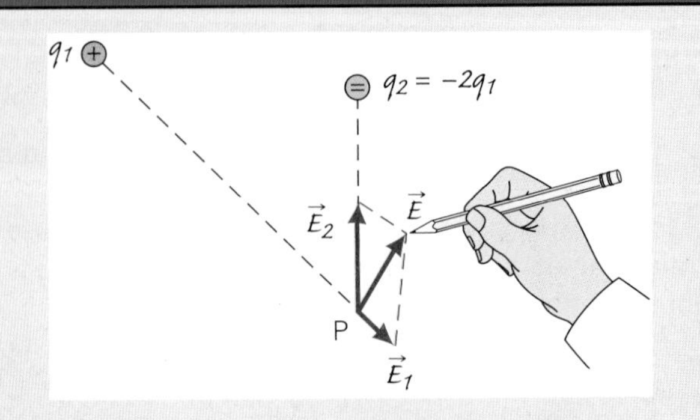

Example 15.6 ■ Electric Fields in One Dimension: Zero Field by Superposition

Two point charges are placed on the x-axis as in ▸Fig. 15.13. Find all locations on the axis where the electric field is zero.

Thinking It Through. Each point charge produces its own field. By the superposition principle, the electric field is the vector sum of the two fields. We are looking for locations where these fields are equal and opposite, so as to cancel and give no (*total* or *net*) electric field.

Solution. Let us specify the location as a distance x from q_1 (located at $x = 0$) and convert charges from microcoulombs to coulombs as usual.

Given: $d = 0.60$ m (distance between charges)
$q_1 = +1.5\ \mu C = +1.5 \times 10^{-6}$ C
$q_2 = +6.0\ \mu C = +6.0 \times 10^{-6}$ C

Find: x [the location(s) of zero E]

Since both charges are positive, their fields point to the right at all locations to the right of q_2. Therefore, the fields cannot cancel in that region. Similarly, to the left of q_1, both fields point to the left and cannot cancel. The only possibility of cancellation is *between* the charges. In that region, the two fields will cancel if their magnitudes are equal, because they are oppositely directed. Setting the magnitudes equal and solving for x:

$$E_1 = E_2 \quad \text{or} \quad \frac{kq_1}{x^2} = \frac{kq_2}{(d-x)^2}$$

Rearranging this expression and canceling the constant k yields

$$\frac{1}{x^2} = \frac{(q_2/q_1)}{(d-x)^2}$$

With $q_2/q_1 = 4$, taking the square root of both sides:

$$\sqrt{\frac{1}{x^2}} = \sqrt{\frac{q_2/q_1}{(d-x)^2}} = \sqrt{\frac{4}{(d-x)^2}} \quad \text{or} \quad \frac{1}{x} = \frac{2}{d-x}$$

Solving, $x = d/3 = 0.60\ \text{m}/3 = 0.20$ m. (Why don't we use the negative square root? Try it.) The result being closer to q_1 makes sense physically. Because q_2 is the larger charge, for the two fields to be equal in magnitude, the location must be closer to q_1.

Follow-Up Exercise. Repeat this Example, changing the only sign of the right-hand charge.

Where is $E = 0$?
$q_1 = +1.5\ \mu C$ $q_2 = +6.0\ \mu C$
0 0.10 0.20 0.30 0.40 0.50 0.60 x (m)

▲ **FIGURE 15.13 Electric field in one dimension** See Example 15.6.

Note: Total electric field: $\vec{E} = \Sigma \vec{E}_i$.

Integrated Example 15.7 ■ Electric Fields in Two Dimensions: Using Vector Components and Superposition

▾ Fig. 15.14a shows a configuration of three point charges. **(a)** In what quadrant is the electric field at the origin: (1) the first quadrant, (2) the second quadrant, or (3) the third quadrant? Explain your reasoning, using the superposition principle. **(b)** Calculate the magnitude and direction of the electric field at the origin due to these charges.

(continues on next page)

y (m)
4.00 $q_3 = -1.50\ \mu C$
$q_2 = +2.00\ \mu C$
$q_1 = -1.00\ \mu C$
x (m)
−5.00 0 3.50
(a)

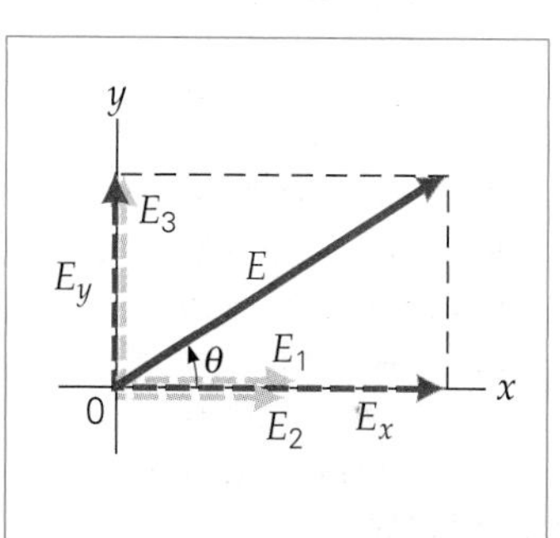

(b)

◂ **FIGURE 15.14 Finding the electric field** See Integrated Example 15.7.

(a) Conceptual Reasoning. The electric field points towards negative point charges and away from positive point charges. Therefore, $\vec{\mathbf{E}}_1$ and $\vec{\mathbf{E}}_2$ point in the positive x-direction and $\vec{\mathbf{E}}_3$ points along the positive y-axis. Because the electric field is the sum of these three fields, both of its components are positive. Therefore $\vec{\mathbf{E}}$ is in the first quadrant (Fig. 15.14b). Thus, the correct answer is (1).

(b) Quantitative Reasoning and Solution. The directions of the individual electric fields are shown in the sketch in part (a). According to the superposition principle, we need to add the fields vectorially to find the electric field ($\vec{\mathbf{E}} = \vec{\mathbf{E}}_1 + \vec{\mathbf{E}}_2 + \vec{\mathbf{E}}_3$).

Listing the data given and converting the charges into coulombs, we have:

Given:
$q_1 = -1.00\ \mu C = -1.00 \times 10^{-6}\ C$
$q_2 = +2.00\ \mu C = +2.00 \times 10^{-6}\ C$
$q_3 = -1.50\ \mu C = -1.50 \times 10^{-6}\ C$
$r_1 = 3.50\ m$
$r_2 = 5.00\ m$
$r_3 = 4.00\ m$

Find: $\vec{\mathbf{E}}$ (electric field at origin)

From the sketch E_y is due to $\vec{\mathbf{E}}_3$ and E_x is the sum of the magnitudes of $\vec{\mathbf{E}}_1$ and $\vec{\mathbf{E}}_2$. The magnitudes of the three fields are determined from Eq. 15.4. These magnitudes are

$$E_1 = \frac{kq_1}{r_1^2} = \frac{(9.00 \times 10^9\ N \cdot m^2/C^2)(1.00 \times 10^{-6}\ C)}{(3.50\ m)^2} = 7.35 \times 10^2\ N/C$$

$$E_2 = \frac{kq_2}{r_2^2} = \frac{(9.00 \times 10^9\ N \cdot m^2/C^2)(2.00 \times 10^{-6}\ C)}{(5.00\ m)^2} = 7.20 \times 10^2\ N/C$$

$$E_3 = \frac{kq_3}{r_3^2} = \frac{(9.00 \times 10^9\ N \cdot m^2/C^2)(1.50 \times 10^{-6}\ C)}{(4.00\ m)^2} = 8.44 \times 10^2\ N/C$$

The x- and y-components of the field are

$$E_x = E_1 + E_2 = +7.35 \times 10^2\ N/C + 7.20 \times 10^2\ N/C = +1.46 \times 10^3\ N/C$$

and

$$E_y = E_3 = +8.44 \times 10^2\ N/C$$

In component form,

$$\vec{\mathbf{E}} = E_x\hat{\mathbf{x}} + E_y\hat{\mathbf{y}} = (1.46 \times 10^3\ N/C)\hat{\mathbf{x}} + (8.44 \times 10^2\ N/C)\hat{\mathbf{y}}$$

You should be able to show that in magnitude–angle form this is

$$E = 1.69 \times 10^3\ N/C \text{ at } \theta = 30.0° \quad (\theta \text{ is in the first quadrant, relative to the } +x\text{-axis})$$

Follow-Up Exercise. In this Example, suppose q_1 was moved to the origin. Find the electric field at its former location.

Electric Lines of Force

Illustration 23.3 Field-Line Representation of Vector Fields

A convenient way of *graphically* representing the electric field is by use of *electric lines of force*, or **electric field lines**. To start, consider the electric field vectors near a positive point charge, as in Fig. 15.12a. In Fig. 15.12b these vectors have been "connected." This constructs the *electric field line pattern* due to a positive point charge. Notice that the field lines come closer together (their spacing decreases) as we near the charge, because the field increases in strength. Also note that at any location on a field line, the electric field *direction* is tangent to the line. (The lines usually have arrows attached to them that indicate the general field direction.) It should be clear that electric field lines can't cross. If they did, it would mean that at the crossing spot there would be two directions for the force on a charge placed there—a physically unreasonable result.

The general rules for sketching and interpreting electric field lines are as follows:

1. The closer together the field lines, the stronger the electric field.
2. At any point, the direction of the electric field is tangent to the field lines.
3. The electric field lines start at positive charges and end at negative charges.
4. The number of lines leaving or entering a charge is proportional to the magnitude of that charge.
5. Electric field lines can never cross.

These rules enable us to "map" the pattern of electric lines of force due to various charge configurations. (See the accompanying Learn by Drawing on Sketching Electric Lines of Force.)

Let's now apply these rules and the superposition principle to map out the electric field line pattern due to an *electric dipole* in Example 15.8. An **electric dipole** consists of two equal, but opposite, electric charges (or "poles," as they were known historically). Even though the net charge on the dipole is zero, it creates an electric field because the charges are separated. If the charges were at the same location, their fields would cancel everywhere.

In addition to using dipoles to learn about electric field sketching, dipoles are important in themselves, because they occur in nature. For example, electric dipoles can serve as a model for important polarized molecules, such as the water molecule. (See Fig. 15.7.) Also see Insight 15.2 on Electric Fields in Law Enforcement and Nature: Stun Guns and Electric Fish on page 524.

LEARN BY DRAWING

Sketching Electric Lines of Force

Sketching Electric Lines of Force

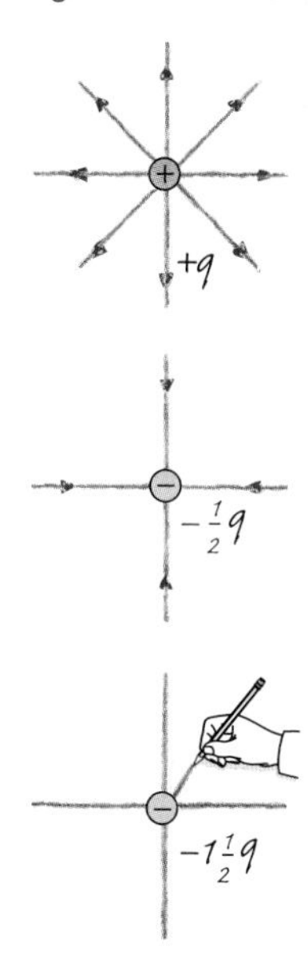

How many lines should be drawn for $-1\frac{1}{2}q$, and what should their direction be?

Example 15.8 ■ Constructing the Electric Field Pattern Due to a Dipole

Use the superposition principle and the electric field line rules to construct a typical electric field line due to an electric dipole.

Thinking It Through. The construction involves vector addition of the individual electric fields from the two opposite ends of the dipole.

Solution.

Given: an electric dipole of two equal and opposite charges separated by a distance, d

Find: a typical electric field line

An electric dipole is shown in ▾Fig. 15.15a. To keep track of the two fields, let's label the positive charge q_+ and the negative charge q_-. Their individual fields, $\vec{\mathbf{E}}_+$ and $\vec{\mathbf{E}}_-$, will be designated by the same subscripts.

Because electric fields (and field lines) start at positive charges, let's begin at location A, near charge q_+. Because this is much closer to q_+ it follows that $E_+ > E_-$. We know that $\vec{\mathbf{E}}_+$ will *always* point away from q_+ and $\vec{\mathbf{E}}_-$ will *always* point toward q_-. Putting these two facts together enables us to qualitatively draw the two fields at A. The parallelogram method determines their vector sum: the electric field at A.

To map the electric field line, the general direction of the electric field at A points us approximately to our next location, B. At B, there is a reduced magnitude (why?) and slight directional change for both $\vec{\mathbf{E}}_+$ and $\vec{\mathbf{E}}_-$. You should now be able to see how the fields at C and D are determined. Location D is special because it is on the perpendicular bisector of the dipole axis (the line that connects the two charges). The electric field points downward anywhere on this line. You should be able to continue the construction at points E, F, and G.

Lastly, to construct the electric field line, start at the positive end of the dipole, because the field lines leave that end. Because the electric field vectors are tangent to the field lines, we draw the line to fulfill this requirement. [You should be able to sketch in the other lines and understand the complete dipole field pattern shown in Fig. 15.15b.]

Follow-Up Exercise. Using the techniques in this Example, construct the field lines that start (a) just above the positive charge, (b) just below the negative charge, and (c) just below the positive charge.

Exploration 23.2 Field Lines and Trajectories

Note: Remember the name "electric lines of force" is a misnomer. These field lines represent the electric field, not the electric force.

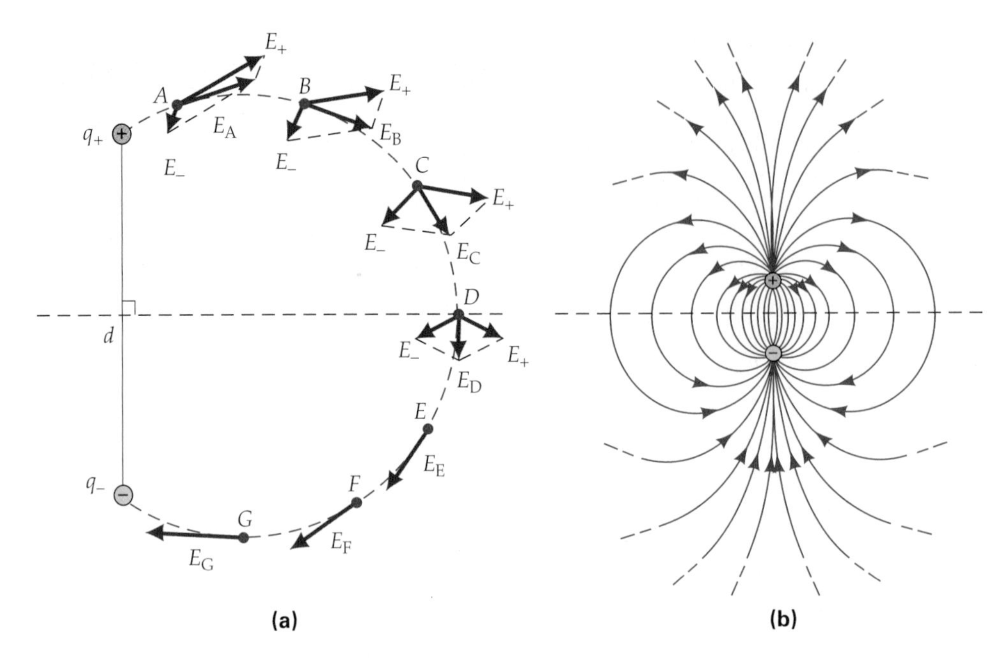

▶ **FIGURE 15.15 Mapping the electric field due to a dipole** **(a)** The construction of one electric field line from a dipole is shown. The electric field is the vector sum of the two fields produced by the two ends of the dipole. (See Example 15.8 for details.) **(b)** The full electric dipole field is determined by following the procedure in part (a) at other locations near the dipole.

▼Figure 15.16a shows the use of the superposition principle to construct the electric field line pattern due to a large positively charged plate. Notice that the field points perpendicularly away from the plate on both sides. Figure 15.16b shows the result if the plate is negatively charged, the only difference being the field direction. Putting these two together, we can find the field between two closely spaced and oppositely charged plates. The result is the pattern in Fig. 15.16c. Due to the cancellation of the horizontal field components (as long as we stay away from the plate edges), the electric field is uniform and points from the positive to negative. (Think of the direction of the force acting on a positive test charge placed between the plates.)

The derivation of the expression for the electric field magnitude between two closely spaced plates is beyond the scope of this text. However, the result is

$$E = \frac{4\pi kQ}{A} \quad \textit{(electric field between parallel plates)} \tag{15.5}$$

where Q is the magnitude of the total charge on *one* of the plates and A is the area of *one* plate. Parallel plates are common in electronic applications. For example, in Chapter 16 we will study an important circuit element called a *capacitor*, which, in its simplest form, is just a set of parallel plates. Capacitors play a crucial role in lifesaving devices such as *heart defibrillators*, as we shall also see in Chapter 16.

Cloud-to-ground lightning can be approximated by closely spaced parallel plates as in the next Example. (See Insight 15.1 on Lightning and Lightning Rods on accompanying page.)

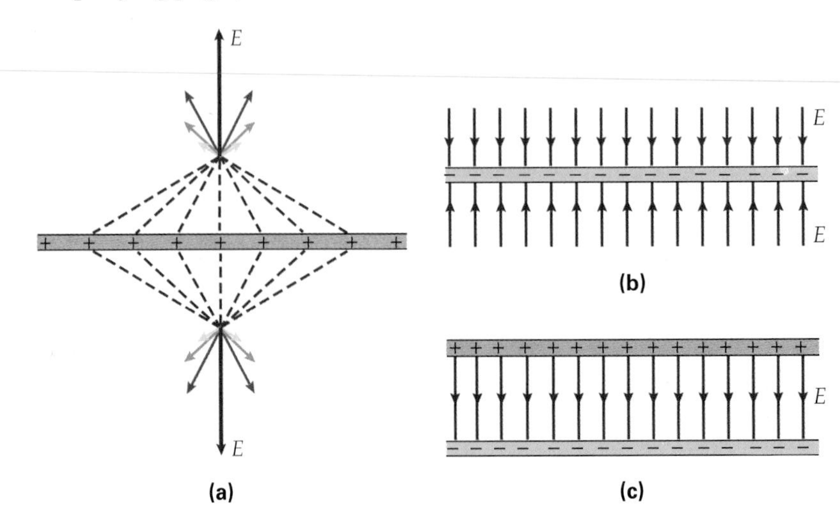

▶ **FIGURE 15.16 Electric field due to very large parallel plates** **(a)** Above a positively charged plate, the net electric field points upward. Here, the horizontal components of the electric fields from various locations on the plate cancel out. Below the plate, $\vec{E}$ points downward. **(b)** For a negatively charged plate, the electric field directions (shown on both sides of the plate) are reversed. **(c)** Superimposing the fields from both plates results in cancellation outside the plates and an approximately uniform field between them.

INSIGHT 15.1 LIGHTNING AND LIGHTNING RODS

Although the violent release of electrical energy in the form of lightning is common, we have a lot to learn about its formation. It is known that during the development of a cumulonimbus (storm) cloud, a separation of charge occurs. How the separation of charge takes place in a cloud is not fully understood, but it must be associated with the rapid vertical movement of air and moisture within storm clouds. Whatever the mechanism, the cloud acquires regions of different charge, with the bottom usually negatively charged.

As a result, an opposite charge is induced on the Earth's surface (Fig. 1a). Eventually, lightning may reduce this charge difference by ionizing the air, allowing a flow of charge between the cloud and the ground. However, air is a good insulator, so the electric field must be quite strong for ionization to occur. (See Example 15.9 for a quantitative estimate of the charge on a cloud.)

Most lightning occurs entirely within a cloud (intracloud discharges), where it cannot be seen. However, visible discharges do take place between clouds (cloud-to-cloud discharges) and between a cloud and the Earth (cloud-to-ground discharges). Special high-speed-camera photographs of cloud-to-ground discharges reveal a nearly invisible downward ionization path. The lightning discharges in a series of jumps or steps and so is called a *stepped leader*. As the leader nears the ground, positively charged ions in the form of a *streamer* rise from trees, tall buildings, or the ground to meet it.

When a streamer and a leader make contact, the electrons along the leader channel begin to flow downward. The initial flow is near the ground. As it continues, electrons positioned successively higher begin to migrate downward. Hence, the path of electron flow is extended upward in a *return stroke*. The surge of charge in the return stroke causes the conductive path to be illuminated, producing the bright flash seen by the eye and recorded in time-exposure photographs (Fig. 1b). Most lightning flashes have a duration of less than 0.50 s. Usually, after the initial discharge, ionization again takes place along the original channel, and another return stroke occurs. Typical lightning events have three or four return strokes.

Ben Franklin is often said to have been the first to demonstrate the electrical nature of lightning. In 1750, he suggested an experiment using a metal rod on a tall building. However, a Frenchman named Thomas François d'Alibard was the one who first set up the experiment using a rod during a thunderstorm (Fig. 1c). Franklin later performed a similar experiment with a kite, also during a thunderstorm.

A practical outcome of Franklin's work was the *lightning rod*, a pointed metal rod connected by a wire to a metal rod driven into the Earth, or "grounded." The elevated rod's tip, with its dense accumulation of induced positive charge and large electric field (see Fig. 15.19b), intercepts the downward, ionized stepped leader, discharging it harmlessly to the ground before the leader reaches a structure or makes contact with an upward streamer. This prevents the formation of a damaging electrical surge associated with a return stroke.

(a)

(b)

(c)

FIGURE 1 Lightning and lightning rods (a) Cloud polarization induces a charge on the Earth's surface. **(b)** When the electric field becomes large enough, an electrical discharge results, which we call lightning. **(c)** A lightning rod mounted atop a tall structure provides a path to ground so as to prevent damage.

INSIGHT 15.2 ELECTRIC FIELDS IN LAW ENFORCEMENT AND NATURE: STUN GUNS AND ELECTRIC FISH

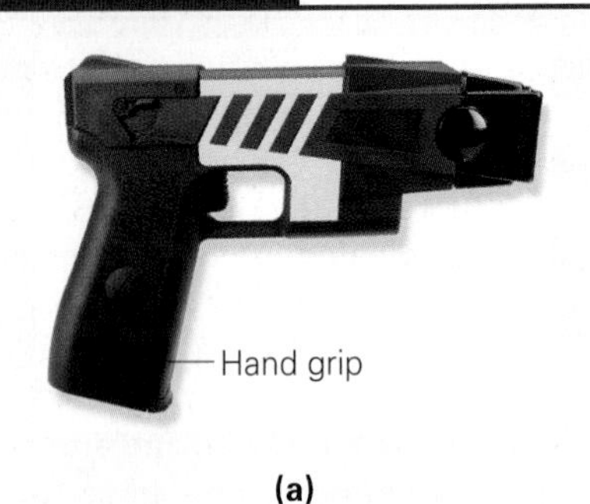

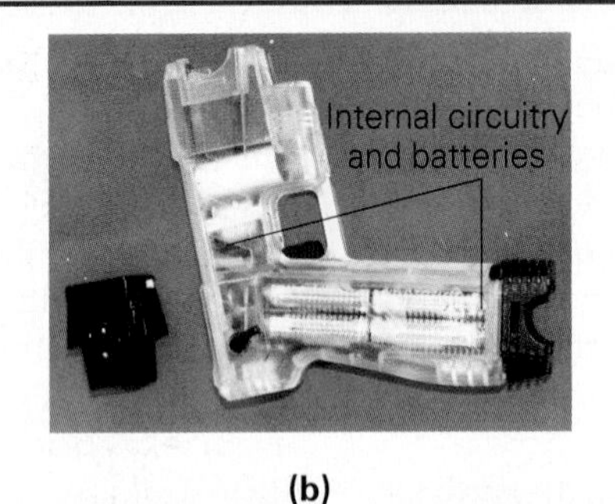

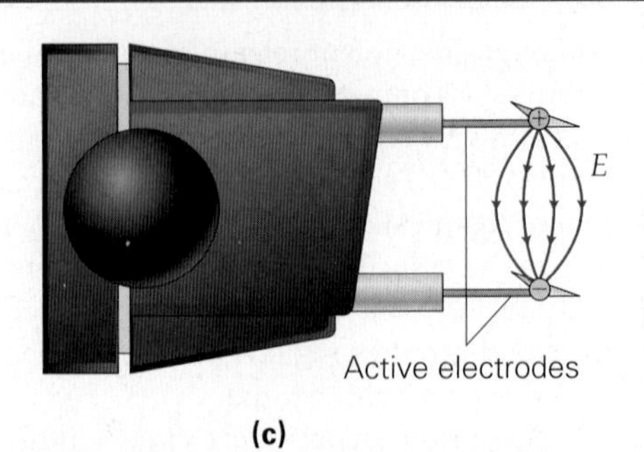

FIGURE 1 The Taser stun gun **(a)** The exterior of a stun gun; notice the grip and two electrodes. **(b)** The interior: the circuitry necessary to increase the electric field and charge separation to the strength required to disrupt nerve communication. **(c)** A sketch of the electric field between the electrodes. (The charges change sign periodically, producing an oscillating electric field.)

Stun guns and electric fish have similar electric field properties. Stun guns (a generic name for several types, the most familiar being the hand-held Taser) generate a charge separation by using batteries and internal circuitry. This circuitry can produce a large charge polarization—that is, equal and opposite charge on the electrodes. Figures 1a and 1b show a typical Taser. The charges on the electrodes oscillate in sign, but at any instant, the field is close to that of a dipole (Fig. 1c). Tasers are used for subduing a criminal, theoretically without permanent harm. A law enforcement officer, holding the grip, applies the electrodes to the body—say, the thigh. The electric field disrupts the electrical signals in the nerves that control the large thigh muscle, rendering the muscle inoperative and making the criminal more easily subdued.

The phrase *electric fish* conjures an image of an electric eel (which is actually an eel-shaped fish). However, there are other fish that are "electric." The electric eel and others such as the electric catfish are *strongly electric fish*. They can generate large electric fields to stun prey but can also use the fields for location and communication. *Weakly electric fish*, such as the elephant nose (Fig. 2a), use their fields (Fig. 2b) for location and communication only. Fish that actively produce electric fields are called *electrogenic fish*.

In electrogenic fish, the charge separation is accomplished by the *electric organ* (shown for the elephant nose fish in Fig. 2b), which is a specialized stack of *electroplates*. Each electroplate is a disklike structure that is normally uncharged. When the brain sends a signal, the disks become polarized through a chemical process similar to that of nerve action, creating the fish's field.

Weakly electric fish are capable of producing electric fields about the same as those produced by batteries. This is good only for *electrocommunication* and *electrolocation*. Strongly electric fish produce fields hundreds of times stronger and can kill prey if the fish touches them simultaneously with the oppositely charged areas. The electric eel has thousands of electroplates stacked in the electric organ, which typically extends from behind its head well into its tail and may take up to 50% of its body length (Fig. 2c).

As an example how these fields are used for electrolocation, consider the change to the elephant nose fish's normal electric field pattern (Fig. 2b) if it approaches a small conducting object (Fig. 3). Notice that the field lines change to bend toward the object; because the object is conducting, the field lines must be oriented at right angles to its surface. This results in a stronger field at the part of the fish's skin surface near the object. Skin sensors detect this increase and send a signal to the brain to that effect. A nonconducting object, such as a rock, would have the opposite effect. In reality electrolocation and electrocommunication are determined by an interplay of the electric field and the sensory organs. However, the basic properties of electrostatic fields provide us with the general idea of how such fish operate.

Example 15.9 ■ Parallel Plates: Estimating the Charge on Storm Clouds

The electric field (magnitude) E required to ionize moist air is about 1.0×10^6 N/C. When the field reaches this value, the least bound electrons are pulled off their molecules (ionization of the molecules), which can lead to a lightning stroke. Assume that the value for E between the negatively charged lower cloud surface and the positively charged ground is 1.00% of this, or 1.0×10^4 N/C. (See Fig. 1a of Insight 15.1 on page 523.) Take the clouds to be squares 10 miles on each side. Estimate the magnitude of the total negative charge on the lower surface.

Thinking It Through. The electric field is given, so Eq. 15.5 can be used to estimate Q. The cloud area A (one of the "plates") must be expressed in square meters.

Solution.

Given: $E = 1.0 \times 10^4$ N/C
$d = 10 \text{ mi} \approx 1.6 \times 10^4$ m

Find: Q (the magnitude of the charge on the lower cloud surface)

(a)

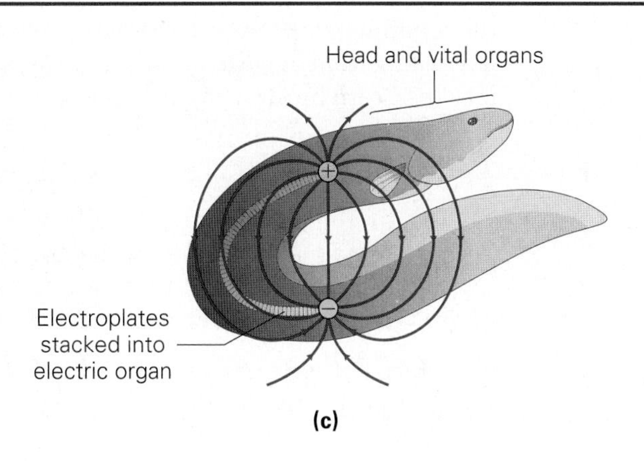

(c)

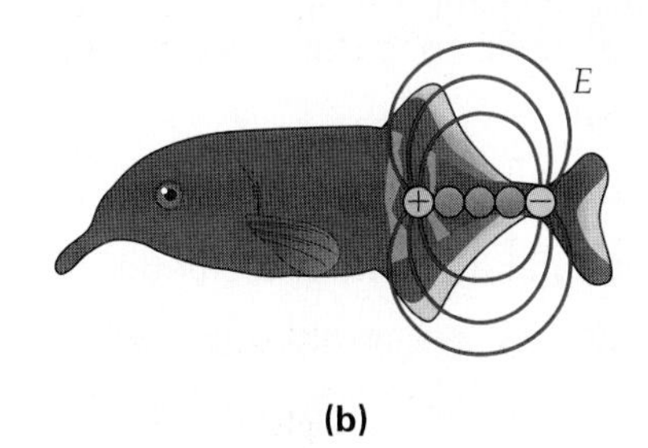

(b)

FIGURE 2 Electric fish **(a)** The elephant nose fish, which is a weakly electric fish—a fish that uses its electric field for electrolocation and communication. **(b)** At an instant in time, the approximate electric field produced by the elephant nose fish's electric organ, located near its tail. (The field actually oscillates.) **(c)** At an instant in time, the approximate electric field produced by an electric eel. The electric organ in the eel is capable of producing fields that can kill and stun as well as locate and communicate.

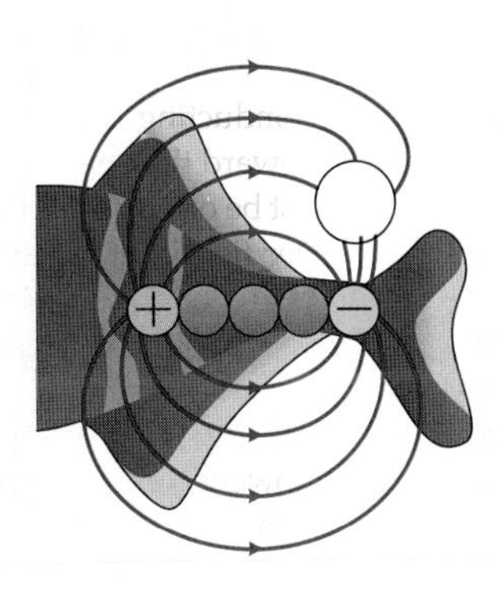

FIGURE 3 Electrolocation The field from an elephant nose fish with a conducting object nearby. Note the decrease in spacing of the field lines as they enter the skin surface. This increase in field strength is picked up by sensory organs on the skin, which send a signal to the fish's brain.

Using $A = d^2$ for the area of a square and solving Eq. 15.5 for the magnitude of the charge (the cloud surface is negative):

$$Q = \frac{EA}{4\pi k} = \frac{(1.0 \times 10^4\ \text{N/C})(1.6 \times 10^4\ \text{m})^2}{4\pi(9.0 \times 10^9\ \text{N}\cdot\text{m}^2/\text{C}^2)} = 23\ \text{C}$$

This expression is justified only if the distance between the clouds and the ground is much less than their size. (Why?) Such an assumption is equivalent to assuming that the 10-mile-long clouds are less than several miles from the Earth's surface.

This amount of charge is huge compared with the frictional static charges developed when shuffling on a carpet. However, because the cloud charge is spread out over a large area, any region of the cloud does not contain a lot of charge.

Follow-Up Exercise. In this Example, (a) what is the direction of the electric field between the cloud and the Earth? (b) How much charge would be required to ionize moist air?

Illustration 23.4 Practical Uses of Charges and Electric Fields

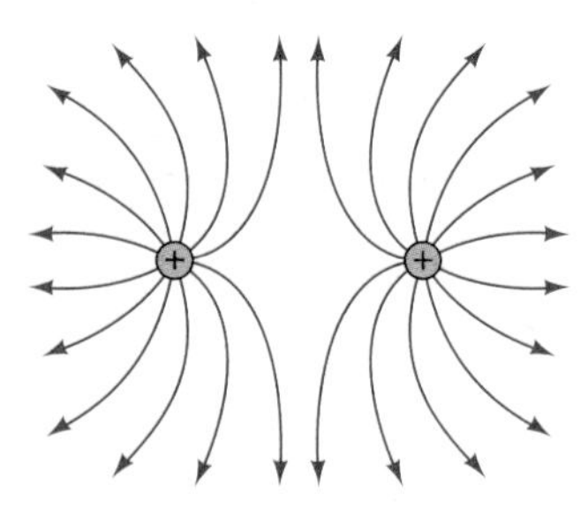

(a) Like point charges

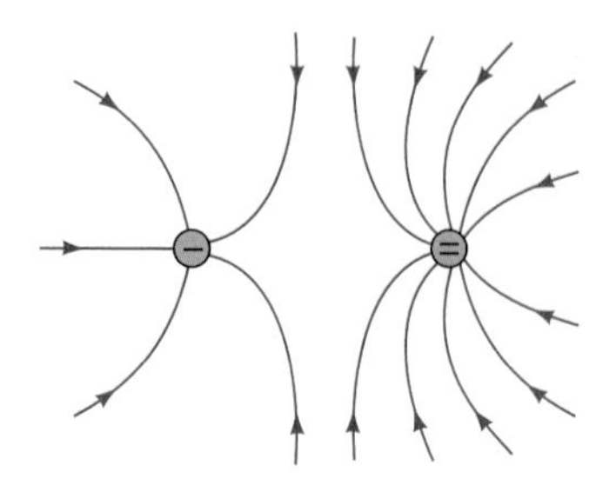

(b) Unequal like point charges

▲ **FIGURE 15.17** Electric fields Electric fields for **(a)** like point charges and **(b)** unequal like point charges.

The electric field patterns for some other common charge configurations are shown in ◂Fig. 15.17. You should be able to see how they are sketched qualitatively. Note that the electric field lines begin on positive charges and end on negative ones (or at infinity when there is no nearby negative charge). Choose the number of lines emanating from or ending at a charge in proportion to the magnitude of that charge. (See the Learn by Drawing on Sketching Electric Lines of Force on page 521.)

15.5 Conductors and Electric Fields

OBJECTIVES: To (a) describe the electric field near the surface and in the interior of a conductor, (b) determine where charge accumulates on a charged conductor, and (c) sketch the electric field line outside a charged conductor.

The electric fields associated with charged conductors have several interesting properties. By definition in electrostatics, the charges are at rest. Because conductors possess electrons that are free to move, and they don't, the electrons must experience no electric force and thus no electric field. Hence we conclude that:

The electric field is zero *inside* a charged conductor.

Excess charges on a conductor tend to get as far away from each other as possible, because they are highly mobile. Then:

Any *excess* charge on an isolated conductor resides entirely on the surface of the conductor.

Another property of static electric fields and conductors is that there cannot be a tangential component of the field at the surface of the conductor. If this were not true, charges would move *along* the surface, contrary to our assumption of a static situation. Thus:

The electric field at the surface of a charged conductor is perpendicular to the surface.

Lastly, the excess charge on a conductor of irregular shape is most closely packed where the surface is highly curved (at the sharpest points). Since the charge is densest there, the electric field will also be the largest at these locations. That is:

Excess charge tends to accumulate at sharp points, or locations of highest curvature, on charged conductors. As a result, the electric field is greatest at such locations.

These last two results are summarized in ▸Fig. 15.18. *Note that they are true only for conductors under static conditions*. Electric fields *can* exist inside nonconducting materials and inside conductors when conditions vary with time.

To understand *why* most of the charge accumulates in the highly curved surface regions, consider the forces acting *between* charges on the surface of the conductor. (See ▸Fig. 15.19a.) Where the surface is fairly flat, these forces will be directed nearly parallel to the surface. The charges will spread out until the parallel forces from neighboring charges in opposite directions cancel out. At a sharp end, the forces between charges will be directed more nearly perpendicular to the surface, and so there will be little tendency for the charges to move parallel to the surface. Therefore one would expect highly curved regions of the surface to accumulate the highest concentration of charge.

An interesting situation occurs if there is a large concentration of charge on a conductor with a sharp point (Fig. 15.19b). The electric field above the point may be high enough to ionize air molecules (to pull or push electrons off the molecules). The freed electrons are then further accelerated by the field and can cause secondary ionizations by striking other molecules. This results in an "avalanche" of electrons, visible as a spark discharge. More charge can be placed

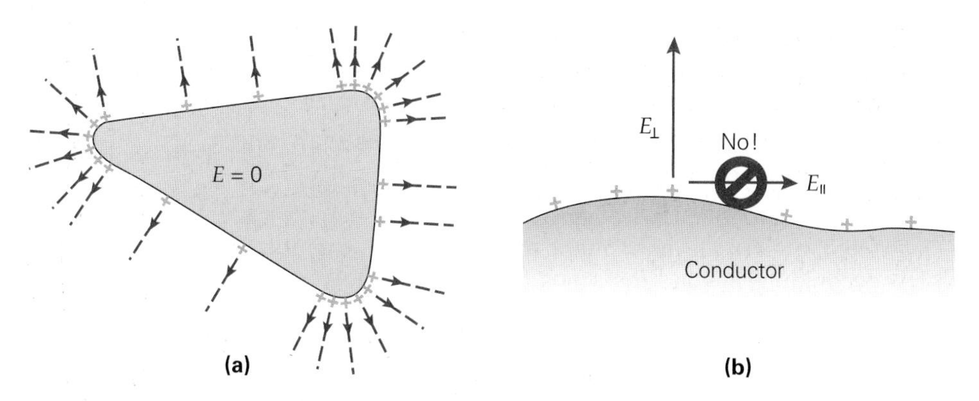

▲ **FIGURE 15.18 Electric fields and conductors (a)** Under static conditions, the electric field is zero inside a conductor. Any excess charge resides on the conductor's surface. For an irregularly shaped conductor, the excess charge accumulates in the regions of highest curvature (the sharpest points), as shown. The electric field near the surface is perpendicular to that surface and strongest where the charge is densest. **(b)** Under static conditions, the electric field must *not* have a component tangential to the conductor's surface.

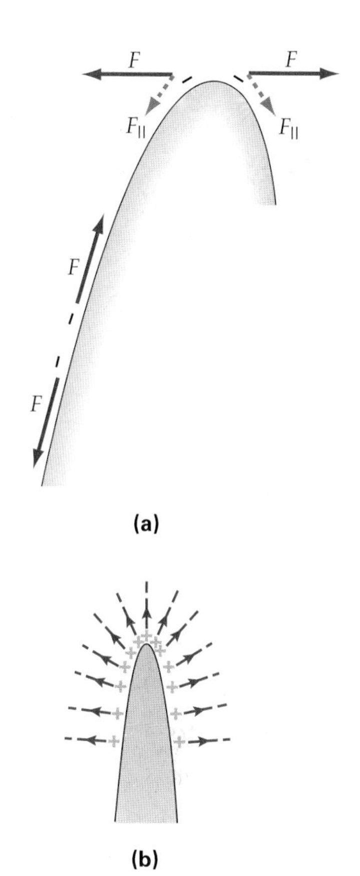

▲ **FIGURE 15.19 Concentration of charge on a curved surface (a)** On a flat surface, the repulsive forces between excess charges are parallel to the surface and tend to push the charges apart. On a sharply curved surface, in contrast, these forces are directed at an angle to the surface. Their components parallel to the surface are smaller, allowing charge to concentrate in such areas. **(b)** Taken to the extreme, a sharply pointed metallic needle has a dense concentration of charge at the tip. This produces a large electric field in the region above the tip, which is the principle of the lightning rod.

on a gently curved conductor, such as a sphere, before a spark discharge will occur. The concentration of charge at the sharp point of a conductor is one reason for the effectiveness of lightning rods. (See Insight 15.1 on Lightning and Lightning Rods on page 523.)

For some law enforcement and biological applications of electric fields and conductors, refer to Insight 15.2 on Stun Guns and Electric Fish on page 524.

As an illustration of an early experiment done on conductors with excess charge, consider the following Example.

Conceptual Example 15.10 ■ The Classic Ice Pail Experiment

A positively charged rod is held inside an isolated metal container that has uncharged electroscopes conductively attached to its inside surface and to its outside surface (▼Fig. 15.20). What will happen to the leaves of the electroscopes? (Justify your answer.) (a) Neither electroscope's leaf will show a deflection. (b) Only the outside-connected electroscope's leaf will show a deflection. (c) Only the inside-connected electroscope's leaf will show a deflection. (d) The leaves of both electroscopes will show deflections.

Reasoning and Answer. The positively charged rod will attract negative charges, causing the inside of the metal container to become negatively charged. The outside electroscope will thus acquire a positive charge. Hence, both electroscopes will be charged (though with opposite signs) and show deflections, so the answer is (d). This experiment was performed by the nineteenth-century English physicist Michael Faraday using ice pails, so this setup is often called *Faraday's ice pail experiment*.

Follow-Up Exercise. Suppose in this Example that the positively charged rod actually *touched* the metal container. How would the electroscopes react now?

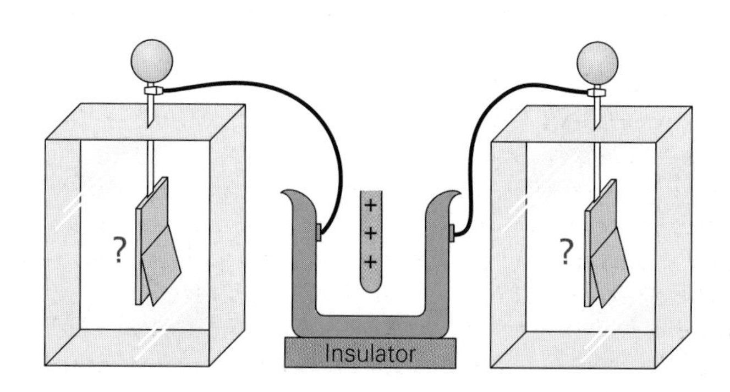

◀ **FIGURE 15.20 An ice pail experiment** See Conceptual Example 15.10.

*15.6 Gauss's Law for Electric Fields: A Qualitative Approach

OBJECTIVES: To (a) state the physical basis of Gauss's law, and (b) use the law to make qualitative predictions.

One of the fundamental laws of electricity was discovered by Karl Friedrich Gauss (1777–1855), a German mathematician. Using it for quantitative calculations involves techniques beyond the scope of this book. However, a conceptual look at this law can teach us some interesting physics.

Consider the single positive charge in ◀Fig. 15.21a. Now picture an *imaginary closed surface* surrounding this charge. Such a surface is called a **Gaussian surface**. Let us designate electric field lines that pass through the surface outwardly as positive and inward pointing ones as negative. If the lines of both types are counted and totaled (that is, subtract the number of negative lines from the number of positive ones), we find that the total is positive, because in this case there are *only* positive lines. This result reflects the fact that there is a net number of outward-pointing electric field lines through the surface. Similarly, for a negative charge (Fig. 15.21b), the count would yield a negative total, indicating a net number of inward-pointing lines passing through the surface. Note that these results would be true for *any* closed surface surrounding the charge, regardless of its shape or size. If we double the magnitude of the negative charge (Fig. 15.21c), our negative field line count would also double. (Why?)

Figure 15.21d shows a dipole with four different imaginary Gaussian surfaces. surface 1 encloses a net positive charge and therefore has a positive line count. Similarly, surface 2 has a negative line count. The more interesting cases are surfaces 3 and 4. Note that both include zero net charge—surface 3 because it includes no charges at all and Surface 4 because it includes equal and opposite charges. Note that both surfaces 3 and 4 have a net line count of zero, correlating with no net charge enclosed.

These situations can be generalized (conceptually) to give us the underlying physical principle of **Gauss's law:***

> The net number of electric field lines passing through an imaginary closed surface is proportional to the amount of net charge enclosed within that surface.

An everyday analogy illustrated in ▼Fig. 15.22 may help you understand this principle. If you surround a lawn sprinkler with an imaginary surface (surface 1), you find that there is a net flow of water out through that surface—because inside is a "source" of water (disregarding the pipe's bringing water into the sprinkler). In an analogous way, a net outward-pointing electric field indicates the presence of a net positive charge inside the surface, because positive charges are "sources" of the electric field. Similarly, a puddle would form inside our imaginary surface 2 because there would be a net inward water flow through the surface. The following Example illustrates the power of Gauss's law in its qualitative form.

Gaussian surface

(a)

Gaussian surface

(b)

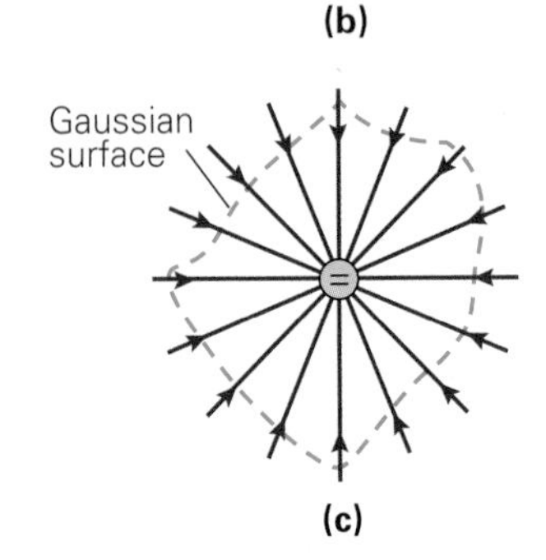

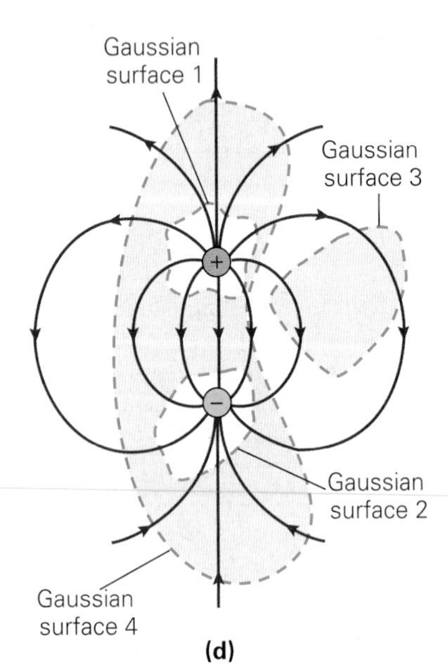

▲ **FIGURE 15.21 Various Gaussian surfaces and lines of force** **(a)** Surrounding a single positive point charge, **(b)** surrounding a single negative point charge, and **(c)** surrounding a larger negative point charge. **(d)** Four different surfaces surrounding various parts of an electric dipole.

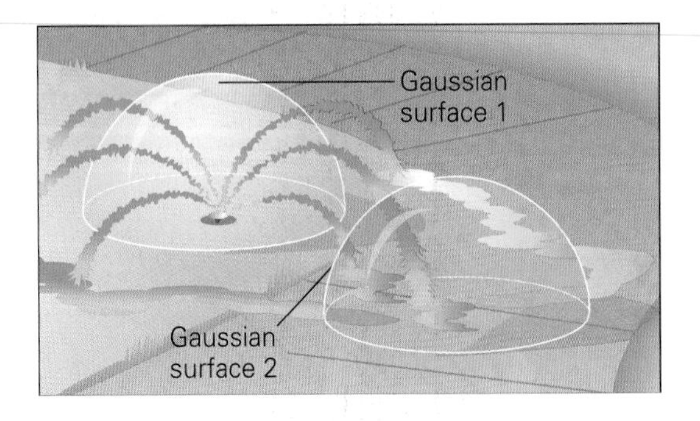

◀ **FIGURE 15.22 Water analogy to Gauss's law** A net outward flow of water indicates a source of water inside closed surface 1. A net inward flow of water indicates a water drain inside closed surface 2.

*Strictly speaking, this is Gauss's law for electric fields. There is also a version of Gauss's law for magnetic fields, which will not be discussed.

Conceptual Example 15.11 ■ Charged Conductors Revisited: Gauss's Law

A net charge Q is placed on a conductor of arbitrary shape (▸Fig. 15.23). Use the qualitative version of Gauss's law to prove that all the charge must lie on the conductor's surface under electrostatic conditions.

Reasoning and Answer. Because the situation is static equilibrium, there can be no electric field inside the volume of the conductor; otherwise, the almost-free electrons would move around. Take a Gaussian surface that follows the shape of the conductor, but is *just barely* inside the actual surface. Because there are no electric field lines inside the conductor, there are also no electric field lines passing through our imaginary surface. Thus zero electric field lines penetrate the Gaussian surface. But by Gauss's law, the net number of field lines is proportional to the amount of charge inside the surface. Therefore, there must be no net charge within the surface.

Because our surface can be made as close as we wish to the conductor surface, it follows that the excess charge, if it cannot be inside the volume of the conductor, must be on the surface.

Follow-Up Exercise. In this Example, if the net charge on the conductor is negative, what is the sign of the net number of lines through a Gaussian surface that completely encloses the conductor? Explain your reasoning.

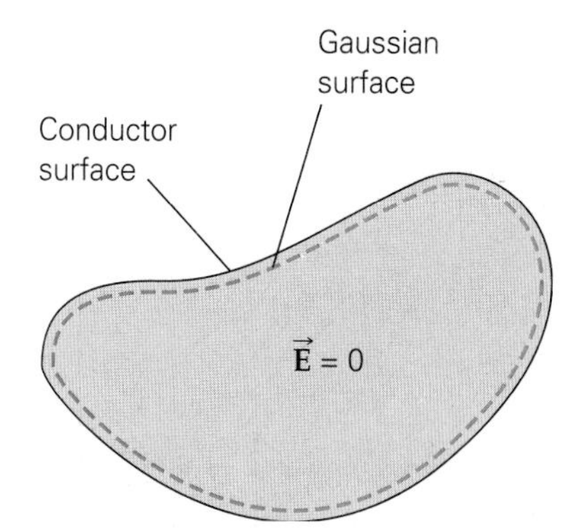

▲ **FIGURE 15.23 Gauss's law: Excess charge on a conductor** See Conceptual Example 15.12.

Chapter Review

- The **law of charges**, or **charge–force law**, states that like charges repel and opposite charges attract.

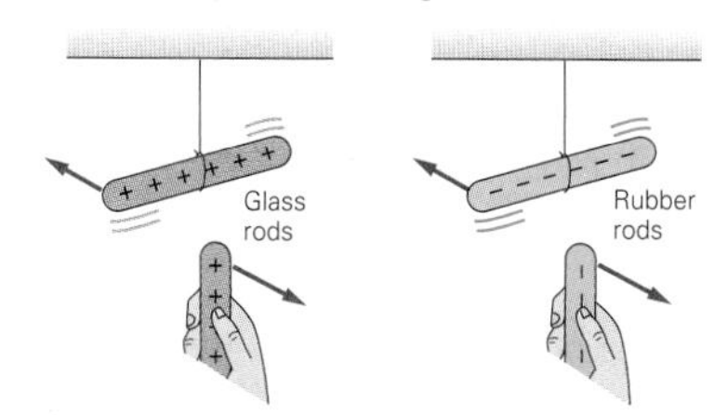

- **Charge conservation** means that the net charge of an isolated system remains constant.
- **Conductors** are materials that conduct electric charge readily because their atoms have one or more loosely bound electrons.
- **Insulators** are materials that do not easily gain, lose, or conduct electric charge.
- **Electrostatic charging** involves processes that enable an object to gain a net charge. Among these processes are charging by friction, contact (conduction), and induction.

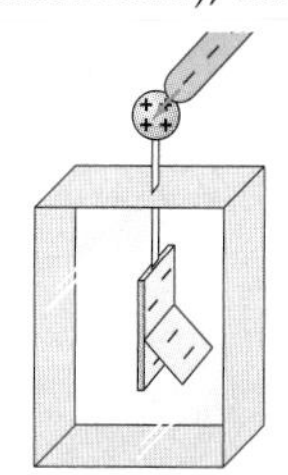

- The **electric polarization** of an object involves creating separate and equal amounts of positive and negative charge in different locations on that object.

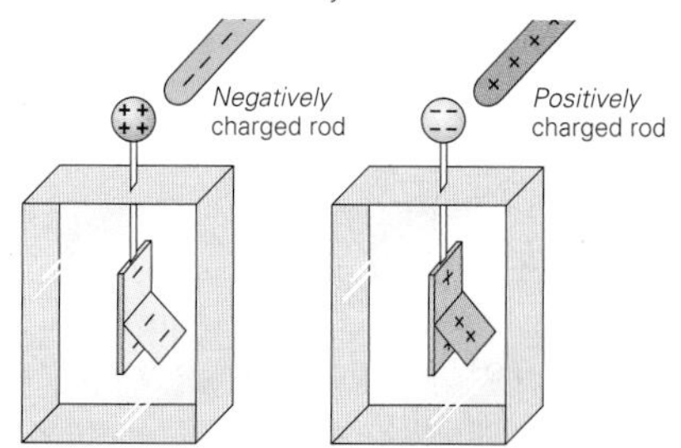

- **Coulomb's law** expresses the magnitude of the force between two point charges:

$$F_e = \frac{kq_1q_2}{r^2} \quad \textit{(two point charges)} \qquad (15.2)$$

where $k \approx 9.00 \times 10^9\ \mathrm{N \cdot m^2/C^2}$.

$$F_{12} = \frac{kq_1q_2}{r^2} \qquad q_1 \qquad\qquad q_2 \qquad F_{21} = \frac{kq_1q_2}{r^2}$$

- The **electric field** is a vector field that describes how charges modify the space around them. It is defined as the electric force per unit positive charge, or

$$\vec{\mathbf{E}} = \frac{\vec{\mathbf{F}}_{\text{on } q_+}}{q_+} \qquad (15.3)$$

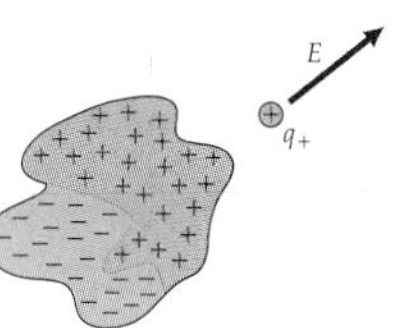

- According to the **superposition principle for electric fields**, the (net) electric field at any location due to a configuration of charges is the vector sum of the individual electric fields from individual charges of that configuration.

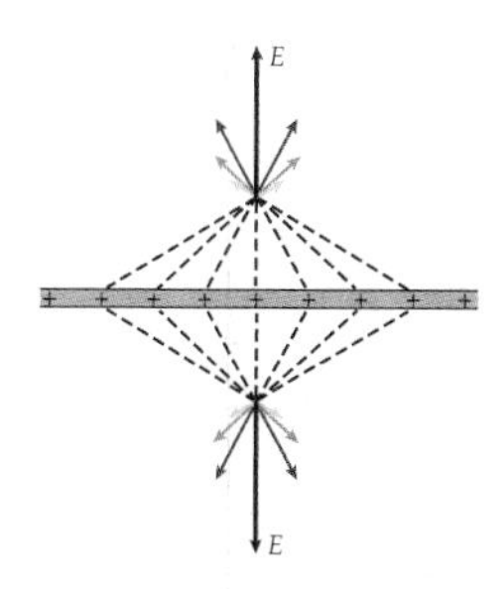

- **Electric field lines** are a visualization of the electric field. The line spacing is inversely related to the field strength, and tangents to the lines give the field direction.

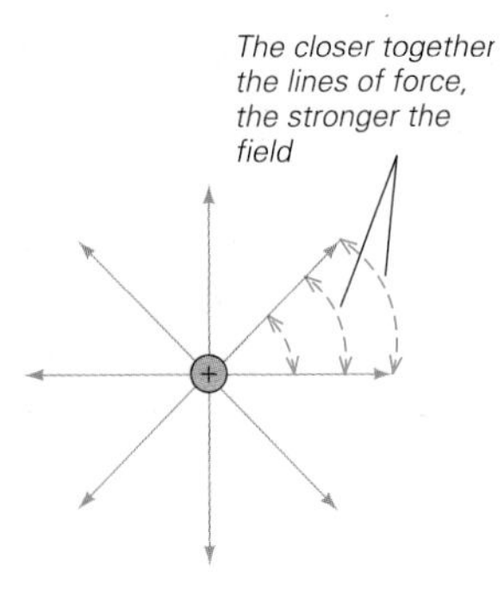
The closer together the lines of force, the stronger the field

- Under electrostatic conditions, the **electric fields associated with conductors** have the following properties:

  The electric field is zero inside a charged conductor.
  Excess charge on a conductor resides entirely on its surface.
  The electric field near the surface of a charged conductor is perpendicular to that surface.
  Excess charge on the surface of a conductor is most dense at locations of highest surface curvature.
  The electric field near the surface of a charged conductor is greatest at locations of highest surface curvature.

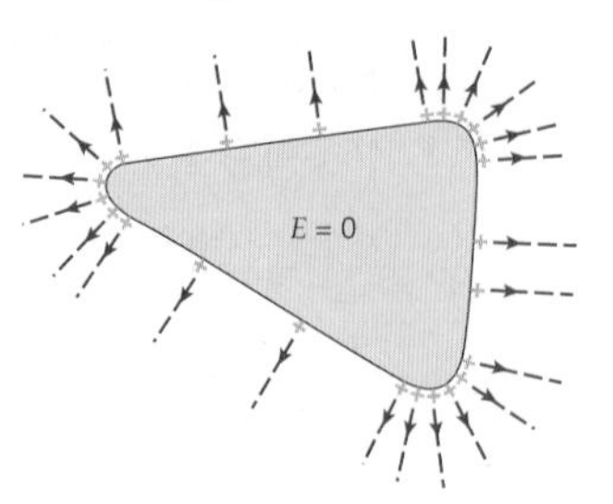
E = 0

Exercises*

MC = *Multiple Choice Question*, **CQ** = *Conceptual Question, and* **IE** = *Integrated Exercise. Throughout the text, many exercise sections will include "paired" exercises. These exercise pairs, identified with* **red numbers**, *are intended to assist you in problem solving and learning. In a pair, the first exercise (even numbered) is worked out in the Study Guide so that you can consult it should you need assistance in solving it. The second exercise (odd numbered) is similar in nature, and its answer is given at the back of the book.*

15.1 Electric Charge

1. **MC** A combination of two electrons and three protons would have a net charge of (a) +1, (b) −1, (c) $+1.6 \times 10^{-19}$ C, (d) -1.6×10^{-19} C.

2. **MC** An electron is just above a fixed proton. The direction of the force on the electric proton is (a) up, (b) down, (c) zero.

3. **MC** In Exercise 2, which one feels the bigger size force: (a) The electron, (b) proton, or (c) both feel the same size force?

4. **CQ** (a) How do we know that there are two types of electric charge? (b) What would be the effect of designating the charge on the electron as positive and the charge on the proton as negative?

5. **CQ** An electrically neutral object can be given a net charge by several means. Does this violate the conservation of charge? Explain.

6. **CQ** If a solid neutral object becomes positively charged, does its mass increase or decrease? What if it becomes negatively charged?

7. **CQ** How can you determine the type of charge on an object using an electroscope that has a net charge of a known sign? Explain.

*Take k to be exact at $9.00 \times 10^9\ \mathrm{N \cdot m^2/C^2}$ and e to be exact at 1.60×10^{-19} C for significant-figure purposes.

8. **CQ** If two objects electrically repel each other, are both necessarily charged? What if they attract each other?

9. ● What is the net charge of an object that has 1.0 million excess electrons?

10. ● In walking across a carpet, you acquire a net negative charge of 50 μC. How many excess electrons do you have?

11. ●● An alpha particle is the nucleus of a helium atom with no electrons. What would be the charge on two alpha particles?

12. **IE** ●● A glass rod rubbed with silk acquires a charge of $+8.0 \times 10^{-10}$ C. (a) Is the charge on the silk (1) positive, (2) zero, or (3) negative? Why? (b) What is the charge on the silk, and how many electrons have been transferred to the silk? (c) How much mass has the glass rod lost?

13. **IE** ●● A rubber rod rubbed with fur acquires a charge of -4.8×10^{-9} C. (a) Is the charge on the fur (1) positive, (2) zero, or (3) negative? Why? (b) What is the charge on the fur, and how much mass is transferred to the rod? (c) How much mass has the rubber rod gained?

15.2 Electrostatic Charging

14. **MC** A rubber rod is rubbed with fur. The fur is then quickly brought near the bulb of an uncharged electroscope. The sign of the charge on the leaves of the electroscope is (a) positive, (b) negative, (c) zero.

15. **MC** A stream of water is deflected toward a nearby electrically charged object that is brought close to it. The sign of the charge on the object (a) is positive, (b) is negative, (c) is zero, (d) can't be determined by the data given.

16. **MC** A balloon is charged and then clings to a wall. The sign of the charge on the balloon (a) is positive, (b) is negative, (c) is zero, (d) can't be determined by the data given.

17. **CQ** Fuel trucks often have metal chains reaching from their frames to the ground. Why is this important?

18. **CQ** Is there a gain or loss of electrons when an object is electrically polarized? Explain.

19. **CQ** Explain carefully the steps you would use to create an electroscope that is positively charged by induction. After you are done, how can you verify that the electroscope is positively (and thus not negatively) charged?

20. **CQ** Two metal spheres mounted on insulated supports are in contact. Bringing a negatively charged object close to the right-hand sphere would enable you to temporarily charge both spheres by induction. Explain clearly how this would work and what the sign of the charge on each sphere would be.

15.3 Electric Force

21. **MC** How does the magnitude of the electric force between two point charges change as the distance between them is increased? The force (a) decreases, (b) increases, (c) stays the same.

22. **MC** Compared with the electric force, the gravitational force between two protons is (a) about the same, (b) somewhat larger, (c) very much larger, (d) very much smaller.

23. **CQ** The Earth attracts us by its gravitational force, but we have seen that the electric force is much greater than the gravitational force. Why don't we experience an electric force from the Earth?

24. **CQ** Two nearby electrons would fly apart if released. How could you prevent this by placing a single charge in their neighborhood? Explain clearly what the sign of the charge and its location would have to be.

25. **CQ** Coulomb's law is an example of an inverse-square law. Use this inverse-square idea to determine the ratio of electric force (final divided by initial) between two charges when the distance between them is cut to one third of its initial value.

26. **IE** ● An electron that is a certain distance from a proton is acted on by an electrical force. (a) If the electron were moved twice that distance away from the proton, would the electrical force be (1) 2, (2) $\frac{1}{2}$, (3) 4, or (4) $\frac{1}{4}$ times the original force? Why? (b) If the initial electric force is F, and the electron were moved to one third the original distance toward the proton, what would be the new electrical force?

27. ● Two identical point charges are a fixed distance apart. By what factor would the magnitude of the electric force between them change if (a) one of their charges were doubled and the other were halved, (b) both their charges were halved, and (c) one charge were halved and the other were left unchanged?

28. ● In a certain organic molecule, the nuclei of two carbon atoms are separated by a distance of 0.25 nm. What is the magnitude of the electric force between them?

29. ● An electron and a proton are separated by 2.0 nm. (a) What is the magnitude of the force on the electron? (b) What is the net force on the system?

30. **IE** ● Two charges originally separated by a certain distance are moved farther apart until the force between them has decreased by a factor of 10. (a) Is the new distance (1) less than 10, (2) equal to 10, or (3) greater than 10 times the original distance? Why? (b) If the original distance was 30 cm, how far apart are the charges?

31. ● Two charges are brought together until they are 100 cm apart, causing the electric force between them to increase by a factor of exactly 5. What was their initial separation distance?

32. ● The distance between neighboring singly charged sodium and chlorine ions in crystals of table salt (NaCl) is 2.82×10^{-10} m. What is the attractive electric force between the ions?

33. ●● Two point charges of $-2.0\ \mu C$ are fixed at opposite ends of a meterstick. Where on the meterstick could (a) a free electron and (b) a free proton be in electrostatic equilibrium?

34. ●● Two point charges of $-1.0\ \mu C$ and $+1.0\ \mu C$ are fixed at opposite ends of a meterstick. Where could (a) a free electron and (b) a free proton be in electrostatic equilibrium?

35. ●● Two charges, q_1 and q_2, are located at the origin and at (0.50 m, 0), respectively. Where on the x-axis must a third charge, q_3, of arbitrary sign be placed to be in electrostatic equilibrium if (a) q_1 and q_2 are like charges of equal magnitude, (b) q_1 and q_2 are unlike charges of equal magnitude, and (c) $q_1 = +3.0\ \mu C$ and $q_2 = -7.0\ \mu C$?

36. ●● Compute the gravitational force and electrical force between the electron and proton in the hydrogen atom (▼Fig. 15.24) assuming they are 5.3×10^{-11} m apart. Then calculate the ratio of the magnitudes of the electric force to the gravitational force.

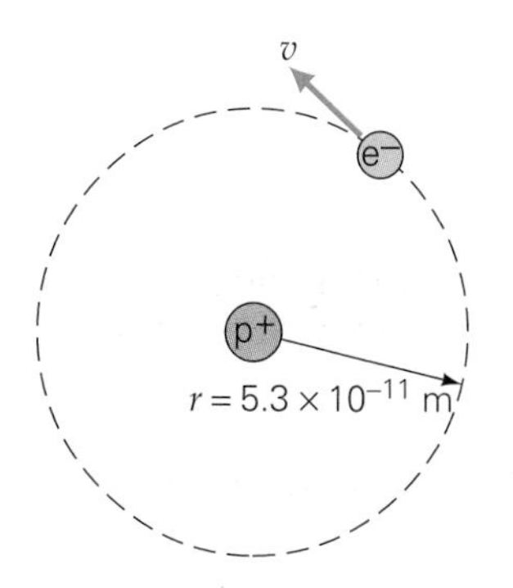

◀ **FIGURE 15.24 Hydrogen atom** See Exercises 36 and 37.

37. ●●● On average, the electron and proton in a hydrogen atom are separated by a distance of 5.3×10^{-11} m (Fig. 15.24). Assuming the orbit of the electron to be circular, (a) what is the electric force on the electron? (b) What is the electron's orbital speed? (c) What is the magnitude of the electron's centripetal acceleration in units of g?

38. ●●● Three charges are located at the corners of an equilateral triangle, as depicted in ▼Fig. 15.25. What are the magnitude and the direction of the force on q_1?

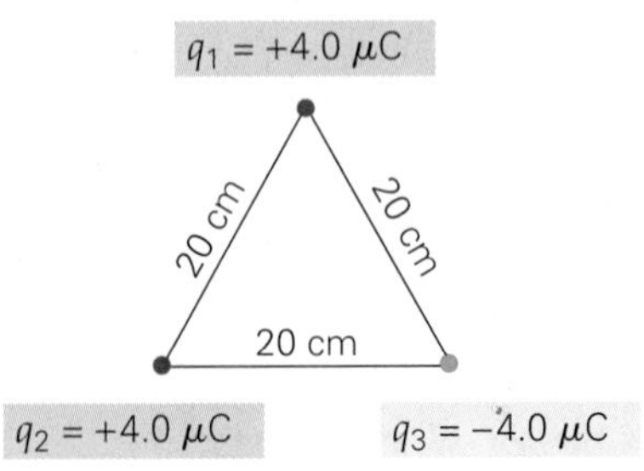

▲ **FIGURE 15.25 Charge triangle** See Exercises 38, 59, and 60.

39. ●●● Four charges are located at the corners of a square, as illustrated in ▼Fig. 15.26. What are the magnitude and the direction of the force (a) on charge q_2 and (b) on charge q_4?

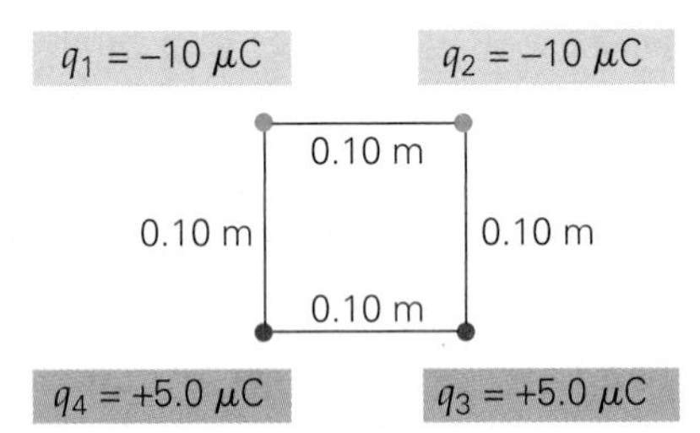

▲ **FIGURE 15.26 Charge rectangle** See Exercises 39, 61, and 65.

40. ●●● Two 0.10-g pith balls are suspended from the same point by threads 30 cm long. (Pith is a light insulating material once used to make helmets worn in tropical climates.) When the balls are given equal charges, they come to rest 18 cm apart, as shown in ▼Fig. 15.27. What is the magnitude of the charge on each ball? (Neglect the mass of the thread.)

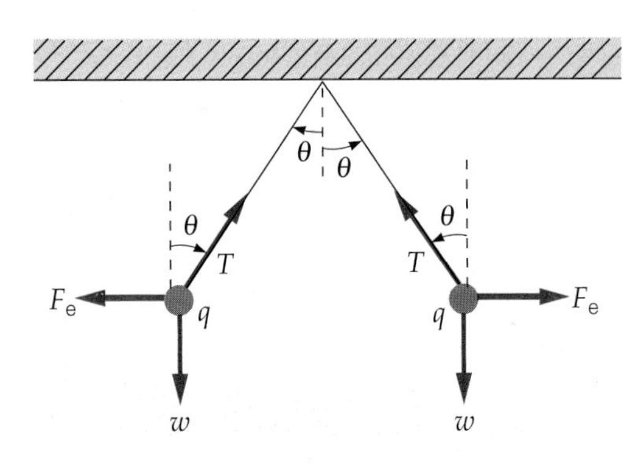

▲ **FIGURE 15.27 Repelling pith balls** See Exercise 40.

15.4 Electric Field

41. **MC** How is the magnitude of the electric field due to a point charge reduced when the distance from that charge is tripled: (a) It stays the same, (b) it is reduced to one third of its original value charge, (c) it is reduced to one ninth of its original value charge, or (d) it is reduced to one twenty-seventh of its original value charge?

42. **MC** The SI units of electric field are (a) C, (b) N/C, (c) N, (d) J.

43. **MC** At a point in space, an electric force acts vertically upward on an electron. The direction of the electric field at that point is (a) down, (b) up, (c) zero, (d) undetermined by the data.

44. **CQ** How is the relative magnitude of the electric field in different regions determined from a field vector diagram?

45. **CQ** How can the relative magnitudes of the field in different regions be determined from an electric field line diagram?

46. **CQ** Explain clearly why electric field lines can never cross.

47. **CQ** A positive charge is inside an isolated metal spherical shell, as shown in ▼Fig. 15.28. Describe the electric field in the three regions: between the charge and the inside surface of the shell, inside the shell itself, and outside the outer shell surface. What is the sign of the charge on the two shell surfaces? How would your answers change if the charge were negative?

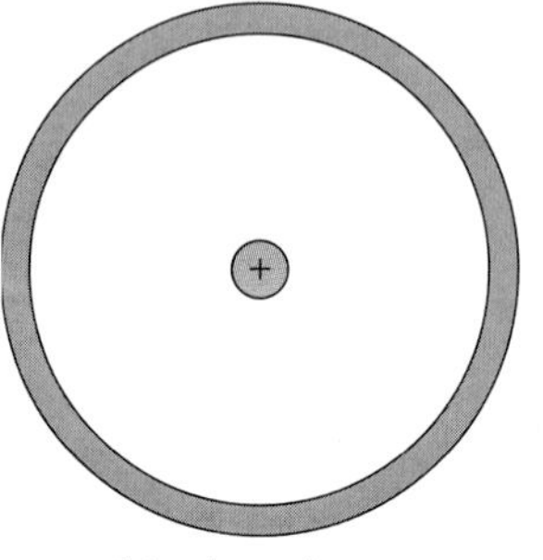

◀ **FIGURE 15.28 A point charge inside a thick metal spherical shell** See Exercise 47.

48. **CQ** At a certain location, the electric field due to the excess charge on the Earth's surface points downward. What is the sign of the charge on the Earth's surface at that location? Why?

49. **CQ** (a) Could the electric field due to two identical negative charges ever be zero at some location(s) nearby? Explain. If yes, describe and sketch the situation. (b) How would your answer change if the charges were equal but oppositely charged? Explain.

50. **IE** ● (a) If the distance from a charge is doubled, is the magnitude of the electric field (1) increased, (2) decreased, or (3) the same compared to the initial value? (b) If the original electric field due to a charge is 1.0×10^{-4} N/C, what is the magnitude of the new electric field at twice the distance from the charge?

51. ● An electron is acted on by an electric force of 3.2×10^{-14} N. What is the magnitude of the electric field at the electron's location?

52. ● What is the magnitude and direction of the electric field at a point 0.75 cm away from a point charge of +2.0 pC?

53. ● At what distance from a proton is the magnitude of its electric field 1.0×10^5 N/C?

54. IE ●● Two fixed charges, $-4.0\ \mu$C and $-5.0\ \mu$C, are separated by a certain distance. (a) Is the net electric field at a location halfway between the two charges (1) directed toward the $-4.0\ \mu$C charge, (2) zero, or (3) directed toward the $-5.0\ \mu$C charge? Why? (b) If the charges are separated by 20 cm, calculate the magnitude of the net electric field halfway between the charges?

55. ●● What would be the magnitude and the direction of an electric field that would just support the weight of a proton near the surface of the Earth? What about an electron?

56. IE ●● Two charges, $-3.0\ \mu$C and $-4.0\ \mu$C, are located at $(-0.50$ m, 0) and (0.50 m, 0), respectively. There is a point on the x-axis between the two charges where the electric field is zero. (a) Is that point (1) left of the origin, (2) at the origin, or (3) right of the origin? (b) Find the location of the point where the electric field is zero.

57. ●● Three charges, $+2.5\ \mu$C, $-4.8\ \mu$C, and $-6.3\ \mu$C, are located at $(-0.20$ m, 0.15 m), (0.50 m, -0.35 m), and $(-0.42$ m, -0.32 m), respectively. What is the electric field at the origin?

58. ●● Two charges of $+4.0\ \mu$C and $+9.0\ \mu$C are 30 cm apart. Where on the line joining the charges is the electric field zero?

59. ●●● What is the electric field at the center of the triangle in Fig. 15.25?

60. ●●● Compute the electric field at a point midway between charges q_1 and q_2 in Fig. 15.25.

61. ●●● What is the electric field at the center of the square in Fig. 15.26?

62. ●●● A particle with a mass of 2.0×10^{-5} kg and a charge of $+2.0\ \mu$C is released in a (parallel-plate) uniform horizontal electric field of 12 N/C. (a) How far horizontally does the particle travel in 0.50 s? (b) What is the horizontal component of its velocity at that point? (c) If the plates are 5.0 cm on each side, how much charge is on each?

63. ●●● Two very large parallel plates are oppositely and uniformly charged. If the field between the plates is 1.7×10^6 N/C, how dense is the charge on each plate (in μC/m^2)?

64. ●●● Two square, oppositely charged conducting plates measure 20 cm on each side. The plates are close together and parallel to each other. Their charges are +4.0 nC and -4.0 nC, respectively. (a) What is the electric field between the plates? (b) What force is exerted on an electron between the plates?

65. ●●● Compute the electric field at a point 4.0 cm from q_2 along a line running toward q_3 in Fig. 15.26.

66. ●●● Two equal and opposite point charges form a dipole, as shown in ▼Fig. 15.29. (a) Add the electric fields due to each end at point P, thus graphically determining the direction of the field there. (b) Derive a symbolic expression for the magnitude of the electric field at point P, in terms of k, q, d, and x. (c) If point P is very far away, use the exact result to show that $E \approx kqd/x^3$. (d) Why is it an inverse-*cube* fall-off instead of inverse-*square*? Explain.

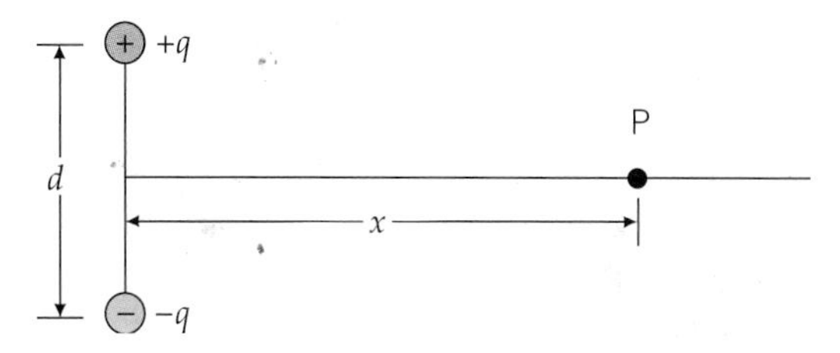

▲ **FIGURE 15.29 Electric dipole field** See Exercise 66.

15.5 Conductors and Electric Fields

67. **MC** In electrostatic equilibrium, is the electric field just below the surface of a charged conductor (a) the same value as the field just above the surface (b) zero, (c) dependent on the amount of charge on the conductor, or (d) given by kq/R^2?

68. **MC** An uncharged thin metal slab is placed in an external electric field that points horizontally to the left. What is the electric field inside the slab: (a) zero, (b) the same value as the original external field but oppositely directed, (c) less than the original external field value but not zero, or (d) depends on the magnitude of the external field?

69. **MC** The direction of the electric field at the surface of a charged conductor under electrostatic conditions (a) is parallel to the surface, (b) is perpendicular to the surface, (c) is at a 45° angle to the surface, or (d) depends on the charge on the conductor.

70. CQ Is it safe to stay in a car in a lightning storm (▼Fig. 15.30)? Explain.

▲ **FIGURE 15.30 Safe inside a car?** See Exercise 70.

71. **CQ** Under electrostatic conditions, the excess charge on a conductor is uniformly spread over its surface. What is the shape of the surface?

72. **CQ** Tall buildings have lightning rods to protect them from lightning strikes. Explain why the rods are pointed in shape and taller than the buildings.

73. **IE** ● A solid conducting sphere is surrounded by a thick, spherical conducting shell. Assume that a total charge $+Q$ is placed at the center of the sphere and released. (a) After equilibrium is reached, the inner surface of the shell will have (1) negative, (2) zero, (3) positive charge. (b) In terms of Q, how much charge is on the interior of the sphere? (c) The surface of the sphere? (d) The inner surface of the shell? (e) The outer surface of the shell?

74. ● In Exercise 73, what is the electric field direction (a) in the interior of the solid sphere, (b) between the sphere and the shell, (c) inside the shell, and (d) outside the shell?

75. ●● In Exercise 73, write expressions for the electric field magnitude (a) in the interior of the solid sphere, (b) between the sphere and the shell, (c) inside the shell, and (d) outside the shell. Your answer should be in terms of Q, r (the distance from the center of the sphere), and k.

76. ●● A flat, triangular piece of metal with rounded corners has a net positive charge on it. Sketch the charge distribution on the surface and the electric field lines near the surface of the metal (including their direction).

77. ●●● Approximate a metal needle as a long cylinder with a very pointed, but slightly rounded, end. Sketch the charge distribution and outside electric field lines if the needle has an excess of electrons on it.

*15.6 Gauss's Law for Electric Fields: A Qualitative Approach

78. **MC** A Gaussian surface surrounds an object with a net charge of $-5.0\ \mu C$. Which of the following is true? (a) More electric field lines will point outward than inward. (b) More electric field lines will point inward than outward. (c) The net number of field lines through the surface is zero. (d) There must be only field lines passing inward through the surface.

79. **MC** What can you say about the net number of electric field lines passing through a Gaussian surface located completely within the region between a set of oppositely charged parallel plates? (a) The net number point outward. (b) The net number point inward. (c) The net number is zero. (d) The net number depends on the amount of charge on each plate.

80. **MC** Two concentric spherical surfaces enclose a charged particle. The radius of the outer sphere is twice that of the inner one. Which sphere will have more electric field lines passing through its surface? (a) The larger one. (b) The smaller one. (c) Both spheres would have the same number of field lines passing through them. (d) The answer depends on the amount of charge on the particle.

81. **CQ** The same Gaussian surface is used to surround two charged objects separately. The net number of field lines penetrating the surface is the same in both cases, but the lines are oppositely directed. What can you say about the net charges on the two objects?

82. **CQ** If a net number of electric field lines point outward from a Gaussian surface, does that necessarily mean there are no negative charges in the interior? Explain with an example.

83. ● Suppose a Gaussian surface encloses both a positive point charge that has six field lines leaving it and a negative point charge with twice the magnitude of charge of the positive one. What is the net number of field lines passing through the Gaussian surface?

84. **IE** ●● A Gaussian surface has 16 field lines leaving it when it surrounds a point charge of $+10.0\ \mu C$ and 75 field lines entering it when it surrounds an unknown point charge. (a) The magnitude of the unknown charge is (1) greater than $10.0\ \mu C$, (2) equal to $10.0\ \mu C$, or (3) less than $10.0\ \mu C$. Why? (b) What is the unknown charge?

85. ●● If 10 field lines leave a Gaussian surface when it completely surrounds the positive end of an electric dipole, what would the count be if the surface surrounded just the other end?

Comprehensive Exercises

86. A negatively charged pith ball (mass 6.00×10^{-3} g, charge -1.50 nC) is suspended vertically from a light nonconducting string of length 15.5 cm. This apparatus is then placed in a horizontal uniform electric field. After being released, the pith ball comes to a stable position at an angle of 12.3° to the left of the vertical. (a) What is the direction of the external electric field? (b) Determine magnitude of the electric field.

87. A positively charged particle with a charge of 9.35 pC is suspended in equilibrium in the electric field between two oppositely charged horizontal parallel plates. The square plates each have a charge 5.50×10^{-5} C, are separated by 6.25 mm, and have an edge length of 11.0 cm. (a) Which plate must be positively charged? (b) Determine the mass of the particle.

88. **CQ** Use the superposition principle and/or symmetry arguments to determine the direction of the electric field (a) at the center of a uniformly positively charged semicircular wire, (b) in the plane of a negatively charged flat plate, just off one of the edges, and (c) on the perpendicular bisector axis of a long thin insulator with more negative charge on one end than positive charge on the other end.

89. An electron starts from one plate of a charged closely spaced (vertical) parallel plate arrangement with a velocity of 1.63×10^4 m/s to the right. Its speed on reaching the other plate, 2.10 cm away, is 4.15×10^4 m/s. (a) What type of charge is on each plate? (b) What is the direction of the electric field between the plates? (c) If the plates are square with an edge length of 25.4 cm, determine the charge on each.

90. Two fixed charges, $-3.0\ \mu C$ and $-5.0\ \mu C$, are 0.40 m apart. (a) Where should a third charge of $-1.0\ \mu C$ be placed to put the system of three charges in electrostatic equilibrium? (b) What if the third charge was $+1.0\ \mu C$ instead?

91. Find the electric field at point O for the charge configuration shown in ▼Fig. 15.31.

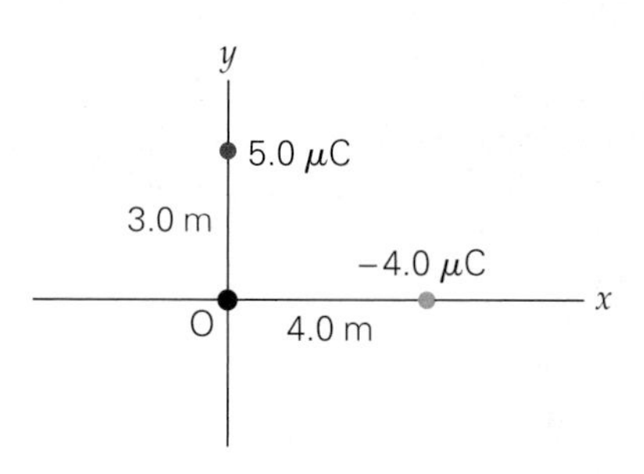

▲ **FIGURE 15.31 Electric field** See Exercise 91.

92. **CQ** A uniform metal slab (less thick than the plate separation distance) is inserted between and parallel to a pair of oppositely charged parallel plates. Sketch the resulting electric field everywhere between the plates including the slab.

93. An electron in a computer monitor enters midway between two parallel oppositely charged plates, as shown in ▼Fig. 15.32. The initial speed of the electron is 6.15×10^7 m/s and its vertical deflection (d) is 4.70 mm. (a) What is the magnitude of the electric field between the plates? (b) Determine the magnitude of the surface charge density on the plates in C/m^2.

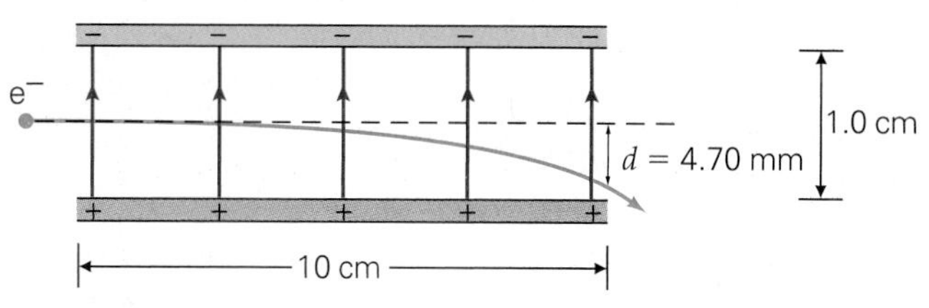

▲ **FIGURE 15.32 Electron in a computer monitor** See Exercise 93.

94. **CQ** For an electric dipole, the product qd is called the *dipole moment* and given the symbol p. The dipole moment is actually a vector $\vec{\mathbf{p}}$ that points from the negative to the positive end. Assuming an electric dipole is free to move and rotate and starts from rest, (a) use a sketch to show that if it is placed in a uniform field it will rotate as it tries to "line up" with the field direction. (b) What is different about the motion of the dipole if the field is not uniform?

The following Physlet Physics Problems can be used with this chapter.
22.1, 22.2, 22.3, 22.4, 22.5, 22.7, 22.8, 22.9, 23.1, 23.2, 23.4, 23.6, 23.7, 23.8

CHAPTER

16 ELECTRIC POTENTIAL, ENERGY, AND CAPACITANCE

16.1 Electric Potential Energy and Electric Potential Difference 537

16.2 Equipotential Surfaces and the Electric Field 543

16.3 Capacitance 549

16.4 Dielectrics 552

16.5 Capacitors in Series and in Parallel 557

PHYSICS FACTS

- The unit of electrical capacitance, the farad, is named for the British scientist Michael Faraday (1791–1867). With little formal education he was appointed (at age 21), a laboratory assistant at the Royal Institution (London). Eventually he became director of the laboratory. He discovered electromagnetic induction, which is the principle behind modern electric generating plants.
- In electrochemistry, an important amount of charge called the faraday is equal to 96 485.341 5 coulombs. The name honors Michael Faraday for his electrochemistry experiments that showed that 1 faraday of charge is required to deposit 1 mole of silver on the negatively charged cathode of his apparatus.
- Count Alessandro Volta was born in Como, Italy, in 1745. Because he did not speak until age 4, his family was convinced he was mentally retarded. However, in 1778 he was the first to isolate methane (the main component of natural gas). Like many chemists of his time, he did significant work with electricity in relationship to chemical reactions. He constructed the first electric battery, and the unit of electromotive force, the volt (V), was named in his honor.
- Electric eels can kill or stun prey by producing potential differences (or voltages) up to 650 volts, more than fifty times as large as that of a car battery. Other electric fish, such as the elephant nose, generate only about 1 volt, useful for electrolocation but not hunting.

The girl in the photo is experiencing some electrical effects as she is electrically charged to several thousand volts. Household circuits operate at 120 volts and can give you a potentially dangerous shock. Yet this girl doesn't seem to be having a problem. What's going on? You'll find the explanation of this and other electrical phenomena in this and the two following chapters. Here the concept of electric potential will be introduced and its properties and usefulness examined.

Although this chapter concentrates on the study of fundamental electrical concepts such as voltage and capacitance, there are discussions involving practical applications. For example, your dentist's X-ray machine works by using high voltage to accelerate electrons. Heart defribillators use capacitors to temporarily store the electrical energy required to stimulate the heart into its correct rhythm. Capacitors are used to store the energy that triggers the flash unit in your camera. Our body's nervous system, its communication network, is capable of sending thousands of electrical voltages per second shuttling back and forth along "cables" we call nerves. These signals are generated by chemical activity. The body uses them to enable us to do many things we take for granted such as muscle movement, thought processes, vision, and hearing. And, in following chapters, we will see even more practical uses of electricity, such as electric appliances, computers, medical instruments, electrical energy distribution systems, and household wiring.

16.1 Electric Potential Energy and Electric Potential Difference

OBJECTIVES: To (a) understand the concept of electric potential difference (voltage) and its relationship to electric potential energy, and (b) calculate electric potential differences.

In Chapter 15, electrical effects were analyzed in terms of electric field vectors and lines of force. Recall that the study of mechanics in early chapters began similarly through the use of Newton's laws, free-body diagrams, and forces (*vectors*). A search for a simpler approach led to *scalar* quantities such as work, kinetic energy, and potential energy. Using these quantities, energy methods were employed to solve problems that are much more difficult when using the vector (force) approach. It turns out to be extremely useful, both conceptually and for problem solving, to extend the energy methods to the study of electric fields.

Electric Potential Energy

To investigate electric potential energy, let's start with one of the simplest electric field patterns: the field between two large, oppositely charged parallel plates. As was learned in Chapter 15, near the center of the plates the field is uniform in magnitude and direction (▼Fig. 16.1a). Suppose a small positive charge q_+ is moved at constant velocity against the electric field, $\vec{\mathbf{E}}$, in a straight line from the negative plate (A) to the positive one (B). An external force ($\vec{\mathbf{F}}_{\text{ext}}$) with the same magnitude as the electric force is required (why?), and so we have $F_{\text{ext}} = q_+E$. The work done by this external force is positive, because the force and displacement are in the same direction. Then the work done by the external force is $W_{\text{ext}} = F_{\text{ext}}(\cos 0°)d = q_+Ed$.

Note: From Eq. 15.3, $\vec{\mathbf{E}} = \vec{\mathbf{F}}/q_+$, and $\vec{\mathbf{F}} = q_+\vec{\mathbf{E}}$.

Suppose the positive charge is now released from the positive plate. It will accelerate toward the negative plate, gaining kinetic energy. This kinetic energy is a result of the work done on the charge, and the initial energy (when there is no kinetic energy at B) must be some type of potential energy. In moving from A to B, the charge's **electric potential energy, U_e,** has increased ($U_B > U_A$) by an amount equal to the external work done on it. So the *change* in electric potential energy of the charge is

$$\Delta U_e = U_B - U_A = q_+Ed$$

The gravitational analogy to the parallel-plate electric field is the gravitational field near the Earth's surface, where it is uniform. When an object is raised a vertical distance h at constant velocity, the change in its potential energy is positive ($U_B > U_A$) and equal to the work done by the external (lifting) force. If it is assumed that there is no acceleration, this force must equal the object's weight, or $F_{\text{ext}} = w = mg$ (Fig. 16.1b). The increase in gravitational potential energy is then

Illustration 25.1 Energy and Voltage

$$\Delta U_g = U_B - U_A = F_{\text{ext}}h = mgh$$

(*Note*: Different distance symbols (h and d) are used to distinguish between the two different situations.)

▼ **FIGURE 16.1 Changes in potential energy in uniform electric and gravitational fields** **(a)** Moving a positive charge q_+ against the electric field requires positive work and increases the electric potential energy. **(b)** Moving a mass m against the gravitational field requires positive work and increases the gravitational potential energy.

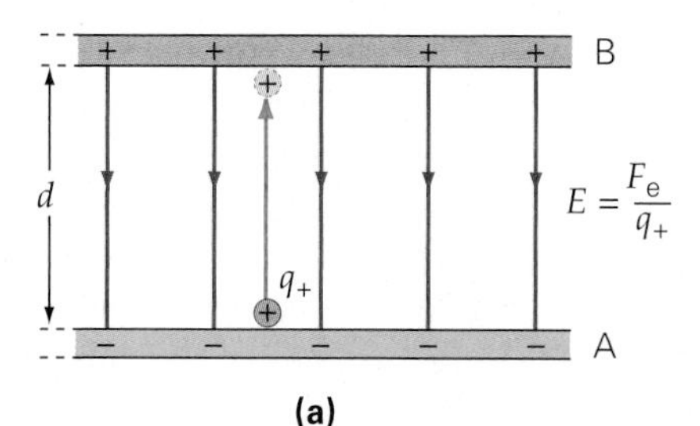

(a)

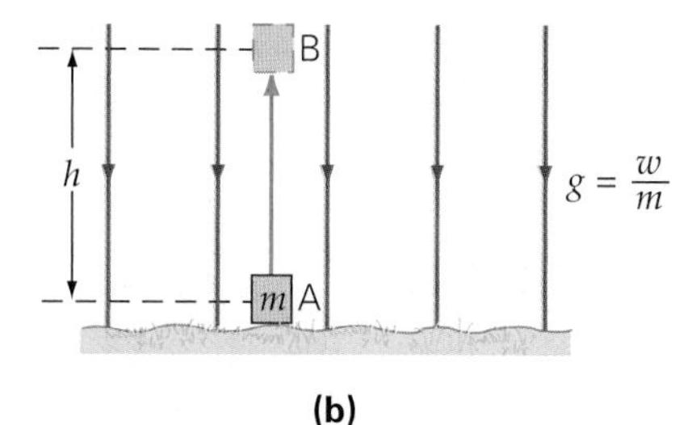

(b)

Electric Potential Difference

Recall that defining the electric field as the electric force *per unit positive test charge* eliminated the dependence on the test charge. Thus knowing the electric field, the force on *any* charge placed could be determined from $F_e = q_+E$. Similarly, the **electric potential difference, ΔV,** between any two points in space is defined as the change in potential energy *per unit positive test charge*:

$$\Delta V = \frac{\Delta U_e}{q_+} \quad \textit{(electric potential difference)} \tag{16.1}$$

SI unit of electric potential difference:
joule/coulomb (J/C), or volt (V)

The SI unit of electric potential difference is the joule per coulomb. This unit is named the **volt (V)** in honor of Alessandro Volta (1745–1827), an Italian scientist who constructed the first battery (Chapter 17), and 1 V = 1 J/C. Potential difference is commonly called **voltage**, and the symbol for potential difference is routinely changed from ΔV to just V, as is done later in this chapter.

Notice a crucial point: Electric potential difference, although based on electric potential-*energy* difference, is *not* the same. Electric potential difference is defined as electric potential energy difference *per unit charge*, and therefore does *not* depend on the amount of charge moved. Like the electric field, potential difference is a very useful quantity. If ΔV is known, we can then calculate ΔU_e for *any* amount of charge moved. To illustrate this, let's calculate the potential difference associated with the uniform field between two parallel plates:

$$\Delta V = \frac{\Delta U_e}{q_+} = \frac{q_+Ed}{q_+} = Ed \quad \begin{array}{l}\textit{potential difference}\\ \textit{(parallel plates only)}\end{array} \tag{16.2}$$

Notice that the amount of charge moved, q_+, cancels out. The potential difference ΔV depends only on the characteristics of the charged plates—that is, the field produced (E) and the separation (d). The language used is as follows:

For a pair of oppositely charged parallel plates, the positively charged plate is at a *higher electric potential* than the negatively charged one by an amount ΔV.

Notice that electric potential *difference* is defined without defining electric potential itself (V). Although this may seem backward, there is a good reason for it. Of the two, electric potential *difference* is the physically meaningful quantity; that is, the quantity we actually measure. (Electric potential differences, or voltages, are measured with voltmeters, see Chapter 18.). The electric potential V, in contrast, isn't definable in an absolute way—it depends entirely on the choice of a reference point. This means that an arbitrary constant can be added to, or subtracted from, potentials, changing them. However this has no affect on the meaningful quantity of potential *difference*.

We encountered this idea during the study of potential energy associated with springs and gravitation (Sections 5.2 and 5.4). Recall that only *changes* in potential energies were important. Specific values of potential energy could be determined, but only after the zero reference point was defined. For example, in the case of gravity, zero gravitational potential energy is sometimes chosen at the Earth's surface. However, it is just as correct (and sometimes more convenient) to define zero at an infinite distance from Earth (Section 7.5).

These ideas also hold for electric potential energy and potential. The electric potential may be chosen as zero at the negative plate of a pair of parallel plates. However, it is sometimes convenient to locate the zero value at infinity, as will be seen in the case of a point charge. Either way, *differences* are unaffected. For a visualization of this, refer to the Learn by Drawing feature on this page. In this figure, with a certain choice of zero electric potential, point A is a potential of +100 V and B at +300 V. With a different zero, the potential at A might be +1100 V, in which case, the electric potential at B would then be +1300 V. Regardless of the zero, B will *always* be 200 V higher in potential than A.

LEARN BY DRAWING

ΔV Is Independent of Reference Point

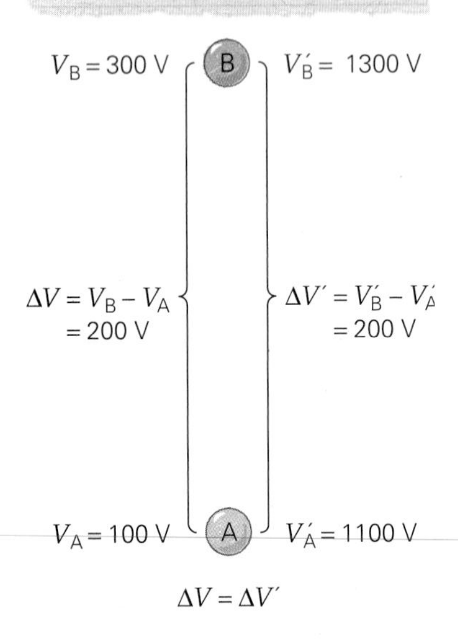

Note: We commonly assign a value of zero for the electric potential of the negative plate, but this is arbitrary. Only potential *differences* are meaningful.

The following Example illustrates the relationship between electric potential energy and electric potential.

Example 16.1 ■ Energy Methods in Moving a Proton: Potential Energy versus Potential

Imagine moving a proton from the negative plate to the positive plate of a parallel-plate arrangement (▸Fig. 16.2a). The plates are 1.50 cm apart, and the field is uniform with a magnitude of 1500 N/C. (a) What is the change in the proton's electric potential energy? (b) What is the electric potential difference (voltage) between the plates? (c) If the proton is released from rest at the positive plate (Fig. 16.2b), what speed will it have just before it hits the negative plate?

Thinking It Through. (a) The change in potential energy can be computed from the work required to move the charge. (b) The electric potential difference between the plates can then be found by dividing the work by the charge moved. (c) When the proton is released, its electric potential energy is converted into kinetic energy. Knowing the proton's mass, we can calculate its speed.

Solution. The magnitude of the electric field, E, is given. Because a proton is involved, we can find its mass and charge (Table 15.1).

Given: $E = 1500\ \text{N/C}$
$q_p = +1.60 \times 10^{-19}\ \text{C}$
$m_p = 1.67 \times 10^{-27}\ \text{kg}$
$d = 1.50\ \text{cm} = 1.50 \times 10^{-2}\ \text{m}$

Find: (a) ΔU_e (potential-energy change)
(b) ΔV (potential difference between plates)
(c) v (speed of released proton just before it reaches negative plate)

(a) Th electric potential energy is increased, so positive work is done to move the proton against the field, toward the positive plate:

$$\Delta U_e = q_p Ed = (+1.60 \times 10^{-19}\ \text{C})(1500\ \text{N/C})(1.50 \times 10^{-2}\ \text{m}) = +3.60 \times 10^{-18}\ \text{J}$$

(b) The potential difference, or voltage, is the potential energy *change* per unit charge (defined by Eq. 16.1):

$$\Delta V = \frac{\Delta U_e}{q_p} = \frac{+3.60 \times 10^{-18}\ \text{J}}{+1.60 \times 10^{-19}\ \text{C}} = +22.5\ \text{V}$$

We would say that the positive plate is 22.5 V higher in electric potential than the negative one.

(c) The total energy of the proton is constant; therefore, $\Delta K + \Delta U_e = 0$. The proton has no initial kinetic energy ($K_o = 0$). Hence, $\Delta K = K - K_o = K$. From this, the speed of the proton can be calculated:

$$\Delta K = K = -\Delta U_e$$

or

$$\tfrac{1}{2} m_p v^2 = -\Delta U_e$$

But on its return to the negative plate, the proton's potential energy change is negative (why?), hence $\Delta U_e = -3.60 \times 10^{-18}$ J. So its speed is

$$v = \sqrt{\frac{2(-\Delta U_e)}{m_p}} = \sqrt{\frac{2[-(-3.60 \times 10^{-18}\ \text{J})]}{1.67 \times 10^{-27}\ \text{kg}}} = 6.57 \times 10^4\ \text{m/s}$$

Notice that even though the kinetic energy gained is very small, the proton acquires a high speed because its mass is extremely small.

Follow-Up Exercise. In this Example, how would your answers change if an alpha particle were moved instead of a proton? (An alpha particle is the nucleus of a helium atom and has a charge of $+2e$ and a mass approximately four times that of a proton.) *(Answers to all Follow-Up Exercises are at the back of the text.)*

The principles in Example 16.1 can be used to show another interesting property of electric potential energy (and potential) changes: Both are *independent of the path* on which the charged particle is taken. Recall from Section 5.5 that this means the *electrostatic force is conservative.* As shown in Fig. 16.2c, the work done in moving the proton from A to B is the same, regardless of the route. The alternative wiggly paths from A to B and A to B′ require the same work as do the straight-line paths. This is fundamentally because movement at right angles to the field requires no work. (Why?)

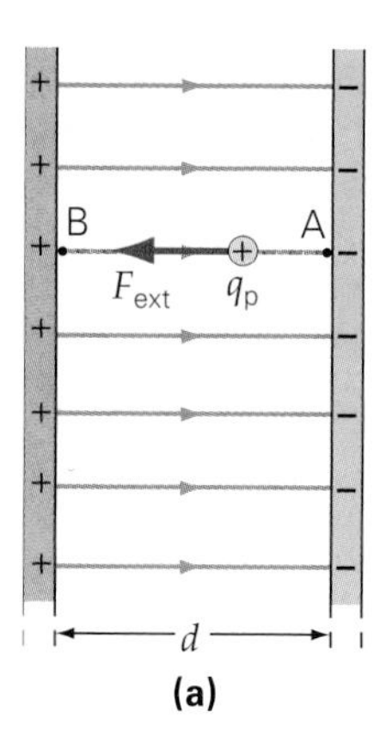

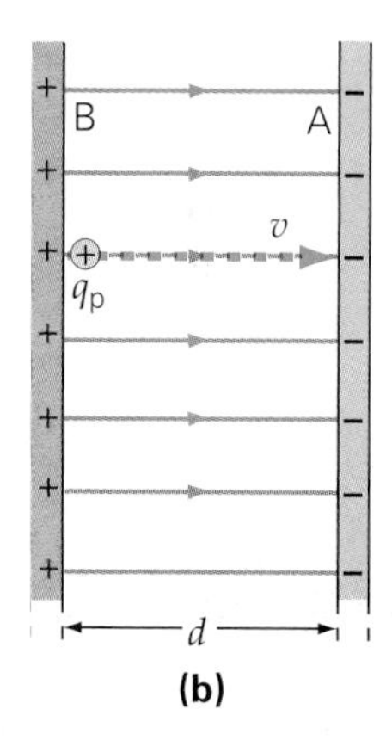

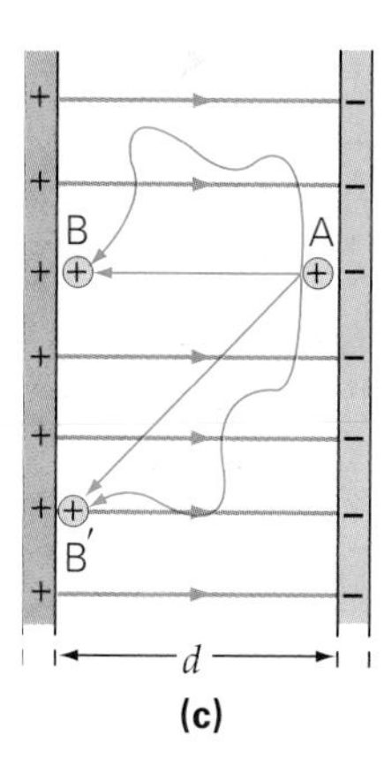

▲ **FIGURE 16.2** Accelerating a charge **(a)** Moving a proton from the negative to the positive plate increases the proton's potential energy. (See Example 16.1.) **(b)** When it is released from the positive plate, the proton accelerates toward the negative plate, gaining kinetic energy at the expense of electric potential energy. **(c)** The work done to move a proton between any two points in an electric field, such as A and B or A and B′, is independent of the path.

Note: Potential differences, like electric fields, are defined in terms of positive charges. Negative charges are subject to the same potential difference, but the opposite potential energy change. To determine whether the potential increases or decreases, decide whether an external force does positive or negative work on a *positive* test charge.

The gravitational analogy to Example 16.1 is that of raising an object in a uniform gravitational field. When the object is raised, its gravitational potential energy increases, because the force of gravity acts downward. However, with electricity we know that there are two types of charge, and the force between them can be repulsive or attractive. At this point, the analogy to gravity breaks down.

To understand why the analogy fails, consider how the discussion in Example 16.1 would differ if instead an electron had been moved. Because an electron is negatively charged, it would be attracted to plate B, and the external force would be *opposite* the electron's displacement (to prevent the electron from accelerating). For an electron, this force would do *negative* work, thereby *decreasing* the electric potential energy. Unlike the proton, the electron would be attracted to the positive plate (the plate with the higher electric potential). If allowed to move freely, electrons would "fall" (accelerate) toward regions of higher potential. Recall that the proton "fell" (accelerated) toward the region of lower potential. Regardless, both the proton and electron ended up losing electric potential energy and gaining kinetic energy. Thus the behavior of charged particles in electric fields can be summarized, in "potential language" as follows:

> Positive charges, when released, accelerate toward regions of lower electric potential.

> Negative charges, when released, accelerate toward regions of higher electric potential.

Consider the following medical application involving the creation of X-rays from fast-moving electrons, accelerated by large electric potential differences (voltages).

(a)

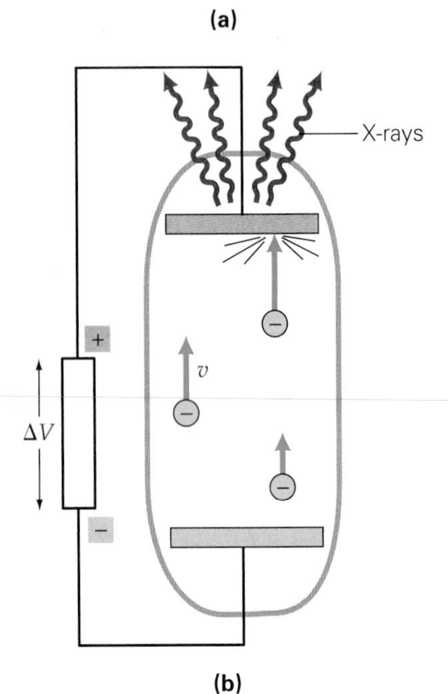

(b)

▲ **FIGURE 16.3** X-ray production **(a)** An illustration of a dental X-ray machine. **(b)** A schematic diagram of the X-ray production.

Example 16.2 ■ Creating X-Rays: Accelerating Electrons

Modern dental offices use X-ray machines for diagnosing hidden dental problems (◂Fig. 16.3a). Typically, electrons are accelerated through electric potential differences (voltages) of 25 000 V. When the electrons hit the positive plate, their kinetic energy is converted into high-energy particles called *X-ray photons* (Fig. 16.3b). (Photons are particles of light discussed in Chapter 27.) Suppose a single electron's kinetic energy is distributed equally among five X-ray photons. How much energy would one photon have?

Thinking It Through. From energy conservation, the kinetic energy gained by one electron is equal in magnitude to the electric potential energy it loses. From the kinetic energy lost by one electron, the energy of one X-ray photon can be calculated.

Solution. The charge of an electron is known (from Table 15.1), and the accelerating voltage is given.

Given: $q = -1.60 \times 10^{-19}\ \text{C}$, $\Delta V = 2.50 \times 10^{4}\ \text{V}$

Find: energy (E) of one X-ray photon

The electron leaves the negatively charged plate and moves toward the region of highest electric potential ("uphill"). Thus, the change in its electric potential energy is

$$\Delta U_e = q\Delta V = (-1.60 \times 10^{-19}\ \text{C})(+2.50 \times 10^{4}\ \text{V}) = -4.00 \times 10^{-15}\ \text{J}$$

The gain in kinetic energy comes from this loss in electric potential energy. Because the electrons have no appreciable kinetic energy when they start,

$$K = |\Delta U_e| = 4.00 \times 10^{-15}\ \text{J}$$

Therefore, if equally shared, one photon will have an energy of

$$E = \frac{K}{5} = 8.00 \times 10^{-16}\ \text{J}$$

Follow-Up Exercise. In this Example, use energy methods to determine the speed of one electron when it is halfway to the positive plate.

Electric Potential Difference Due to a Point Charge

In nonuniform electric fields, the potential difference between two points is determined by applying the fundamental definition (Eq. 16.1). However, in this case the field strength (and thus work done) varies, making the calculation beyond the scope of this text. The only nonuniform field we will consider in any detail is that due to a point charge (▸Fig. 16.4). Here we will simply state the result for the potential difference (voltage) between two points at distances r_A and r_B from a point charge q:

$$\Delta V = \frac{kq}{r_B} - \frac{kq}{r_A} \quad \textit{electric potential difference (point charge only)} \tag{16.3}$$

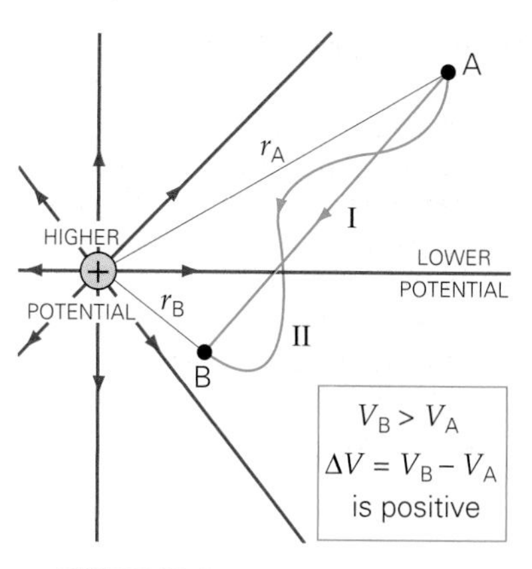

▲ **FIGURE 16.4 Electric field and potential due to a point charge** Electric potential increases as you move closer to a positive charge. Thus, B is at a higher potential than A.

In Fig. 16.4, the point charge is positive. Since point B is closer to the charge than A, the potential difference is positive, that is $V_B - V_A > 0$ or $V_B > V_A$. Thus B is at a higher potential than A. This is fundamentally because changes in potential are determined by visualizing the movement of a positive test charge. Here it takes *positive* work to move such a charge from A to B.

From this we see that electric potential increases as we move nearer to a positive charge. Notice also (in Fig. 16.4) that the work done on path II is the same as that for path I. Because the electric force is conservative, the potential difference is also the same, regardless of path.

Consider what would happen if the point charge were negative. In this case, B would be at a *lower* potential than A because the work required to move a positive test charge closer would be *negative* (why?).

Electric potential values change according to the following:

> Electric potential increases when moving nearer to positive charges *or* farther from negative charges

and

> Electric potential decreases when moving farther from positive charges *or* nearer to negative charges.

The potential at a very large distance from a point charge is usually chosen to be zero (as was done for the gravitational case of a point mass in Chapter 7). With this choice, the *electric potential* V at a distance r from a point charge is

$$V = \frac{kq}{r} \quad \textit{electric potential (point charge only, zero at infinity)} \tag{16.4}$$

Even though this expression is for the electric potential, V, keep in mind that only electric potential *differences* (ΔV) are important, as the next Example illustrates.

Integrated Example 16.3 ■ Describing the Hydrogen Atom: Potential Differences Near a Proton

According to the Bohr model of the hydrogen atom (Chapter 27), the electron in orbit around the proton can exist only in certain sized circular orbits. The smallest has a radius of 0.0529 nm, and the next largest has a radius of 0.212 nm. (a) How do the values of electric potential compare between each orbit: (1) The smaller one is at a higher potential, (2) the larger is at a higher potential, or (3) they both have the same potential? Explain your reasoning. (b) Verify your answer to part (a) by calculating the values of the electric potential at the locations of the two orbits.

(a) Conceptual Reasoning. The electron orbits in the field of a proton, whose charge is positive. Because electric potential increases with decreasing distance from a positive charge, the answer must be (1).

(b) Quantitative Reasoning and Solution. We know the charge on the proton, so Eq. 16.4 can be used to find the potential values. Listing the values,

Given: $q_p = +1.60 \times 10^{-19}$ C
$r_1 = 0.0529$ nm $= 5.29 \times 10^{-11}$ m
$r_2 = 0.212$ nm $= 2.12 \times 10^{-10}$ m (Note: 1 nm $= 10^{-9}$ m)

Find: The value of the electric potential (V) for each orbit

Applying Equation 16.4 we find, for the smaller orbit,

$$V_1 = \frac{kq_p}{r_1} = \frac{(9.00 \times 10^9\ \text{N}\cdot\text{m}^2/\text{C}^2)(+1.60 \times 10^{-19}\ \text{C})}{5.29 \times 10^{-11}\ \text{m}} = +27.2\ \text{V}$$

(continues on next page)

and for the larger orbit, we have

$$V_2 = \frac{kq_p}{r_2} = \frac{(9.00 \times 10^9 \text{ N}\cdot\text{m}^2/\text{C}^2)(+1.60 \times 10^{-19}\text{ C})}{2.12 \times 10^{-10}\text{ m}} = +6.79 \text{ V}$$

Follow-Up Exercise. In this Example, suppose the electron were moved from the smallest to the next orbit. (a) Has it moved to a region of higher or lower electrical potential? (b) What would be the change in the electric potential *energy* of the electron?

Electric Potential Energy of Various Charge Configurations

In Chapter 7 the gravitational potential energy of *systems* of masses was considered in some detail. The expressions for electric force and gravitational force are mathematically similar, and so are those for potential energy, except that charge takes the place of mass (remembering that charge comes in two signs). In the gravitational case of two masses, the mutual gravitational potential energy is negative, because the force is always *attractive*. For electric potential energy, the result can be positive *or* negative, because the electric force can be *repulsive* or *attractive*.

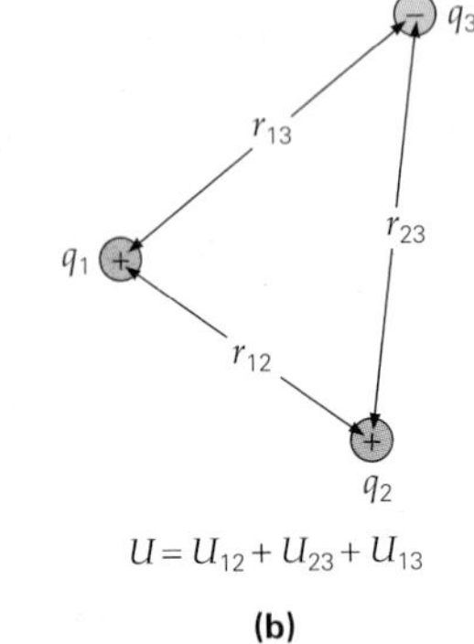

▲ **FIGURE 16.5 Mutual electric potential energy of point charges** **(a)** If a positive charge is moved from a large distance to a distance r_{12} from another positive charge, there is an increase in potential energy because positive work must be done to bring the mutually repelling charges closer. **(b)** For more than two charges, the system's electric potential energy is the sum of the mutual potential energies of each pair.

For example, consider a positive point charge, q_1, fixed in space. Suppose a second positive charge q_2 is brought toward it from a very large distance (that is, let its initial location $r \to \infty$) to a distance r_{12} (◂Fig. 16.5a). In this case, the work required is positive (why?). Therefore, this particular system gains electric potential energy. The potential at a large distance (V_∞) is, as is usual for point charges and masses, chosen as zero. (The zero point is arbitrary.) Thus, from Eq. 16.3, the change in potential energy is

$$\Delta U_e = q_2 \Delta V = q_2(V_1 - V_\infty) = q_2\left(\frac{kq_1}{r_{12}} - 0\right) = \frac{kq_1q_2}{r_{12}}$$

Because the large-distance value of the electric potential energy is chosen as zero, it follows that $\Delta U_e = U_{12} - U_\infty = U_{12}$. With this choice of reference, the potential energy of *any* two-charge system is

$$U_{12} = \frac{kq_1q_2}{r_{12}} \quad \textit{mutual electric potential energy (two charges)} \qquad (16.5)$$

Notice that for *unlike* charges the electric potential energy is negative and for *like* charges, the value is positive. So if the two charges are of the same sign, when released, they will move apart, gaining kinetic energy as they lose potential energy. Conversely, it would take positive work to increase the separation of two opposite charges, such as the proton and the electron, much like stretching a spring. (See the Follow-Up Exercise related to Integrated Example 16.3.)

Because energy is a scalar, for a configuration of any number of point charges, the *total* potential energy (U) is the algebraic sum of the mutual potential energies of all pairs of charges:

$$U = U_{12} + U_{23} + U_{13} + U_{14} \cdots \qquad (16.6)$$

Only the first three terms of Eq. 16.6 would be needed for the configuration shown in Fig. 16.5b. Note that the signs of the charges keep things straight mathematically, as the biomolecular situation in Example 16.4 shows.

Example 16.4 ■ Molecule of Life: The Electric Potential Energy of a Water Molecule

The water molecule is the foundation of life as we know it. Many of its properties (such as the reason it is a liquid on the Earth's surface) are related to the fact that it is a permanent polar molecule (see Section 15.4 on electric dipoles). A simple picture of the water molecule, including the charges, is shown in ▸Fig. 16.6. The distance from each hydrogen atom to the oxygen atom is 9.60×10^{-11} m, and the angle (θ) between the two hydrogen–oxygen bond directions is 104°. What is the total electrostatic energy of the water molecule?

Thinking It Through. The model of this molecule involves three charges. The charges are given, but the distance between the hydrogen atoms must be calculated using trigonometry. The total electrostatic potential energy is the algebraic sum of the potential energies of the three pairs of charges (that is, Eq. 16.6 will have three terms).

Solution. The following data are taken from Fig. 16.6.

Given: $q_1 = q_2 = +5.20 \times 10^{-20}$ C
$q_3 = -10.4 \times 10^{-20}$ C
$r_{13} = r_{23} = 9.60 \times 10^{-11}$ m
$\theta = 104°$

Find: U (total electrostatic potential of energy water molecule)

Notice that $(r_{12}/2)/r_{13} = \sin(\theta/2)$. Hence, we can solve for r_{12}:

$$r_{12} = 2r_{13}\left(\sin\frac{\theta}{2}\right) = 2(9.60 \times 10^{-11}\text{ m})(\sin 52°) = 1.51 \times 10^{-10}\text{ m}$$

Before determining the total potential energy of this system, let's calculate each pair's contribution separately. Note that $U_{13} = U_{23}$. (Why?) Applying Eq. 16.5,

$$U_{12} = \frac{kq_1q_2}{r_{12}} = \frac{(9.00 \times 10^9\text{ N}\cdot\text{m}^2/\text{C}^2)(+5.20 \times 10^{-20}\text{ C})(+5.20 \times 10^{-20}\text{ C})}{1.51 \times 10^{-10}\text{ m}}$$
$$= +1.61 \times 10^{-19}\text{ J}$$

and

$$U_{13} = U_{23} = \frac{kq_2q_3}{r_{23}} = \frac{(9.00 \times 10^9\text{ N}\cdot\text{m}^2/\text{C}^2)(+5.20 \times 10^{-20}\text{ C})(-10.4 \times 10^{-20}\text{ C})}{9.60 \times 10^{-11}\text{ m}}$$
$$= -5.07 \times 10^{-19}\text{ J}$$

Thus the total electrostatic potential energy is

$$U = U_{12} + U_{13} + U_{23} = (+1.61 \times 10^{-19}\text{ J}) + (-5.07 \times 10^{-19}\text{ J}) + (-5.07 \times 10^{-19}\text{ J})$$
$$= -8.53 \times 10^{-19}\text{ J}$$

The negative result indicates that the molecule requires positive work to break it apart. (That is, it must be pulled apart.)

Follow-Up Exercise. Another common polar molecule is carbon monoxide (CO), a toxic gas commonly produced by automobiles when fuel combustion is incomplete. The carbon atom is positively charged and the oxygen atom is negative. The distance between the carbon and oxygen atoms is 1.20×10^{-10} m, and the (average) charge on each is 6.60×10^{-20} C. Determine the electrostatic energy of this molecule. Is it more or less electrically stable than a water molecule in this Example?

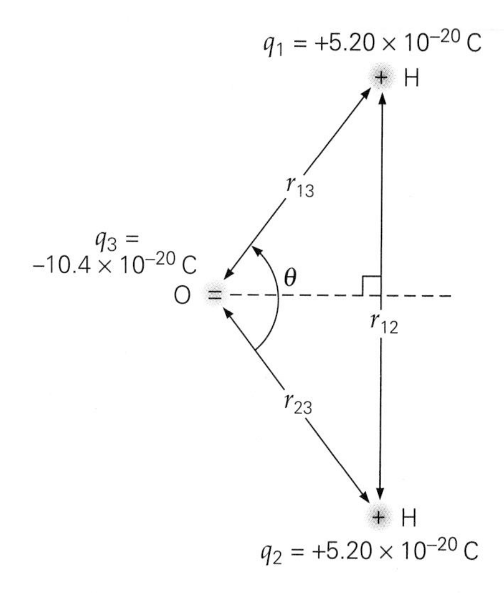

▲ **FIGURE 16.6 Electrostatic potential energy of a water molecule** A charge configuration has electric potential energy, because work is required to bring the charges together from large distances. The charges shown on the water molecule are net average charges because the atoms within the molecule share electrons. So the charges on the ends of the water molecule can be smaller than the charge on the electron or proton. (See Example 16.4 for details.)

16.2 Equipotential Surfaces and the Electric Field

OBJECTIVES: To (a) explain what is meant by an equipotential surface, (b) sketch equipotential surfaces for simple charge configurations, and (c) explain the relationship between equipotential surfaces and electric fields.

Equipotential Surfaces

Suppose a positive charge is moved perpendicularly to an electric field (such as path I of ▼Fig. 16.7a). As the charge moves from A to A′, *no work* is done by the electric field (why?). If no work is done, then the potential energy of the charge does not change, so $\Delta U_{AA'} = 0$. From this, it can be concluded that these two points (A and A′)—and *all* other points on path I—are at the same potential V; that is,

$$\Delta V_{AA'} = V_{A'} - V_A = \frac{\Delta U_{AA'}}{q} = 0 \quad \text{or} \quad V_A = V_{A'}$$

This result actually holds for all points on the *plane* parallel to the plates and containing path I. A surface like this plane, on which the potential is constant, is called an **equipotential surface** (or simply an *equipotential*). The word *equipotential*

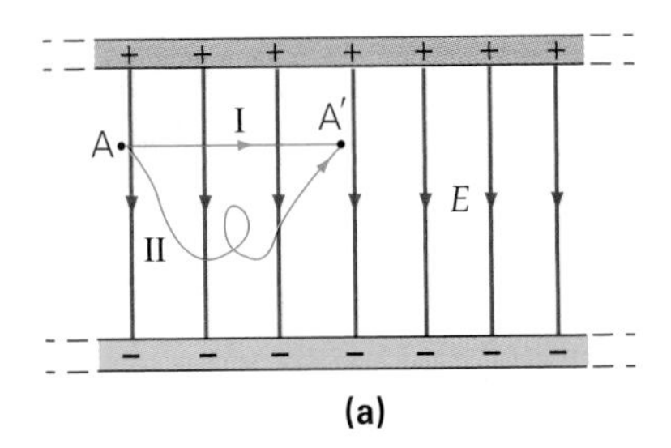

(a)

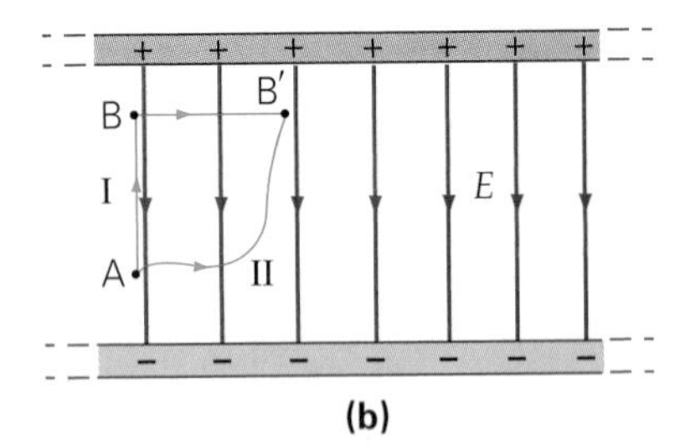

(b)

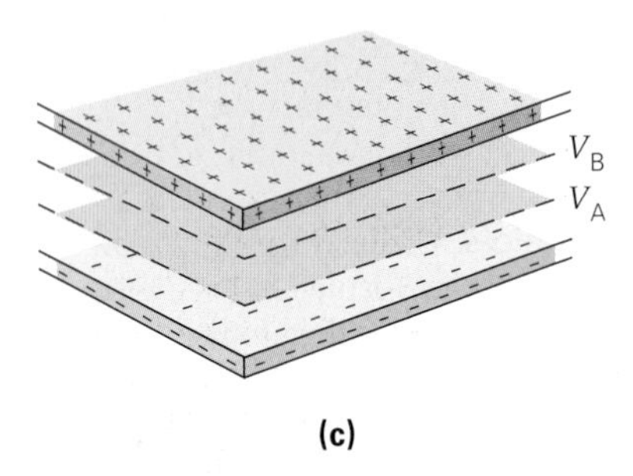

(c)

▲ **FIGURE 16.7 Construction of equipotential surfaces between parallel plates** **(a)** The work done in moving a charge is zero as long as you start and stop on the same equipotential surface. (Compare paths I and II.) **(b)** Once the charge moves to a higher potential (for example, from point A to point B), it can stay on that new equipotential surface by moving perpendicularly to the electric field (B to B′). The change in potential is independent of the path, since the same change occurs whether path I or path II is used. (Why?) **(c)** The actual equipotential surfaces within the parallel plates are planes parallel to those plates. Two such plates are shown, with $V_B > V_A$.

means "same potential." Note that, unlike this special case, an equipotential need not be a flat plane.

Since no work is required to move a charge along an equipotential surface, it must be generally true that

> Equipotential surfaces are always at right angles to the electric field.

Moreover, because the electric field is conservative, the work is the same whether path I, path II, or *any other* path from A to A′ is taken (Fig. 16.7a). As long as the charge returns to the same equipotential surface from which it started, the work done on it is zero and the value of the electric potential is the same.

If the positive charge is moved opposite to $\vec{\mathbf{E}}$ (path I in Fig. 16.7b)—at right angles to the equipotentials—the electric potential energy, and hence the electric potential, increases. (Why?) When B is reached, the charge is on a different equipotential—one of a higher potential than A. If, instead, the charge had been moved from A to B′, the work would be the same as that in moving from A to B. Hence, B and B′ are on the same equipotential surface. For parallel plates, the equipotentials are planes parallel to the plates (Fig. 16.7c).

To help understand the concept of an electric equipotential surface, consider a gravitational analogy. If the gravitational potential energy is designated as zero at ground level and an object is raised a height $h = h_B - h_A$ (from A to B in ▼Fig. 16.8), then the work done by an external force is mgh and is positive. For horizontal movement, the potential energy does not change. This means that the dashed plane at height h_B is a gravitational equipotential surface—and so is the plane at h_A, but it has a lower potential value than the plane at h_B. Therefore, surfaces of constant gravitational potential energy are planes parallel to the Earth's surface. Topographic maps, which display land contours by plotting lines of constant elevation (usually relative to sea level), are actually maps of constant gravitational potential (▸Fig. 16.9a, b). Note how the equipotentials near a point charge (Fig. 16.9c, d) are qualitatively similar to the gravitational contours due to a hill.

It is useful to know how to sketch equipotential surfaces, because they are intimately related to the electric field and to practical aspects such as voltage. The Learn by Drawing on Graphical Relationship between Electric Field Lines and Equipotentials (page 547) summarizes a qualitative method useful for sketching equipotential surfaces, given an electric field line pattern. As this feature shows, the method is also useful for the converse problem: sketching the electric field lines if the equipotential surfaces are given. Can you see how these ideas were used to construct the equipotentials of an electric dipole in ▸Fig. 16.10?

To determine the relationship between the electric field (E) and the electric potential (V), consider the special case of a uniform electric field (▸Fig. 16.11). The potential difference (ΔV) between any two equipotential planes (labeled V_1 and V_2 in the figure) can be calculated with the same technique used to derive Eq. 16.2. The result is

$$\Delta V = V_3 - V_1 = E\Delta x \qquad (16.7)$$

Thus, if you start on equipotential surface 1 and move *perpendicularly away* from it and *opposite to* the electric field to equipotential surface 3, there is a potential *increase* (ΔV) that depends on the electric field strength (E) and the distance (Δx).

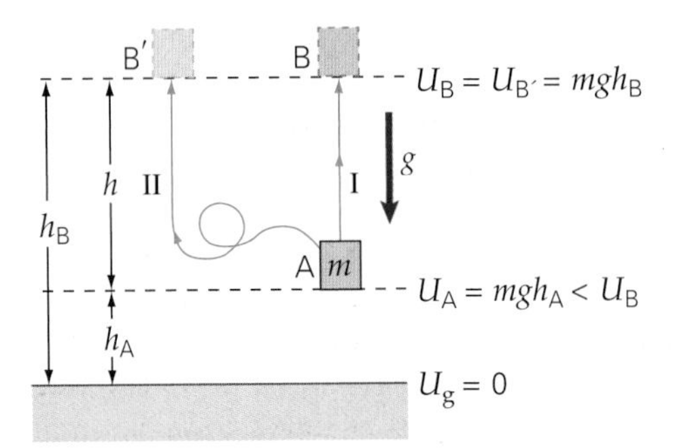

◀ **FIGURE 16.8 Gravitational potential energy analogy** Raising an object in a uniform gravitational field results in an increase in gravitational potential energy, and, $U_B > U_A$. At a given height, the object's potential energy is constant as long as it remains on that (gravitational) equipotential surface. Here, $\vec{\mathbf{g}}$ points downward, like $\vec{\mathbf{E}}$ in Fig. 16.7.

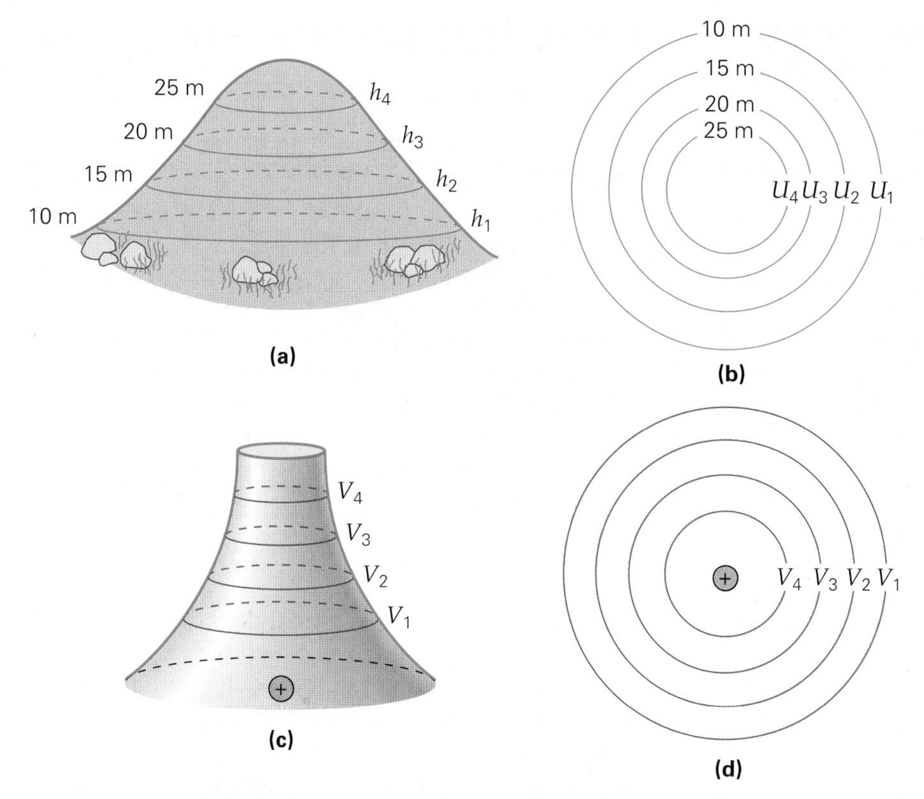

▲ **FIGURE 16.9 Topographic maps—a gravitational analogy to equipotential surfaces** **(a)** A symmetrical hill with slices at different elevations. Each slice is a plane of constant gravitational potential. **(b)** A topographic map of the slices in (a). The contours, where the planes intersect the surface, represent increasingly larger values of gravitational potential as one goes up the hill. **(c)** The electric potential V near a point charge q forms a similar symmetrical hill. V is constant at fixed distances from q. **(d)** Electrical equipotentials around a point charge are spherical (in two dimensions they are circles) centered on the charge. The closer the equipotential to the positive charge, the larger its electric potential.

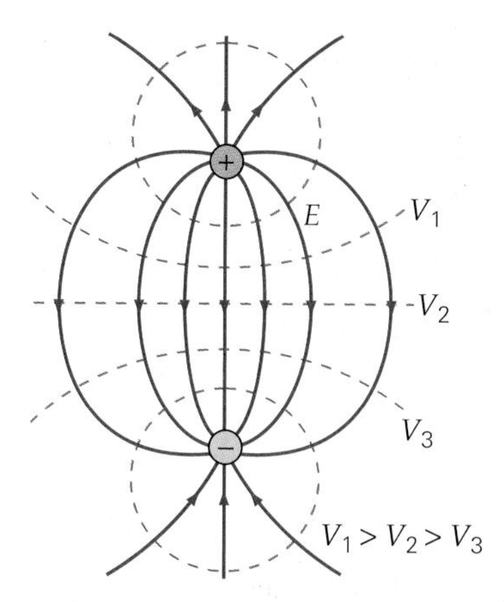

▲ **FIGURE 16.10 Equipotentials of an electric dipole** Equipotentials are perpendicular to electric field lines. $V_1 > V_2$ because equipotential surface 1 is closer to the positive charge than is surface 2. To understand how equipotentials are constructed, see the accompanying Learn by Drawing. (See page 547.)

For a given distance Δx, this perpendicular movement yields the maximum possible gain in potential. Think of taking one step of length Δx in *any* direction, starting from surface 1. The way to maximize the increase would be to step onto surface 3. A step in any direction not perpendicular to surface 1 (for example, ending on surface 2) yields a smaller increase in potential.

By finding the direction of the *maximum* potential increase, we are finding the direction opposite that of $\vec{\mathbf{E}}$. Thus, as a general rule, we can state that:

> The direction of the electric field $\vec{\mathbf{E}}$ is that in which the electric potential decreases the most rapidly.

Then, at any location, the magnitude of the electric field is the maximum rate of change of the potential with distance, or

$$E = \left|\frac{\Delta V}{\Delta x}\right|_{\text{max}} \tag{16.8}$$

The unit of electric field is volts per meter (V/m). Previously, E was expressed in newtons per coulomb (N/C; see Section 15.4). You should show, through dimensional analysis, that 1 V/m = 1 N/C. A graphical interpretation of the relationship between $\vec{\mathbf{E}}$ and V is shown in the Learn by Drawing on page 547.

In most practical situations, it is the potential difference (*voltage*), rather than the electric field, that is specified. For example, a D-cell flashlight battery has a terminal voltage of 1.5 V, meaning that it can maintain a potential difference of 1.5 V between its terminals. Most automotive batteries have a terminal voltage of about 12 V. Some of the common potential differences, or voltages, are listed in Table 16.1.

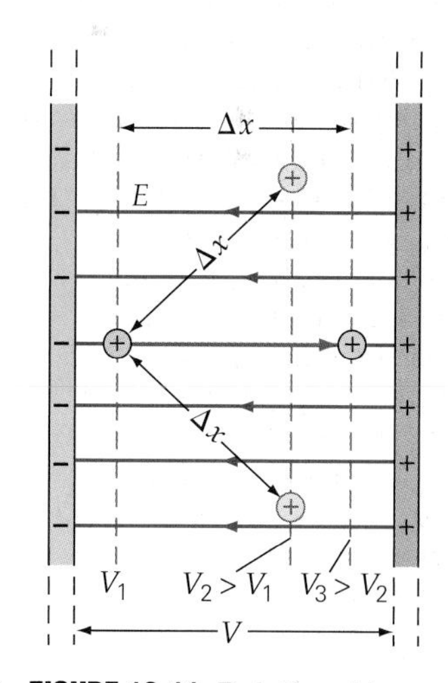

▲ **FIGURE 16.11 Relationship between the potential change (ΔV) and the electric field ($\vec{\mathbf{E}}$)** The electric field direction is that of maximum decrease in potential, or opposite the direction of maximum increase in potential (here, this maximum is in the direction of the solid blue arrow, not the angled ones; why?). The electric field magnitude is given by the maximum rate at which the potential changes over distance (usually in volts per meter).

Illustration 25.2 Work and Equipotentials

Illustration 25.3 Electric Potential of Charged Spheres

TABLE 16.1 Common Electric Potential Differences (Voltages)

Source	*Approximate Voltage* (ΔV)
Across nerve membranes	100 mV
Small-appliance batteries	1.5 to 9.0 V
Automotive batteries	12 V
Household outlet (United States)	110 to 120 V
Houschold outlets (Europe)	220 to 240 V
Automotive ignitions (spark plug firing)	10 000 V
Laboratory generators	25 000 V
High-voltage electric power delivery lines	300 kV or more
Cloud-to-Earth surface during thunderstorm	100 MV or more

Whether you know it or not, you live in an electric field near the Earth's surface. This field varies with weather conditions and, consequently, can be an indicator of approaching storms. Example 16.5 applies the equipotential surface concept to help in understanding the Earth's electric field.

Example 16.5 ■ The Earth's Electric Field and Equipotential Surfaces: Electric Barometers?

Under normal atmospheric conditions, the Earth's surface is electrically charged. This creates an approximately constant electric field of about 150 V/m pointing *down* near the surface. (a) Under these conditions, what is the shape of the equipotential surfaces, and in what direction does the electric potential decrease the most rapidly? (b) How far apart are two equipotential surfaces that have a 1000-V difference between them? Which has a higher potential, the one farther from the Earth or the one closer?

Thinking It Through. (a) Near the Earth's surface, the electric field is approximately uniform, so the equipotentials are similar to those of parallel plates. The discussion of Eqs. 16.7 and 16.8 enables us to determine which way the potential increases. (b) Equation 16.8 can then be used to determine how far apart the equipotential surfaces are.

Solution. Listing the data,

Given: $E = 150$ V/m, downward
$\Delta V = 1000$ V

Find: (a) shape of equipotential surfaces and direction of decrease in potential
(b) Δx (distance between equipotentials)

(a) Uniform electric fields are associated with plane equipotentials; in this case, the planes are parallel to the Earth's surface. The electric field points downward. This is the direction in which the potential decreases most rapidly.

(b) To determine the distance between the two equipotentials, think of moving vertically so that $\Delta V/\Delta x$ has its maximum value. Solving Eq. 16.8 for Δx yields

$$\Delta x = \frac{\Delta V}{E} = \frac{1000\text{ V}}{150\text{ V/m}} = 6.67\text{ m}$$

Because the potential decreases as we move downward (in the direction of $\vec{\mathbf{E}}$), the higher potential is associated with the surface that is 6.67 m *farther* from the ground.

Follow-Up Exercise. Re-examine this Example under storm conditions. During a lightning storm, the electric field can rise to many times the normal value. (a) Under these conditions, if the field is 900 V/m and points *upward*, how far apart are two equipotential surfaces that differ by 2000 V? (b) Which surface is at a higher potential, the one closer to the Earth or the one farther away? (c) Can you tell how far the two surfaces are from the ground? Why or why not?

Exploration 25.1 Investigate Equipotential Lines

Exploration 25.2 Electric Field Lines and Equipotentials

LEARN BY DRAWING

GRAPHICAL RELATIONSHIP BETWEEN ELECTRIC FIELD LINES AND EQUIPOTENTIALS

Because it takes no work to move a charge along an equipotential surface, such surfaces must be perpendicular to the electric field lines. Also, the electric field has a magnitude equal to the change in potential per unit distance (V/m) and points in the direction in which the potential decreases most rapidly. These facts can be used to construct equipotentials if we know the field pattern. The reverse is also true: Given the equipotentials, the electric field lines can be constructed. Furthermore, if the potential (in volts) associated with each equipotential is known, the strength and direction of the field can be estimated from the rate at which the potential changes with distance (Eq. 16.8).

A couple of examples should provide a graphical insight into the connection between equipotential surfaces and their associated electric fields. Consider Fig. 1, in which you are given the electric field lines and want to determine the shape of the equipotentials. Pick any point, such as A, and begin moving at right angles to the field lines. Keep moving so as to maintain this perpendicular orientation to the lines. Between lines you may have to approximate, but plan ahead to the next field line so it is crossed at a right angle. To find another equipotential, start at another point, such as B, and proceed the same way. Sketch as many equipotentials as you need to map the area of interest. The figure shows the result of sketching four equipotentials, from A (at the highest potential—can you tell why?) to D (at the lowest potential).

Now suppose you are given the equipotentials instead of the field lines (Fig. 2). The electric field lines point in the direction of decreasing V and are perpendicular to the equipotential surfaces. Thus, to map the field, start at any point, and move in such a way that your path intersects each equipotential surface at a right angle. The resulting field line is shown in Fig. 2, beginning at point A. Starting at points B, C, and D provides additional field lines that suggest the complete electric field pattern; you need only add the arrows in the direction of decreasing potential.

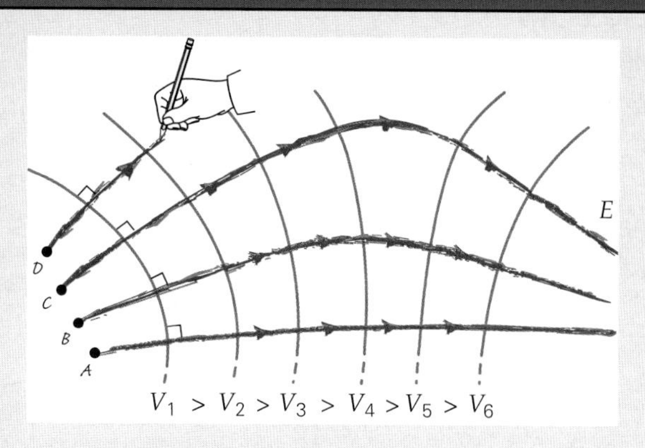

FIGURE 2 Mapping the electric field from equipotentials Start at a convenient point, and trace a line that crosses each equipotential at a right angle. Repeat the process as often as needed to reveal the field pattern, adding arrows to indicate the direction of the field lines from high to low potential. In going from one potential to the next, plan ahead so that each succeeding equipotential is also crossed at right angles.

Lastly, suppose you want to estimate the magnitude of $\vec{\mathbf{E}}$ at some point P (Fig. 3), knowing the values of the equipotentials 1.0 cm on either side of it. From this, you know that the field points roughly from A to B (why?) and its approximate magnitude would be

$$E = \left|\frac{\Delta V}{\Delta x}\right|_{\max} = \frac{(1000\text{ V} - 950\text{ V})}{2.0 \times 10^{-2}\text{ m}} = 2.5 \times 10^3\text{ V/m}$$

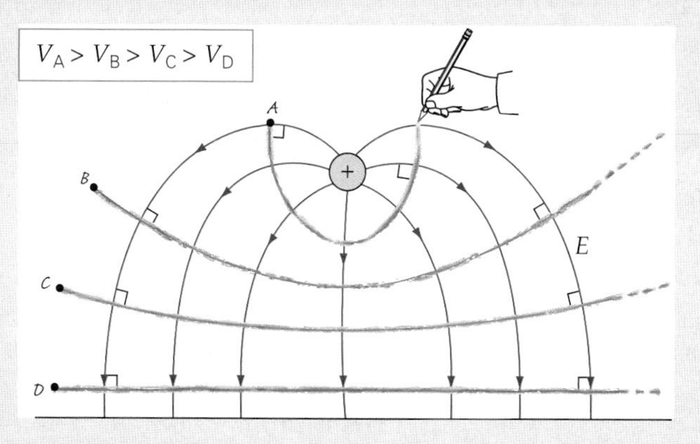

FIGURE 1 Sketching equipotentials from electric field lines If you know the electric field pattern, pick a point in the region of interest and move so that your path is always perpendicular to the next field line. Keep your path as smooth as possible, planning ahead so that each succeeding field line is also crossed at right angles. To map a surface with a higher (or lower) potential, move in the opposite (or the same) direction as the electric field and repeat the process. Here, $V_A > V_B$, and so on.

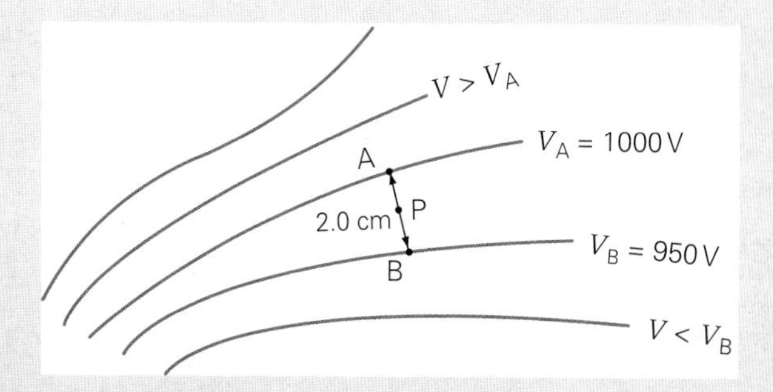

FIGURE 3 Estimating the magnitude of the electric field The magnitude of the potential change per meter at any point gives the strength of the electric field at that point.

Equipotential surfaces can be useful for describing the field near a charged conductor, as the following Conceptual Example shows.

Conceptual Example 16.6 ■ The Equipotential Surfaces Outside a Charged Conductor

A solid conductor with an excess positive charge is shown in ◂Fig. 16.12a. Which of the following best describes the shape of the equipotential surfaces just outside the conductor's surface: (a) flat planes, (b) spheres, or (c) approximately the shape of the conductor's surface? Explain your reasoning.

Reasoning and Answer. Choice (a) can be eliminated immediately, because flat (plane) equipotential surfaces are associated with flat plates. While it might be tempting to pick answer (b), a quick look at the electric field near the surface (Chapter 15), in conjunction with the Learn by Drawing on p. 547, shows that the correct answer is (c). To verify that (c) is the correct answer, recall that near the surface the electric field is perpendicular to that surface. Since the equipotential surfaces are perpendicular to the field lines, they must follow the contour of the conductor's surface (Fig. 16.12b).

Follow-Up Exercise. In this Example, (a) which of the two equipotentials (1 or 2) shown in Fig. 16.12c is at a higher potential? (b) What is the approximate shape of the equipotential surfaces *very far from* this conductor? Explain your reasoning. [*Hint:* What does the conductor look like when you are very far from it?]

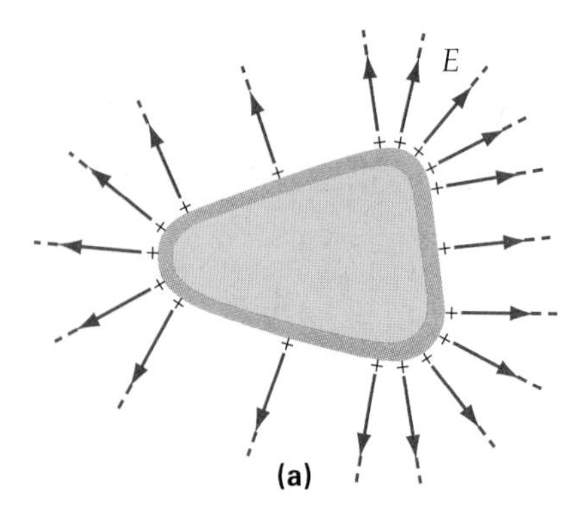

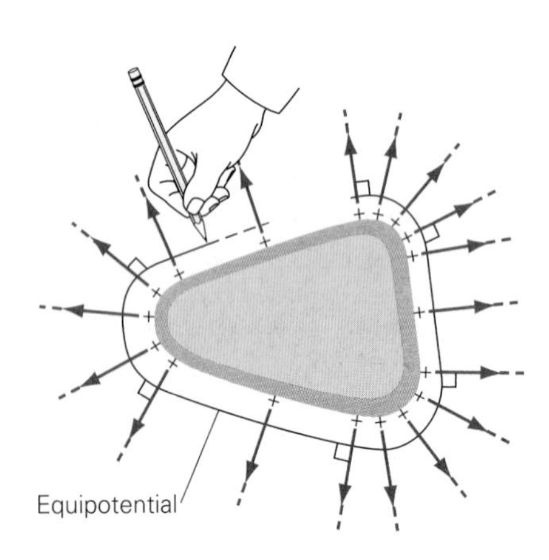

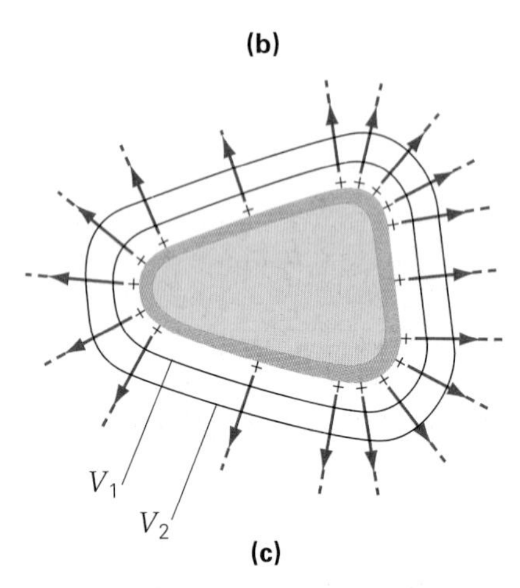

▲ **FIGURE 16.12** Equipotential surfaces near a charged conductor
See Conceptual Example 16.6.

PHYSLET®

Exploration 25.3 Electric Potential around Conductors

Note: The electron-volt is a unit of energy. The volt is not. Do not confuse the two!

The Electron-Volt

The concept of electric potential provides a unit of energy that is particularly useful in molecular, atomic, nuclear, and elementary particle physics. The **electron-volt (eV)** is defined as the kinetic energy acquired by an electron (or proton) accelerated through a potential difference, or voltage, of exactly 1 V. The gain in kinetic energy is equal (but opposite) to the change in electric potential energy. For an electron, its gain in kinetic energy in joules is:

$$\Delta K = -\Delta U_e = -(e\Delta V) = -(-1.60 \times 10^{-19}\,\text{C})(1.00\,\text{V}) = +1.60 \times 10^{-19}\,\text{J}$$

Since this is what is meant by 1 electron-volt, the conversion factor between the electron volt and the joule (to three significant figures) is

$$1\,\text{eV} = 1.60 \times 10^{-19}\,\text{J}$$

The electron-volt is typical of energies on the atomic scale, so it is convenient to express atomic energies in terms of electron-volts instead of joules. The energy of *any* charged particle accelerated through *any* potential difference can be expressed in electron-volts. For example, if an electron is accelerated through a potential difference of 1000 V, its gain in kinetic energy (ΔK) is one thousand times that of a 1-eV electron, or

$$\Delta K = e\Delta V = (1\,\text{e})(1000\,\text{V}) = 1000\,\text{eV} = 1\,\text{keV}$$

The abbreviation *keV* stands for *kiloelectron-volt.*

The electron-volt is defined in terms of a particle with the minimum charge (the electron or proton). However, the energy of a particle with *any* charge can also be expressed in electron-volts. Thus, if a particle with a charge of $+2e$, such as an alpha particle, were accelerated through a potential difference of 1000 volts, it would gain a kinetic energy of $\Delta K = e\Delta V = (2\,\text{e})(1000\,\text{V}) = 2000\,\text{eV} = 2\,\text{keV}$. Note how easy it is to compute the kinetic energy if you work in electron-volts.

Occasionally, larger units than the electron-volt are needed. For example, in nuclear and elementary particle physics, it is not uncommon to find particles with energies of *megaelectron-volts* (MeV) and *gigaelectron-volts* (GeV); $1\,\text{MeV} = 10^6\,\text{eV}$ and $1\,\text{GeV} = 10^9\,\text{eV}$.*

*At one time, a billion electron-volts was referred to as BeV, but this usage was abandoned because confusion arose. In some countries, such as Great Britain and Germany, a billion means 10^{12} (which is called a trillion in the United States).

In working problems, it is important to be aware that the electron-volt (eV) is *not* an SI unit. Hence, when using energies, you must convert from electron-volts to joules. For example, to calculate the speed of an electron accelerated from rest through 10.0 V, first convert the kinetic energy (10.0 eV) to joules:

$$K = (10.0 \text{ eV})(1.60 \times 10^{-19} \text{ J/eV}) = 1.60 \times 10^{-18} \text{ J}$$

Continuing in the SI system, the mass of the electron must be in kilograms. Then the speed is

$$v = \sqrt{2K/m} = \sqrt{2(1.60 \times 10^{-18} \text{ J})/(9.11 \times 10^{-31} \text{ kg})} = 1.87 \times 10^{6} \text{ m/s}$$

16.3 Capacitance

OBJECTIVES: To (a) define capacitance and explain what it means physically, and (b) calculate the charge, voltage, electric field, and energy storage for parallel-plate capacitors.

A pair of parallel plates, if charged, stores electrical energy (▼Fig. 16.13). Such an arrangement of conductors is an example of a **capacitor**. (Any pair of conductors qualifies as a capacitor.) The energy storage occurs because it takes work to transfer the charge from one plate to the other. Imagine that one electron is moved between a pair of initially uncharged plates. Once that is done, transferring a *second* electron would be more difficult, because it is not only repelled by the first electron on the negative plate, but also attracted by a double positive charge on the positive plate. Separating the charges requires more and more work as more and more charge accumulates on the plates. (This is analogous to stretching a spring. The more you stretch it, the harder it is to stretch it further.)

Note: Capacitors store energy in their electric fields.

The work needed to charge parallel plates can be done quickly (usually in a few microseconds) by a battery. Although we won't discuss battery action in detail until the next chapter, all you need to know now is that a battery removes electrons from the positive plate and transfers, or "pumps," them through a wire to the negative plate. In the process of doing work, the battery loses some of its internal chemical potential energy. Of primary interest here is the result: a separation of charge and the creation of an electric field in the capacitor. The battery will continue to charge the capacitor until the potential difference between the plates is equal to the terminal voltage of the battery—for example, 12 V if you use a standard automotive battery. When the capacitor is disconnected from the battery, it becomes a storage "reservoir" of electrical energy.

Note: Recall that our notation for potential difference, or voltage (ΔV), will be replaced by V for convenience.

For a capacitor, the potential difference across the plates is proportional to the charge Q on the plates, or $Q \propto V$.* (Here, Q denotes the magnitude of the charge

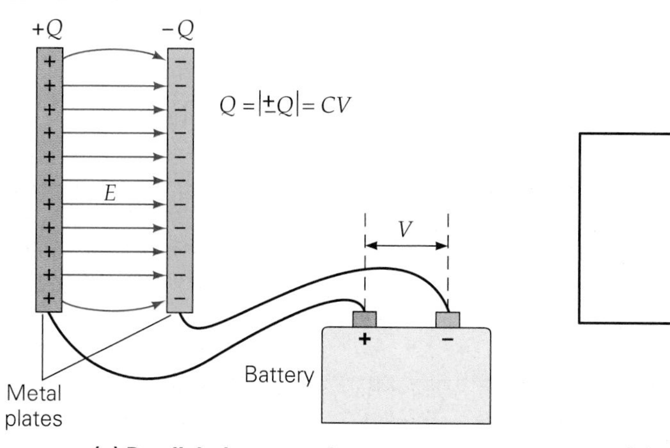

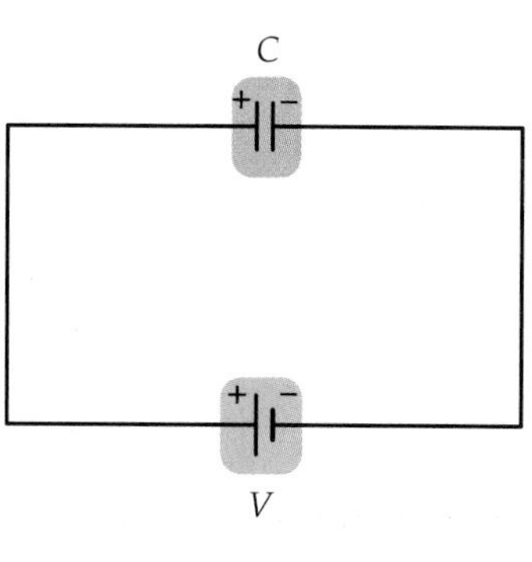

◀ **FIGURE 16.13 Capacitor and circuit diagram** **(a)** Two parallel metal plates are charged by a battery that moves electrons from the positive plate to the negative one through the wire. Work is done while charging the capacitor, and energy is stored in the electric field. **(b)** This diagram represents the charging situation shown in part (a). It also shows the symbols commonly used for a battery (V) and a capacitor (C). The longer line of the battery symbol is the positive terminal, and the shorter line represents the negative terminal. The symbol for a capacitor is similar, but the lines are of equal length.

*At this point, we will begin using V to denote potential differences instead of ΔV. This is a common practice. Always remember that the important quantity is potential difference, ΔV.

Note: The charges on the plates are $+Q$ and $-Q$, but it is customary to refer in general to the magnitude of these charges, Q (meaning $|\pm Q|$), on a capacitor.

Note: The farad was named for the English scientist Michael Faraday (1791–1867), an early investigator of electrical phenomena who first introduced the concept of the electric field.

Note: Generally, we will use the lowercase letter q to represent charges on single particles and the uppercase letter Q for the larger amounts of charges on capacitor plates.

PHYSLET®

Illustration 26.2 A Capacitor Connected to a Battery

PHYSLET®

Exploration 26.1 Energy

on *either* plate, *not* the net charge on the whole capacitor, which is zero.) This proportionality can be made into an equation by using a constant, C, called *capacitance:*

$$Q = CV \quad \text{or} \quad C = \frac{Q}{V} \tag{16.9}$$

SI unit of capacitance: coulomb per volt (C/V), or farad (F)

The coulomb per volt equals a **farad**, $1\ \text{C/V} = 1\ \text{F}$. The farad is a large unit (see Example 16.7), so the *microfarad* ($1\ \mu\text{F} = 10^{-6}\ \text{F}$), the *nanofarad* ($1\ \text{nF} = 10^{-9}\ \text{F}$), and the *picofarad* ($1\ \text{pF} = 10^{-12}\ \text{F}$) are commonly used.

Capacitance represents the charge stored *per volt*. When a capacitor has a large capacitance, it holds a large amount of charge *per volt* compared with one of smaller capacitance. If you connected the same battery to two different capacitors, the one with the larger capacitance would store more charge and more energy.

Capacitance depends *only* on the geometry (size, shape, and spacing) of the plates (and the material between the plates, Section 16.5) and *not* the charge on the plates. Consider the parallel-plate capacitor, which has an electric field given by Eq. 15.5:

$$E = \frac{4\pi kQ}{A}$$

The voltage across the plates can be computed from Eq. 16.2:

$$V = Ed = \frac{4\pi kQd}{A}$$

The capacitance of a parallel-plate arrangement is then

$$C = \frac{Q}{V} = \left(\frac{1}{4\pi k}\right)\frac{A}{d} \quad \textit{(parallel plates only)} \tag{16.10}$$

It is common to replace the expression in the parentheses in Eq. 16.10 with a single quantity called the **permittivity of free space (ε_o)**. The value of this constant (to three significant figures) is

$$\varepsilon_o = \frac{1}{4\pi k} = 8.85 \times 10^{-12}\ \text{C}^2/(\text{N}\cdot\text{m}^2) \quad \textit{permittivity of free space} \tag{16.11}$$

ε_o describes the electrical properties of free space (vacuum), but its value in air is only 0.05% larger. In our calculations, they will be taken to be the same.

It is common to rewrite Eq. 16.10 in terms of ε_o:

$$C = \frac{\varepsilon_o A}{d} \quad \textit{(parallel plates only)} \tag{16.12}$$

Let's use Eq. 16.12 in the next Example to show just how unrealistically large an air-filled capacitor with a capacitance of 1.0 F would be.

Example 16.7 ■ Parallel-Plate Capacitors: How Large Is a Farad?

What would be the plate area of an air-filled 1.0-F parallel-plate capacitor if the plate separation were 1.0 mm? Would it be realistic to consider building such a capacitor?

Thinking It Through. The area can be calculated directly from Eq. 16.12. Remember to keep all quantities in SI units, so that the answer will be in square meters. The vacuum value of ε_o for air can be used without creating a significant error.

Solution.

Given: $C = 1.0\ \text{F}$
$d = 1.0\ \text{mm} = 1.0 \times 10^{-3}\ \text{m}$

Find: A (area of one of the plates)

Solving Eq. 16.12 for the area gives

$$A = \frac{Cd}{\varepsilon_o} = \frac{(1.0\ \text{F})(1.0 \times 10^{-3}\ \text{m})}{8.85 \times 10^{-12}\ \text{C}^2/(\text{N}\cdot\text{m}^2)} = 1.1 \times 10^8\ \text{m}^2$$

This is more than 100 km^2 (40 mi^2), that is, a square more than 10 km (6 mi) on a side. It is unrealistic to build a capacitor that big; 1.0 F is therefore a very large value of capacitance. There are ways, however, to make high-capacity capacitors (Section 16.4).

Follow-Up Exercise. In this Example, what would the plate spacing have to be if you wanted the capacitor to have a plate area of 1 cm^2? Compare your answer with a typical atomic diameter of 10^{-9} to 10^{-10} m. Is it feasible to build this capacitor?

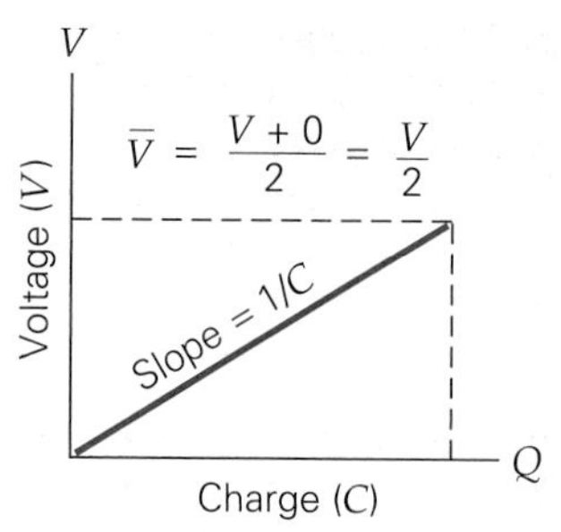

▲ **FIGURE 16.14 Capacitor voltage versus charge** A plot of voltage (V) versus charge (Q) for a capacitor is a straight line with slope $1/C$ (because $V = (1/C)Q$). The average voltage is $\bar{V} = \frac{1}{2}V$, and the total work done is equivalent to transferring the charge through $\bar{V}$. Thus, $U_C = W = Q\bar{V} = \frac{1}{2}QV$, the area under the curve (a triangle).

The expression for the energy stored in a capacitor can be obtained by graphical analysis, since both Q and V vary during charging—for example, as the charge is separated by a battery. A plot of voltage versus charge for charging a capacitor is a straight line with a slope of $1/C$, because $V = (1/C)Q$ (▸Fig. 16.14). The graph represents the charging of an initially uncharged capacitor ($V_o = 0$) to a final voltage (V). The work done is equivalent to transferring the total charge, using an average voltage $\bar{V}$. Because the voltage varies linearly with charge, the average voltage is half the final voltage V:

$$\bar{V} = \frac{V_{\text{final}} + V_{\text{initial}}}{2} = \frac{V+0}{2} = \frac{V}{2}$$

The energy stored in the capacitor (equal to the work done by the battery) is then

$$U_C = W = Q\bar{V} = \tfrac{1}{2}QV$$

Because $Q = CV$, this equation can be written in several equivalent forms:

$$U_C = \tfrac{1}{2}QV = \frac{Q^2}{2C} = \tfrac{1}{2}CV^2 \quad \textit{energy storage in a capacitor} \quad (16.13)$$

Typically, the form $U_C = \frac{1}{2}CV^2$ is the most practical, since the capacitance and the voltage are usually the known quantities. A very important medical application of the capacitor is in the *cardiac defibrillator* discussed in the next Example.

Note: Do not confuse U_C, the energy stored in a capacitor, with ΔU_e, the change in electric potential energy of a charged particle. (See Section 16.1.)

Note: Practice using the various forms of capacitor energy. In a pinch, you need recall only one, together with the definition of capacitance, $C = Q/V$.

Example 16.8 ■ Capacitors to the Rescue: Energy Storage in a Cardiac Defibrillator

During a heart attack, the heart beats in an erratic fashion, called *fibrillation*. One way to get it back to normal rhythm is to shock it with electrical energy supplied by a *cardiac defibrillator* (▸Fig. 16.15). About 300 J of energy is required to produce the desired effect. Typically, a defibrillator stores this energy in a capacitor charged by a 5000-V power supply. (a) What capacitance is required? (b) What is the charge on the capacitor's plates?

Thinking It Through. (a) To find the capacitance, solve for C in Eq. 16.13. (b) The charge then follows from the definition of capacitance (Eq. 16.9).

Solution. We list the given data:

Given: $U_C = 300$ J, $V = 5000$ V

Find: (a) C (the capacitance) (b) Q (charge on capacitor)

(a) The most useful form of Eq. 16.13 is $U_C = \frac{1}{2}CV^2$. Solving for C,

$$C = \frac{2U_C}{V^2} = \frac{2(300\text{ J})}{(5000\text{ V})^2} = 2.40 \times 10^{-5}\text{ F} = 24.0\ \mu\text{F}$$

(b) The charge (magnitude) on either plate is then

$$Q = CV = (2.40 \times 10^{-5}\text{ F})(5000\text{ V}) = 0.120\text{ C}$$

Follow-Up Exercise. For the capacitor in this Example, if the maximum allowable energy for any single defibrillation attempt is 750 J, what is the maximum voltage that should be used?

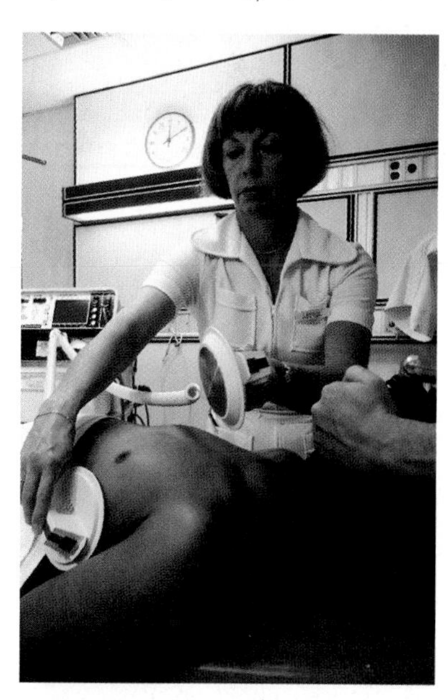

▲ **FIGURE 16.15 Defibrillator** A burst of electric current (flow of charge) from a defibrillator may restore a normal heartbeat in people in cardiac arrest. Capacitors store the electrical energy on which the device depends.

Sometimes capacitors can successfully model real-life phenomena. For example, a lightning storm can be considered to be the discharge of a negatively charged cloud to the positively charged ground—in effect, a "cloud-ground" capacitor. Another interesting application of electric potential treats nerve membranes as cylindrical capacitors to help explain nerve signal transmission. (See Insight 16.1 on Electric Potential and Nerve Signal Transmission on page 552.)

INSIGHT

16.1 ELECTRIC POTENTIAL AND NERVE SIGNAL TRANSMISSION

The human body's nervous system is responsible for the reception of external stimuli through our senses (such as touch) as well as communication between the brain and our organs and muscles. If you touch something hot, nerves in your hand detect the problem and send a signal to your brain; your brain then sends the signal "Pull back!" through other parts of the nervous system to your hand. But what are these signals, and how do they work?

A typical nerve consists of a bundle of nerve cells called *neurons*, much like individual telephone wires bundled into a single cable. The structure of a typical neuron is shown in Fig. 1a. The cell body, or *soma*, has long branchlike extensions called *dendrites*, which receive the input signal. The soma is responsible for processing the signal and transmitting it down a long extension called the *axon*. At the other end of the axon are projections with knobs called *synaptic terminals*. At these knobs, the electrical signal is transmitted to another neuron across a gap called the *synapse*. The human body contains on the order of 100 billion neurons, and each neuron can have several hundred synapses! Running the nervous system costs the body about 25% of its energy intake.

To understand the electrical nature of nerve signal transmission, let us focus on the axon. A vital component of the axon is its cell membrane, which is typically about 10 nm thick and consists of *phospholipids* (electrically polarized hydrocarbon molecules) and embedded protein molecules (Fig. 1b). The membrane has proteins called *ion channels*, which form pores where large protein molecules regulate the flow of ions (primarily sodium) across the membrane. The key to nerve signal transmission is the fact that these ion channels are selective: They allow only certain types of ions to cross the membrane; others cannot.

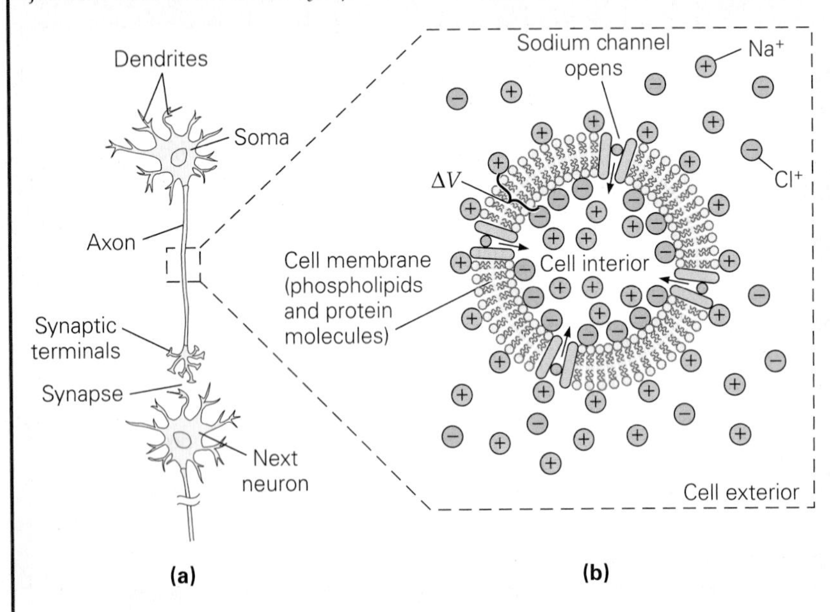

FIGURE 1 **(a)** The structure of a typical neuron. **(b)** An enlargement of the axon membrane, showing the membrane (about 10 nm thick) and the concentration of ions inside and outside the cell. The charge polarization across the membrane leads to a voltage, or membrane potential. When triggered by an external stimulus, the sodium ion channels open, allowing sodium ions into the cell. This influx changes the membrane potential.

16.4 Dielectrics

OBJECTIVES: To (a) understand what a dielectric is, and (b) understand how it affects the physical properties of a capacitor.

PHYSLET®

Illustration 26.1 Microscopic View of a Capacitor

In most capacitors, a sheet of insulating material, such as paper or plastic, is between the plates. Such an insulating material, called a **dielectric**, serves several purposes. For one, it keeps the plates from coming into contact. Contact would allow the electrons to flow back onto the positive plate, neutralizing the charge on the capacitor and the energy stored. A dielectric also allows flexible plates of metallic foil to be rolled into a cylinder, giving the capacitor a more compact (more practical) size. Finally, a dielectric increases the charge storage capacity of the capacitor and therefore, under the right conditions, the energy stored in the capacitor. This capability depends on the type of material and is characterized by the **dielectric constant** (κ). Values of the dielectric constant for some common materials are given in Table 16.2.

The fluid outside the axon, although electrically neutral, contains sodium ions (Na^+) and chlorine ions (Cl^-) in solution. In contrast, the axon's internal fluid is rich in potassium ions (K^+) and negatively charged protein molecules. If it were not for the selective nature of the cell membrane, the Na^+ concentration would be equal on both sides of the membrane. Under normal (or *resting*) conditions, it is difficult for Na^+ to penetrate the interior of the nerve cell. This ion selectivity gives rise to a polarization of charge across the membrane. The exterior is positive (with Na^+ trying to enter the region of lower concentration), attracting the negative proteins to the inner surface of the membrane (Fig. 1b). Thus a cylindrical capacitor-like charge-storage system exists across an axon membrane at rest. The *resting membrane potential* (the voltage across the membrane) is defined as $\Delta V = V_{in} - V_{out}$. Because the outside is positively charged, as defined, the resting potential is a negative quantity; it ranges from about −40 to −90 mV (millivolts), with a typical value of −70 mV in humans.

Signal conduction occurs when the cell membrane receives a stimulus from the dendrites. Only then does the membrane potential change, and this change is propagated down the axon. The stimulus triggers Na^+ channels in the membrane (which are closed while resting, like a gate) to open and temporarily allows sodium ions to enter the cell (Fig. 1b). These positive ions are attracted to the negative charge layer on the interior and are driven by the difference in concentration. In about 0.001 s, enough sodium ions have passed through the gated channel to cause a reversal of polarity, and the membrane potential rises, typically to +30 mV in humans. The time sequence for this change in membrane potential is shown in Fig. 2. When the difference in Na^+ concentration causes the membrane voltage to become positive, the Na^+ channels close. A chemical process involving proteins known as the *Na/K–ATPase molecular pump* then re-establishes the resting potential at about −70 mV by selectively transporting the excess Na^+ back to the cell's exterior.

This temporary change in membrane potential (a total of 100 mV, from −70 mV to +30 mV) is called the cell's *action potential*. The action potential is the signal that is transmitted down the axon. This "voltage wave" travels at speeds of 1 to 100 m/s on its way to triggering another such pulse in the adjacent neuron. This speed, along with other factors such as time delays in the synapse region, is responsible for typical human reaction times totaling a few tenths of a second.

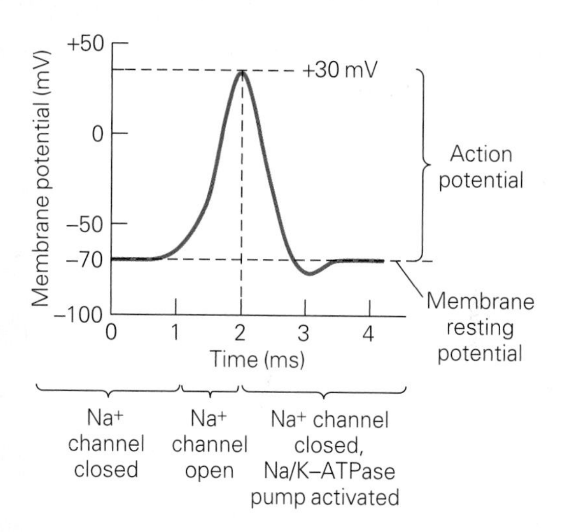

FIGURE 2 As the sodium channels open and sodium ions rush to the cell's interior, the membrane potential changes quickly from its resting value of −70 mV to about +30 mV. The resting potential is restored (about 4 ms later) by a protein "pumping" process that chemically removes the excess Na^+ after the sodium channels have closed (at 2 ms).

TABLE 16.2 Dielectric Constants for Some Materials

Material	*Dielectric Constant* (κ)	*Material*	*Dielectric Constant* (κ)
Vacuum	1.0000	Glass (range)	3–7
Air	1.00059	Pyrex glass	5.6
Paper	3.7	Bakelite	4.9
Polyethylene	2.3	Silicon oil	2.6
Polystyrene	2.6	Water	80
Teflon	2.1	Strontium titanate	233

How a dielectric affects the electrical properties of a capacitor is illustrated in ▾Fig. 16.16. The capacitor is fully charged (creating a field $\vec{\mathbf{E}}_o$) and disconnected from the battery, after which a dielectric is inserted (Fig. 16.16a). In the dielectric material,

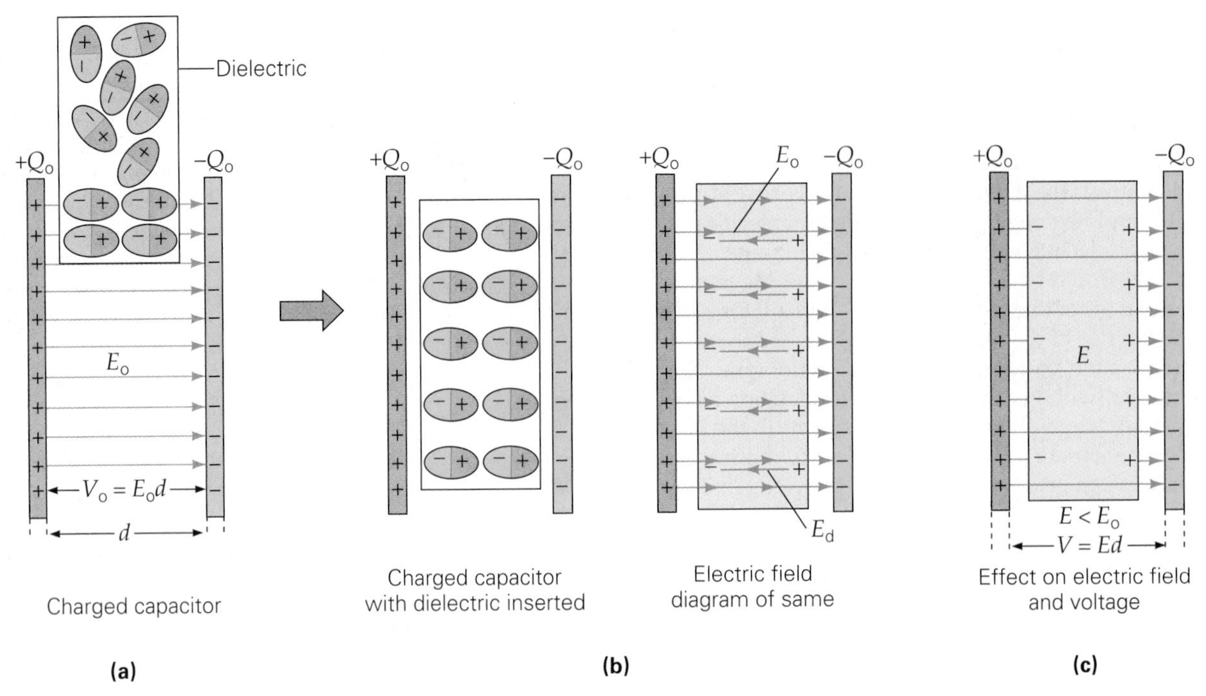

▲ **FIGURE 16.16 The effects of a dielectric on an isolated capacitor** **(a)** A dielectric material with randomly oriented permanent molecular dipoles (or dipoles induced by the electric field) is inserted between the plates of an isolated charged capacitor. As the dielectric is inserted, the capacitor tends to pull it in, thus doing work on it. (Note the attractive forces between the plate charges and those induced on the dielectric surfaces.) **(b)** When the material is in the capacitor's electric field, the dipoles orient themselves with the field, giving rise to an opposing electric field $\vec{\mathbf{E}}_d$. **(c)** The dipole field partially cancels the field due to the plate charges. The net effect is a decrease in both the electric field and the voltage. Because the stored charge remains the same, the capacitance increases.

Illustration 26.3 Capacitor with a Dielectric

work is done on molecular dipoles by the existing electric field, aligning them with that field (Fig. 16.16b). (The molecular polarization may be permanent or temporarily induced by the electric field. In either case, the effect is the same.) Work is also done on the dielectric sheet as a whole, because the charged plates pull it into them.

The result is that the dielectric creates a "reverse" electric field ($\vec{\mathbf{E}}_d$ in Fig. 16.16c) that partially cancels the field between the plates. This means that the *net* field ($\vec{\mathbf{E}}$) between the plates is reduced, and so is the voltage across the plates (because $V = Ed$). The dielectric constant κ of the material is defined as the ratio of the voltage with the material in place (V) to the vacuum voltage (V_o). Because V is proportional to E, this ratio is the same as the electric field ratio:

Note: Equation 16.14 holds only if the battery is disconnected.

$$\kappa = \frac{V_o}{V} = \frac{E_o}{E} \quad \textit{(only when the capacitor charge is constant)} \tag{16.14}$$

Note that κ is dimensionless and is greater than 1, because $V < V_o$. Equation 16.14 shows that the dielectric constant can be determined by measuring the two voltages. (Voltmeters are discussed in detail in Chapter 18.) Because the battery was disconnected and the capacitor isolated, the charge on the plates, Q_o, is unaffected. Because $V = V_o/\kappa$, the value of the capacitance with the dielectric inserted is larger than the vacuum value by a factor of κ. In effect, the same amount of charge is now being stored at a lower voltage, and the result is an increase in capacitance. To understand this effect, apply the definition of capacitance:

$$C = \frac{Q}{V} = \frac{Q_o}{(V_o/\kappa)} = \kappa\left(\frac{Q_o}{V_o}\right) \quad \text{or} \quad C = \kappa C_o \tag{16.15}$$

So inserting a dielectric into an isolated capacitor results in a larger capacitance. But what about energy storage? Because there is no energy input (the battery is disconnected) and the capacitor does work on the dielectric by pulling it into the region

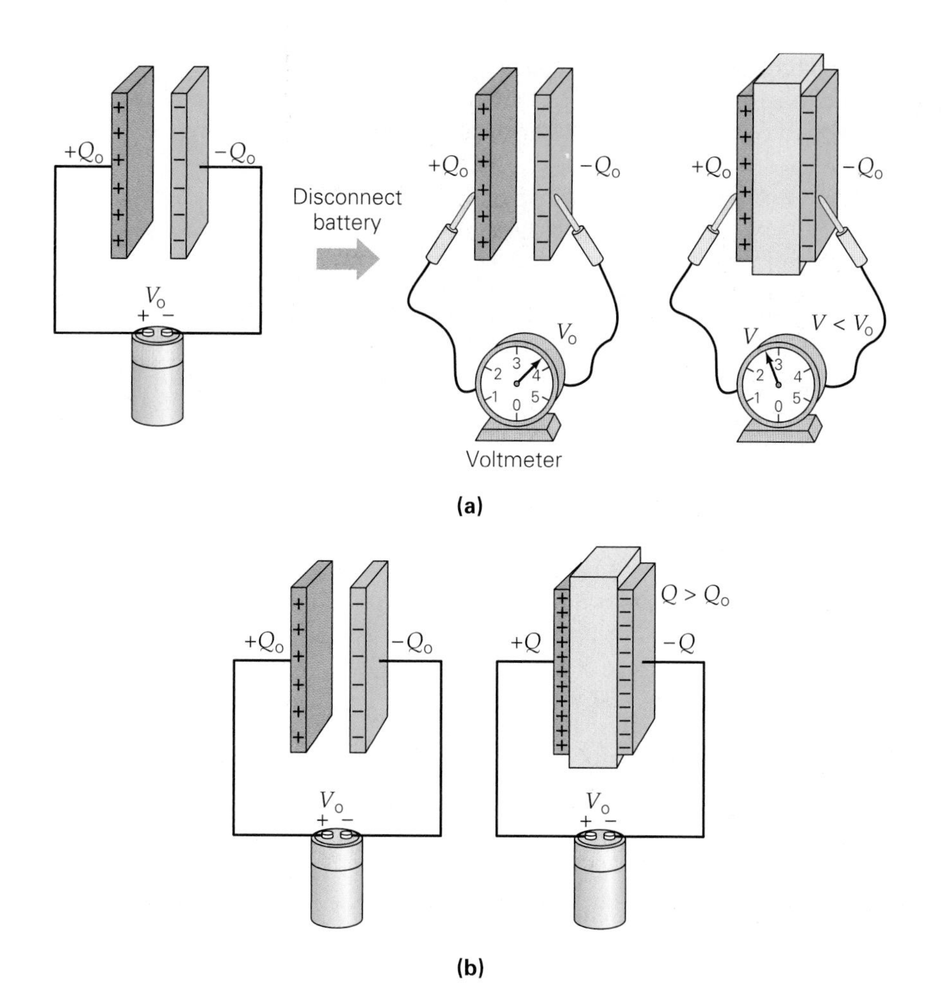

FIGURE 16.17 Dielectrics and capacitance **(a)** A parallel-plate capacitor in air (no dielectric) is charged by a battery to a charge Q_o and a voltage V_o (left). If the battery is disconnected and the potential across the capacitor is measured by a voltmeter, a reading of V_o is obtained (center). But if a dielectric is now inserted between the capacitor plates, the voltage drops to $V = V_o/\kappa$ (right), so the stored energy decreases. (Can you estimate the dielectric constant from the voltage readings?) **(b)** A capacitor is charged as in part (a), but the battery is left connected. When a dielectric is inserted into the capacitor, the voltage is maintained at V_o. (Why?) However, the charge on the plates increases to $Q = \kappa Q_o$. Therefore, more energy is now stored in the capacitor. In both cases, the capacitance increases by a factor of κ.

between the plates, the stored energy *drops* by a factor of κ (▲Fig. 16.17a), as the following calculation shows:

$$U_C = \frac{Q^2}{2C} = \frac{Q_o^2}{2\kappa C_o} = \frac{Q_o^2/2C_o}{\kappa} = \frac{U_o}{\kappa} < U_o \quad \textit{(battery disconnected)}$$

A different situation occurs, however, if the dielectric is inserted *and the battery remains connected*. In this case, the voltage stays constant and the battery supplies (pumps) more charge—and therefore does work (Fig. 16.17b). Because the battery does work, we expect the energy stored in the capacitor to *increase*. With the battery remaining connected, the charge on the plates increases by a factor κ, or $Q = \kappa Q_o$. Once again the capacitance increases, but now it is because more charge is stored at the same voltage. From the definition of capacitance, the result is the same as Eq. 16.15, because $C = Q/V = \kappa Q_o/V_o = \kappa(Q_o/V_o) = \kappa C_o$. Thus,

the effect of a dielectric is to increase the capacitance by a factor of κ, regardless of the conditions under which the dielectric is inserted.

In the case of a capacitor kept at constant voltage, the energy storage of the capacitor increases at the expense of the battery. To see this, let's calculate the energy with the dielectric in place under these conditions:

$$U_C = \tfrac{1}{2}CV^2 = \tfrac{1}{2}\kappa C_o V_o^2 = \kappa\left(\tfrac{1}{2}C_o V_o^2\right) = \kappa U_o > U_o \quad \textit{(battery connected)}$$

For a parallel-plate capacitor with a dielectric, the capacitance is increased over its (air) value in Eq. 16.12 by a factor of κ:

$$C = \kappa C_o = \frac{\kappa \varepsilon_o A}{d} \quad \textit{(parallel plates only)} \qquad (16.16)$$

This relationship is sometimes written as $C = \varepsilon A/d$, where $\varepsilon = \kappa\varepsilon_o$ is called the **dielectric permittivity** of the material, which is always greater than ε_o. (How do you know this?)

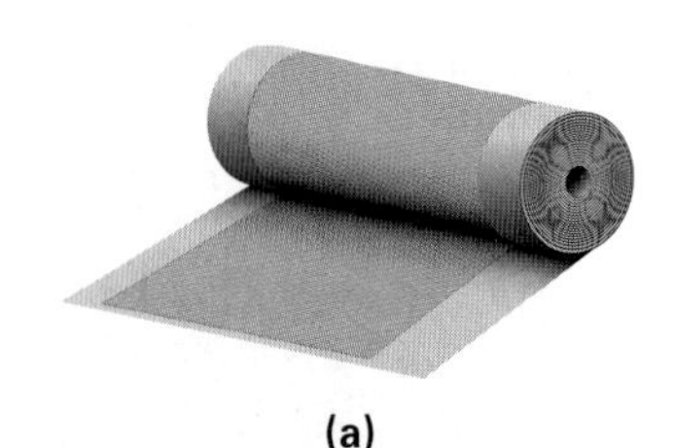

(a)

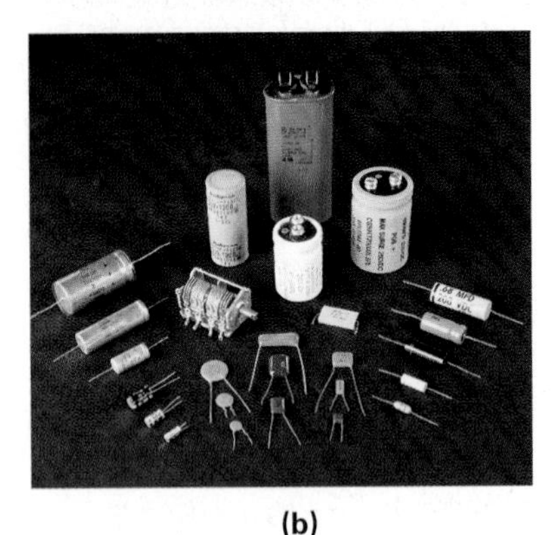

(b)

▲ **FIGURE 16.18** Capacitors in use **(a)** The dielectric material between the capacitor plates enables the plates to be constructed so that they are close together, thus increasing the capacitance. In addition, the plates can then be rolled up into a compact, more practical capacitor. **(b)** Capacitors (flat, brown circles and purple cylinders) among other circuit elements in a microcomputer.

A sketch of the inside of a typical cylindrical capacitor and an assortment of real capacitors is shown in ◂Fig. 16.18. Changes in capacitance can be used to monitor motion in our technological world, as the next Example shows.

Example 16.9 ■ The Capacitor as a Motion Detector: Computer Keyboards

Consider a capacitor (with dielectric) underneath a computer key (▾Fig. 16.19). The capacitor is connected to a 12.0-V battery and has a normal (uncompressed—without a keystroke) plate separation of 3.00 mm and a plate area of 0.750 cm^2. (a) What is the required dielectric constant if the capacitance is 1.10 pF? (b) How much charge is stored on the plates under normal conditions? (c) How much charge flows onto the plates (that is, what is the change in their charge) if they are compressed to a separation of 2.00 mm?

Thinking It Through. (a) The capacitance of air-filled plates can be found from Eq. 16.12, and then the dielectric constant can be determined from Eq. 16.15. (b) The charge follows from Eq. 16.9. (c) The compressed-plate separation distance must be used to recompute the capacitance. Then the new charge can be found as in (b).

Solution. The given data are as follows:

Given:
$V = 12.0\text{ V}$
$d = 3.00\text{ mm} = 3.00 \times 10^{-3}\text{ m}$
$A = 0.750\text{ cm}^2 = 7.50 \times 10^{-5}\text{ m}^2$
$C = 1.10\text{ pF} = 1.10 \times 10^{-12}\text{ F}$
$d' = 2.00\text{ mm} = 2.00 \times 10^{-3}\text{ m}$

Find:
(a) κ (dielectric constant)
(b) Q (initial capacitor charge)
(c) ΔQ (change in capacitor charge)

(a) From Eq. 16.12, the capacitance if the plates were separated by air would be

$$C_o = \frac{\varepsilon_o A}{d} = \frac{(8.85 \times 10^{-12}\text{ C}^2/\text{N}\cdot\text{m}^2)(7.50 \times 10^{-5}\text{ m}^2)}{3.00 \times 10^{-3}\text{ m}} = 2.21 \times 10^{-13}\text{ F}$$

Because the dielectric increases the capacitance, its value is

$$\kappa = \frac{C}{C_o} = \frac{1.10 \times 10^{-12}\text{ F}}{2.21 \times 10^{-13}\text{ F}} = 4.98$$

(b) The initial charge is then

$$Q = CV = (1.10 \times 10^{-12}\text{ F})(12.0\text{ V}) = 1.32 \times 10^{-11}\text{ C}$$

(c) Under compressed conditions, the capacitance is

$$C' = \frac{\kappa\varepsilon_o A}{d'} = \frac{(4.98)(8.85 \times 10^{-12}\text{ C}^2/\text{N}\cdot\text{m}^2)(7.50 \times 10^{-5}\text{ m}^2)}{2.00 \times 10^{-3}\text{ m}} = 1.65 \times 10^{-12}\text{ F}$$

The voltage remains the same, $Q' = C'V = (1.65 \times 10^{-12}\text{ F})(12.0\text{ V}) = 1.98 \times 10^{-11}\text{ C}$. Because the capacitance increased, the charge increased by

$$\Delta Q = Q' - Q = (1.98 \times 10^{-11}\text{ C}) - (1.32 \times 10^{-11}\text{ C}) = +6.60 \times 10^{-12}\text{ C}$$

As the key is depressed, a charge, whose magnitude is related to the displacement, flows onto the capacitor, providing a way of measuring the movement electrically.

Follow-Up Exercise. In this Example, suppose instead that the spacing between the plates were *increased* by 1.00 mm from the normal value of 3.00 mm. Would charge flow onto or away from the capacitor? How much charge would this be?

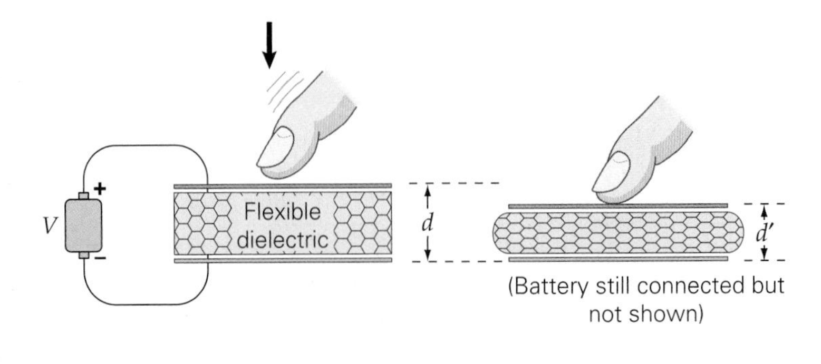

▶ **FIGURE 16.19** Capacitors in use Capacitors can be used to convert movement into electrical signals that can be measured and analyzed by computer. As the distance between the plates changes, so does the capacitance, which causes a change in the charge on the capacitor. Some computer keyboards operate in this way, as do other instruments such as seismographs (Chapter 13). See Example 16.9.

16.5 Capacitors in Series and in Parallel

OBJECTIVES: To (a) find the equivalent capacitance of capacitors in series and in parallel, (b) calculate the charge, voltage, and energy storage of individual capacitors in series and parallel configurations, and (c) analyze capacitor networks that include both series and parallel arrangements.

Capacitors can be connected in two basic ways: *in series* or *in parallel*. In series, the capacitors are connected head to tail (▼Fig. 16.20a). When connected in parallel, all the leads on one side of the capacitors have a common connection. (Think of all the "tails" connected together and all the "heads" connected together; Fig. 16.20b.)

Note: For so-called sandwiched dielectric parallel-plate capacitors, there is no head or tail distinction between the leads. Some types of capacitors do have particular positive and negative sides, and then the distinction must be made.

Illustration 26.4 Microscopic View of Capacitors in Series and Parallel

▼ **FIGURE 16.20** **Capacitors in series and in parallel** **(a)** All capacitors connected in series have the same charge, and the sum of the voltage drops is equal to the voltage of the battery. The total series capacitance is equivalent to the value of C_s. **(b)** When capacitors are connected in parallel, the voltage drops across the capacitors are the same, and the total charge is equal to the sum of the charges on the individual capacitors. The total parallel capacitance is equivalent to the value of C_p. **(c)** In a parallel connection, thinking of the plates makes it easier to see why the total charge is the sum of the individual charges. In effect, this arrangement represents a capacitor with two large plates.

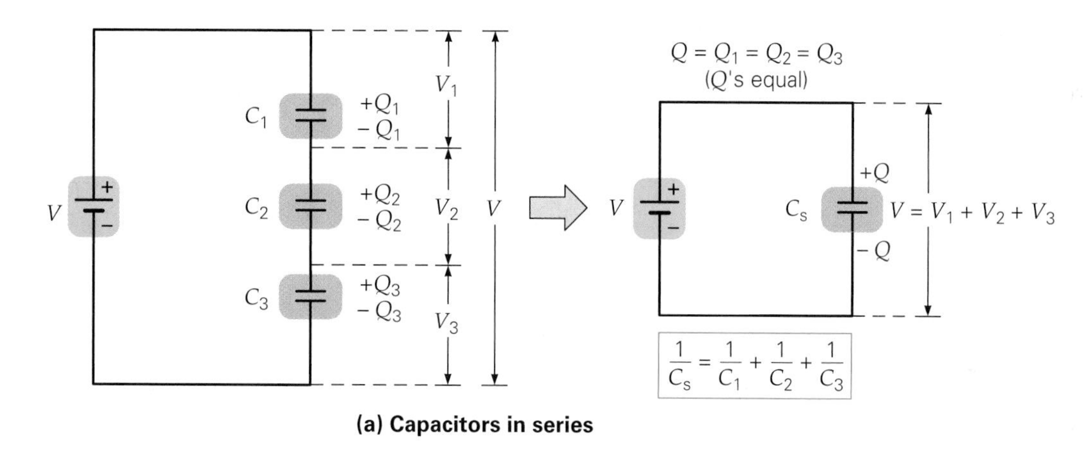

(a) Capacitors in series

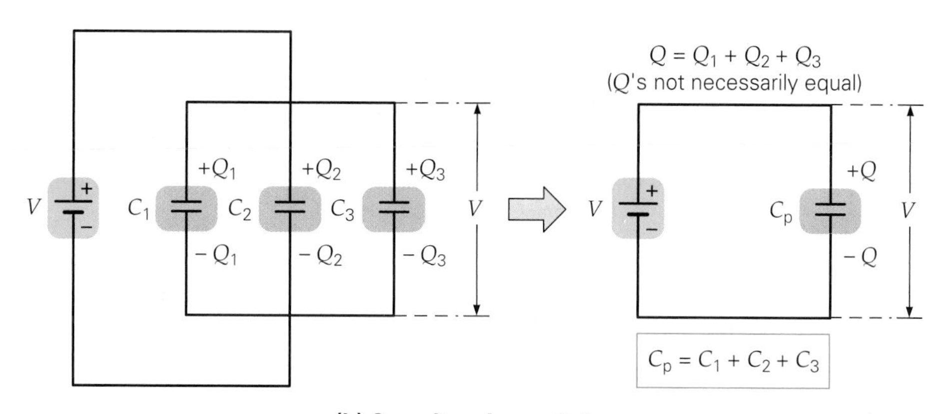

(b) Capacitors in parallel

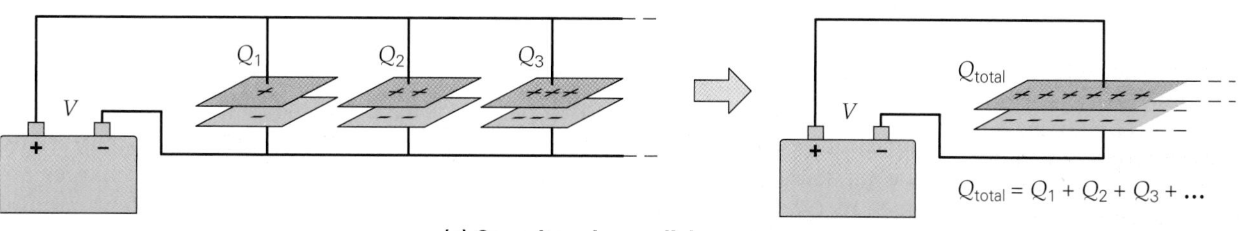

(c) Capacitors in parallel

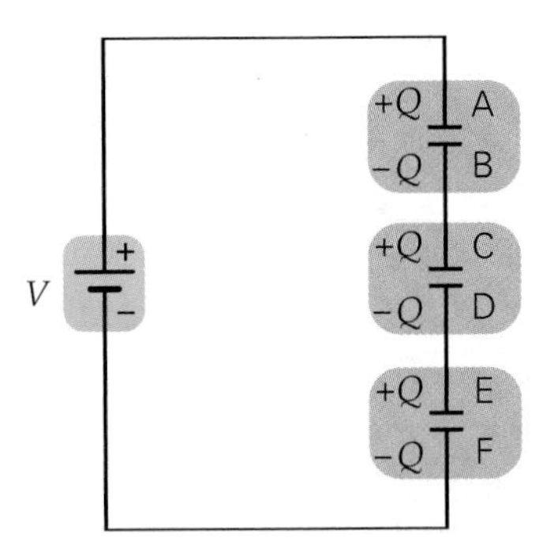

▲ **FIGURE 16.21 Charges on capacitors in series** Plates B and C together had zero net charge to start. When the battery placed $+Q$ on plate A, charge $-Q$ was induced on B; thus, C must have acquired $+Q$ for the BC combination to remain neutral. Continuing this way through the string, we see that all the charges must be the same in magnitude.

Capacitors in Series

When capacitors are wired in series, the charge Q must be the same on all the plates:

$$Q = Q_1 = Q_2 = Q_3 = \cdots$$

To see why this must be true, examine ◂Fig. 16.21. Note that only plates A and F are actually connected to the battery. Because the plates labeled B and C are isolated, the total charge on them must always be zero. So if the battery puts a charge of $+Q$ on plate A, then $-Q$ is induced on B at the expense of plate C, which acquires a charge of $+Q$. This charge in turn induces $-Q$ on D, and so on down the line.

As we have seen, "voltage drop" is just another name for "change in electrical potential energy per unit charge." When we add up all the series capacitor voltage drops (see Fig 16.20a), we must get the same value as the voltage across the battery terminals. The sum of the individual voltage drops across all the capacitors is equal to the voltage of the source:

$$V = V_1 + V_2 + V_3 + \cdots$$

The **equivalent series capacitance, C_s**, is defined as the value of a single capacitor that could replace the series combination and store the same charge at the same voltage. Because the combination of capacitors stores a charge of Q at a voltage of V, it follows that $C_s = Q/V$, or $V = Q/C_s$. However, the individual voltages are related to the individual charges by $V_1 = Q/C_1$, $V_2 = Q/C_2$, $V_3 = Q/C_3$, and so on.

Substituting these expressions into the voltage equation, we have

$$\frac{Q}{C_s} = \frac{Q}{C_1} + \frac{Q}{C_2} + \frac{Q}{C_3} + \cdots$$

Canceling the common Q's, we get

$$\frac{1}{C_s} = \frac{1}{C_1} + \frac{1}{C_2} + \frac{1}{C_3} + \cdots \quad \textit{equivalent series capacitance} \qquad (16.17)$$

This relationship means that C_s is always smaller than the smallest capacitance in the series combination. For example, try Eq. 16.17 with $C_1 = 1.0\ \mu\text{F}$ and $C_2 = 2.0\ \mu\text{F}$. You should be able to show that $C_s = 0.67\ \mu\text{F}$, which is less than $1.0\ \mu\text{F}$ (the general proof will be left to you). Physically, the reasoning goes like this. In series, all the capacitors have the same charge, so the charge stored by this arrangement is $Q = C_iV_i$ (where the subscript i refers to *any* of the individual capacitors in the string). Because $V_i < V$, the series arrangement stores *less* charge than any individual capacitor connected by itself to the same battery.

It makes sense that in series the smallest capacitance receives the largest voltage. A small value of C means less charge stored per volt. In order for the charge on all the capacitors to be the same, the smaller the value of capacitance, the larger the fraction of the total voltage required ($Q = CV$).

Capacitors in Parallel

With a parallel arrangement (Fig. 16.20b), the voltages across the capacitors are the same (why?), and each individual voltage is equal to that of the battery:

$$V = V_1 = V_2 = V_3 = \cdots$$

The total charge is the sum of the charges on each capacitor (Fig 16.20c):

$$Q_{\text{total}} = Q_1 + Q_2 + Q_3 + \cdots$$

The equivalent capacitance in parallel is expected to be larger than the largest capacitance, because more charge per volt can be stored in this way than if any one capacitor were connected to the battery by itself. The individual charges are given by $Q_1 = C_1V$, $Q_2 = C_2V$, and so on. A capacitor with the **equivalent parallel capacitance, C_p**, would hold this same total charge when connected to the battery, so $C_p = Q_{total}/V$, or $Q_{total} = C_pV$. Substituting these expressions into the previous equation gives

$$C_pV = C_1V + C_2V + C_3V + \cdots$$

and canceling the common V we obtain

$$C_p = C_1 + C_2 + C_3 + \cdots \quad \textit{equivalent parallel capacitance} \qquad (16.18)$$

In the parallel case, the equivalent capacitance C_p is the sum of the individual capacitances. In this case, the equivalent capacitance is larger than the largest individual capacitance. Because capacitors in parallel have the same voltage, the largest capacitance will store the most charge. For a comparison of capacitors in series and in parallel, consider the next Example.

Example 16.10 ■ Charging without Credit Cards: Capacitors in Series and in Parallel?

Given two capacitors, one with a capacitance of 2.50 μF and the other of 5.00 μF, what are the charge on each and the total charge stored if they are connected across a 12.0-V battery (a) in series and (b) in parallel?

Exploration 26.4 Equivalent Capacitance

Thinking It Through. (a) Capacitors in series have the same charge. From Eq. 16.17 we can find the equivalent capacitance and then the charge on each capacitor. (b) Capacitors in parallel have the same voltage; from that the charge on each can be easily determined because their individual capacitances are known.

Solution. Listing the data, we have the following:

Given: $C_1 = 2.50\ \mu\text{F} = 2.50 \times 10^{-6}$ F
$C_2 = 5.00\ \mu\text{F} = 5.00 \times 10^{-6}$ F
$V = 12.0$ V

Find: (a) Q on each capacitor in series and Q_{total} (total charge)
(b) Q on each capacitor in parallel and Q_{total} (total charge)

(a) In series, the total (equivalent) capacitance is determined as follows:

$$\frac{1}{C_s} = \frac{1}{2.50 \times 10^{-6}\ \text{F}} + \frac{1}{5.00 \times 10^{-6}\ \text{F}} = \frac{3}{5.00 \times 10^{-6}\ \text{F}}$$

so

$$C_s = 1.67 \times 10^{-6}\ \text{F}$$

(*Note*: C_s is less than the smallest capacitance in the series, as expected.)

Because the charge on each capacitor is the same in series (and the same as the total), we have

$$Q_{total} = Q_1 = Q_2 = C_sV = (1.67 \times 10^{-6}\ \text{F})(12.0\ \text{V}) = 2.00 \times 10^{-5}\ \text{C}$$

(b) Here, the parallel equivalent capacitance relationship is used:

$$C_p = C_1 + C_2 = 2.50 \times 10^{-6}\ \text{F} + 5.00 \times 10^{-6}\ \text{F} = 7.50 \times 10^{-6}\ \text{F}$$

(This result is reasonable because it is greater than the largest individual value in the parallel arrangement.)

Therefore,

$$Q_{total} = C_pV = (7.50 \times 10^{-6}\ \text{F})(12.0\ \text{V}) = 9.00 \times 10^{-5}\ \text{C}$$

In parallel, each capacitor has the full 12.0 V across it; hence,

$$Q_1 = C_1V = (2.50 \times 10^{-6}\ \text{F})(12.0\ \text{V}) = 3.00 \times 10^{-5}\ \text{C}$$
$$Q_2 = C_2V = (5.00 \times 10^{-6}\ \text{F})(12.0\ \text{V}) = 6.00 \times 10^{-5}\ \text{C}$$

As a final double check, notice that the total stored charge is equal to the sum of the charges on both capacitors.

Follow-Up Exercise. In this Example, determine which combination, series or parallel, stores the most energy.

Capacitor arrangements generally can involve *both* series and parallel connections, as shown in the next Example. In this situation, you simplify the circuit, using the equivalent parallel and series capacitance expressions, until you end up with one single, overall equivalent capacitance. To find the results for each individual capacitor, you work backward until you get to the original arrangement.

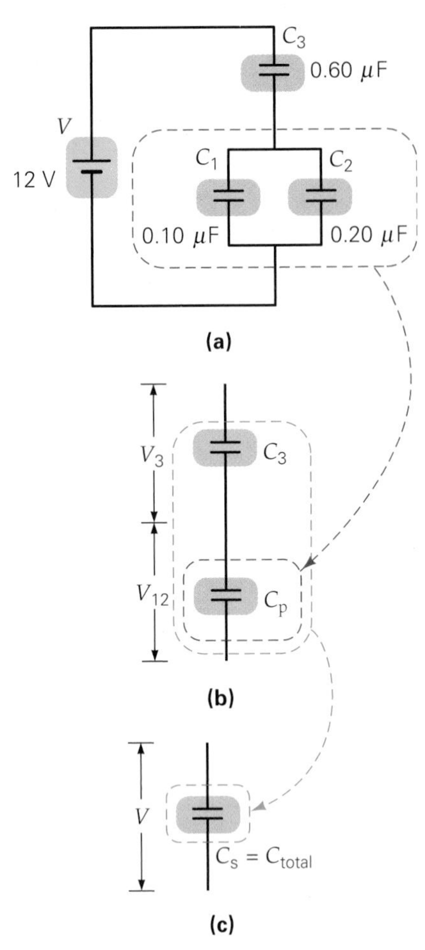

▲ **FIGURE 16.22** Circuit reduction When capacitances are combined, the combination of capacitors is reduced to a single equivalent capacitance. See Example 16.11.

Example 16.11 ■ The Electrical Two-Step—Forward, Then Backward: A Series–Parallel Combination of Capacitors

Three capacitors are connected in a circuit as shown in ◂Fig. 16.22a. What is the voltage across each capacitor?

Thinking It Through. The voltage across each capacitor could be found from $V = Q/C$ *if* the charge on each capacitor were known. The total charge on the capacitors is found by reducing the series–parallel combination to a single equivalent capacitance. Two of the capacitors are in parallel. Their single equivalent capacitance (C_p) is itself in series with the last capacitor—a fact that enables the total capacitance to be found. Working backward will allow the voltage across each capacitor to be found.

Solution.

Given: Values of capacitance and voltage from the figure

Find: V_1, V_2, and V_3 (voltages across capacitors)

Starting with the parallel combination, we have

$$C_p = C_1 + C_2 = 0.10\ \mu\text{F} + 0.20\ \mu\text{F} = 0.30\ \mu\text{F}$$

Now the arrangement is partially reduced, as shown in Fig. 16.22b. Next, considering C_p in series with C_3, we can find the total, or overall, equivalent capacitance of the original arrangement:

$$\frac{1}{C_s} = \frac{1}{C_3} + \frac{1}{C_p} = \frac{1}{0.60\ \mu\text{F}} + \frac{1}{0.30\ \mu\text{F}} = \frac{1}{0.60\ \mu\text{F}} + \frac{2}{0.60\ \mu\text{F}} = \frac{1}{0.20\ \mu\text{F}}$$

Therefore,

$$C_s = 0.20\ \mu\text{F} = 2.0 \times 10^{-7}\ \text{F}$$

This is the total equivalent capacitance of the arrangement (Fig. 16.22c). Treating the problem as one single capacitor, we find the charge on that equivalent capacitance:

$$Q = C_s V = (2.0 \times 10^{-7}\ \text{F})(12\ \text{V}) = 2.4 \times 10^{-6}\ \text{C}$$

This is the charge on C_3 and C_p, because they are in series. We can use this to calculate the voltage across C_3:

$$V_3 = \frac{Q}{C_3} = \frac{2.4 \times 10^{-6}\ \text{C}}{6.0 \times 10^{-7}\ \text{F}} = 4.0\ \text{V}$$

The sum of the voltages across the capacitors equals the voltage across the battery terminals. The voltages across C_1 and C_2 are the same because they are in parallel. Because the voltage across C_1 (or C_2) plus the voltage across C_3 equals the total voltage (the battery voltage), we can write $V = V_{12} + V_3 = 12$ V. (See Fig. 16.22a.) Here, V_{12} represents the voltage across either C_1 or C_2. Solving for V_{12},

$$V_{12} = V - V_3 = 12\ \text{V} - 4.0\ \text{V} = 8.0\ \text{V}$$

Notice that C_p is less than C_3. Because C_p and C_3 are in series, it follows that C_p (and therefore C_1 and C_2) have most of the voltage.

Follow-Up Exercise. In this Example, find (a) the charge stored on each capacitor and (b) the energy stored in each.

Chapter Review

- The **electric potential difference** (or **voltage**) between two points is the work done per unit positive charge between those two points, or the change in electric potential energy per unit positive charge. Expressed in equation form, this relationship is

$$\Delta V = \frac{\Delta U_e}{q_+} = \frac{W}{q_+} \quad (16.1)$$

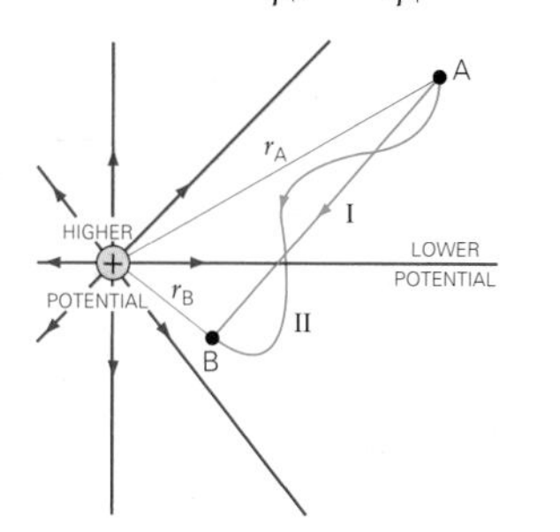

- **Equipotential surfaces** (surfaces of constant electric potential, also called **equipotentials**) are surfaces on which a charge has a constant electric potential energy. These surfaces are everywhere perpendicular to the electric field.

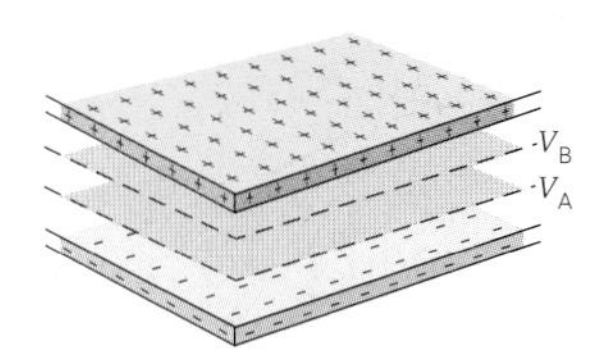

- The expression for the **electric potential due to a point charge** (choosing $V = 0$ at $r = \infty$) is

$$V = \frac{kq}{r} \quad (16.4)$$

- The **electric potential energy for a pair of point charges** is (choosing $U = 0$ at $r = \infty$)

$$U_{12} = \frac{kq_1q_2}{r_{12}} \quad (16.5)$$

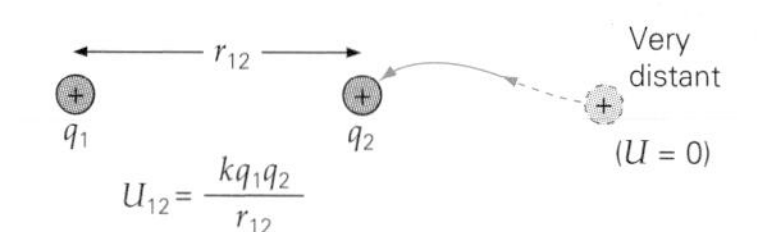

- The **electric potential energy of a configuration of more than two point charges** is the sum of point-charge pair terms from Eq. 16.5:

$$U_{\text{total}} = U_{12} + U_{23} + U_{13} + \cdots \quad (16.6)$$

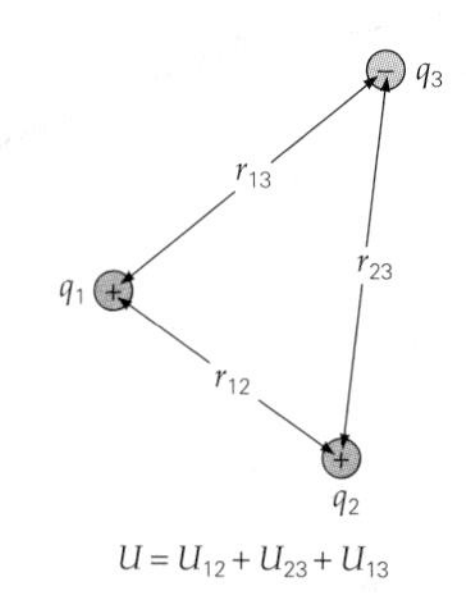

- The electric field is related to the rate of change of electric potential with distance. The electric field ($\vec{\mathbf{E}}$) is in the direction of the most rapid decrease in electric potential (V). The electric field magnitude (E) is the rate of change of the potential with distance, or

$$E = \left|\frac{\Delta V}{\Delta x}\right|_{\max} \quad (16.8)$$

- The **electron-volt (eV)** is the kinetic energy gained by an electron or a proton accelerated through a potential difference of 1 volt.

- A **capacitor** is any arrangement of two metallic plates. Capacitors store charge on their plates, and therefore electric energy.

- **Capacitance** is a quantitative measure of how effective a capacitor is in storing charge. It is the magnitude of the charge stored on either plate per volt, or

$$Q = CV \quad \text{or} \quad C = \frac{Q}{V} \quad (16.9)$$

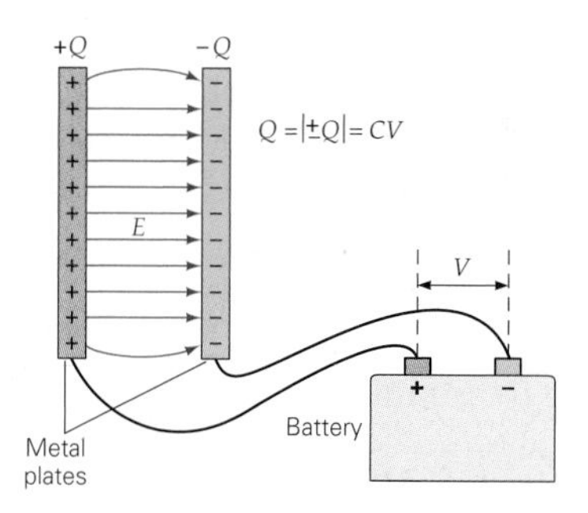

- The **capacitance of a parallel-plate capacitor** (in air) is

$$C = \frac{\varepsilon_o A}{d} \quad (16.12)$$

where $\varepsilon_o = 8.85 \times 10^{-12}\ \text{C}^2/(\text{N}\cdot\text{m}^2)$ is called the **permittivity of free space.**

- The **energy stored in a capacitor** depends on its capacitance and the charge the capacitor stores (or, equivalently, the voltage across its plates). There are three equivalent expressions for this energy:

$$U_C = \tfrac{1}{2}QV = \frac{Q^2}{2C} = \tfrac{1}{2}CV^2 \quad (16.13)$$

- A **dielectric** is a nonconducting material that increases capacitance.

- The **dielectric constant κ** describes the effect of a dielectric on capacitance. A dielectric increases the capacitor's capacitance over its value with air between the plates by a factor of κ

$$C = \kappa C_o \quad (16.15)$$

- Capacitors in series are equivalent to one capacitor, with a capacitance called the **equivalent series capacitance** C_s. The equivalent series capacitance is

$$\frac{1}{C_s} = \frac{1}{C_1} + \frac{1}{C_2} + \frac{1}{C_3} + \cdots \tag{16.17}$$

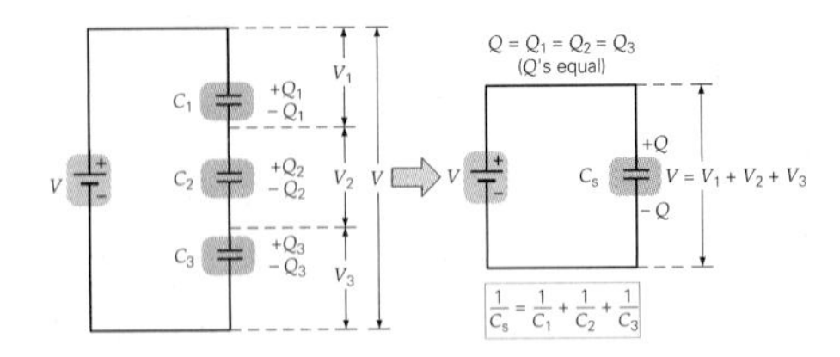

- Capacitors in parallel are equivalent to one capacitor, with a capacitance called the **equivalent parallel capacitance** C_p. In parallel, all the capacitors have the same voltage. The equivalent parallel capacitance is

$$C_p = C_1 + C_2 + C_3 + \cdots \tag{16.18}$$

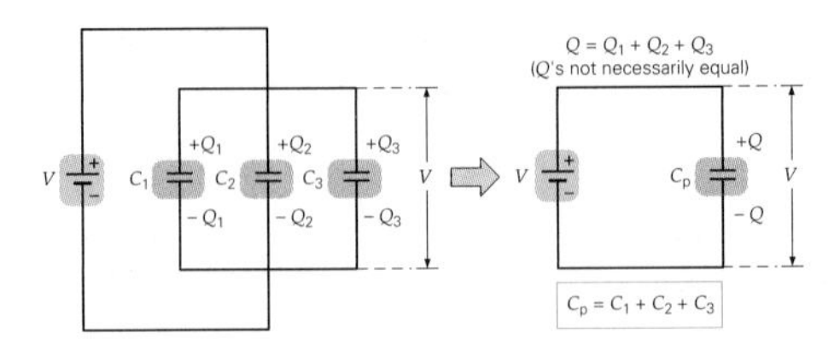

Exercises

MC = Multiple Choice Question, CQ = Conceptual Question, and IE = Integrated Exercise. Throughout the text, many exercise sections will include "paired" exercises. These exercise pairs, identified with **red numbers**, *are intended to assist you in problem solving and learning. In a pair, the first exercise (even numbered) is worked out in the Study Guide so that you can consult it should you need assistance in solving it. The second exercise (odd numbered) is similar in nature, and its answer is given at the back of the book.*

16.1 Electric Potential Energy and Electric Potential Difference

1. **MC** The SI unit of electric potential difference is the (a) joule, (b) newton per coulomb, (c) newton-meter, (d) joule per coulomb.

2. **MC** How does the electrostatic potential energy of two positive point charges change when the distance between them is tripled: (a) It is reduced to one third its original value, (b) it is reduced to one ninth its original value, (c) it is unchanged, or (d) it is increased to three times its original value?

3. **MC** An electron is moved from the positive to negative plate of a charged parallel plate arrangement. How does the sign of the change d in its electrostatic potential energy compare to the sign of the change in electrostatic potential it experiences: (a) Both are positive, (b) the energy change is positive, the potential change is negative, (c) the energy change is negative, the potential change is positive, or (d) both are negative?

4. **CQ** What is the difference (a) between electrostatic potential energy and electric potential and (b) between electric potential difference and voltage?

5. **CQ** When a proton approaches another fixed proton, what happens to (a) the kinetic energy of the approaching proton, (b) the electric potential energy of the system, and (c) the total energy of the system?

6. **CQ** Using the language of electrical potential and energy (not forces), explain why positive charges speed up as they approach negative charges.

7. **CQ** An electron is released in a region where the electric potential decreases to the left. Which way will the electron begin to move? Explain.

8. **CQ** An electron is released in a region where the electric potential is constant. Which way will the electron accelerate? Explain.

9. **CQ** If two locations are at the same electrical potential, how much work does it take to move a charge from the first location to the second? Explain.

10. ● A pair of parallel plates is charged by a 12-V battery. How much work is required to move a particle with a charge of $-4.0\ \mu\text{C}$ from the positive to the negative plate?

11. ● If it takes $+1.6 \times 10^{-5}$ J to move a positively charged particle between two charged parallel plates, (a) what is the charge on the particle if the plates are connected to a 6.0-V battery? (b) Was it moved from the negative to the positive plate or from the positive to the negative plate?

12. ● What are the magnitude and direction of the electric field between the two charged parallel plates in Exercise 11 if the plates are separated by 4.0 mm?

13. ● In a dental X-ray machine, a beam of electrons is accelerated by a potential difference of 10 kV. At the end of the acceleration, how much kinetic energy does each electron have if they start from rest?

14. ● An electron is accelerated by a uniform electric field (1000 V/m) pointing vertically upward. Use Newton's laws to determine the electron's velocity after it moves 0.10 cm from rest.

15. ● (a) Repeat Exercise 14, but find the speed by using energy methods. Get the direction in which the electron is moving by considering electric potential-energy changes. (b) Does the electron gain or lose potential energy?

16. IE ● Consider two points at different distances from a positive point charge. (a) The point closer to the charge is at a (1) higher, (2) equal, (3) lower potential than the point farther away. Why? (b) How much different is the electric potential 20 cm from a charge of 5.5 μC compared to 40 cm from the same charge?

17. IE ●● (a) At one third the original distance from a positive point charge, by what factor is the electric potential changed: (1) 1/3, (2) 3, (3) 1/9, or (4) 9? Why? (b) How far from a $+1.0$-μC charge is a point with an electric potential value of 10 kV? (c) How much of a change in potential would occur if the point were moved to three times that distance?

18. IE ●● In the Bohr model of the hydrogen atom (see Chapter 27), we will learn that the electron can exist only in circular orbits of certain radii about a proton. (a) Will a larger orbit have a (1) a higher, (2) an equal, or (3) a lower electric potential than a smaller orbit? Why? (b) Determine the potential difference between two orbits of radii 0.21 nm and 0.48 nm.

19. ●● In Exercise 18, by how much does the potential energy of the atom change if the electron goes (a) from the lower to the higher orbit, (b) from the higher to the lower orbit, and (c) from the larger orbit to a very large distance?

20. ●● How much work is required to completely separate two charges (each -1.4 μC) and leave them at rest if they were initially 8.0 mm apart?

21. ●● In Exercise 20, if the two charges are released at their initial separation distance, how much kinetic energy would each have when they are very distant from one another?

22. ●● It takes $+6.0$ J of work to move two charges from a large distance apart to 1.0 cm from one another. If the charges have the same magnitude, (a) how large is each charge, and (b) what can you tell about their signs?

23. ●● A $+2.0$-μC charge is initially 0.20 m from a fixed -5.0-μC charge and is then moved to a position 0.50 m from the fixed charge. (a) How much work is required to move the charge? (b) Does the work depend on the path through which the charge is moved?

24. ●● An electron is moved from point A to point B and then to point C along two legs of an equilateral triangle with sides of length 0.25 m (▼Fig. 16.23). If the horizontal electric field is 15 V/m, (a) what is the magnitude of the work required? (b) What is the potential difference between points A and C? (c) Which point is at a higher potential?

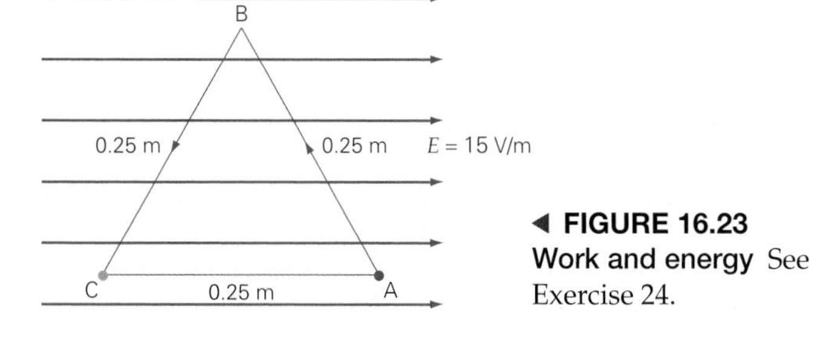

◄ **FIGURE 16.23 Work and energy** See Exercise 24.

25. ●● Compute the energy necessary to bring together (from a very large distance) the charges in the configuration shown in ▼Fig. 16.24.

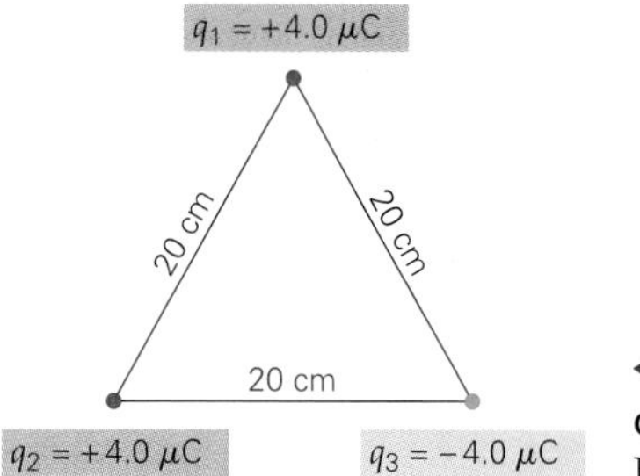

◄ **FIGURE 16.24 A charge triangle** See Exercises 25 and 27.

26. ●● Compute the energy necessary to bring together (from a very large distance) the charges in the configuration shown in ▼Fig. 16.25.

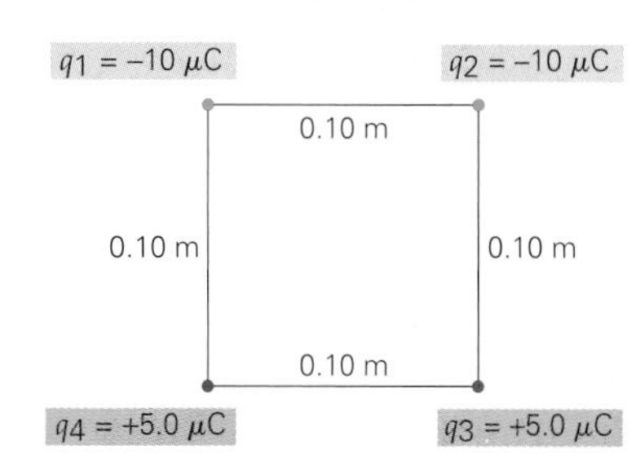

◄ **FIGURE 16.25 A charge rectangle** See Exercises 26 and 28.

27. ●●● What is the value of the electric potential at (a) the center of the triangle and (b) a point midway between q_2 and q_3 in Fig. 16.24?

28. ●●● What is the value of electric potential at (a) the center of the square and (b) a point midway between q_1 and q_4 in Fig. 16.25?

29. IE ●●● In a computer monitor, electrons are accelerated from rest through a potential difference in an "electron gun" arrangement (▼Fig. 16.26). (a) Should the left side of the gun be at (1) a higher, (2) an equal, or (3) a lower potential than the right side? Why? (b) If the potential difference in the gun is 5.0 kV, what is the "muzzle speed" of the electrons emerging from the gun? (c) If the gun is directed at a screen 25 cm away, how long do the electrons take to reach the screen?

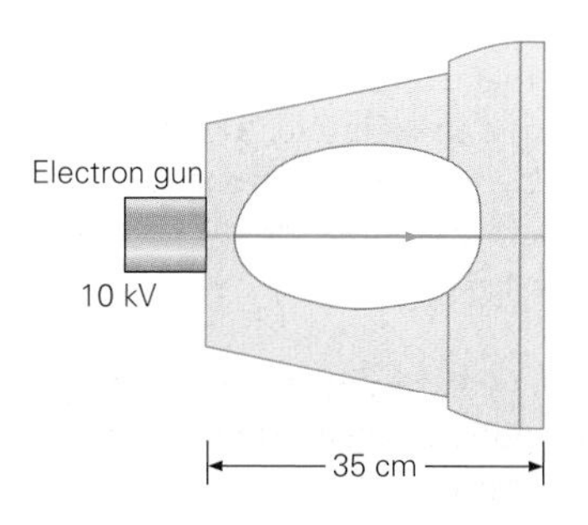

◄ **FIGURE 16.26 Electron speed** See Exercise 29.

16.2 Equipotential Surfaces and the Electric Field

30. **MC** On an equipotential surface (a) the electric potential is constant, (b) the electric field is zero, (c) the electric potential is zero, (d) there must be equal amounts of negative and positive charge.

31. **MC** Equipotential surfaces (a) are parallel to the electric field, (b) are perpendicular to the electric field, (c) can be at any angle with respect to the electric field.

32. **MC** An electron is moved from an equipotential surface at +5.0 V to one at +10.0 V. It is moving generally in a direction (a) parallel to the electric field, (b) opposite to the electric field, (c) in the same direction as the electric field.

33. **CQ** Sketch the topographic map you would expect as you walk away from the ocean up a gently sloping uniform beach. Label the gravitational equipotentials as to relative height and potential value. Show how to predict, from the map, which way a ball would accelerate if it is initially rolled up the beach away from the water.

34. **CQ** Explain why two equipotential surfaces cannot intersect.

35. **CQ** Suppose a charge starts at rest on an equipotential surface, is moved off that surface, and then is eventually returned to the same surface at rest after a round trip. How much work did it take to do this? Explain.

36. **CQ** What geometrical shape are the equipotential surfaces between two charged parallel plates?

37. **CQ** (a) What is the approximate shape of the equipotential surfaces inside the axon cell membrane? (See Fig. 1, p. 552.) (b) Under resting-potential conditions, where is the region of highest electric potential inside the membrane? (c) What about during reversed polarity conditions?

38. **CQ** Near a fixed positive point charge, if you go from one equipotential surface to another one with a smaller radius, (a) what happens to the value of the potential? (b) What was your general direction relative to the electric field?

39. **CQ** (a) If a proton is accelerated from rest by a potential difference of 1 million volts, how much kinetic energy does it gain? (b) How would your answer to part (a) change if the accelerated particle had twice the charge of the proton (same sign) and four times the mass?

40. **CQ** (a) Can the electric field at a point be zero while there is also a nonzero electric potential at that point? (b) Can the electric potential at a point be zero while there is also a nonzero electric field at that point? Explain. If the answer to either part is yes, give an example.

41. ● For a $+3.50\text{-}\mu C$ point charge, what is the radius of the equipotential surface that is at a potential of 2.50 kV?

42. ● A uniform electric field of 10 kV/m points vertically upward. How far apart are the equipotential planes that differ by 100 V?

43. ● In Exercise 42, if the ground is designated as zero potential, how far above the ground is the equipotential surface corresponding to 7.0 kV?

44. ● Determine the potential 2.5 mm from the negative plate of a pair of parallel plates separated by 10 mm and connected to a 24-V battery.

45. ● Relative to the positive plate in Exercise 44, where is the point with a potential of 20 V?

46. ● If the radius of the equipotential surface of point charge is 14.3 m at a potential of 2.20 kV, what is the magnitude of the point charge creating the potential?

47. **IE** ● (a) The equipotential surfaces in the neighborhood of a point charge are (1) concentric spheres, (2) concentric cylinders, (3) planes. (b) Calculate the amount of work (in electron-volts) it would take to move an electron from 12.6 m to 14.3 m away from a $+3.50\text{-}\mu C$ point charge.

48. ● The potential difference involved in a typical lightning discharge may be up to 100 MV (million volts). What is the gain in kinetic energy of an electron accelerated through this potential difference? Give your answer in both electron-volts and joules. (Assume that there are no collisions.)

49. ● In a typical Van de Graaff linear accelerator, protons are accelerated through a potential difference of 20 MV. What is their kinetic energy if they started from rest? Give your answer in (a) eV, (b) keV, (c) MeV, (d) GeV, and (e) joules.

50. ● In Exercise 49, how do your answers change if a doubly charged (+2e) alpha particle is accelerated instead? (An alpha particle consists of two neutrons and two protons.)

51. ●● In Exercises 49 and 50, compute the speed of the proton and alpha particle after being accelerated.

52. ●● Calculate the voltage required to accelerate a beam of protons initially at rest, and calculate their speed if they have a kinetic energy of (a) 3.5 eV, (b) 4.1 keV, and (c) 8.0×10^{-16} J.

53. ●● Repeat the calculation in Exercise 52 for electrons instead of protons.

54. ●●● Two large parallel plates are separated by 3.0 cm and connected to a 12-V battery. Starting at the negative plate and moving 1.0 cm toward the positive plate at a 45° angle (▼Fig. 16.27), (a) what value of potential would be reached,

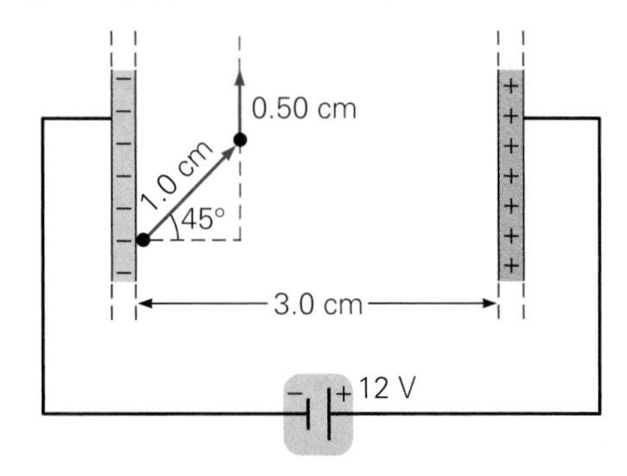

▲ **FIGURE 16.27 Reaching our potential** See Exercises 54 and 55.

assuming the negative plate were defined as zero potential? (b) What would be the value of the potential if you then moved 0.50 cm parallel to the plates?

55. ●●● Consider a point midway between the two large charged plates in Fig. 16.27. Compute the change in electric potential if from there you moved (a) 1.0 mm toward the positive plate, (b) 1.0 mm toward the negative plate, and (c) 1.0 mm parallel to the plates.

56. ●●● Using the results of Exercise 55, determine the electric field (direction and magnitude) at the midway point between the plates.

16.3 Capacitance

57. **MC** A capacitor is first connected to a 6.0-V battery and then disconnected and connected to a 12.0-V battery. How does its capacitance change: (a) It increases, (b) it decreases, or (c) it stays the same?

58. **MC** A capacitor is first connected to a 6.0-V battery and then disconnected and connected to a 12.0-V battery. How does the charge on one of its plates change: (a) It increases, (b) it decreases, or (c) it stays the same?

59. **MC** A capacitor is first connected to a 6.0-V battery and then disconnected and connected to a 12.0-V battery. By how much does the electric field strength between its plates change: (a) two times, (b) four times, or (c) it stays the same.

60. **MC** A capacitor has the distance between its plates cut in half. By what factor does its capacitance change: (a) It is cut in half, (b) it is reduced to one fourth its original value, (c) it is doubled, or (d) it is quadrupled?

61. **MC** A capacitor has the area of its plates reduced. How would you adjust the distance between those plates to keep the capacitance constant: (a) increase it, (b) decrease it, or (c) changing the distance cannot ever make up for the plate area change?

62. **CQ** If the plates of an isolated parallel-plate capacitor are moved closer to each other, does the energy storage increase, decrease, or remain the same? Explain.

63. **CQ** If the potential difference across a capacitor is doubled, what happens to (a) the charge on the capacitor and (b) the energy stored in the capacitor?

64. **CQ** A capacitor is connected to a 12-V battery. If the plate separation is tripled and the capacitor remains connected to the battery, by what factor does the charge on the capacitor change?

65. ● How much charge flows through a 12-V battery when a 2.0-μF capacitor is connected across its terminals?

66. ● A parallel-plate capacitor has a plate area of 0.50 m^2 and a plate separation of 2.0 mm. What is its capacitance?

67. ● What plate separation is required for a parallel-plate capacitor to have a capacitance of 5.0×10^{-9} F if the plate area is 0.40 m^2?

68. **IE** ● (a) For a parallel-plate capacitor, a larger plate area results in (1) a larger, (2) an equal, (3) a smaller capacitance. (b) A 2.5×10^{-9} F parallel-plate capacitor has a plate area of 0.425 m^2. If the capacitance is to double, what is the required plate area?

69. ●● A 12-V battery is connected to a parallel-plate capacitor with a plate area of 0.20 m^2 and a plate separation of 5.0 mm. (a) What is the charge on the capacitor? (b) How much energy is stored in the capacitor?

70. ●● If the plate separation of the capacitor in Exercise 69 changed to 10 mm after the capacitor is disconnected from the battery, how do your answers change?

71. ●●● Current state-of-the-art capacitors are capable of storing many times the energy of older ones. Such a capacitor, with a capacitance of 1.0 F, is able to light a small 0.50-W bulb at steady full power for 5.0 s before it quits. What was the terminal voltage of the battery that charged the capacitor?

72. ●●● A 1.50-F capacitor is connected to a 12.0-V battery for a long time, and then is disconnected. The capacitor briefly runs a 1.00-W toy motor for 2.00 s. After this time, (a) by how much has the energy stored in the capacitor decreased? (b) What is the voltage across the plates? (c) How much charge is stored on the capacitor? (d) How much longer could the capacitor run the motor, assuming the motor ran at full power until the end?

73. ●●● Two parallel plates have a capacitance value of 0.17 μF when they are 1.5 mm apart. They are connected permanently to a 100-V power supply. If you pull the plates out to a distance of 4.5 mm, (a) what is the electric field between them? (b) By how much has the capacitor's charge changed? (c) By how much has its energy storage changed? (d) Repeat these calculations assuming the power supply is disconnected before you pull the plates further apart.

16.4 Dielectrics

74. **MC** Putting a dielectric in a charged parallel-plate capacitor that is not connected to a battery (a) decreases the capacitance, (b) decreases the voltage, (c) increases the charge, (d) causes a discharge because the dielectric is a conductor.

75. **MC** A parallel-plate capacitor is connected to a battery. If a dielectric is inserted between the plates, (a) the capacitance decreases, (b) the voltage increases, (c) the voltage decreases, (d) the charge increases.

76. **MC** A parallel-plate capacitor is connected to a battery and then disconnected. If a dielectric is then inserted between the plates, what happens to the charge on its plates: (a) The charge decreases, (b) the charge increases, or (c) the charge stays the same?

77. **CQ** Give several reasons why a conductor would not be a good choice as a dielectric for a capacitor.

78. **CQ** A parallel-plate capacitor is connected to a battery and then disconnected. If a dielectric is inserted between the plates, what happens to (a) the capacitance and (b) the voltage?

79. CQ Explain clearly why the electric field between two parallel plates of a capacitor decreases when a dielectric is inserted if the capacitor is not connected to a power supply, but remains the same when it is connected to a power supply.

80. ● A capacitor has a capacitance of 50 pF, which increases to 150 pF when a dielectric material is between its plates. What is the dielectric constant of the material?

81. ● A 50-pF capacitor is immersed in silicone oil ($\kappa = 2.6$). When the capacitor is connected to a 24-V battery, what will be the charge on the capacitor and the amount of stored energy?

82. ●● The dielectric of a parallel-plate capacitor is to be constructed from glass that completely fills the volume between the plates. The area of each plate is 0.50 m^2. (a) What thickness should the glass have if the capacitance is to be 0.10 μF? (b) What is the charge on the capacitor if it is connected to a 12-V battery?

83. ●●● A parallel-plate capacitor has a capacitance of 1.5 μF with air between the plates. The capacitor is connected to a 12-V battery and charged. The battery is then removed. When a dielectric is placed between the plates, a potential difference of 5.0 V is measured across the plates. (a) What is the dielectric constant of the material? (b) Did the energy stored in the capacitor increase, decrease, or stay the same? (c) By how much did the energy storage of this capacitor change when the dielectric was inserted?

84. IE ●●● An air-filled parallel-plate capacitor has rectangular plates with dimensions of 6.0 cm × 8.0 cm. It is connected to a 12-V battery. While the battery remains connected, a sheet of 1.5-mm-thick Teflon ($\kappa = 2.1$) is inserted and completely fills the space between the plates. (a) While the dielectric was being inserted, (a) charge flowed onto the capacitor, (2) charge flowed off the capacitor, (3) no charge flowed. (b) Determine the change in the charge storage of this capacitor because of the dielectric insertion.

16.5 Capacitors in Series and in Parallel

85. MC Capacitors in series have the same (a) voltage, (b) charge, (c) energy storage.

86. MC Capacitors in parallel have the same (a) voltage, (b) charge, (c) energy storage.

87. MC Capacitors 1, 2, and 3 have the same capacitance value C. 1 and 2 are in series and their combination is in parallel with 3. What is their effective total capacitance. (a) C, (b) $1.5C$, (c) $3C$, or (d) $C/3$?

88. CQ Under what conditions would two capacitors in series have the same voltage?

89. CQ Under what conditions would two capacitors in parallel have the same charge?

90. CQ If you are given two capacitors, how should you connect them to get (a) maximum equivalent capacitance and (b) minimum equivalent capacitance?

91. CQ You have N (an even number ≥ 2) identical capacitors, each with a capacitance of C. In terms of N and C, what is their total effective capacitance if (a) they are all connected in series? (b) they are all connected in parallel? (c) two halves ($N/2$ each) are connected in series and these two sets are connected in parallel?

92. ● What is the equivalent capacitance of two capacitors with capacitances of 0.40 μF and 0.60 μF when they are connected (a) in series and (b) in parallel?

93. IE ● (a) Two capacitors can be connected to a battery in either a series or parallel combination. The parallel combination will draw (1) more, (2) equal, or (3) less energy from a battery than the series combination. Why? (b) When a series combination of two uncharged capacitors is connected to a 12-V battery, 173 μJ of energy is drawn from the battery. If one of the capacitors has a capacitance of 4.0 μF, what is the capacitance of the other?

94. ●● For the arrangement of three capacitors in ▾Fig. 16.28, what value of C_1 will give a total equivalent capacitance of 1.7 μF?

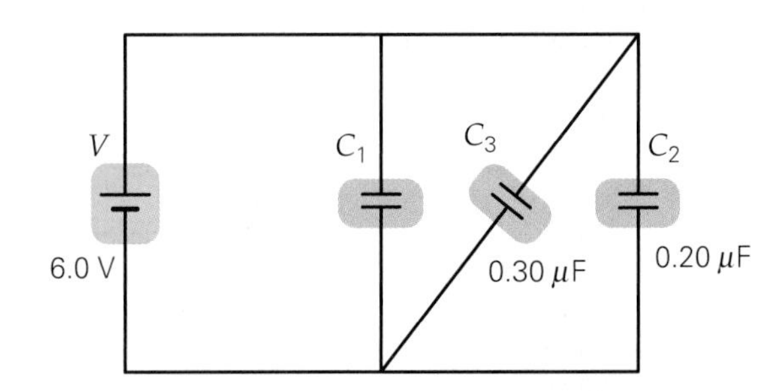

▲ **FIGURE 16.28 A capacitor triad** See Exercises 94 and 98.

95. IE ●● (a) Three capacitors of equal capacitance are connected in parallel to a battery, and together they draw a certain amount of charge Q from that battery. Will the charge on each capacitor be (1) Q, (2) $3Q$, or (3) $Q/3$? (b) Three capacitors of 0.25 μF each are connected in parallel to a 12-V battery. What is the charge on each capacitor? (c) How much charge is drawn from the battery?

96. IE ●● (a) If you are given three identical capacitors, you can obtain (1) three, (2) five, (3) seven different capacitance values. (b) If the three capacitors each have a capacitance of 1.0 μF, what are the different values of equivalent capacitance?

97. ●● What are the maximum and minimum equivalent capacitances that can be obtained by combinations of three capacitors of 1.5 μF, 2.0 μF, and 3.0 μF?

98. ●●● If the capacitance $C_1 = 0.10$ μF, what is the charge on each of the capacitors in the circuit in Fig. 16.28?

99. ●●● Four capacitors are connected in a circuit as illustrated in ▸Fig. 16.29. Find the charge on, and the voltage difference across, each of the capacitors.

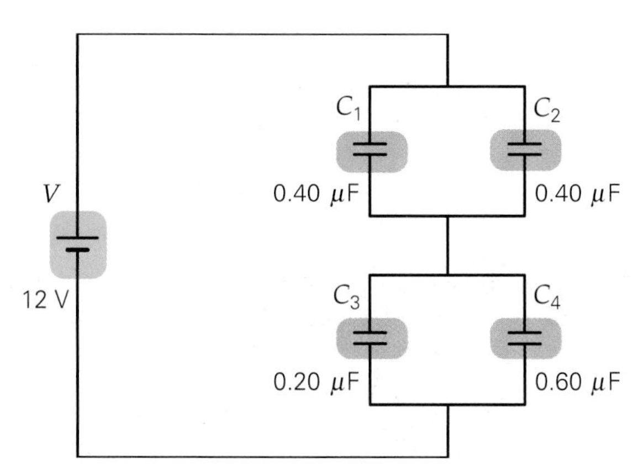

▲ **FIGURE 16.29 Double parallel in series** See Exercise 99.

Comprehensive Exercises

100. IE A tiny dust particle in the form of a long thin needle has charges of ± 7.14 pC on its end. The length of the particle is 3.75 μm. (a) Which location is at a higher potential: (1) 7.65 μm above the positive end, (2) 5.15 μm above the positive end, or (3) both locations are at the same potential? (b) Compute the potential at the two the points in part (a). (c) Use your answer from part (b) to determine the work needed to move an electron from the near point to the far point.

101. A vacuum tube has a vertical height of 50.0 cm. An electron leaves from the top at a speed of 3.2×10^6 m/s downward and is subjected to a "typical" Earth field of 150 V/m downward. (a) Use energy methods to determine if it reaches the bottom surface of the tube. (b) If it does, with what speed does it hit, if not, how close does it come to the bottom surface?

102. CQ Sketch the equipotential surfaces and the electric field line pattern outside a uniformly (negatively) charged long wire. Label the surfaces with relative potential value and indicate the electric field direction.

103. A helium atom with one electron already removed (a positive helium ion) consists of a single orbiting electron and a nucleus of two protons. The electron is in its minimum orbital radius of 0.027 nm. (a) What is the potential energy of the system? (b) What is the centripetal acceleration of the electron? (c) What is the total energy of the system? (d) What is the minimum energy required to ionize this atom so the electron leaves completely?

104. Suppose that the three capacitors in Figure 16.22 have the following values: $C_1 = 0.15\ \mu$F, $C_2 = 0.25\ \mu$F, and $C_3 = 0.30\ \mu$F. (a) What is the equivalent capacitance of this arrangement? (b) How much charge will be drawn from the battery? (c) What is the voltage across each capacitor?

105. IE Two very large horizontal parallel plates are separated by 1.50 cm. An electron is to be suspended in midair between them. (a) The top plate should be at (1) a higher potential, (2) an equal potential, (3) a lower potential compared with the bottom plate. Explain. (b) What voltage across the plates is required? (c) Does the electron have to be positioned midway between the plates, or is any location between the plates just as good?

106. (See the Insight on Electric Potential and Nerve Signal Transmission on p. 552 and the Learn by Drawing on graphical relationships between $\vec{\mathbf{E}}$ and V on p. 547.) Suppose an (axon) cell membrane is experiencing the end of a stimulus event and the voltage across the cell membrane is instantaneously at 30 mV. Assume the membrane is 10 nm thick. At this point the Na/K-ATPase molecular pump starts to move the excess Na^+ ions back to the exterior. (a) How much work does it take for the pump to move the first sodium ion? (b) Estimate the electric field (including direction) in the membrane under these conditions? (c) What is the electric field (including direction) under normal conditions when the voltage across the membrane is -70 mV?

107. In exercise 106, assume that the inside and outside surfaces of the axon membrane act like a parallel plate capacitor with an area of 1.1×10^{-9} m^2. (a) Estimate the capacitance of an axon's membrane, assuming it is filled with lipids with a dielectric constant of 3.0. (b) How much charge would be on each surface under resting potential conditions?

108. Two parallel plates, 9.25 cm on a side, are separated by 5.12 mm. (a) Determine their capacitance if the volume from one plate to midplane is filled with a material of dielectric constant 2.55 and the rest is filled with a different material (dielectric constant 4.10). See ▼Fig. 16.30a. [*Hint*: Do you see two capacitors in series?] (b) Repeat part (a), except fill the volume from one edge to the middle with the same two materials. See Figure 16.30b. (Do you see two capacitors in parallel?)

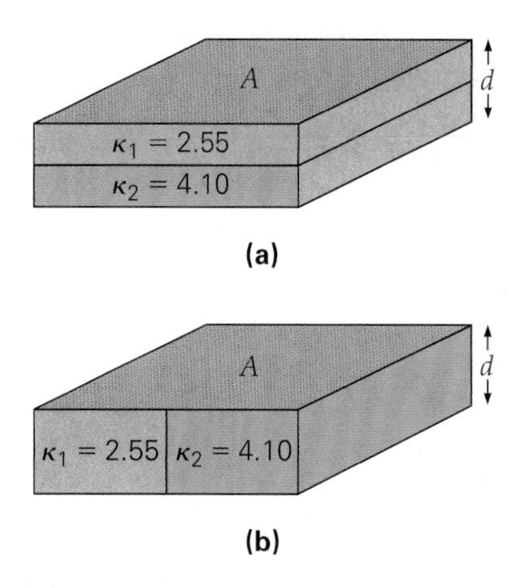

▲ **FIGURE 16.30 Double-stuffed capacitor** See Exercise 108.

The following Physlet Physics Problems can be used with this chapter.
25.1, 25.2, 25.3, 25.4, 25.5, 25.6, 25.7, 26.1, 26.2, 26.3, 26.5, 26.9, 26.11

CHAPTER 17

ELECTRIC CURRENT AND RESISTANCE

17.1 Batteries and Direct Current 569

17.2 Current and Drift Velocity 571

17.3 Resistance and Ohm's Law 573

17.4 Electric Power 580

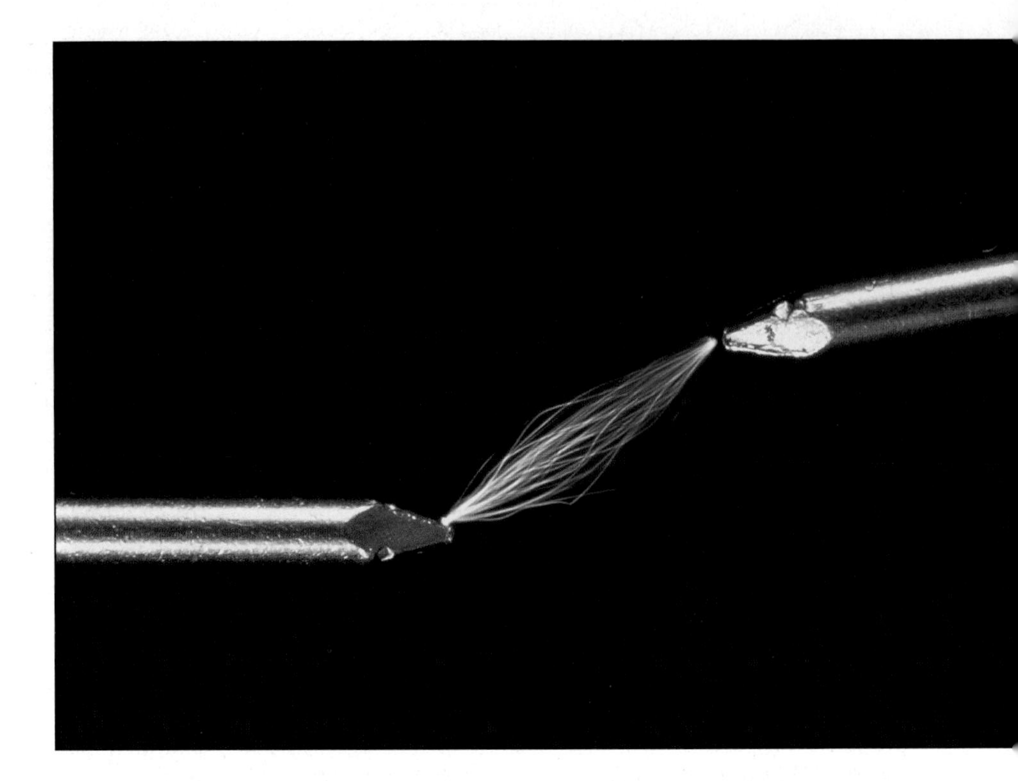

PHYSICS FACTS

- André Marie Ampère (1775–1836) was a French physicist/mathematician known for his work with electric currents. His name is used for the SI unit of current, the ampere (usually shortened to *amp*). He also worked in chemistry, being involved in the classification of the elements and the discovery of fluorine. In physics, Ampère is famous for being one of the first to attempt a combined theory of electricity and magnetism. Ampere's law, which describes the *magnetic* field created by a flow of *electric* charge, is one of the four fundamental equations of classical electromagnetism.
- In a metal wire, the electric *energy* travels at the speed of light (in the wire), which is much faster than the speed of the charge carriers themselves. The speed of the latter is only several millimeters per second.
- The SI unit of electrical resistance, the ohm (Ω), is named after Georg Simon Ohm (1789–1854), a German mathematician and physicist. A quantity called electrical conductivity, proportional to the *inverse* of resistance, is named, appropriately enough, the mho—his last name spelled backwards.
- Using a voltage of up to 600 volts, electric eels and rays can, for brief times, discharge as much as 1 ampere of current through flesh. The energy is delivered at a rate of 600 J/s, or about three-fourths of a horsepower.

If you were asked to think of electricity and its uses, many favorable images would probably come to mind, including such diverse applications as lamps, television remote controls, computers, and electric leaf blowers. You might also think of some unfavorable images, such as dangerous lightning, a shock, or sparks you may have experienced from an overloaded electric outlet.

Common to all of these images is the concept of electric energy. For an electric appliance, energy is supplied by electric current in wires; for lightning or a spark, it is conducted through the air. In either case, the light, heat, or mechanical energy given off is simply electric energy converted to a different form. In the photograph, for example, the light given off by the spark is emitted by air molecules.

In this chapter, we are concerned with the fundamental principles governing electric circuits. These principles will enable us to answer questions such as the following: What is electric current and how does it travel? What causes an electric current to move through an appliance when we flip a switch? Why does the electric current cause the filament in a bulb to glow brightly, but not affect the connecting wires in the same way? We can apply electrical principles to gain an understanding of a wide range of phenomena, from the operation of household appliances to the workings of Nature's spectacular fireworks—lightning.

17.1 Batteries and Direct Current

OBJECTIVES: To (a) introduce the properties of a battery, (b) explain how a battery produces a direct current in a circuit, and (c) learn various circuit symbols for sketching schematic circuit diagrams.

After studying electric force and energy in Chapters 15 and 16, you can probably guess what is required to produce an *electric current*, or a flow of charge. Here are some analogies to help. Water naturally flows downhill, from higher to lower gravitational potential energy—that is, because there is a *difference* in gravitational potential energy. Heat flows naturally because of temperature *differences*. In electricity, a flow of electric charge is caused by an *electric* potential *difference*—what we call "voltage."

In solid conductors, particularly metals, some of the outer electrons of atoms are relatively free to move. (In liquid conductors and charged gases called *plasmas*, positive and negative ions as well as electrons can move.) Energy is required to move electric charge. Electric energy is generated through the conversion of other forms of energy, giving rise to a potential difference, or voltage. Any device that can produce and maintain a potential difference is called by the general name of a *power supply*.

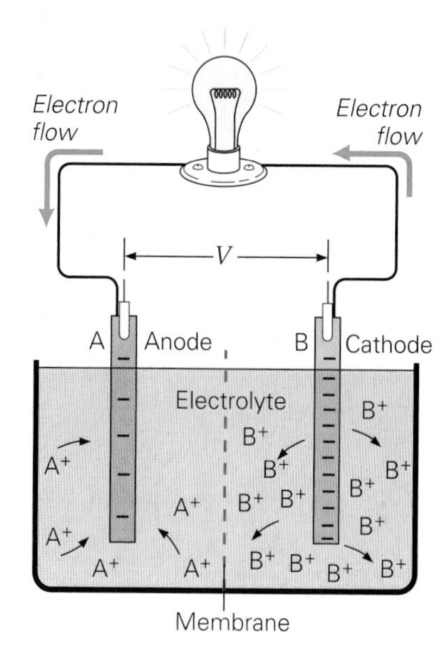

▲ **FIGURE 17.1 Battery action in a chemical battery or cell** Chemical processes involving an electrolyte and two unlike metal electrodes cause ions of both metals to dissolve into the solution at different rates. Thus, one electrode (the cathode) becomes more negatively charged than the other (the anode). The anode is at a higher potential than the cathode. By convention, the anode is designated the positive terminal and the cathode the negative. This potential difference (V) can cause a current, or a flow of charge (electrons), in the wire. The positive ions migrate as shown. (A membrane is necessary to prevent mixing of the two types of ion; why?)

Battery Action

One common type of power supply is the battery. A **battery** converts stored *chemical* potential energy into electrical energy. The Italian scientist Alessandro Volta constructed one of the first practical batteries. A simple battery consists of two unlike metal *electrodes* in an *electrolyte*, a solution that conducts electricity. With the appropriate electrodes and electrolyte, a potential difference develops across the electrodes as a result of chemical action (▸Fig. 17.1).

When a complete circuit is formed, for example, by connecting a lightbulb and wires (Fig. 17.1), electrons from the more negative electrode (B) will move through the wire and bulb to the less negative electrode (A).* The result is a flow of electrons in the wire. As electrons move through the bulb's filament, colliding with and transferring energy to its atoms (typically tungsten), the filament reaches a sufficient temperature to give off visible light (glow). Since electrons move to regions of higher potential, electrode A is at a higher potential than B. Thus the battery action has created a potential *difference* (V) across its terminals. Electrode A is the **anode** and labeled with a plus (+) sign. Electrode B is the **cathode** and labeled as negative (−). It is easy to keep track of this sign convention because the negatively charged electrons will move through the wire from B (−) to A (+).

For the study of circuits, we can just picture a battery as a "black box" that maintains a constant potential difference across its terminals. Inserted into a circuit, a battery can do work on, and transfer energy to, electrons in the wire (at the expense of its own internal chemical energy), which in turn delivers that energy to external circuit elements. In these elements, the energy is converted into other forms, such as mechanical motion (as in electric fans), heat (as in immersion heaters), and light (as in flashlights). Other sources of voltage, such as generators and photocells, will be considered later.

To help better visualize the role of a battery, consider the gravitational analogy in ▸Fig. 17.2. A gasoline-fueled pump (analogous to the battery) does work on the water as it lifts it. The increase in the water's gravitational potential energy comes at the expense of the chemical potential energy of the gasoline molecules. The water then returns to the pump by flowing down the trough (analogous to the wire) into the pond. On the way down, the water does work on the wheel, resulting in rotational kinetic energy, analogous to the electrons transferring energy to a lightbulb.

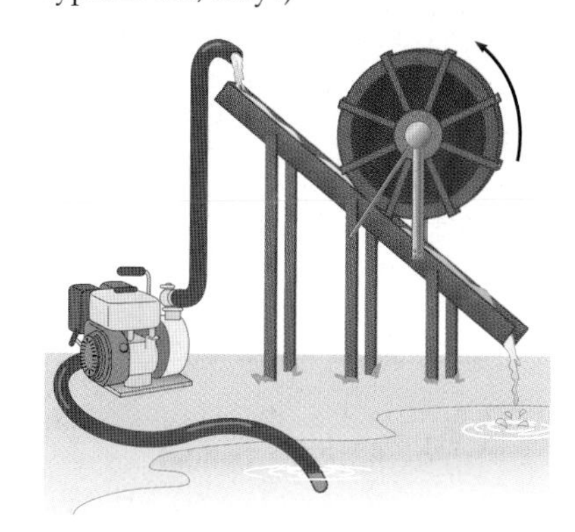

▲ **FIGURE 17.2 Gravitational analogy to a battery and lightbulb** A gasoline-powered pump lifts water from the pond, increasing the potential energy of the water. As the water flows downhill, it transfers energy to (or does work on) a waterwheel, causing the wheel to spin. This action is analogous to the delivery of energy to a lightbulb by an electrical current (for example, as in Fig. 17.1).

*As we shall see soon, a *complete circuit* is any complete loop consisting of wires and electrical devices (such as batteries and lightbulbs).

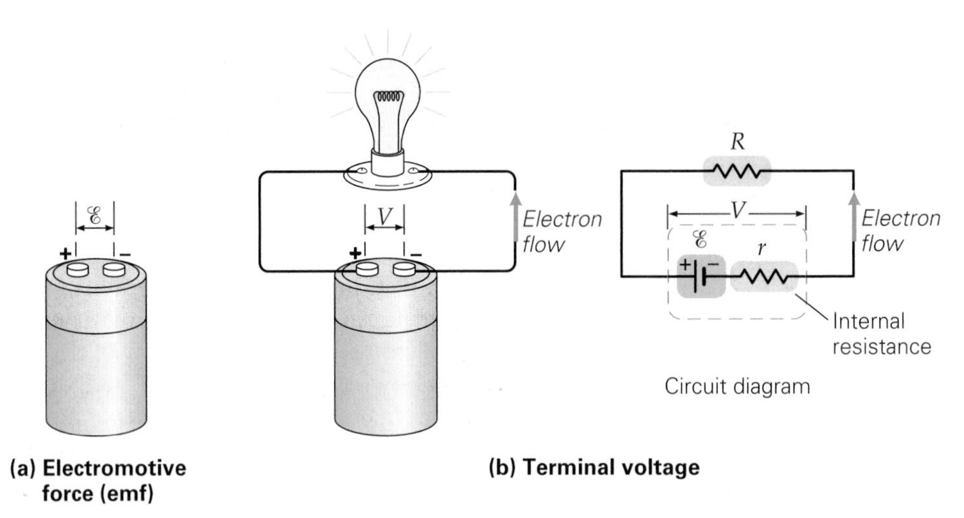

▶ **FIGURE 17.3 Electromotive force (emf) and terminal voltage (a)** The emf (ℰ) of a battery is the maximum potential difference across its terminals. This maximum occurs when the battery is not connected to an external circuit. **(b)** Because of internal resistance (r), the terminal voltage V when the battery is in operation is less than the emf ℰ. Here, R is the resistance of the lightbulb.

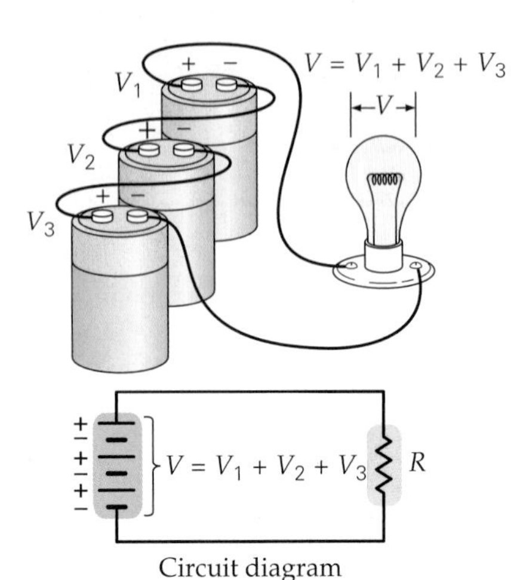

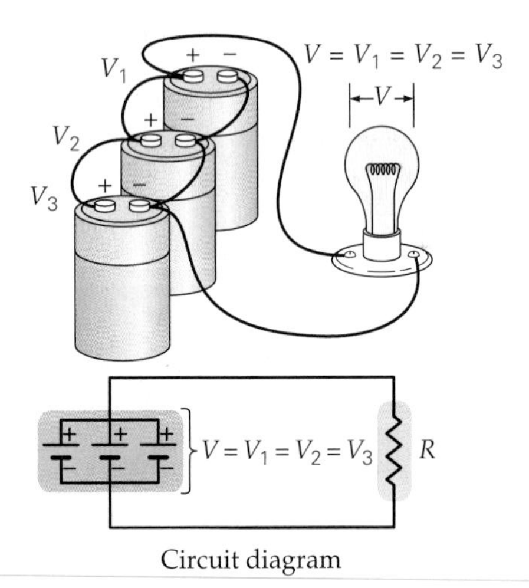

▲ **FIGURE 17.4 Batteries in series and in parallel (a)** When batteries are connected in series, their voltages add, and the voltage across the resistance R is the sum of the voltages. **(b)** When batteries of the same voltage are connected in parallel, the voltage across the resistance is the same, as if only a single battery were present. In this case, each battery supplies part of the total current.

Battery EMF and Terminal Voltage

The potential difference across the terminals of a battery *when it is not connected* to a circuit is called the battery's **electromotive force (emf)**, symbolized by ℰ. The name is misleading, because emf is *not* a force, but a potential difference, or voltage. To avoid confusion with force, we will call electromotive force just *emf*. Thus a battery's emf represents the work done by the battery *per coulomb* of charge that passes through it. If a battery does 1 joule of work on 1 coulomb of charge, then its emf is 1 joule per coulomb (1 J/C), or 1 volt (1 V).

The emf actually represents the maximum potential difference across the terminals (▲Fig. 17.3a). Under practical circumstances, when a battery is in a circuit and charge flows, the voltage across the terminals is always slightly *less* than the emf. This "operating voltage" (V) of a battery (the battery symbol is the pair of unequal-length parallel lines in Fig. 17.3b) is called its **terminal voltage**. Because batteries in actual operation are of most interest, it is the terminal voltage that is important.

Under many conditions, the emf and terminal voltage are essentially the same. Any difference is due to the battery's *internal resistance* (r), shown explicitly in the circuit diagram in Fig. 17.3b. (Resistance, defined in Section 17.3, is a quantitative measure of the opposition to charge flow.) Internal resistances are typically small, so the terminal voltage of a battery is essentially the same as the emf, that is, $V \approx \mathscr{E}$. However, when a battery supplies a large current or when its internal resistance is high (older batteries), the terminal voltage may drop appreciably below the emf. The reason is that it takes some voltage just to produce a current in the internal resistance itself. Mathematically, the terminal voltage is related to the emf, current, and internal resistance by $V = \mathscr{E} - Ir$ where I is the *electric current* (Section 17.2) in the battery.

For example, most modern cars have a battery "voltage readout." Upon startup, the 12-V battery's voltage typically reads only 10 V (this value is normal). Because of the enormous current required at startup, the Ir term (2 V) reduces the emf by about 2 V to the measured terminal voltage of 10 V. When the engine is running and supplying most of the electric energy to run the car's functions, the current required from the battery is essentially zero and the battery readout rises back to normal voltage levels. Thus, the terminal voltage, and not the emf, is a true indication of the state of the battery. Unless otherwise specified, we will assume negligible internal resistance, so that $V \approx \mathscr{E}$.

There is a wide variety of batteries in use. One of the most common is the 12-V automobile battery, consisting of six 2-V cells connected in *series*.* That is, the positive terminal of each cell is connected to the negative terminal of the next (see the three cells in ◀Fig. 17.4a). When batteries or cells are connected in this fashion, their voltages add. If cells are connected in *parallel*, their positive terminals are connected to each other, as are the negative ones (Fig. 17.4b). When identical batteries are connected this way, the potential difference or terminal

*Chemical energy is converted to electrical energy in a chemical *cell*. The term *battery* generally refers to a collection, or "battery," of cells.

voltage is the same for all of them. However, each supplies only a fraction of the current to the circuit. For example, if you have three batteries with equal voltages connected in parallel, each supplies one third of the total current. A parallel connection of two batteries is the main method for "jump-starting" a car. For such a start, the weak (high r) battery is connected in parallel to a normal (low r) battery, which delivers most of the current to start the car.

Note: Recall that *voltage* is used to mean "difference in electric potential."

Circuit Diagrams and Symbols

To help analyze and visualize circuits, it is common to draw *circuit diagrams* that are schematic representations of the wires, batteries, appliances, and so on. Each element in the circuit is represented by its own symbol in such a diagram. As in Fig. 17.3b and Fig. 17.4, the battery symbol is two parallel lines, the longer representing the positive terminal and the shorter the negative terminal. Any element (such as a lightbulb or appliance) that *opposes* the flow of charge is represented by the symbol —\/\/\/—. (Electrical resistance is defined in Section 17.3; here we merely introduce the symbol.) Connecting wires are unbroken lines and assumed, unless stated otherwise, to have negligible resistance. Where lines cross, it is assumed that they do *not* contact one another, unless they have a dot at their intersection. Lastly, switches are shown as "drawbridges," capable of going up (to open the circuit and stop the current) and down (closed to complete the circuit and allow current). These symbols, along with that of the capacitor (from Chapter 16), are summarized in the accompanying Learn by Drawing. An example of the use of these symbols and circuit diagrams to understand circuits conceptually is shown in the next example.

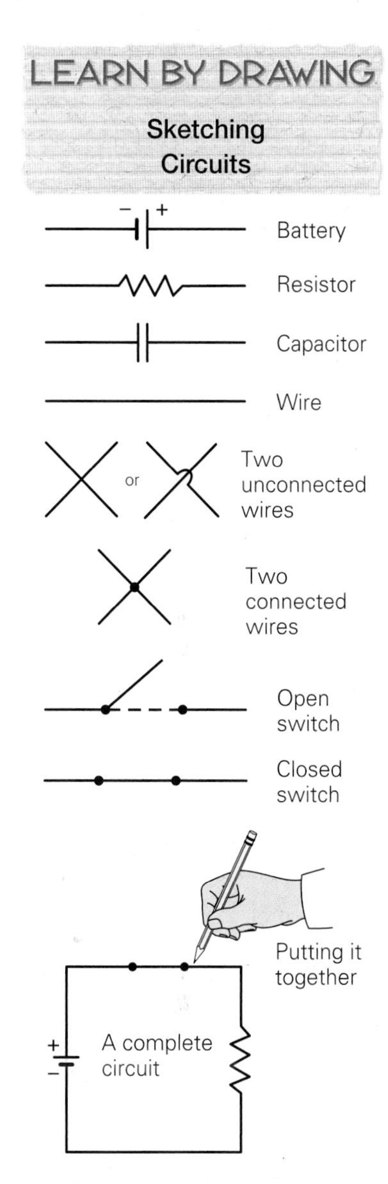

Conceptual Example 17.1 ■ Asleep at the Switch?

▸Fig. 17.5 shows a circuit diagram that represents two identical batteries (each with a terminal voltage V) connected in parallel to a lightbulb (represented by a resistor). Because it is assumed that the wires have no resistance, we know that before switch S_1 is opened, the voltage across the lightbulb equals V (that is, $V_{AB} = V$). What happens to the voltage across the lightbulb when S_1 is opened? (a) The voltage remains the same (V) as before the switch was opened. (b) The voltage drops to $V/2$, because only one battery is now connected to the bulb. (c) The voltage drops to zero.

Reasoning and Answer. It might be tempting to choose answer (b), because there is now just one battery. But look again. The remaining battery is still connected to the lightbulb. This means that there must be *some* voltage across the lightbulb, so the answer certainly cannot be (c). But it also means that the answer cannot be (b), because the remaining battery itself will maintain a voltage of V across the bulb. Hence, the answer is (a).

Follow-Up Exercise. In this Example, what would the correct answer be if, in addition to opening S_1, switch S_2 were also opened? Explain your answer and reasoning. *(Answers to all Follow-Up Exercises are at the back of the text.)*

17.2 Current and Drift Velocity

OBJECTIVES: To (a) define electric current, (b) distinguish between electron flow and conventional current, and (c) explain the concept of drift velocity and electric energy transmission.

As we have just seen, to sustain an electric current requires a voltage source and a **complete circuit**—the name given to a continuous conducting path. Most practical circuits include a switch to "open" or "close" the circuit. An open switch eliminates the continuous part of the path, thereby stopping the flow of charge in the wires.

Electric Current

Because it is the electrons that move in any circuit's wires, the charge flow is away from the negative terminal of the battery. Historically, however, circuit analysis has been done in terms of **conventional current**. The conventional current's direction is

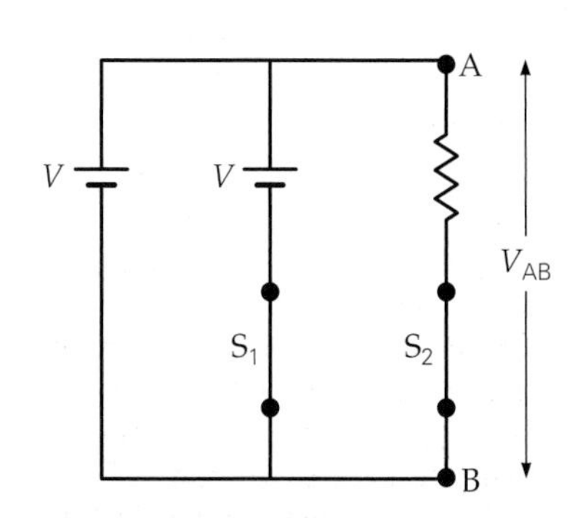

▲ **FIGURE 17.5** What happens to the voltage? See Example 17.1.

that in which positive charges would flow, that is, *opposite* the actual electron flow (◂Fig. 17.6). (Some situations do exist in which a positive charge flow *is* responsible for the current—for example, in semiconductors.)

The battery is said to *deliver* current to a circuit or a component of that circuit (a circuit element). Alternatively, we sometimes say that a circuit (or its components) *draws* current from the battery. The current then returns to the battery. A battery can produce a current in only one direction. One-directional charge flow is called **direct current (dc)**. (Note if the current changes direction and/or magnitude, it is *alternating current*. We will study this type of situation in detail in Chapter 21.)

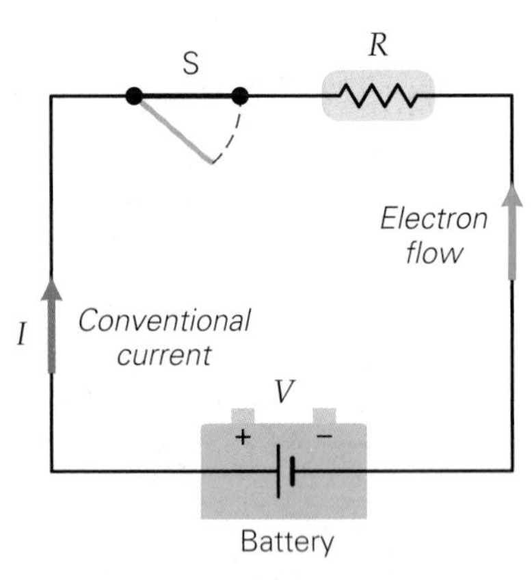

▲ **FIGURE 17.6 Conventional current** For historical reasons, circuit analysis is usually done with conventional current. Conventional current is in the direction in which positive charges would flow, or opposite to the electron flow.

Quantitatively, the **electric current (I)** is the time rate of flow of net charge. We will be concerned primarily with steady charge flow. In this case, if a net charge q passes through a cross-sectional area in a time interval t (◂Fig. 17.7), the electric current is defined as

$$I = \frac{q}{t} \quad \textit{electric current} \tag{17.1}$$

SI unit of current: coulomb per second (C/s) or ampere (A)

The coulomb per second is designated the **ampere (A)** in honor of the French physicist André Ampère (1775–1836), an early investigator of electrical and magnetic phenomena. In everyday usage, the ampere is commonly shortened to *amp*. Thus, a current of 10 A is read as "ten amps." Small currents are routinely expressed in *milliamperes* (mA, or 10^{-3} A), *microamperes* (μA, or 10^{-6} A), or *nanoamperes* (nA, or 10^{-9} A). These are usually shortened to *milliamps, microamps,* and *nanoamps,* respectively. In a typical household circuit, it is not unusual for the wires to carry several amps of current. To understand the relationship between charge and current, consider the next example.

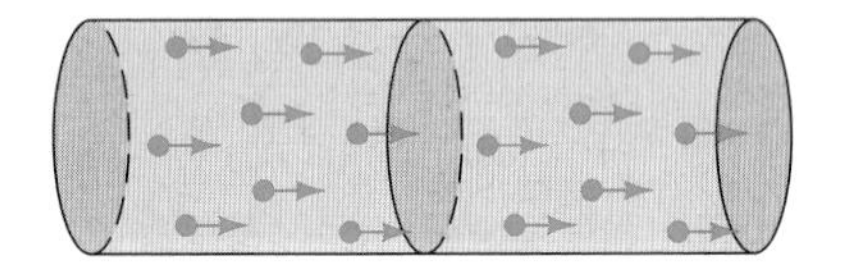

▲ **FIGURE 17.7 Electric current** Electric current (I) in a wire is defined as the rate at which the net charge (q) passes through the wire's cross-sectional area: $I = q/t$. The units of I are amperes (A), or *amps* for short.

Example 17.2 ■ Counting Electrons: Current and Charge

Suppose there is a steady current of 0.50 A in a flashlight bulb lasting for 2.0 min. How much charge passes through the bulb during this time? How many electrons does this represent?

Thinking It Through. The current and time elapsed are given; therefore, the definition of current (Eq. 17.1) allows us to find the charge q. Since each electron carries a charge of magnitude 1.6×10^{-19} C, then q can be converted into the number of electrons.

Solution. Listing the data given and converting the time into seconds:

Given: $I = 0.50$ A, $t = 2.0 \text{ min} = 1.2 \times 10^2$ s

Find: q (amount of charge), n (number of electrons)

By Eq. 17.1, $I = q/t$, so the magnitude of the charge is

$$q = It = (0.50 \text{ A})(1.2 \times 10^2 \text{ s}) = (0.50 \text{ C/s})(1.2 \times 10^2 \text{ s}) = 60 \text{ C}$$

Solving for the number of electrons (n), we have

$$n = \frac{q}{e} = \frac{60 \text{ C}}{1.6 \times 10^{-19} \text{ C/electron}} = 3.8 \times 10^{20} \text{ electrons}$$

(It takes a lot of electrons.)

Follow-Up Exercise. Many sensitive laboratory instruments can measure currents in the nanoamp range or smaller. How long, in years, would it take for 1.0 C of charge to flow past a given point in a wire that carries a current of 1.0 nA?

Drift Velocity, Electron Flow, and Electric Energy Transmission

Although we frequently mention charge flow in analogy to water flow, electric charge traveling in a conductor does not flow the same way that water flows in a pipe. In the absence of a potential difference across the ends of a metal wire, the free electrons move randomly at high speeds, colliding many times per second with the metal atoms. As a result, there is no net flow of charge, because equal amounts of charge pass through a given cross-sectional area in opposite directions during a specific time interval.

However, when a potential difference (voltage) *is* applied across the wire (such as by a battery), an electric field appears in the wire in one direction. A flow of electrons then begins *opposite* that direction (why?). This does *not* mean the electrons move directly from one end of the wire to the other. They still move in all directions as they collide with the atoms of the conductor but there is now an *added* component (in one direction) to their velocities (▸Fig. 17.8). The result is that their velocities are now, on average, more toward the positive terminal of the battery than away.

This net electron flow is characterized by an average velocity called the **drift velocity**. The drift velocity is much smaller than the random (thermal) velocities of the electrons themselves. Typically the magnitude of the drift velocity is on the order of 1 mm/s. At that speed, it would take an electron about 17 min to travel 1 m along a wire. Yet a lamp comes on almost instantaneously when the switch is closed (completing the circuit), and the electronic signals carrying telephone conversations travel almost instantaneously over miles of wire. How can that be?

Evidently, *something* must be moving faster than the "drifting" electrons. Indeed, this something is the electric field. When a potential difference is applied, the associated electric field in the conductor travels at a speed close to that of light (in the material, roughly 10^8 m/s). Thus the electric field influences the electrons *throughout the conductor* almost instantaneously. This means that the current starts everywhere in the circuit essentially simultaneously. You don't have to wait for electrons to "get there" from a distant place (say, near the switch). Thus in the light bulb, the electrons that are *already* in its filament begin to move almost immediately, delivering energy and creating light with no noticeable delay.

This effect is analogous to toppling a row of standing dominos. When you tip a domino at one end, that *signal* (or energy) is transmitted rapidly down the row. Very quickly, at the other end, the last domino topples (and delivers energy). Note that the domino delivering the signal or energy is *not* the one you pushed. It was the energy, not the dominos, that traveled down the row.

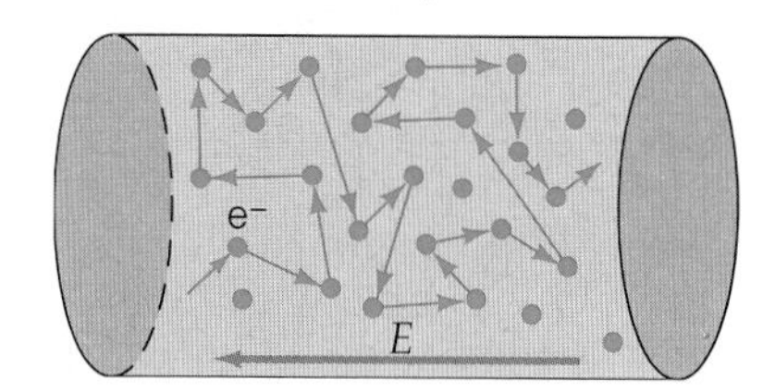

▲ **FIGURE 17.8** Drift velocity Because of collisions with the atoms of the conductor, electron motion is random. However, when the conductor is connected, for example, to a battery to form a complete circuit, there is a small net motion in the direction opposite the electric field [toward the high-potential (positive) terminal, or anode]. The speed and direction of this net motion form the drift velocity of the electrons.

17.3 Resistance and Ohm's Law

OBJECTIVES: To (a) define electrical resistance and explain what is meant by an ohmic resistor, (b) summarize the factors that determine resistance, and (c) calculate the effect of these factors in simple situations.

Note: *Resistor* is a generic term for any object that possesses significant electrical resistance.

If you place a voltage (potential difference) across the ends of any conducting material, what factors determine the current? As might be expected, usually the greater the voltage, the greater the current. However, another factor also influences current. Just as internal friction (viscosity; see Chapter 9) affects fluid flow in pipes, the resistance of the wire's material will affect the flow of charge. Any object that offers significant resistance to electrical current is called a *resistor* and is represented by the zigzag symbol (Section 17.1). This symbol is used to represent all types of resistors, from the cylindrical color-coded ones on printed circuit boards to electrical devices and appliances such as hair dryers and lightbulbs (▸Fig. 17.9).

Note: Remember, V stands for ΔV.

But how is resistance quantified? We know, for example, that if a large voltage applied across an object produces only a small current, then that object has a high resistance. Thus, the **resistance (*R*)** of any object should be related to the ratio of the voltage across the object to the resulting current through that object. Resistance is therefore defined as

$$R = \frac{V}{I} \quad \textit{electrical resistance} \qquad (17.2a)$$

SI unit of resistance: volt per ampere (V/A), or ohm (Ω)

The units of resistance are volts per ampere (V/A), called the **ohm** (Ω) in honor of the German physicist Georg Ohm (1789–1854), who investigated the relationship between current and voltage. Large values of resistance are commonly expressed as kilohms (kΩ) and megohms (MΩ). A schematic circuit diagram

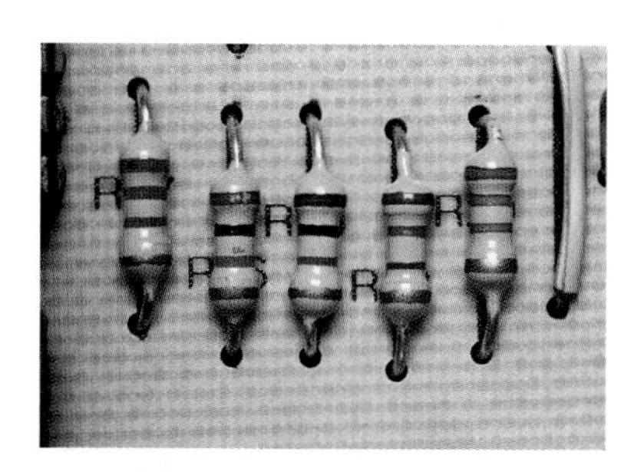

▲ **FIGURE 17.9** Resistors in use A printed circuit board, typically used in computers, includes resistors of different values. The large, striped cylinders are resistors; their four-band color code indicates their resistance in ohms.

Note: *Ohmic* means "having a constant resistance."

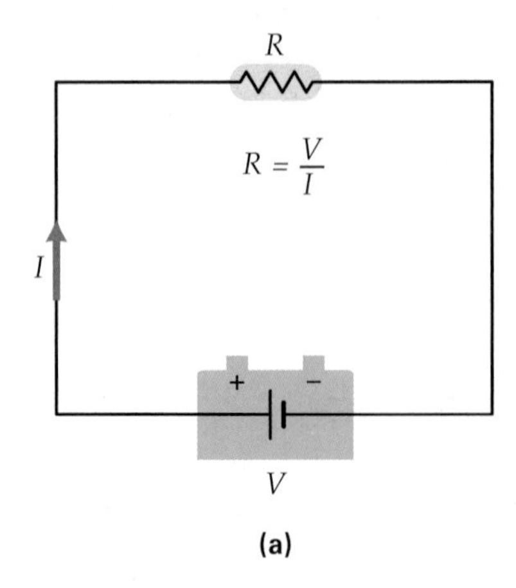

(a)

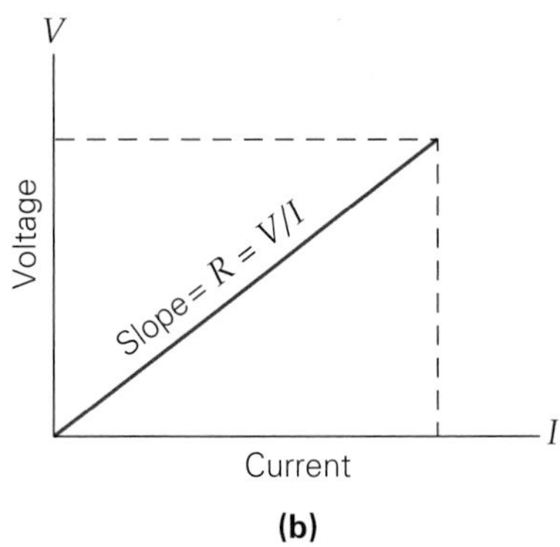

(b)

▲ **FIGURE 17.10 Resistance and Ohm's law (a)** In principle, any object's electrical resistance can be determined by dividing the voltage across it by the current through it. **(b)** If the object obeys Ohm's law (applicable only to a constant resistance), then a plot of voltage versus current is a straight line with a slope equal to R, the element's resistance. (Its resistance does not change with voltage.)

showing how, in principle, resistance is determined is illustrated in ◂Fig. 17.10a. (In Chapter 18, we will study the instruments that are used to measure electrical currents and voltages, called *ammeters* and *voltmeters*, respectively.)

For some materials, the resistance may be constant over a range of voltages. A resistor that exhibits constant resistance is said to obey **Ohm's law**, or to be *ohmic*. The law was named after Ohm, who found materials possessing this property. A plot of voltage versus current for a material with an ohmic resistance gives a straight line with a slope equal to its resistance R (Fig. 17.10b). A common and practical form of Ohm's law is

$$V = IR \quad \text{(Ohm's law)} \tag{17.2b}$$

$$(\text{or } I \propto V, \text{ only when } R = \text{constant})$$

Ohm's law is *not* a fundamental law in the same sense as, for example, the law of conservation of energy. There is no "law" that states that materials *must* have constant resistance. Indeed, many of our advances in electronics are based on materials such as semiconductors, which have *nonlinear* (nonohmic) voltage–current relationships.

Unless specified otherwise, we will assume resistor to be ohmic. Always remember, however, that many materials are nonohmic. For instance, the tungsten filaments in lightbulbs have resistance that increases with temperature, being much larger at their operating temperature than at room temperature. The following Example shows how the resistance of the human body can make the difference between life and death.

Example 17.3 ■ Danger in the House: Human Resistance

Any room in the house that is exposed to water and electrical voltage can present hazards. (See the discussion of electrical safety in Section 18.5.) For example, suppose a person steps out of a shower and inadvertently touches an exposed 120-V wire (perhaps a frayed cord on a hair dryer) with a finger. The human body, when wet, can have an electrical resistance as low as 300 Ω. Using this value, estimate the current in that person's body.

Thinking It Through. The wire is at an electrical potential of 120 V above the floor, which is "ground" and taken to be at 0 V. Therefore the voltage (or potential difference) across the body is 120 V. To determine the current, we can use Eq. 17.2, the definition of resistance.

Solution. Listing the data,

Given: $V = 120$ V, $R = 300\ \Omega$ *Find:* I (current in the body)

From Eq. 17.2, we have

$$I = \frac{V}{R} = \frac{120\text{ V}}{300\ \Omega} = 0.400\text{ A} = 400\text{ mA}$$

While this is a small current by everyday standards, it is a large current for the human body. A current over 10 mA can cause severe muscle contractions, and currents on the order of 100 mA can stop the heart. So this current is potentially deadly. (See Insight 18.2 on Electricity and Personal Safety, Table 1, in Chapter 18 on page 614.)

Follow-Up Exercise. When the human body is dry, its resistance (over its length) can be as high as 100 kΩ. What voltage would be required to produce a current of 1.0 mA (the value that a person can barely feel)?

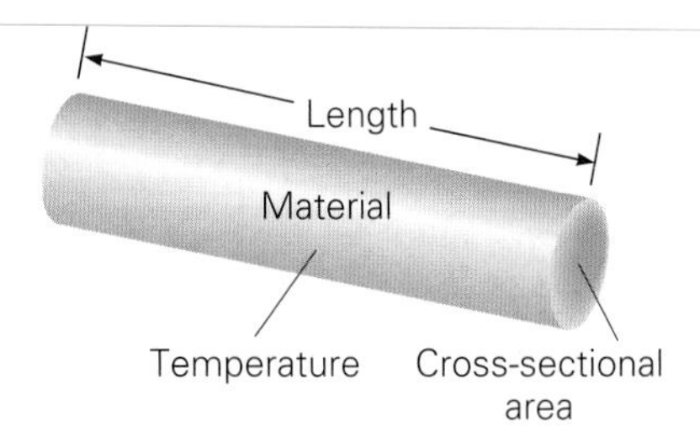

▲ **FIGURE 17.11 Resistance factors** Factors directly affecting the electrical resistance of a cylindrical conductor are the type of material it is made of, its length (L), its cross-sectional area (A), and its temperature (T).

Factors That Influence Resistance

On the atomic level, resistance arises when electrons collide with the atoms that make up a material. Thus, resistance partially depends on the type of material of which an object is composed. However, geometrical factors also influence resistance. In summary, the resistance of an object of uniform cross-section, such as a length of wire, depends on four properties: (1) the type of material, (2) its length, (3) its cross-sectional area, and (4) its temperature (◂Fig. 17.11).

As you might expect, the resistance of an object (such as a piece of wire) is *inversely* proportional to its cross-sectional area (A), and *directly* proportional to its length (L); that is, $R \propto L/A$. For example, a uniform metal wire 4 m long offers twice as much resistance as a similar wire 2.0 m long, but a wire with a cross-sectional area of 0.50 mm^2 has only half the resistance of one with an area of 0.25 mm^2. These geometrical resistance conditions are analogous to those for liquid flow in a pipe. The longer the pipe, the more is its resistance (drag), but the larger its cross-sectional area, the more liquid it can carry per second. To see an interesting use of the dependence of resistance on length and area by living organisms, read Insight 17.1 on The "Bio-Generation" of High Voltage on this page.

Illustration 30.5 Ohm's "Law"

INSIGHT 17.1 THE "BIO-GENERATION" OF HIGH VOLTAGE

From the discussion of battery action in Section 17.1, you know that two different metals in acid can generate a constant separation of charge (a voltage) and thus can produce electric current. However, living organisms can also create voltages by a process sometimes called "bio-generation." Electric eels (see Insight 15.2 on page 527), in particular, can generate 600 V, more than enough to kill humans. But how do they accomplish this? As we shall see, the process has similarities both to regular "dry cells" and to nerve signal transmission.

Eels have three organs related to its electrical activities. The Sachs' organ generates low-voltage pulsations for navigation. The others, named the Hunter organ and the Main organ, are sources of high voltage (Fig. 1). In these organs, cells called *electrocytes*, or *electroplates*, are arranged in a stack. Each cell has a flat, disklike shape. The electroplate stack is a series connection similar to that in a car battery, in which there are six cells at 2 V each, producing a total of 12 V. Each electroplate is capable of producing a voltage of only about 0.15 V, but four or five thousand in series can add up to a voltage of 600 V. The electroplates are electrically similar to muscle cells in that they, like muscle cells, receive nerve impulses by synaptic connection. However, these nerve impulses do not cause movement. Instead they trigger voltage generation by the following mechanism.

Each electroplate has the same structure. The top and bottom membranes behave similarly to nerve membranes (see Insight 16.1 on page 552). Under resting conditions, the Na^+ ions cannot penetrate the membrane. To equilibrate their concentrations on both sides, the ions reside near the outside surface. This, in turn, attracts the (interior) negatively charged proteins to the interior surface. As a result, the interior is at a potential of 0.08 V lower than the outside. Therefore, under resting conditions, the outside top (toward the head, or anterior) surface and the outside bottom (posterior) surface of *all* the electroplates are positive (one is shown in Fig. 2a) and exhibit no voltage ($\Delta V_1 = 0$). Thus, under resting conditions a series stack has no voltage $\left(\Delta V_{\text{total}} = \Sigma \Delta V_i = 0\right)$ from top to bottom (Fig. 2b).

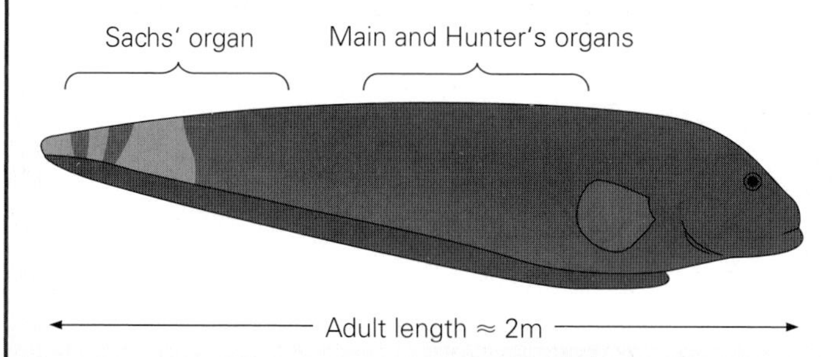

FIGURE 1 Anatomy of an electric eel 80% of an electric eel's body is devoted to voltage generation. Most of that portion contains the two organs (Main and Hunter's) responsible for the high voltage associated with killing of prey. The Sachs' organ produces a lower pulsating voltage used for navigation.

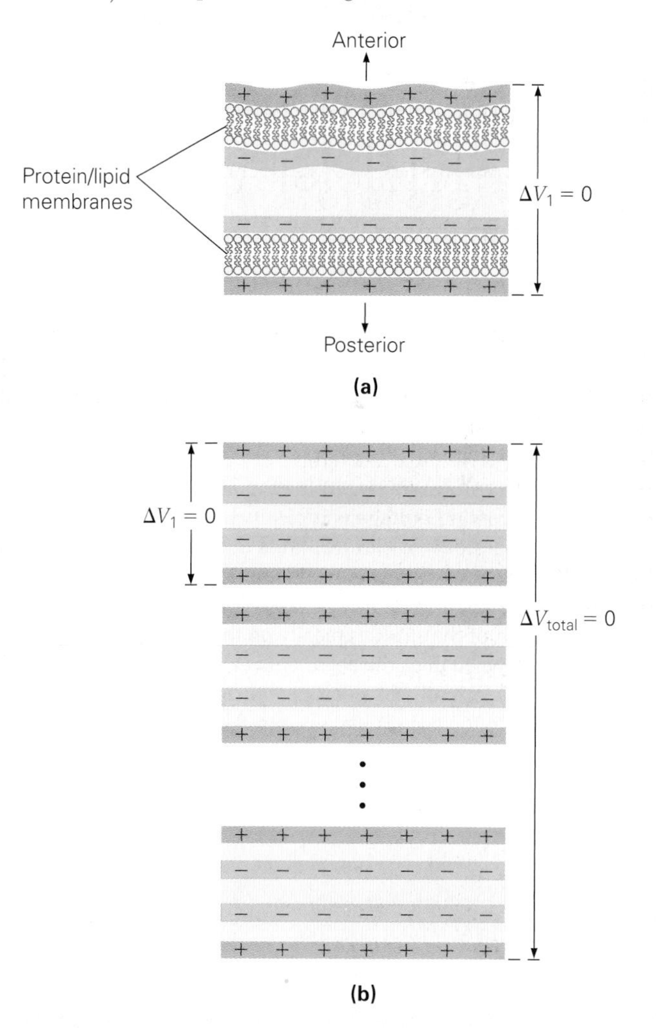

FIGURE 2 (a) A single, resting electroplate One of the thousands of electroplates in the eel's electric organs has, under resting conditions, equal amounts of positive charge at its top and bottom, resulting in no voltage. **(b) Resting electroplates in series** Several thousand electroplates in series under resting conditions have a total voltage of zero.

(continues on next page)

However, when an eel locates prey, the eel's brain sends a signal along a neuron *to only the bottom membrane* of each electroplate (one cell is shown in Fig. 3a). A chemical (*acetylcholine*) diffuses across the synapse onto the membrane, briefly opening the ion channels and allowing in Na^+. *For a few milliseconds the lower membrane polarity is reversed,* creating a voltage across one cell of $\Delta V_1 \approx 0.15$ V. The whole stack does this simultaneously, causing a large voltage across the ends of the stack ($\Delta V_{total} \approx 4000\ \Delta V_1 =$ 600 V; see Fig. 3b). When the eel touches the prey with the stack ends, the resulting current pulse through the prey (about 0.5 A) delivers enough energy to kill or at least stun.

An interesting biological "wiring" arrangement enables all electroplates to be triggered simultaneously—a requirement crucial for generation of the maximum voltage. Since each electroplate is a different distance from the brain, the action potential traveling down the neurons must be carefully timed. To do this, the neurons attached to the top of the stack (closest to the brain) are longer and thinner than those attached to the bottom. From what you know about resistance (see, for example, the discussion of Eq. 17.3 and $R \propto L/A$), it should be clear that both a reduction in area and increase in length of the neurons serve to increase neuron resistance compared with those attached to more distant electroplates. Increased resistance means that the action potential travels slower through the closer neurons, thus enabling the closer electroplates to receive their signal at the same time as the more distant ones—a very interesting and practical use of physics (from the eel's perspective, not the prey's).

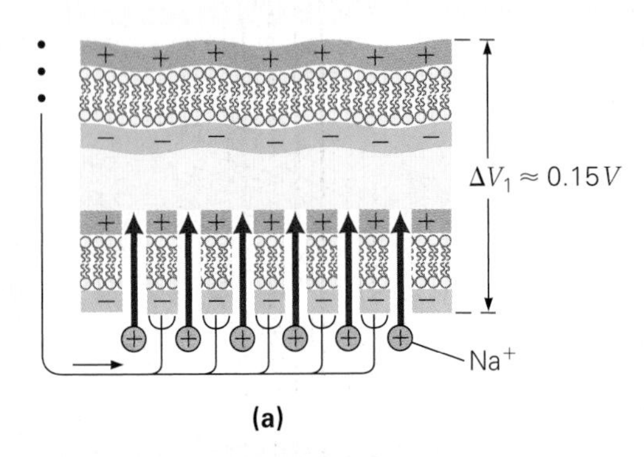

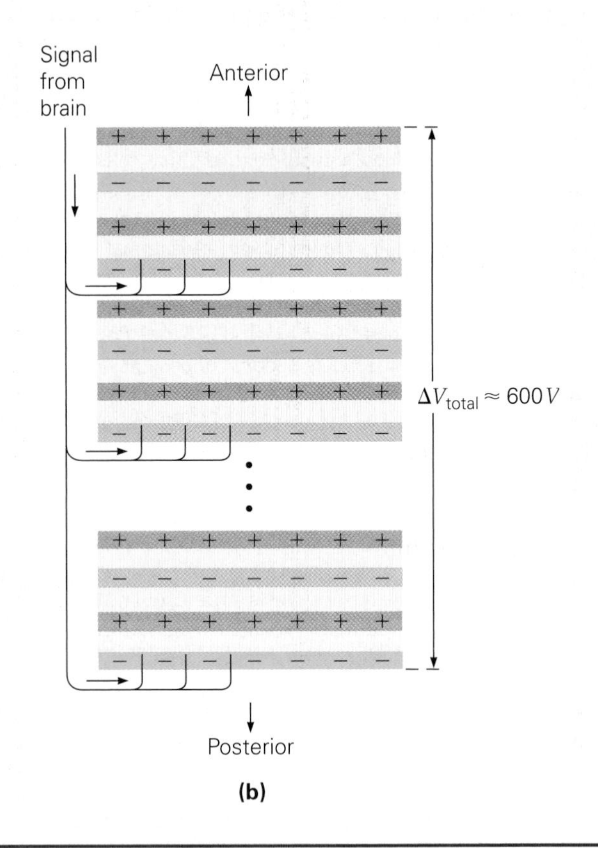

FIGURE 3 (a) An electroplate in action Upon location of prey, a signal is sent from the eel's brain to each electroplate along a neuron attached only to the bottom of the plate. This triggers a brief opening of the ion channel allowing Na^+ ions to the interior, temporarily reversing the polarity of the lower membrane. This creates a temporary electric potential difference (voltage) between the top and bottom membranes. Each electroplate voltage is typically a few tenths of a volt. **(b) A series stack of electroplates in action** When each electroplate in the stack is triggered into action by the lower neuron signal, this results in a large voltage between the top and bottom of the stack, typically on the order of 600 V. This large voltage enables the eel to deliver a pulse of current on the order of a few tenths of an ampere through the prey. The energy deposited in the prey is usually enough to stun or kill it.

Resistivity

Note: Do not confuse resistivity with mass density, which has the same symbol (ρ).

The resistance of an object is partly determined by its material's atomic properties, quantitatively described by that material's **resistivity (ρ)**. The resistance of uniform cross-section object is given by

$$R = \rho\left(\frac{L}{A}\right) \tag{17.3}$$

SI unit of resistivity: ohm-meter ($\Omega \cdot$m)

The units of resistivity (ρ) are ohm-meters ($\Omega \cdot$m). (You should show this.) Thus, knowing its resistivity (the material type) and using Eq. 17.3, the resistance of any

TABLE 17.1 Resistivities (at 20°C) and Temperature Coefficients of Resistivity for Various Materials*

	ρ ($\Omega \cdot m$)	α (1/C°)		ρ ($\Omega \cdot m$)	α (1/C°)
Conductors			*Semiconductors*		
Aluminum	2.82×10^{-8}	4.29×10^{-3}	Carbon	3.6×10^{-5}	-5.0×10^{-4}
Copper	1.70×10^{-8}	6.80×10^{-3}	Germanium	4.6×10^{-1}	-5.0×10^{-2}
Iron	10×10^{-8}	6.51×10^{-3}	Silicon	2.5×10^{2}	-7.0×10^{-2}
Mercury	98.4×10^{-8}	0.89×10^{-3}			
Nichrome (alloy of nickel and chromium)	100×10^{-8}	0.40×10^{-3}	*Insulators*		
Nickel	7.8×10^{-8}	6.0×10^{-3}	Glass	10^{12}	
Platinum	10×10^{-8}	3.93×10^{-3}	Rubber	10^{15}	
Silver	1.59×10^{-8}	4.1×10^{-3}	Wood	10^{10}	
Tungsten	5.6×10^{-8}	4.5×10^{-3}			

*Values for semiconductors are general ones, and resistivities for insulators are typical orders of magnitude.

constant-area object can be calculated, as long as its length and cross-sectional area is known.

The values of the resistivities of some conductors, semiconductors, and insulators are given in Table 17.1. The values apply at 20°C, because resistivity can depend on temperature. Most common wires are composed of copper or aluminum with cross-sectional areas on the order of 10^{-6} m^2 or 1 mm^2. For a length of 1.5 m, you should be able to show that the resistance of a copper wire with this area is about 0.025 Ω (= 25 mΩ). This explains why wire resistances are neglected in circuits—their values are much less than most household devices.

An interesting and potentially important medical application involves the measurement of human body resistance and its relationship to body fat. (See Insight 17.2 on Bioelectrical Impedance Analysis on page 578.) To get a feeling for the magnitudes of these quantities in living tissue, consider the following Example.

Example 17.4 ■ Electric Eels: Cooking with Bio-Electricity?

Suppose an electric eel touches the head and tail of a cylindrically shaped fish and applies a voltage of 600 V across it. (See Insight 17.1 on page 575.) If a current of 0.80 A results (likely killing the prey), estimate the average resistivity of the fish's flesh, assuming it is 20 cm long and 4.0 cm in diameter.

Thinking It Through. With cylindrical geometry, we know its length and can find its cross-sectional area from the given dimensions. From the voltage and current, its resistance can be determined. Lastly, its resistivity can be estimated using Eq. 17.3.

Solution. Listing the data:

Given: $L = 20 \text{ cm} = 0.20 \text{ m}$
$d = 4.0 \text{ cm} = 4.0 \times 10^{-2} \text{ m}$
$V = 600 \text{ V}$
$I = 0.80 \text{ A}$

Find: ρ (resistivity)

The cross-sectional area of the fish is

$$A = \pi r^2 = \pi\left(\frac{d}{2}\right)^2 = \frac{\pi(2.0 \times 10^{-2} \text{ m})^2}{4} = 3.1 \times 10^{-4} \text{ m}^2$$

(continues on next page)

We also know that the fish's overall resistance is $R = \frac{V}{I} = \frac{600\text{ V}}{0.80\text{ A}} = 7.5 \times 10^2\ \Omega$. From Eq. 17.3, we have

$$\rho = \frac{RA}{L} = \frac{(7.5 \times 10^2\ \Omega)(3.1 \times 10^{-4}\text{ m}^2)}{0.20\text{ m}} = 1.2\ \Omega \cdot \text{m or about } 120\ \Omega \cdot \text{cm}$$

Comparing this to the values in Table 17.1, you can see that, as expected, the fish's flesh is much more resistive than metals, but certainly not a great insulator. The value is on the order of the resistivities measured for different human tissues; for example, cardiac muscle has a resistivity of about 175 $\Omega \cdot$cm and liver on the order of 200 $\Omega \cdot$cm. Clearly our answer is an average over the whole fish and tells us nothing about different regions of the fish.

Follow-Up Exercise. Suppose for its next meal, the eel in this Example chooses a different species of fish. The next fish has twice the average resistivity, half the length, and half the diameter of the first fish. What current would be expected in this fish if the eel applied 400 V across its body?

For many materials, especially metals, the temperature dependence of resistivity is nearly linear if the temperature change is not too great. That is, the resistivity at a temperature T after a temperature change $\Delta T = T - T_o$ is given by

$$\rho = \rho_o(1 + \alpha \Delta T) \quad \textit{temperature variation of resistivity} \quad (17.4)$$

where α is a constant (usually only over a certain temperature range) called the **temperature coefficient of resistivity** and ρ_o is a reference resistivity at T_o (usually 20°C). Equation 17.4 can be rewritten as

$$\Delta\rho = \rho_o \alpha \Delta T \quad (17.5)$$

Note: Compare the form of Eq. 17.5 with Eq. 10.10 for the linear expansion of a solid.

where $\Delta\rho = \rho - \rho_o$ is the change in resistivity that occurs when the temperature changes by ΔT. The ratio $\Delta\rho/\rho_o$ is dimensionless, so α has units of inverse Celsius degrees, written as 1/C°. Physically, α represents the fractional change in resistivity $(\Delta\rho/\rho_o)$ per Celsius degree. The temperature coefficients of resistivity for some materials are listed in Table 17.1. These coefficients are assumed to be constant over normal temperature ranges. Notice that for semiconductors and insulators the coefficients are generally orders of magnitude and are usually not constant.

INSIGHT 17.2 BIOELECTRICAL IMPEDANCE ANALYSIS (BIA)

Traditional methods for estimating body-fat percentages involve the use of buoyancy tanks (for density measurements, see Chapter 9) or calipers to pinch the flesh. However, in recent years, electrical resistance experiments have been designed to measure the body fat of the human body.* In theory, these measurements—termed *bioelectrical impedance analysis* (BIA)—have the potential to determine, with more accuracy than traditional methods, a patient's total water content, fat-free mass, and body fat (*adipose tissue*).

The principle of BIA is based on the water content of the human body. Water in the human body is a relatively good conductor of electric current, due to the presence of ions such as potassium (K^+) and sodium (Na^+). Because muscle tissue holds more water per kilogram than fat holds, it is a better conductor than fat. Thus for a given voltage, the difference in currents should be a good indicator of the fat-to-muscle percentage.

In practice, one electrode of a low-voltage power supply is connected to the wrist and the other to the opposite ankle during a BIA test. The current is kept below 1 mA for safety, with typical currents being about 800 μA. The subject cannot feel this small current. Typical resistance values are about 250 Ω. From Ohm's law, the required voltage is $V = IR = (8 \times 10^{-4}\text{ A})(250\ \Omega) = 0.200$ V, or about 200 mV. In actuality, the voltage alternates in polarity at a frequency of 50 kHz, because this frequency is known *not* to trigger electrically excitable tissues, such as nerves and cardiac muscle.

From what has been presented in this chapter (for example, Eq. 17.3), you should be able to understand some of the factors involved in interpreting the results of human resistance measurements. The measured resistance is the total resistance. However, the current travels not through a uniform material but rather through the arm, trunk, and leg. Not only does each of these body parts have a different fat-to-muscle ratio, which affects resistivity (ρ), but also they differ widely in length (L) and cross-sectional area (A). The arm and the leg, usually dominated by muscle and with a small cross-sectional area, offer the most resistance. The trunk, which usually contains a relatively high percentage of fat and has a large cross-sectional area, has low resistance.

By subjecting BIAs to statistical analysis, researchers hope to understand how the wide range of physical and genetic parameters present in humans affects resistance measurements. Among these parameters are height, weight, body type, and ethnicity. Once the correlations are understood, BIA may well become a valuable medical tool in routine physicals and in the diagnosis of various diseases.

*Technically, this technique measures the body's *impedance*, which includes effects of capacitance and magnetic effects as well as resistance. (See Chapter 21.) However, these contributions are about 10% of the total. Hence, the word *resistance* is used here.

Resistance is directly proportional to resistivity (Eq. 17.3). This means that an object's resistance has the same dependence on temperature as its resistivity (Eqs. 17.3 and 17.4). The resistance of an object of uniform cross section varies with temperature:

$$R = R_o(1 + \alpha\Delta T) \quad \text{or} \quad \Delta R = R_o\alpha\Delta T \qquad \textit{temperature variation of resistance} \qquad (17.6)$$

Here, $\Delta R = R - R_o$, the change in resistance relative to its reference value R_o, usually taken to be at 20°C. The variation of resistance with temperature provides a means of measuring temperature in the form of an *electrical resistance thermometer*, as illustrated in the next Example.

Example 17.5 ■ An Electrical Thermometer: Variation of Resistance with Temperature

A platinum wire has a resistance of 0.50 Ω at 0°C. It is placed in a water bath, where its resistance rises to a final value of 0.60 Ω. What is the temperature of the bath?

Thinking It Through. From the temperature coefficient of resistivity for platinum from Table 17.1, ΔT can be found from Eq. 17.6 and added to 0°C, the initial temperature, to find the temperature of the bath.

Solution.

Given: $T_o = 0°\text{C}$
$R_o = 0.50\ \Omega$
$R = 0.60\ \Omega$
$\alpha = 3.93 \times 10^{-3}/\text{C}°$ (Table 17.1)

Find: T (temperature of the bath)

The ratio $\Delta R/R_o$ is the fractional change in the initial resistance R_o (at 0°C). We solve Eq. 17.6 for ΔT, using the given values:

$$\Delta T = \frac{\Delta R}{\alpha R_o} = \frac{R - R_o}{\alpha R_o} = \frac{0.60\ \Omega - 0.50\ \Omega}{(3.93 \times 10^{-3}/\text{C}°)(0.50\ \Omega)} = 51\ \text{C}°$$

Thus, the bath is at $T = T_o + \Delta T = 0°\text{C} + 51\ \text{C}° = 51°\text{C}$.

Follow-Up Exercise. In this Example, if the material had been copper with $R_o = 0.50\ \Omega$, rather than platinum, what would its resistance be at 51°C? From this, you should be able to explain which material makes the more "sensitive" thermometer, one with a high temperature coefficient of resistivity or one with a low value.

Superconductivity

Carbon and other semiconductors have negative temperature coefficients of resistivity. However, many materials, including most metals, have positive temperature coefficients, which means that their resistances increase as temperature increases. You might wonder how far electrical resistance can be reduced by lowering the temperature. In certain cases, the resistance can reach zero—not just close to zero, but, as accurately as can be measured, *exactly* zero. This phenomenon is called **superconductivity** (first discovered in 1911 by Heike Kamerlingh Onnes, a Dutch physicist). Currently the required temperatures are about 100 K or below. Thus its current usage is restricted to high-tech laboratory apparatus and research equipment.

However, it does have the potential for important new everyday applications, especially if materials can be found whose superconducting temperature is near room temperature. Among the applications are superconducting magnets (already in use in labs and small-scale naval propulsion units). In the absence of resistance, high currents and very high magnetic fields are possible (Chapter 19). Used in motors or engines, superconducting electromagnets would be more efficient, providing more power for the same energy input. Superconductors might also be used as electrical transmission lines with no resistive losses. Some envision superfast superconducting computer memories. The absence of electrical resistance opens almost endless possibilities. You're likely to hear more about superconductor applications in the future as new materials are developed.

17.4 Electric Power

OBJECTIVES: To (a) define electric power, (b) calculate the power delivery of simple electric circuits, and (c) explain joule heating and its significance.

When a sustained current exists in a circuit, the electrons are given energy by the voltage source, such as a battery. As these charge carriers pass through circuit components, they collide with the atoms of the material (that is, they encounter resistance) and lose energy. The energy transferred in the collisions can result in an increase in the temperature of the components. In this way, electrical energy can be transformed, at least partially, into thermal energy.

However, electric energy can also be converted into other forms of energy such as light (as in lightbulbs) and mechanical motion (as in electric drills). According to conservation of energy, whatever forms the energy may take, the *total* energy delivered to the charge carriers by the battery must be *completely* transferred to the circuit elements (neglecting losses in the wires). That is, on return to the voltage source or battery, a charge carrier loses all the electric potential energy it gained from that source, and is ready to repeat the process.

The energy gained by an amount of charge q from a voltage source (voltage V) is qV [by units, $C(J/C) = J$]. Over a time interval t, the *rate* at which energy is delivered may not be constant. Thus the average rate of energy delivery, called the average **electric power, $\overline{P}$,** is given by

$$\overline{P} = \frac{W}{t} = \frac{qV}{t}$$

In the special case when the current and voltage are steady with time (as with a battery), then the average power is the same as the power at all times. For steady (dc) currents, $I = q/t$ (Eq. 17.1). Thus, we can rewrite the preceding equation as:

$$P = IV \quad \textit{electric power} \tag{17.7a}$$

Recall from Chapter 5 that the SI unit of power is the watt (W). The ampere (the unit of current I) times the volt (the unit of voltage V) gives the joule per second (J/s), or watt. (You should check this.)

A visual mechanical analogy to help explain Eq. 17.7a is given in ▼Fig. 17.12. The figure depicts a simple electric circuit as a system for transferring energy, in analogy to a conveyor belt delivery system.

Because $R = V/I$, power can be written in three equivalent forms:

$$P = IV = \frac{V^2}{R} = I^2R \quad \textit{electric power} \tag{17.7b}$$

Joule Heat

The thermal energy expended in a current-carrying resistor is referred to as **joule heat**, or **I^2R losses** (pronounced "I squared R" losses). In many instances (such as in electrical transmission lines), joule heating is an undesirable side effect. However, in

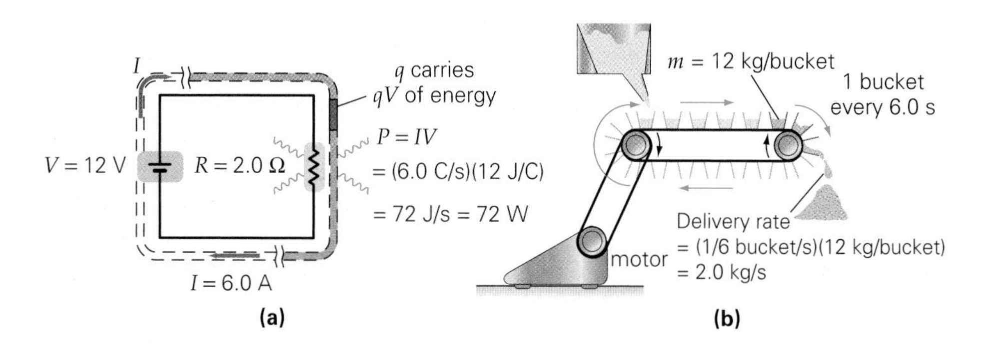

▶ **FIGURE 17.12 Electric power analogy** Electric circuits can be thought of as energy delivery systems much like a conveyor belt. **(a)** Imagine the current as being made up of consecutive segments of charge $q = 1.0$ C, each carrying $qV = 12$ J of energy supplied by the battery. The current is $I = V/R = 6.0$ A, or 6.0 C/s. Then the power (or energy delivery rate) to the resistor is $(6.0\ C/s)(12\ J/C) = 72\ J/s = 72$ W. **(b)** The conveyor belt comprises a series of buckets, each carrying 12 kg of sand (analogous to the energy carried by each charge q), arriving at a rate of one bucket every 6.0 s (analogous to current I). The delivery rate in kg/s is analogous to the power in J/s in part (a).

other situations, the conversion of electrical energy to thermal energy is the main purpose. Heating applications include the heating elements (burners) of electric stoves, hair dryers, immersion heaters, and toasters.

Electric lightbulbs are rated in watts (power)—for example, 60 W (▸Fig. 17.13a). Incandescent lamps are relatively inefficient as light sources. Typically, less than 5% of the electrical energy is converted to visible light; most of the energy produced is invisible infrared radiation and heat.

Electrical appliances are tagged or stamped with their power ratings. Either the voltage and power requirements or the voltage and current requirements are given (Fig. 17.13b). In either case, the current, power, and effective resistance can be calculated. Typical power requirements of some household appliances are given in Table 17.2. Even though most common appliances specify a nominal operating voltage of 120 V, it should be noted that household voltage can vary from 110 V to 120 V and still be considered in the "normal" range.

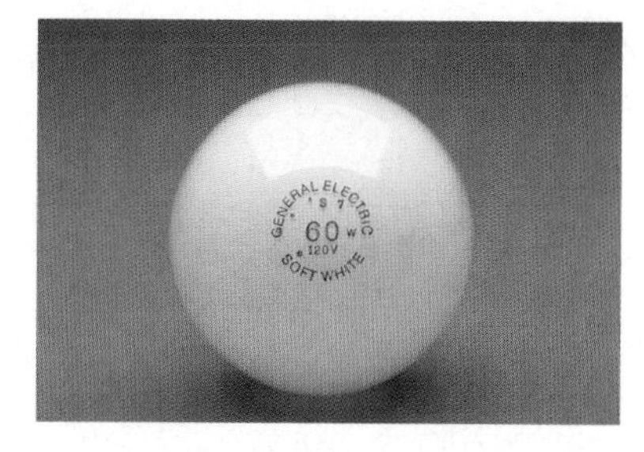

(a)

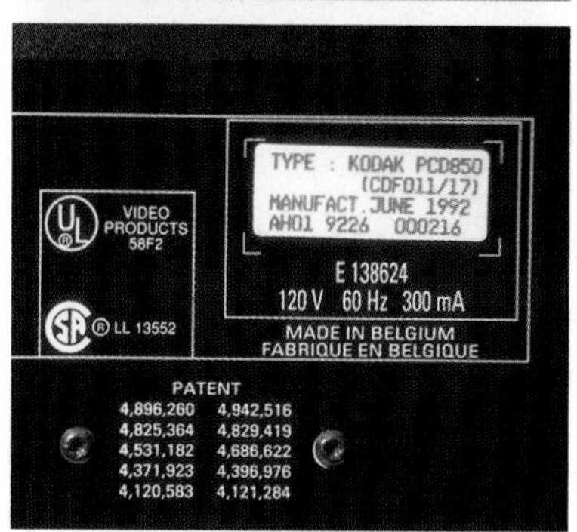

(b)

▲ **FIGURE 17.13 Power ratings** **(a)** Lightbulbs are rated in watts. Operated at 120 V, this 60-W bulb uses 60 J of energy each second. **(b)** Appliance ratings list either voltage and power or voltage and current. From either, the current, power, and effective resistance can be found. Here, one appliance is rated at 120 V and 18 W and the other at 120 V and 300 mA. Can you compute the current and resistance for the former and the power required and resistance for the latter?

Integrated Example 17.6 ■ A Modern Appliance Dilemma: Computing or Eating

(a) Consider two appliances that operate at the same voltage. Appliance A has a higher power rating than appliance B. (a) How does the resistance of A compare with that of B: (1) larger, (2) smaller, or (3) the same? (b) A computer system includes a color monitor with a power requirement of 200 W, whereas a countertop broiler/toaster oven is rated at 1500 W. What is the resistance of each if both are designed to run at 120 V?

(a) Conceptual Reasoning. Power depends on current and voltage. Because the two appliances operate at the same voltage, they can't carry the same current and yet have different power requirements. Therefore, answer (3) cannot be correct. Because both appliances operate at the same voltage, the one with higher power (A) must carry more current. For A to carry more current at the same voltage as B, it must have less resistance than B. Therefore, the correct answer is (2); A has less resistance than B.

(b) Quantitative Reasoning and Solution. The definition of resistance is $R = V/I$ (Eq. 17.2). To use this definition, we need the current, which can be determined from Eq. 17.7 ($P = IV$). This will be done twice, once for the monitor and then for the broiler/toaster. We list the data, using the subscript m for monitor and b for broiler/toaster:

Given: $P_m = 200$ W
$P_b = 1500$ W
$V = 120$ V

Find: R (resistance of each appliance)

The monitor current is (using Eq. 17.7)

$$I_m = \frac{P_m}{V} = \frac{200\text{ W}}{120\text{ V}} = 1.67\text{ A}$$

and that in the broiler/toaster is

$$I_b = \frac{P_b}{V} = \frac{1500\text{ W}}{120\text{ V}} = 12.5\text{ A}$$

Thus their resistances are

$$R_m = \frac{V}{I_m} = \frac{120\text{ V}}{1.67\text{ A}} = 71.9\ \Omega$$

and

$$R_b = \frac{V}{I_b} = \frac{120\text{ V}}{12.5\text{ A}} = 9.60\ \Omega$$

Because they all operate at the same voltage, an appliance's power output is controlled by its resistance. The appliance's resistance is *inversely* related to the appliance's power requirement.

Follow-Up Exercise. An immersion heater is a common "appliance" in most college dorms, useful for heating water for tea, coffee, or soup. Assuming 100% of the heat goes into the water, what must be the heater's resistance (operating at 120 V) to heat a cup of water (mass 250 g) from room temperature (20°C) to boiling in 3.00 minutes?

TABLE 17.2 Typical Power and Current Requirements for Various Household Appliances (120 V)

Appliance	Power	Current	Appliance	Power	Current
Air conditioner, room	1500 W	12.5 A	Heater, portable	1500 W	12.5 A
Air-conditioning, central	5000 W	41.7 A*	Microwave oven	900 W	5.2 A
Blender	800 W	6.7 A	Radio–cassette player	14 W	0.12 A
Clothes dryer	6000 W	50 A*	Refrigerator, frost-free	500 W	4.2 A
Clothes washer	840 W	7.0 A	Stove, top burners	6000 W	50.0 A*
Coffeemaker	1625 W	13.5 A	Stove, oven	4500 W	37.5 A*
Dishwasher	1200 W	10.0 A	Television, color	100 W	0.83 A
Electric blanket	180 W	1.5 A	Toaster	950 W	7.9 A
Hair dryer	1200 W	10.0 A	Water heater	4500 W	37.5 A*

*A high-power appliance such as this is typically wired to a 240-V house supply to reduce the current to half these values (Section 18.5).

Example 17.7 ■ A Potentially Dangerous Repair: Don't Do It Yourself!

A hair dryer is rated at 1200 W for a 115-V operating voltage. The uniform wire filament breaks near one end, and the owner repairs it by removing the section near the break and simply reconnecting it. The filament is then 10.0% shorter than its original length. What will be the heater's power output after this "repair"?

Thinking It Through. The wire always operates at 115 V. Thus shortening the wire, which decreases its resistance, will result in a larger current. With this increase in current, one would expect the power output to go up.

Solution. Let's use a subscript 1 to indicate "before breakage" and a subscript 2 to mean "after the repair." Listing the given values,

Given: $P_1 = 1200\text{ W}$
$V_1 = V_2 = 115\text{ V}$
$L_2 = 0.900L_1$

Find: P_2 (power output after the repair)

After the repair, the wire has 90.0% of its original resistance, because (see Eq. 17.3) a wire's resistance is directly proportional to its length. To show the reduction to 90% explicitly, let's express the resistance after repair ($R_2 = \rho L_2/A$) in terms of the original resistance ($R_1 = \rho L_1/A$):

$$R_2 = \rho\frac{L_2}{A} = \rho\frac{0.900L_1}{A} = 0.900\left(\rho\frac{L_1}{A}\right) = 0.900R_1$$

as expected.

The current will increase, because the voltage is the same ($V_2 = V_1$). This requirement can be written as $V_2 = I_2R_2 = V_1 = I_1R_1$. So the new current in terms of the original current is

$$I_2 = \left(\frac{R_1}{R_2}\right)I_1 = \left(\frac{R_1}{0.900R_1}\right)I_1 = (1.11)I_1$$

which means that the current after the repair is about 11% larger than before.

The original power is $P_1 = I_1V = 1200\text{ W}$. The power after the repair is $P_2 = I_2V$ (note that the voltages have no subscripts because they remained the same and will cancel out). Forming a ratio gives

$$\frac{P_2}{P_1} = \frac{I_2V}{I_1V} = \frac{I_2}{I_1} = 1.11$$

which can be solved for P_2:

$$P_2 = 1.11P_1 = 1.11(1200\text{ W}) = 1.33 \times 10^3\text{ W}$$

The power output of the dryer has been increased by about 120 W. *Do not perform such a repair job!*

Follow-Up Exercise. In this Example, determine the (a) initial and final resistances and (b) initial and final currents.

We often complain about our electric bills, but what do we actually pay for? What is bought is electric *energy* measured in units of the **kilowatt-hour (kWh)**. Power is the rate at which work is done ($P = W/t$ or $W = Pt$), so work has units of watt-seconds (power $\times$ time). Converting this unit to the larger unit of kilowatt-hours (kWh), it can be seen that the kilowatt-hour is a unit of work (or energy), equivalent to 3.6 million joules, because

$$1 \text{ kWh} = (1000 \text{ W})(3600 \text{ s}) = (1000 \text{ J/s})(3600 \text{ s}) = 3.6 \times 10^6 \text{ J}$$

Thus, we pay the "power" company for electrical energy that we use to do work with our appliances.

The cost of electric energy varies with location. In the United States, it ranges from a low of several cents (per kilowatt-hour) to several times that value. Recently, electric energy rates have been deregulated. Coupled with increasing demand (without a corresponding increase in supply), deregulation has given rise to sky-rocketing rates in some areas of the country. Do you know the price of electricity in your locality? Check an electric bill to find out, especially if you are in one of the areas affected by dramatically rising rates. Let's look at the cost of electricity to run a typical appliance in the next Example.

Example 17.8 ■ Electric Energy Cost: The Price of Coolness

If the motor of a frost-free refrigerator runs 15% of the time, how much does it cost to operate per month (to the nearest cent) if the power company charges 11¢ per kilowatt-hour? (Assume 30 days in a month.)

Thinking It Through. From the power and the time the motor is on per day, we can compute the electrical energy the refrigerator requires *daily* and project that to a 30-day month.

Solution. Practical energy amounts are usually written in kilowatt-hours because the joule is a relatively tiny unit. Listing the data,

Given: $P = 500$ W (Table 17.2)
Cost $= \$0.11/\text{kWh}$

Find: operating cost per month

The refrigerator motor operates 15% of the time, so in one day it runs $t = (0.15)(24 \text{ h}) = 3.60$ h. Because $P = W/t$, the electrical energy required *per day* is

$$W = Pt = (500 \text{ W})(3.60 \text{ h/day}) = 1.80 \times 10^3 \text{ Wh} = 1.80 \text{ kWh/day}$$

So the cost per day is

$$\left(\frac{1.80 \text{ kWh}}{\text{day}}\right)\left(\frac{\$0.11}{\text{kWh}}\right) = \frac{\$0.20}{\text{day}}$$

or 20¢ per day. For a 30-day month, the cost would be

$$\left(\frac{\$0.20}{\text{day}}\right)\left(\frac{30 \text{ day}}{\text{month}}\right) \approx \$6 \text{ per month}$$

Follow-Up Exercise. How long would you have to leave a 60-W lightbulb on to use the same amount of electrical energy that the refrigerator motor in this Example uses each hour it is on?

Electrical Efficiency and Natural Resources

About 25% of the electric energy generated in the United States goes to lighting (▸Fig. 17.14). This percentage is roughly equivalent to the output of 100 electric-generating (power) plants. Refrigerators consume about 7% of the electric energy produced in the United States (the output of about twenty-eight such plants).

This huge (and growing) consumption of electric energy has prompted the federal government and many state governments to set minimum efficiency limits for such appliances as refrigerators, freezers, air conditioners, and water heaters

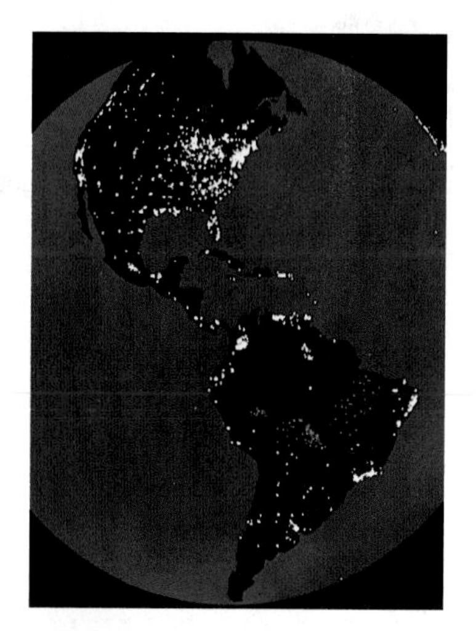

▲ **FIGURE 17.14 All lit up** A satellite image of the Americas at night. Can you identify the major population centers in the United States and elsewhere? The red spots across part of South America indicate large-scale burning of vegetation. The small yellow spot in Central America shows burning gas flares at oil production sites. At the top right edge, you can just glimpse a few of the white city lights of Europe. This image was recorded by a visible–infrared system.

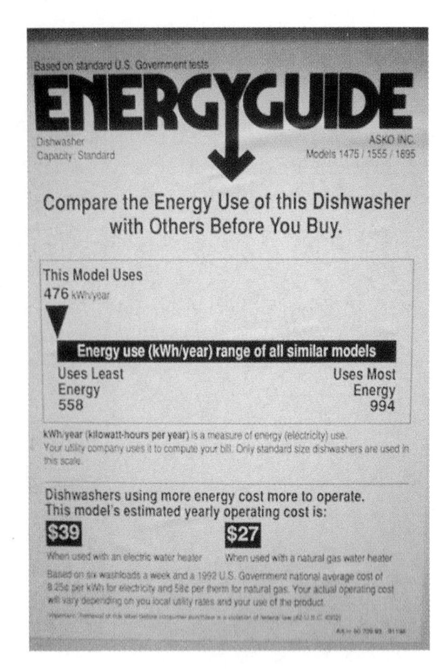

▲ **FIGURE 17.15** Energy guide
Consumers are made aware of the efficiencies of appliances in terms of the average yearly cost of their operation. Sometimes the yearly cost is given for different kilowatt-hour (kWh) rates, which vary around the United States.

(◂Fig. 17.15). Also, efficient fluorescent lighting has been developed and put into more common use. The most efficient fluorescent lamp now in use consumes about 25–30% less energy than the average fluorescent lamp and roughly 75% less energy than incandescent lamps with an equivalent light output.

The result of all these measures has been significant energy savings as new, more efficient appliances gradually replace inefficient models. Energy saved translates directly into savings of fuels and other natural resources, as well as a reduction in environmental hazards such as chemical pollution and global warming. To see what kind of results can be achieved by applying just one energy efficiency standard, consider the next Example.

Example 17.9 ■ What We Can Save: Increasing Electrical Efficiency

Many modern power plants produce electric energy at a rate of about 1.0 GW (gigawatt electric power output). Estimate how many fewer such power plants California would need if all its households switched from the 500-W refrigerators of Example 17.8 to more-efficient 400-W refrigerators. (Assume that there are about 10 million homes in California with an average of 1.2 refrigerators operating per home.)

Thinking It Through. The results from Example 17.8 can be used to calculate the overall effect.

Solution.

Given: Plant rate $= 1.0\ \text{GW} = 1.0 \times 10^6\ \text{kW}$
Energy requirement,
500-watt model $= 1.80\ \text{kWh/day}$
(Example 17.8)
Number of homes $= 10 \times 10^6$
Number of refrigerators per home $= 1.2$

Find: how many fewer power plants are required by switching to more efficient refrigerators

For the entire state, the energy usage per day with the less-efficient refrigerators is

$$\left(\frac{1.80\ \text{kWh/day}}{\text{refrig}}\right)(10 \times 10^6\ \text{homes})\left(\frac{1.2\ \text{refrigerators}}{\text{home}}\right) = 2.2 \times 10^7\ \frac{\text{kWh}}{\text{day}}$$

The more-efficient refrigerators would use only 80% ($400\ \text{W}/500\ \text{W} = 0.80$) of this, or 1.7×10^7 kWh/day. The difference, 5.0×10^6 kWh/day, is the rate at which electric energy is saved. One 1.0-GW power plant produces

$$\left(1.0 \times 10^6\ \frac{\text{kW}}{\text{plant}}\right)\left(24\ \frac{\text{h}}{\text{day}}\right) = 2.4 \times 10^7\ \frac{(\text{kWh/plant})}{\text{day}}$$

So the replacement refrigerators would save about

$$\frac{5.0 \times 10^6\ \text{kWh/day}}{2.4 \times 10^7\ \text{kWh/(plant-day)}} = 0.21\ \text{plant}$$

or about 20% of the output of a typical plant. Note that this saving results from a change in a *single* appliance. Imagine what could be done if all appliances, including lighting, were made more efficient. Developing and using more efficient electrical appliances is one way to avoid having to build new electric energy generating plants.

Follow-Up Exercise. Electric and gas water heaters are often said to be equally efficient—typically, about 95%. In reality, while gas water heaters are capable of 95% efficiency, it might be more accurate to describe electric water heaters as only about 30% efficient, even though approximately 95% of the *electrical* energy they require is transferred to the water in the form of heat. Explain. [*Hint*: What is the source of energy for an electric water heater? Compare this to the energy delivery of natural gas. Recall the discussion of electrical generation in Section 12.4 and Carnot efficiency in Section 12.5.]

Chapter Review

- A **battery** produces an **electromotive force (emf)**, or voltage, across its terminals. The high-voltage terminal is the **anode,** and the low-voltage one is the **cathode**.

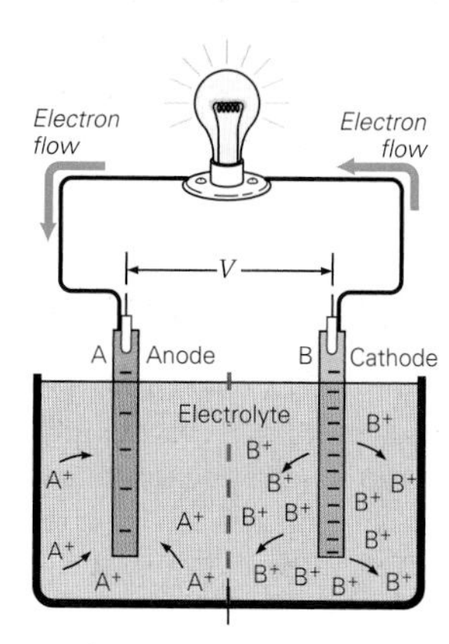

- **Electromotive force (emf $\mathscr{E}$)** is measured in volts and represents the number of joules of energy that a battery (power supply) gives to 1 coulomb of charge passing through it; that is, 1 J/C = 1 V.

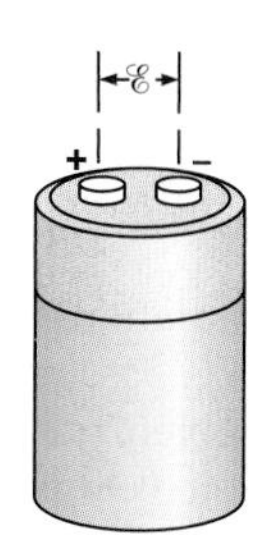

- **Electric current (*I*)** is the rate of charge flow. Its direction is that of **conventional current**, which is the direction in which positive charge actually flows or appears to flow. In metals, because the charge flow is electrons, the current direction is opposite the electron flow direction. Current is measured in **amperes** (1 A = 1 C/s) and defined as

$$I = \frac{q}{t} \tag{17.1}$$

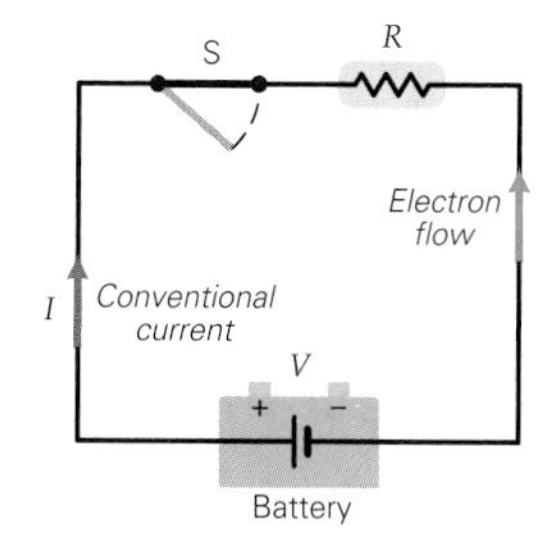

- For an electric current to exist in a circuit, it must be a **complete circuit**—that is, a circuit (set of circuit elements and wires) that connects both terminals of a battery or power supply with no break.

- The electrical **resistance (*R*)** of an object is the voltage across the object divided by the current in it, or

$$R = \frac{V}{I} \quad \text{or} \quad V = IR \tag{17.2}$$

The units of resistance are the **ohm** or the volt per ampere.

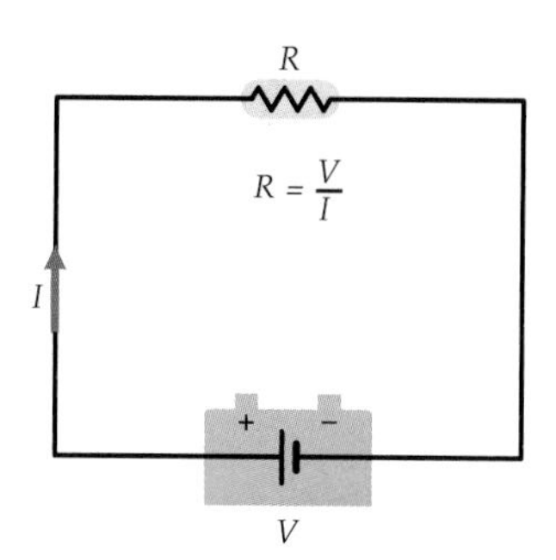

- A circuit element obeys **Ohm's law** if it exhibits constant electrical resistance. Ohm's law is commonly written $V = IR$, where R is constant.

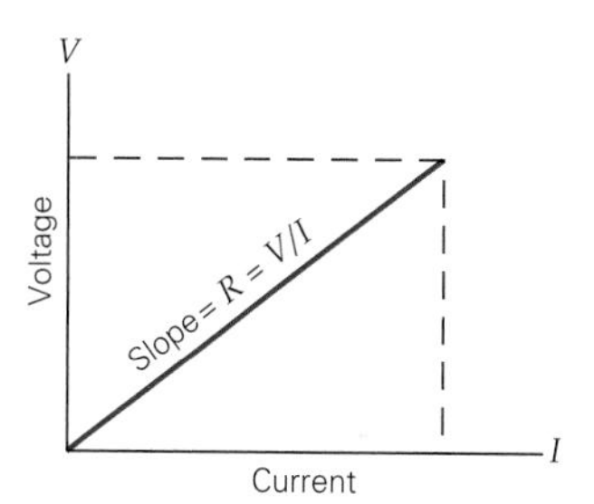

- The resistance of an object depends on the **resistivity (*ρ*)** of the material (based on its atomic properties), the cross-sectional area A, and the length L. For objects of uniform cross section,

$$R = \rho\left(\frac{L}{A}\right) \tag{17.3}$$

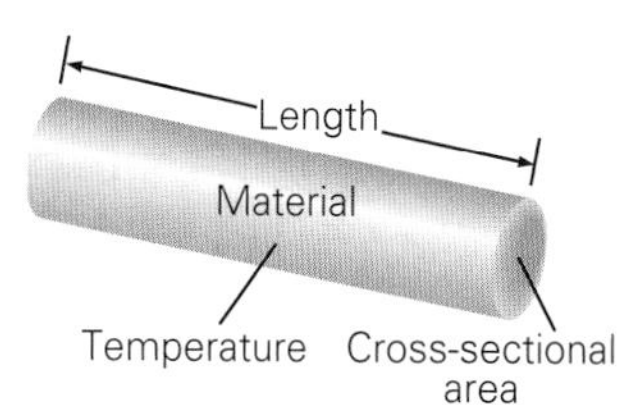

- **Electric power** (P) is the rate at which work is done by a battery (power supply), or the rate at which energy is transferred to a circuit element. The power delivered to a circuit element depends on the element's resistance, the current in it, and the voltage across it. Electrical power can be written in three equivalent ways:

$$P = IV = \frac{V^2}{R} = I^2R \tag{17.7b}$$

Exercises

MC = *Multiple Choice Question,* **CQ** = *Conceptual Question, and* **IE** = *Integrated Exercise. Throughout the text, many exercise sections will include "paired" exercises. These exercise pairs, identified with* **red numbers**, *are intended to assist you in problem solving and learning. In a pair, the first exercise (even numbered) is worked out in the Study Guide so that you can consult it should you need assistance in solving it. The second exercise (odd numbered) is similar in nature, and its answer is given at the back of the book.*

In this chapter, assume that all batteries have negligible internal resistance unless otherwise indicated.

17.1 Batteries and Direct Current

1. **MC** When a battery is part of a complete circuit, the voltage across its terminals is its (a) emf, (b) terminal voltage, (c) power output, (d) all of the preceding.

2. **MC** As a battery ages, its (a) emf increases, (b) emf decreases, (c) terminal voltage increases, (d) terminal voltage decreases.

3. **MC** When four 1.5-V batteries are connected, the output voltage of the combination is 1.5 V. These batteries are connected in (a) series, (b) parallel, (c) a pair in series connected in parallel to the other pair in series, (d) none of the preceding.

4. **MC** When helping someone whose car has a "dead" battery, how should your car's battery be connected in relation to the "dead" battery: (a) series, (b) parallel, or (c) either series or parallel would work fine?

5. **MC** When several 1.5-V batteries are connected in series, the overall output voltage of the combination is measured to be 12 V. How many batteries are needed to achieve this voltage: (a) two, (b) ten, (c) eight, or (d) six?

6. **CQ** Why does the battery design shown in Fig. 17.1 require a chemical membrane?

7. **CQ** A battery has its voltage measured while sitting on the workbench and the technician reads its manufacturer's rating, which is 12 V. Does this mean it will perform as expected when placed in a complete circuit? Explain.

8. **CQ** Sketch the following *complete* circuits, using the symbols show in the Learn by Drawing on page 571. (a) Two ideal 6.0-V batteries in series wired to a capacitor followed by a resistor. (b) Two ideal 12.0-V batteries in parallel, connected as a unit to two identical resistors in series with one another. (c) A nonideal battery (one with internal resistance) wired to two identical capacitors that are in parallel with each other, followed by two resistors in series with one another.

9. • (a) Three 1.5-V dry cells are connected in series. What is the total voltage of the combination? (b) What would be the total voltage if the cells were connected in parallel?

10. • What is the voltage across six 1.5-V batteries when they are connected (a) in series and (b) in parallel?

11. •• Two 6.0-V batteries and one 12-V battery are connected in series. (a) What is the voltage across the whole arrangement? (b) What arrangement of these three batteries would give a total voltage of 12 V?

12. •• Given three batteries with voltages of 1.0 V, 3.0 V, and 12 V, respectively, how many different voltages could be obtained by connecting one or more of the batteries in series or parallel, and what are these voltages?

13. **IE** •• You are given four AA batteries that are rated at 1.5 V each. The batteries are grouped in pairs. In arrangement A, the two batteries in each pair are in series, and then the pairs are connected in parallel. In arrangement B, the two batteries in each pair are in parallel, and then the pairs are connected in series. (a) Compared with arrangement B, will arrangement A have (1) a higher, (2) the same, or (3) a lower total voltage? (b) What are the total voltages for each arrangement?

17.2 Current and Drift Velocity

14. **MC** In which of these situations does more charge flow past a given point on a wire: when the wire has (a) a current of 2.0 A for 1.0 min, (b) 4.0 A for 0.5 min, (c) 1.0 A for 2.0 min, or (d) all are the same charge?

15. **MC** Which of these situations involves the least current: a wire that has (a) 1.5 C passing a given point in 1.5 min, (b) 3.0 C passing a given point in 1.0min, or (c) 0.5 C passing a given point in 0.10 min?

16. **MC** In a dental X-ray machine, the movement of accelerated electrons is east. The current associated with these electrons is in what direction: (a) east, (b) west, or (c) zero?

17. **CQ** In the circuit shown in Fig. 17.4a, what is the direction of (a) the electron flow in the resistor, (b) the current in the resistor, and (c) the current in the battery?

18. **CQ** Drift velocity, the average speed of electrons in a complete circuit, is typically a few mm per second. Yet a lamp 3.0 m away from a light switch comes on instantaneously when you flip the switch. Explain this apparent paradox.

19. • A net charge of 30 C passes through the cross-sectional area of a wire in 2.0 min. What is the current in the wire?

20. • How long would it take for a net charge of 2.5 C to pass a location in a wire if it is to carry a steady current of 5.0 mA?

21. • A small toy car draws a 0.50-mA current from a 3.0-V nicad (nickel–cadmium) battery. In 10 min of operation, (a) how much charge flows through the toy car, and (b) how much energy is lost by the battery?

22. •• A car's starter motor draws 50 A from the car's battery during startup. If the startup time is 1.5 s, how many electrons pass a given location in the circuit during that time?

23. ●● A net charge of 20 C passes a location in a wire in 1.25 min. How long does it take for a net 30-C charge to pass that location if the current in the wire is doubled?

24. ●●● Car batteries are often rated in "ampere-hours" or A·h. (a) Show that A·h has units of charge and that the value of 1 A·h is 3600 C. (b) A fully charged, heavy-duty battery is rated at 100 A·h and can deliver a current of 5.0 A steadily until depleted. What is the maximum time this battery can deliver current, assuming it isn't being recharged? (c) How much charge will the battery deliver in this time?

25. IE ●●● Imagine that some protons are moving to the left at the same time that some electrons are moving to the right past the same location. (a) Will the net current be (1) to the right, (2) to the left, (3) zero, or (4) none of the preceding? (b) In 4.5 s, 6.7 C of electrons flow to the right at the same time that 8.3 C of protons flow to the left. What is the magnitude of the total current?

26. ●●● In a proton linear accelerator, a 9.5-mA proton current hits a target. (a) How many protons hit the target each second? (b) How much energy is delivered to the target each second if the protons each have a kinetic energy of 20 MeV and lose all their energy in the target?

17.3 Resistance and Ohm's Law*

27. **MC** The ohm is just another name for the (a) volt per ampere, (b) ampere per volt (c) watt, or (d) volt.

28. **MC** Two ohmic resistors are placed across a 12-V battery one at a time. The resulting current in resistor A is measured to be twice as that in B. What can you say about their resistance values? (a) $R_A = 2R_B$, (b) $R_A = R_B$, (c) $R_A = R_B/2$, or (d) none of the preceding.

29. **MC** An ohmic resistor is placed across two different batteries. When connected to battery A, the resulting current is measured to be three times the current when the resistor is attached to battery B. What can you say about the battery voltages: (a) $V_A = 3V_B$ (b) $V_A = V_B$ (c) $V_B = 3V_A$ or (d) none of the preceding.

30. **MC** If you double the voltage across an ohmic resistor while at the same time cutting its resistance to one-third its original value, what happens to the current in the resistor: (a) it doubles, (b) it triples, (c) it goes up by six times, or (d) can't tell from the data given?

31. CQ If the voltage (V) were plotted on the same graph versus current (I) for two ohmic conductors with different resistances, how could you tell which is less resistive?

32. CQ Filaments in lightbulbs usually fail just after the bulbs are turned on rather than when they have already been on for a while. Why?

*Assume that the temperature coefficients of resistivity given in Table 17.1 apply over large temperature ranges.

33. CQ A wire is connected across a steady voltage source. (a) If that wire is replaced with one of the same material that is twice as long and has twice the cross-sectional area, how will the current in the wire be affected? (b) How will the current be affected if, instead, the new wire has the same length as the old one but half the diameter?

34. CQ A real battery always has some internal resistance r (▼Fig. 17.16) that increases with the battery's age. Explain why the terminal voltage drops as the internal resistance increases.

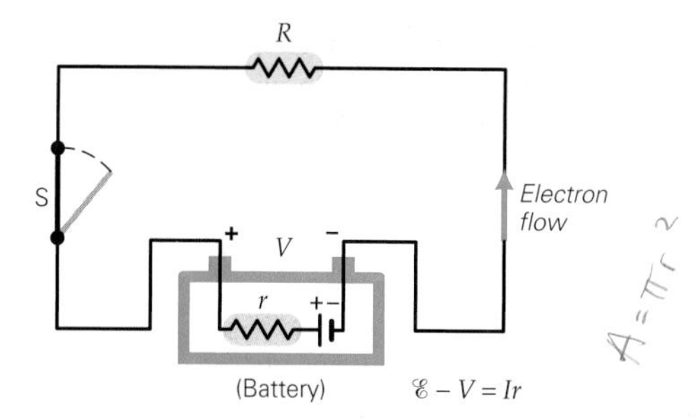

▲ **FIGURE 17.16 Emf and terminal voltage** See Exercises 34 and 35.

35. ● A battery labeled 12.0 V supplies 1.90 A to a 6.00-Ω resistor (Fig. 17.16). (a) What is the terminal voltage of the battery? (b) What is its internal resistance?

36. ● What is the emf of a battery with a 0.15-Ω internal resistance if the battery delivers 1.5 A to an externally connected 5.0-Ω resistor?

37. IE ● Some states allow the use of aluminum wire in houses in place of copper. (a) If you wanted the resistance of your aluminum wires to be the same as that of copper, would the aluminum wire have to have (1) a greater diameter than, (2) a smaller diameter than, or (3) the same diameter as the copper wire? (b) Calculate the thickness ratio of aluminum to that of copper.

38. ● How much current is drawn from a 12-V battery when a 15-Ω resistor is connected across its terminals?

39. ● What voltage must a battery have to produce a 0.50-A current through a 2.0-Ω resistor?

40. ● During a research experiment on the conduction of current in the human body, a medical technician attaches one electrode to the wrist and a second to the shoulder. If 100 mV are applied across the two electrodes and the resulting current is 12.5 mA, what is the overall resistance of the patient's arm?

41. ●● A 0.60-m-long copper wire has a diameter of 0.10 cm. What is the resistance of the wire?

42. ●● A material is formed into a long rod with a square cross-section 0.50 cm on each side. When a 100-V voltage is applied across a 20-m length of the rod, a 5.0-A current is carried. (a) What is the resistivity of the material? (b) Is the material a conductor, an insulator, or a semiconductor?

43. ●● Two copper wires have equal cross-sectional areas and lengths of 2.0 m and 0.50 m, respectively. (a) What is the ratio of the current in the shorter wire to that in the longer one if they are connected to the same power supply? (b) If you wanted the two wires to carry the same current, what would the ratio of their cross-sectional areas have to be? (Give your answer as a ratio of longer to shorter.)

44. IE ●● Two copper wires have equal lengths, but the diameter of one is three times that of the other. (a) The resistance of the thinner wire is (1) 3, (2) $\frac{1}{3}$, (3) 9, (4) $\frac{1}{9}$ times that of the resistance of the thicker wire. Why? (b) If the thicker wire has a resistance of 1.0 Ω, what is the resistance of the thinner wire?

45. ●● The wire in a heating element of an electric stove burner has a 0.75-m effective length and a 2.0×10^{-6}-m^2 cross-sectional area. (a) If the wire is made of iron and operates at a 380°C temperature, what is its operating resistance? (b) What is its resistance when the stove is "off"?

46. ●● (a) What is the percentage variation of the resistivity of copper over the temperature range from room temperature (20°C) to 100°C? (b) Assume a copper wire's resistance changes only due to resistivity changes over this temperature range. Further assume it is connected to the same power supply. By what percentage would its current change? Would it be an increase or decrease?

47. ●● A copper wire has a 25-mΩ resistance at 20°C. When the wire is carrying a current, heat produced by the current causes the temperature of the wire to increase by 27 C°. (a) What is the change in the wire's resistance? (b) If its original current was 10.0 mA, what is its final current?

48. ●● When a resistor is connected to a 12-V source, it draws a 185-mA current. The same resistor connected to a 90-V source draws a 1.25-A current. Is the resistor ohmic? Justify your answer mathematically.

49. ●● A particular application requires a 20-m length of aluminum wire to have a 0.25-mΩ resistance at 20°C. What must be the wire's diameter?

50. ●● If the resistance of the wire in Exercise 49 cannot vary by more than ±5.0%, what is the wire's operating temperature range?

51. IE ●●● As a wire is stretched out so that its length increases, its cross-sectional area decreases, while the total volume of the wire remains constant. (a) Will the resistance after the stretch be (1) greater than, (2) the same as, or (3) less than that before the stretch? (b) A 1.0-m length of copper wire with a 2.0-mm diameter is stretched out; its length increases by 25% while its cross-sectional area decreases, but remains uniform. Compute the resistance ratio (final to initial).

52. ●●● ▼Figure 17.17 shows data on the dependence of the current through a resistor on the voltage across that resistor. (a) Is the resistor ohmic? Explain. (b) What is its resistance? (c) Use the data to predict what voltage would be needed to produce a 4.0-A current in the resistor.

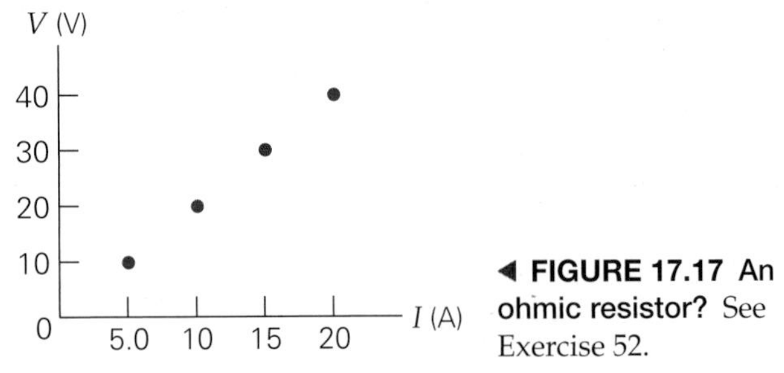

◀ **FIGURE 17.17** **An ohmic resistor?** See Exercise 52.

53. ●●● At 20°C, a silicon rod is connected to a battery with a terminal voltage of 6.0 V and a 0.50-A current results. The temperature of the rod is then increased to 25°C. Assume its temperature coefficient of resistance is constant. (a) What is its new resistance? (b) How much current does it carry? (c) If you wanted to cut the current from its room temperature value of 0.50 A to 0.40 A, at what temperature would the sample have to be?

54. IE ●●● A platinum wire is connected to a battery. (a) If the temperature increases, will the current in the wire (1) increase, (2) remain the same, or (3) decrease? Why? (b) An electrical resistance thermometer is made of platinum wire that has a 5.0-Ω resistance at 20°C. The wire is connected to a 1.5-V battery. When the thermometer is heated to 2020°C, by how much does the current change?

17.4 Electric Power

55. MC The electric power unit, the watt, is equivalent to what combination of SI units: (a) $A^2 \cdot \Omega$, (b) J/s, (c) V^2/Ω, or (d) all of the preceding?

56. MC If the voltage across an ohmic resistor is doubled, the power expended in the resistor (a) increases by a factor of 2, (b) increases by a factor of 4, (c) decreases by half, or (d) none of the preceding.

57. MC If the current through an ohmic resistor is halved, the power expended in the resistor (a) increases by a factor of 2, (b) increases by a factor of 4, (c) decreases by half, (d) decreases by a factor of 4.

58. CQ Assuming that the resistance of your hair dryer obeys Ohm's law, what would happen to its power output if you plugged it directly into a 240-V outlet in Europe if it is designed to be used in the 120-V outlets of the United States?

59. CQ Most lightbulb filaments are made of tungsten and are about the same length. What would be different about the filament in a 60-W bulb compared with that in a 40-W bulb?

60. CQ Which one consumes more power from a 12-V battery, a 5.0-Ω resistor or a10-Ω resistor? Why?

61. • A digital video disk (DVD) player is rated at 100 W at 120 V. What is its resistance?

62. • A freezer of resistance 10 Ω is connected to a 110-V source. What is the power delivered when this freezer is on?

63. • The current through a refrigerator with a resistance of 12 Ω is 13 A (when the refrigerator is on). What is the power delivered to the refrigerator?

64. • Show that the quantity volts squared per ohm (V^2/Ω) has SI units of power.

65. • An electric water heater is designed to produce 50 kW of heat when it is connected to a 240-V source. What must be the resistance of the heater?

66. •• If the heater in Exercise 65 is 90% efficient, how long would it take to heat 50 gal of water from 20°C to 80°C?

67. IE •• An ohmic resistor in a circuit is designed to operate at 120 V. (a) If you connect the resistor to a 60-V power source, will the resistor dissipate heat at (1) 2, (2) 4, (3) $\frac{1}{2}$, or (4) $\frac{1}{4}$ times the designed power? Why? (b) If the designed power is 90 W at 120 V, but the resistor is connected to 40 V, what is the power delivered to the resistor at the lower voltage?

68. •• An electric toy with a resistance of 2.50 Ω is operated by a 1.50-V battery. (a) What current does the toy draw? (b) Assuming that the battery delivers a steady current for its lifetime of 6.00 h, how much charge passed through the toy? (c) How much energy was delivered to the toy?

69. •• A welding machine draws 18 A of current at 240 V. (a) What is its power rating? (b) What is its resistance?

70. •• On average, an electric water heater operates for 2.0 h each day. (a) If the cost of electricity is \$0.15/kWh, what is the cost of operating the heater during a 30-day month? (b) What is the resistance of a typical water heater? [*Hint:* See Table 17.2.]

71. •• (a) What is the resistance of a heating coil if it is to generate 15 kJ of heat per minute when it is connected to a 120-V source? (b) How would you change the resistance if instead you wanted 10 kJ of heat per minute?

72. •• A 200-W computer power supply is on 10 h per day. If the cost of electricity is \$0.15/kWh, what is the cost (to the nearest dollar) of using the computer for a year (365 days)?

73. •• A 120-V air conditioner unit draws 15 A of current. If it operates for 20 min, (a) how much energy in kilowatt-hours does it use? (b) If the cost of electricity is \$0.15/kWh, what is the cost (to the nearest penny) of operating the unit for 20 min?

74. •• Two resistors, 100 Ω and 25 kΩ, are rated for a maximum power output of 1.5 W and 0.25 W, respectively. What is the maximum voltage that can be safely applied to each resistor?

75. •• A wire 5.0 m long and 3.0 mm in diameter has a resistance of 100 Ω. A 15-V potential difference is applied across the wire. Find (a) the current in the wire, (b) the resistivity of its material, and (c) the rate at which heat is being produced in the wire.

76. IE •• When connected to a voltage source, a coil of tungsten wire initially dissipates 500 W of power. In a short time, the temperature of the coil increases by 150 C° because of joule heating. (a) Will the dissipated power (1) increase, (2) remain the same, or (3) decrease? Why? (b) What is the corresponding change in the power?

77. •• A 20-Ω resistor is connected to four 1.5-V batteries. What is the joule heat loss per minute in the resistor if the batteries are connected (a) in series and (b) in parallel?

78. •• A 5.5-kW water heater operates at 240 V. (a) Should the heater circuit have a 20-A or a 30-A circuit breaker? (A circuit breaker is a safety device that opens the circuit at its rated current.) (b) Assuming 85% efficiency, how long will the heater take to heat the water in a 55-gal tank from 20° to 80°C?

79. •• A student uses an immersion heater to heat 0.30 kg of water from 20°C to 80°C for tea. If the heater is 75% efficient and takes 2.5 min, what is its resistance? (Assume 120-V household voltage.)

80. •• An ohmic appliance is rated at 100 W when it is connected to a 120-V source. If the power company cuts the voltage by 5.0% to conserve energy, what is (a) the current in the appliance and (b) the power consumed by the appliance after the voltage drop?

81. •• A lightbulb's output is 60 W when it operates at 120 V. If the voltage is cut in half and the power dropped to 20 W during a brownout, what is the ratio of the bulb's resistance at full power to its resistance during the brownout?

82. •• To empty a flooded basement, a water pump must do work (lift the water) at a rate of 2.00 kW. If the pump is wired to a 240-V source and is 84% efficient, (a) how much current does it draw and (b) what is its resistance?

83. ••• Find the total monthly (30-day) electric bill (to the nearest dollar) for the following household appliance usage if the utility rate is \$0.12/kWh: Central air conditioning runs 30% of the time; a blender is used 0.50 h/month; a dishwasher is used 8.0 h/month; a microwave oven is used 15 min/day; the motor of a frost-free refrigerator runs 15% of the time; a stove (burners plus oven) is used a total of 10 h/month; and a color television is operated 120 h/month. (Use the information given in Table 17.2.)

Comprehensive Exercises

84. IE A piece of carbon and a piece of copper have the same resistance at room temperature. (a) If the temperature of each piece is increased by 10.0 C°, will the copper piece have (1) a higher resistance than, (2) the same resistance as, or (3) a lower resistance than the carbon piece? Why? (b) Calculate the ratio of the resistance of copper to that of carbon at the raised temperature.

85. Two pieces of aluminum and copper wire are identical in length and diameter. At some temperature, one of the wires will have the same resistance that the other has at 20°C. What is that temperature? (Is there more than one temperature?)

86. A battery delivers 2.54 A to an ohmic resistor rated at 4.52 Ω. When it is connected to a 2.21-Ω resistor, it delivers 4.98 A. Determine the battery's (a) internal resistance (assumed constant), (b) emf , and (c) terminal voltage (in both cases).

87. An external resistor is connected to a battery with a variable emf but constant internal resistance. At an emf of 3.00 V, the resistor draws a current of 0.500 A, and at 6.00 V, the resistor draws a current of 1.00 A. Is the external resistor ohmic? Prove your answer.

88. An electric eel delivers a current of 0.75 A to a small pencil-thin prey 15 cm long. If the eel's "bio-battery" was charged to 500 V, and it was constant for 20 ms before dropping to zero, estimate (a) the resistance of the fish, (b) the energy delivered to the fish, and (c) the average electric field (magnitude) in the fish's flesh.

89. Most modern TVs have an "instant warm-up" feature. Even though the set appears to be off, it is "off" only in that there is no picture and audio. To provide a "quick on" feature, the TV's electronics are kept ready. This takes about 10 W of electric power, constantly. Assume that there is one TV with this feature for every two households and estimate how many electric power plants this feature takes to run in the United States.

90. ▾Figure 17.18 shows free-charge carriers that each have a charge q and move with a speed v_d (drift speed) in a conductor of cross-sectional area A. Let n be the number of free charge carriers per unit volume. (a) Prove that the total charge (ΔQ) free to move in the volume element shown is given by $\Delta Q = (nAx)q$. (b) Prove that the current in the conductor is given by $I = nqv_d A$.

91. A copper wire with a cross-sectional area of 13.3 mm^2 (AWG No. 6) carries a 1.2-A current. If the wire contains 8.5×10^{22} free electrons per cubic centimeter, what is the drift velocity of the electrons? [*Hint*: See Exercise 90 and Fig. 17.18.]

92. A computer CD-ROM drive that operates on 120 V is rated at 40 W when it is operating. (a) How much current does the drive draw? (b) What is the drive's resistance?

93. The tungsten filament of an incandescent lamp has a resistance of 200 Ω at room temperature. What would the resistance be at an operating temperature of 1600°C?

94. A common sight in our modern world is high-voltage lines carrying electric energy over long distances from the power plant to populated areas. The delivery voltage of these lines is typically 500 kV, whereas by the time the energy reaches our households it is down to 120 V (see Chapter 20 for how this is done). (a) Explain clearly why electric power is delivered over long distances at high voltages when we know that high voltages can be dangerous. (b) Calculate the ratio of heating loss in a given length of wire (assumed ohmic) carrying current at 500 kV to when it operates at 120 V.

95. In a country setting it is common to see hawks sitting on a single high-voltage electric power line searching for a road-kill meal (▾Fig. 17.19). To understand why this bird isn't electrocuted, let's do a ballpark estimate of the voltage between her feet. Assume dc conditions in a power line that is 1.0 km long, has a resistance of 30 Ω, and is at an electric potential of 250 kV above the other wire (the one the bird is not on), which is at ground or zero volts. (a) If the wires are carrying energy at the rate of 100 MW, what is the current in them? (b) Assuming the bird's feet are 15 cm apart, what is the resistance of that segment of the hot wire? (c) What is the voltage *difference* between the bird's feet? Comment on the size of your answer and whether you think this might be dangerous. (d) What is the voltage difference between her feet if she places one on the ground wire while continuing to hold onto the hot wire? Comment on the size of your answer and whether you think this might be dangerous.

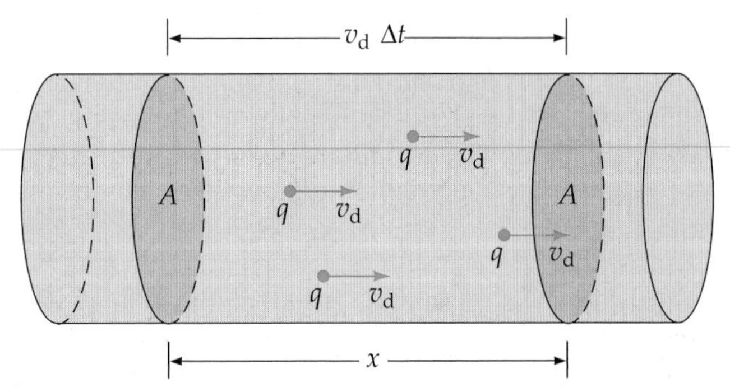

▲ **FIGURE 17.18 Total charge and current** See Exercises 90 and 91.

◀ **FIGURE 17.19 Bird on a wire** See Exercise 95.

PHYSLET® The following Physlet Physics Problem can be used with this chapter. 30.5

CHAPTER

18 BASIC ELECTRIC CIRCUITS

18.1 Resistances in Series, Parallel, and Series–Parallel Combinations 592

18.2 Multiloop Circuits and Kirchhoff's Rules 599

18.3 RC Circuits 604

18.4 Ammeters and Voltmeters 607

18.5 Household Circuits and Electrical Safety 611

PHYSICS FACTS

- Physics for parents of teenagers: Usually more than one hair dryer/blower cannot be on the same household circuit without tripping the circuit breaker. Either two separate circuits are needed in the bathroom, or someone has to go to another room and use a different circuit.
- An ammeter wired *incorrectly* in parallel with a circuit element not only measures the wrong current, but risks being burned out. Hence all ammeters have protective fuses. On the other hand, if a voltmeter is *incorrectly* connected in series with a circuit element, the circuit current drops to zero, and although the voltage measurement is incorrect, there is no damage.
- Less than 0.01 of an amp of current through the human body can trigger muscle paralysis. If the person cannot then let go of the exposed wiring, death could result if the current passes through a vital organ, such as the heart.
- Specialized *pacemaker cells* located in a small region of the heart trigger your heartbeat. Their electrical signals travel across the heart in about 50 ms. If these fail, other parts of the heart's electrical system can take over as a backup. The pacemaker cells can be influenced by the body's nervous system, so the rate at which they tell the heart to beat can vary dramatically—for example, from a calm 60 beats per minute when asleep to more than 100 beats per minute during physical exertion.

We usually think of metallic wires as the "connectors" between resistors in a circuit. However, wires are not the only conductors of electricity, as the photo shows. Because the bulb is lit, the circuit must be complete. It can be concluded, therefore, that the "lead" in a pencil (actually a form of carbon called *graphite*) conducts electricity. The same must be true for the liquid in the beaker—in this case, a solution of water and ordinary table salt.

Electric circuits are of many kinds and can be designed for many purposes, from boiling water to lighting a Christmas tree. Circuits containing "liquid" conductors (as in the photo), have practical applications in the laboratory and in industry; for example, they can be used to synthesize or purify chemical substances and to *electroplate* metals. (Electroplating means to chemically attach metals to surfaces using electrical techniques, such as in making silver plate.) Armed with the principles learned in Chapters 15, 16, and 17, you are now ready to analyze some electric circuits. This analysis will give you an appreciation of how electricity actually works.

Circuit analysis most often deals with voltage, current, and power requirements. A circuit may be analyzed theoretically before being assembled. The analysis might show that the circuit would not function properly as designed or that there could be a safety problem (such as overheating due to joule heat). To help in this chapter's analysis, we will rely heavily on diagrams of circuits to visualize and understand their function. A few of these diagrams were included in Chapter 17.

Let's begin our analysis of circuits by looking at arrangements of resistive elements, such as lightbulbs, toasters, and immersion heaters.

18.1 Resistances in Series, Parallel, and Series–Parallel Combinations

OBJECTIVES: To (a) determine the equivalent resistance of resistors in series, parallel, and series–parallel combinations, and (b) use equivalent resistances to analyze simple circuits.

The resistance symbol –\/\/\/– can represent *any* type of circuit element such as a lightbulb or toaster. Here it is assumed that all the elements are ohmic (constant resistance) unless otherwise stated. (Note that lightbulbs, in particular, are not ohmic because their resistance increases significantly as they warm up.) In addition, as usual, wire resistance will be neglected.

Resistors in Series

In analyzing a circuit, because voltage represents energy per unit charge, to conserve energy, the *sum of the voltages around a complete circuit loop is zero*. Remember that *voltage* means "change in electrical potential," so voltage gains and losses are represented by + and − signs, respectively. For the circuit in ▾Fig. 18.1a, by conservation of energy (per coulomb) the individual voltages (V_i, where $i = 1, 2$, or 3) across the resistors add to equal the voltage (V) across the battery terminals. Each resistor in series must carry the same current (I) because charge can't "pile up" or "leak out" at any location in the circuit. Summing the voltage gains and losses, we have $V - \Sigma V_i = 0$. Finally, we know how the voltage is related to the resistance for each resistor, namely $V_i = IR_i$. Substituting this into the previous equation,

Note: For resistors, $V = IR$.

$$V - \Sigma(IR_i) = 0 \quad \text{or} \quad V = \Sigma(IR_i) \tag{18.1}$$

The elements in Fig. 18.1a are said to be connected in **series**, or end to end. *When resistors are in series, the current must be the same through all the resistors*, as required by the conservation of charge. If this were not true, then charge would build up or disappear, which cannot happen. ▸Figure 18.2 shows the analogous flow of water in a smooth streambed punctuated by a series of rapids (representing "resistance").

Labeling the common current in the resistors as I, then Eq. 18.1 can be written explicitly for three resistors (such as in Fig. 18.1a):

$$\begin{aligned} V &= V_1 + V_2 + V_3 \\ &= IR_1 + IR_2 + IR_3 = I(R_1 + R_2 + R_3) \end{aligned}$$

Note: To better understand resistor connections, review the discussion of capacitors connected in series and parallel arrangements in Chapter 16.

To **equivalent series resistance, (R_s)** is the value of a single resistor that could replace the three resistors by one resistor R_s and maintain the same current means that $V = IR_s$, or $R_s = V/I$. Hence, the resistors in series have an equivalent resistance

$$R_s = \frac{V}{I} = R_1 + R_2 + R_3$$

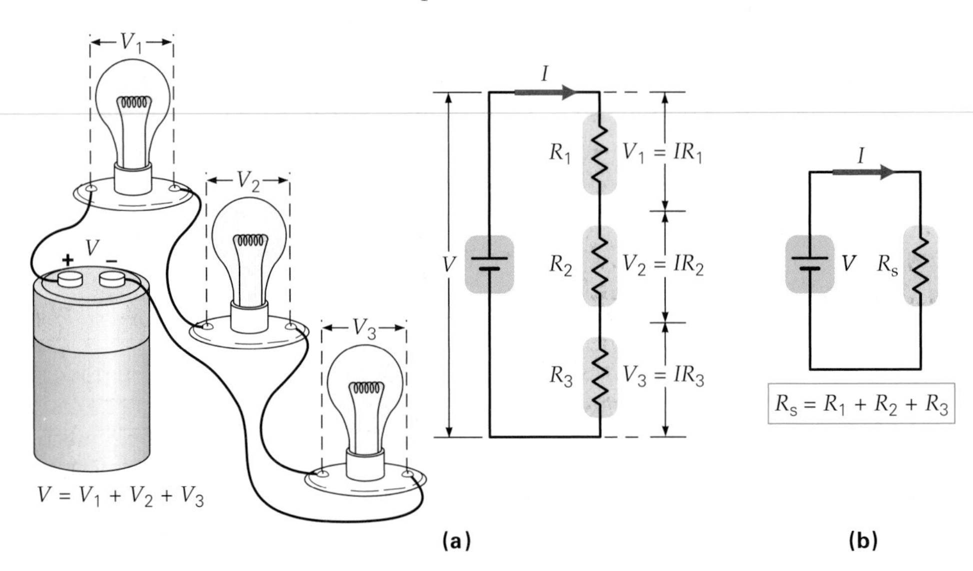

▸ **FIGURE 18.1** Resistors in series **(a)** When resistors (representing the resistances of lightbulbs here) are in series, the current in each is the same. ΣV_i, the sum of the voltage drops across the resistors, is equal to V, the battery voltage. **(b)** The equivalent resistance R_s of the resistors in series is the sum of the resistances.

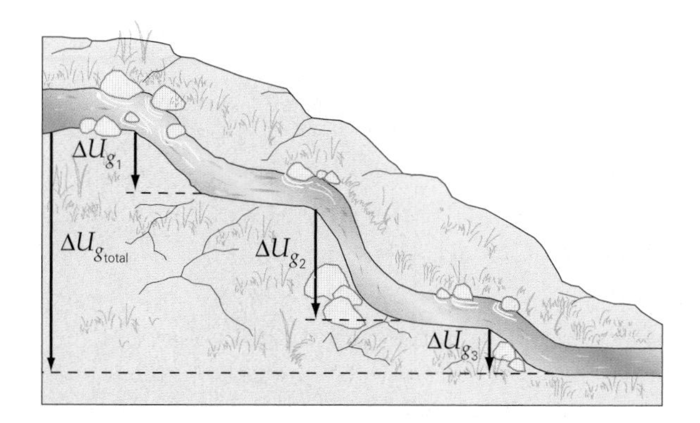

◀ FIGURE 18.2 Water-flow analogy to resistors in series Even though, in general, a different amount of gravitational potential energy (per kilogram) is lost as the water flows down each set of rapids, the *current* of water is the same everywhere. The *total* loss of gravitational potential energy (per kilogram) is the sum of the losses. (To make this a "complete" water circuit, some external agent, such as a pump, would need to continuously do work on the water by returning it to the top of the hill, restoring its original gravitational potential energy.)

That is, the equivalent resistance of resistors in series is the sum of the individual resistances. This means that the three resistors (bulbs) in Fig. 18.1a could be replaced by a single resistor of resistance R_s (Fig. 18.1b) without affecting the current. For example, if each resistor in Fig. 18.1a had a value of 10 Ω, then R_s would be 30 Ω. *Note that the equivalent series resistance is larger than the resistance of the largest resistor in the series.*

This result can be extended to any number of **resistors in series**:

$$R_s = R_1 + R_2 + R_3 + \cdots = \Sigma R_i \quad \textit{equivalent series resistance} \tag{18.2}$$

Note: R_s is larger than the largest resistance in a series arrangement.

Series connections are not common in some circuits such as house wiring, because there are two major disadvantages compared to parallel wiring. The first is clear considering what happens if one of the bulbs in Fig. 18.1a burns out (or you wish to turn only that one bulb off). In this case, *all* of the bulbs go out, because the circuit would no longer be complete, or continuous. In this situation, the circuit is said to be *open*. An *open circuit* has an infinite equivalent resistance, because the current in it is zero, even though the battery voltage is not.

A second disadvantage of series connections is that each resistor operates at less than the battery voltage (V). Consider what would happen if a fourth resistor were added. The result would be that the voltage across each of the original bulbs (and the current) would decrease, resulting in reduced power delivered to all the bulbs. That is, the bulbs would not glow at the rated brightness or light output. Clearly, this situation is *not* acceptable in a household setting.

Resistors in Parallel

Resistors can also be connected to a battery in **parallel** (▼Fig. 18.3a). In this case, all the resistors have common connections—that is, all the leads on one side of the resistors are attached together to one terminal of the battery. All the leads on the other side are attached to the other terminal. *When resistors are connected in parallel to a source of emf, the voltage drop across each resistor is the same.* It may not surprise

(a)

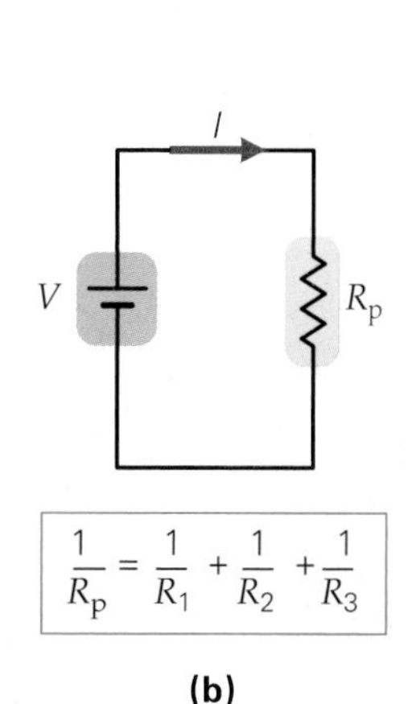

◀ FIGURE 18.3 Resistors in parallel **(a)** When resistors are connected in parallel, the voltage drop across each resistor is the same. The current from the battery divides (generally unequally) among the resistors. **(b)** The equivalent resistance, R_p, of resistors in parallel is given by a reciprocal relationship.

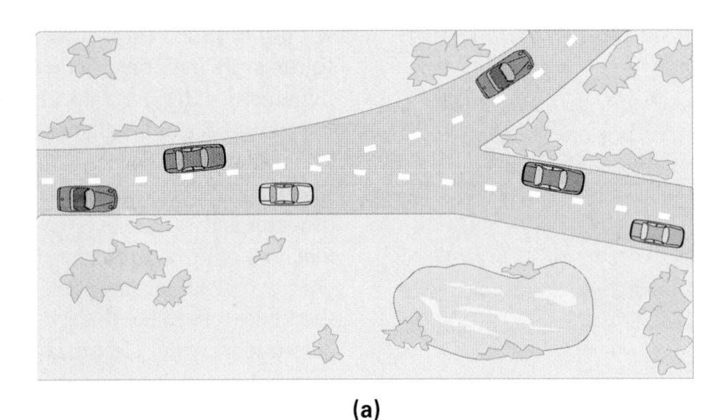
(a)

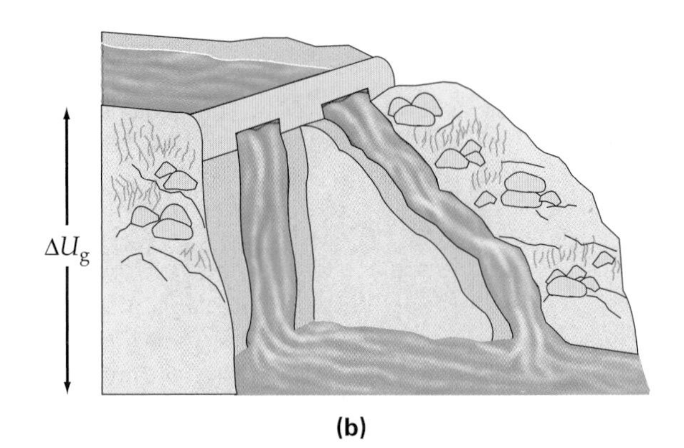

(b)

▲ **FIGURE 18.4 Analogies for resistors in parallel (a)** When a road forks, the total number of cars entering the two branches each minute is equal to the number of cars arriving at the fork each minute. Movement of charge into and then out of a junction can be considered in the same way. **(b)** When water flows from a dam, the gravitational potential energy lost (per kilogram of water) in falling to the stream below is the same regardless of the path. This is analogous to voltages across parallel resistors.

Note: In actuality, the wires of a circuit are not ordinarily arranged in the neat rectangular pattern of a circuit diagram. The rectangular form is a convention that provides a neater and consistent presentation along with an easier visualization of the actual circuit.

you to learn that household circuits are wired in parallel. (See Section 18.5.) This is because when wired in parallel, each appliance operates at full voltage, and turning one appliance off or on does not affect the others.

Unlike resistors in series, the current in a parallel circuit divides into the different paths (Fig. 18.3a). This occurs whenever we have a *junction* (a location where several wires come together), much as traffic divides when it reaches a fork in the road (▲Fig. 18.4a). The total current out of the battery is equal to the sum of these currents. Specifically, for three resistors in parallel, $I = I_1 + I_2 + I_3$. Notice that if the resistances are equal, the current will divide so that each resistor has the same current. However, in general, the resistances will not be equal and the current will divide among the resistors in inverse proportion to their resistances. This means that the largest current will take the path of least resistance. Remember, however, that no one resistor carries the total current.

The **equivalent parallel resistance (R_p)** is the value of a single resistor that could replace all the resistors and maintain the same current. Thus, $R_p = V/I$, or $I = V/R_p$. In addition, the voltage drop (V) must be the same across each resistor. To visualize this situation, imagine a water analogy. Consider two separate water paths, each leading from the top of a dam to the bottom. The water loses the same amount of gravitational potential energy (analogous to V) regardless of the path (Fig. 18.4b). For electricity, a given amount of charge loses the same amount of electrical potential energy, regardless of which parallel resistor it passes through.

The current through each resistor is $I_i = V/R_i$. (The subscript i represents *any* of the resistors: 1, 2, 3, ...) Substituting for each current, we obtain

$$I = I_1 + I_2 + I_3 = \frac{V}{R_1} + \frac{V}{R_2} + \frac{V}{R_3}$$

Therefore,

$$\frac{V}{R_p} = V\left(\frac{1}{R_p}\right) = V\left(\frac{1}{R_1} + \frac{1}{R_2} + \frac{1}{R_3}\right)$$

By equating the two expressions in the parentheses, we see that R_p is related to the individual resistances by a reciprocal equation

$$\frac{1}{R_p} = \frac{1}{R_1} + \frac{1}{R_2} + \frac{1}{R_3}$$

This result can be generalized to include any number of **resistors in parallel**:

$$\frac{1}{R_p} = \frac{1}{R_1} + \frac{1}{R_2} + \frac{1}{R_3} + \cdots = \Sigma\left(\frac{1}{R_i}\right) \quad \textit{equivalent parallel resistance} \qquad (18.3)$$

For the special case when there are just two resistors, the equivalent resistance can be rewritten (using a common denominator) as

$$\frac{1}{R_p} = \frac{1}{R_1} + \frac{1}{R_2} = \frac{R_1 + R_2}{R_1 R_2}$$

or

$$R_p = \frac{R_1 R_2}{R_1 + R_2} \quad \textit{(only for two resistors in parallel)} \qquad (18.3a)$$

Note: R_p is smaller than the smallest resistance in a parallel arrangement.

Problem-Solving Hint

Note that Eq. 18.3 gives $1/R_p$, *not* R_p. At the end of the calculation, the reciprocal must be taken to find R_p. Unit analysis will show that the units are not ohms until inverted. As usual, carrying units along with calculations makes errors of this type less likely to occur.

The equivalent resistance of resistors in parallel is always less than the smallest resistance in the arrangement. For example, two parallel resistors—say, of resistances 6.0 Ω and 12.0 Ω—are equivalent to a single with a resistance of 4.0 Ω (you should show this). But why should we expect this seemingly strange answer?

PHYSLET®

Illustration 30.3 Current and Voltage Dividers

Physically, the reason can be seen by first considering a 12-V battery in a circuit with a *single* 6.0-Ω resistor. The current in the circuit is 2.0 A ($I = V/R$). Now imagine connecting a 12.0-Ω resistor in parallel to the 6.0-Ω resistor. The current through the 6.0-Ω resistor will be unaffected—it will remain at 2.0 A. (Why?) However, the new resistor will have a current of 1.0 A (using $I = V/R$ again). Thus the *total* current in the circuit is 1.0 A + 2.0 A = 3.0 A. Now look at the overall result. When the second resistor is attached to the first *in parallel*, the total current delivered by the battery increases. Since the voltage did not increase, the equivalent resistance of the circuit *must have decreased* (below its initial value of 6.0 Ω) when the 12-Ω resistor was attached. In other words, every time an extra parallel path is added, the result is more total current. Thus the circuit behaves as if its equivalent resistance *decreased*.

Illustration 30.4 Batteries and Switches

Notice that this argument does not depend on the value of the added resistor. All that matters is that another path with some resistance is added. (Try this using a 2-Ω or a 2-MΩ resistor in place of the 12-Ω resistor. A decrease in equivalent resistance happens again. Note, however, that the *value* of the equivalent resistance will be different.)

In general, then, series connections provide a way to increase total resistance and parallel connections provide a way to decrease total resistance. To see how these ideas work, consider Example 18.1.

Example 18.1 ■ Connections Count: Resistors in Series and in Parallel

What is the equivalent resistance of three resistors (1.0 Ω, 2.0 Ω, and 3.0 Ω) when connected (a) in series (Fig. 18.1a) and (b) in parallel (Fig. 18.3a)? (c) How much current will be delivered by a 12-V battery in each of these arrangements?

Thinking It Through. To find the equivalent resistances for parts (a) and (b), apply Eqs. 18.2 and 18.3, respectively. For the series current in part (c), calculate the current through the battery by treating the battery as if it were connected to a single resistor—the series equivalent resistance. For the parallel arrangement, the total current can be determined by using the parallel equivalent resistance. From the knowledge that each resistor in parallel has the same voltage across it, the individual currents can be calculated.

Solution. Listing the data

Given: $R_1 = 1.0\ \Omega$
$R_2 = 2.0\ \Omega$
$R_3 = 3.0\ \Omega$
$V = 12\ \text{V}$

Find: (a) R_s (series resistance)
(b) R_p (parallel resistance)
(c) I (total current for each case)

(a) The equivalent series resistance (Eq. 18.2) is

$$R_s = R_1 + R_2 + R_3 = 1.0\ \Omega + 2.0\ \Omega + 3.0\ \Omega = 6.0\ \Omega$$

Our result is larger than the largest resistance, as expected.

(b) The equivalent parallel resistance is determined from Eq. 18.3

$$\frac{1}{R_p} = \frac{1}{R_1} + \frac{1}{R_2} + \frac{1}{R_3} = \frac{1}{1.0\ \Omega} + \frac{1}{2.0\ \Omega} + \frac{1}{3.0\ \Omega}$$

$$= \frac{6.0}{6.0\ \Omega} + \frac{3.0}{6.0\ \Omega} + \frac{2.0}{6.0\ \Omega} = \frac{11}{6.0\ \Omega}$$

or, after inverting,

$$R_p = \frac{6.0\ \Omega}{11} = 0.55\ \Omega$$

which is less than the least resistance, also as expected.

(continues on next page)

(c) From the equivalent series resistance and the battery voltage:

$$I = \frac{V}{R_s} = \frac{12\text{ V}}{6.0\ \Omega} = 2.0\text{ A}$$

Let's calculate the voltage drop across each resistor:

$$V_1 = IR_1 = (2.0\text{ A})(1.0\ \Omega) = 2.0\text{ V}$$
$$V_2 = IR_2 = (2.0\text{ A})(2.0\ \Omega) = 4.0\text{ V}$$
$$V_3 = IR_3 = (2.0\text{ A})(3.0\ \Omega) = 6.0\text{ V}$$

Notice that to ensure that the current through each resistor is the same, it must be that *in series, the larger resistors require more voltage*. As a check, note that the sum of the resistor voltage drops ($V_1 + V_2 + V_3$) equals the battery voltage.

For the parallel arrangement, the total current is:

$$I = \frac{V}{R_p} = \frac{12\text{ V}}{0.55\ \Omega} = 22\text{ A}$$

Note that the current for the parallel combination is much larger than that for the series combination. (Why?) Now the current through each resistor can be determined, because each has a voltage of 12 V across it. Therefore,

$$I_1 = \frac{V}{R_1} = \frac{12\text{ V}}{1.0\ \Omega} = 12\text{ A}$$

$$I_2 = \frac{V}{R_2} = \frac{12\text{ V}}{2.0\ \Omega} = 6.0\text{ A}$$

$$I_3 = \frac{V}{R_3} = \frac{12\text{ V}}{3.0\ \Omega} = 4.0\text{ A}$$

As a check, note that the sum of the currents is equal to the current through the battery.

As can be seen, for resistors in parallel, the resistor with the smallest resistance gets most of the total current because resistors in parallel experience the same voltage. (Note that for parallel arrangements the least resistance never has *all* the current, just the largest.)

Follow-Up Exercise. (a) Calculate the power delivered to each resistor for both arrangements in this Example. (b) What generalizations can you make? For instance, which resistor gets the most power in series? in parallel? (c) For each arrangement, does the total power delivered to all the resistors equal the power output of the battery? *(Answers to all Follow-Up Exercises are at the back of the text.)*

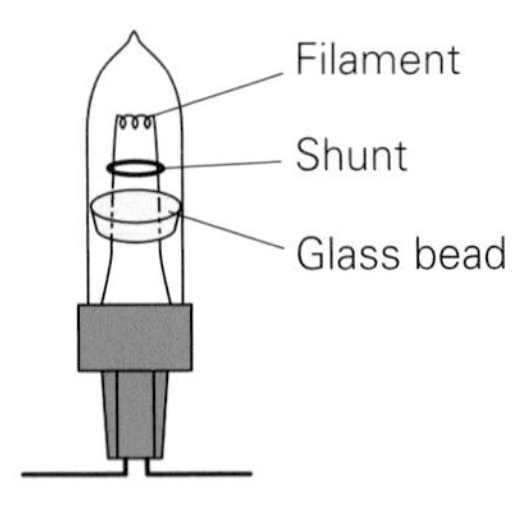

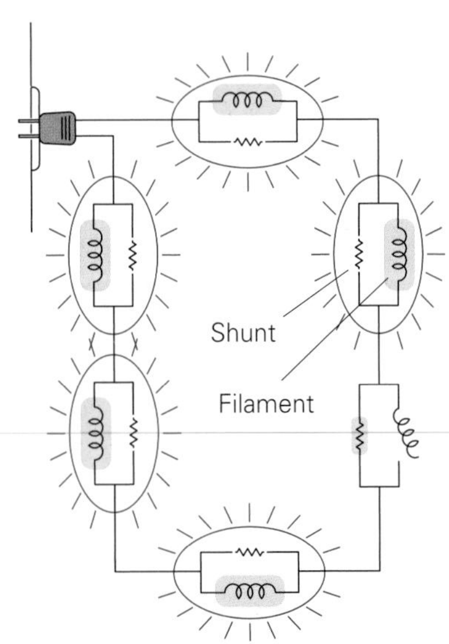

▲ **FIGURE 18.5 Shunt-wired Christmas tree lights** A shunt, or "jumper," in parallel with the bulb filament reestablishes a complete circuit when one of the filaments burns out (lower right bulb). Without the shunt, if one were to burn out, all the bulbs would go out.

As a wiring application, consider strings of Christmas tree lights. In the past, such strings had large bulbs connected in series. When one bulb burned out, all the others in the string went out, leaving you to hunt for the faulty bulb. Now, with newer strings that have smaller bulbs, one or more bulbs may burn out, but the others remain lit. Does this mean that the bulbs are now wired in parallel? No; parallel wiring would give a small resistance and large current, which could be dangerous.

Instead, an insulated jumper, or "shunt," is wired in parallel with each bulb's filament (◂Fig. 18.5). In normal operation, the shunt is insulated from the filament wires and does not carry current. When the filament breaks or "burns out," there is *momentarily* an open circuit, and there is no current anywhere in the string. Thus, the voltage across the open circuit at the broken filament will be the full 120-V household voltage. This voltage causes sparking that burns off the shunt's insulating material. Now the shunt is in electrical contact with the other filament wires, again completing the circuit, and the rest of the lights in the string continue to glow. (The shunt, a wire with little resistance, is indicated by the small resistance symbol in the circuit diagram of Fig. 18.5. Under normal operation, there is a gap—the insulation—between the shunt and the filament wire.) To understand what happens to the remaining bulbs in a string with a burnt-out bulb, consider the following Example.

Conceptual Example 18.2 ■ Oh, Tannenbaum! Christmas Tree Lights Burning Brightly

Consider a string of Christmas tree lights composed of bulbs with jumper shunts. If the filament of one bulb burns out and the shunt completes the circuit, will the other bulbs each (a) glow a little more brightly, (b) glow a little more dimly, or (c) be unaffected?

Reasoning and Answer. If one bulb filament burns out and its shunt completes the circuit, there will be less total resistance in the circuit, because the shunt's resistance is much less than the filament's resistance. (Note that the filaments of the good bulbs and the shunt of the burnt-out bulb are in now series, so the resistances add.)

With less total resistance, there will be more current in the circuit, and the remaining good bulbs will glow a little brighter because the light output of a bulb is directly related to the power delivered to that bulb. (Recall that electrical power is related to the current by $P = I^2R$.) So the answer is (a). For example, suppose the string initially has 18 identical bulbs. Because the total voltage across the string is 120 V, the voltage drop in any one bulb is $(120\text{ V})/18 = 6.7\text{ V}$. If one bulb is out (and thus shunted), the voltage across each of the remaining lighted bulbs becomes $(120\text{ V})/17 = 7.1\text{ V}$. This increased voltage causes the current to increase. Both increases contribute to more power delivered to each bulb, and brighter lights (recall the alternative expression for electric power, $P = IV$).

Follow-Up Exercise. In this Example, if you were to remove one bulb, what would be the voltage across (a) the empty socket and (b) any of the remaining bulbs? Explain.

PHYSLET®

Illustration 30.1 Complete Circuits

Series–Parallel Resistor Combinations

Resistors may be connected in a circuit in a variety of series–parallel combinations. As shown in ▼Fig. 18.6, circuits with only one voltage source can sometimes be reduced to a single equivalent loop, containing just the voltage source and one equivalent resistance, by applying the series and parallel results.

A procedure for analyzing circuits (determining voltage and current for each circuit element) for such combinations is as follows:

Illustration 30.2 Switches, Voltages, and Complete Circuits

1. Determine which groups of resistors are in series and which are in parallel, and reduce all groups to equivalent resistances, using Eqs. 18.2 and 18.3.

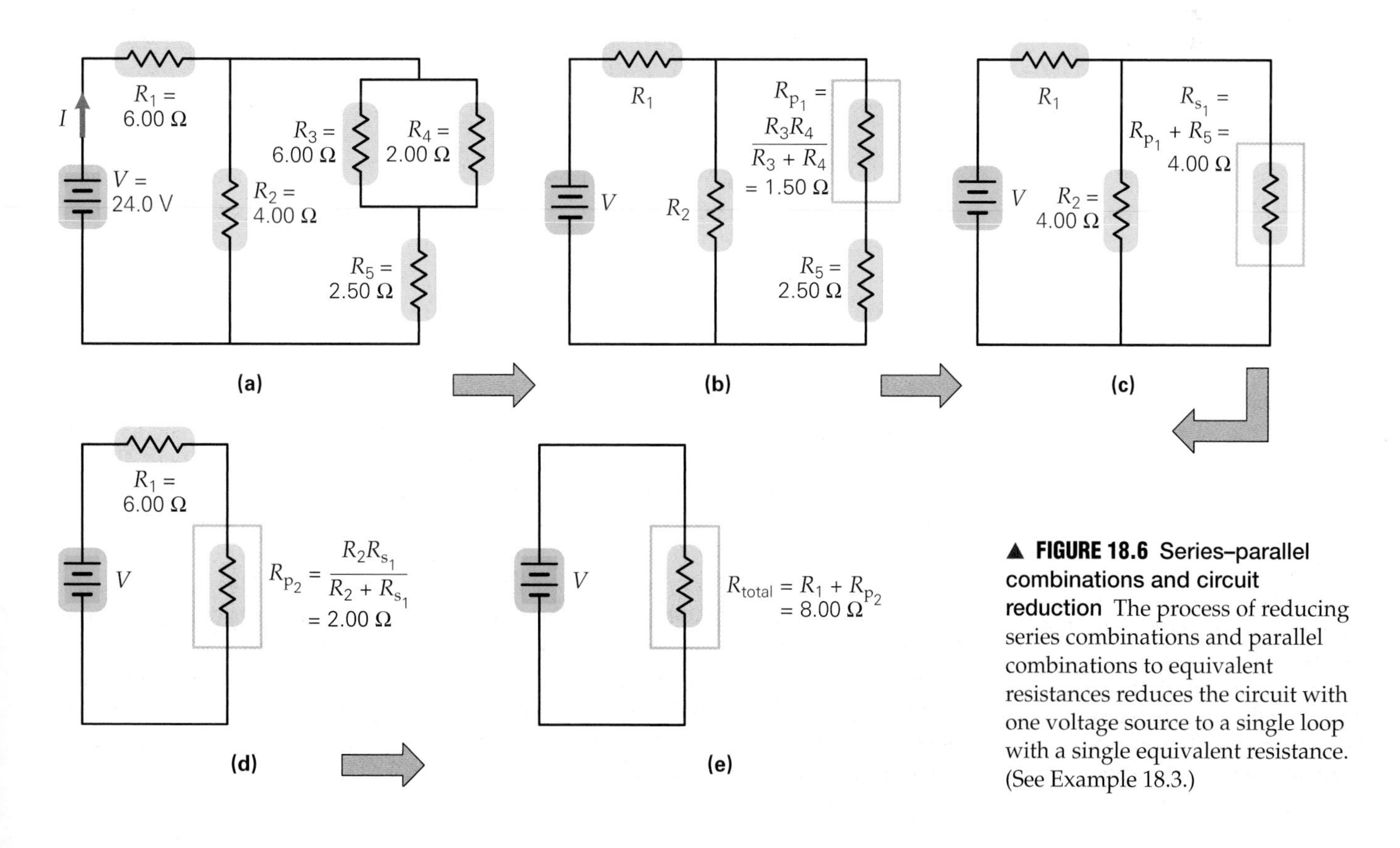

▲ **FIGURE 18.6 Series–parallel combinations and circuit reduction** The process of reducing series combinations and parallel combinations to equivalent resistances reduces the circuit with one voltage source to a single loop with a single equivalent resistance. (See Example 18.3.)

2. Reduce the circuit further by treating the separate equivalent resistances (from Step 1) as individual resistors. Proceed until you get to a single loop with one total (overall or equivalent) resistance value.
3. Find the current delivered to the reduced circuit using $I = V/R_{\text{total}}$.
4. Expand the reduced circuit back to the actual circuit by reversing the reduction steps, one at a time. Use the current of the reduced circuit to find the currents and voltages for the resistors in each step.

To see this procedure in use, consider the next Example.

Example 18.3 ■ Series–Parallel Combination of Resistors: Same Voltage or Same Current?

What are the voltages across and the currents in each of the resistors R_1 through R_5 in Fig. 18.6a?

Thinking It Through. Applying the steps described previously, it is important to identify parallel and series combinations before starting. It should be clear that R_3 is in parallel with R_4 (written $R_3 \| R_4$). This parallel combination is itself in series with R_5. Furthermore, the $(R_3 \| R_4) + R_5$ leg is in parallel with R_2. Lastly, this parallel combination is in series with R_1. Combining the resistors step by step should enable us to determine the total equivalent circuit resistance (Step 2). From that value, the total current can be calculated. Then, working backward, we can find the current in, and voltage across, each resistor.

Solution. To avoid rounding errors, the results will be carried to three significant figures.

Given: Values in Fig. 18.6a *Find:* Current and voltage for each resistor (Fig. 18.6a)

The parallel combination at the right-hand side of the circuit diagram can be reduced to the equivalent resistance R_{p_1} (see Fig. 18.6b), using Eq. 18.3:

$$\frac{1}{R_{p_1}} = \frac{1}{R_3} + \frac{1}{R_4} = \frac{1}{6.00\ \Omega} + \frac{1}{2.00\ \Omega} = \frac{4}{6.00\ \Omega}$$

This expression is equivalent to

$$R_{p_1} = 1.50\ \Omega$$

This operation leaves a series combination of R_{p_1} and R_5 along that side, which is reduced to R_{s_1}, using Eq. 18.2 (Fig. 18.6c):

$$R_{s_1} = R_{p_1} + R_5 = 1.50\ \Omega + 2.50\ \Omega = 4.00\ \Omega$$

Then, R_2 and R_{s_1} are in parallel and can be reduced (again using Eq. 18.3) to R_{p_2} (Fig. 18.6d):

$$\frac{1}{R_{p_2}} = \frac{1}{R_2} + \frac{1}{R_{s_1}} = \frac{1}{4.00\ \Omega} + \frac{1}{4.00\ \Omega} = \frac{2}{4.00\ \Omega}$$

This expression is equivalent to

$$R_{p_2} = 2.00\ \Omega$$

This operation leaves two resistances (R_1 and R_{p_2}) in series. These resistances combine to give the total equivalent resistance (R_{total}) of the circuit (Fig. 18.6e):

$$R_{\text{total}} = R_1 + R_{p_2} = 6.00\ \Omega + 2.00\ \Omega = 8.00\ \Omega$$

Thus, the battery delivers a current of

$$I = \frac{V}{R_{\text{total}}} = \frac{24.0\ \text{V}}{8.00\ \Omega} = 3.00\ \text{A}$$

Now let's work backward and "rebuild" the actual circuit. Note that the battery current is the same as the current through R_1 and R_{p_2}, because they are all in series. (In Fig. 18.6d, $I = I_1 = 3.00$ A and $I = I_{p_2} = 3.00$ A.) Therefore, the voltages across these resistors are

$$V_1 = I_1 R_1 = (3.00\ \text{A})(6.00\ \Omega) = 18.0\ \text{V}$$

and

$$V_{p_2} = I_{p_2} R_{p_2} = (3.00\ \text{A})(2.00\ \Omega) = 6.00\ \text{V}$$

Because R_{p_2} is made up of R_2 and R_{s_1} (Fig. 18.6c and d), there must be a 6.00-V drop across both of these resistors. We can use this to calculate the current in each.

$$I_2 = \frac{V_2}{R_2} = \frac{6.00\ \text{V}}{4.00\ \Omega} = 1.50\ \text{A} \quad \text{and} \quad I_{s_1} = \frac{V_{s_1}}{R_{s_1}} = \frac{6.00\ \text{V}}{4.00\ \Omega} = 1.50\ \text{A}$$

Next, notice that I_{s_1} is also the current in R_{p_1} and R_5, because they are in series. (In Fig. 18.6b, $I_{s_1} = I_{p_1} = I_5 = 1.50$ A.)

The resistors' individual voltages are therefore

$$V_{P_1} = I_{s_1} R_{P_1} = (1.50\ \text{A})(1.50\ \Omega) = 2.25\ \text{V}$$

and

$$V_5 = I_{s_1} R_5 = (1.50\ \text{A})(2.50\ \Omega) = 3.75\ \text{V}$$

respectively. (As a check, note that the voltages do, in fact, add to 6.00 V.)

Finally, the voltage across R_3 and R_4 is the same as V_{P_1} (why?), and

$$V_{P_1} = V_3 = V_4 = 2.25\ \text{V}$$

With these voltages and known resistances, the last two currents, I_3 and I_4, are

$$I_3 = \frac{V_3}{R_3} = \frac{2.25\ \text{V}}{6.00\ \Omega} = 0.38\ \text{A}$$

and

$$I_4 = \frac{V_4}{R_4} = \frac{2.25\ \text{V}}{2.00\ \Omega} = 1.13\ \text{A}$$

The current (I_{s_1}) is expected to divide at the R_3–R_4 junction. Thus, a check is available: $I_3 + I_4$ does equal I_{s_1}, within rounding errors.

Follow-Up Exercise. In this Example, verify that the total power delivered to all of the resistors is the same as the power output of the battery.

18.2 Multiloop Circuits and Kirchhoff's Rules

OBJECTIVES: To (a) understand the physical principles that underlie Kirchhoff's circuit rules, and (b) apply these rules in the analysis of actual circuits.

Series–parallel circuits with a single voltage source can always be reduced to a single loop, as we have seen in Example 18.3. However, circuits may contain several loops, each one having several voltage sources, resistances, or both. In many cases, resistors may not be connected either in series or in parallel. A multiloop circuit, which does not lend itself to the methods of Section 18.1, is shown in ▾Fig. 18.7a. Even though some groups of resistors may be replaced by their equivalent resistances (Fig. 18.7b), this circuit can be reduced only so far by using parallel and series procedures.

Analyzing these types of circuits requires a more general approach—that is, the application of **Kirchhoff's rules.*** These embody conservation of charge and energy. (Although not stated specifically, Kirchhoff's rules were applied to the parallel and series arrangements in Section 18.1.) First, it is useful to introduce some terminology that will help us describe more complex circuits:

- A point where three or more wires are joined is called a **junction**, or **node**—for example, point A in Fig. 18.7b.
- A path connecting two junctions is called a **branch**. A branch may contain one or more circuit elements and there may be more than two branches between two junctions.

Note: Kirchhoff's rules were developed by the German physicist Gustav Kirchhoff (1824–1887).

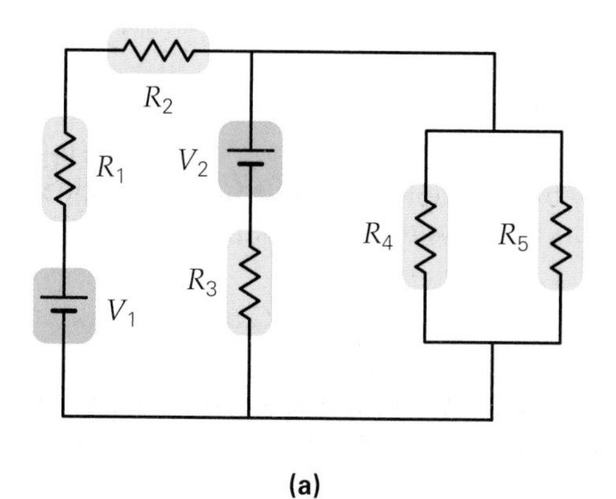

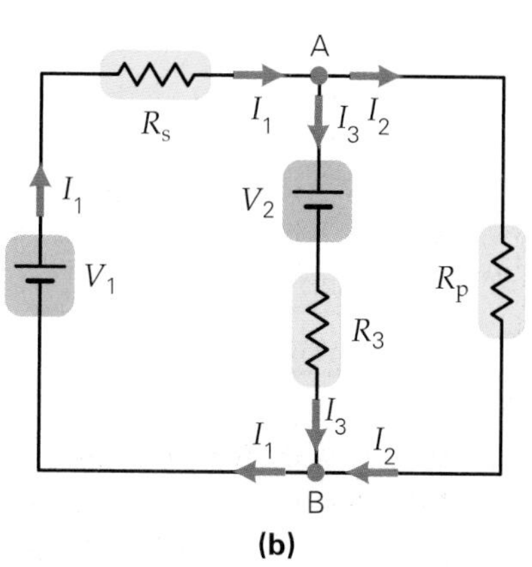

◂ **FIGURE 18.7 Multiloop circuit** In general, a circuit that contains voltage sources in more than one loop cannot be completely reduced by series and parallel methods alone. However, some reductions within each loop may be possible, such as from part **(a)** to part **(b)**. At a junction the current divides or comes together, as at junctions A and B in part (b), respectively. Any path between two junctions is called a *branch*. In part (b), there are three branches—that is, there are three different ways to get from junction A to junction B.

*Gustav Robert Kirchhoff (1824–1887) was a German scientist who made important contributions to electrical circuit theory and light spectroscopy. He invented the spectroscope, a device that separated light in its colors and studied the emitted light "signature" of various elements (see Chapter 27).

Kirchhoff's Junction Theorem

Kirchhoff's first rule, or **junction theorem**, states that the algebraic sum of the currents at any junction is zero:

$$\Sigma I_i = 0 \quad \textit{sum of currents at a junction} \tag{18.4}$$

This means that the sum of the currents entering at a junction (taken as positive) and the currents leaving the junction (taken as negative) is zero. This rule is just a statement of charge conservation—no charge can pile up at a junction (why?). For the junction at point A in Fig. 18.7b, for example, the algebraic sum of the currents is $I_1 - I_2 - I_3 = 0$; equivalently,

$$I_1 = I_2 + I_3$$
$$\textit{current in} = \textit{current out}$$

(This rule was applied in analyzing parallel resistances in Section 18.1.)

Illustration 30.7 The Loop Rule

Exploration 30.1 Circuit Analysis

Problem-Solving Hint

Sometimes it is not evident whether a particular current is directed into or out of a junction just by looking at a circuit diagram. In this case, a direction is simply *assumed*. Then the currents are calculated, without worry about their directions. If some of the assumptions turn out to be opposite to the actual directions, then negative answers for these currents will result. This outcome means that the directions of these currents are opposite to the directions initially chosen (or guessed).

Kirchhoff's Loop Theorem

Kirchhoff's second rule, or **loop theorem**, states that the algebraic sum of the potential differences (voltages) across all of the elements of any *closed loop* is zero:

$$\Sigma V_i = 0 \quad \textit{sum of voltages around a closed loop} \tag{18.5}$$

This expression means that the sum of the voltage rises (an increase in potential) equals the sum of the voltage drops (a decrease in potential) around a closed loop, which must be true if energy is conserved. (This rule was used in analyzing series resistances in Section 18.1.)

Notice that traversing a circuit loop in different directions will yield either a voltage rise or a voltage drop across each circuit element. Thus, it is important to establish a sign convention for voltages. We will use the convention illustrated in Fig. 18.8. The voltage across a battery is taken as positive (a voltage rise) if it is traversed from the negative terminal to the positive (Fig. 18.8a). Thus voltage across a battery will be negative if it is traversed in the opposite direction, from the positive to negative. (Note that the direction of the *current* through the battery has *nothing* to do with the sign of the battery voltage. The sign of this voltage depends only on the direction we choose to cross the battery.)

The voltage across a resistor is taken to be negative (a decrease) if the resistor is traversed in the same direction as the assigned current, in essence going "downhill" potential-wise (Fig. 18.8b). Clearly the voltage will be positive if the resistor is traversed in the opposite direction (going counter to the current direction, gaining electrical potential). Used together, these sign conventions allow the summation of the voltages around a closed loop, regardless of the direction chosen to do that sum. It should be clear that Eq. 18.5 is the same in either case. To see this, note that reversing the chosen loop direction simply amounts to multiplying Eq. 18.5 (from the original direction) by -1. This operation, of course, does not change the equation.

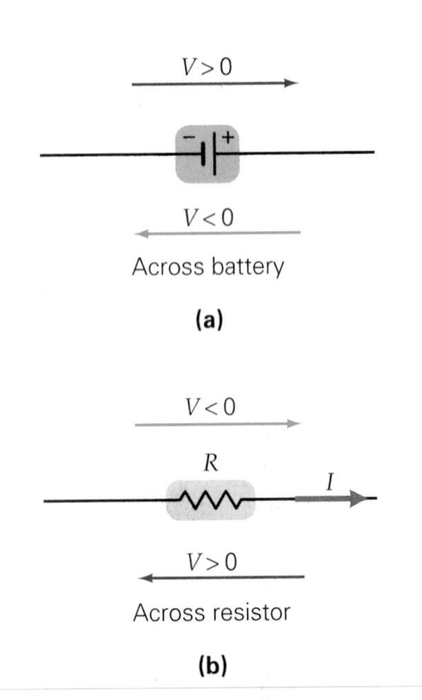

▲ **FIGURE 18.8 Sign convention for Kirchhoff's loop theorem (a)** The battery voltage is taken as positive if it is traversed from the negative to the positive terminal. It is assigned a negative value if traversed from the positive to the negative. **(b)** The voltage across a resistor is taken as negative if the resistor is traversed in the direction of the assigned current ("downstream"). It is taken as positive if the resistance is traversed opposite that of the assigned branch current ("upstream").

Problem-Solving Hint

In applying Kirchhoff's loop theorem, the sign of a voltage across a resistor is determined by the direction of the current in that resistor. However, there can be situations in which the current direction is not obvious. How do you handle the voltage signs in such cases? The answer is simple: After assuming a direction for the current, follow the voltage sign convention *based on this assumed direction*. This guarantees that the two sign conventions are mathematically consistent. Thus, if it turns out that the actual current direction is opposite your choice, the voltage drops will automatically reflect that.

A graphical interpretation of Kirchhoff's loop theorem is presented in the Learn by Drawing future on page 602. Integrated Example 18.4, in which a simple parallel circuit is reexamined using Kirchhoff's rules, shows that our previous series–parallel considerations were consistent with these circuit rules. In this Example, take care to notice how important it is draw a correct circuit diagram—it can guide you as to how to proceed.

Integrated Example 18.4 ■ A Simple Circuit: Using Kirchhoff's Rules

Two resistors R_1 and R_2 are connected in parallel. This combination is in series with a third resistor R_3 that has the largest resistance of the three. A battery completes the circuit, with one electrode connected to the beginning and the other to the end of this network. (a) Which resistor will carry the most current, (1) R_1, (2) R_2, or (3) R_3? Explain. (b) In the actual circuit, assume $R_1 = 6.0\ \Omega$, $R_2 = 3.0\ \Omega$, $R_3 = 10.0\ \Omega$, and the battery's terminal voltage is 12.0 V. Apply Kirchhoff's rules to determine the current in each resistor and the voltage across each resistor.

(a) Conceptual Reasoning. It is best to first look at a schematic circuit diagram based on the word description of the network (◂Fig. 18.9). You might think that the resistor with the least resistance would carry the most current. But be careful; this holds only if *all* the resistors are in parallel. This is *not* true here. The two parallel resistors each carry only a portion of the total current. However, because the total of their two currents is in R_3, that resistor carries the total, and therefore the most, current. Thus, the correct answer is (3).

(b) Quantitative Reasoning and Solution.

Given: $R_1 = 6.0\ \Omega$, $R_2 = 3.0\ \Omega$, $R_3 = 10.0\ \Omega$, $V = 12.0$ V

Find: the current in each resistor and the voltage across each resistor

There are three unknown currents: the total current (I) and the currents in each of the parallel resistors (labeled as I_1 and I_2). Since there is only one battery, the current must be clockwise (shown in the figure). Applying Kirchhoff's junction theorem to the first junction (J in Fig.18.9a), we have

$$\Sigma I_i = 0 \quad \text{or} \quad I - I_1 - I_2 = 0 \qquad (1)$$

Using the loop theorem in the clockwise direction in Fig. 18.9b, we cross the battery from the negative to the positive terminal and then traverse R_1 and R_3 to complete the loop. The resulting equation (showing the voltage signs explicitly) is

$$\Sigma V_i = 0 \quad \text{or} \quad +V + (-I_1R_1) + (-IR_3) = 0 \qquad (2)$$

A third equation can be obtained by applying the loop theorem but this time going through R_2 instead of R_1 (Fig. 18.9c). This yields

$$\Sigma V_i = 0 \quad \text{or} \quad +V + (-I_2R_2) + (-IR_3) = 0 \qquad (3)$$

Putting in the battery voltage (in volts) and resistances (in ohms) and rearranging:

$$I = I_1 + I_2 \qquad (1a)$$

$$12 - 6I_1 - 10I = 0 \quad \text{or} \quad 6 - 3I_1 - 5I = 0 \qquad (2a)$$

$$12 - 3I_2 - 10I = 0 \qquad (3a)$$

Adding Eqs. (2a) and (3a) yields $18 - 3(I_1 + I_2) - 15I = 0$. However, from Eq. (1a), $I = I_1 + I_2$. Therefore, this equation becomes

$$18 - 3I - 15I = 0 \quad \text{or} \quad 18I = 18$$

and solving for the total current yields $I = 1.00$ A.

Eqs. (3a) and (1a) can then be solved for the remaining currents:

$$I_2 = \tfrac{2}{3}\text{A} \quad \text{and} \quad I_1 = \tfrac{1}{3}\text{A}$$

These answers are consistent with our circuit-diagram reasoning in part (a).

Because the currents are now known, the voltages can be obtained from Ohm's law, $V = IR$. Thus,

$$V_1 = I_1R_1 = \left(\tfrac{1}{3}\text{A}\right)(6.0\ \Omega) = 2.0\text{ V}$$

$$V_2 = I_2R_2 = \left(\tfrac{2}{3}\text{A}\right)(3.0\ \Omega) = 2.0\text{ V}$$

$$V_3 = I_3R_3 = (1.0\text{ A})(10.0\ \Omega) = 10.0\text{ V}$$

Take a quick look at the results to see if they are reasonable. As expected, the voltage drops across the parallel resistors are equal. Because of that, two thirds of the total current is in the resistor with the least resistance. Also, the total voltage across the network is 12.0 V, as it must be.

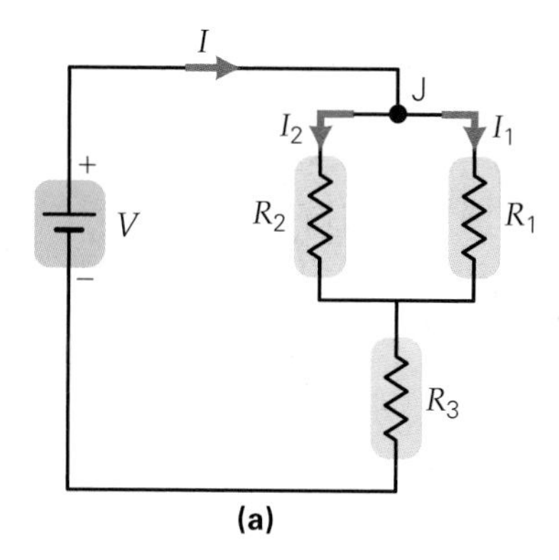

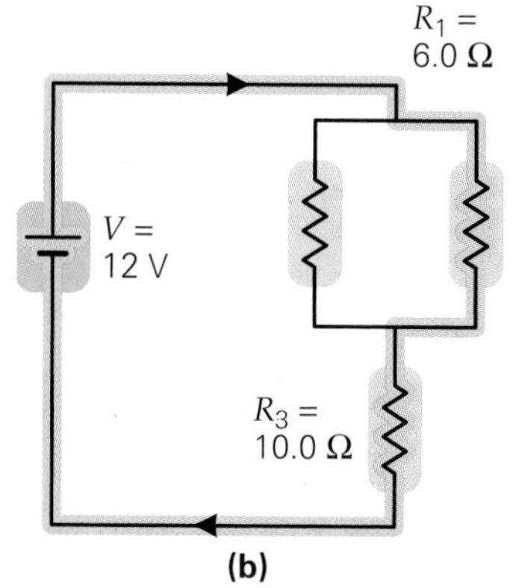

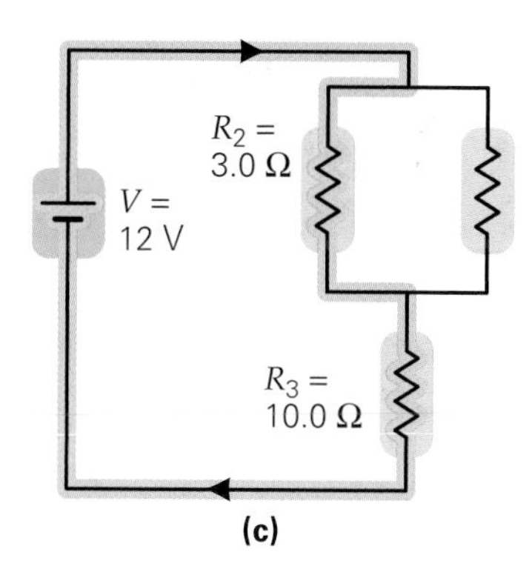

▲ **FIGURE 18.9** Sketching circuit diagrams and Kirchhoff's rules **(a)** The circuit diagram resulting from the written description in Integrated Example 18.4. **(b)** and **(c)** The two loops used in the analysis of Integrated Example 18.4.

Exploration 30.2 Lightbulbs

(continues on page 603)

LEARN BY DRAWING KIRCHHOFF PLOTS: A GRAPHICAL INTERPRETATION OF KIRCHHOFF'S LOOP THEOREM

The equation form of Kirchhoff's loop theorem has a geometrical visualization that may help you develop better insight into its meaning. This graphical approach allows us to visualize the potential changes in a circuit, either to anticipate the results of mathematical analysis or to qualitatively confirm the results. (Don't forget that a complete analysis usually also includes the junction theorem—see Example 18.5.)

The idea is to make a three-dimensional plot based on the circuit diagram. The wires and elements of the circuit form the basis for the x–y plane, or the diagram's "floor." Plotted perpendicularly to this plane, along the z-axis, is the electric potential (V), with an appropriate choice for zero. Such a diagram is called a *Kirchhoff plot* (Fig. 1).

The rules for constructing a Kirchhoff plot are simple: Start at a known potential value, and go around a complete loop, finishing where you started. Because you come back to the same location, the sum of all the rises in potential (positive voltages) must be balanced by the sum of the drops (negative voltages). This requirement is the geometrical expression of energy conservation, embodied mathematically by Kirchhoff's loop theorem.

Thus, if the potential increases (say, in traversing a battery from negative terminal to positive terminal), draw a rise in the z-direction. In this instance, the rise represents the terminal voltage of the battery. Similarly, if the potential decreases (for example, in traversing a resistor in the direction of the current), make sure the potential drops. If possible, try to draw the rises and drops (the voltages) to scale. That is, if there is a large rise in potential (such as across a high-voltage battery), then draw that rise to be large in proportion to the others on the diagram.

For elaborate circuits, this graphical method may prove to be too complicated for practical use. Nevertheless, it is always good to keep this concept in mind, as it illustrates the fundamental physics behind the loop theorem.

As an example of the power of this method, consider the circuit in Fig. 1: a battery with internal resistance r wired to a single external resistor R. The direction of the current is from the anode to the cathode through the external resistor. The potential of the battery's cathode is chosen as zero. Starting there and traversing the circuit in the direction of the current, there is a rise in potential from the battery cathode to the anode. From there the potential remains constant as the current travels through the wires to the external resistor. That is, no significant voltage *drop* should be indicated along connecting wires (why?).

At the resistor, there must be a drop in potential. However, it must not drop to zero, because there must be some voltage left to produce current through the internal resistance. Thus it can be reasoned visually why the terminal voltage of the battery, V, must be less than its emf (the rise between a and b).

Figure 2 shows two resistors in series, and that combination in parallel with a third resistor. For simplicity, all three resistors have the same resistance (R) and the battery's internal resistance is assumed to be zero. Starting at point a, there is a rise in potential corresponding to the battery voltage. Then, as the loop is traced, it goes through the single resistor, so there must be one drop in potential equal in magnitude to $\mathscr{E}$.

Following the loop that includes the two resistors, each must have half the total drop (why?). So each will carry half the current of the single resistor. Recall that in parallel circuits, the largest resistance carries the least current. Notice how nicely the geometrical approach helps develop your intuition and allows anticipation of quantitative results.

As an exercise, try redrawing Fig. 2 if, instead, the series resistors had resistance values of R and $2R$. Which of these two now has the largest voltage? How do the currents in the resistors compare to the previous situation? Lastly, analyze the circuit mathematically, to see whether your expectations are confirmed.

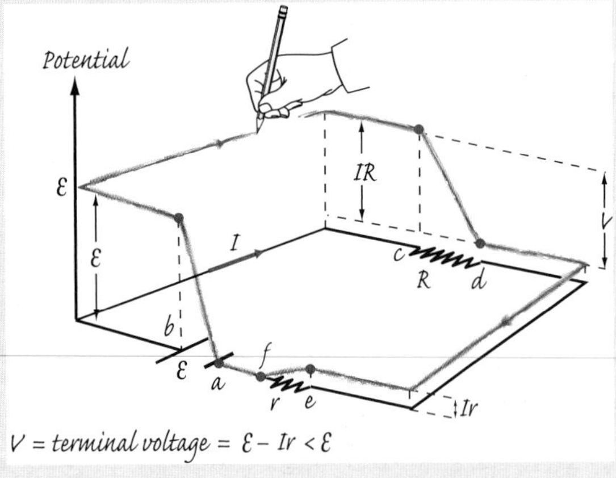

FIGURE 1 Kirchhoff plots: A graphical problem-solving strategy The schematic of the circuit is laid out in the x–y plane, and the electric potential is plotted perpendicularly along the z-axis. Usually, the zero of the potential is taken to be the negative terminal of the battery. A direction for current is assigned, and the value of the potential is plotted around the circuit, following the rules for gains and losses. This particular plot shows a rise in potential when the battery is traversed from cathode to anode, followed by a drop in potential across the external resistor, and a smaller drop in potential across the battery's internal resistance.

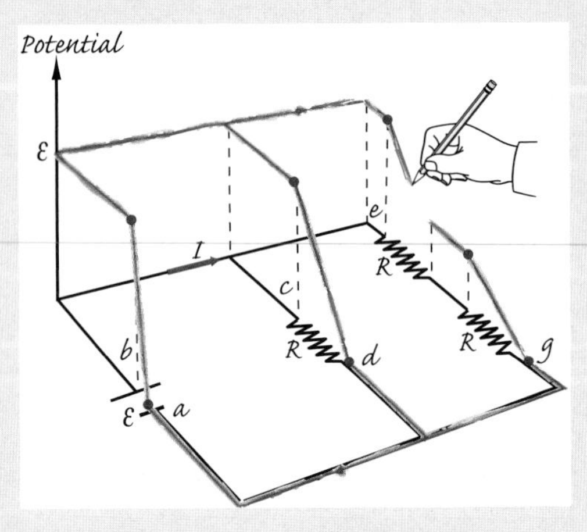

FIGURE 2 Kirchhoff plot of a more complex circuit Imagine how the plot would change if you were to vary the values of the three resistors. Then analyze the circuit mathematically to see whether your plot allowed you to anticipate the voltages and currents.

A special note before leaving this Example: We know that the answers must be in amperes and volts because we used amps, volts, and ohms were used consistently throughout. If you stay within this system (that is, express quantities in volts, amps and ohms), you don't *need* to carry units; the answers will automatically be in these units. (Of course, it is always a good idea to check your units if there is a question.)

Follow-Up Exercise. (a) In this Example, predict what will happen to each of the currents if R_2 is increased. Explain your reasoning. (b) Rework part (b) of this Example, changing R_2 to 8.0 Ω, and see if your reasoning is correct.

Application of Kirchhoff's Rules

Integrated Example 18.4 could have been worked using the expressions for equivalent resistances. However, more complicated, multiloop circuits (which may have resistors neither in parallel or series) require a more structured approach. In this book, the following general steps will be used when applying Kirchhoff's rules:

1. Assign a current and direction of current for each branch in the circuit. This assignment is done most conveniently at junctions.
2. Indicate the loops and the directions in which they are to be traversed (▸Fig. 18.10). Every branch *must* be in at least one loop.
3. Apply Kirchhoff's first rule (junction rule) at each junction that gives a unique equation. (This step gives a set of equations that includes *all* currents, but you may have redundant equations from two different junctions.)
4. Traverse the number of loops necessary to include all branches. In traversing a loop, apply Kirchhoff's second rule, the loop theorem (using $V = IR$ for each resistor), and write the equations, using our sign conventions.

If this procedure is applied properly, Steps 3 and 4 give a set of N equations if there are N unknown currents. These equations may then be solved for the currents. If more loops are traversed than necessary, you might also have redundant loop equations. Only the number of loops that includes each branch *once* is needed.

This procedure may seem complicated, but it's generally straightforward, as the following Example shows.

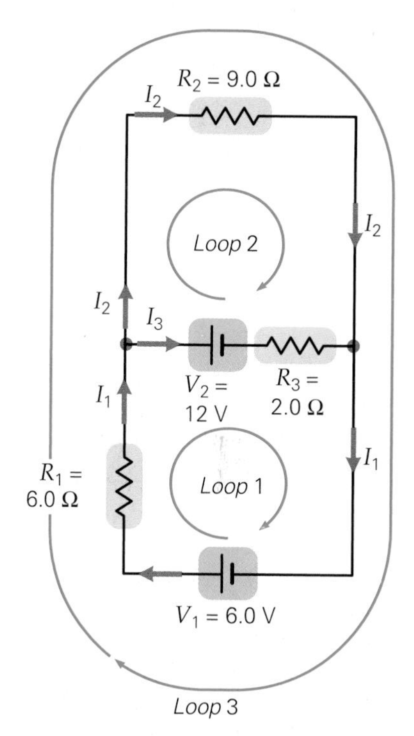

▲ **FIGURE 18.10 Application of Kirchhoff's rules** To analyze a circuit such as the one shown for Example 18.5, assign a current and its direction for each branch in the circuit (most conveniently done at junctions). Identify each loop and the direction of traversal. Then write current equations for each independent junction (using Kirchhoff's junction theorem). Also write voltage equations for as many loops as needed to include every branch (using Kirchhoff's loop theorem). Be careful to observe sign conventions.

Example 18.5 ■ Branch Currents: Using Kirchhoff's Rules

For the circuit diagrammed in Fig. 18.10, find the current in each branch.

Thinking It Through. Series or parallel calculations cannot be used here. (Why?) Instead, the solution is begun by assigning current directions ("best guesses") in each loop, and then the junction theorem and the loop theorem are used (twice—once for each inner loop) to generate three equations, because there are three currents.

Solution.

Given: Values in Fig. 18.10 *Find:* The current in each of the three branches

The chosen current directions and loop traversal directions are shown in the figure. (Remember, these directions are not unique; choose them, work the problem, and check the final current signs to see if your choices were correct.) There is a current in every branch, and every branch is in at least one loop. (Some branches are in more than one loop, which is acceptable.)

Applying Kirchhoff's first rule at the left-hand junction gives

$$I_1 - I_2 - I_3 = 0$$

or, after rearranging,

$$I_1 = I_2 + I_3 \qquad (1)$$

(For the other junction, $I_2 + I_3 - I_1 = 0$ but this is equivalent to Eq. (1), so we are done with the junctions.)

Going around loop 1 as in Fig. 18.10 and applying Kirchhoff's loop theorem with the sign conventions gives

$$\Sigma V_i = +V_1 + (-I_1R_1) + (-V_2) + (-I_3R_3) = 0 \qquad (2)$$

Putting in the numerical values gives

$$+6 - 6I_1 - 12 - 2I_3 = 0$$

(continues on next page)

Rearranging this equation and dividing both sides by 2, we have

$$3I_1 + I_3 = -3$$

For convenience, units are omitted (they are all in amps and ohms, and thus are self-consistent).

For loop 2, the loop theorem yields

$$\Sigma V_i = +V_2 + (-I_2R_2) + (+I_3R_3) = 0 \quad (3)$$

Again, after substituting in the values and rearranging, we have

$$9I_2 - 2I_3 = 12 \quad (3a)$$

Equations (1), (2a), and (3a) form a set of three equations with three unknowns. You can solve for the currents in many ways. For example, substitute Eq. (1) into Eq. (2a) to eliminate I_1:

$$3(I_2 + I_3) + I_3 = -3$$

which, after rearranging and dividing by 3, simplifies to

$$I_2 = -1 - \tfrac{4}{3}I_3 \quad (4)$$

Then, substituting Eq. (4) into Eq. (3a) eliminates I_2:

$$9\left(-1 - \tfrac{4}{3}I_3\right) - 2I_3 = 12$$

Finishing the algebra and solving for I_3, we have

$$-14I_3 = 21 \quad \text{or} \quad I_3 = -1.5\ \text{A}$$

The minus sign tells us that the wrong direction was assumed for I_3.

Putting the value of I_3 into Eq. (4) gives I_2:

$$I_2 = -1 - \tfrac{4}{3}(-1.5\ \text{A}) = 1.0\ \text{A}$$

Then, from Eq. (1),

$$I_1 = I_2 + I_3 = 1.0\ \text{A} - 1.5\ \text{A} = -0.5\ \text{A}$$

Again, the minus sign indicates that the direction of I_1 was also initially chosen incorrectly.

Note that this analysis did not have to use loop 3. The equation for this loop would be redundant, containing no new information (can you show this?).

Follow-Up Exercise. Rework this Example, using the junction theorem and loops 3 and 1 instead of loops 1 and 2.

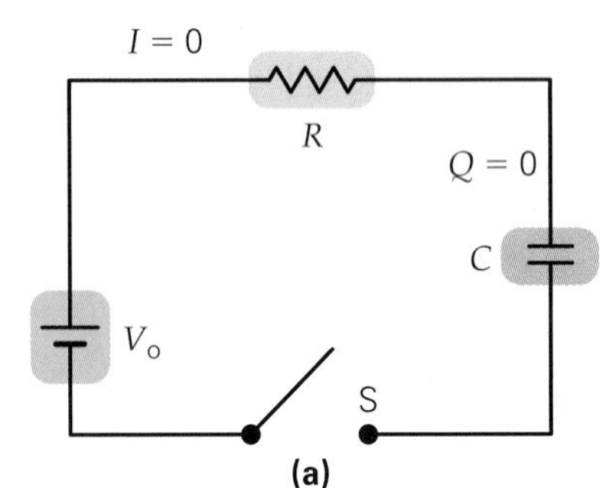

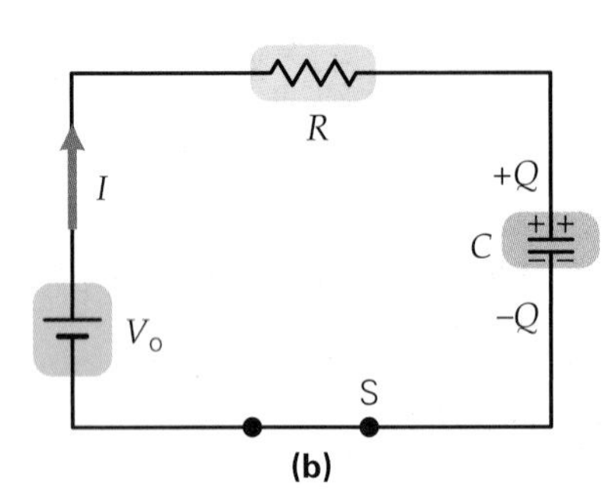

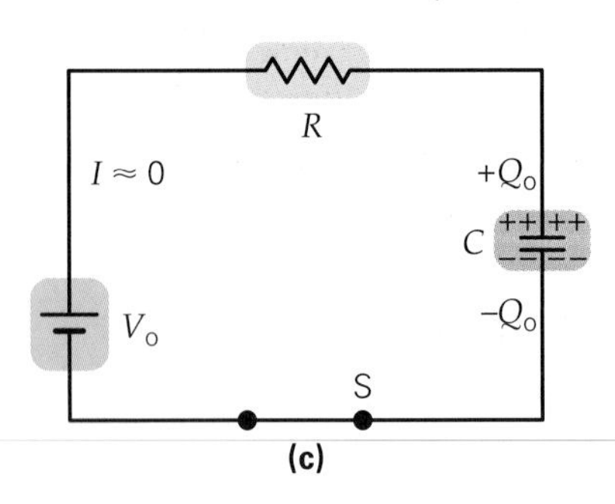

▲ **FIGURE 18.11 Charging a capacitor in a series RC circuit** **(a)** Initially there is no current and no charge on the capacitor. **(b)** When the switch is closed, there is a current in the circuit until the capacitor is charged to its maximum value. The rate of charging depends on the circuit's time constant, τ (= RC). **(c)** For times much larger than τ, the current is very close to zero, and the capacitor is said to be fully charged.

18.3 RC Circuits

OBJECTIVES: To (a) understand the charging and discharging of a capacitor through a resistor, and (b) calculate the current and voltage at specific times during these processes.

Until now, only circuits that have constant currents have been considered. In some direct-current (dc) circuits, the current can *vary with time* while maintaining a constant direction (it still is "dc"). Such is the case in **RC circuits**, which most generally consist of several resistors and capacitors.

Charging a Capacitor through a Resistor

The charging of an uncharged capacitor by a battery is depicted in ◂Fig. 18.11. After the switch is closed, even though there is a gap (the capacitor plates), charge *must* flow while the capacitor is charging.

The maximum charge (Q_o) that the capacitor can attain depends on its capacitance (C) and the battery voltage (V_o). To determine the value of Q_o and understand how both the current and capacitor charge vary with time, consider the following argument. At $t = 0$, there is no charge on the capacitor and thus no voltage across it. By Kirchhoff's loop theorem, this means that the full battery voltage must appear across the resistor, resulting in an initial (maximum) current $I_o = V_o/R$. As charge on the capacitor increases, so must the voltage across its plates, thereby reducing the resistor's voltage and current. Eventually, when the capacitor is charged to its maximum, the current becomes zero. At this time, the resistor's voltage is zero and the capacitor's voltage must be V_o. Because of the relationship between the charge on a capacitor and its voltage (Ch. 16, Eq 16.19), the maximum capacitor charge is given by $Q_o = CV_o$. (This sequence is depicted in Fig. 18.11.)

The resistance is one of two factors that determines how fast the capacitor is charged, because the larger its value, the greater the resistance to charge flow. The capacitance is the other factor that influences the charging speed—it simply takes longer to charge a larger capacitor. Analysis of this type of circuit requires mathematics beyond the level of this book. However, it can be shown that the voltage across the capacitor increases exponentially with time according to

$$V_C = V_o[1 - e^{-t/(RC)}] \quad \textit{(charging capacitor voltage in an RC circuit)} \tag{18.6}$$

where e has an approximate value of 2.718. (Recall that the irrational number e is the base of the system of *natural logarithms*.) A graph of V_C versus t is shown in ▸Fig. 18.12a. As expected, V_C approaches V_o, the capacitor's maximum voltage, after a "long" time.

A graph of I versus t is given in Fig. 18.12b. The current varies with time according to

$$I = I_o e^{-t/(RC)} \tag{18.7}$$

Thus the current decreases exponentially with time and has its largest value initially, as expected.

According to Eq. 18.6, it would take an infinite time for the capacitor to become fully charged. However, in practice, we know that most capacitors become essentially completely charged in relatively short times. It is therefore customary to use a special value to express the "charging time." This value, called the **time constant (τ)**, is

$$\tau = RC \quad \textit{time constant for RC circuits} \tag{18.8}$$

(You should be able to show that RC has units of seconds.) After an elapsed time of one time constant, that is $t = \tau = RC$, the voltage across the charging capacitor has risen to 63% of the maximum possible. This can be seen by evaluating V_C (Eq. 18.6), replacing t with τ $(= RC)$:

$$V_C = V_o(1 - e^{-\tau/\tau}) = V_o(1 - e^{-1})$$
$$\approx V_o\left(1 - \frac{1}{2.718}\right) = 0.63V_o$$

Because $Q \propto V_C$, this result means that the capacitor has 63% of its maximum possible charge after one time constant has elapsed. You should be able to show that after one time constant, the current has dropped to 37% of its initial (maximum) value, I_o.

After a time of two time constants has elapsed ($t = 2\tau = 2RC$), you should be able to show that the capacitor is charged to more than 86% of its maximum value; at $t = 3\tau = 3RC$, the capacitor is charged to 95% of its maximum value; and so on. As a general rule of thumb, a capacitor is considered to be "fully charged" after "several time constants" have elapsed.

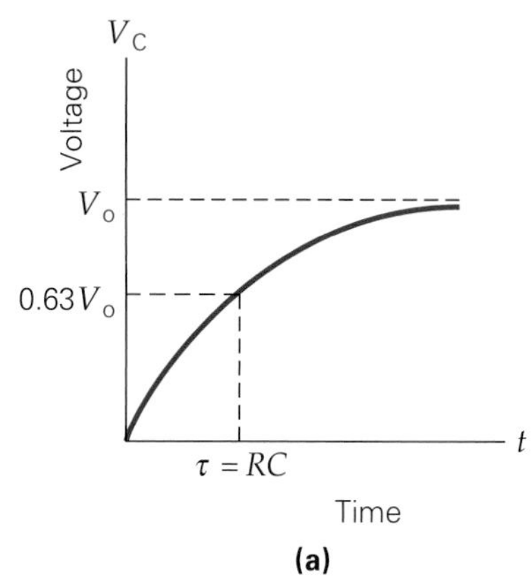

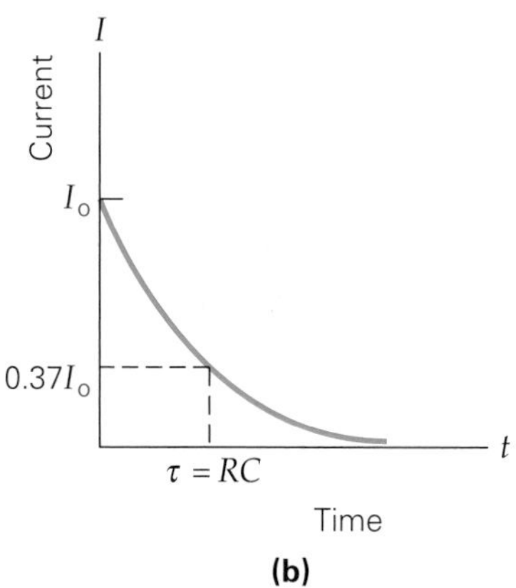

▲ **FIGURE 18.12 Capacitor charging in a series RC circuit**
(a) In a series RC circuit, as the capacitor charges, the voltage across it increases nonlinearly, reaching 63% of its maximum voltage (V_o) in one time constant, τ. **(b)** The current in this circuit is initially a maximum ($I_o = V_o/R$) and decays exponentially, falling to 37% of its initial value in one time constant, τ.

PHYSLET® Illustration 30.6 RC Circuit

PHYSLET® Exploration 30.6 RC Time Constant

Note: Most calculators now have an e^x button. For exponential calculations, practice using it. For example, make sure your calculator gives you $e^{-1} \approx 0.37$.

Discharging a Capacitor through a Resistor

▸Figure 18.13a shows a capacitor being *discharged* through a resistor. In this case, the voltage across the capacitor *decreases* exponentially with time, as does the current. The expression for the decay of the capacitor's voltage (from its maximum voltage of V_o) is

$$V_C = V_o e^{-t/(RC)} = V_o e^{-t/\tau} \quad \textit{(discharging capacitor voltage in an RC circuit)} \tag{18.9}$$

Then after one time constant, the capacitor voltage is at 37% of its original value (Fig. 18.13b). The current in the circuit decays exponentially also, following Eq. 18.7. This is also the behavior of a capacitor in a heart defibrillator as it discharges its stored energy (as a flow of charge or current) through the heart (resistance R) in a discharge time of about 0.1 sec. RC circuits are also an integral part of cardiac pacemakers, which alternately charge a capacitor, transfer the energy to the heart, and repeat this at a rate determined by the time constant. For details on these interesting and important instruments, refer to Insight 18.11 on Applications of RC Circuits to Cardiac Medicine on page 608. Aspects of RC circuits in medical settings are also covered in Exercises 107 and 108. As a practical application, consider their use in modern cameras in Example 18.6.

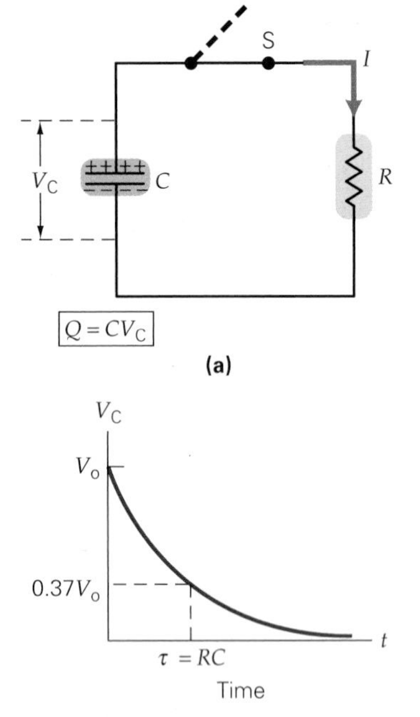

▲ **FIGURE 18.13** Capacitor discharging in a series RC circuit **(a)** The capacitor is initially fully charged. When the switch is closed, current appears in the circuit as the capacitor begins to discharge. **(b)** In this case, the voltage across the capacitor (and the current in the circuit) decays exponentially with time, falling to 37% of its initial value in one time constant, τ.

Example 18.6 ■ RC Circuits in Cameras: Flash Photography Is as Easy as Falling Off a Log(arithm)

In many cameras, the built-in flash gets its energy from that stored in a capacitor. The capacitor is charged using long-life batteries with voltages of typically 9.00 V. Once the bulb is fired, the capacitor must recharge quickly through an internal RC circuit. If the capacitor has a value of 0.100 F, what must the resistance be so the capacitor is charged to 80% of its maximum charge (the minimum charge to fire the bulb again) in 5.00 s?

Thinking It Through. After one time constant, the capacitor will be charged to 63% of its maximum voltage and charge. Because the capacitor needs 80%, the time constant must be *less* than 5.00 s. Eq. 18.6 can be used (along with a calculator) to determine the time constant. From that the required value of resistance can be determined.

Solution. The data given include the final voltage across the capacitor, V_C, which is 80% of the battery's voltage, which means that Q is 80% of the maximum charge.

Given: $C = 0.100$ F
$V_B = V_o = 9.00$ V
$V_C = 0.80V_o = 7.20$ V
$t = 5.00$ s

Find: R (the resistance required so that the capacitor is 80% charged in 5.00 s)

Putting the data into Eq. 18.6, $V_C = V_o(1 - e^{-t/\tau})$, we have

$$7.20 = 9.00(1 - e^{-5.00/\tau})$$

Rearranging this equation yields $e^{-5.00/\tau} = 0.20$, and the reciprocal of this expression (to make the exponent positive) is

$$e^{5.00/\tau} = 5.00$$

To solve for the time constant, recall that if $e^a = b$, then a is the *natural logarithm* (ln) of b. Thus, in our case, $5.00/\tau$ is the natural logarithm of 5.00. Using a calculator, we find that $\ln 5.00 = 1.61$. Therefore

$$\frac{5.00}{\tau} = \ln 5.00 = 1.61$$

or

$$\tau = RC = \frac{5.00}{1.61} = 3.11 \text{ s}$$

Solving for R yields

$$R = \frac{3.11 \text{ s}}{C} = \frac{3.11 \text{ s}}{0.10 \text{ F}} = 31 \ \Omega$$

As expected, the time constant is less than 5.0 s, because achieving 80% of the maximum voltage requires a time interval longer than one time constant.

Follow-Up Exercise. (a) In this Example, how does the energy stored in the capacitor (after 5.00 s) compare with the maximum energy storage? Explain why it isn't 80%. (b) If you waited 10.00 s to charge the capacitor, what would its voltage be? Why isn't it twice the voltage that exists across the capacitor after 5.00 s?

An application of an RC circuit is diagrammed in ▼Fig. 18.14a. This circuit is called a *blinker circuit* (or a *neon-tube relaxation oscillator*). The resistor and capacitor are initially wired in series, and then a miniature neon tube is connected in parallel with the capacitor.

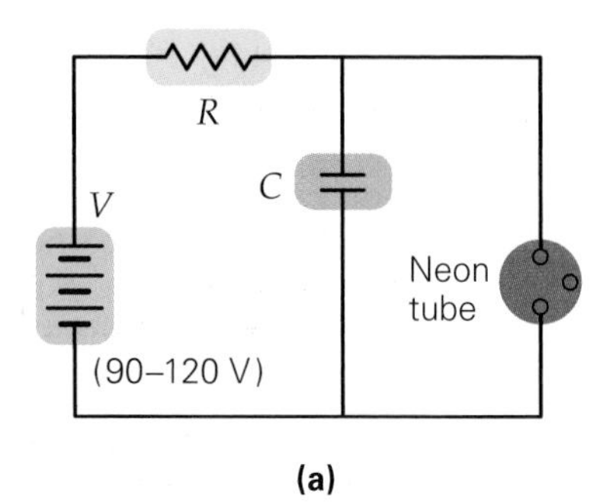

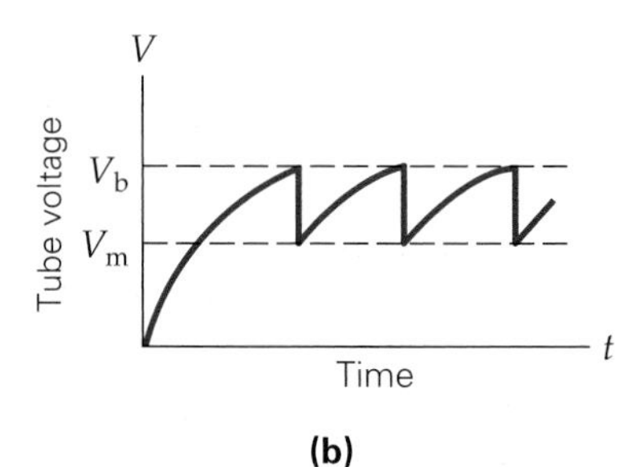

▶ **FIGURE 18.14** Blinker circuit **(a)** When a neon tube is connected across the capacitor in a series RC circuit that has the proper voltage source, the voltage across the tube will oscillate with time. As a result, the tube periodically flashes or blinks. **(b)** A graph of tube voltage versus time shows the voltage oscillating between V_b, the "breakdown" voltage, and V_m, the "maintaining" voltage. See text for detailed discussion.

When the circuit is closed, the voltage across the capacitor (and the neon tube) rises from 0 to V_b, which is the *breakdown voltage* of the neon gas in the tube (about 80 V). At that voltage, the gas becomes ionized (that is, electrons are freed from atoms, creating positive and negative charges that are free to move). Thus the gas begins to conduct electricity, and the tube lights. When the tube is in this conducting state, the capacitor discharges through it, and the voltage falls rapidly (Figure 18.14b). When the voltage drops below V_m, called the *maintaining voltage*, the ionization in the tube cannot be sustained, and the tube stops conducting. The capacitor begins charging again, the voltage rises from V_m to V_b, and the cycle repeats. This continual repetition causes the tube to blink on and off.

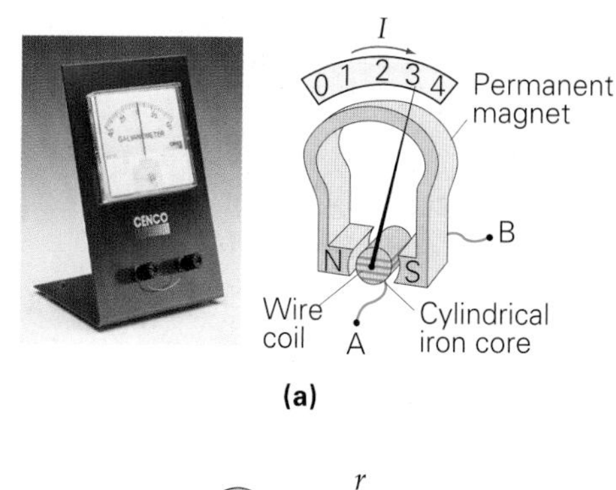

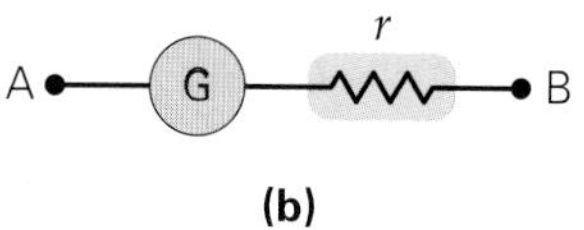

▲ **FIGURE 18.15 The galvanometer** **(a)** A galvanometer is a current-sensitive device whose needle deflection is proportional to the current in its coil. **(b)** The circuit symbol for a galvanometer is a circle containing a G. The internal resistance (r) of the meter is indicated explicitly as r.

18.4 Ammeters and Voltmeters

OBJECTIVES: To understand (a) how galvanometers are used to make ammeters and voltmeters, (b) how multirange versions of these devices are constructed, and (c) how they are connected to measure current and voltage in real circuits.

As the names imply, an **ammeter** measures current *through* circuit elements and a **voltmeter** measures voltages *across* circuit elements. The basic component common to both of these meters is a **galvanometer** (▸Fig. 18.15a). The galvanometer operates on magnetic principles covered in Chapter 19. In this chapter, it will be treated simply as a circuit element with an internal resistance r (typically about 50 Ω) whose needle deflection is proportional to the current in it (Fig. 18.15b).

The Ammeter

A galvanometer measures current, but because of its small resistance, only currents in the microampere range can be measured without burning out its wires. However, there is a way to construct an ammeter to measure larger currents with a galvanometer. To do this, a small *shunt resistor* (with a resistance of R_s) is employed in parallel with a galvanometer. The job of the shunt resistor (or "shunt" for short) is to take most of the current (▾Fig. 18.16). This requires the shunt to have much less resistance than the galvanometer ($R_s \ll r$). The following Example illustrates how the resistance of the shunt is determined in the design of an ammeter.

Note: Ammeters are connected in series with the element whose current they are measuring (Fig. 18.16b).

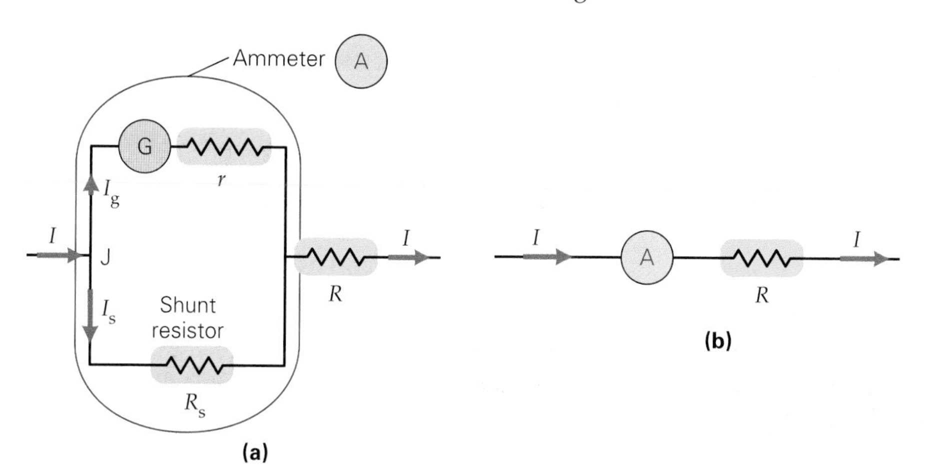

◂ **FIGURE 18.16 A dc ammeter** Here, R is the resistance of the resistor whose current is being measured. **(a)** A galvanometer in parallel with a shunt resistor (R_s) creates an ammeter capable of measuring various ranges of current, depending on the value of R_s. **(b)** The circuit symbol for an ammeter is a circle with an A inside it. (See Example 18.7 for a detailed discussion of ammeter design.)

Example 18.7 ■ Ammeter Design Using Kirchhoff's Rules: Choosing a Shunt Resistor

Suppose you have a galvanometer that can safely carry a maximum coil current of 200 μA (called its *full-scale sensitivity*) and that has a coil resistance of 50 Ω. It is to be used in an ammeter designed to measure currents up to 3.0 A (at full scale). What is the required shunt resistance? (See Fig. 18.16a).

Thinking It Through. The galvanometer itself can carry only a small current, so most of the current will have to be diverted, or "shunted," through the shunt. Thus, the shunt resistance

Exploration 30.4 *Galvanometers and Ammeters*

(continues on next page)

will have to be much less than the galvanometer's internal resistance. Because the shunt and coil resistance are really two resistors in parallel, they have the same voltage across them. This reasoning along with Kirchhoff's laws should enable us to determine the value of R_s.

Solution. Listing the data, we have

Given: $I_g = 200\ \mu A = 2.00 \times 10^{-4}$ A   *Find:* R_s (shunt resistance)
$r = 50\ \Omega$
$I_{max} = 3.0$ A

Because the voltages across the galvanometer and the shunt resistor are equal, we can write (using subscripts "g" for galvanometer and "s" for shunt—see Fig. 18.16a)

$$V_g = V_s \quad \text{or} \quad I_g r = I_s R_s$$

Using Kirchhoff's junction rule at J, the current I in the external circuit is $I = I_g + I_s$, or $I_s = I - I_g$. Substituting this into the previous equation, we have

$$I_g r = (I - I_g)R_s$$

INSIGHT 18.1 APPLICATIONS OF RC CIRCUITS TO CARDIAC MEDICINE

The normal human heart beats between 60 and 70 times per minute, with each beat delivering about 70 mL of blood—about a gallon per minute. Your heart is essentially a pump composed of specialized muscle cells. The cells are triggered to beat when they receive electrical signals (Fig. 1). These signal (see Insight 16.1 on Nerve Transmission in Chapter 16) are sent by special *pacemaker cells* located in the *sinotrial node* (*SA node* for short) in one of the upper chambers of the heart.

During a heart attack or after an electrical shock, the heart may go into an unregulated beat pattern. If left untreated, this condition would be fatal in minutes. Fortunately, it is possible to return the heart to its normal pattern by passing an electrical current through it. The instrument that does this is called a *cardiac defibrillator*. The main component of a defibrillator is, for our purposes, a capacitor charged to a high voltage.*

Several hundred joules of electric energy are needed to restart the heart. The high-voltage and low-voltage plates of the capacitor are attached to the patient's skin by two "paddles" placed just above the two sides of the heart (Fig. 2a and Fig. 2b). When a switch is thrown, current flows through the heart, thus transferring the capacitor's energy to the heart in an attempt to cause it to beat correctly.

This discharge through the heart is essentially that of an RC circuit. Typically the capacitor has a value of 10 μF and is charged to 1000 V. (Recent advances in dielectrics have produced compact capacitors at 1 F or more. This reduces the need for high voltage because the stored energy is proportional to the capacitance, $U_C = CV^2/2$.) The resistance of the heart (R_h) is typically about 1000 Ω, giving a time constant (for discharge) of $\tau = R_h C = 10^{-2}$ s = 10 ms.

Because of this 10 ms discharge time constant, the capacitor is essentially fully discharged after 50 ms. If needed, a second charge can be applied. To be able to do this, the capacitor should be able to recharge in about 5 s (Fig. 2c). Thus the *charging* time constant should be about 1 s. This means the charging resistor should have an approximate value (R_c) of $R_c = \tau/C \approx 10^5\ \Omega$.

In some forms of heart disease, the heartbeats are irregular due to problems with the pacemaker cells. The heart can be triggered to beat correctly by using an electronic (implanted) *cardiac pacemaker*. These units are the size of a book of matches, powered by a long-life battery, and usually inserted surgically near the SA node.

Most pacemakers are controlled by a sophisticated triggering circuit that allows the pacemaker to send signals to the heart only if they are needed ("on demand" pacemakers; see Figs. 3a and b). The triggering circuit sends a signal to the pacemaker to "fire" only if the heart does not beat. If the heart is beating normally, the capacitor switch is left in the fully charged position, waiting for the signal to fire.

For our purposes, the pacemaker is simply an RC circuit. The capacitor (typically 10 μF) is kept charged by the battery, and must be ready to release its energy as rapidly as 70 times per minute—the worst-case scenario if the heart's own pacemaker cells are not operating at all. Typically, the resistance of the heart muscle between the pacemaker leads is about 100 Ω meaning the pacemaker discharge time constant is $\tau \approx 1$ ms. Thus it is effectively fully discharged in 5 ms.

To operate at 70 times per second, the capacitor has to charge, fire, and recharge in $1/70 \approx 14$ ms. Since it takes about 5 ms to discharge, it has about 9 ms to recharge, meaning a recharge time constant of about 2 ms. This requires a recharge resistor R_C (the one in the circuit through which the capacitor is charged) to be, at most, about 200 Ω (Fig. 3c).

*Because portable batteries aren't capable of high voltages, the charging uses a phenomenon called electromagnetic induction, which will be studied in Chapter 20.

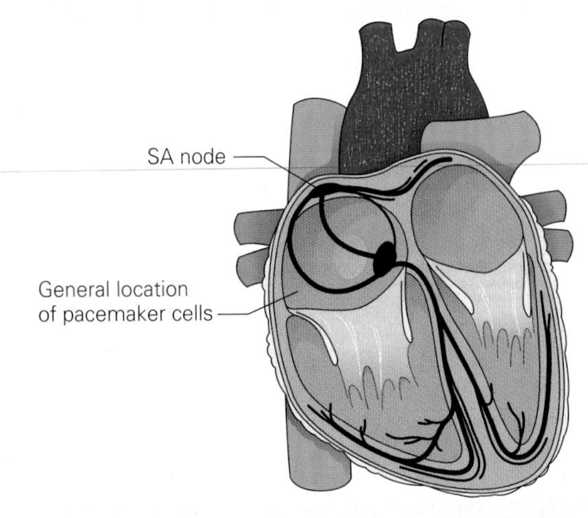

FIGURE 1 The heart The pacemaker cells are located primarily in the SA node. Electrical signals that trigger a heartbeat reach the lower areas of the heart in about 50 milliseconds.

Thus the shunt's resistance R_s is

$$R_s = \frac{I_g r}{I_{max} - I_g} = \frac{(2.00 \times 10^{-4}\,A)(50\,\Omega)}{3.0\,A - 2.00 \times 10^{-4}\,A} = 3.3 \times 10^{-3}\,\Omega = 3.3\,m\Omega$$

The shunt resistance is very small compared with the coil's resistance. This allows most of the current (2.9998 A at full scale) to pass through the shunt. This ammeter will be able to read currents linearly up to 3.0 A. For example, if a current of 1.5 A were to flow into the ammeter, there would be 100 μA (half the maximum) in the coil, which would give a half-scale reading, or 1.5 A.

Follow-Up Exercise. In this Example, if we had used a shunt resistance of 1.0 mΩ, what would be the full-scale reading (maximum current reading) of the ammeter?

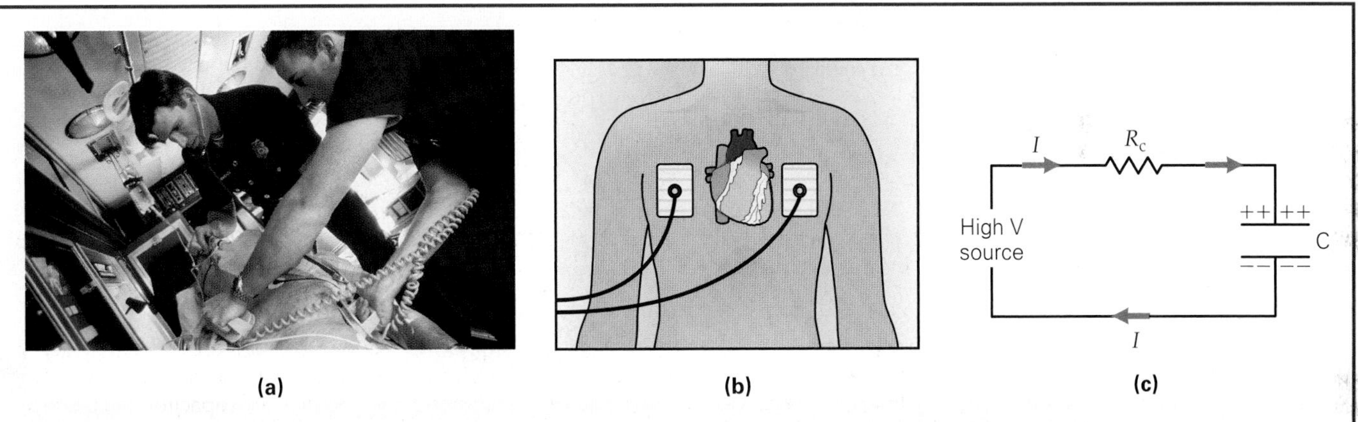

FIGURE 2 Restart the heart! (a) Paddles are placed externally to either side of the heart, and energy from a charged capacitor passes through it, hopefully triggering it into a normal beating pattern. **(b)** This shows the schematic diagram for correct defibrillator use. The discharge is that of an RC circuit. **(c)** Recharging the defibrillator's capacitor, getting it ready to go again, through a (charging) resistor $R_C \approx 10^5\,\Omega$.

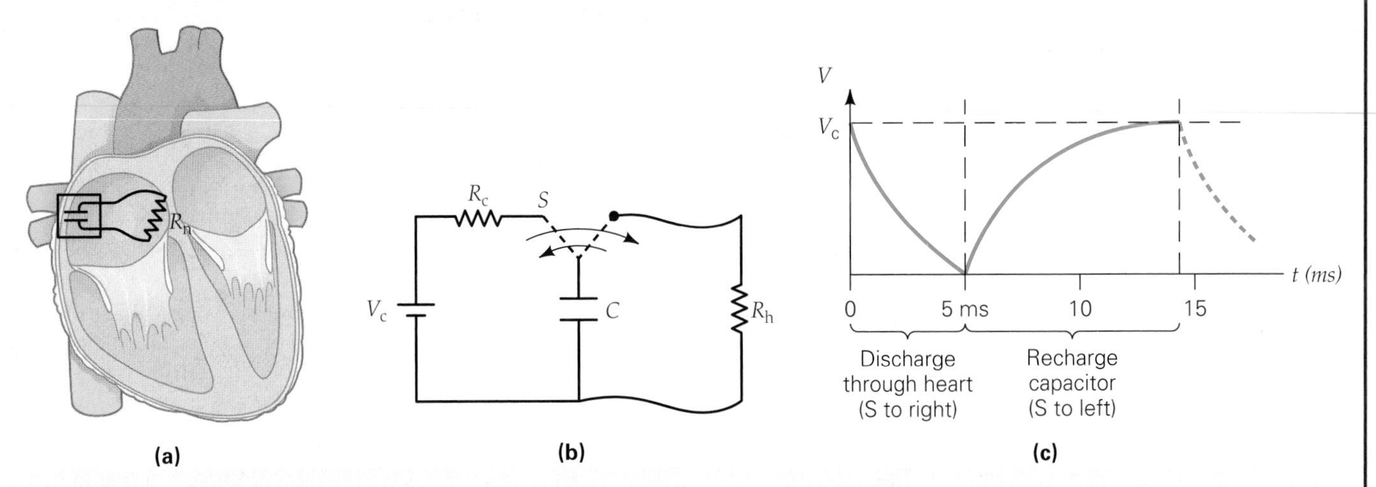

FIGURE 3 Cardiac pacemaker (a) The typical pacemaker (shown as a capacitor in a box) is implanted surgically on or near the heart surface, with its leads attached to the heart muscle (resistance R_h). (The capacitor's charging circuit is not shown.) Other leads (not shown) receive signals from the heart to determine whether the pacemaker needs to "fire." **(b)** The sensing circuit determines the position of the capacitor's "switch." If the heart is not beating, the sensing circuit flips the switch to the right, initiating energy discharge through the heart muscle. If the heart is beating properly, the sensing circuit sets the switch to the left, keeping the capacitor fully charged. **(c)** If the pacemaker is in operation, one full cycle requires about 15 ms. About 5 ms is for discharge through the heart muscle, and 10 ms is required to recharge the capacitor. The recharge is accomplished using a long-life battery, V_c.

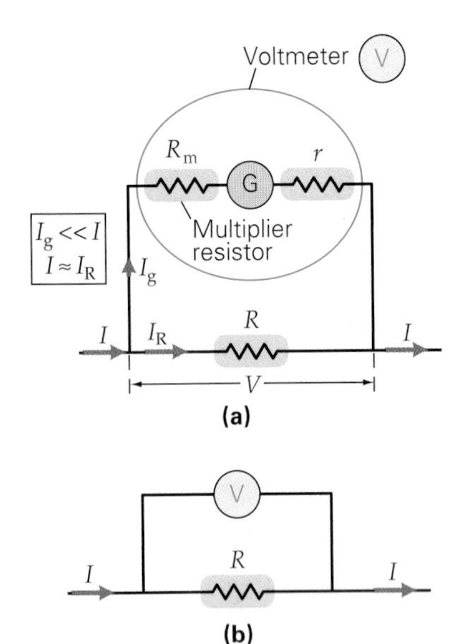

▲ **FIGURE 18.17** A dc voltmeter Here, R is the resistance of the resistor whose voltage is being measured. **(a)** A galvanometer in series with a multiplier resistor (R_m) is a voltmeter capable of measuring various ranges of voltage, depending on the value of R_m. **(b)** The circuit symbol for a voltmeter is a circle with a V inside it. (See Example 18.8 for a detailed discussion of voltmeter design.)

The Voltmeter

A voltmeter that is capable of reading voltages higher than the microvolt range (anything higher than a microvolt would burn out the galvanometer if it were alone) is constructed by connecting a large *multiplier resistor* in *series* with a galvanometer (◂Fig. 18.17). Because the voltmeter has a large resistance, due to the multiplier resistor, it draws little current from the circuit element whose voltage it measures. However, the current that exists in the voltmeter is proportional to the voltage across the circuit element. Thus, the voltmeter can be calibrated in volts. To better understand this configuration, consider Example 18.8.

Example 18.8 ■ Voltmeter Design: Using Kirchhoff's Rules to Choose a Multiplier Resistor

Suppose that the galvanometer in Example 18.7 is to be used instead in a voltmeter with a full-scale reading of 3.0 V. What is the required multiplier resistance?

Thinking It Through. To turn a galvanometer into a voltmeter, we need a reduction in current—accomplished by adding a large "multiplier resistor" in series. All the data necessary to calculate the multiplier resistance are given here and in Example 18.7.

Solution. First, we list the data:

Given: $I_g = 200\ \mu\text{A}$
$= 2.00 \times 10^{-4}$ A (from Example 18.7)
$r = 50\ \Omega$ (from Example 18.7)
$V_{max} = 3.0$ V

Find: R_m (multiplier resistance)

The resistances of the galvanometer and multiplier are in series. This combination is itself in parallel with the external circuit element (R). Therefore, the voltage across the external circuit element is the sum of the voltages across the galvanometer and multiplier (Fig. 18.17):

$$V = V_g + V_m$$

The voltages across the galvanometer and multiplier resistors are

$$V_g = I_g r \quad \text{and} \quad V_m = I_g R_m$$

Combining these three equations, we have

$$V = V_g + V_m = I_g r + I_g R_m = I_g(r + R_m)$$

Solving for the resistance of the multiplier, we have

$$R_m = \frac{V - I_g r}{I_g} = \frac{3.0\ \text{V} - (2.00 \times 10^{-4}\ \text{A})(50\ \Omega)}{2.00 \times 10^{-4}\ \text{A}} = 1.5 \times 10^4\ \Omega = 15\ \text{k}\Omega$$

Notice that the second term in the numerator ($I_g r$) is negligible compared with the full-scale reading of 3.0 V. Thus, to a good approximation, $R_m \approx V/I_g$, or $V \propto I_g$, and the measured voltage is proportional to the current in the galvanometer.

Follow-Up Exercise. The voltmeter in this Example is used to measure the voltage of a resistor in a circuit. A current of 3.00 A flows through the resistor (1.00 Ω) *before* the voltmeter is connected. Assuming that the total *incoming* current (I in Fig. 18.17b) remains the same after the voltmeter is connected, calculate the current in the galvanometer.

Note: Voltmeters are connected in parallel with, or across, the element whose voltage they are measuring (Fig. 18.17b).

Exploration 30.5 Voltmeters

For versatility, ammeters and voltmeters may be constructed with a multi-range feature. This is accomplished by providing a choice of shunt or multiplier resistors (▸Fig. 18.18a and b). Combinations of these meters are manufactured and sold as *multimeters*, which are capable of measuring voltage, current, and often resistance. Electronic digital multimeters are now commonplace (Fig. 18.18c). In place of mechanical galvanometers, these use electronic circuits to analyze digital signals and calculate voltages, currents, and resistances, which are then displayed.

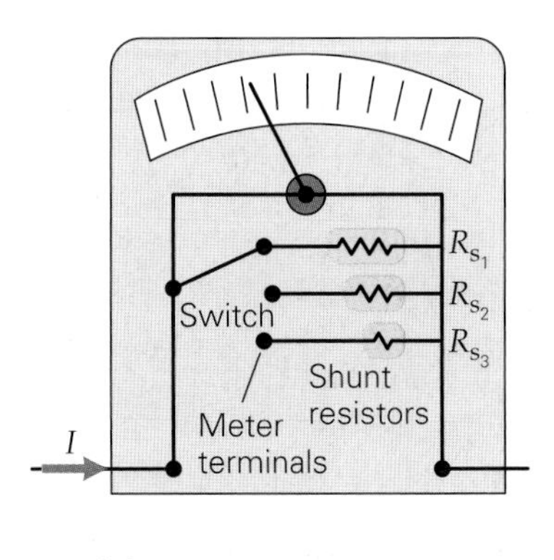

(a) **Multirange ammeter**

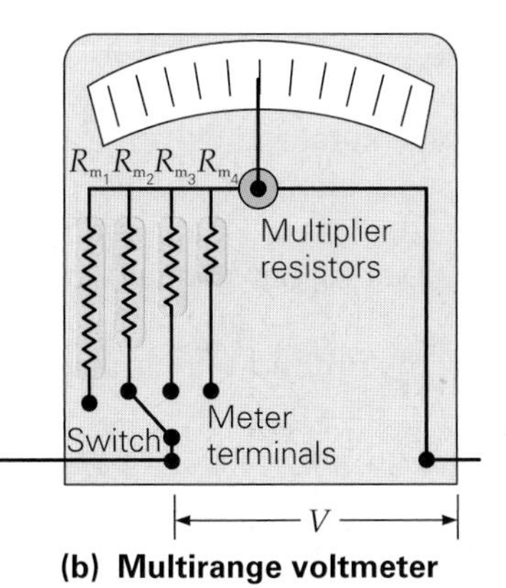

(b) **Multirange voltmeter**

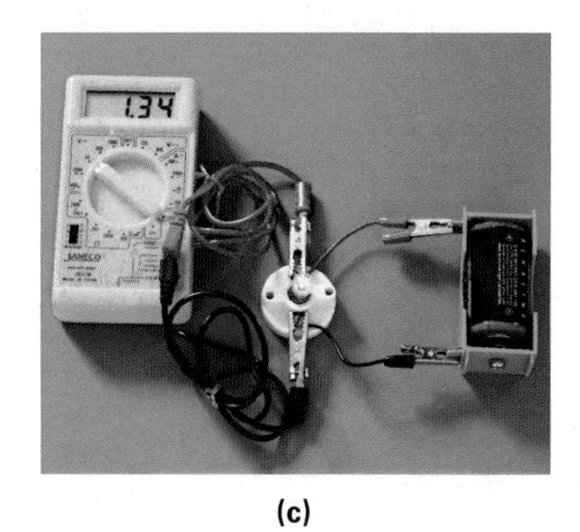

(c)

▲ **FIGURE 18.18 Multirange meters** **(a)** An ammeter or **(b)** a voltmeter can measure different ranges of current and voltage by switching among different shunt or multiplier resistors, respectively. (Instead of a switch, there may be an exterior terminal for each range.) **(c)** Both functions can be combined in a multimeter, shown here on the left measuring the voltage across a lightbulb. (How can you tell it is not measuring the current?).

18.5 Household Circuits and Electrical Safety

OBJECTIVES: To understand (a) how household circuits are wired, and (b) the underlying principles that govern electrical safety devices.

Although household circuits use alternating current, which has not yet been discussed, you can understand their operation (and many practical applications) using the circuit principles just studied.

For example, would you expect the elements (lamps, appliances, and so on) in a household circuit to be in series or parallel? From the discussion of Christmas tree lights (Section 18.1), it should be apparent that they must be connected in parallel. For example, when a bulb in a lamp in your kitchen burns out, other appliances on that circuit, such as the coffee maker, must continue to work. Moreover, household appliances and lamps are generally rated to work at 120 V. If these elements were in series, none of the individual appliances would have a voltage of 120 V across it.

Electrical power is supplied to a house by a three-wire system (▼Fig. 18.19). There is a difference in potential of 240 V between the two "hot," or high-potential,

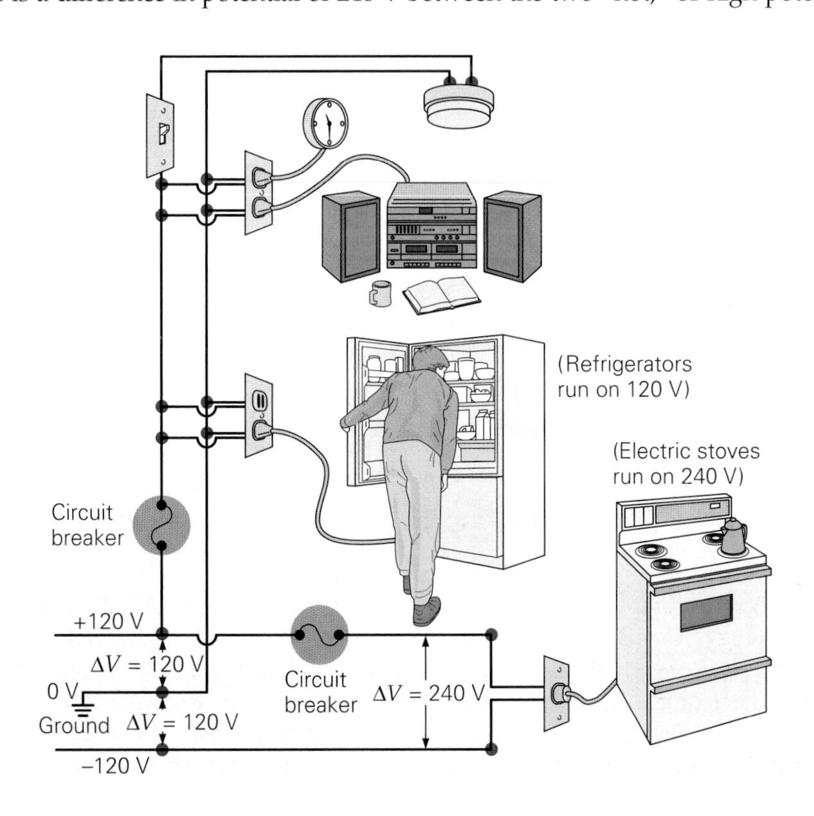

◀ **FIGURE 18.19 Household wiring schematic** A 120-V circuit is obtained by connecting between either of the "hot" lines and the ground line. A voltage of 240 V (for appliances that require a lot of power such as electric stoves) can be obtained by connecting between the two "hot" lines of opposite polarity. (*Note:* For clarity, the dedicated ground wire [the third line that takes the rounded prong] is not shown.)

Note: Household voltage can fluctuate, under normal conditions, between 110 and 120 V. Similarly, 240-V connections can be as low as 220 V and still be considered normal.

wires. Each of these "hot" wires has a 120-V difference in potential with respect to the ground. The third wire is grounded at the point where the wires enter the house, usually by a metal rod driven into the ground. This wire is defined to be at zero potential and is called the *ground*, or *neutral, wire*.

The 120 V needed for most appliances is obtained by connecting them between the ground and either high-potential wire. The result is the same in either case, because $\Delta V = 120\text{ V} - 0\text{ V} = 120\text{ V}$ or $\Delta V = 0\text{ V} - (-120\text{ V}) = 120\text{ V}$. (See Fig. 18.19.)

Even though the ground wire is at zero potential, it *is* a current-carrying wire, because it is part of the complete circuit. Large appliances such as central air conditioners, ovens, and water heaters operate at 240 V. This value is obtained by connecting them between the two hot wires: $\Delta V = 120\text{ V} - (-120\text{ V}) = 240\text{ V}$. The current in an appliance (under 120-V operating conditions) may be given on a rating tag. If not, it can usually be determined from the power rating on the tag (from $I = P/V$). Thus a stereo rated at 180 W would draw a current of 1.50 A (because $I = P/V = 180\text{ W}/120\text{ V} = 1.50\text{ A}$).

There are limitations on the number of appliances in a circuit because of a limitation on the *total* current in the wires of that circuit. Specifically, joule heating (or I^2R loss) of the wires must be carefully considered. Remember that the more elements in parallel, the smaller their equivalent resistance. Thus adding appliances (in their "on" position) increases the total current. Real wires have resistance and could be subject to significant joule heating if the current is large. Therefore, by adding too many appliances, it is possible to overload a household circuit and produce too much heat *in the wires*. This could melt insulation and perhaps even start a fire.

This potential overloading is prevented by limiting the current. Two types of devices are commonly employed as limiters: fuses and circuit breakers. **Fuses** are still fairly common in older homes (◂Fig. 18.20). An Edison-base fuse has threads like those on the base of a lightbulb. (See Fig. 18.20b.) Inside the fuse is a metal strip that melts when the current is larger than the rated value (typically 15 A for a 120-V circuit). The melting of the strip opens the circuit, and the current drops to zero.

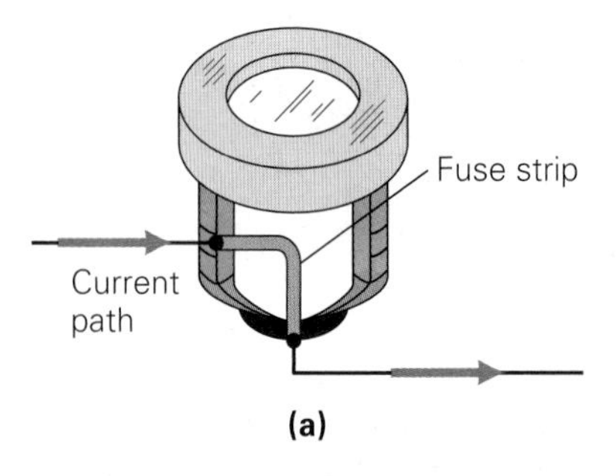

(a)

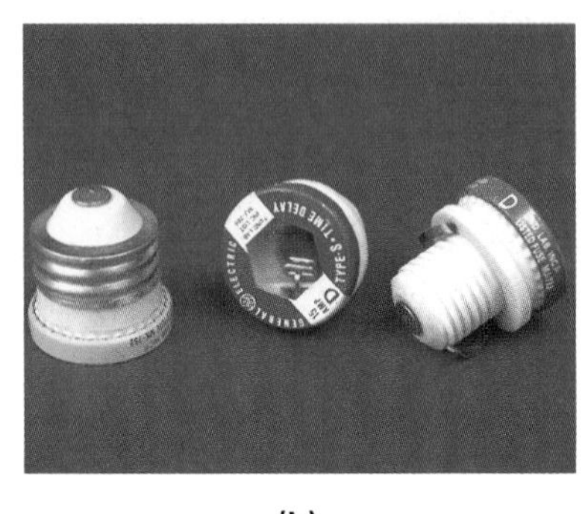

(b)

▲ **FIGURE 18.20 Fuses (a)** A fuse contains a metallic strip that melts when the current exceeds a rated value. This action opens the circuit and prevents overheating. **(b)** Edison-base fuses (left) have threads similar to lightbulbs. The threads are identical in this type of fuse; thus, different ampere-rated fuses can be interchanged—something that is not desirable (why?). Type-S fuses (right) have different threads for different ratings and thus cannot be interchanged.

Circuit breakers are used exclusively in newer homes. One type (▾Fig. 18.21) uses a bimetallic strip (see Chapter 10). As the current in the strip increases, the strip becomes warmer and bends. At the rated value of current, the strip will bend sufficiently to open the circuit. The strip then cools, so the breaker can be reset. However, a blown fuse or a tripped circuit breaker indicates that the circuit is attempting to draw too much current! *Find and correct the problem before replacing the fuse or resetting the circuit breaker.* Also, under no circumstances should the blown fuse be even temporarily replaced by one of a higher current rating (why?). If a fuse of the correct current rating is not available, for safety purposes it is better to leave that circuit open (unless it controls items needed for emergency or crucial for living) until the correct fuse is found.

Switches, fuses, and circuit breakers are placed in the "hot" (high-potential) side of the circuit. They would, of course, also work if placed in the grounded side. To see why they aren't, consider the following. If they were placed there, even if the switch were open, the fuse blown, or the breaker tripped, the appliances would all remain connected to a high voltage—which could be potentially dangerous if a person made electrical contact (▸Fig. 18.22a).

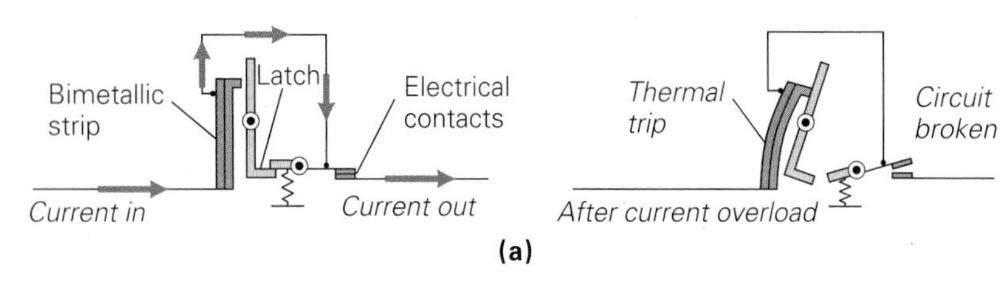

(a)

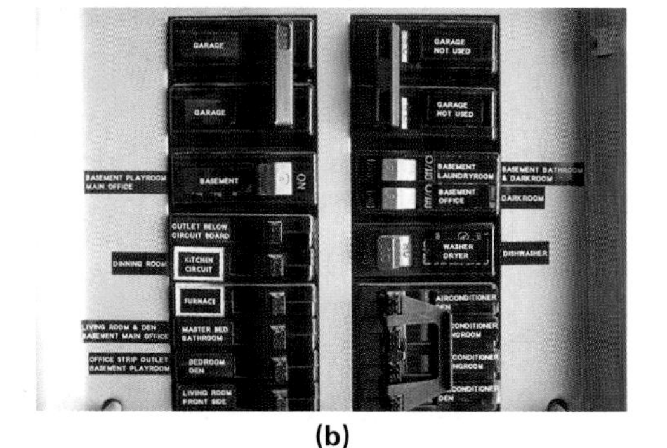

(b)

▲ **FIGURE 18.21 Circuit breakers (a)** A diagram of a thermal trip element. With increased current and joule heating, the element bends until it opens the circuit at some preset current value. Trip elements using magnetic principles also exist. **(b)** A typical bank of household circuit breakers.

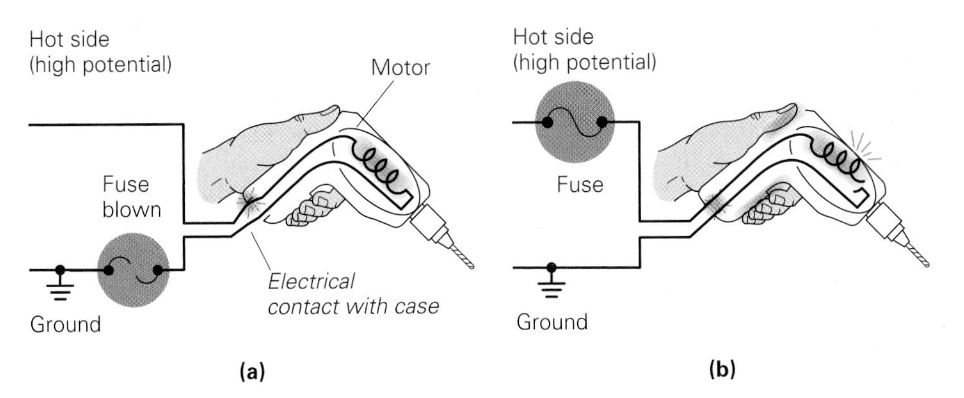

▲ **FIGURE 18.22 Electrical safety (a)** Switches and fuses or circuit breakers should always be wired in the hot side of the line, *not* in the grounded side as shown. If these elements are wired in the grounded side, the line (and potentially the metallic case of the appliance) remains at a high voltage even when the fuse is blown or a switch is open. **(b)** Even if the fuse or circuit breaker is wired in the hot side, a potentially dangerous situation exists. If an internal wire comes in contact with the metal casing of an appliance or power tool, a person touching the casing, which is at high voltage, can get a shock. To prevent this possibility, a third dedicated ground line runs from the case to ground (see Fig. 18.23).

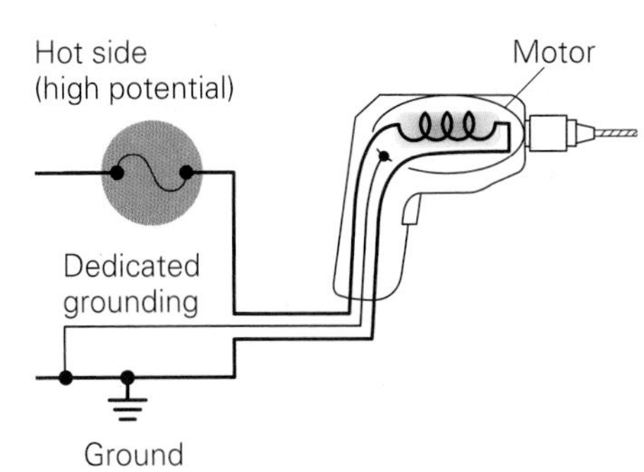

▲ **FIGURE 18.23 Dedicated grounding** For safety, a third wire is connected from an appliance or power tool to ground. This dedicated grounding wire normally carries no current (as opposed to the grounded wire of the circuit). If the hot wire should come in contact with the metal case, the current will follow the ground wire (path of least resistance) rather than go through the body of the operator holding the case. The plug used for this is shown in Fig. 18.24.

Even with fuses or circuit breakers wired correctly into the "hot" side of the circuit, there is still a possibility of an electrical shock from a defective appliance that has a metal casing, such as a hand drill. For example, if a wire comes loose inside, it could make contact with the casing, which would then be at a high voltage (Fig. 18.22b). A person's body could then provide a path for current to ground, thus receiving a shock. For a discussion of the effects of electric shock, see Insight 18.2 on Electricity and Personal Safety on page 614.

To prevent a shock, a third, dedicated grounding wire is added to the circuit that grounds the metal casing of appliances or power tools (▸Fig. 18.23). This wire provides a path of very low resistance, bypassing the tool. This wire does not normally carry current. If a hot wire comes in contact with the casing, the circuit is completed through this grounding wire. Then the fuse is blown or the circuit breaker tripped. Most of the current would be in the ground wire rather than in you. Most likely you would not be harmed. Remember, however, that the breaker, if reset, will continue to trip unless you find the source of the problem and fix it.

On **three-prong grounded plugs**, the large, round prong connects with the grounding wire. Adapters can be used between a three-prong plug and a two-prong socket. Such an adapter has a grounding lug or grounding wire (▾Fig. 18.24a) that should be fastened to the receptacle box by the plate-fastening screw. The receptacle box is itself attached to the grounding wire. If the adapter lug or wire is *not* connected, the system is left unprotected, which defeats the purpose of the dedicated grounding safety feature.

You may have noticed another type of plug, a two-prong plug that fits in the socket only in one orientation, because one prong is wider than the other and one

(a)

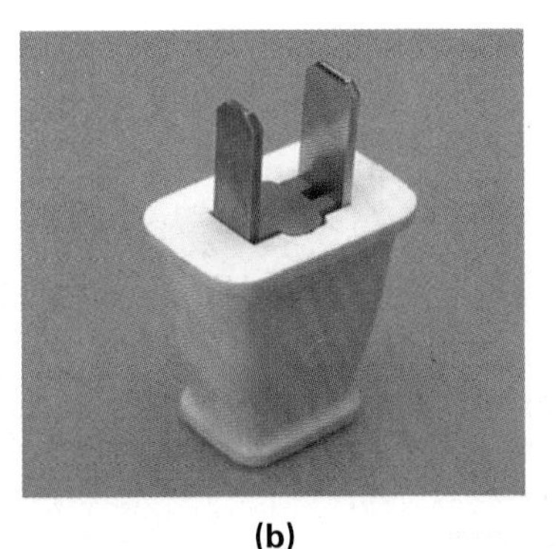

(b)

◂ **FIGURE 18.24 Plugging into ground (a)** To accommodate the dedicated ground wire (Fig. 18.23) a three-prong plug is used. The adapter shown enables a three-prong plug to be used in a two-prong socket. The grounding lug (loop) on the adapter should be connected to the plate-fastening screw on the grounded receptacle box—otherwise, this safety feature is lost. **(b)** A polarized plug. The differently sized prongs permit prewired identification of the high and ground sides of the line. See text for details.

of the slits of the receptacle is also larger (Fig. 18.24b). This type is called a **polarized plug**. *Polarizing* in the electrical circuit sense is a method of identifying the hot and grounded sides of the line so that particular connections can be made.

Such polarized plugs and sockets are now a common safety feature. Wall receptacles are wired so that the small slit connects to the hot side and the large slit connects to the neutral, or ground, side. Having the hot side identified in this way makes two safeguards possible. First, the manufacturer of an electrical appliance can design it so that the switch is always in the hot side of the line. Thus, all of the wiring of the appliance beyond the switch is safely neutral when the switch is open and the appliance off. Moreover, the casing of an appliance is connected by the manufacturer to the ground side by means of a polarized plug. Should a hot wire inside the appliance come loose and contact the metal casing, the effect would be similar to that with a dedicated grounding system. The hot side of the line would be shorted to the ground, which would blow a fuse or trip a circuit breaker. Once again, you would be spared.

Another type of electrical safety device, the ground fault circuit interrupter, or GFCI, is discussed in Chapter 20.

INSIGHT 18.2 ELECTRICITY AND PERSONAL SAFETY

Safety precautions are necessary to prevent injuries when people work with electrical devices or wiring. Electrical conductors (wires) are coated with insulating materials so they can be handled safely. However, if a person comes in contact with a charged conductor, a difference in potential could exist across part of the person's body. A bird can sit on a high-voltage line without any problem because both of its feet are at the same potential—thus there is *no difference in potential* to generate a current in the bird. But if a person carrying an aluminum (conducting) ladder touches it to a bare electrical line, a difference in potential exists between the line and the ground. Thus the ladder and person become part of a current-carrying circuit.

The extent of personal injury in such cases depends on the size of the current in the body and on its path through the body. We know that the current in the body is given by $I = V/R_{\text{body}}$. Clearly, the current depends on the body's resistance.

The body's resistance can vary. If the skin is dry, the total body resistance can be $0.50\ \text{M}\Omega\ (0.50 \times 10^6\ \Omega)$ or more. For a voltage of 120 V, there would be a current of about one-quarter of a milliamp, because

$$I = \frac{V}{R_{\text{body}}} = \frac{120\ \text{V}}{0.50 \times 10^6\ \Omega} = 0.24 \times 10^{-3}\ \text{A} = 0.24\ \text{mA}$$

This current is almost too weak to be felt (see Table 1). But if the skin is wet, then R_{body} can drop to as low as $5.0\ \text{k}\Omega (5.0 \times 10^3\ \Omega)$ and the current will be 24 mA (you should show this), a value which could potentially be dangerous. (See Table 1 again.)

A basic precaution is to avoid contact with any exposed electrical conductor that might cause a voltage across any part of your body. The physical damage resulting from such contact depends on the current path through the body. If that path is from a finger to the thumb on one hand, a large current probably would result only in a burn. However, if the path is from hand to hand through the chest (and therefore likely through the heart), the effect could be much worse. Some of the possible effects of this type of path are also given in Table 1.

Injury results because the current interferes with muscle function and/or causes burns. Muscle function is regulated by electrical nerve impulses (see Chapter 16), which can be influenced by external currents. Muscle reaction and pain can occur from a current of just a few milliamperes. At about 10 mA, muscle paralysis can prevent a person from releasing the conductor. At about 20 mA, contraction of the chest muscles occurs, which can cause impairment or stoppage of breathing. Death can occur in a few minutes. At 100 mA, rapid uncoordinated movements of the heart muscles (called *ventricular fibrillation*) prevent the proper heart pumping action and can be fatal in seconds. Working safely with electricity requires a knowledge of fundamental electrical principles *and* common sense. Electricity must be treated with respect.

Related Exercises: 94 and 95

TABLE 1 Effects of Electric Current on the Human Body*

Current (approximate)	*Effect*
2.0 mA (0.002 A)	Mild shock or heating
10 mA (0.01 A)	Paralysis of motor muscles
20 mA (0.02 A)	Paralysis of chest muscles, causing respiratory arrest; fatal in a few minutes
100 mA (0.1 A)	Ventricular fibrillation, preventing coordination of the heart's beating; fatal in a few seconds
1000 mA (1 A)	Serious burns; fatal almost instantly

*The effect of a given amount of current depends on a variety of conditions. This table gives only general and relative descriptions and assumes a circuit path that includes the upper chest.

Chapter Review

- When resistors are wired in **series**, the current through each of them is the same. The **equivalent resistance** of resistors in series is

$$R_s = R_1 + R_2 + R_3 + \cdots = \Sigma R_i \qquad (18.2)$$

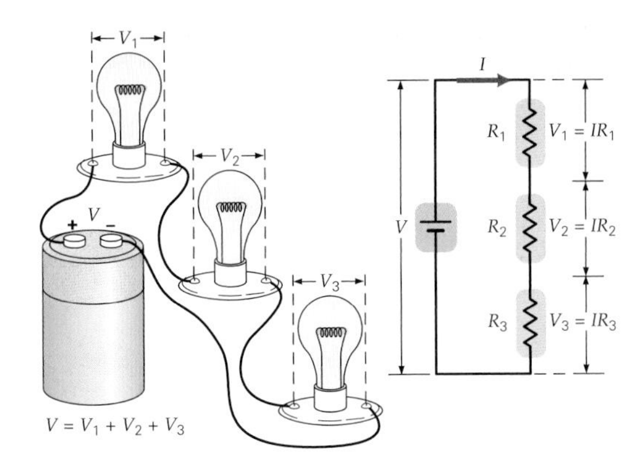

- When resistors are wired in **parallel**, the voltage across each of them is the same. The **equivalent resistance** is

$$\frac{1}{R_p} = \frac{1}{R_1} + \frac{1}{R_2} + \frac{1}{R_3} + \cdots = \Sigma \frac{1}{R_i} \qquad (18.3)$$

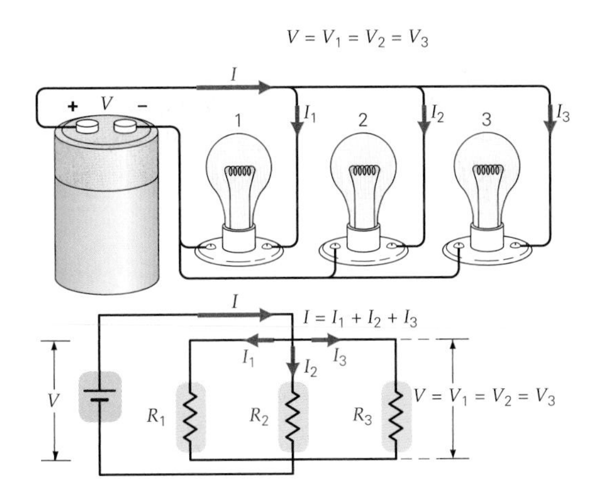

- **Kirchhoff's junction theorem** states that the total current into any **junction** equals the total current out of that junction (conservation of electric charge).

$$\Sigma I_i = 0 \quad \textit{sum of currents at a junction} \qquad (18.4)$$

- **Kirchhoff's loop theorem** states that in traversing a complete circuit loop, the algebraic sum of the voltage gains and losses is zero, or the sum of the voltage gains equals the sum of the voltage losses (conservation of energy in an electric circuit). In terms of voltages, this can be written as

$$\Sigma V_i = 0 \quad \textit{sum of voltage around a closed loop} \qquad (18.5)$$

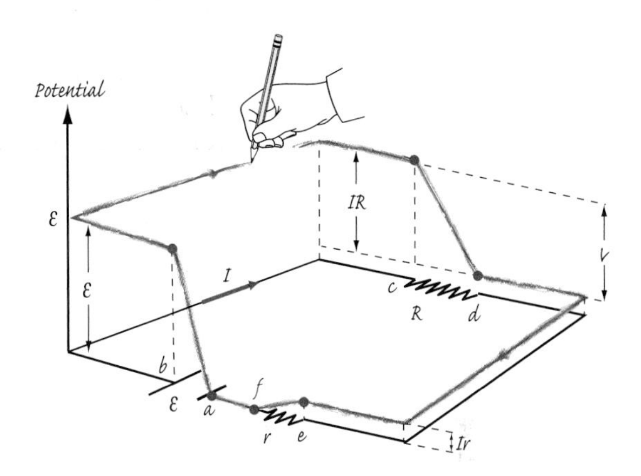

- The **time constant (τ)** for an RC circuit is a characteristic time by which we measure the capacitor's charging and discharging rate. τ is given by

$$\tau = RC \qquad (18.8)$$

- An **ammeter** is a device for measuring current; it consists of a galvanometer and a shunt resistor in parallel. Ammeters are connected in series with the circuit element carrying the current to be measured, and have very little resistance.

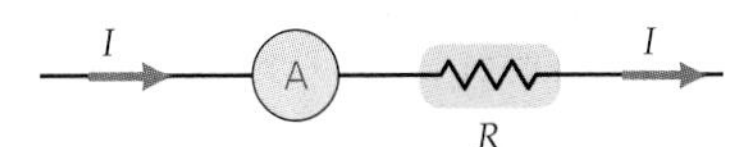

- A **voltmeter** is a device for measuring voltage; it consists of a galvanometer and a multiplier resistor wired in series. Voltmeters are connected in parallel, with the circuit element experiencing the voltage to be measured, and have large resistance.

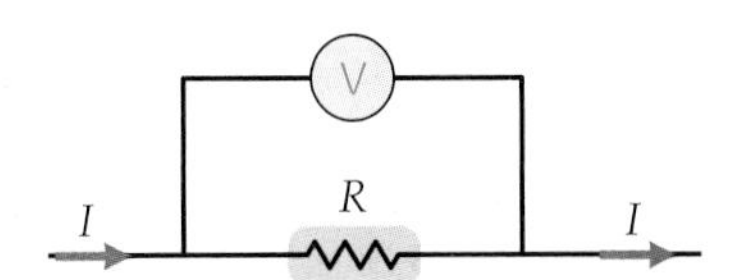

Exercises

MC = *Multiple Choice Question,* **CQ** = *Conceptual Question, and* **IE** = *Integrated Exercise. Throughout the text, many exercise sections will include "paired" exercises. These exercise pairs, identified with* **red numbers**, *are intended to assist you in problem solving and learning. In a pair, the first exercise (even numbered) is worked out in the Study Guide so that you can consult it should you need assistance in solving it. The second exercise (odd numbered) is similar in nature, and its answer is given at the back of the book.*

Assume that all resistors are ohmic unless otherwise stated.

18.1 Resistances in Series, Parallel, and Series–Parallel Combinations

1. **MC** Which of the following quantities must be the same for resistors in series? (a) voltage, (b) current, (c) power, (d) energy.

2. **MC** Which of the following quantities must be the same for resistors in parallel? (a) voltage, (b) current, (c) power, (d) energy.

3. **MC** Two resistors (A and B) are connected in series to a 12-V battery. Resistor A has 9 V across it. Which resistor has the least resistance? (a) A, (b) B, (c) both have the same, (d) can't tell from the data given.

4. **MC** Two resistors (A and B) are connected in parallel to a 12-V battery. Resistor A has 2.0 A in it and the total current in the battery is 3.0 A. Which resistor has the most resistance? (a) A, (b) B, (c) both have the same, (d) can't tell from the data given.

5. **MC** Two resistors (one with a resistance of 2.0 Ω and the other with 6.0 Ω resistance) are connected in parallel to a battery. Which one produces the most joule heating? (a) 2.0 Ω, (b) 6.0 Ω, (c) both produce the same, (d) can't tell from the data given.

6. **MC** Two lightbulbs (bulb A has a rating of 100 W *at 120 V* and B has a rating of 60 W *at 120 V*) are connected in series to a wall socket at 120 V. Which one produces the most light? (a) A, (b) B, (c) both produce the same, (d) can't tell from the data given.

7. **CQ** Are the voltage drops across resistors in series generally the same? If not, under what circumstance(s) could they be the same?

8. **CQ** Are the currents in resistors in parallel generally the same? If not, under what circumstance(s) could they be the same?

9. **CQ** If a large resistor and a small resistor are connected in series, will the effective resistance be closer in value to that of the large resistance or the small one? What if they are connected in parallel?

10. **CQ** Lightbulbs have a power output marked on them by the manufacturer. For example, a 60-W light bulb assumes the bulb is connected to a 120-V source. Suppose you have two bulbs. One labeled 60 W is followed by a 40 W bulb in series to a 120-V source. Which one glows the brightest? Why? What happens if you switch the order of the bulbs? Are either of them at full power rating? Explain.

11. **CQ** Three identical resistors are connected to a battery. Two are wired in parallel, and that combination is followed in series by the third resistor. Which resistor(s) has (a) the largest current, (b) the largest voltage, and (c) the largest power output?

12. **CQ** Three resistors have values of 5 Ω, 2 Ω, and 1 Ω. The first one is followed in series by the last two wired in parallel. When this arrangement is connected to a battery, which resistor(s) has (a) the largest current, (b) the largest voltage, and (c) the largest power output?

13. ● Three resistors that have values of 10 Ω, 20 Ω, and 30 Ω, respectively, are to be connected. (a) How should you connect them to get the maximum equivalent resistance, and what is this maximum value? (b) How should you connect them to get the minimum equivalent resistance, and what is this minimum value?

14. ● Two identical resistors (R) are connected in series and then wired in parallel to a 20-Ω resistor. If the total equivalent resistance is 10 Ω, what is the value of R?

15. ● Two identical resistors (R) are connected in parallel and then wired in series to a 40-Ω resistor. If the total equivalent resistance is 55 Ω, what is the value of R?

16. **IE** ● (a) In how many different ways can three 4.0-Ω resistors, be wired: (1) three, (2) five, or (3) seven? (b) Sketch the different ways you found in part (a) and determine the equivalent resistance for each.

17. ● Three resistors with values of 5.0 Ω, 10 Ω, and 15 Ω, respectively, are connected in series in a circuit with a 9.0-V battery. (a) What is the total equivalent resistance? (b) What is the current in each resistor? (c) At what rate is energy delivered to the 15-Ω resistor?

18. ● Find the equivalent resistances for all possible combinations of two or more of the three resistors in Exercise 17.

19. ● Three resistors with values 1.0 Ω, 2.0 Ω, and 4.0 Ω, respectively, are connected in parallel in a circuit with a 6.0-V battery. What are (a) the total equivalent resistance, (b) the voltage across each resistor, and (c) the power delivered to the 4.0-Ω resistor?

20. **IE** ●● (a) If you had only an infinite supply of 1.0-Ω resistors, what is the minimum number of resistors required to make an equivalent resistance of 1.5 Ω: (1) two, (2) three,

or (3) four? (b) Describe or show by a sketch how the resistors should be connected.

21. IE ●● A length of wire with a resistance R is cut into two equal segments. The segments are then twisted together to form a conductor half as long as the original wire. (a) The resistance of the shortened conductor is (1) $R/4$, (2) $R/2$, (3) R. (b) If the resistance of the original wire is 27 $\mu\Omega$ and the wire is cut into three equal segments, what is the resistance of the shortened conductor?

22. IE ●● You have four 5.00-Ω resistors. (a) Can you connect all the resistors to produce an effective total resistance of 3.75 Ω? (b) Describe how you would connect them.

23. ●● Three resistors with values of 2.0 Ω, 4.0 Ω, and 6.0 Ω, respectively, are connected in series in a circuit with a 12-V battery. (a) How much current is delivered to the circuit by the battery? (b) What is the current in each resistor? (c) How much power is delivered to each resistor? (d) How does this power compare with the power delivered to the total equivalent resistance?

24. ●● Suppose that the resistors in Exercise 23 are connected in parallel. (a) How much current is delivered to the circuit by the battery? (b) What is the current in each resistor? (c) How much power is delivered to each resistor? (d) How does this power compare with the power delivered to the total equivalent resistance?

25. ●● Two 8.0-Ω resistors are connected in parallel, as are two 4.0-Ω resistors. These two combinations are then connected in series in a circuit with a 12-V battery. What is the current in each resistor and the voltage across each resistor?

26. ●● What is the equivalent resistance of the resistors in ▾Fig. 18.25?

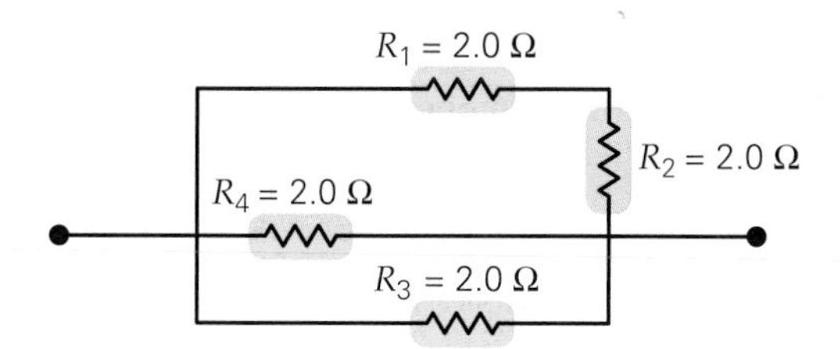

▲ **FIGURE 18.25 Series–parallel combination** See Exercises 26 and 34.

27. ●● What is the equivalent resistance between points A and B in ▾Fig. 18.26?

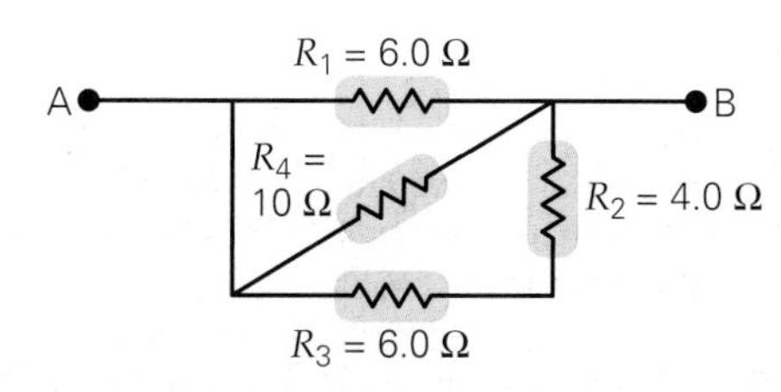

▲ **FIGURE 18.26 Series–parallel combination** See Exercises 27 and 36.

28. ●● What is the equivalent resistance of the arrangement of resistors shown in ▾Fig. 18.27?

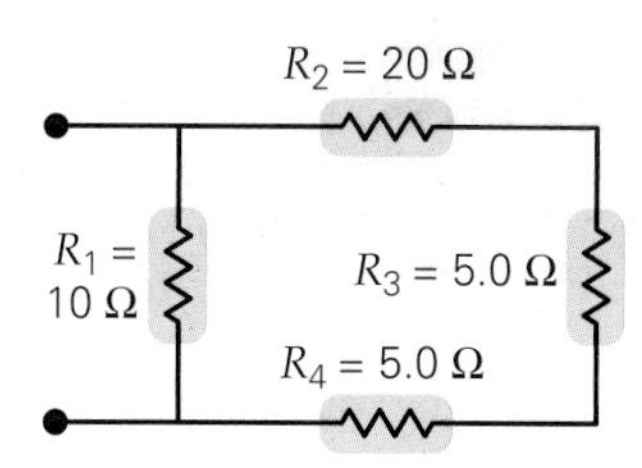

▲ **FIGURE 18.27 Series–parallel combination** See Exercise 28.

29. ●● Several 60-W lightbulbs are connected in parallel with a 120-V source. The very last bulb blows a 15-A fuse in the circuit. (a) Sketch a schematic circuit diagram to show the fuse in relation to the bulbs. (b) How many lightbulbs are in the circuit (including the very last bulb)?

30. ●● Find the current in and voltage across the 10-Ω resistor shown in ● ▾Fig. 18.28.

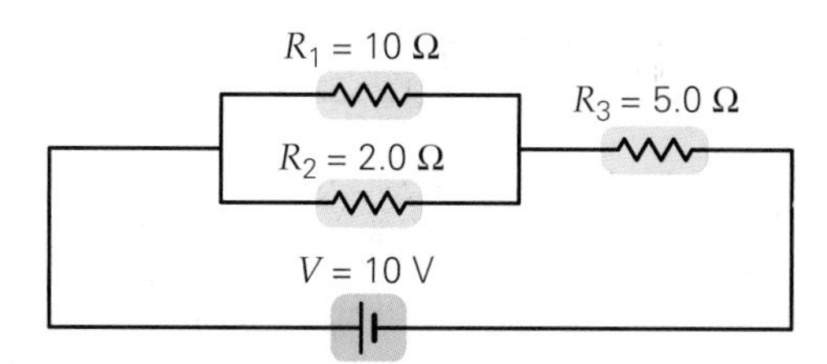

▲ **FIGURE 18.28 Current and voltage drop of a resistor** See Exercises 30 and 52.

31. ●● For the circuit shown in ▾Fig. 18.29, find (a) the current in each resistor, (b) the voltage across each resistor, and (c) the total power delivered.

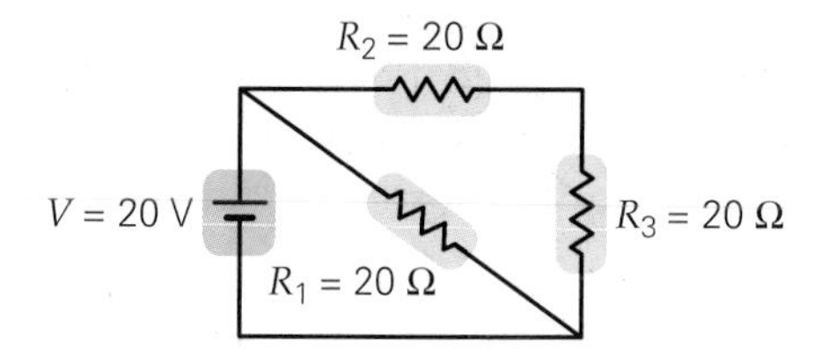

▲ **FIGURE 18.29 Circuit reduction** See Exercises 31 and 53.

32. ●● A 120-V circuit has a circuit breaker rated to trip (to create an open circuit) at 15 A. How many 300-Ω resistors could be connected in parallel without tripping the breaker?

33. ●● In your dorm room, you have two 100-W lights, a 150-W color TV, a 300-W refrigerator, a 900-W hairdryer, and a 200-W computer (including the monitor). If there is a 15-A circuit breaker in the 120-V power line, will the breaker trip (create an open circuit)?

34. ●● Suppose that the resistor arrangement in Fig. 18.25 is connected to a 12-V battery. What will be (a) the current in each resistor, (b) the voltage drop across each resistor, and (c) the total power delivered?

35. ●● To make hot tea, you use a 500-W heater connected to a 120-V line to heat 0.20 kg of water from 20°C to 80°C. Assuming that there is no heat loss other than that delivered to the water, how long does this process take?

36. ●●● The terminals of a 6.0-V battery are connected to points A and B in Fig. 18.26. (a) How much current is in each resistor? (b) How much power is delivered to each? (c) Compare the sum of the individual powers with the power delivered to the equivalent resistance for the circuit.

37. ●●● Lightbulbs with the power ratings (expressed in watts) given in ▾Fig. 18.30 are connected in a circuit as shown. (a) What current does the voltage source deliver to the circuit? (b) Find the power delivered to each bulb. (Take the bulbs' resistances to be the same as at their normal operating voltage.)

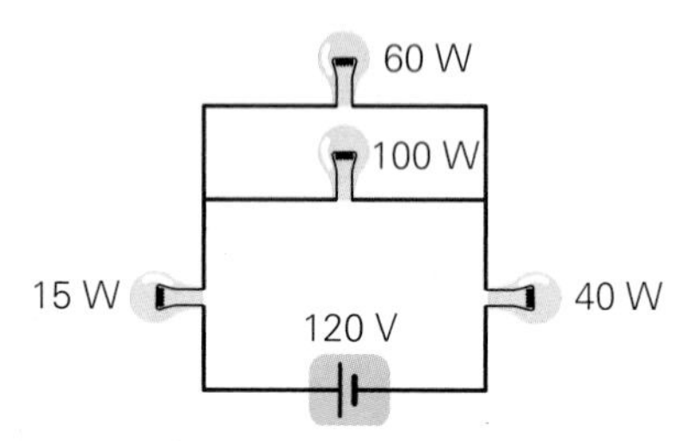

▲ **FIGURE 18.30 Watt's up?** See Exercise 37.

38. ●●● Two resistors R_1 and R_2 are in series with a 7.0-V battery. If R_1 has a resistance of 2.0 Ω and R_2 receives energy at the rate of 6.0 W, what is (are) the value(s) for the circuit's current(s)? (There may be more than one answer.)

39. ●●● For the circuit in ▾Fig. 18.31, find (a) the current in each resistor and (b) the voltage across each resistor.

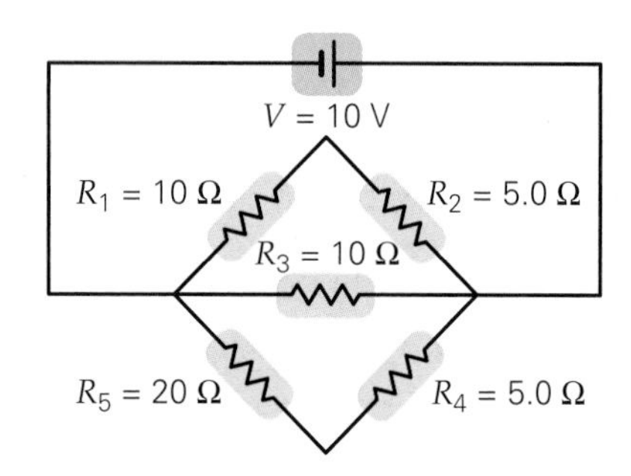

▲ **FIGURE 18.31 Resistors and currents** See Exercise 39.

40. ●●● What is the total power delivered to the circuit shown in ▾Fig. 18.32?

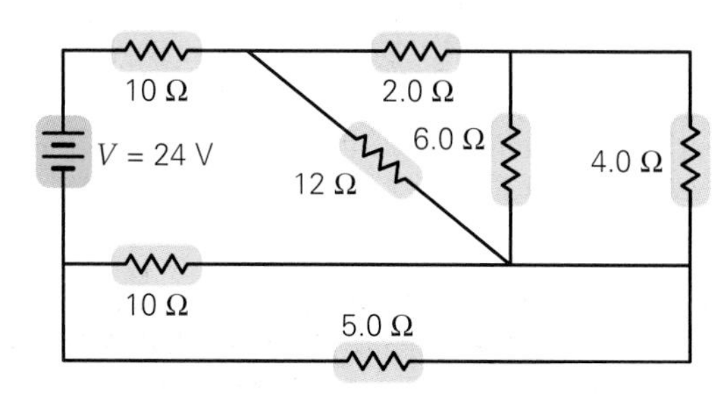

▲ **FIGURE 18.32 Power dissipation** See Exercise 40.

41. ●●● What is the equivalent resistance of the arrangement shown in ▾Fig. 18.33?

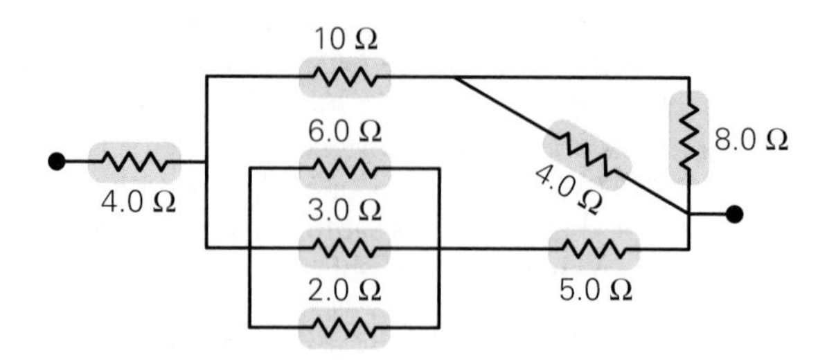

▲ **FIGURE 18.33 Equivalent resistance** See Exercise 41.

42. ●●● The circuit in ▾Fig. 18.34, named the *Wheatstone bridge* after Sir Charles Wheatstone (1802–1875), is used to measure resistance without the corrections sometimes necessary when using ammeter–voltmeter measurements. (See, for example, Exercises 88 and 89.) The resistances R_1, R_2, and R_s are known, and R_x is the unknown. R_s is variable and is adjusted until the bridge circuit is balanced—that is, until the galvanometer (G) reads zero (no current). Show that when the bridge is balanced, R_x can be determined from the following relationship:

$$R_x = \left(\frac{R_2}{R_1}\right)R_s.$$

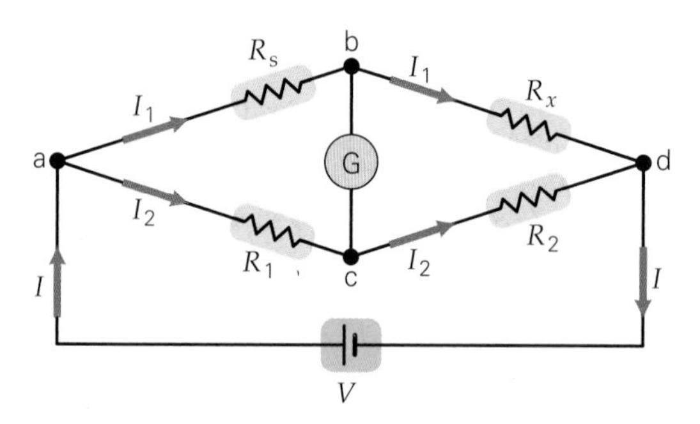

▲ **FIGURE 18.34 Wheatstone bridge** See Exercise 42.

18.2 Multiloop Circuits and Kirchhoff's Rules

43. **MC** You have a multiloop circuit with one battery. After leaving the battery, the current encounters a junction into two wires. One wire carries 1.5 A and the other 1.0 A. What is the current in the battery? (a) 2.5 A, (b) 1.5 A, (c) 1.0 A, (d) 5.0 A, (e) can't be determined from the given data.

44. **MC** By our sign convention, if a resistor is traversed in the actual direction of the current in it, what can you say about the sign of the change in electric potential (the voltage): (a) it is negative, (b) it is positive, (c) it is zero, or (d) you can't tell from the data given?

45. **MC** By our sign convention, if a battery is traversed in the actual direction of the current in it, what can you say about the sign of the change in electric potential (the battery's terminal voltage): (a) It is negative, (b) it is positive, (c) it is zero, or (d) you can't tell from the data given.

46. **MC** You have a multiloop circuit with one battery that has a terminal voltage of 12 V. After leaving the positive terminal

of the battery, a short wire takes you to a junction where the current splits into three wires. From that point until you return to the negative terminal of the battery, what can you say about the sum of the voltages in each wire: (a) they total +12 V, (b) they total −12 V, (c) their magnitude is less than 12 V, or (d) their magnitude is greater than 12 V.

47. **CQ** Must the current in a battery (in a complete circuit) always travel from its negative terminal to its positive terminal? Explain. If not, give an example.

48. **CQ** Use Kirchhoff's junction theorem to explain why the total equivalent resistance of a circuit is reduced by connecting a second resistor in parallel to another resistor.

49. **CQ** Use Kirchhoff's loop theorem to explain why a 60-W lightbulb produces more light than one rated at 100 W when they are connected in series to a 120-V source. [*Hint*: Recall that the power ratings are meaningful only at 120 V.]

50. ● Traverse loop 3 *opposite* to the direction shown in Fig. 18.10, and demonstrate that the resulting equation is the same as if you had followed the direction of the arrows.

51. ● For the circuit shown in Fig. 18.10, reverse the directions of loops 1 and 2, and demonstrate that equations equivalent to those in Example 18.5 are obtained.

52. ●● Use Kirchhoff's loop theorem to find the current in each resistor in Fig. 18.28.

53. ●● Apply Kirchhoff's rules to the circuit in Fig. 18.29 to find the current in each resistor.

54. **IE** ●● Two batteries with terminal voltages of 10 V and 4 V, respectively, are connected with their positive terminals together. A 12-Ω resistor is wired between their negative terminals. (a) The current in the resistor is (1) 0 A, (2) between 0 A and 1.0 A, (3) greater than 1.0 A. Why? (b) Use Kirchhoff's loop theorem to find the current in the circuit and the power delivered to the resistor. (c) Compare this result with the power output of each battery.

55. ●● Using Kirchhoff's rules, find the current in each resistor in ▾Fig. 18.35.

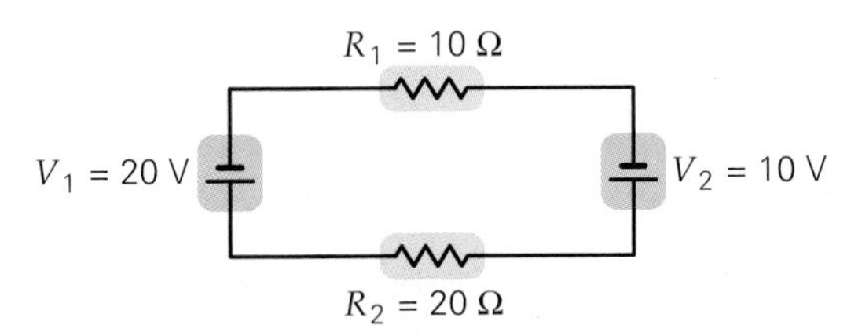

▲ **FIGURE 18.35 Single-loop circuit** See Exercise 55.

56. ●● Apply Kirchhoff's rules to the circuit in ▸Fig. 18.36, and find (a) the current in each resistor and (b) the rate at which energy is being delivered to the 8.0-Ω resistor.

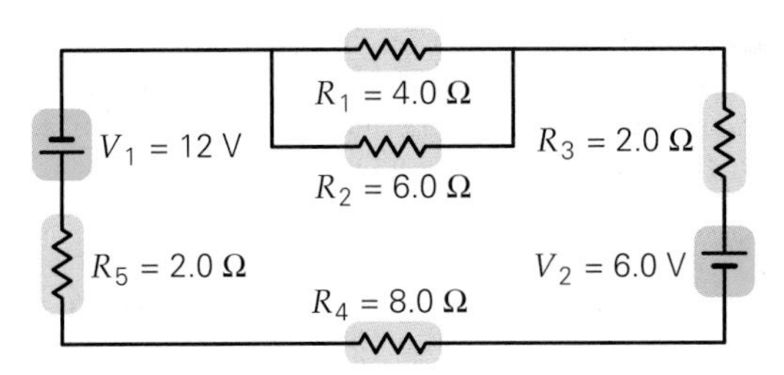

▲ **FIGURE 18.36 A loop in a loop** See Exercise 56.

57. ●●● Find the current in each resistor in the circuit shown in ▾Fig. 18.37.

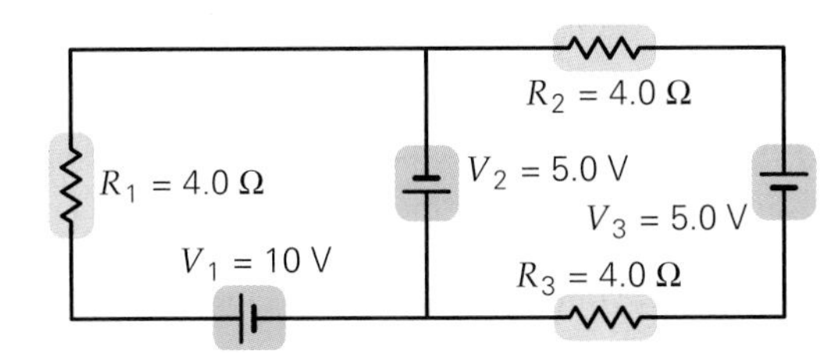

▲ **FIGURE 18.37 Double-loop circuit** See Exercise 57.

58. ●●● Find the currents in the circuit branches in ▾Fig. 18.38.

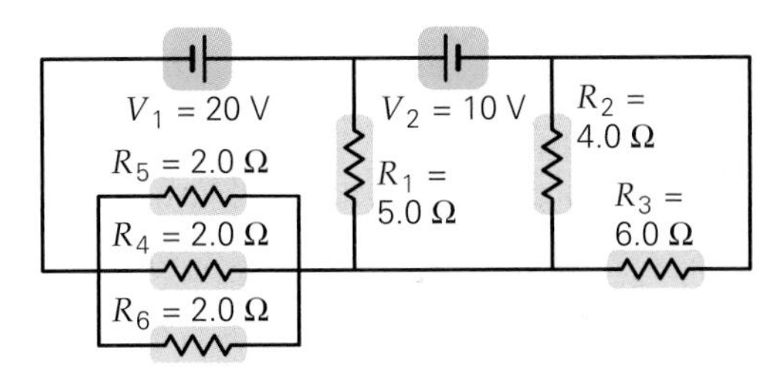

▲ **FIGURE 18.38 How many loops?** See Exercise 58.

59. ●●● For the multiloop circuit shown in ▾Fig. 18.39, what is the current in each branch?

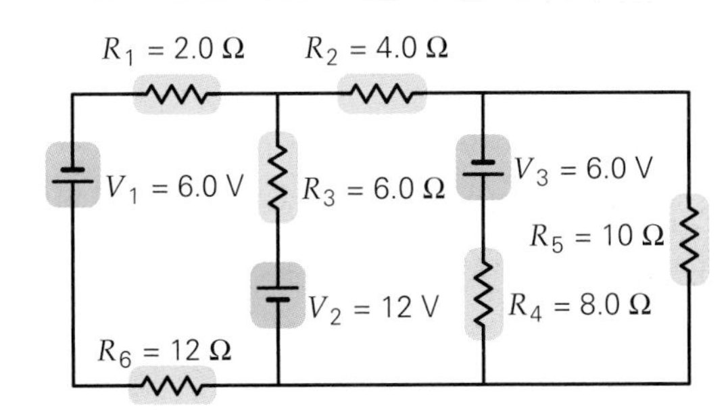

▲ **FIGURE 18.39 Triple-loop circuit** See Exercise 59.

18.3 RC Circuits

60. **MC** As a capacitor discharges through a resistor, the voltage across the resistor is a maximum (a) at the beginning of the process, (b) near the middle of the process, (c) at the end of the process, (d) after one time constant.

61. **MC** When a capacitor discharges through a resistor, the current in the circuit is a minimum (a) at the beginning of the process, (b) near the middle of the process, (c) at the end of the process, (d) after one time constant.

62. **MC** A charged capacitor discharges through a resistor (call this #1). If the value of the resistor is then doubled and the identical capacitor allowed to discharge again (call this #2), how do the time constants compare? (a) $\tau_1 = 2\tau_2$, (b) $\tau_1 = \tau_2$, (c) $\tau_1 = \frac{1}{2}\tau_2$, (d) $\tau_2 = 4\tau_1$.

63. **MC** A capacitor discharges through a resistor (call this #1). The capacitor is then recharged to twice the initial charge in #1, and the discharge occurs through the same resistor (call this #2). How do the time constants compare? (a) $\tau_1 = 2\tau_2$, (b) $\tau_1 = \tau_2$, (c) $\tau_1 = \frac{1}{2}\tau_2$, (d) can't tell from the data given.

64. **CQ** Another way to describe the discharge time of an RC circuit is to use a time interval called the *half-life*, which is defined as the time for the capacitor to lose half its initial charge. Is the time constant longer or shorter than the half-life? Explain your reasoning.

65. **CQ** Does charging a capacitor in an RC circuit to 25% its maximum value take longer or shorter than one time constant? Explain.

66. **CQ** Explain why the current in a charging RC circuit decreases as the capacitor is being charged.

67. ● In Fig. 18.11b, the switch is closed at $t = 0$, and the capacitor begins to charge. What is the voltage across the resistor and across the capacitor, expressed as fractions of V_o (to two significant figures), (a) just after the switch is closed, (b) after two time constants have elapsed, and (c) after many time constants have elapsed?

68. ● A capacitor in a single-loop RC circuit is charged to 63% of its final voltage in 1.5 s. Find (a) the time constant for the circuit and (b) the percentage of the circuit's final voltage after 3.5 s.

69. **IE** ● In a flashing neon sign display, a certain time constant is desired. (a) To increase this time constant, you should (1) increase the capacitance, (2) decrease the capacitance, or (3) eliminate the capacitor. Why? (b) If a 2.0-s time constant is desired and you have a 1.0-μF capacitor, what resistance should you use in the circuit?

70. ●● How many time constants will it take for an initially charged capacitor to be discharged to half of its initial voltage?

71. ●● A 1.00-μF capacitor, initially charged to 12 V, discharges when it is connected in series with a resistor. (a) What resistance is necessary to cause the capacitor to have only 37% of its initial charge 1.50 s after starting? (b) What is the voltage across the capacitor at $t = 3\tau$ if the capacitor is instead *charged* by the same battery through the same resistor?

72. ●● A series RC circuit with $C = 40\ \mu$F and $R = 6.0\ \Omega$ has a 24-V source in it. With the capacitor initially uncharged, an open switch in the circuit is closed. (a) What is the voltage across the resistor immediately afterward? (b) What is the voltage across the capacitor at that time? (c) What is the current in the resistor at that time?

73. ●● (a) For the circuit in Exercise 72, after the switch has been closed for $t = 4\tau$, what is the charge on the capacitor? (b) After a long time has passed, what are the voltages across the capacitor and the resistor?

74. ●●● An RC circuit with a 5.0-MΩ resistor and a 0.40-μF capacitor is connected to a 12-V source. If the capacitor is initially uncharged, what is the change in voltage across it between $t = 2\tau$ and $t = 4\tau$?

75. ●●● A 3.0-MΩ resistor is connected in series with a 0.28-μF capacitor. This arrangement is then connected across four 1.5-V batteries (also in series). (a) What is the maximum current in the circuit and when does it occur? (b) What percentage of the maximum current is in the circuit after 4.0 s? (c) What is the maximum charge on the capacitor and when does it occur? (d) What percentage of the maximum charge is on the capacitor after 4.0 s?

18.4 Ammeters and Voltmeters

76. **MC** To accurately measure the voltage across a 1-kΩ resistor, the voltmeter should have a resistance that is (a) much larger than 1 kΩ, (b) much smaller than 1 kΩ, (c) about the same as 1 kΩ, (d) zero.

77. **MC** To accurately measure the current in a 1-kΩ resistor, the ammeter should have a resistance that is (a) much larger than 1 kΩ, (b) much smaller than 1 kΩ, (c) about the same as 1 kΩ, (d) as large as possible—essentially infinite if possible.

78. **MC** To correctly measure the voltage across a circuit element, a voltmeter should be connected (a) in series with it, (b) in parallel with it, (c) between the high potential side of the element and ground, (d) none of the preceding.

79. **CQ** (a) What would happen if an ammeter were connected in parallel with a current-carrying circuit element? (b) What would happen if a voltmeter were connected in series with a current-carrying circuit element?

80. **CQ** Explain clearly, using Kirchhoff's laws, why the resistance of an ideal voltmeter is infinite.

81. **CQ** If designed properly, a good ammeter should have a very small resistance. Why? Explain clearly, using Kirchhoff's laws.

82. **IE** ● A galvanometer with a full-scale sensitivity of 2000 μA has a coil resistance of 100 Ω. It is to be used in an ammeter with a full-scale reading of 30 A. (a) Should you use (1) a shunt resistor, (2) a zero-resistor, or (3) a multiplier resistor? Why? (b) What is the necessary resistance?

83. **IE** ● The galvanometer in Exercise 82 is to be used in a voltmeter with a full-scale reading of 15 V. (a) Should you use (1) a shunt resistor, (2) a zero-resistor, or (3) a multiplier resistor? Why? (b) What is the required resistance?

84. ● A galvanometer with a full-scale sensitivity of 600 μA and a coil resistance of 50 Ω is to be used to build an am-

meter designed to read 5.0 A at full scale. What is the required shunt resistance?

85. ●● A galvanometer has a coil resistance of 20 Ω. A current of 200 μA deflects the needle through 10 divisions at full scale. What resistance is needed to convert the galvanometer to a full-scale 10-V voltmeter?

86. ●● An ammeter has a resistance of 1.0 mΩ. Find the current in the ammeter when it is properly connected to a 10-Ω resistor and a 6.0-V source. (Express your answer to five significant figures to show how it differs from 0.60 A.)

87. ●● A voltmeter has a resistance of 30 kΩ. What is the current in the meter when it is properly connected across a 10-Ω resistor that is hooked to a 6.0-V source?

88. IE ●●● An ammeter and a voltmeter can measure the value of a resistor. Suppose that the ammeter is connected in series with the resistor and that the voltmeter is placed across the resistor only. (a) For accurate measurement, the internal resistance of the voltmeter should be (1) zero, (2) equal to the resistance to be measured, (3) infinite. Why? (b) Explain why the correct resistance is *not* given by $R = \frac{V}{I}$. (c) Show that the correct resistance is actually larger than the result in part (b) and is given by $R = \frac{V}{I - (V/R_V)}$ where V is the voltage measured by the voltmeter, I is the current measured by the ammeter, and R_V is the resistance of the voltmeter. (d) Show that the result in part (c) reduces to $R = \frac{V}{I}$ for an ideal voltmeter.

89. IE ●●● An ammeter and a voltmeter can measure the value of a resistor. Suppose that the ammeter is connected in series with the resistor and that the voltmeter is placed across both the ammeter and the resistor. (a) For accurate measurement, the internal resistance of the ammeter should be (1) zero, (2) equal to the resistance to be measured, (3) infinite. Why? (b) Explain why the correct resistance is *not* given by $R = \frac{V}{I}$. (c) Show that the correct resistance is actually smaller than the result in part (b) and is given by $R = (V/I) - R_A$ where V is the voltage measured by the voltmeter, I is the current measured by the ammeter, and R_A is the resistance of the ammeter. (d) Show that the result in part (c) reduces to $R = \frac{V}{I}$ for an ideal ammeter.

18.5 Household Circuits and Electrical Safety

90. MC The ground wire in household wiring (a) is a current-carrying wire, (b) is at a voltage of 240 V from one of the "hot" wires, (c) carries no current, (d) none of the preceding.

91. MC A dedicated grounding wire (a) is the basis for the polarized plug, (b) is necessary for a circuit breaker, (c) normally carries no current, (d) none of the preceding.

92. CQ In terms of electrical safety, explain clearly what is wrong with the circuit in ▸Fig. 18.40, and why?

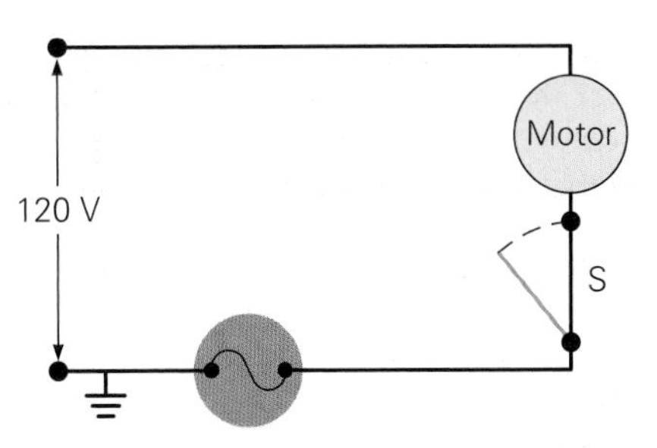

▲ **FIGURE 18.40 A safety problem?** See Exercise 92.

93. CQ The severity of bodily injury from electrocution depends on the magnitude of the current and its path, yet you commonly see signs that warn "Danger: High Voltage" (▾Fig. 18.41). Shouldn't such signs be changed to refer to high current? Explain.

▲ **FIGURE 18.41 Danger—high voltage** Shouldn't it be "high current" instead of "high voltage?" See Exercise 93.

94. CQ Explain why it is perfectly safe for birds to perch with both feet on the same high-voltage wire, even if the insulation is worn through.

95. CQ After a collision with a power pole, you are trapped in your car, with a high-voltage line (frayed insulation) in contact with the hood of the car. Is it safer to step out of the car one foot at a time or to jump with both feet leaving the car at the same time? Explain your reasoning.

96. CQ Most electrical codes require the metal case of an electric clothes dryer to have a wire running from the case to a nearby faucet (or any metal plumbing). Explain why.

Comprehensive Exercises

97. Find the current in each resistor in the circuit in ▾Fig. 18.42.

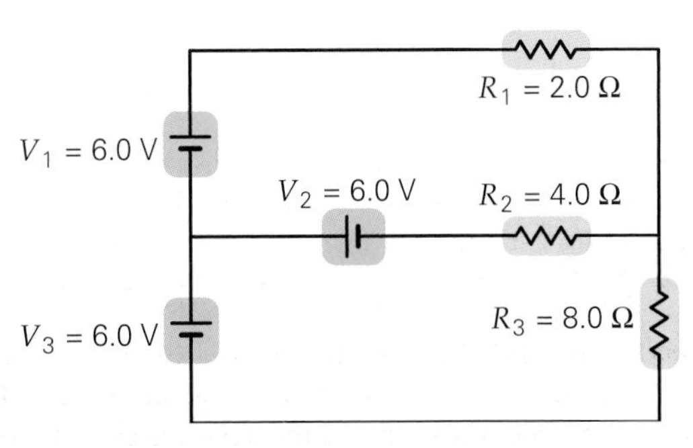

▲ **FIGURE 18.42 Kirchhoff's rules** See Exercise 97.

98. Four resistors are connected to a 90-V source, as shown in ▼Fig. 18.43. (a) Which resistor(s) receive(s) the most power and how much is that? (b) What is the total power delivered to the circuit by the power supply?

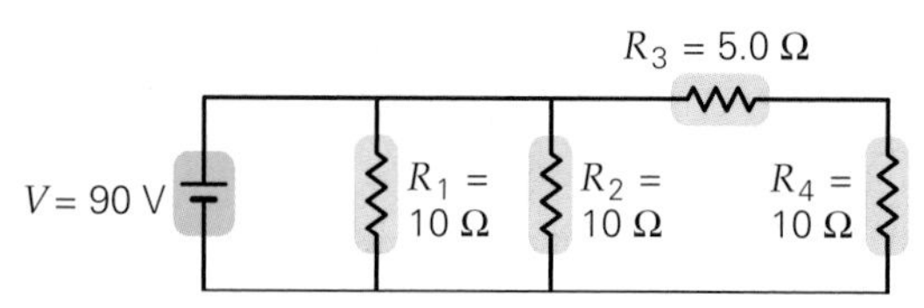

▲ **FIGURE 18.43 How much power is delivered?** See Exercise 98.

99. Four resistors are connected in a circuit with a 110-V source, as shown in ▼Fig. 18.44. (a) What is the current in each resistor? (b) How much power is delivered to each resistor?

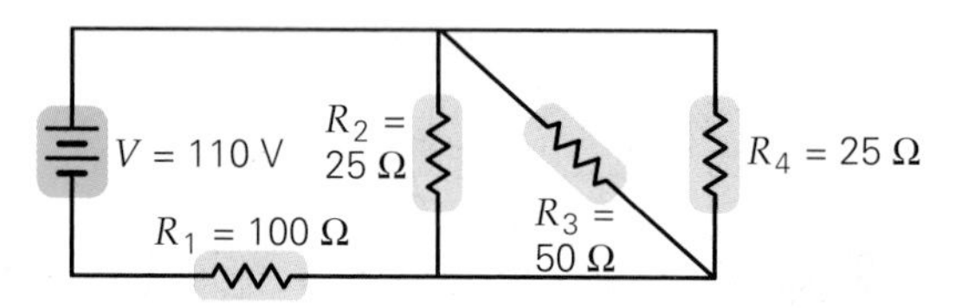

▲ **FIGURE 18.44 Joule-heat losses** See Exercise 99.

100. Nine resistors, each of value R, are connected in a "ladder" fashion, as shown in ▼Fig. 18.45. (a) What is the effective resistance of this network between points A and B? (b) If $R = 10\ \Omega$ and a 12.0-V battery is connected from point A to point B, how much current is in each resistor?

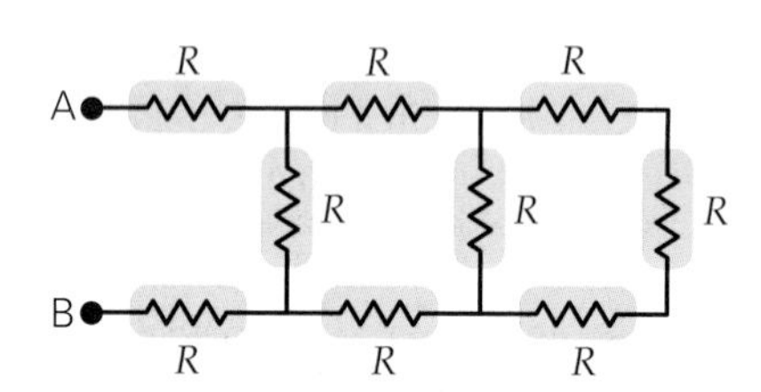

▲ **FIGURE 18.45 A resistance ladder** See Exercise 100.

101. A 4.0-Ω resistor and a 6.0-Ω resistor are connected in series. A third resistor is connected in parallel with the 6.0-Ω resistor. The whole arrangement gives a total equivalent resistance of 7.0 Ω. What is the value of the third resistor?

102. A galvanometer with an internal resistance of 50 Ω and a full-scale sensitivity of 200 μA is used to construct a multirange voltmeter. What values of multiplier resistors allow for three full-scale voltage readings of 20 V, 100 V, and 200 V? (See Fig. 18.18b.)

103. A galvanometer with an internal resistance of 100 Ω and a full-scale sensitivity of 100 μA is used to construct a multirange ammeter. What values of shunt resistors allow for three full-scale current readings of 1.0 A, 5.0 A and 10 A? (See Fig. 18.18a.)

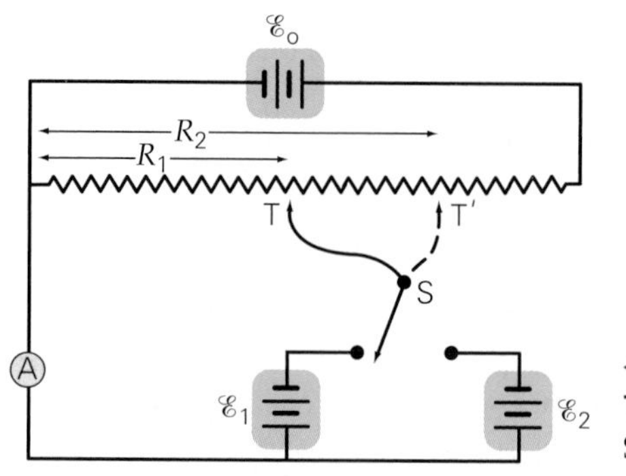

◀ **FIGURE 18.46 The potentiometer** See Exercise 104.

104. ▲Fig. 18.46 shows a schematic circuit of an instrument called a *potentiometer* that is a very accurate device for determining emf values of power supplies. It consists of three batteries, an ammeter, several resistors, and a uniform wire that can be "tapped" for a specific fraction of its total resistance. $\mathscr{E}_o$ is the emf of a working battery, $\mathscr{E}_1$ designates a battery with a precisely known emf, and $\mathscr{E}_2$ designates a battery whose emf is unknown. The switch S is thrown toward battery 1, and the point T (for "tapped") is moved along the resistor until the ammeter reads zero. Let's call resistance of this arrangement R_1. This procedure is repeated with the switch thrown toward battery 2, and the point T is moved to T′ until the ammeter again reads zero. Let's designate the resistance of this arrangement as R_2. Show that the unknown emf can be determined from the following relationship: $\mathscr{E}_2 = \frac{R_2}{R_1}\mathscr{E}_1$.

105. If a combination of three 30-Ω resistors receives energy at the rate of 3.2 W when connected to a 12-V battery, how are the resistors connected?

106. A battery has three cells, each with an internal resistance of 0.020 Ω and an emf of 1.50 V. The battery is connected in parallel with a 10.0-Ω resistor. (a) Determine the voltage across the resistor. (b) How much current is in each cell? (The cells in a battery are in series.)

107. A 10.0-μF capacitor in a heart defibrillator unit is charged fully by a 10 000-V power supply. Each capacitor plate is connected to the chest of a patient by wires and flat "paddles," one on either side of the heart. The energy stored in the capacitor is delivered through an RC circuit, where R is the resistance of the body between the two paddles. Data indicates that it takes 75.1 ms for the voltage to drop to 20.0 V. (a) Find the time constant. (b) Determine the resistance, R. (c) How much time does it take for the capacitor to lose 90% of its stored energy?

108. During an operation, one of the electrical instruments has its metal case shorted to the 120-V "hot" wire that powers it. The attending physician is isolated from ground because of rubber-soled shoes, and inadvertently touches the case with his elbow, while simultaneously having his opposite hand in contact with the patient's chest. The patient, lying on a metal table, is well grounded. If the patient's head-to-ground resistance is 2200 Ω, what is the minimum resistance for the physician so that they both feel, at most, a "mild shock"?

PHYSLET® The following Physlet Physics Problems can be used with this chapter.
30.1, 30.2, 30.3, 30.4, 30.6, 30.7, 30.8, 30.11, 30.12

CHAPTER 19

MAGNETISM

19.1 Magnets, Magnetic Poles, and Magnetic Field Direction 624
19.2 Magnetic Field Strength and Magnetic Force 626
19.3 Applications: Charged Particles in Magnetic Fields 629
19.4 Magnetic Forces on Current-Carrying Wires 632
19.5 Applications: Current-Carrying Wires in Magnetic Fields 635
19.6 Electromagnetism: The Source of Magnetic Fields 637
19.7 Magnetic Materials 641
*19.8 Geomagnetism: The Earth's Magnetic Field 644

PHYSICS FACTS

- The SI unit of current, the ampere or the coulomb per second, is officially defined in terms of the magnetic field it creates and the magnetic force that field can exert on another current.
- Nikola Tesla (1856–1943) was a Serbian-American researcher perhaps best known for the Tesla coil, which is capable of producing high voltages (see Chapter 20) and is a common sight at high school science fairs. Tesla's name became the SI unit of magnetic field. When Westinghouse secured the patent rights to his alternating-current designs, this resulted in a battle between Edison's direct-current system and the Tesla-Westinghouse alternating-current system. The latter eventually won out and has become the primary means of delivering electric energy throughout the world.
- Pierre Curie (1859–1906) pioneered in widely varying areas ranging from magnetism to radioactivity. He discovered that ferromagnetic substances exhibited a temperature transition above which they lost their ferromagnetic behavior. This is now known as the Curie temperature.

When thinking about magnetism, most people tend to think of an attraction because it is well known that certain things can be picked up with a magnet. You have probably encountered magnetic latches that hold cabinet doors shut, or used magnets to stick notes to the refrigerator. It is less likely that one thinks of repulsion. Yet repulsive magnetic forces exist—and they can be just as useful as attractive ones.

In this regard, the photo shows an interesting example. At first glance, the vehicle looks like an ordinary train. But where are the wheels? In fact, it isn't a conventional train at all, but a high-speed, *magnetically levitated* one. It doesn't physically touch the rails. Rather it "floats" above them, supported by repulsive forces produced by powerful magnets. The advantages are obvious: with no wheels, there is no rolling friction and no bearings to lubricate—in fact, there are few moving parts of any kind.

Where do these useful magnetic forces come from? For centuries, the properties of magnets were attributed to the supernatural. The original "natural" magnets were called lodestone. Magnetism is now associated with electricity, because physicists discovered that both are really different aspects of a single force: the electromagnetic force. Electromagnetism is used in motors, generators, radios, and many other familiar applications. In the future, the development of high-temperature superconductors (Chapter 17) may open the way for the practical application of many devices now found only in the laboratory.

Although electricity and magnetism are fundamentally different manifestations of the same force, it is instructive to consider them individually and then put them together, so to speak, as electromagnetism. This chapter and the next will investigate magnetism and its intimate relationship to electricity.

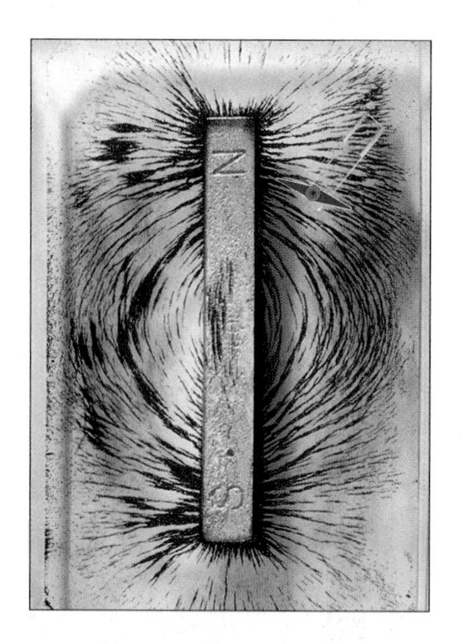

▲ **FIGURE 19.1 Bar magnet** The iron filings indicate the poles, or centers of force, of a common bar magnet. The compass direction designates these poles as north (N) and south (S). (See Fig. 19.3.)

Illustration 27.2 Earth's Magnetic Field

19.1 Magnets, Magnetic Poles, and Magnetic Field Direction

OBJECTIVES: To (a) learn the force rule between magnetic poles, and (b) explain how the direction of a magnetic field is determined with a compass.

One of the features of a common bar magnet is that it has two "centers" of force, called *poles*, near each end (◂Fig. 19.1). To avoid confusion with the plus-minus designation used for electric charge, these poles are instead called north (N) and south (S). This terminology stems from the early use of the magnetic compass to determine direction. The north pole of a compass magnet was historically defined as the *north-seeking* end—that is, the end that points *north* on the Earth. The other end of the compass was labeled as south, or a south pole.

By using two bar magnets, the nature of the forces acting between the magnetic poles can be determined. Each pole of a bar magnet is attracted to the opposite pole of the other and repelled by the same pole of the other. Thus we have the **pole–force law**, or **law of poles**:

> Like magnetic poles repel each other, and unlike magnetic poles attract each other (▾Fig. 19.2).

One immediate (and sometimes confusing) result of the historical definition of a magnet's north pole has to do with the Earth's magnetic field. Because the north pole of a bar magnet is attracted to the Earth's north *polar* region (that is, *geographic* north), that area must be acting, or magnetically speaking, as a south (magnetic) pole. (See Section 19.8 for more details on the geophysics of the Earth's magnetic field.) So the Earth's south magnetic pole is in the general vicinity of its north geographic pole.

Two opposite magnetic poles, such as those of a bar magnet, form a *magnetic dipole*. At first glance, a bar magnet 's field might appear to be the magnetic analog of the electric dipole. There are, however, fundamental differences between the two. For example, permanent magnets always have two poles, never one. You might think that breaking a bar magnet in half would yield two isolated poles. However, the resulting pieces of the magnet always turn out to be two shorter magnets, *each with its own set of north and south poles*. While a single magnetic pole (a *magnetic monopole*) could exist in theory, it has yet to be found experimentally.

The fact that there is no magnetic analog to electric charge provides a strong hint about the differences between electric and magnetic fields. For example, the source of magnetism is electric charge, just like the electric field. However, as we will see in Sections 19.6 and 19.7, magnetic fields are produced only when electric charges are *in motion*, such as electric currents in circuits and orbiting (or spinning) atomic electrons. The latter is actually the source of the bar magnet's field.

Magnetic Field Direction

The historical approach to analyzing a bar magnet's field was to try to express the magnetic force between poles in a mathematical form similar to Coulomb's law for the electric force (Chapter 15). In fact, Coulomb developed such a law, using magnetic pole strengths in place of electric charge. However, this approach is rarely used, because it does not fit our modern understanding—that is, single magnetic poles do not exist. Instead, the modern description uses the concept of the *magnetic field*.

Recall that electric charges produce electric fields, which can be represented by electric field lines. The electric field (vector) is defined as the force per unit charge at

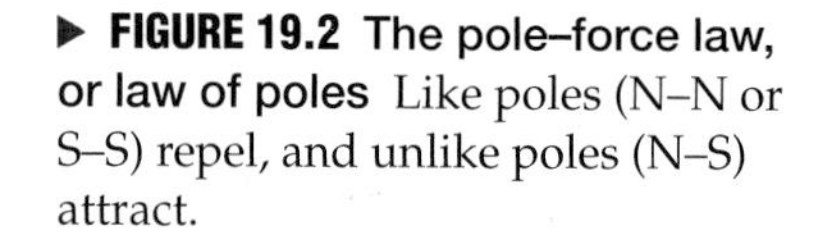

▸ **FIGURE 19.2 The pole–force law, or law of poles** Like poles (N–N or S–S) repel, and unlike poles (N–S) attract.

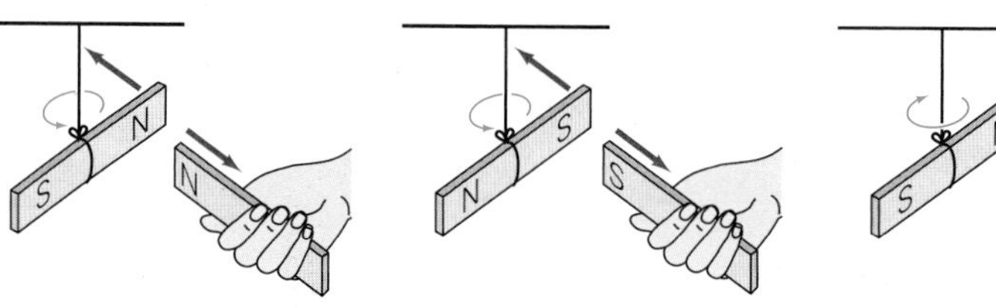

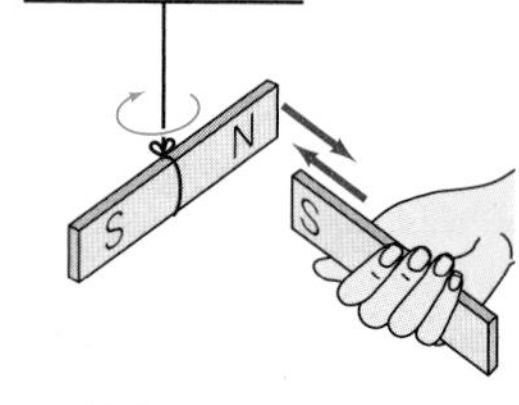

any location in space, or $\vec{\mathbf{E}} = \vec{\mathbf{F}}_e/q_o$. Similarly, magnetic interactions can be described in terms of the **magnetic field**, a vector quantity represented by the symbol $\vec{\mathbf{B}}$. Just as electric fields exist near electric charges, we know that magnetic fields occur near permanent magnets. The magnetic field pattern surrounding a magnet can be made visible by sprinkling iron filings over a magnet covered with paper or glass (Fig. 19.1). Because of the magnetic field, the iron filings themselves become magnetized into little magnets (basically compass needles) and, behaving like compasses, line up in the direction of $\vec{\mathbf{B}}$.

Because the magnetic field is a vector field, both magnitude (or "strength") and direction must be specified. The direction of a magnetic field (usually called a "B field") is defined in terms of a compass that has been calibrated (for direction) using the Earth's magnetic field:

> The direction of a magnetic field ($\vec{\mathbf{B}}$) at any location is the direction that the north pole of a compass would point if placed at that location.

Illustration 27.1 Magnets and Compass Needles

This provides a method for mapping a magnetic field by moving a small compass to various locations in the field. At any location, the compass needle will line up in the direction of the B field that exists there. If the compass is then moved in the direction in which its needle (the north end) points, the path of the needle traces out a *magnetic field line*, as illustrated in ▼Fig. 19.3a.

Because the north end of a compass points away from the north pole of a bar magnet, the field lines of a bar magnet point away from that pole and point toward its south pole. The rules that govern the interpretation of the magnetic field lines are the same that apply to electric field lines:

> The closer together (that is, the denser) the B field lines, the stronger the magnetic field. At any location, the direction of the magnetic field is tangent to the field line, or equivalently, the way the north end of a compass points.

Note the concentration of iron filings in the pole regions (Figs. 19.3b and c). This indicates closely spaced field lines and therefore a relatively strong magnetic field compared with other locations. As for field direction, note that just outside the middle of the magnet the field points vertically downward, tangent to the field line at that point (Fig. 19.3a, point P).

You might think that the magnitude of $\vec{\mathbf{B}}$ would be defined as the magnetic force per unit pole strength, in analogy to $\vec{\mathbf{E}}$. However, because magnetic monopoles don't exist, the magnitude of $\vec{\mathbf{B}}$ is defined in terms of the magnetic force exerted on a moving electric charge, which is discussed next.

▼ **FIGURE 19.3 Magnetic fields (a)** Magnetic field lines can be traced and outlined by using iron filings or a compass, as shown in the case of the magnetic field due to a bar magnet. The filings behave like tiny compasses and line up with the field. The closer together the field lines, the stronger the magnetic field. **(b)** Iron filing pattern for the magnetic field between unlike poles; the field lines converge. **(c)** Iron filing pattern for the magnetic field between like poles; the field lines diverge.

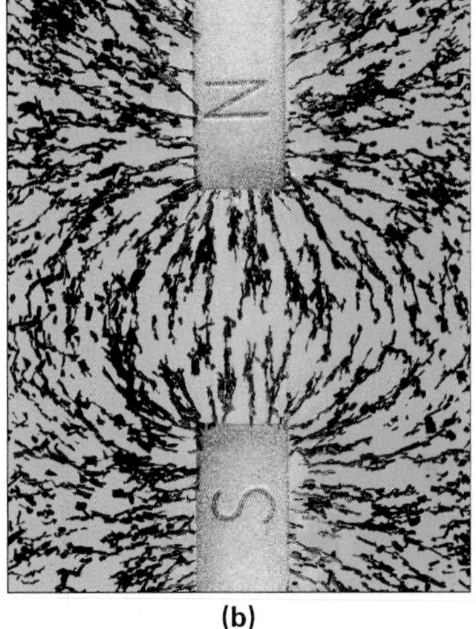

(b)

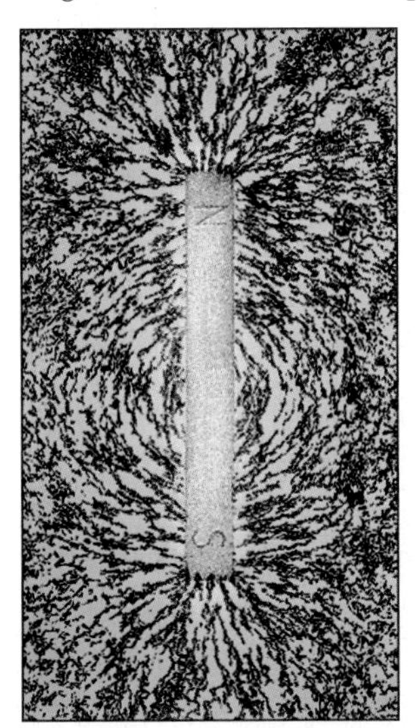

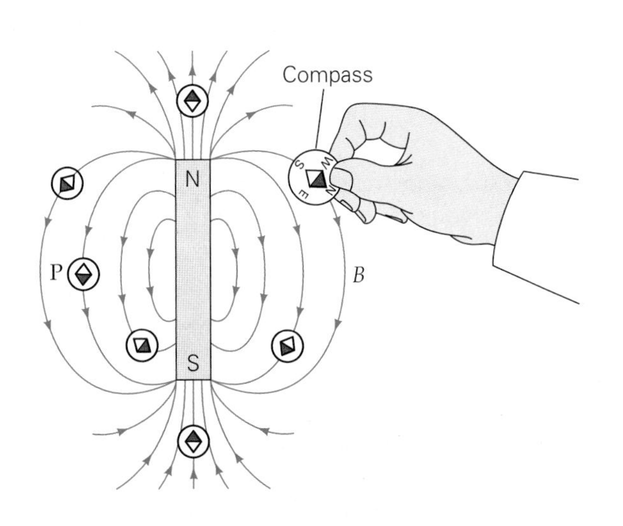

(a)

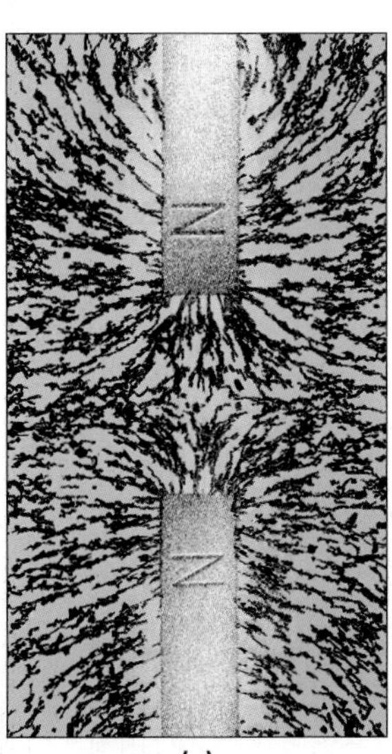

(c)

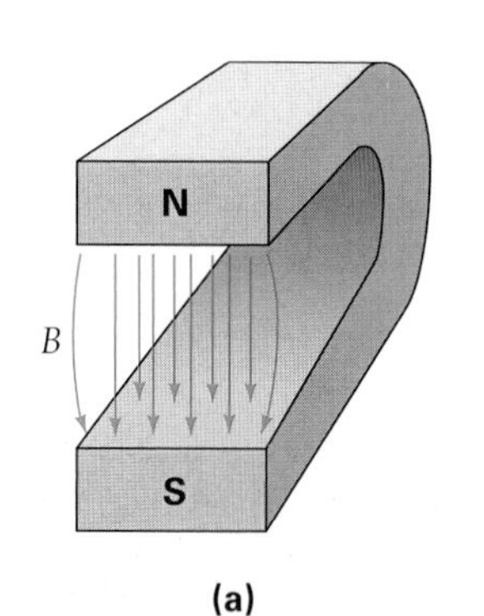

(a)

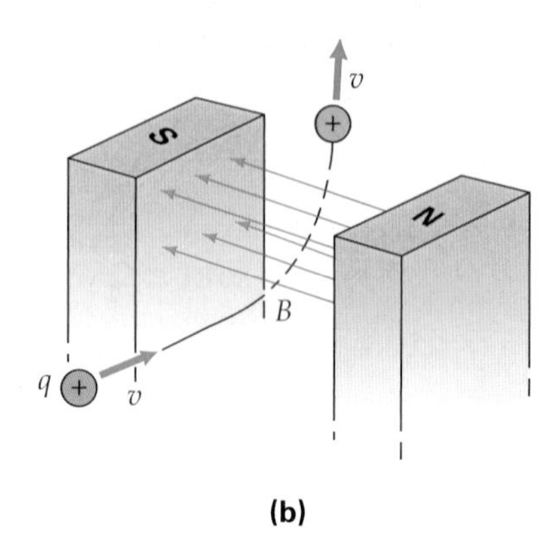

(b)

▲ **FIGURE 19.4 Force on a moving charged particle** **(a)** A horseshoe magnet, created by bending a permanent bar magnet, produces a fairly uniform field between its poles. **(b)** When a charged particle enters a magnetic field, the particle is acted on by a force whose direction is obvious by the deflection of the particle from its original path.

Note: The magnetic field plays a vital role in magnetic resonance imaging (MRI), a technique widely used in medical diagnostics. (See Chapter 28.)

19.2 Magnetic Field Strength and Magnetic Force

OBJECTIVES: To (a) define magnetic field strength, and (b) determine the magnetic force exerted by a magnetic field on a moving charged particle.

Experiments indicate that one quantity that determines the *magnetic* force on a particle is its *electric* charge. That is, there is a connection between *electrical* properties of objects and how they respond to *magnetic* fields. The study of these interactions is called **electromagnetism**. Consider the following electromagnetic interaction. Suppose a positively charged particle is moving at a constant velocity as it enters a uniform magnetic field. For simplicity, let us also assume that its velocity is perpendicular to the field. (A fairly uniform B field exists between the poles of a "horseshoe" magnet, as shown in ◂Fig. 19.4a.) When the charged particle enters the field, it is *deflected* into an upward curved path, which is actually part of a circular path (if the B field is uniform), as shown in Fig. 19.4b.

From our study of circular motion (Section 7.3), for a particle to move in a circular arc, a centripetal force must exist that is always perpendicular to its velocity. But what provides this force here? No electric field is present. The gravitational force, besides being too weak to cause such a deflection, would deflect the particle into a downward parabolic arc, not an upward circular one. Evidently, the force is a magnetic one, due to the interaction of the moving charge and the magnetic field. *Thus a magnetic field can exert a force on a moving charged particle.*

From detailed measurements, the magnitude of this force is found to be proportional to the particle's charge and its speed. When the particle's velocity ($\vec{\mathbf{v}}$) is perpendicular to the magnetic field ($\vec{\mathbf{B}}$), the magnitude of the field, or the field strength B, is defined as

$$B = \frac{F}{qv} \quad \textit{(valid only when } \vec{\mathbf{v}} \textit{ is perpendicular to } \vec{\mathbf{B}}) \tag{19.1}$$

SI unit of magnetic field:
newton per ampere-meter [$\mathrm{N/(A\cdot m)}$, or tesla (T)]

Physically, B means the magnetic force exerted on a charged particle *per unit charge* (coulomb) and *per unit speed* (m/s). From this relationship, the units of B are $\mathrm{N/(C\cdot m/s)}$ or $\mathrm{N/(A\cdot m)}$, because $1\ \mathrm{A} = 1\ \mathrm{C/s}$. This combination of units is named the **tesla (T)** after Nikola Tesla (1856–1943), an early researcher in magnetism, and $1\ \mathrm{T} = 1\ \mathrm{N/(A\cdot m)}$. Most everyday magnetic field strengths, such as those from permanent magnets, are much smaller than 1 T. In such situations, it is common to express magnetic field strengths in milliteslas ($1\ \mathrm{mT} = 10^{-3}\ \mathrm{T}$) or microteslas ($1\ \mu\mathrm{T} = 10^{-6}\ \mathrm{T}$). A non-SI unit commonly used by geologists and geophysicists, called the *gauss* (G), is defined as one ten-thousandth of a tesla ($1\ \mathrm{G} = 10^{-4}\ \mathrm{T} = 0.1\ \mathrm{mT}$). For example, the Earth's magnetic field is on the order of several tenths of a gauss or several hundredths of a millitesla. On the other hand, conventional laboratory magnets can produce fields as high as 3 T, and superconducting magnets up to 25 T or higher.

Once the magnetic field strength has been determined (Eq. 19.1), the force on any charged particle moving at any speed can be found.* Equation 19.1 can be solved to find the force's magnitude:

$$F = qvB \quad \textit{(valid only if } \vec{\mathbf{v}} \textit{ is perpendicular to } \vec{\mathbf{B}}) \tag{19.2}$$

In the more general case, a particle's velocity may *not* be perpendicular to the field. Then the magnitude of the force depends on the sine of the angle (θ) between the velocity vector and the magnetic field vector. In general, the magnitude of the magnetic force is

$$F = qvB \sin\theta \quad \textit{magnetic force on a charged particle} \tag{19.3}$$

Note this means that the magnetic force is zero when $\vec{\mathbf{v}}$ and $\vec{\mathbf{B}}$ are parallel ($\theta = 0°$) or oppositely directed ($\theta = 180°$), because $\sin 0° = \sin 180° = 0$. The force has maximum value when these two vectors are perpendicular. With $\theta = 90°$ ($\sin 90° = 1$), this maximum value is $F = qvB \sin 90° = qvB$.

*Strictly speaking, the speeds must be considerably less than the speed of light to avoid relativistic complications (Chapter 26).

The Right-Hand Force Rule for Moving Charges

The *direction* of the magnetic force on any moving charged particle is determined by the orientation of the particle's velocity relative to the magnetic field. Experiment shows that the magnetic force direction is given by the **right-hand force rule** (▾Fig. 19.5a):

> When the fingers of the right hand are pointed in the direction of a charged particle's velocity $\vec{\mathbf{v}}$ and then curled (through the smallest angle) toward the vector $\vec{\mathbf{B}}$, the extended thumb points in the direction of the magnetic force $\vec{\mathbf{F}}$ that acts on a *positive* charge. If the particle has a negative charge, the magnetic force is in the direction opposite to that of the thumb.

Note: *B* fields that point into the plane of the page are designated by ×. *B* fields that point out of the plane are indicated by •. Visualize these symbols as the feathered end and the tip of an arrow, respectively.

You might imagine the fingers of the right hand to be physically turning or rotating the vector $\vec{\mathbf{v}}$ into $\vec{\mathbf{B}}$ so that $\vec{\mathbf{v}}$ and $\vec{\mathbf{B}}$ are aligned.

Notice that the magnetic force is always *perpendicular to the plane formed by* $\vec{\mathbf{v}}$ *and* $\vec{\mathbf{B}}$ (Fig. 19.5b). Because the force is perpendicular to the particle's direction of motion ($\vec{\mathbf{v}}$), it cannot do any work on the particle. (This follows from the definition of work in Chapter 5, with a right angle between the force and displacement, $W = Fd \cos 90° = 0$.) Therefore, a magnetic field does not change the speed (that is, kinetic energy) of a particle—only its direction.

Several common alternative (and physically equivalent) right-hand force rules are shown in Fig. 19.5c. For negative charges, it is suggested that you start by assuming that the charge is positive. Next, determine the force direction, using the right-hand force rule. Lastly, *reverse* this direction to find the actual force direction on the negative charge. To see how this rule is applied to both charge signs, consider the following Conceptual Example.

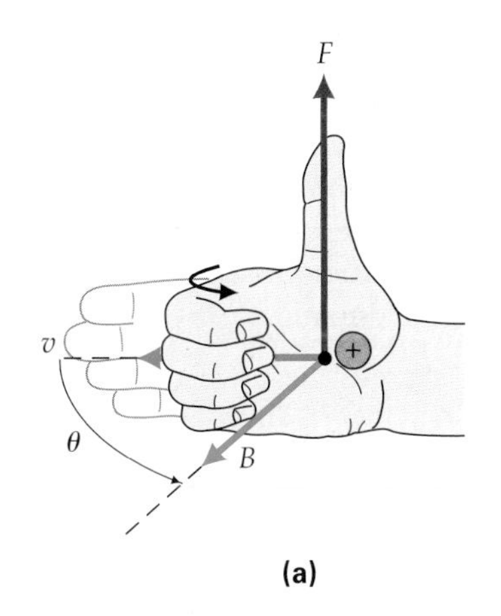

◀ **FIGURE 19.5 Right-hand rules for magnetic force** **(a)** When the fingers of the right hand are pointed in the direction of $\vec{\mathbf{v}}$ and then curled toward the direction of $\vec{\mathbf{B}}$, the extended thumb points in the direction of the force $\vec{\mathbf{F}}$ on a *positive* charge. **(b)** The magnetic force is always perpendicular to the plane of $\vec{\mathbf{B}}$ and $\vec{\mathbf{v}}$ and thus is always perpendicular to the direction of the particle's motion. **(c)** When the extended forefinger of the right hand points in the direction of $\vec{\mathbf{v}}$ and the middle finger points in the direction of $\vec{\mathbf{B}}$, the extended right thumb points in the direction of $\vec{\mathbf{F}}$ on a *positive* charge. **(d)** When the fingers of the right hand are pointed in the direction of $\vec{\mathbf{B}}$ and the thumb in the direction of $\vec{\mathbf{v}}$, the palm points in the direction of the force $\vec{\mathbf{F}}$ on a *positive* charge. (Regardless of the rule you employ, remember to use your right hand and reverse the direction for a negative charge.)

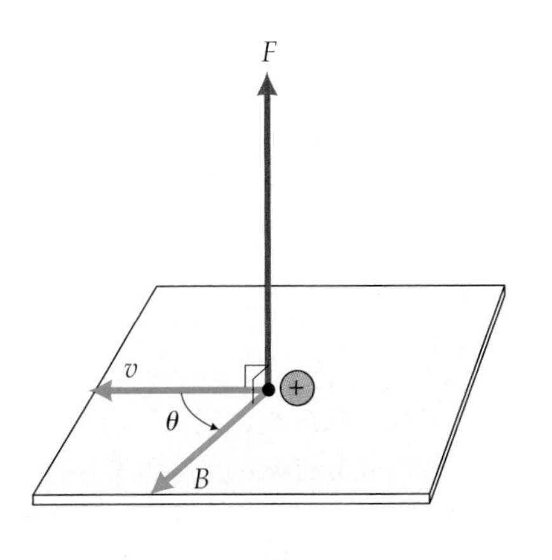

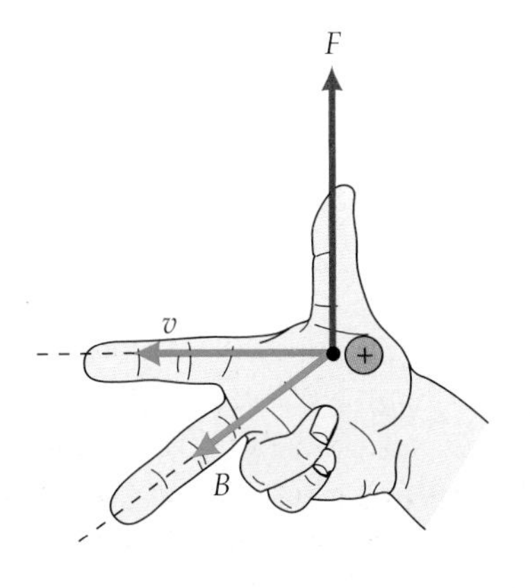

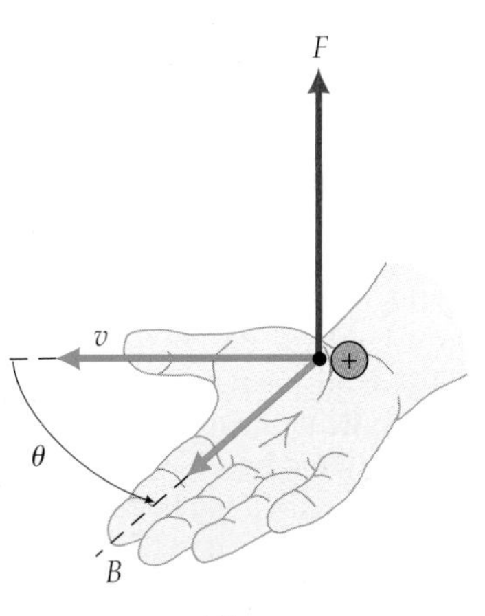

Conceptual Example 19.1 ■ Even "Lefties" Use the Right-Hand Rule

In a linear particle accelerator, a beam of protons travels horizontally northward. To deflect the protons eastward with a uniform magnetic field, the field should point in which direction? (a) vertically downward, (b) west, (c) vertically upward, or (d) south.

Reasoning and Answer. Because the force is perpendicular to the plane formed by $\vec{\mathbf{v}}$ and $\vec{\mathbf{B}}$, the field *cannot* be horizontal. If it were, it would deflect the protons down or up. Thus, we can eliminate choices (b) and (d). Use the right-hand force rule (assuming a positive charge) to see if $\vec{\mathbf{B}}$ could be downward (answer a). You should verify (make a sketch) that for a downward magnetic field, the force would be to the west. Hence, the answer must be (c). The magnetic field must point upward to deflect the protons to the east. You should verify that this answer is correct, using the right-hand force rule.

Follow-Up Exercise. What direction would the particles in this Example deflect if they were electrons moving southward? *(Answers to all Follow-Up Exercises are at the back of the text.)*

According to our previous discussion, charged particles traveling in uniform magnetic fields have circular arcs as their trajectory. To see this in more detail, consider Example 19.2 carefully.

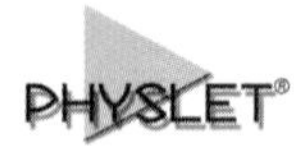

Illustration 27.3 A Mass Spectrometer

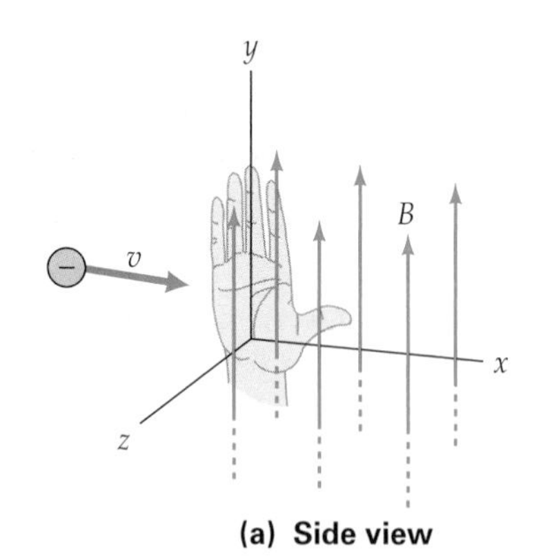

(a) Side view

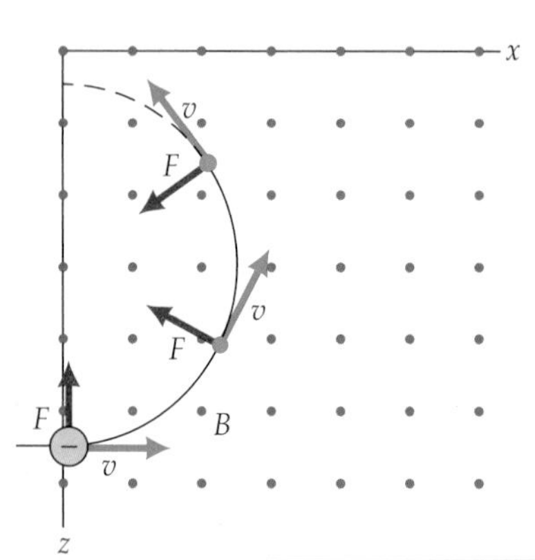

(b) Top view

▲ **FIGURE 19.6 Path of a charged particle in a magnetic field (a)** A charged particle entering a uniform magnetic field will be deflected—here toward the xy-plane by the right-hand rule, because the charge is negative. **(b)** In the field, the force is always perpendicular to the particle's velocity. The particle moves in a circular path if the field is constant and the particle enters the field perpendicularly to its direction. (See Example 19.2.)

Example 19.2 ■ Going Around in Circles: Force on a Moving Charge

A particle with a charge of -5.0×10^{-4} C and a mass of 2.0×10^{-9} kg moves at 1.0×10^3 m/s in the $+x$-direction. It enters a uniform magnetic field of 0.20 T that points in the $+y$-direction (see ◀Fig. 19.6a). (a) Which way will the particle deflect just as it enters the field? (b) What is the magnitude of the force on the particle just as it enters the field? (c) What is the radius of the circular arc that the particle will travel while in the field?

Thinking It Through. The initial deflection is, of course, in the direction of the initial magnetic force. Since the particle is negative, care must be taken in applying the right-hand force rule. A circular arc is expected, as the magnetic force is always perpendicular to the particle's velocity. The magnitude of the magnetic force on a single charge is given by Eq. 19.3. This force is the only significant force on the electron; therefore it is also the net force. From this, Newton's second law should enable us to determine the circular orbit radius.

Solution. We list the given data.

Given: $q = -5.0 \times 10^{-4}$ C
$v = 1.0 \times 10^3$ m/s ($+x$-direction)
$m = 2.0 \times 10^{-9}$ kg
$B = 0.20$ T ($+y$-direction)

Find: (a) Initial deflection direction
(b) Magnitude of initial magnetic force F
(c) Radius r of the orbit

(a) By the right-hand force rule, the force on a positive charge would be in the $+z$-direction (see direction of palm, Fig 19.6a). Because the charge is negative, the force is actually opposite this; thus, the particle will begin to deflect in the $-z$-direction.

(b) The force's magnitude can be determined from Eq. 19.3. Because only the magnitude is of interest, the sign of q can be dropped.

$$F = qvB\sin\theta$$
$$= (5.0 \times 10^{-4}\,\text{C})(1.0 \times 10^3\,\text{m/s})(0.20\,\text{T})(\sin 90°) = 0.10\,\text{N}$$

(c) Because the magnetic force is the only force acting on the particle, it is also the net force (Fig. 19.6b). This net force points toward the circle's center and is called the centripetal force ($\vec{\mathbf{F}}_{\text{net}} = \vec{\mathbf{F}}_{\text{c}}$; see Chapter 7). Therefore, in describing circular motion, Newton's second law becomes

$$\vec{\mathbf{F}}_{\text{c}} = m\vec{\mathbf{a}}_{\text{c}}$$

Now substitute the magnetic force (from Eq. 19.2, because $\theta = 90°$) for the net force and the expression for centripetal acceleration ($a_c = v^2/r$; see Section 7.3) to obtain:

$$qvB = \frac{mv^2}{r} \quad \text{or} \quad r = \frac{mv}{qB}$$

Finally, inserting the numerical values,

$$r = \frac{mv}{qB} = \frac{(2.0 \times 10^{-9}\,\text{kg})(1.0 \times 10^3\,\text{m/s})}{(5.0 \times 10^{-4}\,\text{C})(0.20\,\text{T})} = 2.0 \times 10^{-2}\,\text{m} = 2.0\,\text{cm}$$

Follow-Up Exercise. In this Example, if the particle had been a proton traveling initially in the $+z$-direction, (a) in what direction would it be initially deflected? (b) If the radius of its circular path were 10 cm and its speed were 1.0×10^6 m/s, what would be the magnetic field strength?

19.3 Applications: Charged Particles in Magnetic Fields

OBJECTIVE: To understand how the magnetic force on charged particles is employed in various practical applications.

We have seen that a charged particle moving in a magnetic field usually experiences a magnetic force. This force deflects the particle by an amount that depends on its mass, charge, and velocity (speed and direction), as well as the field strength. Let's take a look at the important role this force plays in some common appliances, machines, and instruments.

The Cathode-Ray Tube (CRT): Oscilloscope Screens, Television Sets, and Computer Monitors*

The **cathode-ray tube (CRT)** is a vacuum tube that is commonly used as a display screen for a laboratory instrument called an *oscilloscope* (▸Fig. 19.7). The basic operation of the oscilloscope and television picture tube is similar, and is shown in ▾Fig. 19.8. Electrons are emitted from a hot metallic filament and accelerated by a voltage applied between the cathode (−) and the anode (+) in an "electron gun" arrangement. In one kind of design, these tubes use current-carrying coils to produce a magnetic field (Section 19.6), which in turn controls the deflection of the electron beam. As the field strength is quickly varied, the electron beam scans the fluorescent screen in a fraction of a second. When the electrons hit the fluorescent material, they cause the atoms to emit light (Section 27.4). For a black-and-white TV, the signals reproduce an image on the screen as a mosaic of light and dark dots, depending on whether the beam is on or off at a particular instant.

Producing the images on a color TV or a color computer monitor is somewhat more involved. A common color picture tube has three beams, one for each of the primary colors (red, green, and blue; Chapter 25). Phosphor dots on the screen are arranged in groups of three (triads), with one dot for each primary color. The excitation of the appropriate dots and the resulting emission (fluorescence) of a combination of colors produce a color picture.

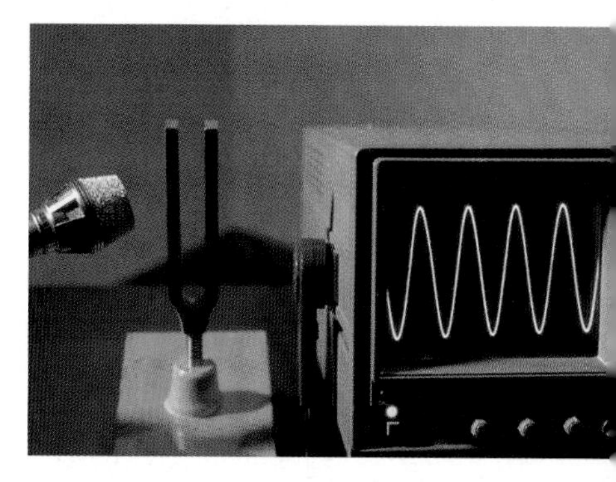

▲ **FIGURE 19.7 Cathode-ray tube (CRT)** The motion of the deflected beam traces a pattern on a fluorescent screen.

The Velocity Selector and the Mass Spectrometer

Have you ever thought about how the mass of an atom or molecule is measured? Electric and magnetic fields provide a way in a **mass spectrometer** ("mass spec" for short). Mass spectrometers perform many functions in modern laboratories. For example, they can be used to track short-lived molecules in studies of the biochemistry of living organisms. They can also determine the structure of large organic molecules and to analyze the composition of complex mixtures, such as a sample of smog-laden air.

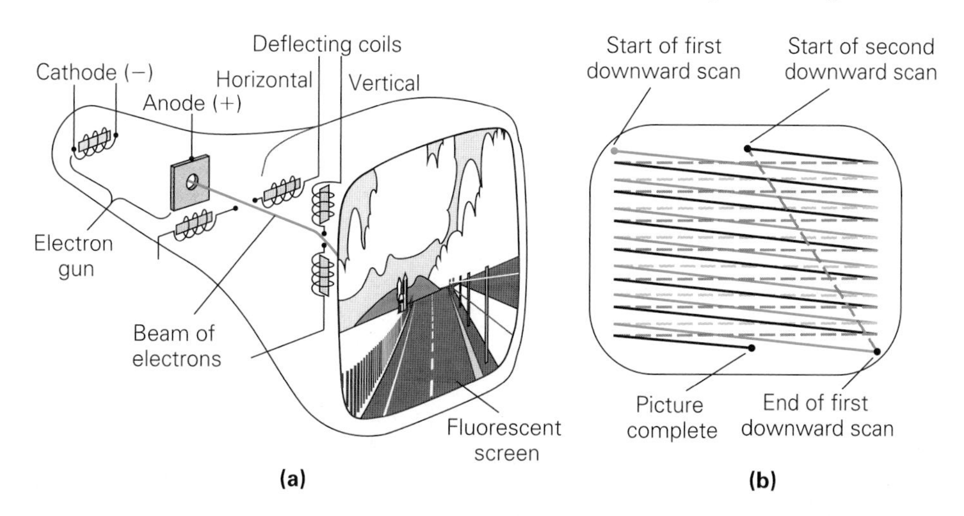

◀ **FIGURE 19.8 Television tube** **(a)** A television picture tube is a cathode-ray (electron) tube, or CRT. The electrons are accelerated between the cathode and anode and are then deflected to the proper location on a fluorescent screen by magnetic fields produced by current-carrying coils. **(b)** In this design the beam scans every other line on the screen in one downward pass that takes $\frac{1}{60}$ s and then scans the lines in between in a second pass of $\frac{1}{60}$ s. This yields a complete picture of 525 lines in $\frac{1}{30}$ s.

*More and more, vacuum tube displays are being replaced by flat-screen displays that employ materials such as liquid crystals (LCD TVs and flat-screen computer monitors). These do not use magnetic forces in their operation.

In criminal cases, forensic chemists use the mass spectrometer to identify traces of materials, such as a streak of paint from a car accident. In other fields such as archaeology and paleontology, these instruments can be used to separate atoms to establish the age of ancient rocks and human artifacts. In modern hospitals, mass spectrometers are essential for measuring and maintaining the proper balance of gaseous medications, such as anesthetic gases administered during an operation.

Note: The difference in mass between a neutral atom or molecule and its charged counterparts (ions) amounts to only the mass of an electron or two and is therefore negligible under most circumstances.

In actuality, it is the masses of *ions*, or charged molecules, that are measured in mass spectrometer.* Ions with a known charge $(+q)$ are produced by removing electrons from atoms and molecules. At this point, the ion beam would have a distribution of speeds, rather than a single speed. If these entered the mass spectrometer, then different-speed ions would take different paths in the mass spectrometer. Thus, before they enter the spectrometer, a specific ion velocity must be selected for analysis. This can be accomplished by using a *velocity selector*. This instrument consists of an electric field and magnetic field at right angles.

This arrangement allows particles traveling only at a unique velocity to pass through undeflected. To see this, consider a positive ion approaching the crossed-field arrangement at right angles to both fields. The electric field produces a downward force $(F_e = qE)$, and the magnetic field produces an upward force $(F_m = qvB_1)$. (You should verify each force's direction in ▾Fig. 19.9.)

If the beam is not to be deflected, the resultant, or net, force on each particle must be zero. In other words, these two forces cancel. Therefore they are equal in magnitude and oppositely directed. Equating the two force magnitudes,

$$F_e = F_m \quad \text{or} \quad qE = qvB_1$$

which can be solved for the "selected" speed:

$$v = \frac{E}{B_1}$$

If the plates are parallel, the electric field between them is given by $E = V/d$, where V is the voltage across the plates and d is the distance between them. A more practical version of the previous equation is

PHYSLET®

Exploration 27.2 *Velocity Selector*

$$v = \frac{V}{B_1 d} \quad \textit{speed selected by a velocity selector} \tag{19.4}$$

The desired speed can be selected by varying V, as usually B_1 and d are difficult to change.

Once through the velocity selector, the beam passes through a slit into another magnetic field $(\vec{\mathbf{B}}_2)$ that is perpendicular to the beam direction. At this point, the particles are bent into a circular arc. The analysis is identical to that in Example 19.2. Therefore,

$$F_c = ma_c \quad \text{or} \quad qvB_2 = m\frac{v^2}{r}$$

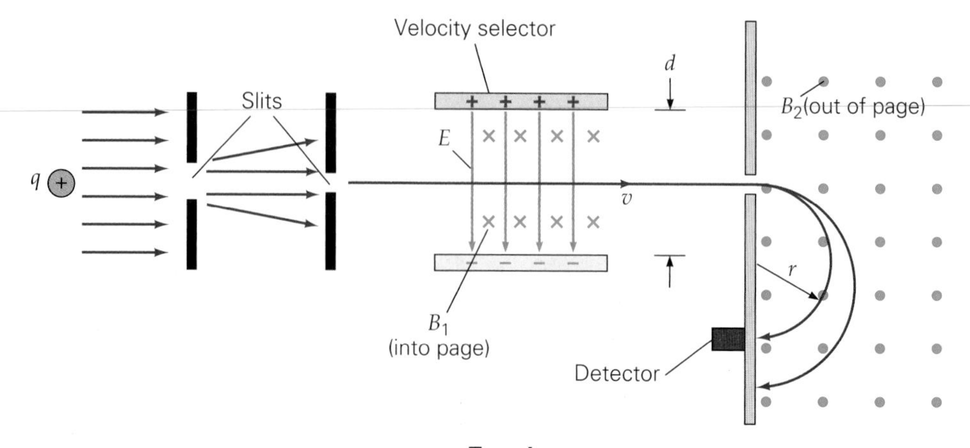

▶ **FIGURE 19.9 Principle of the mass spectrometer** Ions pass through the velocity selector; only those with a particular velocity $(v = E/B_1)$ then enter a magnetic field (B_2). These ions are deflected, with the radius of the circular path depending on the mass and charge of the ion. Paths of two different radii indicate that the beam contains ions of two different masses (assuming that they have the same charge).

*Recall that removing electrons from or adding electrons to an atom or molecule produces an ion. However, an ion's mass is negligibly different from that of its neutral atom, because the electron's mass is very small compared to the masses of the protons and neutrons in the atomic nuclei.

Using Eq. 19.4, the mass of the particle is given by

$$m = \left(\frac{qdB_1B_2}{V}\right)r \quad \textit{(mass determined with a mass spectrometer)} \tag{19.5}$$

The quantity inside the parentheses is a constant (assuming all ions have the same charge). Hence, the greater the mass of an ion, the greater its circular-path radius. Two paths of different radii are shown in Fig. 19.9. This indicates that the beam actually contains ions of two different masses. If the radius is measured (say, by recording the position where the ions hit a detector), the ion mass can be determined using Eq. 19.5.

In a mass spectrometer of slightly different design, the detector is kept at a fixed position. This design employs a time-varying magnetic field (B_2), and a computer records and stores the detector reading as a function of time. Notice that in this design, m is proportional to B_2. To see this, rewrite Eq. 19.5 as $m = (qdB_1r/V)B_2$. Because the quantity in the parentheses is a constant, then $m \propto B_2$. Thus as B_2 is varied, the detector data in connection with a high-speed computer enables us to determine the masses and relative number (that is, percentage) of ion of each mass. Regardless of design, the result—called a *mass spectrum* (the number of ions plotted against their mass)—is typically displayed on an oscilloscope or computer screen and digitized for storage and analysis (▸Fig. 19.10). Consider the following Example of a mass spectrometer arrangement.

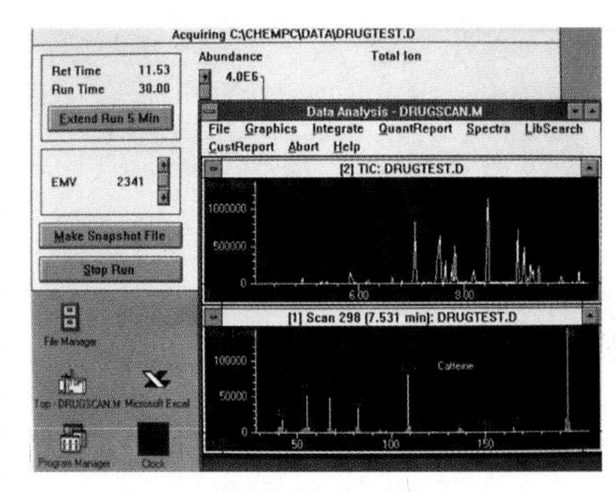

▲ **FIGURE 19.10 Mass spectrometer** Display of a mass spectrometer, with the number of molecules plotted vertically and the molecular mass horizontally. The molecule being analyzed is myoglobin, a protein that stores oxygen in muscle tissue. For myoglobin, each peak on the display represents the mass of an ionized fragment. Such patterns, *mass spectra*, help determine the composition and structure of large molecules. The mass spectrometer can also be used to identify tiny amounts of a molecule in a complex mixture.

PHYSLET®

Exploration 27.3 Mass Spectrometer

Example 19.3 ■ The Mass of a Molecule: a Mass Spectrometer

One electron is removed from a methane molecule before it enters the mass spectrometer in Fig. 19.9. After passing through the velocity selector, the ion has a speed of 1.00×10^3 m/s. It then enters the main magnetic field region, in which the field strength is 6.70×10^{-3} T. From there it follows a circular path and lands 5.00 cm from the field entrance. Determine the mass of this molecule. (Neglect the mass of the electron that is removed.)

Thinking It Through. The centripetal force is provided by the magnetic force on the ion. Because the velocity and magnetic field are at right angles, the magnetic force is given by Eq. 19.2. Applying Newton's second law to circular motion, we can solve for the molecule's mass.

Solution. First, list the given data.

Given: $q = 1.60 \times 10^{-19}\,\text{C}$ (electron)
$r = d/2 = (5.00\text{ cm})/2 = 0.0250$ m
$B_2 = 6.70 \times 10^{-3}$ T
$v = 1.00 \times 10^3$ m/s

Find: m (mass of a methane molecule)

The centripetal force on the ion ($F_c = mv^2/r$) is provided by the magnetic force ($F_m = qvB_2$):

$$\frac{mv^2}{r} = qvB_2$$

Solving this equation for m and putting in the numerical values, we have:

$$m = \frac{qB_2r}{v} = \frac{(1.60 \times 10^{-19}\,\text{C})(6.70 \times 10^{-3}\,\text{T})(0.0250\text{ m})}{1.00 \times 10^3\text{ m/s}} = 2.68 \times 10^{-26}\text{ kg}$$

Follow-Up Exercise. In this Example, if the magnetic field between the velocity selector's parallel plates, which are 10.0 mm apart, is 5.00×10^{-2} T, what voltage must be applied to the plates?

Silent Propulsion: Magnetohydrodynamics

In a search for quiet and efficient methods of propulsion at sea, engineers have invented a system based on *magnetohydrodynamics*—the study of the interactions of moving fluids and magnetic fields. This method of propulsion relies on the magnetic force and does not require moving parts, such as motors, bearings, and shafts, that are common to most ships and submarines. To avoid detection, this "silent-running" feature is of particular importance to modern submarine design.

Basically, seawater enters the front of the unit and is expelled at high speeds out the rear (▸Fig. 19.11). A superconducting electromagnet (see Section 19.7) is used to produce a large magnetic field. At the same time, an electric generator produces a large dc voltage, sending a current through the seawater. [Recall that seawater is a

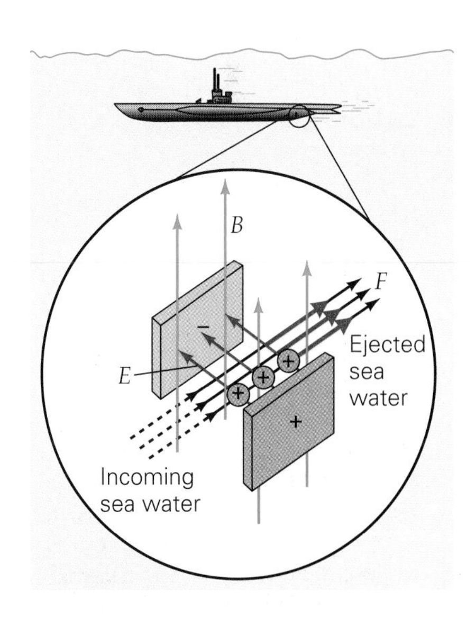

▲ **FIGURE 19.11 Propulsion via magnetohydrodynamics** In magnetohydrodynamic propulsion, seawater has an electric current passed through it by a dc voltage. A magnetic field exerts a force on the current, pushing the water out of the submarine or boat. The reaction force pushes the vessel in the opposite direction.

good conductor because it has a large concentration of sodium (Na^+) and chlorine (Cl^-) ions.] The magnetic force on the electric current pushes the water backward, expelling it as a jet of water. By Newton's third law, a reaction force acts forward on the submarine, enabling it to accelerate silently.

19.4 Magnetic Forces on Current-Carrying Wires

OBJECTIVES: To (a) calculate the magnetic force on a current-carrying wire and the torque on a current-carrying loop, and (b) explain the concept of the magnetic moment of a loop or coil.

Any charged particle moving in a magnetic field will generally experience a magnetic force. Because an electric current is composed of moving charges, we should expect a current-carrying wire, when placed in a magnetic field, to experience such a force as well. The sum of the individual magnetic forces on the charges that make up the current yield the total magnetic force on the wire.

Note: Remember that thinking of a current in terms of a positive charge is simply a useful convention. In reality, negative electrons are the charge carriers in ordinary electric current.

Recall that the direction of the "conventional current" assumes that electric current in a wire is due to the motion of positive charges, as depicted in ▾Fig. 19.12.* In this orientation, the magnetic force is a maximum, because $\theta = 90°$. In a time t, any one charge q_i (that is, an electron) would move on the average a length $L = vt$, where v is the average drift speed. Because all the moving charges (total charge $= \Sigma q_i$) in this length of wire are acted on by a magnetic force in the same direction, the magnitude of the total force on this length of wire is (assuming the field to be uniform), from Eq. 19.2,

$$F = (\Sigma q_i)vB$$

Substituting L/t for v and rearranging gives

$$F = (\Sigma q_i)\left(\frac{L}{t}\right)B = \left(\frac{\Sigma q_i}{t}\right)LB$$

Illustration 27.4 Magnetic Forces on Currents

But $\Sigma q_i/t$ is just the current (I). Therefore, in terms of the circuit current, we can write

$$F = ILB \quad \textit{(valid only if current and magnetic field are perpendicular)} \tag{19.6}$$

This result yields the maximum force on the wire. If the current makes an angle θ with respect to the field direction, then the force on the wire will be less. In general, the force on a length of current-carrying wire in a uniform magnetic field is

$$F = ILB \sin\theta \quad \textit{magnetic force on a current-carrying wire} \tag{19.7}$$

If the current is parallel to or directly opposite to the field, then the force on the wire is zero.

The direction of the magnetic force on a current-carrying wire is also given by a right-hand rule. As was the case for individual charged particles, there are several equivalent versions of the **right-hand force rule for a current-carrying wire**, the most common being:

Exploration 27.1 Map Field Lines and Determine Forces

When the fingers of the right hand are pointed in the direction of the conventional current I and then curled toward the vector $\vec{\mathbf{B}}$, the extended thumb points in the direction of the magnetic force on the wire (see ▸Figs. 19.13a and b).

▶ **FIGURE 19.12 Force on a wire segment** Magnetic fields exert forces on wires that carry currents, because electric current is composed of moving charged particles. The maximum magnetic force on a current-carrying wire is shown, because the angle between the charge velocity and the field is 90°.

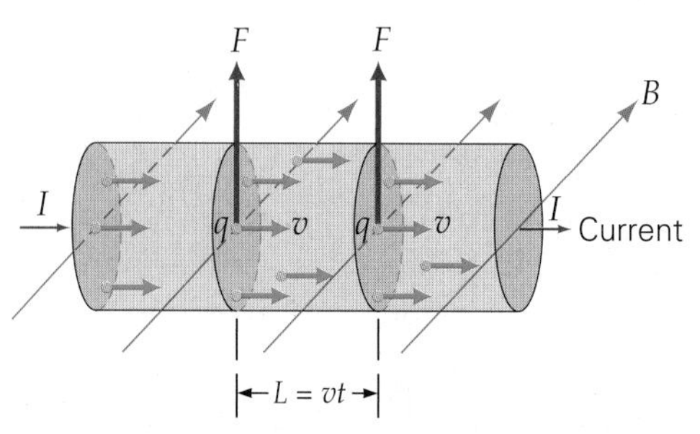

*Use the right-hand force rule to convince yourself that electrons traveling to the left would give the same magnetic force direction.

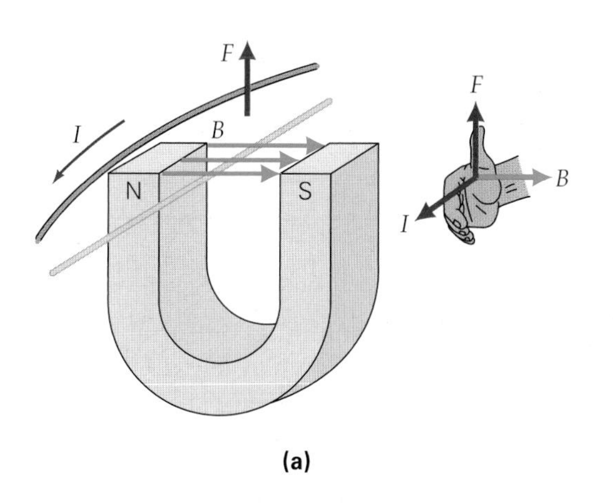

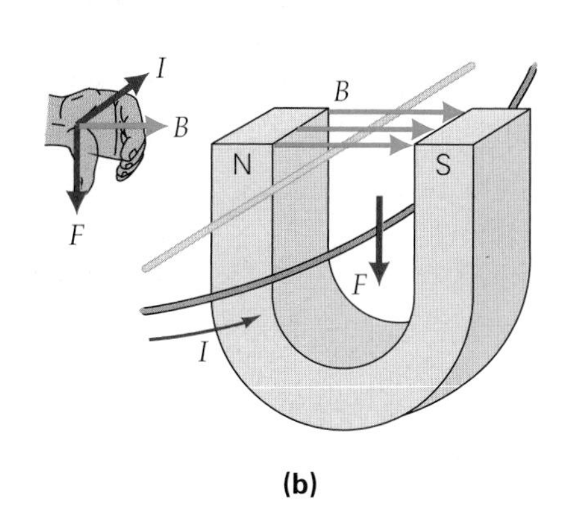

◀ **FIGURE 19.13 A right-hand force rule for current-carrying wires** The direction of the force is given by pointing the fingers of the right hand in the direction of the conventional current I and then curling them toward $\vec{\mathbf{B}}$. The extended thumb points in the direction of $\vec{\mathbf{F}}$. The force is **(a)** upward and **(b)** downward.

An equivalent alternative is shown in ▸Fig. 19.14.

> When the fingers of the right hand are extended in the direction of the magnetic field and the thumb pointed in the direction of the conventional current I carried by the wire, the palm of the right hand points in the direction of the magnetic force on the wire.

Both rules give the same direction, because they are extensions of the right-hand force rules for individual charges. To see how two current-carrying wires interact magnetically, consider the next example.

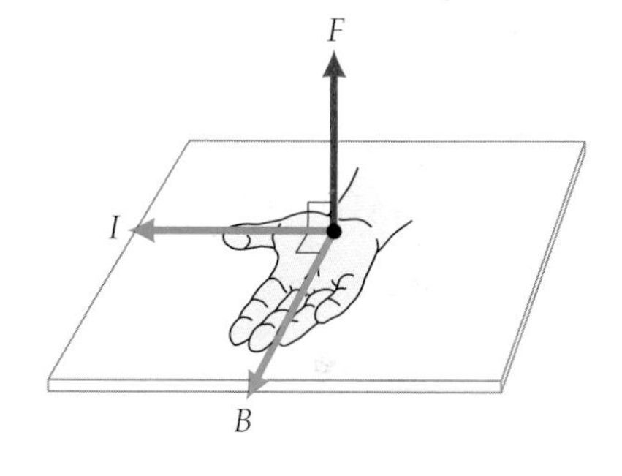

▲ **FIGURE 19.14 An alternative right-hand force rule** When the fingers of the right hand are extended in the direction of the magnetic field $\vec{\mathbf{B}}$ and the thumb is pointed in the direction of the conventional current I, then the palm points in the direction of $\vec{\mathbf{F}}$. You should check that this gives the same direction as the equivalent rule shown in Fig. 19.13.

Integrated Example 19.4 ■ Magnetic Forces on Wires Suspended at the Equator

Because a current-carrying wire is acted on by a magnetic force, it would seem possible to suspend such a wire at rest above the ground using the Earth's magnetic field. (a) A long, straight wire is located at the equator. In what direction would the current in the wire have to be to perform such a feat: (1) up, (2) down, (3) east, or (4) west? Draw a sketch to help you decide. (b) Calculate the current required to suspend the wire, assuming that the Earth's magnetic field is 0.40 G (gauss) at the equator and the wire is 1.0 m long with a mass of 30 g.

(a) Conceptual Reasoning. The magnetic force direction must be upward, because gravity acts downward (▸Fig. 19.15). The Earth's magnetic field at the equator is parallel to the ground and points north. Because the magnetic force is perpendicular to both the current and the field, the current cannot be up or down, which eliminates the first two choices. To decide between east and west, simply choose one and see if it works (or doesn't work). Suppose the current is to the west. Using the right-hand force rule shows that the magnetic force acts downward. Because this is incorrect, the only correct answer is (3), eastward. You should verify that this is correct by using the force right-hand rule directly.

(b) Quantitative Reasoning and Solution. The mass of the wire is known, and therefore its weight can be calculated. This must be equal and opposite to the magnetic force. The current and the field are at right angles to each other; hence, the magnetic force is given by Eq. 19.6, and from that, the current can be determined.

First we list the data (and convert to SI units at the same time):

Given: $m = 30\text{ g} = 3.0 \times 10^{-2}\text{ kg}$
$B = (0.40\text{ G})(10^{-4}\text{ T/G}) = 4.0 \times 10^{-5}\text{ T}$
$L = 1.0\text{ m}$

Find: I (current required to suspend the wire)

The wire's weight is $w = mg = (3.0 \times 10^{-2}\text{ kg})(9.8\text{ m/s}^2) = 0.29\text{ N}$. With the wire suspended, this would equal the magnetic force, or

$$w = ILB$$

Therefore,

$$I = \frac{w}{LB} = \frac{0.29\text{ N}}{(1.0\text{ m})(4.0 \times 10^{-5}\text{ T})} = 7.4 \times 10^{3}\text{ A}$$

This is a huge current, so suspending the wire in this manner is probably not a practical idea.

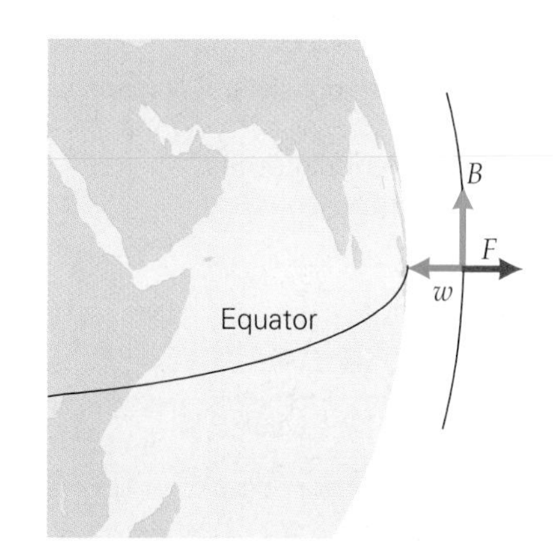

▲ **FIGURE 19.15 Defying gravity by using a magnetic field?** Near the Earth's equator it is theoretically possible to cancel the pull of gravity with an upward magnetic force on a wire. What must be the direction and magnitude of the current? (See Integrated Example 19.4.)

(continues on next page)

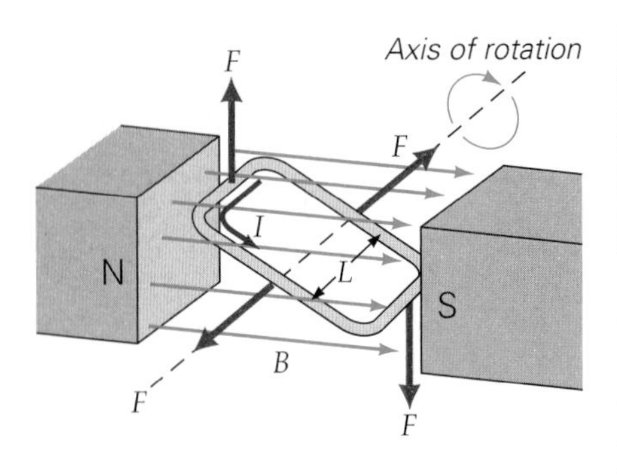

(a)

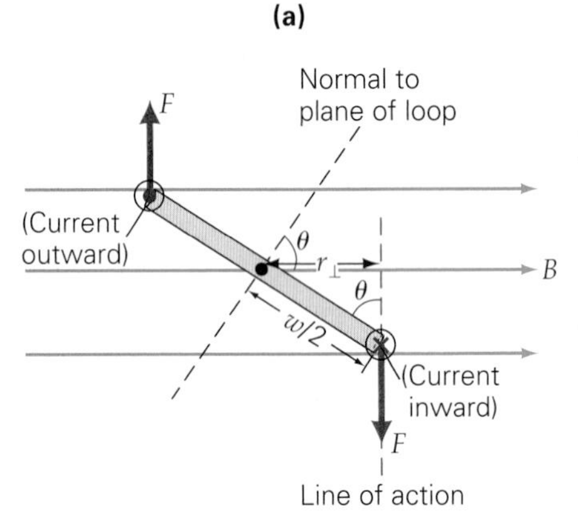

(b) Side view (pivot side)

▲ **FIGURE 19.16** **Force and torque on a current-carrying pivoted loop** **(a)** A current-carrying rectangular loop oriented in a magnetic field as shown is acted on by a force on each of its sides. Only the forces on the sides parallel to the axis of rotation produce a torque that causes the loop to rotate. **(b)** A side view shows the geometry for determining the torque. (See text for details.)

Note: A *coil* consists of *N* loops of the same size, all carrying the same current *I* in series.

Follow-Up Exercise. (a) Using the right-hand force rule, show that the idea of suspending a wire, as in this Example, could not work at either the south or north magnetic poles of the Earth. (b) In this Example, what would the wire's mass have to be for it to be suspended when carrying a more reasonable current of 10 A? Does this seem like a reasonable mass for a 1-m length of wire?

Torque on a Current-Carrying Loop

Another important use of magnetism is to exert forces and torques on current-carrying loops, such as the rectangular loop in a uniform field shown in ◂Fig. 19.16a. (The wires connecting the loop to a voltage source are not shown.) Suppose the loop is free to rotate about an axis passing through opposite sides, as shown. There is no net force or torque due to the forces acting on its pivoted sides (the sides through which the rotation axis passes). The forces on these sides are equal and opposite and in the plane of the loop. Therefore they produce no net torque or force. However, the equal and opposite forces on the two sides of the loop *parallel* to the axis of rotation, while not creating a net force, *do* create a net torque (see Chapter 8).

To see how this works, consider Fig. 19.16b, which is a side view of Fig.19.16a. The magnitude of the magnetic force F on each of the non-pivoted sides (length L) is given by $F = ILB$. The torque produced by this force (Section 8.2) is $\tau = r_\perp F$, where $r_\perp$ is the perpendicular distance (lever arm) from the axis of rotation to the line of action of the force. From Fig. 19.16b, $r_\perp = \frac{1}{2}w \sin\theta$, where w is the width of the loop and θ is the angle between the normal to the loop's plane and the direction of the field. The net torque τ is due to the torques from both forces and is their sum, or twice one of them (why?)

$$\tau = 2r_\perp F = 2\left(\tfrac{1}{2}w \sin\theta\right)F = wF\sin\theta$$
$$= w(ILB)\sin\theta$$

Then, because wL is the area (A) of the loop, the torque on a single pivoted, current-carrying loop can be rewritten as

$$\tau = IAB\sin\theta \quad \textit{torque on one current-carrying loop} \tag{19.8}$$

Although derived for a rectangular loop, Eq. 19.8 is valid for a flat loop of *any* shape and area. A *coil* is composed of N loops connected in series (where $N = 2, 3, \ldots$). Thus, for a coil, the torque is N times that on one loop (because each loop carries the same current). Therefore the torque on a coil is

$$\tau = NIAB\sin\theta \quad \textit{torque on current-carrying coil of N loops} \tag{19.9}$$

The magnitude of a coil's **magnetic moment** vector, m, is defined as

$$m = NIA \quad \textit{magnetic moment of a coil} \tag{19.10}$$

(SI units of magnetic moment: ampere · meter² or A · m²)

The direction of the magnetic moment vector $\vec{\mathbf{m}}$ is determined by circling the fingers of the right hand in the direction of the (conventional) current. The thumb gives the direction of this vector. Note that $\vec{\mathbf{m}}$ is always perpendicular to the plane of the coil (▸Fig. 19.17a). Equation 19.10 can be rewritten in terms of the magnetic moment:

$$\tau = mB\sin\theta \tag{19.11}$$

The magnetic torque tends to align the magnetic moment vector ($\vec{\mathbf{m}}$) with the magnetic field direction. To see this, notice that a loop or coil in a magnetic field is subject to a torque until $\sin\theta = 0$ (that is, $\theta = 0°$), at which point the forces producing the torque are parallel to the plane of the loop (see Fig. 19.17b). This situation exists when the plane of the loop is perpendicular to the field. If the loop is started from rest with its magnetic moment making some angle with the field, the loop will undergo an angular acceleration that will rotate it to the zero angle position. Rotational inertia will carry it through the equilibrium position (zero angle, Fig. 19.17c) to the other side. On that side, the torque will slow the loop, stop it, and then reaccelerate it back toward equilibrium. In other words, the torque on the loop is *restoring* and tends to cause the magnetic moment to oscillate about the field direction, much like a compass needle as it settles down to point north.

Example 19.5 ■ Magnetic Torque: Doing the Twist?

A laboratory technician makes a circular coil out of 100 loops of thin copper wire with a resistance of 0.50 Ω. The coil diameter is 10 cm and the coil is connected to a 6.0-V battery. (a) Determine the magnetic moment (magnitude) of the coil. (b) Determine the maximum torque (magnitude) on the coil if it were placed between the pole faces of a magnet where the magnetic field strength was 0.40 T.

Thinking It Through. The magnetic moment includes not only the number of loops and the area of the coil, but also the current in the wires. Ohm's law can be used to find the current. The maximum torque occurs when the angle between the magnetic moment vector and the B field is 90°, as given by Eq. 19.11.

Solution. Listing the given data with the radius of the circle expressed in SI units:

Given: $N = 100$ loops
$r = d/2 = 5.0\text{ cm} = 5.0 \times 10^{-2}\text{ m}$
$R = 0.50\ \Omega$
$V = 6.0\text{ V}$

Find: (a) m (coil magnetic moment)
(b) τ (maximum torque on the coil)

(a) The magnetic moment is given by Eq. 19.10, so the area and current are needed:

$$A = \pi r^2 = (3.14)(5.0 \times 10^{-2}\text{ m})^2 = 7.9 \times 10^{-3}\text{ m}^2$$

and

$$I = \frac{V}{R} = \frac{6.0\text{ V}}{0.50\ \Omega} = 12\text{ A}$$

Therefore the magnetic moment magnitude is

$$m = NIA = (100)(12\text{ A})(7.9 \times 10^{-3}\text{ m}^2) = 9.5\text{ A}\cdot\text{m}^2$$

(b) The magnitude of the maximum torque (using $\theta = 90°$ in Eq. 19.11) is:

$$\tau = mB\sin\theta = (9.5\text{ A}\cdot\text{m}^2)(0.40\text{ T})(\sin 90°) = 3.8\text{ m}\cdot\text{N}$$

Follow-Up Exercise. In this Example, (a) show that if the coil were rotated so that its magnetic moment vector was at 45°, the torque would *not* be half the maximum torque. (b) At what angle would the torque be half the maximum torque?

19.5 Applications: Current-Carrying Wires in Magnetic Fields

OBJECTIVE: To explain the operation of various instruments whose functions depend on electromagnetic interactions between currents and magnetic fields.

With the principles of electromagnetic interactions learned so far, you can understand the operation of some familiar applications such as the ones that follow in this section.

The Galvanometer: The Foundation of the Ammeter and Voltmeter

Recall that ammeters and voltmeters use the *galvanometer* as the heart of their design.* Now you can understand exactly how the galvanometer works. As ▼Fig. 19.18a shows, a galvanometer consists of a coil of wire loops on an iron core that pivots between the pole faces of a permanent magnet. When a current is in the coil, a torque is exerted on the coil. A small spring supplies a countertorque, and when the two torques cancel (equilibrium) a pointer indicates a deflection angle ϕ that is proportional to the coil's current.

A problem arises if the galvanometer's magnetic field is not shaped correctly. If the coil rotated from its position of maximum torque ($\theta = 90°$), the torque would become less, and the pointer deflection ϕ would *not* then be proportional to the current. This problem is avoided by making the pole faces curved and by wrapping the coil on a cylindrical iron core. The core tends to concentrate the field lines such that $\vec{\mathbf{B}}$ is always perpendicular to the nonpivoted side of the coil (Fig. 19.18b). With this design, the deflection angle is proportional to the current through the galvanometer ($\phi \propto I$), as required.

*Even though most voltmeters and ammeters are now digital, it is useful to understand how the mechanical versions use magnetic forces to make electrical measurements.

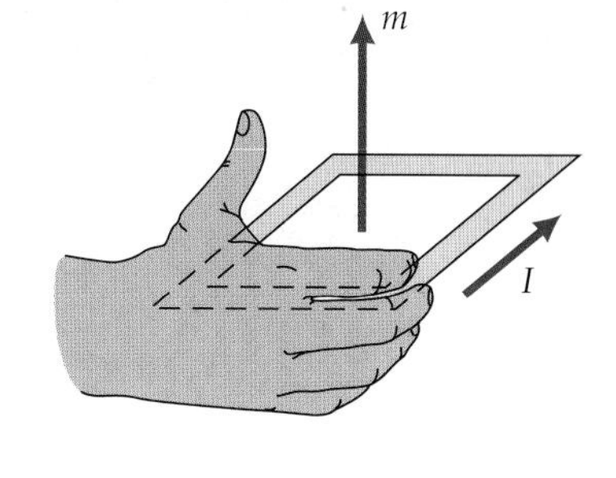

(a)

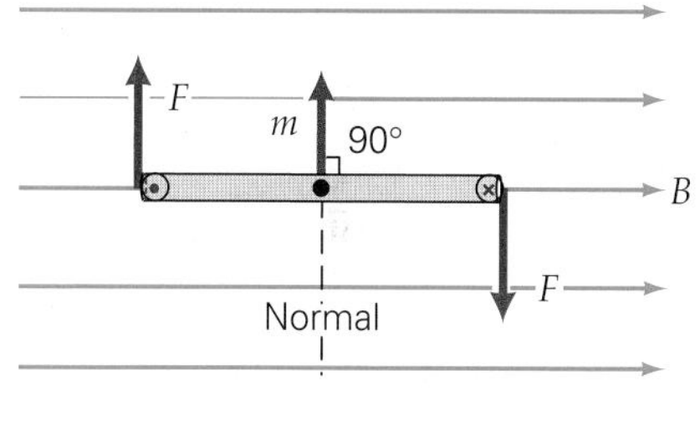

(b)

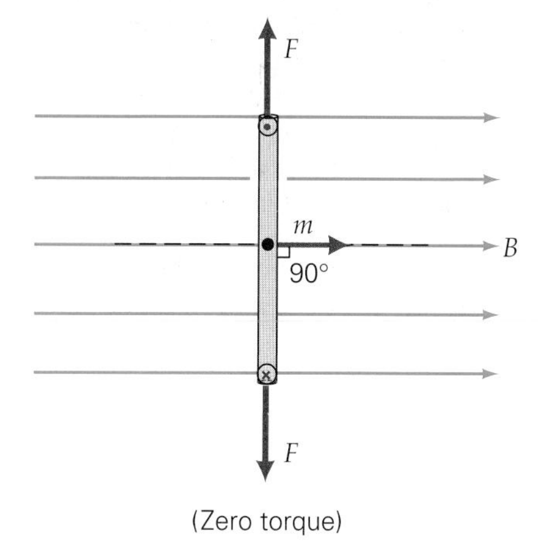

(c)

▲ **FIGURE 19.17 Magnetic moment of a current-carrying loop (a)** A right-hand rule determines the direction of the loop's magnetic moment vector $\vec{\mathbf{m}}$. The fingers wrap around the loop in the direction of the current, and the thumb gives the direction of $\vec{\mathbf{m}}$. **(b)** Condition of maximum torque. **(c)** Condition of zero torque. If the loop is free to rotate, the magnetic moment vector of the loop will tend to align with the direction of the external magnetic field.

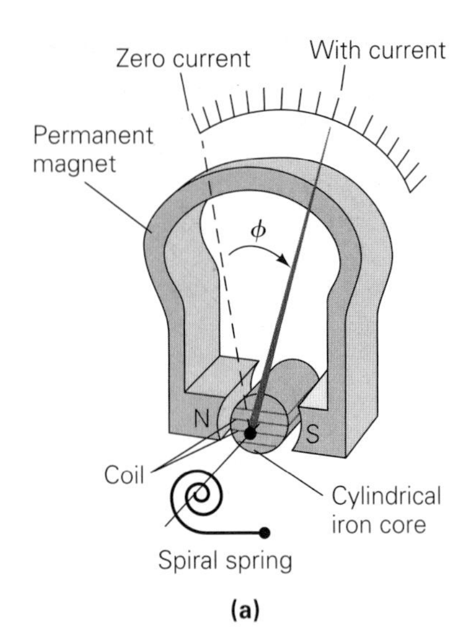

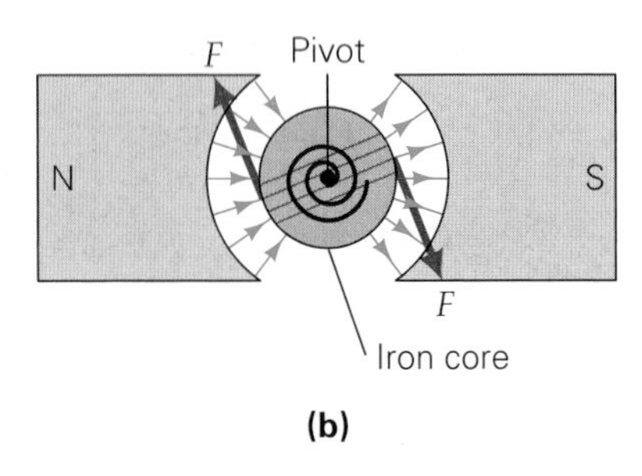

▲ **FIGURE 19.18 The galvanometer** **(a)** The deflection (ϕ) of the needle from its zero-current position is proportional to the current in the coil. A galvanometer can therefore detect and measure currents. **(b)** A magnet with curved pole faces is used so that the field lines are always perpendicular to the core surface and the torque does not vary with ϕ.

The dc Motor

In general, *an electric motor is a device that converts electrical energy into mechanical energy*. Such a conversion actually occurs during the movement of a galvanometer needle. However, a galvanometer is not considered a motor, because a practical **dc motor** must have continuous rotation for continuous energy output.

A pivoted, current-carrying coil in a magnetic field will rotate, but for only a half-cycle. When the magnetic field and the coil's magnetic moment are lined up ($\sin\theta = 0$), the torque on the coil is zero and it is in equilibrium.

To provide continuous rotation, the current is reversed every half-turn so that the torque-producing forces are reversed. This is done by using a *split-ring commutator*, which is an arrangement of two metal half-rings insulated from each other (▾Fig. 19.19a). The ends of the wire of the coil are fixed to the half-rings so that they rotate together. The current is supplied to the coil through the commutator by means of contact brushes. Then, with one half-ring electrically positive and the other negative, the coil and ring will rotate. When they have gone through half a rotation, the half-rings come in contact with the opposite brushes. Because their polarity is now reversed, the current in the coil also reverses. In turn, this action reverses the directions of the magnetic forces, keeping the torque in the same direction (clockwise in Fig. 19.19b). Even though the torque is zero at the equilibrium position, the coil is in unstable equilibrium and has enough rotational momentum to continue through the equilibrium point, whereupon the torque takes over and the coil rotates another half-cycle. The process repeats in continuous operation. In a real motor, the rotating shaft is called the *armature*.

The Electronic Balance

Traditional laboratory balances measure mass by balancing the weight of an unknown mass against a known one. Digital electronic balances (▸Fig. 19.20a) work on a different principle. In one type of design, there is still a suspended beam with a pan on one end that holds the object to be weighed, but no known mass is needed. Instead the balancing downward force is supplied by current-carrying coils of wire in the field of a permanent magnet (Fig. 19.20b). The coils move up and down in the cylindrical gap of the magnet, and the downward force is proportional to the current in the coils. The weight of the object in the pan is determined from the coil current, which produces a force just sufficient to balance the beam. From the weight, the balance determines the object's mass using the local value for g from $m = w/g$.

▾ **FIGURE 19.19 A dc motor** **(a)** A split-ring commutator reverses the polarity and current each half-cycle, so the coil rotates continuously. **(b)** An end view shows the forces on the coil and its orientation during a half-cycle. [For simplicity, we depict a single loop, but the coil actually has many (N) loops.] Notice the current reversal (shown by the dot and cross notation) between situations (3) and (4).

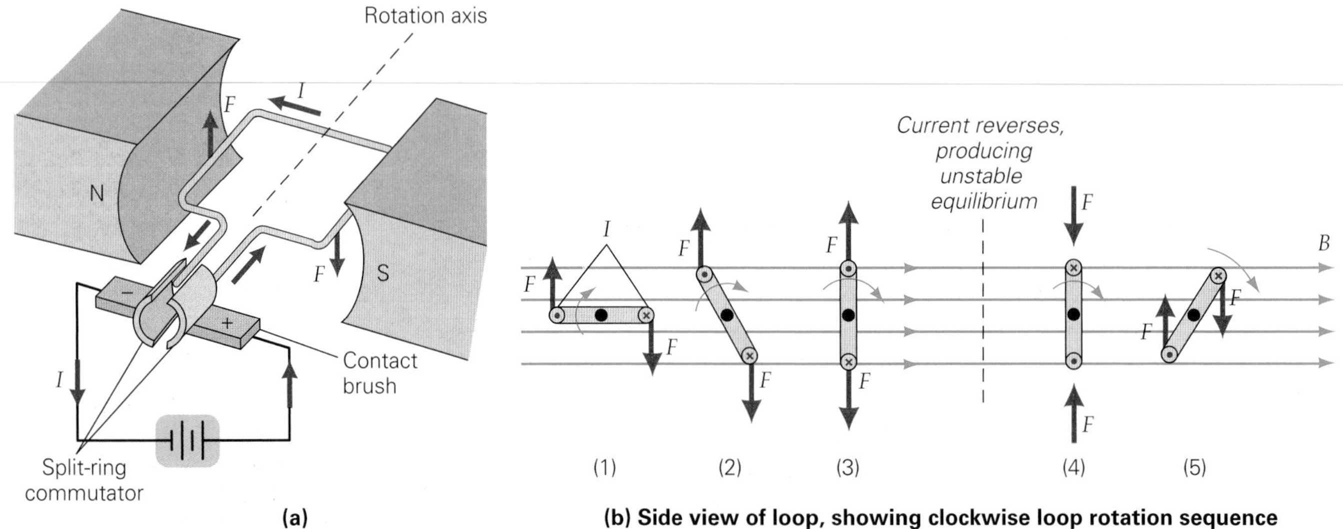

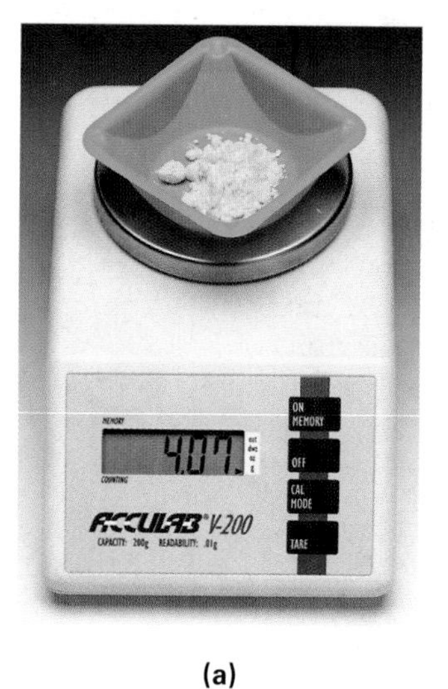

(a)

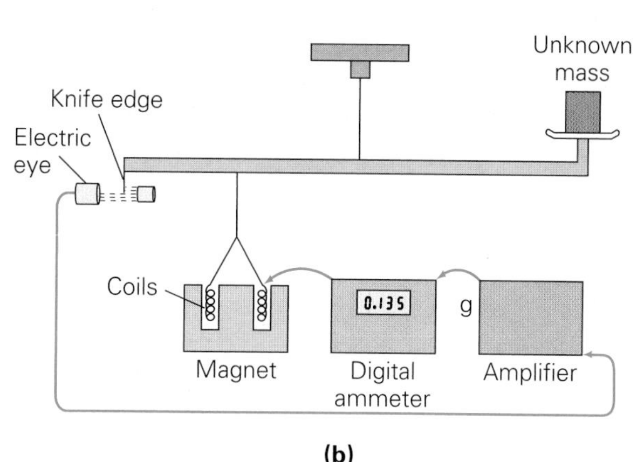

(b)

◀ **FIGURE 19.20 Electronic balance** **(a)** A digital electronic balance. **(b)** Diagram of the principle of an electronic balance. The balance force is supplied by electromagnetism.

The current required to produce balance is controlled automatically by photosensors and an electronic feedback loop. When the beam is balanced and horizontal, a knife-edge obstruction cuts off part of the light from a source that falls on a photosensitive "electric eye," the resistance of which depends on the amount of light falling on it. This resistance controls the current that an amplifier sends through the coil. For example, if the beam tilts so that the knife edge rises and more light strikes the eye, the current in the coil is increased to counterbalance the tilting. In this manner, the beam is electronically maintained in nearly horizontal equilibrium. Lastly, the current that keeps the beam in the horizontal position is read out on a digital ammeter calibrated in grams or milligrams instead of amperes.

19.6 Electromagnetism: The Source of Magnetic Fields

OBJECTIVES: To (a) understand the production of a magnetic field by electric currents, (b) calculate the strength of the magnetic field in simple cases, and (c) use the right-hand source rule to determine the direction of the magnetic field from the direction of the current that produces it.

Electric and magnetic phenomena, although clearly different, are closely and fundamentally related. As we have seen, the *magnetic* force on a particle depends on the particle's *electric* charge. But what is the source of the magnetic field? Danish physicist Hans Christian Oersted discovered the answer in 1820, when he found that *electric currents produce magnetic fields*. His studies marked the beginnings of the discipline called **electromagnetism**, which involves the relationship between electric currents and magnetic fields.

In particular, Oersted first noted that an electric current could produce a deflection of a compass needle. This property can be demonstrated with an arrangement like that in ▸Fig. 19.21. When the circuit is open and there is no current, the compass needle points, as usual, in the northerly direction. However, when the switch is closed and there is current in the circuit, the compass needle points in a different direction, indicating that an additional magnetic field (due to the current) must be affecting the needle.

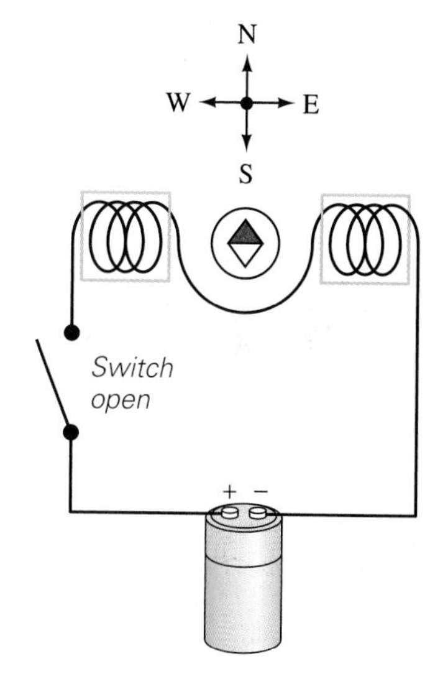

(a) No current

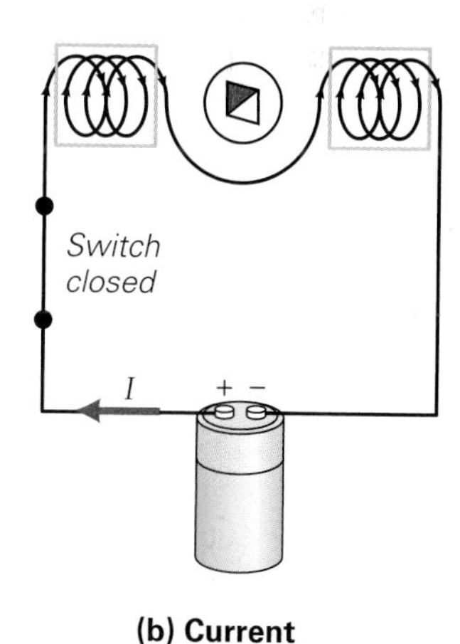

(b) Current

▲ **FIGURE 19.21 Electric current and magnetic field** **(a)** With no current in the wire, the compass needle points north. **(b)** With a current in the wire, the compass needle is deflected, indicating the presence of an additional magnetic field superimposed on that of the Earth. In this case, the strength of the additional field is roughly equal in magnitude to that of the Earth. How can you tell?

Developing expressions for the magnetic field created by various configurations of current-carrying wires requires mathematics beyond the scope of this book. This section will present the results for the magnetic fields for several common current configurations.

Magnetic Field Near a Long, Straight, Current-Carrying Wire

At a perpendicular distance d from a long, straight wire carrying a current I (▾Fig. 19.22), the magnitude of $\vec{\mathbf{B}}$ is given by

$$B = \frac{\mu_o I}{2\pi d} \quad \textit{magnetic field due to a long, straight wire} \tag{19.12}$$

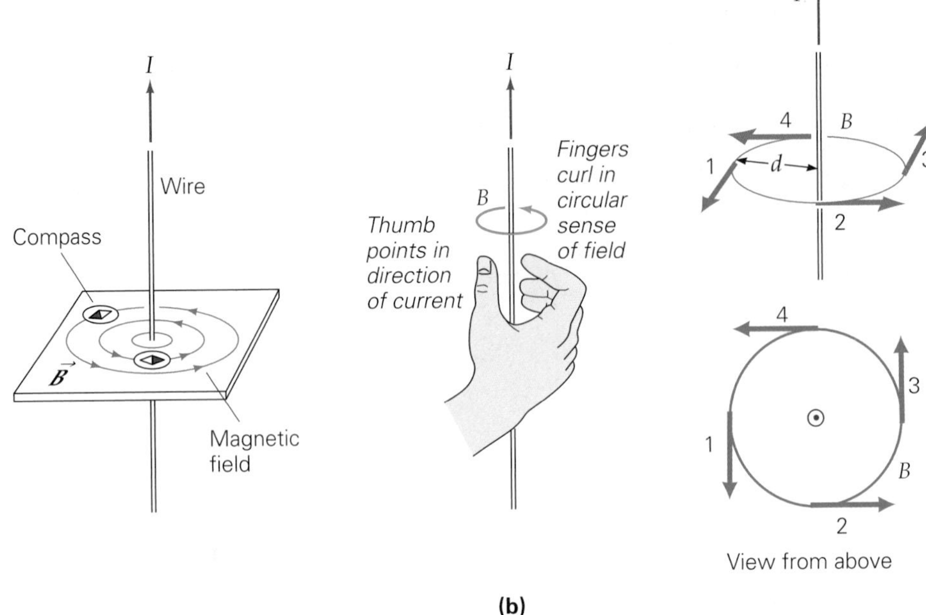

▲ **FIGURE 19.22 Magnetic field around a long, straight current-carrying wire** **(a)** The field lines form concentric circles around the wire, as revealed by the pattern of iron filings. **(b)** The circular sense of the field lines is given by the right-hand source rule, and the magnetic field vector is tangent to the circular field line at any point.

where $\mu_o = 4\pi \times 10^{-7}\,\mathrm{T \cdot m/A}$ is a proportionality constant called the **magnetic permeability of free space**. For long, straight wires, the field lines are closed circles centered on the wire (Fig. 19.22a). Note in Fig. 19.22b that the direction of $\vec{\mathbf{B}}$ due to a current in a long straight wire is given by a **right-hand source rule**:

> If a long straight current-carrying wire is grasped with the right hand and the extended thumb points in the direction of the current (I), then the curled fingers indicate the circular sense of the magnetic field lines.

Example 19.6 ■ Common Fields: Magnetic Field from a Current-Carrying Wire

The maximum household current in a wire is about 15 A. Assume that this current exists in a long straight wire in a west-to-east direction (◂Fig. 19.23). What are the magnitude and direction of the magnetic field the current produces 1.0 cm directly below the wire?

Thinking It Through. To find the magnitude of the field, we should use Eq. 19.12 for a long, straight wire. The direction of the field is given by the right-hand source rule.

Solution. Listing the data:

Given: $I = 15\,\mathrm{A}$, $d = 1.0\,\mathrm{cm} = 0.010\,\mathrm{m}$ *Find:* $\vec{\mathbf{B}}$ (magnitude and direction)

From Eq. 19.12, the magnitude of the field 1.0 cm below the wire is

$$B = \frac{\mu_o I}{2\pi d} = \frac{(4\pi \times 10^{-7}\,\mathrm{T \cdot m/A})(15\,\mathrm{A})}{2\pi(0.010\,\mathrm{m})} = 3.0 \times 10^{-4}\,\mathrm{T}$$

By the source rule (Fig. 19.23), the field direction directly below the wire is north.

Follow-Up Exercise. (a) In this Example, what is the field direction 5.0 cm above the wire? (b) What current is needed to produce a magnetic field at this location with one-half the strength of the field in the Example?

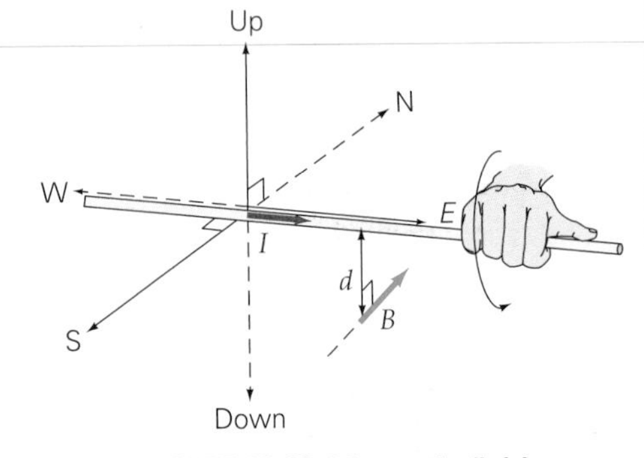

▲ **FIGURE 19.23 Magnetic field** Finding the magnitude and direction of the magnetic field produced by a straight current-carrying wire. (See Example 19.6.)

Magnetic Field at the Center of a Circular Current-Carrying Wire Loop

At the *center* of a circular coil of wire consisting of N loops, each of radius r and each carrying the same current I (▸Fig. 19.24a shows one such loop), the magnitude of $\vec{\mathbf{B}}$ is

$$B = \frac{\mu_o N I}{2r} \quad \textit{magnetic field at center of circular coil of N loops} \qquad (19.13)$$

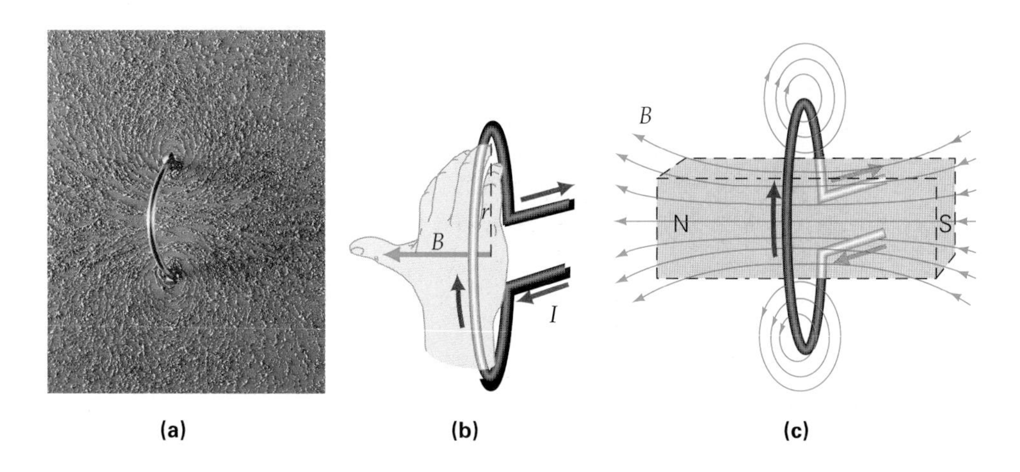

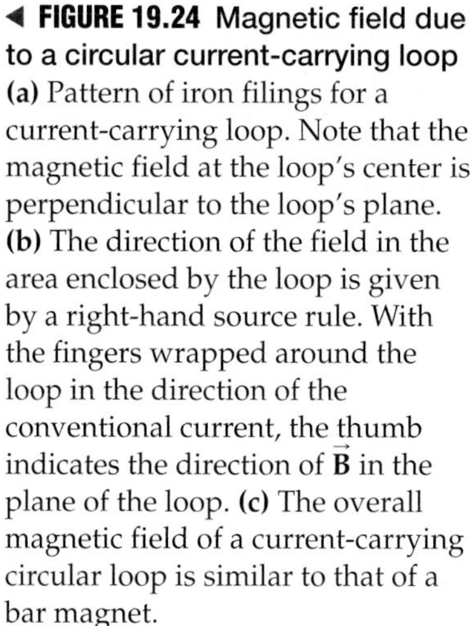

◀ **FIGURE 19.24 Magnetic field due to a circular current-carrying loop** **(a)** Pattern of iron filings for a current-carrying loop. Note that the magnetic field at the loop's center is perpendicular to the loop's plane. **(b)** The direction of the field in the area enclosed by the loop is given by a right-hand source rule. With the fingers wrapped around the loop in the direction of the conventional current, the thumb indicates the direction of $\vec{\mathbf{B}}$ in the plane of the loop. **(c)** The overall magnetic field of a current-carrying circular loop is similar to that of a bar magnet.

Illustration 28.1 Field from Wires and Loops

In this case (and all circular current arrangements, such as solenoids, discussed next), it is convenient to determine the magnetic field direction using the right-hand source rule that is slightly different from (but equivalent to) the one for straight wires:

> If a circular loop of current-carrying wires is grasped with the right hand so the fingers are curled in the direction of the current, the magnetic field direction *inside* the circular area formed by the loop is the direction in which the extended thumb points. (see Fig. 19.24b)

In all cases, the magnetic field lines form closed loops, the direction of which is determined by the right-hand source rule. Recall however, the direction of $\vec{\mathbf{B}}$ is tangent to the field line, and therefore depends on the location (Fig. 19.24c). Notice that the overall field pattern of the loop is geometrically similar to that of a bar magnet. More about this later.

Magnetic Field in a Current-Carrying Solenoid

A *solenoid* is constructed by winding a long wire into a coil, or *helix*, with many circular loops, as shown in ▼Fig. 19.25. If the solenoid's radius is small compared with its length (L), the interior magnetic field is parallel to the solenoid's longitudinal axis and constant in magnitude. The longer the solenoid, the more uniform the internal field. Notice how the solenoid's field (Fig. 19.25) closely resembles that of a permanent bar magnet.

As usual, the direction of the interior field is given by the right-hand source rule for circular geometry. If the solenoid has N turns and each carries a current I, the magnitude of the magnetic field near its center is given by

$$B = \frac{\mu_o N I}{L} \quad \textit{magnetic field near the center of a solenoid} \tag{19.14}$$

Notice that the solenoid field depends on how closely packed (note the N/L) or how densely the turns are wound. Thus n is defined as $n = N/L$ to quantify this. Its units are turns per meter, and it is called the *linear turn density*. In terms of this, Eq. 19.14 is sometimes rewritten as $B = \mu_o n I$.

To see why the solenoid might be best suited for magnetic applications requiring a large magnetic field, consider the next Example.

▼ **FIGURE 19.25 Magnetic field of a solenoid** **(a)** The magnetic field of a current-carrying solenoid is fairly uniform near the central axis of the solenoid, as seen in this pattern of iron filings. **(b)** The direction of the field in the interior can be determined by applying the right-hand source rule to any of the loops. Notice its resemblance to the field near a bar magnet.

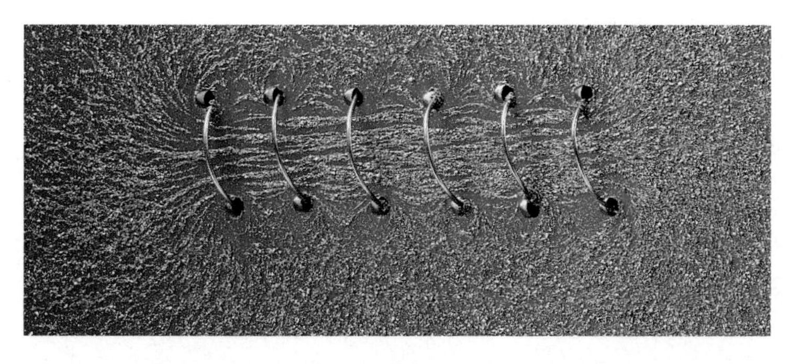

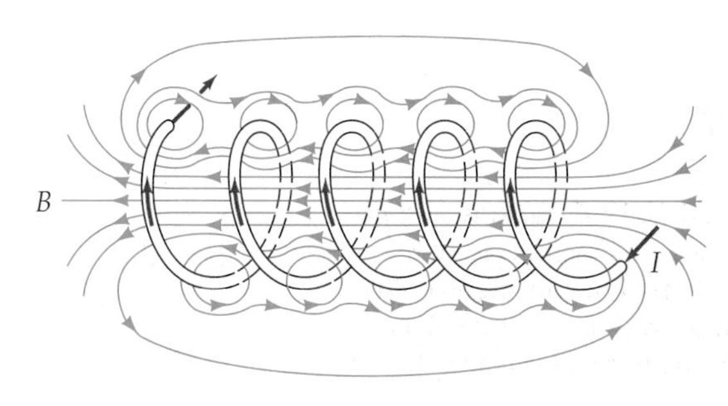

(a) (b)

Example 19.7 ■ Wire versus Solenoid: Concentrating the Magnetic Field

A solenoid is 0.30 m long with 300 turns and carries a current of 15.0 A. (a) What is the magnitude of the magnetic field at the center of this solenoid? (b) Compare your result with the field near the single wire in Example 19.6, which carries the same current, and comment.

Thinking It Through. The B field depends on the number of turns (N), the solenoid length (L), and the current (I). This is a direct application of Eq. 19.14.

Solution.

Given: $I = 15.0\text{ A}$, $N = 300\text{ turns}$, $L = 0.30\text{ m}$

Find: (a) B (magnitude of magnetic field near the solenoid center)
(b) Compare the answer in part (a) with that from a long straight wire in Example 19.6

(a) From Eq. 19.14,

$$B = \frac{\mu_o NI}{L} = \frac{(4\pi \times 10^{-7}\text{ T}\cdot\text{m/A})(300)(15.0\text{ A})}{0.30\text{ m}} = 6\pi \times 10^{-3}\text{ T} \approx 18.8\text{ mT}$$

(b) Notice that this is more than *60 times as large as* the field near the wire in Example 19.6. Winding many loops close together in a helix fashion increases the field while enabling the use of the same current. The fundamental reason is that the solenoid's field is the vector sum of the fields from 300 loops, and the individual magnetic field directions are all approximately the same.

Follow-Up Exercise. In this Example, if the current were reduced to 1.0 A and the solenoid shortened to 0.10 m, how many turns would be needed to create the same magnetic field?

Illustration 28.2 Forces Between Wires

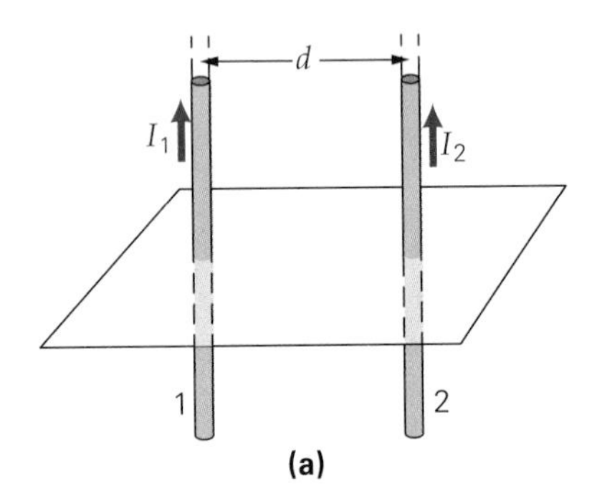

(a)

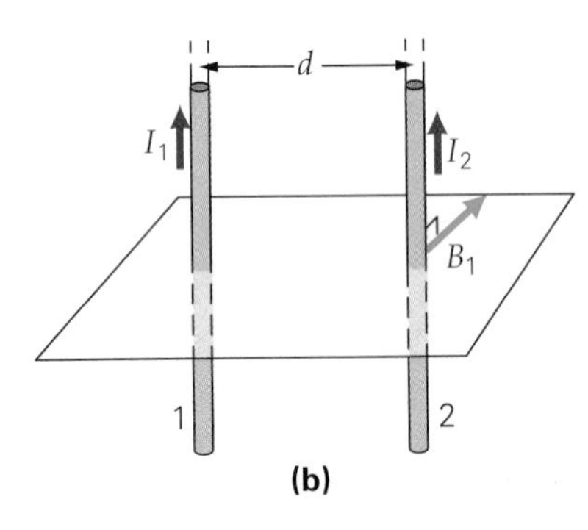

(b)

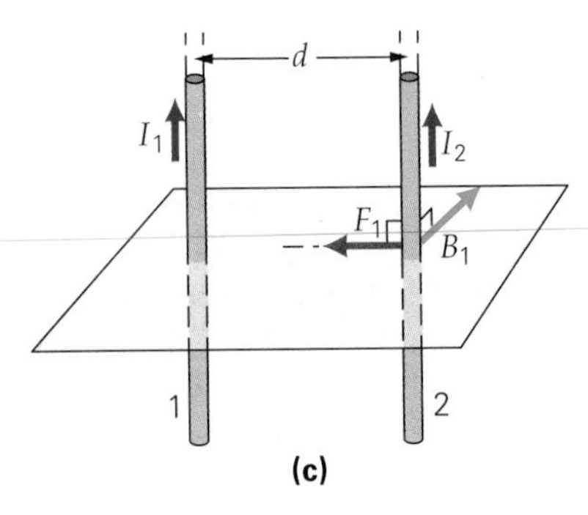

(c)

▲ **FIGURE 19.26** Mutual interaction of parallel current-carrying wires **(a)** Two parallel wires carry current in the same direction. **(b)** Wire 1 creates a magnetic field at the site of wire 2. **(c)** Wire 2 is pulled toward wire 1 by a force. (See Integrated Example 19.8 for details.)

In the next Integrated Example, all aspects of electromagnetism are involved: forces on electric currents and the production of magnetic fields by electric currents. Study the example carefully, especially the use of the appropriate right-hand rule.

Integrated Example 19.8 ■ Attraction or Repulsion: Magnetic Force between Two Parallel Wires

Two long, parallel wires carry currents in the same direction, as illustrated in ◀Fig. 19.26a. (a) Is the magnetic force between these wires (1) attractive or (2) repulsive? Make a sketch to show how you obtained your result. (b) If both wires carry the same current of 5.0 A, have a length of 50 cm, and are separated by 3.0 mm, determine the magnitude of the force on each wire.

(a) Conceptual Reasoning. Choose one wire and determine the direction of the magnetic field it produces at the location of the other wire. (Note that the field of interest is determined first using the right-hand source rule.) In Fig. 19.26b, wire 1 was chosen. The field produced by the current in wire 1 is the field in which wire 2 is located. Next, use the right-hand force rule on wire 2 to determine the direction of the force on it. The result, shown in Fig. 19.26c, is an attractive force, so (1) is the correct answer. You should be able to show that, at the same time, wire 2 exerts an attractive force on wire 1, in keeping with Newton's third law.

(b) Quantitative Reasoning and Solution. To find the magnetic field strength produced by wire 1, use Eq. 19.12. Because the field is at right angles to the current in wire 2, the magnitude of the force on it is ILB. Be careful to use the appropriate field and current. The symbols are shown in Fig. 19.26. We list the given data and convert to SI units.

Given: $I_1 = I_2 = 5.0\text{ A}$, $d = 3.0\text{ mm} = 3.0 \times 10^{-3}\text{ m}$, $L = 50\text{ cm} = 5.0 \times 10^{-1}\text{ m}$

Find: F (the magnitude of the magnetic force between the wires)

The magnetic field due to I_1 at the location of wire 2 is

$$B_1 = \frac{\mu_o I_1}{2\pi d} = \frac{(4\pi \times 10^{-7}\text{ T}\cdot\text{m/A})(5.0\text{ A})}{2\pi(3.0 \times 10^{-3}\text{ m})} = 3.3 \times 10^{-4}\text{ T}$$

The magnitude of the magnetic force on wire 2 due to the field created by wire 1 is

$$F_2 = I_2 LB_1 = (5.0\text{ A})(0.50\text{ m})(3.3 \times 10^{-4}\text{ T}) = 8.3 \times 10^{-4}\text{ N}$$

Follow-Up Exercise. (a) In this Example, determine the force direction if the current in either one of them is reversed. (b) If the magnitude of the force between the wires is kept the same as in the Example, but the current tripled, how far apart are the wires?

The magnetic force between parallel wires in an arrangement like that analyzed in Integrated Example 19.8 provides the modern basis for the definition of the ampere. The National Institute of Standards and Technology (NIST) defines the ampere as

> that current which, if maintained in each of two long parallel wires separated by a distance of exactly 1 m in free space, would produce a magnetic force between the wires of exactly 2×10^{-7} N for each meter of wire.

This definition was chosen as a universal standard, in part because it is easier to measure forces than to count electrons (that is, measure charge) over time.

19.7 Magnetic Materials

OBJECTIVES: To (a) explain how ferromagnetic materials enhance external magnetic fields, (b) understand the concept of a material's magnetic permeability, (c) explain how "permanent" magnets are produced, and (d) explain how "permanent" magnetism can be destroyed.

Why are some materials magnetic or easily magnetized, whereas others are not? How can a bar magnet create a magnetic field when it carries no obvious current? To answer these questions, let's start with some basics. It is known that a current is needed to produce a magnetic field. If the magnetic fields of a bar magnet and a long solenoid are compared (see Figs. 19.1 and 19.25), it seems that the magnetic field of the bar magnet might be due to *internal* currents. Perhaps these "invisible" currents are due to electrons orbiting the atomic nucleus or electron spin. However, detailed analysis of atomic structure shows that the net magnetic field produced by orbital motion is zero or very small.

What then *is* the source of the magnetism produced by magnetic materials? Modern atomic quantum theory tells us that the permanent type of magnetism, like that exhibited by an iron bar magnet, is produced by *electron spin*. Classical physics likens a spinning electron to the Earth rotating on its axis. However, this mechanical analog is *not* actually the case. Electron spin is a quantum mechanical effect with no direct classical analog. Nonetheless, the picture of spinning electrons creating magnetic fields is useful for qualitative thinking and reasoning. In effect, each "spinning" electron produces a field similar to a current loop (Fig. 19.24c). This pattern, resembling that from a small bar magnet, enables us to treat electrons, magnetically speaking, as tiny compass needles.

In multielectron atoms, the electrons *usually* are arranged in pairs with their spins oppositely aligned (that is, one with "spin up" and one with "spin down" in chemistry parlance). In this case, their magnetic fields will effectively cancel, and the material cannot be magnetic. Aluminum is such a material.

However, in certain materials, known as **ferromagnetic materials**, the fields due to electron spins in individual atoms do not cancel. Thus each atom possesses a magnetic moment. There is a strong interaction between these neighboring moments that leads to the formation of regions called **magnetic domains**. In a given domain, the electron spin moments are aligned in approximately the same direction, producing a relatively strong (net) magnetic field. Not many ferromagnetic materials occur naturally. The most common are iron, nickel, and cobalt. Gadolinium and certain manufactured alloys, such as neodymium and other rare earth alloys, are also ferromagnetic.

In an unmagnetized ferromagnetic material, the domains are randomly oriented and there is no net magnetization (▾Fig. 19.27a). But when a ferromagnetic material

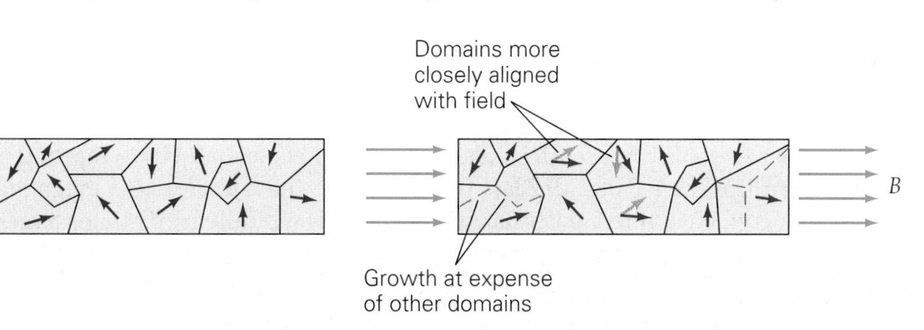

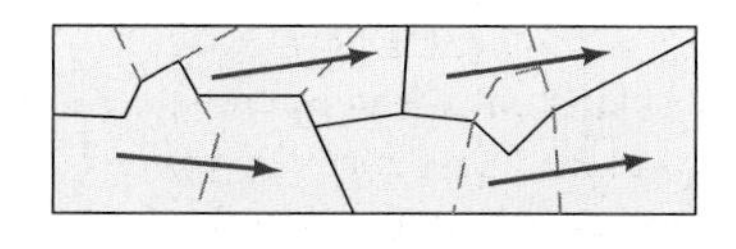

▾ **FIGURE 19.27 Magnetic domains** **(a)** With no external magnetic field, the magnetic domains of a ferromagnetic material are randomly oriented and the material is unmagnetized. **(b)** In an external magnetic field, domains with orientations parallel to the field may grow at the expense of other domains, and the orientations of some domains may become more aligned with the field. **(c)** As a result, the material becomes magnetized, or exhibits magnetic properties.

INSIGHT 19.1 THE MAGNETIC FORCE IN FUTURE MEDICINE

From ancient times to present, humans have sought healing power in magnetism. Claims that magnetism can heal bunions, eliminate tennis elbow, and cure cancer have been made, but none have ever been substantiated. However, there are many magnetic applications in modern medicine, such as the well-known magnetic resonance imaging (MRI) system (see Chapter 28).

Many new ideas are on the horizon as well. For example, certain types of bacteria can create nanometer-sized permanent magnets inside themselves (see Insight 19.2 on Magnetism in Nature on page 645). Scientists have proposed harvesting these tiny magnets, which are just small enough to fit through a hypodermic needle. These magnets could then be attached to drug molecules. By placing a magnetic field near the site of interest, the molecules could be attracted and held there. Holding drug molecules in place would increase their effectiveness and reduce side effects that can occur when drugs circulate into other parts of the body.

A major problem with this proposal is the need to develop techniques for extracting the tiny bacterial magnets and growing them in large enough quantities. Alternative proposals include creating nano-sized unmagnetized pieces of iron by chemical means, attaching them to the drug molecules, and moving them around with magnetic fields in a microscopic version of iron filings. Both proposals present dangers, such as the drug molecules attracting each other, thus clumping and potentially blocking blood flow.

Instead, perhaps, tiny magnetic microspheres could be filled with drugs or radioactive material and steered to the site and held in place by external magnetic fields. One application would be to the treatment of diabetic foot ulcers—open lesions that are difficult to heal due to diabetes-related circulation problems. The wound would be covered by thin, strong magnets in a bandage. An injection of microspheres filled with slow-release drugs, such as an antibiotic, would follow. The magnets would attract the microspheres to the ulcer site and hold them in place. As the microspheres would break down over the course of several weeks, they would release the drugs slowly, helping the body repair the wound. Microspheres filled with radioactive material could potentially treat liver, lung, brain, and other deep tumors.

Another experimental therapy uses magnetically induced heating (*hyperthermic* techniques) to target breast cancer. This therapy could be especially important in destroying the smaller tumors now being found by modern imaging techniques. For these tumors, fluid magnetite (Fe_3O_4) would be injected directly into the tumor. For tumors larger than a few cubic millimeters, the iron particles could be delivered via the circulatory system if they were first attached to biomolecules that specifically target cancer cells. Regardless of the method, the treatment would be the same.

In the presence of an external alternating magnetic field, the iron particles would heat up due to induced currents (see Chapter 20 on electromagnetic induction). A rise in temperature of only a few degrees Celsius above normal body temperature has been shown to kill cancer cells. In theory, this heating would occur locally only at the tumor sites and be minimally invasive. Initial experiments have had positive results, and the outlook is promising.

Illustration 27.5 Permanent Magnets and Ferromagnetism

(such as an iron bar) is placed in an external magnetic field, the domains change their orientation and size (Fig. 19.27b). Remember the picture of the electron as a small compass; the electrons begin to "line up" in the external field. As the external field and the iron bar begin to interact, the iron exhibits the following two effects:

1. Domain boundaries change, and the domains with magnetic orientations in the direction of the external field grow at the expense of the others.
2. The magnetic orientation of some domains may change slightly so as to be more aligned with the field.

On removal of the external fields, the iron domains remain more or less aligned in the original external field direction, thus creating an overall "permanent" magnetic field of their own.

Now you can also understand why an unmagnetized piece of iron is attracted to a magnet and why iron filings line up with a magnetic field. Essentially, the pieces of iron become induced magnets (Fig. 19.27c). Some uses of permanent magnets and magnetic forces in modern medicine are discussed in Insight 19.1 on The Magnetic Force in Future Medicine on this page.

Electromagnets and Magnetic Permeability

Ferromagnetic materials are used to make electromagnets, usually by wrapping a wire around an iron core (▸Fig. 19.28a). The current in the coil creates a magnetic field in the iron, which in turn creates its own field that is typically many times larger than the coil's field. By turning the current on and off, we can turn the magnetic field on and off at will. When the current is on, it induces magnetism in ferromagnetic materials (for example, the iron sliver in Fig. 19.28b) and, if the forces are large enough, it can be used to pick up large amounts of scrap iron (Fig. 19.28c).

The iron used in an electromagnet is called *soft iron*. In this type of iron, when the external field is removed, the magnetic domains become unaligned and the iron reverts to its demagnetized state. The adjective *soft* refers not to the metal's mechanical hardness, but to its magnetic properties.

When an electromagnet is on (lower drawing in Fig. 19.28a), the iron core is magnetized and adds to the field of the solenoid. The total field is expressed as

$$B = \frac{\mu N I}{L} \quad \textit{magnetic field at the center of the iron–core solenoid} \tag{19.15}$$

Notice that this equation is identical to that for the magnetic field of an air-core solenoid (Eq. 19.14), except that it contains μ instead of μ_o, the permeability of free space. Here, μ represents the **magnetic permeability** *of the core material*, not free space. The role permeability plays in magnetism is similar to that of the permittivity ε in electricity (Chapter 16). For magnetic materials, the magnetic permeability is defined in terms of its value in free space; thus,

$$\mu = \kappa_m \mu_o \tag{19.16}$$

where κ_m is called the *relative* permeability (dimensionless) and is the magnetic analog of the dielectric constant κ.

The value of κ_m for a vacuum is equal to unity. (Why?) Because, for ferromagnetic materials, the total magnetic field far exceeds that from the wire wrapping, it follows that $\mu \gg \mu_o$ and $\kappa_m \gg 1$. A core of a ferromagnetic material with a large permeability in an electromagnet can enhance its field thousands of times compared with an air core. In other words, ferromagnetic materials have values of κ_m on the order of thousands. To see the effect of ferromagnetic materials, refer to the next Example.

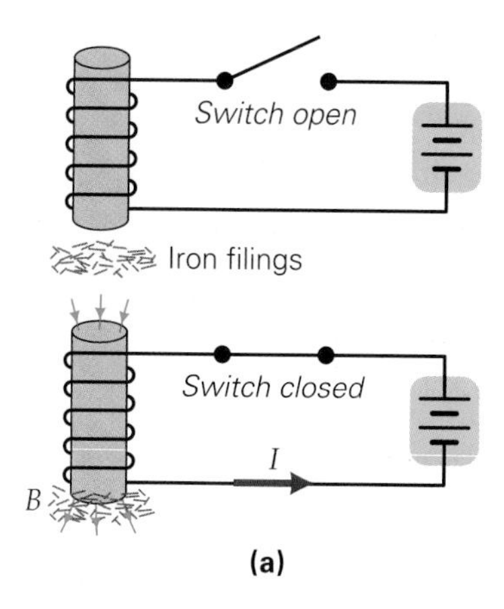

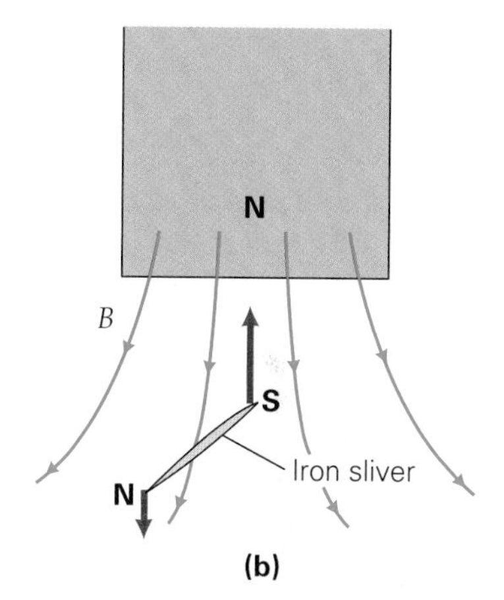

▲ **FIGURE 19.28 Electromagnet** **(a)** *(top)* With no current in the circuit, there is no magnetic force. *(bottom)* However, with a current in the coil, there is a magnetic field and the iron core becomes magnetized. **(b)** Detail of the lower end of the electromagnet in part (a). The sliver of iron is attracted to the end of the electromagnet. **(c)** An electromagnet picking up scrap metal.

Example 19.9 ■ Magnetic Advantage: Using Ferromagnetic Materials

A laboratory solenoid with 200 turns in a length of 30 cm is limited to carrying a maximum current of 2.0 A. The scientists need an interior magnetic field strength of at least 2.0 T and are debating whether they need to employ a ferromagnetic core. (a) Is their field possible if no material fills its core? (b) If not, determine the minimum magnetic permeability of the ferromagnetic material that would comprise the core.

Thinking It Through. The *B* field depends on the number of turns (*N*), the solenoid length (*L*), the current (*I*), and the permeability of the core material (μ). This is a direct application of Eqs. 19.14 and 19.15.

Solution.

Given: $I_{max} = 2.0$ A
$N = 200$ turns
$L = 0.30$ m

Find: (a) Is $B = 2.0$ T possible with no core material?
(b) Magnetic permeability required to attain $B = 2.0$ T

(a) From Eq. 19.14, without any core material, the interior field clearly would not be large enough:

$$B = \frac{\mu_o N I}{L} = \frac{(4\pi \times 10^{-7}\,\text{T}\cdot\text{m/A})(200)(2.0\,\text{A})}{0.30\,\text{m}} = 1.7 \times 10^{-3}\,\text{T} = 1.7\,\text{mT}$$

(b) The required field is about $2.0\,\text{T}/1.7 \times 10^{-3}\,\text{T}$ or about 1200 times as strong as the answer to part (a). Thus, because $B \propto \mu$ if everything else is constant, to attain a value of 2.0 T requires a permeability $\mu \geq 1200\,\mu_o$ or $\mu \geq 1.5 \times 10^{-3}\,\text{T}\cdot\text{m/A}$.

Follow-Up Exercise. In this example, if the scientists found a way for the solenoid to handle up to 5.0 A, what would be the new required permeability?

From Eq. 19.15, the electromagnet's field strength also depends on its current. Large currents produce large fields, but this is accompanied by much greater joule heating (I^2R losses) in the wires, which then require water cooling. The problem can be alleviated by using a superconducting wire. This is because superconducting materials have zero resistance and there is no joule heating. This technology currently exists in laboratories. For commercial use, superconducting magnets are not yet practical because of the energy required in the form of cooling to keep the conductors at their low temperatures and thus in their superconducting state. If (and when) near–room temperature superconductors are found, high-strength magnetic fields will become commonplace in many appliances and applications.

The type of iron that retains some of its magnetism after being in an external magnetic field is called *hard iron* and is used to make so-called permanent magnets. You may have noticed that a paper clip or a screwdriver blade becomes slightly magnetized after being near a magnet. Permanent magnets are produced by heating pieces of some ferromagnetic material in an oven and then cooling them in a strong magnetic field to get the maximum effect. In permanent magnets, the domains do *not* become unaligned when the external magnetic field is removed.

A *permanent* magnet is not necessarily truly permanent, because its magnetism can be destroyed. Hitting such a magnet with a hard object or dropping it on the floor can cause a loss of some domain alignment, reducing the magnet's magnetic field. Also, it can cause a loss of magnetism, because the increase in random (thermal) motions of atoms tends to disrupt the domain alignment. One of the worst things you can do to a magnetic VHS or audiocassette tape is to leave it on the dashboard of a car on a hot day. The increased thermal motion of the electron spins can partially destroy the magnetic voice or video signal imprinted on the tape. Above a certain critical temperature, called the **Curie temperature** (or Curie point), domain coupling is destroyed by these increased thermal oscillations, and a ferromagnetic material loses its ferromagnetism. This effect was discovered by the French physicist Pierre Curie (1859–1906), husband of Madame Marie Curie. The Curie temperature for iron is 770°C.

Ferromagnetic domain alignment plays an important role in geology and geophysics. For example, it is well known that when cooled, lava flows that initially contain iron above its Curie temperature can retain some magnetism due to the Earth's field as it existed when the lava cooled below the Curie temperature and hardened. Measuring the strength and orientation of older lava flows at various locations has enabled geophysicists to map the changes in the Earth's magnetic field and polarity over time.

Some of the first evidence in support of plate tectonic motion came from measuring the direction of the magnetic polarity of seafloor samples containing iron.* The seafloor near the mid-Atlantic ridge, for example, is composed of lava flows from underwater volcanoes. These solidified flows were found to exhibit permanent magnetism, but the polarity varied with time (older samples are farther out from the ridge) as the Earth's magnetic polarity changed.

*19.8 Geomagnetism: The Earth's Magnetic Field

OBJECTIVES: To (a) state some of the general characteristics of the Earth's magnetic field, (b) explain some theories about its possible source, and (c) discuss some of the ways in which the Earth's magnetic field affects our planet's local environment.

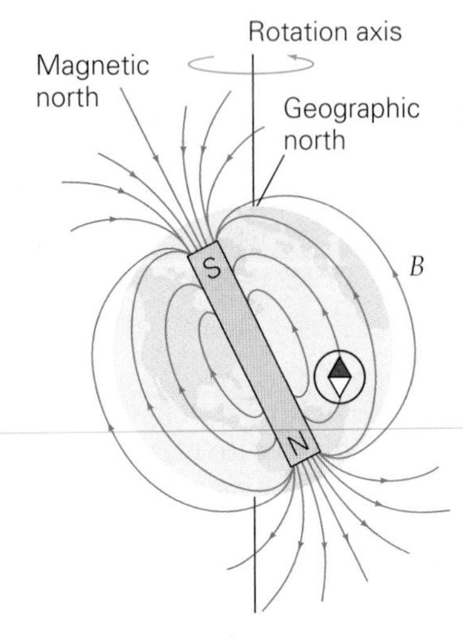

▲ **FIGURE 19.29** Geomagnetic field The Earth's magnetic field is similar to that of a bar magnet. However, a permanent solid magnet could not exist within the Earth because of the high temperatures there. The Earth's magnetic field is believed to be associated with motions in the liquid outer core deep within the planet.

The magnetic field of the Earth was used for centuries before people had any clues about its origin. In ancient times, navigators used lodestones or magnetized needles to locate north. Some other forms of life, including certain bacteria and homing pigeons, also use the Earth's magnetic field for navigation. (See Insight 19.2 on Magnetism in Nature.)

An early study of magnetism was carried out by the English scientist Sir William Gilbert about 1600. In investigating the magnetic field of a specially cut *lodestone* (the name for naturally occurring magnetized rocks) that simulated that of the Earth, he concluded that the Earth as a whole acts as a magnet. Gilbert thought that a large body of permanently magnetized material within the Earth might produce its field.

In fact, the Earth's external magnetic field, or the *geomagnetic field*, as it is called, does have a configuration similar to that which would be produced by a large interior bar magnet with the south pole of the magnet pointing north (◂Fig. 19.29).

*The outermost solid layer of the Earth's crust is known to be composed of sections or "plates." These plates are in constant, but very slow, motion—a centimeter per year is a typical speed. At certain intersections, such as that where the Pacific plate meets the North American plate along the southern coastline of Alaska, the plates collide, giving rise to volcanism and earthquakes. At other intersections, such as the mid-Atlantic ridge, the plates are receding from one another, with new material coming up from the Earth's interior in the form of hot lava.

INSIGHT 19.2 MAGNETISM IN NATURE

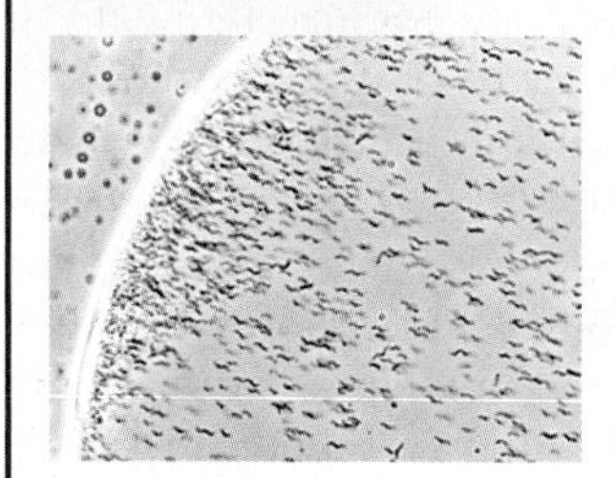

FIGURE 1 Magnetotactic bacteria in migration Bacteria in a drop of muddy water, as viewed under a microscope, align along the direction of the applied magnetic field (north to the left) and accumulate at the edge. When the field is reversed, so is the direction of migration.

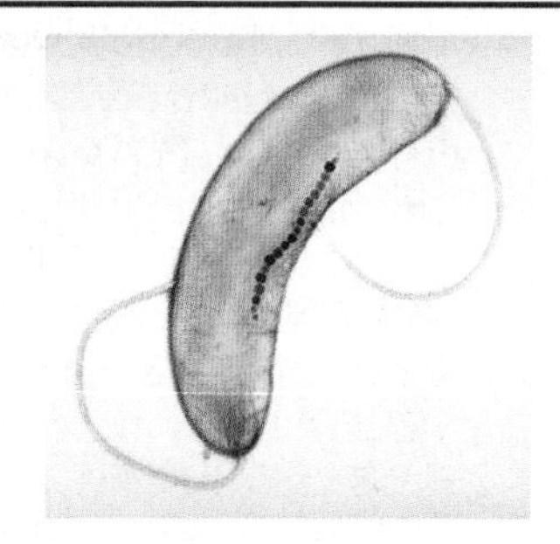

FIGURE 2 A magnetotactic bacterium A freshwater magnetotactic bacterium, shown in an electron micrograph. Two whiplike appendages, or flagella, are clearly visible, along with a chain of magnetite particles.

For centuries, humans have relied on compasses to provide directional information (Fig. 19.29). Studies indicate that certain organisms seem to have their own built-in directional sensors. For example, some species of bacteria are *magnetotactic*—that is, able to sense the presence and direction of the Earth's magnetic field.

In the 1980s, experiments were performed on bacteria commonly found in bogs, swamps, and ponds.* In a laboratory magnetic field, when a droplet of muddy water was viewed under a microscope, one species of bacteria always aligned and migrated in the direction of the field (Fig. 1)—just as these bacteria do in their natural environment with the Earth's field. Furthermore, when these bacteria died and therefore could no longer migrate, they maintained their alignment with the field even when its direction changed. It became apparent that members of this species act like biological magnetic dipoles, or biological compasses. Once aligned with the field, they migrate along magnetic field lines by moving their *flagella* (whiplike appendages), as shown in Fig. 2.

What makes these bacteria act as living compasses? Even among known magnetotactic species, "new" bacteria (formed by cell division) do not initially have this magnetotactic sense. However, if they live in a solution containing a minimum concentration of iron, they are able to synthesize a chain of small magnetic particles (Fig. 2). Oddly enough, these internal compasses have the same chemical composition as the original slivers of naturally occurring ore used as compasses by ancient sailors: magnetite (chemical symbol Fe_3O_4). The individual particles in the chain are approximately 50 nm across, and the chain of a mature bacterium typically contains about 20 such particles, each of which is a single magnetic domain.

In essence, these bacteria are passively steered by their internal compasses. But why is it biologically important for these bacteria to follow the Earth's magnetic field? A piece of the puzzle was found while investigators were studying the same species from Southern Hemisphere waters. These bacteria migrate *opposite* the direction of the Earth's field, unlike their Northern Hemisphere counterparts. Recall that in the Northern Hemisphere the Earth's magnetic field inclines downward and that the reverse is true in the Southern Hemisphere. This discovery lead scientists to believe that the bacteria are using the field direction for survival. Oxygen is toxic to them, so they are most likely to survive in the muddy, oxygen-poor depths of their bog, swamp, or pond, and following the Earth's magnetic field direction enables them to head that way (Fig. 3). This directional sense also aids them near the equator. There it does not direct them downward but instead keeps them at a constant depth, thus avoiding an upward migration to the deadly oxygen-rich surface waters.

Evidence of magnetic field navigation has been found not only in bacteria, but also in such diverse organisms as bees, butterflies, homing pigeons, and dolphins.

*See, for example, R. P. Blakemore and R. B. Frankel, "Magnetic Navigation in Bacteria," *Scientific American*, December 1981. We are indebted to Professor Frankel for several interesting discussions of this topic.

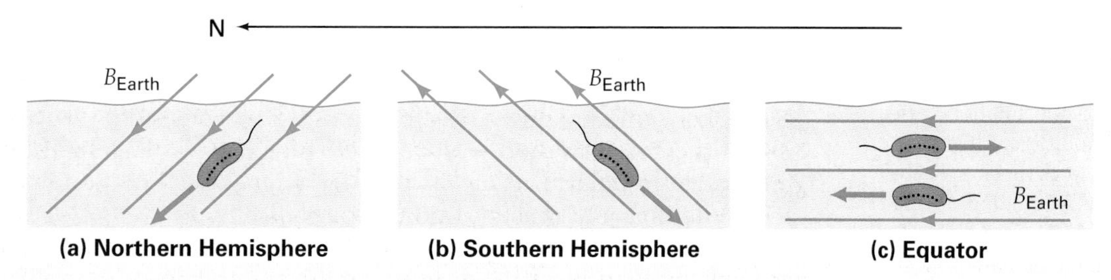

FIGURE 3 Survival of the fittest? (a) In the Northern Hemisphere, where the Earth's magnetic field inclines downward, magnetotactic bacteria follow the field to the oxygen-poor depths. **(b)** In the Southern Hemisphere, the Earth's field is inclined upward, but the bacteria migrate opposite the field and therefore are also able to head for deep waters, like their Northern Hemisphere cousins. **(c)** Near the equator, the bacteria move parallel to the water surface and thus are kept away from shallow, oxygen-rich (and hence toxic to them) waters.

The magnitude of the horizontal component of the Earth's magnetic field at the magnetic equator is on the order of 10^{-5} T (about 0.4 G), and the vertical component at the geomagnetic poles is about 10^{-4} T (roughly, 1 G). It has been calculated that for a ferromagnetic material of maximum magnetization to produce this field, it would have to occupy only about 0.01% of the Earth's volume.

The idea of a ferromagnet of this size within the Earth may not seem unreasonable at first, but this cannot be a correct model. It is known that the interior temperatures deep inside the Earth are well above the Curie temperatures of iron and nickel, the ferromagnetic materials believed to be the most abundant in the Earth's interior. For example, the Curie temperature for iron is attained at a depth of only 100 km below the Earth's surface. The temperatures are even higher at greater depths. So the existence of a permanent internal Earth magnet is not possible.

Knowing that electric currents produce magnetic fields has led scientists to speculate that the Earth's field is associated with motions in the liquid outer core, which, in turn, may be connected some way with the Earth's rotation. We know that Jupiter, a planet that is largely gaseous and rotates very rapidly, has a magnetic field much larger than that of Earth. Mercury and Venus have very weak magnetic fields; these planets are more like Earth and rotate relatively slowly.

Several theoretical models have been proposed to explain the Earth's magnetic field. For example, it has been suggested that the field arises from currents associated with thermal convection cycles in the liquid outer core caused by heat from the inner core. But the details of this mechanism are still not clear.

It is well known that the axis of the Earth's magnetic field does *not* coincide with the planet's rotational axis, which defines the geographic poles. Hence, the Earth's (south) magnetic pole and the geographic North Pole do not coincide (see Fig. 19.29). The magnetic pole is several thousand kilometers south of the geographic North Pole (true north). The Earth's north magnetic pole is displaced even more from its south geographic pole, meaning that the magnetic axis does not even pass through the center of the Earth.

A compass indicates the direction of *magnetic north*, not "true," or geographic, north. The angular difference in these two directions is called the *magnetic declination* (◂Fig. 19.30). The magnetic declination varies with location. Knowing these variations has historically been particularly important in the accurate navigation of airplanes and ships, as you can imagine. Most recently, with the advent of super-accurate GPSs (Global Positioning Systems), high-tech travelers no longer depend on compasses as much as before.

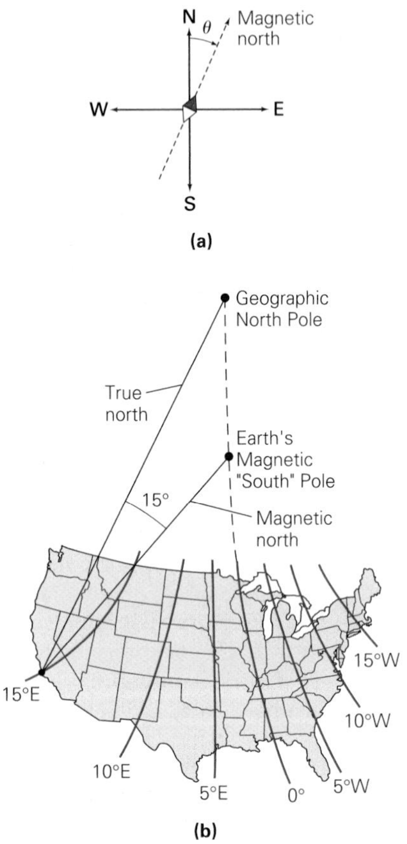

N θ Magnetic north W E S (a)

Geographic North Pole True north Earth's Magnetic "South" Pole 15° Magnetic north 15°W 15°E 10°W 10°E 5°E 0° 5°W (b)

▲ **FIGURE 19.30 Magnetic declination (a)** The angular difference between magnetic north and "true," or geographic, north is called the magnetic declination. **(b)** The magnetic declination varies with location and time. The map shows *isogonic* lines (lines with the same magnetic declination) for the continental United States. For locations on the 0° line, magnetic north is in the same direction as true (geographic) north. On either side of this line, a compass has an easterly or westerly variation. For example, on a 15°E line, a compass has an easterly declination of 15°. (Magnetic north is 15° east of true north.)

The Earth's magnetic field also exhibits a variety of fluctuations with time. As we have discussed, the permanent magnetism created in iron-rich rocks as they cooled in the Earth's magnetic field has provided us with much evidence of these fluctuations over long time scales. For example, the Earth's magnetic poles have switched polarity at various times in the past, most recently about 700 000 years ago. During a period of reversed polarity, the south magnetic pole is near the south geographic pole—the opposite of today's polarity. The mechanism for this periodic magnetic polarity reversal is not clearly understood and scientists are still investigating it.

On a shorter time scale, the magnetic poles also tend to "wander," or change location. For example, the Earth's south magnetic pole (near the north geographic pole) has recently been moving about 1° in latitude (roughly 110 km or 70 mi) per decade. For some unknown reason, it has moved consistently northward from its 1904 latitude of 69°N, and westward, crossing the 100°W longitudinal meridian. This long-term polar drift means that the magnetic declination map (Fig. 19.30b) varies with time and must be updated periodically.

On a still shorter time scale, there are sometimes dramatic daily shifts of magnetic pole location by as much as 80 km (50 mi), followed by a return to the starting position. These shifts are thought to be caused by charged particles from the Sun that reach the Earth's upper atmosphere and set up currents that change the planet's overall magnetic field.

Charged particles from the Sun entering the Earth's magnetic field give rise to another phenomenon. A charged particle that enters a uniform magnetic field at an angle that is *not* perpendicular to the field spirals in a helix (▸Fig. 19.31a). This

▲ **FIGURE 19.31 Magnetic confinement** **(a)** A charged particle entering a uniform magnetic field at an angle other than 90° moves in a spiraling path. **(b)** In a nonuniform, bulging magnetic field, particles spiral back and forth as though confined in a magnetic bottle. **(c)** Charged particles are trapped in the Earth's magnetic field, and the regions where they are concentrated are called Van Allen belts.

is because the component of the particle's velocity parallel to the field does not change. (Recall that a magnetic field acts only on the perpendicular component of the velocity.) The motions of charged particles in a nonuniform field are quite complex. However, for a bulging field such as that depicted in Fig. 19.31b, the particles spiral back and forth as though in a "magnetic bottle."

An analogous phenomenon occurs in the Earth's magnetic field, giving rise to regions with concentrations of charged particles. Two large donut-shaped regions at altitudes of several thousand kilometers are called the *Van Allen radiation belts* (Fig. 19.31c). In the lower Van Allen belt, light emissions called *auroras* occur—the aurora borealis, or northern lights, in the Northern Hemisphere and the aurora australis, or southern lights, in the Southern Hemisphere. These eerie, flickering lights are most commonly observed in the Earth's polar regions, but have been seen at lower latitudes (▸Fig. 19.32).

An aurora is created when charged solar particles become trapped in the Earth's magnetic field. Maximum aurora activity occurs after a solar disturbance, such as a solar flare—a violent magnetic storm on the Sun that spews out enormous quantities of charged particles. Trapped in the Earth's magnetic field, these particles are guided toward the polar regions, where they excite or ionize oxygen and nitrogen atoms in the atmosphere. When the excited atoms return to their normal state and the ions regain their normal number of electrons, light is emitted (Chapter 27), producing the glow of the aurora.

▲ **FIGURE 19.32 Aurora borealis: the northern lights** This spectacular display is caused by energetic solar particles trapped in the Earth's magnetic field. The particles excite or ionize air atoms; on de-excitation (or recombination) of the atoms, light is emitted.

Chapter Review

- The **pole–force law,** or **law of poles:** opposite magnetic poles attract and like poles repel.

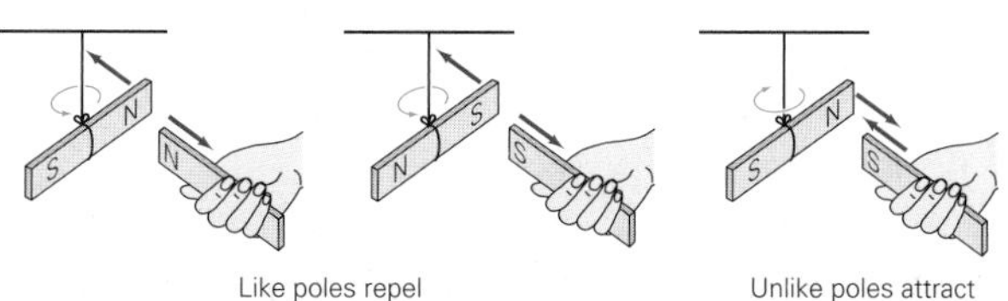

- The **magnetic field** ($\vec{\mathbf{B}}$) has SI units of the **tesla (T)**, where $1\text{ T} = 1\text{ N}/(\text{A}\cdot\text{m})$. Magnetic fields can exert forces on moving charged particles and electric currents. The magnitude of the magnetic force on a charged particle is

$$F = qvB\sin\theta \tag{19.3}$$

The magnitude of the magnetic force on a current-carrying wire is

$$F = ILB\sin\theta \tag{19.7}$$

- **Right-hand force rules** determine the direction of a *magnetic force* on moving charged particles and current-carrying wires.

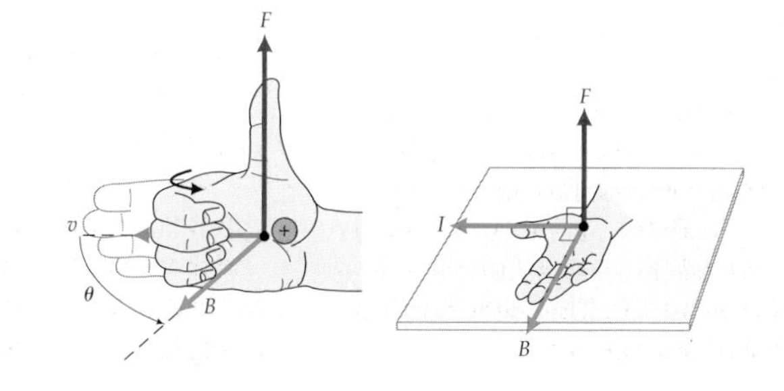

• A series of N current-carrying circular loops, each with a plane area A and carrying a current I, can experience a **magnetic torque** when placed in a magnetic field. The magnitude of the torque on such an arrangement is

$$\tau = NIAB \sin\theta \quad (19.9)$$

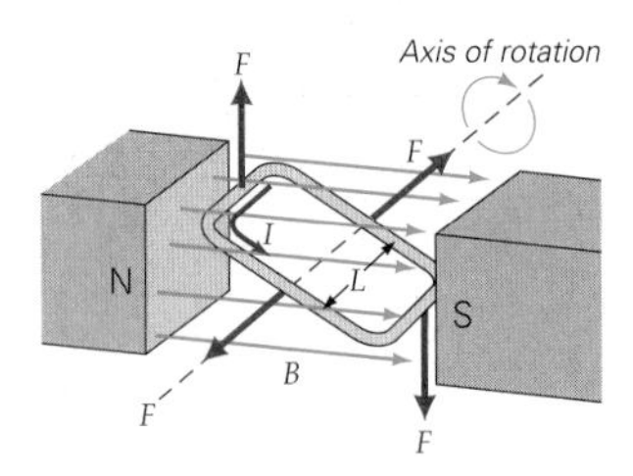

• The magnitude of the **magnetic field** produced by a long, straight wire is

$$B = \frac{\mu_o I}{2\pi d} \quad (19.12)$$

where $\mu_o = 4\pi \times 10^{-7}\,\mathrm{T \cdot m/A}$ is the **magnetic permeability of free space**. For long, straight wires, the field lines are closed circles centered on the wire.

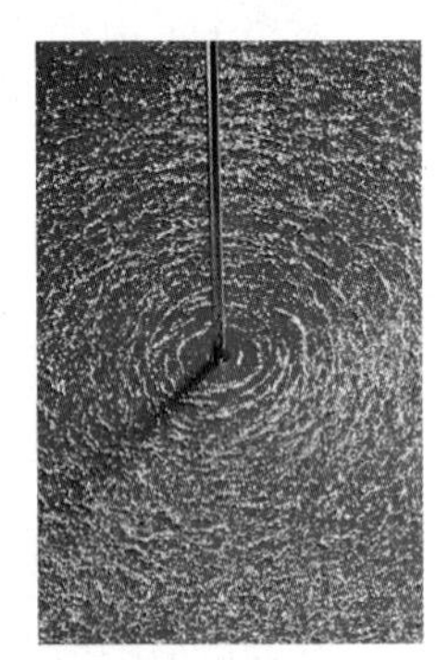

• The magnitude of the magnetic field produced at the center of an arrangement of N circular current-carrying loops of radius r is

$$B = \frac{\mu_o NI}{2r} \quad (19.13)$$

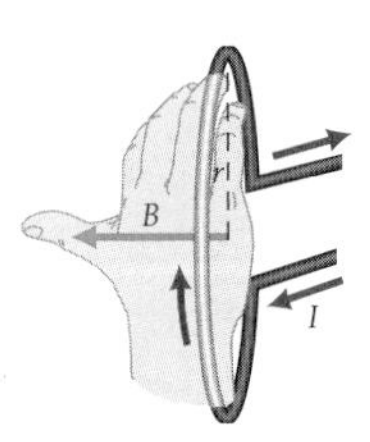

• The magnitude of the magnetic field produced near the center of the interior of a *solenoid* with N windings and a length L is

$$B = \frac{\mu_o NI}{L} \quad (19.14)$$

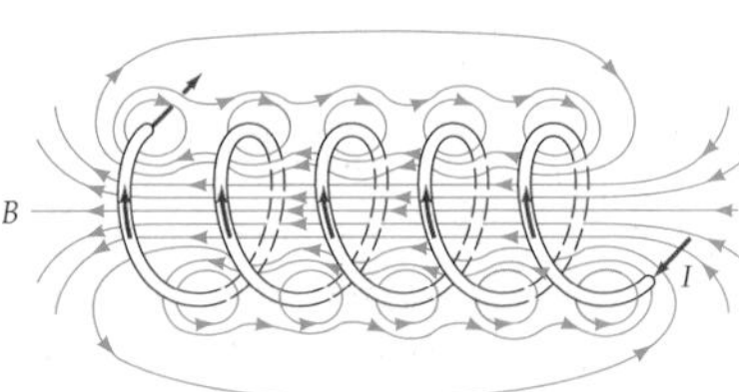

• **Right-hand source rules** are used to determine the direction of the magnetic field from various current configurations.

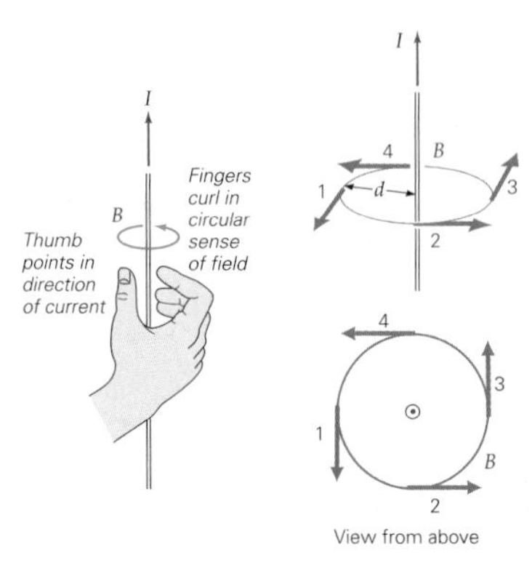

• In **ferromagnetic materials,** the electron spins align, creating **domains.** When an external field is applied, the effect is to increase the size of those domains that already point in the direction of the field at the expense of the others. When the external magnetic field is removed, a **permanent magnet** remains.

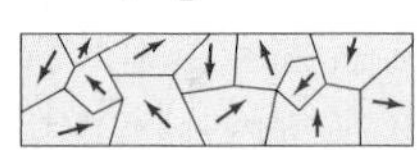

(a) No external magnetic field

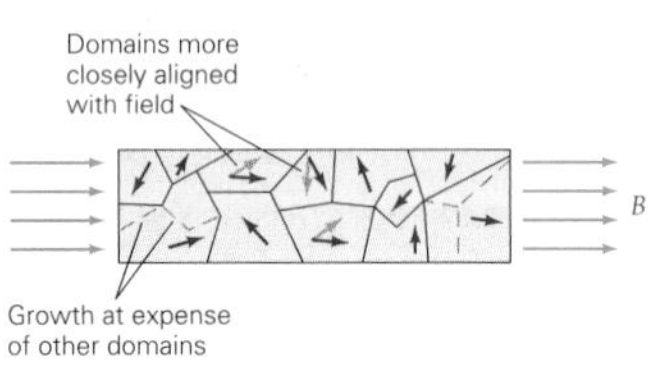

(b) With external magnetic field

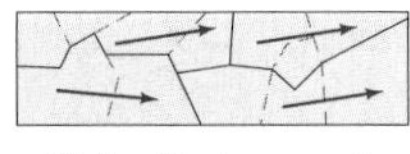

(c) Resulting bar magnet

Exercises

MC = *Multiple Choice Question,* **CQ** = *Conceptual Question, and* **IE** = *Integrated Exercise. Throughout the text, many exercise sections will include "paired" exercises. These exercise pairs, identified with* **red numbers**, *are intended to assist you in problem solving and learning. In a pair, the first exercise (even numbered) is worked out in the Study Guide so that you can consult it should you need assistance in solving it. The second exercise (odd numbered) is similar in nature, and its answer is given at the back of the book.*

19.1 Magnets, Magnetic Poles, and Magnetic Field Direction

1. **MC** When the ends of two bar magnets are near each other, they attract one another. The ends must be (a) one north, the other south, (b) both north, (c) both south, (d) either a or b.

2. **MC** A compass that has been calibrated in the Earth's magnetic field is placed near the end of a permanent bar magnet, and points away from that end of the magnet. It can be concluded that this end of the magnet (a) acts as a north magnetic pole, (b) acts as a south magnetic pole, (c) you can't conclude anything about the magnetic properties of the permanent magnet.

3. **MC** If you look directly at the south pole of a bar magnet, its magnetic field points (a) to the right, (b) to the left, (c) away from you, or (d) toward you.

4. **CQ** Given two identical iron bars, one of which is a permanent magnet and the other unmagnetized, how could you tell which is which by using only the two bars?

5. **CQ** The direction of any magnetic field is taken to be in the direction that an Earth-calibrated compass points. Explain why this means that magnetic field lines must leave from the north pole of a permanent bar magnet and enter its south pole.

19.2 Magnetic Field Strength and Magnetic Force

6. **MC** A proton moves vertically upward perpendicularly to a uniform magnetic field and deflects to the right as you watch it. What is the magnetic field direction: (a) directly away from you, (b) directly toward you, (c) to the right, or (d) to the left?

7. **MC** An electron is moving horizontally to the east in a uniform magnetic field that is vertical. It is found to deflect north. What direction is the magnetic field? (a) up (b) down or (c) the direction can't be determined from the given data.

8. **MC** If a negatively charged particle were moving downward along the right edge of this page, which way should a magnetic field (known to be perpendicular to the plane of the paper) be oriented so that the particle would initially be deflected to the left: (a) out of the page, (b) in the plane of the page, or (c) into the page?

9. **MC** An electron passes through a magnetic field without being deflected. What do you conclude about the angle between the magnetic field direction and that of the electron's velocity, assuming that no other forces act: (a) They could be in the same direction, (b) they could be perpendicular, (c) they could be opposite, or (d) both a and c are possible?

10. **CQ** A proton and an electron are moving at the same velocity perpendicularly to a constant magnetic field. (a) How do the magnitudes of the magnetic forces on them compare? (b) What about the magnitudes of their accelerations?

11. **CQ** If a charged particle moves in a straight line and there are no other forces on it except possibly from a magnetic field, can you say with certainty that no magnetic field is present? Explain.

12. **CQ** Three particles enter a uniform magnetic field as shown in ▸Fig. 19.33a. Particles 1 and 3 have equal speeds and charges of the same magnitude. What can you say about (a) the charges of the particles and (b) their masses?

13. **CQ** You want to deflect a positively charged particle in an S-shaped path, as shown in Fig. 19.33b, using only magnetic fields. (a) Explain how this could be done by using magnetic fields perpendicular to the plane of the page. (b) How does the magnitude of an emerging particle's velocity compare with the particle's initial velocity?

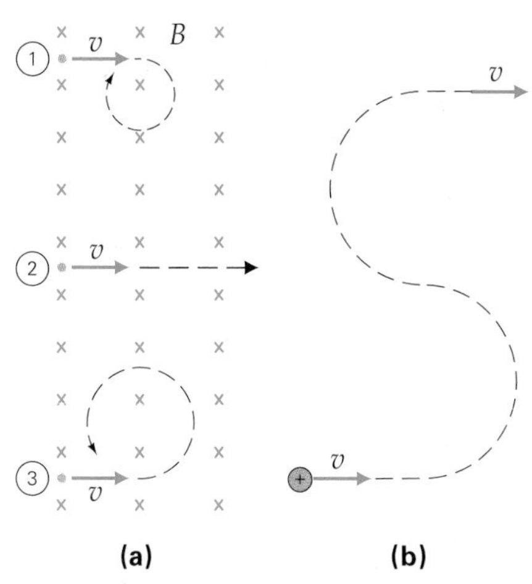

▲ **FIGURE 19.33 Charges in motion** See Exercises 12 and 13.

14. **IE** ● A positive charge moves horizontally to the right across this page and enters a magnetic field directed vertically downward in the plane of the page. (a) What is the direction of the magnetic force on the charge: (1) into the page, (2) out of the page, (3) downward in the plane of the page, or (4) upward in the plane of the page? Explain. (b) If the charge is 0.25 C, its speed is 2.0×10^2 m/s, and it is acted on by a force of 20 N, what is the magnetic field strength?

15. ● A charge of 0.050 C moves vertically in a field of 0.080 T that is oriented 45° from the vertical. What speed must the charge have such that the force acting on it is 10 N?

16. ●● A magnetic field can be used to determine the sign of charge carriers in a current-carrying wire. Consider a wide conducting strip in a magnetic field oriented as shown in ▾Fig. 19.34. The charge carriers are deflected by the magnetic force and accumulate on one side of the strip, giving rise to a measurable voltage across it. (This phenomenon is known as the *Hall effect*.) If the sign of the charge carriers is unknown (they are either positive charges moving as indicated by the arrows in the figure or negative charges moving in the opposite direction), how does polarity or sign of the measured voltage allow the sign of the charge to be determined? Assume that only one type of charge carrier is responsible for the current.

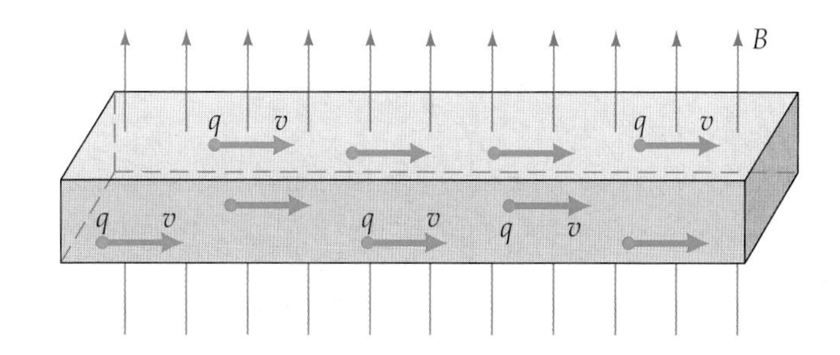

▲ **FIGURE 19.34 The Hall effect** See Exercise 16.

17. ●● A beam of protons is accelerated to a speed of 5.0×10^6 m/s in a particle accelerator and emerges horizontally from the accelerator into a uniform magnetic field. What $\vec{\mathbf{B}}$ field perpendicular to the velocity of the proton would cancel the force of gravity and keep the beam moving exactly horizontally?

18. **IE ●●** An electron travels in the $+x$-direction in a magnetic field and is acted on by a magnetic force in the $-y$-direction. (a) In which of the following directions could the magnetic field be oriented: (1) $-x$-, (2) $+y$-, (3) $+z$-, or (4) $-z$-direction? Explain. (b) If the electron speed is 3.0×10^6 m/s and the magnitude of the force is 5.0×10^{-19} N, what is the magnetic field strength?

19. **●●** An electron travels at a speed of 2.0×10^4 m/s through a uniform magnetic field whose magnitude is 1.2×10^{-3} T. What is the magnitude of the magnetic force on the electron if its velocity and the magnetic field (a) are perpendicular, (b) make an angle of 45°, (c) are parallel, and (d) are exactly opposite?

20. **●●** What angle(s) does a particle's velocity have to make with the magnetic field direction for the particle to be subjected to half the maximum possible magnetic force?

21. **●●●** A proton beam is first accelerated due east to a speed of 3.0×10^5 m/s in a particle accelerator. The beam then enters a uniform magnetic field of 0.50 T, which is oriented at an angle of 37° above the horizontal relative to the direction of the beam. (a) What is the initial acceleration of a proton in the accelerated beam? (b) What if the magnetic field were angled at 37° below the horizontal instead? (c) If the beam were made of electrons rather than protons and the field angle upward at 37°, what would be the difference in the force on the particles as the beam entered the magnetic field?

19.3 Applications: Charged Particles in Magnetic Fields

22. **MC** In a mass spectrometer two ions with identical charge and speed are accelerated into two different semicircular arcs. Ion A's arc has a radius of 25.0 cm and ion B's arc's radius is 50.0 cm. What can you say about their relative masses? (a) $m_A = m_B$ (b) $m_A = 2m_B$ (c) $m_A = \frac{1}{2}m_B$ (d) you can't say anything given just this data.

23. **MC** In a mass spectrometer two ions with identical mass and speed are accelerated into two different semicircular arcs. Ion A's arc has a radius of 25.0 cm and ion B's arc's radius is 50.0 cm. What can you say about their net charges? (a) $q_A = q_B$ (b) $q_A = 2q_B$ (c) $q_A = \frac{1}{2}q_B$ (d) you can't say anything given just this data.

24. **MC** In the velocity selector shown in Fig. 19.9, which way will an ion be deflected if its velocity is larger than E/B_1? (a) up (b) down (c) There will be no deflection.

25. **CQ** Explain how a nearby magnet can distort the display of a computer monitor or television picture tube (▼Fig. 19.35).

◀ **FIGURE 19.35 Magnetic disturbance** See Exercise 25.

26. **CQ** The enlarged circular inset in Fig. 19.11 shows how the positive (Na^+) ions in seawater are accelerated out the rear of the submarine to provide a propulsive force. But what about the negative (Cl^-) ions in the seawater? Because they have charge of the opposite sign, aren't they accelerated *forward* resulting in a net force of zero on the submarine? Explain.

27. **CQ** Explain clearly why the speed selected in a velocity selector setup (similar to that in Fig. 19.9) does not depend on the charges of any of the ions passing through.

28. **●** An ionized deuteron (a particle with a $+e$ charge) passes through a velocity selector whose perpendicular magnetic and electric fields have magnitudes of 40 mT and 8.0 kV/m, respectively. Find the speed of the ion.

29. **●** In a velocity selector, the uniform magnetic field of 1.5 T is produced by a large magnet. Two parallel plates with a separation of 1.5 cm produce the perpendicular electric field. What voltage should be applied across the plates so that (a) a singly charged ion traveling at 8.0×10^4 m/s will pass through undeflected or (b) a doubly charged ion traveling at the same speed will pass through undeflected?

30. **●** A charged particle travels undeflected through perpendicular electric and magnetic fields whose magnitudes are 3000 N/C and 30 mT, respectively. Find the speed of the particle if it is (a) a proton or (b) an alpha particle. (An alpha particle is a helium nucleus—a positive ion with a double positive charge.)

31. **●●** In an experimental technique for treating deep tumors, unstable positively charged pions (π^+, elementary particles with a mass of 2.25×10^{-28} kg) are aimed to penetrate the flesh and to disintegrate at the tumor site, releasing energy to kill cancer cells. If pions with a kinetic energy of 10 keV are required and if a velocity selector with an electric field strength of 2.0×10^3 V/m is used, what must be the magnetic field strength?

32. **●●** In a mass spectrometer, a singly charged ion having a particular velocity is selected by using a magnetic field of 0.10 T perpendicular to an electric field of 1.0×10^3 V/m. This same magnetic field is then used to deflect the ion, which moves in a circular path with a radius of 1.2 cm. What is the mass of the ion?

33. **●●** In a mass spectrometer, a doubly charged ion having a particular velocity is selected by using a magnetic field of 100 mT perpendicular to an electric field of 1.0 kV/m. This same magnetic field is then used to deflect the ion in a circular path with a radius of 15 mm. Find (a) the mass of the ion and (b) the kinetic energy of the ion. (c) Does the kinetic energy of the ion increase in the circular path? Explain.

34. **●●●** In a mass spectrometer, a beam of protons enters a magnetic field. Some protons make exactly a one-quarter circular arc of radius 0.50 m. If the field is always perpendicular to the proton's velocity, what is the field's magnitude if exiting protons have a kinetic energy of 10 keV?

19.4 Magnetic Forces on Current-Carrying Wires *and* 19.5 Applications: Current-Carrying Wires in Magnetic Fields

35. **MC** A long, straight, horizontal wire located on the equator carries a current directed toward the east. What is the direction of the force on the wire due to the Earth's magnetic field: (a) east, (b) west, (c) south, or (d) upward?

36. **MC** A long, straight, horizontal wire located on the equator carries a current. In what direction should the current be if the object is to attempt to balance the wire's weight with the magnetic force on it: (a) east, (b) west, (c) south, or (d) upward?

37. **MC** You are looking horizontally due west directly at the circular plane of a current-carrying coil. The coil is in a uniform vertically upward magnetic field. When released, the top of the coil starts to rotate away from you as the bottom rotates toward you. Which direction is the current in the coil: (a) clockwise, (b) counterclockwise, or (c) can't tell from the data given?

38. **CQ** Two straight wires are parallel to each other and carry different currents in the same direction. Do they attract or repel each other? How do the magnitudes of these forces on each wire compare?

39. **CQ** Predict what should happen to the length of a spring when a large current passes through it. [*Hint*: Consider the current direction in the neighboring spring coils.]

40. **CQ** Is it possible to orient a current loop in a uniform magnetic field so that there is no torque on the loop? If yes, describe the possible orientation(s).

41. **CQ** Explain the operation of the doorbell and door chimes illustrated in ▼Fig. 19.36.

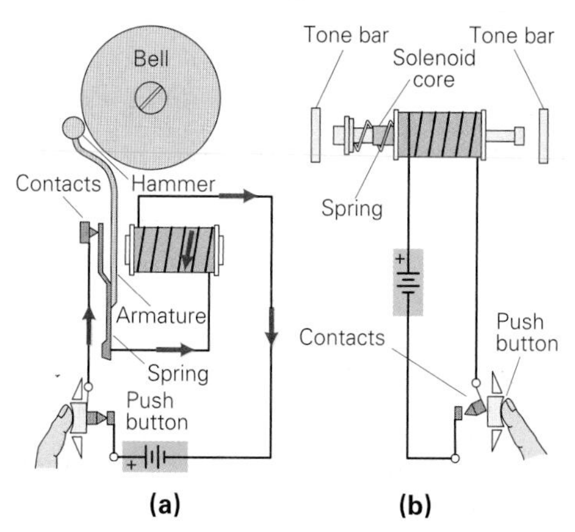

▲ **FIGURE 19.36 Electromagnetic applications** Both **(a)** a doorbell and **(b)** door chimes have electromagnets. See Exercise 41.

42. **IE** ● A straight, horizontal segment of wire carries a current in the $+x$-direction in a magnetic field that is directed in the $-z$-direction. (a) Is the magnetic force on the wire directed in the (1) $-x$-, (2) $+z$-, (3) $+y$-, or (4) $-y$-direction? Explain. (b) If the wire is 1.0 m long and carries a current of 5.0 A and the magnitude of the magnetic field is 0.30 T, what is the magnitude of the force on the wire?

43. ● A 2.0-m length of straight wire carries a current of 20 A in a uniform magnetic field of 50 mT whose direction is at an angle of 37° from the direction of the current. Find the force on the wire.

44. ●● Show how you can use a right-hand force rule to find the direction of the current in a wire in a magnetic field if you know the force on the wire. The forces on some specific wires are shown in ▼Fig. 19.37.

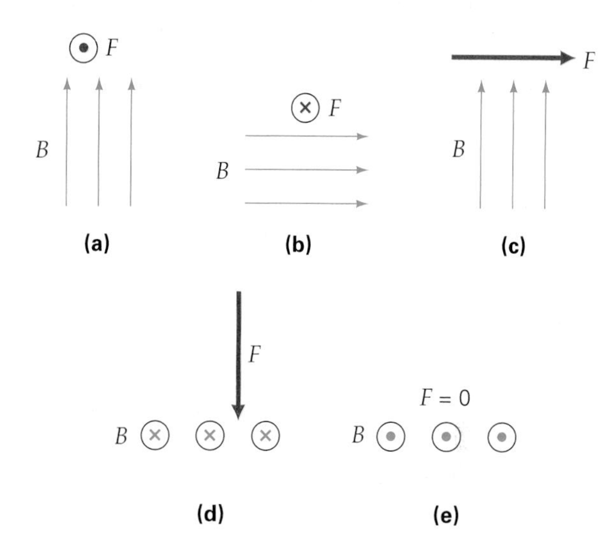

▲ **FIGURE 19.37 The right-hand force rule** See Exercise 44.

45. ●● A straight wire 50 cm long conducts a current of 4.0 A directed vertically upward. If the wire is acted on by a force of 1.0×10^{-2} N in the eastward direction due to a magnetic field at right angles to the length of the wire, what are the magnitude and direction of the magnetic field?

46. ●● A horizontal magnetic field of 1.0×10^{-4} T is at an angle of 30° to the direction of the current in a straight, horizontal wire 75 cm long. If the wire carries a current of 15 A, what is the magnitude of the force on the wire?

47. ●● A wire carries a current of 10 A in the $+x$-direction in a uniform magnetic field of 0.40 T. Find the magnitude of the force per unit length and the direction of the force on the wire if the magnetic field points in (a) the $+x$-direction, (b) the $+y$-direction, (c) the $+z$-direction, (d) the $-y$-direction, and (e) the $-z$-direction.

48. ●● A straight wire 25 cm long is oriented vertically in a uniform horizontal magnetic field of 0.30 T pointing in the $-x$-direction. What current (including direction) would cause the wire to be subject to a force of 0.050 N in the $+y$-direction?

49. ●● A wire carries a current of 10 A in the $+x$-direction. Find the force per unit length on the wire if it is in a magnetic field that has components of $B_x = 0.020$ T, $B_y = 0.040$ T, and $B_z = 0$ T.

50. ●● A set of jumper cables used to start a car from another car's battery is connected to the terminals of both batteries. If 15 A of current exists in the cables during the starting procedure and the cables are parallel and 15 cm apart, what is the force per unit length on the cables?

51. IE ●● Two long, straight, parallel wires carry current in the same direction. (a) Use the right-hand source and force rules to determine whether the forces on the wires are (1) attractive or (2) repulsive. (b) If the wires are 24 cm apart and carry currents of 2.0 A and 4.0 A, respectively, find the force per unit length on each wire.

52. IE ●● Two long, straight, parallel wires 10 cm apart carry currents in opposite directions. (a) Use the right-hand source and force rules to determine whether the forces on the wires are (1) attractive or (2) repulsive. (b) If the wires carry equal currents of 3.0 A, what is the force per unit length on the wires?

53. ●● A nearly horizontal dc power line in the midlatitudes of North America carries a current of 1000 A directly eastward. If the Earth's magnetic field at the location of the power line is northward with a magnitude of 5.0×10^{-5} T at an angle of 45° below the horizontal, what are the magnitude and direction of the magnetic force on a 15-m section of the line?

54. ●● What is the force (including direction) per unit length on wire 1 in ▼Fig. 19.38?

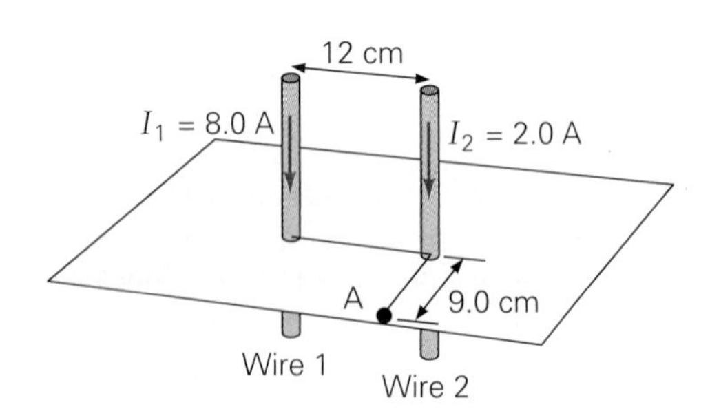

▲ **FIGURE 19.38 Parallel current-carrying wires** See Exercises 54, 55, 73, and 76.

55. ●● What is the force (including direction) per unit length on wire 2 in Fig. 19.38?

56. IE ●● A long wire is placed 2.0 cm directly below a rigidly mounted second wire (▼Fig. 19.39). (a) Use the right-hand source and force rules to determine whether the currents in the wires should be in (1) the same or (2) the opposite direction so that the lower wire is in equilibrium. (It "floats.") (b) If the lower wire has a linear mass density of 1.5×10^{-3} kg/m and the wires carry the same current, what should be the current?

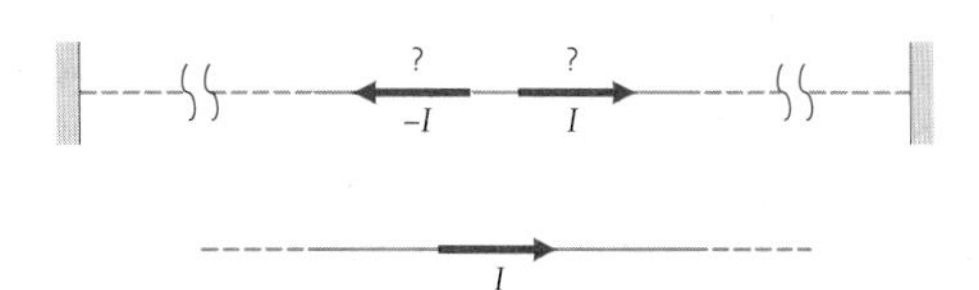

▲ **FIGURE 19.39 Magnetic suspension** The bottom wire is magnetically attracted to the top (rigidly fixed) wire. See Exercise 56.

57. ●● A wire is bent as shown in ▼Fig. 19.40 and placed in a magnetic field with a magnitude of 1.0 T in the indicated direction. Find the net force on the wire if x = 50 cm and it carries a current of 5.0 A in the direction shown.

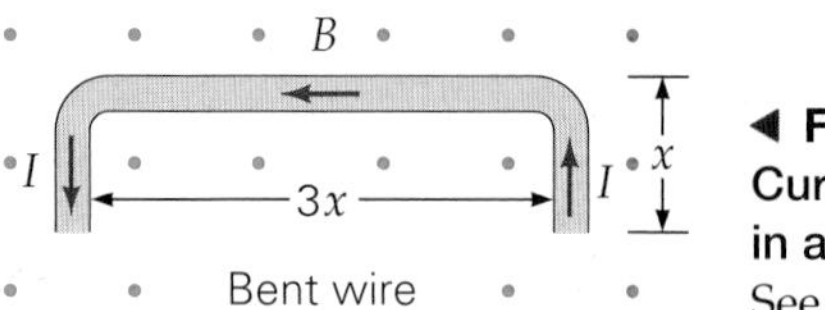

◀ **FIGURE 19.40 Current-carrying wire in a magnetic field** See Exercise 57.

58. IE ●●● A loop of current-carrying wire is in a 1.6-T magnetic field. (a) For the magnetic torque on the loop to be at maximum, should the plane of the coil be (1) parallel, (2) perpendicular, or (3) at a 45° angle to the magnetic field? Explain. (b) If the loop is rectangular with dimensions 20 cm by 30 cm and carries a current of 1.5 A, what is the magnitude of the magnetic moment of the loop, and what is the maximum torque? (c) What would be the angle(s) between the magnetic moment vector and the magnetic field direction if the loop felt only 20% of its maximum torque?

59. ●●● Two straight wires are positioned at right angles to each other as in ▼Fig. 19.41. What is the net force on each wire? Is there a net torque on each wire?

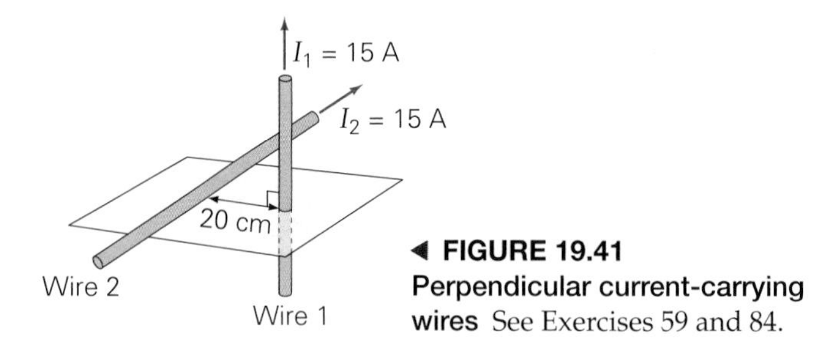

◀ **FIGURE 19.41 Perpendicular current-carrying wires** See Exercises 59 and 84.

60. ●●● A rectangular wire loop with a cross-sectional area of 0.20 m² carries a current of 0.25 A. The loop is free to rotate about an axis that is perpendicular to a uniform magnetic field with strength 0.30 T. The plane of the loop is at an angle of 30° to the direction of the magnetic field. (a) What is the magnitude of the torque on the loop? (b) How would you change the magnetic field to double the magnitude of the torque in part (a)? (c) Could you double the torque in part (a) only by changing the angle? Explain. If so, find that angle.

19.6 Electromagnetism: The Source of Magnetic Fields

61. **MC** A long, straight wire is parallel to the ground and carries a steady current to the east. At a point directly below the wire, what is the direction of the magnetic field the wire produces: (a) north, (b) east, (c) south, or (d) west?

62. **MC** You are looking directly into one end of a long solenoid. The magnetic field at its center points at you. What is the direction of the current in the solenoid, as viewed by you: (a) clockwise, (b) counterclockwise, (c) directly toward you, or (d) directly away from you?

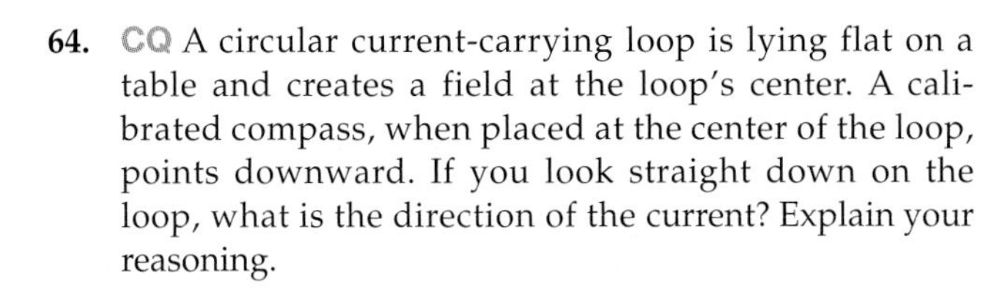

63. **MC** A current-carrying loop of wire is in the plane of this paper. Outside the loop, its magnetic field points into the paper. What is the direction of the current in the loop: (a) clockwise, (b) counterclockwise, or (c) can't tell from the data given?

64. **CQ** A circular current-carrying loop is lying flat on a table and creates a field at the loop's center. A calibrated compass, when placed at the center of the loop, points downward. If you look straight down on the loop, what is the direction of the current? Explain your reasoning.

65. **CQ** If you doubled your distance from a long current-carrying wire, what would you have to do to the current to keep the magnetic field strength the same as at the close position but reverse its direction? Explain.

66. **CQ** There are two solenoids, one with 100 turns and the other with 200 turns. If both carry the same current, will the one with more turns necessarily produce a stronger magnetic field at its center? Explain.

67. **CQ** To minimize the effects of the magnetic field, most appliance wires are placed close together. Explain how this works to reduce the external field from the current in the wire.

68. **CQ** Two circular wire loops are coplanar (that is, their areas are in the same plane) and have a common center. The outer one carries a current of 10 A in the clockwise direction. To create a zero magnetic field at their center, what should be the direction of the current in the inner loop? Should its current be 10 A, larger than 10 A, or smaller than 10 A? Explain your reasoning.

69. ● The magnetic field at the center of a 50-turn coil of radius 15 cm is 0.80 mT. Find the current in the coil.

70. ● A long, straight wire carries a current of 2.5 A. Find the magnitude of the magnetic field 25 cm from the wire.

71. ● In a physics lab, a student discovers that the magnitude of the magnetic field at a certain distance from a long wire is 4.0 μT. If the wire carries a current of 5.0 A, what is the distance of the magnetic field from the wire?

72. ● A solenoid is 0.20 m long and consists of 100 turns of wire. At its center, the solenoid produces a magnetic field with a strength of 1.5 mT. Find the current in the coil.

73. ●● Two long, parallel wires carry currents of 8.0 A and 2.0 A (Fig. 19.38). (a) What is the magnitude of the magnetic field midway between the wires? (b) Where on a line perpendicular to and joining the wires is the magnetic field zero?

74. ●● Two long, parallel wires separated by 50 cm each carry currents of 4.0 A in a horizontal direction. Find the magnetic field midway between the wires if the currents are (a) in the same direction and (b) in opposite directions.

75. ●● Two long, parallel wires separated by 0.20 m carry equal currents of 1.5 A in the same direction. Find the magnitude of the magnetic field 0.15 m away from each wire on the side opposite the other wire (▼Fig. 19.42).

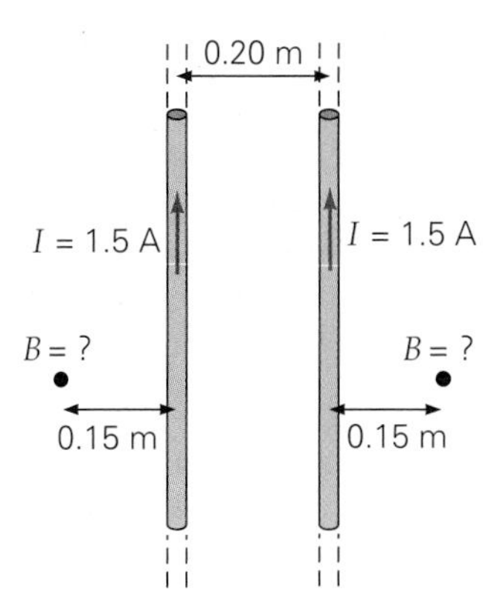

▲ **FIGURE 19.42 Magnetic field summation** See Exercise 75.

76. ●● In Fig. 19.38, find the magnetic field (magnitude and direction) at point A, which is located 9.0 cm away from wire 2 on a line perpendicular to the line joining the wires.

77. ●● Suppose that the current in wire 1 in Fig. 19.38 were in the opposite direction. What would be the magnetic field midway between the wires?

78. ●● How much current must flow in a circular loop of radius 10 cm to produce a magnetic field at the center of the loop that is the same magnitude as the horizontal component of the Earth's magnetic field at the equator (about 0.40 G)?

79. ●● A coil of four circular loops of radius 5.0 cm carries a current of 2.0 A clockwise, as viewed from above the coil's plane. What is the magnetic field at the center of the coil?

80. **IE** ●● A circular loop of wire in the horizontal plane carries a counterclockwise current, as viewed from above. (a) Use the right-hand source rule to determine whether the direction of the magnetic field at the center of the loop is (1) toward or (2) away from the observer. (b) If the diameter of the loop is 12 cm and the current is 1.8 A, what is the magnitude of the magnetic field at the center of the loop?

81. ●● A circular loop of wire with a radius of 5.0 cm carries a current of 1.0 A. Another circular loop of wire is concentric with (that is, has a common center with) the first and has a radius of 10 cm. The magnetic field at the center of the loops is double what the field would be from the first one alone, but oppositely directed. What is the current in the second loop?

82. ●●● A current-carrying solenoid is 10 cm long and is wound with 1000 turns of wire. It produces a magnetic field of 4.0×10^{-4} T at the solenoid's center. (a) How long would you make the solenoid in order to produce a field of 6.0×10^{-4} T at its center? (b) Adjusting only the windings, what number would be needed to produce a field of 8.0×10^{-4} T at the center? (c) What current in the solenoid would be needed to produce a field of 9.0×10^{-4} T but in the opposite direction?

83. ●●● A solenoid is wound with 200 turns per centimeter. An outer layer of insulated wire with 180 turns per centimeter is wound over the solenoid's first layer of wire. When the solenoid is operating, the inner coil carries a current of 10 A and the outer coil carries a current of 15 A in the direction opposite to that of the current in the inner coil (▼Fig. 19.43). (a) What is the magnitude of the magnetic field at the center of the doubly wound solenoid? (b) What is the direction of the magnetic field at the center for this configuration?

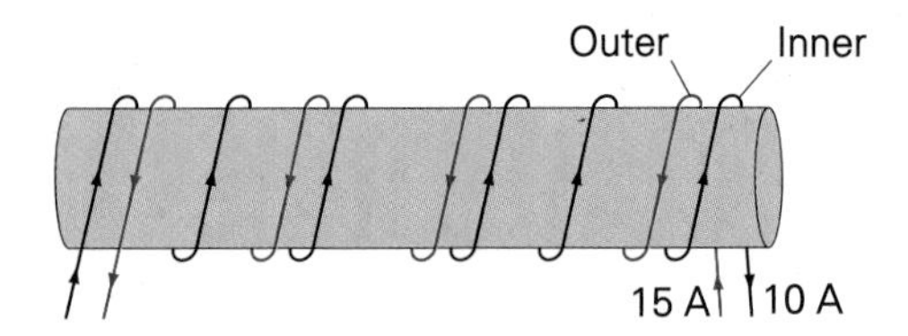

▲ **FIGURE 19.43 Double it up?** See Exercise 83.

84. ●●● Two long, perpendicular wires carry currents of 15 A, as illustrated in Fig. 19.41. What is the magnitude of the magnetic field at the midpoint of the line joining the wires?

85. ●●● Four wires running through the corners of a square with sides of length a, as shown in ▼Fig. 19.44, carry equal currents I. Calculate the magnetic field at the center of the square in terms of these parameters.

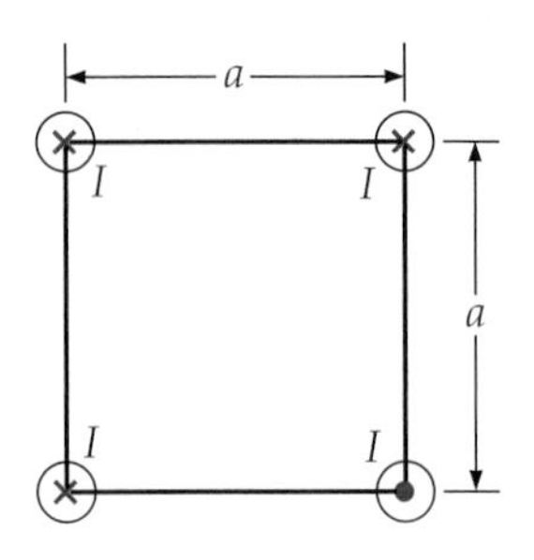

◀ **FIGURE 19.44 Current-carrying wires in a square array** See Exercise 85.

86. ●●● A particle with charge q and mass m moves in a horizontal plane at right angles to a uniform vertical magnetic field B. (a) What is the frequency f of the particle's circular motion in terms of q, B, and m? (This frequency is called the *cyclotron frequency*.) (b) Show that the time required for any charged particle to make one complete revolution is independent of its speed and radius. (c) Compute the path radius and the cyclotron frequency if the particle is an electron with speed $v = 1.0 \times 10^5$ m/s and the field strength is $B = 1.0 \times 10^{-4}$ T.

19.7 Magnetic Materials

87. **MC** The main source of magnetism in magnetic materials is from (a) electron orbits, (b) electron spin, (c) magnetic poles, (d) nuclear properties.

88. **MC** When a ferromagnetic material is placed in an external magnetic field, (a) the domain orientation may change, (b) the domain boundaries may change, (c) new domains are created, (d) both a and b.

89. **CQ** If you are looking down on the orbital plane of the electron in a hydrogen atom and the electron orbits counterclockwise, what is the direction of the magnetic field the electron produces at the proton?

90. **CQ** What is the purpose of the iron core often used at the center of a solenoid?

91. **CQ** Discuss several ways you destroy or reduce the magnetic field of a permanent magnet.

92. ●● A solenoid with 100 turns per centimeter has an iron core with a relative permeability of 2000. The solenoid carries a current of 0.040 A. (a) What is the magnetic field at the center of the solenoid? (b) How much greater is the magnetic field with the iron core than it would be without it?

93. ●●● What is the magnetic field (due to the electron only) at the center of the circular orbit of the electron in a hydrogen atom? The orbital radius is 0.0529 nm. [*Hint*: Find the electron's period by considering the centripetal force.]

*19.8 Geomagnetism: The Earth's Magnetic Field

94. **MC** The Earth's magnetic field (a) has poles that coincide with the geographic poles, (b) only exists at the poles, (c) reverses polarity every few hundred years, (d) none of these.

95. **MC** Auroras (see Fig. 19.32) (a) occur only in the Northern Hemisphere, (b) are related to the lower Van Allen belt, (c) occur because of Earth's magnetic pole reversals, (d) happen predominantly when there are no solar disturbances.

96. **MC** If the direction of your calibrated compass pointed straight down, where would you be: (a) near the Earth's north geographic pole, (b) near the equator, or (c) near the Earth's south geographic pole?

97. **MC** If a proton was orbiting above the Earth's equator in the Van Allen belt, which way would it have to be orbiting: (a) to the west, (b) to the east, or (c) either direction would work?

98. **CQ** Determine the direction of the force on an electron due to the Earth's magnetic field on an electron near the equator for each of the following situations. The electron's velocity is directed (a) due south, (b) northwest, or (c) upward.

99. **CQ** In a relatively short time geologically speaking, it is conjectured that the Earth's magnetic field direction will reverse. After that, what would be the polarity of the magnetic pole near the Earth's geographic North Pole?

Comprehensive Exercises

100. A beam of protons is accelerated from rest through a potential difference of 3.0 kV. It enters a region where its velocity is initially perpendicular to an electric field. The field is created by two parallel plates separated by 10 cm with a potential difference of 250 V across them. Find the magnitude of the magnetic field (perpendicular to $\vec{E}$) needed so the beam passes undeflected through the plates.

101. A solenoid 10 cm long has 3000 turns of wire and carries a current of 5.0 A. A 2000-turn coil of wire of the same length as the solenoid surrounds it and is concentric (shares a common central axis) with it. The outer coil carries a current of 10 A in the same direction as that of the current in the solenoid. Find the magnetic field at the common center.

102. IE A horizontal beam of electrons travels from north to south in a discharge tube located in the Northern Hemisphere. (a) Is the magnetic force on the electron directed (1) west, (2) east, (3) south, or (4) north? Explain. (b) If their speed is 1.0×10^3 m/s and the vertical component of the Earth's magnetic field at that location is known to be 5.0×10^{-5} T, what is the magnitude of the force on each electron?

103. A proton enters a uniform magnetic field that is at a right angles to its velocity. The field strength is 0.80 T and the proton follows a circular path with a radius of 4.6 cm. What is its (a) momentum and (b) kinetic energy?

104. Exiting a linear accelerator, a narrow horizontal beam of protons travels due north. If 1.75×10^{13} protons pass a given point per second, determine the magnetic field direction and strength at a location of 2.40 m east of the beam. Does it seem likely this would interfere with a magnetic strip on an ATM card, in comparison to the Earth's field?

105. A 200-turn circular coil of wire has a radius of 10.0 cm and a total resistance of 0.115 Ω. At its center the magnetic field strength is 7.45 mT. Determine the voltage of the power supply creating the current in the coil.

106. A 100-turn circular coil of wire has a radius of 20.0 cm and carries a current of 0.400 A. The normal to the coil area points due east. A compass, when placed at the center of the coil, does not point east, but instead makes an angle of 60° north of east. Using this data, determine (a) the magnitude of the horizontal component of the Earth's field at that location and (b) the magnitude of the Earth's field at that location if it makes an angle of 55° below the horizontal.

107. IE Two long straight wires are oriented perpendicularly to the page. Wire 1 carries current of 20.0 A into the page and 15.0 cm to its left, wire 2 carries a 5.00 A current. Somewhere on the line joining the two wires there is to be a zero magnetic field. (a) What is the direction of the current in wire 2: (1) out of the paper, (2) into the paper, or (3) can't tell from the data given? (b) Find the location where the zero magnetic field exists.

108. IE A circular coil of wire has the normal to its area pointing upward. A second smaller concentric coil carries a current in the opposite direction. (a) Where, in the plane of these coils, could the magnetic field be zero: (1) only inside the smaller one, (2) only between the inner and outer one, (3) only outside the larger one, (4) or inside the smaller one and outside the larger one? (b) The larger one is a 200-turn coil of wire with a radius of 9.50 cm and carries a current of 11.5 A. The second one is a 100-turn coil with a radius of 2.50 cm. Determine the current in the inner coil so the magnetic field at their common center is zero. Neglect the Earth's field.

109. A 50-cm-long solenoid has 100 turns of wire and carries a current of 0.95 A. It has a ferromagnetic core completely filling its interior where the field is 0.71 T. Determine the (a) magnetic permeability and (b) relative magnetic permeability of the material.

The following Physlet Physics Problems can be used with this chapter.
27.1, 27.2, 27.3, 27.4, 27.5, 27.6, 27.7, 27.8, 27.9, 27.10, 28.1, 28.6

CHAPTER

20 ELECTROMAGNETIC INDUCTION AND WAVES

20.1 Induced emf: Faraday's Law and Lenz's Law 657

20.2 Electric Generators and Back emf 663

20.3 Transformers and Power Transmission 668

20.4 Electromagnetic Waves 672

PHYSICS FACTS

- Nikola Tesla (1856–1943), the Serbian-American scientist-inventor whose last name is the SI unit of magnetic field strength, invented ac dynamos, transformers, and motors. He sold the patent rights to these to George Westinghouse. This eventually led to a struggle between Thomas Edison's dc systems and Westinghouse's ac version of power generation and distribution. The latter eventually won, with the installation of the first large-scale electric generator at Niagara Falls.
- To prove the safety of electric energy to a skeptical public at the turn of the twentieth century, Tesla gave exhibitions of lighting lamps by allowing electricity to flow through his body. Westinghouse used his system to light the World's Columbian Exposition at Chicago in 1893. Tesla proved that the Earth could be used as a conductor and lighted 200 lamps without wires at a distance of 25 miles. With his giant transformer (a Tesla coil) Westinghouse created artificial lightning, producing bolts measuring over 100 feet long.
- Radio waves, radar, visible light, and X-rays are all electromagnetic waves. Better known as light, they all obey the same mathematical relationships. The only difference is their frequency and wavelength. In a vacuum, they all travel at exactly the same speed, c (3.00×10^8 m/s).
- The Scottish physicist James Clerk Maxwell (1831–1879) fully developed and integrated the equations of electricity and magnetism. This set became known as Maxwell's equations, and his interpretation of them was one of the great achievements of nineteenth-century physics.

As we saw in Chapter 19, an electric current produces a magnetic field. But the relationship between electricity and magnetism does not stop there. In this chapter, you will learn that under the right conditions, a magnetic field can produce an electric current. How is this done? Chapter 19 considered only *constant* magnetic fields. No current is produced in a loop of wire that is stationary in a constant magnetic field. However, if the magnetic field changes with time, or if the wire loop moves into or out of, or is rotated in, the field, a current *is* produced in the wire.

The uses of this interrelationship of electricity and magnetism are many. One example happens during the playing of a videotape, which is actually a magnetic tape that has information encoded on it as variations in its magnetism. These variations can be used to produce electrical currents, which, in turn, are amplified and the signal sent for replay to the television set. Similar processes are involved when information is stored on or retrieved from a magnetic disk in your computer.

On a larger scale, consider the generation of the electric energy that provides the basis for our modern civilization. At hydroelectric plants such as that in this photo, one of the oldest and simplest energy sources on Earth—falling water—is used to generate electric energy. The gravitational potential energy of the water is converted into kinetic energy, and some of this kinetic energy is transformed, eventually, into electric energy. But how does this last step take place? Regardless of the ultimate source of the energy—the burning of oil, coal, or gas; a nuclear reactor; or falling water—the actual conversion to electric energy is accomplished by means of magnetic fields and electromagnetic induction. This chapter not only examines the underlying principles that make such conversion possible, but also discusses several practical applications. Moreover, we will also see that the creation and propagation of electromagnetic radiation is intimately related to electromagnetic induction.

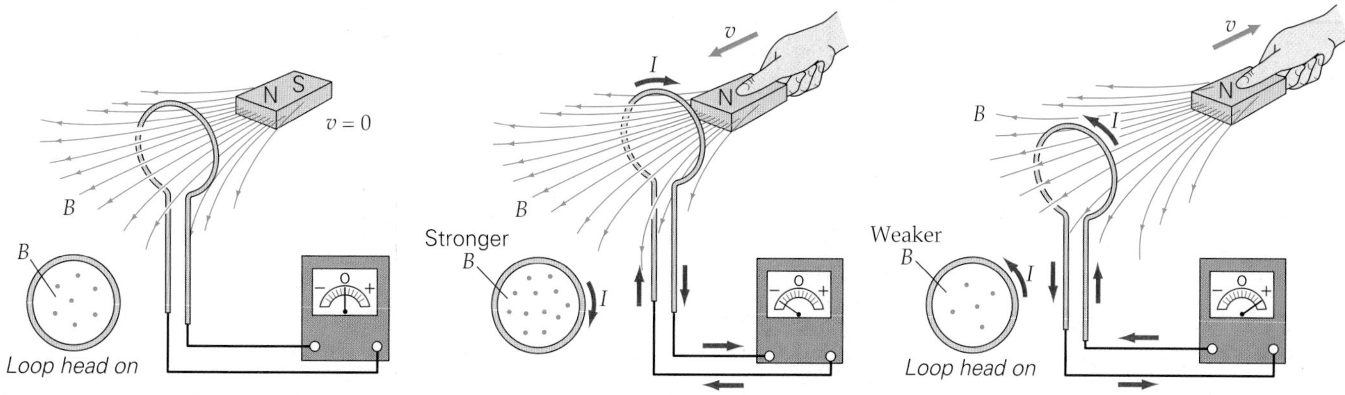

(a) No motion between magnet and loop (b) Magnet is moved toward loop (c) Magnet is moved away from loop

▲ **FIGURE 20.1 Electromagnetic induction** **(a)** When there is no relative motion between the magnet and the wire loop, the number of field lines through the loop (in this case, 7) is constant, and the galvanometer shows no deflection. **(b)** Moving the magnet toward the loop increases the number of field lines passing through the loop (now 12), and an induced current is detected. **(c)** Moving the magnet away from the loop decreases the number of field lines passing through the loop (to 5). The induced current is now in the opposite direction. (Note the needle deflection.)

20.1 Induced emf: Faraday's Law and Lenz's Law

OBJECTIVES: To (a) define magnetic flux and explain how an induced emf is created, and (b) determine induced emfs and currents.

Recall from Chapter 17 that the term *emf* stands for *e*lectro*m*otive *f*orce, which is a voltage or electric potential difference capable of creating an electric current. It is observed experimentally that a magnet held stationary near a conducting wire loop does *not* induce an emf (and therefore produces no current) in that loop (▲Fig. 20.1a). If the magnet is moved toward the loop, however, as shown in Fig. 20.1b, the deflection of the galvanometer needle indicates that current exists in the loop, but only during the motion. Furthermore, if the magnet is moved away from the loop, as shown in Fig. 20.1c, the galvanometer needle is deflected in the opposite direction, which indicates a reversal of the current's direction, but, again, only during the motion.

Deflections of the galvanometer needle, indicating the presence of *induced currents*, also occur if the loop is moved toward or away from the stationary magnet. The effect thus depends on *relative* motion of the loop and magnet. It also turns out that the magnitude of the induced current depends on the speed of that motion. However, experimentally, there is a noteworthy exception. If a loop is moved (but not rotated) in a *uniform* magnetic field, as shown in ▸Fig. 20.2, no current is induced. We will see why this is so later in this section.

Yet another way to induce a current in a stationary wire loop is to vary the current in another, nearby loop. When the switch in the battery-powered circuit in ▾Fig. 20.3a is closed, the current in the loop on the right goes from zero to some constant value in a short time. Only during the buildup time does the magnetic field caused by the current in this loop increase in the region of the loop on the left. During the buildup, the galvanometer needle deflects, indicating current in the left loop. When the current in the right loop attains its steady value, the field it produces becomes constant, and the current in the left loop drops to zero. Similarly, when the switch in the right loop is opened (Fig. 20.3b), its current and field decrease to zero, and the galvanometer deflects in the opposite direction, indicating a reversal in direction of the current induced in the left loop. The important fact to note is that *induced current in a loop occurs only when the magnetic field through that loop changes.*

In Fig. 20.1, moving the magnet changed the magnetic environment in a loop, caused an induced emf that, in turn, caused an induced current. For the case of *two* stationary loops (Fig. 20.3), a changing current in the right loop produced a changing magnetic environment in the left loop, thereby inducing an emf and a current in the left loop.* There is a convenient way of summarizing what is happening in both Fig. 20.1 and Fig. 20.3: To induce currents in a loop or complete circuit, a process called **electromagnetic induction**, all that matters is whether the magnetic field through the loop or circuit is changing.

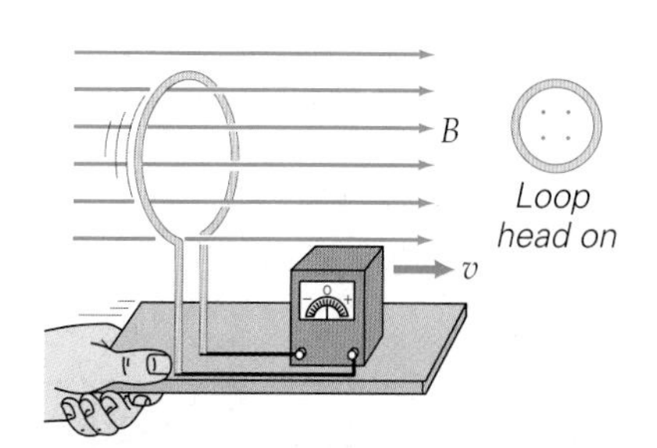

▲ **FIGURE 20.2 Relative motion and no induction** When a loop is moved parallel to a uniform magnetic field, there is no change in the number of field lines passing through the loop, and there is no induced current.

*The term *mutual induction* is used to describe the situation in which emfs and currents are induced between two (or more) loops.

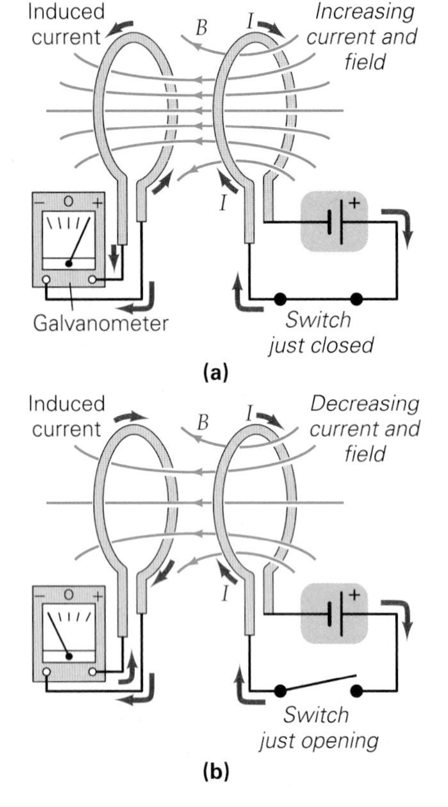

▲ **FIGURE 20.3 Mutual induction** **(a)** When the switch is closing in the right-loop circuit, the buildup of current produces a changing magnetic field in the other loop, inducing a current in it. **(b)** When the switch is opened, the magnetic field collapses, and the magnetic field in the left loop decreases. The induced current in this loop is then in the opposite direction. The induced currents occur only when the magnetic field passing through a loop changes and vanish when the field reaches a constant value.

Detailed experiments on electromagnetic induction were done independently by Michael Faraday in England and Joseph Henry in the United States around 1830. Faraday found that the important factor in electromagnetic induction was the time rate of change of the number of magnetic field lines passing through the loop or circuit area. That is, he discovered that

> an induced emf is produced in a loop or complete circuit whenever the number of magnetic field lines passing through the plane of the loop or circuit changes.

Magnetic Flux

Because the induced emf in a loop depends on the rate of change of the number of magnetic field lines passing through it, determining induced emf demands that the number of field lines through the loop be quantified. Consider a loop of wire in a uniform magnetic field (▼Fig. 20.4a). The number of field lines through the loop depends on the loop's area, its orientation relative to the field, and the strength of that field. To describe the loop's orientation, the concept of an *area vector* ($\vec{\mathbf{A}}$) is employed. Its direction is normal to the loop's plane, and its magnitude is equal to the loop area. θ, the angle between the magnetic field ($\vec{\mathbf{B}}$) and the area vector ($\vec{\mathbf{A}}$), is a measure of their relative orientation. For example, in Fig. 20.4a, $\theta = 0°$, meaning that the vectors are in the same direction, or, alternatively, the area plane is perpendicular to the field.

For the case of a magnetic field that does not vary over the area, the number of magnetic field lines passing through a particular area (the area inside a loop in our case) is proportional to the **magnetic flux (Φ)**, which is defined as

$$\Phi = BA\cos\theta \quad \textit{magnetic flux (in a constant magnetic field)} \tag{20.1}$$

SI unit of magnetic flux: tesla-meter squared ($\text{T}\cdot\text{m}^2$), or weber (Wb)*

The SI unit of the magnetic field is the tesla, and magnetic flux has SI units of $\text{T}\cdot\text{m}^2$. This combination is sometimes expressed as the weber, defined as $1\ \text{Wb} = 1\ \text{T}\cdot\text{m}^2$. The orientation of the loop with respect to the magnetic field affects the number of field lines passing through it, and this factor is accounted for by the cosine term in Eq. 20.1. Let us consider several possible orientations:

- If $\vec{\mathbf{B}}$ and $\vec{\mathbf{A}}$ are parallel ($\theta = 0°$), then the magnetic flux is positive and has a maximum value of $\Phi_{\text{max}} = BA\cos 0° = +BA$. The maximum possible number of magnetic field lines pass through the loop in this orientation (Fig. 20.4b).
- If $\vec{\mathbf{B}}$ and $\vec{\mathbf{A}}$ are oppositely directed ($\theta = 180°$), then the magnitude of the magnetic flux is a maximum again, but of opposite sign: $\Phi_{180°} = BA\cos 180° = -BA = -\Phi_{\text{max}}$ (Fig. 20.4c).

▼ **FIGURE 20.4 Magnetic flux** **(a)** Magnetic flux (Φ) is a measure of the number of field lines passing through an area (A). The area can be represented by a vector $\vec{\mathbf{A}}$ perpendicular to the plane of the area. **(b)** When the plane of a loop is perpendicular to the field and $\theta = 0°$, then $\Phi = \Phi_{\text{max}} = +BA$. **(c)** When $\theta = 180°$, the magnetic flux has the same magnitude, but is opposite in direction: $\Phi = -\Phi_{\text{max}} = -BA$. **(d)** When $\theta = 90°$, then $\Phi = 0$. **(e)** As the loop's plane is changed from being perpendicular to the field to one more parallel to the field, less area is available to the field lines, and therefore the flux decreases. In general, $\Phi = BA\cos\theta$.

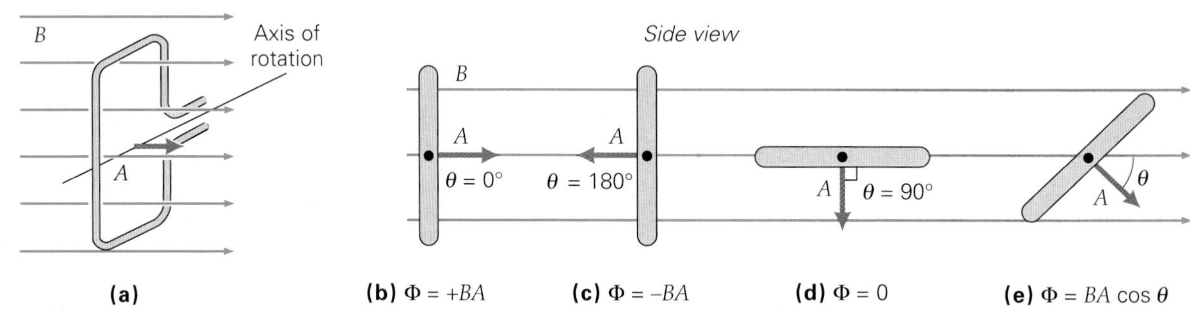

*Wilhelm Eduard Weber (1804–1891), a German physicist, was noted for his work in magnetism and electricity, particularly terrestrial magnetism. The unit name *weber* was introduced as the SI unit of magnetic flux in 1935.

- If $\vec{\mathbf{B}}$ and $\vec{\mathbf{A}}$ are perpendicular, then there are no field lines passing through the plane of the loop, and the flux is zero: $\Phi_{90°} = BA\cos 90° = 0$ (Fig. 20.4d).
- For situations at intermediate angles, the flux is less than the maximum value, but nonzero (Fig. 20.4e). $A\cos\theta$ can be interpreted as the effective area of the loop perpendicular to the field lines (▸Fig. 20.5a). Alternatively, $B\cos\theta$ can be viewed as the perpendicular component of the field through the full area of the loop, A, as shown in Fig. 20.5b. Thus, Eq. 20.1 can be thought of as either $\Phi = (B\cos\theta)A$ or $\Phi = B(A\cos\theta)$, depending on the interpretation. In either case, the answer is the same.

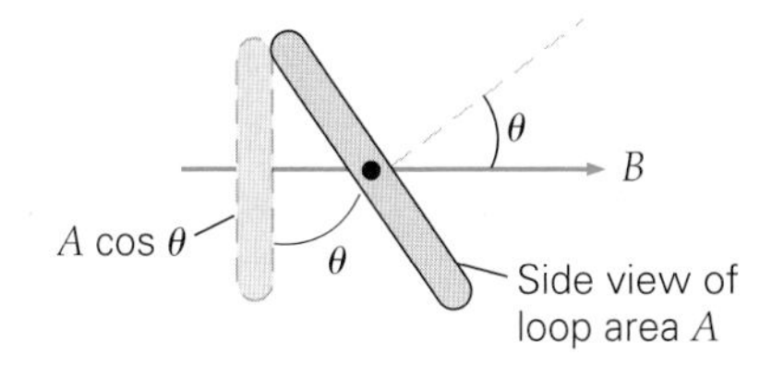

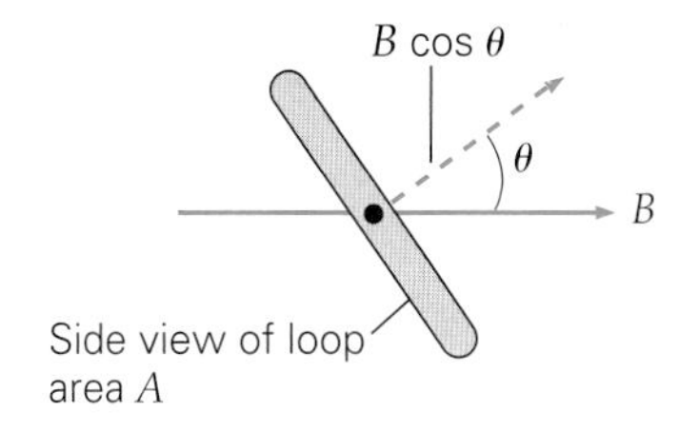

▲ **FIGURE 20.5 Magnetic flux through a loop: An alternative interpretation** Instead of defining the flux (Φ) **(a)** in terms of the magnetic field magnitude (B) passing through a reduced area ($A\cos\theta$), we can define it **(b)** in terms of the perpendicular component of the magnetic field ($B\cos\theta$) passing through A. Either way, Φ is a measure of the number of field lines passing through A and is given by $\Phi = BA\cos\theta$ (Eq. 20.1).

Faraday's Law of Induction and Lenz's Law

From quantitative experiments, Faraday determined that the emf ($\mathscr{E}$) induced in a coil (a coil, by definition, consists of a series connection of N individual loops) depends on the time rate of change of the number of magnetic field lines through all the loops, or the *time rate of change of the magnetic flux through all the loops (total flux)*. This dependence, known as **Faraday's law of induction**, is expressed mathematically as

$$\mathscr{E} = -N\frac{\Delta\Phi}{\Delta t} = -\frac{\Delta(N\Phi)}{\Delta t} \quad \textit{Faraday's law for induced emf} \qquad (20.2)$$

where $\Delta\Phi$ is the change in flux through one loop. In a coil consisting of N loops, the total change in flux is $N\Delta\Phi$. Note that the induced emf in Eq. 20.2 is an average value over the time interval Δt (why?).

The minus sign is included in Eq. 20.2 to give an indication of the *direction* of the induced emf, which we haven't discussed as yet. The Russian physicist, Heinrich Lenz (1804–1865), discovered the law that governs the direction of the induced emf. **Lenz's law** is stated as follows:

> An induced emf in a wire loop or coil has a direction such that the current it creates produces its own magnetic field that opposes the *change* in magnetic flux through that loop or coil.

This law means that the magnetic field *due to the induced current* is in a direction that tries to keep the flux through the loop from changing. For example, if the flux increases in the $+x$-direction, the magnetic field due to the induced current will be in the $-x$-direction (▾Fig. 20.6a). This effect tends to cancel the increase in the flux, or *oppose the change*. Essentially, the magnetic field due to the induced current tries to maintain the existing magnetic flux. This effect is sometimes called "electromagnetic inertia," by analogy to the tendency of objects to resist changes in their velocity. In the long run, the induced current cannot prevent the magnetic flux from changing. However, during the time that the flux is changing, the induced magnetic field will oppose that change.

Illustration 29.2 Loop in a Changing Magnetic Field

The direction of the induced current is given by the **induced-current right-hand rule**:

> With the thumb of the right hand pointing in the direction of the induced field, the fingers curl in the direction of the induced current.

PHYSLET®

Exploration 29.1 Lenz's Law

(See Fig. 20.6b and Integrated Example 20.1.) You might recognize this rule as a version of the right-hand rules used to find the direction of a magnetic field produced

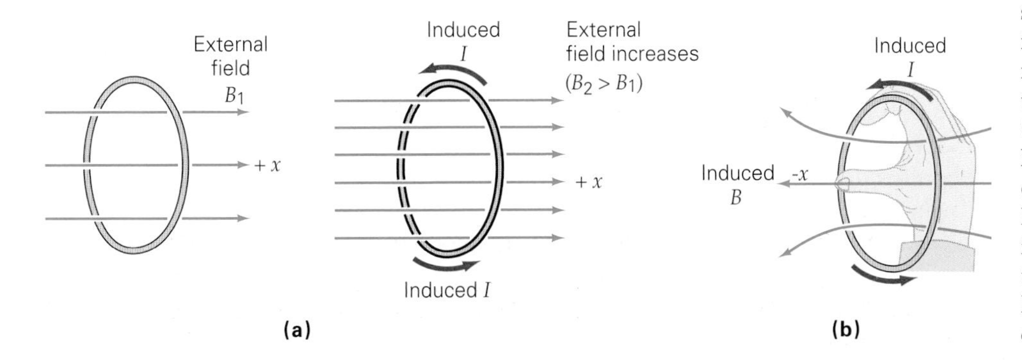

◀ **FIGURE 20.6 Finding the direction of the induced current** **(a)** An external magnetic field is shown increasing to the right. The induced current creates its own magnetic field to try to counteract the flux change that is occurring. **(b)** The (induced) current right-hand (source) rule determines the direction of the induced current. Here the direction of the induced field must be to the left. With the thumb of the right hand pointing left, the fingers give the induced current direction.

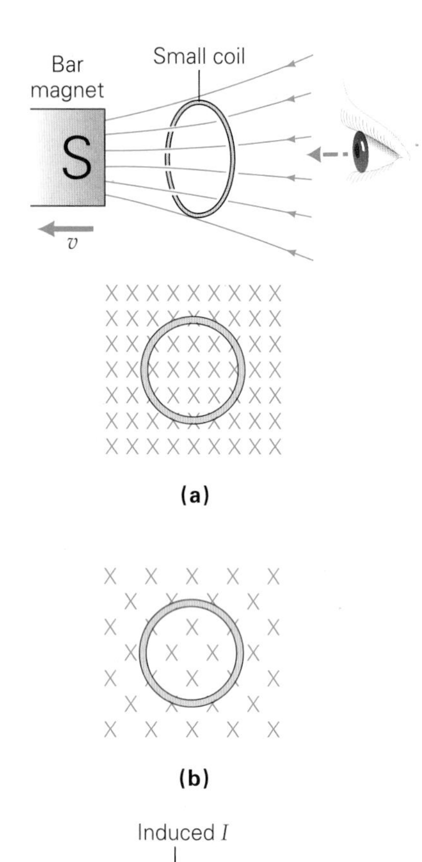

▲ **FIGURE 20.7 Using a bar magnet to induce currents (a)** The south end of a bar magnet is pulled away from a wire loop. **(b)** The view from the right of the loop shows the magnetic field pointing away from the observer, or into the page, and decreasing. **(c)** To counteract this loss of flux into the page, current is induced in the clockwise direction, so as to provide its own field into the page. See Integrated Example 20.1.

Exploration 29.3 Loop Near a Wire

by a current (Chapter 19). Here it is used in reverse. Typically, you know the induced field direction (for example, $-x$ in Fig. 20.6b) and want the direction of the current that produces it. An application of Lenz's law is illustrated in Integrated Example 20.1.

Integrated Example 20.1 ■ Lenz's Law and Induced Currents

(a) The south end of a bar magnet is pulled far away from a small wire coil. (See ◂Fig. 20.7a.) Looking from behind the coil toward the south end of the magnet (Fig. 20.7b), what is the direction of the induced current: (1) counterclockwise, (2) clockwise, or (3) there is no induced current? (b) Suppose that the magnetic field over the area of the coil is initially constant at 40 mT, the coil's radius is 2.0 mm, and there are 100 loops in the coil. Determine the magnitude of the average induced emf in the coil if the bar magnet is removed in 0.75 s.

(a) Conceptual Reasoning. There is initially magnetic flux *into* the plane of the coil (Fig. 20.7b), and later, when the magnet is far away from the coil, there is no flux; and the flux has changed. Therefore, there must be an induced emf, so answer (3) cannot be correct. As the bar magnet is pulled away, the field weakens, but maintains the same direction. The induced emf will produce an (induced) current that, in turn, will produce a magnetic field into the page so as to try to prevent this decrease in flux. Therefore, the induced emf and current are in the clockwise direction, as found using the induced-current right-hand rule (Fig. 20.7c) and the correct answer is (2), clockwise.

(b) Quantitative Reasoning and Solution. This example is a straightforward application of Eq. 20.2. The initial flux is the maximum possible. The following data are given and converted to SI units:

Given: $B_i = 40\text{ mT} = 0.040\text{ T}$
$r = 2.00\text{ mm} = 2.00 \times 10^{-3}\text{ m}$
$N = 100\text{ loops}$
$\Delta t = 0.75\text{ s}$

Find: Average induced emf $\mathscr{E}$ (magnitude)

To find the initial magnetic flux through one loop of the coil, use Eq. 20.1 with an angle of $\theta = 0°$. (Why?) The area is $A = \pi r^2 = \pi(2.00 \times 10^{-3}\text{ m})^2 = 1.26 \times 10^{-5}\text{ m}^2$. Therefore, the initial flux, Φ_i, through one loop is positive (why?) and given by

$$\Phi_i = B_i A \cos\theta = (0.040\text{ T})(1.26 \times 10^{-5}\text{ m}^2)\cos 0° = +5.03 \times 10^{-7}\text{ T}\cdot\text{m}^2$$

Because the final flux is zero, $\Delta\Phi = \Phi_f - \Phi_i = 0 - \Phi_i = -\Phi_i$. Therefore, the absolute value of the average induced emf is

$$|\mathscr{E}| = N\frac{|\Delta\Phi|}{\Delta t} = (100\text{ loops})\frac{(5.03 \times 10^{-7}\text{ T}\cdot\text{m}^2\text{ loop})}{(0.75\text{ s})} = 6.70 \times 10^{-5}\text{ V}$$

Follow-Up Exercise. In this Example, (a) in which direction is the induced current if instead a north magnetic pole approaches the coil quickly? Explain. (b) In this example, what would be the average induced current if the coil had a total resistance of 0.20 Ω? *(Answers to all Follow-Up Exercises are at the back of the text.)*

Lenz's law incorporates the principle of energy conservation. Consider a situation in which a wire loop has an increasing magnetic flux through its area. Contrary to Lenz's law, suppose instead that the magnetic field from the induced current *added* to the flux instead of keeping it at its original value. This increased flux would then lead to an even greater induced current. In turn, this greater induced current would produce a still greater magnetic flux, which in turn would give a greater induced current, and so on. Such a something-for-nothing energy situation would violate conservation of energy.

To understand the direction of the induced emf in a loop in terms of forces, consider the case of the moving magnet (for example, Fig. 20.1b). Recall that a current-carrying loop creates its own magnetic field similar to that of a bar magnet. (See Figs. 19.3 and 19.25.) The induced current sets up a magnetic field in the loop, and that loop acts like a bar magnet with a polarity that will oppose the motion of the real bar magnet (▸Fig. 20.8). You should be able to show that if the bar magnet is pulled away from the loop, the loop exerts a magnetic *attraction* to try to keep the magnet from leaving—electromagnetic inertia in action.

Substituting the expression for the magnetic flux (Φ) given by Eq. 20.1 into Eq. 20.2, we have

$$\mathscr{E} = -N\frac{\Delta\Phi}{\Delta t} = -\frac{N\Delta(BA\cos\theta)}{\Delta t} \qquad (20.3)$$

Thus, an induced emf results if

1. the strength of the magnetic field changes,
2. the loop area changes, and/or
3. the orientation between the loop area and the field direction changes.

In situation (1), a flux change is created by a time-varying field, such as that from a time-varying current in a nearby circuit or created by moving a magnet near a coil, as in Fig. 20.1 (or by moving the coil near the magnet).

In situation (2), a flux change results because of a varying loop area. This situation might occur if a loop had an adjustable circumference (such as the loop around an inflatable balloon, as in Exercise 23 at the end of the chapter).

Finally, in situation (3), a change in flux can result from a *change in orientation of the loop*. This situation can occur when a coil is rotated in a magnetic field. The change in the number of field lines through a single loop is evident in the sequential views in Fig. 20.4. Rotating a coil in a field is a common way of inducing an emf and will be considered separately in Section 20.2. The emfs that result from changing the field strength and loop area are analyzed in the next two Examples. Also, see Insight 20.1 on Electromagnetic Induction at Work: Flashlights and Antiterrorism on page 664 for ways in which electromagnetic induction helps make our everyday lives safer and easier.

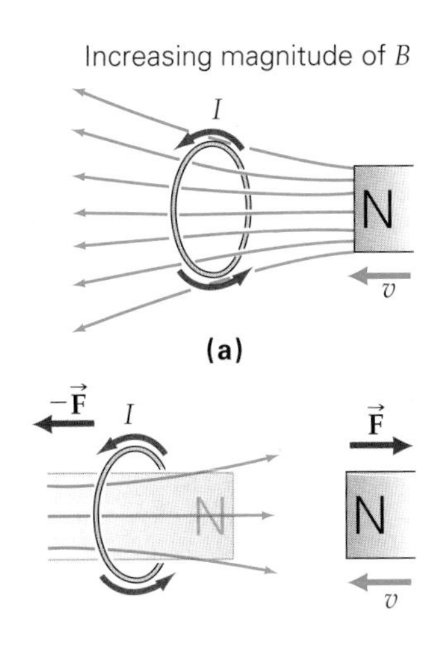

▲ **FIGURE 20.8 Lenz's law in terms of forces** **(a)** If the north end of a bar magnet is moved rapidly toward a wire loop, current is induced in the direction shown. **(b)** While the induced current exists, the loop then acts like a bar magnet with its "north end" close to the north end of the real bar magnet. Thus there is a magnetic repulsion. This is an alternative way of viewing Lenz's law: Induce a current so as to try to keep the flux from changing—in this case, to try to keep the bar magnet away and maintain the initial value of flux, zero.

Conceptual Example 20.2 ■ Fields in the Fields: Electromagnetic Induction

In rural areas where electric power lines carry electricity to big cities, it is possible to generate small electric currents by means of induction in a conducting loop. The overhead power lines carry alternating currents that periodically reverse direction 60 times per second. How would you orient the plane of the loop to maximize the induced current if the power lines run north to south: (a) parallel to the Earth's surface, (b) perpendicular to the Earth's surface in the north–south direction, or (c) perpendicular to the Earth's surface in the east–west direction? (See ▸Fig. 20.9a.)

Reasoning and Answer. Magnetic field lines from long wires are circular. (See Fig. 19.23.) By the source right-hand rule, the magnetic field direction at ground level is parallel to the Earth's surface and alternates in direction. The orientation choices are shown in Fig. 20.9b. Neither answer (a) nor (c) can be correct, because in these orientations there would *never* be any magnetic flux passing through the loop. In this situation, the flux would be constant and there would be no induced emf. Hence, the answer is (b). If the loop is oriented perpendicular to the Earth's surface with its plane in the north–south direction, the flux through it would vary from zero to its maximum value and back sixty times per second, and this would maximize the induced emf and current in the loop.

Follow-Up Exercise. Suggest possible ways of increasing the induced current in this Example by changing only properties of the loop and not of the overhead wires.

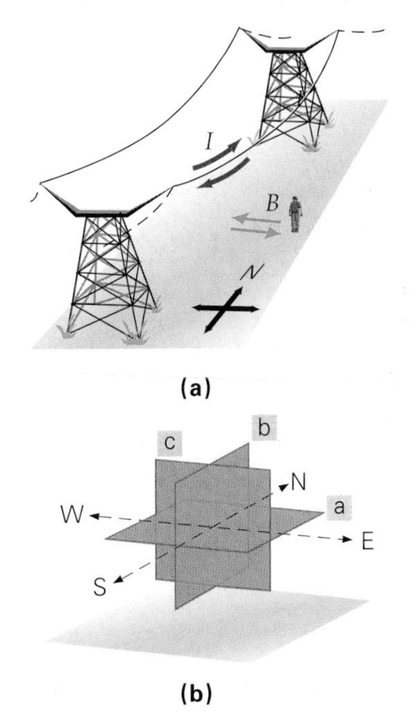

▲ **FIGURE 20.9 Induced emfs below power lines** **(a)** If current-carrying wires run in the north– south direction, then directly below the alternating current produces a magnetic field that oscillates between pointing east and west. **(b)** These are the three choices for loop orientation in Conceptual Example 20.2.

Example 20.3 ■ Induced Currents: A Potential Hazard to Equipment?

Electrical instruments can be damaged or destroyed if they are in a rapidly changing magnetic field. This can occur if an instrument is located near an electromagnet operating under ac conditions; the electromagnet's external field could produce a changing flux within a nearby instrument. If the induced currents are large enough, they could damage the instrument. Consider a computer speaker that is near such an electromagnet (▾Fig. 20.10, p. 662). Suppose an electromagnet exposes the speaker to a maximum magnetic field of 1.00 mT that reverses direction every 1/120 s.

Assume that the speaker's coil consists of 100 circular loops (each with a radius of 3.00 cm) and has a total resistance of 1.00 Ω. According to the manufacturer of the speaker, the current in the coil should not exceed 25.0 mA. (a) Calculate the magnitude of the average induced emf in the coil during the 1/120 s interval. (b) Is the induced current likely to damage the speaker coil?

Thinking It Through. (a) The flux goes from a (maximum) positive to a (maximum) negative value in 1/120 s. The magnetic flux change can be determined from Eq. 20.1 with $\theta = 0°$ and $\theta = 180°$. The average induced emf can then be calculated from Eq. 20.2. (b) Once we know the emf, the induced current can be calculated from $I = \mathscr{E}/R$.

(continues on next page)

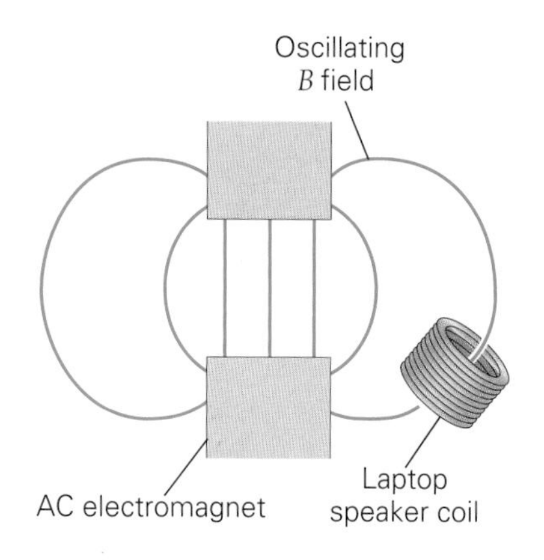

▲ **FIGURE 20.10 Instrument hazard?** The coil of a computer speaker system is close to an alternating-current electromagnet. The changing flux in the coil produces an induced emf and, thus, an induced current that depends on the resistance of the coil. See Example 20.3.

Solution. Listing the data and converting to SI units,

Given:
$B_i = +1.00\text{ mT} = +1.00 \times 10^{-3}\text{ T}$ (+ pointing one way)
$B_f = -1.00\text{ mT} = -1.00 \times 10^{-3}\text{ T}$ (pointing the opposite way)
$\Delta t = 1/120\text{ s} = 8.33 \times 10^{-3}\text{ s}$
$N = 100$ loops
$R = 1.00\ \Omega$
$r = 3.00\text{ cm} = 3.00 \times 10^{-2}\text{ m}$
$I_{max} = 25.0\text{ mA} = 2.50 \times 10^{-2}\text{ A}$

Find: (a) $\mathscr{E}$ (magnitude of average induced emf)
(b) I (magnitude of average induced current)

(a) The circular loop area is $A = \pi r^2 = \pi(3.00 \times 10^{-2}\text{ m})^2 = 2.83 \times 10^{-3}\text{ m}^2$. Thus the initial flux through *one* loop is (see Eq. 20.1):

$$\Phi_i = B_i A \cos\theta = (1.00 \times 10^{-3}\text{ T})(2.83 \times 10^{-3}\text{ m}^2/\text{loop})(\cos 0°) = 2.83 \times 10^{-6}\text{ T}\cdot\text{m}^2/\text{loop}$$

Because the final flux is the negative of this, the change in flux through one loop is

$$\Delta\Phi = \Phi_f - \Phi_i = -\Phi_i - \Phi_i = -2\Phi_i = -5.66 \times 10^{-6}\text{ T}\cdot\text{m}^2/\text{loop}$$

Therefore, the average induced emf is (using Eq. 20.2)

$$\mathscr{E} = N\frac{|\Delta\Phi|}{\Delta t} = (100\text{ loops})\left(\frac{5.66 \times 10^{-6}\text{ T}\cdot\text{m}^2/\text{loop}}{8.33 \times 10^{-3}\text{ s}}\right) = 6.79 \times 10^{-2}\text{ V}$$

(b) This voltage is small by everyday standards, but keep in mind that the speaker coil's resistance is also small. To determine the induced current in the coil, use the relationship between voltage, resistance, and current:

$$I = \frac{\mathscr{E}}{R} = \frac{6.79 \times 10^{-2}\text{ V}}{1.00\ \Omega} = 6.79 \times 10^{-2}\text{ A} = 67.9\text{ mA}$$

This value exceeds the allowed speaker current of 25.0 mA and therefore the speaker coil is possibly subject to damage.

Follow-Up Exercise. In this example, if the speaker coil were moved farther from the electromagnet, it could reach a point where the induced average current would be below the "dangerous" level of 25 mA. Determine the magnetic field strength B_{max} at this point.

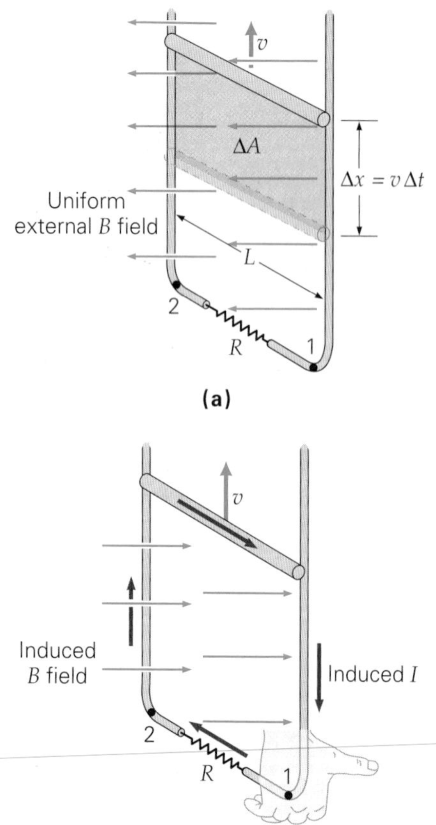

▲ **FIGURE 20.11 Motional emf** **(a)** As the metal rod is pulled on the metal frame, the area of the rectangular loop varies with time. A current is induced in the loop as a result of the changing flux. **(b)** To counteract the increase of flux to the left, an induced current creates its own magnetic field to the right. See Integrated Example 20.4.

As a special case, emfs and currents can be induced in conductors as they are moved through a magnetic field. In this situation, the induced emf is called a *motional emf*. To see how this works, consider the situation in ◂Fig. 20.11a. As the bar moves upward, the circuit area increases by $\Delta A = L\Delta x$ (Fig. 20.11a.) At constant speed, the distance traveled by the bar in a time Δt is $\Delta x = v\Delta t$. Therefore $\Delta A = Lv\Delta t$. The angle between the magnetic field and the normal to the area (θ) is always 0°. However, the area is changing, so the flux varies. However, we know that $\Phi = BA\cos 0° = BA$; hence we can write $\Delta\Phi = B\Delta A$, or $\Delta\Phi = BLv\Delta t$. Therefore, from Faraday's law, the magnitude of this "motional" (induced) emf, $\mathscr{E}$, is $|\mathscr{E}| = |\Delta\Phi|/\Delta t = BLv\Delta t/\Delta t = BLv$. This is the fundamental idea behind electric energy generation: Move a conductor in a magnetic field, and convert the work done on it into electrical energy. To see some of the details, consider the following Integrated Example.

Integrated Example 20.4 ■ The Essence of Electric-Energy Generation: Mechanical Work into Electrical Current

Consider the situation in Fig. 20.11a. An external force does work as the movable bar moves, and this work is converted to electrical energy. Because the "circuit" (wires, resistor, and bar) is in a magnetic field, the flux through it changes with time, inducing a current. (a) What is the direction of the induced current in the resistor, (1) from 1 to 2 or (2) from 2 to 1? (b) If the bar is 20 cm long and is pulled at a steady speed of 10 cm/s, what is the induced current if the resistor has a value of 5.0 Ω and the circuit is in a uniform magnetic field of 0.25 T?

(a) Conceptual Reasoning. In Fig. 20.11a, the magnetic flux points left and increases. According to Lenz's law, the field due to the induced current must then be to the right. Using the induced-current right-hand rule, we find that the direction of the induced current is from 1 to 2 (Fig. 20.11b), and the correct answer is (1).

(b) Quantitative Reasoning and Solution. The flux change is due to an area change as the bar is pulled upward. The analysis for motional emfs has been done in the preceding text. Lastly, once the motional emf has been found, the induced current can be determined using Ohm's law.

Listing the data and converting to SI units, we have:

Given: $B = 0.25\ \text{T}$
$L = 20\ \text{cm} = 0.20\ \text{m}$
$v = 10\ \text{cm/s} = 0.10\ \text{m/s}$
$R = 5.0\ \Omega$

Find: Induced current in the resistor

In the preceding text, it was shown that the magnitude of the induced emf $\mathscr{E}$ is given by BLv, so numerically we have:

$$|\mathscr{E}| = BLv = (0.25\ \text{T})(0.20\ \text{m})(0.10\ \text{m/s}) = 5.0 \times 10^{-3}\ \text{V}$$

Hence the induced current is

$$I = \frac{\mathscr{E}}{R} = \frac{5.0 \times 10^{-3}\ \text{V}}{5.0\ \Omega} = 1.0 \times 10^{-3}\ \text{A}$$

Clearly this arrangement isn't a practical way to generate large amounts of electrical energy. Here the power dissipated in the resistor is only 5.0×10^{-6} W. (You should verify this.)

Follow-Up Exercise. In this Example, if the field were increased by three times and the bar's width changed to 45 cm, what would the bar's speed be to induce a current of 0.1 A?

20.2 Electric Generators and Back emf

OBJECTIVES: To (a) understand the operation of electrical generators and calculate the emf produced by an ac generator, and (b) explain the origin of back emf and its effect on the behavior of motors.

One way to induce an emf in a loop is through a change in the loop's orientation in its magnetic field (Fig. 20.4). This is the operational principle behind electric generators.

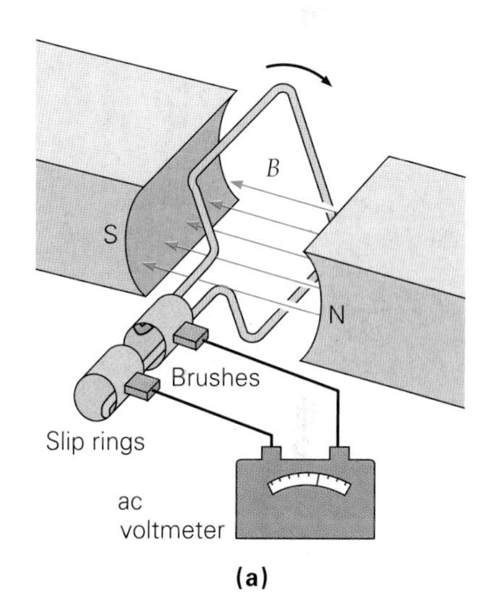

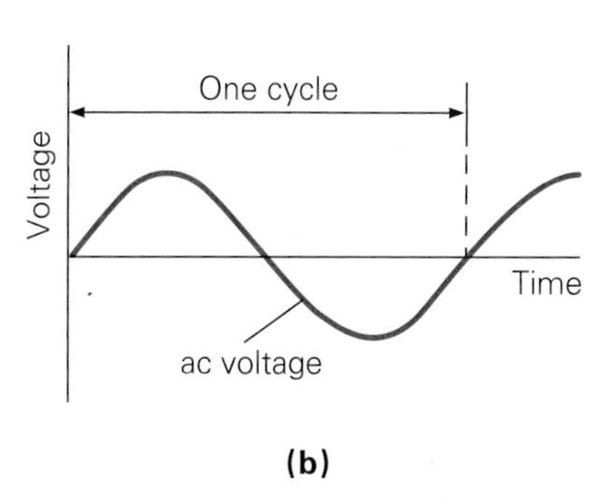

▲ **FIGURE 20.12** A simple ac generator **(a)** The rotation of a wire loop in a magnetic field produces **(b)** a voltage output whose polarity reverses with each half-cycle. This alternating voltage is picked up by a brush/slip ring arrangement as shown.

Electric Generators

An *electric generator* is a device that converts mechanical energy into electrical energy. Basically, the function of a generator is the reverse of that of a motor.

Recall that a battery supplies direct current (dc). That is, the voltage polarity (and therefore the current direction) do not change. However, most generators produce *alternating current* (ac), named because the polarity of the voltage (and therefore the current direction) change periodically. Thus, the electric energy used in homes and industry is delivered in the form of alternating voltage and current. (See Chapter 21 for analysis of ac circuits and Chapter 18 for household wiring diagrams.)

An **ac generator** is sometimes called an *alternator*. The elements of a simple ac generator are shown in ▸Fig. 20.12. A wire loop called an *armature* is mechanically rotated in a magnetic field by some external means, such as water flow or steam hitting turbine blades. The rotation of the blades in turn causes a rotation of the loop. This results in a change in the loop's magnetic flux and an induced emf in the loop. The ends of the loop are connected to an external circuit by means of slip rings and brushes. In this case, the induced currents will be delivered to that circuit. In practice, generators have many loops, or windings, on their armatures.

When the loop is rotated at a constant angular speed (ω), the angle (θ) between the magnetic-field vector and the area vector of the loop changes with time: $\theta = \omega t$ (assuming that $\theta = 0°$ at $t = 0$). As a result, the number of field lines through the loop changes with time, causing an induced emf. From Eq. 20.1, the flux (for one loop) varies as

$$\Phi = BA\cos\theta = BA\cos\omega t$$

From this it can be seen that the induced emf will also vary with time. For a rotating coil of N loops, Faraday's law yields

$$\mathscr{E} = -N\frac{\Delta\Phi}{\Delta t} = -NBA\left(\frac{\Delta(\cos\omega t)}{\Delta t}\right)$$

INSIGHT 20.1 ELECTROMAGNETIC INDUCTION AT WORK: FLASHLIGHTS AND ANTITERRORISM

We use electromagnetic induction in many ways in our daily lives, in most cases without realizing it. One recent invention is a flashlight that works without a battery (Fig. 1a). As the flashlight is shaken, a strong permanent magnet in it oscillates through induction coils, inducing an oscillating emf and current. To charge a capacitor, the ac current must be *rectified* into dc current and not change direction. The schematic of this flashlight is shown in Fig. 1b. Here a solid-state *rectifier circuit* (triangular symbol) acts as a "one-way current valve." In this diagram, only clockwise dc current goes to the capacitor and charges it. After about a minute, the capacitor is fully charged. When the switch S is thrown, the capacitor discharges through an efficient *light-emitting diode* (LED). The resulting beam of light lasts for several minutes before the flashlight needs to be reshaken. This device could, at the least, play an important backup role to the more traditional flashlights that rely on batteries.

(a)

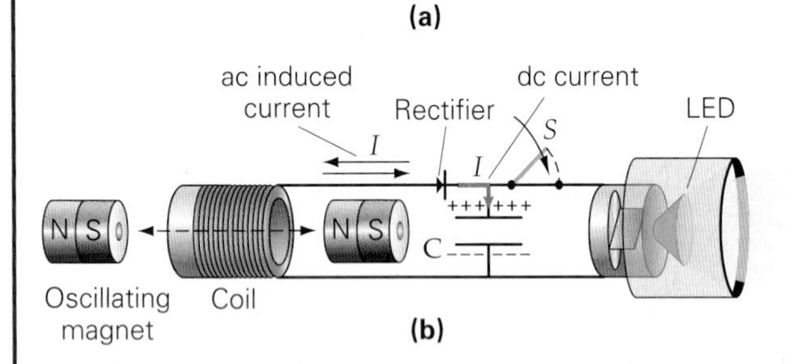

(b)

FIGURE 1 A batteryless flashlight (a) A photo of a relatively new type of flashlight that produces light using electric energy generated by shaking (induction). **(b)** A schematic diagram of the flashlight shown in part (a). As the flashlight is shaken, its internal permanent magnet passes through a coil, inducing a current. This current alternates in direction (why?) and thus needs to be turned into dc ("rectified") before it can charge a capacitor. Once the capacitor is fully charged, it can be used to create a current through a light-emitting diode (LED), which in turn gives off light, typically for several minutes.

In air travel safety, induction is used to prevent dangerous metallic objects (such as knives and guns) from being carried onto airplanes. As a passenger walks through the arch of an airport metal detector (see Fig. 2), a series of large "spiked" currents is periodically delivered to a coil (solenoid) in one of the nonmagnetic sides. In the most common system, called PI (for *pulsed induction*), these current spikes occur hundreds of times per second. As the current rises and falls, a changing magnetic field is created in the passenger. If the passenger is carrying nothing metallic, there will be no significant induced current and no induced magnetic field. However, if the passenger has a metal object, a current will be induced in that object, which in turn will produce its own (induced) magnetic field that can be sensed by the emitting coil—that is, a "magnetic echo." Sophisticated electronics measure the echo-induced emf and trigger a warning light to suggest that further inspection of that passenger is warranted.

FIGURE 2 Screening at the airport As passengers walk through the arch, they are subjected to a series of magnetic field pulses. If they have a metal object on their person, the currents induced in that object create their own magnetic field "echo" that, when detected, gives the safety inspectors reason to check the passenger more closely.

Here, we have removed B and A from the time rate of change, because they are constant. By using methods beyond the scope of this book, it can be shown that the induced emf expression can be rewritten as

$$\mathscr{E} = (NBA\omega) \sin \omega t$$

Notice that the product of terms, $NBA\omega$, represents the magnitude of the maximum emf, which occurs whenever $\sin \omega t = \pm 1$. If $NBA\omega$ is called $\mathscr{E}_o$, the maximum value of the emf, then the previous equation can be rewritten compactly as

$$\mathscr{E} = \mathscr{E}_o \sin \omega t \tag{20.4}$$

Because the sine function varies between ± 1, the polarity of the emf changes with time (▸Fig. 20.13). Note that the emf has its maximum value $\mathscr{E}_o$ when $\theta = 90°$ or $\theta = 270°$. That is, at the instants when the plane of the loop is parallel to the field, and the magnetic flux is zero, the emf will be at its largest (magnitude). The

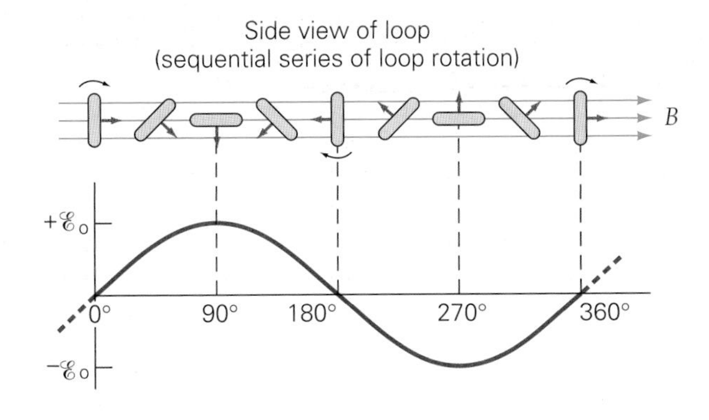

◀ **FIGURE 20.13 An ac generator output** A graph of the sinusoidal output of a generator, with a side view of the corresponding loop orientations during a cycle, showing the flux variation with time. Note that the emf is a maximum when the flux changes most rapidly, as it passes through zero and changes in sign.

change in flux is greatest at these angles, because although the flux is momentarily zero, it is changing rapidly due to a *sign* change. Near the angles that produce the flux's largest value ($\theta = 0°$ and $\theta = 180°$), the flux is approximately constant and thus the induced emf is zero at those angles.

Illustration 29.3 Electric Generator

Because the induced current is produced by this alternating induced emf, the current also changes direction periodically. In everyday applications, it is common to refer to the frequency (f) of the armature [in hertz (Hz) or rotations per second], rather than the angular frequency (ω). Because they are related by $\omega = 2\pi f$, Eq. 20.4 can be rewritten as

$$\mathscr{E} = \mathscr{E}_o \sin(2\pi f t) \quad \textit{alternator emf} \tag{20.5}$$

The ac frequency in the United States and most of the western hemisphere is 60 Hz. A frequency of 50 Hz is common in Europe and other areas.

Keep in mind that Eqs. 20.4 and 20.5 give the instantaneous value of the emf and that $\mathscr{E}$ varies between $+\mathscr{E}_o$ and $-\mathscr{E}_o$ over half of an armature rotational period (1/120 of a second in the United States). For practical ac electrical circuits, time-averaged values for ac voltage and current are more important. This concept will be developed in Chapter 21. To see how various factors influence the generator's output, examine the next Example closely. Also, see Insight 20.2 on Electromagnetic Induction at Play on page 666 for ways in which electromagnetic induction makes for an interesting hobby and helps generate the electric energy needed to power hybrid automobiles for more fuel-efficient transportation.

Example 20.5 ■ An ac Generator: Renewable Electric Energy

A farmer decides to use a waterfall to create a small hydroelectric power plant for his farm. He builds a coil consisting of 1500 circular loops of wire with a radius of 20 cm, which rotates on the generator's armature at 60 Hz in a magnetic field. To generate an rms voltage of 120 V, he needs to generate a maximum emf of 170 V (we will learn more about ac voltages in Chapter 21). What is the magnitude of the generator's magnetic field necessary for this to happen?

Thinking It Through. We can determine the magnetic field from the expression for $\mathscr{E}_o$.

Solution.

Given: $\mathscr{E}_o = 170$ V
$N = 1500$ loops
$r = 20$ cm $= 0.20$ m
$f = 60$ Hz

Find: magnitude of the magnetic field (B)

The generator's maximum (or peak) emf is given by $\mathscr{E}_o = NBA\omega$. Because $\omega = 2\pi f$ and, for a circle, $A = \pi r^2$, this can be rewritten as

$$\mathscr{E}_o = NB(\pi r^2)(2\pi f) = 2\pi^2 NBr^2 f$$

Solving for B,

$$B = \frac{\mathscr{E}_o}{2\pi^2 Nr^2 f} = \frac{170 \text{ V}}{2\pi^2(1500)(0.20 \text{ m})^2(60 \text{ Hz})} = 2.4 \times 10^{-3} \text{ T}$$

Follow-Up Exercise. In this Example, suppose that the farmer wanted to generate an emf with an rms value of 240 V, which requires a maximum emf of 340 V. If he chose to do so by changing the size of the coils, what would their new radius have to be?

INSIGHT 20.2 ELECTROMAGNETIC INDUCTION AT PLAY: HOBBIES AND TRANSPORTATION

FIGURE 1 A two-coil metal detector Note both the transmitter (larger outer) and receiver (smaller inner) coils.

Electromagnetic induction plays an important part in our leisure and transportation activities. For example, some hobbyists use metal detectors to hunt for "buried treasure" of a metallic kind. A common design consists of two coils of wire at the end of a shaft used for sweeping just above the ground (see Fig. 1). At the hand-held end are electronics for displaying information about any detected items. The outer, or *transmitter*, coil contains a current oscillating at several thousand hertz, creating an ever-changing magnetic field in the ground below it. (Usually it can penetrate a foot or more below the surface, depending on soil type and condition.) If no metallic objects are within range of this oscillating field, then no significant currents will be induced. Therefore no induced magnetic field "echo" will be detected by the inner, or *receiver*, coil. However, if a metallic object is present, the current induced in it will create a magnetic echo (field) that the receiver will detect as an induced emf and current. By means of sophisticated computer software to evaluate the strength of the induced signal, the object's depth and chemical makeup can be estimated.

The increasing price of gasoline has many drivers turning to *gas-electric hybrid* automobiles, in which the gasoline engine is considerably smaller than conventional ones. In addition to helping power the car, at least part of the hybrid engine's job is to supply electric energy (through induction in a generator) to batteries and an electric motor, which in turn supply power to the wheels. In this way, more work can be extracted from a gallon of gasoline then in a conventional engine.

A cutaway of a typical hybrid car is shown in Fig. 2a. Hybrid cars currently come in two basic designs: parallel and series. In the *parallel hybrid* arrangement (Fig. 2b), the gasoline engine is connected to the wheels via a standard transmission. However, it also turns a generator that, through induction, creates and supplies electric energy to charge the batteries and/or to operate the electric motor. Sophisticated power electronics monitor the charge on the batteries and divert current to where it is needed. The electric motor is connected to the wheels through its own separate transmission, hence the name *parallel hybrid*—as the gasoline and electric motor work together, in parallel. *Fully hybrid* models are capable of moving the car with either engine alone (for maximum fuel economy—say, while cruising on a freeway) or both simultaneously (when more power is needed— say, while accelerating on to a freeway).

Alternatively the engines/motors can be connected in series—in the *series hybrid* automobile. Here the electric motor is what actually powers the wheels (Fig. 2c). The job of the gasoline engine is to supply electric energy (through induction in its generator) to the batteries and electric motor. If the batteries are fully charged and the motor is running well, the gasoline engine can idle down or power off. With frequent accelerations, when the electric motor is called on for a high-power output, the batteries may drain quickly. Under these conditions, the power electronics direct the gasoline engine to begin generating electric energy to recharge the batteries.

Regardless of the design, the object is the same: higher efficiency—that is, more miles per gallon. Hybrids, unlike purely electric cars, are never "plugged in"; they derive all their energy from burning gasoline. However, they are much more efficient, and thus considerably less polluting, than conventional cars. Some recent car models employ hybrid engines that are capable of more horsepower than their gasoline counterparts. For these reasons, the hybrid engine is increasingly likely to be the engine of choice for many drivers in the near future.

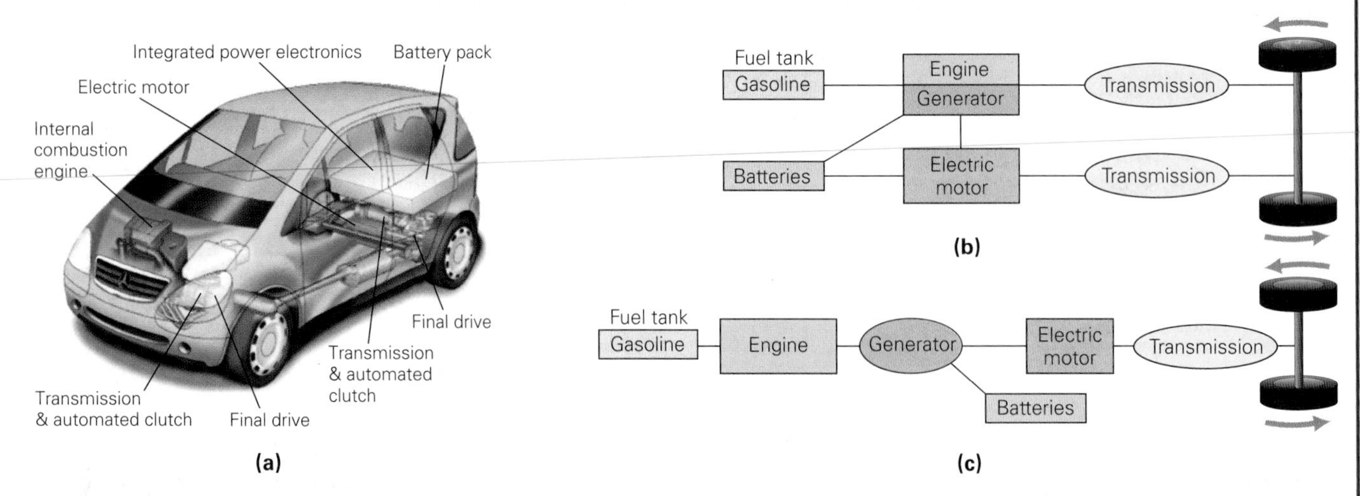

FIGURE 2 Hybrid automobiles **(a)** A cutaway of a typical modern hybrid vehicle. **(b)** A schematic of the main systems in a parallel hybrid. **(c)** A schematic of the main systems in a series hybrid.

(a)

(b)

◀ **FIGURE 20.14 Electrical generation** **(a)** Turbines such as those depicted here generate electric energy in much larger quantities than can the "small hydro" plant in the chapter-opening photo. **(b)** Gravitational potential energy of water, here trapped behind the Glen Canyon dam on the Colorado River in Arizona, is converted into electric energy.

In most large-scale ac generators (power plants), the armature is actually stationary, and magnets revolve about it. The revolving magnetic field produces a time-varying flux through the coils of the armature and thus an ac output. A turbine supplies the mechanical energy required to spin the magnets in the generator (▲Fig. 20.14a). Turbines are typically powered by steam generated from the heat of combustion of fossil fuels or by heat generated from nuclear fission (see Chapters 29 and 30), but they can also be rotated by falling water (*hydroelectricity*), as in Fig. 20.14b. Thus the basic difference between the various types of power plants is the source of the energy that turns the turbines.

Back emf

Although their main job is to convert electric energy into mechanical energy, motors also generate (induced) emfs at the same time. Like a generator, a motor has a rotating armature in a magnetic field. For motors, the induced emf is called a **back emf** (or *counter emf*), $\mathscr{E}_b$, because its direction is opposite that of the line voltage and tends to reduce the current in the armature coils.

If V is the line voltage, then the net voltage driving the motor is less than V (because the line voltage and the back emf are of opposite polarity). Therefore the net voltage is $V_{net} = V - \mathscr{E}_b$. If the motor's armature has a resistance of R, the currentthe motor draws while in operation is $I = V_{net}/R = (V - \mathscr{E}_b)/R$ or, solving for the back emf,

$$\mathscr{E}_b = V - IR \quad \textit{(back emf of a motor)} \tag{20.6}$$

where V is the line voltage.

The back emf of a motor depends on the rotational speed of the armature and increases from zero to some maximum value as the armature goes from rest to its normal operating speed. On startup, the back emf is zero (why?). Therefore the starting current is a maximum (Eq. 20.6 with $\mathscr{E}_b = 0$). Ordinarily, a motor turns something, such as a drill bit; that is, it has a mechanical load. Without a load, the armature speed will increase until the back emf almost equals the line voltage. The result is a small current in the coils, just enough to overcome friction and joule heat losses. Under normal load conditions, the back emf is less than the line voltage. The larger the load, the slower the motor rotates and the smaller the back emf. If a motor is overloaded and turns very slowly, the back emf may be reduced so much that the current becomes very large (note that V_{net} increases as $\mathscr{E}_b$ decreases) and may burn out the coils. The back emf plays a vital role in the regulation of a motor's operation by limiting the current in it.

Schematically, a back emf in a dc motor circuit can be represented as an "induced battery" with polarity opposite that of the driving voltage (▶Fig. 20.15). To see how the back emf affects the current in a motor, consider the next Example.

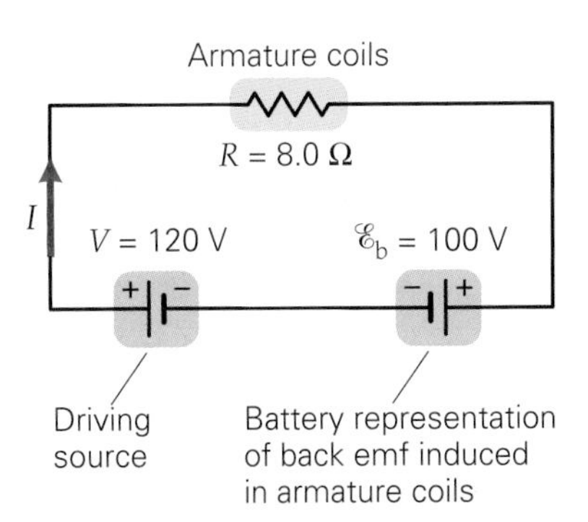

▲ **FIGURE 20.15 Back emf** The back emf in the armature of a dc motor can be represented as a battery with polarity opposite that of the driving voltage.

Example 20.6 ■ Getting up to Speed: Back emf in a dc Motor

A dc motor is built with windings that have a resistance of 8.00 Ω and operates at a line voltage of 120 V. With a normal load, there is a back emf of 100 V when the motor reaches full speed. (See Fig. 20.15.) Determine (a) the starting current drawn by the motor and (b) the armature current at operating speed under a normal load.

Thinking It Through. (a) The only difference between startup and full speed is that there is no back emf at startup. The net voltage and resistance determine the current, so Eq. 20.6 can be applied. (b) At operating speed, the back emf increases and is opposite in polarity to the line voltage. Equation 20.6 can again be used to determine the current.

Solution. Let's list the data as usual:

Given: $R = 8.00\ \Omega$, $V = 120\ \text{V}$, $\mathscr{E}_b = 100\ \text{V}$

Find: (a) I_s (starting current) (b) I (operating current)

(a) From Eq. 20.6, the current in the windings is

$$I_s = \frac{V}{R} = \frac{120\ \text{V}}{8.00\ \Omega} = 15.0\ \text{A}$$

(b) When the motor is at full speed, the back emf is 100 V; thus, the current is less.

$$I = \frac{V - \mathscr{E}_b}{R} = \frac{120\ \text{V} - 100\ \text{V}}{8.00\ \Omega} = 2.50\ \text{A}$$

With little or no back emf, the *starting* current is relatively large. When a big motor, such as that of a central air-conditioning unit, starts up, the lights in the building might momentarily dim, because of the large starting current that the motor draws. In some designs, resistors are temporarily connected in series with a motor's coil to protect the windings from burning out as a result of large starting currents.

Follow-Up Exercise. In this Example, (a) how much energy is required to bring the motor to operating speed if it takes 10 s and the back emf averages 50 V during that time? (b) Compare this amount with the amount of energy required to keep the motor running for 10 s once it reaches its operating conditions.

Because motors and generators are opposites, so to speak, and a back emf develops in a motor, you may be wondering whether a back force develops in a generator. The answer is yes. When an operating generator is not connected to an external circuit, no current exists, and therefore there is no magnetic force on the armature coils. However, when the generator delivers energy to an external circuit and current *is* in the coils, the magnetic force on the armature coils produces a *countertorque* that opposes the rotation of the armature. As more current is drawn, the countertorque increases and a greater driving force is needed to turn the armature. Therefore, the higher the generator's current output, the greater the energy expended (that is, fuel consumed) in overcoming the countertorque.

20.3 Transformers and Power Transmission

OBJECTIVES: To (a) explain transformer action in terms of Faraday's law, (b) calculate the output of step-up and step-down transformers, and (c) understand the importance of transformers in electric energy delivery systems.

Electric energy is transmitted by power lines over long distances. It is desirable to minimize I^2R losses (joule heat) that can occur in these transmission lines. Because the resistance of a line is fixed, reducing I^2R losses means reducing current. However, the power output of a generator is determined by its outputs of current and voltage ($P = IV$), and for a fixed voltage, such as 120 V, a reduction in current would mean a reduced power output. It might appear that there is no way to reduce the current while maintaining the power level. Fortunately, electromagnetic induction enables us to reduce power-transmission losses by increasing voltage while simultaneously reducing current in such a way that the delivered *power* is essentially unchanged. This is done using a device called a **transformer**.

A simple transformer consists of two coils of insulated wire wound on the same iron core (▸Fig. 20.16a). When ac voltage is applied to the input coil, or *primary coil*, the alternating current produces an alternating magnetic flux concentrated in the iron core, without any significant leakage of flux outside the core. Under these conditions, the same changing flux also passes through the output coil, or *secondary coil*, inducing an alternating voltage and current in it. (Note that it is common in transformer design to refer to emfs as "voltages," as was done in Chapter 18. We will use this language here also.)

The ratio of the induced voltage in the secondary coil to that of the voltage in the primary coil depends on the ratio of the numbers of turns in the two coils. By Faraday's law, the induced voltage in the secondary coil is

$$V_s = -N_s \frac{\Delta\Phi}{\Delta t},$$

where N_s is the number of turns in the secondary coil. The changing flux in the primary coil produces a back emf of

$$V_p = -N_p \frac{\Delta\Phi}{\Delta t},$$

where N_p is the number of turns in the primary coil. If the resistance of the primary coil is neglected, this back emf is equal in magnitude to the external voltage applied to the primary coil (why?). Forming a ratio of output voltage (secondary) to input voltage (primary) yields

$$\frac{V_s}{V_p} = \frac{-N_s(\Delta\Phi/\Delta t)}{-N_p(\Delta\Phi/\Delta t)}$$

or

$$\frac{V_s}{V_p} = \frac{N_s}{N_p} \quad \textit{(voltage ratio in a transformer)} \qquad (20.7)$$

If the transformer is 100% efficient (that is, there are no energy losses), then the power input is equal to the power output. Because $P = IV$, we have

$$I_p V_p = I_s V_s \qquad (20.8)$$

Although some energy is always lost, this equation is a good approximation, as a well-designed transformer will have an efficiency greater than 95%. (The sources of energy losses will be discussed shortly.) Assuming this ideal case, from Eq. 20.8, the transformer currents and voltages are related to the turn ratio by

$$\frac{I_p}{I_s} = \frac{V_s}{V_p} = \frac{N_s}{N_p} \qquad (20.9)$$

To summarize the transformer action in terms of voltage and current output, we have

$$V_s = \left(\frac{N_s}{N_p}\right)V_p \qquad (20.10a)$$

and *(ideal relationship between transformer voltage and current)*

$$I_s = \left(\frac{N_p}{N_s}\right)I_p \qquad (20.10b)$$

If the secondary coil has more windings than the primary coil does (that is, $N_s/N_p > 1$), as in Fig. 20.16a, the voltage is "stepped up," because $V_s > V_p$. This is called a *step-up transformer*. Notice that because of this there is *less* current in the secondary than in the primary ($N_p/N_s < 1$ and $I_s < I_p$).

If the secondary coil has fewer turns than the primary does, we have a *step-down transformer* (Fig. 20.16b). In the usual transformer language, this means that the voltage is "stepped down," and the current, therefore, is increased. Depending on the design details, a step-up transformer may be used as a step-down transformer by simply reversing output and input connections.

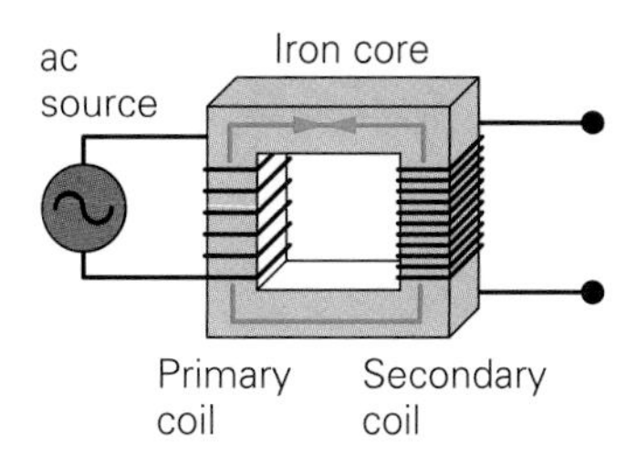

(a) Step-up transformer: high-voltage (low-current) output

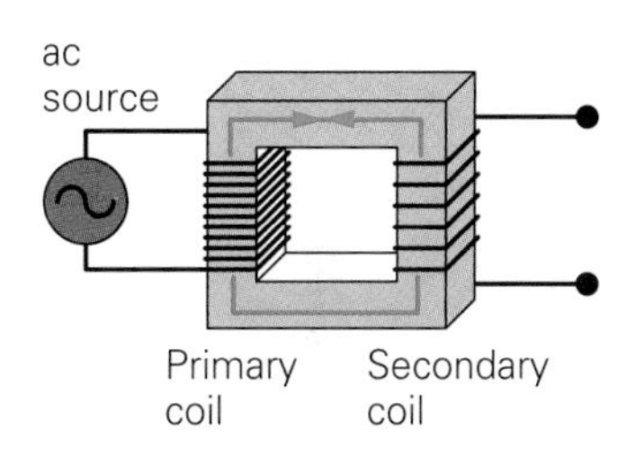

(b) Step-down transformer: low-voltage (high-current) output

▲ **FIGURE 20.16** Transformers **(a)** A step-up transformer has more turns in the secondary coil than in the primary coil. **(b)** A step-down transformer has more turns in the primary coil than in the secondary coil.

Integrated Example 20.7 ■ Transformer Orientation: Step-Up or Step-Down Configuration?

An ideal 600-W transformer has 50 turns on its primary coil and 100 turns on its secondary coil. (a) Is this transformer a (1) a step-up or (2) step-down arrangement? (b) If the primary coil is connected to a 120-V source, what are the output voltage and current of this transformer?

(a) Conceptual Reasoning. Step-up or step-down refers to what happens to the voltage, not the current. Because the voltage is proportional to the number of turns, in this case the secondary voltage is greater than the primary voltage. Thus the correct answer is (1) a step-up transformer.

(b) Quantitative Reasoning and Solution. The output voltage can be determined from Eq. 20.10a, once the turn ratio is established. From the power, the current can be determined.

Given: $N_p = 50$
$N_s = 100$
$V_p = 120$ V

Find: V_s and I_s (secondary voltage and current)

The secondary voltage can be found using Eq. 20.10a with a turn ratio of 2, because $N_s = 2N_p$:

$$V_s = \left(\frac{N_s}{N_p}\right)V_p = (2)(120\text{ V}) = 240\text{ V}$$

If the transformer is ideal, then the input power equals the output power. On the primary side, the input power is $P_p = I_pV_p = 600$ W, so the input current must be

$$I_p = \frac{600\text{ W}}{V_p} = \frac{600\text{ W}}{120\text{ V}} = 5.00\text{ A}$$

Because the voltage is stepped up by a factor of two, the output current should be stepped down by a factor of two. From Eq. 20.10b,

$$I_s = \left(\frac{N_p}{N_s}\right)I_p = \left(\frac{1}{2}\right)(5.00\text{ A}) = 2.50\text{ A}$$

Follow-Up Exercise. (a) When a European visitor (the average ac voltages are 240 V in Europe) visits the United States, what type of transformer should be used to enable her hair dryer to work properly? Explain. (b) For a 1500-W hair dryer (assumed ohmic), what would be the transformer's input current in the United States, assuming it to be ideal?

The preceding relationships strictly apply only to ideal (or "lossless") transformers; actual transformers have energy losses. Well-designed transformers generally have a loss of less than 5%. Consequently there is no such thing as an ideal transformer. Many factors combine to determine how close a real transformer comes to performing like an ideal one.

First, there is flux leakage; that is, not all of the flux passes through the secondary coil. In some transformer designs, one of the insulated coils is wound directly on top of the other (interlocking) rather than having two separate coils. This configuration helps minimize flux leakage while reducing transformer size.

Second, the ac current in the primary means there is a changing magnetic flux through those coils. In turn this gives rise to an induced emf in the primary. This is called *self-induction*. By Lenz's law, the self-induced emf will oppose the change in current and thus limit the primary current (this is a similar effect to that of the back emf in a motor).

A third reason that transformers are less than ideal is joule heating (I^2R losses) due to the resistance of the wires. Usually this loss is small because the wires have little resistance.

Lastly, consider the effect of induction in the core material. To increase magnetic flux, the core is made of a highly permeable material (such as iron), but such materials are also good conductors. The changing magnetic flux in the core induces emfs there, which in turn create *eddy (or "swirling") currents* in the core material. These eddy currents can cause energy loss between the primary and secondary by heating the core (I^2R losses again).

To reduce the loss of energy due to eddy currents, transformer cores are made of thin sheets of material (usually iron) laminated with an insulating glue between them. The insulating layers between the sheets break up the eddy currents or confine them to the thin sheets, greatly reducing energy loss.

The effects of eddy currents can be demonstrated by allowing a plate made of a conductive, but nonmagnetic, metal, such as aluminum, to swing through a magnetic

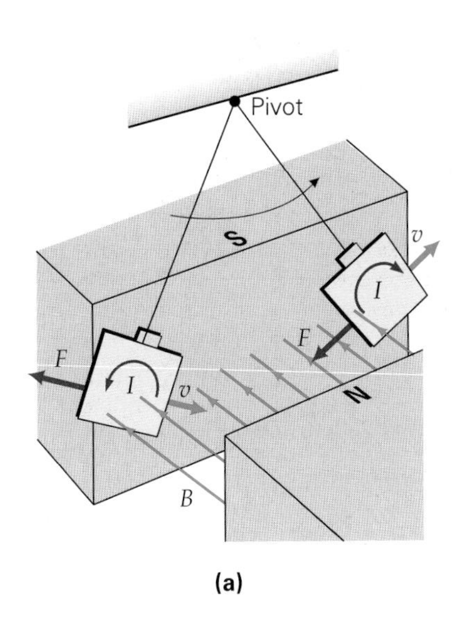

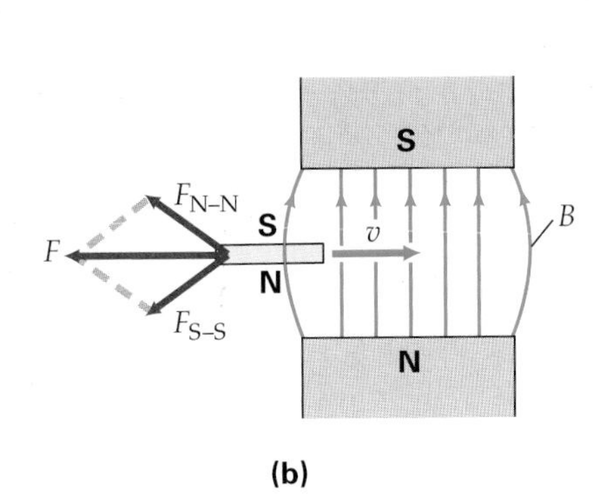

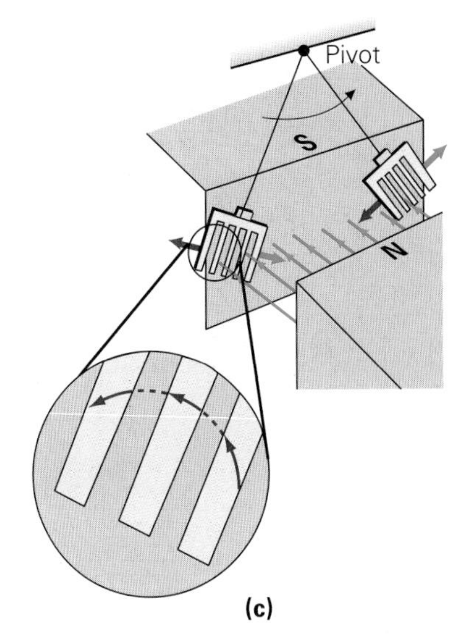

▲ FIGURE 20.17 Eddy currents **(a)** Eddy currents are induced in a metal plate moving in a magnetic field. The induced currents oppose the change in flux. These currents then experience a retarding magnetic force to oppose the motion first into, then out of, the field region. To see this, note that the currents reverse direction as the plate leaves the field. **(b)** An overhead view as the plate swings toward the field from the left. The retarding force $\vec{\mathbf{F}}$ (to slow it from entering the field) results from the two repulsive forces ($\vec{\mathbf{F}}_{N-N}$ and $\vec{\mathbf{F}}_{S-S}$) acting between magnetic poles. This is because the side of the plate closest to the north pole of the permanent magnet acts as a north pole, and the other side acts as a south pole. **(c)** If the plate has slits, the eddy currents, and thus magnetic forces, are drastically reduced and the plate will swing more freely.

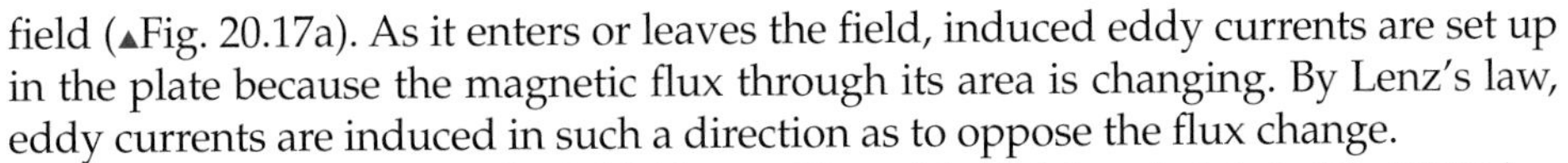

field (▲Fig. 20.17a). As it enters or leaves the field, induced eddy currents are set up in the plate because the magnetic flux through its area is changing. By Lenz's law, eddy currents are induced in such a direction as to oppose the flux change.

When the plate enters the field (the position of the left-hand plate in Fig. 20.17a), a counterclockwise current is induced. (You should apply Lenz's law to show this.) The induced current produces its own magnetic field, which means that, in effect, the plate has a north magnetic pole near the permanent magnet's north pole and a south magnetic pole near the permanent magnet's south pole (Fig. 20.17b). Two repulsive magnetic forces act on the plate. The effect of the net force is to slow the plate down as it enters the field. The plate's eddy currents are reversed in direction as it leaves the field, producing a net attractive magnetic force, thus tending to slow the plate from leaving the field. In both cases, the induced emfs act to slow the plate's motion.

The reduction of eddy currents (similar to how the laminated layers in a transformer work) can be demonstrated by using a plate with slits cut into it (Fig. 20.17c). When this plate swings between the magnet's poles, it swings relatively freely, because the eddy currents are greatly reduced by the air gaps (slits). Consequently, the magnetic force on the plate is also reduced.

The damping effect of eddy currents has been applied in the braking systems of rapid-transit railcars. When an electromagnet (housed in the car) is turned on, it applies a magnetic field to a rail. The repulsive force due to the induced eddy currents in the rail acts as a braking force (▼Fig. 20.18). As the car slows, the eddy currents in the rail decrease, allowing a smooth braking action.

Power Transmission and Transformers

For power transmission over long distances, transformers provide a way to increase the voltage and reduce the current of an electric generator, thus cutting down the joule heating (I^2R) losses in the transmission wires that are carrying current. A schematic diagram of an ac power distribution system is shown in ▼Fig. 20.19. The voltage output of the generator is stepped up, reducing the current. The energy is transmitted over long distances to an area substation near the consumers. There,

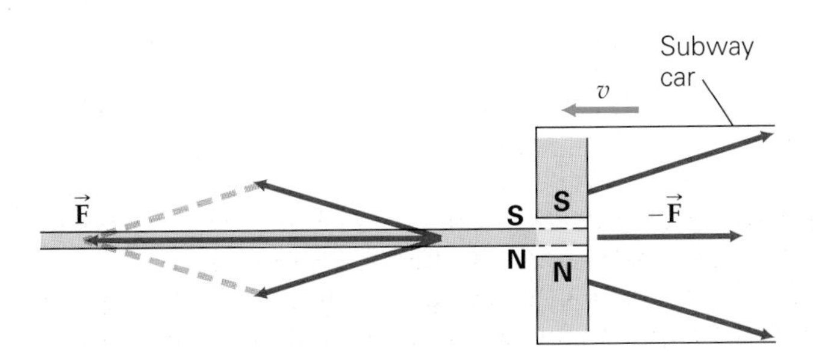

◀ FIGURE 20.18 Electromagnetic braking and mass transit When braking, a train energizes an electromagnet onboard. This electromagnet straddles a long metal rail. The induced currents in the rail produce a mutually repulsive force between the rail and the train, thereby slowing the train.

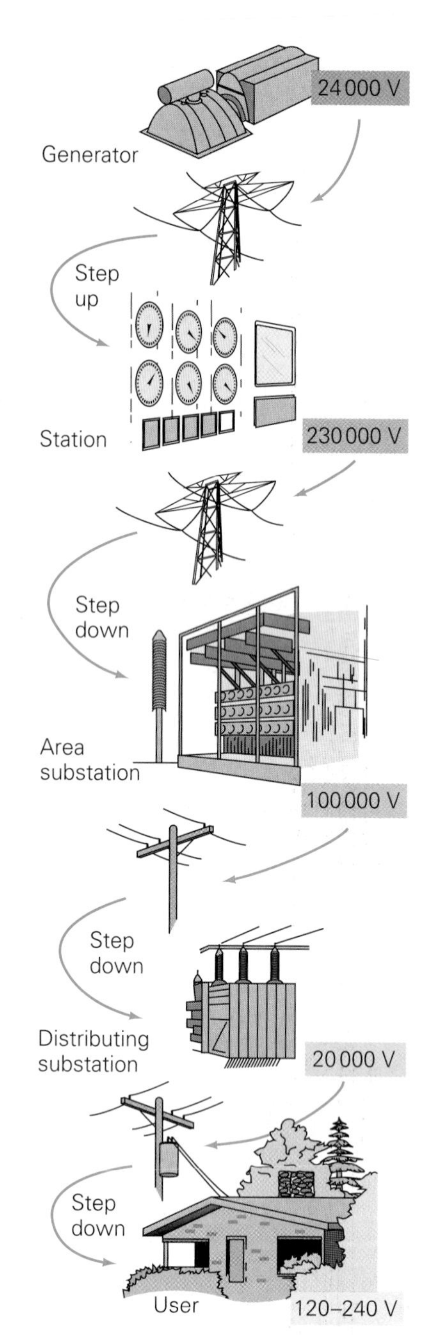

▲ **FIGURE 20.19 Power transmission** A diagram of a typical electrical-power distribution system.

Illustration 31.3 Transformers

the voltage is stepped down, increasing the current. There are further step-downs at distributing substations and utility poles before the electricity is supplied to homes and businesses at the normal voltage and current.

The following Example illustrates the benefits of being able to step up the voltage (and step down the current) for electrical power transmission.

Example 20.8 ■ Cutting Your Losses: Power Transmission at High Voltage

A small hydroelectric power plant produces energy in the form of electric current at 10 A and a voltage of 440 V. The voltage is stepped up to 4400 V (by an ideal transformer) for transmission over 40 km of power line, which has a total resistance of 20 Ω. (a) What percentage of the original energy would have been lost in transmission if the voltage had not been stepped up? (b) What percentage of the original energy is actually lost when the voltage is stepped up?

Thinking It Through. (a) The power output can be computed from $P = IV$ and compared with the power lost in the wire, $P = I^2R$. (b) Equations 20.10a and 20.10b should be used to determine the stepped-up voltage and stepped-down currents, respectively. Then the calculation is repeated, and the results are compared with those of part (a).

Solution.

Given: $I_p = 10\text{ A}$, $V_p = 440\text{ V}$, $V_s = 4400\text{ V}$, $R = 20\ \Omega$

Find: (a) percentage energy loss without voltage step-up
(b) percentage energy loss with voltage step-up

(a) The power output by the generator is

$$P = I_pV_p = (10\text{ A})(440\text{ V}) = 4400\text{ W}$$

The rate of energy loss of the wire (joules per second, or watts) in transmitting a current of 10 A is very high, because

$$P_{\text{loss}} = I^2R = (10\text{ A})^2(20\ \Omega) = 2000\text{ W}$$

Thus, the percentage of the produced energy lost to joule heat in the wires is nearly 50% because

$$\%\text{ loss} = \frac{P_{\text{loss}}}{P} \times 100\% = \frac{2000\text{ W}}{4400\text{ W}} \times 100\% = 45\%$$

(b) When the voltage is stepped up to 4400 V, this allows for transmission of energy at a current that is reduced by a factor of 10 from its value in part (a). We have

$$I_s = \left(\frac{V_p}{V_s}\right)I_p = \left(\frac{440\text{ V}}{4400\text{ V}}\right)(10\text{ A}) = 1.0\text{ A}$$

The power is thus reduced by a factor of 100, because it varies as the square of the current:

$$P_{\text{loss}} = I^2R = (1.0\text{ A})^2(20\ \Omega) = 20\text{ W}$$

Therefore, the percentage of power lost is also reduced by a factor of 100 to a much more acceptable level:

$$\%\text{ loss} = \frac{P_{\text{loss}}}{P} \times 100\% = \frac{20\text{ W}}{4400\text{ W}} \times 100\% = 0.45\%$$

Follow-Up Exercise. Some heavy-duty electrical appliances, such as water pumps, can be wired to 240 V or 120 V. Their power rating is the same regardless of the voltage at which they run. (a) Explain the efficiency advantage of operating such appliances at the higher voltage. (b) For a 1.00-hp pump (746 W), estimate the ratio of the power lost in the wires at 240 V to the power lost at 120 V (assuming that all resistances are ohmic and the connecting wires are the same).

20.4 Electromagnetic Waves

OBJECTIVES: To (a) explain the nature, origin, and propagation of electromagnetic waves, and (b) describe some of the properties and uses of the different types of electromagnetic waves.

Electromagnetic waves (or *electromagnetic radiation*) were considered as a means of heat transfer in Section 11.4. You can now understand the production and characteristics of electromagnetic radiation, because these waves are composed of electric and magnetic fields.

Scottish physicist James Clerk Maxwell (1831–1879) is credited with first bringing together, or *unifying*, electric and magnetic phenomena. Using mathematics beyond the scope of this book, he took the equations that governed each field and predicted the existence of electromagnetic waves. In fact, he went further and calculated their speed in a vacuum, and his prediction agreed with experiment. Because of these contributions, the set of equations is known as *Maxwell's equations*, although they were, for the most part, developed by others (for example, Faraday's law of induction).

Essentially, Maxwell showed how the electric field and the magnetic field could be thought of as a single electromagnetic field. The apparently separate fields are symmetrically related in the sense that either one can create the other under the proper conditions. This symmetry is evident by looking at the equations (not shown). A qualitative summary of the results is sufficient:

> A time-varying magnetic field produces a time-varying electric field.
> A time-varying electric field produces a time-varying magnetic field.

PHYSLET® Illustration 32.1 *Creation of Electromagnetic Waves*

The first statement summarizes our observations in Section 20.1: A changing magnetic flux gives rise to an induced emf, which in turn can cause a current. The second statement (which we will not study in detail) is crucial to the self-propagating characteristic of electromagnetic waves. Together, these two phenomena enable these waves to travel through a vacuum, whereas all other waves, such as string waves, require a supporting medium.

According to Maxwell's theory, *accelerating* electric charges, such as an oscillating electron, produce electromagnetic waves. The electron in question could, for example, be one of the many electrons in the metal antenna of a radio transmitter, driven by an electrical (voltage) oscillator at a frequency of 10^6 Hz (1 MHz). As each electron oscillates, it continually accelerates and decelerates and thus radiates an electromagnetic wave (▾Fig. 20.20a). The driven oscillations of many electrons produce time-varying electric and magnetic fields in the vicinity of the antenna. The electric field, shown in red in Fig. 20.20a, is in the plane of the page and continually changes direction, as does the magnetic field (shown in blue and pointing into and out of the paper).

Both the electric and the magnetic fields carry energy and propagate outward at the speed of light. This speed is symbolized by the letter c. To three significant figures, $c = 3.00 \times 10^8$ m/s. Maxwell's results showed that at large distances from the source, these electromagnetic waves become plane waves. (Figure 20.20b shows a wave at an instant in time.) Here, the electric field ($\vec{\mathbf{E}}$) is perpendicular to the magnetic field ($\vec{\mathbf{B}}$), and each varies sinusoidally with time. Both $\vec{\mathbf{E}}$ and $\vec{\mathbf{B}}$ are perpendicular to the direction of wave propagation. Thus, electromagnetic waves are *transverse* waves, with the *fields* oscillating perpendicularly to the direction of propagation. According to Maxwell's theory, as one field changes, it creates the other. This process, repeated again and again, gives rise to the traveling electromagnetic wave we call light. An important result of all this is as follows:

> In a vacuum, all electromagnetic waves, regardless of frequency or wavelength, travel at the same speed, $c = 3.00 \times 10^8$ m/s.

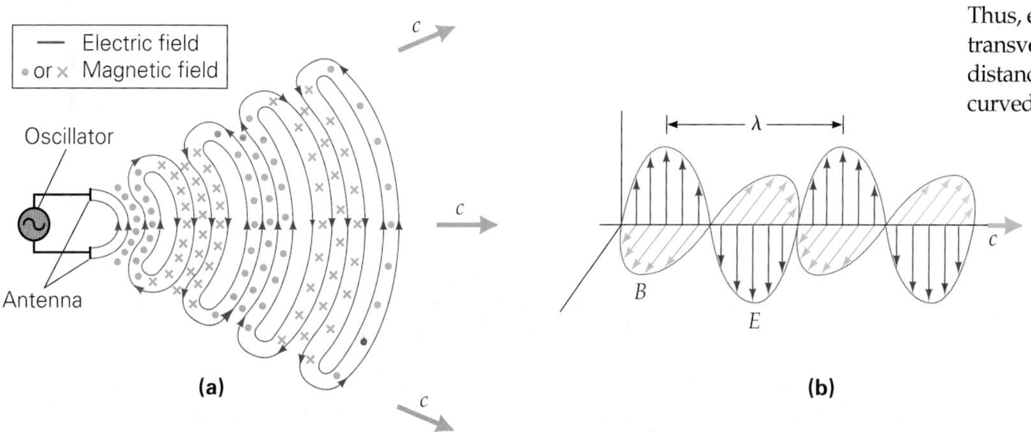

▾ **FIGURE 20.20** **Source of electromagnetic waves** Electromagnetic waves are produced, fundamentally, by accelerating electric charges. **(a)** Here charges (electrons) in a metal antenna are driven by an oscillating voltage source. As the antenna polarity and current direction periodically change, alternating electric and magnetic fields propagate outward. The electric and magnetic fields are perpendicular to the direction of wave propagation. Thus, electromagnetic waves are transverse waves. **(b)** At large distances from the source, the initially curved wavefronts become planer.

For everyday distances, the time delay due to the speed of light can usually be neglected. However, for interplanetary trips, this delay can be a problem. Consider the following Example.

Example 20.9 ■ Long-Distance Guidance: The Speed of Electromagnetic Waves in a Vacuum

The first successful Mars landings were the *Viking* probes in 1976. They sent radio and TV signals (both are electromagnetic waves) back to the Earth. How much longer would it take for a signal to reach us when Mars was farthest from the Earth than when it was closest to us? The average distances of Mars and the Earth from the Sun are 229 million km (d_M) and 150 million km (d_E), respectively. Assume that both planets have circular orbits, and use the average distances as the radii of the circles.

Thinking It Through. This situation calls for a time–distance calculation. The planets are farthest apart when they are on opposite sides of the Sun and separated by a distance of $d_M + d_E$. (This arrangement requires signals to be sent through the Sun, which, of course, is not possible. However, it does serve to determine the upper limit on transmission times.) The planets are closest when they are aligned on the same side of the Sun. In this case, their separation distance is at a minimum value of $d_M - d_E$. (Draw a diagram to help visualize this.) Because the speed of electromagnetic waves in a vacuum is known, the times can be found from $t = d/c$.

Solution. Listing the data and converting the distances to meters,

Given: $d_M = 229 \times 10^6\ \text{km} = 2.29 \times 10^{11}\ \text{m}$
$d_E = 150 \times 10^6\ \text{km} = 1.50 \times 10^{11}\ \text{m}$

Find: Δt (difference in time for light to travel the longest and shortest distances)

Radio and TV waves travel at speed c. The longest travel time t_L is

$$t_L = \frac{d_M + d_E}{c} = \frac{3.79 \times 10^{11}\ \text{m}}{3.00 \times 10^8\ \text{m/s}} = 1.26 \times 10^3\ \text{s} \quad (\text{or } 21.1\ \text{min})$$

For the shortest distance the shortest travel time t_s is

$$t_S = \frac{d_M - d_E}{c} = \frac{7.90 \times 10^{10}\ \text{m}}{3.00 \times 10^8\ \text{m/s}} = 2.63 \times 10^2\ \text{s} \quad (\text{or } 4.39\ \text{min})$$

The difference in the times is thus $\Delta t = t_L - t_s = 1.00 \times 10^3$ s (or 16.7 min).

Follow-Up Exercise. Assume that a *Rover* Martian vehicle (see Example 2.1 on page 34) is heading for a collision with a rock 2.0 m ahead of it. When it is at that distance, the vehicle sends a picture of the rock to controllers on Earth. If Mars is at the closest point to the Earth, what is the maximum speed that the *Rover* could have and still avoid a collision? Assume that the video signal from the *Rover* reaches the Earth and that the signal for it to stop is sent back immediately.

Radiation Pressure

An electromagnetic wave carries energy. Consequently, it can do work and can exert a force on a material it strikes. Consider light striking an electron at rest on a surface (◂Fig. 20.21). The electric field of the wave exerts a force on the electron, giving it a downward velocity ($\vec{v}$) as shown in the figure. Because a charged particle moving in a magnetic field experiences a force, there is a magnetic force on the electron, due to the magnetic-field component of the light wave. By the force right-hand rule, this force is in the direction that the wave is propagating (Fig. 20.21a). Therefore, because the electromagnetic wave will produce the same force on many electrons in a material, it exerts a force on the surface as a whole, and that force is in the direction in which it is traveling.

The radiation force per area is called **radiation pressure**. Radiation pressure is negligibly small for most everyday situations, but it can be important in atmospheric and astronomical phenomena, as well as in atomic and nuclear

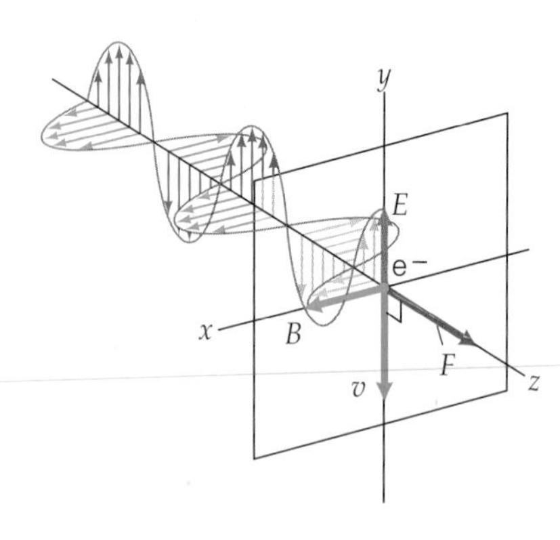

▲ **FIGURE 20.21 Radiation pressure** The electric field of an electromagnetic wave that strikes a surface acts on an electron, giving it a velocity. The magnetic field then exerts a force on that moving charge in the direction of propagation of the incident light. (Verify this direction, using the magnetic right-hand force rule.)

physics, where masses are small and there is no friction. For example, radiation pressure plays a key role in determining the direction in which the tail of a comet points. Sunlight delivers energy to the comet's "head," which consists of ice and dust. Some of this material evaporates as the comet nears the Sun, and the evaporated gases are pushed away from the Sun by radiation pressure. Thus, the tail generally points away from the Sun, no matter whether the comet is approaching or leaving the Sun's vicinity.

Another potential use of radiation pressure from sunlight is to propel interplanetary "sailing" satellites outward from the Sun toward the outer planets in an ever enlarging, spiraling orbit (▸Fig. 20.22a). To create enough force, given the extremely low pressure of the sunlight, the sails would have to be very large in area, and the satellite would have to have as little mass as possible. The payoff is that no fuel (except for small amounts for course corrections) would be needed once the satellite was launched. Consider the following Conceptual Example, which concerns radiation pressure and space travel.

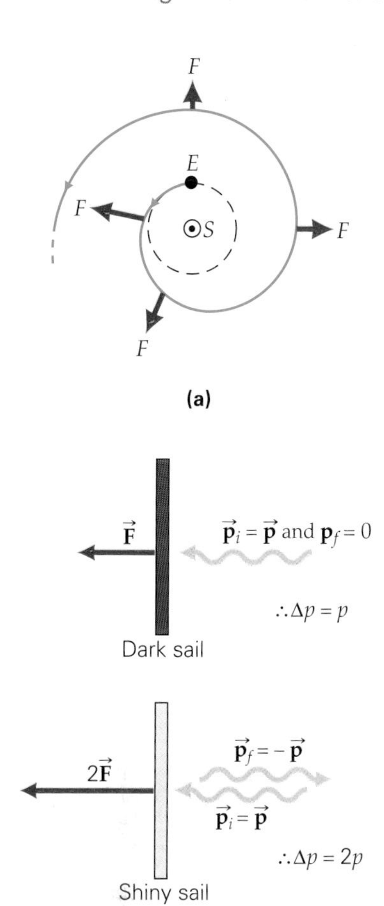

▲ **FIGURE 20.22 "Sailing" the solar system** **(a)** A space probe launched from the Earth (E) equipped with a large sail would be acted on by radiation pressure from sunlight (Sun is at S). This cost-free force would cause the satellite to spiral outward. With proper planning, the craft could get to outer planets with little or no extra fuel. Note the reduction in force with distance. **(b)** Is it better for the sail to be dark or shiny? See Conceptual Example 20.10 and momentum conservation.

Conceptual Example 20.10 ■ Sailing the Sea of Space: Radiation Pressure in Action

Consider the design of a relatively light spacecraft with a huge "sail" to be used as an interplanetary probe. With little or no power of its own, it would be designed to use the pressure of sunlight to propel it to the outer planets. To get the maximum propulsive force, what kind of surface should the sail have: (a) shiny and reflective, (b) dark and absorptive, or (c) surface characteristics would not matter?

Reasoning and Answer. At first glance, you might think that the answer is (c). However, as we have seen, radiation is capable of exerting force and can transfer momentum to whatever it strikes. The interaction between the radiation and the sail can be described in terms of conservation of momentum, as shown in Fig. 20.22b. (See Section 6.3.) If the radiation is absorbed, the situation is analogous to a completely inelastic collision (such as a putty wad sticking to a door), and the sail would acquire all of the momentum ($\vec{p}$) originally possessed by the radiation.

However, if the radiation is *reflected*, the situation is analogous to a completely elastic collision, like a Superball bouncing off a wall (see Section 6.1). Because the momentum of the radiation after the collision would be equal to its original momentum in magnitude, but opposite in direction, its momentum would thus be reversed (from $\vec{p}$ to $-\vec{p}$). To conserve momentum, the momentum transferred to the shiny sail would be twice as great ($2\vec{p}$) as that for the dark sail. Because force is the rate of change of momentum, reflective sails would experience, on average, twice as much force as absorptive ones. So the answer is (a).

Follow-Up Exercise. (a) Would the sail in this Example provide less or more acceleration as the interplanetary sailing ship moves farther from the Sun? (b) Explain how a change in sail area could counteract this change.

Types of Electromagnetic Waves

Electromagnetic waves are classified by the range of frequencies or wavelengths they encompass. Recall from Chapter 13 that frequency and wavelength are inversely related by the traveling-wave relationship $\lambda = c/f$, where the speed of light, c, has been substituted for the general wave speed v. The higher the frequency, the shorter the wavelength, and vice versa. The electromagnetic spectrum is continuous, so the limits of the various types of radiation are approximate (▾Fig. 20.23). Table 20.1 (p. 676) lists these ranges for the general types of electromagnetic waves.

Power Waves Electromagnetic waves with a frequency of 60 Hz result from alternating currents in electric power lines. These power waves have a wavelength of 5.0×10^6 m, or 5000 km (more than 3000 mi). Waves of such low frequency are of little practical use. They may occasionally produce a so-called 60-Hz hum on your stereo or introduce, via induction, unwanted electrical noise in delicate instruments. More serious concerns have been expressed about the possible effects of these waves

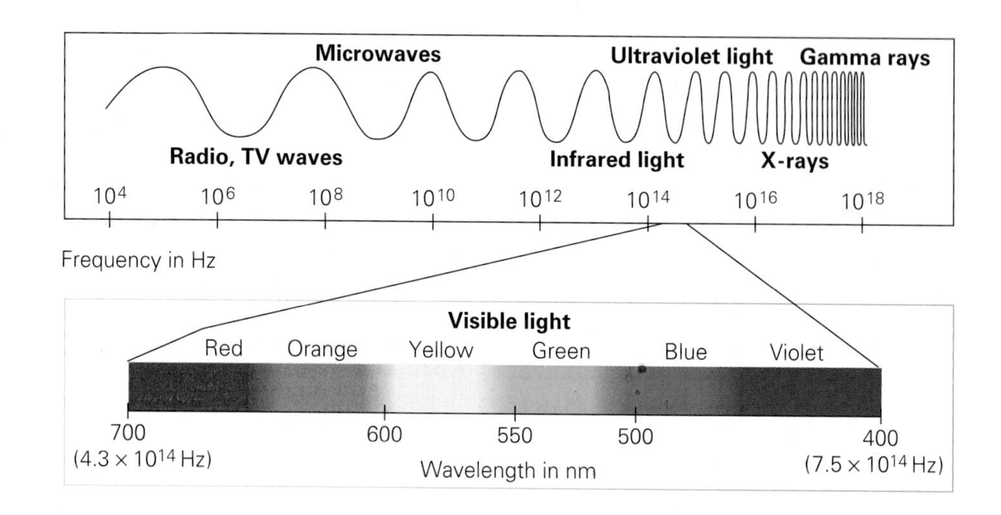

▶ **FIGURE 20.23 The electromagnetic spectrum** The spectrum of frequencies or wavelengths is divided into various regions, or ranges. The visible-light region is a very small part of the total spectrum. For visible light, wavelengths are usually expressed in nanometers ($1 \text{ nm} = 10^{-9} \text{ m}$). (The relative sizes of the wavelengths at the top of the figure are not to scale.)

TABLE 20.1 Classification of Electromagnetic Waves

Type of Wave	*Approximate Frequency Range (Hz)*	*Approximate Wavelength Range (m)*	*Some Typical Sources*
Power waves	60	5.0×10^6	Electric currents
Radio waves—AM	$(0.53 \times 10^6)-(1.7 \times 10^6)$	570–186	Electric circuits/antennae
Radio waves—FM	$(88 \times 10^6)-(108 \times 10^6)$	3.4–2.8	Electric circuits/antennae
TV	$(54 \times 10^6)-(890 \times 10^6)$	5.6–0.34	Electric circuits/antennae
Microwaves	10^9-10^{11}	$10^{-1}-10^{-3}$	Special vacuum tubes
Infrared radiation	$10^{11}-10^{14}$	$10^{-3}-10^{-7}$	Warm and hot bodies, stars
Visible light	$(4.0 \times 10^{14})-(7.0 \times 10^{14})$	10^{-7}	The Sun and other stars, and lamps
Ultraviolet radiation	$10^{14}-10^{17}$	$10^{-7}-10^{-10}$	Very hot bodies, stars, and special lamps
X-rays	$10^{17}-10^{19}$	$10^{-10}-10^{-12}$	High-speed electron collisions and atomic processes
Gamma rays	above 10^{19}	below 10^{-12}	Nuclear reactions and nuclear decay processes

on health. Some early research tended to suggest that very low-frequency fields may have potentially harmful biological effects on cells and tissues. However, recent surveys indicate that this is not the case.

Radio and TV Waves Radio and TV waves are generally in the frequency range from 500 kHz to about 1000 MHz. The AM (*amplitude-modulated*) band runs from 530 to 1710 kHz (1.71 MHz). Higher frequencies, up to 54 MHz, are used for "shortwave" bands. TV bands range from 54 MHz to 890 MHz. The FM (*frequency-modulated*) radio band runs from 88 to 108 MHz, which lies in a gap between channels 6 and 7 of the range of TV bands. Cellular phones use radio waves to transmit voice communication in the ultrahigh-frequency (UHF) band, with frequencies similar to those used for TV channels 13 and higher.

Early global communications used the "shortwave" bands, as do amateur (ham) radio operators today. But how are the normally straight-line radio waves transmitted around the curvature of the Earth? This feat is accomplished by reflection off ionic layers in the upper atmosphere. Energetic particles from the Sun ionize gas molecules, giving rise to several ion layers. Certain of these layers reflect radio waves. By "bouncing" radio waves off these layers, radio transmissions can be sent beyond the horizon, to any region of the Earth.

It turns out that such reflection requires the ionic layers to have fairly uniform density. When, from time to time, a solar disturbance produces a larger-than-average shower of energetic charged particles that upsets this uniformity, a communications "blackout" can occur as the radio waves are scattered in various directions rather than reflected in straight lines. To avoid such disruptions, global communications have, in the past, relied largely on transoceanic cables. Now we also have communications satellites, which can provide line-of-sight transmission to any point on the globe.

Microwaves Microwaves, with frequencies in the gigahertz (GHz) range, are produced by special vacuum tubes (called *klystrons* and *magnetrons*). Microwaves are commonly used in communications and radar applications. In addition to its roles in navigation and guidance, radar provides the basis for the speed guns used to time such things as baseball pitches and motorists through the use of the Doppler effect (see Section 14.5). When the waves are reflected off an object of interest, the magnitude and sign of the frequency shift allow determination of the object's velocity.

Infrared (IR) Radiation The infrared region of the electromagnetic spectrum lies adjacent to the low-frequency, or long-wavelength, end of the visible spectrum. A warm body emits IR radiation, which depends on that body's temperature. (See Chapter 27.) An object at or near room temperature emits radiation in the far infrared region. ("Far" means relative to the visible region.)

Recall from Section 11.4 that infrared radiation is sometimes referred to as "heat rays." This is because water molecules, which are present in most materials and which possess electrical permanent polarization, readily absorb electromagnetic radiation at frequencies in the infrared-wavelength region. When they do, their random thermal motion is increased—and the molecules "heat up," as well as their surroundings. Infrared lamps are used in therapeutic applications, such as easing pain in strained muscles, and to keep food warm. IR is also associated with maintaining the Earth's temperature through the *greenhouse effect*. In this effect, incoming visible light (which passes relatively easily through the atmosphere) is absorbed by the Earth's surface and reradiated as infrared radiation. This IR radiation is in turn trapped by "greenhouse gases," such as carbon dioxide and water vapor, which are opaque to IR radiation. Its name comes from actual glass-enclosed greenhouse, where the glass rather than atmospheric gases traps the IR energy.

Visible Light The region of visible light occupies only a small portion of the electromagnetic spectrum and covers a frequency range from 4×10^{14} Hz to about 7×10^{14} Hz. In terms of wavelengths, the range is from about 700 to 400 nm (Fig. 20.23). Recall that 1 nanometer (nm) $= 10^{-9}$ m. Only the radiation in this region activates the receptors on the retina of human eyes. Visible light emitted or reflected from objects around us provides us with visual information about our world. Visible light and optics will be discussed in Chapters 22 to 25.

It is interesting to note that not all animals are sensitive to the same range of wavelengths. For example, snakes can visually detect infrared radiation, and the visible range of many insects extends well into the ultraviolet range. The sensitivity range of the human eye conforms closely to the spectrum of wavelengths emitted by the Sun. The human eye's maximum sensitivity is in the same yellow-green region where the Sun's energy output is at its maximum (wavelengths of about 550 nm).

Ultraviolet (UV) Radiation The Sun's spectrum has a small component of ultraviolet (UV) light, whose frequency range lies beyond the violet end of the visible region. UV is also produced artificially by special lamps and very hot objects. In addition to causing tanning of the skin, UV radiation can cause sunburn and/or skin cancer if exposure to it is too high.

Upon its arrival at the Earth, most of the Sun's ultraviolet emission is absorbed in the ozone (O_3) layer in the atmosphere, at an altitude of about 30 to 50 km. Because the ozone layer plays a protective role, there is concern about its depletion by chlorofluorocarbon gases (such as Freon, once commonly used in refrigerators) that drift upward and react with the ozone.

Most UV radiation is absorbed by ordinary glass. Therefore, you cannot get much of a tan through glass windows. Sunglasses are labeled to indicate which UV protection standards they meet in shielding the eyes from this potentially harmful radiation. Certain types of high-tech glass (called "photogray" glass) darken when exposed to UV. These materials are used to create "transition" sunglasses that darken when

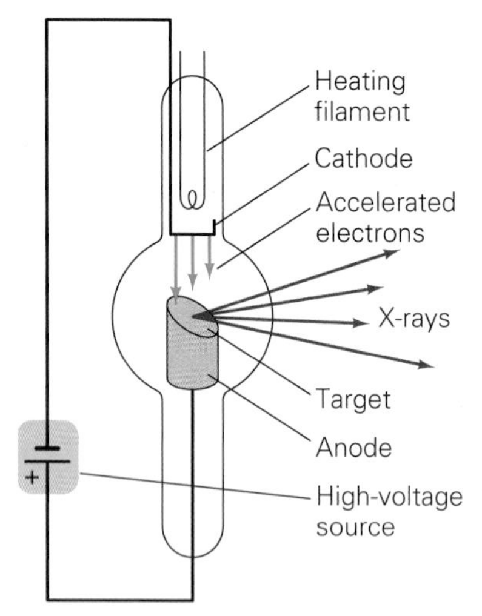

▲ **FIGURE 20.24** The X-ray tube Electrons accelerated through a large voltage strike a target electrode. There they slow down and interact with the electrons of the target material. Energy is emitted in the form of X-rays during this "braking" (deceleration) process.

exposed to sunlight. Of course, these sunglasses aren't very useful while driving a car. (Why?) Welders wear special glass goggles to protect their eyes from large amounts of UV produced by the arcs of welding torches. Similarly, it is important to shield your eyes from sunlamps or snow-covered surfaces. The ultraviolet component of sunlight reflected from snow-covered surfaces can produce snowblindness in unprotected eyes.

X-Rays Beyond the ultraviolet region of the electromagnetic spectrum is the important X-ray region. We are familiar with X-rays primarily through medical applications. X-rays were discovered accidentally in 1895 by the German physicist Wilhelm Roentgen (1845–1923) when he noted the glow of a piece of fluorescent paper, evidently caused by some mysterious radiation coming from a cathode-ray tube. Because of the apparent mystery involved, this radiation was named *x-radiation* or *X-rays* for short.

The basic elements of an X-ray tube are shown in ◂Fig. 20.24. An accelerating voltage, typically several thousand volts, is applied across the electrodes in a sealed, evacuated tube. Electrons emitted from the heated negative electrode (cathode) are accelerated toward the positive electrode (anode). When they strike the anode, some of their kinetic-energy loss is converted to electromagnetic energy in the form of X-rays.

A similar process takes place in color TV picture tubes, which use high voltages and electron beams. When the high-speed electrons hit the screen, they can emit X-rays. Fortunately, all modern televisions have the shielding necessary to protect viewers from exposure to this radiation. In the early days of color television, this was not always the case—hence the warning that came with the set: "Do not sit too close to the screen."

As you will learn in Chapter 27, the energy carried by electromagnetic radiation depends on its frequency. High-frequency X-rays have very high energies and can cause cancer, skin burns, and other harmful effects. However, at low intensities, X-rays can be used with relative safety to view the internal structure of the human body and other opaque objects.* X-rays can pass through materials that are opaque to other types of radiation. The denser the material, the greater is its absorption of X-rays and the less intense the transmitted radiation will be. For example, as X-rays pass through the human body, many more are absorbed or scattered by bone than by tissue. If the transmitted radiation is directed onto a photographic plate or film, the exposed areas show variations in intensity—a picture of internal structures.

The combination of the computer with modern X-ray machines permits the formation of three-dimensional images by means of a technique called *computerized tomography*, or *CT* (▾Fig. 20.25).

Gamma Rays The electromagnetic waves of the uppermost frequency range of the known electromagnetic spectrum are called *gamma rays* (γ-rays). This high-frequency radiation is produced in nuclear reactions, in particle accelerators, and in certain types of nuclear decay (radioactivity). Gamma rays will be discussed in more detail in Chapter 29.

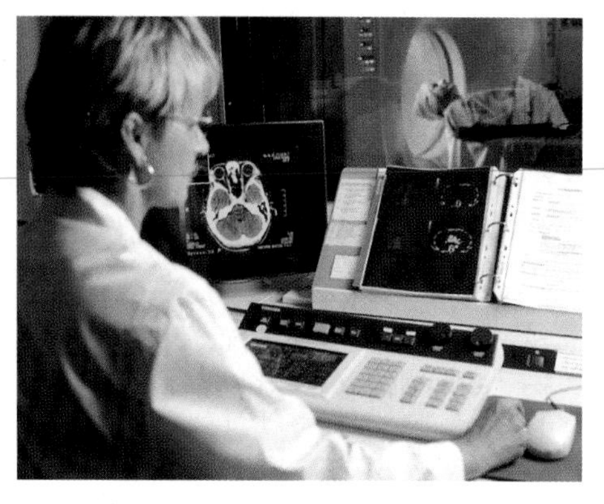

(a)

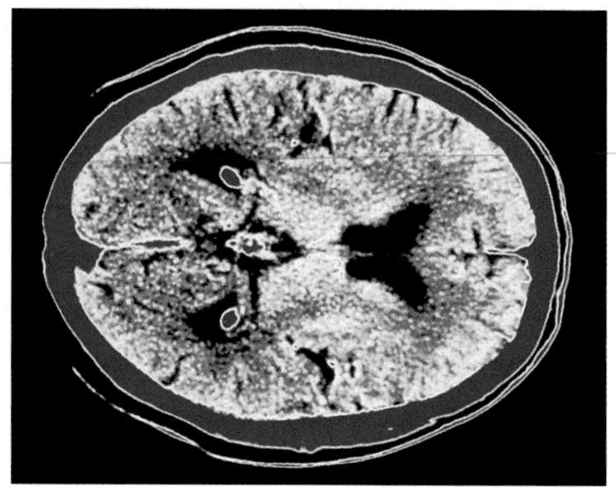

(b)

▸ **FIGURE 20.25** CT scan In an ordinary X-ray image, the entire thickness of the body is projected onto the film and internal structures often overlap, making details hard to distinguish. In CT—computerized tomography (from the Greek *tomo*, meaning "slice," and *graph*, meaning "picture")—X-ray beams scan a slice of the body. **(a)** The transmitted radiation is recorded by a series of detectors and processed by a computer. Using information from multiple slices, the computer constructs a three-dimensional image. Any single slice can be displayed for further study. **(b)** CT image of a brain with a benign tumor.

*Many health scientists believe that there is no safe "threshold" level for X-rays or other energetic radiation—that is, no level of exposure that is completely risk-free—and that some of the dangerous effects are cumulative over a lifetime. People should therefore avoid unnecessary medical X-rays or any other unwarranted exposure to "hard" radiation (Chapter 29). However, when properly used, X-rays can be an extremely useful diagnostic tool capable of saving lives.

Chapter Review

- **Magnetic flux (Φ)** is a measure of the number of magnetic field lines that pass through an area. For a single wire loop of area A, it is defined as

$$\Phi = BA\cos\theta \qquad (20.1)$$

where B is the magnetic field strength (assumed constant), A is the loop area, and θ is the angle between the direction of the magnetic field and the normal to the area's plane.

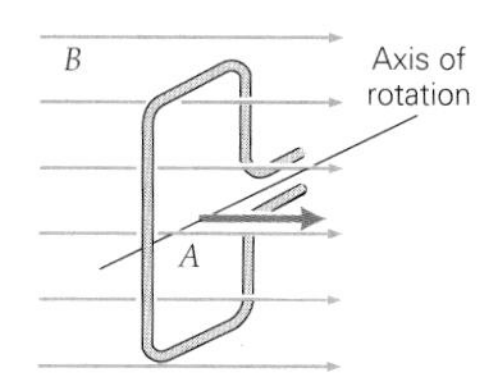

- **Faraday's law of induction** relates the induced emf in a loop (or coil composed of N loops in series) to the time rate of change of the magnetic flux through that loop (or coil).

$$\mathscr{E} = -N\frac{\Delta\Phi}{\Delta t} \qquad (20.2)$$

where $\Delta\Phi$ is the change in flux through *one loop* and there are N total loops.

Increasing magnitude of B

- **Lenz's law** states that when a change in magnetic flux induces an emf in a coil, loop, or circuit, the resulting, or induced, current direction is such as to create a magnetic field to oppose the change in flux.

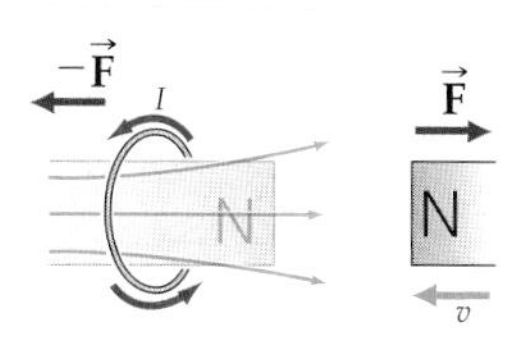

- An **ac generator** converts mechanical energy into electrical energy. The generator's emf as a function of time is

$$\mathscr{E} = \mathscr{E}_o \sin\omega t \qquad (20.4)$$

where $\mathscr{E}_o$ is the maximum emf.

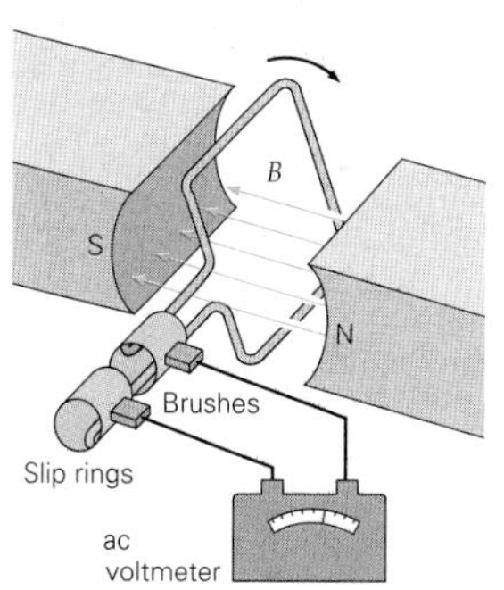

- A **transformer** is a device that changes the voltage supplied to it by means of induction. The voltage applied to the input, or primary (p), side of the transformer is changed into the output, or secondary (s), voltage. The current and voltage relationships for a transformer are

$$V_s = \left(\frac{N_s}{N_p}\right)V_p \qquad (20.10a)$$

$$I_s = \left(\frac{N_p}{N_s}\right)I_p \qquad (20.10b)$$

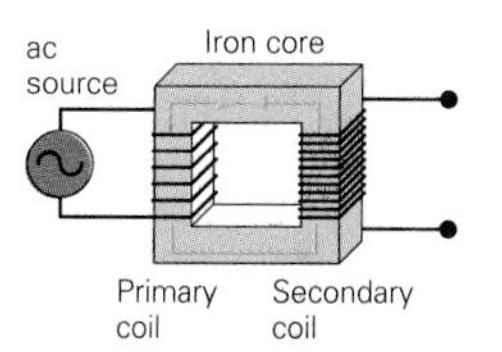

- An **electromagnetic wave** (light) consists of time-varying electric and magnetic fields that propagate at a speed of c (3.00×10^8 m/s) in a vacuum. The different types of electromagnetic radiation (such as UV, radio waves, and visible light) differ in frequency and wavelength.

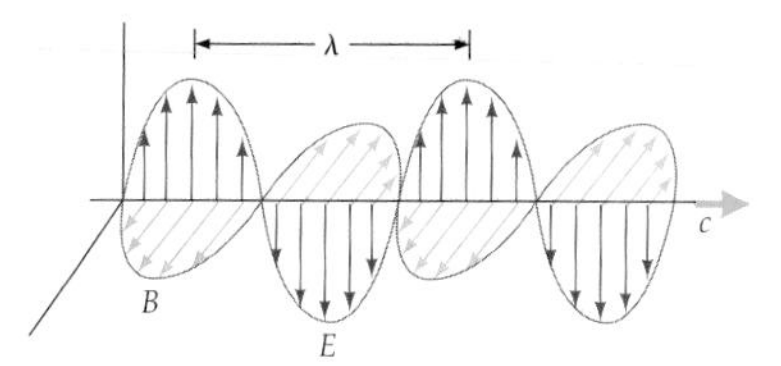

Exercises

MC = *Multiple Choice Question,* **CQ** = *Conceptual Question, and* **IE** = *Integrated Exercise. Throughout the text, many exercise sections will include "paired" exercises. These exercise pairs, identified with* **red numbers**, *are intended to assist you in problem solving and learning. In a pair, the first exercise (even numbered) is worked out in the Study Guide so that you can consult it should you need assistance in solving it. The second exercise (odd numbered) is similar in nature, and its answer is given at the back of the book.*

20.1 Induced emf: Faraday's Law and Lenz's Law

1. **MC** A unit of magnetic flux is (a) Wb, (b) $T \cdot m^2$, (c) $T \cdot m/A$, or (d) both a and b.

2. **MC** The magnetic flux through a loop can change due to a change in (a) the area of the coil, (b) the strength of the magnetic field, (c) the orientation of the loop, or (d) all of the preceding.

3. **MC** For a current to be induced in a wire loop, (a) there must be a large magnetic flux through the loop, (b) the loop's plane must be parallel to the magnetic field, (c) the loop's plane must be perpendicular to the magnetic field, or (d) the magnetic flux through the loop must vary with time.

4. **MC** Identical single loops A and B are oriented so they initially have the maximum amount of flux in a magnetic field. Loop A is then quickly rotated so its normal is perpendicular to the magnetic field, and in the same time, B is rotated so its normal makes an angle of only 45° with the field. How do their induced emfs compare? (a) They are the same (b) A's is larger than B's (c) B's is larger than A's or (d) you can't tell the relative emf magnitudes from the data given.

5. **MC** Identical single loops A and B are oriented so they have the maximum amount of flux when placed in a magnetic field. Both loops maintain their orientation relative to the field, but in the same amount of time A is moved to a region of stronger field, while B is moved to a region of weaker field. How do their induced emfs compare? (a) They are the same (b) A's is larger than B's (c) B's is larger than A's or (d) you can't tell the relative emf magnitudes from the data given.

6. **CQ** A bar magnet is dropped through a coil of wire as shown in ▼Fig. 20.26. (a) Describe what is observed on the galvanometer by sketching a graph of induced emf versus t. (b) Does the magnet fall freely? Explain.

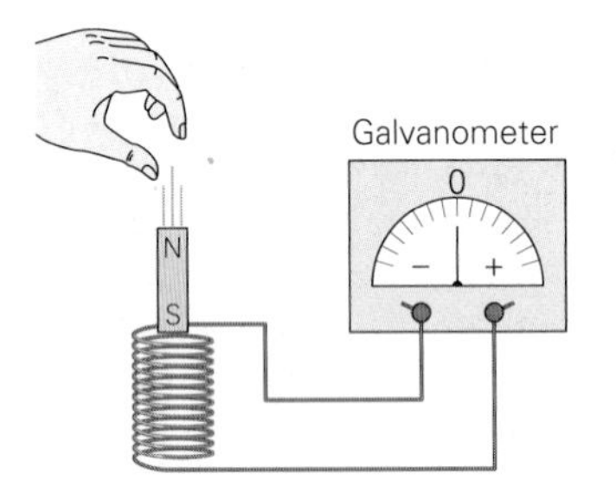

◀ **FIGURE 20.26** **A time-varying magnetic field** What will the galvanometer measure? See Exercise 6.

7. **CQ** In Fig. 20.1b, what would be the direction of the induced current in the loop if the south pole of the magnet were approaching instead of the north pole?

8. **CQ** In Fig. 20.7a, how would you move the coil so to prevent any current from being induced in it? Explain.

9. **CQ** Does the induced emf in a closed loop depend on the value of the magnetic flux in the loop? Explain.

10. **CQ** Two identical strong magnets are dropped simultaneously by two students into two vertical tubes of the same dimensions (▼Fig. 20.27). One tube is made of copper, and the other is made of plastic. From which tube will the magnet emerge first? Why?

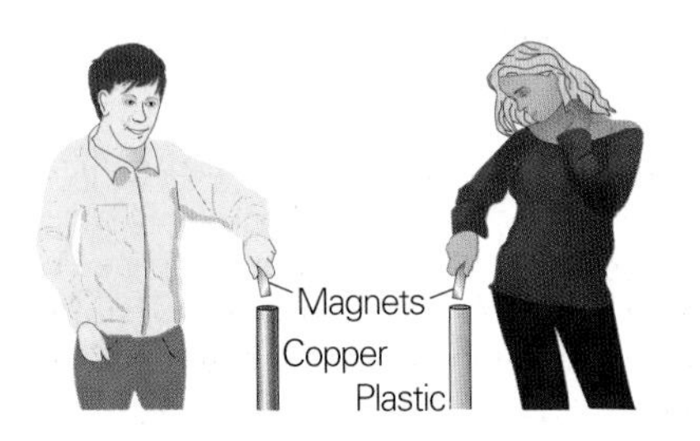

◀ **FIGURE 20.27** **Free fall?** See Exercise 10.

11. **CQ** A basic telephone has both a speaker-transmitter and a receiver (▼Fig. 20.28). Until the advent of digital phones in the 1990s, the transmitter had a diaphragm coupled to a carbon chamber (called the *button*), which contained loosely packed granules of carbon. As the diaphragm vibrated because of incident sound waves, the pressure on the granules varied, causing them to be more or less closely packed. As a result, the resistance of the button changed. The receiver converted these electrical impulses to sound. Applying the principles of electricity and magnetism that you have learned, explain the basic operation of this type of telephone.

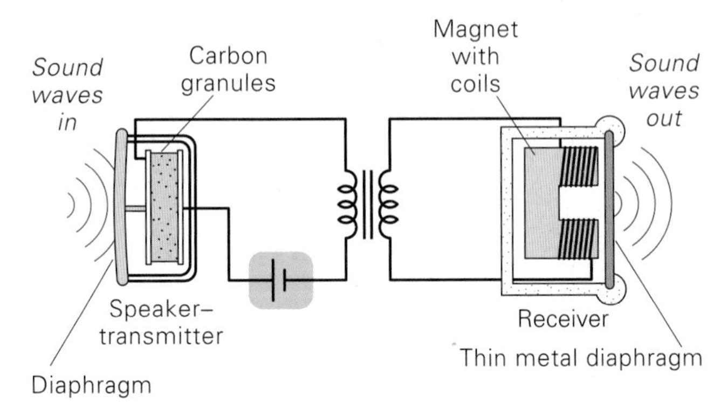

▲ **FIGURE 20.28 Telephone operation** See Exercise 11.

12. ● A circular loop with an area of 0.015 m^2 is in a uniform magnetic field of 0.30 T. What is the flux through the loop's plane if it is (a) parallel to the field, (b) at an angle of 37° to the field, and (c) perpendicular to the field?

13. ● A circular loop (radius of 20 cm) is in a uniform magnetic field of 0.15 T. What angle(s) between the normal to the plane of the loop and the field would result in a flux with a magnitude of $1.4 \times 10^{-2}\ T \cdot m^2$?

14. ● The plane of a conductive loop with an area of 0.020 m^2 is perpendicular to a uniform magnetic field of 0.30 T. If the field drops to zero in 0.0045 s, what is the magnitude of the average emf induced in the loop?

15. ● A loop in the form of a right triangle with one side of 40.0 cm and a hypotenuse of 50.0 cm lies in a plane perpendicular to a uniform magnetic field of 550 mT. What is the flux through the loop?

16. ● A square coil of wire with 10 turns is in a magnetic field of 0.25 T. The total flux through the coil is $0.50\ T \cdot m^2$. Find the area of one turn if the field (a) is perpendicular to the plane of the coil and (b) makes an angle of 60° with the plane of the coil.

17. ●● An ideal solenoid with a current of 1.5 A has a radius of 3.0 cm and a turn density of 250 turns/m. What is the magnetic flux (due to its own field) through only one of its loops at its center?

18. ●● A uniform magnetic field is at right angles to the plane of a wire loop. If the field decreases by 0.20 T in 1.0×10^{-3} s and the magnitude of the average emf induced in the loop is 80 V, what is the area of the loop?

19. ●● A square loop of wire with sides of length 40 cm is in a uniform magnetic field perpendicular to its area. If the field's strength is initially 100 mT and it decays to zero in 0.010 s, what is the magnitude of the average emf induced in the loop?

20. ●● The magnetic flux through one loop of wire is reduced from 0.35 Wb to 0.15 Wb in 0.20 s. The average induced current in the coil is 10 A. Find the resistance of the wire.

21. ●● When the magnetic flux through a single loop of wire increases by $30\ \mathrm{T \cdot m^2}$, an average current of 40 A is induced in the wire. Assuming that the wire has a resistance of $2.5\ \Omega$, over what period of time did the flux increase?

22. ●● In 0.20 s, a coil of wire with 50 loops experiences an average induced emf of 9.0 V, due to a changing magnetic field perpendicular to the plane of the coil. The radius of the coil is 10 cm, and the initial strength of the magnetic field is 1.5 T. Assuming that the strength of the field decreased with time, what is the final strength of the field?

23. IE ●● A single strand of wire of adjustable length is wound around the circumference of a round balloon. A uniform magnetic field is perpendicular to the plane of the loop (▾Fig. 20.29). (a) If the balloon is inflated, in which direction is the induced current, looking down from above: (1) counterclockwise, (2) clockwise, or (3) there is no induced current? (b) If the magnitude of the magnetic field is 0.15 T and the wire's diameter increases from 20 cm to 40 cm in 0.040 s, what is the magnitude of the average value of the emf induced in the loop?

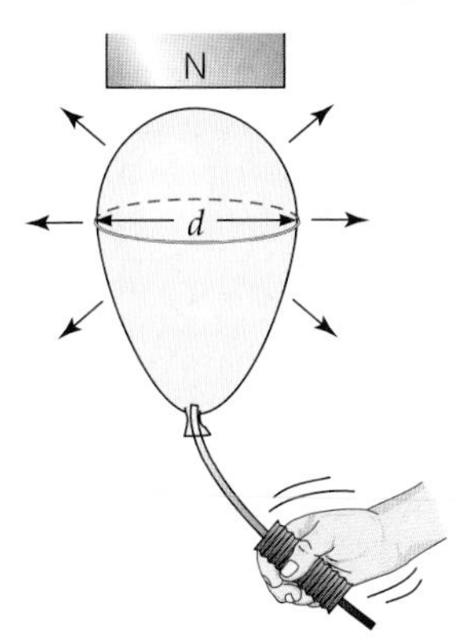

◂ FIGURE 20.29 Pumping energy See Exercise 23.

24. ●● The magnetic field perpendicular to the plane of a wire loop with an area of $0.10\ \mathrm{m^2}$ changes with time as shown in ▸Fig. 20.30. What is the magnitude of the average emf induced in the loop for each segment of the graph (for example, from 0 to 2.0 ms)?

25. IE ●● A boy is traveling due north at a constant speed while carrying a metal rod. The rod's length is oriented in the east–west direction and is parallel to the ground. (a) There will be no induced emf when the rod is (1) at the equator, (2) near the Earth's magnetic poles, (3) somewhere between the equator and the poles. Why? (b) Assume that the Earth's magnetic field is 1.0×10^{-4} T near the North Pole and 1.0×10^{-5} T near the Equator. If the boy runs with a speed of 5.0 m/s northward near each location, and the rod is 1.0 m long, calculate the induced emf in the rod in each location.

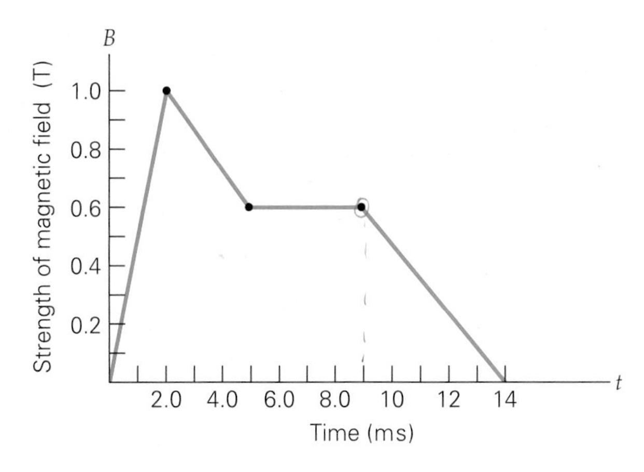

▴ FIGURE 20.30 Magnetic field versus time See Exercise 24.

26. ●● A metal airplane with a wingspan of 30 m flies horizontally at a constant speed of 320 km/h in a region where the vertical component of the Earth's magnetic field is 5.0×10^{-5} T. What is the induced emf between the tips of the plane's wings?

27. ●● Suppose that the metal rod in Fig. 20.11 is 20 cm long and is moving at a speed of 10 m/s in a magnetic field of 0.30 T and that the metal frame is covered with an insulating material. Find (a) the magnitude of the induced emf across the rod and (b) the current in the rod.

28. ●●● The flux through a loop of wire changes uniformly from +40 Wb to −20 Wb in 1.5 ms. (a) What is the significance of the negative flux? (b) What is the average induced emf in the loop? (c) If you wanted to double the average induced emf by changing only the time, what would the new time interval be? (d) If you wanted to double the average induced emf by changing only the final flux value, what would it be?

29. ●●● A coil of wire with 10 turns and an area of $0.055\ \mathrm{m^2}$ is placed in a magnetic field of 1.8 T and oriented so that the area is perpendicular to the field. The coil is then flipped by 90° in 0.25 s and ends up with the area parallel to the field (▾Fig. 20.31). What is the magnitude of the average emf induced in the coil?

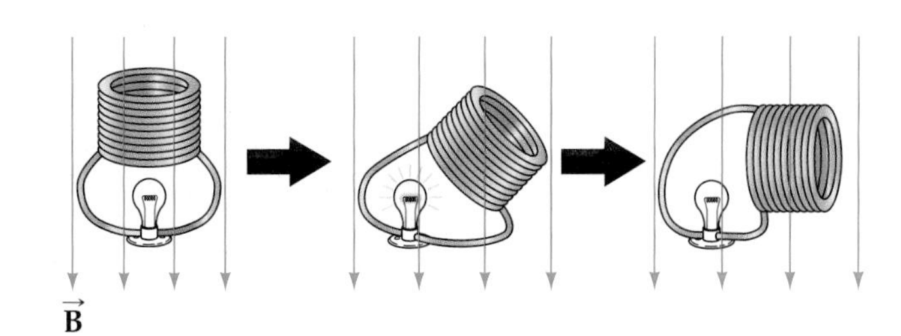

▴ FIGURE 20.31 Flipping the coil See Exercises 29 and 30.

30. IE ●●● In Fig 20.31, the coil is flipped by 180° in the same time interval as in Exercise 29. (a) How does the magnitude of the average emf compare with that in Exercise 29, where the coil was flipped only 90°: (1) higher, (2) the same, or (3) lower? Why? (b) What is the magnitude of the average emf in this case?

31. ●●● A uniform magnetic field of 0.50 T penetrates a double-incline block as shown in ▼Fig. 20.32. (a) Determine the magnetic flux through each inclined surface of the block. (b) Determine the flux through the vertical back surface of the block. (c) Determine the flux through the flat horizontal surface of the block. (d) What is the total flux through all the outside surfaces? Explain the meaning of your answer.

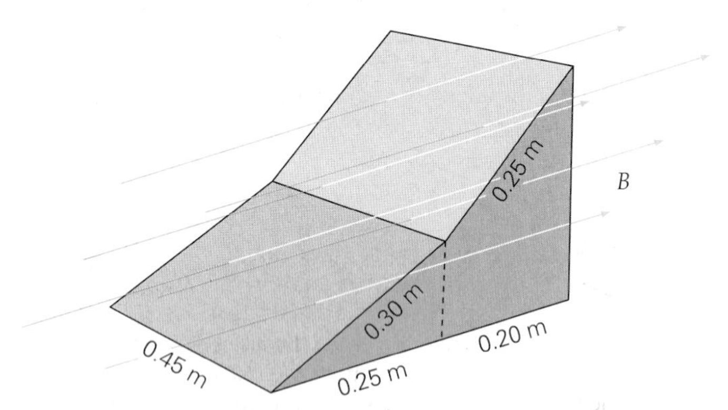

▲ **FIGURE 20.32 Magnetic flux** See Exercise 31. (Not drawn to scale.)

20.2 Electric Generators and Back emf

32. **MC** Doing nothing but increasing the coil area in an ac generator would result in (a) an increase in the frequency of rotation, (b) a decrease in the maximum induced emf, (c) an increase in the maximum induced emf.

33. **MC** The back emf of a motor depends on (a) the input voltage, (b) the input current, (c) the armature's rotational speed, or (d) none of the preceding.

34. **CQ** What is the orientation of the armature loop in a simple ac generator when the value of (a) the emf is a maximum and (b) the magnetic flux is a maximum? Explain why the maximum emf does not occur when the flux is a maximum.

35. **CQ** A student has a bright idea for a generator: For the arrangement shown in ▼Fig. 20.33, the magnet is pulled down and released. With a highly elastic spring, the student thinks that there should be a relatively continuous electrical output. What is wrong with this idea?

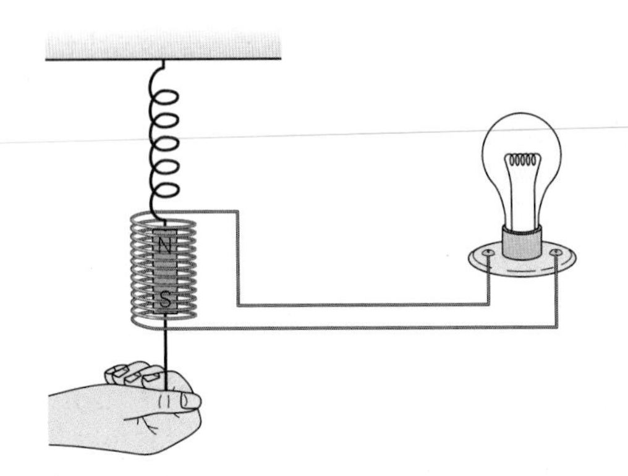

▲ **FIGURE 20.33 Inventive genius?** See Exercise 35.

36. **CQ** In a dc motor, if the armature is jammed or turns very slowly under a heavy load, the coils in the motor may burn out. Why?

37. **CQ** If you wanted to make a more compact ac generator by reducing the area of the coils, how would could you compensate with other factors in order to maintain the same output as before?

38. ● A hospital emergency room ac generator operates at a rotation frequency of 60 Hz. If the output voltage is a maximum (in magnitude) at $t = 0$, when is it next (a) a maximum (in magnitude), (b) zero, and (c) at its initial value?

39. ● A student makes a simple generator by using a single square loop 10 cm on each side. The loop is then rotated at a frequency of 60 Hz in a magnetic field of 0.015 T. (a) What is the maximum emf output? (b) What would be the maximum emf output if 10 such loops were used instead?

40. ●● A simple ac generator consists of a coil with 10 turns (each turn has an area of 50 cm^2). The coil rotates in a uniform magnetic field of 350 mT with a frequency of 60 Hz. (a) Write an expression in the form of Eq. 20.5 for the generator's emf variation with time. (b) Compute the maximum emf.

41. **IE** ●● A 60-Hz ac voltage source has a maximum voltage of 120 V. A student wants to determine the voltage 1/180 s after it is zero. (a) How many possible voltages are there: (1) one, (2) two, or (3) three? Why? (b) Determine all possible voltages.

42. ●● An ac generator with a maximum emf of 400 V is to be constructed from wire loops (radius 0.15 m). It will operate at a frequency of 60 Hz and use a magnetic field of 200 mT. How many loops will be needed?

43. ●● An ac generator operates at a rotational frequency of 60 Hz and has a maximum emf of 100 V. Assume that it has zero emf at startup. What is the instantaneous emf (a) 1/240 s after startup and (b) 1/120 s after the emf passes through zero as it begins to reverse polarity?

44. ●● The armature of a simple ac generator has 20 circular loops of wire, each with a radius of 10 cm. It is rotated with a frequency of 60 Hz in a uniform magnetic field of 800 mT. What is the maximum emf induced in the loops, and how often is this value attained?

45. ●● The armature of an ac generator has 100 turns. Each turn is a rectangular loop measuring 8.0 cm by 12 cm. The generator has a sinusoidal voltage output with an amplitude of 24 V. If the magnetic field of the generator is 250 mT, with what frequency does the armature turn?

46. **IE** ●● (a) To increase the output of an ac generator, a student has the choice of doubling *either* the generator's magnetic field *or* its frequency. To maximize the increase in emf output, (1) he should double the magnetic field, (2) he should double the frequency, (3) it does not matter which one he doubles. Explain. (b) Two students display their ac generators at a science fair. The generator made by student A has a loop area of 100 cm^2 rotating in a magnetic field of 20 mT at 60 Hz. The one made by student B has a loop area of 75 cm^2 rotating in a magnetic field of 200 mT at 120 Hz. Which one generates the largest maximum emf? Justify your answer mathematically.

47. IE ●● A motor has a resistance of 2.50 Ω and is connected to a 110-V line. (a) Is the operating current of the motor (1) higher than 44 A, (2) 44 A, or (3) lower than 44 A? Why? (b) If the back emf of the motor at operating speed is 100 V, what is its operating current?

48. ●●● The starter motor in an automobile has a resistance of 0.40 Ω in its armature windings. The motor operates on 12 V and has a back emf of 10 V when running at normal operating speed. How much current does the motor draw (a) when running at its operating speed, (b) when the motor is running at half its final rotational speed, and (c) when starting up?

49. ●●● A 240-V dc motor has an armature whose resistance is 1.50 Ω. When running at its operating speed, it draws a current of 16.0 A. (a) What is the back emf of the motor when it is operating normally? (b) What is the starting current? (Assume that there is no additional resistance in the circuit.) (c) What series resistance would be required to limit the starting current to 25 A?

20.3 Transformers and Power Transmission

50. MC A transformer in the electrical energy delivery system just before your house has (a) more windings in the primary coil, (b) more windings in the secondary coil, or (c) the same number of windings in the primary and secondary coils.

51. MC The output power delivered by a realistic step-down transformer is (a) greater than the input power, (b) less than the input power, or (c) the same as the input power.

52. CQ Explain why electric energy delivery systems operate at such high voltages when such voltages can be dangerous.

53. CQ In your automotive workshop emergency repair, you need a step-down transformer, but have only step-up transformers on the shelves. Can you use a step-up transformer as a step-down one? If so, explain how you would wire it.

54. IE ● The secondary coil of an ideal transformer has 450 turns, and the primary coil has 75 turns. (a) Is this transformer a (1) step-up or (2) step-down transformer? Why? (b) What is the ratio of the current in the primary coil to the current in the secondary coil? (c) What is the ratio of the voltage across the primary coil to the voltage in the secondary coil?

55. ● An ideal transformer steps 8.0 V up to 2000 V, and the 4000-turn secondary coil carries 2.0 A. (a) Find the number of turns in the primary coil. (b) Find the current in the primary coil.

56. ● The primary coil of an ideal transformer has 720 turns, and the secondary coil has 180 turns. If the primary coil carries 15 A at a voltage of 120 V, what are (a) the voltage and (b) the output current of the secondary coil?

57. ● The transformer in the power supply for a computer's 250-MB Zip drive changes a 120-V input to a 5.0-V output. Find the ratio of the number of turns in the primary coil to the number of turns in the secondary coil.

58. ● The primary coil of an ideal transformer is connected to a 120-V source and draws 10 A. The secondary coil has 800 turns and a current of 4.0 A. (a) What is the voltage across the secondary coil? (b) How many turns are in the primary coil?

59. ● An ideal transformer has 840 turns in its primary coil and 120 turns in its secondary coil. If the primary coil draws 2.50 A at 110 V, what are (a) the current and (b) the output voltage of the secondary coil?

60. ●● The efficiency e of a transformer is defined as the ratio of the power output to the power input:

$$e = \frac{P_s}{P_p} = \frac{I_s V_s}{I_p V_p}.$$

(a) Show that in terms of the ratios of currents and voltages given in Eq. 20.10 for an ideal transformer, an efficiency of 100% is obtained for that situation. (b) Suppose a step-up transformer increased the line voltage from 120 to 240 V, while at the same time the output current was reduced to 5.0 A from 12 A. What is the transformer's efficiency? Is it ideal?

61. IE ●● The specifications of a transformer used with a small appliance read as follows: Input, 120 V, 6.0 W; Output, 9.0 V, 300 mA. (a) Is this transformer (1) an ideal or (2) a nonideal transformer? Why? (b) What is its efficiency? (See Exercise 60.)

62. ●● A circuit component operates at 20 V and 0.50 A. A transformer with 300 turns in its primary coil is used to convert 120-V household electricity to the proper voltage. (a) How many turns must the secondary coil have? (b) How much current is in the primary coil?

63. ●● A transformer in a door chime steps down the voltage from 120 V to 6.0 V and supplies a current of 0.50 A to the chime mechanism. (a) What is the turn ratio of the transformer? (b) What is the current input to the transformer?

64. ●● An ac generator supplies 20 A at 440 V to a 10 000-V power line. If the step-up transformer has 150 turns in its primary coil, how many turns are in the secondary coil?

65. ●● The electricity supplied in Exercise 64 is transmitted over a line 80.0 km long with a resistance of 0.80 Ω/km. (a) How many kilowatt-hours are saved in 5.00 h by stepping up the voltage? (b) At $0.10/kWh, how much of a savings (to the nearest $10) is this to the consumer in a 30-day month, assuming that the energy is supplied continuously?

66. ●● At an area substation, the power-line voltage is stepped down from 100 000 V to 20 000 V. If 10 MW of power is delivered to the 20 000-V circuit, what are the primary and secondary currents in the transformer?

67. ●● A voltage of 200 000 V in a transmission line is reduced to 100 000 V at an area substation, to 7200 V at a distributing substation, and finally to 240 V at a utility pole outside a house. (a) What turn ratio N_s/N_p is required for each reduction step? (b) By what factor is the transmission-line current stepped up in each voltage step-down? (c) What is the overall factor by which the current is stepped up from transmission line to utility pole?

68. ●● A plant produces electric energy at 50 A and 20 kV. The energy is transmitted 25 km over transmission lines whose resistance is 1.2 Ω/km. (a) What is the power loss in the lines if the energy is transmitted at 20 kV? (b) What should be the output voltage of the generator to decrease the power loss by a factor of 15?

69. ●● Electrical power is transmitted through a power line 175 km long with a resistance of 1.2 Ω/km. The generator's output is 50 A at its operating voltage of 440 V. This voltage has a single step-up for transmission at 44 kV. (a) How much power is lost as joule heat during the transmission? (b) What must be the turn ratio of a transformer at the delivery point in order to provide an output voltage of 220 V? (Neglect the voltage drop in the line.)

20.4 Electromagnetic Waves

70. **MC** Relative to the blue end of the visible spectrum, the yellow and green regions have (a) higher frequencies, (b) longer wavelengths, (c) shorter wavelengths, or (d) both a and c.

71. **MC** Which of the following electromagnetic waves has the lowest frequency? (a) UV (b) IR (c) X-ray or (d) microwave

72. **MC** Which of the following electromagnetic waves travels fastest in a vacuum? (a) green light (b) infrared light (c) gamma rays (d) radiowaves or (e) they all have the same speed

73. **MC** If you doubled the frequency of a blue light source, what kind of light would it then put out? (a) red (b) blue (c) violet or (d) UV (e) X-ray

74. CQ An antenna is connected to a car battery. Will the antenna emit electromagnetic radiation? Why or why not? Explain.

75. CQ On a cloudy summer day, you work outside and feel cool, yet that evening you find that you are sunburned. Explain how this is possible.

76. CQ Radiation exerts pressure on surfaces on which it falls (*radiation pressure*). Will this pressure be greater on a shiny surface or a dark surface? Will it be greater using a bright source, or one of the same color but fainter? Explain both of your answers.

77. CQ Radar operates at wavelengths on the order of centimeters, whereas FM radio operates at wavelengths of several meters. How do radar frequencies compare with the frequencies of the FM band on a radio? How do their speeds in a vacuum compare?

78. ● Find the frequencies of electromagnetic waves with wavelengths of (a) 3.0 cm, (b) 650 nm, and (c) 1.2 fm. (d) Classify the type of light in each case.

79. ● In a small town there are only two AM radio stations, one at 920 kHz and one at 1280 kHz. What are the wavelengths of the radio waves transmitted by each station?

80. ● A meteorologist in a TV station is using radar to determine the distance to a cloud. He notes that a time of 0.24 ms elapses between the sending and the return of a radar pulse. How far away is the cloud?

81. ● How long does a laser beam take to travel from the Earth to a reflector on the Moon and back? Take the distance from the Earth to the Moon to be 2.4×10^5 mi. (This experiment was done when the *Apollo* flights of the early 1970s left laser reflectors on the lunar surface.)

82. ●● Orange light has a wavelength of 600 nm, and green light has a wavelength of 510 nm. (a) What is the difference in frequency between the two types of light? (b) If you doubled the wavelength of both of these, what type of light would they become?

83. ●● A certain type of radio antenna is called a *quarter-wavelength antenna*, because its length is equal to one quarter of the wavelength to be received. If you were going to make such antennae for the AM and FM radio bands by using the middle frequencies of each band, what lengths of wire would you use?

84. IE ●●● Microwave ovens can have cold spots and hot spots, due to standing electromagnetic waves, analogous to standing wave nodes and antinodes in strings (▼Fig. 20.34). (a) The longer the distance between the cold spots, (1) the higher the frequency, (2) the lower the frequency, (3) the frequency is independent of this distance. Why? (b) In your microwave the cold spots (nodes) are approximately every 5.0 cm, and your neighbor's microwave produces them every 6.0 cm. Which microwave operates at a higher frequency and by how much?

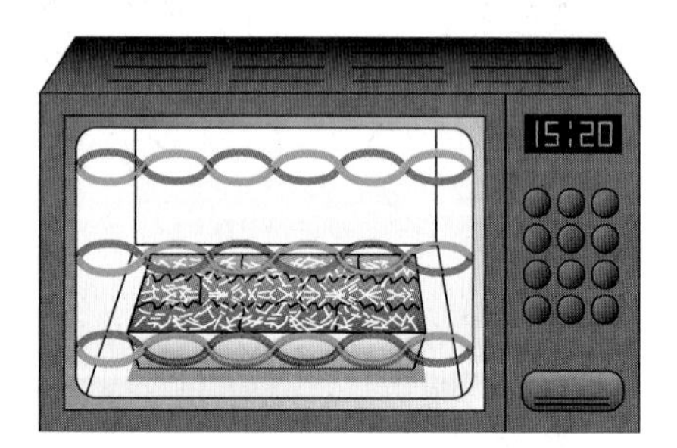

▲ **FIGURE 20.34 Cold spots?** See Exercise 84.

Comprehensive Exercises

85. IE In ▼Fig. 20.35, a metal bar of length L moves in a region of constant magnetic field. That field is directed into the page. (a) The direction of the induced current through the resistor is (1) up, (2) down, (3) there is no current. Why? (b) If the magnitude of the magnetic field is 250 mT, what is the current?

86. Suppose a modern power plant has a 1.00-GW electric power output at 500 V. It is stepped up to a transmission voltage of 750 kV in a series of five identical transformers. (a) What is the output current at the plant? (b) What is the turn ratio in each of the transformers? (c) What is the current in the high voltage delivery lines?

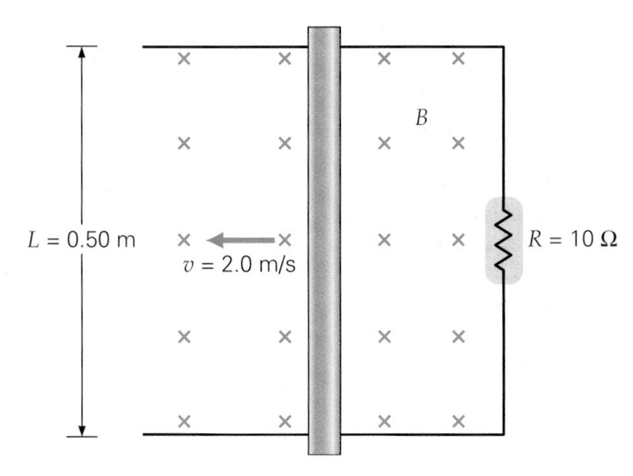

▲ **FIGURE 20.35 Motional emf** See Exercise 85.

87. A transformer is used by a European traveler while she is visiting the United States. Primarily she uses it to run a 1200-watt hair dryer/blower she brought with her. When plugged in to her hotel room outlet in Los Angeles, she notices it runs *exactly* as it does at home. The input voltage and current are 120 V and 11.0 A, respectively. (a) Is this an ideal transformer? Explain how you came to your conclusion. (b) If it is not an ideal transformer, what is its efficiency?

88. A solenoid of length 20.0 cm is made of 5000 circular coils. It carries a steady current of 10.0 A. Near its center is placed a very flat and small coil made of 100 circular loops, each with a radius of 3.00 mm. This small coil is oriented so that its area receives the maximum magnetic flux. A switch is opened in the solenoid circuit and its current drops to zero in 15.0 ms. (a) What was the initial magnetic flux through the inner coil? (b) Determine the average induced emf in the small coil during the 15.0 ms. (c) If you look along the long axis of the solenoid so the initial 10.0 A current is clockwise, determine the direction of the induced current in the small inner coil during the time the current drops to zero.

89. IE A flat coil of copper wire consists of 100 loops and has a total resistance of 0.500 Ω. The coil diameter is 4.00 cm and it is in a uniform magnetic field pointing away from you (into the page). Its orientation is in the plane of the page. It is then pulled to the right and completely out of the field. (a) What is the direction of the induced current in the coil: (a) clockwise, (b) counterclockwise, or (c) there is no induced current? (b) During the time the coil leaves the field, an average induced current of 10.0 mA is measured. What is the average induced emf in the coil? (c) If the field strength is 3.50 mT, how much time did it take to pull the coil out?

90. The transformer on a utility pole steps the voltage down from 20 000 V to 220 V for use in a college science building. During the day, the transformer delivers electric energy at the rate of 6.60 kW. (a) Assuming the transformer to be ideal, during that time what are the primary and secondary currents in the transformer? (b) If the transformer is only 95% efficient (but still delivers energy at a rate of 6.60 kW at 220 V), how does its input current compare to the ideal case? (c) At what rate is heat lost in the nonideal transformer?

91. Suppose you wanted to build an electric generator using the Earth's magnetic field. Assume it has a strength of 0.040 mT at your location. Your generator design calls for a coil of 1000 windings rotated at exactly 60 Hz. The coil is oriented so that the normal to the area lines up with the Earth's field at the end of each cycle. What must the coil diameter be to generate a maximum voltage of 170 V (required in order to average 120 V)? Does this seem like a practical way to generate electric energy?

92. A radio signal is sent to a deep space probe traveling in the plane of the solar system. 3.5 days later, Earth receives a response. Assuming the probe computers took 4.5 hours to process the signal instructions and to send out the return, is the probe within the Solar System? (Assume a Solar System radius of about 40 times the Earth–Sun distance.)

93. A 100-loop coil of wire with a diameter of 2.50 cm is oriented in a 0.250-T constant magnetic field so that it initially has no magnetic flux. In 0.115 sec, the coil is turned so that its normal makes an angle of 45° with the field direction. If a 4.75-mA average current is induced in the coil during this rotation, what is the coil's resistance?

PHYSLET® The following Physlet Physics Problems can be used with this chapter.
29.1, 29.2, 29.5, 29.6, 29.7, 29.8

CHAPTER

21 AC CIRCUITS

21.1 Resistance in an AC Circuit 687

21.2 Capacitive Reactance 689

21.3 Inductive Reactance 691

21.4 Impedance: RLC Circuits 693

21.5 Circuit Resonance 697

PHYSICS FACTS

- Under ac conditions (alternating voltage direction), a capacitor, even with the gap between the plates, allows current in the circuit during the charging and discharging stages. Under dc conditions (steady voltage across the plates), there is no current.
- Under dc voltages, an inductor offers no impedance to the flow of charge and thus can readily conduct current. However, under ac conditions, an inductor impedes the change in current by producing a reverse emf in accordance with Faraday's law of induction.
- A capacitor, inductor, and resistor connected in series to an ac power supply is analogous to a mechanical damped, driven spring–mass system. Driven at its natural frequency, the circuit "resonates," that is, exhibits a current maximum, just as the mechanical system has its largest amplitude under the same condition. The capacitor stores electric potential energy in analogy to the spring's elastic potential energy. The inductor stores magnetic energy (associated with moving charges), in analogy to the mass on the spring having kinetic (moving) energy. The resistor will dissipate the system's energy, as air resistance might do to the mechanical system.

Direct-current (dc) circuits have many uses, but the nuclear reactor control panel in the photo operates many devices that use alternating current (ac). The electric power delivered to our homes and offices is also ac, and most everyday devices and appliances require alternating current.

There are several reasons for our reliance on alternating current. For one thing, almost all electric energy generators produce electric energy using electromagnetic induction, and thus produce ac outputs (Chapter 20). Furthermore, electrical energy produced in ac fashion can be transmitted economically over long distances through the use of transformers. But perhaps the most important reason is that ac currents produce electromagnetic effects that can be exploited in a variety of devices. For example, when you tune a radio to a station, you take advantage of a special *resonance* property of ac circuits (studied in this chapter).

To determine currents in dc circuits, resistance values were of main concern. There is, of course, resistance present in ac circuits as well, but additional factors can affect the flow of charge. For instance, a capacitor in a dc circuit is equivalent to an infinite resistance (an open circuit). However, in an ac circuit the alternating voltage continually charges and discharges a capacitor. Under such conditions, current can exist in a circuit even if it contains a capacitor. Moreover, wrapped coils of wire can oppose an ac current through the principles electromagnetic induction (Lenz's law; Section 20.1).

In this chapter, the principles of ac circuits will be studied. More generalized forms of Ohm's law and expressions for power, applicable to ac circuits, will be developed. Finally, the phenomenon and uses of *circuit resonance* will be explored.

21.1 Resistance in an AC Circuit

OBJECTIVES: To (a) specify how voltage, current, and power vary with time in an ac circuit, (b) understand the concepts of rms and peak values, and (c) learn how resistors respond under ac conditions.

An ac circuit contains an ac voltage source (such as a small generator or simply a household outlet) and one or more elements. An ac circuit with a single resistive element is shown in ▸Fig. 21.1. If the source's output voltage varies sinusoidally, as that from a generator (see Section 20.2), the voltage across the resistor varies with time in accordance with the equation

$$V = V_o \sin \omega t = V_o \sin 2\pi f t \tag{21.1}$$

where ω is the angular frequency of the voltage (in rad/s) and is related to its frequency f (in Hz) by $\omega = 2\pi f$. The voltage oscillates between $+V_o$ and $-V_o$ as $\sin 2\pi f t$ oscillates between ± 1. The voltage V_o, called the **peak** (or *maximum*) **voltage**, represents the amplitude of the voltage oscillations.

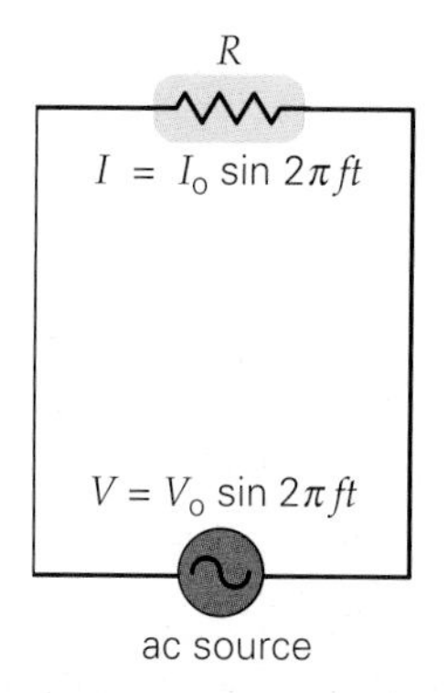

▲ **FIGURE 21.1 A purely resistive circuit** The ac source delivers a sinusoidal voltage to a circuit consisting of a single resistor. The voltage across, and current in, the resistor vary sinusoidally at the frequency of the applied ac voltage.

AC Current and Power

Under ac conditions, the current through the resistor oscillates in direction and magnitude. From Ohm's law, the ac current in the resistor, as a function of time, is given by

$$I = \frac{V}{R} = \left(\frac{V_o}{R}\right) \sin 2\pi f t$$

Because V_o represents the peak voltage across the resistor, the expression in the parentheses represents the maximum current in the resistor. Thus, this expression can be rewritten as

$$I = I_o \sin 2\pi f t \tag{21.2}$$

where the amplitude of the current is $I_o = V_o/R$ and is called the **peak** (or *maximum*) **current**.

▸Figure 21.2 shows both current and voltage as functions of time for a resistor. Note that they are in step, or *in phase*. That is, both reach their zero, minimum, and maximum values at the same time. The current oscillates and takes on both positive and negative values, indicating its directional changes during each cycle. Because the current spends equal time in both directions, *the average current is zero.* Mathematically, this is because the time-averaged value of the sine function over one or more *complete* (360°) cycles is zero. Using overbars to denote a time-averaged value, we have $\overline{\sin \theta} = \overline{\sin 2\pi f t} = 0$. Similarly, $\overline{\cos \theta} = 0$.

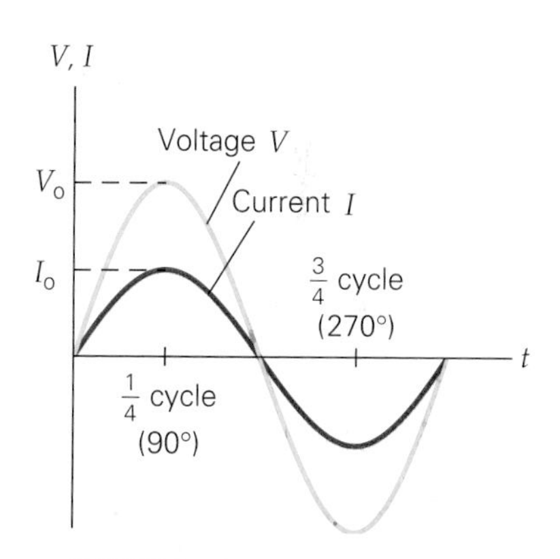

▲ **FIGURE 21.2 Voltage and current in phase** In a purely resistive ac circuit, the voltage and current are in step, or in phase.

Even though the *average* current is zero, this does not mean that there is no joule heating (I^2R losses). This is because the dissipation of electrical energy in a resistor does not depend on the current's direction. The instantaneous power as a function of time is obtained from the instantaneous current (Eq. 21.2). Thus,

$$P = I^2R = (I_o^2 R) \sin^2 2\pi f t \tag{21.3}$$

Even though the current changes sign, the *square* of the current, I^2, is always positive. Thus the average value of I^2R is *not* zero. The average, or mean, value of I^2 is

$$\overline{I^2} = \overline{I_o^2 \sin^2 2\pi f t} = I_o^2\, \overline{\sin^2 2\pi f t}$$

Using the trigonometric identity $\sin^2 \theta = \frac{1}{2}(1 - \cos 2\theta)$, we obtain $\overline{\sin^2 \theta} = \frac{1}{2}(1 - \overline{\cos 2\theta})$. Because $\overline{\cos 2\theta} = 0$ (just as $\overline{\cos \theta} = 0$), it follows that $\overline{\sin^2 \theta} = \frac{1}{2}$. Thus the previous expression for $\overline{I^2}$ can be rewritten as

$$\overline{I^2} = I_o^2\, \overline{\sin^2 2\pi f t} = \tfrac{1}{2} I_o^2 \tag{21.4}$$

The average power is therefore

$$\overline{P} = \overline{I^2} R = \tfrac{1}{2} I_o^2 R \tag{21.5}$$

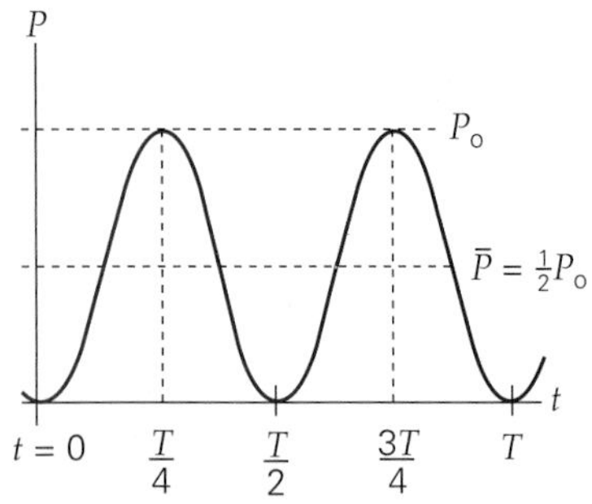

▲ **FIGURE 21.3 Power variation with time in a resistor** Although both current and voltage oscillate in direction (sign), their product (power) is always a positive oscillating quantity. The average power is one-half the peak power.

It should be emphasized that ac power has the same form as dc power ($P = I^2R$) and is *valid at all times*. It is customary, however, to work with average power and a special kind of "average" current defined as follows:

$$I_{rms} = \sqrt{\overline{I^2}} = \sqrt{\tfrac{1}{2}I_o^2} = \frac{I_o}{\sqrt{2}} = \frac{\sqrt{2}}{2}I_o = 0.707I_o \tag{21.6}$$

I_{rms} is called the **rms current**, or **effective current**. (Here, *rms* stands for *root-mean-square*, indicating the square *root* of the *mean* value of the *square* of the current.) The rms current represents the value of a steady (dc) current required to produce the same power as its ac current counterpart, hence the name *effective* current.

Using $I_{rms}^2 = (I_o/\sqrt{2})^2 = \frac{1}{2}I_o^2$, we can rewrite the average power (Eq. 21.5) as

$$\bar{P} = \tfrac{1}{2}I_o^2R = I_{rms}^2R \quad \textit{(time-averaged power)} \tag{21.7}$$

The average power is just the time-varying (oscillating) power averaged over time (◂Fig. 21.3).

AC Voltage

The peak values of voltage and current for a resistor are related by $V_o = I_oR$. Using a development similar to that for rms current, we define the **rms voltage**, or **effective voltage**, as

$$V_{rms} = \frac{V_o}{\sqrt{2}} = \frac{\sqrt{2}}{2}V_o = 0.707V_o \tag{21.8}$$

For resistors under ac conditions, then, dc relationships can be used—as long as we realize that the quantities represent rms values. Thus for ac situations involving only a resistor, the relationship between rms values of current and voltage is

$$V_{rms} = I_{rms}R \quad \textit{voltage across a resistor} \tag{21.9}$$

Combining Eqs. 21.9 and 21.7, we have several physically equivalent expressions for ac power:

$$\bar{P} = I_{rms}^2R = I_{rms}V_{rms} = \frac{V_{rms}^2}{R} \quad \textit{ac power} \tag{21.10}$$

It is customary to measure and specify rms values when dealing with ac quantities. For example, the household line voltage of 120 V, as you may have guessed, is really the rms value of the voltage. It actually has a peak value of

$$V_o = \sqrt{2}\,V_{rms} = 1.414(120\text{ V}) = 170\text{ V}$$

Visual interpretations of peak and rms values of current and voltage are shown in ◂Fig. 21.4.

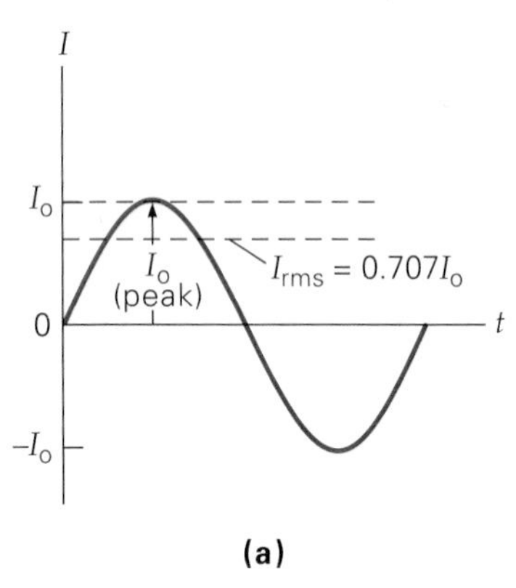

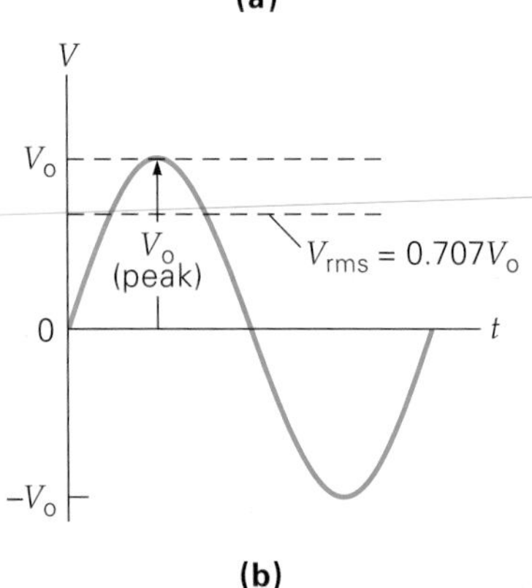

▲ **FIGURE 21.4 Root-mean-square (rms) current and voltage** The rms values of **(a)** current and **(b)** voltage are 0.707, or $1/\sqrt{2}$, times the peak (maximum) values.

Example 21.1 ■ A Bright Lightbulb: Its RMS and Peak Values

A lamp with a 60-W bulb is plugged into a 120-V outlet. (a) What are the rms and peak currents through the lamp? (b) What is the resistance of the bulb under these conditions?

Thinking It Through. (a) Because the average power and rms voltage are known, we can find the rms current from Eq. 21.10. From the rms current, Eq. 21.6 can be used to calculate the peak current. (b) The resistance is found from Eq. 21.9.

Solution. The average power and the rms voltage of the source are given.

Given: $\bar{P} = 60$ W
$V_{rms} = 120$ V

Find: (a) I_{rms} and I_o (rms and peak currents)
(b) R (bulb resistance)

(a) The rms current is

$$I_{rms} = \frac{\overline{P}}{V_{rms}} = \frac{60\ W}{120\ V} = 0.50\ A$$

and the peak current is determined by rearranging Eq. 21.6:

$$I_o = \sqrt{2}I_{rms} = \sqrt{2}(0.50\ A) = 0.71\ A$$

(b) The resistance of the bulb is

$$R = \frac{V_{rms}}{I_{rms}} = \frac{120\ V}{0.50\ A} = 240\ \Omega$$

Follow-Up Exercise. What would be the (a) rms current and (b) peak current in a 60-W lightbulb in Great Britain, where the house rms voltage is 240 V at 50 Hz? (c) How would the resistance of a 60-W bulb in Great Britain compare with one designed for operation at 120 V? Why are the two resistances different? *(Answers to all Follow-Up Exercises are at the back of the text.)*

Illustration 31.2 AC Voltage and Current

Conceptual Example 21.2 ■ Across the Pond: British versus American Electrical Systems

In many countries, the line voltage is 240 V. If a British tourist visiting in the United States plugged in a hair dryer from home (where the voltage is 240 V), you would expect it (a) not to operate, (b) to operate normally, (c) to operate poorly, or (d) to burn out.

Reasoning and Answer. British appliances operate at 240 V. At a decreased voltage (namely 120 V), there would be decreased current ($I = V/R$) and reduced joule heating (because $P = IV$). If the resistance of the appliance were constant, then, at half the voltage, there would be only one-fourth the power output. Thus the heating element of the hair dryer might get warm, but it would not work as expected, so the answer is (c). In addition, the decreased current could cause the motor to run slower than normal.

When on foreign travel, most people do not make this mistake. This is because plugs and sockets vary from country to country. If you are traveling with appliances from home, a converter/adapter kit can be useful (▸Fig. 21.5). This kit contains a selection of plugs for adapting to the foreign sockets, as well as a voltage converter. The converter is a device that converts 240 V to 120 V for U.S. travelers and vice versa for tourists visiting the United States who have 240 V electrical supplies at home. (There are appliances that can be switched between 120 V and 240 V.)

Follow-Up Exercise. What happens if an American tourist inadvertently plugs a 120-V appliance into a British 240-V outlet without a converter? Explain.

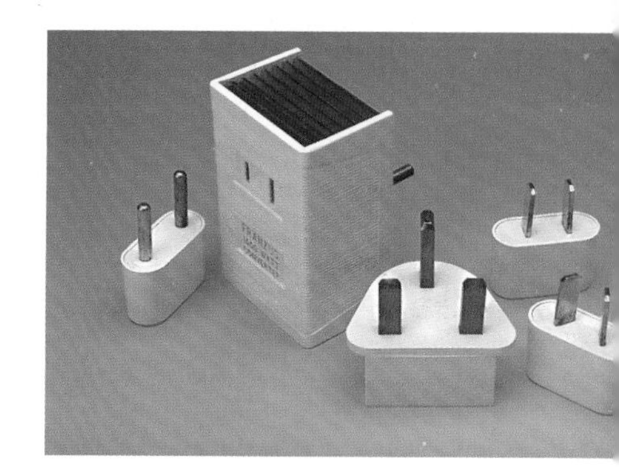

▲ **FIGURE 21.5 Converter and adapters** In countries that have 240-V line voltages, U.S. tourists need conversion to 120 V to operate normal U.S. appliances properly. Note the different types of plugs for different countries. The small plugs go into the foreign sockets, and the converter prongs fit into the back of a socket. A U.S. standard two-prong plug fits into the converter, which has a 120-V output. (See Conceptual Example 21.2.)

21.2 Capacitive Reactance

OBJECTIVES: To (a) explain the behavior of capacitors in ac circuits, and (b) calculate the effect of a capacitor on ac current (capacitive reactance).

In Chapter 16, we learned that when a capacitor is connected to a dc voltage source, current exists only for the short time required to charge the capacitor. As charge accumulates on the capacitor's plates, the voltage across them increases, opposing the external voltage and reducing the current. When the capacitor is fully charged, the current drops to zero.

Things are different when a capacitor is driven by an ac voltage source (▸Fig. 21.6a). Under these conditions, the capacitor limits the current, but doesn't completely prevent the flow of charge. This is because the capacitor is alternately charged and discharged as the current and voltage reverse each half-cycle.

Plots of ac current and voltage versus time for a circuit with just a capacitor are shown in Fig. 21.6b. Let's look at the changing conditions of the capacitor with time (▾Fig. 21.7).

- In Fig. 21.7a, $t = 0$ is arbitrarily chosen as the time of maximum voltage ($V = V_o$)*. At the start, the capacitor is assumed fully charged ($Q_o = CV_o$) with the polarity shown. Because the plates cannot accommodate more charge, there is no current in the circuit.

*We have arbitrarily chosen the polarity of initial capacitor voltage as positive (Figs. 21.6b and 21.7a).

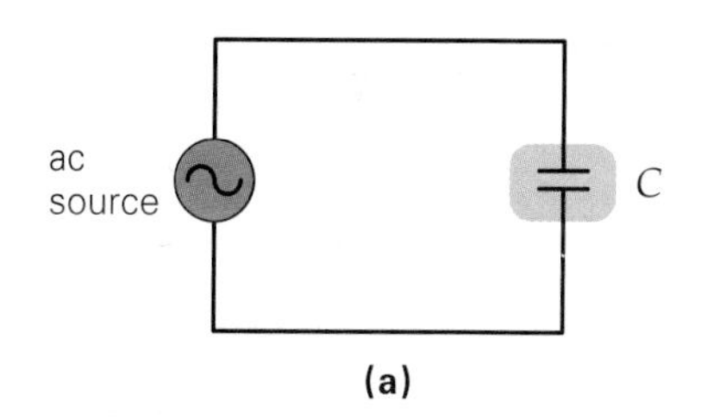

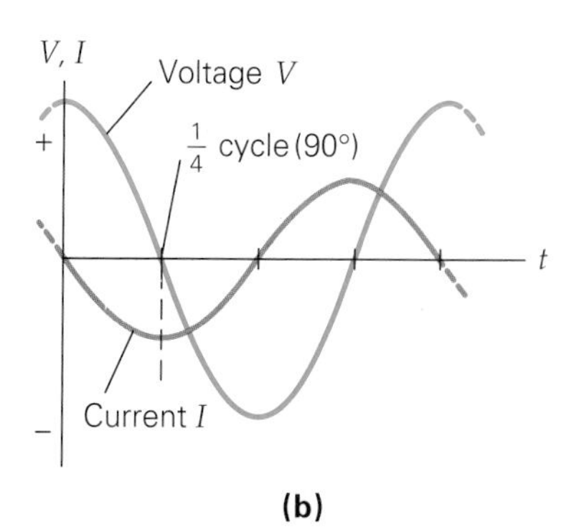

▲ **FIGURE 21.6 A purely capacitive circuit (a)** In a circuit with only capacitance, **(b)** the current leads the voltage by 90°, or one quarter cycle. Half of a cycle of voltage and current, shown, corresponds to Fig. 21.7.

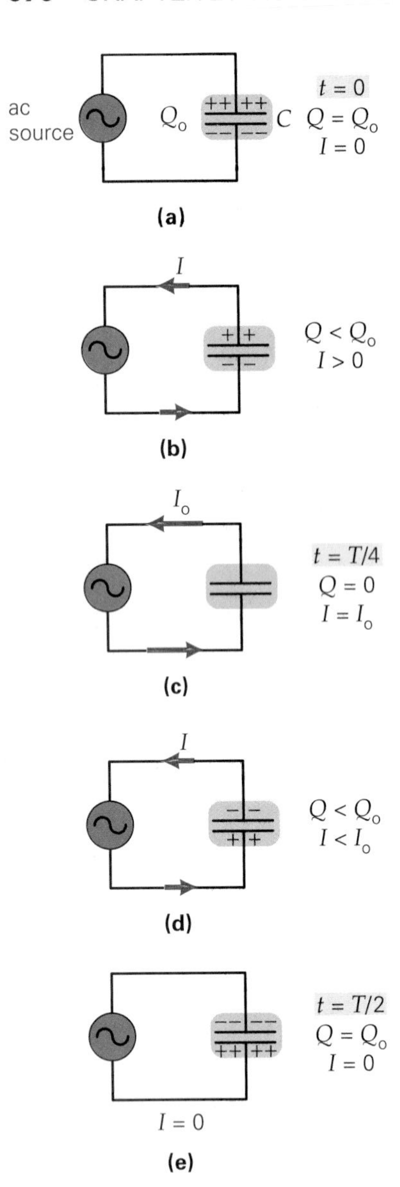

▲ **FIGURE 21.7 A capacitor under ac conditions** This sequence shows the voltage, charge, and current in a circuit containing only a capacitor and an ac voltage source. All five circuit diagrams taken together represent physically what is plotted in the first half of the cycle (from $t = 0$ to $t = T/2$) in the graph shown in Fig. 21.6b.

- As the voltage decreases so that $0 < V < V_o$, the capacitor begins to discharge, giving rise to a counterclockwise current (labeled negative, compare Fig. 21.6b to Fig. 21.7b).
- The current reaches its maximum value when the voltage drops to zero and the capacitor plates are completely discharged (Fig. 21.7c). This occurs one quarter of the way through the cycle ($t = T/4$).
- The ac voltage source now reverses polarity and starts to increase in magnitude, so that $-V_o < V < 0$. The capacitor begins to charge, this time with the opposite polarity (Fig. 21.7d). With the plates uncharged, there is no opposition to the current, so the current is at its maximum value. However, as the plates accumulate charge, they begin to inhibit the current, and the current decreases in magnitude.
- Halfway through the cycle ($t = T/2$), the capacitor is fully charged, but opposite in polarity to its starting condition (Fig. 21.7e). The current is zero, and the voltage is at its maximum magnitude, but opposite the initial polarity ($V = -V_o$).

During the next half-cycle (not shown in Fig. 21.7), the process is reversed and the circuit returns to its initial condition.

Note that the current and voltage are *not* in step (that is, *not* in phase). The current reaches its maximum a quarter-cycle *ahead* of the voltage. The phase relationship between the current and the voltage for a capacitor is commonly stated this way:

> In a purely capacitive ac circuit, the *current* leads the *voltage* by 90°, or one-quarter $\left(\frac{1}{4}\right)$ cycle.

Thus in an ac situation, a capacitor provides opposition to the charging process, but it is not totally limiting. (Recall that under dc conditions it behaves as an open circuit). The quantitative measure of this "capacitive opposition" to current is called the capacitor's **capacitive reactance (X_C)**. In an ac circuit, capacitive reactance is given by

$$X_C = \frac{1}{\omega C} = \frac{1}{2\pi f C} \quad \textit{capacitive reactance} \quad (21.11)$$

SI unit of capacitive reactance:
ohm (Ω), or second per farad (s/F)

where, as usual, $\omega = 2\pi f$, C is the capacitance (in farads), and f is the frequency (in Hz). Like resistance, reactance is measured in ohms (Ω). Using unit analysis, you should be able to show that the ohm is equivalent to the second per farad.

Equation 21.11 shows that the reactance is inversely proportional to both the capacitance (C) and the voltage frequency (f). Both of these dependencies can be understood physically as follows.

Recall that *capacitance* means "charge stored per volt" ($C = Q/V$). Therefore, for a particular voltage, the greater the capacitance, the more charge the capacitor can accommodate. This requires a larger charge flow rate, or current. Increasing the capacitance offers less opposition to charge flow (that is, a reduced capacitive reactance) at a given frequency.

To see the frequency dependence, consider the fact that the greater the frequency of the voltage, the *shorter* the time for charging each cycle. A shorter charging time means less charge will be able to accumulate on the plates and there will be less opposition to the current. Increasing the frequency results in a decrease in capacitive reactance. Hence capacitive reactance is inversely proportional to *both* frequency and capacitance.

It is always good to check a general relationship to see if it gives a result you know to be true in a special case [or several special case(s)]. As a special case for the capacitor, note that if $f = 0$ (that is, nonoscillating dc conditions), the capacitive reactance is infinite. As expected under such conditions, there would be no current.

Capacitive reactance is related to the voltage across the capacitor and the current by an equation that has the same form as $V = IR$ for pure resistances:

$$V_{rms} = I_{rms} X_C \quad \textit{voltage across a capacitor} \quad (21.12)$$

Consider the next Example: a capacitor connected to an ac voltage source.

Example 21.3 ■ Current under ac Conditions: Capacitive Reactance

A 15.0-μF capacitor is connected to a 120-V, 60-Hz source. What are (a) the capacitive reactance and (b) the current (rms and peak) in the circuit?

Thinking It Through. The capacitive reactance can be obtained from the capacitance and the frequency by using Eq. 21.11. (b) We can then calculate the rms current from the reactance and rms voltage via Eq. 21.12. Finally, Eq. 21.6 gives the peak current.

Solution. Assuming the 60-Hz frequency to be exact, our answers are to three significant figures. We list the data:

Given: $C = 15.0\ \mu\text{F} = 15.0 \times 10^{-6}\ \text{F}$
$V_{\text{rms}} = 120\ \text{V}$
$f = 60\ \text{Hz}$

Find: (a) X_C (capacitive reactance)
(b) I_o (peak current), I_{rms} (rms current)

(a) The capacitive reactance is

$$X_C = \frac{1}{2\pi f C} = \frac{1}{2\pi(60\ \text{Hz})(15.0 \times 10^{-6}\ \text{F})} = 177\ \Omega$$

(b) Then, the rms current is

$$I_{\text{rms}} = \frac{V_{\text{rms}}}{X_C} = \frac{120\ \text{V}}{177\ \Omega} = 0.678\ \text{A}$$

and therefore, the peak current is

$$I_o = \sqrt{2} I_{\text{rms}} = \sqrt{2}(0.678\ \text{A}) = 0.959\ \text{A}$$

The current oscillates at 60 cycles per second with a magnitude of 0.959 A.

Follow-Up Exercise. In this Example, (a) what is the peak voltage and (b) what frequency would give the same current if the capacitance were reduced by half?

21.3 Inductive Reactance

OBJECTIVES: To (a) explain what an inductor is, (b) explain the behavior of inductors in ac circuits, and (c) calculate the effect of inductors on ac current (inductive reactance).

Inductance is a measure of the opposition a circuit element presents to a time-varying current (by Lenz's law). In principle, all circuit elements (even resistors) have some inductance. However, a coil of wire with negligible resistance has, in effect, only inductance. When placed in a circuit with a time-varying current, such a coil, called an **inductor**, exhibits a reverse voltage, or back emf, in opposition to the changing current. The changing current through the coil produces a changing magnetic field and flux. The back emf is the induced emf in opposition to this changing flux. Because the back emf is induced in the inductor as a result of its own changing magnetic field, this phenomenon is called *self-induction.*

Note: Lenz's law is given in Section 20.1.

The self-induced emf (for a coil consisting of N loops) is given by Faraday's law (Eq. 20.2): $\mathcal{E} = -N\Delta\Phi/\Delta t$. The time rate of change of the total flux through the coil, $N\Delta\Phi/\Delta t$, is proportional to the rate of change of the current in the coil, $\Delta I/\Delta t$. This is because the current produces the magnetic field responsible for the changing flux. Thus the back emf is proportional to, and oppositely directed to, the rate of current change. This relationship is expressed using a proportionality constant, L:

$$\mathcal{E} = -L\left(\frac{\Delta I}{\Delta t}\right) \qquad (21.13)$$

where L is the *inductance* of the coil (more properly, its *self*-inductance). You should be able to show, using unit analysis, that the units of inductance are volt-seconds per ampere (V · s/A). This combination is called a **henry** (**H**, 1 H = 1 V · s/A), in honor of Joseph Henry (1797–1878), an American physicist and early investigator of electromagnetic induction. Smaller units, such as the millihenry (mH), are commonly used (1 mH = 10^{-3} H).

The opposition presented to current by an inductor under ac conditions depends on the inductance and the voltage frequency. This is expressed quantitatively by the circuit's **inductive reactance (X_L)**, which is

$$X_L = \omega L = 2\pi f L \quad \textit{inductive reactance} \tag{21.14}$$

SI unit of inductive reactance:
ohm (Ω), or henry per second (H/s)

where f is the frequency of the driving voltage, $\omega = 2\pi f$, and L is the inductance. Like capacitive reactance, inductive reactance is measured in ohms (Ω), which you should be able to show are equivalent to henrys per second.

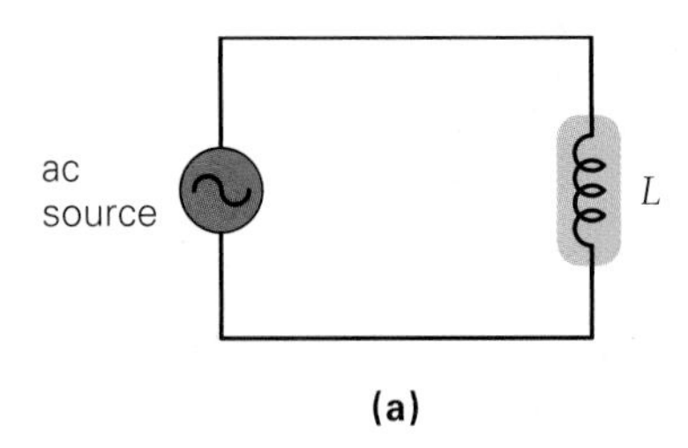

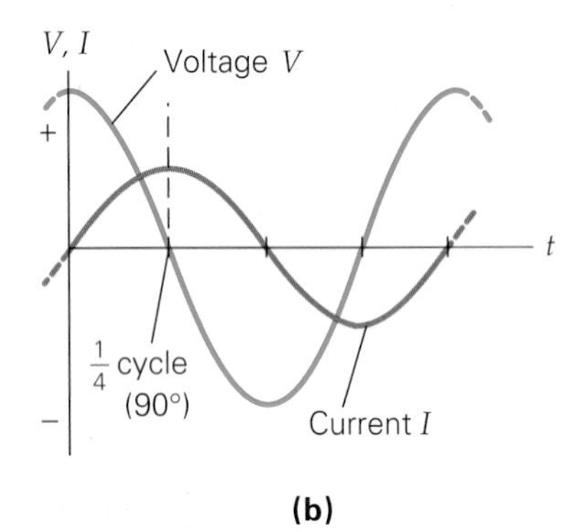

▲ **FIGURE 21.8 A purely inductive circuit** **(a)** In a circuit with only inductance, **(b)** the voltage leads the current by 90°, or one quarter cycle.

Note that the inductive reactance is proportional to both the coil inductance (L) and the voltage frequency (f). The inductance is a property of the coil that depends on the number of turns, the coil's diameter and length, and the material of the core (if any). The frequency of the voltage plays a role because the more rapidly the current in the coil changes, the greater the rate of change of its magnetic flux. This implies a larger self-induced (back or reverse) emf to oppose the changes in current.

In terms of X_L, the voltage across an inductor is related to the current and inductive reactance by:

$$V_{rms} = I_{rms} X_L \quad \textit{voltage across an inductor} \tag{21.15}$$

The circuit symbol for an inductor and the graphs of the voltage across the inductor and the current in the circuit are shown in ◂Fig. 21.8. When an inductor is connected to an ac voltage source, maximum voltage corresponds to zero current. When the voltage drops to zero, the current is maximum. This happens because, as the voltage changes polarity (causing the magnetic flux through the inductor to drop to zero), the inductor acts to prevent the change in accordance with Lenz's law, so the induced emf creates a current. In an inductor, the current *lags* one quarter cycle behind the voltage, a relationship commonly expressed as follows:

> In a purely inductive ac circuit, the *voltage* leads the *current* by 90°, or one quarter $\left(\frac{1}{4}\right)$ cycle.

Illustration 31.4 Phase Shift

Exploration 31.2 Reactance

Because the phase relationships between current and voltage for purely inductive and purely capacitive circuits are opposite, there is a phrase that may help you remember the difference: *E<u>L</u>I* the *I<u>C</u>E* man. Here *E* represents voltage (for *e*mf) and *I* represents current. The three letters *ELI* indicate that for inductance (L), the voltage leads the current (I)—reading the acronym from left to right. Similarly, *ICE* means that for capacitance (C), the current leads the voltage.

Example 21.4 ■ Current Opposition without Resistance: Inductive Reactance

A 125-mH inductor is connected to a 120-V, 60-Hz source. What are (a) the inductive reactance and (b) the rms current in the circuit?

Thinking It Through. Because the inductance and frequency are known, we can compute the inductive reactance from Eq. 21.14 and the current from Eq. 21.15.

Solution. Listing the given data,

Given: $L = 125\text{ mH} = 0.125\text{ H}$
$V_{rms} = 120\text{ V}$
$f = 60\text{ Hz}$

Find: (a) X_L (inductive reactance)
(b) I_{rms}

(a) The inductive reactance is

$$X_L = 2\pi f L = 2\pi(60\text{ Hz})(0.125\text{ H}) = 47.1\ \Omega$$

(b) The rms current is then

$$I_{rms} = \frac{V_{rms}}{X_L} = \frac{120\text{ V}}{47.1\ \Omega} = 2.55\text{ A}$$

Follow-Up Exercise. In this Example, (a) what is the peak current? (b) What voltage frequency would yield the same current if the inductance were reduced to one third the value in this Example?

21.4 Impedance: RLC Circuits

OBJECTIVES: To (a) calculate currents and voltages when a combination of reactive and resistive circuit elements are present in ac circuits, (b) use phase diagrams to calculate overall impedance and rms currents, and (c) understand and use the concept of the power factor in ac circuits.

In the previous sections purely capacitive or purely inductive circuits were considered separately and without resistance present. However, in the real world, it is impossible to have purely reactive circuits, because there is always some resistance—at a minimum, that from the connecting wires. Thus resistances, capacitive reactances, and inductive reactances *combine* to impede the current in ac circuits. An analysis of some combination circuits illustrates these effects.

Series RC Circuit

Suppose an ac circuit consists of a voltage source, a resistor, and a capacitor connected in series (▸Fig. 21.9a). The phase relationship between the current and the voltage is different for each circuit element. As a result, a special graphical method is needed to find the overall opposition to the current in the circuit. This method employs a *phase diagram.*

In a phase diagram, such as in Fig. 21.9b for an RC circuit, the resistance and reactance in the circuit are endowed with vectorlike properties and their magnitudes represented by arrows called *phasors*. On a set of x–y coordinate axes, the resistance is plotted on the positive x-axis (that is, at 0°), because the voltage–current phase difference for a resistor is zero. The capacitive reactance is plotted along the negative y-axis, to reflect a phase difference (ϕ) of $-90°$ because for a capacitor, the voltage lags behind the current by one quarter of a cycle.

The phasor sum is the effective, or net, opposition to the current, which we call the **impedance (Z)**. Phasors must be added in the same way as vectors because the effects of the resistor and capacitor are not in phase. For the series RC circuit,

$$Z = \sqrt{R^2 + X_C^2} \quad \textit{series RC circuit impedance} \tag{21.16}$$

The unit of impedance is the ohm.

The generalization of Ohm's law to circuits containing capacitors and inductors along with resistors is

$$V_{rms} = I_{rms}Z \quad \textit{Ohm's law for ac circuits} \tag{21.17}$$

To illustrate how phasors can be used to analyze an RC circuit, consider the next Example. Take particular note of part (b), in which there is an *apparent* violation of Kirchhoff's loop theorem—explained by phase differences between the voltages across the two elements in the circuit.

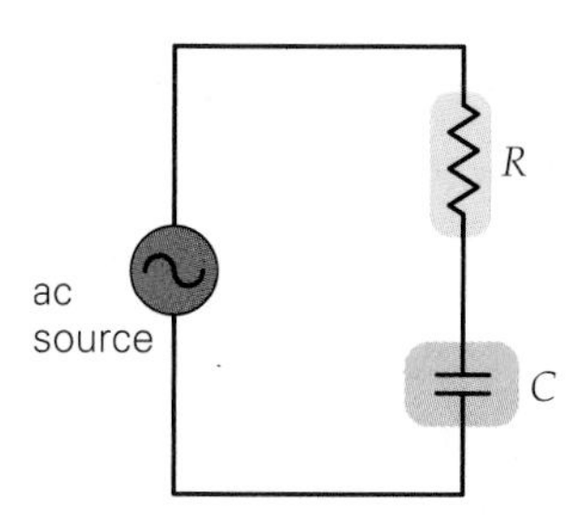

(a) RC circuit diagram

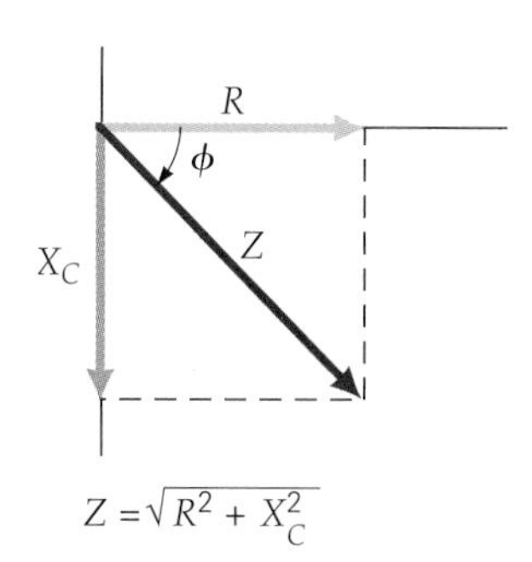

$Z = \sqrt{R^2 + X_C^2}$

(b) Phase diagram

▲ **FIGURE 21.9** A series RC circuit **(a)** In a series RC circuit, **(b)** the impedance Z is the phasor sum of the resistance R and the capacitive reactance X_C.

Note: *Impedance* (denoted by Z) refers to the overall circuit opposition to current. We reserve *reactance* and *resistance* for opposition in individual elements.

Note: $V_{rms} = I_{rms}Z$ can be applied to any circuit, as long as the impedance Z is properly calculated, using phasors.

Example 21.5 ■ RC Impedance and Kirchhoff's Loop Theorem

A series RC circuit has a resistance of 100 Ω and a capacitance of 15.0 μF. (a) What is the (rms) current in the circuit when it is driven by a 120-V, 60-Hz source? (b) Compute the (rms) voltage across each circuit element and the two elements combined. Compare it with that of the voltage source. Is Kirchhoff's loop theorem satisfied? Comment and explain your reasoning.

Thinking It Through. (a) Note that the voltage and capacitor values are the same as those in Example 21.3 and a resistor has been added in series. This will help in reducing the necessary calculation. Then, using phasors, the capacitive reactance and the resistance can be combined to determine the overall impedance (Eq. 21.16). From Eq. 21.17, the impedance and voltage can be used to find the current. (b) Because the current is the same everywhere at any given time in a series circuit, the result of part (a) can be used to calculate the voltages. The rms voltage across both elements together is found by recalling that the individual voltages are out of phase by 90°. What this means physically is that they reach their peak values *not* at the same time, but rather one fourth of a period apart. Thus, we *cannot* simply add the voltages.

(continues on next page)

Solution.

Given: $R = 100\ \Omega$
$C = 15.0\ \mu\text{F} = 15.0 \times 10^{-6}\ \text{F}$
$V_{\text{rms}} = 120\ \text{V}$
$f = 60\ \text{Hz}$

Find: (a) I (rms current)
(b) V_C (rms voltage across capacitor)
V_R (rms voltage across resistor)
$V_{(R+C)}$ (combined rms voltage)

(a) In Example 21.3, we found that the reactance for this capacitor at this frequency was $X_C = 177\ \Omega$. Now Eq. 21.16 can be used to calculate the circuit impedance:

$$Z = \sqrt{R^2 + X_C^2} = \sqrt{(100\ \Omega)^2 + (177\ \Omega)^2} = 203\ \Omega$$

Because $V_{\text{rms}} = I_{\text{rms}}Z$, the rms current is

$$I_{\text{rms}} = \frac{V_{\text{rms}}}{Z} = \frac{120\ \text{V}}{203\ \Omega} = 0.591\ \text{A}$$

(b) Using Eq. 21.17 first for the rms voltage across the resistor alone ($Z = R$),

$$V_R = I_{\text{rms}}R = (0.591\ \text{A})(100\ \Omega) = 59.1\ \text{V}$$

For the capacitor alone ($Z = X_C$), the rms voltage across the capacitor is

$$V_C = I_{\text{rms}}X_C = (0.591\ \text{A})(177\ \Omega) = 105\ \text{V}$$

Illustration 31.6 Voltage and Current Phasors

PHYSLET®

Illustration 31.7 RC Circuits and Phasors

The algebraic sum of these two rms voltages is 164 V, which is *not* the same as the rms value of the voltage source (120 V). This does *not* mean that Kirchhoff's loop theorem has been violated. In fact, the source voltage does equal the combined voltages across the capacitor and resistor *if you account for phase differences*. The combined voltage must be calculated properly to take into account the 90° phase difference between the two voltages. Using the Pythagorean theorem to get the total voltage, we have

$$V_{(R+C)} = \sqrt{V_R^2 + V_C^2} = \sqrt{(59.1\ \text{V})^2 + (105\ \text{V})^2} = 120\ \text{V}$$

Thus when the individual voltages are combined properly (taking into account that the voltages do not peak at the same time), Kirchhoff's law are still valid. Here it has been shown that the total rms voltage across both elements is equal to the rms voltage of the source. We must take care to add voltages this way because they are out of phase in general. *Thus, Kirchhoff's laws are valid at any instant of time, not just for rms values, but care must be taken to account for phase differences.*

Follow-Up Exercise. (a) How would the result in part (a) of this Example change if the circuit were driven by a voltage source with the same rms voltage, but oscillating at 120 Hz? (b) Is the resistor or the capacitor responsible for the change?

Series RL Circuit

The analysis of a series RL circuit (◂Fig. 21.10) is similar to that of a series RC circuit. However, the inductive reactance is plotted along the *positive y*-axis in the phase diagram, to reflect a phase difference of +90° with respect to the resistance. Remember that a positive phase angle means that the voltage *leads* the current, as is true for an inductor.

Thus the impedance in an RL series circuit is

$$Z = \sqrt{R^2 + X_L^2} \quad \textit{series RL circuit impedance} \tag{21.18}$$

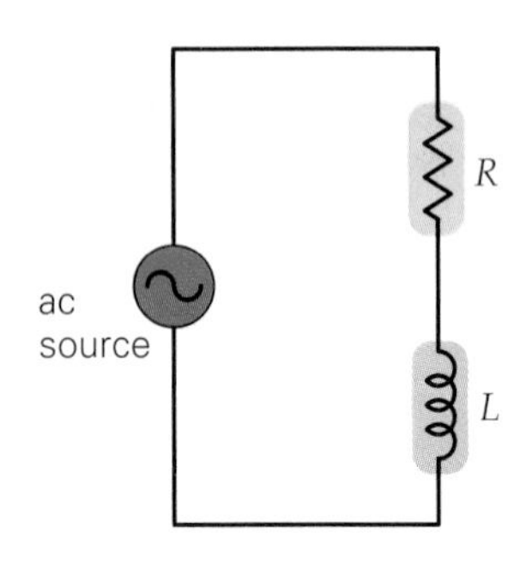

(a) RL circuit diagram

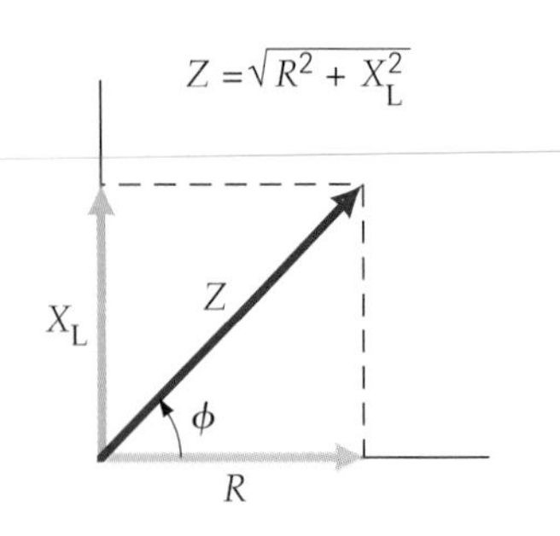

(b) Phase diagram

▲ **FIGURE 21.10** A series RL circuit **(a)** In a series RL circuit, **(b)** the impedance Z is the phasor sum of the resistance R and the inductive reactance X_L.

Series RLC Circuit

More generally, an ac circuit may contain all three circuit elements—a resistor, an inductor, and a capacitor—as shown in series in ▸Fig. 21.11. Again, phasor addition must be used to determine the overall circuit impedance. Combining the vertical components (that is, inductive and capacitive reactances) gives the *total reactance*, $X_L - X_C$. Subtraction is used because the phase difference between X_L and X_C is 180°. The overall circuit impedance is the phasor sum of the resistance and the total reactance. Employing the Pythagorean theorem once more on the phasor diagram, we have

$$Z = \sqrt{R^2 + (X_L - X_C)^2} \quad \textit{series RLC circuit impedance} \tag{21.19}$$

The **phase angle (ϕ)** between the source voltage and the current in the circuit is the angle between the overall impedance phasor (Z) and the $+x$-axis (Fig. 21.11b), or

$$\tan\phi = \frac{X_L - X_C}{R} \quad \textit{phase angle in series RLC circuit} \tag{21.20}$$

TABLE 21.1 Impedances and Phase Angles for Series Circuits

Circuit Element(s)	Impedance Z (in W)	Phase Angle ϕ
R	R	0°
C	X_C	−90°
L	X_L	+90°
RC	$\sqrt{R^2 + X_C^2}$	negative (meaning that ϕ is between 0° and −90°)
RL	$\sqrt{R^2 + X_L^2}$	positive (meaning that ϕ is between 0° and +90°)
RLC	$\sqrt{R^2 + (X_L - X_C)^2}$	positive if $X_L > X_C$ negative if $X_C > X_L$

Notice that if X_L is greater than X_C (as in Fig. 21.11b), the phase angle is positive ($+\phi$), and the circuit is said to be *inductive*, because the nonresistive part of the impedance (that is, the reactance) is dominated by the inductor. If X_C is greater than X_L, the phase angle is negative ($-\phi$), and the circuit is said to be *capacitive*, because capacitive reactance dominates over inductive reactance.

A summary of impedances and phase angles for the three circuit elements and various combinations is given in Table 21.1. Example 21.6 analyzes an RLC circuit.

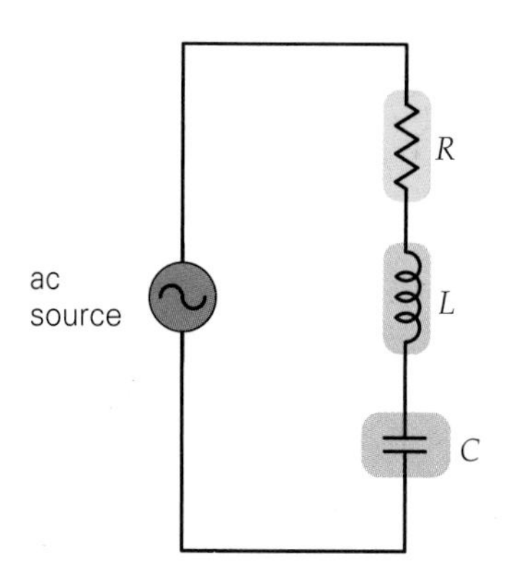

(a) RLC circuit diagram

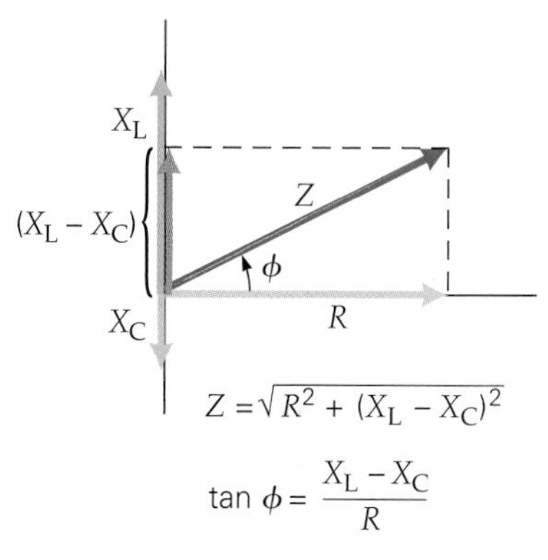

(b) Phase diagram

▲ **FIGURE 21.11 A series RLC circuit (a)** In a series RLC circuit, **(b)** the impedance Z is the phasor sum of the resistance R and the total (or net) reactance $(X_L - X_C)$. Note that the phasor diagram is drawn for the case of $X_L > X_C$.

Example 21.6 ■ All Together Now: Impedance in an RLC Circuit

A series RLC circuit has a resistance of 25.0 Ω, a capacitance of 50.0 μF, and an inductance of 0.300 H. If the circuit is driven by a 120-V, 60-Hz source, what are (a) the total impedance of the circuit, (b) the rms current in the circuit, and (c) the phase angle between the current and the voltage?

Thinking It Through. (a) To calculate the overall impedance from Eq. 21.19, the individual reactances must first be determined. (b) The current is computed from the generalization of Ohm's law, $V_{rms} = I_{rms}Z$ (Eq. 21.17). (c) The phase angle is calculated from Eq. 21.20.

Solution. We are given all the necessary data:

Given:
$R = 25.0\ \Omega$
$C = 50.0\ \mu F = 5.00 \times 10^{-5}\ F$
$L = 0.300\ H$
$V_{rms} = 120\ V$
$f = 60\ Hz$

Find:
(a) Z (overall circuit impedance)
(b) I_{rms}
(c) ϕ (phase angle)

(a) The individual reactances are

$$X_C = \frac{1}{2\pi f C} = \frac{1}{2\pi(60\ Hz)(5.00 \times 10^{-5}\ F)} = 53.1\ \Omega$$

and

$$X_L = 2\pi f L = 2\pi(60\ Hz)(0.300\ H) = 113\ \Omega$$

Then,

$$Z = \sqrt{R^2 + (X_L - X_C)^2} = \sqrt{(25.0\ \Omega)^2 + (113\ \Omega - 53.1\ \Omega)^2} = 64.9\ \Omega$$

(b) Because $V_{rms} = I_{rms}Z$, we have

$$I_{rms} = \frac{V_{rms}}{Z} = \frac{120\ V}{64.9\ \Omega} = 1.85\ A$$

(c) Solving $\tan \phi = (X_L - X_C)/R$ for the phase angle gives

$$\phi = \tan^{-1}\left(\frac{X_L - X_C}{R}\right) = \tan^{-1}\left(\frac{113\ \Omega - 53.1\ \Omega}{25.0\ \Omega}\right) = +67.3°$$

We should have expected a positive phase angle, because the inductive reactance is greater than the capacitive reactance [see part (a)]. Thus this circuit is *inductive* in nature.

(continues on next page)

Follow-Up Exercise. (a) Consider the RLC circuit in this Example, but with the driving frequency doubled. Reasoning conceptually, should the phase angle ϕ be greater or less than the $+67.3°$ after the increase? (b) Compute the new phase angle to show that your reasoning is correct.

By now, you should appreciate the usefulness of phasor diagrams in determining impedances, voltages, and currents in ac circuits. However, you might still be wondering what the use and meaning are of the phase angle ϕ. To illustrate its importance, let's examine the power loss in an RLC circuit. Note that this power analysis also depends on the use of phasor diagrams.

Power Factor for a Series RLC Circuit

In considering an RLC circuit, a crucial thing to realize is that any circuit power loss (joule heating) can take place only in the resistor. *There are no power losses associated with capacitors and inductors.* Capacitors and inductors simply store energy and give it back, without loss. Ideally, neither has any resistance, and thus any joule heating attributed to them is zero.

The average (rms) power dissipated by a resistor is $P_{rms} = I_{rms}^2 R$. This rms power can also be expressed in terms of the rms current and voltage, but the voltage *must be that across the resistor* (V_R), because it is the only dissipative element. The average power dissipated in a series RLC circuit can be alternatively expressed as

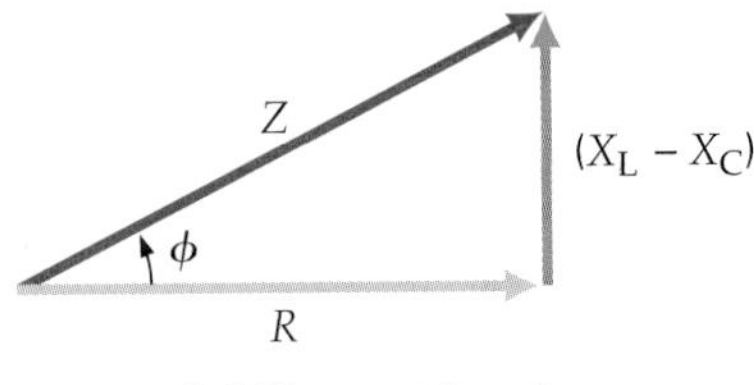

(a) Phasor triangle

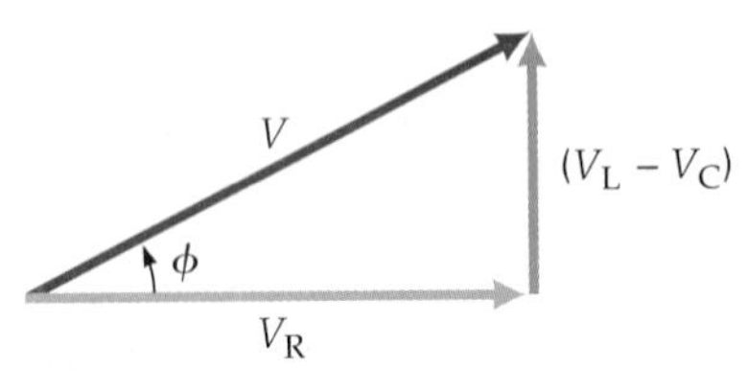

(b) Equivalent voltage triangle

▲ **FIGURE 21.12 Phasor and voltage triangles** The rms voltages across the components of a series RLC circuit are given by $V_R = I_{rms}R$, $V_L = I_{rms}X_L$, and $V_C = I_{rms}X_C$. Because the current is the same through each, **(a)** the phasor triangle can be converted to **(b)** a voltage triangle. Note that $V_R = V\cos\phi$. Both phasor diagrams are drawn for the case of $X_L > X_C$.

$$\bar{P} = P_R = I_{rms}V_R$$

The voltage across the resistor can be found from a voltage triangle that corresponds to the phasor triangle (◀Fig. 21.12). The rms voltages across the individual components in an RLC circuit are $V_R = I_{rms}R$, $V_L = I_{rms}X_L$, and $V_C = I_{rms}X_C$. Combining the last two voltages, we can write $(V_L - V_C) = I_{rms}(X_L - X_C)$. If each leg of the phasor triangle (Fig. 21.12a) is multiplied by the rms current, an equivalent voltage triangle results (Fig. 21.12b). As this figure shows, the voltage across the resistor is

$$V_R = V_{rms}\cos\phi \quad (21.21)$$

The term $\cos\phi$ is called the **power factor**. From Fig. 21.11,

$$\cos\phi = \frac{R}{Z} \quad \textit{series RLC power factor} \quad (21.22)$$

The average power, rewritten in terms of the power factor, is

$$\bar{P} = I_{rms}V_{rms}\cos\phi \quad \textit{series RLC power} \quad (21.23)$$

Because power is dissipated only in the resistance ($\bar{P} = I_{rms}^2 R$), Eq. 21.22 enables us to express the average power as

$$\bar{P} = I_{rms}^2 Z\cos\phi \quad \textit{series RLC power} \quad (21.24)$$

Note that $\cos\phi$ varies from a maximum of $+1$ (when $\phi = 0°$) to a minimum of zero (when $\phi = \pm 90°$). When $\phi = 0°$, the circuit is said to be *completely resistive.* That is, there is maximum power dissipation (as though the circuit contained only a resistor). The power factor decreases as the phase angle increases in either direction [because $\cos(-\phi) = \cos\phi$]—in other words, as the circuit becomes inductive or capacitive. At $\phi = +90°$, the circuit is *completely inductive*; at $\phi = -90°$, it is *completely capacitive*. In these cases, the circuit contains only an inductor or a capacitor, respectively, so no power is dissipated. In practice, because there is always some resistance, a circuit can never be completely inductive or capacitive. It is possible, however, for an RLC circuit to *appear* to be completely resistive even if it contains a capacitor and an inductor, as we shall see in Section 21.5. Let's look at our previous RLC example with an emphasis on power.

Illustration 31.5 Power and Reactance

Exploration 31.4 Phase Angle and Power

Example 21.7 ■ Power Factor Revisited

What is the average power dissipated in the circuit described in Example 21.6?

Thinking It Through. The power factor can be determined because the resistance (R) and impedance (Z) are known. Once the power factor is known, the actual power can be calculated.

Solution.

Given: See Example 21.6 *Find:* $\bar{P}$ (average power)

In Example 21.6, it was determined that the circuit had an impedance of $Z = 64.9\ \Omega$, and its resistance was $R = 25.0\ \Omega$. Therefore its power factor is

$$\cos\phi = \frac{R}{Z} = \frac{25.0\ \Omega}{64.9\ \Omega} = 0.385$$

Using the other data from Example 21.6 and Eq. 21.23 gives

$$\bar{P} = I_{rms}V_{rms}\cos\phi = (1.85\ \text{A})(120\ \text{V})(0.385) = 85.5\ \text{W}$$

This is less than the power that would be dissipated without a capacitor and an inductor. (Can you show this to be true? Why is it true?)

Follow-Up Exercise. If the frequency were doubled and the capacitor removed from this Example, what would be the rms power?

21.5 Circuit Resonance

OBJECTIVES: To (a) understand the concept of resonance in ac circuits, and (b) calculate the resonance frequency of an RLC circuit.

From the previous discussion, it can be seen that when the power factor ($\cos\phi$) of an RLC series circuit is equal to unity, maximum power is transferred to the circuit. In this situation, the current in the circuit must be maximum, because the impedance is at its minimum. This occurs because at this unique frequency, the inductive and capacitive reactances *effectively cancel*—that is, they are equal in magnitude and 180° out of phase, or opposite. This situation can happen in any RLC circuit if the appropriate source frequency is chosen.

Illustration 31.8 Impedance and Resonance, RLC Circuit

The key to finding this frequency is to realize that because inductive and capacitive reactances are frequency dependent, so is the overall impedance. From the expression for the RLC series impedance, $Z = \sqrt{R^2 + (X_L - X_C)^2}$, it can be seen that the impedance is a minimum when $X_L - X_C = 0$. This occurs at a frequency f_o, found by setting $X_L = X_C$. Using the expressions for the reactances, this means that $2\pi f_o L = 1/2\pi f_o C$. Solving for f_o yields

Exploration 31.7 RLC Circuit

$$f_o = \frac{1}{2\pi\sqrt{LC}} \quad \textit{series RLC resonance frequency} \tag{21.25}$$

This frequency satisfies the condition of minimum impedance—and therefore maximizes the current in the circuit. In analogy to pumping a swing at just the right frequency or having a violin string in one of its normal modes, f_o is called the circuit's **resonance frequency**. A plot of capacitive and inductive reactances versus frequency is shown in ▼Fig. 21.13a. The curves X_C and X_L intersect at f_o, the frequency at which their values are equal.

A Physical Explanation of Resonance

The physical explanation of resonance in a series RLC circuit is worth exploring. We have seen that the capacitor and inductor voltages are *always* 180° out of phase, or have opposite polarity. In other words, they tend to cancel out but usually don't completely do so because their values are not equal. If this is the case, then the voltage across the resistor is less than that of the source voltage because there is a net voltage across the

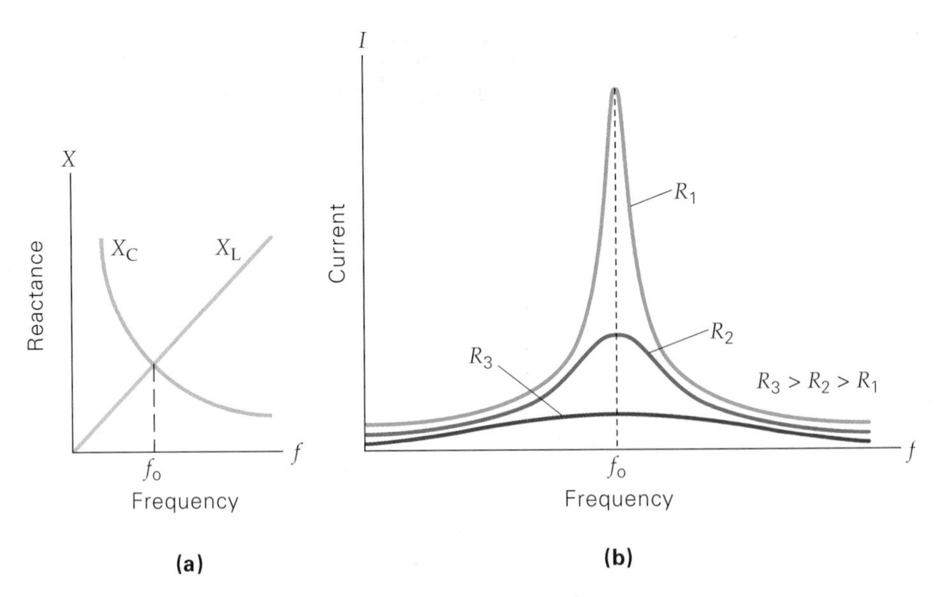

▲ **FIGURE 21.13 Resonance frequency for a series RLC circuit** **(a)** At the resonance frequency (f_o), the capacitive and inductive reactances are equal ($X_L = X_C$). On a graph of X versus f, this is the frequency at which the curves of X_C and X_L intersect. **(b)** On a graph of I versus f, the current is a maximum at f_o. The curve becomes sharper and narrower as the resistance in the circuit decreases.

combination of capacitor and inductor. This means that the power dissipated in the resistor is less than its maximum value. However, in the special situation when the capacitive and inductive voltages do cancel, the full source voltage appears across the resistor, the power factor becomes 1, and the resistor dissipates the maximum possible power. This is what we mean by the circuit being driven "at resonance."

Applications of Resonance

As we have seen, when a series RLC circuit is driven at its resonance frequency, both the current in the circuit and the power transfer to the circuit are at a maximum. A graph of rms current versus driving frequency is shown in Fig. 21.13b for several different values of resistance. As expected, the maximum current occurs at frequency f_o. Notice also that the curve becomes sharper and narrower as the resistance decreases.

Resonant circuits have a variety of applications. One common application is in the tuning mechanism of a radio. Each radio station has an assigned broadcast frequency at which its radio waves are transmitted (see Insight 21.1 on Oscillator Circuits: Broadcasters of Electromagnetic Radiation). When the waves are received at the antenna, their oscillating electric and magnetic fields set the electrons in the antenna into regular back-and-forth motion. In other words, they produce an alternating current in the receiver circuit, just as a regular ac voltage source would do.

▲ **FIGURE 21.14 Variable air capacitor** Rotating the movable plates between the fixed plates changes the overlap area and thus the capacitance. Such capacitors were common in tuning circuits in older radios.

In a given area, each radio station is assigned its own broadcast frequency. Usually several different radio signals reach an antenna together, but a good receiver circuit selectively picks up only the one with a frequency at or near its resonance frequency. Most radios allow you to alter this resonance frequency to "tune in" different stations. In the early days of radio, variable air capacitors were used for this purpose (▸Fig. 21.14). Today, more compact variable capacitors in smaller radios have a polymer dielectric between thin plates. The polymer sheets help maintain the plate separation and increase the capacitance, thus allowing manufacturers to use plates of a much smaller area. (Recall from Chapter 16 that $C = \kappa\varepsilon_o A/d$.) In most modern radios, solid-state devices replace variable capacitors.

INSIGHT 21.1 OSCILLATOR CIRCUITS: BROADCASTERS OF ELECTROMAGNETIC RADIATION

To generate the high-frequency electromagnetic waves used in radio communications and television (Fig. 1), electric current must be made to oscillate at high frequencies in electronic circuits. This can be accomplished with RLC circuits. Such circuits are called *oscillator circuits*, because the current in them oscillates at a frequency determined by their inductive and capacitive elements.

When the resistance in an RLC circuit is very small, the circuit is essentially an LC circuit. The current in such a circuit oscillates at a frequency f, which is the circuit's "natural" frequency and also its resonance frequency (Eq. 21.25). Any small resistance in the circuit would dissipate energy. However, in an ideal LC circuit (which we are considering here) with no resistance, the oscillation would continue indefinitely.

To understand this oscillation, consider the energy oscillations in an ideal (resistanceless) parallel LC circuit, shown in Fig. 2a. Let's assume that the capacitor is initially charged and the switch is then closed ($t = 0$). The following sequence of events occurs:

1. The capacitor would discharge instantaneously (because $RC = 0$) if it were not for the current having to pass through the coil. At $t = 0$, the current in the coil is zero (Fig. 2a). As the current builds, so does the magnetic field in the coil. By Lenz's law, the increasing magnetic field and change of flux in the coil induce a back emf to oppose this current increase. Because of this back emf, the capacitor does takes time to discharge.
2. When the capacitor is fully discharged (Fig. 2b), all of its energy (in its electric field) has been transferred to the inductor in the form of its magnetic field. (Because in this circuit it is assumed that $R = 0$, no energy is lost to joule heating.) At this time (one-quarter of a period; $T/4$), the magnetic field and the current in the coil are maximum, and all the energy is stored in the inductor. (Refer to the energy "histograms" accompanying the circuit diagrams in Figs. 2a, 2b, and 2c to visualize the energy trades as the cycle proceeds.)
3. As the magnetic field collapses from its maximum value, an emf that opposes the collapse is induced in the coil. This emf acts in a direction that tends to continue the current in the coil even as it is decreasing (Lenz's law again). The polarity of the emf is now opposite that in step 1. Thus current continues to charge to the capacitor, but the result is a polarity reversed from its initial polarity.
4. When the capacitor is again fully charged (but at reverse polarity), it has its initial energy again (Fig. 2c). This occurs halfway through the cycle, or half a period from the start ($T/2$). The magnetic field in the coil is zero, as is the circuit current.
5. The capacitor again begins to discharge, and these four steps are repeated over and over. Thus we have a current and energy oscillation in the circuit. In an ideal case of a circuit without resistance, the oscillations would continue indefinitely.

FIGURE 1 A broadcast antenna

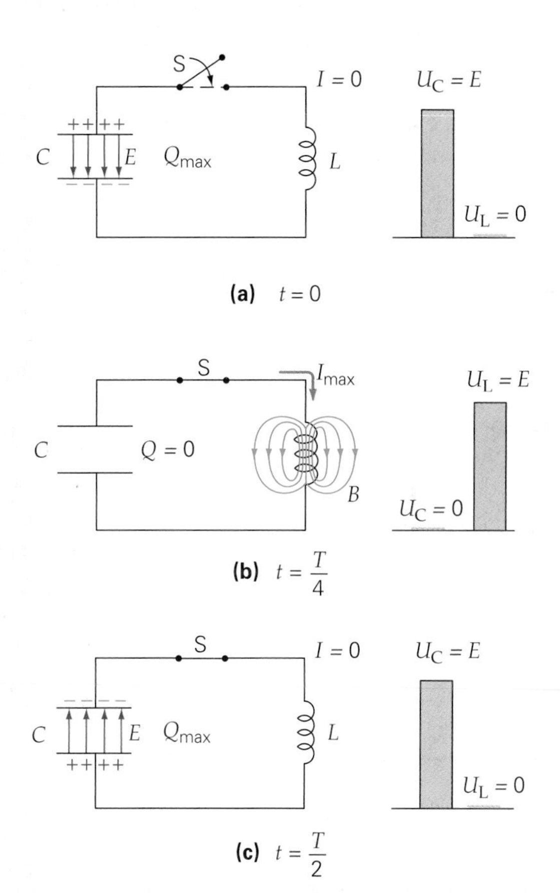

FIGURE 2 An oscillating LC circuit If the resistance is negligible, this circuit will oscillate indefinitely. Half a full cycle is shown between $t = 0$ and $t = T/2$. Energy is transferred back and forth between magnetic and electric types of energy (as shown by the energy histograms to the right). The oscillating electrons in the wire will give off electromagnetic radiation at the circuit's oscillating frequency.

Integrated Example 21.8 ■ AM versus FM: Resonance in Radio Reception

(a) When you switch from an AM station (on the "AM band"—the word *band* refers to a specific range of frequencies) to one on the FM band, you are effectively changing the capacitance of the receiving circuit, assuming constant inductance. Is the capacitance (1) increased or (2) decreased when you make this change? (b) Suppose you were listening to news on an AM station at 920 kHz and switched to a music on the FM band at 99.7 MHz. By what factor would you have changed the capacitance of the receiving circuit in the radio, assuming constant inductance?

(a) Conceptual Reasoning. Because FM stations broadcast at significantly higher frequencies than do AM stations (see Table 20.1, p. 676), the resonance frequency of the receiver must be increased to receive signals on the FM band. An increase of the resonance frequency requires reducing the capacitance since the inductance is fixed. Thus the correct answer is (2).

(b) Quantitative Reasoning and Solution. The resonance frequency (Eq. 21.25) depends on the inductance and the capacitance. Because the question asks for a "factor," it is clearly asking for a ratio of the new capacitance to the original capacitance. The frequencies have to be expressed in the same units, so convert MHz into kHz and use unprimed quantities for AM and primed quantities for FM.

Given: $f_o = 920$ kHz, $f'_o = 99.7 \text{ MHz} = 99.7 \times 10^3$ kHz

Find: C'/C (ratio of FM capacitance to AM capacitance)

From Eq. 21.25, the two resonant frequencies are given by

$$f_o = \frac{1}{2\pi\sqrt{LC}} \quad \text{and} \quad f'_o = \frac{1}{2\pi\sqrt{LC'}}$$

Dividing the first of these equations by the second gives

$$\frac{f_o}{f'_o} = \frac{2\pi\sqrt{LC'}}{2\pi\sqrt{LC}} = \sqrt{\frac{C'}{C}}$$

Solving for the capacitance ratio by squaring, and substituting the numbers, we have

$$\frac{C'}{C} = \left(\frac{f_o}{f'_o}\right)^2 = \left(\frac{920 \text{ kHz}}{99.7 \times 10^3 \text{ kHz}}\right)^2 = 8.51 \times 10^{-5}$$

Thus, $C' = 8.51 \times 10^{-5}\, C$ and the capacitance was decreased by a factor of almost one ten-thousandth ($8.51 \times 10^{-5} \approx 10^{-4}$).

Follow-Up Exercise. (a) Based on the resonance curves shown in Fig. 21.13b, can you explain how it is possible to pick up two radio stations *simultaneously* on your radio? (You may have encountered this phenomenon, particularly between two cities located far apart from one another. Two stations are sometimes granted licenses for broadcasting at closely spaced frequencies under the assumption that they won't both be received by the same radio. However, under certain atmospheric conditions, this may not be true.) (b) In part (b) of this Example, if you next increased the capacitance by a factor of two (starting with the news at 920 kHz) to listen to a hockey game, to what new frequency on the AM band would you now be tuned?

Chapter Review

- An **ac voltage** is described by

$$V = V_o \sin \omega t = V_o \sin 2\pi f t \qquad (21.1)$$

- For a sinusoidally varying current, called **ac current,** the **peak current I_o** and the **rms** (root-mean-square or effective) **current I_{rms}** are related by

$$I_{rms} = \frac{I_o}{\sqrt{2}} = 0.707 I_o \qquad (21.6)$$

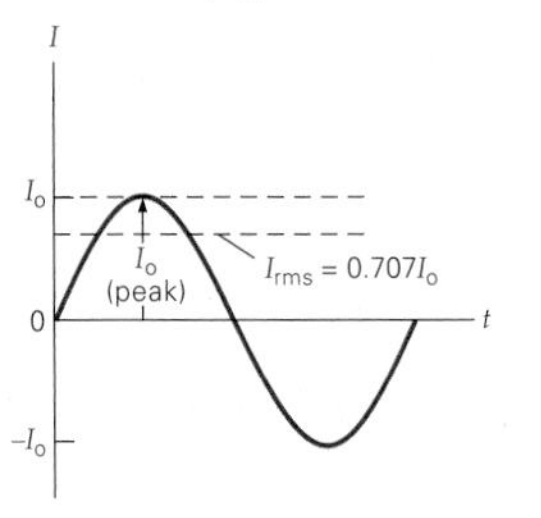

- For an ac voltage, the **peak voltage V_o** is related to its the **rms** (root-mean-square) **voltage V_{rms}** by

$$V_{rms} = \frac{V_o}{\sqrt{2}} = 0.707 V_o \qquad (21.8)$$

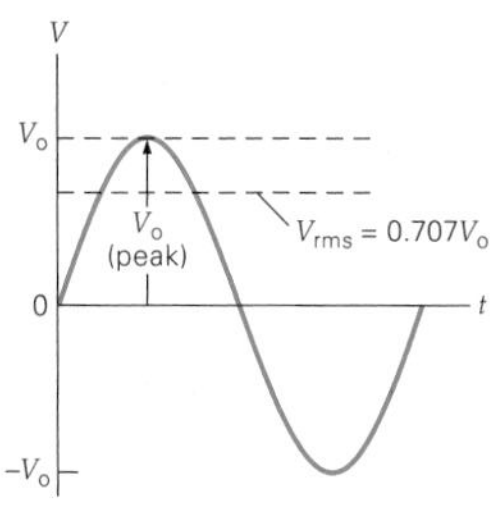

- The current in a resistor is in phase with the voltage across it. For a capacitor, the current is 90° (one quarter of a cycle) ahead of the voltage. For an inductor, the current lags the voltage by 90°.

- In ac circuits, joule heating is due entirely to the resistive elements, and the time-averaged **power dissipation** is

$$\bar{P} = I_{rms}^2 R \quad (21.10)$$

- In an ac circuit, capacitors and inductors allow current and create opposition to current. This opposition is characterized by **capacitive reactance (X_C)** and **inductive reactance (X_L)**, respectively. The capacitive reactance is given by

$$X_C = \frac{1}{\omega C} = \frac{1}{2\pi f C} \quad (21.11)$$

The inductive reactance is given by

$$X_L = \omega L = 2\pi f L \quad (21.14)$$

- Ohm's law, as applied to each type of circuit element, is a generalization of the version from dc circuits. The relationship between rms current and the rms voltage for a resistor is:

$$V_{rms} = I_{rms} R \quad (21.9)$$

The relationship between rms current and the rms voltage for a capacitor is

$$V_{rms} = I_{rms} X_C \quad (21.12)$$

The relationship between rms current and the rms voltage for an inductor is

$$V_{rms} = I_{rms} X_L \quad (21.15)$$

- **Phasors** are vectorlike quantities that allow resistances and reactances to be represented graphically.

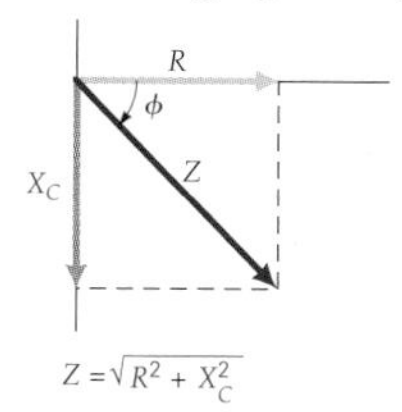

- **Impedance (Z)** is the total, or effective, opposition to current that takes into account both resistances and reactances. Impedance is related to current and circuit voltage by a generalization of Ohm's law:

$$V_{rms} = I_{rms} Z \quad (21.17)$$

- The **impedance for a series RLC circuit** is

$$Z = \sqrt{R^2 + (X_L - X_C)^2} \quad (21.19)$$

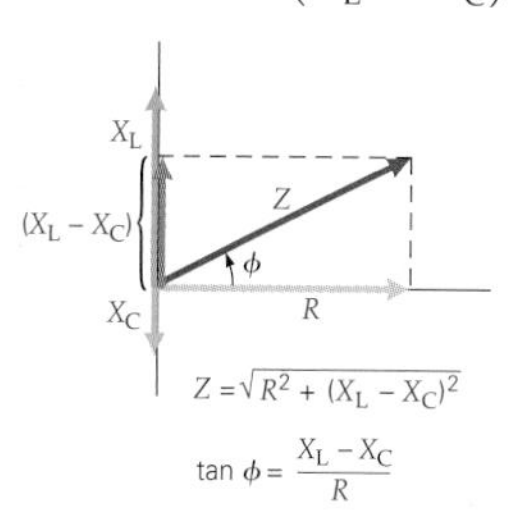

- The **phase angle (ϕ)** between the rms voltage and the rms current in a series RLC circuit is

$$\tan\phi = \frac{X_L - X_C}{R} \quad (21.20)$$

- The **power factor ($\cos\phi$)** for a series RLC circuit is a measure of how close to the maximum power dissipation the circuit is. The power factor is

$$\cos\phi = \frac{R}{Z} \quad (21.22)$$

The average power dissipated (joule heating in the resistor) is

$$\bar{P} = I_{rms} V_{rms} \cos\phi \quad (21.23)$$

or

$$\bar{P} = I_{rms}^2 Z \cos\phi \quad (21.24)$$

- The **resonance frequency (f_o)** of an RLC circuit is the frequency at which the circuit dissipates maximum power. This frequency is

$$f_o = \frac{1}{2\pi\sqrt{LC}} \quad (21.25)$$

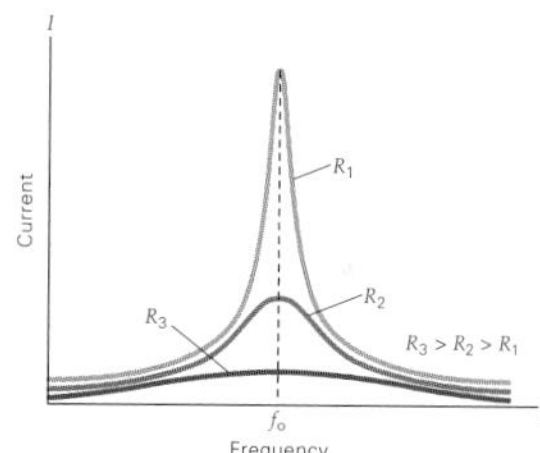

Exercises

MC = *Multiple Choice Question,* **CQ** = *Conceptual Question, and* **IE** = *Integrated Exercise. Throughout the text, many exercise sections will include "paired" exercises. These exercise pairs, identified with* **red numbers,** *are intended to assist you in problem solving and learning. In a pair, the first exercise (even numbered) is worked out in the Study Guide so that you can consult it should you need assistance in solving it. The second exercise (odd numbered) is similar in nature, and its answer is given at the back of the book.*

21.1 Resistance in an AC Circuit

1. **MC** Which of the following voltages is larger for a sinusoidally varying ac voltage: (a) V_o, (b) V_{rms}, or (c) they have the same value?

2. **MC** During the course of one ac voltage cycle (United States) how long does the direction of the current stay constant in a resistor: (a) 1/60 s, (b) 1/120 s, or (c) 1/30 s?

3. **MC** During seven complete ac voltage cycles (United States) what is the average voltage: (a) 0 V, (b) 60 V, (c) 120 V, or (d) 170 V?

4. **CQ** The average current in a resistor in an ac circuit is zero. Explain why the average power delivered to a resistor isn't zero.

5. **CQ** The voltage and current associated with a resistor in an ac circuit are *in phase*. What does that mean?

6. **CQ** A 60-W lightbulb designed to work at 240 V in England is instead connected to a 120-V source. Discuss the changes in the bulb's rms current and power when it is at 120 V compared with 240 V. Assume the bulb is ohmic.

7. **CQ** If the ac voltage and current for a particular circuit element are given respectively by $V = 120\sin(120\pi t)$ and $I = 30\sin(120\pi t + \pi/2)$, could the circuit element be a resistor? Is the frequency 60 Hz? Explain.

8. ● What are the peak voltages of a 120-V ac line and a 240-V ac line?

9. ● An ac circuit has an rms current of 5.0 A. What is the peak current?

10. ● The maximum voltage across a resistor in an ac circuit is 156 V. Find the resistor's rms voltage.

11. ● How much ac rms current must be in a 10-Ω resistor to produce an average power of 15 W?

12. ● An ac circuit contains a resistor with a resistance of 5.0 Ω. The resistor has an rms current of 0.75 A. (a) Find its rms voltage and peak voltage. (b) Find the average power delivered to the resistor.

13. ● A hair dryer is rated at 1200 W when plugged into a 120-V outlet. Find (a) its rms current, (b) its peak current, and (c) its resistance.

14. IE ●● The voltage across a 10-Ω resistor varies as $V = (170\text{ V})\sin(100\pi t)$. (a) Will the current in the resistor (1) be in phase with the voltage, (2) lead the voltage by 90°, or (3) lag the voltage by 90°? (b) Write the expression for the current in the resistor as a function of time and determine the voltage frequency.

15. ●● An ac voltage is applied to a 25.0-Ω resistor so that it dissipates 500 W of power. Find the resistor's (a) rms and peak currents and (b) rms and peak voltages.

16. IE ●● An ac voltage source has a peak voltage of 85 V and a frequency of 60 Hz. The voltage at $t = 0$ is zero. (a) A student wants to calculate the voltage at $t = 1/240$ s. How many possible answers are there: (1) one, (2) two, or (3) three? Why? (b) Determine all possible answers.

17. ●● An ac voltage source has an rms voltage of 120 V. Its voltage goes from zero to its maximum value in 4.20 ms. Write an expression for the voltage as a function of time.

18. ●● What are the resistance and rms current of a 100-W, 120-V computer monitor?

19. ●● Find the rms and peak currents in a 40-W, 120-V lightbulb.

20. ●● A 50-kW heater is designed to run using a 240-V ac source. Find its (a) peak current and (b) peak voltage.

21. ●● The current in a resistor is given by $I = (8.0\text{ A})\sin(40\pi t)$ when a voltage given by $V = (60\text{ V})\sin(40\pi t)$ is applied to it. (a) What is the frequency and period of the voltage source? (b) What is the average power delivered to the resistor?

22. ●● The current and voltage outputs of an operating ac generator have peak values of 2.5 A and 16 V, respectively. (a) What is the average power output of the generator? (b) What is the effective resistance of the circuit it is in?

23. ●●● The current in a 60-Ω resistor is given by $I = (2.0\text{ A})\sin(380t)$. (a) What is the frequency of the current? (b) What is the rms current? (c) How much average power is delivered to the resistor? (d) Write an equation for the voltage across the resistor as a function of time. (e) Write an equation for the power delivered to the resistor as a function of time. (f) Show that the rms power obtained in part (e) is the same as your answer to part (c).

21.2 Capacitive Reactance *and* 21.3 Inductive Reactance

24. MC In a purely capacitive ac circuit, (a) the current and voltage are in phase, (b) the current leads the voltage, (c) the current lags the voltage, or (d) none of the preceding.

25. MC A single capacitor is connected to an ac voltage source. When the voltage across the capacitor is at a maximum, then the charge on it is (a) zero (b) at a maximum or (c) neither of the preceding, but somewhere in between.

26. MC A single inductor is connected to an ac voltage source. When the voltage across the inductor is at a maximum, then the current in it is not changing. (a) true (b) false or (c) cannot be determined from the given information.

27. CQ Explain why, under very low frequency ac conditions, a capacitor acts almost as an open circuit while an inductor acts almost as a short circuit.

28. CQ Can an inductor oppose dc current? What about a capacitor? Explain each and why they are different.

29. CQ If the current on a 10-μF capacitor is described by $I = (120\text{ A})\sin(120\pi t + \pi/2)$, explain why the instantaneous voltage across it at $t = 0$ is zero whereas the current at that time is not.

30. ● Find the frequency at which a 25-μF capacitor has a reactance of 25 Ω.

31. ● A single 2.0-μF capacitor is connected across the terminals of a 60-Hz voltage source, and a current of 2.0 mA is measured on an ac ammeter. What is the capacitive reactance of the capacitor?

32. ● What capacitance would have a reactance of 100 Ω in a 60-Hz ac circuit?

33. ● A single 50-mH inductor forms a complete circuit when connected to an ac voltage source at 120 V and 60 Hz. (a) What is the inductive reactance of the circuit? (b) How much current is in the circuit? (c) What is the phase angle between the current and the applied voltage? (Assume negligible resistance.)

34. ● How much current is in a circuit containing only a 50-μF capacitor connected to an ac generator with an output of 120 V and 60 Hz?

35. ●● A variable capacitor in a circuit with a 120-V, 60-Hz source initially has a capacitance of 0.25 μF. The capacitance is then increased to 0.40 μF. What is the percentage change in the current in the circuit?

36. ●● An inductor has a reactance of 90 Ω in a 60-Hz ac circuit. What is its inductance?

37. ●● Find the frequency at which a 250-mH inductor has a reactance of 400 Ω.

38. **IE** ●● A capacitor is connected to a variable-frequency ac voltage source. (a) If the frequency increases by a factor of 3, the capacitive reactance will be (1) 3, (2) $\frac{1}{3}$, (3) 9, (4) $\frac{1}{9}$ times the original reactance. Why? (b) If the capacitive reactance of a capacitor at 120 Hz is 100 Ω, what is its reactance if the frequency is changed to 60 Hz?

39. ●● With a single 150-mH inductor in a circuit with a 60-Hz voltage source, a current of 1.6 A is measured on an ac ammeter. (a) What is the rms voltage of the source? (b) What is the phase angle between the current and that voltage?

40. ●● What inductance has the same reactance in a 120-V, 60-Hz circuit as a capacitance of 10 μF?

41. ●● A circuit with a single capacitor is connected to a 120-V, 60-Hz source. What is its capacitance if there is a current of 0.20 A in the circuit?

42. **IE** ●● An inductor is connected to a variable-frequency ac voltage source. (a) If the frequency decreases by a factor of 2, the rms current will be (1) 2, (2) $\frac{1}{2}$, (3) 4, (4) $\frac{1}{4}$ times the original rms current. Why? (b) If the rms current in an inductor at 40 Hz is 9.0 A, what is its rms current if the frequency is changed to 120 Hz?

21.4 Impedance: RLC Circuits *and* 21.5 Circuit Resonance

43. **MC** The impedance of an RLC circuit depends on (a) frequency, (b) inductance, (c) capacitance, or (d) all of the preceding.

44. **MC** If the capacitance of a series RLC circuit is decreased, (a) the capacitive reactance increases, (b) the inductive reactance increases, (c) the current remains constant, or (d) the power factor remains constant.

45. **MC** When a series RLC circuit is driven at its resonance frequency, (a) energy is dissipated only by the resistive element, (b) the power factor has a value of one, (c) there is maximum power delivered to the circuit, or (d) all of the preceding.

46. **CQ** What is the impedance of an RLC circuit at resonance and why?

47. **CQ** Is any power ever delivered to capacitors or inductors in ac circuits? Why or why not?

48. **CQ** What are the factors that determine the resonant frequency of an RLC circuit? Is resistance a factor? Explain?

49. ● A coil in a 60-Hz circuit has a resistance of 100 Ω and an inductance of 0.45 H. Calculate (a) the coil's reactance and (b) the circuit's impedance.

50. ● A series RC circuit has a resistance of 200 Ω and a capacitance of 25 μF and is driven by a 120-V, 60-Hz source. (a) Find the capacitive reactance and impedance of the circuit. (b) How much current is drawn from the source?

51. ● A series RL circuit has a resistance of 100 Ω and an inductance of 100 mH and is driven by a 120-V, 60-Hz source. (a) Find the inductive reactance and the impedance of the circuit. (b) How much current is drawn from the source?

52. ● An RC circuit has a resistance of 250 Ω and a capacitance of 6.0 μF. If the circuit is driven by a 60-Hz source, find (a) the capacitive reactance and (b) the impedance of the circuit.

53. **IE** ● An RC circuit has a resistance of 100 Ω and a capacitive reactance of 50 Ω. (a) Will the phase angle be (1) positive, (2) zero, or (3) negative? Why? (b) What is the phase angle of this circuit?

54. ●● A series RLC circuit has a resistance of 25 Ω, an inductance of 0.30 H, and a capacitance of 8.0 μF. (a) At what frequency should the circuit be driven for the maximum power to be transferred from the driving source? (b) What is the impedance at that frequency?

55. **IE** ●● In a series RLC circuit, $R = X_C = X_L = 40\ \Omega$ for a particular driving frequency. (a) This circuit is (1) inductive, (2) capacitive, (3) in resonance. Why? (b) If the driving frequency is doubled, what will be the impedance of the circuit?

56. **IE** ●● (a) An RLC series circuit is in resonance. Which one of the following can you change without upsetting the resonance: (1) resistance, (2) capacitance, (3) inductance, or (4) frequency? Why? (b) A resistor, an inductor, and a capacitor have values of 500 Ω, 500 mH, and 3.5 μF, respectively. They are connected in series to a power supply of 240 V with a frequency of 60 Hz. What values of resistance and inductance would be required for this circuit to be in resonance (without changing the capacitor)?

57. ●● How much power is dissipated in the circuit described in Exercise 56b using the initial values of resistance, inductance and capacitance?

58. ●● What is the resonant frequency of an RLC circuit with a resistance of 100 Ω, an inductance of 100 mH, and a capacitance of 5.00 μF?

59. ●● A tuning circuit in a radio receiver has a fixed inductance of 0.50 mH and a variable capacitor. If the circuit is tuned to a radio station broadcasting at 980 kHz on the AM dial, what is the capacitance of the capacitor?

60. ●● What would be the range of the variable capacitor in Exercise 59 required for tuning over the complete AM band? [*Hint*: See Table 20.1.]

61. ●● Find the currents supplied by the ac source for all possible connections in ▼Fig. 21.15.

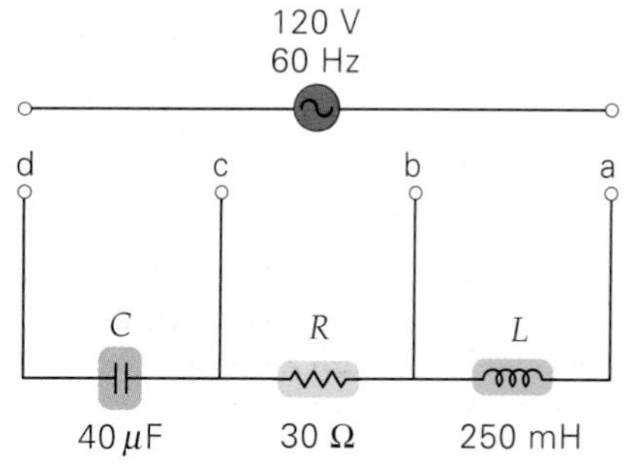

◀ **FIGURE 21.15** **A series RLC circuit** See Exercise 61.

62. IE ●● A coil with a resistance of 30 Ω and an inductance of 0.15 H is connected to a 120-V, 60-Hz source. (a) Is the phase angle of this circuit (1) positive, (2) zero, or (3) negative? Why? (b) What is the phase angle of the circuit? (c) How much rms current is in the circuit? (d) What is the average power delivered to the circuit?

63. ●● A small welder uses a voltage source of 120 V at 60 Hz. When the source is operating, it requires 1200 W of power, and the power factor is 0.75. Find the rms current in the welder.

64. ●● A series circuit is connected to a 220-V, 60-Hz power supply. The circuit has the following components: a 10-Ω resistor, a coil with an inductive reactance of 120 Ω, and a capacitor with a reactance of 120 Ω. Compute the rms voltage across (a) the resistor, (b) the inductor, and (c) the capacitor.

65. ●● A series RLC circuit has a resistance of 25 Ω, a capacitance of 0.80 μF, and an inductance of 250 mH. The circuit is connected to a variable-frequency source with a fixed rms voltage output of 12 V. If the frequency that is supplied is set at the circuit's resonance frequency, what is the rms voltage across each of the circuit elements?

66. ●● (a) In Exercises 64 and 65, determine the numerical (scalar) sum of the rms voltages across the three circuit elements and explain why it is much larger than the source voltage. (b) Determine the sum of these voltages using the proper phasor techniques and show that your result is equal to the source voltage.

67. IE ●● (a) If the circuit in ▾Fig. 21.16 is in resonance, the impedance of the circuit is (1) greater than 25 Ω, (2) equal to 25 Ω, (3) less than 25 Ω. Why? (b) If the driving frequency is 60 Hz, what is the circuit's impedance?

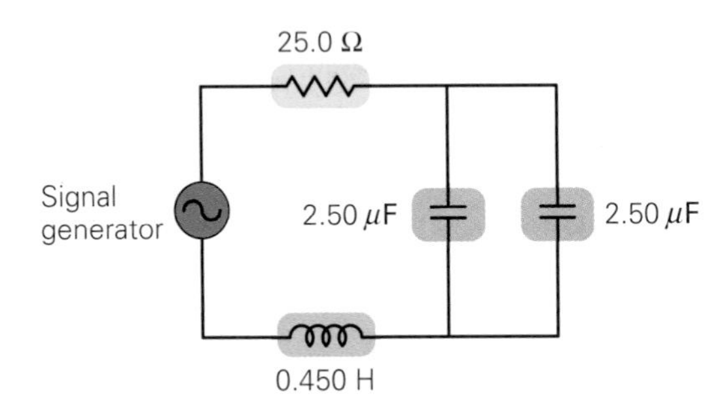

▲ **FIGURE 21.16** Tune to resonance See Exercise 67.

68. ●●● A series RLC circuit with a resistance of 400 Ω has capacitive and inductive reactances of 300 Ω and 500 Ω, respectively. (a) What is the power factor of the circuit? (b) If the circuit operates at 60 Hz, what additional capacitance should be connected to the original capacitance to give a power factor of unity, and how should the capacitors be connected?

69. ●●● A series RLC circuit has components with $R = 50\ \Omega$, $L = 0.15$ H, and $C = 20\ \mu$F. The circuit is driven by a 120-V, 60-Hz source. What is the power delivered to the circuit, expressed as a percentage of the power delivered when the circuit is in resonance?

Comprehensive Exercises

70. A series RLC radio receiver circuit with an inductance of 1.50 μH is tuned to an FM station at 98.9 MHz by adjusting a variable capacitor. When the circuit is tuned to this station, (a) what is its inductive reactance? (b) What is its capacitive reactance? (c) What is its capacitance?

71. A circuit connected to a 110-V, 60-Hz source contains a 50-Ω resistor and a coil with an inductance of 100 mH. Find (a) the reactance of the coil, (b) the impedance of the circuit, (c) the current in the circuit, and (d) the power dissipated by the coil, and (e) calculate the phase angle between the current and the applied voltage.

72. A 1.0-μF capacitor is connected to a 120-V, 60-Hz source. (a) What is the capacitive reactance of the circuit? (b) How much current is in the circuit? (c) What is the phase angle between the current and the applied voltage?

73. IE (a) If an RLC circuit is in resonance, the phase angle of the circuit is (1) positive, (2) zero, (3) negative. Why? (b) A circuit has an inductive reactance of 280 Ω at 60 Hz. What value of capacitance would set this circuit into resonance?

74. The circuit in ▾Fig. 21.17a is called a *low-pass filter* because a large current and voltage (and thus a lot of power) is delivered to the load resistor (R_L) only by a low-frequency source. The circuit in Fig. 21.17b is called a *high-pass filter* because a large current and voltage (and thus a lot of power) is delivered to the load only by a high-frequency source. Describe conceptually why the circuits have these characteristics.

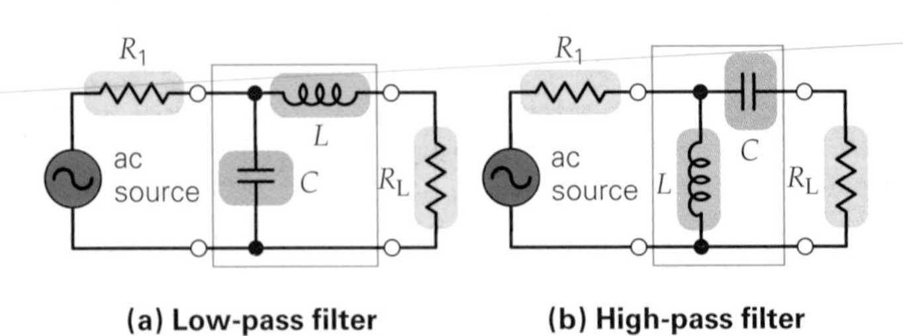

▲ **FIGURE 21.17** Low-pass and high-pass filters See Exercise 74.

The following Physlet Physics Problems can be used with this chapter.
31.1, 31.3, 31.4, 31.5, 31.8, 31.9, 31.11, 31.12, 31.14

CHAPTER

22 REFLECTION AND REFRACTION OF LIGHT

22.1 Wave Fronts and Rays 706

22.2 Reflection 707

22.3 Refraction 708

22.4 Total Internal Reflection and Fiber Optics 717

22.5 Dispersion 721

PHYSICS FACTS

- Due to total internal reflection, optical fibers allow signals to travel for long distances without repeaters (amplifiers), to compensate for reductions in signal strength. Fiber-optic repeaters are currently about 100 km (about 62 mi) apart, compared to about 1.5 km (about 1 mi) for electrical (wire-based) systems.
- Every day installers lay enough new fiber-optic cables for computer networks to circle the Earth three times. Optical fibers can be drawn to smaller diameters than copper wire. Fibers can be as small as 10 microns in diameter. In comparison, the average human hair is about 25 microns in diameter.
- Most camera lenses are coated with a thin film to reduce light loss due to reflection. For a typical seven-element camera lens, about 50% of the light would be lost due to reflection if the lens were not coated with thin films.
- In 1998, scientists at MIT made a perfect mirror, a mirror with 100% reflection. A tube lined with this type of mirror would transmit light over long distances better than optical fibers.

We live in a visual world, surrounded by eye-catching images such as that refractive image of the turtle shown in the photo. How these images are formed is something taken largely for granted—until we see something that can't be easily explained. *Optics* is the study of light and vision. Human vision requires *visible light* of wavelength from 400 nm to 700 nm (see Fig. 20.23). Optical properties, such as reflection and refraction, are shared by all electromagnetic waves. Light acts as a wave in its propagation (Chapter 24) and as a particle (photon) when it interacts with matter (Chapter 27–30).

In this chapter, we will investigate the basic optical phenomena of reflection, refraction, total internal reflection, and dispersion. The principles that govern reflection explain the behavior of mirrors, while those that govern refraction explain the properties of lenses. With the aid of these and other optical principles, we can understand many optical phenomena experienced every day—why a glass prism spreads light into a spectrum of colors, what causes mirages, how rainbows are formed, and why the legs of a person standing in a lake or swimming pool seem to shorten. Some less familiar but increasingly useful territory, including the fascinating field of fiber optics will also be explored.

A simple geometrical approach involving straight lines and angles can be used to investigate many aspects of the properties of light, especially how light propagates. For these purposes, we need not be concerned with the physical (wave) nature of electromagnetic waves described in Chapter 20. The principles of geometrical optics will be introduced here and applied in greater detail in the study of mirrors and lenses in Chapter 23.

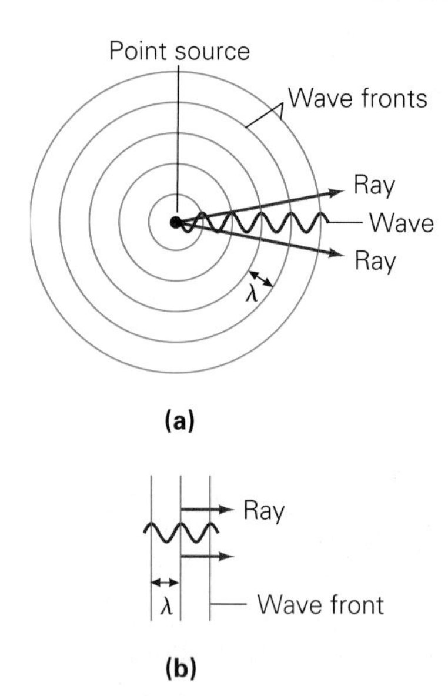

▲ **FIGURE 22.1 Wave fronts and rays** A wave front is defined by adjacent points on a wave that are in phase, such as those along wave crests or troughs. A line perpendicular to a wave front in the direction of the wave's propagation is called a ray. **(a)** Near a point source, the wave fronts are circular in two dimensions and spherical in three dimensions. **(b)** Very far from a point source, the wave fronts are approximately linear or planar and the rays nearly parallel.

22.1 Wave Fronts and Rays

OBJECTIVE: To define and explain the concepts of wave fronts and rays.

Waves, electromagnetic or otherwise, are conveniently described in terms of wave fronts. A **wave front** is the line or surface defined by adjacent portions of a wave that are in phase. If an arc is drawn along one of the crests of a circular water wave moving out from a point source, all the particles on the arc will be in phase (◂Fig. 22.1a). An arc along a wave trough would work equally well. For a three-dimensional spherical wave, such as a sound or light wave emitted from a point source, the wave front is a spherical surface rather than a circle.

Very far from the source, the curvature of a short segment of a circular or spherical wave front is extremely small. Such a segment may be approximated as a *linear wave front* (in two dimensions) or a **plane wave front** (in three dimensions), just as we take the surface of the Earth to be locally flat (Fig. 22.1b). A plane wave front can also be produced directly by a large luminous flat surface. In a uniform medium, wave fronts propagate outward from the source at a speed characteristic of the medium. This was seen for sound waves in Chapter 14, and the same occurs for light, although at a much faster speed. The speed of light is greatest in a vacuum: $c = 3.00 \times 10^8$ m/s. The speed of light in air, for all practical purposes, is the same as that in vacuum.

The geometrical description of a wave in terms of wave fronts tends to neglect the fact that the wave is actually oscillating, like those studied in Chapter 13. This simplification is carried a step further with the concept of a ray. As illustrated in Fig. 22.1, a line drawn perpendicular to a series of wave fronts and pointing in the direction of propagation is called a **ray**. Note that a ray points in the direction of the energy flow of a wave. A plane wave is assumed to travel in a straight line in a medium in the direction of its rays, perpendicular to its plane wave fronts. A beam of light can be represented by a group of rays or simply as a single ray (◂Fig. 22.2). The representation of light as rays is adequate and convenient for describing many optical phenomena.

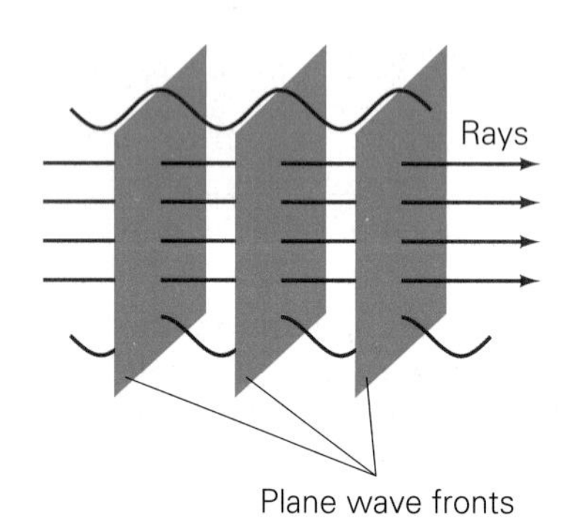

▲ **FIGURE 22.2 Light rays** A plane wave travels in a direction perpendicular to its wave fronts. A beam of light can be represented by a group of parallel rays (or by a single ray).

How do we see things and objects around us? We see them because rays from the objects, or rays that appear to come from the objects, enter our eyes (▾Fig. 22.3). In the eyes rays form images of the objects on the retina. The rays could be coming directly from the objects as in the case of light sources or could be reflected or refracted by the objects or other optical systems. Our eyes and brain working together, however, cannot tell whether the rays actually come from the objects or only *appear* to come from the objects. This is one way magicians can fool our eyes with seemingly impossible illusions.

The use of the geometrical representations of wave fronts and rays to explain phenomena such as the reflection and refraction of light is called **geometrical optics**. However, certain other phenomena, such as the interference of light, cannot be treated in this manner and must be explained in terms of actual wave characteristics. These phenomena will be considered in Chapter 24.

▸ **FIGURE 22.3 How we see things** We see things because **(a)** rays from the objects or **(b)** rays appearing to come from the objects enter our eyes.

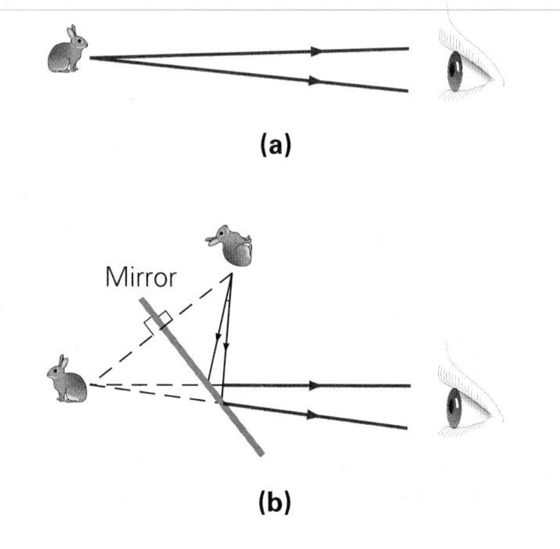

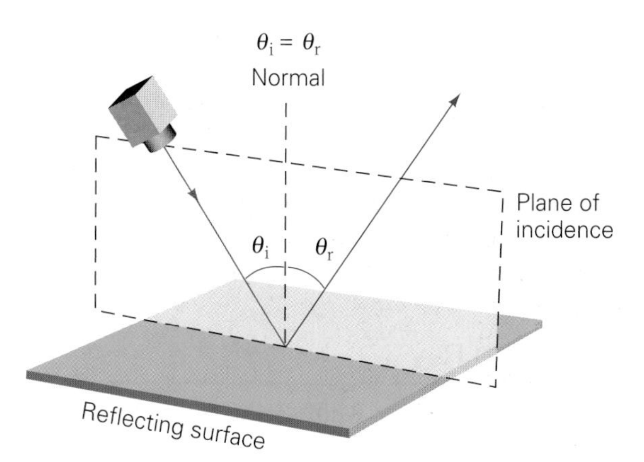

◀ **FIGURE 22.4 The law of reflection** According to the law of reflection, the angle of incidence (θ_i) is equal to the angle of reflection (θ_r). Note that the angles are measured relative to a normal (a line perpendicular to the reflecting surface). The normal and the incident and reflected rays always lie in the same plane.

22.2 Reflection

OBJECTIVES: To (a) explain the law of reflection, and (b) distinguish between regular (specular) and irregular (diffuse) reflection.

The reflection of light is an optical phenomenon of enormous importance: If light were not reflected to our eyes by objects around us, we wouldn't see the objects at all. **Reflection** involves the absorption and re-emission of light by means of complex electromagnetic vibrations in the atoms of the reflecting medium. However, the phenomenon is easily described by using rays.

A light ray incident on a surface is described by an **angle of incidence (θ_i)**. This angle is measured relative to a *normal*—a line perpendicular to the reflecting surface (▲Fig. 22.4). Similarly, the reflected ray is described by an **angle of reflection (θ_r)**, also measured from the normal. The relationship between these angles is given by the **law of reflection**: The angle of incidence is equal to the angle of reflection, or

$$\theta_i = \theta_r \quad \textit{law of reflection} \tag{22.1}$$

Two other attributes of reflection are that the incident ray, the reflected ray, and the normal all lie in the same plane, which is sometimes called the plane of incidence, and that the incident and the reflected rays are on opposite sides of the normal.

When the reflecting surface is smooth and flat, the reflected rays from parallel incident rays are also parallel (▸Fig. 22.5a). This type of reflection is called **specular**, or **regular**, **reflection**. The reflection from a highly polished flat mirror is an example of specular (regular) reflection (Fig. 22.5b). If the reflecting surface is rough, however, the reflected rays are not parallel, because of the irregular nature of the surface (▸Fig. 22.6). This type of reflection is termed **diffuse**, or **irregular**, **reflection**. The reflection of light from this page is an example of diffuse reflection because the paper is microscopically rough. Insight 22.1 on A Dark, Rainy Night on page 709, discusses more about the difference between specular and diffuse reflection in a real-life situation.

Note in Fig. 22.5a and Fig. 22.6 that the law of reflection still applies locally to both specular and diffuse reflection. However, the type of reflection involved determines whether we see images from a reflecting surface. In specular reflection, the reflected, parallel rays produce an image when they are viewed by an optical system such as an eye or a camera. Diffuse reflection does not produce an image, because the light is reflected in various directions.

Experience with friction and direct investigations show that all surfaces are rough on a microscopic scale. What, then, determines whether reflection is specular or diffuse? In general, if the dimensions of the surface irregularities are greater than the wavelength of the light, the reflection is diffuse. Therefore, to make a good mirror, glass (with a metal coating) or metal must be polished at least until the surface

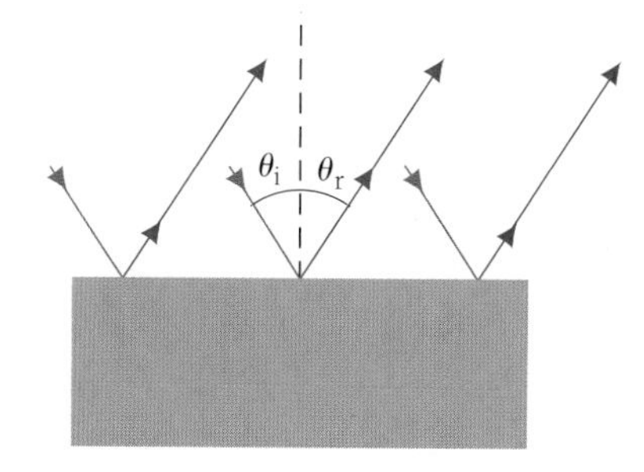

(a) Specular (regular) reflection (diagram)

(b) Specular (regular) reflection (photo)

▲ **FIGURE 22.5 Specular (regular) reflection** **(a)** When a light beam is reflected from a smooth surface and the reflected rays are parallel, the reflection is said to be specular or regular. **(b)** Specular (regular) reflection from a smooth water surface produces an almost perfect mirror image of salt mounds at this Australian salt mine.

▲ **FIGURE 22.6 Diffuse (irregular) reflection** Reflected rays from a relatively rough surface, such as this page, are not parallel; the reflection is said to be diffuse or irregular. (Note that the law of reflection still applies locally to each individual ray.)

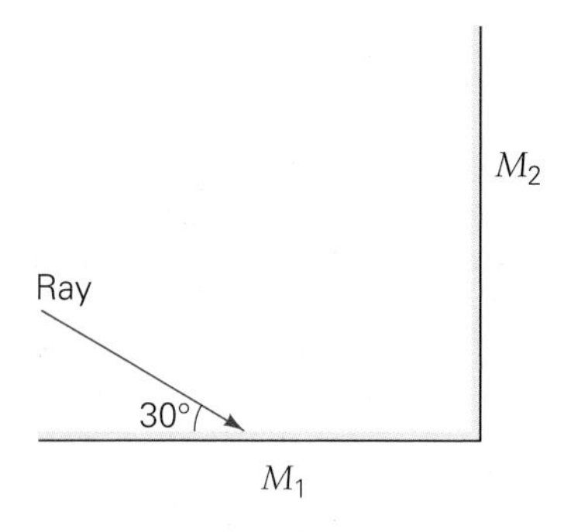

▲ **FIGURE 22.7** Trace the ray See Example 22.1.

Note: Drawing diagrams like these is extremely important in the study of geometrical optics.

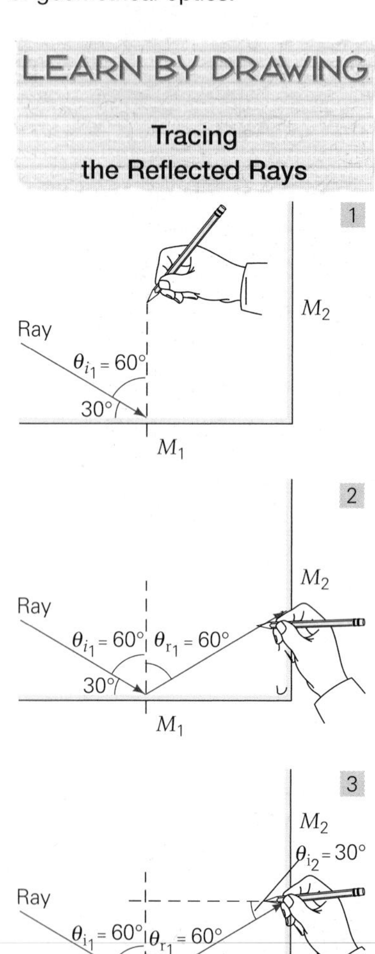

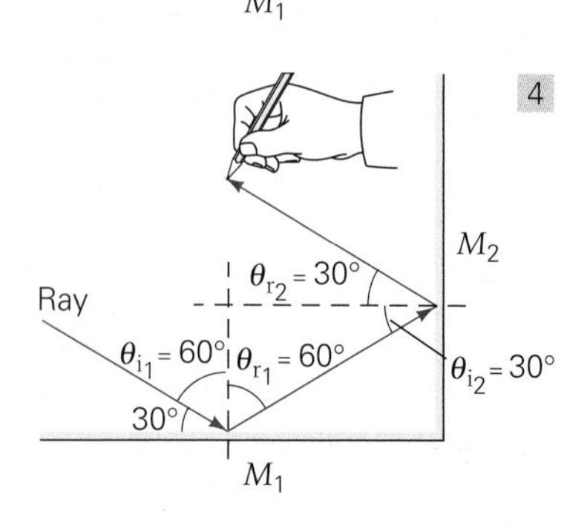

irregularities are about the same size as the wavelength of light. Recall from Chapter 20 that the wavelength of visible light is on the order of 10^{-7} m. (You will learn more about reflection from a mirror in the Learn by Drawing presented in Example 22.1.)

Diffuse reflection enables us to see illuminated objects, such as the Moon. If the Moon's spherical surface were smooth, only the reflected sunlight from a small region would come to an observer on the Earth, and only that small illuminated area would be seen. Also, you can see the beam of light from a flashlight or spotlight because of diffuse reflection from dust and particles in the air.

Example 22.1 ■ Tracing the Reflected Ray

Two mirrors, M_1 and M_2, are perpendicular to each other, with a light ray incident on one of the mirrors as shown in ◂Fig. 22.7. (a) Sketch a diagram to trace the path of the reflected light ray. (b) Find the direction of the ray after it is reflected by M_2.

Thinking It Through. The law of reflection can be used to determine the direction of the ray after it leaves the first and then the second mirror.

Solution.

Given: $\theta = 30°$ (angle relative to M_1)

Find: (a) Sketch a diagram tracing the light ray
(b) θ_{r_2} (angle of reflection from M_2)

Follow steps 1–4 in Learn by Drawing:

(a) 1. Since the incident and reflected rays are measured from the normal (a line perpendicular to the reflecting surface), we draw the normal to mirror M_1 at the point the incident ray hits M_1. From geometry, it can be seen that the angle of incidence on M_1 is $\theta_{i_1} = 60°$.

2. According to the law of reflection, the angle of reflection from M_1 is also $\theta_{r_1} = 60°$. Next, draw this reflected ray with an angle of reflection of 60°, and extend it until it hits M_2.

3. Draw another normal to M_2 at the point where the ray hits M_2. Also from geometry (focus on the triangle in the diagram), the angle of incidence on M_2 is $\theta_{i_2} = 30°$. (Why?)

(b) 4. The angle of reflection off M_2 is $\theta_{r_2} = \theta_{i_2} = 30°$. This is the final direction of the ray reflected after both mirrors.

What if the directions of the rays are reversed? In other words, if a ray is first incident on M_2, in the direction opposite that of the one drawn for part (b), will all the rays reverse their directions? Draw another diagram to prove that this is indeed the case. Light rays are reversible.

Follow-Up Exercise. When following an eighteen-wheel truck, you may see a sign on the back stating, "If you can't see my mirror, I can't see you." What does this mean? *(Answers to all Follow-Up Exercises are at the back of the text.)*

22.3 Refraction

OBJECTIVES: To (a) explain refraction in terms of Snell's law and the index of refraction, and (b) give examples of refractive phenomena.

Refraction refers to the change in direction of a wave at a boundary where the wave passes from one transparent medium into another. In general, when a wave is incident on a boundary between media, some of the wave's energy is reflected and some is transmitted. For example, when light traveling in air is incident on a transparent material such as glass, it is partially reflected and partially transmitted (▸Fig. 22.8). But the direction of the transmitted light is different from the direction of the incident light, so the light is said to have been refracted; in other words, it has changed direction.

This change in direction is caused by the fact that light travels with different speeds in different media. Intuitively, you might expect the passage of light to take longer through a medium with more atoms per volume, and the

INSIGHT 22.1 A DARK, RAINY NIGHT

When you drive on a dry night, you can clearly see the road and the street signs ahead of you from your headlights. However, here is a familiar scene on a dark, rainy night: Even with headlights, you can hardly see the road ahead. When a car approaches, the situation becomes even worse. You see the reflections of the approaching car's headlights from the surface of the road, and they appear brighter than usual. Often nothing can be seen except the reflective glare of the oncoming headlights.

What causes these conditions? When the road surface is dry, the reflection of light off the road is diffuse (irregular), because the surface is rough. Light from your headlights hits the road in front of you and reflects in all directions. Some of it reflects back, and you can see the road clearly (just as you can read this page because the paper is microscopically rough). However, when the road surface is wet, water fills the crevices, turning the road into a relatively smooth reflecting surface (Fig. 1a). Light from the headlights then reflects ahead. The normally diffuse reflection is gone and is replaced by specular reflection. Reflected images of lighted buildings and road lights form, blurring the view of the surface, and the specular reflection of oncoming cars' headlights may make it difficult for you to see the road (Fig. 1b).

Besides wet, slippery surfaces, specular reflection is a major cause of accidents on rainy nights. Thus, extra caution is advised under such conditions.

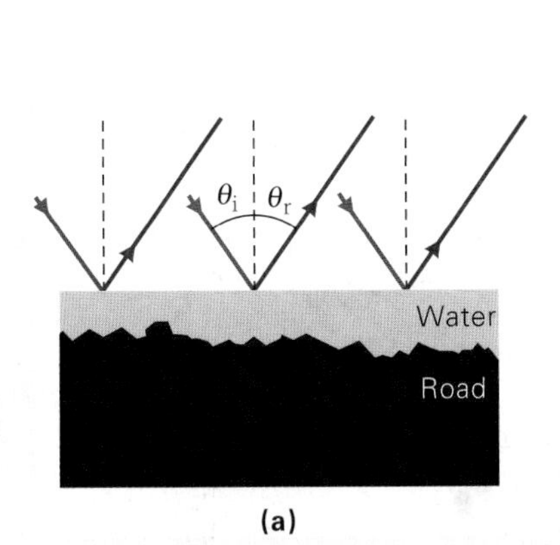

(a)

(b)

FIGURE 1 Diffuse to specular (a) The diffuse reflection from a dry road is turned into specular reflection by water on the road's surface. **(b)** Instead of seeing the road, a driver sees the reflected images of lights, buildings, and so on.

speed of light is, in fact, generally less in denser media. For example, the speed of light in water is about 75% of that in air or a vacuum. ▼Fig. 22.9a shows the refraction of light at an air–water boundary.

The change in the direction of wave propagation is described by the **angle of refraction**. In Fig. 22.9b, θ_1 is the angle of incidence and θ_2 is the angle of refraction. We use notations of θ_1 and θ_2 for the angles of incidence and refraction to avoid confusion with θ_i and θ_r for the angles of incidence and reflection. Willebrord Snell (1580–1626), a Dutch physicist, discovered a relationship between the angles (θ) and the speeds (v) of light in two media (Fig. 22.9b):

$$\frac{\sin\theta_1}{\sin\theta_2} = \frac{v_1}{v_2} \quad \textit{Snell's law} \tag{22.2}$$

This expression is known as **Snell's law**. Note that θ_1 and θ_2 are always taken with respect to the normal.

Thus, light is refracted when passing from one medium into another because the speed of light is different in the two media. The speed of light is greatest in a vacuum, and it is therefore convenient to compare the speed of light in other

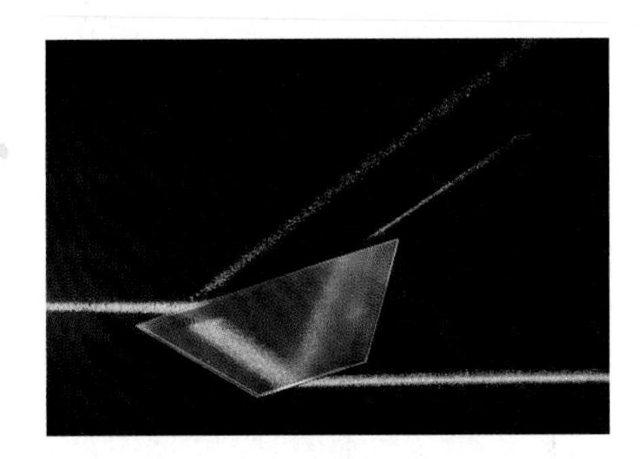

▲ **FIGURE 22.8 Reflection and refraction** A beam of light is incident on a trapezoidal prism from the left. Part of the beam is reflected, and part is refracted. The refracted beam is partially reflected and partially refracted at the bottom glass–air surface.

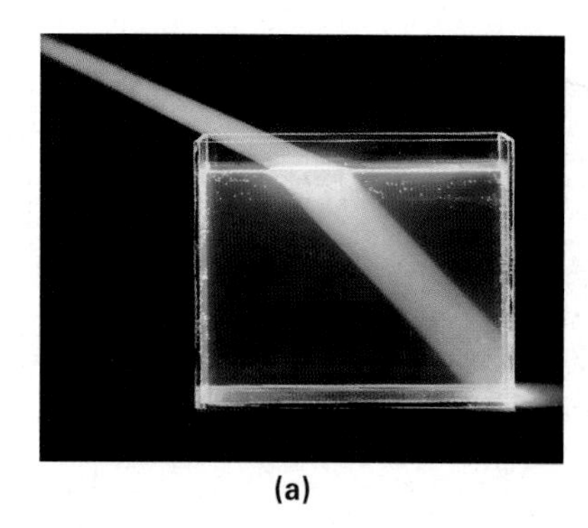

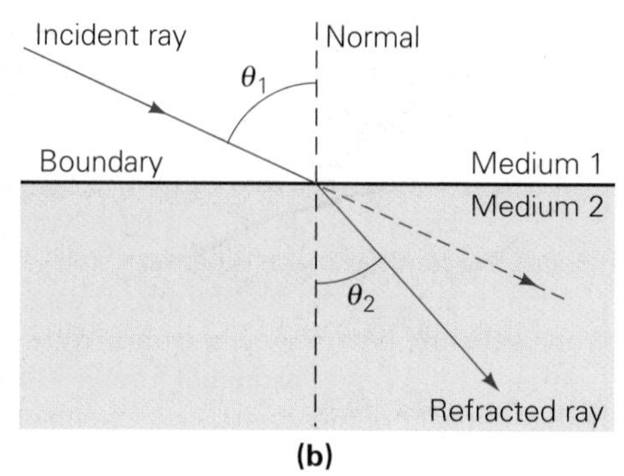

(a) (b)

▶ **FIGURE 22.9 Refraction** **(a)** Light changes direction on entering a different medium. **(b)** The refracted ray is described by the angle of refraction, θ_2, measured from the normal.

Illustration 34.1 Huygen's Principle and Refraction

Note: When light is refracted,
- its speed and wavelength are changed;
- its frequency remains unchanged.

media with this constant value (c). This is done by defining a ratio called the **index of refraction (n)**:

$$n = \frac{c}{v} \left(\frac{\text{speed of light in a vacuum}}{\text{speed of light in a medium}} \right) \tag{22.3}$$

As a ratio of speeds, the index of refraction is a unitless quantity. The indices of refraction of several substances are given in Table 22.1. Note that these values are for a specific wavelength of light. The wavelength is specified because v, and consequently n, are slightly different for different wavelengths. (This is the cause of dispersion, to be discussed later in the chapter.) The values of n given in the table will be used in examples and exercises in this chapter for all wavelengths of light in the visible region, unless otherwise noted. Observe that n is always greater than 1, because the speed of light in a vacuum is greater than the speed of light in any material ($c > v$).

The frequency (f) of light does not change when the light enters another medium, but the wavelength of light in a material (λ_m) differs from the wavelength of that light in a vacuum (λ), as can be easily shown:

$$n = \frac{c}{v} = \frac{\lambda f}{\lambda_m f}$$

or

$$n = \frac{\lambda}{\lambda_m} \tag{22.4}$$

The wavelength of light in the medium is then $\lambda_m = \lambda/n$. Since $n > 1$, it follows that $\lambda_m < \lambda$.

TABLE 22.1

Indices of Refraction (at λ = 590 nm)*

Substance	*n*
Air	1.000 29
Water	1.33
Ice	1.31
Ethyl alcohol	1.36
Fused quartz	1.46
Human eye	1.336–1.406
Polystyrene	1.49
Oil (typical value)	1.50
Glass (by type)†	1.45–1.70
crown	1.52
flint	1.66
Zircon	1.92
Diamond	2.42

*One nanometer (nm) is 10^{-9} m.

†Crown glass is a soda–lime silicate glass; flint glass is a lead–alkali silicate glass. Flint glass is more dispersive than crown glass (Section 22.5).

Example 22.2 ■ The Speed of Light in Water: Index of Refraction

Light from a laser with a wavelength of 632.8 nm travels from air into water. What are the speed and wavelength of the laser light in water?

Thinking It Through. If we know the index of refraction (n) of a medium, the speed and wavelength of light in the medium can be obtained from Eq. 22.3 and Eq. 22.4.

Solution.

Given: $n = 1.33$ (from Table 22.1)
$\lambda = 632.8$ nm
$c = 3.00 \times 10^8$ m/s (speed of light in air)

Find: v and λ_m (speed and wavelength of light in water)

Since $n = c/v$,

$$v = \frac{c}{n} = \frac{3.00 \times 10^8 \text{ m/s}}{1.33} = 2.26 \times 10^8 \text{ m/s}$$

Note that $1/n = v/c = 1/1.33 = 0.75$; therefore, v is 75% of the speed of light in a vacuum. Also, $n = \lambda/\lambda_m$, so

$$\lambda_m = \frac{\lambda}{n} = \frac{632.8 \text{ nm}}{1.33} = 475.8 \text{ nm}$$

Follow-Up Exercise. The speed of light of wavelength 500 nm (in air) in a particular liquid is 2.40×10^8 m/s. What is the index of refraction of the liquid and the wavelength of light in the liquid?

The index of refraction, n, is a measure of the speed of light in a transparent material, or technically, a measure of the *optical density* of the material. For example, the speed of light in water is less than that in air, so water is said to be optically denser than air. (Optical density in general correlates with mass density. However, in some instances, a material with a greater optical density than another can have a lower mass density.) Thus, the greater the index of refraction of a material, the greater is the material's optical density and the smaller is the speed of light in the material.

For practical purposes, the index of refraction is measured in air rather than in a vacuum, since the speed of light in air is very close to c, and

$$n_{\text{air}} = \frac{c}{v_{\text{air}}} \approx \frac{c}{c} = 1$$

(From Table 22.1, $n_{\text{air}} = 1.000\,29$, and we will usually assume $n_{\text{air}} = 1$.)

A more practical form of Snell's law can be rewritten as

$$\frac{\sin\theta_1}{\sin\theta_2} = \frac{v_1}{v_2} = \frac{c/n_1}{c/n_2} = \frac{n_2}{n_1}$$

or

$$n_1 \sin\theta_1 = n_2 \sin\theta_2 \qquad \textit{Snell's law (another form)} \qquad (22.5)$$

where n_1 and n_2 are the indices of refraction for the first and second media, respectively.

Note that Eq. 22.5 can be used to measure the index of refraction. If the first medium is air, then $n_1 \approx 1$ and $n_2 \approx \sin\theta_1/\sin\theta_2$. Thus, only the angles of incidence and refraction need to be measured to determine the index of refraction of a material experimentally. On the other hand, if the index of refraction of a material is known, it can be used in Snell's law to find the angle of refraction for any angle of incidence.

Note also that the sine of the refraction angle is inversely proportional to the index of refraction: $\sin\theta_2 \approx \sin\theta_1/n_2$. Hence, for a given angle of incidence, the greater the index of refraction, the smaller is $\sin\theta_2$ and the smaller is the angle of refraction, θ_2.

More generally, the following relationships hold:

- If the second medium is more optically dense than the first medium ($n_2 > n_1$), the ray is refracted *toward* the normal ($\theta_2 < \theta_1$), as illustrated in ▼Fig. 22.10a.
- If the second medium is less optically dense than the first medium ($n_2 < n_1$), the ray is refracted *away from* the normal ($\theta_2 > \theta_1$), as illustrated in Fig. 22.10b.

PHYSLET®

Exploration 34.4 Fermat's Principle and Snell's Law

Note: During refraction the product of $n \sin\theta$ remains a constant from medium to medium.

▼ **FIGURE 22.10 Index of refraction and ray deviation (a)** When the second medium is more optically dense than the first ($n_2 > n_1$), the ray is refracted toward the normal, as in the case of light entering water from air. **(b)** When the second medium is less optically dense than the first ($n_2 < n_1$), the ray is refracted away from the normal. [This is the case if the ray in part (a) is traced in reverse, going from medium 2 to medium 1.]

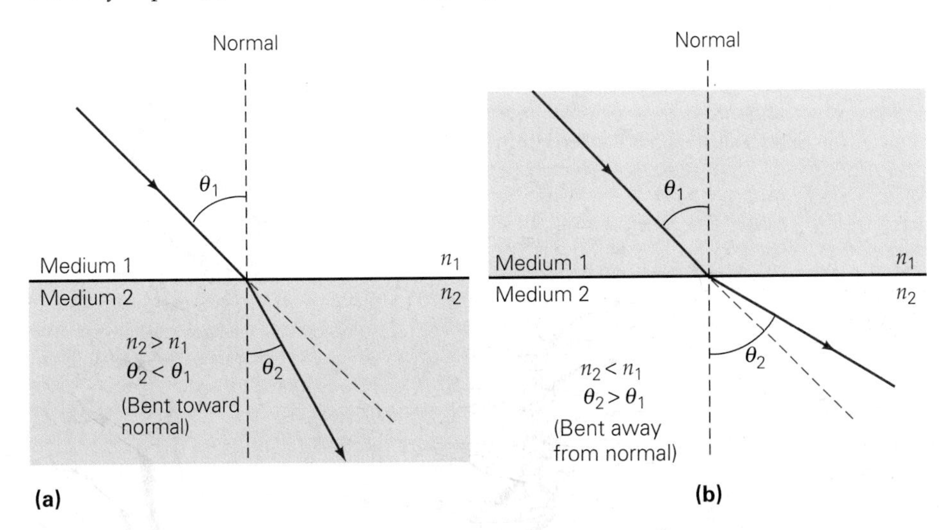

Integrated Example 22.3 ■ Angle of Refraction: Snell's Law

Light in water is incident on a piece of crown glass at an angle of 37° (relative to the normal). (a) Will the transmitted ray be (1) bent toward the normal, (2) bent away from the normal, or (3) not bent at all? Use a diagram to illustrate. (b) What is the angle of refraction?

(a) Conceptual Reasoning. We can use Table 22.1 to look up the indices of refraction of water and crown glass. According to the alternative form of Snell's law (Eq. 22.5), $n_1 \sin\theta_1 = n_2 \sin\theta_2$, (1) is the correct answer. Since $n_2 > n_1$, the angle of refraction must be smaller than the angle of incidence ($\theta_2 < \theta_1$). Because both θ_1 and θ_2 are measured from the normal, the refracted ray will bend toward the normal. The ray diagram in this case is identical to Fig. 22.10a.

(b) Quantitative Reasoning and Solution. Again, the alternative form of Snell's law (Eq. 22.5) is most practical in this case. (Why?) Listing the given quantities,

Given: $\theta_1 = 37°$
$n_1 = 1.33$ (water, from Table 22.1)
$n_2 = 1.52$ (crown glass, from Table 22.1)

Find: (b) θ_2 (angle of refraction)

The angle of refraction is found by using Eq. 22.5,

$$\sin\theta_2 = \frac{n_1 \sin\theta_1}{n_2} = \frac{(1.33)(\sin 37°)}{1.52} = 0.53$$

and

$$\theta_2 = \sin^{-1}(0.53) = 32°$$

Follow-Up Exercise. It is found experimentally that a beam of light entering a liquid from air at an angle of incidence of 37° exhibits an angle of refraction of 29° in the liquid. What is the speed of light in the liquid?

Example 22.4 ■ A Glass Tabletop: More about Refraction

A beam of light traveling in air strikes the glass top of a coffee table at an angle of incidence of 45° (▼Fig. 22.11). The glass has an index of refraction of 1.5. (a) What is the angle of refraction for the light transmitted into the glass? (b) Prove that the emergent beam is parallel to the incident beam—that is, that $\theta_4 = \theta_1$. (c) If the glass is 2.0 cm thick, what is the lateral displacement between the ray entering and the ray emerging from the glass (the perpendicular distance between the two rays—d in the figure)?

Thinking It Through. Since two refractions are involved in this example, we use Snell's law in parts (a) and (b), and then some geometry and trigonometry in part (c).

Solution. Listing the data:

Given: $\theta_1 = 45°$
$n_1 = 1.0$ (air)
$n_2 = 1.5$
$y = 2.0$ cm

Find: (a) θ_2 (angle of refraction)
(b) Show that $\theta_4 = \theta_1$
(c) d (lateral displacement)

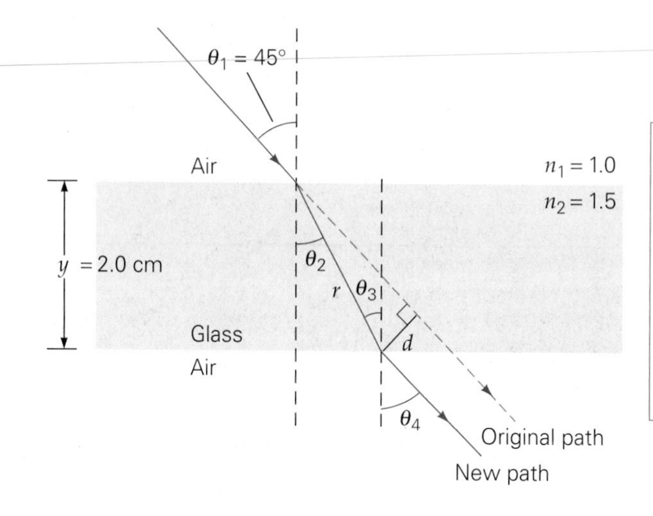

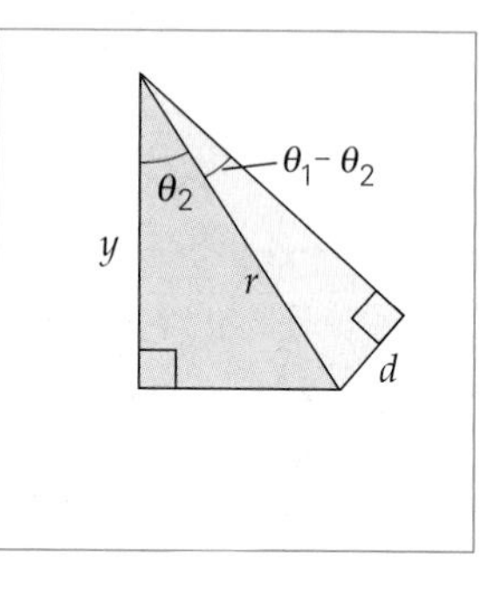

▶ **FIGURE 22.11 Two refractions** In the glass, the refracted ray is displaced laterally (sideways) a distance d from the incident ray, and the emergent ray is parallel to the original ray. (See Example 22.4.)

(a) Using the practical form of Snell's law, Eq. 22.5, with $n_1 = 1.0$ for air gives

$$\sin\theta_2 = \frac{n_1 \sin\theta_1}{n_2} = \frac{(1.0)\sin 45°}{1.5} = \frac{0.707}{1.5} = 0.47$$

Thus,

$$\theta_2 = \sin^{-1}(0.47) = 28°$$

Note that the beam is refracted toward the normal.

(b) If $\theta_1 = \theta_4$, then the emergent ray is parallel to the incident ray. Then applying Snell's law to the beam at both surfaces,

$$n_1 \sin\theta_1 = n_2 \sin\theta_2$$

and

$$n_2 \sin\theta_3 = n_1 \sin\theta_4$$

From the figure, $\theta_2 = \theta_3$. Therefore,

$$n_1 \sin\theta_1 = n_1 \sin\theta_4$$

or

$$\theta_1 = \theta_4$$

Thus, the emergent beam is parallel to the incident beam but displaced laterally or perpendicularly to the incident direction at a distance d.

(c) It can be seen from the inset in Fig. 22.11 that, to find d, we need to first find r from the known information in the pink right triangle. Then,

$$\frac{y}{r} = \cos\theta_2 \quad \text{or} \quad r = \frac{y}{\cos\theta_2}$$

In the yellow right triangle, $d = r\sin(\theta_1 - \theta_2)$. Substituting r from the previous step yields

$$d = \frac{y\sin(\theta_1 - \theta_2)}{\cos\theta_2} = \frac{(2.0\text{ cm})\sin(45° - 28°)}{\cos 28°} = 0.66\text{ cm}$$

Follow-Up Exercise. If the glass in this Example had $n = 1.6$, would the lateral displacement be the same, larger, or smaller? Explain your answer conceptually, and then calculate the actual value to verify your reasoning.

Conceptual Example 22.5 ■ The Human Eye: Refraction and Wavelength

A simplified representation of the crystalline lens in a human eye shows it to have a cortex (an outer layer) of $n_{\text{cortex}} = 1.386$ and a nucleus (core) of $n_{\text{nucleus}} = 1.406$. (See Fig. 25.1b.) Note that both refraction indices are within the range listed for the human eye in Table 22.1. If a beam of monochromatic (single-frequency or -wavelength) light of wavelength 590 nm is directed from air through the front of the eye and into the crystalline lens, qualitatively compare and list the frequency, speed, and wavelength of light in air, the cortex, and the nucleus. First do the comparison without numbers, and then calculate the actual values to verify your reasoning.

Reasoning and Answer. First, the relative magnitudes of the indices of refraction are needed, where $n_{\text{air}} < n_{\text{cortex}} < n_{\text{nucleus}}$.

As learned earlier in this section, the frequency (f) of light is the same in all three media: air, the cortex, and the nucleus. Thus, the frequency can be calculated by using the speed and the wavelength of light in any of these materials, but it is easiest in air. (Why?) From the wave relationship $c = \lambda f$ (Eq. 13.17),

$$f = f_{\text{air}} = f_{\text{cortex}} = f_{\text{nucleus}} = \frac{c}{\lambda} = \frac{3.00 \times 10^8\text{ m/s}}{590 \times 10^{-9}\text{ m}} = 5.08 \times 10^{14}\text{ Hz}$$

The speed of light in a medium depends on its index of refraction, since $v = c/n$. The smaller the index of refraction, the higher the speed. Therefore, the speed of light is the highest in air ($n = 1.00$) and lowest in the nucleus ($n = 1.406$).

The speed of light in the cortex is

$$v_{\text{cortex}} = \frac{c}{n_{\text{cortex}}} = \frac{3.00 \times 10^8\text{ m/s}}{1.386} = 2.16 \times 10^8\text{ m/s}$$

(continues on next page)

and the speed of light in the nucleus is

$$v_{\text{nucleus}} = \frac{3.00 \times 10^8 \text{ m/s}}{1.406} = 2.13 \times 10^8 \text{ m/s}$$

We also know that the wavelength of light in a medium depends on the index of refraction of the medium ($\lambda_m = \lambda/n$). The smaller the index of refraction, the longer the wavelength. Therefore, the wavelength of light is the longest in air ($n = 1.00$ and $\lambda = 590$ nm) and shortest in the nucleus ($n = 1.406$).

The wavelength in the cortex can be calculated from Eq. 22.4:

$$\lambda_{\text{cortex}} = \frac{\lambda}{n_{\text{cortex}}} = \frac{590 \text{ nm}}{1.386} = 426 \text{ nm}$$

and the wavelength in the nucleus is

$$\lambda_{\text{nucleus}} = \frac{590 \text{ nm}}{1.406} = 420 \text{ nm}$$

Finally, a table can be constructed to compare more easily the index of refraction, frequency, speed, and wavelength of light in the three media:

	Index of refraction	*Frequency (Hz)*	*Speed (m/s)*	*Wavelength (nm)*
Air	1.00	5.08×10^{14}	3.00×10^8	590
Cortex	1.386	5.08×10^{14}	2.16×10^8	426
Nucleus	1.406	5.08×10^{14}	2.13×10^8	420

Follow-Up Exercise. A light source of a single frequency is submerged in water in a special fish tank. The beam travels in the water, through double glass panes at the side of the tank (each glass pane has a different n), and into air. In general, what happens to (a) the frequency and (b) the wavelength of the light when it emerges into the outside air?

Refraction is common in everyday life and explains many things we observe. Let's look at refraction in action.

Mirage: A common example of this phenomenon sometimes occurs on a highway on a hot summer day. The refraction of light is caused by layers of air that are at different temperatures (the layer closer to the road is at a higher temperature, lower density, and lower index of refraction). This variation in indices of refraction gives rise to the observed "wet" spot and an inverted image of an object such as a car (▾Fig. 22.12a). The

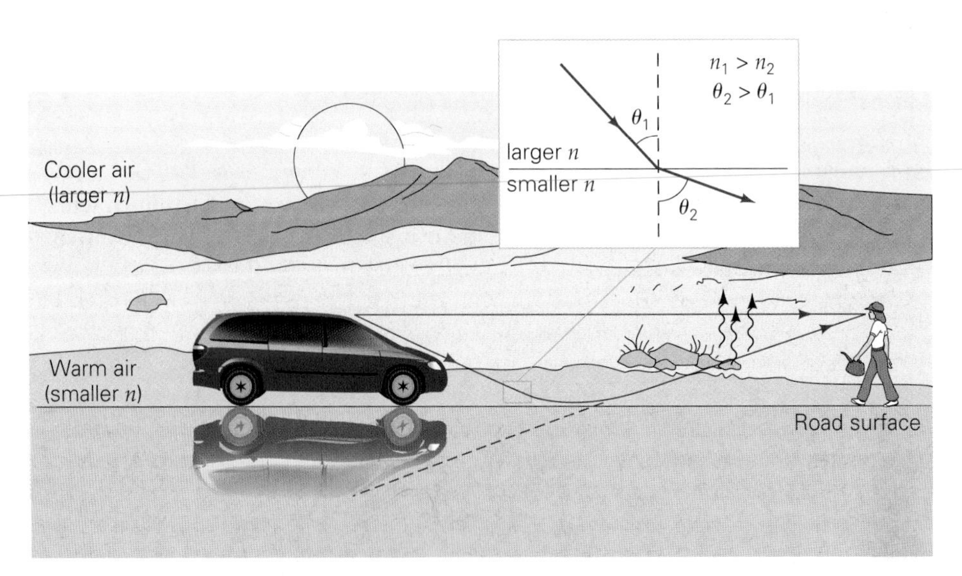

(a) (b)

▾ **FIGURE 22.12 Refraction in action** **(a)** An inverted car on a "wet" road, a mirage. **(b)** The mirage is formed when light from the object is refracted by layers of air at different temperatures near the surface of the road.

INSIGHT 22.2 NEGATIVE INDEX OF REFRACTION AND THE "PERFECT" LENS

In 1968 physicists predicted the existence of a material with a negative index of refraction. They expected that, in the presence of such a negative-index material, nearly all known wave propagation and optical phenomena would be substantially altered. Negative-index materials were not known to exist at the time, however.

At the beginning of the twenty-first century, a new class of artificially structured materials was created that were found to have negative indices of refraction. Moreover, a natural ferroelastic material called a *"twinned" alloy* (containing yttrium, vanadium, and oxygen) has also displayed a negative index of refraction (Fig. 1).

Figure 2 illustrates the difference between materials with positive and negative indices of refraction. In Fig. 2a, light incident on a positive-index material is refracted to the other side of the normal to the interface. However, if the material has a negative index of refraction, the same incident light is refracted to the same side of the normal to the interface (Fig. 2b). Due to this "abnormal" refraction, negative-index slab materials with flat surfaces can even focus light as shown in Fig. 2c, resulting in a new class of lenses (to be discussed in Chapter 23). If a light source is placed on one side of a slab with a refractive index $n = -1$, the light rays are refracted in such a way as to produce a focal point inside the material and then another just outside the material. The "focal length" of such a lens would depend on both the object distance and the thickness of the slab.

The undesirable characteristics of lenses made of materials with a positive index of refraction are energy loss due to reflection, aberrations, and low resolution due to diffraction limit (more on this in Chapter 24). The latest experiments provide strong evidence that negative-index materials have an important future in imaging. This is because negative-index lenses offer a new degree of flexibility that could lead to more compact lenses with reduced lens aberration. The diffraction limit—which is the most fundamental limitation to image resolution—may be circumvented by negative-index materials. Furthermore, total negative refraction—that is, absence of a reflection—has been observed in materials with a negative index of refraction. Such a lens would truly be a "perfect lens."

FIGURE 1 Material with a negative index of refraction This artificial material made from grids of rings and wires has a negative index of refraction.

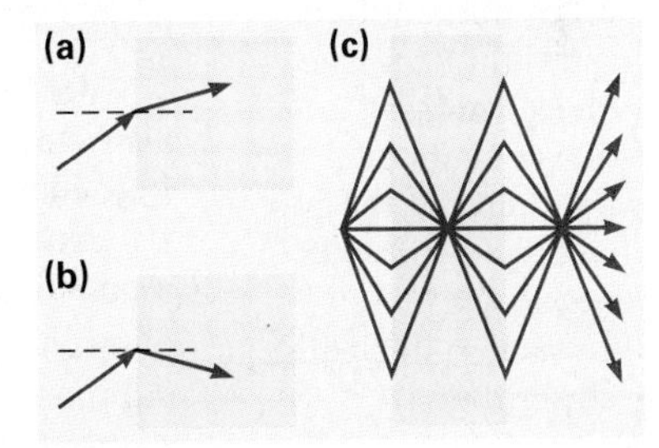

FIGURE 2 Reflection in positive-index versus negative-index materials **(a)** Light incident on the interface between air and a positive-index material is bent toward the other side of the normal, **(b)** whereas in a negative-index material light is bent toward the same side of the normal. **(c)** If a light source is placed on one side of a slab with a refractive index of $n = -1$, the waves are refracted in such a way as to produce a focus inside the material and then another just outside the material.

term *mirage* generally brings to mind a thirsty person in the desert "seeing" a pool of water that really isn't there. This optical illusion plays tricks on the mind, with the image usually seen as in a pool of water and our eye's past experience unconsciously leading us to conclude that there is water on the road.

In Fig. 22.12b, there are two ways for light to get to our eyes from the car. First, the horizontal rays come directly from the car to our eyes, so we see the car above the ground. Also, the rays from the car that travel toward the road surface will be gradually refracted by the layered air. After hitting the surface, these rays will be refracted again and travel toward our eyes. (See the inset in the figure.) Cooler air has a higher density and so a higher index of refraction. A ray traveling toward the road surface will be gradually refracted with increasing angle of refraction until it hits the surface. It will then be refracted again with decreasing angle of refraction, going toward our eyes. As a consequence, we also see an inverted image of the car, below the road surface. In other words, the surface of the road acts almost as a mirror.

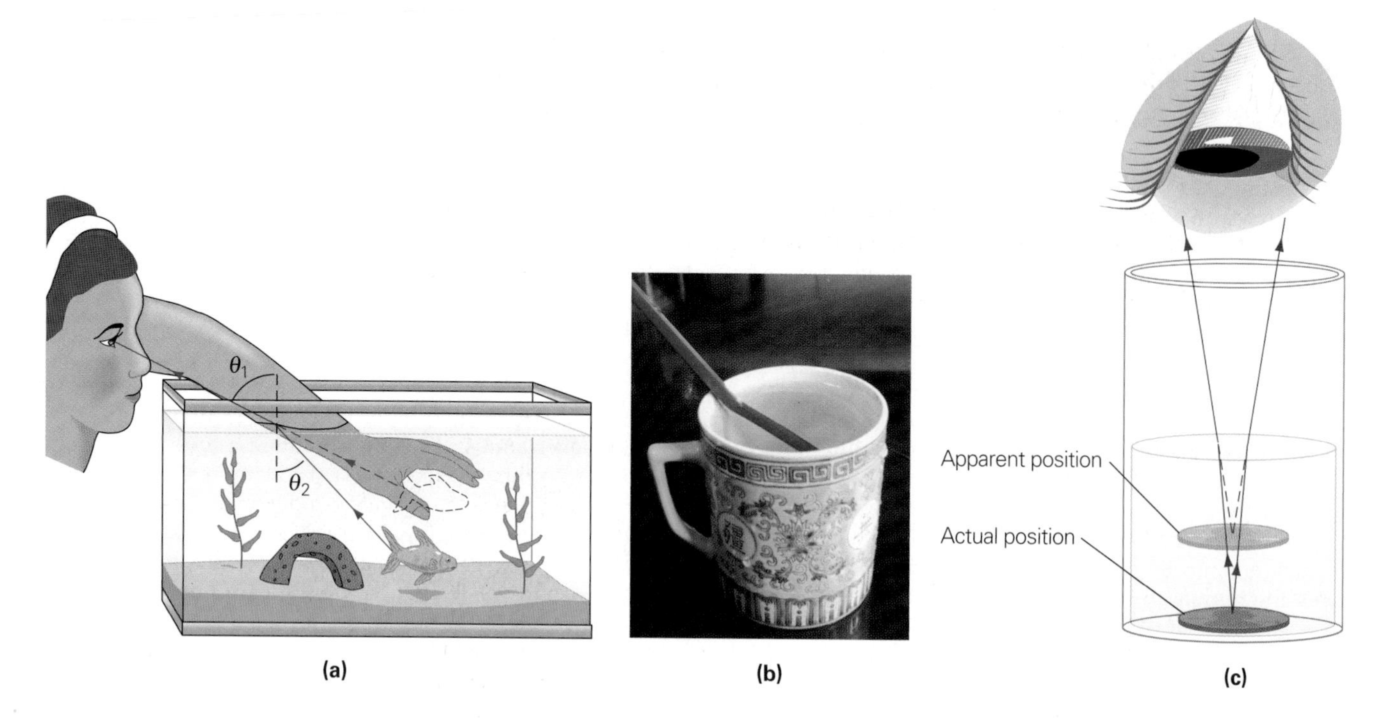

▲ **FIGURE 22.13** Refractive effects **(a)** The light is refracted, and because we tend to think of light as traveling in straight lines, the fish is below where we think it is. **(b)** The chopstick appears bent at the air–water boundary. If the cup is transparent, we see a different refraction. (See Exercise 21.) **(c)** Because of refraction, the coin appears to be closer than it actually is.

The "pool of water" is actually sky light being refracted—an image of the sky. This layering of air of different temperatures, creating different indices of refraction, causes us to "see" the rising hot air as a result of continually changing refraction.

The opposite of this is the mirage at sea (looming effect). At the sea, the air above is warmer than below. This causes the light to be refracted opposite as in Fig. 22.12b, causing objects to be seen in the air above sea.

Not where it should be: You may have experienced a refractive effect while trying to reach for something underwater, such as a fish (▲Fig. 22.13a). We are used to light traveling in straight lines from objects to our eyes, but the light reaching our eyes from a submerged object has a directional change at the air–water interface. (Note in the figure that the ray is refracted away from the normal.) As a result, the object appears closer to the surface than it actually is, and therefore we tend to miss the object when reaching for it. For the same reason, a chopstick in a cup appears bent (Fig. 22.13b), a coin in a glass of water will appear closer than it really is (Fig. 22.13c), and the legs of a person standing in water seem shorter than their actual length. The relationship between the true depth and the apparent depth can be calculated. (See Exercise 37.)

Atmospheric effects: The Sun on the horizon sometimes appears flattened, with its horizontal dimension greater than its vertical dimension (▼Fig. 22.14a). This effect is the result of temperature and density variations in the denser air along the horizon. These variations occur predominantly vertically, so light from the top and bottom portions of the Sun are refracted differently as the two sets of beams pass through different atmospheric densities with different indices of refraction.

Atmospheric refraction lengthens the day, so to speak, by allowing us to see the Sun (or the Moon, for that matter) just before it actually rises above the horizon and

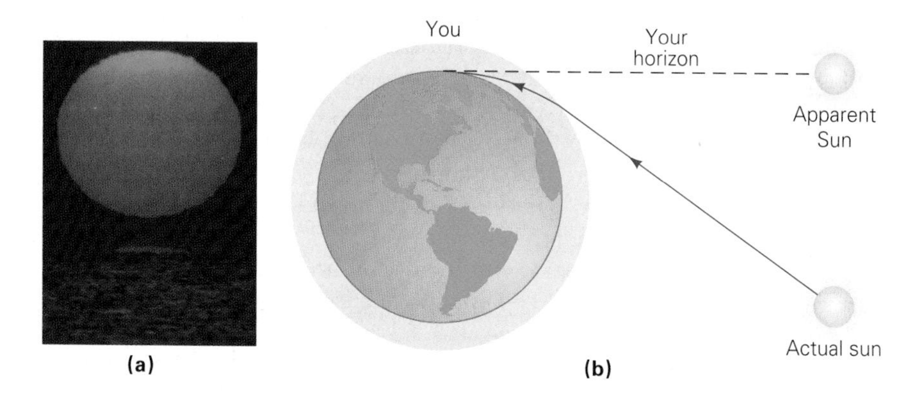

▶ **FIGURE 22.14** Atmospheric effects **(a)** The Sun on the horizon commonly appears flattened as a result of atmospheric refraction. **(b)** Before rising and after setting, the Sun can be seen briefly also because of atmospheric refraction. (Exaggerated for illustration).

just after it actually sets below the horizon (as much as 20 min on both ends). The denser air near the Earth refracts the light over the horizon toward us (Fig. 22.14b).

The twinkling of stars is due to atmospheric turbulence, which distorts the light from the stars. The turbulences refract light in random directions and cause the stars to appear to "twinkle." Stars on the horizon will appear to twinkle more than stars directly overhead, because the light has to pass through more of the Earth's atmosphere. However, planets do not "twinkle" as much. This is because stars are much farther away than planets, so they appear as point sources. Outside the Earth's atmosphere, the stars don't twinkle.

22.4 Total Internal Reflection and Fiber Optics

OBJECTIVES: To (a) describe total internal reflection, and (b) understand fiber-optic applications.

An interesting phenomenon occurs when light travels from a more optically dense medium into a less optically dense one, such as when light goes *from* water *into* air. As you know, in such a case a ray will be refracted away from the normal. (The angle of refraction is larger than the angle of incidence.) Furthermore, Snell's law states that the greater the angle of incidence, the greater the angle of refraction. That is, as the angle of incidence increases, the farther the refracted ray diverges from the normal.

However, there is a limit. For a certain angle of incidence called the **critical angle (θ_c)**, the angle of refraction is 90°, and the refracted ray is directed along the boundary between the media. But what happens if the angle of incidence is even larger? If the angle of incidence is greater than the critical angle ($\theta_1 > \theta_c$), the light isn't refracted at all, but is internally reflected (▼Fig. 22.15). This condition is called **total internal reflection**. The reflection process is about 100% efficient. (There is always some absorption of light *in* the materials.) Because of total internal reflection, glass prisms can be used as mirrors (▶Fig. 22.16). In summary, where $n_1 > n_2$, reflection and refraction occur at all angles for $\theta_1 \leq \theta_c$, but the refracted or transmitted ray disappears at $\theta_1 > \theta_c$.

PHYSLET® Exploration 34.2 Snell's Law and Total Internal Reflection

An expression for the critical angle can be obtained from Snell's law. If $\theta_1 = \theta_c$ in the optically denser medium, $\theta_2 = 90°$, and it follows that

$$n_1 \sin\theta_1 = n_2 \sin\theta_2 \qquad \text{or} \qquad n_1 \sin\theta_c = n_2 \sin 90°$$

Since $\sin 90° = 1$,

$$\sin\theta_c = \frac{n_2}{n_1} \quad \textit{where } n_1 > n_2 \qquad (22.6)$$

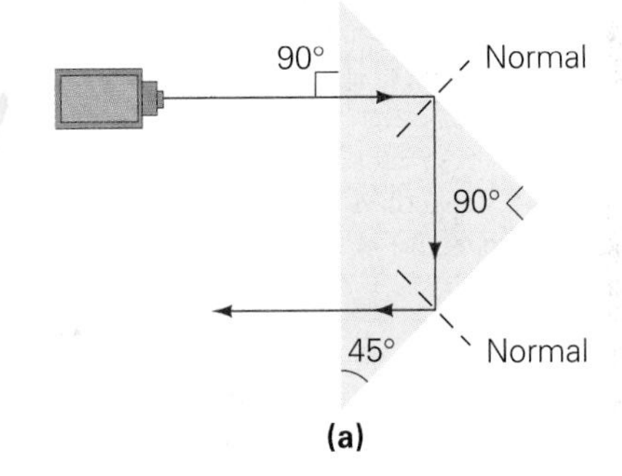

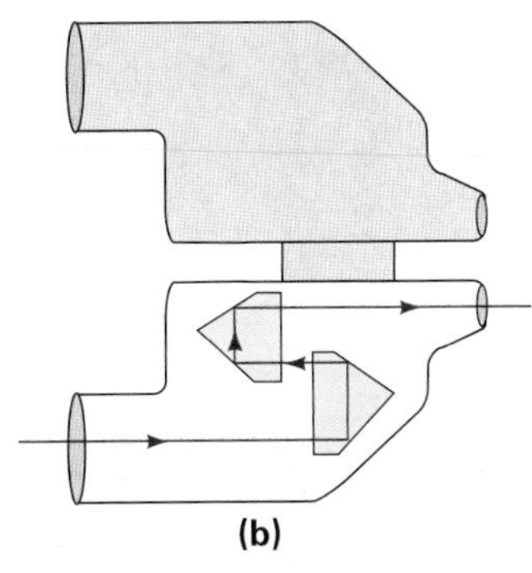

▲ **FIGURE 22.16 Internal reflection in a prism (a)** Because the critical angle of glass is less than 45°, prisms with 45° and 90° angles can be used to reflect light through 180°. **(b)** Internal reflection of light by prisms in binoculars makes this instrument much shorter than a telescope because the rays are "folded" by the prisms.

▼ **FIGURE 22.15 Internal reflection (a)** When light enters a less optically dense medium, it is refracted away from the normal. At a critical angle (θ_c), the light is refracted along the interface (common boundary) of the media. At an angle greater than the critical angle ($\theta_1 > \theta_c$), there is total internal reflection. **(b)** Can you estimate the critical angle in the photograph?

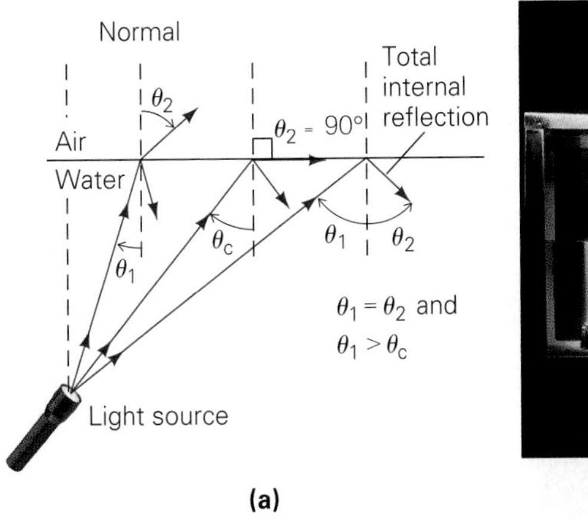

If the second medium is air, $n_2 \approx 1$, and the critical angle at the boundary from a medium into air is given by $\sin\theta_c = 1/n$, where n is the index of refraction of the medium. This is another method that can be used to measure the index of refraction in laboratories.

Example 22.6 ■ A View from the Pool: Critical Angle

(a) What is the critical angle for light traveling in water and incident on a water–air boundary? (b) If a diver submerged in a pool looked up at the surface of the water at an angle of $\theta < \theta_c$, what would she see? (Neglect any thermal or motional effects.)

Thinking It Through. (a) The critical angle is given by Eq. 22.6. (b) As shown in Fig. 22.15a, θ_c forms a cone of vision for viewing from below the water.

Solution.

Given: $n_1 = 1.33$ (for water, from Table 22.1)
$n_2 \approx 1$ (why?)

Find: (a) θ_c (critical angle)
(b) view for $\theta < \theta_c$

(a) The critical angle is

$$\theta_c = \sin^{-1}\left(\frac{n_2}{n_1}\right) = \sin^{-1}\left(\frac{1}{1.33}\right) = 48.8°$$

(b) Using Fig. 22.15a, trace the rays in reverse for light coming from all angles outside the pool. Light coming from the above-water 180° panorama could be viewed only in a cone with a half-angle of 48.8°. As a result, objects above the surface would also appear distorted. An underwater panoramic view is seen in ◂Fig. 22.17. Now can you explain why wading birds like herons usually keep their bodies low while trying to catch a fish?

Follow-Up Exercise. What would the diver see when looking up at the water surface at an angle of $\theta > \theta_c$?

▲ **FIGURE 22.17 Panoramic and distorted** An underwater view of the surface of a swimming pool in Hawaii. (See Example 22.6.)

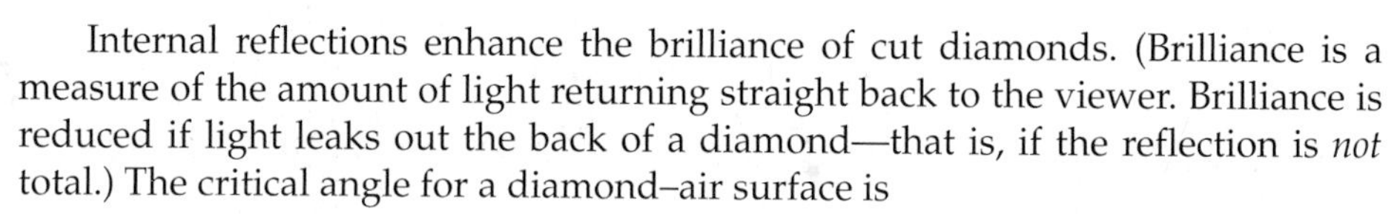

Internal reflections enhance the brilliance of cut diamonds. (Brilliance is a measure of the amount of light returning straight back to the viewer. Brilliance is reduced if light leaks out the back of a diamond—that is, if the reflection is *not* total.) The critical angle for a diamond–air surface is

$$\theta_c = \sin^{-1}\left(\frac{1}{n}\right) = \sin^{-1}\left(\frac{1}{2.42}\right) = 24.4°$$

A so-called brilliant-cut diamond has many facets, or faces (58 in all—33 on the upper face and 25 on the lower). Light from above hitting the lower facets at angles greater than the critical angle is internally reflected in the diamond. The light then emerges from the upper facets, giving rise to the diamond's brilliance (▾Fig. 22.18).

Fiber Optics

When a fountain is illuminated from below, the light is transmitted along the curved streams of water. This phenomenon was first demonstrated in 1870 by the British scientist John Tyndall (1820–1893), who showed that light was "conducted"

(a)

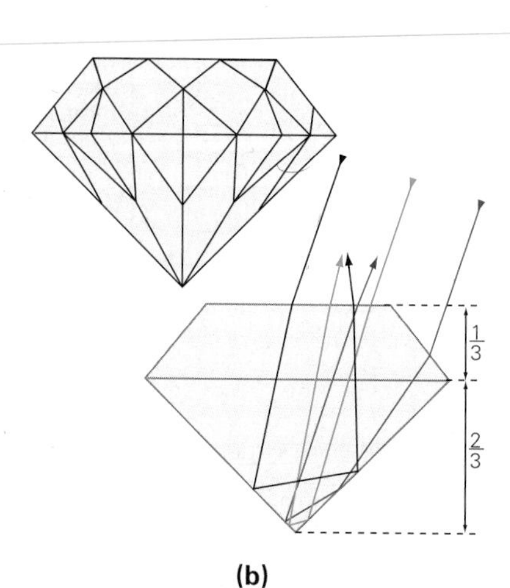

(b)

▸ **FIGURE 22.18 Diamond brilliance** **(a)** Internal reflection gives rise to a diamond's brilliance. **(b)** The "cut," or the depth proportions, of the facets is critical. If a stone is too shallow or too deep, light will be lost (refracted out) through the lower facets.

along the curved path of a stream of water flowing from a hole in the side of a container. The phenomenon is observed because light undergoes total internal reflection along the stream.

Total internal reflection forms the basis of **fiber optics**, a fascinating modern technology field centered on the use of transparent fibers to transmit light. Multiple total internal reflections make it possible to "pipe" light along a transparent rod (as in streams of water), even if the rod is curved (▸Fig. 22.19). Note from the figure that the smaller the diameter of the light pipe, the more total internal reflections it has. A small fiber can produce as many as several hundred total internal reflections per centimeter.

Total internal reflection is an exceptionally efficient process. Optical fibers can be used to transmit light over very long distances with losses of only about 25% per kilometer. These losses are due primarily to impurities in the fiber, which scatter the light. Transparent materials have different degrees of transmission. Fibers are made of special plastics and glasses for maximum transmission efficiency. The greatest efficiency is achieved with infrared radiation, because there is less scattering, as will be learned in Section 24.5.

The greater efficiency of multiple total internal reflections compared with multiple mirror reflections can be illustrated by a good reflecting plane mirror, which has at best a reflectivity of about 95%. After each reflection, the beam intensity is 95% of that of the incident beam from the preceding reflection ($I_1 = 0.95\, I_o$; $I_2 = 0.95\, I_1 = 0.95^2\, I_o; \ldots$). Therefore, the intensity I of the reflected beam after n reflections is given by

$$I = 0.95^n\, I_o$$

where I_o is the initial intensity of the beam before the first reflection. Thus, after 14 reflections,

$$I = 0.95^{14}\, I_o = 0.49\, I_o$$

In other words, after 14 reflections, the intensity is reduced to less than half (49%). For 100 reflections, $I = 0.006\, I_o$, and the intensity is only 0.6% of the initial intensity! Compare this to about 75% of the initial intensity in optical fibers over a kilometer in length with *thousands* of reflections, so you can see the advantage of total internal reflection.

Fibers whose diameters are about 10 μm (10^{-5} m) are grouped together in flexible bundles that are 4 to 10 mm in diameter and up to several meters in length, depending on the application (▾Fig. 22.20). A fiber bundle with a

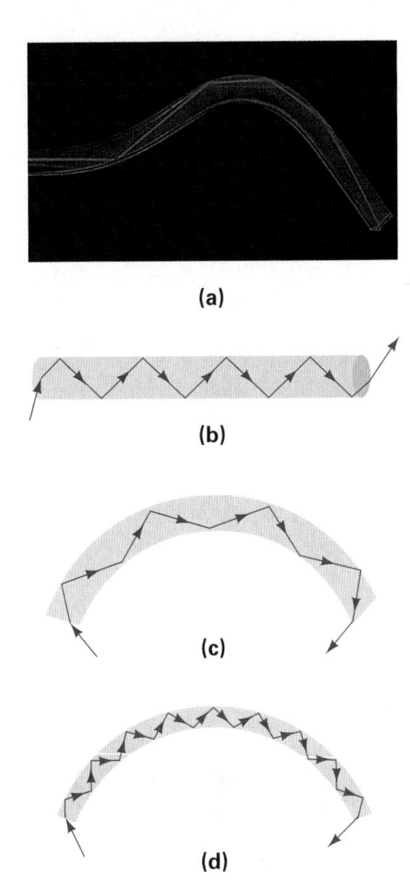

▲ **FIGURE 22.19** Light pipes
(a) Total internal reflection in an optical fiber. **(b)** When light is incident on the end of a cylindrical form of transparent material such that the internal angle of incidence is greater than the critical angle of the material, the light undergoes total internal reflection down the length of the light pipe. **(c)** Light is also transmitted along curved light pipes by total internal reflection. **(d)** As the diameter of the rod or fiber becomes smaller, the number of reflections per unit length increases.

PHYSLET®
Illustration 34.2 Fiber Optics

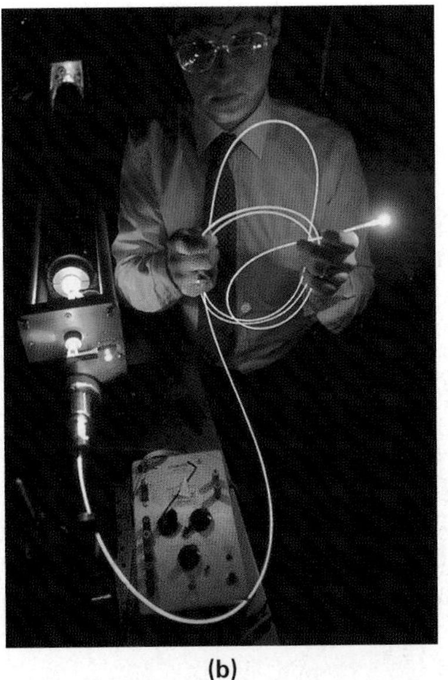

◀ **FIGURE 22.20** Fiber-optic bundle
(a) Hundreds or even thousands of extremely thin fibers are grouped together **(b)** to make an optical fiber, here colored blue by a laser.

INSIGHT 22.3 FIBER OPTICS: MEDICAL APPLICATIONS

Before fiber optics, *endoscopes*—instruments used to view internal portions of the human body—consisted of lens systems in long, narrow tubes. Some endoscopes contained a dozen or more lenses and produced relatively poor images. Because the lenses had to be aligned in certain ways, the tubes had to have rigid sections, which limited the endoscope's maneuverability. Such an endoscope could be inserted down the patient's throat into the stomach to observe the stomach lining. However, there were blind spots due to the curvature of the stomach and the inflexibility of the instrument.

Fiber-optic bundles have eliminated these problems. Lenses placed at the end of the fiber bundles focus the light, and a prism is used to change the direction for its return. The incident light is usually transmitted by an outer layer of fiber bundles, and the image is returned through a central core of fibers. Mechanical linkages allow maneuverability. The end of a fiber endoscope can be equipped with devices to obtain specimens of the viewed tissues for biopsy (diagnostic examination) or even to perform surgical procedures. For example, arthroscopic surgery is performed on injured joints (Fig. 1). The *arthroscope* that is now routinely used for inspecting *and* repairing damaged joints is simply a fiber endoscope fitted with appropriate surgical implements.

A fiber-optic *cardioscope* (for direct observation of heart valves) typically is a fiber bundle about 4 mm in diameter and 30 cm long. Such a cardioscope passes easily to the heart through the jugular vein, which is about 15 mm in diameter, in the neck. To displace the blood and provide a clear field of view for observing and photographing, a transparent balloon at the tip of the cardioscope is inflated with saline (saltwater) solution.

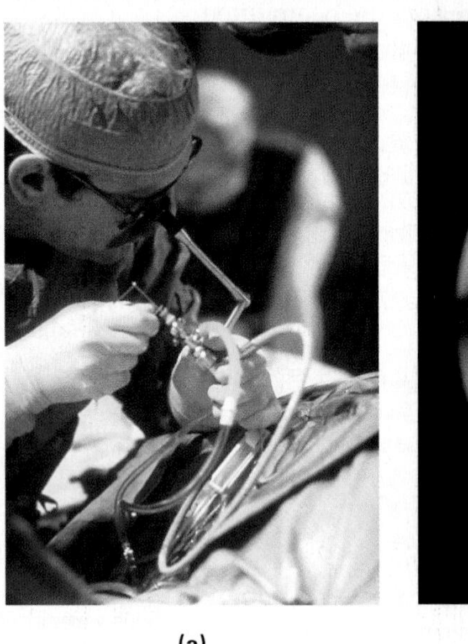

(a)

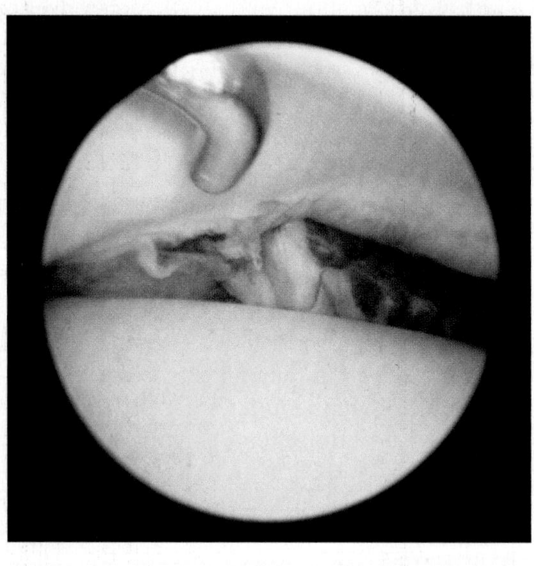

(b)

FIGURE 1 Arthroscopy (a) A fiber-optic arthroscope used to perform surgery. **(b)** An arthroscopic view of a torn knee meniscus.

cross-sectional area of 1 cm^2 can contain as many as 50 000 individual fibers. (A coating on each fiber is needed to keep the fibers from touching each other.)

There are many important and interesting applications of fiber optics, including communications, computer networking, and medical applications. (See Insight 22.3 on Fiber Optics: Medical Applications.) Light signals, converted from electrical signals, are transmitted through optical telephone lines and computer networks. At the other end, they are converted back to electrical signals. Optical fibers have lower energy losses than electric current–carrying wires, particularly at higher frequencies, and can carry far more data. Also, optical fibers are lighter than metal wires, have greater flexibility, and are not affected by electromagnetic disturbances (electric and magnetic fields), because they are made of materials that are electrical insulators.

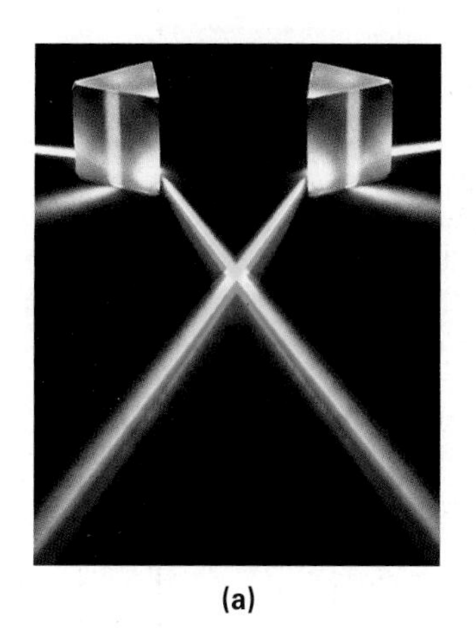

(a)

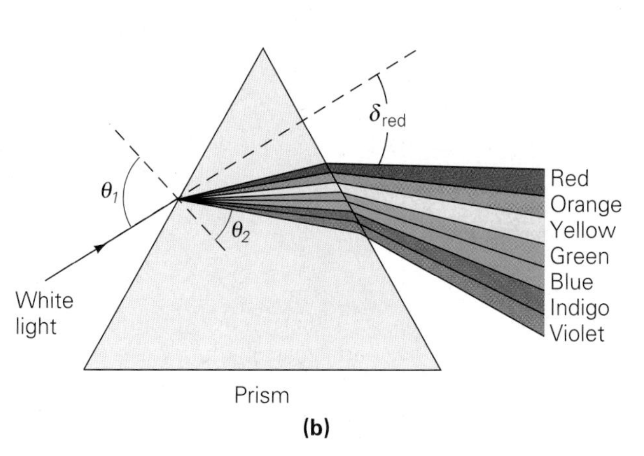

(b)

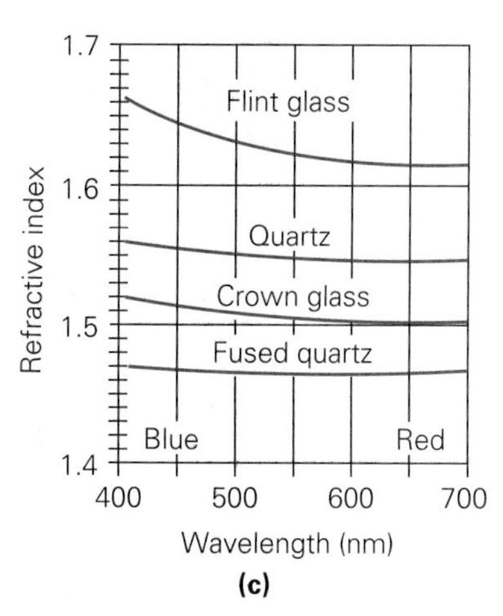

(c)

▲ **FIGURE 22.21 Dispersion (a)** White light is dispersed into a spectrum of colors by glass prisms. **(b)** In a dispersive medium, the index of refraction varies slightly with wavelength. Red light, longest in wavelength, has the smallest index of refraction and is refracted least. The angle between the incident beam and an emergent ray is the angle of deviation (δ) for that ray. (The angles are exaggerated for clarity.) **(c)** Variation in the index of refraction with wavelength for some common transparent media.

22.5 Dispersion

OBJECTIVE: To explain dispersion and some of its effects.

Light of a single frequency, and consequently a single wavelength, is called *monochromatic light* (from the Greek *mono*, meaning "one," and *chroma*, meaning "color"). Visible light that contains all the component frequencies, or colors, at about the same intensities (such as sunlight) is termed *white light*. When a beam of white light passes through a glass prism, as shown in ▲Fig. 22.21a, it is spread out, or dispersed, into a spectrum of colors. This phenomenon led Newton to believe that sunlight is a mixture of colors. When the beam enters the prism, the component colors, corresponding to different wavelengths of light, are refracted at slightly different angles, so they spread out into a spectrum (Fig. 22.21b).

Illustration 34.3 Prisms and Dispersion

The emergence of a spectrum indicates that the index of refraction of glass is slightly different for different wavelengths, which is true of many transparent media (Fig. 22.21c). The reason has to do with the fact that in a dispersive medium the speed of light is slightly different for different wavelengths. Since the index of refraction n of a medium is a function of the speed of light in that medium ($n = c/v$), the index of refraction is different for different wavelengths. It follows from Snell's law that light of different wavelengths will be refracted at different angles.

Note: You can remember the sequence of the colors of the visible spectrum (from the long-wavelength end to the short-wavelength end) by using the name ROY G. BIV, which is an acronym for *r*ed, *o*range, *y*ellow, *g*reen, *b*lue, *i*ndigo, and *v*iolet.

We can summarize the preceding discussion by saying that in a transparent material with different indices of refraction for different wavelengths of light, refraction causes a separation of light according to wavelength, and the material is said to be *dispersive* and exhibit **dispersion**. Dispersion varies with different media (Fig. 22.21c). Also, because the differences in the indices of refraction for different wavelengths are small, a representative value at some specified wavelength can be used for general purposes. (See Table 22.1.)

A good example of a dispersive material is diamond, which is about five times as dispersive as glass. In addition to revealing the brilliance resulting from internal reflections off many facets, a cut diamond shows a display of colors, or "fire," resulting from the dispersion of the refracted light.

Dispersion is a cause of chromatic aberration in lenses, which is described more fully in Chapter 23. Optical systems in cameras often consist of several lenses to minimize this problem (see Section 23.4).

Another dramatic example of dispersion is the production of a rainbow, as discussed in Insight 22.4 on The Rainbow on page 722.

INSIGHT 22.4 THE RAINBOW

Everyone has been fascinated by the beautiful array of colors of a rainbow. With the optical principles learned in this chapter, we are now in a position to understand the formation of this spectacular display.

A rainbow is formed by refraction, dispersion, and internal reflection of light within water droplets. When sunlight shines on millions of water droplets in the air during and after a rain, a multicolored arc is seen whose colors run from violet along the lower part of the bow (in order of wavelength) to red along the upper part. Occasionally, more than one rainbow is seen: The main, or primary, rainbow is sometimes accompanied by a fainter and higher secondary rainbow (Fig. 1) or even a third rainbow. These higher-order rainbows are caused by more than one total internal reflection within the water droplets.

FIGURE 1 Rainbow The colors of the primary rainbow run vertically from red (top) to violet (bottom).

The light that forms the primary rainbow is first refracted and dispersed in each water droplet, then totally reflected once at the back surface of each droplet. Finally, it is refracted and dispersed again upon exiting each droplet, resulting in the light being spread out in different directions into a spectrum of colors (Fig. 2a). However, because of the conditions for refraction and total internal reflection in water, the angles between incoming and outgoing rays for violet to red light lie within a narrow range of 40° to 42°. This means that you can see a rainbow only when the Sun is behind you, so that the dispersed light travels to you only at these angles.

Red appears on the top of the rainbow because light of shorter wavelengths from those water droplets will pass over our eyes (Fig. 2b). Similarly, violet is at the bottom of the rainbow because light of longer wavelengths passes under our eyes.

The secondary rainbow has reversed color orders because of the extra reflection.

Rainbows are generally seen only as arcs, because their formation is cut off at the ground. When on a cliff or on an airplane, you might see a complete circular rainbow (Fig. 2b). Also, the higher the Sun is in the sky, the less of a rainbow will be seen from the ground. In fact, a primary rainbow would not be seen if the Sun's angle above the horizon is greater than 42°. The primary rainbow can still be seen from a height, however. As an observer's elevation increases, more of the arc becomes visible. You may also have seen a circular rainbow in the spray from a garden hose.

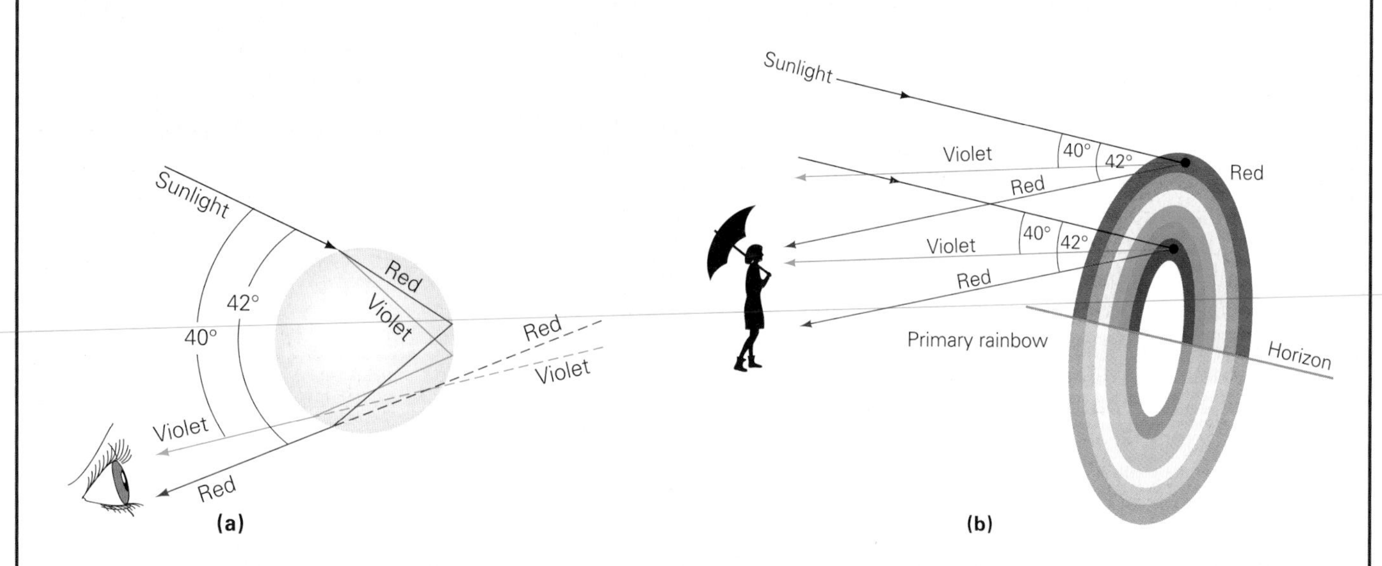

FIGURE 2 The rainbow Rainbows are created by the refraction, dispersion, and internal reflection of sunlight within water droplets. **(a)** Light of different colors emerges from the droplet in different directions. **(b)** An observer sees red light at the top of the rainbow and violet at the bottom.

Integrated Example 22.7 ■ Forming a Spectrum: Dispersion

The index of refraction of a particular transparent material is 1.4503 for the red end ($\lambda_r = 700$ nm) of the visible spectrum and 1.4698 for the blue end ($\lambda_b = 400$ nm). White light is incident on a prism of this material, as in Fig. 22.21b, at an angle of incidence θ_i of 45°. (a) Inside the prism, the angle of refraction of the red light is (a) larger than, (2) smaller than, or (3) the same as the angle of refraction of the blue light. Explain. (b) What is the angular separation of the visible spectrum inside the prism?

(a) Conceptual Reasoning. The angle of refraction is given by Snell's law, $n_1 \sin\theta_1 = n_2 \sin\theta_2$. Since red light has a smaller index of refraction than blue light, the angle of refraction of red light is larger than that of blue light, for the same angle of incidence. Sometimes, we also say that the red light is "refracted less" than blue light because the larger angle of refraction of the red light means it is closer to the direction of the original incident ray. So the answer is (a).

(b) Quantitative Reasoning and Solution. Again, we use Snell's law to compute the angle of refraction for the red and blue ends of the visible spectrum. The angular separation of the two colors inside the prism is the difference between these angles of refraction.

Given: (red) $n_r = 1.4503$ for $\lambda_r = 700$ nm
(blue) $n_b = 1.4698$ for $\lambda_b = 400$ nm
$\theta_1 = 45°$

Find: $\Delta\theta_2$ (angular separation)

Using Eq. 22.5 with $n_1 = 1.00$ (air),

$$\sin\theta_{2_r} = \frac{\sin\theta_1}{n_{2_r}} = \frac{\sin 45°}{1.4503} = 0.48756 \quad \text{and} \quad \theta_{2_r} = 29.180°$$

Similarly,

$$\sin\theta_{2_b} = \frac{\sin\theta_1}{n_{2_b}} = \frac{\sin 45°}{1.4698} = 0.48109 \quad \text{and} \quad \theta_{2_b} = 28.757°$$

So

$$\Delta\theta_2 = \theta_{2_r} - \theta_{2_b} = 29.180° - 28.757° = 0.423°$$

This is not much of a deviation, but as the light travels to the other side of the prism, it is refracted and dispersed again by the second boundary. Thus the colors spread out even farther. When the light emerges from the prism, the dispersion becomes evident (Fig. 22.21a).

Follow-Up Exercise. In the prism in this Example, if the green light exhibits an angular separation of 0.156° from the red light, what is the index of refraction for green light in the material? Will the green light refract more or less than the red light? Explain.

Chapter Review

- **Law of reflection:** The angle of incidence equals the angle of reflection (as measured from the normal to the reflecting surface):

$$\theta_i = \theta_r \qquad (22.1)$$

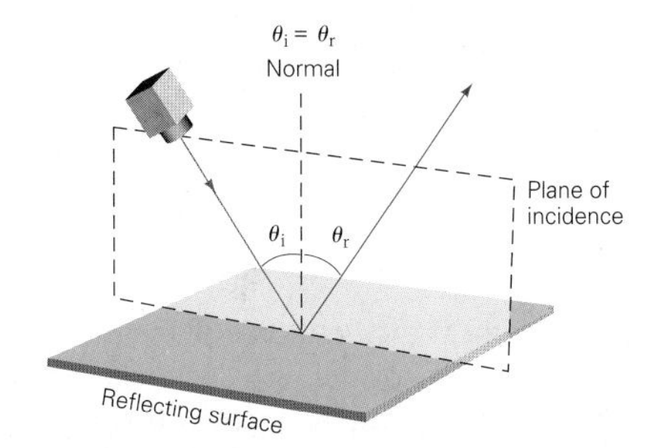

- The **index of refraction (*n*)** of any medium is the ratio of the speed of light in a vacuum to its speed in that medium:

$$n = \frac{c}{v} = \frac{\lambda}{\lambda_m} \qquad (22.3, 22.4)$$

- The refraction of light as it enters one medium from another is given by **Snell's law**. If the second medium is more optically dense, the ray is refracted toward the normal; if the medium is less dense, the ray is refracted away from the normal. Snell's law is

$$\frac{\sin\theta_1}{\sin\theta_2} = \frac{v_1}{v_2} \qquad (22.2)$$

$$n_1 \sin\theta_1 = n_2 \sin\theta_2 \qquad (22.5)$$

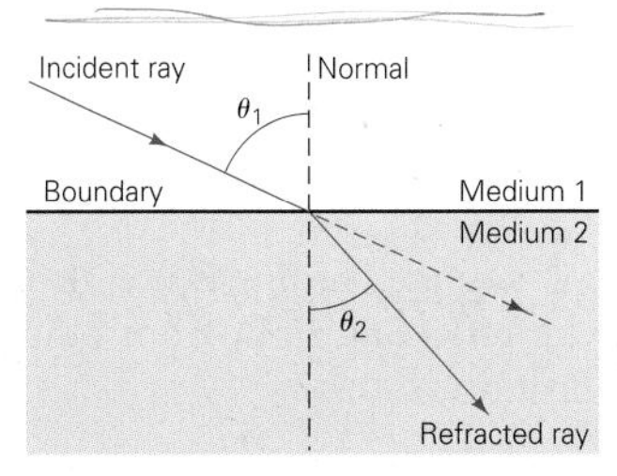

- **Total internal reflection** occurs if the second medium is less dense than the first and the angle of incidence exceeds the critical angle, which is given by

$$\sin\theta_c = \frac{n_2}{n_1} \quad (n_1 > n_2) \qquad (22.6)$$

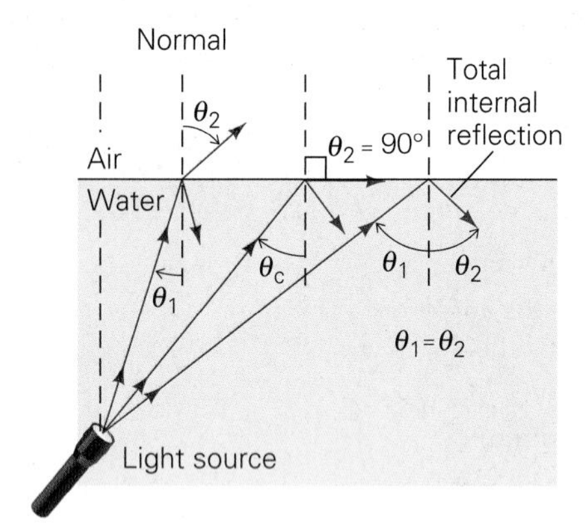

- **Dispersion** of light occurs in some media because different wavelengths have slightly different indices of refraction and hence different speeds. This results in slightly different refraction angles for different wavelengths.

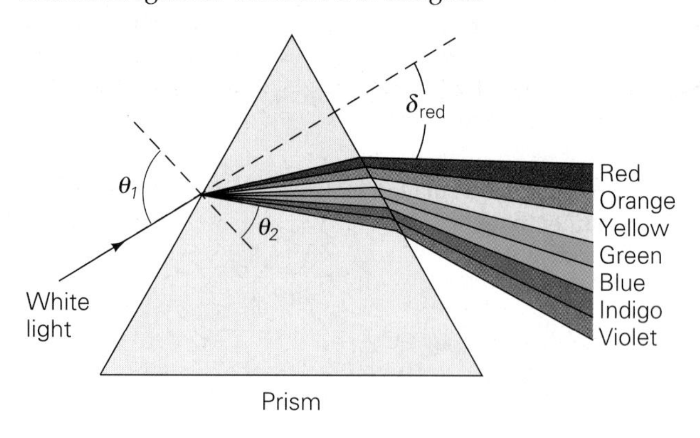

Exercises*

MC = *Multiple Choice Question,* **CQ** = *Conceptual Question, and* **IE** = *Integrated Exercise. Throughout the text, many exercise sections will include "paired" exercises. These exercise pairs, identified with* **red numbers**, *are intended to assist you in problem solving and learning. In a pair, the first exercise (even numbered) is worked out in the Study Guide so that you can consult it should you need assistance in solving it. The second exercise (odd numbered) is similar in nature, and its answer is given at the back of the book.*

22.1 Wave Fronts and Rays *and* 22.2 Reflection

1. **MC** A ray (a) is perpendicular to the direction of energy flow, (b) is always parallel to other rays, (c) is perpendicular to a series of wave fronts, (d) illustrates the wave nature of light.

2. **MC** The angle of incidence is the angle between (a) the incident ray and the reflecting surface, (b) the incident ray and the normal to the surface, (c) the incident ray and the reflected ray, (d) the reflected ray and the normal to the surface.

3. **MC** For both specular (regular) and diffuse (irregular) reflections, (a) the angle of incidence equals the angle of reflection, (b) the incident and reflected rays are on opposite sides of the normal, (c) the incident ray, the reflected ray, and the local normal lie in the same plane, (d) all of the preceding.

4. **CQ** Under what circumstances will the angle of reflection be smaller than the angle of incidence?

5. **CQ** The book you are reading is not, itself, a light source, so it must be reflecting light from other sources. What type of reflection is this?

*Assume angles to be exact.

6. **CQ** When you see the Sun over a lake or the ocean, you often observe a long swath of light (▼Fig. 22.22). What causes this effect, sometimes called a "glitter path"?

◀ **FIGURE 22.22** **A glitter path** See Exercise 6.

7. ● The angle of incidence of a light ray on a mirrored surface is 35°. What is the angle between the incident and reflected rays?

8. ● A beam of light is incident on a plane mirror at an angle of 32° relative to the normal. What is the angle between the reflected rays and the surface of the mirror?

9. **IE** ● A beam of light is incident on a plane mirror at an angle α relative to the surface of the mirror. (a) Will the angle between the reflected ray and the normal be (1) α, (2) $90° - \alpha$, or (3) 2α? (b) If $\alpha = 43°$, what is the angle between the reflected ray and the normal?

10. IE ●● Two upright plane mirrors touch along one edge, where their planes make an angle of α. A beam of light is directed onto one of the mirrors at an angle of incidence $\beta < \alpha$ and is reflected onto the other mirror. (a) Will the angle of reflection of the beam from the second mirror be (1) α, (2) β, (3) $\alpha + \beta$, or (4) $\alpha - \beta$? (b) If $\alpha = 60°$ and $\beta = 40°$, what will be the angle of reflection of the beam from the second mirror?

11. IE ●● Two identical plane mirrors of width w are placed a distance d apart with their mirrored surfaces parallel and facing each other. (a) A beam of light is incident at one end of one mirror so that the light just strikes the far end of the other mirror after reflection. Will the angle of incidence be (1) $\sin^{-1}(w/d)$, (2) $\cos^{-1}(w/d)$, or (3) $\tan^{-1}(w/d)$? (b) If $d = 50$ cm and $w = 25$ cm, what is the angle of incidence?

12. ●● Two people stand 3.0 m away from a large plane mirror and spaced 5.0 m apart in a dark room. At what angle of incidence should one of them shine a flashlight on the mirror so that the reflected beam directly strikes the other person?

13. ●● A beam of light is incident on a plane mirror at an angle of incidence of 35°. If the mirror rotates through a small angle of θ, through what angle will the reflected ray rotate?

14. ●●● Two plane mirrors, M_1 and M_2, are placed together as illustrated in ▼Fig. 22.23. (a) If the angle α between the mirrors is 70° and the angle of incidence, θ_{i_1}, of a light ray incident on M_1 is 35°, what is the angle of reflection, θ_{r_2}, from M_2? (b) If $\alpha = 115°$ and $\theta_{i_1} = 60°$, what is θ_{r_2}?

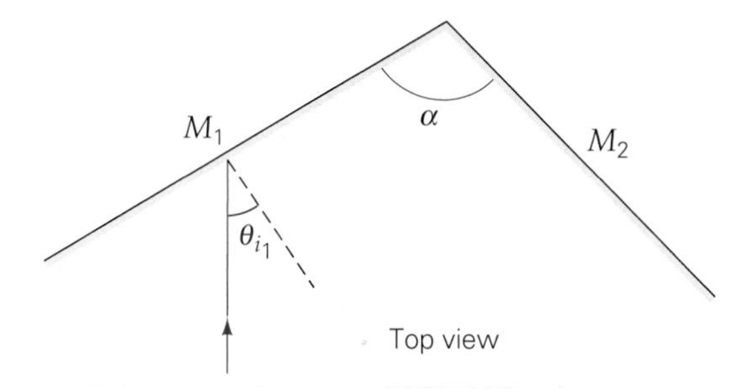

▲ **FIGURE 22.23 Plane mirrors together** See Exercises 14 and 15.

15. ●●● For the plane mirrors in Fig. 22.23, what angles α and θ_{i_1} would allow a ray to be reflected back in the direction from which it came (parallel to the incident ray)?

22.3 Refraction *and* 22.4 Total Internal Reflection and Fiber Optics

16. **MC** Light refracted at the boundary of two different media (a) is bent toward the normal when $n_1 > n_2$, (b) is bent away from the normal when $n_1 > n_2$, (c) is bent away from the normal when $n_1 < n_2$, (d) has the same angle of refraction as the angle of incidence.

17. **MC** The index of refraction (a) is always greater than or equal to 1, (b) is inversely proportional to the speed of light in a medium, (c) is inversely proportional to the wavelength of light in the medium, (d) all of the preceding.

18. **MC** Which of the following must be satisfied for total internal reflection to occur: (a) $n_1 > n_2$, (b) $n_2 > n_1$, (c) $\theta_1 > \theta_c$, or (d) $\theta_1 < \theta_c$.

19. CQ Explain the fundamental physical reason for refraction.

20. CQ As light travels from one medium to another, does its wavelength change? Its frequency? Its speed?

21. CQ Explain why the pencil in ▼Fig. 22.24 appears almost severed. Also, compare this figure with Fig. 22.13b and explain the difference.

◀ **FIGURE 22.24 Refraction effect** See Exercise 21.

22. CQ The photos in ▼Fig. 22.25 were taken with a camera on a tripod at a fixed angle. There is a penny in the container, but only its tip is seen initially. However, when water is added, more of the coin is seen. Why? Use a diagram to explain.

▲ **FIGURE 22.25 You barely see it, but then you do** See Exercises 22 and 52.

23. CQ Two hunters, one with bow and arrow and the other with a laser gun, see a fish under water. They both aim directly where they see it. Which one, the arrow or the laser beam, has a better chance of hitting the fish? Explain.

24. ● The speed of light in the core of the crystalline lens in a human eye is 2.13×10^8 m/s. What is the index of refraction of the core?

25. IE ● The indices of refraction for diamond and zircon can be found in Table 22.1. (a) The speed of light in zircon is (1) greater than, (2) less than, (3) the same as the speed of light in diamond. Explain. (b) Compute the ratio of the speed of light in zircon to that in diamond.

26. IE ● A beam of light enters water from air. (a) Will the angle of refraction be (1) greater than, (2) equal to, or (3) less than the angle of incidence? Explain. (b) If the beam enters the water at an angle of 60° relative to the normal of the surface, find the angle of refraction.

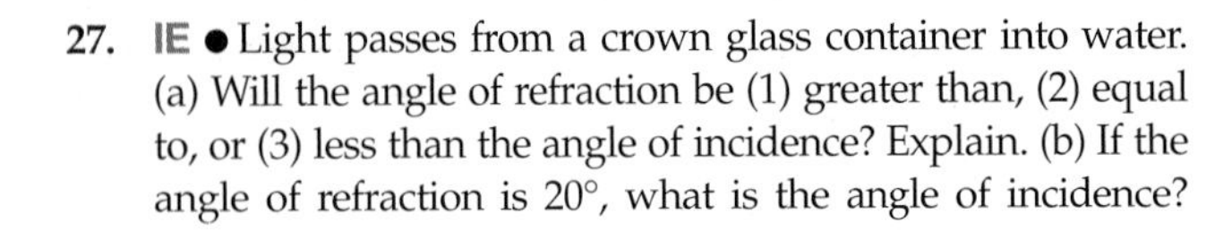

27. IE ● Light passes from a crown glass container into water. (a) Will the angle of refraction be (1) greater than, (2) equal to, or (3) less than the angle of incidence? Explain. (b) If the angle of refraction is 20°, what is the angle of incidence?

28. ● A beam of light traveling in air is incident on a transparent plastic material at an angle of incidence of 50°. The angle of refraction is 35°. What is the index of refraction of the plastic?

29. IE ● (a) For total internal reflection to occur, should the light be directed from (1) air to a diamond, (2) a diamond to air? Explain. (b) What is the critical angle of the diamond in air?

30. ● The critical angle for a certain type of glass in air is 41.8°. What is the index of refraction of the glass?

31. ●● A beam of light in air is incident on the surface of a slab of fused quartz. Part of the beam is transmitted into the quartz at an angle of refraction of 30° relative to a normal to the surface, and part is reflected. What is the angle of reflection?

32. ●● A beam of light is incident from air onto a flat piece of polystyrene at an angle of 55° relative to a normal to the surface. What angle does the refracted ray make with the plane of the surface?

33. ●● Monochromatic blue light that has a frequency of 6.5×10^{14} Hz enters a piece of flint glass. What are the frequency and wavelength of the light in the glass?

34. IE ●● Light passes from material A, which has an index of refraction of $\frac{4}{3}$, into material B, which has an index of refraction of $\frac{5}{4}$. (a) The speed of light in material A is (1) greater than, (2) the same as, (3) less than the speed of light in material B. Explain. (b) Find the ratio of the speed of light in material A to the speed of light in material B.

35. IE ●● In Exercise 34, (a) the wavelength of light in material A is (1) greater than, (2) the same as, (3) less than the wavelength of light in material B. Explain. (b) What is the ratio of the light's wavelength in material A to that in material B?

36. ●● The laser used in cornea surgery to treat corneal disease is the *excimer laser*, which emits ultraviolet light at a wavelength of 193 nm in air. The index of refraction of the cornea is 1.376. What are the wavelength and frequency of the light in the cornea?

37. ●● (a) An object immersed in water appears closer to the surface than it actually is. What is the cause of this illusion? (b) Using ▼Fig. 22.26, show that the apparent depth for small angles of refraction is d/n, where n is the index of refraction of the water. [*Hint*: Recall that for small angles, $\tan \theta \approx \sin \theta$.]

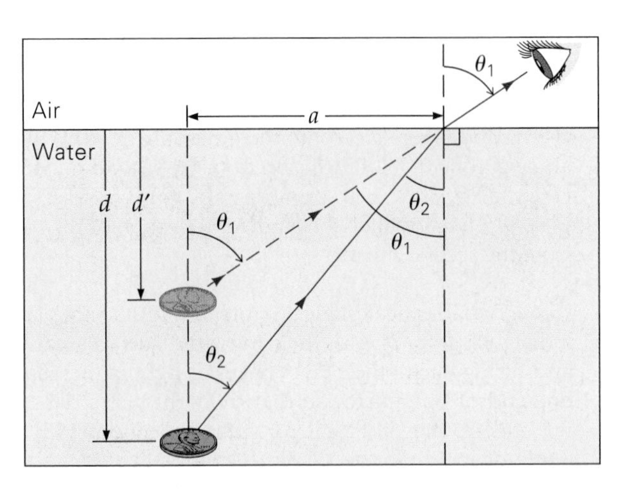

▲ **FIGURE 22.26 Apparent depth?** See Exercise 37. (For small angles only; angles enlarged for clarity.)

38. ●● A person lying at poolside looks over the edge of the pool and sees a bottle cap on the bottom directly below, where the depth is 3.2 m. How far below the water surface does the bottle cap appear to be? (See Exercise 37b.)

39. ●● What percentage of the actual depth is the apparent depth of an object submerged in water if the observer is looking almost straight downward? (See Exercise 37b.)

40. ●● A light ray in air is incident on a glass plate 10.0 cm thick at an angle of incidence of 40°. The glass has an index of refraction of 1.65. The emerging ray on the other side of the plate is parallel to the incident ray, but is laterally displaced. What is the perpendicular distance between the original direction of the ray and the direction of the emerging ray? [*Hint*: See Example 22.4.]

41. IE ●● To a submerged diver looking upward through the water, the altitude of the Sun (the angle between the Sun and the horizon) appears to be 45°. (a) The actual altitude of the Sun is (1) greater than, (2) the same as, (3) less than 45°. Explain. (b) What is the Sun's actual altitude?

42. ●● At what angle to the surface must a diver submerged in a lake look toward the surface to see the setting Sun just along the horizon?

43. ●● A submerged diver shines a light toward the surface of a body of water at angles of incidence of 40° and 50°. Can a person on the shore see a beam of light emerging from the surface in either case? Justify your answer mathematically.

44. IE ●● A beam of light is to undergo total internal reflection through a 45°–90°–45° prism (►Fig. 22.27). (a) Will this arrangement depend on (1) the index of refraction of the prism, (2) the index of refraction of the surrounding medium, or (3) the indices of refraction of both? Explain. (b) Calculate the minimum index of refraction of the prism if the surrounding medium is air. Repeat if it is water.

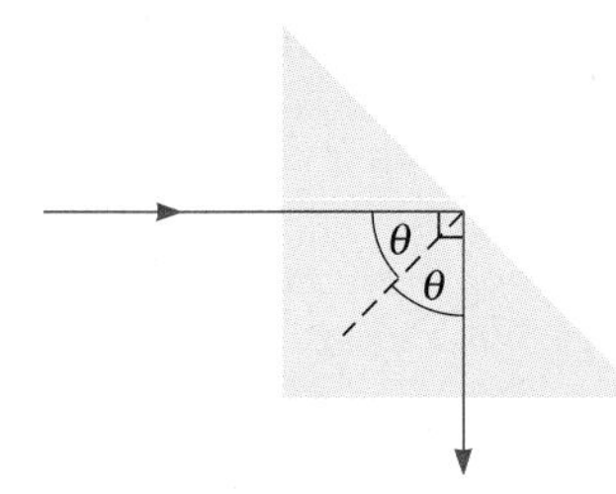

◀ **FIGURE 22.27 Total internal reflection in a prism** See Exercises 44 and 45.

45. ●● A 45°–90°–45° prism (Fig. 22.27) is made of a material with an index of refraction of 1.85. Can the prism be used to deflect a beam of light by 90° (a) in air? (b) What about in water?

46. ●● A coin lies on the bottom of a pool under 1.5 m of water and 0.90 m from the sidewall (▼Fig. 22.28). If a light beam is incident on the water surface at the wall, at what angle θ relative to the wall must the beam be directed so that it will illuminate the coin?

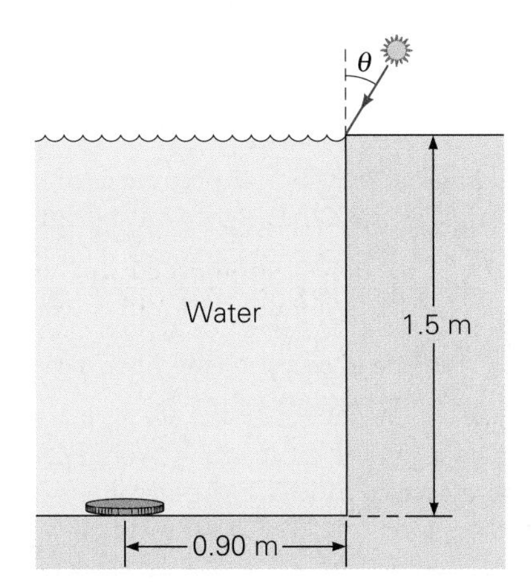

▲ **FIGURE 22.28 Find the coin** See Exercise 46 (not drawn to scale).

47. ●● Can you determine the index of refraction of the fluid in air as shown in Fig. 22.9a? If yes, what is its value?

48. ●● A crown-glass plate 2.5 cm thick is placed over a newspaper. How far beneath the top surface of the plate would the print appear to be if you were looking almost vertically downward through the plate? (See Exercise 37b.)

49. ●● A beam of light traveling in water strikes a surface of a transparent material at an angle of incidence of 45°. If the angle of refraction in the material is 35°, what is the index of refraction of the transparent material?

50. ●● Yellow-green light of wavelength 550 nm is incident on the surface of a flat piece of crown glass at an angle of 40°. What is (a) the angle of refraction of the light? (b) the speed of the light in the glass? (c) the wavelength of the light in the glass?

51. IE ●●● A light beam traveling upward in a plastic material with an index of refraction of 1.60 is incident on an upper horizontal air interface. (a) At certain angles of incidence, the light is not transmitted into air. The cause of this is (1) reflection, (2) refraction, (3) total internal reflection. Explain. (b) If the angle of incidence is 45°, is some of the beam transmitted into air? (c) Suppose the upper surface of the plastic material is covered with a layer of liquid with an index of refraction of 1.20. What happens in this case?

52. ●●● A 15-cm-deep opaque container is empty except for a single coin resting on its bottom surface. When looking into the container at a viewing angle of 50° relative to the vertical side of the container, you see nothing on the bottom. When the container is filled with water, you see the coin (from the same viewing angle) on the bottom of, and just beyond, the side of the container. (See Fig. 22.25.) How far is the coin from the side of the container?

53. ●●● An outdoor circular fish pond has a diameter of 4.00 m and a uniform full depth of 1.50 m. A fish halfway down in the pond and 0.50 m from the near side can just see the full height of a 1.80-m-tall person. How far away from the edge of the pond is the person?

54. ●●● A cube of flint glass sits on a newspaper on a table. The bottom half of the vertical sides of the cube is painted so that portion is opaque but the top half is transparent. By looking into one of the *vertical* sides of the cube, is it possible to see the portion of the newspaper covered by the center of the glass? Prove your answer. [*Hint*: Sketch the light leaving the location of interest.]

55. ●●● Two glass prisms are placed together (▼Fig. 22.29). (a) If a beam of light strikes the face of one of the prisms at normal incidence as shown, at what angle θ does the beam emerge from the other prism? (b) At what angle of incidence would the beam be refracted along the interface of the prisms?

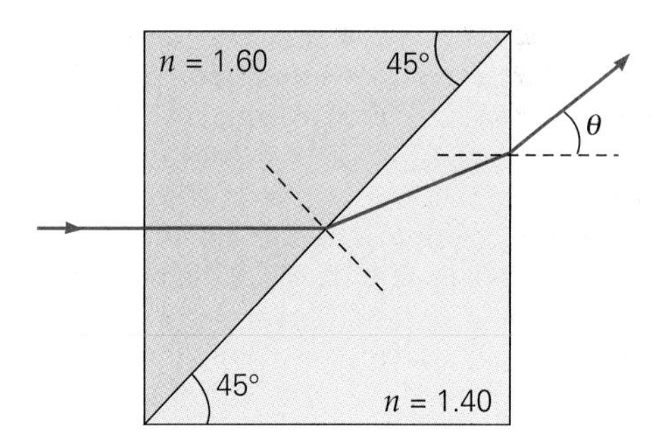

▲ **FIGURE 22.29 Joined prisms** See Exercise 55.

22.5 Dispersion

56. **MC** Dispersion can occur only if the light is (a) monochromatic, (b) polychromatic, (c) white light, (d) both b and c.

57. **MC** Dispersion can occur only during (a) reflection, (b) refraction, (c) total internal reflection, (d) all of the preceding.

58. **MC** Dispersion is caused by (a) the difference in the speed of light in different media, (b) the difference in the speed of light for different wavelengths of light in a given medium, (c) the difference in the angle of incidence for different wavelengths of light in a given medium, (d) the difference in the indices of refraction of light in different media.

59. CQ Why is dispersion more prominent when using a triangular prism rather than a square block?

60. CQ A glass prism disperses white light into a spectrum. Can a second glass prism be used to recombine the spectral components? Explain.

61. CQ You can never walk under a rainbow. Explain why.

62. CQ A light beam consisting of two colors, A and B, is sent through a prism. Color A is refracted more than color B. Which color has a longer wavelength? Explain.

63. CQ (a) If glass is dispersive, why don't we normally see a spectrum of colors when sunlight passes through a windowpane? (b) Does dispersion occur for polychromatic light incident on a dispersive medium at an angle of 0°? Explain. (Are the speeds of each color of light the same in the medium?)

64. IE ●● The index of refraction of crown glass is 1.515 for red light and 1.523 for blue light. (a) If light is incident on crown glass from air, which color, red or blue, will be refracted more? Explain? (b) Find the angle separating rays of the two colors in a piece of crown glass if their angle of incidence is 37°.

65. ●● A beam of light with red and blue components of wavelengths 670 nm and 425 nm, respectively, strikes a slab of fused quartz at an incident angle of 30°. On refraction, the different components are separated by an angle of 0.001 31 rad. If the index of refraction of the red light is 1.4925, what is the index of refraction of the blue light?

66. ●● White light passes through a piece of crown glass and strikes an interface with air at an angle of 41.15°. Assume the indices of refraction of crown glass are the same as given in Exercise 64. Which color(s) of light will be refracted out into the air?

67. ●●● A beam of red light is incident on an equilateral prism as shown in ▼Fig. 22.30. (a) If the index of refraction of red light of the prism is 1.400, at what angle θ does the beam emerge from the other face of the prism? (b) Suppose the incident beam were white light. What would be the angular separation of the red and blue components in the emergent beam if the index of refraction of blue light were 1.403? (c) If the index of refraction of blue light were 1.405?

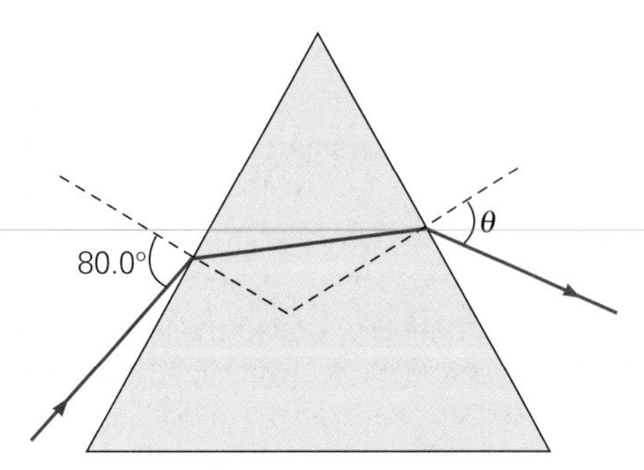

▲ **FIGURE 22.30 Prism revisited** See Exercise 67.

Comprehensive Exercises

68. In Fig. 22.21b, if the glass prism has an index of refraction of 1.5 and the experiment is done in water rather than in air, what happens to the spectrum emerging from the prism? How about in a liquid that also has an index of refraction of 1.5? Explain.

69. Light passes from medium A into medium B at an angle of incidence of 30°. The index of refraction of A is 1.5 times that of B. (a) What is the angle of refraction? (b) What is the ratio of the speed of light in B to the speed of light in A? (c) What is the ratio of the frequency of the light in B to the frequency of light in A? (d) At what angle of incidence would the light be internally reflected?

70. For total internal reflection to occur inside an optic fiber as shown in ▼Fig. 22.31, the angle θ must be greater than the critical angle for the fiber–air interface. At the end of the fiber, the incident light undergoes a refraction to enter the fiber. If total internal reflection is to occur for *any* angle of incidence, θ_i, outside the end of the fiber, what is the minimum index of refraction of the fiber?

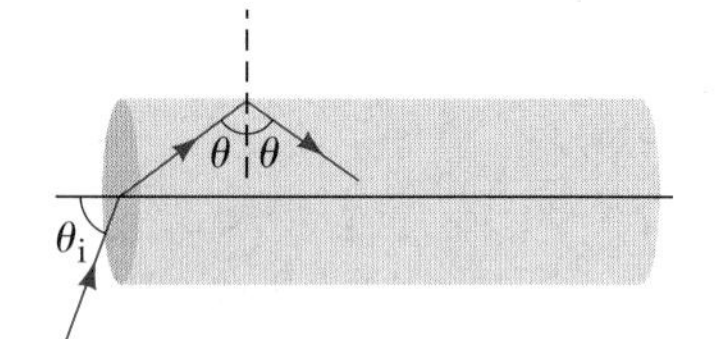

◀ **FIGURE 22.31 Optic fiber** See Exercise 70.

71. IE The critical angle for a glass–air interface is 41.11° for red light and 41.04° for blue light. (a) During the time the blue light travels 1.000 m, the red light will travel (1) more than (2) less than, (3) exactly 1.000 m. Explain. (b) Calculate the difference in distance traveled by the two colors.

72. In Exercise 67, if the angle of incidence is too small, light will not emerge from the other side of the prism. How could this happen? Calculate the minimum angle of incidence for the red light so that it does not emerge from the other side of the prism.

73. Light in air is incident on a transparent material. It is found that the angle of reflection is twice the angle of refraction. What is the *range* of the index of refraction of the material?

PHYSLET® The following Physlet Physics Problems can be used with this chapter.
34.1, 34.2, 34.3, 34.5, 34.6, 34.7, 34.8, 34.10

CHAPTER

23 MIRRORS AND LENSES

23.1 Plane Mirrors 730

23.2 Spherical Mirrors 732

23.3 Lenses 740

23.4 The Lens Maker's Equation 750

*23.5 Lens Aberrations 752

PHYSICS FACTS

- The largest refracting optical lens in the world measures 1.827 m (5.99 ft) in diameter. It was constructed by a team at the Optics Shop of the Optical Sciences Center of the University of Arizona in Tucson, Arizona, and completed in January 2000.
- The largest mirror under development for the European Space Agency's Herschel Space Observatory is 3.5 m (11.5 ft) in diameter. It is made from silicon carbide, which reduces its mass by a factor of 5 compared with traditional materials.
- A typical camera lens actually has more than one element (lens) inside. Many camera lenses have seven or more compensating elements to reduce or eliminate various types of lens aberrations. A single lens would produce distorted images.

What would life be like if there were no mirrors in bathrooms or cars, and if eyeglasses did not exist? Imagine a world without optical images of any kind—no photographs, no movies, no TV. Think about how little we'd know about the universe if there were no telescopes to observe distant planets and stars—or how little we'd know about biology and medicine if there were no microscopes to see bacteria and cells. It is often forgotten how dependent we are on mirrors and lenses.

The first mirror was probably the reflecting surface of a pool of water. Later, people discovered that polished metals and glass also have reflective properties. They must also have noticed that when they looked at things through glass, the objects looked different than when viewed directly, depending on the shape of the glass. In some cases, the objects appeared to be reduced or inverted, as the flower in the photo. In time, people learned to shape glass purposefully into lenses, paving the way for the eventual development of the many optical devices we now take for granted.

The optical properties of mirrors and lenses are based on the principles of reflection and refraction of light, as introduced in Chapter 22. In this chapter, you'll learn the principles of mirrors and lenses. Among other things, you'll discover why the images in the photo are upside down and reduced, whereas your image in an ordinary flat mirror is right side up—but the image doesn't seem to comb your hair with the same hand you use!

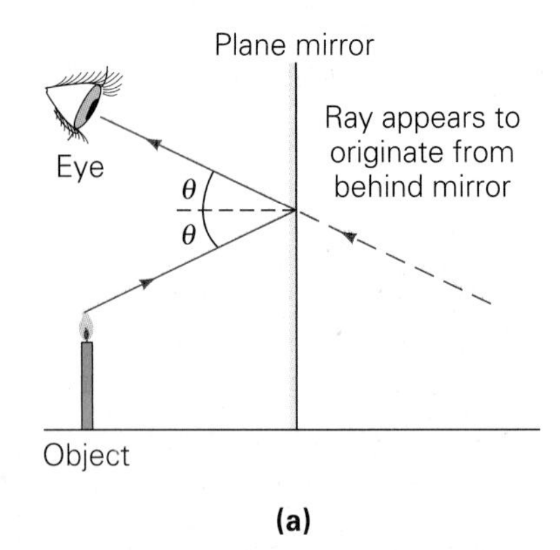

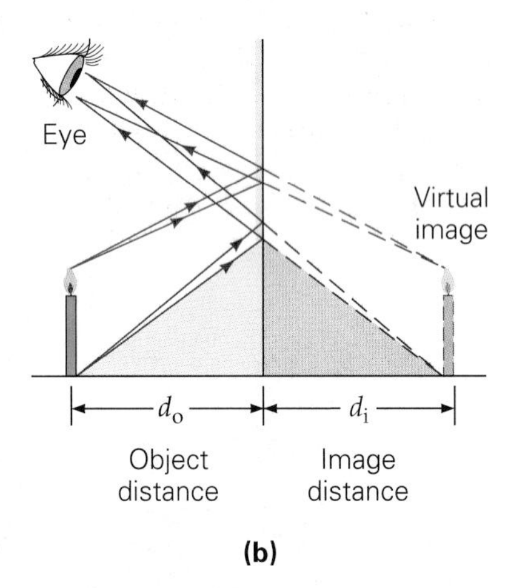

▲ **FIGURE 23.1 Image formed by a plane mirror (a)** A ray from a point on the object is reflected in the mirror according to the law of reflection. **(b)** Rays from various points on the object produce an image. Because the two shaded triangles are identical, the image distance d_i (the distance of the image from the mirror) is equal to the object distance d_o. That is, the image appears to be the same distance behind the mirror as the object is in front of the mirror. The rays appear to diverge from the image position. In this case, the image is said to be virtual.

Illustration 33.2 Flat Mirrors

23.1 Plane Mirrors

OBJECTIVES: To (a) understand how images are formed, and (b) describe the characteristics of images formed by plane mirrors.

Mirrors are smooth reflecting surfaces, usually made of polished metal or glass that has been coated with some metallic substance. As you know, even an uncoated piece of glass, such as a window pane, can act as a mirror. However, when one side of a piece of glass is coated with a compound of tin, mercury, aluminum or silver, the reflectivity of the glass is increased, as light is not transmitted through the coating. A mirror may be front coated or back coated. Most mirrors are backcoated.

When you look directly into a mirror, you see the images of yourself and objects around you (apparently on the other side of the surface of the mirror). The geometry of a mirror's surface affects the size, orientation, and type of image. In general, an *image* is the visual counterpart of an object, produced by reflection (mirrors) or refraction (lenses).

A mirror with a flat surface is called a **plane mirror**. How images are formed by a plane mirror is illustrated by the ray diagram in ◂Fig. 23.1. An image appears to be behind or "inside" the mirror. This is because when the mirror reflects a ray of light from the object to the eye (Fig. 23.1a), the ray appears to originate from behind the mirror. Reflected rays from the top and bottom of an object are shown in Fig. 23.1b. In actuality, light rays coming from all points on the side of the object facing the mirror are reflected, and an image of the complete object is observed.

The image formed in this way *appears* to be behind the mirror. Such an image is called a **virtual image**. Light rays appear to diverge from virtual images, but do not actually do so. No light energy actually comes from or passes through the image. However, spherical mirrors (discussed in Section 23.2) can project images in front of the mirror where light actually passes through the image. This type of image is called a **real image**. An example of a real image is the image produced by an overhead projector in a classroom.

Notice in Fig. 23.1b the distances of the object and image from the mirror. Quite logically, the distance of an object from a mirror is called the *object distance* (d_o), and the distance its image appears to be behind the mirror is called the *image distance* (d_i). By geometry of identical triangles and the law of reflection, $\theta_i = \theta_r$, it can be shown that $d_o = d_i$, which means that *the image formed by a plane mirror appears to be at a distance behind the mirror that is equal to the distance between the object and the front of the mirror*. (See Exercise 17.)

We are interested in various characteristics of images. Two of these features are the height and orientation of an image compared with those of its object. Both are expressed in terms of the **lateral magnification factor (*M*)**, which is defined as a ratio of heights of the image (h_i) and object (h_o):

$$M = \frac{\text{image height}}{\text{object height}} = \frac{h_i}{h_o} \tag{23.1}$$

A lighted candle used as an object allows us to address an important image characteristic: orientation—that is, whether the image is upright or inverted with respect to the orientation of the object. (In sketching ray diagrams, an arrow is a convenient object for this purpose.) For a plane mirror, the image is always upright (or erect). This means that the image is oriented in the same direction as the object. We say that h_i and h_o have *the same sign* (both positive or both negative), so M is positive. Note that M is a dimensionless quantity, as it is a ratio of heights.

In ▸Fig. 23.2, you should also be able to see that the image and object have the same sizes (heights), so $h_i = h_o$. Therefore, $M = +1$ for a plane mirror, the image is upright, and there is no magnification. That is, you and your image in a plane mirror are the same size.

With other types of mirrors, such as spherical mirrors (which we will consider shortly), it is possible to have inverted images where M is negative. In summary, the sign of M tells us the orientation of the image relative to the object, and the absolute value of M gives the magnification.

TABLE 23.1 Characteristics of Images Formed by Plane Mirrors

$d_i = d_o$	The image distance is equal to the object distance; that is, the image appears to be as far behind the mirror as the object is in front.
$M = +1$	The image is virtual, upright, and unmagnified.

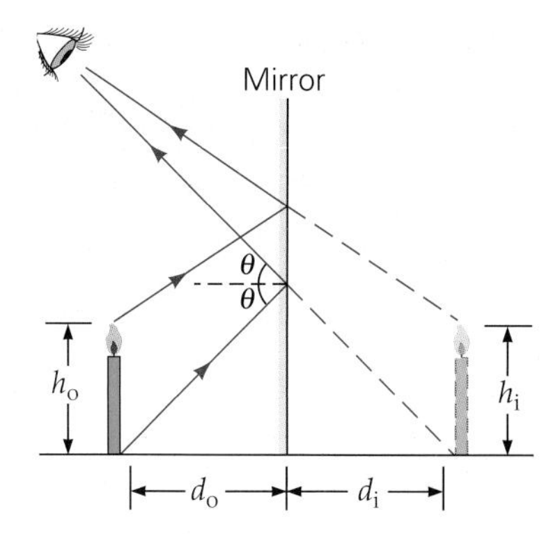

▲ **FIGURE 23.2** Magnification The lateral, or height magnification factor is given by $M = h_i/h_o$. For a plane mirror, $M = +1$, which means that $h_i = h_o$, that is, the image is the same height as the object, and that means the image is upright.

Another characteristic of reflected images of plane mirrors is the so-called *right–left reversal*. When you look at yourself in a mirror and raise your right hand, it appears that your image raises its left hand. However, this right–left reversal is actually caused by the front–back reversal. For example, if your front faces south, then your back "faces" north. Your image, on the other hand, has its front to the north and back to the south—a front–back reversal. You can demonstrate this reversal by asking one of your friends to stand facing you (without a mirror). If your friend raises his right hand, you can see that that hand is actually on your left side.

The main characteristics of an image formed by a plane mirror are summarized in Table 23.1. See also Insight 23.1, It's All Done with Mirrors, on page 733.

Example 23.1 ■ All of Me: Minimum Mirror Length

What is the minimum vertical length of a plane mirror needed for a person to be able to see a complete (head-to-toe) image of himself or herself (▼Fig. 23.3)?

Thinking It Through. Applying the law of reflection, we see in the figure that two triangles are formed by the rays needed for the image to be complete. These triangles relate the person's height to the minimum mirror length.

Solution. To determine this length, consider the situation shown in Fig. 23.3. With a mirror of minimum length, a ray from the top of the person's head would be reflected at the top of the mirror, and a ray from the person's feet would be reflected at the bottom of the mirror, to the eyes. The length L of the mirror is then the distance between the dashed horizontal lines perpendicular to the mirror at its top and bottom.

However, these lines are also the normals for the ray reflections. By the law of reflection, they bisect the angles between incident and reflected rays; that is, $\theta_i = \theta_r$. Then, because their respective triangles on each side of the dashed normal are identical, the length of the mirror from its bottom to a point even with the person's eyes is $h_1/2$, where h_1 is the person's height from the feet to the eyes. Similarly, the small upper length of the mirror is $h_2/2$ (the vertical distance between the person's eyes and the top of mirror). Then,

$$L = \frac{h_1}{2} + \frac{h_2}{2} = \frac{h_1 + h_2}{2} = \frac{h}{2}$$

where h is the person's total height.

Exploration 33.1 Image in a Flat Mirror

(continues on next page)

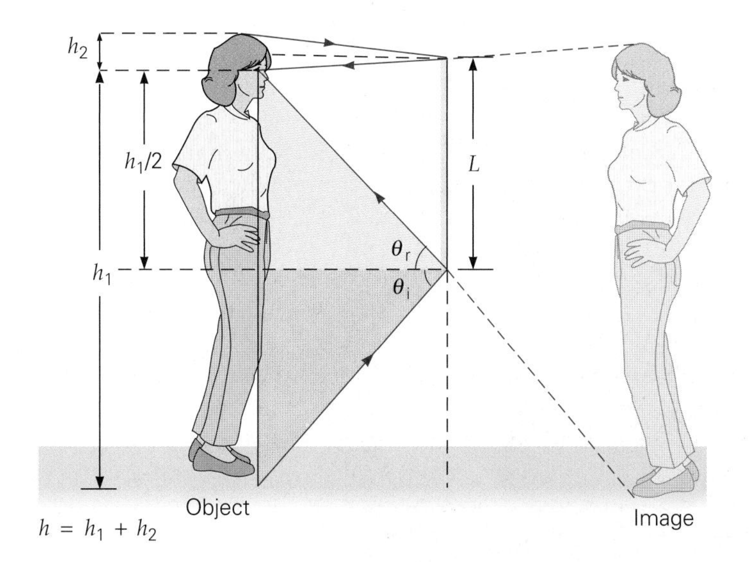

◀ **FIGURE 23.3** Seeing it all The minimum height, or vertical length, of a plane mirror needed for a person to see his or her complete (head-to-toe) image turns out to be half the person's height. See Example 23.1.

Hence, for a person to see his or her complete image in a plane mirror, the minimum height, or vertical length, of the mirror must be half the height of the person.

You can do a simple experiment to prove this conclusion. Get some newspaper and tape, and find a full-length mirror. Gradually cover parts of the mirror with the newspaper until you cannot see your complete image. You will find you need a mirror length that is only half your height to see a complete image.

Follow-Up Exercise. What effect does a person's distance from the mirror have on the minimum mirror length required to produce his or her complete image? *(Answers to all Follow-Up Exercises are at the back of the text.)*

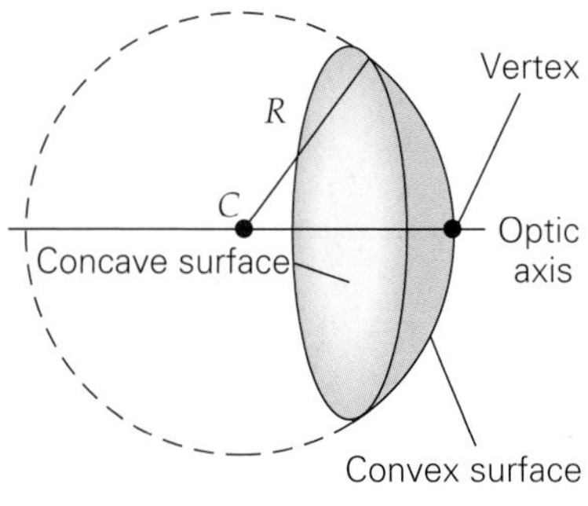

▲ **FIGURE 23.4** Spherical mirrors A spherical mirror is a section of a sphere. Either the outside (convex) surface or the inside (concave) surface of the spherical section may be the reflecting surface.

23.2 Spherical Mirrors

OBJECTIVES: To (a) distinguish between converging and diverging spherical mirrors, (b) describe images and their characteristics, and (c) determine image characteristics from ray diagrams and the spherical-mirror equation.

As the name implies, a **spherical mirror** is a reflecting surface with spherical geometry. ◂Figure 23.4 shows that if a portion of a sphere of radius R is sliced off along a plane, the severed section has the shape of a spherical mirror. Either the inside or outside of such a section can be the reflecting surface. For reflections on the inside surface, the section behaves as a **concave mirror**. (Think of looking into a *cave* in order to help yourself remember that a con*cave* mirror has a recessed surface.) For reflections from the outside surface, the section behaves as a **convex mirror**.

The radial line through the center of the spherical mirror that intersects the surface of the mirror at the *vertex* of the spherical section (Fig. 23.4) is called the *optic axis*. The point on the optic axis that corresponds to the center of the sphere of which the mirror forms a section is called the **center of curvature (*C*)**. The distance between the vertex and the center of curvature is equal to the radius of the sphere and is called the **radius of curvature (*R*)**.

When rays parallel and close to the optic axis are incident on a concave mirror, the reflected rays intersect, or converge, at a common point called the **focal point (*F*)**. As a result, a concave mirror acts as a **converging mirror** (◂Fig. 23.5a). Note that the law of reflection, $\theta_i = \theta_r$, is satisfied for each ray.

Similarly, rays parallel and close to the optic axis of a convex mirror diverges on reflection, as though the reflected rays came from a focal point behind the mirror's surface (Fig. 23.5b). Thus, a convex mirror acts as a **diverging mirror** (▾Fig. 23.6). When you see diverging rays, your brain interprets the image to mean that there is an object from which the rays *appear* to diverge, even though no such object is actually there.

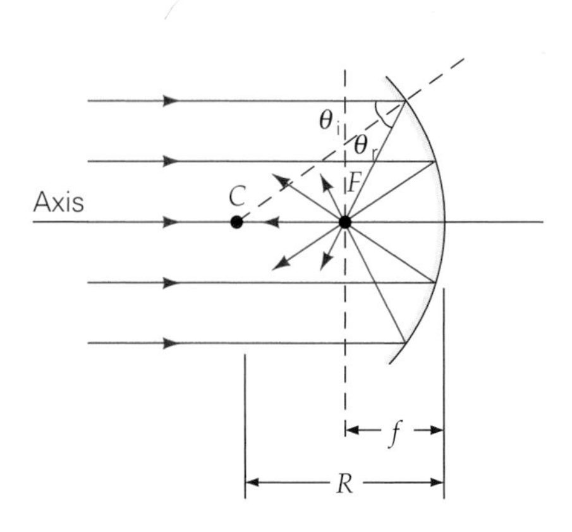

(a) Concave, or converging, mirror

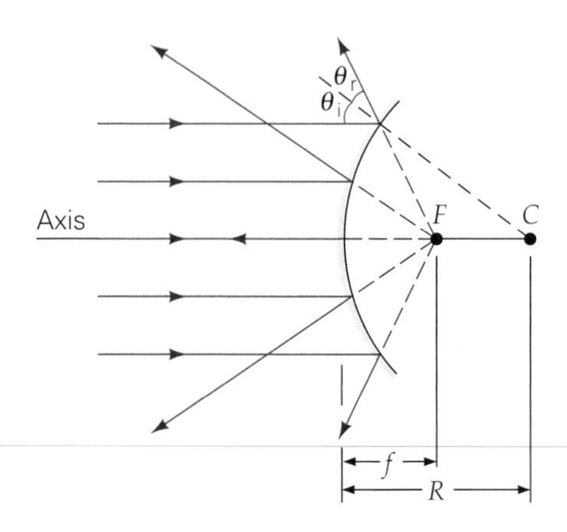

(b) Convex, or diverging, mirror

▲ **FIGURE 23.5** Focal point **(a)** Rays parallel and close to the optic axis of a concave spherical mirror converge at the focal point F. **(b)** Rays parallel and close to the optic axis of a convex spherical mirror are reflected along paths as though they diverge from a focal point behind the mirror. Note that the law of reflection, $\theta_i = \theta_r$, is satisfied for each ray in each diagram.

◂ **FIGURE 23.6** Diverging mirror Note by reverse-ray tracing in Fig 23.5b that a diverging (convex) spherical mirror gives an expanded, although distorted, field of view, as can be seen in this store-monitoring mirror.

INSIGHT 23.1 IT'S ALL DONE WITH MIRRORS

FIGURE 1 The Sphinx, an illustration of Tobin's sensational illusion The body was concealed by two plane mirrors.

FIGURE 2 Houdini and Jennie, the disappearing elephant The elephant vanished from view when Houdini fired a pistol.

Most of us are fascinated by a stage magician's sensational tricks that appear to make objects and animals suddenly appear or disappear. Of course they do not really appear or disappear. The magician requires special skills to make the performance quick and smooth so as to "fool" the audience. It's all done with mirrors, as they say.

The very first mirror illusion, "The Sphinx," was invented for stage magicians by Thomas William Tobin in 1876. His invention used mirrors to conceal a person or an object, as shown in Fig. 1, and it was used as the front cover of *Modern Magic* in 1876. Two plane mirrors are placed between the legs of the three-legged table to hide the person's body.

Harry Houdini, the world-famous master of illusion, felt it was too easy to make a pigeon fly out of a hat or a rabbit disappear into thin air. In 1918, Houdini made a 10 000-lb elephant named Jennie "disappear" on the stage of the Hippodrome Theater in New York City (Fig. 2). The act was called "The Vanishing Elephant."

When the time for the elephant's disappearance came, two large plane mirrors at right angles to each other were slid quickly into place. When properly aligned, the mirrors reflected light from the side walls of the stage to form virtual images that matched the pattern of the stage backdrop. Thus the audience apparently saw the stage with no elephant visible (Fig. 3). A strobe light was used to conceal the brief motion of the mirrors. Unseen by the audience, the elephant was quickly led off stage.

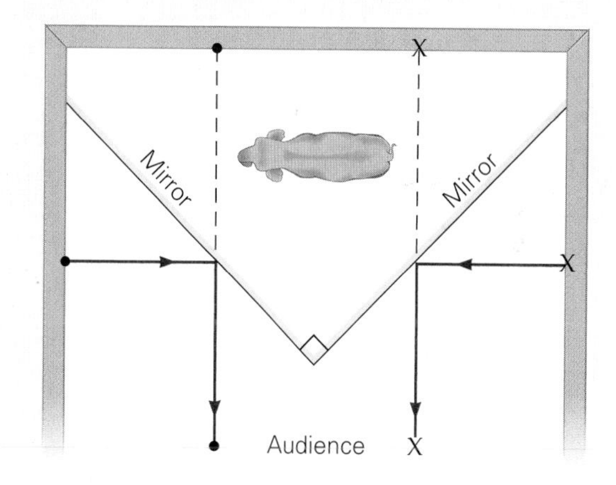

FIGURE 3 The disappearing elephant Two large mirrors at right angles to each other were used to conceal the elephant.

The distance from the vertex to the focal point (F) of parallel rays near the axis of a spherical mirror is called the **focal length** f. (See Fig. 23.5.) The focal length is related to the radius of curvature by the following simple equation:

$$f = \frac{R}{2} \quad \textit{focal length of spherical mirror} \tag{23.2}$$

Note:
Concave mirror = converging mirror

Convex mirror = diverging mirror

The preceding result is valid only when the rays are close to the optic axis—that is, for small-angle approximation. Rays far away from the optic axis will focus at different focal points, resulting in some image distortion. In optics, this distortion is an example of an *aberration*. Some telescope mirrors are parabolic in shape, rather than spherical, so *all* rays parallel to the optic axis are focused at the focal point, thus eliminating *spherical aberration*.

LEARN BY DRAWING

A Mirror Ray Diagram (see Example 23.2)

Converging (concave) mirrior

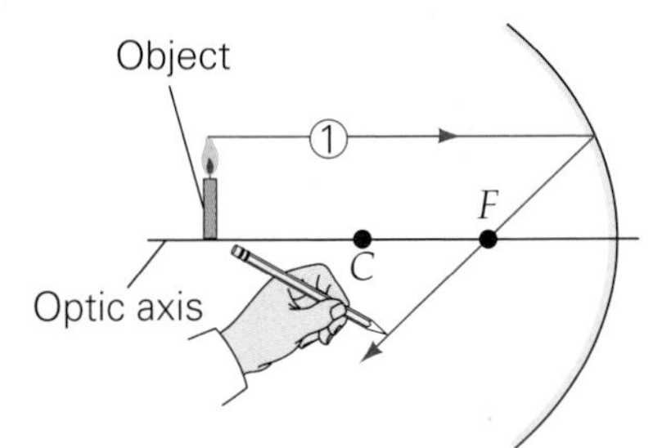

1 **Parallel ray**

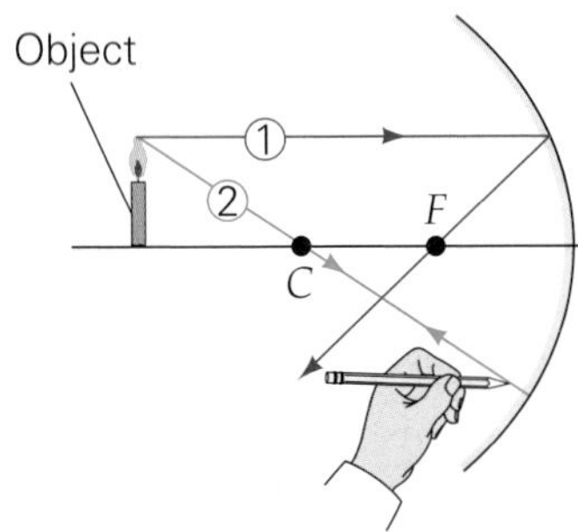

2 **Chief (radial) ray**

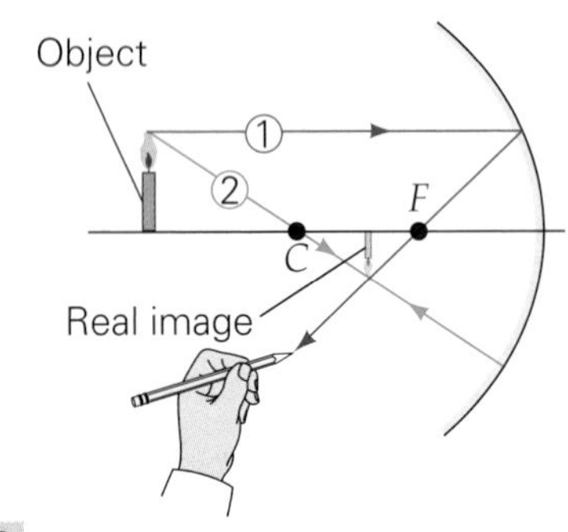

3 **Locating image**

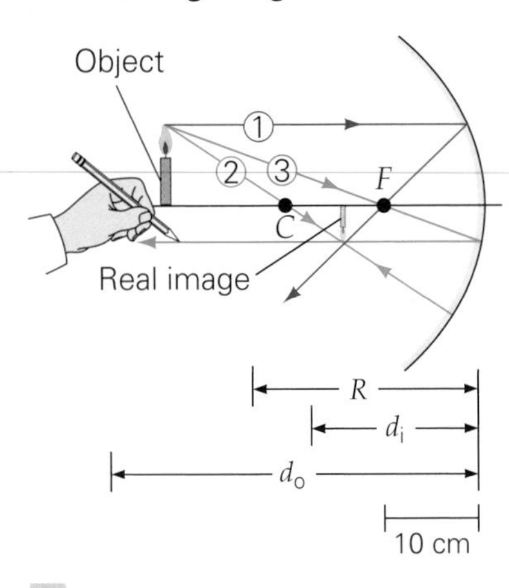

4 **Can also use focal ray to confirm image**

Ray Diagrams

The characteristics of images formed by spherical mirrors can be determined from geometrical optics (Chapter 22). The method involves drawing rays emanating from one or more points on an object. The law of reflection ($\theta_i = \theta_r$) applies, and three key rays are defined with respect to the mirror's geometry as follows:

1. A **parallel ray** is a ray that is incident along a path parallel to the optic axis and is reflected through (or appears to go through) the focal point F (as do all rays near and parallel to the axis).
2. A **chief ray**, or **radial ray**, is a ray that is incident through the center of curvature (C) of the spherical mirror. Since the chief ray is incident normal to the mirror's surface, this ray is reflected back along its incident path, through point C.
3. A **focal ray** is a ray that passes through (or appears to go through) the focal point and is reflected parallel to the optic axis. (It is a reversed parallel ray, so to speak.)

Using any two of these three key rays, we can locate the image (image distance) and determine its size (magnified or reduced), orientation (upright or inverted), and type (real or virtual). It is customary to use the tip of an asymmetrical object (for example, the head of an arrow or the flame of a candle) as the origin point of the rays. The corresponding point of the image is at the point of intersection of the rays. This makes it easy to see whether the image is upright or inverted.

Keep in mind, however, that *properly traced rays from any point on the object can be used to find the image.* Every point on a visible object acts as an emitter of light. For example, for a candle, the flame emits its own light, and many other points on the candle surface reflect light.

Example 23.2 ■ Learn by Drawing: A Mirror Ray Diagram

An object is placed 39.0 cm in front of a concave spherical mirror of radius 24.0 cm. (a) Use a ray diagram to locate the image formed by this mirror. (b) Discuss the characteristics of the image.

Thinking It Through. A ray diagram, drawn accurately, can by itself provide "quantitative" information about image location and image characteristics that might otherwise be determined mathematically.

Solution.

Given: $R = 24.0$ cm, $d_o = 39.0$ cm

Find: (a) image location (b) image characteristics

(a) Since we have been asked to use a ray diagram (drawing) to locate the image, the first thing we need to decide on is a scale for the drawing. If a scale of 1 cm (on the drawing) represents 10 cm, the object would be drawn 3.90 cm in front of the mirror.

First we draw the optic axis, the mirror, the object (a lighted candle), and the center of curvature (C). From Eq. 23.2, $f = 24.0\text{ cm}/2 = 12.0$ cm, and the focal point (F) is halfway from the vertex to the center of curvature.

To locate the image, follow steps 1–4 in the accompanying Learn by Drawing:

1. The first ray drawn is the parallel ray (① in the drawing). From the tip of the flame, draw a ray parallel to the optic axis. After reflecting, this ray goes through the focal point, F.

2. Then draw the chief ray (② in the drawing). From the tip of the flame, draw a ray going through the center of curvature, C. This ray will be reflected back along the original direction. (Why?)

3. It can be seen that these two rays intersect. The point of intersection is the tip of the *image* of the candle. From this point, draw the image by extending the tip of the flame to the optic axis. The image distance $d_i = 17$ cm, as measured from the diagram.

4. Only two rays are needed to locate the image. However, if the third ray is drawn as a double check, the focal ray in this case (③ in the drawing), it must go through the same point on the image at which the other two rays intersected (if drawn carefully). The focal ray from the tip of the flame going through the focal point, F, after reflection, will travel out parallel to the optic axis.

(b) From the ray diagram drawn in part (a), it can clearly be seen that the image is real (because the reflected rays intersect *in front* of the mirror). They converge and pass through the image. As a result, the real image could be seen on a screen (for example, a piece of white paper) that is positioned at a distance d_i from the concave mirror. The image is also inverted (the image of the candle points downward) and is smaller than the object.

Follow-Up Exercise. In this example, what would the characteristics of the image be if the object were 15.0 cm in front of the mirror? Locate the image and discuss its characteristics.

Exploration 33.3 Ray Diagrams

An example of a ray diagram using the same three rays for a convex (diverging) mirror will be shown in Integrated Example 23.4.

A converging mirror does *not* always form a real image. For a converging spherical mirror, the characteristics of the image change with the distance of the object from the mirror. Dramatic changes take place at two points: C (the center of curvature) and F (the focal point). These points divide the optic axis into three regions (▼Fig. 23.7a): $d_o > R$, $R > d_o > f$, and $d_o < f$.

▼ **FIGURE 23.7 Concave mirrors (a)** For a concave, or converging, mirror, the object is located within one of three regions defined by the center of curvature (C) and the focal point (F), or at one of these two points. For $d_o > R$, the image is real, inverted, and smaller than the object, as shown by the ray diagrams in Example 23.2. **(b)** For $R > d_o > f$, the image will also be real and inverted but enlarged, or magnified. **(c)** For an object at the focal point F, or $d_o = f$, the image is said to be formed at infinity. **(d)** For $d_o < f$, the image will be virtual, upright, and enlarged.

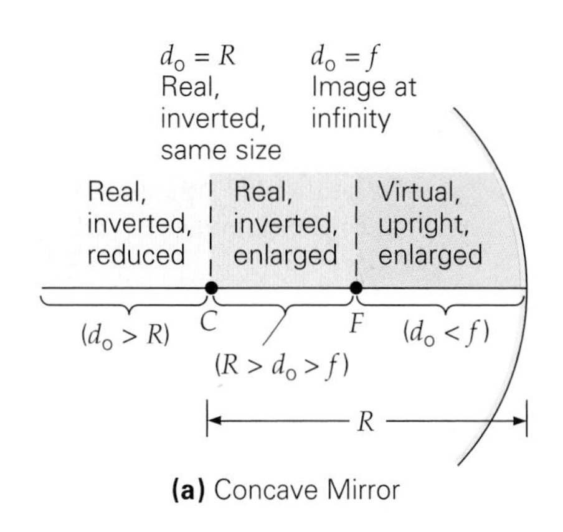

(a) Concave Mirror

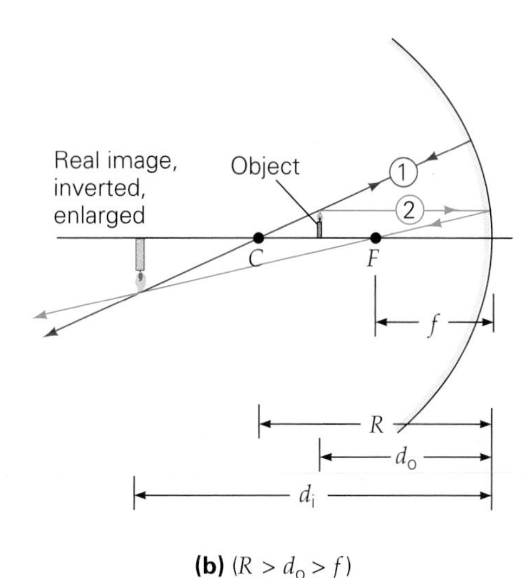

(b) $(R > d_o > f)$

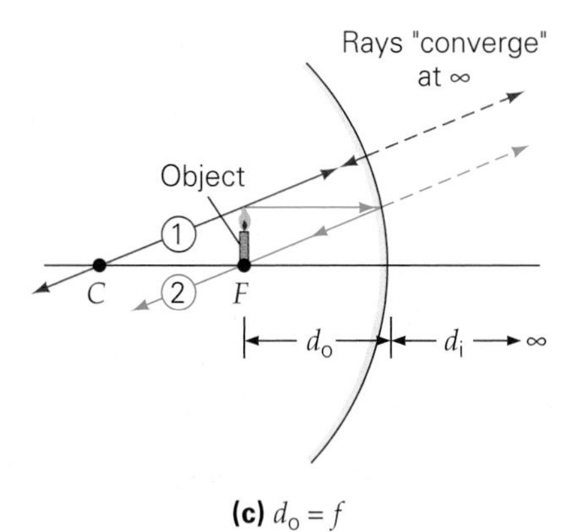

(c) $d_o = f$

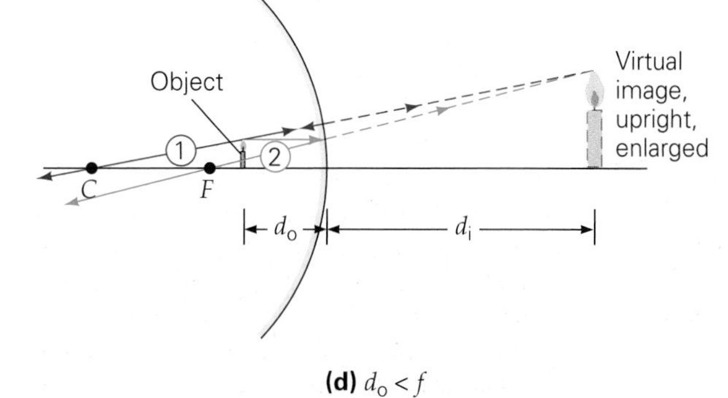

(d) $d_o < f$

Let's start with an object in the region farthest from the mirror ($d_o > R$) and move toward the mirror:

- The case of $d_o > R$ was shown in Example 23.2.
- When $d_o = R = 2f$, the image is real, inverted, and the same size as the object.
- When $R > d_o > f$, an enlarged, inverted, real image is formed (Fig. 23.7b). The image is magnified when the object is inside the center of curvature, C.
- When $d_o = f$, the object is at the focal point (Fig. 23.7c). The reflected rays are parallel, and the image is said to be "formed at infinity." The focal point F is a special "crossover" point, as it divides the space in front of the mirror into two regions.
- When $d_o < f$, the object is inside the focal point (between the focal point and the mirror's surface). A virtual, enlarged, and upright image is formed (Fig. 23.7d).

PHYSLET®

Exploration 33.2 Looking at Curved Mirrors

When $d_o > f$, the image is real; when $d_o < f$, the image is virtual. For $d_o = f$, we say that the image is formed at infinity (Fig. 23.7c). That is, when an object is at "infinity"—it is so far away that rays emanating from it and falling on the mirror are essentially parallel—its image is formed at the focal plane. This fact provides an easy method for determining the focal length of a concave mirror.

As we have seen, the position, orientation, and size of the image can be approximately determined graphically from ray diagrams drawn to scale. However, these characteristics can be determined more accurately by analytical methods. It can be shown by means of geometry that the object distance (d_o), the image distance (d_i), and the focal length (f) are related. That relationship is known as the **spherical-mirror equation**:

$$\frac{1}{d_o} + \frac{1}{d_i} = \frac{1}{f} = \frac{2}{R} \quad \textit{spherical-mirror equation} \tag{23.3}$$

Note that this equation can be written in terms of either the radius of curvature, R, or the focal distance, f, since by Eq. 23.2, $f = R/2$. Both R and f can be either positive or negative, as we will discuss shortly.

If d_i is the quantity to be found for a spherical mirror, it may be convenient to use an alternative form of the spherical-mirror equation:

$$d_i = \frac{d_o f}{d_o - f} \tag{23.3a}$$

But, you can always use the reciprocal form of Eq. 23.3.

The signs of the various quantities are very important in the application of Eqs. 23.3. We will use the sign conventions summarized in Table 23.2. For example,

Exploration 33.4 Focal Point and Image Point

TABLE 23.2 Sign Conventions for Spherical Mirrors

Focal length (f)	
Concave (converging) mirror	f (or R) is positive
Convex (diverging) mirror	f (or R) is negative
Object distance (d_o)	
Object is in front of the mirror (real object)	d_o is positive
Object is behind the mirror (virtual object)*	d_o is negative
Image distance (d_i) and *image type*	
Image is formed in front of the mirror (real image)	d_i is positive
Image is formed behind the mirror (virtual image)	d_i is negative
Image orientation (M)	
Image is upright with respect to the object	M is positive
Image is inverted with respect to the object	M is negative

*In a combination of two (or more) mirrors, the image formed by the first mirror is the object of the second mirror (and so on). If this image–object falls behind the second mirror, it is referred to as a *virtual* object, and the object distance is taken to be negative. This concept is more important for lens combinations, as we will see in Section 23.3, and is mentioned here only for completeness.

for a real object, a positive d_i indicates a real image and a negative d_i corresponds to a virtual image.

The **lateral magnification factor** M defined in Eq. 23.1 can also be found analytically for a spherical mirror. Again, by using geometry, it can be expressed in terms of the image and object distances:

$$M = -\frac{d_i}{d_o} \quad \textit{magnification equation} \qquad (23.4)$$

Note: A helpful hint for remembering that the magnification is d_i over d_o is that the ratio is in alphabetical order ("i" over "o").

The minus sign is added by convention to indicate the orientation of the image: A positive value for M indicates an upright image, whereas a negative M implies an inverted image. Also, if $|M| > 1$, the image is magnified, or larger than the object. If $|M| < 1$, the image is reduced, or smaller than the object. Note that for mirrors, the lateral magnification M, also called the *magnification factor*, or simply *magnification*, is conveniently expressed in terms of the image distance d_i and the object distance d_o rather than in terms of the image and object heights used in Eq. 23.1. (A description of the origin of Eqs. 23.3 and 23.4 follows as optional content.)

Example 23.3 and Integrated Example 23.4 show how these equations and sign conventions are used for spherical mirrors. In general, this approach usually involves finding the image of an object; you will be asked where the image is formed (d_i) and what the image characteristics are (M). These characteristics tell whether the image is real or virtual, upright or inverted, and larger or smaller than the object (magnified or reduced).

Note: $|M|$ is the absolute value of M: its magnitude without regard to sign. For example, $|+2| = |-2| = 2$.

***(Optional) Derivation of the Spherical-Mirror Equation** You might wonder from where Equations 23.3 and 23.4 originate. The spherical-mirror equation can be derived with the aid of a little geometry. Consider the ray diagram in ▼Fig. 23.8. The object and image distances (d_o and d_i) and the heights of the object and image (h_o and h_i) are shown. Note that these lengths make up the bases and heights of triangles formed by the ray reflected at the vertex (V). These triangles ($O'VO$ and $I'VI$) are similar, since, by the law of reflection, their angles at V are equal. Hence, we can write

$$\frac{h_i}{h_o} = -\frac{d_i}{d_o} \qquad (1)$$

This equation is Eq. 23.4, from the definition of Eq. 23.1. The negative sign inserted here signifies the fact that the image is inverted, so h_i is negative.

The (focal) ray through F also forms similar triangles, $O'FO$ and VFA in the approximation that the mirror is small compared with its radius. (Why are the triangles similar?) The bases of these triangles are $VF = f$ and $OF = d_o - f$. Then, if VA is taken to be h_i,

$$\frac{h_i}{h_o} = -\frac{VF}{OF} = -\frac{f}{d_o - f} \qquad (2)$$

Again, the negative sign inserted here signifies the fact that the image is inverted, so h_i is negative.

Equating Eqs. 1 and 2,

$$\frac{d_i}{d_o} = \frac{f}{d_o - f} \qquad (3)$$

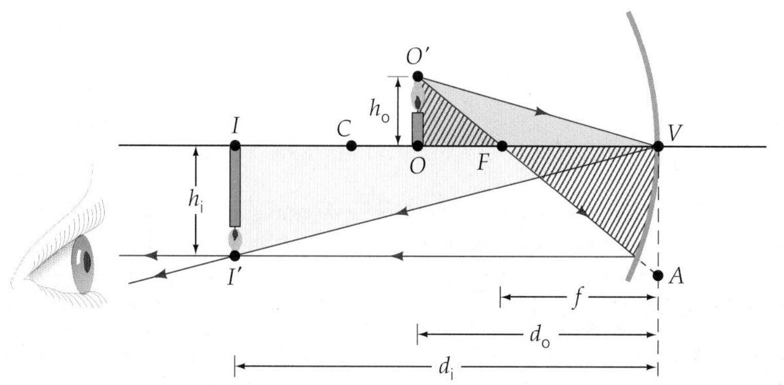

◀ **FIGURE 23.8 Spherical-mirror equation** The rays provide the geometry, through similar triangles, for the derivation of the spherical-mirror equation.

Algebraic manipulation yields

$$\frac{1}{d_o} + \frac{1}{d_i} = \frac{1}{f}$$

which is the spherical-mirror equation (Eq. 23.3).

Example 23.3 ■ What Kind of Image? Characteristics of a Concave Mirror

A concave mirror has a radius of curvature of 30 cm. If an object is placed (a) 45 cm, (b) 20 cm, and (c) 10 cm from the mirror, where is the image formed, and what are its characteristics? (Specify whether each image is real or virtual, upright or inverted, and magnified or reduced.)

Thinking It Through. Here we are given R, from which we can compute the focal length $f = R/2$. Also given are three different object distances, which can be applied in Eqs. 23.3 and 23.4 to determine the image location and characteristics.

Solution.

Given: $R = 30$ cm, so $f = R/2 = 15$ cm
(a) $d_o = 45$ cm
(b) $d_o = 20$ cm
(c) $d_o = 10$ cm

Find: d_i, M, and image characteristics for given object distances

Note that the given object distances correspond to the regions shown in Fig. 23.7a. There is no need to convert the distances to meters as long as all distances are expressed in the same unit (centimeters in this case). You could draw representative ray diagrams for each of these cases in order to find the characteristics of each image.

(a) In this case, the object distance is greater than the radius of curvature ($d_o > R$), and

$$\frac{1}{d_o} + \frac{1}{d_i} = \frac{1}{f} \quad \text{or} \quad \frac{1}{d_i} = \frac{1}{f} - \frac{1}{d_o} = \frac{1}{15\text{ cm}} - \frac{1}{45\text{ cm}} = \frac{2}{45\text{ cm}}$$

Then

$$d_i = \frac{45\text{ cm}}{2} = +22.5\text{ cm} \quad \text{and} \quad M = -\frac{d_i}{d_o} = -\frac{22.5\text{ cm}}{45\text{ cm}} = -\frac{1}{2}$$

Thus, the image is real (positive d_i), inverted (negative M), and half as large as the object $\left(|M| = \frac{1}{2}\right)$.

(b) Here, $R > d_o > f$, and the object is between the focal point and the center of curvature:

$$\frac{1}{d_i} = \frac{1}{15\text{ cm}} - \frac{1}{20\text{ cm}} = \frac{1}{60\text{ cm}}$$

Thus,

$$d_i = +60\text{ cm} \quad \text{and} \quad M = -\frac{60\text{ cm}}{20\text{ cm}} = -3.0$$

In this case, the image is real (positive d_i), inverted (negative M), and three times the size of the object ($|M| = 3$).

(c) For this case, $d_o < f$, and the object is inside the focal point.
Using the alternate form of Eq. 23.3 for illustration:

$$d_i = \frac{d_o f}{d_o - f} = \frac{(10\text{ cm})(15\text{ cm})}{10\text{ cm} - 15\text{ cm}} = -30\text{ cm}$$

Then

$$M = -\frac{d_i}{d_o} = -\frac{(-30\text{ cm})}{10\text{ cm}} = +3.0$$

In this case, the image is virtual (negative d_i), upright (positive M), and three times the size of the object ($|M| = 3$).

From the denominator of the expression for d_i, you can see that d_i will always be negative when d_o is less than f. Therefore, a virtual image is always formed for an object inside the focal point of a converging mirror.

Follow-Up Exercise. For the converging mirror in this Example, where is the image formed and what are its characteristics if the object is at 30 cm, or $d_o = R$?

Problem-Solving Hint

When using the spherical-mirror equations to find image characteristics, it is helpful to first make a quick sketch (approximate, not necessarily to scale) of the ray diagram for the situation. This sketch shows you the image characteristics and helps you avoid making mistakes when applying the sign conventions. *The ray diagram and the mathematical solution must agree.*

Integrated Example 23.4 ■ Similarities and Differences: Characteristics of a Convex Mirror

An object (in this case, a candle) is 20 cm in front of a diverging mirror that has a focal length of −15 cm (see the sign conventions in Table 23.2). (a) Use a ray diagram to determine whether the image formed is (1) real, upright, magnified, (2) virtual, upright, magnified, (3) real, upright, reduced, (4) virtual, upright, reduced, (5) real, inverted, magnified, or (6) virtual, inverted, reduced. (b) Find the location and characteristics of the image by using the mirror equations.

Exploration 35.5 *Convex Mirrors, Focal Point, and Radius of Curvature*

(a) Conceptual Reasoning. Since we know the object distance and the focal length of the convex mirror, a ray diagram can be drawn and the image characteristics can be determined. The first thing to decide on is a scale for the ray diagram. In this example, a scale of 1 cm (on the drawing) could be used to represent 10 cm. That way, the object would be 2.0 cm in front of the mirror in our drawing. Draw the optic axis, the mirror, the object (a lighted candle), and the focal point (F). Since this mirror is convex, the focal point (F) and the center of curvature (C) are behind the mirror. From Eq. 23.2, $R = 2f = 2(-15\text{ cm}) = -30\text{ cm}$. So C is drawn at twice the distance of F from the vertex.

Only two out of the three key rays are necessary to locate the image (▾Fig. 23.9). The parallel ray ① starts from the tip of the flame, travels parallel to the optic axis, and then diverges from the mirror after reflection, appearing to come from F. The chief ray ② originates from the tip of the flame, appears to go through C, and then reflects straight back, but appears to come from C. It is clearly seen that these two rays, after reflection, diverge from each other, and there is no chance for them to intersect. However, they appear to start from a common point behind the mirror: the image point of the tip of the flame. We can also draw the focal ray ③ to verify that all three rays appear to emanate from the same image point.

The image is virtual (the reflecting rays don't actually come from a point behind the mirror), upright, and smaller than the object. Therefore, the answer is (4) virtual, upright, reduced. Measuring from the diagram (keep in mind the drawing scale we are using), we find that $d_i \approx -9.0\text{ cm}$, and the magnification $M = \dfrac{h_i}{h_o} \approx \dfrac{0.5\text{ cm}}{1.2\text{ cm}} = +0.4$.

(continues on next page)

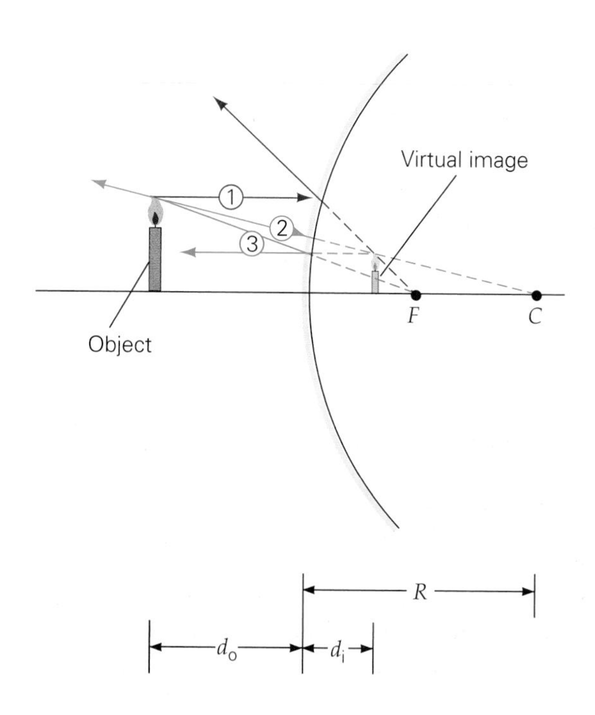

◂ **FIGURE 23.9 Diverging mirror** Ray diagram of a diverging mirror. See Integrated Example 23.4.

(b) Quantitative Reasoning and Solution. The object distance and focal length are given. The image position and characteristics can be calculated by using the mirror equations.

Given: $d_o = 20$ cm, $f = -15$ cm *Find:* d_i, M, and image characteristics

Note that the focal length is negative for a convex mirror. (See Table 23.2.) Using Eq. 23.3, we have

$$\frac{1}{20\text{ cm}} + \frac{1}{d_i} = \frac{1}{-15\text{ cm}}$$

so

$$d_i = -\frac{60\text{ cm}}{7} = -8.6\text{ cm}$$

Then

$$M = -\frac{d_i}{d_o} = -\frac{(-8.6\text{ cm})}{20\text{ cm}} = +0.43$$

Thus, the image is virtual (d_i is negative), upright (M is positive), and 0.43 times the size (height) of the object. These results agree well with those from the ray diagram. The image of an object is always virtual for a diverging (convex) mirror. (Can you prove this using either a ray diagram or the mirror equation?)

Follow-Up Exercise. As has been pointed out, a diverging mirror always forms a virtual image of a real object. What about the other characteristics of the image—its orientation and magnification? Can any general statements be made about them?

PHYSLET® Illustration 33.1 Mirrors and the Small-Angle Approximation

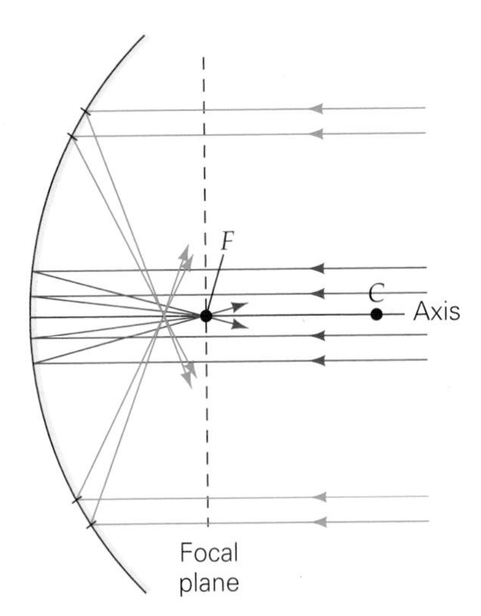

▲ **FIGURE 23.10 Spherical aberration for a mirror** According to the small-angle approximation, rays parallel to and near the mirror's axis converge at the focal point. However, when parallel rays not near the axis are reflected, they converge in front of the focal point. This effect, called *spherical aberration*, gives rise to blurred images.

Spherical-Mirror Aberrations

Technically, our descriptions of image characteristics for spherical mirrors are true only for objects near the optic axis—that is, only for small angles of incidence and reflection. If these conditions do not hold, the images will be blurred (out of focus) or distorted, because not all of the rays will converge in the same plane. As illustrated in ◂Fig. 23.10, incident parallel rays far from the optic axis do not converge at the focal point. The farther the incident ray is from the axis, the farther is the reflected ray from the focal point. This effect is called **spherical aberration.**

Spherical aberration does not occur with a parabolic mirror. (As the name *parabolic mirror* implies, a parabolic mirror has the form of a paraboloid.) *All* of the incident rays parallel to the optic axis of such a mirror have a common focal point. For this reason, parabolic mirrors are used in most astronomical telescopes (Chapter 24). However, these mirrors are more difficult to make than spherical mirrors and are therefore more expensive.

23.3 Lenses

OBJECTIVES: To (a) distinguish between converging and diverging lenses, (b) describe their images and their characteristics, and (c) find image locations and characteristics by using ray diagrams and the thin-lens equation.

The word *lens* is from the Latin *lentil*, which is a round, flattened, edible seed of a pea-like plant. Its shape is similar to that of a lens. An optical **lens** is made from transparent material (most commonly glass, but sometimes plastic or crystal). One or both surfaces usually have a spherical contour. *Biconvex* spherical lenses (with both surfaces convex) and *biconcave* spherical lenses (with both surfaces concave) are illustrated in ▾Fig. 23.11. Lenses can form images by refracting the light that passes through them.

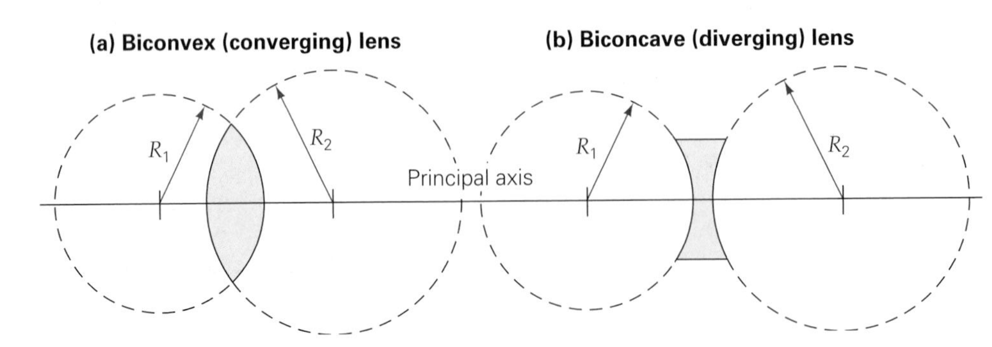

▸ **FIGURE 23.11 Spherical lenses** Spherical lenses have surfaces defined by two spheres, and the surfaces are either convex or concave. **(a)** Biconvex and **(b)** biconcave lenses are shown here. If $R_1 = R_2$, a lens is spherically symmetric.

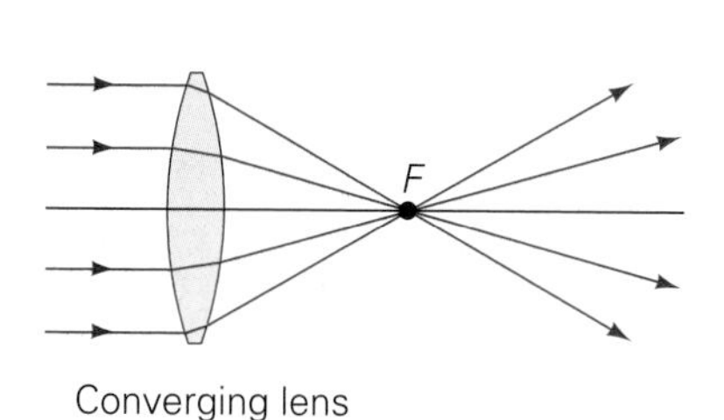

(a) Biconvex (converging) lens

(b)

◀ FIGURE 23.12 Converging lens **(a)** For a thin biconvex lens, rays parallel to the axis converge at the focal point *F*. **(b)** A magnifying glass (converging lens) can be used to focus the Sun's rays to a spot—with incendiary results. Do not try this at home!

Illustration 35.1 Lenses and the Thin-Lens Approximation

A biconvex lens is an example of a **converging lens**. Incident light rays parallel to the axis of the lens converge at a focal point (F) on the opposite side of the lens (▲Fig. 23.12a). This fact provides a way to experimentally determine the focal length of a converging lens. You may have focused the Sun's rays with a magnifying glass (a biconvex, or converging, lens) and thereby witnessed the concentration of radiant energy that results (Fig. 23.12b).

A biconcave lens is an example of a **diverging lens**. Incident parallel rays emerge from the lens as though they emanated from a focal point on the incident side of the lens (▸Fig. 23.13).

There are several types of converging and diverging lenses (▸Fig. 23.14). Convex and concave meniscus lenses are the type most commonly used for corrective eyeglasses. In general, a converging lens is thicker at its center than at its periphery, and a diverging lens is thinner at its center than at its periphery. This discussion will be limited to spherically symmetric biconvex and biconcave lenses, for which both surfaces have the same radius of curvature.

When light passes through a lens, it is refracted and displaced laterally (Example 22.4 and Fig. 22.11). If a lens is thick, this displacement may be fairly large and can complicate analysis of the lens's characteristics. This problem does not arise with thin lenses, for which the refractive displacement of transmitted light is negligible. Our discussion will be limited to thin lenses. A thin lens is a lens for which the thickness of the lens is assumed to be negligible compared with the lens's focal length.

A lens with spherical geometry has, *for each lens surface*, a center of curvature (C), a radius of curvature (R), a focal point (F), and a focal length (f). The focal points are at equal distances on either side of a thin lens. However, for a spherical lens, the focal length is *not* simply related to R by $f = R/2$ as it is for spherical mirrors. Because the focal length also depends on the lens's index of refraction, the focal length of a lens is usually specified, rather than its radius of curvature. This will be discussed in Section 23.4.

The general rules for drawing ray diagrams for lenses are similar to those for spherical mirrors. But some modifications are necessary, since light passes through a lens. Opposite sides of a lens are generally distinguished as the *object side* and the *image side*. The object side is the side on which an object is positioned, and the image side is the *opposite* side of the lens (where a real image would be formed). The three rays from a point on an object are drawn as follows (see Learn by Drawing for Example 23.5 on page 743):

1. A **parallel ray** is a ray that is parallel to the lens's optic axis on incidence and, after refraction, either (a) passes through the focal point on the image side of a converging lens *or* (b) appears to diverge from the focal point on the object side of a diverging lens.

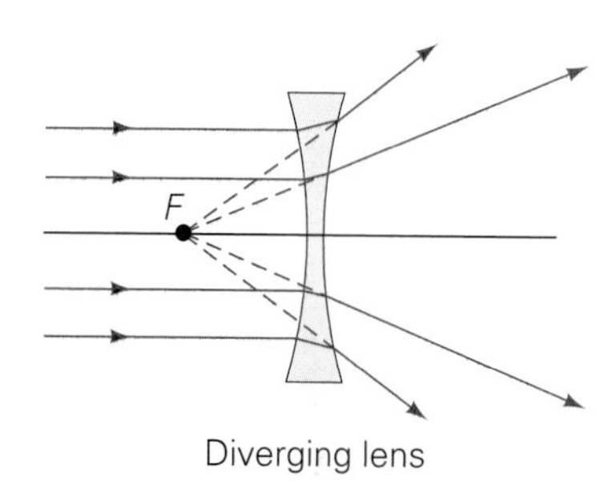

Biconcave (diverging) lens

▲ FIGURE 23.13 Diverging lens Rays parallel to the axis of a biconcave, or diverging, lens appear to diverge from a focal point on the incident side of the lens.

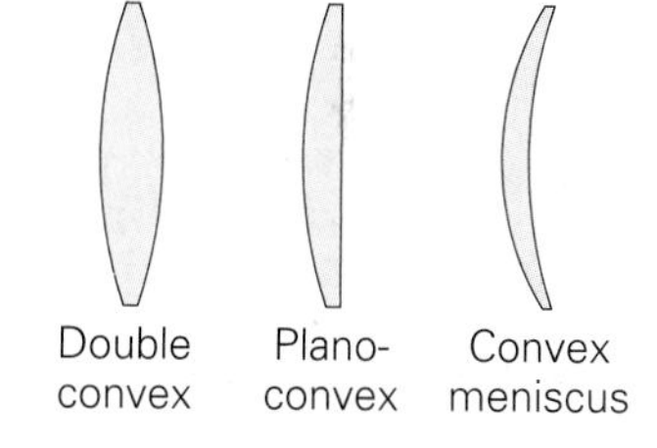

Converging lenses

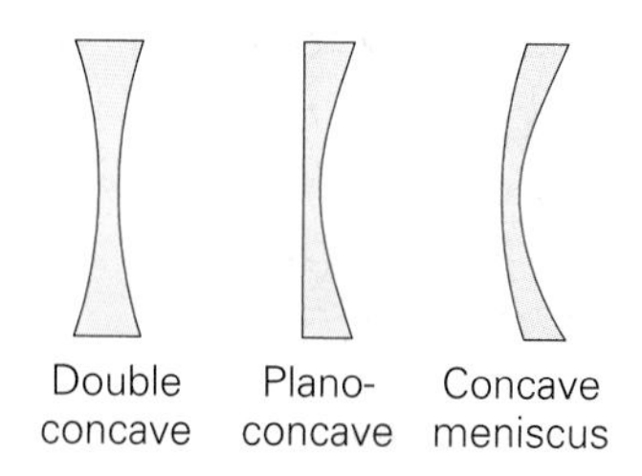

Diverging lenses

▲ FIGURE 23.14 Lens shapes Lens shapes vary widely and are normally categorized as converging or diverging. In general, a converging lens is thicker at its center than at the periphery, and a diverging lens is thinner at its center than at the periphery.

Exploration 35.2 Ray Diagrams

2. A **chief ray**, or **central ray**, is a ray that passes through the center of the lens and is undeviated because the lens is "thin."
3. A **focal ray** is a ray that (a) passes through the focal point on the object side of a converging lens *or* (b) appears to pass through the focal point on the image side of a diverging lens and, after refraction, is parallel to the lens's optic axis.

As with spherical mirrors, only two rays are needed to determine the image; we will normally use the parallel and chief rays. (As in the case of mirrors, however, it is generally a good idea to include a third ray, the focal ray, in your diagrams as a check.)

Example 23.5 ■ Learn by Drawing: A Lens Diagram

An object is placed 30 cm in front of a thin biconvex lens of focal length 20 cm. (a) Use a ray diagram to locate the image. (b) Discuss the characteristics of the image.

Thinking It Through. Follow the steps for lens ray diagrams, as given previously.

Solution.

Given: $d_o = 30$ cm, $f = 20$ cm

Find: (a) location of the image (using a ray diagram)
(b) the image's characteristics

(a) Since we have been asked to use a ray diagram to locate the image (see the accompanying Learn by Drawing), the first thing to decide is a scale for the drawing. In this example, a scale of 1 cm to represent 10 cm is used. That way, the object would be 3.00 cm in front of the mirror in our drawing.

First the optic axis, the lens, the object (a lighted candle), and the focal points (F) are drawn. A vertical dashed line through the center of the lens is drawn because, for simplicity, the refraction is depicted as if it occurs at the center of each lens. In reality, it would occur at the air–glass and glass–air surfaces of each lens.

Follow steps 1–4 in the accompanying Learn by Drawing:

1. The first ray drawn is the parallel ray (① in the drawing). From the tip of the flame, draw a horizontal ray (parallel to the optic axis). After passing through the lens, this ray goes through the focal point F on the image side.

2. Then draw the chief ray (② in the drawing). From the tip of the flame, draw a ray passing through the center of the lens. This ray will go undeviated through the thin lens to the image side.

3. It can be clearly seen that these two rays intersect on the image side. The point of intersection is the image point of the tip of the candle. From this point, draw the image by extending the tip of the flame to the optic axis.

4. Only two rays are needed to locate the image. However, if you draw the third ray, in this case the focal ray (③ in the drawing), it must go through the same point on the image at which the other two rays intersect (if you are drawing the diagram carefully). The ray from the tip of the flame passing through the focal point F on the object side will travel parallel to the optic axis on the image side.

(b) From the ray diagram in part (a), the image is real (because the rays intersect or converge on the image side). As a result, this real image could be seen on a screen (for example, a piece of white paper) that is positioned at a distance d_i from the converging lens. The image is also inverted (the image of the candle points downward) and is larger than the object.

In this case, $d_o = 30$ cm and $f = 20$ cm, so $2f > d_o > f$. Using similar ray diagrams, you can prove that for any d_o in this range, the image is always real, enlarged, and inverted. Actually, the overhead projector in your classroom uses this particular arrangement.

Follow-Up Exercise. In this Example, what does the image look like if the object is 10 cm in front of the lens? Locate the image graphically and discuss the characteristics of the image.

LEARN BY DRAWING

A LENS RAY DIAGRAM (SEE EXAMPLE 23.5)

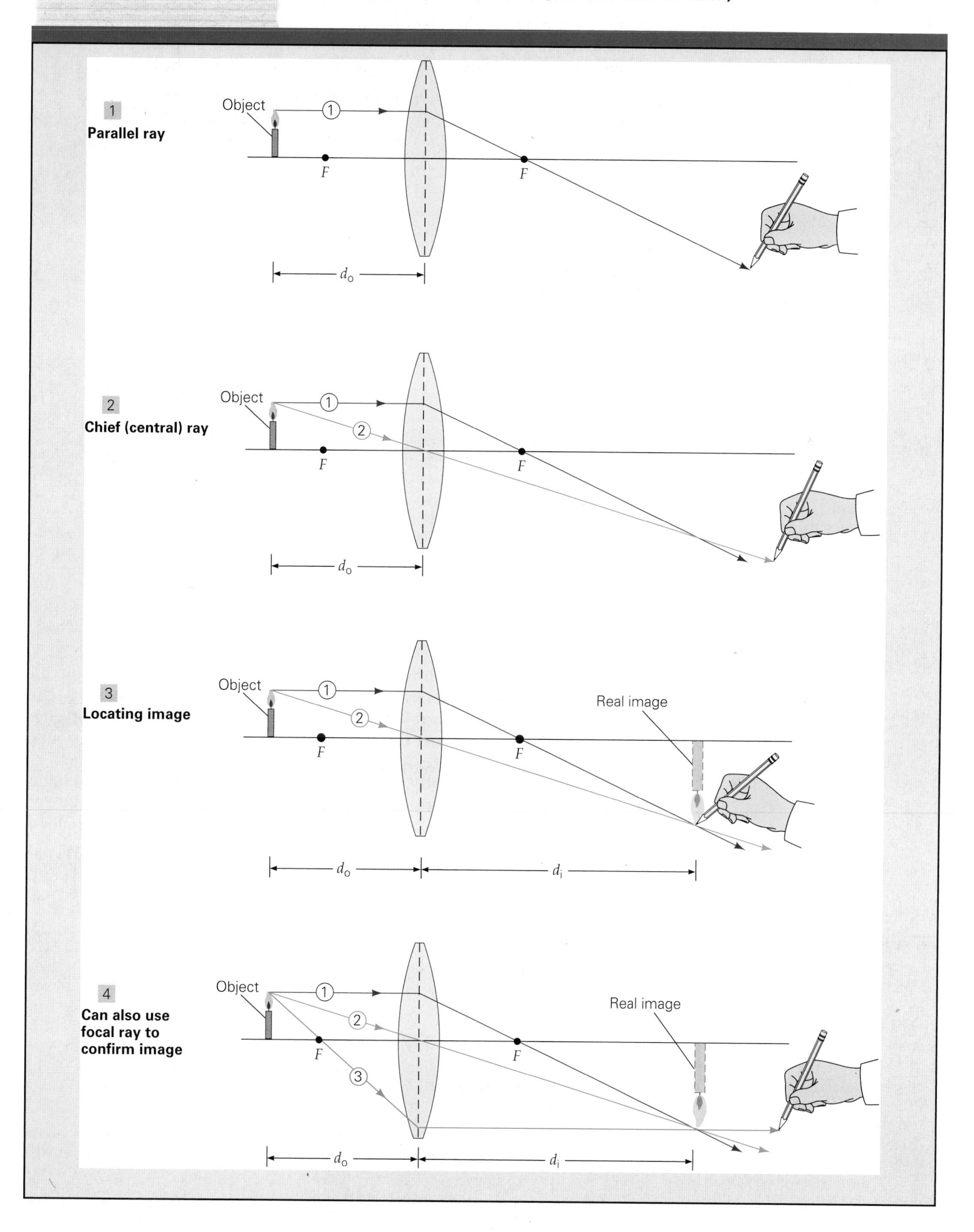

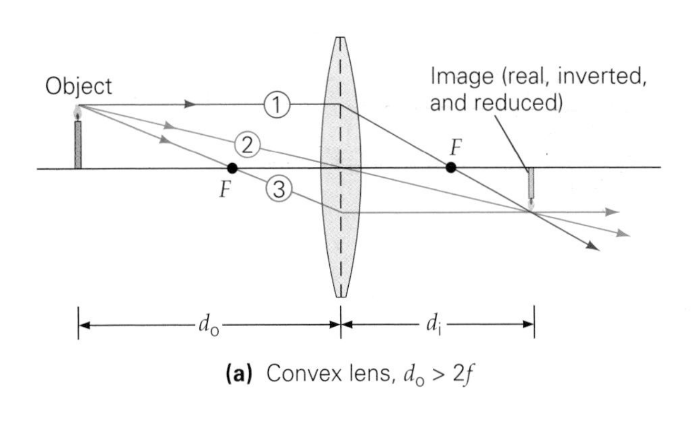

(a) Convex lens, $d_o > 2f$

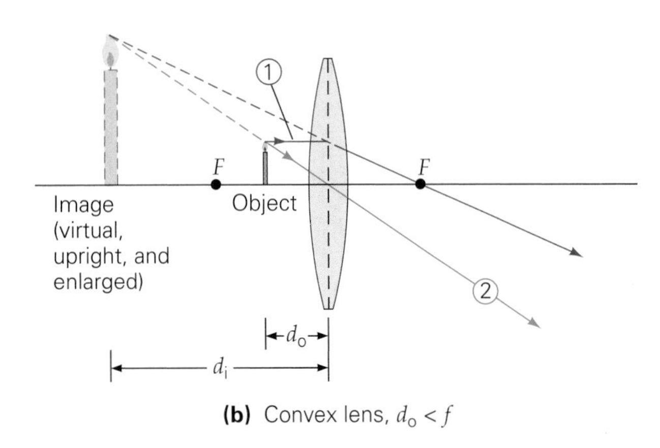

(b) Convex lens, $d_o < f$

▲ **FIGURE 23.15 Ray diagrams for lenses** **(a)** A converging biconvex lens forms a real object when $d_o > 2f$. The image is real, inverted, and reduced. **(b)** Ray diagram for a converging lens with $d_o < f$. The image is virtual, upright, and magnified. Practical examples are shown for both cases.

To illustrate these procedures, ▲Fig. 23.15 shows other ray diagrams with different object distances for a converging lens, along with real-life applications. The image of an object is real when it is formed on the side of the lens *opposite* the object's side (see Fig. 23.15a). A virtual image is said to be formed on the same side of the lens as the object (see Fig. 23.15b).

Regions could be similarly defined for the object distance for a converging lens as was done for a converging mirror in Fig. 23.7a. Here, an object distance of $d_o = 2f$ for a converging lens has significance similar to that of $d_o = R = 2f$ for a converging mirror (▼Fig. 23.16).

The ray diagram for a diverging lens will be discussed shortly. Like diverging mirrors, diverging lenses can form only virtual images of real objects.

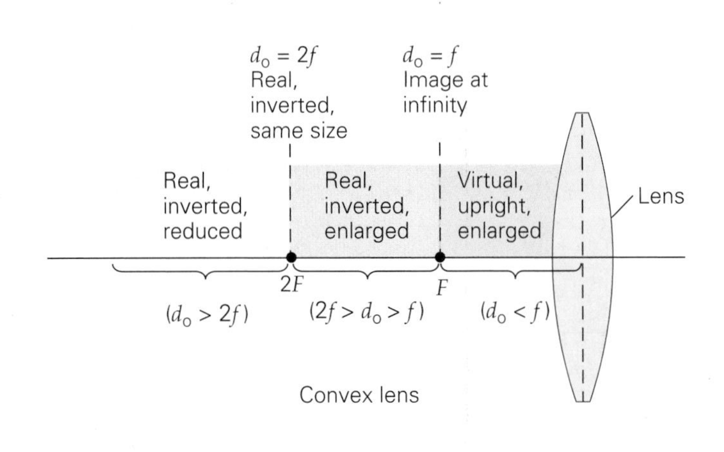

▶ **FIGURE 23.16 Convex lens** For a convex, or converging, lens, the object is located within one of three regions defined by the focal distance (f) and twice the focal distance ($2f$) or at one of these two points. For $d_o > 2f$, the image is real, inverted, and reduced (Fig. 23.15a). For $2f > d_o > f$, the image will also be real and inverted, but enlarged, or magnified, as shown by the ray diagrams in Example 23.5. For $d_o < f$, the image will be virtual, upright, and enlarged (Fig. 23.15b).

TABLE 23.3 Sign Conventions for Thin Lenses

Focal length (f)	
Converging lens (sometimes called a *positive* lens)	f is positive
Diverging lens (sometimes called a *negative* lens)	f is negative
Object distance (d_o)	
Object is in front of the lens (real object)	d_o is positive
Object is behind the lens (virtual object)*	d_o is negative
Image distance (d_i) and *Image type*	
Image is formed on the image side of the lens—opposite to the object (real image)	d_i is positive
Image is formed on the object side of the lens—same side as the object (virtual image)	d_i is negative
Image orientation (M)	
Image is upright with respect to the object	M is positive
Image is inverted with respect to the object	M is negative

*In a combination of two (or more) lenses, the image formed by the first lens is taken as the object of the second lens (and so on). If this image–object falls behind the second lens, it is referred to as a virtual object, and the object distance is taken to be negative (−).

PHYSLET®
Exploration 35.3 Moving a Lens

The image distances and characteristics for a spherical lens can also be found analytically. The equations for thin lenses are identical to those for spherical mirrors. The **thin-lens equation** is

$$\frac{1}{d_o} + \frac{1}{d_i} = \frac{1}{f} \quad \textit{thin-lens equation} \qquad (23.5)$$

As in the case for spherical mirrors, an alternative form of the thin-lens equation,

$$d_i = \frac{d_o f}{d_o - f} \qquad (23.5a)$$

gives a quick and easy way to find d_i.

The **magnification factor**, like that for spherical mirrors, is given by

$$M = -\frac{d_i}{d_o} \qquad (23.6)$$

The sign conventions for these thin-lens equations are given in Table 23.3.

Just as when you are working with mirrors, it is helpful to sketch a ray diagram before working a lens problem analytically.

Example 23.6 ■ Three Images: Behavior of a Converging Lens

A biconvex lens has a focal length of 12 cm. For an object (a) 60 cm, (b) 15 cm, and (c) 8.0 cm from the lens, where is the image formed, and what are its characteristics?

Thinking It Through. With the focal length (f) and the object distances (d_o), we can apply Eq. 23.5 to find the image distances (d_i) and Eq. 23.6 to determine the image characteristics. Sketch ray diagrams first to get an idea of the image characteristics. The diagrams should be in good agreement with the calculations.

Solution.

Given: $f = 12$ cm
(a) $d_o = 60$ cm
(b) $d_o = 15$ cm
(c) $d_o = 8.0$ cm

Find: d_i and the image characteristics for all three cases

(continues on next page)

(a) The object distance is greater than twice the focal length ($d_o > 2f$). Using Eq. 23.5,

$$\frac{1}{d_o} + \frac{1}{d_i} = \frac{1}{f}$$

or

$$\frac{1}{d_i} = \frac{1}{f} - \frac{1}{d_o} = \frac{1}{12\text{ cm}} - \frac{1}{60\text{ cm}} = \frac{5}{60\text{ cm}} - \frac{1}{60\text{ cm}} = \frac{4}{60\text{ cm}} = \frac{1}{15\text{ cm}}$$

Then

$$d_i = 15\text{ cm} \quad \text{and} \quad M = -\frac{d_i}{d_o} = -\frac{15\text{ cm}}{60\text{ cm}} = -0.25$$

The image is real (positive d_i), inverted (negative M), and one-fourth the object's size $(|M| = 0.25)$. A camera uses this arrangement when the object distance is greater than $2f(d_o > 2f)$.

(b) Here, $2f > d_o > f$. Using Eq. 23.5,

$$\frac{1}{d_i} = \frac{1}{12\text{ cm}} - \frac{1}{15\text{ cm}} = \frac{5}{60\text{ cm}} - \frac{4}{60\text{ cm}} = \frac{1}{60\text{ cm}}$$

Then

$$d_i = 60\text{ cm} \quad \text{and} \quad M = -\frac{d_i}{d_o} = -\frac{60\text{ cm}}{15\text{ cm}} = -4.0$$

The image is real (positive d_i), inverted (negative M), and four times the object's size $(|M| = 4.0)$. This situation applies to the overhead projector and slide projector $(2f > d_o > f)$.

(c) For this case, $d_o < f$. Using the alternative form (Eq. 23.5a),

$$d_i = \frac{d_o f}{d_o - f} = \frac{(8.0\text{ cm})(12\text{ cm})}{8.0\text{ cm} - 12\text{ cm}} = -24\text{ cm}$$

Then

$$M = -\frac{d_i}{d_o} = -\frac{(-24\text{ cm})}{8.0\text{ cm}} = +3.0$$

The image is virtual (negative d_i), upright (positive M), and three times the object's size $(|M| = 3.0)$. This situation is an example of a simple microscope or magnifying glass $(d_o < f)$.

As you can see, a converging lens is versatile. Depending on the object distance (relative to the focal length), the lens can be used as a camera, projector, or magnifying glass.

Follow-Up Exercise. If the object distance of a convex lens is allowed to vary, at what object distance does the real image change from being reduced to being magnified?

Conceptual Example 23.7 ■ Half an Image?

Exploration 35.1 Image Formation

A converging lens forms an image on a screen, as shown in ▶Fig. 23.17a. Then the lower half of the lens is blocked, as shown in Fig. 23.17b. As a result, (a) only the top half of the original image will be visible on the screen; (b) only the bottom half of the original image will be visible on the screen; or (c) the entire image will be visible.

Reasoning and Answer. At first thought, you might imagine that blocking off half of the lens would eliminate half of the image. However, rays from *every* point on the object pass through *all parts* of the lens. Thus, the upper half of the lens can form a total image (as could the lower half), so the answer is (c).

You might confirm this conclusion by drawing a chief ray in Fig. 23.17b. Or you might use the scientific method and experiment—particularly if you wear eyeglasses. Block off the bottom part of your glasses, and you will find that you can still read through the top part (unless you wear bifocals).

Follow-Up Exercise. Can you think of any property of the image that *would* be affected by blocking off half of the lens? Explain.

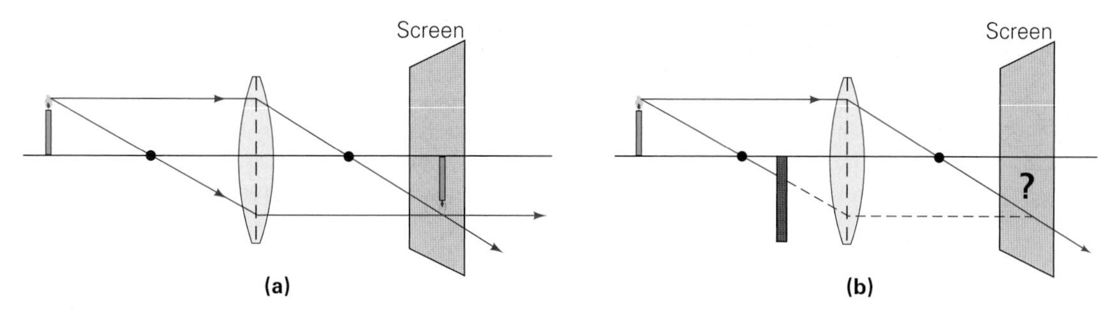

▲ **FIGURE 23.17 Half a lens, half an image? (a)** A converging lens forms an image on a screen. **(b)** The lower half of the lens is blocked. What happens to the image? See Conceptual Example 23.7.

Integrated Example 23.8 ■ Time for a Change: Behavior of a Diverging Lens

An object is 24 cm in front of a diverging lens that has a focal length of −15 cm. (a) Use a ray diagram to determine whether the image is (1) real and magnified, (2) virtual and reduced, (3) real and upright, or (4) upright and magnified. (b) Find the location and characteristics of the image with the thin-lens equations.

(a) Conceptual Reasoning. (See the sign conventions in Table 23.3.) Use a scale of 1 cm (in our drawing of ▸Fig. 23.18) to represent 10 cm. The object will be 2.4 cm in front of the lens in our drawing. Draw the optic axis, the lens, the object (in this case, a lighted candle), the focal point (F), and a vertical dashed line through the center of the lens.

The parallel ray ① starts from the tip of the flame, travels parallel to the optic axis, diverges from the lens after refraction, and appears to diverge from the F on the object side. The chief ray ② originates from the tip of the flame and goes through the center of the lens, with no direction change. We see that these two rays, after refraction, diverge and do not intersect. However, they appear to come from in front of the lens (object side), and that apparent intersection is the image point of the tip of the flame. We can also draw the focal ray ③ to verify that these rays appear to come from the same image point. The focal ray appears to go through the focal point on the image side and travels parallel to the optic axis after refraction from the lens.

This image is virtual (why?), upright, and smaller than the object, so the answer is (2): virtual and reduced. Measuring from the diagram (keeping in mind the drawing scale we are using), we find that $d_i \approx -9$ cm (virtual image) and $M = \frac{h_i}{h_o} \approx \frac{0.5\text{ cm}}{1.4\text{ cm}} = +0.4$.

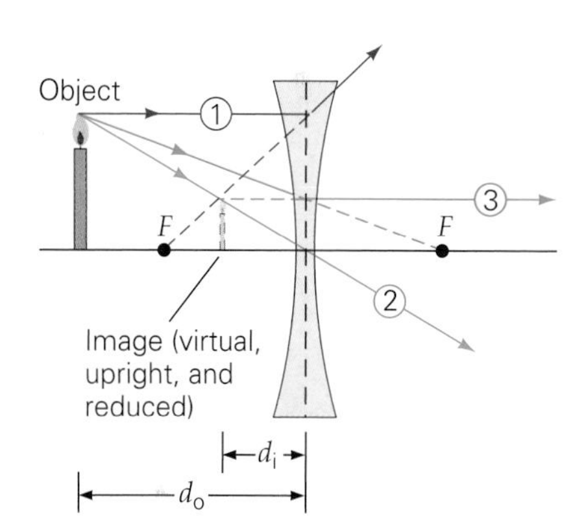

▲ **FIGURE 23.18 Diverging lens** Ray diagram of a diverging lens. Here, the image is virtual and in front of the lens, upright, and smaller than the object. See Integrated Example 23.8.

Illustration 35.2 Image from a Diverging Lens

(b) Quantitative Reasoning and Solution.

Given: $d_o = 24$ cm
$f = -15$ cm (diverging lens)

Find: d_i, M, and image characteristics

Note that the focal length is negative for a diverging lens. (See Table 23.3.) From Eq. 23.5,

$$\frac{1}{24\text{ cm}} + \frac{1}{d_i} = \frac{1}{-15\text{ cm}} \quad \text{or} \quad \frac{1}{d_i} = \frac{1}{-15\text{ cm}} - \frac{1}{24\text{ cm}} = -\frac{13}{120\text{ cm}}$$

so

$$d_i = -\frac{120\text{ cm}}{13} = -9.2\text{ cm}$$

Then

$$M = -\frac{d_i}{d_o} = -\frac{(-9.2\text{ cm})}{24\text{ cm}} = +0.38$$

Thus, the image is virtual (d_i is negative) and upright (M is positive), and it is 0.38 times the height of the object. Due to the fact that f is negative for a diverging lens, d_i is always negative for any positive value of d_o, so the image of an object is always virtual.

Follow-Up Exercise. A diverging lens always forms a virtual image of a real object. What general statements can be made about the image's orientation and magnification?

A special type of lens that you may have encountered is discussed in Insight 23.2 on Fresnel Lenses, on page 748.

INSIGHT 23.2 FRESNEL LENSES

To focus parallel light or to produce a large beam of parallel light rays, a sizable converging lens is sometimes necessary. The large mass of glass necessary to form such a lens is bulky and heavy. Moreover, a thick lens absorbs some of the light and is likely to show distortions. A French physicist named Augustin Fresnel (Fray-nel', 1788–1827) developed a solution to this problem for the lenses used in lighthouses. Fresnel recognized that the refraction of light takes place at the surfaces of a lens. Hence, a lens could be made thinner—and almost flat—by removing glass from the interior, as long as the refracting properties of the surfaces were not changed.

This can be accomplished by cutting a series of concentric grooves in the surface of the lens (Fig. 1a). Note that the surface of each remaining curved segment is nearly parallel to the corresponding surface of the original lens. Together, the concentric segments refract light as does the original biconvex converging lens (Fig. 1b). In effect, the lens has simply been slimmed down by the removal of unnecessary glass between the refracting surfaces.

A lens with a series of concentric curved surfaces is called a *Fresnel lens*. Such lenses are widely used in overhead projectors and in beacons (Fig. 1c). A Fresnel lens is very thin and therefore much lighter in weight than a conventional biconvex lens with the same optical properties. Also, Fresnel lenses are easily molded from plastic—often with one flat side (plano-convex) so that the lens can be attached to a glass surface.

One disadvantage of Fresnel lenses is that concentric circles are visible when an observer is looking through such a lens and when an image produced by the lens is projected on a screen, as when we use an overhead projector.

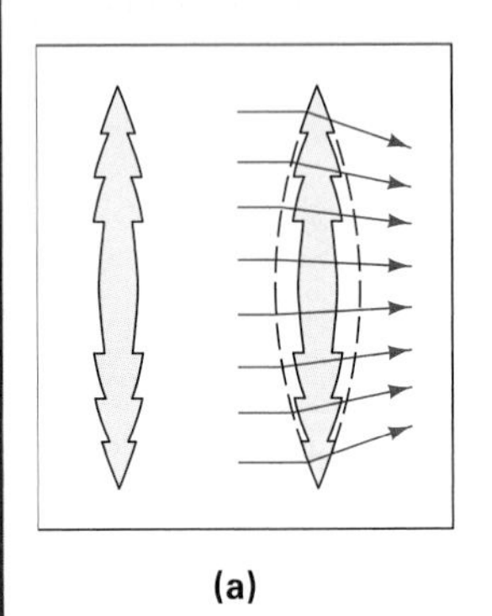
(a)

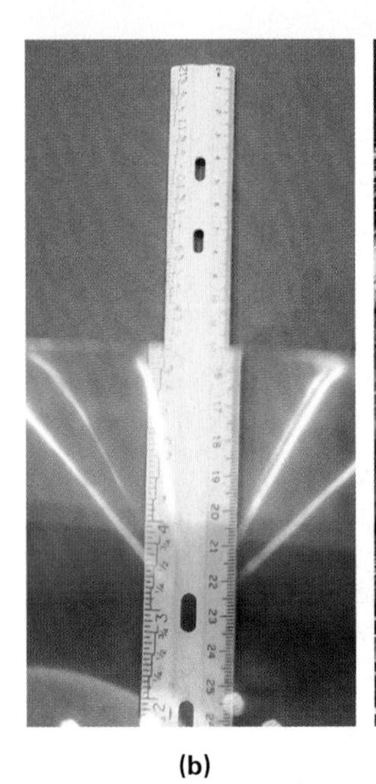
(b)

(c)

FIGURE 1 Fresnel lens **(a)** The focusing action of a lens comes from refraction at its surfaces. It is therefore possible to reduce the thickness of a lens by cutting away glass in concentric grooves, leaving a set of curved surfaces with the same refractive properties as the lens from which they were derived. **(b)** A flat Fresnel lens with concentric curved surfaces magnifies like a biconvex converging lens. **(c)** An array of Fresnel lenses produces focused beams in this Boston Harbor light. (Fresnel lenses were, in fact, developed for use in lighthouses.)

Combinations of Lenses

Exploration 35.4 What is Behind the Curtain?

Many optical instruments, such as microscopes and telescopes (Chapter 25), use a combination of lenses, or a compound-lens system. When two or more lenses are used in combination, we can determine the overall image produced by considering the lenses individually in sequence. That is, the image formed by the first lens becomes the object for the second lens, and so on. For this reason, we present the principles of lens combinations before considering the specifics of their real-life applications.

If the first lens produces an image in front of the second lens, that image is treated as a real object (d_o is positive) for the second lens (▸Fig. 23.19a). If, however, the lenses are close enough so that the image from the first lens is *not* formed before the rays pass through the second lens (Fig. 23.19b), then a modification must be made in the sign conventions. In this case, the image from the first lens is treated as a *virtual* object for the second lens. The virtual object distance is taken to be *negative* in the lens equation (Table 23.3).

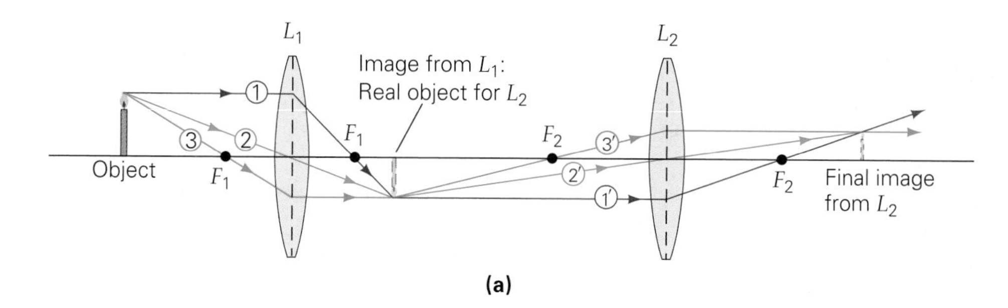

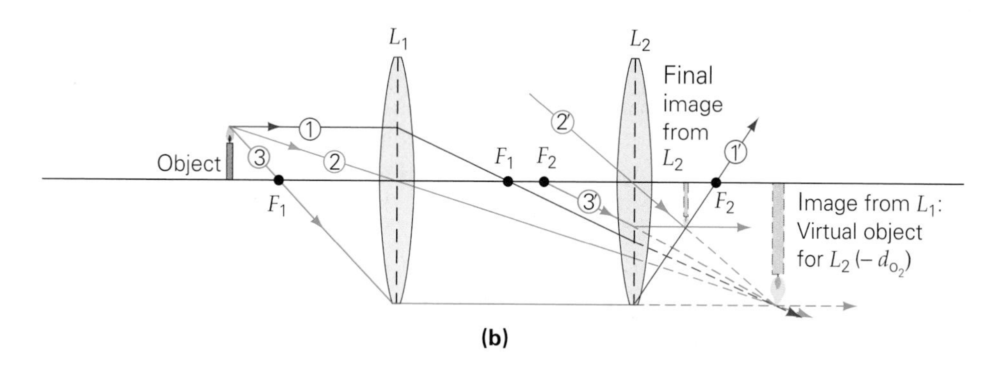

◀ **FIGURE 23.19 Lens combinations** The final image produced by a compound-lens system can be found by treating the image of one lens as the object for the adjacent lens. **(a)** If the image of the first lens (L_1) is formed in front of the second lens (L_2), the object for the second lens is said to be real. (Note that rays 1′, 2′, and 3′ are the parallel, chief, and focal rays, respectively, for L_2. They are *not* continuations of rays 1, 2, and 3—the parallel, chief, and focal rays, respectively, for L_1.) **(b)** If the rays pass through the second lens before the image is formed, the object for the second lens is said to be virtual, and the object distance for the second lens is taken to be negative.

It can be shown that the total magnification (M_{total}) of a compound-lens system is the product of the individual magnification factors of the component lenses. For example, for a two-lens system, as in Fig. 23.19,

$$M_{\text{total}} = M_1 M_2 \tag{23.7}$$

The conventional signs for M_1 and M_2 carry through the product to indicate, from the sign of M_{total}, whether the final image is upright or inverted. (See Exercise 83.)

Example 23.9 ■ A Special Offer: A Lens Combo and a Virtual Object

Consider two lenses similar to those illustrated in Fig. 23.19b. Suppose the object is 20 cm in front of lens L_1, which has a focal length of 15 cm. Lens L_2, with a focal length of 12 cm, is 26 cm from L_1. What is the location of the final image, and what are its characteristics?

Thinking It Through. This is a double application of the thin-lens equation. The lenses are treated successively. The image of lens L_1 becomes the object of lens L_2. We must keep the quantities distinctly labeled and the distances appropriately referenced (with signs!).

Solution. We have

Given: $d_{o_1} = +20$ cm
$f_1 = +15$ cm
$f_2 = +12$ cm
$D = 26$ cm (distance between lenses)

Find: d_{i_2} and image characteristics

The first step is to apply the thin-lens equation (Eq. 23.5) and the magnification factor for thin lenses (Eq. 23.6) to L_1:

$$\frac{1}{d_{i_1}} = \frac{1}{f_1} - \frac{1}{d_{o_1}} = \frac{1}{15\text{ cm}} - \frac{1}{20\text{ cm}} = \frac{4}{60\text{ cm}} - \frac{3}{60\text{ cm}} = \frac{1}{60\text{ cm}}$$

or

$$d_{i_1} = 60\text{ cm (real image from } L_1)$$

and

$$M_1 = -\frac{d_{i_1}}{d_{o_1}} = -\frac{60\text{ cm}}{20\text{ cm}} = -3.0 \text{ (inverted and magnified)}$$

The image from lens L_1 becomes the object for lens L_2. This image is then $d_{i_1} - D = 60\text{ cm} - 26\text{ cm} = 34$ cm on the right, or image, side of L_2. Therefore, it is a *virtual* object (see Table 23.3), and $d_{o_2} = -34$ cm. (Remember that d_o for virtual objects is taken to be negative.)

(continues on next page)

Then applying the equations to the second lens, L_2:

$$\frac{1}{d_{i_1}} = \frac{1}{f_2} - \frac{1}{d_{o_1}} = \frac{1}{12\text{ cm}} - \frac{1}{(-34\text{ cm})} = \frac{23}{204\text{ cm}}$$

or

$$d_{i_2} = 8.9\text{ cm (real image)}$$

and

$$M_2 = -\frac{d_{i_1}}{d_{o_2}} = -\frac{8.9\text{ cm}}{(-34\text{ cm})} = 0.26\text{ (upright and reduced)}$$

(*Note*: The virtual object for L_2 was inverted, and thus the term *upright* means that the *final* image is also inverted.) The total magnification M_{total} is then

$$M_{total} = M_1 M_2 = (-3.0)(0.26) = -0.78$$

The sign is carried through with the magnifications. We determine that the final real image is located at 8.9 cm on the right (image) side of L_2 and that it is inverted (negative sign) relative to the initial object and reduced.

Follow-Up Exercise. Suppose the object in Fig. 23.19b were located 30 cm in front of L_1. Where would the final image be formed in this case, and what would be its characteristics?

23.4 The Lens Maker's Equation

OBJECTIVES: To (a) describe the lens maker's equation, (b) explain how it differs from the thin-lens equation, and (c) understand lens "power" in diopters.

PHYSLET® Exploration 35.5 Lens Maker's Equation

The biconvex and biconcave thin lenses considered so far in this chapter have been relatively easy to analyze. However, there are a variety of other shapes of lenses, as illustrated in Fig. 23.14. For them, the analysis becomes more involved, but it is important to know the focal lengths of such lenses for optical considerations, because lenses are ground for specific purposes or applications.

Lens refraction depends on the shapes of the lens's surfaces and on the index of refraction of the lens. These properties together determine the focal length of a thin lens. The thin-lens focal length is given by the **lens maker's equation**, which enables us to calculate the focal length of a thin lens *in air* ($n_{air} = 1$) as

$$\frac{1}{f} = (n-1)\left(\frac{1}{R_1} + \frac{1}{R_2}\right) \quad \textit{(for thin lens in air)} \qquad (23.8)$$

where n is the index of refraction of the lens material and R_1 and R_2 are the radii of curvature of the first (front side) and second (back side) lens surfaces, respectively. (The first surface is the one on which light from an object is first incident.)

A sign convention is required for the lens maker's equation, and a common one is summarized in Table 23.4. The signs depend only on the shape of the surface, that is, convex or concave (▸Fig. 23.20). For the biconvex lens in Fig. 23.20a, both R_1 and R_2 are positive (both surfaces are convex) and for the biconcave lens in Fig. 23.20b, both R_1 and R_2 are negative (both surfaces are concave).

PHYSLET® Exploration 34.1 Lens and a Changing Index of Refraction

If the lens is surrounded by a medium other than air, then the first term in parentheses in Eq. 23.8 becomes $(n/n_m) - 1$, where n and n_m are the indices of refraction of the lens material and the surrounding medium, respectively. Now we can see why some converging lenses in air become diverging when submerged in water: If $n_m > n$, then f is negative, and the lens is diverging.

TABLE 23.4 Sign Conventions for Lens Maker's Equation

Convex surface	R is positive
Concave surface	R is negative
Plane (flat) surface	$R = \infty$
Converging (positive) lens	f is positive
Diverging (negative) lens	f is negative

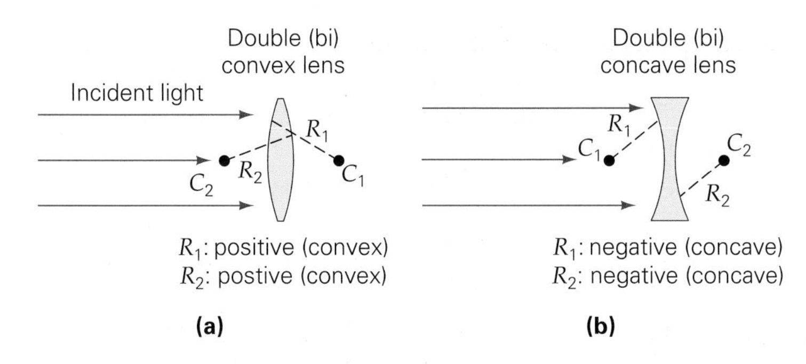

◀ **FIGURE 23.20 Centers of curvature** Lenses, such as **(a)** a biconvex lens and **(b)** a biconcave lens, have two centers of curvature, which define the signs of the radii of curvature. See Table 23.4.

Lens Power: Diopters

Notice that the lens maker's equation (Eq. 23.8) gives the inverse focal length $1/f$. Optometrists use this inverse relationship to express the *lens power* (P) of a lens in units called **diopters** (abbreviated as D). The lens power is the reciprocal of the focal length of the lens expressed in *meters*:

$$P(\text{expressed in diopters}) = \frac{1}{f(\text{expressed in meters})} \tag{23.9}$$

So, $1\text{ D} = 1\text{ m}^{-1}$. The lens maker's equation gives a lens's power $(1/f)$ in diopters if the radii of curvature are expressed in meters.

If you wear glasses, you may have noticed that the prescription the optometrist gave you for your eyeglass lenses was written in terms of diopters. Converging and diverging lenses are referred to as positive (+) and negative (−) lenses, respectively. Thus, if an optometrist prescribes a corrective lens with a power of +2 diopters, it is a converging lens with a focal length of

$$f = \frac{1}{P} = \frac{1}{+2\text{ D}} = \frac{1}{2\text{ m}^{-1}} = 0.50\text{ m} = +50\text{ cm}$$

The greater the power of the lens in diopters, the shorter its focal length is and the more strongly converging or diverging it is. Thus, a "stronger" prescription lens (greater lens power) has a smaller f than does a "weaker" prescription lens (lesser lens power).

Integrated Example 23.10 ■ A Convex Meniscus Lens: Converging or Diverging

The convex meniscus lens shown in Fig. 23.14 has a 15-cm radius for the convex surface and 20 cm for the concave surface. The lens is made of crown glass and is surrounded by air. (a) Is this lens a (1) converging or (2) diverging lens? Explain. (b) What are the focal length and the power of the lens?

(a) Conceptual Reasoning. The index of refraction of crown glass can be obtained from Table 22.1: $n = 1.52$. For a convex meniscus, the first surface is convex, so R_1 is positive; the second surface is concave, so R_2 is negative. Since $R_1 = 15\text{ cm} < |R_2| = 20\text{ cm}$, $1/R_1 + 1/R_2$ will be positive. Therefore, the lens is a converging (positive) lens, according to Eq. 23.8. Thus the answer is (1) converging.

(b) Quantitative Reasoning and Solution.

Given: $R_1 = 15\text{ cm} = 0.15\text{ m}$
$R_2 = -20\text{ cm} = -0.20\text{ m}$
$n = 1.52$ (from Table 22.1 for crown glass)

Find: f and P

From Eq. 23.8, we have

$$\frac{1}{f} = (n-1)\left(\frac{1}{R_1} + \frac{1}{R_2}\right) = (1.52 - 1)\left(\frac{1}{0.15\text{ m}} + \frac{1}{-0.20\text{ m}}\right) = 0.867\text{ m}^{-1}$$

So $f = \dfrac{1}{0.867\text{ m}^{-1}} = +1.15\text{ m}.$

The power of the lens is $P = \dfrac{1}{f} = +0.867\text{ D}.$

Follow-Up Exercise. In this Example, if this lens is immersed in water, what would your answers be?

*23.5 Lens Aberrations

OBJECTIVES: To (a) describe some common lens aberrations, and (b) explain how they can be reduced or corrected.

Lenses, like mirrors, can also have aberrations. We now discuss some common aberrations.

Spherical Aberration

The discussion of lenses thus far has concentrated on rays that are near the optic axis. Like spherical mirrors, however, converging lenses may show **spherical aberration**, an effect that occurs when parallel rays passing through different regions of a lens do not come together on a common focal plane. In general, rays close to the axis of a converging lens are refracted less and come together at a point farther from the lens than do rays passing through the periphery of the lens (▼Fig. 23.21a).

Spherical aberration can be minimized by using an aperture to reduce the effective area of the lens, so that only light rays near the axis are transmitted. Also, combinations of converging and diverging lenses can be used, as the aberration of one lens can be compensated for by the optical properties of another lens.

Chromatic Aberration

PHYSLET®

Exploration 34.5 Index of Refraction and Wavelength

Chromatic aberration is an effect that occurs because the index of refraction of the lens material is *not* the same for all wavelengths of light (that is, the material is dispersive). When white light is incident on a lens, the transmitted rays of different wavelengths (colors) do not have a common focal point, and images of different colors are produced at different locations (Fig. 23.21b).

This dispersive aberration can be minimized, but not eliminated, by using a compound-lens system consisting of lenses of different materials, such as crown glass and flint glass. The lenses are chosen so that the dispersion produced by one is approximately compensated by the opposite dispersion produced by the other. With a properly constructed two-component lens system, called an *achromatic doublet* (*achromatic* means "without color"), the images of any two selected colors can be made to coincide.

▼ **FIGURE 23.21 Lens aberrations (a)** Spherical aberration. In general, rays closer to the axis of a lens are refracted less and come together at a point farther from the lens than do rays passing through the periphery of the lens. **(b)** Chromatic aberration. Because of dispersion, different wavelengths (colors) of light are focused in different planes, which results in distortion of the overall image.

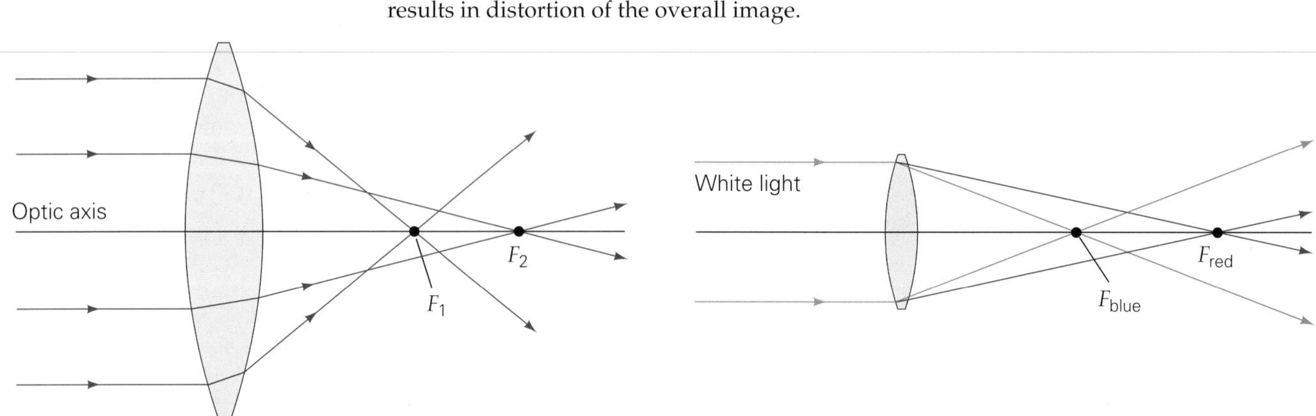

Astigmatism

A circular beam of light along the lens axis forms a circular illuminated area on the lens. When incident on a converging lens, the parallel beam converges at the focal point. However, when a circular beam of light from an off-axis source falls on the convex spherical surface of a lens some distance away, the light forms an *elliptical* illuminated area on the lens. The rays entering along the major and minor axes of the ellipse then focus at different points after passing through the lens. This condition is called **astigmatism**.

With different focal points in different planes, the images in both planes are blurred. For example, the image of a point is no longer a point, but rather two separated short-line images (blurred points). Astigmatism can be reduced by decreasing the effective area of the lens with an aperture or by adding a cylindrical lens to compensate.

Chapter Review

- **Plane mirrors** form virtual, upright, and unmagnified images. The object distance is equal to the image distance ($d_o = d_i$).
- The **lateral magnification factor** for all mirrors and lenses is

$$M = -\frac{d_i}{d_o} \quad (23.4, 23.6)$$

- **Spherical mirrors** are either concave (converging) or convex (diverging). Diverging spherical mirrors always form upright, reduced, virtual images.

Focal length of a spherical mirror:

$$f = \frac{R}{2} \quad (23.2)$$

Spherical-mirror equation:

$$\frac{1}{d_o} + \frac{1}{d_i} = \frac{1}{f} = \frac{2}{R} \quad (23.3)$$

Alternative form:

$$d_i = \frac{d_o f}{d_o - f} \quad (23.3a)$$

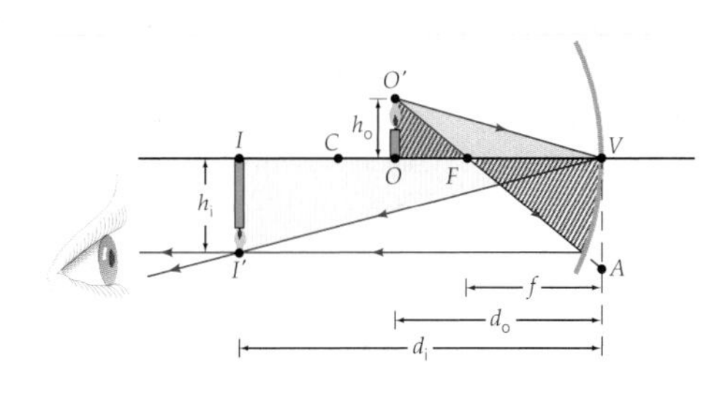

- Bispherical lenses are either convex (converging) or concave (diverging). Diverging spherical lenses always form upright, reduced, virtual images.

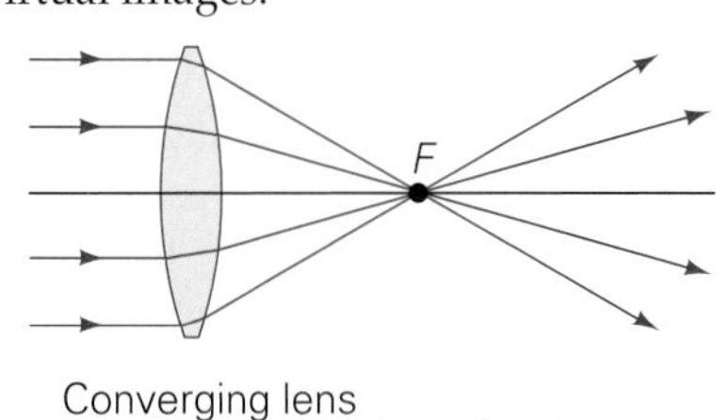

Converging lens

- The **thin-lens equation** relates focal length, object distance, and image distance:

$$\frac{1}{d_o} + \frac{1}{d_i} = \frac{1}{f} \quad (23.5)$$

Alternative form:

$$d_i = \frac{d_o f}{d_o - f} \quad (23.5a)$$

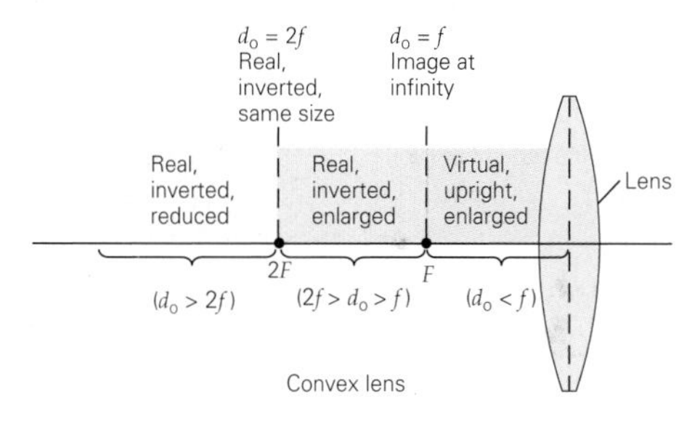

- The **lens maker's equation** is used to compute the grinding radii for the desired focal length of a lens:

$$\frac{1}{f} = (n - 1)\left(\frac{1}{R_1} + \frac{1}{R_2}\right) \quad \text{(thin lens in air only)} \quad (23.8)$$

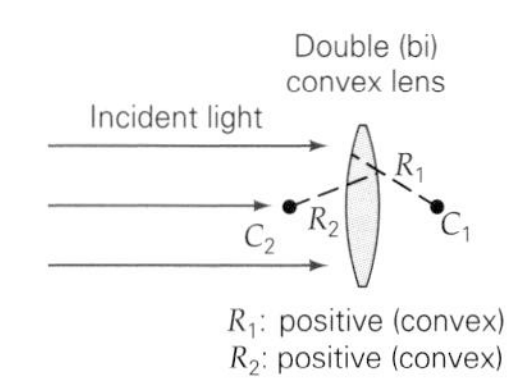

- **Lens power in diopters** (where f is in meters) is given by

$$P = \frac{1}{f} \quad (23.9)$$

Exercises

MC = *Multiple Choice Question,* **CQ** = *Conceptual Question, and* **IE** = *Integrated Exercise. Throughout the text, many exercise sections will include "paired" exercises. These exercise pairs, identified with* **red numbers**, *are intended to assist you in problem solving and learning. In a pair, the first exercise (even numbered) is worked out in the Study Guide so that you can consult it should you need assistance in solving it. The second exercise (odd numbered) is similar in nature, and its answer is given at the back of the book.*

23.1 Plane Mirrors

1. **MC** A plane mirror (a) has a greater image distance than object distance; (b) produces a virtual, upright, unmagnified image; (c) changes the vertical orientation of an object; (d) reverses an object's top and bottom.

2. **MC** A plane mirror (a) produces both real and virtual images, (b) always produces a virtual image, (c) always produces a real image, (d) forms images by diffuse reflection.

3. **MC** The lateral magnification of a plane mirror is (a) greater than 1, (b) less than 1, (c) equal to +1, (d) equal to −1.

4. **CQ** What is the focal length of a plane mirror? Why?

5. **CQ** Day–night rearview mirrors are common in cars. At night, you tilt the mirror backward, and the intensity and glare of headlights behind you are reduced (▼Fig. 23.22). The mirror is wedge shaped and is silvered on the back. The effect has to do with front-surface and back-surface reflections. The unsilvered front surface reflects about 5% of incident light; the silvered back surface reflects about 90% of the incident light. Explain how the day–night mirror works.

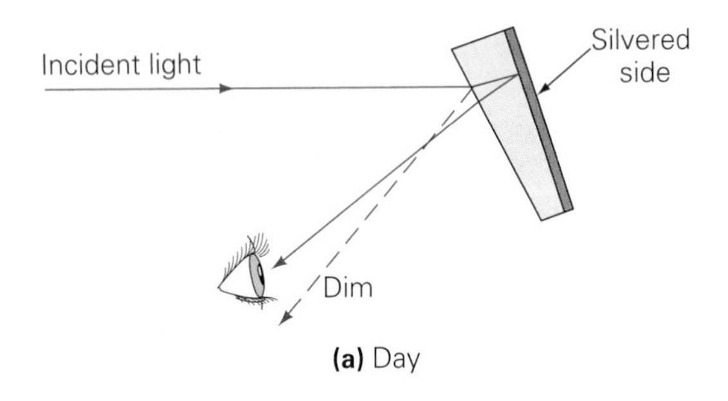

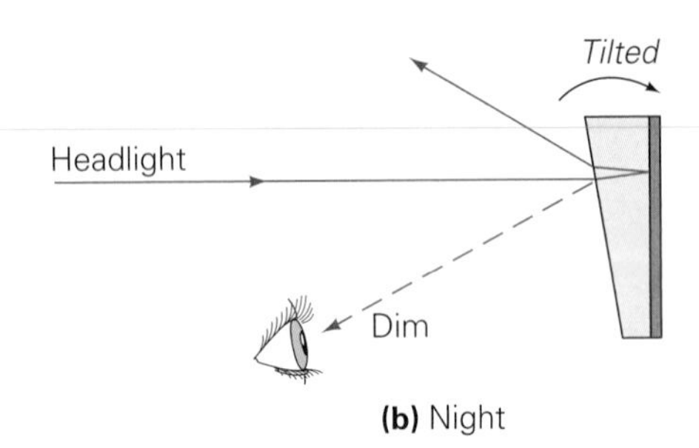

▲ **FIGURE 23.22 Automobile day–night mirror** See Exercise 5.

6. **CQ** When you stand in front of a plane mirror, there is a right–left reversal. (a) Why is there not a top–bottom reversal of your body? (b) Could you affect an apparent top–bottom reversal by positioning your body differently?

7. **CQ** Why do some emergency vehicles have ƎƆИA⅃UᙠMA (▼Fig. 23.23) printed on the front?

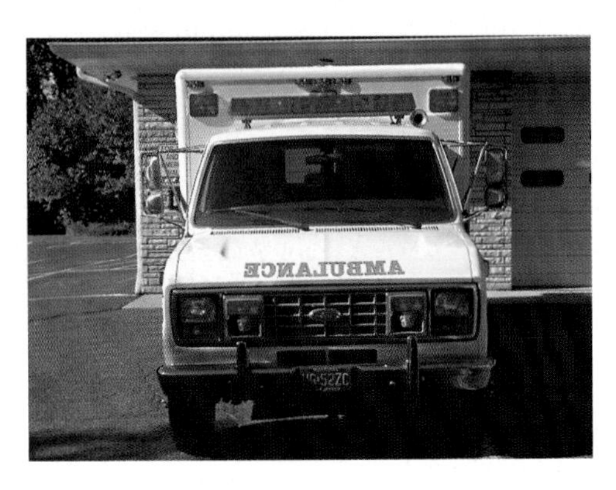

▲ **FIGURE 23.23 Backward and reversed** See Exercise 7.

8. **CQ** Can a virtual image be projected onto a screen? Why or why not?

9. ● A person stands 2.0 m away from the reflecting surface of a plane mirror. (a) What is the distance between the person and his or her image? (b) What are the image characteristics?

10. ● An object 5.0 cm tall is placed 40 cm from a plane mirror. Find (a) the distance from the object to the image, (b) the height of the image, and (c) the image's magnification.

11. ● Standing 2.5 m in front of a plane mirror with your camera, you decide to take a picture of yourself. To what distance should the camera be focused to get a sharp image?

12. ●● If you hold a 900-cm^2 square plane mirror 45 cm from your eyes and can just see the full length of an 8.5-m flagpole behind you, how far are you from the pole? [*Hint*: A diagram is helpful.]

13. ●● A small dog sits 1.5 m in front of a plane mirror. (a) Where is the dog's image in relation to the mirror? (b) If the dog jumps at the mirror at a speed of 0.50 m/s, how fast does the dog approach its image?

14. **IE** ●● A woman fixing the hair on the back of her head holds a plane mirror 30 cm in front of her face so as to look into a plane mirror on the bathroom wall behind her. She is 90 cm from the wall mirror. (a) The image of the back of her head will be from (1) only the front mirror, (2) only the wall mirror, or (3) both mirrors. (b) Approximately how far does the image of the back of her head appear in front of her?

15. IE ●● (a) When you stand between two plane mirrors on opposite walls in a dance studio, you observe (1) one, (2) two, or (3) multiple images. Explain. (b) If you stand 3.0 m from the mirror on the north wall and 5.0 m from the mirror on the south wall, what are the image distances for the first two images in both mirrors?

16. ●● A woman 1.7 m tall stands 3.0 m in front of a plane mirror. (a) What is the minimum height the mirror must be to allow the woman to view her complete image from head to foot? Assume that her eyes are 10 cm below the top of her head. (b) What would be the required minimum height of the mirror if she were to stand 5.0 m away?

17. ●● Prove that $d_o = d_i$ (equal magnitude) for a plane mirror. [*Hint*: Refer to Fig. 23.2 and use similar and identical triangles.]

18. ●●● Draw ray diagrams that show how three images of an object are formed in two plane mirrors at right angles, as shown in ▼Fig. 23.24a. [*Hint*: Consider rays from both ends of the object in the drawing for each image.] Figure 23.24b shows a similar situation from a different point of view that gives four images. Explain the extra image in this case.

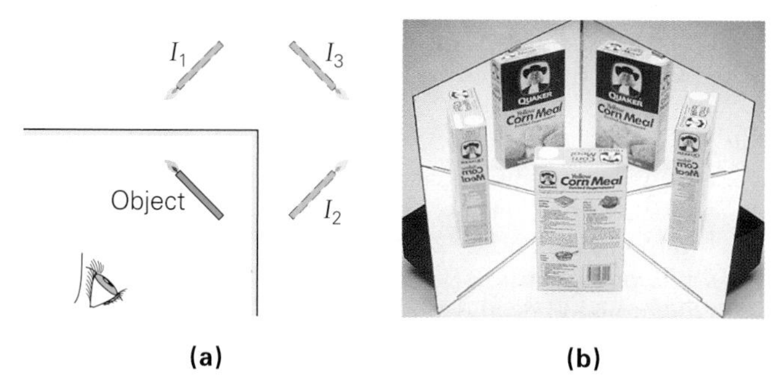

▲ **FIGURE 23.24 Two mirrors—multiple images** See Exercise 18.

23.2 Spherical Mirrors

19. **MC** Which of the following statements concerning spherical mirrors is correct? (a) A converging mirror alone can produce an inverted virtual image. (b) A diverging mirror alone can produce an inverted virtual image. (c) A diverging mirror can produce an inverted real image. (d) A converging mirror can produce an inverted real image.

20. **MC** The image produced by a convex mirror is always (a) virtual and upright, (b) real and upright, (c) virtual and inverted, (d) real and inverted.

21. **MC** A shaving/makeup mirror is used to form an image that is larger than the object, so it is a (a) concave, (b) convex, (c) plane mirror.

22. CQ (a) What is the purpose of using a dual mirror on a car or truck, such as the one shown in ▸Fig. 23.25? (b) Some rearview mirrors on the passenger side of automobiles have the warning "OBJECTS IN MIRROR ARE CLOSER THAN THEY APPEAR." Explain why. (c) Could a TV satellite dish be considered a converging mirror? Explain.

▲ **FIGURE 23.25 Mirror applications** See Exercise 22.

23. CQ (a) If you look into a shiny spoon, you see an inverted image on one side and an upright image on the other (▼Fig. 23.26). (Try it.) Why? (b) Could you see upright images on both sides? Explain.

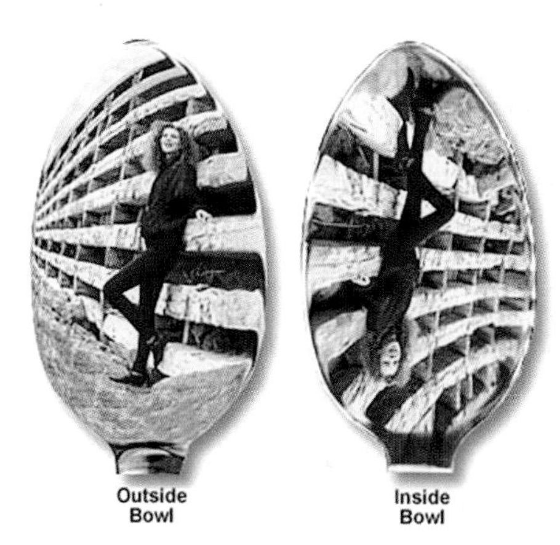

▲ **FIGURE 23.26 Reflections from concave and convex surfaces** See Exercise 23.

24. CQ (a) A 10-cm-tall mirror bears the following advertisement: "Full-view mini mirror. See your full body in 10 cm." How can this be? (b) A popular novelty item consists of a concave mirror with a ball suspended at or slightly inside the center of curvature (▼Fig. 23.27). When the ball swings toward the mirror, its image grows larger and suddenly fills the whole mirror. The image appears to be jumping out of the mirror. Explain what is happening.

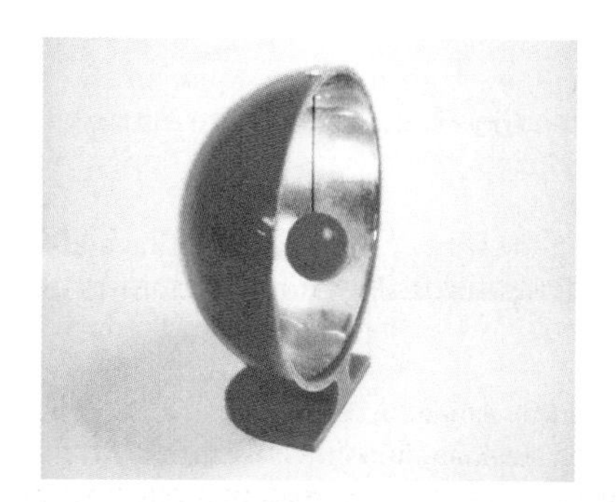

▲ **FIGURE 23.27 Spherical-mirror toy** See Exercise 24.

25. CQ How can the focal length be quickly determined experimentally for a concave mirror? Can you do the same thing for a convex mirror?

26. CQ Can a convex mirror produce an image that is taller than the object? Why or why not?

27. IE ● An object is 30 cm in front of a convex mirror that has a focal length of 60 cm. (a) Use a ray diagram to determine whether the image is (1) real or virtual, (2) upright or inverted, and (3) magnified or smaller than the object. (b) Calculate the image distance and image height.

28. ● An object 3.0 cm tall is placed 20 cm from the front of a concave mirror with a radius of curvature of 30 cm. Where is the image formed, and how tall is it?

29. ● If the object in Exercise 28 is moved to a position 10 cm from the front of the mirror, what will be the characteristics of the image?

30. ● A candle with a flame 1.5 cm tall is placed 5.0 cm from the front of a concave mirror. A virtual image is produced that is 10 cm from the vertex of the mirror. (a) Find the focal length and radius of curvature of the mirror. (b) How tall is the image of the flame?

31. ●● Use the mirror equation and the magnification factor to show that when $d_o = R = 2f$ for a concave mirror, the image is real, inverted, and the same size as the object.

32. ●● An object 3.0 cm tall is placed at different locations in front of a concave mirror whose radius of curvature is 30 cm. Determine the location of the image and its characteristics when the object distance is 40 cm, 30 cm, 15 cm, and 5.0 cm, using (a) a ray diagram and (b) the mirror equation.

33. IE ●● A virtual image of magnification +0.50 is produced when an object is placed in front of a spherical mirror. (a) The mirror is (1) convex, (2) concave, (3) flat. Explain. (b) Find the radius of curvature of the mirror if the object is 7.0 cm in front of it.

34. ●● A bottle 6.0 cm tall is located 75 cm from the concave surface of a mirror with a radius of curvature of 50 cm. Where is the image located, and what are its characteristics?

35. IE ●● A shaving mirror has a magnification of +4.00. (a) The mirror is (1) convex, (2) concave, (3) flat. Explain. (b) What is the focal length of the mirror if your face is 10.0 cm in front of the mirror?

36. ●● Using the spherical-mirror equation and the magnification factor, show that for a concave mirror with $d_o < f$, the image of an object is always virtual, upright, and magnified.

37. ●● Using the spherical-mirror equation and the magnification factor, show that for a convex mirror, the image of an object is always virtual, upright, and reduced.

38. ●● A concave makeup mirror produces a virtual image 1.5 times the size of a person whose face is 20 cm from the mirror. (a) Draw a ray diagram of this situation. (b) What is the focal length of the mirror?

39. IE ●● The image of an object located 30 cm from a mirror is formed on a screen located 20 cm from the mirror. (a) The mirror is (1) convex, (2) concave, (3) flat. Explain. (b) What is the mirror's radius of curvature?

40. IE ●● The erect image of an object 18 cm in front of a mirror is half the size of the object. (a) The mirror is (1) convex, (2) concave, (3) flat. Explain. (b) What is the focal length of the mirror?

41. IE ●● A concave mirror has a magnification of +3.0 for an object placed 50 cm in front of it. (a) The type of image produced is (1) virtual and upright, (2) real and upright, (3) virtual and inverted, (4) real and inverted. Explain. (b) Find the radius of curvature of the mirror.

42. ●● A concave shaving mirror is constructed so that a man at a distance of 20 cm from the mirror sees his image magnified 1.5 times. What is the radius of curvature of the mirror?

43. ●● A child looks at a reflective Christmas tree ball ornament that has a diameter of 9.0 cm and sees an image of her face that is half the real size. How far is the child's face from the ball?

44. IE ●● A dentist uses a spherical mirror that produces an upright image of a tooth that is magnified four times. (a) The mirror is (1) converging, (2) diverging, (3) flat. Explain. (b) What is the mirror's focal length in terms of the object distance?

45. ●● A 15-cm-long pencil is placed with its eraser on the optic axis of a concave mirror and its point directed upward at a distance of 20 cm in front of the mirror. The radius of curvature of the mirror is 30 cm. Use (a) a ray diagram and (b) the mirror equation to locate the image and determine the image characteristics.

46. IE ●● A pill bottle 3.0 cm tall is placed 12 cm in front of a mirror. A 9.0-cm-tall upright image is formed. (a) The mirror is (1) convex, (2) concave, (3) flat. Explain. (b) What is its radius of curvature?

47. ●● A spherical mirror at an amusement park shows anyone who stands 2.5 m in front of it an upright image two times the person's height. What is the mirror's radius of curvature?

48. ●●● For values of d_o from 0 to ∞, (a) sketch graphs of (1) d_i versus d_o and (2) M versus d_o for a converging mirror, and (b) sketch similar graphs for a diverging mirror.

49. ●●● The front surface of a glass cube 5.00 cm on each side is placed a distance of 30.0 cm in front of a converging mirror that has a focal length of 20.0 cm. (a) Where is the image of the front and back surface of the cube located, and what are the image characteristics? (b) Is the image of the cube still a cube?

50. ●●● A section of a sphere is mirrored on both sides. If the magnification of an object is +1.8 when the section is used as a concave mirror, what is the magnification of an object at the same distance in front of the convex side?

51. IE ●●● A concave mirror of radius of curvature of 20 cm forms an image of an object that is twice the height of the object. (a) There could be (1) one, (2) two, (3) three object distance(s) that satisfy the image characteristics. Explain. (b) What are the object distances?

52. ●●● A convex mirror is on the exterior of the passenger side of many trucks (Exercise 22a). If the focal length of such a mirror is −40.0 cm, what will be the location and height of the image of a car that is 2.0 m high and (a) 100 m and (b) 10.0 m behind the truck mirror?

53. ●●● Two students in a physics laboratory each have a concave mirror with the same radius of curvature, 40 cm. Each student places an object in front of her mirror. The image in both mirrors is three times the size of the object. However, when the students compare notes, they find that the object distances are not the same. Is this possible? If so, what are the object distances?

23.3 Lenses

54. **MC** The image produced by a diverging lens is always (a) virtual and magnified, (b) real and magnified, (c) virtual and reduced, (d) real and reduced.

55. **MC** A converging lens (a) must have at least one convex surface, (b) cannot produce a virtual and reduced image, (c) is thicker at its center than at the periphery, (d) all of the preceding.

56. **MC** If an object is placed at the focal point of a converging lens, the image is (a) at zero, (b) also at the focal point, (c) at a distance equal to twice the focal length, (d) at infinity.

57. CQ Explain why a fish in a spherical fish bowl, viewed from the side, appears larger than it really is.

58. CQ Can a converging lens ever form a virtual image of a real object? If yes, under what conditions?

59. CQ How can you quickly determine the focal length of a converging lens? Will the same method work for a diverging lens?

60. CQ If you want to use a converging lens to design a simple overhead projector so as to project the magnified image of some small writing onto a screen on a wall, how far should you place the object in front of the lens?

61. ● An object is placed 50.0 cm in front of a converging lens of focal length 10.0 cm. What are the image distance and the lateral magnification?

62. ● An object placed 30 cm in front of a converging lens forms an image 15 cm behind the lens. What is the focal length of the lens?

63. ● A converging lens with a focal length of 20 cm is used to produce an image on a screen that is 2.0 m from the lens. What is the object distance?

64. IE ●● An object 4.0 cm tall is in front of a converging lens of focal length 22 cm. The object is 15 cm away from the lens. (a) Use a ray diagram to determine whether the image is (1) real or virtual, (2) upright or inverted, and (3) magnified or smaller than the object. (b) Calculate the image distance and lateral magnification.

65. ●● (a) Design the lens in a single-lens slide projector that will form a sharp image on a screen 4.0 m away with the transparent slides 6.0 cm from the lens. (b) If the object on a slide is 1.0 cm tall, how tall will the image on the screen be, and how should the slide be placed in the projector?

66. ●● Using the thin-lens equation and the magnification factor, show that for a spherical diverging lens, the image of a real object is always virtual, upright, and reduced.

67. ●● A biconvex lens has a focal length of 0.12 m. Where on the lens axis should an object be placed in order to get (a) a real, enlarged image with a magnification of 2.0 and (b) a virtual, enlarged image with a magnification of 2.0?

68. ●● An object is placed in front of a biconcave lens whose focal length is −18 cm. Where is the image located and what are its characteristics, if the object distance is (a) 10 cm and (b) 25 cm? Sketch ray diagrams for each case.

69. ●● A biconvex lens produces a real, inverted image of an object that is magnified 2.5 times when the object is 20 cm from the lens. What is the focal length of the lens?

70. ●● A simple single-lens camera (biconvex lens) is used to photograph a man 1.7 m tall who is standing 4.0 m from the camera. If the man's image fills the height of a frame of film (35 mm), what is the focal length of the lens?

71. ●● To photograph a full Moon, a photographer uses a single-lens camera having a focal length of 60 mm. What will be the diameter of the Moon's image on the film? (*Note*: Data about the Moon are inside the back cover of this text.)

72. ●● (a) For values of d_o from 0 to ∞, sketch graphs of (1) d_i versus d_o and (2) M versus d_o for a converging lens. (b) Sketch similar graphs for a diverging lens. (Compare to Exercise 48.)

73. ●● An object is placed 40 cm from a screen. (a) At what point between the object and the screen should a converging lens with a focal length of 10 cm be placed so that it will produce a sharp image on the screen? (b) What is the lens's magnification?

74. ●● An object 5.0 cm tall is 10 cm from a concave lens. The resulting image is one fifth as large as the object. What is the focal length of the lens?

75. ●● (a) For a biconvex lens, what is the *minimum* distance between an object and its image if the image is real? (b) What is the distance if the image is virtual?

76. ●● Using ▾Fig. 23.28, derive (a) the thin-lens equation and (b) the magnification equation for a thin lens. [*Hint*: Use similar triangles.]

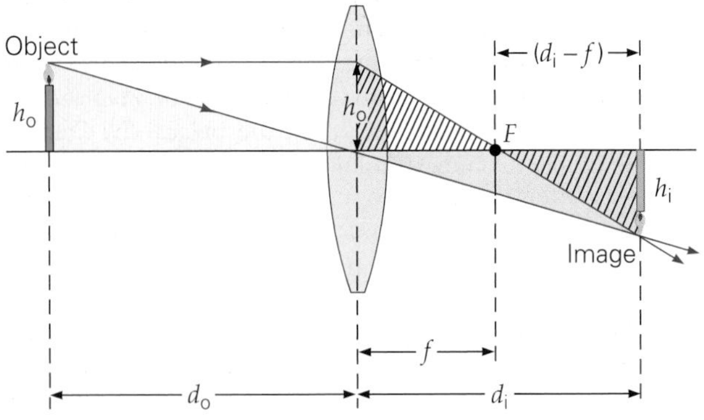

▲ **FIGURE 23.28 The thin-lens equation** The geometry for deriving the thin-lens equation (and magnification factor). Note the two sets of similar triangles. See Exercise 76.

77. ●● (a) If a book is held 30 cm from an eyeglass lens with a focal length of −45 cm, where is the image of the print formed? (b) If an eyeglass lens with a focal length of +57 cm is used, where is the image formed?

78. IE ●● A biology student wants to examine a bug at a magnification of +5.00 (a) The lens should be (1) convex, (2) concave, (3) flat. Explain. (b) If the bug is 5.00 cm from the lens, what is the focal length of the lens?

79. ●● With a magnifying glass, a biology student on a field trip views a small insect. If she sees the insect magnified by a factor of 3.5 when the glass is held 3.0 cm from it, what is the focal length of the lens?

80. ●● The human eye is a complex multiple-lens system. However, it can be approximated to an equivalent single converging lens with an average focal length about 1.7 cm when the eye is relaxed. If an eye is viewing a 2.0-m-tall tree located 15 m in front of the eye, what are the height and orientation of the image of the tree on the retina?

81. ●●● The geometry of a compound microscope, which consists of two converging lenses, is shown in ▼Fig. 23.29. (More detail on microscopes is given in Chapter 25.) The objective lens and the eyepiece lens have focal lengths of 2.8 mm and 3.3 cm, respectively. If an object is located 3.0 mm from the objective lens, where is the final image located, and what type of image is it?

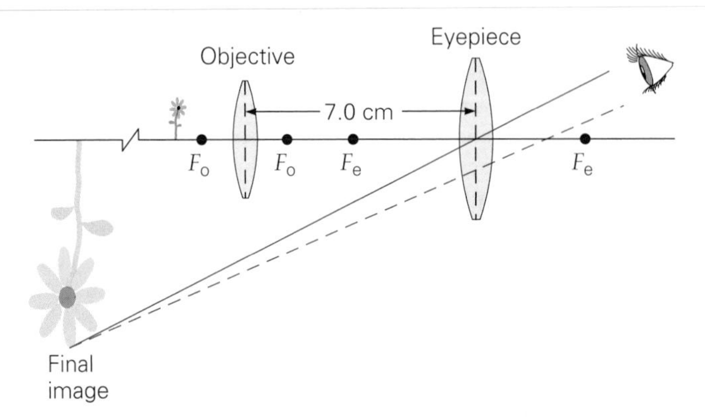

▲ **FIGURE 23.29 Compound microscope** See Exercise 81.

82. ●●● Two converging lenses L_1 and L_2 have focal lengths of 30 cm and 20 cm, respectively. The lenses are placed 60 cm apart along the same axis, and an object is placed 50 cm from L_1 on the side opposite L_2. Where is the image formed relative to L_2, and what are its characteristics?

83. ●●● For a lens combination, show that the total magnification $M_{total} = M_1 M_2$. [*Hint*: Think about the definition of magnification.]

84. ●●● Show that for thin lenses that have focal lengths f_1 and f_2 and are in contact, the effective focal length (f) is given by

$$\frac{1}{f} = \frac{1}{f_1} + \frac{1}{f_2}$$

23.4 The Lens Maker's Equation *and* *23.5 Lens Aberrations

85. **MC** The power of a lens is expressed in units of (a) watts, (b) diopters, (c) meters, (d) both b and c.

86. **MC** A lens aberration that is caused by dispersion is called (a) spherical aberration, (b) chromatic aberration, (c) refractive aberration, (d) none of the preceding.

87. **MC** The focal length of a rectangular glass block is (a) zero, (b) infinity, (c) not defined.

88. CQ Determine the signs of R_1 and R_2 for each lens shown in Fig. 23.14.

89. CQ When you open your eyes underwater, everything is blurry. However, when you wear goggles, you can see clearly. Explain.

90. CQ A lens that is converging in air is submerged in a fluid whose index of refraction is greater than that of the lens. Is the lens still converging?

91. CQ (a) When a lens with $n = 1.60$ is immersed in water, is there a change in the focal length of the lens? If so, which way? (b) What would be the case for a submerged lens whose index of refraction is less than that of the fluid?

92. ● An optometrist prescribed glasses with a power of −2.0 D for a nearsighted student. What is the focal length of the glass lenses?

93. ● A farsighted senior citizen needs glasses with a focal length of 25 cm. What is the power of the lens?

94. ●● An optometrist prescribes a corrective lens with a power of +1.5 D. The lens maker will start with a glass blank that has an index of refraction of 1.6 and a convex front surface whose radius of curvature is 20 cm. To what radius of curvature should the other surface be ground?

95. ●● A plastic plano-concave lens has a radius of curvature of 50 cm for its concave surface. If the index of refraction of the plastic is 1.35, what is the power of the lens?

96. IE ●● A plastic convex-meniscus (Fig. 23.14) contact lens is made of plastic of index of refraction 1.55. The lens has a front radius of 2.50 cm and a back radius of 3.00 cm. (a) The signs of R_1 and R_2 are (1) +, +, (2) +, −, (3) −, +, (4) −, −. Explain. (b) What is the focal length of the lens?

97. ●● A converging glass lens with an index of refraction of 1.62 has a focal length of 30 cm in air. What is the focal length when the lens is submerged in water?

98. IE ●●● A biconvex lens is made of glass whose index of refraction is 1.6. The lens has a radius of curvature of 30 cm for one surface and 40 cm for the other. (a) Will the focal length of this lens (1) increase, (2) remain the same, or (3) decrease if the lens is moved from air to under water? Why? (b) Calculate the focal length of this lens as used in air and under water.

Comprehensive Exercises

99. A method of determining the focal length of a diverging lens is called *autocollimation*. As ▼Fig. 23.30 shows, first a sharp image of a light source is projected on a screen by a converging lens. Second, the screen is replaced with a plane mirror. Third, a diverging lens is placed between the converging lens and the mirror. Light will then be reflected by the mirror back through the compound-lens system, and an image will be formed on a screen near the light source. This image is made sharp by adjusting the distance between the diverging lens and the mirror. The distance at which the image is clearest is equal to the focal length of the lens. Explain why this method works.

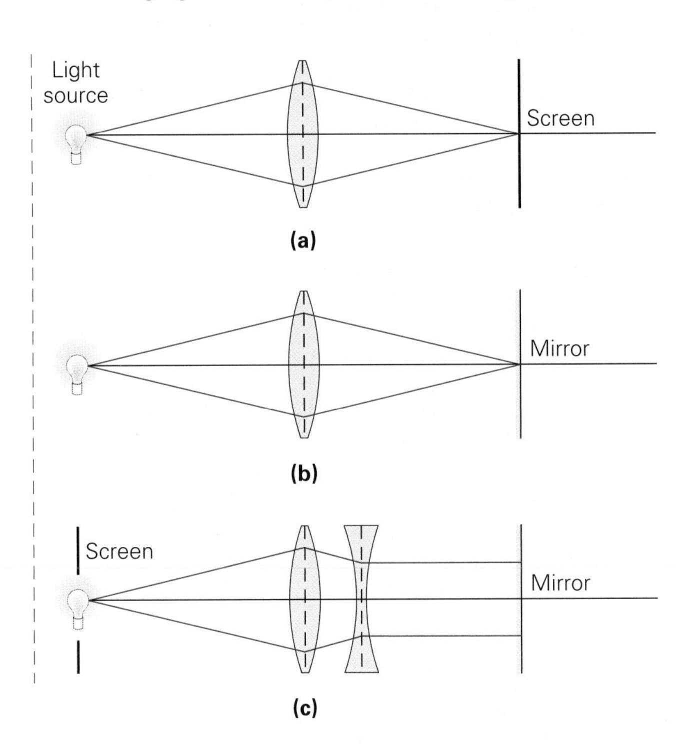

▲ **FIGURE 23.30 Autocollimation** See Exercise 99.

100. For the arrangement shown in ▼Fig. 23.31, an object is placed 0.40 m in front of the converging lens, which has a focal length of 0.15 m. If the concave mirror has a focal length of 0.13 m, where is the final image formed, and what are its characteristics?

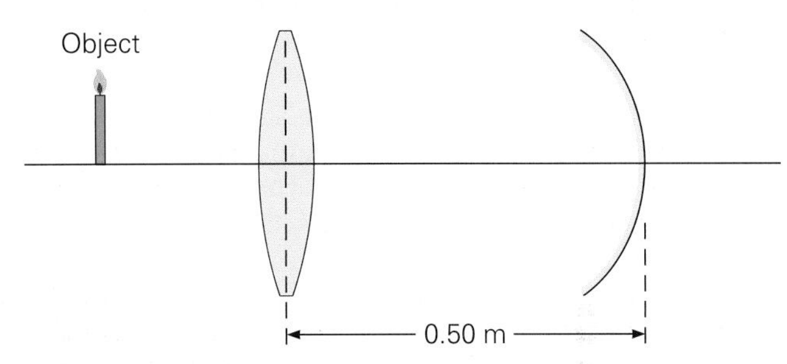

▲ **FIGURE 23.31 Lens–mirror combination** See Exercise 100.

101. Two lenses, each having a power of +10 D, are placed 20 cm apart along the same axis. If an object is 60 cm from the first lens on the side opposite the second lens, where is the final image relative to the first lens, and what are its characteristics?

102. Show that the magnification for objects near the optic axis of a convex mirror is given by $|M| = d_i/d_o$. [*Hint*: Use a ray diagram with rays reflected at the mirror's vertex.]

103. An object is 15 cm from a converging lens whose focal length is 10 cm. On the opposite side of that lens, at a distance of 60 cm, is a converging lens with a focal length of 20 cm. Where is the final image formed, and what are its characteristics?

104. (a) Use ray diagrams to show that a ray parallel to the optic axis of a biconvex lens is refracted toward the axis at the incident surface and again at the exit surface. (b) Show this also holds for a biconcave lens, but with both refractions away from the axis.

The following Physlet Physics Problems can be used with this chapter.
33.1, 33.2, 33.3, 33.4, 33.5, 33.6, 33.7, 33.8, 34.4, 35.1, 35.2, 35.3, 35.4, 35.5, 35.6, 35.7, 35.8, 35.9, 35.10
Optics Appendix: What's Behind the Curtain? Problems 1–15

CHAPTER

24 PHYSICAL OPTICS: THE WAVE NATURE OF LIGHT

24.1 Young's Double-Slit Experiment 761

24.2 Thin-Film Interference 764

24.3 Diffraction 768

24.4 Polarization 775

*24.5 Atmospheric Scattering of Light 782

PHYSICS FACTS

- Some sources say Thomas Young, who first demonstrated the wave nature of light, could read by age 2 and read the Bible twice in his early years.
- The track-to-track distance is 0.74 μm on a DVD-ROM and 1.6 μm on a CD-ROM. In comparison, the diameter of human hair is on the order of 50–150 μm. DVD-ROM and CD-ROM tracks really split hairs.
- AM radio can be heard better in some areas than FM radio. This is because the longer AM waves are more easily diffracted around buildings and other obstacles.
- Skylight is partially polarized. It is believed that some insects, such as bees, use polarized skylight to determine navigational directions relative to the Sun.
- To an observer on Earth, the "red planet" Mars appears reddish because the surface material contains iron oxide. The rusting of iron on Earth produces iron oxide.

It's always intriguing to see brilliant colors produced by objects that we know don't have any color of their own. The glass of a prism, for example, which is clear and transparent by itself, nevertheless gives rise to a whole array of colors when white light passes through it. Prisms, like the water droplets that produce rainbows, don't create color. They merely separate the different colors that make up white light.

The phenomena of reflection and refraction are conveniently analyzed by using geometrical optics (Chapter 22). Ray diagrams (Chapter 23) show what happens when light is reflected from a mirror or passed through a lens. However, other phenomena involving light, such as the interference patterns of the soap bubble in the photo, cannot be adequately explained or described using the ray concept, since this technique ignores the wave nature of light. Other wave phenomena include diffraction and polarization.

Physical optics, or **wave optics**, takes into account wave properties that geometrical optics ignores. The wave theory of light leads to satisfactory explanations of those phenomena that cannot be analyzed with rays. Thus, in this chapter, the wave nature of light must be used to analyze phenomena such as interference and diffraction.

Wave optics must be used to explain how light propagates around small objects or through small openings. We see this in our everyday life with the narrow grooves in CDs, DVDs, and other items. An object or opening is considered small if it is in the order of magnitude of the wavelength of light.

24.1 Young's Double-Slit Experiment

OBJECTIVES: To (a) explain how Young's experiment demonstrated the wave nature of light, and (b) compute the wavelength of light from experimental results.

It has been stated that light behaves like a wave, but no proof of this assertion has been discussed. How would you go about demonstrating the wave nature of light? One method that involves the use of interference was first devised in 1801 by the English scientist Thomas Young (1773–1829). **Young's double-slit experiment** not only demonstrates the wave nature of light, but also allows the measurement of its wavelengths. Essentially, light can be shown to be a wave if it exhibits wave properties such as interference and diffraction.

Recall from the discussion of wave interference in Sections 13.4 and 14.4 that superimposed waves may interfere constructively or destructively. Constructive interference occurs when two crests are superimposed. If a crest and a trough are superimposed, then destructive interference occurs. Interference can be easily observed with water waves, for which constructive and destructive interference produce obvious interference patterns (▸Fig. 24.1).

▲ **FIGURE 24.1 Water-wave interference** The constructive and destructive interference of water waves from two coherent sources in a ripple tank produces interference patterns.

Note: Compare Fig. 24.1 with Fig. 14.8a.

The interference of (visible) light waves is not as easily observed, because of their relatively short wavelengths ($\approx 10^{-7}$ m) and the fact that they usually are not really monochromatic (single frequency). Also, stationary interference patterns are produced only with *coherent sources*—sources that produce light waves having a constant phase relationship to one another. For example, for constructive interference to occur at some point, the waves meeting at that point must be in phase. As the waves meet, a crest must *always* overlap a crest, and a trough must *always* overlap a trough. If a phase difference develops between the waves over time, the interference pattern changes, and a stable or stationary pattern will not be established.

In an ordinary light source, the atoms are excited randomly, and the emitted light waves fluctuate in amplitude and frequency. Thus, light from two such sources is *incoherent* and cannot produce a stationary interference pattern. Interference does occur, but the phase difference between the interfering waves changes so fast that the interference effects are not discernible. To obtain the equivalent of two coherent sources, a barrier with one narrow slit is placed in front of a single light source, and a barrier with two very narrow slits is positioned symmetrically in front of the first barrier (▾Fig. 24.2a).

Illustration 37.1 Ripple Tank

Waves propagating out from the single slit are in phase, and the double slits then act as two coherent sources by separating each wave into two parts. Any random changes in the light from the original source will thus occur for the light passing through both slits, and the phase difference will be constant. The modern laser beam, a coherent light source, makes the observation of a stable interference pattern much

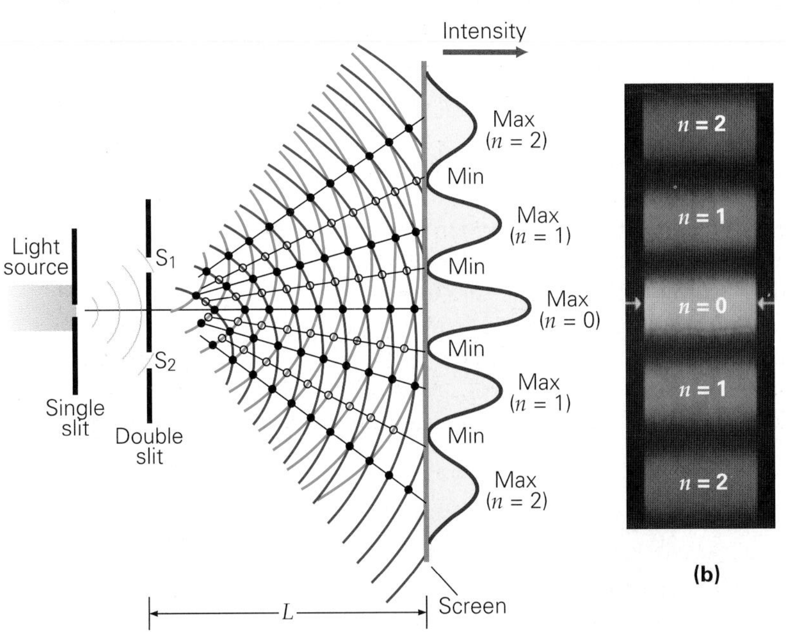

◂ **FIGURE 24.2 Double-slit interference (a)** The coherent waves from two slits are shown in blue (top slit) and red (bottom slit). The waves spread out as a result of diffraction from narrow slits. The waves interfere, producing alternating maxima and minima, on the screen. **(b)** An interference pattern. Note the symmetry of the pattern about the central maximum (n = 0).

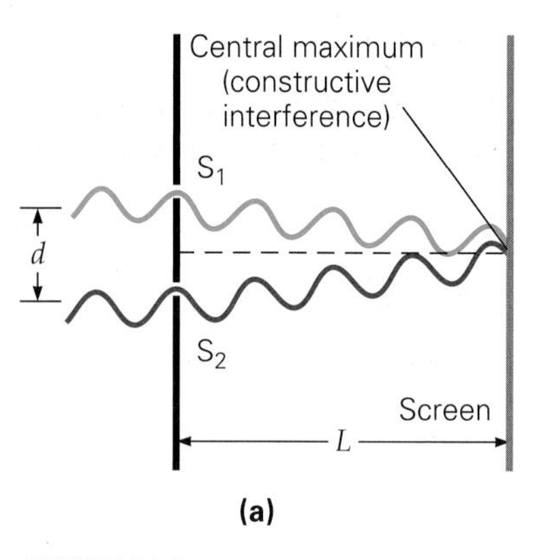

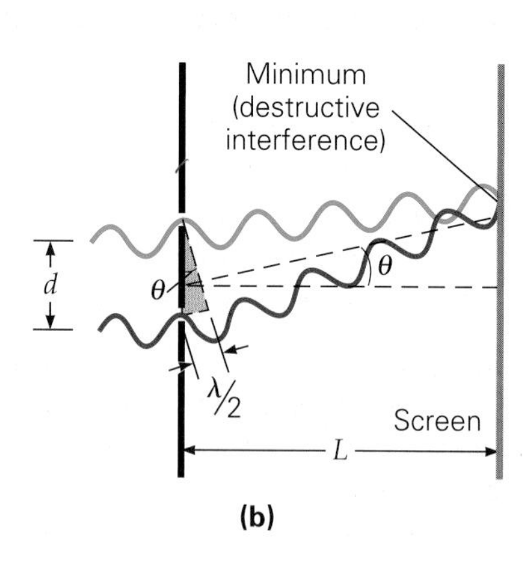

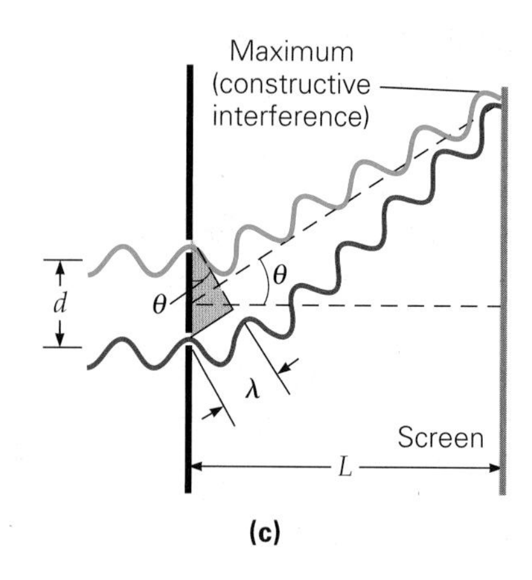

▲ **FIGURE 24.3 Interference** The interference that produces a maximum or minimum depends on the difference in the path lengths of the light from the two slits. **(a)** The path-length difference at the position of the central maximum is zero, so the waves arrive in phase and interfere constructively. **(b)** At the position of the first minimum, the path-length difference is $\lambda/2$, and the waves interfere destructively. **(c)** At the position of the first maximum, the path-length difference is λ, and the interference is constructive.

easier. A series of maxima or bright positions can be observed on a screen placed relatively far from the slits (Fig. 24.2b).

To help analyze Young's experiment, let's imagine that light with a single wavelength (monochromatic light) is used. Because of diffraction (see Sections 13.4 and 14.4 and, in this chapter, Section 24.3), or the spreading of light as it passes through a slit, the waves spread out and interfere as illustrated in Fig. 24.2a. Coming from two coherent "sources," the interfering waves produce a stable interference pattern on the screen. The pattern consists of a bright central maximum (▲Fig. 24.3a) and a series of symmetrical side minima (Fig. 24.3b) and maxima (Fig. 24.3c), which mark the positions at which destructive and constructive interference occur. The existence of this interference pattern clearly demonstrates the wave nature of light. The intensities of each side maxima decrease with distance from the central maximum.

Measuring the wavelength of light requires us to look at the geometry of Young's experiment, as shown in ▼Fig. 24.4. Let the screen be a distance L from the slits and P be an arbitrary point on the screen. P is located a distance y from the center of the central maximum and at an angle θ relative to a normal line between the slits. The slits S_1 and S_2 are separated by a distance d. Note that the light path from slit S_2 to P is longer than the path from slit S_1 to P. As the figure shows, the path-length difference (ΔL) is approximately

$$\Delta L = d \sin\theta$$

The fact that the angle in the small shaded triangle is almost equal to θ can be shown by a simple geometrical argument involving similar triangles when $d \ll L$, as described in the caption of Fig. 24.4.

The relationship of the phase difference of two waves to their path-length difference was discussed in Chapter 14 for sound waves. These conditions hold for any wave, including light. Constructive interference occurs at any point where the path-length difference between the two waves is an integral number of wavelengths:

$$\Delta L = n\lambda \quad \text{for } n = 0, 1, 2, 3, \ldots \quad \textit{condition for constructive interference} \tag{24.1}$$

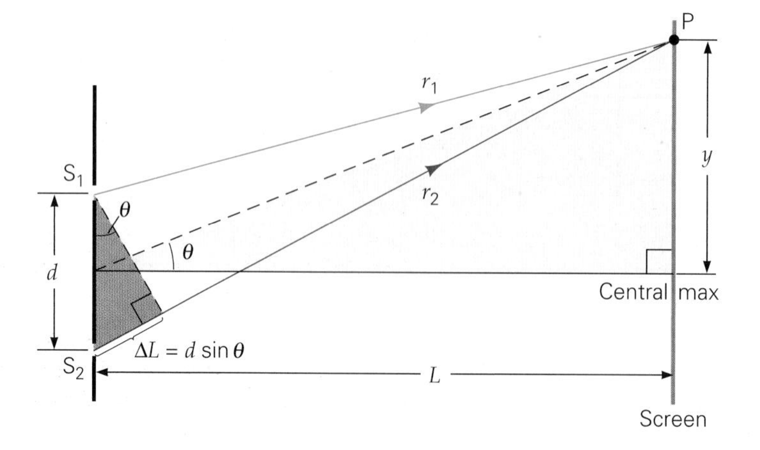

▶ **FIGURE 24.4 Geometry of Young's double-slit experiment** The difference in the path lengths for light traveling from the two slits to a point P is $r_2 - r_1 = \Delta L$, which forms a side of the small shaded triangle. Because the barrier with the slits is parallel to the screen, the angle between r_2 and the barrier (at S_2, in the small shaded triangle) is equal to the angle between r_2 and the screen. When L is much greater than y, that angle is almost identical to the angle between the screen and the dashed line, which is an angle in the large shaded triangle. The two shaded triangles are then almost exactly similar, and the angle at S_1 in the small triangle is almost exactly equal to θ. Thus, $\Delta L = d \sin\theta$. (Not drawn to scale. Assume that $d \ll L$.)

Similarly, for destructive interference, the path-length difference is an odd number of half-wavelengths:

$$\Delta L = \frac{m\lambda}{2} \quad \text{for } m = 1, 3, 5, \ldots \quad \textit{condition for destructive interference} \tag{24.2}$$

PHYSLET®

Exploration 37.1 Varying Numbers and Orientations of Sources

Thus, in Fig. 24.4, the maxima (constructive interference) satisfy

$$d \sin\theta = n\lambda \quad \text{for } n = 0, 1, 2, 3, \ldots \quad \textit{condition for interference maxima} \tag{24.3}$$

Exploration 37.2 Changing the Separation Between Sources

where n is called the *order number*. The zeroth order ($n = 0$) corresponds to the central maximum; the first order ($n = 1$) is the first maximum on either side of the central maximum; and so on. As the path-length difference varies from point to point, so does the phase difference and the resulting type of interference (constructive or destructive).

The wavelength can therefore be determined by measuring d and θ for a maximum of a particular order (other than the central maximum), because Eq. 24.3 can be solved as $\lambda = (d \sin\theta)/n$.

The angle θ locates a maximum relative to the central maximum. This can be measured from a photograph of the interference pattern, such as shown in Fig. 24.2b. If θ is small ($y \ll L$), $\sin\theta \approx \tan\theta = y/L$. Substituting this y/L for $\sin\theta$ into Eq. 24.3 and solving for y gives a good approximation of the distance of the nth maximum (y_n) from the central maximum on either side:

$$y_n \approx \frac{nL\lambda}{d} \quad \text{for } n = 0, 1, 2, 3, \ldots \quad \textit{lateral distance to maxima for small } \theta \textit{ only} \tag{24.4}$$

A similar analysis gives the locations of the minima. (See Exercise 12a.)

From Eq. 24.3, we see that, except for the zeroth order, $n = 0$ (the central maximum), the positions of the maxima depend on wavelength—different wavelengths (λ) give different values of $\sin\theta$ and therefore of θ and y. Hence, when we use white light, the central maximum is white because all wavelengths are at the same location, but the other orders become a "spread out" spectrum of colors. Because y is proportional to λ ($y \propto \lambda$), we observed, in a given order, that red is farther out than blue or that red has a longer wavelength than blue.

By measuring the positions of the color maxima within a particular order, Young was able to determine the wavelengths of the colors of visible light. Note also that the size or "spread" of the interference pattern, y_n, depends inversely on the slit separation d. The smaller d, the more spread out the pattern. For large d, the interference pattern is so compressed that it appears to us as a single white spot (all maxima together at center).

In this analysis, the word *destructive* does *not* imply that energy is destroyed. Destructive interference is simply a description of a physical fact—that if light energy is not present at a particular location, by energy conservation, it *must* be somewhere else. The mathematical description of Young's double-slit experiment tells you that there is no light energy at the minima. The light energy is redistributed and located at the maxima. This is also observed with sound waves as well.

Integrated Example 24.1 ■ Measuring the Wavelength of Light: Young's Double-Slit Experiment

In a lab experiment similar to the one shown in Fig. 24.4, monochromatic light (only one wavelength or frequency) passes through two narrow slits that are 0.050 mm apart. The interference pattern is observed on a white wall 1.0 m from the slits, and the second-order maximum is at an angle of $\theta_2 = 1.5°$. (a) If the slit separation decreases, the second-order maximum will be seen at an angle of (1) greater than 1.5°, (2) 1.5°, (3) less than 1.5°. Explain. (b) What is the wavelength of the light and what is the distance between the second-order and third-order maxima? (c) if $d = 0.040$ mm, what is θ_2?

(a) Conceptual Reasoning. According to the condition for constructive interference, $d \sin\theta = n\lambda$, the product of d and $\sin\theta$ is a constant, for a given wavelength λ and order number n. Therefore, if d decreases, $\sin\theta$ will increase and so will θ. Thus the answer is (1).

(continues on next page)

(b) and (c) Quantitative Reasoning and Solution. Eq. 24.3 can be used to find the wavelength. Since $L \gg d$, that is, 1.0 m $\gg$ 0.050 mm, then θ is small. We could compute y_2 and y_3 from Eq. 24.4 and determine the distance between the second-order and third-order maxima $(y_3 - y_2)$. However, the maxima for a given wavelength of light are evenly spaced (for a small θ). That is, the distance between adjacent maxima is a constant.

Given: $L = 1.0$ m
$n = 2$
(b) $\theta_2 = 1.5°$
$d = 0.050 \text{ mm} = 5.0 \times 10^{-5}$ m
(c) $d = 4.0 \times 10^{-5}$ m

Find: (b) λ (wavelength) and $y_3 - y_2$ (distance between $n = 2$ and $n = 3$)
(c) θ_2 if $d = 0.040$ mm

(b) Using Eq. 24.3 gives

$$\lambda = \frac{d \sin\theta}{n} = \frac{(5.0 \times 10^{-5}\text{ m})\sin 1.5°}{2} = 6.5 \times 10^{-7}\text{ m} = 650\text{ nm}$$

This value is 650 nm, which is the wavelength of orange-red light (see Fig. 20.23). Using a general approach for n and $n + 1$, we get

$$y_{n+1} - y_n = \frac{(n+1)L\lambda}{d} - \frac{nL\lambda}{d} = \frac{L\lambda}{d}$$

In this case, the distance between successive fringes is

$$y_3 - y_2 = \frac{L\lambda}{d} = \frac{(1.0\text{ m})(6.5 \times 10^{-7}\text{ m})}{5.0 \times 10^{-5}\text{ m}} = 1.3 \times 10^{-2}\text{ m} = 1.3\text{ cm}$$

(c) $\sin\theta_2 = \dfrac{n\lambda}{d} = \dfrac{(2)(650 \times 10^{-9}\text{ m})}{(4.0 \times 10^{-5}\text{ m})} = 0.0325$ so $\theta_2 = \sin^{-1}(0.0325) = 1.9° > 1.5°$.

Follow-Up Exercise. Suppose white light were used instead of monochromatic light in this Example. What would be the separation distance of the red ($\lambda = 700$ nm) and blue ($\lambda = 400$ nm) components in the second-order maximum? *(Answers to all Follow-Up Exercises are at the back of the text.)*

24.2 Thin-Film Interference

OBJECTIVES: To (a) describe how thin films can produce colorful displays, and (b) give some examples of practical applications of thin-film interference.

Have you ever wondered what causes the rainbowlike colors when white light is reflected from a thin film of oil or a soap bubble? This effect—known as *thin-film interference*—is a result of the interference of light reflected from opposite surfaces of the film and may be understood in terms of wave interference.

First, however, you need to know how the phase of a light wave is affected by reflection. Recall from Chapter 13 that a wave pulse on a rope undergoes a 180° phase change [or a *half wave shift* ($\lambda/2$)], when reflected from a rigid support and no phase shift when reflected from a free support (▼Fig. 24.5). Similarly, as the figure shows, the phase change for the reflection of light waves at a boundary depends on the indices of refraction (n) of the two materials:

- A light wave undergoes a 180° phase change on reflection if $n_1 < n_2$.
- There is no phase change on reflection if $n_1 > n_2$.

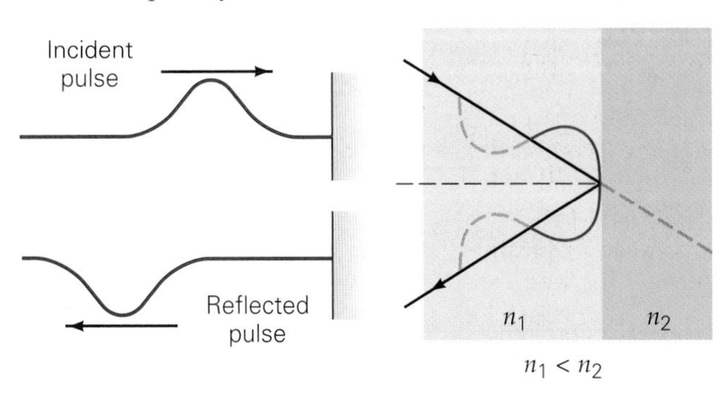

(a) Fixed end: 180° phase shift

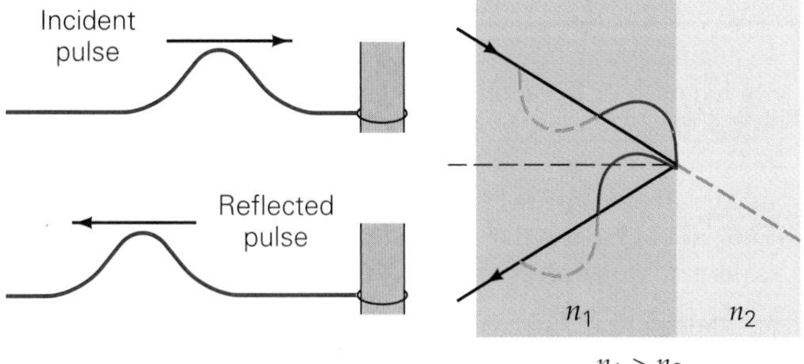

(b) Free end: zero phase shift

▼ **FIGURE 24.5 Reflection and phase shifts** The phase changes that light waves undergo on reflection are analogous to those for pulses in strings. **(a)** The phase of a pulse in a string is shifted by 180° on reflection from a fixed end, and so is the phase of a light wave when it is reflected from a more optically dense medium. **(b)** A pulse in a string has a phase shift of zero (it is not shifted) when reflected from a free end. Analogously, a light wave is not phase shifted when reflected from a less optically dense medium.

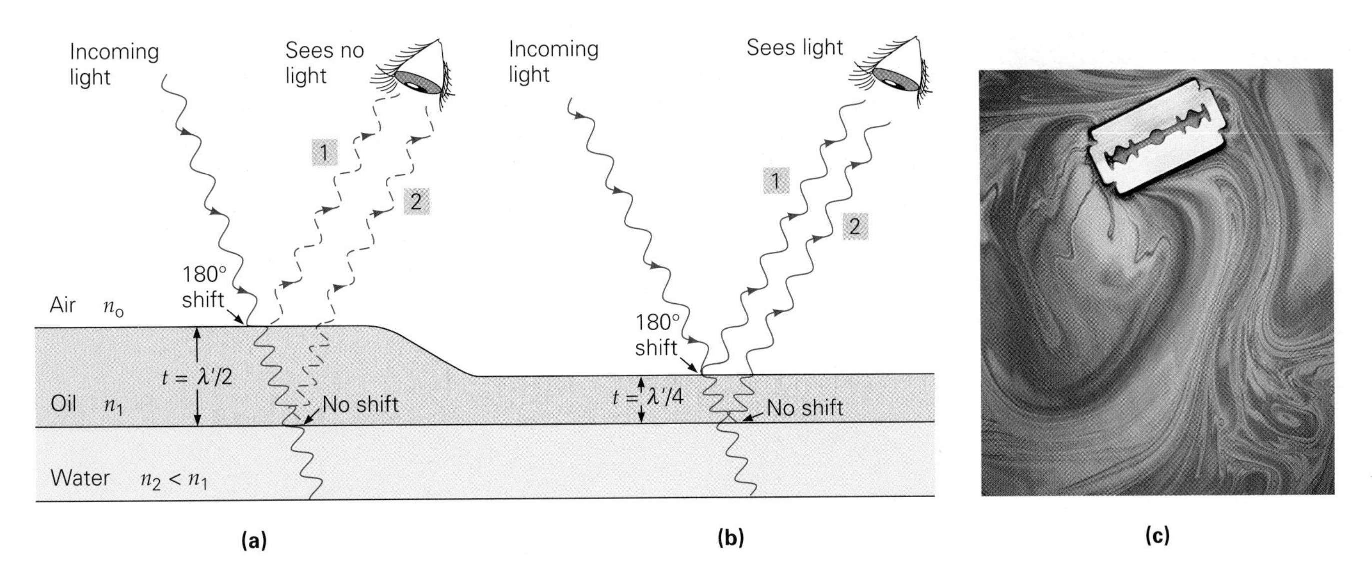

▲ **FIGURE 24.6 Thin-film interference** For an oil film on water, there is a 180° phase shift for light reflected from the air–oil interface and a zero phase shift at the oil–water interface. λ' is the wavelength in the oil. **(a)** Destructive interference occurs if the oil film has a minimum thickness of $\lambda'/2$ for normal incidence. (Waves are displaced and angled for clarity.) **(b)** Constructive interference occurs with a minimum film thickness of $\lambda'/4$. **(c)** Thin-film interference in an oil slick. Different film thicknesses give rise to the reflections of different colors.

To understand why you see colors from a soap bubble or an oil film (for example, floating on water or on a wet road), consider the reflection of monochromatic light from a thin film in ▲Fig. 24.6. The path length of the wave in the film depends on the angle of incidence (why?), but for simplicity, we will assume normal (perpendicular) incidence for the light, even though the rays are drawn at an angle in the figure, for clarity.

The oil film has a greater index of refraction than that of air, and the light reflected from the air–oil interface (wave 1 in the figure) undergoes a 180° phase shift. The transmitted waves pass through the oil film and are reflected at the oil–water interface. In general, the index of refraction of oil is greater than that of water (see Table 22.1)—that is, $n_1 > n_2$—so a reflected wave in this instance (wave 2) does *not* have the phase shift.

You might think that if the path length of the wave in the oil film ($2t$, twice the thickness—down and back) were an integral number of wavelengths—for example, if $2t = 2(\lambda'/2)$ in Fig. 24.6a, where $\lambda' = \lambda/n$ is the wavelength in the oil—then the waves reflected from the two surfaces would interfere constructively. But keep in mind that the wave reflected from the top surface (wave 1) undergoes a 180° phase shift. The reflected waves from the two surfaces are therefore actually *out of phase* and would interfere destructively for this condition. This means that no reflected light for this wavelength would be observed. (The light would be transmitted.)

Similarly, if the path length of the waves in the film were an odd number of half-wavelengths [$2t = 2(\lambda'/4) = \lambda'/2$] in Fig. 24.6b, again where λ' is the wavelength in the oil, then the reflected waves would actually be *in phase* (as a result of a 180° phase shift of wave 1) and would interfere constructively. Reflected light for this wavelength would be observed from above the oil film.

Because oil and soap films generally have different thicknesses in different regions, particular wavelengths (colors) of white light interfere constructively in different regions after reflection. As a result, a vivid display of various colors appears (Fig. 24.6c). This may also change if the film thickness changes with time. Thin-film interference may be seen when two glass slides are stuck together with an air film between them (▼Fig. 24.7a). The bright colors of a peacock, an example of colorful interference in nature, are a result of layers of fibers in its feathers. Light reflected from successive layers interferes constructively, giving bright colors, even though the feather has no pigment of its own. Since the condition for constructive interference depends on the angle of incidence, the color pattern changes somewhat with the viewing angle and motion of the bird (Fig. 24.7b).

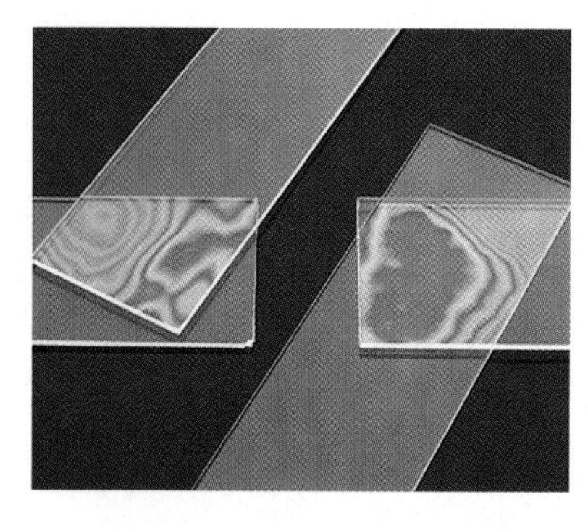

(a)

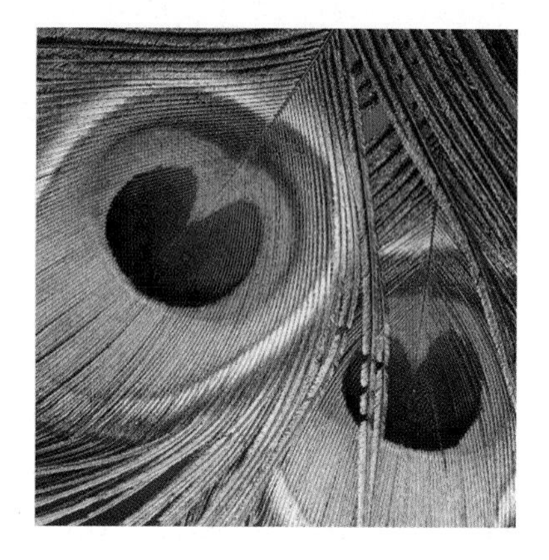

(b)

▲ **FIGURE 24.7 Thin-film interference (a)** A thin air film between microscope slides gives colorful patterns. **(b)** Multilayer interference in a peacock's feathers gives rise to bright colors. The brilliant throat colors of hummingbirds are produced in the same way.

A practical application of thin-film interference is nonreflective coatings for lenses. (See Insight 24.1, on page 768, on Nonreflecting Lenses.) In this situation, a film coating is used to create destructive interference between the reflected waves so as to *increase the light transmission* into the glass (▼Fig. 24.8). The index of refraction of the film has a value between that of air and glass ($n_o < n_1 < n_2$). Consequently, phase shifts of incident light take place at the surfaces of both the film and the glass.

In such a case, the condition for constructive interference of the reflected light is

$$\Delta L = 2t = m\lambda' \text{ or } t = \frac{m\lambda'}{2} = \frac{m\lambda}{2n_1} \quad m = 1, 2, \ldots \quad \begin{matrix}\textit{condition for con-}\\ \textit{structive interference}\\ \textit{when } n_o < n_1 < n_2\end{matrix} \quad (24.5)$$

and the condition for destructive interference is

$$\Delta L = 2t = \frac{m\lambda'}{2} \text{ or } t = \frac{m\lambda'}{4} = \frac{m\lambda}{4n_1} \quad m = 1, 3, 5, \ldots \quad \begin{matrix}\textit{condition for de-}\\ \textit{structive interference}\\ \textit{when } n_o < n_1 < n_2\end{matrix} \quad (24.6)$$

The *minimum* film thickness for destructive interference occurs when $m = 1$, so

$$t_{\text{min}} = \frac{\lambda}{4n_1} \quad \begin{matrix}\textit{minimum film thickness}\\ (\textit{for } n_o < n_1 < n_2)\end{matrix} \quad (24.7)$$

If the index of refraction of the film is greater than that of air and glass, then only the reflection at the air–film interface has the 180° phase shift. Therefore, $2t = m\lambda'$ will actually create destructive interference and $2t = m\lambda'/2$ constructive interference. (Why?)

Example 24.2 ■ Nonreflective Coatings: Thin-Film Interference

A glass lens ($n = 1.60$) is coated with a thin, transparent film of magnesium fluoride ($n = 1.38$) to make the lens nonreflecting. (a) What is the minimum film thickness for the lens to be nonreflecting for normally incident light of wavelength 550 nm? (b) Will a film thickness of 996 nm make the lens nonreflecting?

Thinking It Through. (a) Equation 24.7 can be used directly to get an idea of the minimum film thickness for a nonreflective coating. (b) We need to determine whether 996 nm satisfies the condition in Eq. 24.6.

Solution.

Given: $n_o = 1.00$ (air)
$n_1 = 1.38$ (for film)
$n_2 = 1.60$ (for lens)
$\lambda = 550$ nm

Find: (a) t_{min} (minimum film thickness)
(b) determine whether $t = 996$ nm gives a nonreflecting lens

(a) Because $n_2 > n_1 > n_o$,

$$t_{\text{min}} = \frac{\lambda}{4n_1} = \frac{550 \text{ nm}}{4(1.38)} = 99.6 \text{ nm}$$

which is quite thin ($\approx 10^{-5}$ cm). In terms of atoms, which have diameters on the order of 10^{-10} m, or 10^{-1} nm, the film is 10^3 atoms thick.

PHYSLET®

Illustration 37.2 Dielectric Mirrors

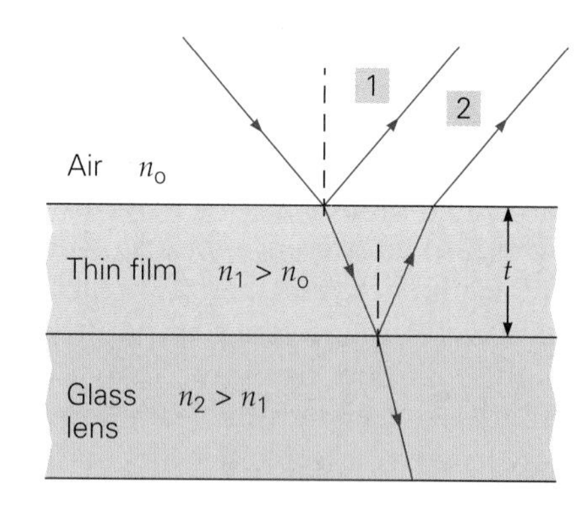

▶ **FIGURE 24.8 Thin-film interference** For a thin film on a glass lens, there is a 180° phase shift at each interface when the index of refraction of the film is less than that of the glass. The waves reflected off the top and bottom surfaces of the film interfere. For clarity, the angle of incidence is drawn to be large, but, in reality, it is almost zero.

(b)

$$t = 996 \text{ nm} = 10(99.6 \text{ nm}) = 10t_{\min} = 10\left(\frac{\lambda}{4n_1}\right) = 5\left(\frac{\lambda}{2n_1}\right)$$

This means that this film thickness does *not* satisfy the nonreflective condition (destructive interference). Actually, it satisfies the requirement for constructive interference (Eq. 24.5) with $m = 5$. Such a coating specific for infrared radiation on car and house windows could be useful in hot climates, because it maximizes reflection and minimizes transmission.

Follow-Up Exercise. For the glass lens in this Example to reflect, rather than transmit, the incident light through the lens, what would be the minimum film thickness?

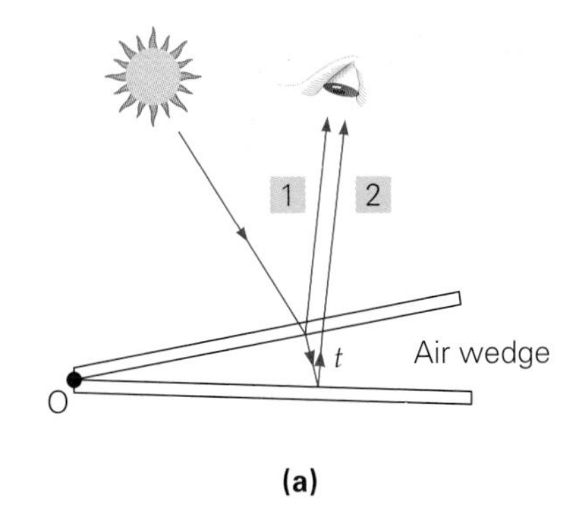

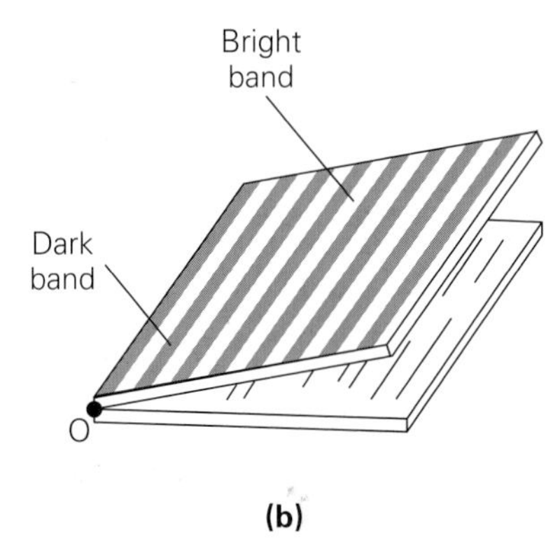

▲ **FIGURE 24.9 Optical flatness** **(a)** An optical flat is used to check the smoothness of a reflecting surface. The flat is placed so that there is an air wedge between it and the surface. The waves reflected from the two plates interfere, and the thickness of the air wedge at certain points determines whether bright or dark bands are seen. **(b)** If the surface is smooth, a regular or symmetrical interference pattern is seen. Note that a dark band is at point O where $t = 0$.

Optical Flats and Newton's Rings

The phenomenon of thin-film interference can be used to check the smoothness and uniformity of optical components such as mirrors and lenses. *Optical flats* are made by grinding and polishing glass plates until they are as flat and smooth as possible. (The surface roughness is usually in the order of $\lambda/20$.) The degree of flatness can be checked by putting two such plates together at a slight angle so that a very thin air wedge is between them (▸Fig. 24.9a). The reflected waves off the bottom of the top plate (wave 1) and top of the bottom plate (wave 2) interfere. Note that wave 2 has a 180° phase shift as it is reflected from an air–plate interface, whereas wave 1 does not. Therefore, at certain points from where the plates touch (point O), the condition for constructive interference is $2t = m\lambda/2$ ($m = 1, 3, 5, \dots$), and the condition for destructive interference is $2t = m\lambda$ ($m = 0, 1, 2, \dots$). The thickness t determines the type of interference (constructive or destructive). If the plates are smooth and flat, a regular interference pattern of bright and dark bands appears (Fig. 24.9b). This pattern is a result of the uniformly varying differences in path lengths between the plates. Any irregularity in the pattern indicates an irregularity in at least one plate. Once a good optical flat is verified, it can be used to check the flatness of a reflecting surface, such as that of a precision mirror.

Direct evidence of the 180° phase shift can be clearly seen in Fig. 24.9. At the point where the two plates touch ($t = 0$), we see a *dark* band. If there were no phase shift, $t = 0$ would correspond to $\Delta L = 0$, and we would expect a bright band to appear. The fact that it is a dark band proves that there is a phase shift in reflection from a more optically dense material.

A similar technique is used to check the smoothness and symmetry of lenses. When a curved lens is placed on an optical flat, a radially symmetric air wedge is formed between the lens and the optical flat (▾Fig. 24.10a). Since the thickness of the air wedge again determines the condition for constructive and destructive interference, the regular interference pattern in this case is a set of concentric circular bright and dark rings (Fig. 24.10b). They are called *Newton's rings*, after Isaac Newton, who first

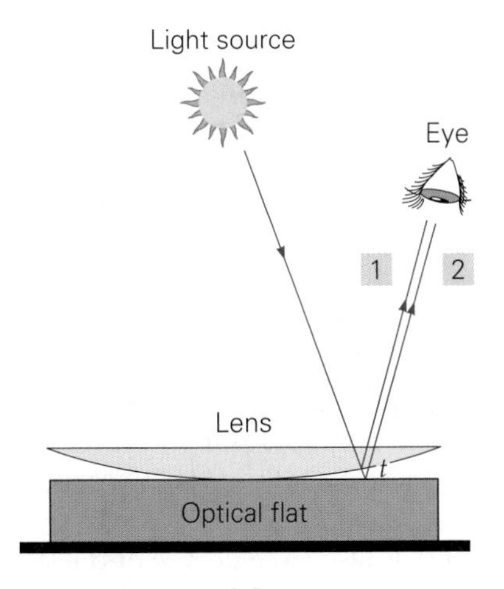

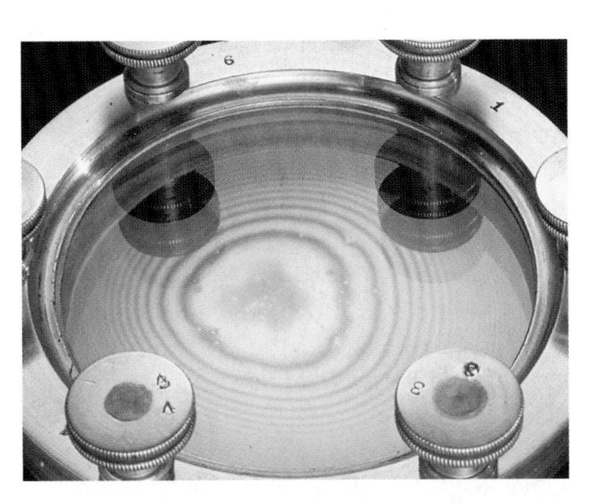

◂ **FIGURE 24.10 Newton's rings** **(a)** A lens placed on an optical flat forms a ring-shaped air wedge, which gives rise to interference of the waves reflected from the top (wave 1) and the bottom (wave 2) of the air wedge. **(b)** The resulting interference pattern is a set of concentric rings called *Newton's rings*. Note that at the center of the pattern is a dark spot. Lens irregularities produce a distorted pattern.

INSIGHT 24.1 NONREFLECTING LENSES

You may have noticed the blue-purple tint of the coated optical lenses used in cameras and binoculars. The coating makes the lenses almost "nonreflecting." If a lens is a nonreflecting type, then the incident light is mostly transmitted through the lens. Maximum transmission of light is desirable for the exposure of photographic film and for viewing objects with binoculars.

For a typical (reflecting) air–glass interface, about 4% of the light is reflected and 96% is transmitted. A modern camera lens is actually made up of a group of lenses (elements) in order to improve image quality. For instance, a 35-mm–70-mm zoom lens might consist of up to 13 elements, thus having 26 reflective surfaces.

After one reflection, $0.96 = 96\%$ of the light is transmitted. After two reflections, or one element, the transmitted light is only $0.96 \times 0.96 = 0.96^2 = 0.92$, or 92%, of the incident light. Thus after 26 reflections, the transmitted light is only $0.96^{26} = 0.35$, or 35%, of the incident light, if lenses are not coated. Therefore, almost all modern lenses are coated with nonreflecting film.

A lens is made nonreflecting by coating it with a thin-film material with an index of refraction between the indices of refraction of air and glass (Fig. 24.8). If the coating is a quarter-wavelength ($\lambda'/4$) thick, the difference in path length between the reflected rays is $\lambda'/2$, where λ' is the wavelength of light in the coating. In this case, both reflected waves undergo a phase shift, and therefore they are out of phase for a path-length difference of $\lambda'/2$ and interfere destructively. That is, the incident light is transmitted, and the coated lens is nonreflecting.

Note that the actual thickness of a quarter-wavelength thickness of film is specific to the particular wavelength of light. The thickness is usually chosen to be a quarter-wavelength of yellow-green light ($\lambda \approx 550$ nm), to which the human eye is most sensitive. The wavelengths at the red and blue ends of the visible region are still partially reflected, giving the coated lens its bluish-purple tint (Fig. 1). Sometimes other quarter-wavelength thicknesses are chosen, giving rise to other hues, such as amber or reddish purple, depending on the application of the lens.

Nonreflective coatings are also applied to the surfaces of solar cells, which convert light into electrical energy (Chapter 27). Because the thickness of such a coating is wavelength dependent, the losses due to reflection can be decreased from around 30% to only 10%. Even so, the process improves the cell's efficiency.

FIGURE 1 Coated lenses The nonreflective coating on binocular and camera lenses generally produces a characteristic bluish-purple hue. (Why?)

described this interference effect. Note that at the point where the lens and the optical flat touch ($t = 0$), there is, once again, a dark spot. (Why?) Lens irregularities give rise to a distorted fringe pattern, and the radii of these rings can be used to calculate the radius of curvature of the lens.

24.3 Diffraction

OBJECTIVES: To (a) define diffraction, and (b) give examples of diffractive effects.

In geometrical optics, light is represented by rays and pictured as traveling in straight lines. If this model were to represent the real nature of light, however, there would be no interference effects in Young's double-slit experiment. Instead, there would be only two bright images of slits on the screen, with a well-defined shadow area where no light enters. But we *do* see interference patterns, which means that the light must deviate from a straight-line path and enter the regions that would otherwise be in shadow. The waves actually "spread out" as they pass through the slits. This spreading is called **diffraction**. Diffraction generally occurs when waves pass through small openings or around sharp edges or corners. The diffraction of water waves is shown in ◂Fig. 24.11. (See also Fig. 13.18.)

▼ **FIGURE 24.11 Water-wave diffraction** This photograph of a beach dramatically shows single-slit diffraction of ocean waves through the barrier openings. Note that the beach has been shaped by the circular wave fronts.

As Fig. 13.18 shows, the amount of diffraction depends on the wavelength in relation to the size of the opening or object. In general, *the longer the wavelength compared to the width of the opening or object, the greater the diffraction.* This effect is also shown in ▸Fig. 24.12. For example, in Fig. 24.12a, the width of the opening w is much greater than the wavelength ($w \gg \lambda$), and there is little diffraction—the wave keeps traveling without much spreading. (There is also *some* degree of diffraction due to the edges of the opening.) In Fig. 24.12b, with the wavelength and opening

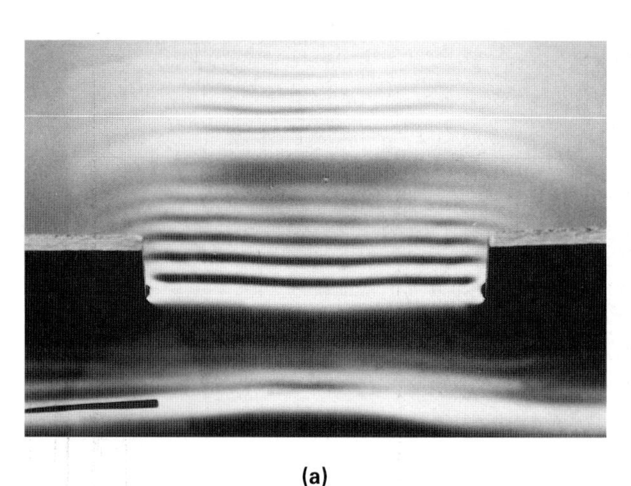

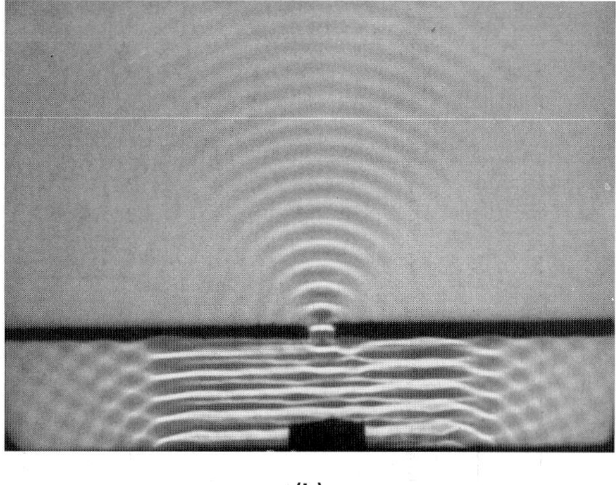

(a) (b)

▲ **FIGURE 24.12 Wavelength and opening dimensions** In general, the narrower the opening compared to the wavelength, the greater the diffraction. **(a)** Without much diffraction ($w \gg \lambda$), the wave would keep traveling in its original direction. **(b)** With noticeable diffraction ($w \approx \lambda$), the wave bends around the opening and spreads out.

width the same order of magnitude ($w \approx \lambda$), there is noticeable diffraction—the wave spreads out and deviates from its original direction. Part of the wave keeps traveling in its original direction but the rest bends *around* the opening and clearly spreads out.

Illustration 38.1 Single Slit Diffraction

The diffraction of sound is quite evident (Chapter 14). Someone can talk to you from another room or around the corner of a building, and even in the absence of reflections, you can easily hear the person. Recall that audible sound wavelengths are on the order of centimeters to meters. Thus, the widths of ordinary objects and openings are about the same as or narrower than the wavelengths of sound, and diffraction will readily occur under these conditions.

Visible light waves, however, have wavelengths on the order of 10^{-7} m. Therefore diffraction phenomena for these waves often go unnoticed, especially through large openings such as doors where sound readily diffracts. However, close inspection of the area around a sharp razor blade will show a pattern of bright and dark bands (▸Fig. 24.13). Diffraction can lead to interference, and thus these interference patterns are evidence of the diffraction of the light around the edge of the blade.

As an illustration of "single-slit" diffraction, consider a slit in a barrier (▾Fig. 24.14). Suppose that the slit (width w) is illuminated with monochromatic light. A diffraction pattern consisting of a bright central maximum and a symmetrical array of maxima (regions of constructive interference) on both sides is observed on a screen at a distance L from the slit (we will assume $L \gg w$).

Thus a diffraction pattern results from the fact that various points on the wave front passing through the slit can be considered to be small point sources of light. The interference of those waves gives rise to the *diffraction* maxima and minima.

The fairly complex analysis is not done here; however, from geometry, it can be proven that the minima (regions of destructive interference) satisfy the relationship

$$w \sin\theta = m\lambda \qquad \text{for } m = 1, 2, 3, \ldots \quad \textit{condition for minima} \qquad (24.8)$$

where θ is the angle of a particular minimum, designated by $m = 1, 2, 3, \ldots$, on either side of the central maximum and m is called the order number. (There is no $m = 0$. Why?)

Although this result is similar in form to that for Young's double-slit experiment (Eq. 24.3), it is extremely important to realize that for the single-slit experiment, minima, rather than maxima, are analyzed. Also, note that the width of the slit (w) is used in diffraction. Physically, this is diffraction from a single slit, *not* interference from two slits.

The small-angle approximation, $\sin\theta \approx y/L$, can be made when $y \ll L$. In this case, the distances of the minima relative to the center of the central maximum are given by

$$y_m = m\left(\frac{L\lambda}{w}\right) \qquad \text{for } m = 1, 2, 3, \ldots \quad \textit{location for minima} \qquad (24.9)$$

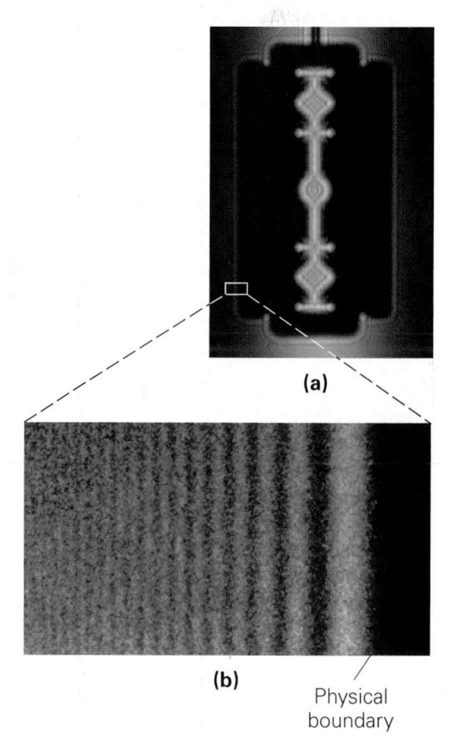

▲ **FIGURE 24.13 Diffraction in action** **(a)** Diffraction patterns produced by a razor blade. **(b)** A close-up view of the diffraction pattern formed at the edge of the blade.

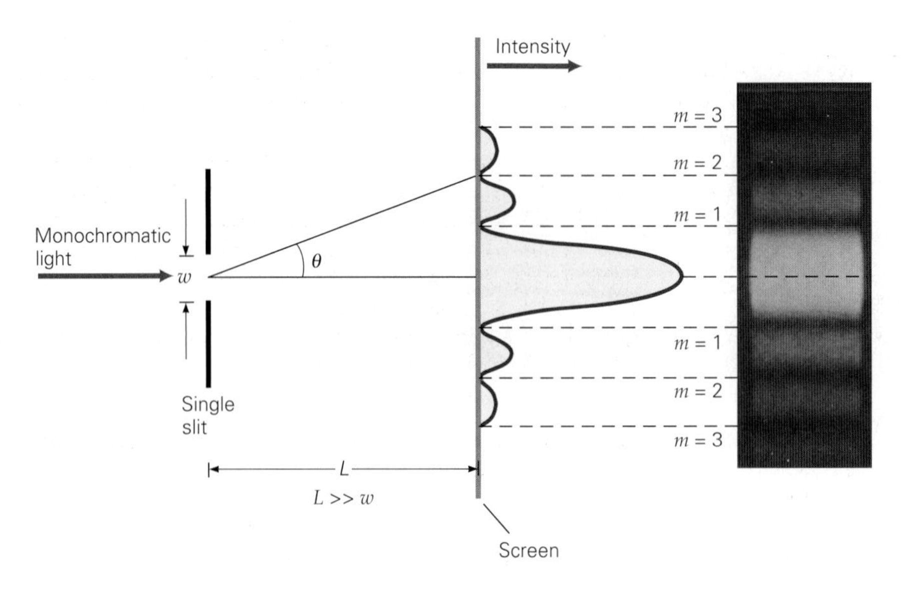

▶ **FIGURE 24.14 Single-slit diffraction** The diffraction of light by a single slit gives rise to a diffraction pattern consisting of a large and bright central maximum and a symmetric array of side maxima. The order number m corresponds to the minima or dark positions. (See text for description.)

The qualitative predictions from Eq. 24.9 are interesting and instructive:

- For a given slit width (w), the longer the wavelength (λ), the wider (more "spread out") the diffraction pattern.
- For a given wavelength (λ) the narrower the slit width (w), the wider the diffraction pattern.
- The width of the central maximum is twice the width of any side maximum.

Let's look in detail at these results. As the slit is made narrower, the central maximum and the side maxima spread out and become larger. Equation 24.9 is not applicable to very small slit widths (because of the small-angle approximation). If the slit width is decreased until it is of the same order of magnitude as the wavelength of the light, then the central maximum spreads out over the whole screen. That is, diffraction becomes dramatically evident when the width of the slit is about the same as the wavelength of the light used. Diffraction effects are most easily observed when $\lambda/w \approx 1$, or $w \approx \lambda$.

Conversely, if the slit is made wider for a given wavelength, then the diffraction pattern becomes less spread out. The maxima move closer together and eventually become difficult to distinguish when w is much wider than λ ($w \gg \lambda$). The pattern then appears as a fuzzy shadow around the central maximum, which is the illuminated image of the slit. This type of pattern is observed for the image produced by sunlight entering a dark room through a hole in a curtain. Such an observation led early experimenters to investigate the wave nature of light. The acceptance of this concept was, in large part, due to the explanation of diffraction offered by physical optics.

The central maximum is twice as wide as any of the side maxima. Taking the width of the central maximum to be the distance between the bounding minima on each side ($m = 1$), or a value of $2y_1$, we obtain, from Eq. 24.9 with $y_1 = L\lambda/w$,

$$2y_1 = \frac{2L\lambda}{w} \quad \textit{width of central maximum} \tag{24.10}$$

Similarly, the width of the side maximum on the sides is given by

$$y_{m+1} - y_m = (m+1)\left(\frac{L\lambda}{w}\right) - m\left(\frac{L\lambda}{w}\right) = \frac{L\lambda}{w} = y_1 \tag{24.11}$$

Thus, the width of the central maximum is twice that of the side maxima.

Conceptual Example 24.3 ■ Diffraction and Radio Reception

While you drive through a city or mountainous areas, the quality of your radio reception varies sharply from place to place, with stations seeming to fade out and reappear. Could diffraction be a cause of this? Which of the following bands would you expect to be least affected by it: (a) Weather (162 MHz); (b) FM (88–108 MHz); or (c) AM (525–1610 kHz)?

Reasoning and Answer. Radio waves, like visible light, are electromagnetic waves and so tend to travel in straight lines when they are long distances from their sources. They can be blocked by objects in their path—especially if the objects are massive (such as hills and buildings).

However, because of diffraction, radio waves can also "wrap around" obstacles or "fan out" as they pass through obstacles and openings, *provided* their wavelength is at least roughly the size of the obstacle or opening. The longer the wavelength, the greater the amount of diffraction, and so the *less likely* the radio waves are to be obstructed.

To determine which band benefits most by such diffraction, we need the wavelengths that correspond to the given frequencies, as given by $c = \lambda f$. AM radio waves, with $\lambda = 186–571$ m, are the longest of the three bands (by a factor of about 100). Thus AM broadcasts are more likely to be diffracted around such objects as buildings or mountains or through the openings between them, and the answer is (c).

Follow-Up Exercise. Woodwind instruments, such as the clarinet and the flute, usually have smaller openings than brass instruments, such as the trumpet and trombone. During halftime at a football game, when a marching band faces you, you can easily hear both the woodwind instruments and the brass instruments. Yet when the band marches away from you, the brass instruments sound muted, but you can hear the woodwinds quite well. Why?

Integrated Example 24.4 ■ Width of the Central Maximum: Single-Slit Diffraction

Monochromatic light passes through a slit whose width is 0.050 mm. (a) The general spreadout of the diffraction pattern, in general, is (1) larger for longer wavelengths, (2) larger for shorter wavelengths, (3) the same for all wavelengths. Explain. (b) At what angle will the third minimum be seen and what is the width of the central maximum on a screen located 1.0 m from the slit, for $\lambda = 400$ nm and 550 nm, respectively?

(a) Conceptual Reasoning. The general size of the diffraction pattern can be characterized by the position of a particular maximum or minimum. From Eq. 24.8, it can be seen that for a given width w and order number m, the position of a minimum $\sin\theta$ is directly proportional to the wavelength λ. Therefore, a longer wavelength will correspond to a greater $\sin\theta$ or a greater θ, and the answer is (1).

(b) Quantitative Reasoning and Solution. This part is a direct application of Eq. 24.8 and Eq. 24.10.

Given: $\lambda_1 = 400\text{ nm} = 4.00 \times 10^{-7}\text{ m}$
$\lambda_2 = 550\text{ nm} = 5.50 \times 10^{-7}\text{ m}$
$w = 0.050\text{ mm} = 5.0 \times 10^{-5}\text{ m}$
$m = 3$
$L = 1.0\text{ m}$

Find: θ_3 and $2y_1$ (width of central maximum)

For $\lambda = 400$ nm:

From Eq. 24.8, we have

$$\sin\theta_3 = \frac{m\lambda}{w} = \frac{3(4.00 \times 10^{-7}\text{ m})}{5.0 \times 10^{-5}\text{ m}} = 0.024 \quad \text{so} \quad \theta_3 = \sin^{-1} 0.024 = 1.4°$$

Equation 24.10 gives

$$2y_1 = \frac{2L\lambda}{w} = \frac{2(1.0\text{ m})(4.00 \times 10^{-7}\text{ m})}{5.0 \times 10^{-5}\text{ m}} = 1.6 \times 10^{-2}\text{ m} = 1.6\text{ cm}$$

For $\lambda = 700$ nm:

$$\sin\theta_3 = \frac{m\lambda}{w} = \frac{3(5.50 \times 10^{-7}\text{ m})}{5.0 \times 10^{-5}\text{ m}} = 0.033 \quad \text{so} \quad \theta_3 = \sin^{-1} 0.033 = 1.9°$$

$$2y_1 = \frac{2L\lambda}{w} = \frac{2(1.0\text{ m})(5.50 \times 10^{-7}\text{ m})}{5.0 \times 10^{-5}\text{ m}} = 2.2 \times 10^{-2}\text{ m} = 2.2\text{ cm}$$

Follow-Up Exercise. By what factor would the width of the central maximum change if red light ($\lambda = 700$ nm) were used instead of 550 nm?

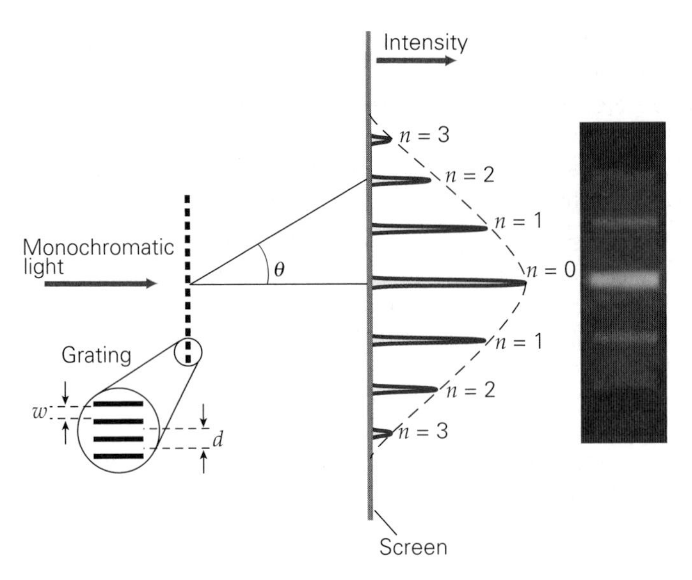

▶ **FIGURE 24.15 Diffraction grating** A diffraction grating produces a sharply defined interference/diffraction pattern. Two parameters define a grating: the slit separation d and the slit width w. The combination of multiple-slit interference and single-slit diffraction determine the intensity distribution of the various orders of maxima.

Illustration 38.2 Application of Diffraction Gratings

Diffraction Gratings

We have seen that maxima and minima result from diffraction followed by interference when monochromatic light passes through a set of double slits. As the number of slits is increased, the maxima become sharper (narrower) and the minima wider. The sharp maxima are very useful in optical analysis of light sources and other applications. ▲Fig. 24.15 shows a typical experiment with monochromatic light incident on a **diffraction grating**, which consists of large numbers of parallel, closely spaced slits. Two parameters define a diffraction grating: the slit separation between successive slits, d, and the individual slit width, w. The resulting pattern of interference and diffraction is shown in ▼Fig. 24.16.

Diffraction gratings were first made of fine strands of wire. They produce effects similar to what can be seen by viewing a candle flame through a feather held close to the eye. Better gratings have a large number of fine lines or grooves on glass or metal surfaces. If light is transmitted through a grating, it is called a *transmission grating*. However, *reflection gratings* are also common. The closely spaced grooves of a compact disc or a DVD act as a reflection grating, giving rise to their familiar iridescent sheen (▶Fig. 24.17). Commercial master gratings are made by depositing a thin film of aluminum on an optically flat surface and then removing some of the reflecting metal by cutting regularly spaced, parallel lines. Precision diffraction gratings

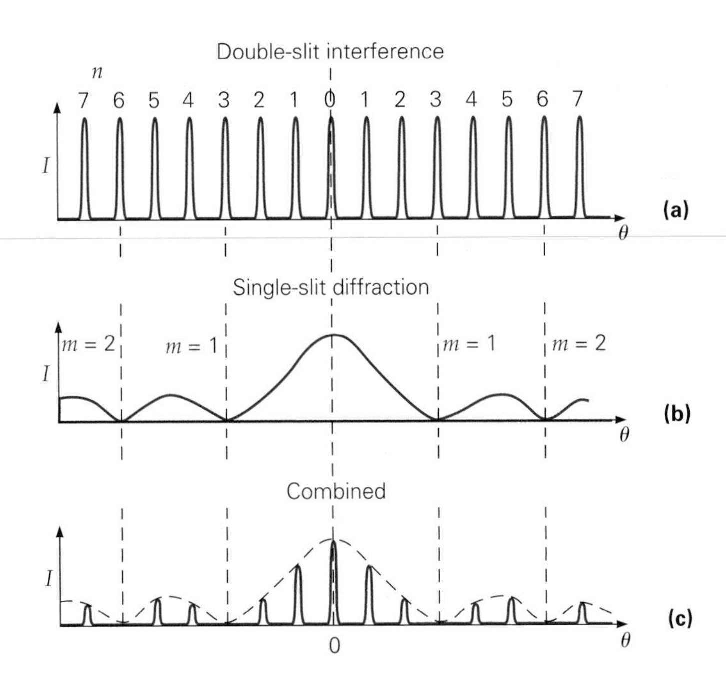

▶ **FIGURE 24.16 Intensity distribution of interference and diffraction** **(a)** Interference determines the positions of the interference maxima: $d \sin\theta = n\lambda, n = 0, 1, 2, 3, \ldots$. **(b)** Diffraction locates the positions of the diffraction minima: $w \sin\theta = m\lambda, m = 1, 2, 3, \ldots$, and the relative intensity of the maxima. **(c)** The combination (product) of interference and diffraction determine the overall intensity distribution.

are made using two coherent laser beams that intersect at an angle. The beams expose a layer of photosensitive material, which is then etched. The spacing of the grating lines is determined by the intersection angle of the beams. Precision gratings may have 30 000 lines per centimeter or more and are therefore expensive and difficult to fabricate. Most gratings used in laboratory instruments are *replica gratings*, which are plastic castings of high-precision master gratings.

It can be shown that the condition for interference maxima for a grating illuminated with monochromatic light is identical to that for a double slit. The expression is

$$d \sin \theta = n\lambda \quad \text{for } n = 0, 1, 2, 3, \ldots \quad \textit{interference maxima} \tag{24.12}$$

where n is called the *order-of-interference maximum* and θ is the angle at which that maximum occurs for a particular wavelength. The zeroth-order maximum is coincident with the central maximum. The spacing between adjacent slits (d) is obtained from the number of lines or slits per unit length of the grating: $d = 1/N$. For example, if $N = 5000$ lines/cm, then

$$d = \frac{1}{N} = \frac{1}{5000/\text{cm}} = 2.0 \times 10^{-4}\ \text{cm}$$

If the light incident on a grating is white light (polychromatic), then the maxima are multicolored (▼Fig. 24.18a). There is no deviation of the components of the light for the zeroth order ($\sin \theta = 0$ for all wavelengths), so the central maximum is white. However, the colors separate for higher orders, since the position of the maximum depends on wavelength (Eq. 24.12). With the longer wavelength having a larger θ, this produces a spectrum. Note that it is possible for higher orders produced by a diffraction grating to overlap. That is, the angles for different orders may be the same for two different wavelengths.

Only a limited number of spectral orders can be obtained using a diffraction grating. The number depends on the wavelength of the light and on the grating's spacing (d). From Eq. 24.12, because θ cannot exceed 90° (that is, $\sin \theta \leq 1$), we have

$$\sin \theta = \frac{n\lambda}{d} \leq 1 \quad \text{or} \quad n_{\text{max}} \leq \frac{d}{\lambda}$$

Diffraction gratings have almost completely replaced prisms in spectroscopy. The creation of a spectrum and the measurement of wavelengths by a grating depend only on geometrical measurements such as lengths and/or angles. Wavelength determination using a prism, in contrast, depends on the dispersive characteristics of the material of which the prism is made. Thus, it is crucial to know precisely how the index of refraction depends on the wavelength of light. In contrast to a prism, which deviates red light least and violet light most, a diffraction grating produces the smallest angle for violet light

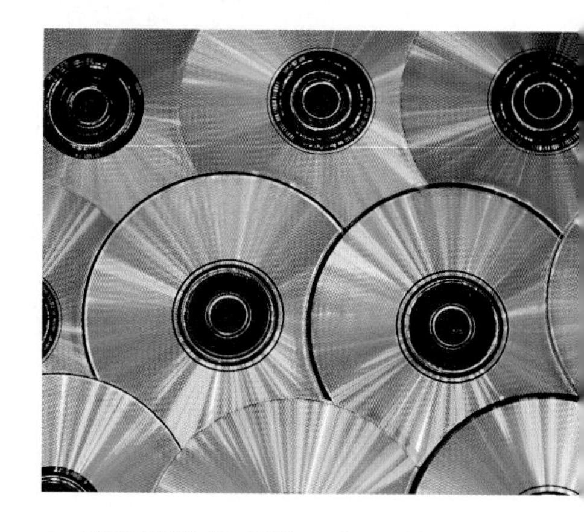

▲ **FIGURE 24.17 Diffraction effects** The narrow grooves of compact discs (CDs) act as reflection diffraction gratings, producing colorful displays.

Note: d is the distance between adjacent slits.

Exploration 38.2 Diffraction Grating

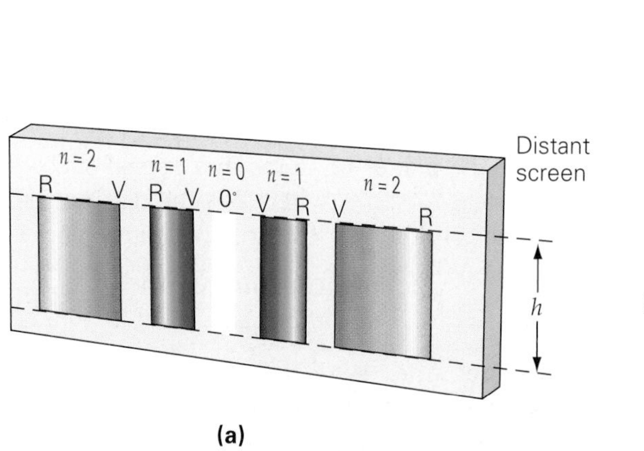

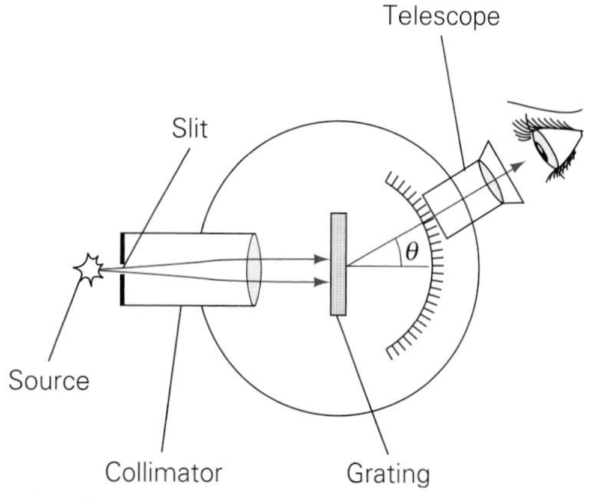

▼ **FIGURE 24.18 Spectroscopy** **(a)** In each side maximum, components of different wavelengths (R = red and V = violet) are separated, because the deviation depends on wavelength: $\theta = \sin^{-1}(n\lambda/d)$. **(b)** As a result, gratings are used in spectrometers to determine the wavelengths present in a beam of light by measuring their angles of diffraction and to separate the various wavelengths for further analysis.

(short λ) and the greatest for red light (long λ). Notice that a prism disperses white light into a single spectrum. A diffraction grating, however, produces a number of spectra, one for each order other than $n = 0$, and the higher the order, the more spread out.

The sharp spectra produced by gratings are used in instruments called *spectrometers* (Fig. 24.18b). Using a spectrometer, materials can be illuminated with light of various wavelengths to find which wavelengths are strongly transmitted or reflected. Their absorption can then be measured and material characteristics determined.

Example 24.5 ■ A Diffraction Grating: Line Spacing and Spectral Orders

A particular diffraction grating produces an $n = 2$ spectral order at an angle of 30° for light with a wavelength of 500 nm. (a) How many lines per centimeter does the grating have? (b) At what angle can the $n = 3$ spectral order be seen?

Thinking It Through. (a) To find the number of lines per centimeter (N) the grating has, we need to know the grating spacing (d), since $N = 1/d$. With the given data, we can find d from Eq. 24.12. (b) Using Eq. 24.12 again, we can find θ for $n = 3$.

Solution.

Given: $\lambda = 500 \text{ nm} = 5.00 \times 10^{-7}\text{ m}$
$n = 2$
$\theta = 30°$ for $n = 2$

Find: (a) N (lines/cm)
(b) θ for $n = 3$

(a) Using Eq. 24.12, we get the grating spacing:

$$d = \frac{n\lambda}{\sin\theta} = \frac{2(5.00 \times 10^{-7}\text{ m})}{\sin 30°} = 2.00 \times 10^{-6}\text{ m} = 2.00 \times 10^{-4}\text{ cm}$$

Then

$$N = \frac{1}{d} = \frac{1}{2.00 \times 10^{-4}\text{ cm}} = 5000 \text{ lines/cm}$$

(b)

$$\sin\theta = \frac{n\lambda}{d} = \frac{3(5.00 \times 10^{-7}\text{ m})}{2.00 \times 10^{-6}\text{ m}} = 0.75$$

so

$$\theta = \sin^{-1} 0.75 = 48.6°$$

Follow-Up Exercise. If white light of wavelength ranging from 400 to 700 nm were used, what would be the angular width of the spectrum for the second order?

X-Ray Diffraction

In principle, the wavelength of any electromagnetic wave can be determined by using a diffraction grating with the appropriate spacing. Diffraction was used to determine the wavelengths of X-rays early in the twentieth century. Experimental evidence indicated that the wavelengths of X-rays were probably around 10^{-10} m or 0.1 nm, but it is impossible to construct a diffraction grating with this spacing. Around 1913, Max von Laue (1879–1960), a German physicist, suggested that the regular spacing of the atoms in a crystalline solid might make the crystal act as a diffraction grating for X-rays, since the atomic spacing is on the order of 0.1 nm (▸Fig. 24.19). When X-rays were directed at crystals, diffraction patterns were indeed observed. (See Fig. 24.19b.)

Figure 24.19a illustrates diffraction by the planes of atoms in a crystal such as sodium chloride. The path-length difference is $2d \sin\theta$, where d is the distance between the crystal's internal planes. Thus, the condition for constructive interference is

$$2d\sin\theta = n\lambda \qquad \text{for } n = 1, 2, 3, \ldots \qquad \textit{constructive interference X-ray diffraction} \qquad (24.13)$$

This relationship is known as **Bragg's law**, after W. L. Bragg (1890–1971), the British physicist who first derived it. Note that θ is *not* measured from the normal, as is the convention in optics.

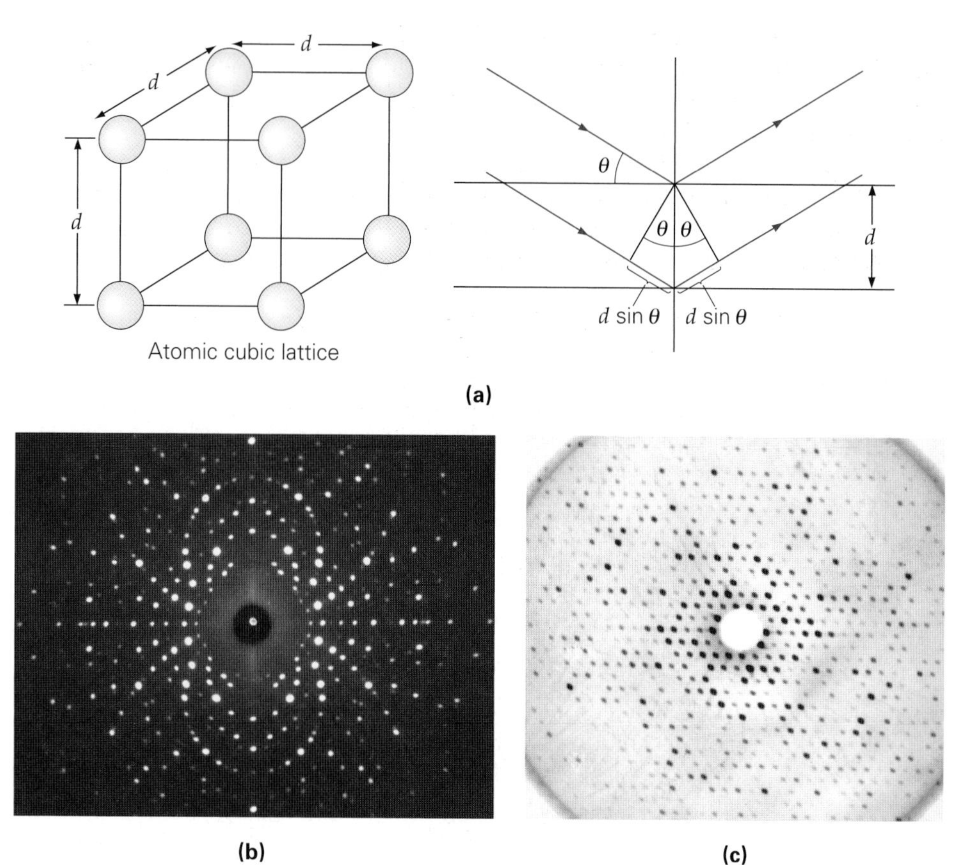

◀ **FIGURE 24.19** **Crystal diffraction** **(a)** The array of atoms in a crystal-lattice structure acts as a diffraction grating, and X-rays are diffracted from the planes of atoms. With a lattice spacing of d, the path-length difference for the X-rays diffracted from adjacent planes is $2d \sin \theta$. **(b)** X-ray diffraction pattern of a crystal of potassium sulfate. By analyzing the geometry of such patterns, investigators can deduce the structure of the crystal and the position of its various atoms. **(c)** X-ray diffraction pattern of the protein hemoglobin, which carries oxygen in blood.

X-ray diffraction is now routinely used to investigate the internal structure not only of simple crystals, but also of large, complex biological molecules such as proteins and DNA (Fig. 24.19c). Because of their short wavelengths, which are comparable with interatomic distances *within* a molecule, X-rays provide a method for investigating atomic structures within molecules.

24.4 Polarization

OBJECTIVES: To (a) explain light polarization, and (b) give examples of polarization, both in the environment and in commercial applications.

When you think of polarized light, you may visualize polarizing (or Polaroid) sunglasses, since this is one of the more common applications of polarization. When something is polarized, it has a preferential direction, or orientation. In terms of light waves, **polarization** refers to the orientation of the transverse wave (electric field) oscillations.

Recall from Chapter 20 that light is an electromagnetic wave with oscillating electric and magnetic field vectors ($\vec{E}$ and $\vec{B}$, respectively) perpendicular (transverse) to the direction of propagation. Light from most sources consists of a large number of electromagnetic waves emitted by the atoms of the source. Each atom produces a wave with a particular orientation, corresponding to the direction of the atomic vibration. However, since electromagnetic waves from a typical source are produced by many atoms, many random orientations of the $\vec{E}$ and $\vec{B}$ fields are in the emitted composite light. When the field vectors are randomly oriented, the light is said to be *unpolarized*. This situation is commonly represented schematically in terms of the electric field vector as shown in ▼Fig. 24.20a. As viewed along the direction of propagation, the electric field is equally distributed in all directions. However, as viewed parallel to the direction of propagation, this random or equal distribution can be represented by two directions (such as the x- and y-directions in a two-dimensional coordinate system). Here, the vertical arrows denote the electric

Note: In many figures, dots represent a direction perpendicular to the paper of the electric field and arrows denote a direction along the arrow of the electric field.

Illustration 39.2 Polarized Electromagnetic Waves

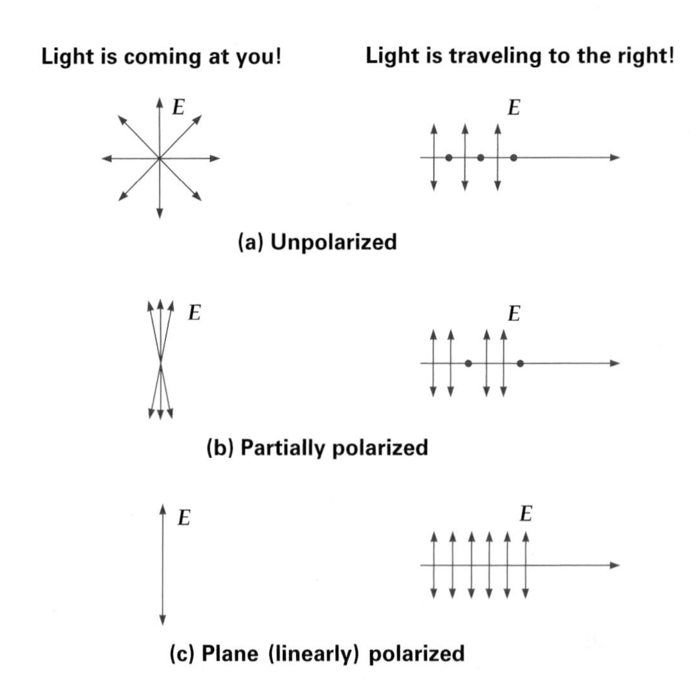

▶ **FIGURE 24.20** Polarization Polarization is represented by the orientation of the plane of vibration of the electric field vectors. **(a)** When the vectors are randomly oriented, the light is unpolarized. The dots represent a direction perpendicular to the paper of the electric field, and the vertical arrows denote an up-and-down direction of the electric field. Equal numbers of dots and arrows are used to represent unpolarized light. **(b)** With preferential orientation of the field vectors, the light is partially polarized. Here, there are fewer dots than arrows. **(c)** When the vectors are in one plane, the light is plane polarized, or linearly polarized. No dots are seen here.

field components in that direction, and the dots represent the component going in and out of the paper. This notation will be used throughout this section.

If there is some preferential orientation of the field vectors, the light is said to be *partially polarized*. Both representations in Fig. 24.20b show that there are more electric field vectors in the vertical direction than in the horizontal direction. If the field vectors oscillate in only *one* plane the light is *plane polarized* or *linearly polarized* (Fig. 24.20c). Note that polarization is evidence that light is a transverse wave. True longitudinal waves, such as sound waves, cannot be polarized, because the molecules of the media do not vibrate perpendicular to the direction of propagation.

Light can be polarized in many ways. However, polarization by selective absorption, reflection, and double refraction will be discussed here. Polarization by scattering will be considered in Section 24.5.

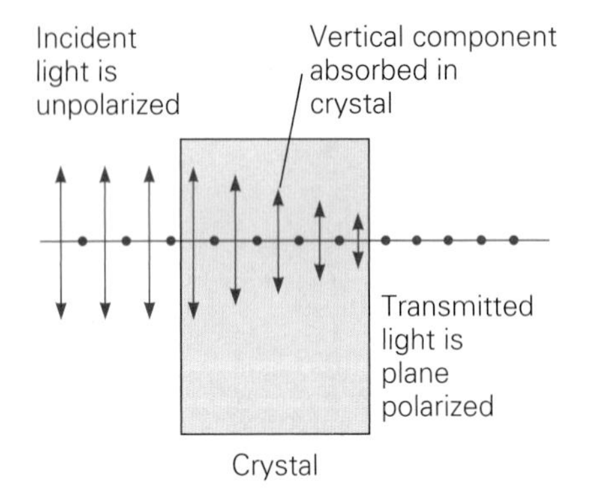

▲ **FIGURE 24.21** Selective absorption (dichroism) Dichroic crystals selectively absorb one polarized component (the vertical component) more than the other. If the crystal is thick enough, the emerging beam is linearly polarized.

Polarization by Selective Absorption (Dichroism)

Some crystals, such as those of the mineral tourmaline, exhibit the interesting property of absorbing one of the electric field components more than the other. This property is called **dichroism**. If a dichroic crystal is sufficiently thick, the more strongly absorbed component may be completely absorbed. In that case, the emerging beam is plane polarized (◂Fig. 24.21).

Another dichroic crystal is quinine sulfide periodide (commonly called *herapathite*, after W. Herapath, an English physician who discovered its polarizing properties in 1852). This crystal was of great practical importance in the development of modern polarizers. Around 1930, Edwin H. Land (1909–1991), an American scientist, found a way to align tiny, needle-shaped dichroic crystals in sheets of transparent celluloid. The result was a thin sheet of polarizing material that was given the commercial name *Polaroid*.

Better polarizing films have been developed that use synthetic polymer materials instead of celluloid. During the manufacturing process, this kind of film is stretched to align the long molecular chains of the polymer. With proper treatment, the outer (valence) electrons of the molecules can move along the oriented chains. As a result, light with $\vec{\mathbf{E}}$ vectors parallel to the oriented chains is readily absorbed, but light with $\vec{\mathbf{E}}$ vectors perpendicular to the chains is transmitted. The direction *perpendicular* to the orientation of the molecular chains is called the **transmission axis**, or the **polarization direction**. Thus, when unpolarized light falls on a polarizing sheet, the sheet acts as a polarizer and transmits polarized light (▸Fig. 24.22). Since one of the two electric field components is absorbed, the

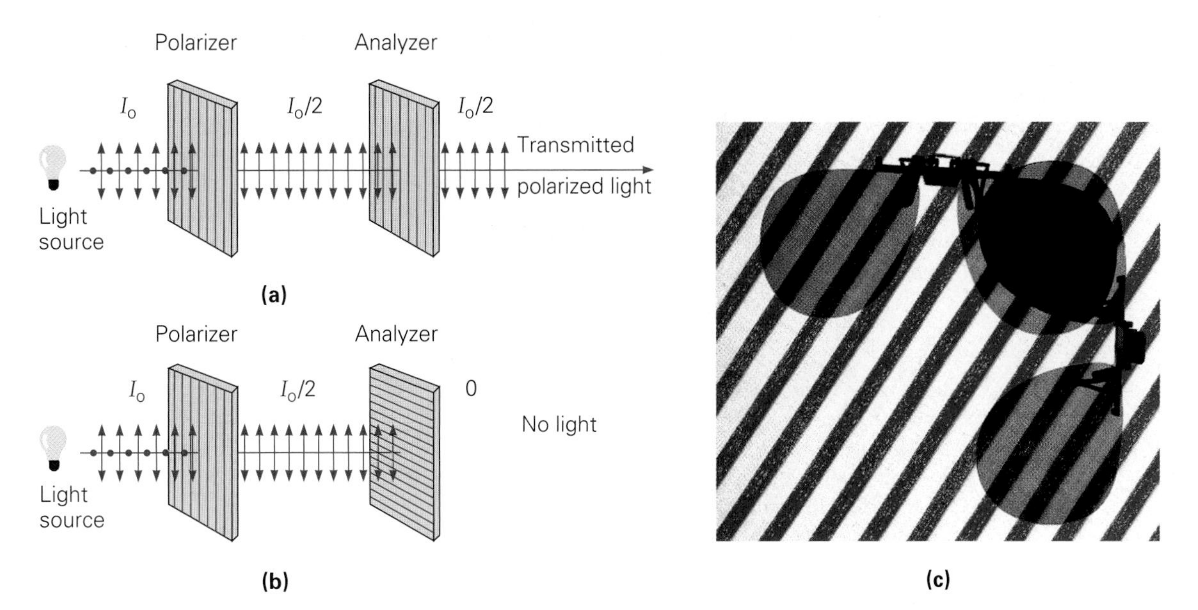

▲ **FIGURE 24.22 Polarizing sheets (a)** When polarizing sheets are oriented so that their transmission axes are the same, the emerging light is polarized. The first sheet acts as a polarizer, and the second acts as an analyzer. **(b)** When one of the sheets is rotated 90° and the transmission axes are perpendicular (crossed polarizers), little light (ideally, none) is transmitted. **(c)** Crossed polarizers made using polarizing sunglasses.

light intensity after the polarizer is half of the intensity incident on it ($I_o/2$). The human eye cannot distinguish between polarized and unpolarized light. To tell whether light is polarized, we must use an *analyzer*, which can simply be another sheet of polarizing film. As shown in Fig. 24.22a, if the transmission axis of an analyzer is parallel to the plane of polarization of polarized light, there is maximum transmission. If the transmission axis of the analyzer is perpendicular to the plane of polarization, little light (ideally, none) will be transmitted.

In general, the intensity of the transmitted light is given by

$$I = I_o \cos^2 \theta \qquad \textit{Malus's law} \qquad (24.14)$$

where θ is the angle between the transmission axes of the polarizer and analyzer. This expression is known as *Malus's law*, after its discoverer, French physicist E. L. Malus (1775–1812).

Polarizing glasses whose lenses have different transmission axes are used to view some 3D movies. The pictures are projected on the screen by two projectors that transmit slightly different images, photographed by two cameras a short distance apart. The projected light from each projector is linearly polarized, but in mutually perpendicular directions. The lenses of the 3D glasses also have transmission axes that are perpendicular. Thus, one eye sees the image from one projector, and the other eye sees the image from the other projector. The brain receives a slight difference in perspective (or "viewing angle") from the two images, and interprets the image as having depth, or a third dimension, just as in normal vision.

Integrated Example 24.6 ■ Make Something Out of Nothing: Three Polarizers

In Figs. 24.22b and c, no light is transmitted through the analyzer, because the transmission axes of the polarizer and analyzer are perpendicular. Assume that the unpolarized light incident on the first polarizer has an intensity of I_o. A second polarizer is now inserted between the first polarizer and analyzer, and the transmission axis of the second polarizer makes an angle of θ with the first polarizer. (a) Is it possible for some light to go through this arrangement? If yes, does it occur at (1) $\theta = 0°$, (2) $\theta = 30°$, (3) $\theta = 45°$, or (4) $\theta = 90°$? Explain. What happens if the second polarizer is rotated? (b) When $\theta = 30°$, what is the intensity of the transmitted light in terms of the incident intensity?

Exploration 39.2 Polarizers

(continues on next page)

(a) Conceptual Reasoning. Yes, it is possible for some light to go through this arrangement at any angle other than 0° or 90°. The accompanying Learn by Drawing can help us understand this situation.

With just the first polarizer and the analyzer, no light is transmitted, according to Malus's law (Eq. 24.14), because the angle between the transmission axes is 90°. However, when a second polarizer is inserted in between the first polarizer and the analyzer, some light can actually pass through the system. For example, if the transmission axis of the second polarizer makes an angle of θ with that of the first polarizer, then the angle between the transmission axes of the second polarizer and the analyzer will be $90° - \theta$. (Why?)

When unpolarized light of intensity I_o is incident on the first polarizer, the transmitted light after the first polarizer is $I_o/2$, because only one of the two electric field components is transmitted. After the second polarizer, the intensity is decreased by a factor of $\cos^2\theta$. After the analyzer, the intensity is decreased further by a factor of $\cos^2(90° - \theta) = \sin^2\theta$. So the transmitted intensity is $I = (I_o/2)(\cos^2\theta)(\sin^2\theta)$. Therefore, as long as θ is not 0° or 90°, some light will be transmitted through the system.

Since the transmitted light depends on the angle θ, rotating the second polarizer will change the intensity transmitted.

(b) Quantitative Reasoning and Solution. Once this situation is understood, part (b) is a straightforward calculation.

Given: $\theta = 30°$ *Find:* (b) I after three polarizers in terms of I_o

When $\theta = 30°$, $I = \frac{I_o}{2}(\cos^2 30°)(\sin^2 30°) = \frac{I_o}{2}\cdot\left(\frac{\sqrt{3}}{2}\right)^2\cdot\left(\frac{1}{2}\right)^2 = \frac{3I_o}{32}$

Follow-Up Exercise. For what value of θ will the transmitted intensity be a maximum in this example?

Polarization by Reflection

When a beam of unpolarized light strikes a smooth, transparent medium such as glass, the beam is partially reflected and partially transmitted. The reflected light may be completely polarized, partially polarized, or unpolarized, depending on the angle of incidence. The unpolarized case occurs for 0°, or normal incidence. As the angle of incidence is changed from 0°, both the reflected and refracted light become partially polarized. For example, the electric field components normal to the surface are reflected more strongly, producing partial polarization (▸Fig. 24.23a).

LEARN BY DRAWING **THREE POLARIZERS (SEE INTEGRATED EXAMPLE 24.6.)**

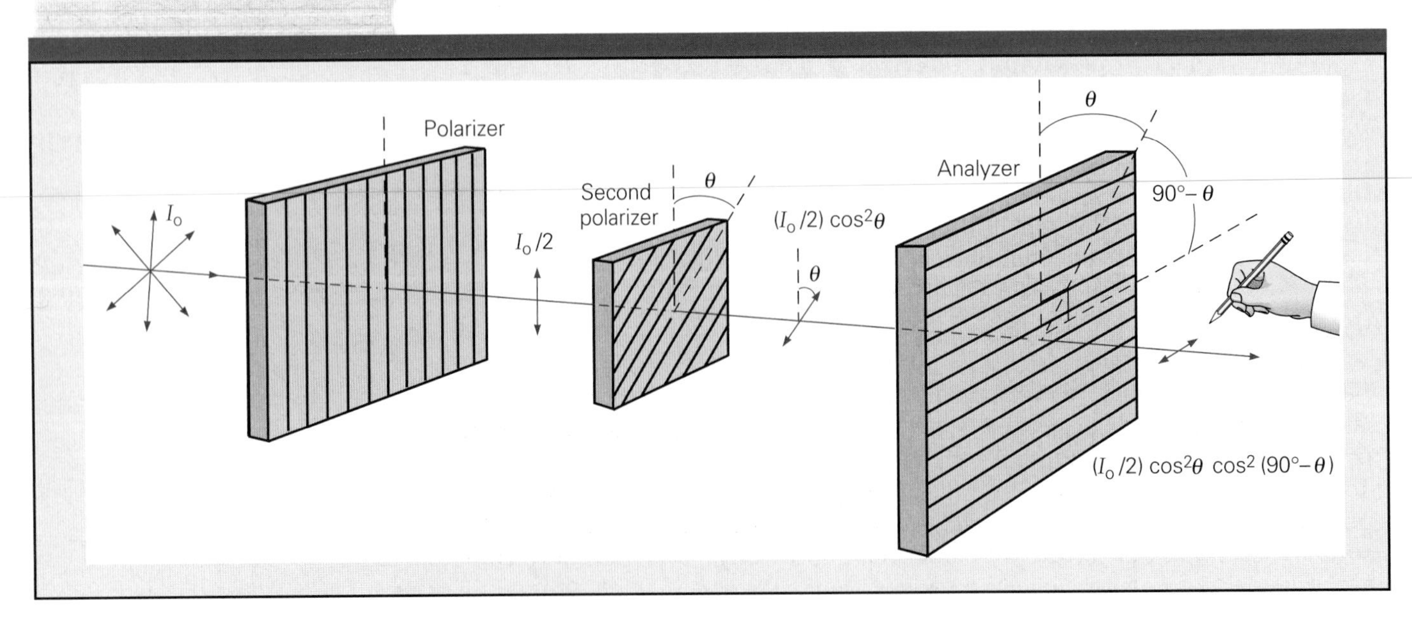

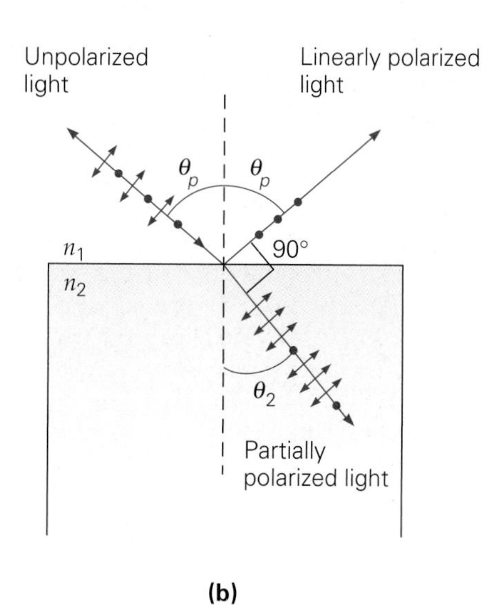

▲ **FIGURE 24.23 Polarization by reflection** **(a)** When a beam of light is incident on a boundary, the reflected and refracted beams are normally partially polarized. **(b)** When the reflected and refracted beams are 90° apart, the reflected beam is linearly polarized, and the refracted beam is partially polarized. This situation occurs when $\theta_1 = \theta_p = \tan^{-1}\left(\frac{n_2}{n_1}\right)$.

However, at one particular angle of incidence, the reflected beam is completely polarized on reflection (Fig. 24.23b). (At this angle, though, the refracted beam is still only partially polarized.)

David Brewster (1781–1868), a Scottish physicist, found that the complete polarization of the reflected beam occurs when the reflected and refracted beams are perpendicular. The incident angle at which complete polarization occurs is the **polarizing angle (θ_p)**, or the **Brewster angle**, and it depends on the indices of refraction of the two media. In Fig. 24.23b, the reflected and refracted beams are at 90° and the incident angle θ_1 is thus the polarizing angle θ_p: $\theta_1 = \theta_p$, so

$$\theta_1 + 90^\circ + \theta_2 = 180^\circ \quad \text{or} \quad \theta_2 = 90^\circ - \theta_1$$

By Snell's law (Chapter 22),

$$n_1 \sin\theta_1 = n_2 \sin\theta_2$$

Hence, we have $\sin\theta_2 = \sin(90^\circ - \theta_1) = \cos\theta_1$. Therefore,

$$\frac{\sin\theta_1}{\sin\theta_2} = \frac{\sin\theta_1}{\cos\theta_1} = \tan\theta_1 = \frac{n_2}{n_1}$$

With $\theta_1 = \theta_p$, we get

$$\tan\theta_p = \frac{n_2}{n_1} \quad \text{or} \quad \theta_p = \tan^{-1}\left(\frac{n_2}{n_1}\right) \tag{24.15}$$

If the first medium is air ($n_1 = 1$), then $\tan\theta_p = \frac{n_2}{1} = n_2 = n$, where n is the index of refraction of the second medium.

Now you can understand the principle behind polarizing sunglasses. Light reflected from a smooth surface is partially polarized. The direction of polarization is mostly parallel to the surface. (See Fig. 24.23b.) Light reflected from the surface of a road or water can be so intense that it gives rise to visual glare (▼Fig. 24.24a). To reduce this effect, the polarizing lenses of glasses are oriented with their transmission axes vertical so that some of the partially polarized light from reflective surfaces is absorbed. Polarizing filters also enable cameras to take "clean" pictures without interference from glare (Fig. 24.24b).

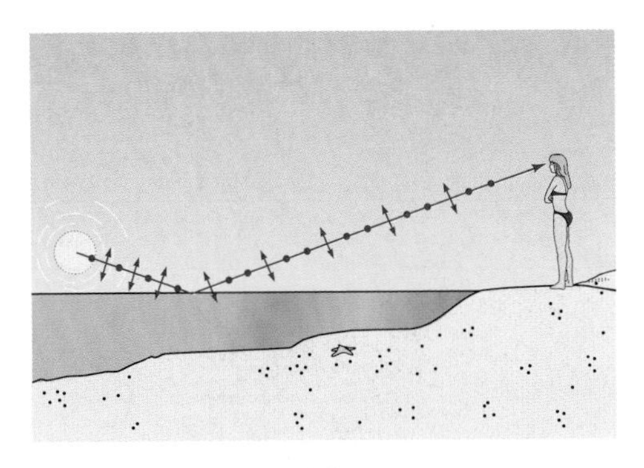

(a)

(b)

▲ **FIGURE 24.24 Glare reduction (a)** Light reflected from a horizontal surface is partially polarized in the horizontal plane. When sunglasses are oriented so that their transmission axis is vertical, the horizontally polarized component of such light is not transmitted, so glare is reduced. **(b)** Polarizing filters for cameras use the same principle. The photo at right was taken with such a filter. Note the reduction in reflections from the store window.

Example 24.7 ■ Sunlight on a Pond: Polarization by Reflection

Sunlight is reflected from the smooth surface of a pond. What is the Sun's altitude (the angle between the Sun and the horizon) when the polarization of the reflected light is greatest?

Thinking It Through. Since the angle of incidence is measured from the normal and the altitude angle is measured from the horizon, the angle of incidence is the angle complementary to the altitude angle (draw a sketch to help yourself visualize this situation). Incident light at the Brewster angle has the greatest polarization upon reflection, so the Sun needs to be at $90° - \theta_p$ from the horizon.

Solution. The index of refraction of water is listed in Table 22.1.

Given: $n_1 = 1$
$n_2 = 1.33$ (Table 22.1)

Find: θ (altitude angle for greatest polarization)

The Sun needs to be at an angle of $\theta = 90° - \theta_p$, where θ_p is the Brewster angle. Using Eq. 24.15, we find that

$$\theta_p = \tan^{-1}\left(\frac{n_2}{n_1}\right) = \tan^{-1}\left(\frac{1.33}{1}\right) = 53.1°$$

So

$$\theta = 90° - \theta_p = 90° - 53.1° = 36.9°$$

Follow-Up Exercise. Light is incident on a flat, transparent material with an index of refraction of 1.52. At what angle of refraction would the transmitted light have the greatest polarization if the transparent material is in water?

Polarization by Double Refraction (Birefringence)

When monochromatic light travels through glass, its speed is the same in all directions and is characterized by a single index of refraction. Any material that has such a property is said to be *isotropic*, meaning that it has the same optical charac-

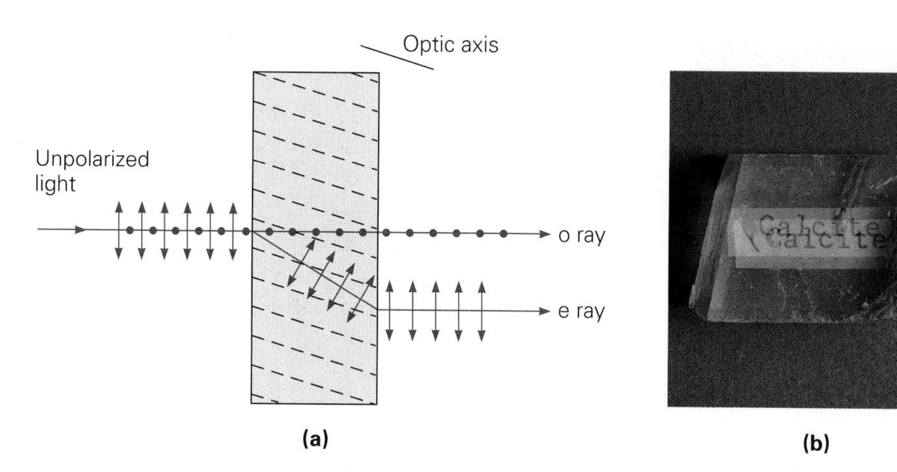

▲ **FIGURE 24.25 Double refraction or birefringence** **(a)** Unpolarized light incident normal to the surface of a birefringent crystal and at an angle to a particular direction in the crystal (dashed lines) is separated into two components. The ordinary (o) ray and the extraordinary (e) ray are plane polarized in mutually perpendicular directions. **(b)** Double refraction seen through a calcite crystal.

teristics in all directions. Some crystalline materials, such as quartz, calcite, and ice, are *anisotropic*; that is, the speed of light, and therefore the index of refraction, is different for different directions within the material. Anisotropy gives rise to some interesting optical properties. These anisotropic materials are said to be doubly refracting, or to exhibit **birefringence**, and polarization is involved.

For example, a beam of unpolarized light incident on a birefringent crystal of calcite ($CaCO_3$, calcium carbonate) is illustrated in ▲Fig. 24.25. When the beam propagates at an angle to a particular crystal axis, the beam is doubly refracted and separated into two components, or rays, upon refraction. These two rays are linearly polarized in mutually perpendicular directions. One ray, called the *ordinary* (o) *ray*, passes straight through the crystal and is characterized by an index of refraction n_o. The second ray, called the *extraordinary* (e) *ray*, is refracted and is characterized by an index of refraction n_e. The particular axis direction indicated by dashed lines in Fig. 24.25a is called the *optic axis*. Along this direction, $n_o = n_e$, and nothing extraordinary is noted about the transmitted light.

Some transparent materials have the ability to *rotate* the plane of polarization of linearly polarized light. This property, called **optical activity**, is due to the molecular structure of the material (▼Fig. 24.26a). The rotation may be clockwise or

▼ **FIGURE 24.26 Optical activity and stress detection** **(a)** Some substances have the property of rotating the polarization direction of linearly polarized light. This ability, which depends on the molecular structure of the substance, is called *optical activity*. **(b)** Glasses and plastics become optically active under stress, and the points of greatest stress are apparent when the material is viewed through crossed polarizers. Engineers can thus test plastic models of structural elements to see where the greatest stresses will occur when the models are "loaded." Here, a model of a suspension-bridge strut is being analyzed.

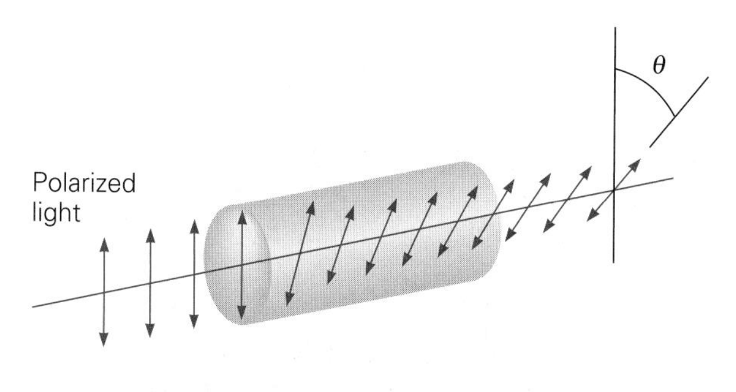

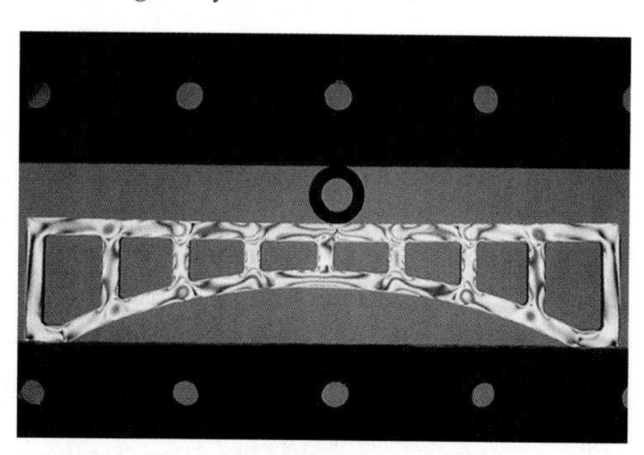

counterclockwise, depending on the molecular orientation. Optically active molecules include those of certain proteins, amino acids, and sugars.

Glasses and plastics become optically active under stress. The greatest rotation of the direction of polarization of the transmitted light occurs in the regions where the stress is the greatest. Viewing the stressed piece of material through crossed polarizers allows the points of greatest stress to be identified. This determination is called *optical stress analysis* (Fig. 24.26b). Another use of polarizing films, the liquid crystal display (LCD), is described in accompanying Insight 24.2, LCDs and Polarized Light.

*24.5 Atmospheric Scattering of Light

OBJECTIVES: To (a) discuss scattering, and (b) explain why the sky is blue and sunsets are red.

When light is incident on a suspension of particles, such as the molecules in air, some of the light may be absorbed and reradiated in all directions. This process is called *scattering*. The scattering of sunlight in the atmosphere produces some interesting effects, including the polarization of skylight (that is, sunlight that has been scattered by the atmosphere), the blueness of the sky, and the redness of sunsets and sunrises.

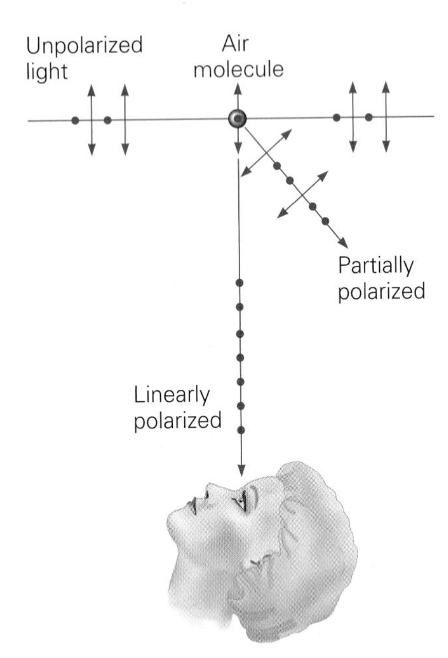

▲ **FIGURE 24.27 Polarization by scattering** When incident unpolarized sunlight is scattered by a gas molecule in the air, the light perpendicular to the direction of the incident ray is linearly polarized. Light scattered at some arbitrary angle is partially polarized. An observer at a right angle (90°) to the direction of the incident sunlight receives linearly polarized light.

Atmospheric scattering causes the skylight to be polarized. When unpolarized sunlight is incident on air molecules, the electric field of the light wave sets electrons of the molecules into vibration. The vibrations are complex, but these accelerated charges emit radiation, like the vibrating electrons in the antenna of a radio broadcast station (see Section 20.4). The intensity of this emitted radiation is strongest along a line perpendicular to the oscillation, and, as illustrated in ◂Fig. 24.27, an observer viewing from an angle of 90° with respect to the direction of the sunlight will receive linearly polarized light, because of the charge oscillations normal to the surface. At other viewing angles, both components are present, and skylight seen through a polarizing filter appears partially polarized, because of the stronger component.

Since the scattering of light with the greatest degree of polarization occurs at a right angle to the direction of the Sun, at sunrise and sunset the scattered light from directly overhead has the greatest degree of polarization. The polarization of skylight can be observed by viewing the sky through a polarizing filter (or a polarizing sunglass lens) and rotating the filter. Light from different regions of the sky will be transmitted in different degrees, depending on its degree of polarization. It is believed that some insects, such as bees, use polarized skylight to determine navigational directions relative to the Sun.

Why the Sky Is Blue

The scattering of sunlight by air molecules is the reason why the sky looks blue. This effect is not due to polarization, but is caused by the selective absorption of light. As oscillators, air molecules have resonant frequencies (at which they scatter most efficiently) in the blue-violet region. Consequently, when sunlight is scattered, the blue end of the visible spectrum is scattered more than in the red end.

For particles such as air molecules, which are much smaller than the wavelength of light, the intensity of the scattered light is inversely proportional to the wavelength to the fourth power ($1/\lambda^4$). This relationship between wavelength and scattering intensity is called **Rayleigh scattering**, after Lord Rayleigh (1842–1919), a British physicist who derived it. This inverse relationship predicts that light of the shorter-wavelength, or blue, end of the spectrum will be scattered much more than light of the longer-wavelength, or red, end. The scattered blue light is rescattered in the atmosphere and eventually is directed toward the ground. This is why the sky appears blue.

INSIGHT 24.2 LCDS AND POLARIZED LIGHT

Today, *liquid crystal displays* (LCDs) are commonplace in items such as watches, calculators, televisions, and computer screens. The name "liquid crystal" may seem self-contradictory. Normally, when a crystalline solid melts, the resulting liquid no longer has an orderly atomic or molecular arrangement. Some organic compounds, however, pass through an intermediate state in which the molecules may rearrange somewhat but still maintain the overall order that is characteristic of a crystal.

A common type of LCD, called a *twisted-nematic display*, makes use of the effect of a liquid crystal on polarized light (Fig. 1). These special liquid crystals are optically active and will rotate the direction of polarization of light by 90° if no voltage is applied across them. However, if voltage is applied, the crystals will lose this optical activity.

The liquid crystals are then placed between crossed polarizing sheets and backed with a mirrored surface. With voltage off, light entering and passing through the LCD is polarized, rotated 90°, reflected, and again rotated 90°. After the return trip through the liquid crystal, the direction of polarization of the light is the same as that of the initial polarizer. Thus, the light is transmitted and leaves the display unit. Because of the reflection and transmission, the display appears to be of a light color (usually light gray) when illuminated with white, unpolarized light.

With voltage on, the polarized light passing through the liquid crystal is absorbed by the second polarizer. Thus, the liquid crystal is opaque and appears dark. Transparent, electrically conductive film coatings arranged in a seven-block pattern are applied to the liquid crystal. Each block, or display segment, has a separate electrical connection. The dark numbers or letters on an LCD are formed by applying an electric voltage to certain blocks of the liquid crystal. Note that all the numerals 0 through 9 can be formed out of pieces of the segmented display.

By using an analyzer, we can readily show that the light from the LCD is polarized (Fig. 2). You can either see or not see the display by rotating the analyzer over the watch. You may have noticed this effect if, while wearing polarizing sunglasses, you have ever tried to see the time on the LCD of a wristwatch.

One of the major advantages of LCDs is their low power consumption. Similar displays, such as those using light-emitting diodes (LEDs), produce light themselves, using relatively large amounts of power. LCDs produce no light but instead use reflected light.

Color flat-panel computer and TV screens, which rely on LCD technology, are popular today. They are about one quarter the size, consume less than half the energy, and are easier on the eyes than CRT monitors and traditional TV screens of the same size. Computer displays and TVs are usually specified in *pixels*, much like the smallest square on graph paper. To produce color, three LCD segments (red, green, and blue) are grouped on each pixel. By controlling the intensities of the three colors, each pixel can generate every color in the visible spectrum.

Mirror Polarizer Liquid crystal Polarizer
Incident light
Reflected light
Voltage off: optical activity allowed
Incident light
No reflected light
Voltage on: no optical activity
Voltage off: no display appears
Voltage on: affected segments darken, number appears

FIGURE 1 Liquid crystal display (LCD) A twisted-nematic display is an application involving the optical activity of a liquid crystal and crossed polarizing sheets. When the crystalline order is disoriented by an electric field from an applied voltage, the liquid crystal loses its optical activity in that region, and light is not transmitted or reflected. Numerals and letters are formed by applying voltages to segments of a block display.

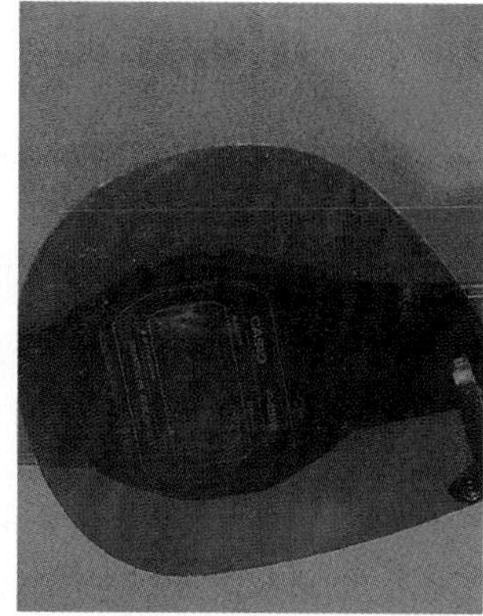

FIGURE 2 Polarized light The light from an LCD is polarized, as can be shown by using polarizing sunglasses as an analyzer.

Example 24.8 ■ The Red and the Blue: Rayleigh Scattering

How much more is light at the blue end of the visible spectrum scattered by air molecules than is light at the red end?

Thinking It Through. We know that Rayleigh scattering is proportional to $1/\lambda^4$ and that light from the blue end of the spectrum (shorter wavelength) is scattered more than light from the red end. The wording "how much more" implies a factor or ratio.

Solution. The Rayleigh scattering relationship is $I \propto 1/\lambda^4$, where I is the amount, or intensity, of scattering for a particular wavelength. Thus, you can form the ratio

$$\frac{I_{\text{blue}}}{I_{\text{red}}} = \left(\frac{\lambda_{\text{red}}}{\lambda_{\text{blue}}}\right)^4$$

The blue end of the spectrum (violet light) has a wavelength of about $\lambda_{\text{blue}} = 400$ nm, and red light has a wavelength of about $\lambda_{\text{red}} = 700$ nm. Inserting these values gives

$$\frac{I_{\text{blue}}}{I_{\text{red}}} = \left(\frac{\lambda_{\text{red}}}{\lambda_{\text{blue}}}\right)^4 = \left(\frac{700\text{ nm}}{400\text{ nm}}\right)^4 = 9.4 \quad \text{or} \quad I_{\text{blue}} = 9.4 I_{\text{red}}$$

Thus, blue light is scattered almost 10 times as much as red light.

Follow-Up Exercise. What wavelength of light is scattered twice as much as red light? What color light is this?

Why Sunsets and Sunrises Are Red

Beautiful red sunsets and sunrises are sometimes observed. When the Sun is near the horizon, sunlight travels a greater distance through the denser air near the Earth's surface. Since the light therefore undergoes a great deal of scattering, you might think that only the least scattered light, the red light, would reach observers on the Earth's surface. This would explain red sunsets. However, it has been shown that the dominant color of white light after only molecular scattering is orange. Thus, other types of scattering must shift the light from the setting (or rising) Sun toward the red end of the spectrum (◂Fig. 24.28).

▲ **FIGURE 24.28** Red sky at night A spectacular red sunset over a mountaintop observatory in Chile. The red sky results from the scattering of sunlight by atmospheric gases and small solid particles. A directly observed reddening Sun is due to the scattering of the wavelengths toward the blue end of the spectrum in the direct line of sight.

Red sunsets have been found to result from the scattering of sunlight by atmospheric gases *and* by small dust particles. These particles are not necessary for the blueness of the sky, but are compulsory for deep-red sunsets and sunrises. (This is why spectacular red sunsets are observed in the months after a large volcanic eruption that can put tons of particulate matter into the atmosphere.) Red sunsets occur most often when there is a high-pressure air mass to the west, since the concentration of dust particles is generally greater in high-pressure air masses than in low-pressure air masses. Similarly, red sunrises occur most often when there is a high-pressure air mass to the east.

Now you can understand the old saying "Red sky at night, sailors' delight; red sky in the morning, sailors take warning." Fair weather generally accompanies high-pressure air masses, because they are associated with reduced cloud formation. Most of the United States lies in the westerlies wind zone, in which air masses generally move from west to east. A red sky at night is thus likely to indicate a fair-weather, high-pressure air mass to the west that will be coming your way. A red sky in the morning means that the high-pressure air mass has passed and poor weather may set in.

As a final note, how would you like a sky that is normally *red*? Then try Mars, the "red planet." The thin Martian atmosphere is about 95% carbon dioxide (CO_2). The CO_2 molecule is more massive than an oxygen (O_2) or a nitrogen (N_2) molecule. As a result, CO_2 molecules have a lower resonant frequency (longer wavelength) and preferentially scatter the red end of the visible spectrum. Hence, the Martian sky is red during the day. And what of the color of sunrises and sunsets on Mars? Think about it. . . .

And finally, the use of light in biomedical application in Insight 24.3, Optical Biopsy.

INSIGHT 24.3 OPTICAL BIOPSY

One of the most reliable ways to detect disease is to perform a biopsy—the removal of tissue samples—and then look for abnormal *changes* in the samples. "Optical biopsy," or biomedical scattering, appears to be a promising tool for diagnosing and monitoring diseases such as cancer *without* such surgery.

Optical biopsies are based on the following physical principle. The particles in the tissues absorb and re-emit light; thus, the scattered light contains information about the makeup of the tissue. Scattering from a tissue depends on the internal structures such as the presence of collagen fibers, and the status of hydration in the tissue. The measurement of the scattered light as a function of wavelength, polarization, or angle can be an important diagnostic tool.

An example of an optical biopsy is the diagnosis and measurement of collagen fibers. A major component of skin and bone, collagen is a fibrous protein found in animal cells. The fibers in the dormant form of collagen (about 2–3 μm in diameter) are composed of bundles of smaller collagen fibrils, about 0.3 μm in diameter, as shown in Fig. 1. The fibrils are made up of entwined tropocollagen molecules and present a banded pattern of striations with 70-nm periodicity due to the staggered alignment of the tropocollagen molecules. Each of these molecules has an electron-dense "head group" that appears dark in the electron micrograph. This periodic variation in refractive index at this level scatters light strongly in the visible and ultraviolet regions. The information contained in the scattered light can reveal abnormal conditions in the collagen fibers.

FIGURE 1 An electron micrograph of collagen fibers The details of collagen fibers show the presence of collagen fibrils and tropocollagen molecules.

Chapter Review

- **Young's double-slit experiment** provides evidence of the wave nature of light and a way to measure the wavelength of light ($\approx 10^{-7}$ m).

  The angular position (θ) of the **maxima** satisfies the condition

$$d \sin \theta = n\lambda \qquad \text{for } n = 0, 1, 2, 3, \ldots \qquad (24.3)$$

where d is the slit separation.

For small θ, the distance between the nth maximum and the central maximum is

$$y_n \approx \frac{nL\lambda}{d} \qquad \text{for } n = 0, 1, 2, 3, \ldots \qquad (24.4)$$

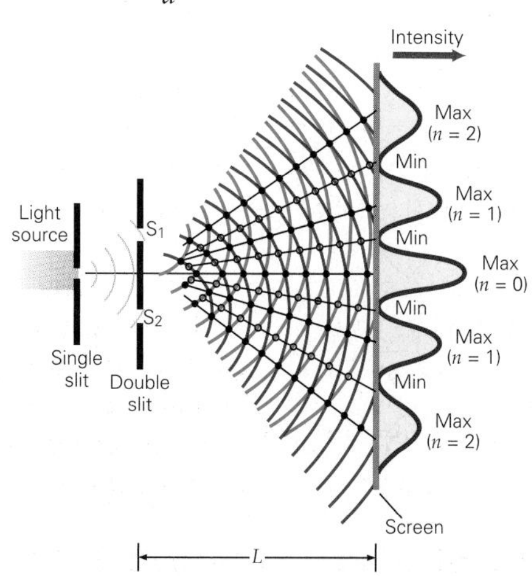

- Light reflected at a media boundary for which $n_2 > n_1$ undergoes a 180° **phase change**. If $n_2 < n_1$, there is no phase change on reflection. The phase changes affect thin-film interference, which also depends on film thickness and index of refraction.

  The **minimum thickness for a nonreflecting film** is

$$t_{\text{min}} = \frac{\lambda}{4n_1} \text{ (for } n_2 > n_1 > n_o) \qquad (24.7)$$

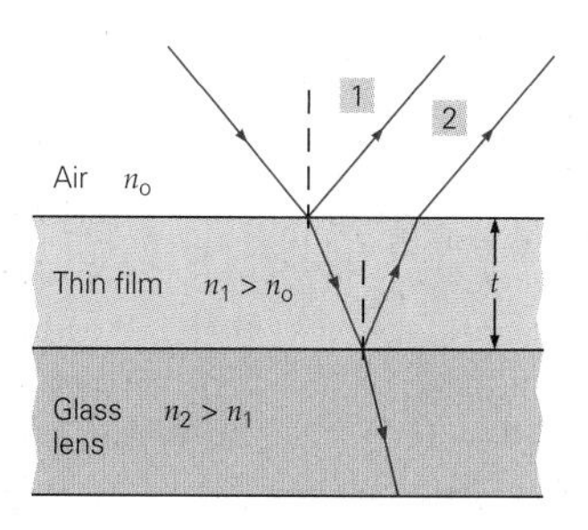

- In a **single-slit diffraction** experiment, the **minima** at location θ satisfy

$$w \sin \theta = m\lambda \quad \text{for } m = 1, 2, 3, \ldots \tag{24.8}$$

where w is the slit width. In general, the longer the wavelength as compared with the width of an opening or object, the greater the diffraction.

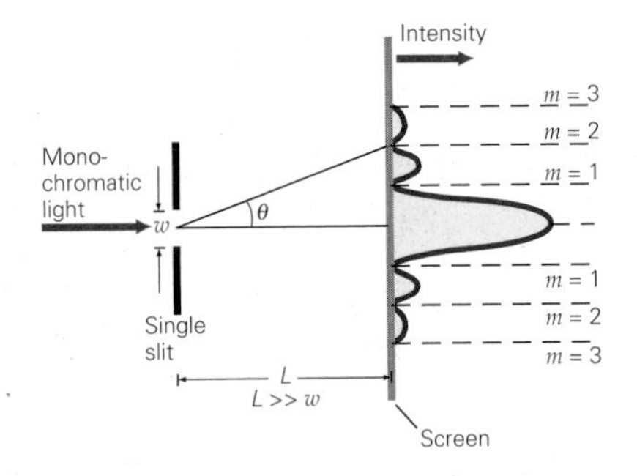

- For a **diffraction grating**, the maxima (bright fringes) satisfy

$$d \sin \theta = n\lambda \quad \text{for } n = 0, 1, 2, \ldots \tag{24.12}$$

where $d = 1/N$ and N is the number of lines per unit length.

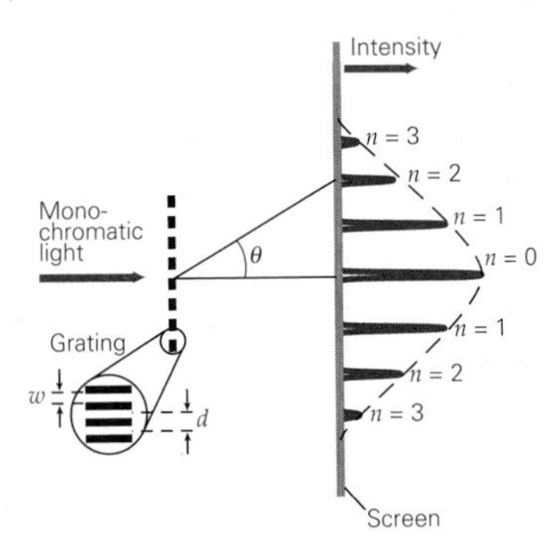

- **Polarization** is the preferential orientation of the electric field vectors that make up a light wave and is evidence that light is a transverse wave. Light can be polarized by selective absorption, reflection, double refraction (**birefringence**), and scattering.

When the transmission axes of a polarizer and an analyzer make an angle of θ, the intensity of the transmitted light is given by **Malus's law**:

$$I = I_o \cos^2 \theta \tag{24.14}$$

In reflection, if the angle of incidence is equal to the **Brewster (polarizing) angle** θ_p, then the reflected light is linearly polarized:

$$\tan \theta_p = \frac{n_2}{n_1} \quad \text{or} \quad \theta_p = \tan^{-1}\left(\frac{n_2}{n_1}\right) \tag{24.15}$$

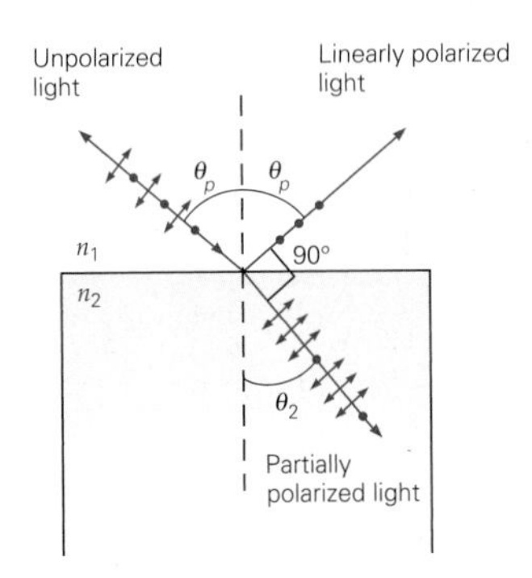

- The intensity of **Rayleigh scattering** is inversely proportional to the fourth power of the wavelength of the light. The blueness of the Earth's sky results from the preferential scattering of sunlight by air molecules.

Exercises

MC = Multiple Choice Question, CQ = Conceptual Question, and IE = Integrated Exercise. Throughout the text, many exercise sections will include "paired" exercises. These exercise pairs, identified with **red numbers**, *are intended to assist you in problem solving and learning. In a pair, the first exercise (even numbered) is worked out in the Study Guide so that you can consult it should you need assistance in solving it. The second exercise (odd numbered) is similar in nature, and its answer is given at the back of the book.*

24.1 Young's Double-Slit Experiment

1. **MC** In a Young's double-slit experiment using monochromatic light, if the slit spacing d decreases, the interference maxima spacing will (a) decrease, (b) increase, (c) remain unchanged, (d) disappear.

2. **MC** If the path-length difference between two identical and coherent beams is 2.5λ when they arrive at a point on a screen, the point will be (a) bright, (b) dark, (c) multicolored, (d) gray.

3. **MC** When white light is used in Young's double-slit experiment, many maxima with a spectrum of colors are seen. In a given maximum, the color closest to the central maximum is (a) red, (b) blue, (c) all colors.

4. **CQ** Non–cable television pictures often flutter when an airplane passes by (▸Fig. 24.29). Explain a possible cause of this fluttering, based on interference effects.

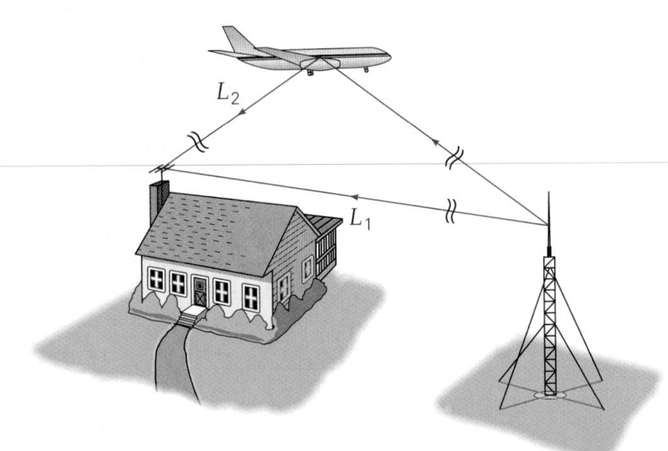

▲ **FIGURE 24.29 Interference** See Exercise 4.

5. **CQ** Describe what would happen to the interference pattern in Young's double-slit experiment if the wavelength of the monochromatic light were to increase.

6. **CQ** The intensity of the central maximum in the interference pattern of a Young's double-slit experiment is about four times that of either light wave. Is this a violation of the conservation of energy? Explain.

7. ● In the development of Young's double-slit experiment, a small-angle approximation ($\tan \theta \approx \sin \theta$) was used to find the lateral displacements of the bright and dark fringes. How good is this approximation? For example, what is the percentage error for $\theta = 15°$?

8. ● To study wave interference, a student uses two speakers driven by the same sound wave of wavelength 0.50 m. If the distances from a point to the speakers differ by 0.75 m, will the waves interfere constructively or destructively at that point? What if the distances differ by 1.0 m?

9. ● Two parallel slits 0.075 mm apart are illuminated with monochromatic light of wavelength 480 nm. Find the angle between the center of the central maximum and the center of the first side maximum.

10. ●● (a) Derive a relationship that gives the locations of the minima in a Young's double-slit experiment. What is the distance between adjacent minima? (b) For a third-order minimum (the third dark-side position from the central maximum), what is the path-length difference between that location and the two slits?

11. ●● In a double-slit experiment that uses monochromatic light, the angular separation between the central maximum and the second-order maximum is 0.160°. What is the wavelength of the light if the distance between the slits is 0.350 mm?

12. **IE** ●● Monochromatic light passes through two narrow slits and forms an interference pattern on a screen. (a) If the wavelength of light used increases, will the distance between the maxima (1) increase, (2) remain the same, or (3) decrease? Explain. (b) If the slit separation is 0.25 mm, the screen is 1.5 m away from the slits, and light of wavelength 550 nm is used, what is the distance from the center of the central maximum to the center of the third-order maximum? (c) What if the wavelength is 680 nm?

13. ●● In a double-slit experiment using monochromatic light, a screen is placed 1.25 m away from the slits, which have a separation distance of 0.0250 mm. The position of the third-order maximum is 6.60 cm from the center of the central maximum. Find (a) the wavelength of the light and (b) the position of the second-order maximum.

14. **IE** ●● (a) If the wavelength used in a double-slit experiment is decreased, the distance between adjacent maxima will (1) increase, (2) also decrease, (3) remain the same. Explain. (b) If the separation between the two slits is 0.20 mm and the adjacent maxima of the interference pattern on a screen 1.5 m away from the slits are 0.45 cm apart, what is the wavelength and color of the light? (c) If the wavelength is 550 nm, what is the distance between adjacent maxima?

15. **IE** ●● Two parallel slits are illuminated with monochromatic light, and an interference pattern is observed on a screen. (a) If the distance between the slits were decreased, would the distance between the maxima (1) increase, (2) remain the same, or (3) decrease? Explain. (b) If the slit separation is 1.0 mm, the wavelength is 640 nm, and the distance from the slits to the screen is 3.00 m, what is the separation between adjacent interference maxima? (c) What if the slit separation is 0.80 mm?

16. **IE** ●● (a) In a double-slit experiment, if the distance from the double slits to the screen is increased, the separation between the adiacent maxima will (1) increase, (2) decrease, (3) remain the same. Explain. (b) Yellow-green light ($\lambda = 550$ nm) is illuminated on a double-slit separated by 1.75×10^{-4} m. If the screen is located 2.00 m from the slits, determine the separation between the adjacent maxima. (c) What if the screen is located 3.00 m from the slits.

17. ●● In a double-slit experiment with monochromatic light and a screen at a distance of 1.50 m from the slits, the angle between the second-order maximum and the central maximum is 0.0230 rad. If the separation distance of the slits is 0.0350 mm, what are (a) the wavelength and color of the light and (b) the lateral displacement of this maximum?

18. **IE** ●●● (a) If the apparatus for a Young's double-slit experiment is completely immersed in water, will the spacing of the interference maxima (1) increase, (2) remain the same, or (3) decrease? Explain. (b) What would the lateral displacements in Exercise 12 be if the entire system were immersed in still water?

19. ●●● Light of two different wavelengths is used in a double-slit experiment. The location of the third-order maximum for the first light, yellow-orange light ($\lambda = 600$ nm), coincides with the location of the fourth-order maximum for the other color's light. What is the wavelength of the other light?

24.2 Thin-Film Interference

20. **MC** For a thin film with $n_1 > n_o$ and $n_1 > n_2$, where n_1 is the index of refraction of the film, a film thickness for constructive interference of the reflected light is (a) $\lambda'/4$, (b) $\lambda'/2$, (c) λ', (d) both a and b.

21. **MC** For a thin film with $n_o < n_1 < n_2$, where n_1 is the index of refraction of the film, the minimum film thickness for destructive interference of the reflected light is (a) $\lambda'/4$, (b) $\lambda'/2$, (c) λ'.

22. **MC** When a thin film of kerosene spreads out on water, the thinnest part looks bright. The index of refraction of kerosene is (a) greater than, (b) less than, (c) the same as that of water.

23. **CQ** Most lenses used in cameras are coated with thin films and appear bluish-purple when viewed with reflected light. What wavelengths are not visible in the reflected light?

24. CQ When destructive interference of two waves occurs at a certain location, there is no energy at that location. Is this situation a violation of the conservation of energy? Explain.

25. CQ At the center of a Newton's rings arrangement (Fig. 24.10a), the air wedge has a thickness of zero. Why is this area always dark?

26. IE ● A film on a lens with an index of refraction of 1.5 is 1.0×10^{-7} m thick and is illuminated with white light. The index of refraction of the film is 1.4. (a) The number of waves that experience the 180° phase shift is (1) zero, (2) one, (3) two. Explain. (b) For what wavelength of visible light will the lens be nonreflecting?

27. ● Light of wavelength 550 nm in air is normally incident on a glass plate ($n = 1.5$) whose thickness is 1.1×10^{-5} m. (a) What is the thickness of the glass in terms of the wavelength of light in glass? (b) Will the reflected light interfere constructively or destructively?

28. ● A lens with an index of refraction of 1.60 is to be coated with a material ($n = 1.40$) that will make the lens nonreflecting for red light ($\lambda = 700$ nm) normally incident on the lens. What is the minimum required thickness of the coating?

29. ●● Magnesium fluoride ($n = 1.38$) is frequently used as a lens coating to make nonreflecting lenses. What is the difference in the minimum film thickness required for maximum transmission of blue light ($\lambda = 400$ nm) and of red light ($\lambda = 700$ nm)?

30. ●● A solar cell is designed to have a nonreflective coating of a transparent material. (a) Will the thickness of the coating depend on the index of refraction of the underlying material in the solar cell? Discuss the possible scenarios. (b) If $n_{solar} > n_{film}$ and $n_{film} = 1.22$, what is the minimum thickness of the film for light with a wavelength of 550 nm?

31. IE ●● A thin layer of oil ($n = 1.50$) floats on water. Destructive interference is observed for light of wavelengths 480 nm and 600 nm, each at a different location. (a) If the order number is the same for both wavelength, which wavelength is at a greater thickness: (1) 480 nm, (2) 600 nm, or (3) both? Explain. (b) Find the two minimum thicknesses of the oil film, assuming normal incidence.

32. ●● A camera lens ($n = 1.50$) is coated with a thin layer of a material that has an index of refraction of 1.35. This coating makes the lens nonreflecting for light of wavelength 450 nm (in air) that is normally incident on the lens. What is the thickness of the thinnest film that will make the lens nonreflecting?

33. ●● Two parallel plates are separated by a small distance as illustrated in ▸Fig. 24.30. If the top plate is illuminated with light from a He–Ne laser ($\lambda = 632.8$ nm), for what minimum separation distances will the light be (a) constructively reflected and (b) destructively reflected? [*Note*: $t = 0$ is *not* an answer for part (b).]

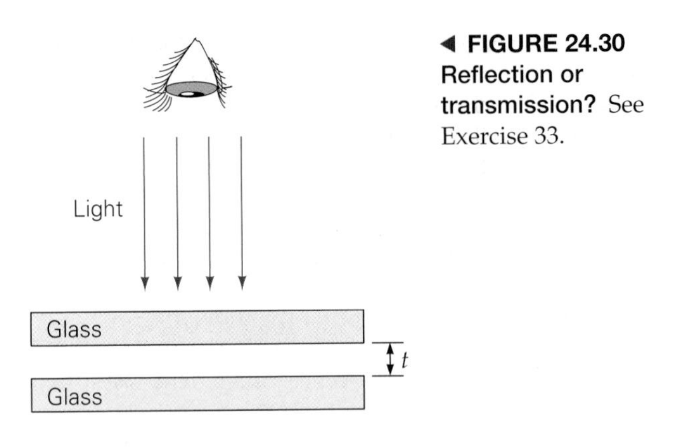

◂ **FIGURE 24.30 Reflection or transmission?** See Exercise 33.

34. IE ●●● An air wedge such as that shown in ▾Fig. 24.31 can be used to measure small dimensions, such as the diameter of a thin wire. (a) If the top glass plate is illuminated with monochromatic light, the interference pattern observed will be (1) bright, (2) dark, (3) bright and dark lines based on the air film thickness. Explain. (b) Express the locations of the bright interference maxima in terms of wedge thickness measured from the apex of the wedge.

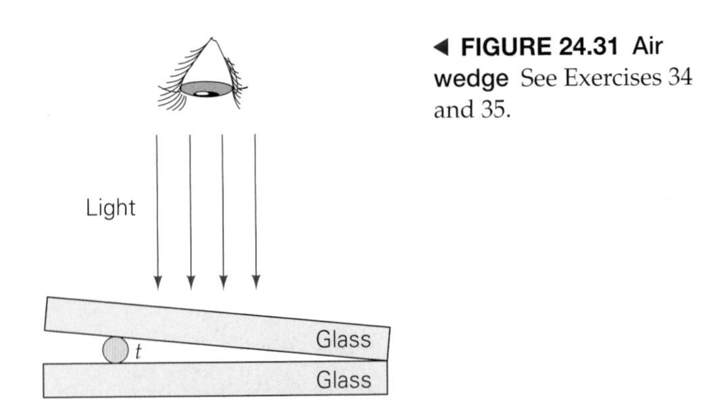

◂ **FIGURE 24.31 Air wedge** See Exercises 34 and 35.

35. ●●● The glass plates in Fig. 24.31 are separated by a thin, round filament. When the top plate is illuminated normally with light of wavelength 550 nm, the filament lies directly below the sixth bright maximum. What is the diameter of the filament?

24.3 Diffraction

36. MC In a single-slit diffraction pattern, (a) all maxima have the same width, (b) the central maximum is twice as wide as the side maxima, (c) the side maxima are twice as wide as the central maximum, (d) none of the preceding.

37. MC As the number of lines per unit length of a diffraction grating increases, the spacing between the maxima (a) increases, (b) decreases, (c) remains unchanged.

38. MC In a single-slit diffraction pattern, if the wavelength of light increases, the width of the central maximum will (a) increase, (2) decrease, (c) remain the same.

39. CQ From Eq. 24.8, can the $m = 2$ minimum be seen if $w = \lambda$? How about the $m = 1$ minimum?

40. CQ In our discussion of single-slit diffraction, the length of the slit was assumed to be much greater than the width. What changes would be observed in the diffraction pattern if the length were comparable with the width of the slit?

41. CQ In a diffraction grating, the slits are very closely spaced. What is the advantage of this design?

42. ● A slit of width 0.20 mm is illuminated with monochromatic light of wavelength 480 nm, and a diffraction pattern is formed on a screen 1.0 m away from the slit. (a) What is the width of the central maximum? (b) What are the widths of the second and third-order maxima?

43. ● A slit 0.025 mm wide is illuminated with red light ($\lambda = 680$ nm). How wide are (a) the central maximum and (b) the side maxima of the diffraction pattern formed on a screen 1.0 m from the slit?

44. ● At what angle will the second-order diffraction maximum be seen from a diffraction grating of spacing 1.25 μm when illuminated by light of wavelength 550 nm?

45. ● A venetian blind is essentially a diffraction grating—not for visible light, but for waves with longer wavelengths. If the spacing between the slats of a blind is 2.5 cm, (a) for what wavelength is there a first-order maximum at an angle of 10°, and (b) what type of radiation is this?

46. IE ●● A single slit is illuminated with monochromatic light, and a screen is placed behind the slit to observe the diffraction pattern. (a) If the width of the slit is increased, will the width of the central maximum (1) increase, (2) remain the same, or (3) decrease? Why? (b) If the width of the slit is 0.50 mm, the wavelength is 680 nm, and the screen is 1.80 m from the slit, what is the width of the central maximum? (c) What if the width of the slit is 0.60 mm?

47. ●● A diffraction grating is designed to have the second-order maxima at 10° from the central maximum for the red end ($\lambda = 700$ nm) of the visible spectrum. How many lines per centimeter does the grating have?

48. ●● A certain crystal gives a deflection angle of 25° for the first-order maximum of monochromatic X-rays with a frequency of 5.0×10^{17} Hz. What is the lattice spacing of the crystal?

49. ●● Find the angles of the blue ($\lambda = 420$ nm) and red ($\lambda = 680$ nm) components of the first- and second-order maxima in a pattern produced by a diffraction grating with 7500 lines/cm.

50. IE ●● (a) Only a limited number of maxima can be observed with a diffraction grating. The factor(s) that limits the number of maxima seen is (a) (1) the wavelength, (2) the grating spacing, (3) both. Explain. (b) How many maxima appear when monochromatic light of wavelength 560 nm illuminates a diffraction grating that has 10 000 lines/cm, and what are their order numbers?

51. ●● In a particular diffraction pattern, the red component (700 nm) in the second-order maximum is deviated at an angle of 20°. (a) How many lines per centimeter does the grating have? (b) If the grating is illuminated with white light, how many maxima of the complete visible spectrum are produced?

52. ●● White light whose components have wavelengths from 400 to 700 nm illuminates a diffraction grating with 4000 lines/cm. Do the first- and second-order spectra overlap? Justify your answer.

53. IE ●● White light ranging from blue (400 nm) to red (700 nm) illuminates a diffraction grating with 8000 lines/cm. (a) For the first maxima, the (1) blue (2) red color is closer to the central maximum. Explain. (b) What are the angles of the first-order maximum for blue and red?

54. ●● A diffraction grating with 8000 lines/cm is illuminated with a red light from a He–Ne laser ($\lambda = 632.8$ nm). How many side maxima are formed in the diffraction pattern, and at what angles are they observed?

55. ●●● Show that for a diffraction grating, the violet ($\lambda = 400$ nm) portion of the third-order maximum overlaps the yellow-orange ($\lambda = 600$ nm) portion of the second-order maximum, regardless of the grating's spacing.

56. IE ●●● A teacher standing in a doorway 1.0 m wide blows a whistle with a frequency of 1000 Hz to summon children from the playground (▼Fig. 24.32). Two boys are playing on the swings 100 m away from the school building. One boy is at an angle of 0° and another one at 19.6° from a line normal to the doorway. (a) (1) Only the boy at 0°, (2) only the boy at 19.6°, or (3) both boys may not hear the whistle. Explain. (b) Taking the speed of sound in air to be 335 m/s, does the boy 19.6° hear the whistle?

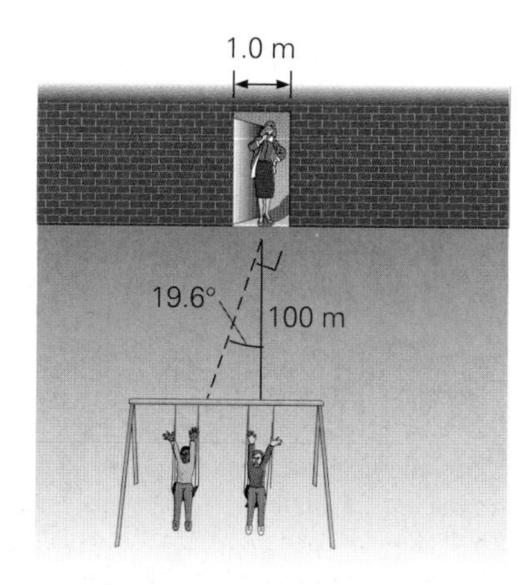

▲ **FIGURE 24.32 Moment of truth** See Exercise 56. (Not drawn to scale.)

24.4 Polarization

57. **MC** Light can be polarized by (a) reflection, (b) refraction, (c) absorption, (d) all of the preceding.

58. **MC** The Brewster angle depends on (a) the indices of refraction of materials, (b) Bragg's law, (c) internal reflection, (d) interference.

59. **MC** A sound wave cannot be polarized. This is because sound is (a) not a light wave, (b) a transverse wave, (c) a longitudinal wave, (d) none of the preceding.

60. **CQ** Given two pairs of sunglasses, could you tell whether one or both were polarizing?

61. **CQ** Suppose that you held two polarizing sheets in front of you and looked through both of them. How many times would you see the sheets lighten and darken (a) if one were rotated through one complete rotation, (b) if both were rotated through one complete rotation at the same rate in opposite directions, (c) if both were rotated through one complete rotation at the same rate in the same direction, and (d) if one rotates twice as fast as the other and the slower one rotates through one complete rotation?

62. **CQ** How does selective absorption produce polarized light?

63. **CQ** If you place a pair of polarizing sunglasses in front of your calculator's LCD display and rotate them, what do you observe?

64. ● Some types of glass have a range of indices of refraction of about 1.4 to 1.7. What is the range of the polarizing (Brewster) angle for these glasses when light is incident on them from air?

65. **IE** ● Light is incident on a certain material in air. (a) If the index of refraction of the material increases, the polarizing (Brewster) angle will (1) also increase, (2) decrease, (3) remain the same. Explain. (b) What are the polarizing angles if the index of refraction is 1.6 and 1.8?

66. ● A polarizer–analyzer pair can have their transmission axes at either 30° or 45° angles. Which angle will allow more light to be transmitted?

67. **IE** ●● Unpolarized light of intensity I_o is incident on a polarizer–analyzer pair. (a) If the angle between the polarizer and analyzer increases in the range of 0° to 90°, the transmitted light intensity will (1) also increase, (2) decrease, (3) remain the same. Explain. (b) If the angle between the polarizer and analyzer is 30°, what light intensity is transmitted through the polarizer and the analyzer, respectively? (c) What if the angle is 60°?

68. ●● A beam of light is incident on a glass plate ($n = 1.62$) in air and the reflected ray is completely polarized. What is the angle of refraction for the beam?

69. ●● The critical angle for internal reflection in a certain medium is 45°. What is the polarizing (Brewster) angle for light externally incident on the medium?

70. ●● The angle of incidence is adjusted so there is maximum linear polarization for the reflection of light from a transparent piece of plastic in air. (a) There is (1) no, (2) maximum, or (3) some light transmitted through the plastic. Explain. (b) If the index of refraction of the plastic is 1.22, what is the angle of refraction in the plastic?

71. ●● Sunlight is reflected off a vertical plate-glass window ($n = 1.55$). What would the Sun's altitude (angle above the horizon) have to be for the reflected light to be completely polarized?

72. **IE** ●● A piece of glass ($n = 1.60$) could be in air or submerged in water. (a) The polarizing (Brewster) angle in water is (1) greater than, (2) less than, (3) the same as that in air. Explain. (b) What is the polarizing angle when it is in air and submerged in water?

73. ●●● A plate of crown glass is covered with a layer of water. A beam of light traveling in air is incident on the water and partially transmitted. Is there any angle of incidence for which the light reflected from the water–glass interface will have maximum linear polarization? Justify your answer mathematically.

*24.5 Atmospheric Scattering of Light

74. **MC** Which of the following colors is scattered the most in the atmosphere: (a) blue, (b) yellow, (c) red, or (d) color makes no difference?

75. **MC** Scattering involves (a) the reflection of light off particles, (b) the refraction of light off particles, (c) the absorption and reradiation of light by particles, (d) none of the preceding.

76. **CQ** Explain why the sky is red in the morning and evening and blue during the day.

77. **CQ** (a) On a clear cloudless day why does the sky not have a uniform blueness? (b) What color would an astronaut on the Moon see when looking at the sky or into space?

Comprehensive Exercises

78. A thin air wedge between two flat glass plates forms bright and dark interference bands when illuminated with normally incident monochromatic light. (See Fig. 24.9.) (a) Show that the thickness of the air wedge changes by $\lambda/2$ from one bright fringe to the next, where λ is the wavelength of the light. (b) What would be the change in the thickness of the wedge between bright fringes if the space were filled with a liquid with an index of refraction n?

79. A salesman tries to sell you an optic fiber that claims to give linearly polarized light when light is totally internally reflected off the fiber–air interface. (a) Would you buy it? Explain. (b) If total internal reflection occurs at an angle of 35°, what is the polarizing (Brewster) angle?

80. Three parallel slits of width w have a slit separation of d, where $d = 3w$. (a) Will you be able to see all the interference maxima? Explain. (b) If not, which interference maxima will be missing? [*Hint*: See Fig. 24.16.]

81. If the slit width in a single-slit experiment were doubled, the distance to the screen reduced by one third, and the wavelength of the light changed from 600 nm to 450 nm, how would the width of the maxima be affected?

82. Show that when the reflected light is completely polarized, the sum of the angle of incidence and the angle of refraction is equal to 90°.

83. What is the highest spectral order that can be seen in a diffraction grating with 9000 lines/cm when the grating is illuminated by white light?

PHYSLET® The following Physlet Physics Problems can be used with this chapter. 37.2, 37.4, 37.7, 37.9, 37.10, 38.1, 38.2, 38.4, 38.5, 38.6, 39.9, 39.10

CHAPTER

25 VISION AND OPTICAL INSTRUMENTS

25.1 The Human Eye 793

25.2 Microscopes 799

25.3 Telescopes 803

25.4 Diffraction and Resolution 807

***25.5** Color 810

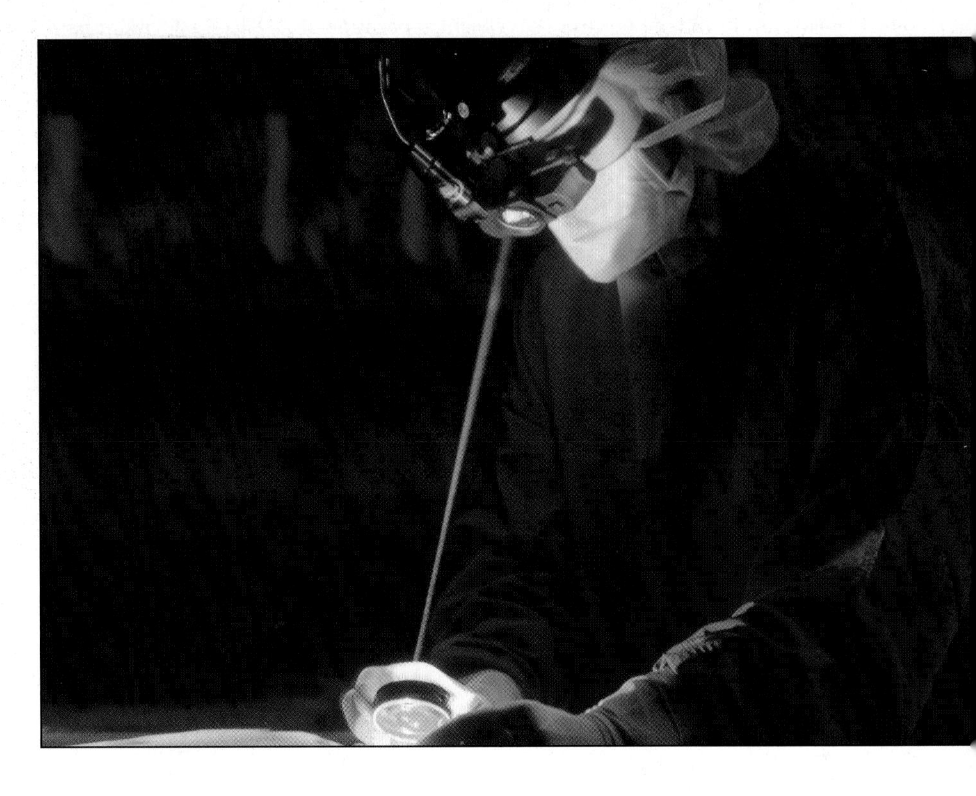

PHYSICS FACTS

- About 80% of the refracting power of a human eye comes from the cornea while the other 20% comes from the crystalline lens. The crystalline lens can change shape to accommodate close or far focusing by means of the ciliary muscles.
- The human eye collects a lot of information. If it were compared to a digital camera, the human eye would be equivalent to 500 megapixels. A common digital camera has 2 to 10 megapixels.
- A red blood cell has a diameter of about 7 μm (7×10^{-6} m). When viewed with a compound microscope at 1000×, it appears to be 7 mm (7×10^{-3} m).
- Some cameras on satellites have excellent resolution. From space they can read the license plates on cars.

Vision is one of our chief means of acquiring information about the world around us. However, the images seen by many eyes are not always clear or in focus, and glasses or some other remedy are needed. Great progress has been made in the last decade in contact lens therapy and surgical correction of vision defects. A popular procedure is laser surgery, as shown in the photo. Laser surgery can be used for such procedures as repairing torn retinas, destroying eye tumors, and stopping abnormal growth of blood vessels that can endanger vision.

Optical instruments, the basic function of which is to improve and extend the power of observation beyond that of the human eye, augment our vision. Mirrors and lenses are used in a variety of optical instruments, including microscopes and telescopes.

The earliest magnifying lenses were drops of water captured in a small hole. By the seventeenth century, artisans were able to grind fair-quality lenses for simple microscopes or magnifying glasses, which were used primarily for botanical studies. (These early lenses also found a use in spectacles.) Soon, the basic compound microscope, which uses two lenses, was developed. Modern compound microscopes, which can magnify an object up to 200 times, extended our vision into the microbe world.

Around 1609, Galileo used lenses to construct an astronomical telescope that allowed him to observe valleys and mountains on the Moon, sunspots, and the four largest moons of Jupiter. Today, huge telescopes that use lenses and mirrors have extended our vision far into the past as we look at farther, and therefore, younger, galaxies.

What would our knowledge be if these instruments had never been invented? Bacteria would still be unknown, and planets, stars, and galaxies would have remained nothing but mysterious points of light.

Mirrors and lenses were discussed in terms of geometrical optics in Chapter 23, and the wave nature of light was investigated in Chapter 24. These principles can be applied to the study of vision and optical instruments. In this chapter, you will learn about our fundamental optical instrument—the human eye, without which all others would be of little use. Also microscopes and telescopes will be discussed, along with the factors that limit their viewing.

25.1 The Human Eye

OBJECTIVES: To (a) describe the optical workings of the eye, and (b) explain some common vision defects and how they are corrected.

The human eye is the most important of all optical instruments, because without it we would know little about our world and the study of optics would not exist. The human eye is analogous to a simple camera in several respects (▾Fig. 25.1). A simple camera consists of a converging lens, which is used to focus images on light-sensitive film (traditional camera) or *charge coupled device*, CCD, (in digital cameras) at the back of the camera's interior chamber. (Recall from Chapter 23 that for relatively distant objects, a converging lens produces a small, inverted, real image.) There is an adjustable diaphragm opening, or aperture, and a shutter to control the amount of light entering the camera.

Note: Image formation by a converging lens is discussed in Section 23.3; see Figure 23.15a.

The eye, too, focuses images onto a light-sensitive lining (the retina) on the rear surface of the eyeball. The eyelid might be thought of as a shutter; however, the shutter of a camera, which controls the exposure time, is generally opened only for a fraction of a second, while the eyelid is normally open for continuous exposure. The human nervous system actually performs a function analogous to a shutter by analyzing image signals from the eye at a rate of 20 to 30 times per second.

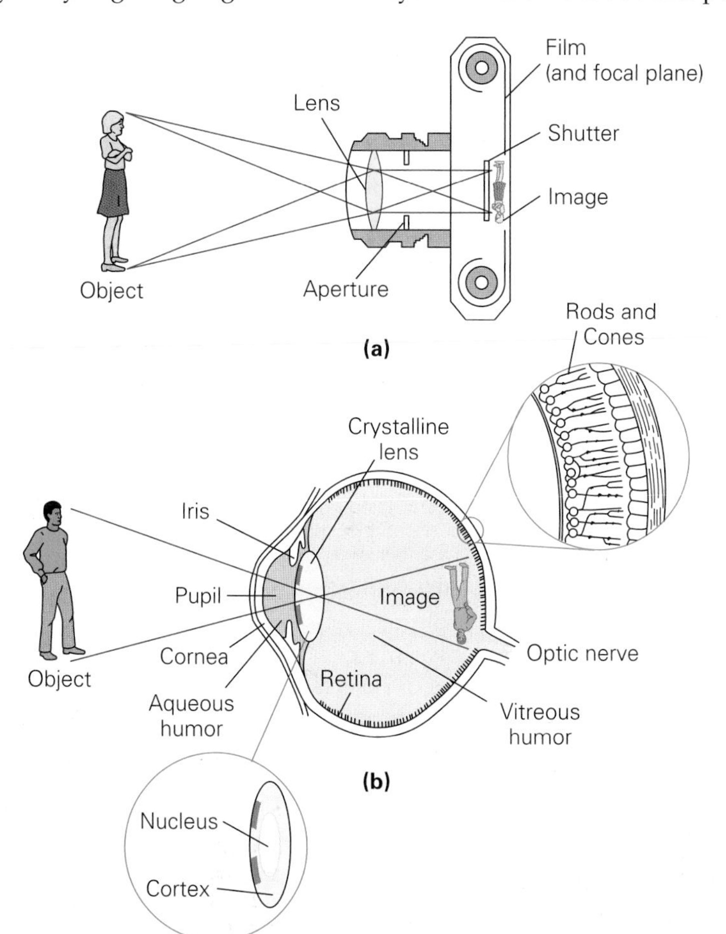

◀ **FIGURE 25.1 Camera and eye analogy** In some respects, **(a)** a simple camera is similar to **(b)** the human eye. An image is formed on the film in a camera and on the retina of the eye. (The complex refractive properties of the eye are not shown here, because multiple refractive media are involved.) See text for a comparative description.

Illustration 36.2 *Camera*

Exploration 36.1 *Camera*

The eye might therefore be better likened to a movie or video camera, which exposes a similar number of frames (images) per second.

Although the optical functions of the eye are relatively simple, its physiological functions are quite complex. As Fig. 25.1b shows, the eyeball is a nearly spherical chamber. It has an internal diameter of about 1.5 cm and is filled with a transparent jellylike substance called the *vitreous humor*. The eyeball has a white outer covering called the *sclera*, part of which is visible as the "white" of the eye. Light enters the eye through a curved, transparent tissue called the *cornea* and passes into a clear fluid known as the *aqueous humor*. Behind the cornea is a circular diaphragm, the *iris*, whose central opening is the *pupil*. The iris contains the pigment that determines eye color. Through muscle action, the area of the pupil can change (from 2 to 8 mm in diameter), thereby controlling the amount of light entering the eye.

Behind the iris is a *crystalline lens*, a converging lens composed of microscopic glassy fibers. (See Conceptual Example 22.5 on page 713 about the internal elements, the nucleous and cortex, inside the crystalline lens.) When tension is exerted on the lens by attached muscles, the glassy fibers slide over each other, causing the shape, and therefore focal length, of the lens to change, to help in focusing the image on the retina properly. Notice that this is an *inverted* image (Fig. 25.1b). We do not see an inverted image, however, because the brain reinterprets this image as being right-side-up.

On the back interior wall of the eyeball is a light-sensitive surface called the **retina**. From the retina, the optic nerve relays retinal signals to the brain. The retina is composed of nerves and two types of light receptors, or photosensitive cells, called **rods** and **cones**, because of their shapes. The rods are more sensitive to light than are the cones and distinguish light from dark in low light intensities (twilight vision). The cones can distinguish frequency ranges but require brighter light. The brain interprets these different frequencies as colors (color vision). Most of the cones are clustered in a central region of the retina called the *macula*. The rods, which are more numerous than the cones, are outside this region and are distributed nonuniformly over the retina.

The focusing adjustment of the eye differs from that of a simple camera. A camera lens has a constant focal length, and the image distance is varied by moving the lens relative to the film to produce sharp images for different object distances. In the eye, the image distance is constant, and the focal length of the lens is varied (as the attached muscles change the lens's shape) to produce sharp images on the retina, regardless of object distance. When the eye is focused on distant objects, the muscles are relaxed, and the crystalline lens is thinnest with a power of about 20 D (diopters). Recall from Chapter 23 that the power (P) of a lens in diopters (D), which is the reciprocal of its focal length *in meters*. So 20 D corresponds to a focal length of $f = 1/(20\text{ D}) = 0.050\text{ m} = 5.0\text{ cm}$. When the eye is focused on closer objects, the lens becomes thicker. Then its radius of curvature and hence its focal length are decreased. For close-up vision, the lens power may increase to 30 D $(f = 0.033\text{ m})$, or even more in young children. The adjustment of the focal length of the crystalline lens is called *accommodation*. (Look at a nearby object and then at an object in the distance, and notice how fast accommodation takes place. It's practically instantaneous.)

The distance extremes over which sharp focus is possible are known as the *far point* and the *near point*. The *far point* is the greatest distance at which the eye can see objects clearly and is infinity for a normal eye. The *near point* is the position closest to the eye at which objects can be seen clearly. This position depends on the extent to which the lens can be deformed (thickened) by accommodation. The range of accommodation gradually diminishes with age as the crystalline lens loses its elasticity. Generally, in the normal eye the near point gradually recedes with age. The approximate positions of the near point at various ages are listed in Table 25.1.

Children can see sharp images of objects that are within 10 cm of their eyes, and the crystalline lens of a normal young-adult eye can do the same for objects as close as 12 to 15 cm. However, adults at about age 40 normally experience a shift in the near point to beyond 25 cm. You may have noticed middle-aged people holding reading material fairly far from their eyes so as to move it out to be within the range of accommodation. When the print becomes too small (or the arms too short), corrective reading glasses are one solution. The recession of the near point with age is not considered an abnormal defect. Since it proceeds at about the same rate in most normal eyes, it is considered mainly a part of the normal "aging" process.

TABLE 25.1

Approximate Near Points of the Normal Eye at Different Ages

Age (years)	*Near Point (centimeters)*
10	10
20	12
30	15
40	25
50	40
60	100

Note: The relationship between lens power in diopters and focal length is presented in Eq. 23.9, in Section 23.4.

Note: The eye sees clearly between its far point and near point.

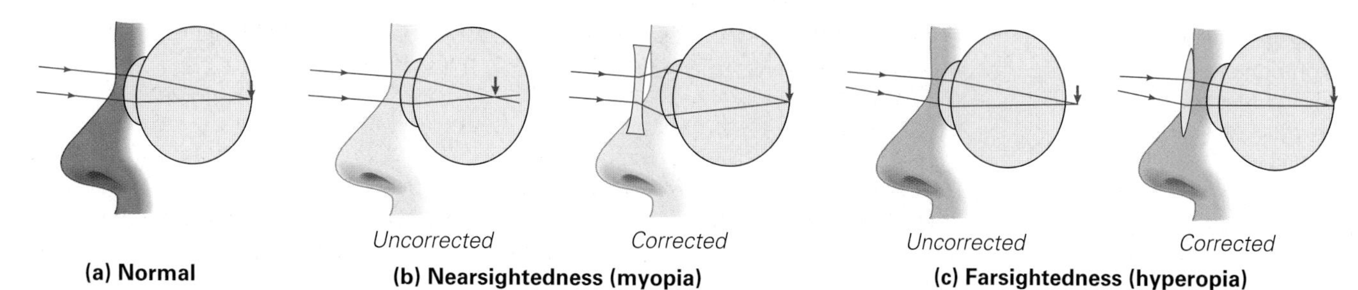

▲ **FIGURE 25.2 Nearsightedness and farsightedness**
(a) The normal eye produces sharp images on the retina for objects located between its near point and its far point. The image is real, inverted, and always smaller than the object. (Why?) Here, the object is a distant, upward-pointing arrow (not shown) and the light rays come from its tip. **(b)** In a nearsighted eye, the image of a *distant* object is focused *in front of* the retina. This defect is corrected with a diverging lens. **(c)** In a farsighted eye, the image of a *nearby* object is focused *behind* the retina. This defect is corrected with a converging lens. (Not drawn to scale.)

Vision Defects

The existence of a "normal" eye (▲Fig. 25.2a) implies that some eyes must have defects. This is indeed the case, as is quite apparent from the number of people who wear corrective glasses or contact lenses. Many people have eyes that cannot accommodate within the normal range (25 cm to infinity). These people usually have one of the two most common visual defects: nearsightedness (myopia) or farsightedness (hyperopia). Both of these conditions can usually be corrected with glasses, contact lenses, or surgery.

Nearsightedness (or *myopia*) is the ability to see nearby objects clearly, but not distant objects. That is, the far point is less than infinity. When an object is beyond the far point, the rays focus in *front* of the retina (Fig. 25.2b). As a result, the image on the retina is blurred, or out of focus. As the object is moved closer, its image moves back toward the retina. When the object reaches the far point for that eye, a sharp image is formed on the retina.

Nearsightedness usually arises because the eyeball is too long or the curvature of the cornea is too great. Whatever the reason, the eyeball overconverges the light from distant objects to a spot in front of the retina. Appropriate diverging lenses correct this condition. Such a lens causes the rays to diverge before reaching the cornea. The eye thus focuses the image farther back on the retina.

Farsightedness (or *hyperopia*) is the ability to see distant objects clearly, but not nearby ones. That is, the near point is farther from the eye than normal. The image of an object that is closer than the near point is formed behind the retina (Fig. 25.2c). Farsightedness arises because the eyeball is too short, because of insufficient curvature of the cornea or because of insufficient elasticity of the crystalline lens. If this occurs as part of the aging process as previously discussed, it is called *presbyopia*.

Farsightedness is usually corrected with appropriate converging lenses. Such a lens causes the rays to converge on the retina, and the eye is then able to focus the image on the retina. Converging lenses are also used in middle-aged people to correct presbyopia, a vision condition in which the crystalline lens of the eye loses its flexibility, which makes it difficult to focus on close objects.

PHYSLET® Illustration 36.1 The Human Eye

Integrated Example 25.1 ■ Correcting Nearsightedness: Use of Diverging Lenses

(a) An optometrist has a choice to give a patient either regular glasses or contact lenses to correct nearsightedness (▼Fig. 25.3). Usually, regular glasses sit a few centimeters in front of the eye and contact lenses right on the eye. Should the power of the contact lenses prescribed be (1) the same as, (2) greater than, or (3) less than that of the regular glasses? Explain. (b) A certain nearsighted person cannot see objects clearly when they are more than 78.0 cm from either eye. What power must corrective lenses have, for both regular glasses and contact lenses, if this person is to see distant objects clearly? Assume that the glasses are 3.00 cm in front of the eye.

(a) Conceptual Reasoning. For nearsightedness, the corrective lens is a diverging one (Fig. 25.3). The lens must effectively put the image of a distant object ($d_o = \infty$) at the far point of the eye, that is, d_f from the eye. The image, which acts as an object for the eye, is then within the range of accommodation. Because the image distance is *measured from the lens*, a contact lens will have a *longer* image distance. For a contact lens, $d_i = -(d_f)$. For regular glasses, $d_i = -|d_f - d|$, where d is the distance between the regular glasses and the eye. A minus sign and absolute values are used for the image distance because the image is virtual, being on the object side of the lens. (You may recall from Chapter 23 that diverging lenses can form only virtual images.)

Note: Review Example 23.6 and 23.8.

Note: Image formation by a diverging lens is discussed in Section 23.3; see Figure 23.18.

(continues on next page)

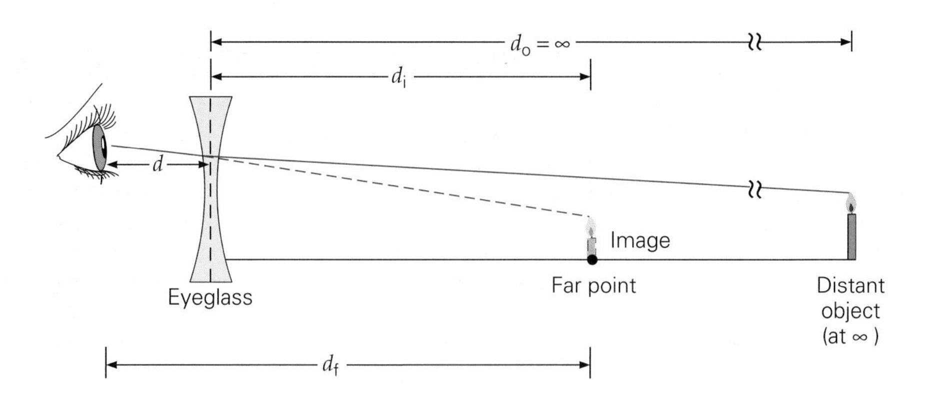

▶ **FIGURE 25.3 Correcting nearsightedness** A diverging lens is used. See Integrated Example 25.1. Only regular glasses are shown. For contact lenses, the lens is immediately in front of the eye ($d = 0$).

Note that d_i is negative. Recall that the power of a lens is $P = 1/f$ (Eq. 23.9). We can use the thin-lens equation (Eq. 23.5) to find P if we can determine the object and image distances, d_o and d_i:

$$P = \frac{1}{f} = \frac{1}{d_o} + \frac{1}{d_i} = \frac{1}{\infty} + \frac{1}{d_i} = \frac{1}{d_i} = -\frac{1}{|d_i|}$$

That is, a longer $|d_i|$ will yield a smaller P, so the contact lenses should have a lower power than the regular glasses. Thus, the answer is (3).

(b) Quantitative Reasoning and Solution. Once we understand how corrective lenses work, the calculation for part (b) is straightforward.

Given: $d_f = 78\text{ cm} = 0.780\text{ m}$ (far point), $d = 3.0\text{ cm} = 0.0300\text{ m}$

Find: P (in diopters) for regular glasses; P (in diopters) for contact lenses

For regular glasses,

$$|d_i| = |d_f - d| = 0.780\text{ m} - 0.0300\text{ m} = 0.750\text{ m}$$

(See Fig 25.3, which is not drawn to scale.) So, $d_i = -0.750$ m.

Then, using the thin-lens equation, we get

$$P = \frac{1}{f} = \frac{1}{d_o} + \frac{1}{d_i} = \frac{1}{\infty} + \frac{1}{-0.750\text{ m}} = -\frac{1}{0.750\text{ m}} = -1.33\text{ D}$$

A negative, or diverging, lens with a power of 1.33 D is needed.

For contact lenses,

$$|d_i| = |d_f| = 0.780\text{ m}$$

($d = 0$.) So, $d_i = -0.78$ m.

Then, using the thin-lens equation, we get

$$P = \frac{1}{\infty} + \frac{1}{-0.780\text{ m}} = -\frac{1}{0.780\text{ m}} = -1.28\text{ D}$$

Follow-Up Exercise. Suppose a mistake was made for regular glasses in this Example such that a "corrective" lens of +1.33 D were used. What happens to the image of objects at infinity? *(Answers to all Follow-Up Exercises are at the back of the text.)*

If the far point for a nearsighted person is changed using diverging lenses (see Integrated Example 25.1), the near point will be affected as well. This causes the close-up vision to worsen, but *bifocal lenses* can be used in this situation to address the problem. Bifocals were invented by Benjamin Franklin, who glued two lenses together. They are now made by grinding or molding lenses with different curvatures in two different regions. Both nearsightedness and farsightedness can be treated at the same time with bifocals. Trifocals are also available, with lenses having three different curvatures. The top lens is for far vision and the bottom lens for near vision. The middle lens is for intermediate vision and is sometimes referred to as a lens for "computer" vision.

More modern techniques involve contact lens therapy or the use of a laser to correct nearsightedness. These are discussed in detail in Insight 25.1, Cornea "Orthodontics" and Surgery, on accompanying page. The purpose of either technique is to change the shape the exposed surface of the cornea, which changes its refractive characteristics. The result, for the nearsighted case, is to make the image of a distant object fall on the retina.

INSIGHT 25.1 CORNEA "ORTHODONTICS" AND SURGERY

The imperfect shape of the cornea of the human eye often causes refractive errors that result in vision defects. For example, a cornea that is curved too much can cause nearsightedness; a flatter-than-normal cornea can cause farsightedness. A cornea that is not spherical can cause astigmatism (Section 23.4).

Recently, a nonsurgical contact lens treatment to improve vision in a matter of hours was developed. This procedure, called *orthokeratology*, or *Ortho-K*, is achieved in a unique way, with the wearing of custom-designed contact lenses. These contact lenses slowly change the shape of the cornea by means of gentle pressure to improve vision safely and quickly. The best analogy to describe Ortho-K is by calling it "orthodontics for the eye."

Laser surgery is also used to reshape the cornea. The surgical procedure corrects the defective shape or irregular surface of the cornea so that it can better focus light on the retina, thereby reducing or even eliminating vision defects (Fig. 1).

In corneal laser surgery, first, a very precise instrument called a *microkeratome* is used to create a thin corneal flap with a hinge on one side of the cornea (Fig. 2a). Once the flap is folded back, a tightly focused ultraviolet pulsed laser is used to reshape the cornea. Each laser pulse accurately removes a microscopic layer of the inner cornea in the targeted area, thus reshaping the cornea to correct vision defects (Fig. 2b). The flap is then placed back in its original position without the need for stitches (Fig. 2c). The procedure is usually painless, and patients typically have only minimal discomfort. Some patients achieve corrected vision within a day after this procedure.

Even more exciting advances in vision correction are on the horizon. For example, researchers have developed techniques for replacing a damaged cornea with freshly bioengineered tissues. If the patient has one healthy eye, stem cells are harvested from it. The cells grow into a sturdy layer of tissue that surgeons can use to replace the bad corneal tissues of the other eye by stitching the new tissue onto that damaged eye. If both of the patient's eyes are damaged, donor tissues may be collected from a close relative.

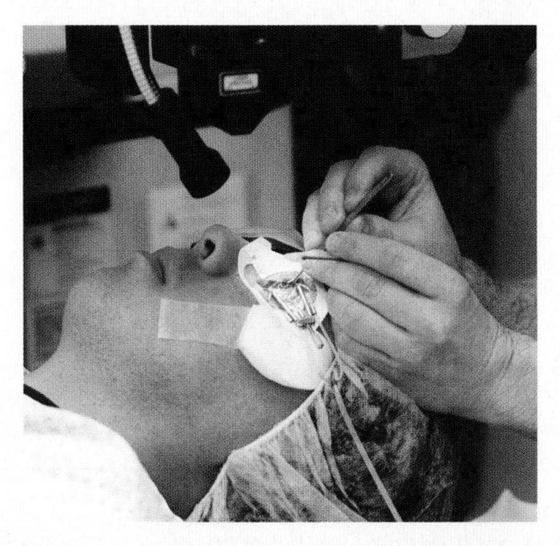

FIGURE 1 Eye surgery Laser surgery is performed to reshape the cornea. Notice that the surgeon is wearing no latex gloves. The fine chalk dust used as a lubricant on the gloves could contaminate the eye.

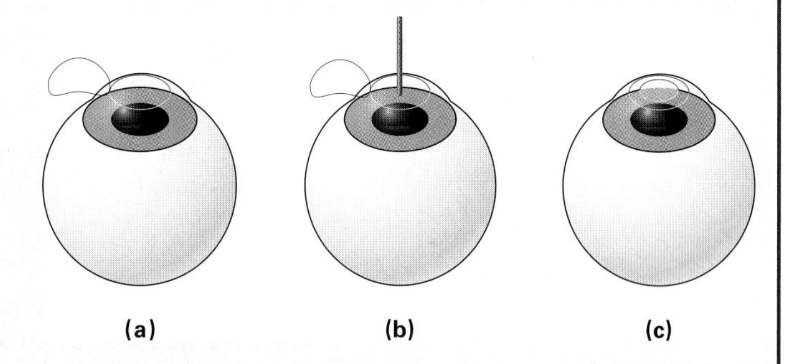

FIGURE 2 Cornea reshaping (a) A flap is made on the corneal surface. **(b)** A laser beam is used to reshape the cornea. **(c)** The flap is placed back.

Integrated Example 25.2 ■ Correcting Farsightedness: Use of a Converging Lens

A farsighted person has a near point of 75 cm for the left eye and a near point of 100 cm for the right one. (a) If the person is prescribed contact lenses, the power of the left lens should be (1) greater than, (2) the same as, (3) less than the power of the right lens. Explain. (b) What powers should contact lenses have to allow the person to see an object clearly at a distance of 25 cm?

(a) Conceptual Reasoning. The normal eye's near point is 25 cm. For farsightedness, the corrective lens must be converging and form the image at its eye's near point of an object at the normal eye's near point. Since the near point of the left eye (75 cm) is closer to the 25-cm normal position than the right eye, the left lens should have less power so the answer is (3).

(b) Quantitative Reasoning and Solution. Let us label the two different eyes as L (left) and R (right). The image distances are negative. (Why?)

Given: $d_{i_L} = -75\text{ cm} = -0.75\text{ m}$
$d_{i_R} = -100\text{ cm} = -1.0\text{ m}$
$d_o = 25\text{ cm} = 0.25\text{ m}$

Find: P_L and P_R (lens power for each eye)

(continues on next page)

Image

Object

F

Near point

$d_o = 25$ cm

d_i

Eyeglass

(a)

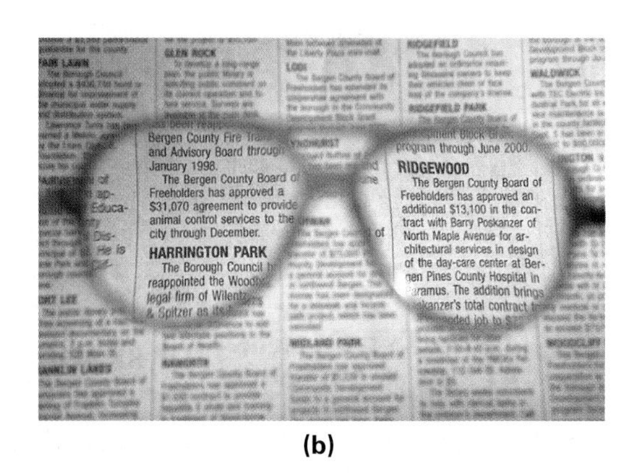
Bergen County Fire
and Advisory Board through
January 1998.
The Bergen County Board of
Freeholders has approved a
$31,070 agreement to provide
animal control services to the
city through December.
HARRINGTON PARK
The Borough Council
reappointed the Wood
legal firm of Wilent
& Spitzer as

RIDGEWOOD
The Bergen County Board of
Freeholders has approved an
additional $13,100 in the con-
tract with Barry Poskanzer of
North Maple Avenue for architectural services in design
of the day-care center at Bergen Pines County Hospital in
ramus. The addition brings
kanzer's total contract

(b)

▲ **FIGURE 25.4 Reading glasses and correcting farsightedness** **(a)** When an object at the normal near point (25 cm) is viewed through reading glasses with converging lenses, the image is formed farther away, but within the eye's range of accommodation (beyond the receded near point). See Integrated Example 25.2. **(b)** Small print as viewed through the lens of reading glasses. The camera used to take this picture is focused past this page onto where the virtual image is.

The optics of these two eyes are usually different (as is typical), as in this problem, and a different lens prescription is usually required for each eye. In this case, each lens is to form an image at its eye's near point of an object that is at a distance (d_o) of 0.25 m. The image will then act as an object within the eye's range of accommodation. This situation is similar to a person wearing reading glasses (▲ Fig. 25.4). (For the sake of clarity, the lens in Fig. 25.4a is not in contact with the eye.)

The image distances are negative, because the images are virtual (that is, the image is on the same side as the object). With contact lenses, the distance from the eye to the object and the distance from the lens to the object are the same. Then

$$P_L = \frac{1}{f_L} = \frac{1}{d_o} + \frac{1}{d_{i_L}} = \frac{1}{0.25\text{ m}} - \frac{1}{0.75\text{ m}} = \frac{2}{0.75\text{ m}} = +2.7\text{ D}$$

and

$$P_R = \frac{1}{f_R} = \frac{1}{d_o} + \frac{1}{d_{i_R}} = \frac{1}{0.25\text{ m}} - \frac{1}{1.0\text{ m}} = \frac{3}{1.0\text{ m}} = +3.0\text{ D}$$

Note that the left lens has less power than the right lens, as expected.

Follow-Up Exercise. A mistake is made in grinding or molding the corrective lenses in this Example such that the left lens is made to the prescription intended for the right eye, and vice versa. Discuss what happens to the images of an object at a distance of 25 cm.

Another common defect of vision is **astigmatism**, which is usually due to a refractive surface, normally the cornea or crystalline lens, being out of round (nonspherical). As a result, the eye has different focal lengths in different planes (▼Fig. 25.5a). Points may appear as lines, and the image of a line may be distinct in one direction and blurred in another or blurred in both directions. A test for astigmatism is given in Fig. 25.5b.

Astigmatism can be corrected with lenses that have greater curvature in the plane in which the cornea or crystalline lens has deficient curvature (Fig. 25.5c). Astigmatism is lessened in bright light, because the pupil of the eye becomes smaller, so only rays near the axis are entering the eye, thus avoiding the outer edges of the cornea.

You have probably heard of *20/20 vision*. But what is it? *Visual acuity* is a measure of how vision is affected by object distance. This quantity is commonly determined by using a chart of letters placed at a given distance from the eyes. The result is usually expressed as a fraction: The *numerator* is the distance at which the test eye sees a standard symbol, such as the letter E, clearly; the *denominator* is the distance at which

▼ **FIGURE 25.5 Astigmatism** When one of the eye's refracting components is not spherical, the eye has different focal lengths in different planes. **(a)** The effect occurs because rays in the vertical plane (red) and horizontal plane (blue) are focused at different points: F_v and F_h, respectively. **(b)** To someone with eyes that are astigmatic, some or all of the lines in this diagram will appear blurred. **(c)** Nonspherical lenses, such as plano-convex cylindrical lenses, are used to correct astigmatism.

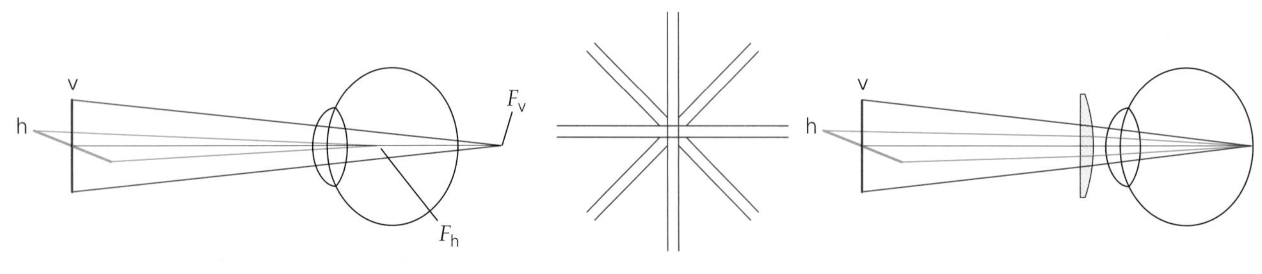

(a) Uncorrected astigmatism **(b) Test for astigmatism** **(c) Corrected by lens**

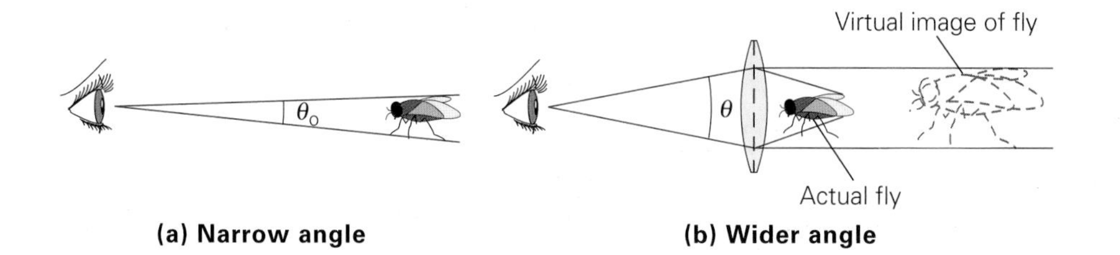

◀ **FIGURE 25.6 Magnification and angle** **(a)** How large an object appears is related to the angle subtended by the object. **(b)** The angle and the size of the virtual image of an object are increased with a converging lens.

the letter is seen clearly by a *normal* eye. A 20/20 (test/normal) rating, which is sometimes called "perfect" vision, means that at a distance of 20 ft, the eye being tested can see standard-sized letters as clearly as can a normal eye.

25.2 Microscopes

OBJECTIVES: To (a) distinguish between lateral and angular magnification, and (b) describe simple and compound microscopes and their magnifications.

Microscopes are used to magnify objects so that we can see more detail or see features that are normally indiscernible. Two basic types of microscopes will be considered here.

The Magnifying Glass (A Simple Microscope)

When we look at a distant object, it appears very small. As it moves closer to our eyes, it appears larger. How large an object appears depends on the size of the image on the retina. This size is related to the angle subtended by the object (▲Fig. 25.6): the greater the angle, the larger the image.

When we want to examine detail or look at something closely, we bring it close to our eyes so that it subtends a greater angle. For example, you may examine the detail of a figure in this book by bringing it closer to your eyes. You'll see the greatest amount of detail when the book is at your near point. If your eyes were able to accommodate to shorter distances, an object brought very close to them would appear even larger. However, as you can easily prove by bringing this book very close to your eyes, images are blurred when objects are inside the near point.

A **magnifying glass**, which is simply a single convex lens (sometimes called a *simple microscope*), forms a clear image of an object when it is closer than the near point (Fig. 23.15b). In such a position, the image of an object subtends a greater angle and therefore appears larger, or magnified (▼Fig. 25.7). The lens produces a virtual image beyond the near point on which the eye focuses. If a handheld magnifying glass is used, its position is usually adjusted until this image is seen clearly.

As illustrated in Fig. 25.7, the angle subtended by the virtual image of an object is much greater when a magnifying glass is used. The magnification of an object *viewed through a magnifying glass* is expressed in terms of this angle. This **angular magnification**, or *magnifying power*, is designated by the symbol m. The angular magnification is defined as the ratio of the angular size of the object as viewed

▼ **FIGURE 25.7 Angular magnification** The angular magnification (m) of a lens is defined as the ratio of the angular size of an object viewed through the lens to the angular size of the object viewed without the lens: $m = \theta/\theta_o$.

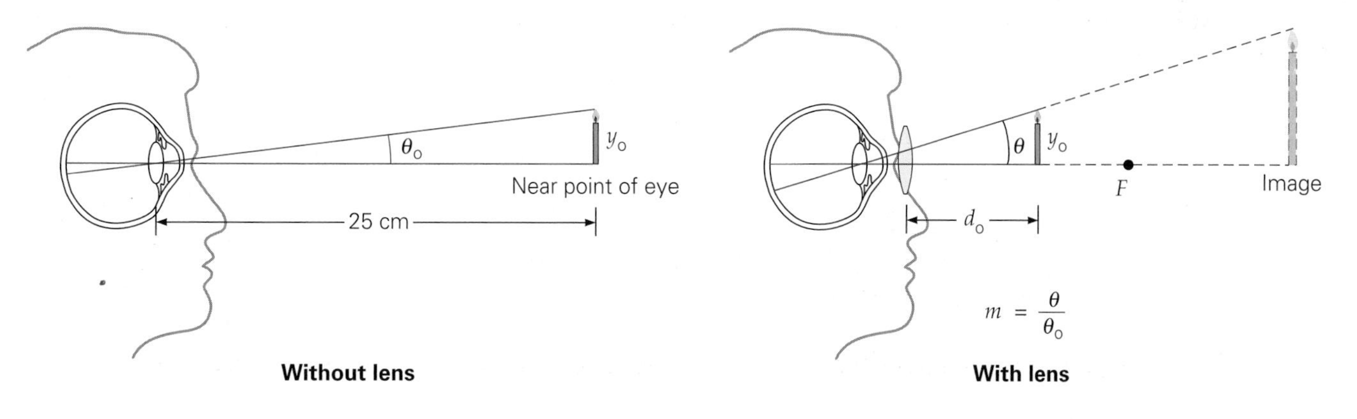

Note: Angular magnification is not the same as lateral magnification, which is discussed in Section 23.1. (See Eq. 23.1.)

through the magnifying glass (θ) to the angular size of the object as viewed without the magnifying glass (θ_o):

$$m = \frac{\theta}{\theta_o} \quad \textit{angular magnification} \tag{25.1}$$

(This m is not the same as M, the lateral magnification, which is a ratio of heights: $M = h_i/h_o$.)

The maximum angular magnification occurs when the image is at the eye's near point, $d_i = -25$ cm, since this position is as close as it can be seen clearly. (A value of 25 cm will be assumed to be typical for the near point of the normal eye. The minus sign is used because the image is virtual; see Chapter 23.) The corresponding object distance can be calculated from the thin-lens equation, Eq. 23.5, as

$$d_o = \frac{d_i f}{d_i - f} = \frac{(-25\text{ cm})f}{-25\text{ cm} - f}$$

or

$$d_o = \frac{(25\text{ cm})f}{25\text{ cm} + f} \tag{25.2}$$

where f must be in centimeters.

The angular sizes of the object are related to its height by

$$\tan\theta_o = \frac{y_o}{25} \quad \text{and} \quad \tan\theta = \frac{y_o}{d_o}$$

(See Fig. 25.7.) Assuming that a small-angle approximation ($\tan\theta \approx \theta$) is valid,

$$\theta_o \approx \frac{y_o}{25} \quad \text{and} \quad \theta \approx \frac{y_o}{d_o}$$

Then the maximum angular magnification can be expressed as

$$m = \frac{\theta}{\theta_o} = \frac{y_o/d_o}{y_o/25} = \frac{25}{d_o}$$

Substituting for d_o from Eq. 25.2 gives

$$m = \frac{25}{25f/(25 + f)}$$

which simplifies to

$$m = 1 + \frac{25\text{ cm}}{f} \quad \textit{angular magnification for image at near point (25 cm)} \tag{25.3}$$

where f is in centimeters. Lenses with shorter focal lengths give greater angular magnifications.

In the derivation of Eq. 25.3, the object being viewed by the unaided eye was taken to be at the near point, as was the image viewed through the lens. Actually, the normal eye can focus on an image located anywhere between the near point and infinity. At the extreme at which the image is at infinity, the eye is more relaxed—the muscles attached to the crystalline lens are relaxed, and the lens is thin. For the image to be at infinity, the object must be at the focal point of the lens. In this case,

$$\theta \approx \frac{y_o}{f}$$

and the angular magnification is

$$m = \frac{25\text{ cm}}{f} \quad \textit{angular magnification for image at infinity} \tag{25.4}$$

Mathematically, it seems that the magnifying power can be increased to any desired value by using lenses that have sufficiently short focal lengths. Physically, however, lens aberrations limit the practical range of a single magnifying glass to about 3× or 4×, or a sharp image magnification of three or four times the size of the object when used normally.

Example 25.3 ■ Elementary: Angular Magnification of a Magnifying Glass

Sherlock Holmes uses a converging lens with a focal length of 12 cm to examine the fine detail of some cloth fibers found at the scene of a crime. (a) What is the maximum magnification given by the lens? (b) What is the magnification for relaxed-eye viewing?

Thinking It Through. Equations 25.3 and 25.4 apply here. Part (a) asks for the maximum magnification, which is discussed in the derivation of Eq. 25.3 and occurs when the image formed by the lens is at the near point of the eye. For part (b), note that the eye is most relaxed when viewing distant objects.

Solution.

Given: $f = 12$ cm   *Find:* (a) $m\ (d_i = \text{near point})$
(b) $m\ (d_i = \infty)$

(a) For Equation 25.3, the near point was taken to be 25 cm:

$$m = 1 + \frac{25\text{ cm}}{f} = 1 + \frac{25\text{ cm}}{12\text{ cm}} = 3.1\times$$

(b) Equation 25.4 gives the magnification for the image formed by the lens at infinity:

$$m = \frac{25\text{ cm}}{f} = \frac{25\text{ cm}}{12\text{ cm}} = 2.1\times$$

Follow-Up Exercise. Taking the maximum practical magnification of a magnifying glass to be 4×, which would have the longer focal length, a glass for near-point viewing or one for distant viewing, and how much longer?

The Compound Microscope

A compound microscope provides greater magnification than is attained with a single lens, or a simple microscope. A basic **compound microscope** consists of a pair of converging lenses, each of which contributes to the magnification (▼Fig. 25.8a). The converging lens with a relatively short focal length ($f_o < 1$ cm) is known as the **objective**. It produces a real, inverted, and enlarged image of an object positioned slightly beyond its focal point. The other lens, called the **eyepiece**, or **ocular**, has a longer focal length (f_e is a few centimeters) and is positioned so that the image formed by the objective falls just *inside* its focal point. This lens forms a magnified, inverted, and virtual image that is viewed by the observer. In essence, the objective gives a magnified real image, and the eyepiece is a simple magnifying glass.

The **total magnification** (m_{total}) of a lens combination is the *product* of the magnifications produced by the two lenses. The image formed by the objective is larger than its object by a factor M_o equal to the lateral magnification ($M_o = -d_i/d_o$). In Fig. 25.8a, note that the image distance for the objective lens is approximately equal to L, the distance between the lenses—that is, $d_i \approx L$. (The image I_o is formed by the

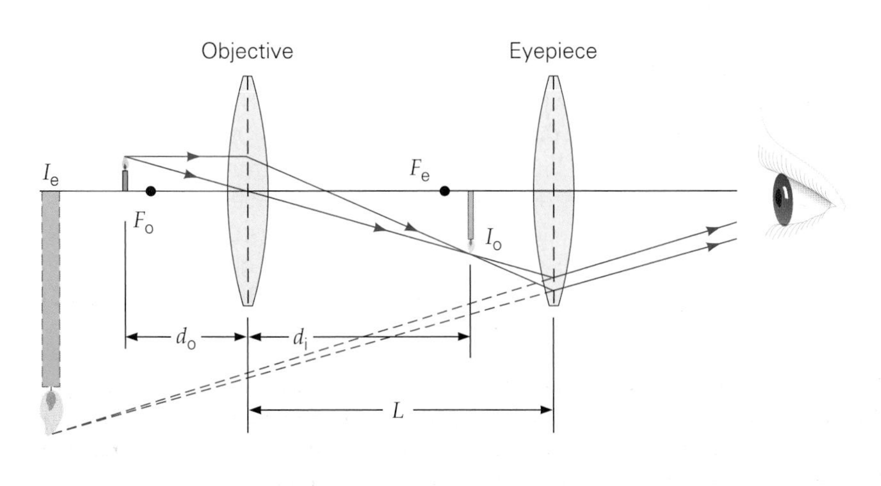

(a)

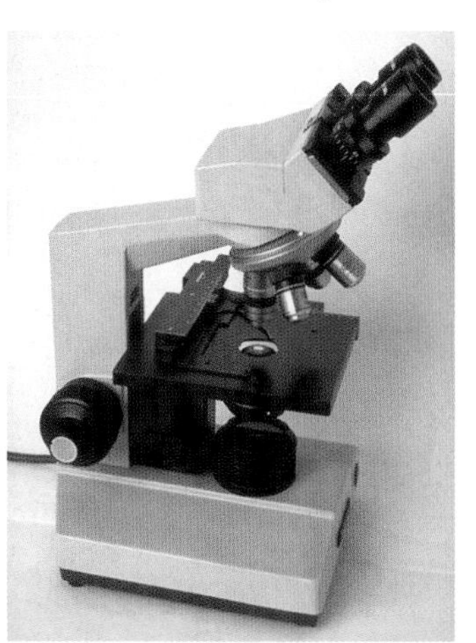

(b)

▼ **FIGURE 25.8 The compound microscope** **(a)** In the optical system of a compound microscope, the real image formed by the objective falls just within the focal point of the eyepiece (F_e) and acts as an object for this lens. An observer looking through the eyepiece sees an enlarged image. **(b)** A compound microscope.

Note: It might be useful to review Section 23.3 and Fig. 23.15.

objective just inside the focal point of the eyepiece.) Also, because the object is very close to the focal point of the objective, $d_o \approx f_o$. With these approximations,

$$M_o \approx -\frac{L}{f_o}$$

Equation 25.4 gives the angular magnification of the eyepiece for an image at infinity.

$$m_e = \frac{25 \text{ cm}}{f_e}$$

Since the object for the eyepiece (the image formed by the objective) is very near the focal point of the eyepiece, a good approximation is given by

$$m_{\text{total}} = M_o m_e = -\left(\frac{L}{f_o}\right)\left(\frac{25 \text{ cm}}{f_e}\right)$$

or

$$m_{\text{total}} = -\frac{(25 \text{ cm})L}{f_o f_e} \quad \textit{angular magnification of compound microscope} \tag{25.5}$$

where f_o, f_e, and L are in centimeters.

The angular magnification of a compound microscope is negative, indicating that the final image is inverted compared to the initial orientation of the object. However, we often state only the magnification (100×, not −100×).

Example 25.4 ■ A Compound Microscope: Finding the Magnification

A microscope has an objective with a focal length of 10 mm and an eyepiece with a focal length of 4.0 cm. The lenses are positioned 20 cm apart in the barrel. Determine the approximate total magnification of the microscope.

Thinking It Through. This is a direct application of Eq. 25.5.

Solution.

Given: $f_o = 10 \text{ mm} = 1.0 \text{ cm}$
$f_e = 4.0 \text{ cm}$
$L = 20 \text{ cm}$

Find: m_{total} (total magnification)

Using Eq. 25.5, we get

$$m_{\text{total}} = -\frac{(25 \text{ cm})L}{f_o f_e} = -\frac{(25 \text{ cm})(20 \text{ cm})}{(1.0 \text{ cm})(4.0 \text{ cm})} = -125\times$$

Note the relatively short focal length of the objective. The negative sign indicates that the final image is inverted.

Follow-Up Exercise. If the focal length of the eyepiece in this Example were doubled, how would the length of the microscope be affected for the same magnification? (Express the change as a percentage.)

A modern compound microscope is shown in Fig. 25.8b. Interchangeable eyepieces with magnifications from about 5× to more than 100× are available. For standard microscopic work in biology or medical laboratories, 5× and 10× eyepieces are normally used. Microscopes are often equipped with rotating turrets, which usually contain three objectives for different magnifications, such as 10×, 43×, and 97×. These objectives and the 5× and 10× eyepieces can be used in various combinations to provide magnifying powers from 50× to 970×. The maximum magnification obtained from a compound microscope is about 2000×.

Opaque objects are usually illuminated with a light source placed above them. Specimens that are transparent, such as cells or thin sections of tissues on glass slides, are illuminated with a light source beneath the microscope stage so that light passes through the specimen. A modern microscope is usually equipped with a light condenser (converging lens) and diaphragm below the stage, which are used to concentrate the light and control its intensity. A microscope may have an internal light source. The light is reflected into the condenser from a mirror. Older microscopes have two mirrors with reflecting surfaces; one is a plane mirror for reflecting light from a high-intensity external source, and the other is a concave mirror for converging low-intensity light such as skylight.

25.3 Telescopes

OBJECTIVES: To (a) distinguish between refractive and reflective telescopes, and (b) describe the advantages of each.

Telescopes apply the optical principles of mirrors and lenses to improve our ability to see distant objects. Used for both terrestrial and astronomical observations, telescopes allow some objects to be viewed in greater detail and other fainter or more distant objects simply to be seen. Basically, there are two types of telescopes—refracting and reflecting—characterized by the gathering and converging of light by lenses or mirrors, respectively.

Refracting Telescope

The principle underlying one type of **refracting telescope** is similar to that of a compound microscope. The major components of a refracting telescope are objective and eyepiece lenses, as illustrated in ▾Fig. 25.9. The objective is a large converging lens with a long focal length, and the movable eyepiece has a relatively short focal length. Rays from a distant object are essentially parallel and form an image (I_o) at the focal point (F_o) of the objective. This image acts as an object for the eyepiece, which is moved until the image lies just inside its focal point (F_e). A large, inverted, virtual image (I_e) is seen by an observer.

For relaxed viewing, the eyepiece is adjusted so that its image (I_e) is at infinity, which means that the objective image (I_o) is at the focal point of the eyepiece (f_e). As Fig. 25.9 shows, the distance between the lenses is then the sum of the focal lengths ($f_o + f_e$), which is the length of the telescope tube. The **magnification of a refracting telescope** focused for the final image at infinity can be shown to be

Exploration 36.2 Telescope

$$m = -\frac{f_o}{f_e} \quad \textit{angular magnification of refracting telescope} \tag{25.6}$$

where the minus sign is inserted to indicate that the image is inverted, as in our lens sign convention described in Section 23.3. Thus, to achieve the greatest magnification, the focal length of the objective should be made as long as possible and the focal length of the eyepiece as short as possible.

The telescope illustrated in Fig. 25.9 is called an **astronomical telescope**. The final image produced by an astronomical telescope is inverted, but this condition poses little

Note: Astronomical telescopes give an inverted image.

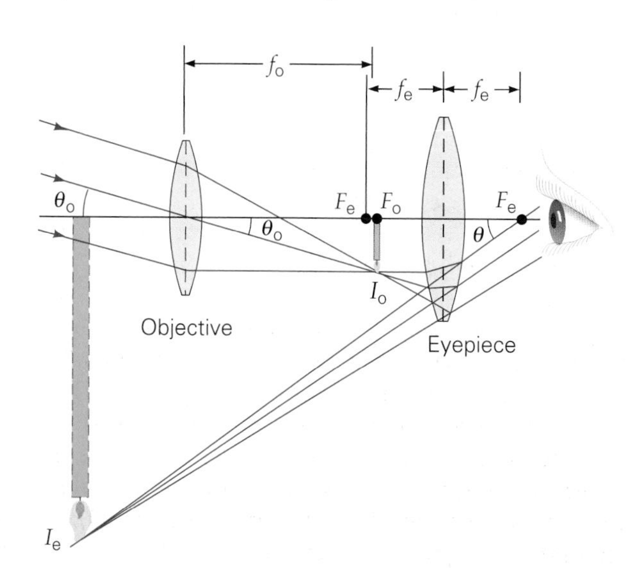

◀ **FIGURE 25.9 The refracting astronomical telescope** In an astronomical telescope, rays from a distant object form an intermediate image (I_o) at the focal point of the objective (F_o). The eyepiece is moved so that the image is at or slightly inside its focal point (F_e). An observer sees an enlarged image at infinity (I_e, shown at a finite distance here for illustration).

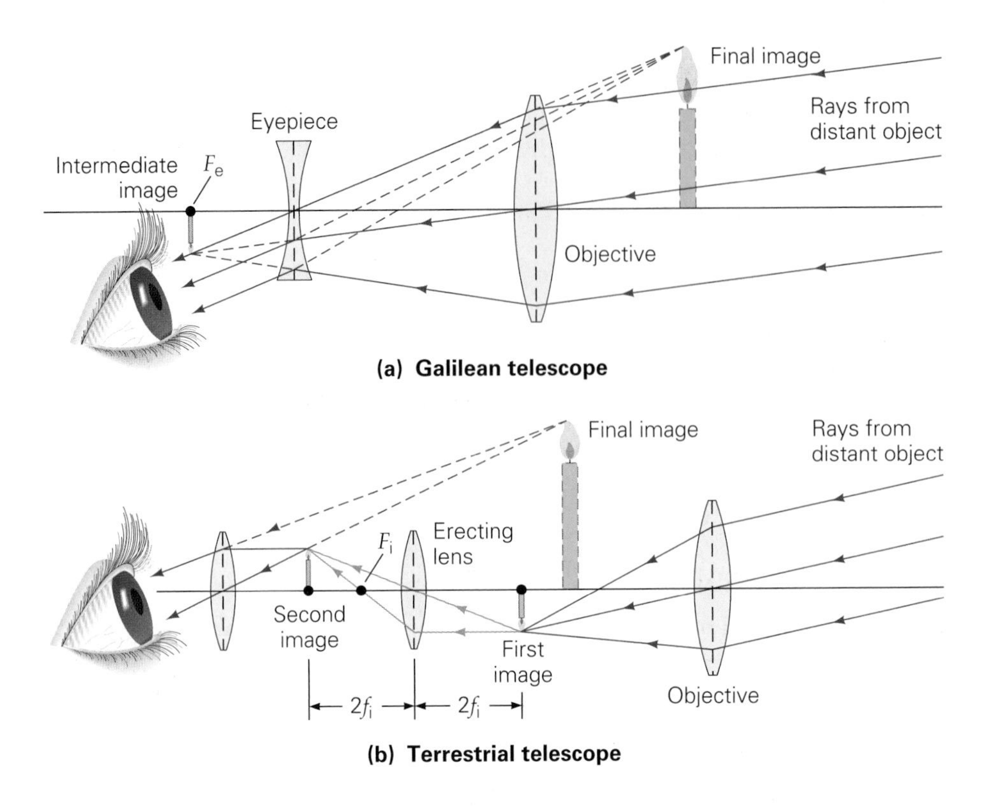

▶ **FIGURE 25.10 Terrestrial telescopes** **(a)** A Galilean telescope uses a diverging lens as an eyepiece, producing upright, virtual images. **(b)** Another way to produce upright images is to use a converging "erecting" lens (focal length f_i) between the objective and eyepiece in an astronomical telescope. This addition elongates the telescope, but the length can be shortened by using internally reflecting prisms.

problem to astronomers. (Why?) However, someone viewing an object on Earth through a telescope finds it more convenient to have an upright image. A telescope in which the final image is upright is called a **terrestrial telescope**. An upright final image can be obtained in several ways; two are illustrated in ▲Fig. 25.10.

In the telescope diagrammed in Fig. 25.10a, a diverging lens is used as an eyepiece. This type of terrestrial telescope is referred to as a *Galilean telescope*, because Galileo built one in 1609. A real image is formed by the objective to the left of the eyepiece, and this image acts as a "virtual" object for the eyepiece. (See Section 23.3.) An observer sees a magnified, upright, virtual image. (Note that with a diverging lens and negative focal length, Eq. 25.6 gives a positive m, indicating an upright image.)

Galilean telescopes have several disadvantages, most notably very narrow fields of view and limited magnification. A better type of terrestrial telescope, illustrated in Fig. 25.10b, uses a third lens, called the *erecting lens*, or *inverting lens*, between converging objective and eyepiece lenses. If the image is formed by the objective at a distance that is twice the focal length of the intermediate erecting lens ($2f_i$), then the lens merely inverts the image without magnification, and the telescope magnification is still given by Eq. 25.6.

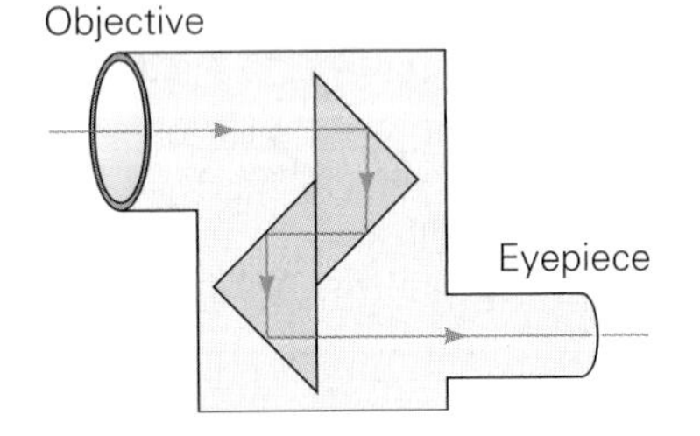

▲ **FIGURE 25.11 Prism binoculars** A schematic cutaway view of one ocular (one half of a pair of prism binoculars), showing the internal reflections in the prisms, which reduce the overall physical length.

However, to achieve the upright image in this way requires a greater telescope length. Using the intermediate erecting lens to invert the image increases the length of the telescope by four times the focal length of the erecting lens ($2f_i$ on each side). The inconvenient length can be avoided by using internally reflecting prisms. This is the principle behind prism binoculars, which are really double telescopes—one for each eye (◀Fig. 25.11).

Example 25.5 ■ An Astronomical Telescope—and a Longer Terrestrial Telescope

An astronomical telescope has an objective lens with a focal length of 30 cm and an eyepiece with a focal length of 9.0 cm. (a) What is the magnification of the telescope? (b) If an erecting lens with a focal length of 7.5 cm is used to convert the telescope to a terrestrial type, what is the overall length of the telescope tube?

Thinking It Through. Equation 25.6 applies directly in part (a). In part (b), the erecting lens elongates the telescope by four times the focal length of the lens ($4f_i$) to the length of the scope (Fig. 25.10b).

Solution. Listing the data, we have

Given: $f_o = 30$ cm
$f_e = 9.0$ cm
$f_i = 7.5$ cm (intermediate erecting lens)

Find: (a) m (magnification)
(b) L (length of telescope tube)

(a) The magnification is given by Eq. 25.6 as

$$m = -\frac{f_o}{f_e} = -\frac{30 \text{ cm}}{9.0 \text{ cm}} = -3.3\times$$

where the minus sign indicates that the final image is inverted.

(b) Taking the length of the astronomical tube to be the distance between the lenses, we find that this length is just the sum of the lenses' focal lengths:

$$L_1 = f_o + f_e = 30 \text{ cm} + 9.0 \text{ cm} = 39 \text{ cm}$$

The overall length is then

$$L = L_1 + L_2 = 39 \text{ cm} + 4f_i = 39 \text{ cm} + 4(7.5 \text{ cm}) = 69 \text{ cm}$$

Hence, the telescope length is more than two-thirds of a meter, with an upright image, but the same magnification, 3.3× (why?).

Follow-Up Exercise. A terrestrial telescope 66 cm in length has an intermediate erecting lens with a focal length of 12 cm. What is the focal length of an erecting lens that would reduce the telescope length to a more manageable 50 cm?

Conceptual Example 25.6 ■ Constructing a Telescope

A student is given two converging lenses, one with a focal length of 5.0 cm and the other with a focal length of 20 cm. To construct a telescope to best view distant objects with these lenses, the student should hold the lenses (a) more than 25 cm apart; (b) less than 25 cm, but more than 20 cm, apart; (c) less than 20 cm, but more than 5.0 cm, apart; (d) less than 5.0 cm apart. Specify which lens should be used as the eyepiece.

Reasoning and Answer. First let's see which lens should be used as the eyepiece. The only type of telescope that can be constructed with two converging lenses is an astronomical telescope. In this type of telescope, the lens with the longer focal length is used as an objective lens to produce a real image of a distant object. That image is then viewed with the lens with the shorter focal length, the eyepiece, used as a simple magnifier.

If the object is at a great distance, a real image is formed by the objective lens in the focal plane of the lens (Fig. 25.9). This image acts as the object for the eyepiece, which is positioned so that the image/object lies just inside its focal point so as to produce a large, inverted second image.

The two lenses must be *slightly* less than 25 cm apart, so answer (a) is not correct. Answers (c) and (d) are also not correct, because the eyepiece would be too close to the objective to produce the large secondary image needed for optimal viewing of a distant object. In these cases, the rays would pass through the second lens before the image was formed, and a *reduced* image might be produced. (See Section 23.3.) Thus, answer (b), with the objective image just inside the eyepiece's focal point, is the correct answer.

Follow-Up Exercise. A third converging lens with a focal length of 4.0 cm is used with the aforementioned two lenses to produce a terrestrial telescope in which the third lens does nothing more than invert the image. How should the lenses be positioned and how far apart should they be for the final image to be of maximum size and upright?

Reflecting Telescope

For viewing the Sun, Moon, and nearby planets, large magnifications are important to see details. However, even with the highest feasible magnification, stars appear only as faint points of light. For distant stars and galaxies, it is more important to gather more light than to increase the magnification, so that the object can be seen and its spectrum analyzed more quickly. The intensity of light from a distant source is sometimes very low. In many instances, such a source can be detected only when the light is gathered and focused on a photographic plate over a long period of time.

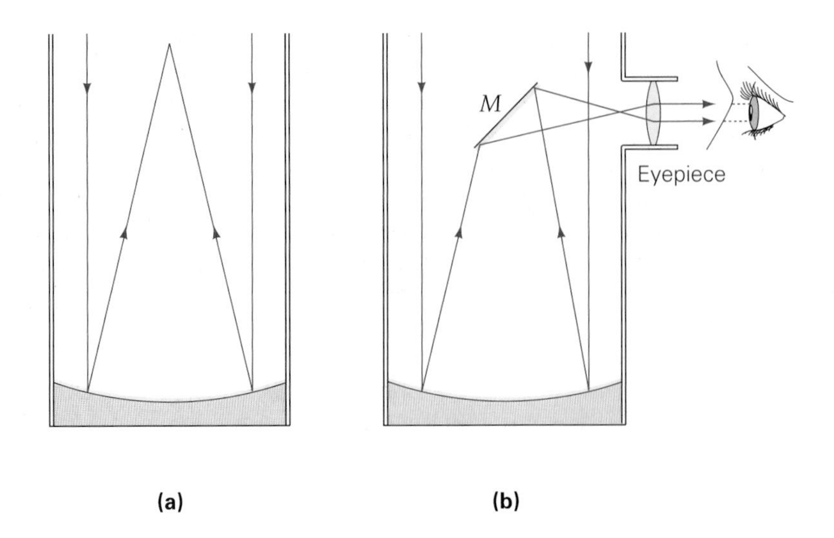

▶ **FIGURE 25.12 Reflecting telescopes** A concave mirror can be used in a telescope to converge light to form an image of a distant object. **(a)** The image may be at the prime focus, or **(b)** a small mirror and lens can be used to focus the image outside the telescope, a configuration called a *Newtonian focus*.

Recall from Section 14.3 that intensity is energy per unit time per unit *area*. Thus, more light energy can be gathered if the size of the objective is increased. This increases the distance at which the telescope can detect faint objects, such as distant galaxies. (Recall that the light intensity of a point source is inversely proportional to the *square* of the distance between the source and the observer.) However, producing a large lens involves difficulties associated with glass quality, grinding, and polishing. Compound-lens systems are required to reduce aberrations, and a very large lens may sag under its own weight, producing further aberrations. The largest objective lens in use has a diameter of 40 in. (102 cm) and is part of the refracting telescope of the Yerkes Observatory at Williams Bay, Wisconsin.

These problems can be reduced by using a **reflecting telescope**, which uses a large, concave, parabolic mirror (▲Fig. 25.12). A parabolic mirror does not exhibit spherical aberration, and a mirror has no inherent chromatic aberration. (Why?) High-quality glass is not needed, because the light is reflected by a mirrored surface. Only one surface has to be ground, polished, and silvered.

The largest single-mirror telescope, with a mirror 8.2 m (323 in.) in diameter, is at the European Southern Observatory in Chile (▼Fig 25.13a). The largest reflecting telescope in the United States, with a mirror 5.1 m (200 in.) in diameter, is the Hale Observatory's reflecting telescope on Mount Palomar in California.

Even though reflecting telescopes have advantages over refracting telescopes, they do have problems. Like a large lens, a large mirror may sag under its own weight, and the weight necessarily increases with the size of the mirror. The weight factor also increases the costs of construction, because the supporting elements for a heavier mirror must be more massive.

These problems are being addressed by new technologies. One approach is to use an array of small mirrors, coordinated to function as a single large mirror. Examples include

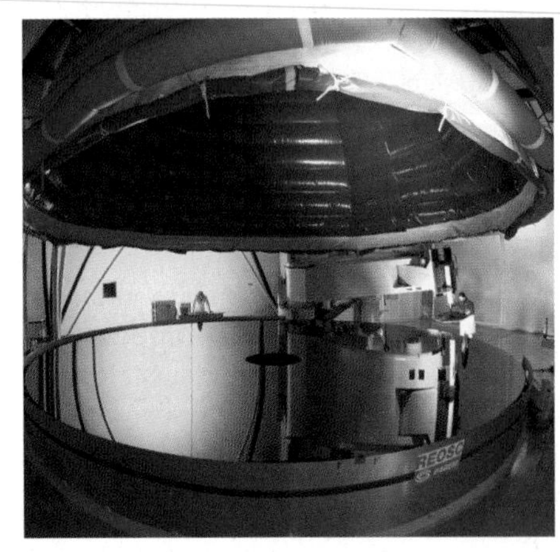

(a) (b)

▶ **FIGURE 25.13 European Southern Observatory, near Paranal, Chile** **(a)** An 8.2-m-diameter mirror is undergoing the final phase of polishing. **(b)** Four 8.2-m telescopes will form a VLT (Very Large Telescope) with an equivalent diameter of 16 m.

the twin Keck telescopes at Mauna Kea in Hawaii. Each has a mirror consisting of 36 hexagonal segments that are computer positioned to be equivalent to a 10-m mirror. The European Southern Observatory plans to have four 8.2-m-diameter mirrors linked to form a VLT (Very Large Telescope,) with an equivalent diameter of 16 m (Fig. 25.13b).

Another way of extending our view into space is to put telescopes into orbit around the Earth. Above the atmosphere, the view is unaffected by the twinkling effect of atmospheric turbulence and refraction, and there is no background problem from city lights. In 1990, the optical Hubble Space Telescope (HST) was launched into orbit (▸Fig. 25.14). Even with a mirror diameter of only 2.4 m, its privileged position has allowed the HST to produce images seven times as clear as those formed by Earth-bound telescopes.

Lastly, you may know that not all telescopes are in the visible region. To see some of these, read Insight 25.2 on Telescopes Using Nonvisible Radiation, on page 808.

▲ **FIGURE 25.14 Hubble Space Telescope (HST)** Late in 1993, astronauts from the space shuttle *Endeavor* visited the HST in orbit. They installed corrective equipment that compensated for many of the telescope's optical flaws and repaired or replaced other malfunctioning systems. The HST is currently in need of repairs.

25.4 Diffraction and Resolution

OBJECTIVES: To (a) describe the relationship of diffraction and resolution, and (b) state and explain Rayleigh's criterion.

The diffraction of light places a limitation on our ability to distinguish objects that are close together when we use microscopes or telescopes. This effect can be understood by considering two point sources located far from a narrow slit of width w (▾Fig. 25.15). The sources could represent distant stars, for example. In the absence of diffraction, two bright spots, or images, would be observed on a screen. As you know from Section 24.3, however, the slit diffracts the light, and each image consists of a central maximum with a pattern of weaker bright and dark fringes on either side. If the sources are close together, the two central maxima may overlap. In this case, the images cannot be distinguished, or are said to be *unresolved*. For the images to be *resolved*, the central maxima must not overlap appreciably.

In general, images of two sources can be resolved if the central maximum of one falls at or beyond the first minimum of the other. This limiting condition for the **resolution** of two images—that is, the ability to distinguish them as separate—was first proposed by Lord Rayleigh (1842–1919), a British physicist. The condition is known as the **Rayleigh criterion**:

> Two images are said to be just resolved when the central maximum of one image falls on the first minimum of the diffraction pattern of the other image.

The Rayleigh criterion can be expressed in terms of the angular separation (θ) of the sources. (See Fig. 25.15.) The first minimum ($m = 1$) for a single-slit diffraction pattern satisfies this relationship:

$$w \sin\theta = m\lambda = \lambda \qquad \text{or} \qquad \sin\theta = \frac{\lambda}{w}$$

According to Fig. 25.15, this is the minimum angular separation for two images to be just resolved according to the Rayleigh criterion. In general, for visible

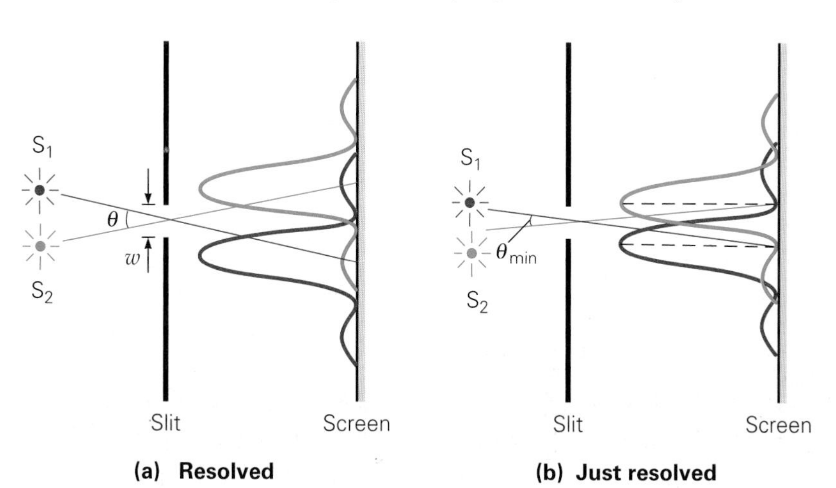

◂ **FIGURE 25.15 Resolution** Two light sources in front of a slit produce diffraction patterns. **(a)** When the angle subtended by the sources at the slit is large enough for the diffraction patterns to be distinguishable, the images are said to be resolved. **(b)** At smaller angles, the central maxima are closer together. At θ_{min}, the central maximum of one image's diffraction pattern falls on the first dark fringe of the other image's pattern, and the images are said to be just resolved. For smaller angles, the patterns are unresolved.

INSIGHT 25.2 TELESCOPES USING NONVISIBLE RADIATION

The word *telescope* usually brings to mind visual observations. However, the visible region is only a very small part of the electromagnetic spectrum, and celestial objects emit many other types of radiation, including radio waves. This fact was discovered accidentally in 1931 by an American electrical engineer named Carl Jansky while he was working on static interference with intercontinental radio communications. Jansky found an annoying static hiss that came from a fixed direction in space, apparently from a celestial source. It was soon clear that radio waves could be another valuable source of astronomical information, and radio telescopes were built to investigate this source.

A radio telescope operates similarly to a reflecting visible light telescope. A reflector with a large area collects and focuses the radio waves at a point where a detector picks up the signal (Fig. 1). The parabolic collector, called a *dish*, is covered with metal wire mesh or metal plates. Since the wavelengths of radio waves range from a few millimeters to several meters, wire mesh is "smooth" enough and a good reflecting surface for such waves.

Radio telescopes supplement optical telescopes and provide some definite advantages over them. For instance, radio waves pass freely through the huge clouds of dust that hide a large part of our galaxy from visual observation. Also, radio waves easily penetrate the Earth's atmosphere, which reflects and scatters a large percentage of the incoming visible light.

Infrared light is also affected by the Earth's atmosphere. For example, water vapor is a strong absorber of infrared radiation. Thus, observations with infrared telescopes are sometimes made from high-flying aircraft or from orbiting spacecraft, beyond the influence of atmospheric water vapor. The first *orbiting* infrared observatory was launched in 1983. Not only is atmospheric interference eliminated in space, but the telescope may be cooled to a very low temperature without becoming coated with condensed water vapor. Cooling the telescope helps eliminate the interference of infrared radiation generated by the telescope itself. The orbiting infrared telescope launched in 1983 was cooled with liquid helium to about 10 K; it carried out an infrared survey of the entire sky.

FIGURE 1 Radio telescopes Several of the dish antennae that make up the Very Large Array (VLA) radio telescope near Socorro, New Mexico. There are 27 movable dishes, each 25 m in diameter, forming the array along a Y-shaped railway network. The data from all the antennae are combined to produce a single radio image. In this way, it is possible to attain a resolution equivalent to that of one giant radio dish (a couple hundred feet in diameter).

The atmosphere is virtually opaque to ultraviolet radiation, X-rays, and gamma rays, so telescopes that detect these types of radiation cannot be Earth based. Orbiting satellites with telescopes sensitive to these types of radiation have mapped out portions of the sky, and other surveys are planned. Observatories by orbiting satellites in the visible region are not affected by air turbulence or refraction. Perhaps in the not-too-distant future, a permanently staffed orbiting observatory carrying a variety of telescopes will replace the uncrewed Hubble Telescope and help expand our knowledge of the universe.

light, the wavelength is much smaller than the slit width ($\lambda < w$), so θ is small and $\sin\theta \approx \theta$. In this case, the limiting, or **minimum angle of resolution (θ_{min})** for a slit of width w is

$$\theta_{min} = \frac{\lambda}{w} \quad \textit{minimum angle of resolution (for a slit)} \tag{25.7}$$

(Note that θ_{min} is a pure number and is therefore in radians.) Thus, the images of two sources will be *distinctly* resolved if the angular separation of the sources is greater than λ/w.

The apertures (openings) of cameras, microscopes, and telescopes are generally circular. Thus, there is a *circular* diffraction pattern around the central maximum, in the form of a bright circular disk (▸Fig. 25.16). Detailed analysis shows that the **minimum angle of resolution for a circular aperture** for the images of two objects to be just resolved is similar to, but slightly different from, Eq. 25.7. It is

$$\theta_{min} = \frac{1.22\lambda}{D} \quad \textit{minimum angle of resolution (for a circular aperture)} \tag{25.8}$$

where D is the diameter of the aperture and θ_{min} is in radians.

Equation 25.8 applies to the objective lens of a microscope or telescope, or the iris of the eye, all of which may be considered to be circular apertures for light. According to Eqs. 25.7 and 25.8, the smaller θ_{min}, the better the resolution. The minimum angle of resolution, $\theta_{\min}$, should be small so that objects close together can be resolved; therefore, the aperture should be as *large* as possible. This is yet another reason for using large lenses (and mirrors) in telescopes.

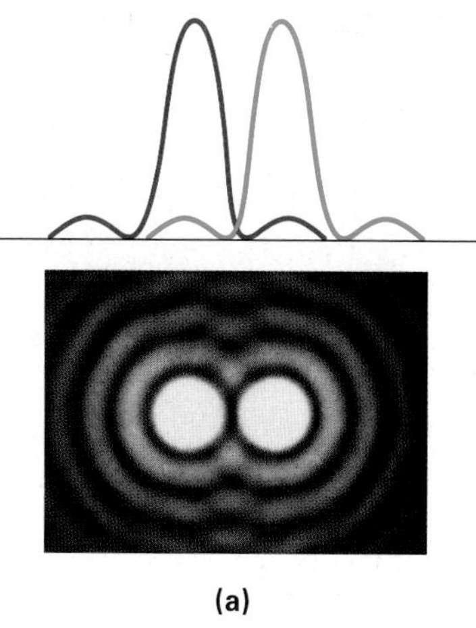

(a)

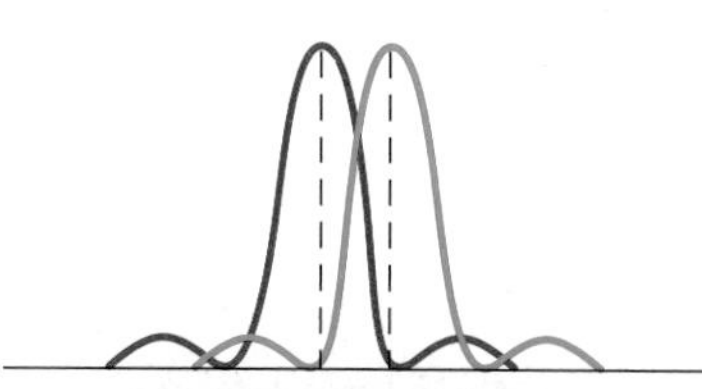

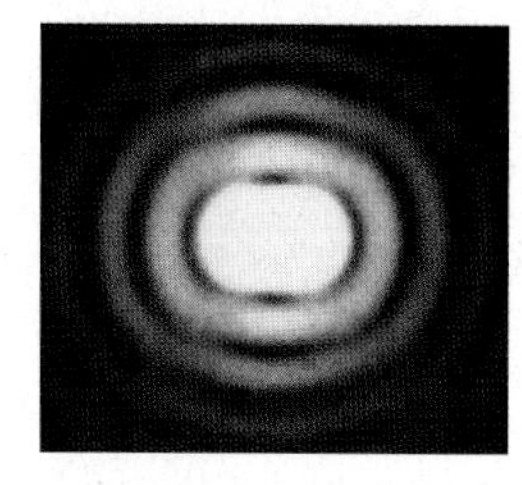

(b)

▲ **FIGURE 25.16 Circular-aperture resolution (a)** When the angular separation of two objects is large enough, the images are well resolved. (Compare with Fig. 25.15a.) **(b)** Rayleigh criterion: The central maximum of the diffraction pattern of one image falls on the first minimum of the diffraction pattern of the other image. (Compare with Fig. 25.15b.) The images of objects with smaller angular separations cannot be clearly distinguished as individual images.

Example 25.7 ■ The Eye and Telescope: Evaluating Resolution with the Rayleigh Criterion

Determine the minimum angle of resolution by the Rayleigh criterion for (a) the pupil of the eye (daytime diameter of about 4.0 mm) for visible light with a wavelength of 660 nm; (b) the European Southern Observatory refracting telescope (diameter of 8.2 m), for visible light of the same wavelength as in part (a); and (c) a radio telescope 25 m in diameter for radiation with a wavelength of 21 cm.

Thinking It Through. This is a comparison of $\theta_{\min}$ for apertures with different diameters—a direct application of Eq. 25.8.

Solution.

Given: (a) $D = 4.0\ \text{mm} = 4.0 \times 10^{-3}\ \text{m}$
$\lambda = 660\ \text{nm} = 6.60 \times 10^{-7}\ \text{m}$
(b) $D = 8.2\ \text{m}$
$\lambda = 660\ \text{nm} = 6.60 \times 10^{-7}\ \text{m}$
(c) $D = 25\ \text{m}$
$\lambda = 21\ \text{cm} = 0.21\ \text{m}$

Find: (a) $\theta_{\min}$ (minimum angles of resolution)
(b) $\theta_{\min}$
(c) $\theta_{\min}$

(a) For the eye,

$$\theta_{\min} = \frac{1.22\lambda}{D} = \frac{1.22(6.60 \times 10^{-7}\ \text{m})}{4.0 \times 10^{-3}\ \text{m}} = 2.0 \times 10^{-4}\ \text{rad}$$

(b) For the light telescope,

$$\theta_{\min} = \frac{1.22(6.60 \times 10^{-7}\ \text{m})}{8.2\ \text{m}} = 9.8 \times 10^{-8}\ \text{rad}$$

(*Note*: The resolution of Earth-bound telescopes with large-diameter objectives is usually not limited by diffraction, but rather by other effects, such as atmospheric turbulence. Thus, in actuality, Earth-bound telescopes have a $\theta_{\min}$ on the order of 10^{-6} rad, or resolution one tenth as good as without the atmosphere.)

(c) For the radio telescope,

$$\theta_{\min} = \frac{1.22(0.21\ \text{m})}{25\ \text{m}} = 0.010\ \text{rad}$$

The smaller the angular separation, the better the resolution. What do the results tell you?

Follow-Up Exercise. As noted in Section 25.3, the Hubble Space Telescope has a mirror diameter of 2.4 m. How does its resolution compare with that of the largest Earth-bound telescopes? (See the note in part (b) of this Example.)

For a microscope, it is more convenient to specify the actual separation (s) between two point sources. Since the objects are usually near the focal point of the objective, to a good approximation,

$$\theta_{\min} = \frac{s}{f} \quad \text{or} \quad s = f\theta_{\min}$$

where f is the focal length of the lens and $\theta_{\min}$ is expressed in radians. (Here, s is taken as the arc length subtended by $\theta_{\min}$, and $s = r\theta_{\min} = f\theta_{\min}$.) Then, using Eq. 25.8, we get

$$s = f\theta_{\min} = \frac{1.22\lambda f}{D} \quad \textit{resolving power of a microscope} \quad (25.9)$$

This minimum distance between two points whose images can be just resolved is called the **resolving power** of the microscope. Note that s is directly proportional to λ, so shorter wavelength gives better resolution. In practice, the resolving power of a microscope indicates the ability of the objective to distinguish fine detail in specimens' structures. For another real-life example of resolution, see ▾Fig. 25.17.

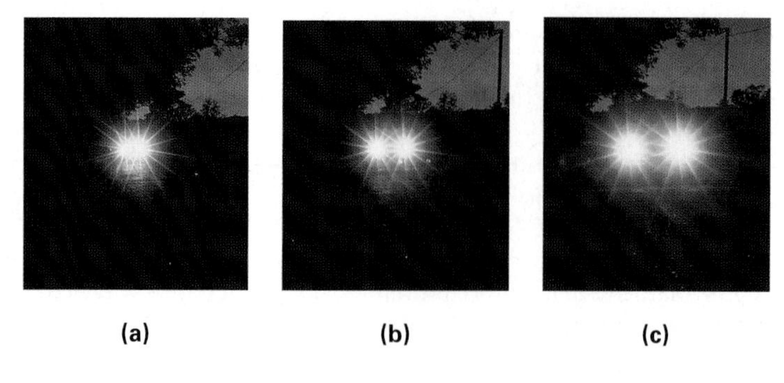

▶ **FIGURE 25.17** Real-life resolution (a), (b), (c) A sequence of an approaching automobile's headlights. In (a), the headlights are almost unresolved through the circular aperture of the camera (or your eye). As the automobile moves closer, the headlights are resolved.

Conceptual Example 25.8 ■ Viewing from Space: The Great Wall of China

The Great Wall of China was originally about 2400 km (1500 mi) long, with a base width of about 6.0 m and a top width of about 3.7 m. Several hundred kilometers of the wall remain intact (◂Fig. 25.18). It is sometimes said that the wall is the only human construction that can be seen with the unaided eye by an astronaut orbiting the Earth. Using the result from part (a) of Example 25.7, see if it is visible. (Neglect any atmospheric effects.)

Reasoning and Answer. Despite the length of the wall, it would not be visible from space unless its *width* subtends the minimum angle of resolution for the eye of an observing astronaut ($\theta_{\text{min}} = 2.0 \times 10^{-4}$ rad from Example 25.7). Actually, guard towers with roofs as wide as 7.0 m were located every 180 m along the Wall. Let's take the maximum observable width to be 7.0 m. (Actually, it is the circular arc length that subtends the angle, but at such a long radius, the chord length is very nearly equal to the circular arc length. Refer to Example 7.2, and make yourself a sketch.)

Let's assume that the astronaut is just able to distinguish the guard roof. Recall that $s = r\theta$ (Eq. 7.3), where s is the maximum observable width of the wall and r is the radial (height) distance. Then, the astronaut would have to be at, or closer than, a distance of

$$r = \frac{s}{\theta} = \frac{7.0 \text{ m}}{2.0 \times 10^{-4} \text{ (rad)}} = 3.5 \times 10^{4} \text{ m} = 35 \text{ km} \; (= 22 \text{ mi})$$

So above 35 km, the wall would *not* be able to be seen with the unaided eye. Orbiting satellites are about 300 km (190 mi) or more above the Earth. The statement about the ability to see the wall from space is false.

Follow-Up Exercise. What would be the minimum diameter of the objective of a telescope that would allow an astronaut orbiting the Earth at an altitude of 300 km to actually see the Great Wall? (Take all conditions to be the same as stated in this Example, and assume that the wavelength of light is 550 nm.)

▲ **FIGURE 25.18** The Great Wall The walkway of the Great Wall of China, which was built as a fortification along China's northern border.

Note: The relationship between wavelength and index of refraction is given in Section 22.3; see Eq. 22.4.

Note from Eq. 25.8 that higher resolution can be gained by using radiation of a shorter wavelength. Thus, a telescope with an objective of a given size will have greater resolution with violet light than with red light. For microscopes, it is possible to increase resolving power by shortening the wavelengths of the light used to create the image. This can be done with a specialized objective called an *oil immersion lens*. When such a lens is used, a drop of transparent oil fills the space between the objective and the specimen. Recall that the wavelength of light in oil is $\lambda' = \lambda/n$, where n is the index of refraction of the oil and λ is the wavelength of light in air. For values of n about 1.50 or higher, the wavelength is significantly reduced, and the resolving power is increased proportionally.

*25.5 Color

OBJECTIVE: To relate color vision and light.

In general, physical properties are fixed or absolute. For example, a particular type of electromagnetic radiation has a certain frequency or wavelength. However, visual *perception* of this radiation may vary from person to person. How we "see" (or our brain "interprets") radiation gives rise to what we call *color vision*.

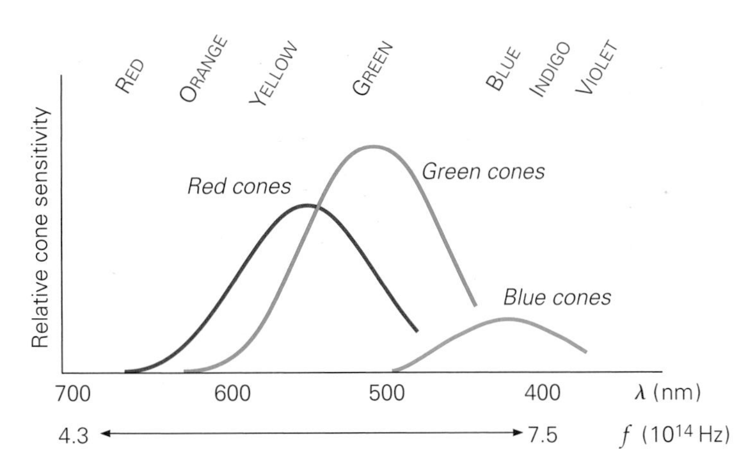

◀ **FIGURE 25.19 Sensitivity of cones** Different types of cones in the retina of the human eye may respond to different frequencies of light to give three general color responses: red, green, and blue.

Color Vision

Color is perceived because of a physiological response to excitation by light of the cone receptors in the retina of the human eye. (Many animals have no cone cells and thus live in a black-and-white world.) The cones are sensitive to light with frequencies approximately between 7.5×10^{14} Hz and 4.3×10^{14} Hz (wavelengths between 400 and 700 nm). The signals representing different frequencies of light are perceived by the brain as different colors. The association of a color with a particular frequency is subjective and may vary from person to person. The concept of pitch is to sound and hearing as color is to light and vision.

The details of color vision are not well understood. It is known that there are three types of cones responding to different parts of the visible spectrum: the red, green, and blue regions (▲Fig. 25.19). Presumably, each cone absorbs light in a specific range of frequencies and all three functionally overlap to form combinations that are interpreted by the brain as the various colors of the spectrum. For example, when red and green cones are stimulated equally, the brain interprets the two superimposed signals as yellow. But when the red cones are stimulated more strongly than the green cones, the brain senses orange. (That is "yellow" but dominated by red.) *Color blindness* results when one or more type of cone is missing or nonfunctional.

As Fig. 25.19 shows, the human eye is not equally sensitive to all colors. Some colors evoke a greater response than others and therefore appear brighter at the same intensity. The wavelength of maximum visual sensitivity is about 550 nm, in the yellow-green region.

The foregoing theory of color vision (mixing or combining) is based on the experimental fact that beams of varying intensities of red, green, and blue light can be arranged to produce most other colors. The red, blue, and green from which we interpret a full spectrum of colors are called the **additive primary colors**. When light beams of the additive primaries are projected and overlapped on a white screen, other colors are produced, as illustrated in ◀Fig. 25.20. This technique is called the **additive method of color production**. Triad dots consisting of three phosphors that emit the additive primary colors are used in television picture tubes to produce colored images.

Note in Fig. 25.20 that a certain combination of the primary colors appears white to the eye. Also, many *pairs* of colors appear white to the eye when combined. The colors of such pairs are said to be **complementary colors**. The complement of blue is yellow, that of red is cyan, and that of green is magenta. As the figure also shows, the complementary color of a particular primary is the combination, or sum, of the other two primaries. Hence, the primary and its complement together appear white.

Edwin H. Land (the developer of Polaroid film) showed that when the proper mixtures of only two wavelengths (colors) of light are passed through black and white transparencies (no color), the wavelengths produce images of various colors. Land wrote, "In this experiment we are forced to the astonishing conclusion that the rays are not in themselves color-making. Rather they are bearers of information that the eye uses to assign appropriate colors to various objects in an image."*

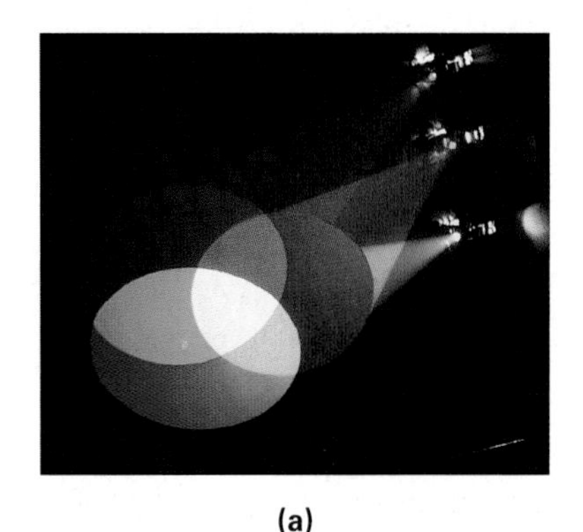

(a)

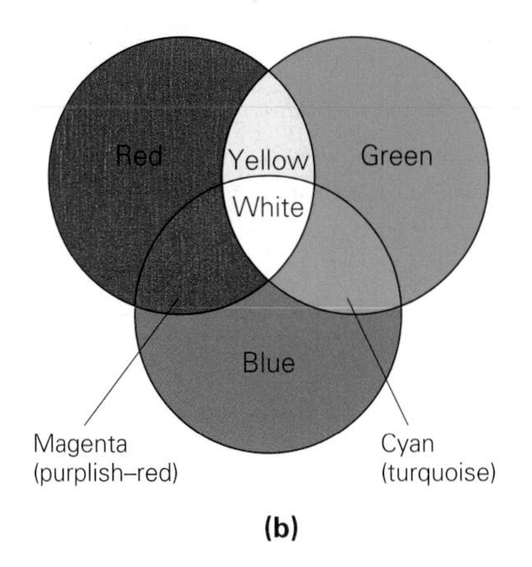

(b)

▲ **FIGURE 25.20 Additive method of color production** When light beams of the primary colors (red, blue, and green) are projected onto a white screen, mixtures of them produce other colors. Varying the intensities of the beams allows most colors to be produced.

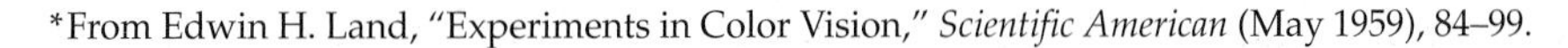

*From Edwin H. Land, "Experiments in Color Vision," *Scientific American* (May 1959), 84–99.

Note: Recall from Chapter 22 that the colors of the visible spectrum may be remembered by the acronym ROY G. BIV.

Objects exhibit a color when they are illuminated with white light because they reflect (scatter) or transmit light predominantly in the frequency range of that color. The other frequencies of the white light are mostly absorbed. For example, when white light strikes a red apple, most of the energy in the red portion of the spectrum is reflected—most of all the others (and thus all other colors) are absorbed. Similarly, when white light passes through a piece of transparent red glass, or a *filter*, mostly the light associated with red is transmitted. This occurs because the color pigments (additives) in the glass are selective absorbers.

Pigments are mixed to form various colors, such as in the production of paints and dyes. You are probably aware that mixing yellow and blue paints produces green. This is because the yellow pigment absorbs most of the wavelengths except those in the yellow and nearby regions (green plus orange) of the visible spectrum, and the blue pigment absorbs most of the wavelengths except those in the blue and nearby regions (violet plus green). The wavelengths in the intermediate (overlap) green region, between the yellow and blue range, are *not* strongly absorbed by either pigment, and therefore the mixture appears green. The same effect can be accomplished by passing white light through stacked yellow and blue filters. The light coming through both filters appears green.

Mixing pigments results in the *subtraction* of colors. The resultant color is created by whatever is *not* absorbed by the pigment—that is, *not* subtracted from the original beam. This is the principle of the **subtractive method of color production**. Three particular pigments—cyan, magenta, and yellow—are the **subtractive primary pigments**. Various combinations of two of the three subtractive primaries produce the three additive primary colors (red, blue, and green), as illustrated in ▼Fig. 25.21. When the subtractive primaries are mixed in the proper proportions, the mixture appears black (because all wavelengths are absorbed). Painters often refer to the subtractive primaries as red, yellow, and blue. They are

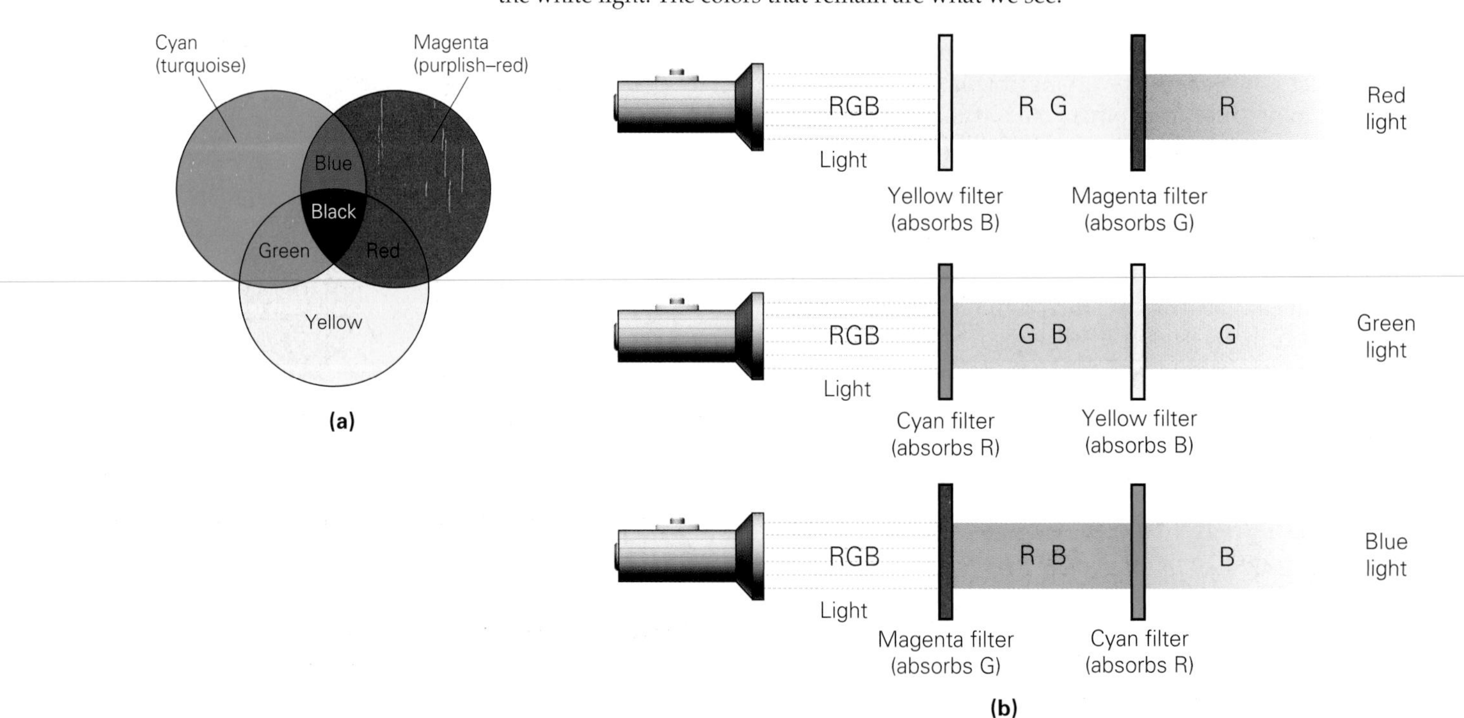

▼ **FIGURE 25.21 Subtractive method of color production (a)** When the primary pigments (cyan, magenta, and yellow) are mixed, different colors are produced by subtractive absorption; for example, the mixing of yellow and magenta produces red. When all three pigments are mixed and all the wavelengths of visible light are absorbed, the mixture appears black. **(b)** Subtractive color mixing, using filters. The principle is the same as in part (a). Each pigment selectively absorbs certain colors, removing them from the white light. The colors that remain are what we see.

loosely referring to magenta (purplish-red), yellow, and cyan ("true" blue). Mixing these paints in the proper proportions produces a broad spectrum of colors.

Note in Fig. 25.21 that the magenta pigment essentially subtracts the color green where it overlaps with cyan and yellow. As a result, magenta is sometimes referred to as "minus green." If a magenta filter were placed in front of a green light, no light would be transmitted. Similarly, cyan is called "minus red," and yellow is called "minus blue." An example of subtractive color mixing is a photographer's use of a yellow filter to bring out white clouds on black and white film. This filter absorbs blue from the sky, darkening it relative to the clouds, which reflect white light. Hence, the contrast between the two is enhanced. What type of filter would you use to darken green vegetation on black and white film? To lighten it?

Chapter Review

- Nearsighted people cannot see distant objects clearly. Farsighted people cannot see nearby objects clearly. These conditions may be corrected by diverging and converging lenses, respectively.

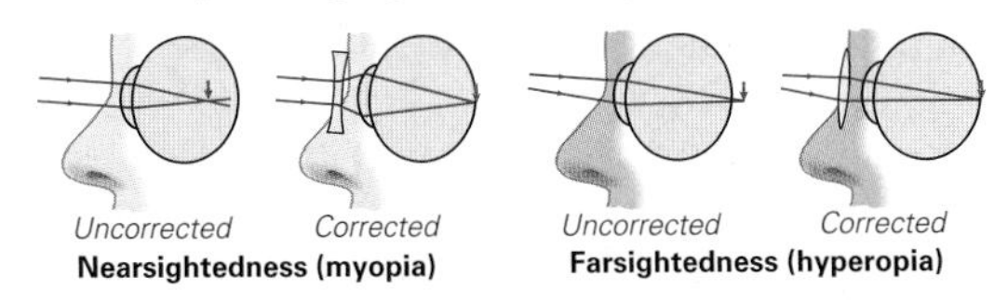

- The magnification of a magnifying glass (or simple microscope) is expressed in terms of **angular magnification (m)**, as distinguished from the lateral magnification (M; see Chapter 23):

$$m = \frac{\theta}{\theta_o} \quad (25.1)$$

Magnification of a magnifying glass with the image at the near point (25 cm) is expressed as

$$m = 1 + \frac{25\text{ cm}}{f} \quad (25.3)$$

Magnification of a magnifying glass with the image at infinity is expressed as

$$m = \frac{25\text{ cm}}{f} \quad (25.4)$$

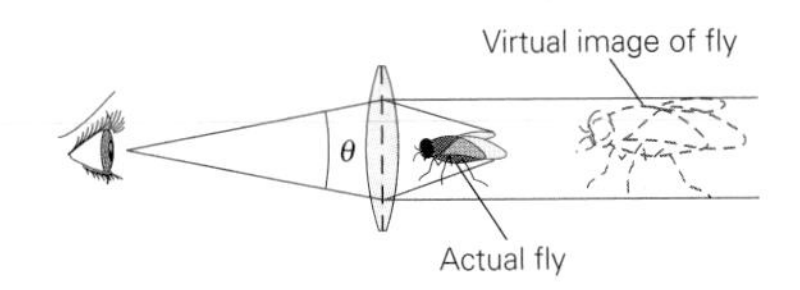

- The objective of a compound microscope has a relatively short focal length, and the eyepiece, or ocular, has a longer focal length. Both contribute to the **total magnification**, m_{total}, given by

$$m_{\text{total}} = M_o m_e = -\frac{(25\text{ cm})L}{f_o f_e} \quad (25.5)$$

where L, f_o, and f_e are in centimeters.

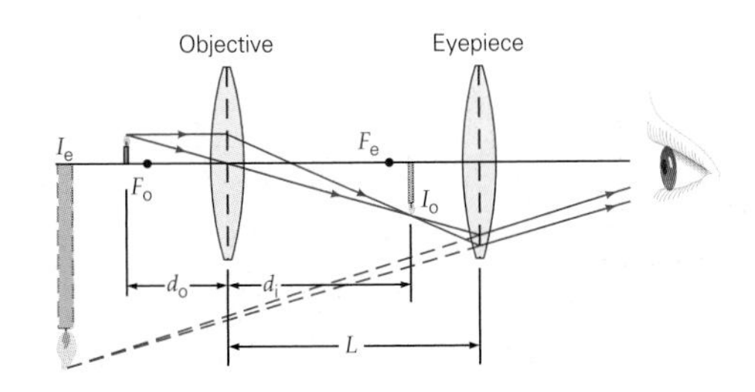

- A refracting telescope uses a converging lens to gather light, and a reflecting telescope uses a converging mirror. The image created by either one is magnified by the eyepiece. The **magnification of a refracting telescope** is

$$m = -\frac{f_o}{f_e} \quad (25.6)$$

- Diffraction places a limit on **resolution**—the ability to resolve, or distinguish, objects that are close together. Two images are said to be just resolved when the central maximum of one image falls on the first minimum of the diffraction pattern of the other image (**Rayleigh criterion**).

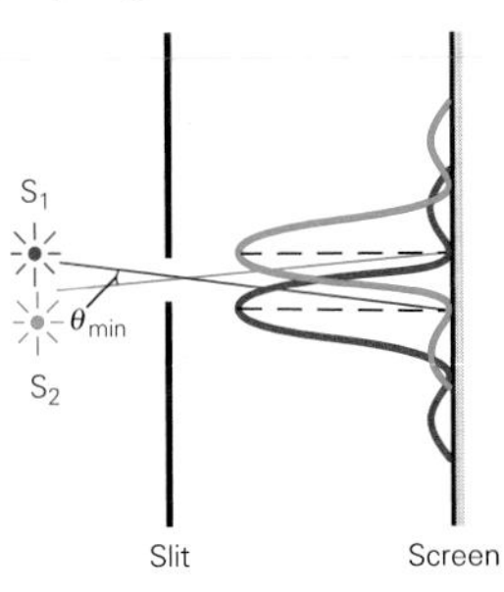

- For a rectangular slit, the **minimum angle of resolution** is

$$\theta_{\text{min}} = \frac{\lambda}{w} \quad (25.7)$$

The **minimum angle of resolution for a circular aperture** of diameter D is

$$\theta_{\text{min}} = \frac{1.22\lambda}{D} \quad (25.8)$$

The **resolving power** of a microscope is

$$s = f\theta_{\text{min}} = \frac{1.22\lambda f}{D} \quad (25.9)$$

Exercises

MC = *Multiple Choice Question,* **CQ** = *Conceptual Question, and* **IE** = *Integrated Exercise. Throughout the text, many exercise sections will include "paired" exercises. These exercise pairs, identified with* **red numbers**, *are intended to assist you in problem solving and learning. In a pair, the first exercise (even numbered) is worked out in the Study Guide so that you can consult it should you need assistance in solving it. The second exercise (odd numbered) is similar in nature, and its answer is given at the back of the book.*

25.1 The Human Eye*

1. **MC** The rods of the retina (a) are responsible for 20/20 vision, (b) are responsible for black-and-white twilight vision, (c) are responsible for color vision, (d) focus light.

2. **MC** An imperfect cornea can cause (a) astigmatism, (b) nearsightedness, (c) farsightedness, (d) all of the preceding.

3. **MC** The image of an object formed on the retina is (a) inverted, (b) upright, (c) the same size as the object, (d) all of the preceding.

4. **MC** The focal length of the crystalline lens of the human eye varies with muscle action. When a distant object is viewed, the radius of the lens is (a) large, (b) small, (c) flat, (d) none of the preceding.

5. **CQ** People and other animals often exhibit "red eye" when photographed with a flash camera. Light reflected from the retina is red because of blood vessels near the surface. Some cameras have an anti-red-eye option, which, when activated, gives a quick flash before the longer picture-taking flash. Explain how this option reduces red eye.

6. **CQ** Which parts of the camera correspond to the iris, crystalline lens, and retina of the eye?

7. **CQ** (a) If an eye has a far point of 15 m and a near point of 25 cm, is that eye nearsighted or farsighted? (b) How about an eye with a far point at infinity and a near point at 50 cm? (c) What type of corrective lenses (converging or diverging) would you use to correct the vision defects in parts (a) and (b)?

8. **CQ** Will wearing glasses to correct nearsightedness and farsightedness, respectively, affect the size of the image on the retina? Explain.

9. ● What are the powers of (a) a converging lens of focal length 20 cm and (b) a diverging lens of focal length -50 cm?

10. **IE** ● The far point of a certain nearsighted person is 90 cm. (a) Which type of contact lens, (1) converging, (2) diverging, or (3) bifocal, should be prescribed to enable the person to see more distant objects clearly? Explain. (b) What would the power of the lens be, in diopters?

11. **IE** ● A certain farsighted person has a near point of 50 cm. (a) Which type of contact lens, (1) converging, (2) diverging, or (3) bifocal, should an optometrist prescribe to enable the person to see objects clearly as close as 25 cm? Explain. (b) What is the power of the lens, in diopters?

12. ●● A woman cannot see objects clearly when they are farther than 12.5 m away. (a) Does she have (1) nearsightedness, (2) farsightedness, or (3) astigmatism? Explain. (b) Which type of lens will allow her to see distant objects clearly, and of what power should the lens be?

13. ●● A nearsighted woman has an uncorrected far point of 200 cm. Which type of contact lens would correct this condition, and of what power should it be?

14. ●● A person can *just* see the print in a book clearly when she holds the book at arm's length (0.80 m from the eyes). (a) Does she have (1) nearsightedness, (2) farsightedness, or (3) astigmatism? Explain. (b) Which type of lens will allow her to read the text at the normal near point and what is the lens's focal length?

15. ●● To correct a case of hyperopia, an optometrist prescribes positive contact lenses that effectively move the patient's near point from 100 cm to 25 cm. (a) To see distant objects clearly, the patient should wear the contact lenses or take them out. Explain. (b) What is the power of the lenses?

16. ●● A farsighted person with a near point of 0.95 m gets contact lenses and can then read a newspaper held at a distance of 25 cm. What is the power of the lenses? (Assume that the lenses are the same for both eyes.)

17. **IE** ●● A farsighted man is unable to focus on objects nearer than 1.5 m. (a) The type of contact lens that allows him to focus on the print of a book held 25 cm from his eyes should be (1) converging, (2) diverging, (3) flat. Explain. (b) What should be the power of the lens?

18. ●● A nearsighted student wears contact lenses to correct for a far point that is 4.00 m from her eyes. When she is not wearing her contact lenses, her near point is 20 cm. What is her near point when she is wearing her contacts?

19. ●● A nearsighted woman has a far point located 2.00 m from one eye. (a) If a corrective lens is worn 2.00 cm from the eye, what would be the necessary power of the lens for her to see distant objects? (b) What would be the necessary power if a contact lens were used?

20. ●● A college professor can see objects clearly only if they are between 70 and 500 cm from her eyes. Her optometrist prescribes bifocals (▾Fig. 25.22) that enable her to see distant objects through the top half of the lenses and read students' papers at a distance of 25 cm through the lower half. What are the respective powers of the top and bottom lenses? [Assume that both lenses (right and left) are the same.]

*Assume that corrective lenses are in contact with the eye (contact lenses) unless otherwise stated.

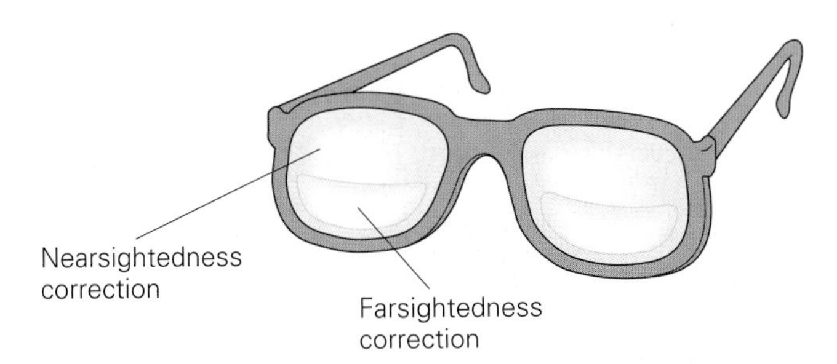

▲ **FIGURE 25.22 Bifocals** See Exercises 20 and 25.

21. ●● A nearsighted man wears eyeglasses whose lenses have a power of −0.15 D. How far away is his far point?

22. ●● An eyeglass lens with a power of +2.8 D allows a farsighted person to read a book held at a distance of 25 cm from her eyes. At what distance must she hold the book to read it without glasses?

23. ●●● A certain myopic man has a far point of 150 cm. (a) What power must a contact lens have to allow him to see distant objects clearly? (b) If he is able to read print at 25 cm while wearing his contacts, is his near point less than 25 cm? If so, what is it? (c) Give an approximation of the man's age, based on the normal rate of recession of the near point.

24. ●●● A middle-aged man starts to wear eyeglasses with lenses of +2.0 D that allow him to read a book held as closely as 25 cm. Several years later, he finds that he must hold a book no closer than 33 cm to read it clearly with the same glasses, so he gets new glasses. What is the power of the new lenses? (Assume that both lenses are the same.)

25. ●●● Bifocal glasses are used to correct both nearsightedness and farsightedness at the same time (Fig. 25.22). If the near points in the right and left eyes are 35.0 cm and 45.0 cm, respectively, and the far point is 220 cm for both eyes, what are the powers of the lenses prescribed for the glasses? (Assume that the glasses are worn 3.00 cm from the eyes.)

25.2 Microscopes*

26. **MC** A magnifying glass (a) is a concave lens, (b) forms virtual images, (c) magnifies by effectively increasing the angle the object subtends, (d) both b and c.

27. **MC** A compound microscope has (a) unlimited magnification, (b) two lenses of the same focal length, (c) a diverging objective lens, (d) an objective of relatively short focal length.

28. **CQ** With an object at the focal point of a magnifying glass, the magnification is given by $m = (25\text{ cm})/f$ (Eq. 25.4). According to this equation, the magnification could be increased indefinitely by using lenses with shorter focal lengths. Why, then, do we need compound microscopes?

29. **CQ** When you use a simple convex lens as a magnifying glass, where should you put the object, farther away than the focal length or inside the focal length? Explain.

*The normal near point should be taken as 25 cm unless otherwise specified.

30. ● Using the small-angle approximation, compare the angular sizes of a car 1.0 m in height when at distances of (a) 500 m and (b) 1025 m.

31. ● An object is placed 10 cm in front of a converging lens with a focal length of 18 cm. What are (a) the lateral magnification and (b) the angular magnification?

32. ● A biology student uses a converging lens to examine the details of a small insect. If the focal length of the lens is 12 cm, what is the maximum angular magnification?

33. ● When viewing an object with a magnifying glass whose focal length is 10 cm, a student positions the lens so that there is minimum eyestrain. What is the observed magnification?

34. **IE** ● A physics student uses a converging lens with a focal length of 15 cm to read a small measurement scale. (a) Maximum magnification is achieved if the image is at (1) the near point, (2) infinity, (3) either place. Explain. (b) What are the magnifications when the image is at the near point and for viewing with the relaxed eye?

35. **IE** ●● A detective wants to achieve maximum magnification when looking at a fingerprint with a magnifying glass. (a) He should use a (1) high-powered, (2) low-powered, or (3) small converging lens. Explain. (b) If he uses lenses of power +3.5 D and +2.5 D, what are the maximum magnifications of the print?

36. ●● What is the maximum magnification of a magnifying glass with a power of +3.0 D for (a) a person with a near point of 25 cm and (b) a person with a near point of 10 cm?

37. ●● A compound microscope has an objective with a focal length of 4.00 mm and an eyepiece with a magnification of 10.0×. If the objective and eyepiece are 15.0 cm apart, what is the total magnification of the microscope?

38. ●● A compound microscope has a distance of 15 cm between lenses and an ocular with a focal length of 8.0 mm. What power should the objective have to give a total magnification of −360×?

39. **IE** ●● Two lenses of focal length 0.45 cm and 0.35 cm are available for a compound microscope using an eyepiece of focal length of 3.0 cm, and the distance between the lenses has to be 15 cm. (a) Which lens, (1) the one with the longer focal length, (2) the one with the shorter focal length, or (3) either should be used as the objective? (b) What are the two possible total magnifications of the microscope.

40. ●● The focal length of the objective lens of a compound microscope is 4.5 mm. The eyepiece has a focal length of 3.0 cm. If the distance between the lenses is 18 cm, what is the magnifications of a viewed image?

41. ●● A compound microscope has an objective lens with a focal length of 0.50 cm and an eyepiece with a focal length of 3.25 cm. The separation distance between the lenses is 22 cm. A student with a normal near point uses the microscope. (a) What is the total magnification? (b) Compare the total magnification (as a percentage) with the magnification of the eyepiece alone as a simple magnifying glass.

42. ●● A $-150\times$ microscope has an objective whose focal length is 0.75 cm. If the distance between the lenses is 20 cm, find the focal length of the eyepiece.

43. ●● A specimen is 5.0 mm from the objective of a compound microscope that has a power of +250 D. What must be the magnifying power of the eyepiece if the total magnification of the specimen is $-100\times$?

44. ●●● A lens with a power of +10 D is used as a simple microscope. (a) For the image of an object to be seen clearly, can the object be placed infinitely close to the lens, or is there a limit on how close it can be? Explain. (b) Calculate how close an object can be brought to the lens. (c) What is the angular magnification at this point?

45. IE ●●● A modern microscope is equipped with a turret that has three objectives with focal lengths of 16 mm, 4.0 mm, and 1.6 mm and interchangeable eyepieces of $5.0\times$ and $10\times$. A specimen is positioned such that each objective produces an image 150 mm from the objective. (a) Which objective-and-eyepiece combination would you use if you want to have the greatest magnification? How about the least magnification? Explain. (b) What are the greatest and least magnifications possible?

25.3 Telescopes

46. **MC** An astronomical telescope has (a) unlimited magnification, (b) two lenses of the same focal length, (c) an objective of relatively long focal length, (d) an objective of relatively short focal length.

47. **MC** An inverted image is produced by (a) a terrestrial telescope, (b) an astronomical telescope, (c) a Galilean telescope, (d) all of the preceding.

48. **MC** Compared with large refracting telescopes, large reflecting telescopes have the advantage of (a) greater light-gathering capability, (b) freedom from chromatic aberration, (c) lower cost, (d) all of the preceding.

49. CQ In Fig. 25.12b, part of the light entering the concave mirror is obstructed by a small plane mirror that is used to redirect the rays to a viewer. Does this mean that only a portion of a star can be seen? How does the size of the obstruction affect the image?

50. CQ Why is chromatic aberration an important factor in refracting telescopes, but not in reflecting telescopes?

51. CQ If you are given two lenses with different focal lengths, which one should you use as the eyepiece for a telescope? Explain.

52. ● Find the magnification and length of a telescope whose objective has a focal length of 50 cm and whose eyepiece has a focal length of 2.0 cm.

53. ● An astronomical telescope has an objective and an eyepiece whose focal lengths are 60 cm and 15 cm, respectively. What are the telescope's (a) magnifying power and (b) length?

54. ●● A astronomical telescope has an eyepiece with a focal length of 10.0 mm. If the length of the tube is 1.50 m, what is the angular magnification of the telescope when it is focused for an object at infinity?

55. A telescope has an angular magnification of $-50\times$ and a barrel 1.02 m long. What are the focal lengths of the objective and the eyepiece?

56. IE ●● A terrestrial telescope has three lenses: an objective, an erecting lens, and an eyepiece. (a) Does the erecting lens (1) increase the magnification, (2) increase the physical length of the telescope, (3) decrease the magnification, or (4) decrease the physical length of the telescope? Explain. (b) The three lenses of this terrestrial telescope have focal lengths of 40 cm, 20 cm, and 15 cm for the objective, erecting lens, and eyepiece, respectively. What is the magnification of the telescope for an object at infinity? (c) What is the length of the telescope barrel?

57. ●● A terrestrial telescope uses an objective and eyepiece with focal lengths of 45 cm and 15 cm, respectively. What should the focal length of the erecting lens be if the overall length of the telescope is to be 0.80 m?

58. ●● An astronomical telescope uses an objective of power +2.0 D. If the length of the telescope is 52 cm, what is the angular magnification of the telescope?

59. IE ●● You are given two objectives and two eyepieces and are instructed to make a telescope with them. The focal lengths of the objectives are 60.0 cm and 40.0 cm, and the focal lengths of the eyepieces are 0.90 cm and 0.80 cm, respectively. (a) Which lens combination would you pick if you want to have maximum magnification? How about minimum magnification? Explain. (b) Calculate the maximum and minimum magnifications.

25.4 Diffraction and Resolution*

60. **MC** The images of two sources are said to be resolved when (a) the central maxima of the diffraction patterns fall on each other, (b) the first maximum of the diffraction patterns fall on each other, (c) the central maximum of one diffraction pattern falls on the first minimum of the other, (d) none of the preceding.

61. **MC** For a telescope with a circular aperture, the minimum angle of resolution is (a) greater for red light than for blue light, (b) independent of the frequency of the light, (c) directly proportional to the radius of the aperture, (d) independent of the area of the aperture.

62. **MC** The purpose of using oil immersion lenses on microscopes is to (a) reduce the size of the microscope, (b) increase the magnification, (c) increase the wavelength of light so as to increase the resolving power, (d) reduce the wavelength of light so as to increase the resolving power.

*Ignore atmospheric blurring unless otherwise stated.

63. CQ When an optical instrument is designed, high resolution is often desired so that the instrument may be used to observe fine details. Does higher resolution mean a smaller or larger minimum angle of resolution? Explain.

64. CQ A reflecting telescope with a large objective mirror can collect more light from stars than a reflecting telescope with a smaller objective mirror. What other advantage is gained with a large mirror? Explain.

65. CQ Modern digital cameras are getting smaller and smaller. Discuss the image resolution of these small cameras.

66. IE ● (a) For a given wavelength, a wider single slit will give a (1) greater, (2) smaller, (3) the same minimum angle of resolution as a narrower slit, according to the Rayleigh criterion. (b) What are the minimum angles of resolution for two point sources of red light ($\lambda = 680$ nm) in the diffraction pattern produced by single slits with a width of 0.55 mm and 0.45 mm?

67. ● The minimum angular separation of the images of two identical monochromatic point sources in a single-slit diffraction pattern is 0.0055 rad. If a slit width of 0.10 mm is used, what is the wavelength of the sources?

68. ● What is the resolution limit due to diffraction for the European Southern Observatory reflecting telescope (8.20-m, or 323-in., diameter) for light with a wavelength of 550 nm?

69. ● What is the resolution due to diffraction for the Hale telescope at Mount Palomar, with its 200-in.-diameter mirror, for light with a wavelength of 550 nm? Compare this value with the resolution limit for the European Southern Observatory telescope found in Exercise 68.

70. ●● From a spacecraft in orbit 150 km above the Earth's surface, an astronaut wishes to observe her hometown as she passes over it. What size features will she be able to identify with the unaided eye, neglecting atmospheric effects? [*Hint*: Estimate the diameter of the human iris.]

71. IE ●● A human eye views small objects of different colors, and the eye's resolution is measured. (a) The eye obtains the maximum resolution and sees the finest details for objects of which color: (1) red, (2) yellow, (3) blue, or (4) it does not matter? Explain. (b) The maximum diameter of the eye's pupil at night is about 7.0 mm. What are the minimum angles of separation for sources with wavelengths 550 nm and 650 nm?

72. ●● Some African tribes people claim to be able to see the moons of Jupiter with the unaided eye. If two moons of Jupiter are at a minimum distance of 3.1×10^8 km away from Earth and at a maximum separation distance of 3.0×10^6 km, is this possible in theory? Explain. Assume that the moons reflect sufficient light and that their observation is not restricted by Jupiter. [*Hint*: See Exercise 71b.]

73. ●● Assuming that the headlights of a car are point sources 1.7 m apart, what is the maximum distance from an observer to the car at which the headlights are distinguishable from each other? [*Hint*: See Exercise 71b.]

74. ●● A refracting telescope with a lens whose diameter is 30.0 cm is used to view a binary star system that emits light in the visible region. (a) What is the minimum angular separation of the two stars for them to be barely resolved? (b) If the binary star is a distance of 6.00×10^{20} km from the Earth, what is the distance between the two stars? (Assume that a line joining the stars is perpendicular to our line of sight.)

75. ●● A radio telescope of diameter 300 m uses a wavelength of 4.0 m to observe a binary star system that is about 2.5×10^{18} km from the Earth. What is the minimum distance of two stars that can be distinguished by the telescope?

76. ●● The objective of a microscope is 2.50 cm in diameter and has a focal length of 30.0 mm. (a) If yellow light with a wavelength of 570 nm is used to illuminate a specimen, what is the minimum angular separation of two fine details of the specimen for them to be just resolved? (b) What is the resolving power of the lens?

77. ●●● A microscope with an objective 1.20 cm in diameter is used to view a specimen via light from a mercury source with a wavelength of 546.1 nm. (a) What is the limiting angle of resolution? (b) If details finer than those observable in part (a) are to be observed, what color of light in the visible spectrum would have to be used? (c) If an oil immersion lens were used ($n_{oil} = 1.50$), what would be the change (expressed as a percentage) in the resolving power?

*25.5 Color

78. MC An additive primary color is (a) blue, (b) green, (c) red, (d) all of the preceding.

79. MC A subtractive primary color is (a) cyan, (b) yellow, (c) magenta, (d) all of the preceding.

80. MC White light is incident on two filters as shown in ▼Fig. 25.23. The color of light that emerges from the yellow filter is (a) blue, (b) yellow, (c) red, (d) green.

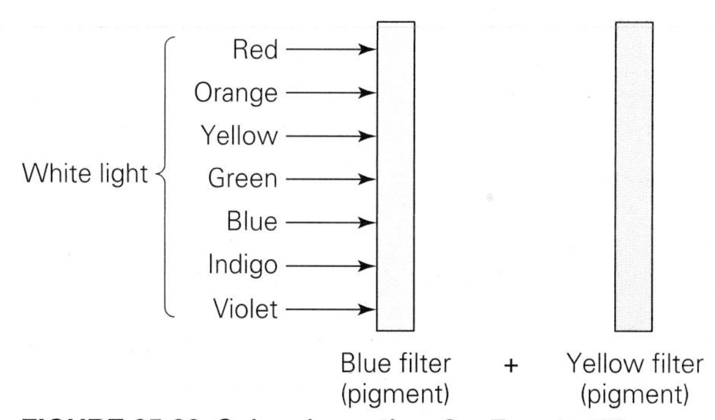

▲ **FIGURE 25.23 Color absorption** See Exercise 80.

81. CQ Describe how the American flag would appear if it were illuminated with light of each of the primary colors.

82. CQ Can white be obtained by the subtractive method of color production? Explain. It is sometimes said that black is the absence of all color or that a black object absorbs all incident light. If so, why do we see black objects?

83. CQ Several beverages, such as root beer, develop a "head" of foam when poured into a glass. Why is the foam generally white or light colored, whereas the liquid is dark?

Comprehensive Exercises

84. A student uses a magnifying glass to examine the details of a microcircuit in the lab. If the lens has a power of 12.5 D and a virtual image is formed at the student's near point (25 cm), (a) how far from the circuit is the lens held, and (b) what is the angular magnification?

85. Referring to ▾Fig. 25.24, show that the magnifying power of a magnifying glass held at a distance d from the eye is given by

$$m = \left(\frac{25}{f}\right)\left(1 - \frac{d}{D}\right) + \frac{25}{D}$$

when the actual object is located at the near point (25 cm). [*Hint*: Use a small-angle approximation, and note that $y_i/y_o = -d_i/d_o$, by similar triangles.]

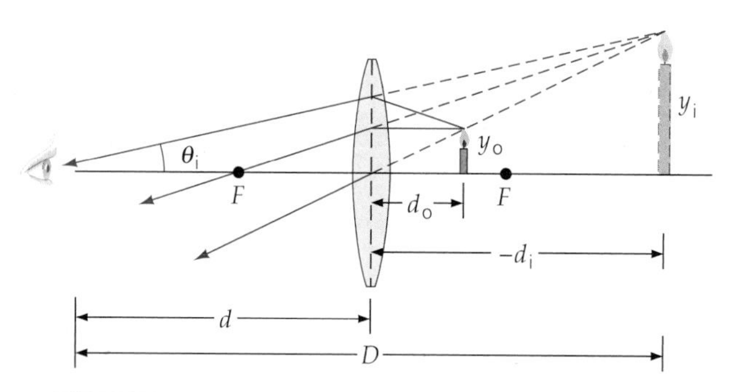

▲ **FIGURE 25.24 Power of a magnifying glass** See Exercise 85.

86. Referring to ▸Fig. 25.25, show that the angular magnification of a refracting telescope focused for the final image at infinity is $m = -f_o/f_e$. (Because telescopes are designed for viewing distant objects, the angular size of an object viewed with the unaided eye is the angular size of the object at its actual location rather than at the near point, as is true for a microscope.)

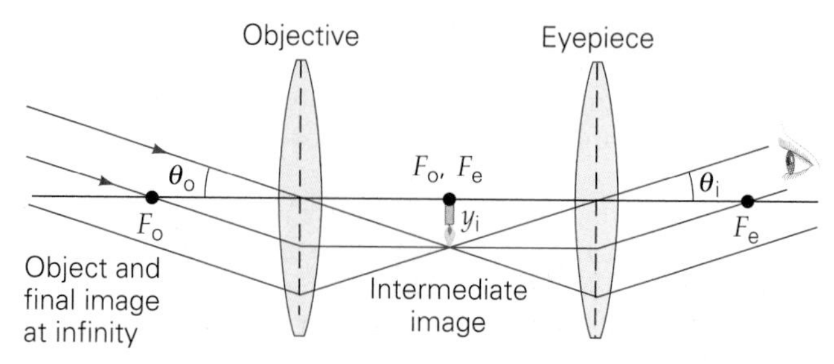

▲ **FIGURE 25.25 Angular modification of a refracting telescope** See Exercise 86.

87. Two astronomical telescopes have the characteristics shown in the following table:

Telescope	Objective Focal Length (cm)	Eyepiece Focal Length (cm)	Objective Diameter (cm)
A	90.0	0.840	75.0
B	85.0	0.770	60.0

(a) Which telescope would you choose (1) for best magnification? (2) for best resolution? Explain. (b) Calculate the maximum magnification and the minimum resolving angle for a wavelength of 550 nm.

88. A refracting telescope has an objective with a focal length of 50 cm and an eyepiece with a focal length of 15 mm. The telescope is used to view an object that is 10 cm high and located 50 m away. What is the apparent angular height of the object as viewed through the telescope?

89. The amount of light that reaches the film in a camera depends on the lens aperture (the effective area) as controlled by the diaphragm. The f-number is the ratio of the focal length of the lens to its effective diameter. For example, an f/8 setting means that the diameter of the aperture is one eighth of the focal length of the lens. The lens setting is commonly referred to as the *f-stop*. (a) Determine how much light each of the following lens settings admits to the camera as compared with f/8: (1) f/3.2 and (2) f/16. (b) The exposure time of a camera is controlled by the shutter speed. If a photographer correctly uses a lens setting of f/8 with a film exposure time of 1/60 s, what exposure time should he use to get the same amount of light exposure if he sets the f-stop at f/5.6?

PHYSLET® The following Physlet Physics Problems can be used with this chapter.
36.1, 36.2, 36.3, 36.4, 36.5

CHAPTER 26

RELATIVITY

26.1 Classical Relativity and the Michelson–Morley Experiment 820

26.2 The Postulates of Special Relativity and the Relativity of Simultaneity 822

26.3 The Relativity of Length and Time: Time Dilation and Length Contraction 825

26.4 Relativistic Kinetic Energy, Momentum, Total Energy, and Mass–Energy Equivalence 833

26.5 The General Theory of Relativity 837

***26.6** Relativistic Velocity Addition 841

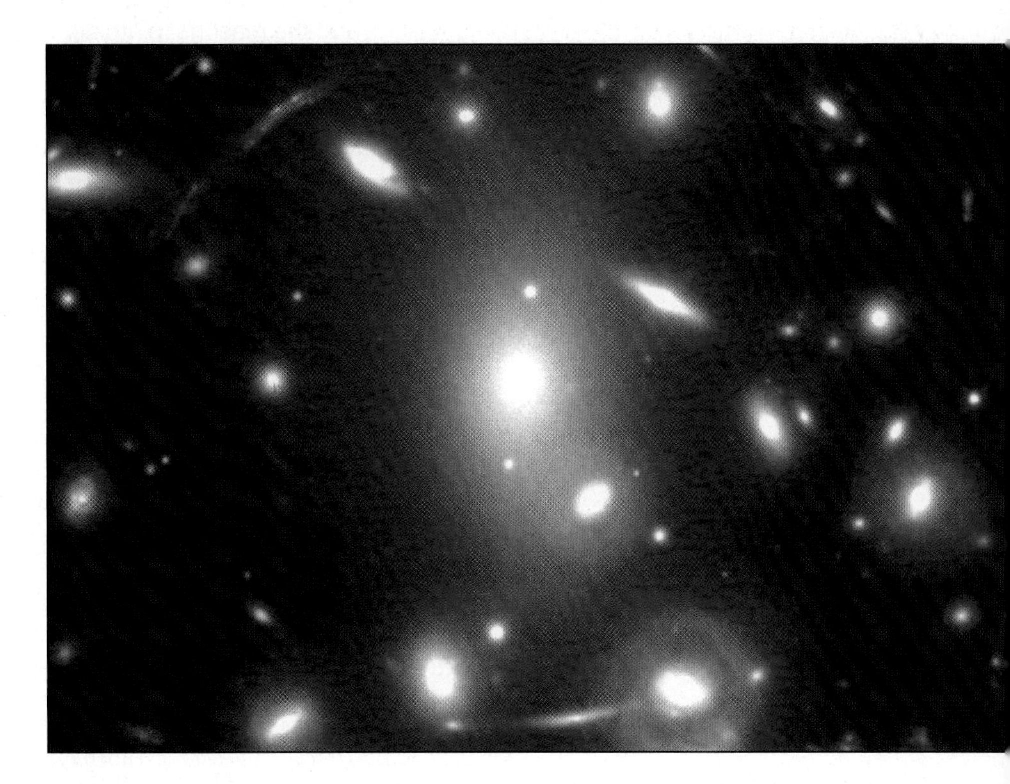

PHYSICS FACTS

- In one nanosecond (10^{-9} s), light travels about 1 ft.
- The atomic clocks in orbit in the Global Positioning System (GPS) can be used to determine the location of an object on the surface of the Earth to within several meters. To achieve this accuracy, their frequencies must be adjusted for special and general relativistic effects due to their speed and the fact that they are in a weaker gravitational field compared to the clocks at the Earth's surface.
- Near the boundary of a black hole, the gravitational field is so strong that just outside the boundary, light can orbit the black hole similar to satellites orbiting the Earth.
- It is currently believed that at the centers of many galaxies may reside a huge black hole, absorbing nearby stars and growing in size.
- Accurate atomic clocks taken on airplane trips around the world show time differences with respect to ones that "stayed at home," in agreement with predictions of relativity theory.
- The U.S. standard atomic clock is about 1 mile above sea level in Boulder, CO. It gains about 5 microseconds per year compared to identical clocks at sea level because of differences in the Earth's gravitational field between their two locations.

You might not think so, but the photograph tells us something very remarkable about our universe. The bright shapes are galaxies, each consisting of billions or trillions of stars. They are very far from us—billions of light-years away. The faint arcs that make the photograph resemble a spider's web are from galaxies even more distant.

What is remarkable about these wisps of light, however, is not the billions of years they took to reach us, but the paths they have taken. We usually think of light traveling in straight lines. Yet the light from these distant galaxies has had its direction changed by the gravitational fields of the galaxies in the foreground, creating the arcs in the photo.

The fact that light can be affected by gravity was predicted by Albert Einstein as a consequence of his theory of general relativity. Relativity originated from the analysis of physical phenomena involving speeds approaching that of light. Indeed, modern relativity caused us to rethink our understanding of space, time, and gravitation. It successfully challenged Newtonian concepts that had dominated science for 250 years.

The impact of relativity has been especially significant in the branches of science concerned with two extremes of physical reality: the subatomic realm of nuclear and particle physics, in which time intervals and distances are inconceivably small (Chapters 29 and 30); and the cosmic realm, in which time intervals and distances are unimaginably large. All modern theories about the birth, evolution, and ultimate fate of our universe are inextricably linked to our understanding of relativity.

In this chapter, you will learn how Einstein's relativity explains the changes in length and time that are observed for rapidly moving objects, the equivalence of energy and mass, and the bending of light by gravitational fields—phenomena that seem strange from the classical Newtonian view.

26.1 Classical Relativity and the Michelson–Morley Experiment

OBJECTIVES: To (a) summarize the concepts of classical relativity, (b) define inertial and noninertial reference frames, and (c) explain the ether hypothesis and the reasons for its demise.

Physics is concerned with the description of the world around us and depends on observations and measurements (Chapter 1). We expect some aspects of nature to be consistent and unvarying; that is, the ground rules by which nature plays should be consistent, and physical principles should not change from observation to observation. This consistency is emphasized by referring to such principles as *laws*—for example, the laws of motion. Not only have physical laws proved valid over time, but they are the same for all observers.

The last sentence means that a physical principle or law should *not* depend on the observer's frame of reference. When a measurement is made or an experiment performed, reference is usually made to a particular frame or coordinate system—most often the laboratory, which is considered to be "at rest." Now envision the same experiment observed by a passerby (moving relative to the laboratory). Upon comparing notes, the experimenter and observer should find the results of the experiment and the physical principles involved to be the same. Physicists believe that the laws of nature are the same regardless of the observer. Measured quantities may vary and descriptions may be different, but the *laws* that these quantities obey must be the same for all observers.

Suppose that you are at rest and observe two cars traveling in the same direction on a straight road at speeds of 60 km/h and 90 km/h. Even though we rarely say it, it is assumed that these speeds are measured relative to your reference frame—the ground. However, a woman in the car traveling at 60 km/h observes the other car traveling at 30 km/h *relative to her reference frame*—the car in which she is riding. (What does someone in the car traveling at 90 km/h observe?) That is, each person observes a *relative* velocity—the one relative to his or her own reference frame. (See Section 3.4.)

Note: Review the discussion of relative velocities in Section 3.4.

In measuring relative velocities, there seems to be no "true" rest frame. Any reference frame can be considered at rest if the observer moves with it. We can, however, make a distinction between *inertial* and *noninertial* reference frames. An **inertial reference frame** is a reference frame in which Newton's first law of motion holds. That is, in an inertial frame, an object on which there is no net force does not accelerate. Since Newton's first law holds in this frame, the second law of motion ($\vec{\mathbf{F}}_{\text{net}} = m\vec{\mathbf{a}}$) also holds.

Note: The relationship between Newton's first and second laws is discussed in a Section 4.3 footnote.

Conversely, in a **noninertial reference frame** (one that is accelerating as measured from an inertial frame), an object with no net force acting on it would *appear* to accelerate. Note, however, that the frame is accelerating, not the object. If observations are made from a noninertial frame, Newton's second law will not correctly describe motion. One example of a noninertial reference frame is an automobile accelerating forward from rest. A cup on the (frictionless) dashboard may appear, when viewed from the car's (noninertial) reference frame, to accelerate backward without being acted on by any force. In fact, the noninertial observer would have to invoke a *fictitious* backward force to explain the cup's apparent acceleration. From the inertial frame of a sidewalk observer, however, the cup stays put (in accordance with the first law, since, with no appreciable friction, no net force acts on it) as the car accelerates away from it.

Any reference frame moving with a constant velocity relative to an inertial reference frame is itself an inertial frame. Given a constant relative velocity, no acceleration effects are introduced in comparing one frame with another. In such cases, $\vec{\mathbf{F}}_{\text{net}} = m\vec{\mathbf{a}}$ can be used by observers in *either* frame to analyze a situation, and both observers will come to the same conclusions. That is, Newton's second law holds in both frames. Thus, at least with respect to the laws of mechanics, no inertial frame is preferred over another. This is called the **principle of classical, or Newtonian, relativity**:

The laws of *mechanics* are the same in all inertial reference frames.

Note: Newtonian relativity refers only to laws of mechanics.

The "Absolute" Reference Frame: The Ether

With the development of the theories of electricity and magnetism in the 1800s, some serious questions arose. Maxwell's equations predicted light to be an electromagnetic wave that travels with a speed of $c = 3.00 \times 10^8$ m/s in a vacuum. But relative to what reference frame does light have this speed? Classically, this speed would be expected to be different when measured from different reference frames. For example, you might expect it to be greater than 3.00×10^8 m/s if you were approaching the beam of light, as in the relative case of another approaching you on a highway.

Consider the situation in ▸Fig. 26.1: A person in a reference frame (truck) moving relative to another frame (ground) with a constant velocity $\vec{\mathbf{v}}'$ throws a ball with a velocity $\vec{\mathbf{v}}_b$ relative to the truck. Then the so-called stationary observer (ground) would say the ball had a velocity of $\vec{\mathbf{v}} = \vec{\mathbf{v}}' + \vec{\mathbf{v}}_b$ relative to the ground. Suppose the truck were moving at 20 m/s east relative to the ground and a ball were thrown at 10 m/s (relative to the truck), also easterly. The ball would have a speed of 20 m/s + 10 m/s = 30 m/s to the east when observed by someone on the ground.

Now suppose that the person on the truck turned on a flashlight, projecting a beam of light to the east. According to Newtonian relativity, $\vec{\mathbf{v}} = \vec{\mathbf{v}}' + \vec{\mathbf{c}}$, and the speed of light measured by the ground observer would be greater than 3.00×10^8 m/s. According to classical relativity then, the speed of light can have *any* value, depending on the observer's reference frame.

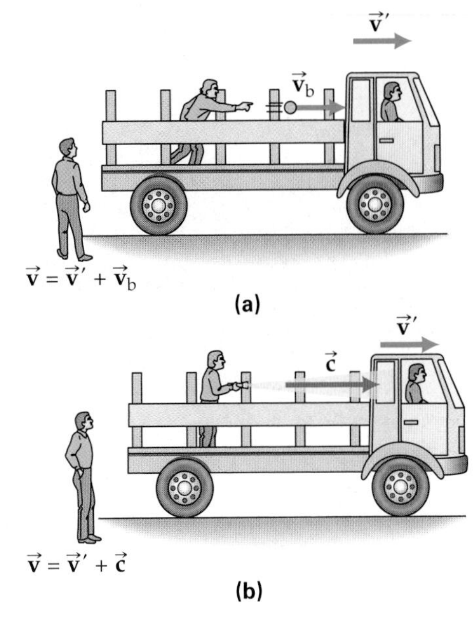

▲ **FIGURE 26.1** Relative velocity **(a)** According to a stationary observer on the ground, $\vec{\mathbf{v}} = \vec{\mathbf{v}}' + \vec{\mathbf{v}}_b$. **(b)** Similarly, the velocity of light would classically be measured to be $\vec{\mathbf{v}} = \vec{\mathbf{v}}' + \vec{\mathbf{c}}$, with a magnitude greater than c. (Velocity vectors are not drawn to scale—can you tell why?)

Assuming Newtonian relativity to be true, it followed that the particular light speed value of 3.00×10^8 m/s must be referenced to some unique frame, in analogy to other waves. Thus, the assumption of a unique reference frame for light seemed quite natural. Since the Earth receives light from the Sun and from distant stars it was thought that a light-transporting medium must permeate all space. This medium was called the *luminiferous ether*, or simply, **ether**.

The idea of an undetected ether became popular in the latter part of the nineteenth century. Maxwell, whose work laid the foundations for our understanding of electromagnetic waves (Chapter 20), believed in the existence of an etherlike substance, as evidenced by a quote from his writings:

> Whatever difficulties we may have in forming a consistent idea of the constitution of the ether, there can be no doubt that the interplanetary and interstellar spaces are not empty, but are occupied by a material substance or body which is certainly the largest, and probably the most uniform body of which we have any knowledge.

Note: Maxwell's equations are discussed in Section 20.4.

It would seem, then, that Maxwell's equations (the basis of electromagnetic theory that describes the propagation of light; see Section 20.4) did *not* satisfy the Newtonian relativity principle, as did the laws of mechanics. On the basis of the preceding discussion, a preferential reference frame would appear to exist—one that could be considered absolutely at rest—the ether frame.

This was the state of affairs toward the end of the nineteenth century, when scientists set out to investigate whether they had come upon a new dimension of physics or perhaps a flaw in what were considered established principles. One of the first attempts to resolve the situation was to prove that the ether existed. Then, presumably, a truly absolute rest frame could finally be identified. This was the purpose of the famous *Michelson–Morley experiment*.

During the 1880s, two American scientists, A. A. Michelson and E. W. Morley,* carried out a series of experiments designed to measure the Earth's velocity relative to the ether. They sought to do this, in effect, by measuring differences in the speed of light due to the Earth's orbital velocity. According to the ether theory, if you were moving relative to the ether (the absolute reference frame), then you would measure the speed of light to be different from c. Their experimental apparatus, while crude by today's standards, was capable of making such a measurement, and yielded a speed of c, *regardless* of the Earth's velocity—a *null* result.

Note: More precisely, the speed of light has been measured as $2.997\,924\,58 \times 10^8$ m/s. We will assume c to be exact at 3.00×10^8 m/s for simplicity.

*Albert Abraham Michelson (1852–1931) was a German-born physicist who devised the Michelson interferometer in an attempt to detect the motion of the Earth through the ether. Edward Morley (1838–1923) was an American chemist who collaborated with Michelson.

▲ **FIGURE 26.2 Einstein and Michelson** A 1931 photo shows Michelson (left) with Einstein during a meeting in Pasadena, California.

26.2 The Postulates of Special Relativity and the Relativity of Simultaneity

OBJECTIVES: To explain (a) the two postulates of relativity, and (b) how they lead to the relativity of simultaneity.

The failure of the Michelson–Morley experiment to detect the ether left the scientific community in a quandary. The inconsistencies between Newtonian mechanics and electromagnetic theory remained unexplained. Many physicists were convinced that the experiment needed to be more accurate; they could not believe that light did not need a medium in which to propagate. These problems were resolved in a theory advanced by Albert Einstein in 1905 (◂Fig. 26.2). Interestingly, Einstein was apparently *not* motivated by the Michelson–Morley experiment in the development of his theory of relativity. When asked later, Einstein could not recall whether he had even known about the experiment when formulating his theory.

In fact, Einstein's insight was based on his intuition that the laws of mechanics should not be the only ones to obey the relativity principle. He reasoned that nature should be symmetrical. *All* physical laws should be encompassed by the relativity principle. In Einstein's view, the inconsistencies in electromagnetic theory were because of the assumption that an absolute rest frame existed. His theory did away with the need for such a frame (and the "ether") by placing all laws of physics on an equal footing, thus eliminating any way of measuring the absolute speed of an inertial reference frame.

The first of the two postulates on which relativity is based is thus an extension (generalization) of Newtonian relativity. Einstein's **principle of relativity** applies to *all* the laws of physics, including those of electricity and magnetism:

> Postulate I (principle of relativity): All the laws of physics are the same in all inertial reference frames.

This postulate means that all inertial reference frames are physically equivalent. That is, all physical laws, not just those of mechanics, are the same in all inertial frames. As a consequence, no experiment performed entirely within an inertial reference frame could enable an observer in that frame to detect its motion. That is, *there is no absolute reference frame*. In hindsight, the first postulate seems reasonable: We have no reason to think that nature would play favorites by picking the laws of mechanics over laws governing other phenomena.

Einstein's second postulate involves the speed of light. Recall that, according to Newtonian relativity, the speed of light can have any value. In fact, if a system were traveling in the same direction as the light and with the same speed, the speed of light in that frame would be zero. This possibility was the source of Einstein's question, "What would I see if I rode a beam of light?" In the same frame as the electromagnetic wave, the electric and magnetic field vectors would not vary with time. However, according to Maxwell's theory of light, the time variation of these two fields is *crucial* to the propagation of light (recall our discussion of time-varying fields and electromagnetic waves in Section 20.4). Hence, *static* fields are *inconsistent* with the propagation or travel of light.

To avoid this inconsistency, Einstein put forth his second postulate, called the **constancy of the speed of light**, as follows:

> Postulate II (constancy of the speed of light): The speed of light in a vacuum has the same value in all inertial systems.

These two postulates form the basis of Einstein's **special theory of relativity**. The "special" designation indicates that the theory deals only with the special case of inertial reference frames. The *general* theory of relativity, discussed later in the chapter, deals with the general case of noninertial, or accelerating, frames.

The second postulate is perhaps more difficult to accept than the first. It means that two observers in different inertial reference frames measure the speed of light to be *c, independent of the speed of the source or the observer*. For example, if a person moving toward you at a constant velocity turned on a flashlight, you both would measure the speed of the emitted light to be c, regardless of the relative velocity (▸Fig. 26.3).

The second postulate is essential to the validity of the first and consistent with the null result of the Michelson–Morley experiment. By doing away with an absolute reference frame, Einstein could reconcile the apparently fundamental differences between mechanics and electromagnetism. Even so, the second postulate seems to go against "common sense." However, keep in mind that most people have no experience dealing with speeds near that of light. The ultimate test of any theory is provided by the scientific method. What does Einstein's theory predict, and can it be experimentally verified? The answer to the last part of the question is a resounding "yes" for every experiment performed, and we will discuss some of these experiments in the following sections.

With the two postulates in place, let's explore some of their implications. The special relativity postulates can be better understood by imagining simple situations. These situations can often tell us what phenomena the postulates predict. Einstein used this method by employing what he termed *gedanken*, or "thought," experiments—"experiments" done in the mind. Let us begin with a series of famous Einstein *gedanken* experiments related to simultaneity and how we measure length and time. We will then look at some experimental evidence that supports the theory and its predictions.

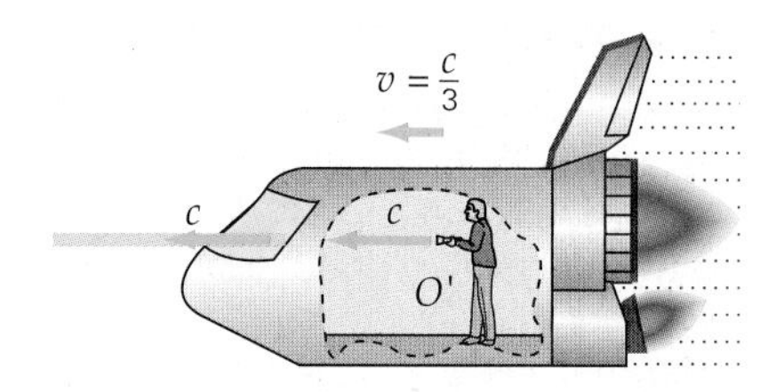

▲ **FIGURE 26.3 Constancy of the speed of light** Two observers in different inertial frames measure the speed of the same beam of light. The observer in frame O' measures a speed of c inside the ship. According to Newtonian mechanics, the observer in frame O would measure a speed of $c + c/3 = 4c/3$ as the beam passes her, but instead, according to Einsteinian mechanics, she measures c.

The Relativity of Simultaneity

In everyday life, we think of two events that are simultaneous to one person as being simultaneous to everyone. That is, we believe the concept of simultaneity is absolute—the same for everyone. What could be more obvious? Simultaneous events occur at the same time, and isn't that the same for all observers? The answer is no—but this result is obvious only for relative velocities near that of light. Thus for all intents and purposes, at everyday speeds this lack of agreement about simultaneity is too small to be observed.

Think of an inertial reference frame (called O) in which two events are *designed* to be simultaneous. For example, two firecrackers (located at points A and B on the x-axis) could be arranged to explode when a switch controlling a voltage source, placed midway between them, is flipped to the "on" position (▾Fig. 26.4a). Let's equip the observer in this frame with a light receptor at point R (for receptor), exactly midway between the firecrackers. This detector is capable of detecting whether two light flashes from the exploded firecrackers arrive at the same time, that is whether they are simultaneous (detected "in coincidence") or not. (Actually, the receptor could be placed anywhere, but the results would need to be corrected due to unequal travel distances. To avoid these complications, all simultaneity detectors will be placed midway between the two events.)

After detonation, the light receptor records that the two explosions went off simultaneously *in the O frame*. But consider the same two explosions as seen by an observer in a different inertial frame, O'. As viewed from O, the other frame O' is moving to the right at a speed v. The observer in O' has equipped himself with a series of light receptors on his x'-axis, because he is not sure which one will end up midway between the explosion points.

After the explosions, there are burn marks on both the x-axis (at A and B) and the x'-axis (at A' and B'), as shown in Fig. 26.4b. These marks can be used to identify the particular O' light receptor (call it R') that was, in fact, located midway between A' and B'. But when the observer in O' reviews the data from this receptor, he finds that it did *not* record the explosions simultaneously. This result from O' does not cause the observer in O to doubt her conclusion, however. She has an explanation of what happened. As she sees the situation, during the time it took for the light to get to R', that receptor had moved toward A and away from B. Consequently, the receptor in O' received the flash from A before that from B.

The question then appears to be: Which observer is correct? It's hard to find any objection to the conclusion reached by the observer in O—so isn't the observer in O' making a mistake? It should be obvious to him that he is moving with respect to the firecrackers. Why doesn't he realize this and take his motion into account? After all, wasn't his light receptor moving toward A and away from B? If so, then it should not surprise him that it recorded the flash from A before the flash

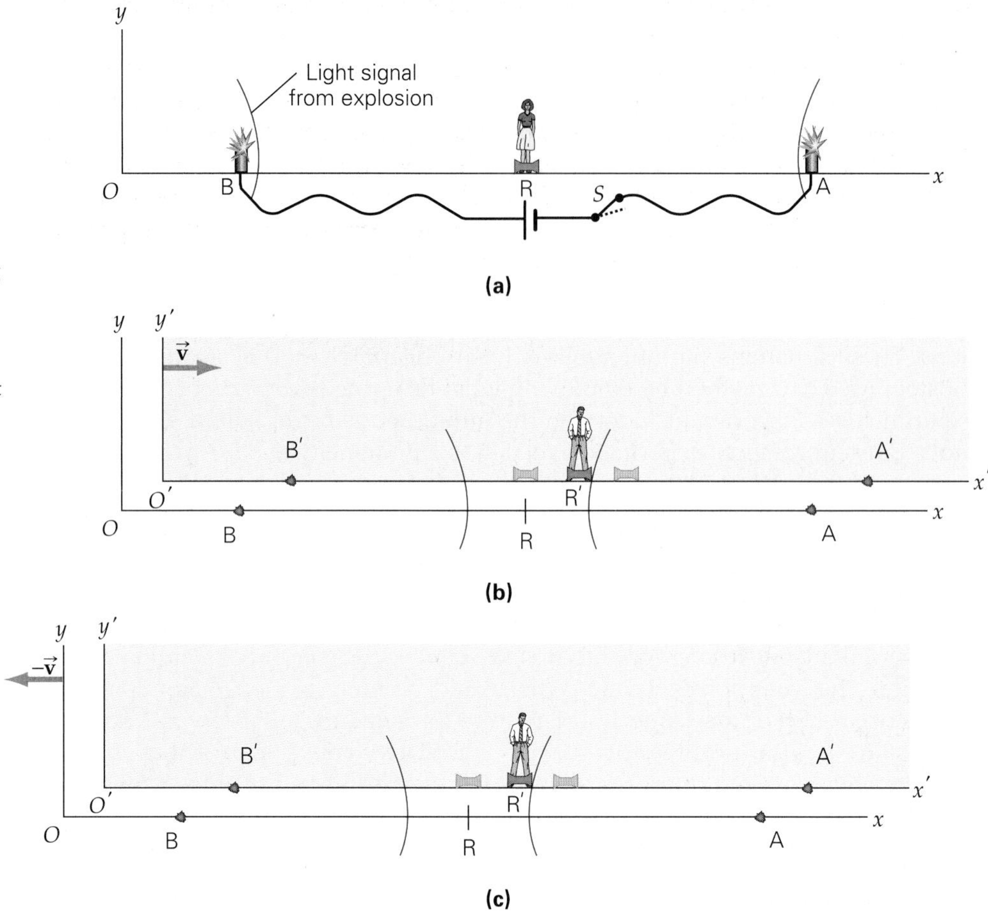

▶ **FIGURE 26.4 The relativity of simultaneity (a)** An observer in reference frame O triggers two explosions (at A and B) simultaneously. A light receptor R, located midway between them, records the two light signals as arriving at the same time. **(b)** An observer midway between the two explosions, but in frame O', moving with respect to O, sees the burn marks made by the two explosions on the x'-axis, but sees the explosion at A happen before that at B. **(c)** The situation as viewed from O'. The observer in O' sees the explosion at A before that at B. To him, O is moving to the left.

from B. All he has to do is to allow for this motion in his calculations, and he will conclude that the flashes "really" were simultaneous.

But if we reason this way, we are ignoring the postulates of relativity. This line of "logic" assumes that when the situation is viewed from the O frame (as in Figs. 26.4a and b), we are looking at what "really" happened from the vantage point of the frame that is "really" at rest. But according to the first postulate, no inertial reference frame is more valid than any other, and none can be considered absolutely at rest. The observer in O' doesn't think of himself as moving. To this observer, *the O frame is moving, and O' is the rest frame*. He observes the firecrackers moving at a speed v to the left (Fig. 26.4c), but this motion would not affect his conclusions. To him, the explosions, equally distant from R', arrived at R' at different times; therefore they were not simultaneous.

You might wonder whether it could be arranged so that the observer in O' would agree that the explosions were simultaneous. However, to accomplish this, the observer in O would have to *delay* the firing of firecracker A relative to B so that R' would receive the two signals at the same time. This does not change the result, because the two observers would still disagree as to whether the events were simultaneous—because now they would no longer be simultaneous in O.

What are we to make of this curious situation? In a nonrelativistic world, one of the observers would have to be wrong. But as we have seen, both observers performed the measurements correctly. Neither one used faulty instruments, or made any errors in logic. So the conclusion must be that *both* are correct.

Notice that there is nothing special about a firecracker explosion. Any "happening" at a particular point in space at a particular time—a karate kick, a soap bubble bursting, a heartbeat—would have done just as well. Such a happening is called an *event* (specified by a location in space *and* a time of occurrence) in the language of relativity. Based on the postulates of relativity, we have the following result:

> Events that are simultaneous in one inertial reference frame may not be simultaneous in a different inertial frame.

This kind of *gedanken* experiment convinced Einstein to give up on simultaneity as an absolute concept.

Note that if the relative speed of the reference frames is slow compared to that of light (as is true for everyday speeds), this lack of simultaneity is completely undetectable. This is why we conclude, erroneously, that simultaneity is absolute. *Most relativistic effects have this property*. That is, their departure from familiar experience is *not* apparent when the speeds involved are much less than the speed of light. Since we have no experience with such high speeds, it is hardly surprising that the predictions of special relativity seem strange.

Conceptual Example 26.1 ■ Agreeing to Disagree: The Relativity of Simultaneity

(a) In Fig. 26.4, estimate the relative speed of the two observers. (b) If the relative speed were only 10 m/s, would there be better agreement on simultaneity? Why?

Reasoning and Answer. (a) To estimate their relative speed, compare the distance between B and B′ in Fig. 26.4b with the distance the light has traveled from B. The figure indicates that the O' frame has moved about 25% as far as the light has. Therefore, the relative speed between the two reference frames is approximately 25% of the speed of light, or $v \approx 0.25c$.
(b) At a relative speed of 10 m/s, the two frames would *not* have moved a noticeable distance; thus, both observers would agree on simultaneity.

Follow-Up Exercise. Show that two events that occur simultaneously on the y-axis of O are perceived as simultaneous by an observer in O', regardless of the relative speed, as long as the relative motion is along their common x–x'-axes. *(Answers to all Follow-Up Exercises are at the back of the text.)*

To grasp the importance of the relativity of simultaneity, think about how important simultaneity is when you are trying to measure the length of a moving object. To measure the length of an object properly, the positions of both ends of it must be marked *simultaneously*. However, two different inertial observers will, in general, *disagree* on simultaneity. Thus, they will also disagree on the object's length.

26.3 The Relativity of Length and Time: Time Dilation and Length Contraction

OBJECTIVES: To (a) understand the origins of time dilation and length contraction, and (b) determine the relationship between time intervals and lengths observed in different inertial frames.

Time Dilation

Another of Einstein's *gedanken* experiments pertained to the measurement of time intervals in different inertial frames. To compare time intervals in different inertial reference frames, he envisioned a *light pulse clock*, illustrated in ▼Fig. 26.5a. A tick (time interval) on the clock corresponds to the time a light pulse would take to make a round trip between the source and the mirror. Let's assume that the observers in the two inertial frames, O and O', have identical light clocks, and the clocks run at the same rate when they are at rest relative to one another. For an observer at rest with respect to one of these clocks, the time interval (Δt_o) for a round trip of a light pulse is the total distance traveled, divided by the speed of light, or

$$\Delta t_o = \frac{2L}{c} \tag{26.1}$$

Now, suppose O' is moving relative to O with a constant velocity $\vec{\mathbf{v}}$ to the right. With his clock (at rest in O') the observer in O' measures the same time interval for

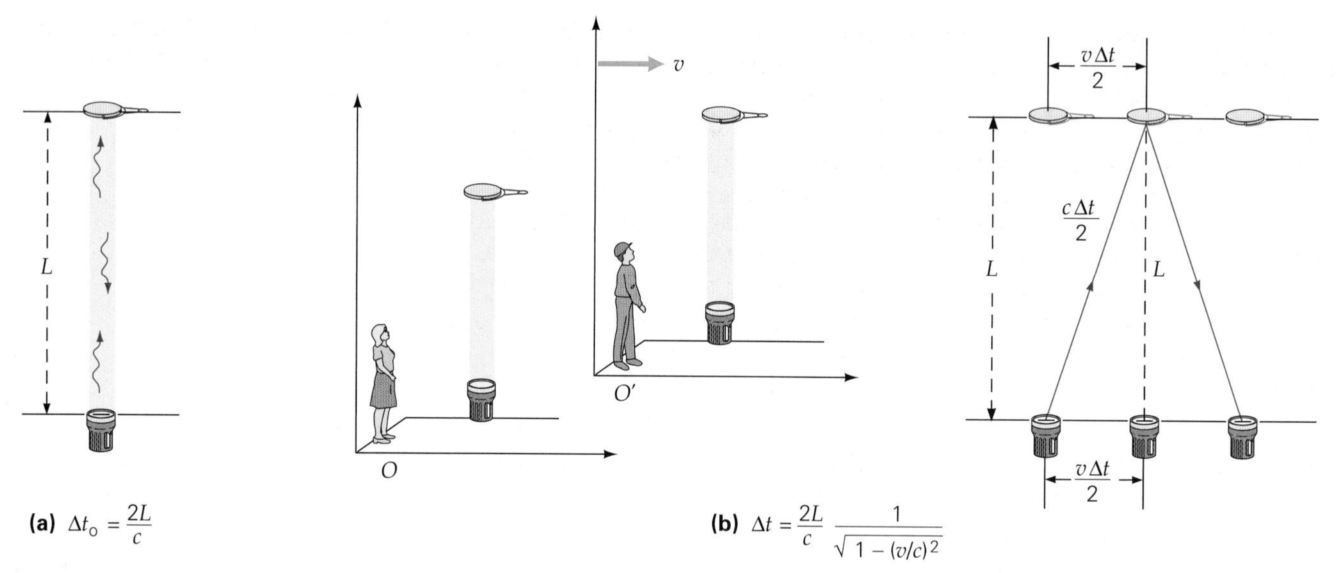

▲ **FIGURE 26.5 Time dilation** **(a)** A light clock that measures time in units of round-trip reflections of light pulses. The time for light to travel up and back is $\Delta t_o = 2L/c$. **(b)** An observer in O measures a time interval of $\Delta t = (2L/c)\left[1/\sqrt{1-(v/c)^2}\right]$ on the clock in the O' frame. Thus, the moving clock appears to run slowly to the observer in O.

his clock, Δt_o. However, according to the observer in O, the clock in the O' system is moving, and the path of its light pulse forms the sides of two right triangles (Fig. 26.5b). Thus, the observer in O sees the light pulse from the clock in O' take a longer path (and therefore a longer time interval Δt) than the light from her own clock. From the O frame, we can apply the Pythagorean theorem:

$$\left(\frac{c\Delta t}{2}\right)^2 = \left(\frac{v\Delta t}{2}\right)^2 + L^2$$

(Here, Δt is the time interval of the O' clock as measured by the observer in O.) Since the speed of light is the same for all observers, the light in the "moving" clock's reference frame takes a longer time to cover the path, *according to the observer in O*. That is, for the observer in the O frame, the "moving" clock runs slowly, that is, the ticks occur at a slower rate. To determine the relationship between the time intervals, we can solve the preceding equation for Δt:

$$\Delta t = \frac{2L}{c}\left[\frac{1}{\sqrt{1-(v/c)^2}}\right] \tag{26.2}$$

But the time interval measured by an observer at rest with respect to a clock is $\Delta t_o = 2L/c$ (Eq. 26.1). Combining the equations, we obtain

$$\Delta t = \frac{\Delta t_o}{\sqrt{1-(v/c)^2}} \quad \textit{relativistic time dilation} \tag{26.3}$$

Since $\sqrt{1-(v/c)^2}$ is less than 1 (why?), then $\Delta t > \Delta t_o$. Thus, an observer in O measures a longer time interval (Δt) *on the O′ clock* than does the observer in O' on the same clock (Δt_o). This effect is called **time dilation**. With longer time between ticks, the O' clock appears, to an observer in O, to run more slowly than the O clock. The situation is symmetric and relative: The observer in O' would say that the clock in the O frame ran slowly relative to the O' clock:

> Moving clocks are observed to run more slowly than clocks that are at rest in the observer's own frame of reference.

This effect, like all relativistic effects, is significant only if the relative speeds are close to that of light.

To distinguish between the two time intervals, the term **proper time interval** is used. As with most measurements, it is usually "proper" or normal to be at rest with respect to a clock when a time interval is measured. In the preceding development, the proper time interval is Δt_o. Stated another way, the proper time interval between two events is the interval measured by an observer at rest relative to the two events and who sees them occur *at the same location in space*. (What are the two events for the light-pulse clock? Are they at the same location in O' for the O' clock?) In Fig. 26.5, the observer in O sees the events by which the time interval of the O' clock is measured at *different* locations. Because the clock is moving, the starting event (the light pulse leaving) occurs at a different location in O from that of the ending event (the light pulse returning). Thus, Δt, the time interval measured by the observer in O, is *not* the proper time interval.

Many of the equations of relativity can be written more compactly if the expression $1/\sqrt{1-(v/c)^2}$ is replaced by γ (Greek letter "gamma"), defined as

$$\gamma \equiv \frac{1}{\sqrt{1-(v/c)^2}} \qquad (26.4)$$

Note that γ is always greater than or equal to 1. (When is it equal to 1?) Also notice that as v approaches c, then γ approaches infinity. Since an infinite time interval is not physically possible, *relative speeds equal to or greater than that of light are not possible*. The values of γ for several values of v (expressed as fractions of c) are listed in Table 26.1. Notice that speeds must be an appreciable fraction of c before relativistic effects can be observed. At $v = 0.10c$, for example, γ differs from 1.00 by only 1%. Using Eq. 26.4, the time dilation relationship (Eq. 26.3) becomes

$$\Delta t = \gamma \Delta t_o \qquad (26.5)$$

Suppose you observed a clock at rest in a system that is moving relative to you at a constant velocity of $v = 0.60c$. For that speed, $\gamma = 1.25$. Thus when 20 min have elapsed on that clock, you would observe an interval of $\Delta t = \gamma\Delta t_o = (1.25)(20\text{ min}) = 25$ min on *your* clock. The 20 min is the proper time, since the events defining the 20-min interval took place at the same location (that of the "moving" clock—for example, for readings at 8:00 A.M. and then at 8:20 A.M.). Thus, the "moving" clock runs more slowly (20 min elapsed, as opposed to 25 min on your clock) when viewed by an observer (you) moving relative to it.

Finally, note that the time-dilation effect cannot apply just to our artificial light-pulse clock. *It must be true for all clocks and hence all time intervals* (that is, anything that keeps a rhythm or frequency, including the heart). If this were not the case—if a mechanical watch, for example, did not exhibit time dilation—then that watch and a light-pulse clock would run at different rates *in the same inertial frame*. This would mean that observers in that frame would be able to tell whether they were moving by making a comparison *solely within their frame*. Since this violates the first postulate of special relativity, it follows that *all* moving clocks, regardless of their nature, must exhibit time dilation. Let's take a look at an actual situation in nature.

Note: The proper time interval Δt_o is always less than the dilated time interval Δt.

TABLE 26.1

Some Values of $\gamma = \dfrac{1}{\sqrt{1-(v/c)^2}}$

v	γ*
0	1.00
$0.100c$	1.01
$0.200c$	1.02
$0.300c$	1.05
$0.400c$	1.09
$0.500c$	1.15
$0.600c$	1.25
$0.700c$	1.40
$0.800c$	1.67
$0.900c$	2.29
$0.950c$	3.20
$0.990c$	7.09
$0.995c$	10.0
$0.999c$	22.4
c	∞

*To illustrate the dependence of γ as v/c approaches 1.

Example 26.2 ■ Muon Decay Viewed from the Ground: Time Dilation Verified by Experiment

Subatomic particles, called *muons*, can be created in the Earth's atmosphere when cosmic rays (mostly protons) collide with the nuclei of the atoms that compose air molecules. Once created, they approach the Earth's surface with speeds near c (typically, about $0.998c$). However, muons are known to be unstable and decay into other particles. The average lifetime of a muon *at rest* has been measured to be 2.20×10^{-6} s. During this time, the muon would travel a distance of $d = v_o\Delta t = (0.998c)(2.20 \times 10^{-6}\text{ s}) = [0.998(3.00 \times 10^8\text{ m/s})](2.20 \times 10^{-6}\text{ s}) = 659$ m. This is 0.659 km, or less than half a mile. Since muons are created at altitudes of 5 to 15 km, we should therefore expect very few of them to reach the Earth's surface. However, an appreciable number *actually do* reach the surface. Using time dilation, explain this apparent paradox.

(continues on next page)

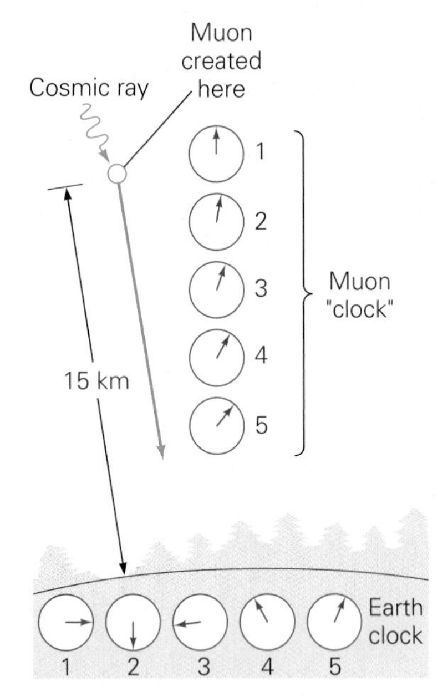

▲ **FIGURE 26.6 Experimental evidence of time dilation** Muons are observed at the surface of the Earth, as predicted by special relativity. The sequence of integers (1 through 5 on each clock) refers to the order of events. Note that from the Earth's viewpoint, the muon's "clock" runs slower than Earth clocks. (See Example 26.2.)

Thinking It Through. The paradox arises because the preceding calculation does not take time dilation into account. That is, a muon decays by its own "internal clock," as measured in its own reference frame. Its "lifetime" of 2.20×10^{-6} s is a proper time interval, and so is its "proper" lifetime. This is because in the muon's rest frame and "birth" and "death" events take place at the same location. Thus, to an observer on the Earth, any "clock" in the muon's reference frame would appear to run more slowly than a clock on the Earth (◂Fig. 26.6). If the muon's lifetime is actually dilated enough, then perhaps its travel distance will be long enough to explain its presence at the Earth's surface.

Solution. We list the given quantities:

Given: $v = 0.998c$
$\Delta t_o = 2.20 \times 10^{-6}$ s (muon proper lifetime)

Find: An explanation, based on time dilation, of why muons make it to the Earth's surface

Instead of the proper time interval $\Delta t_o = 2.20 \times 10^{-6}$ s, an observer on the Earth would measure a time interval longer by a factor of γ. From Eq. 26.4,

$$\gamma = \frac{1}{\sqrt{1 - (v/c)^2}} = \frac{1}{\sqrt{1 - (0.998c/c)^2}} = 15.8$$

From Eq. 26.5, the lifetime of the muon, *according to an observer on the Earth*, is

$$\Delta t = \gamma \Delta t_o = (15.8)(2.20 \times 10^{-6}\text{ s}) = 3.48 \times 10^{-5}\text{ s}$$

The distance the muon travels, *according to an observer on the Earth* (using the dilated time interval), is

$$d = v\Delta t = 0.998(3.0 \times 10^8\text{ m/s})(3.48 \times 10^{-5}\text{ s})$$
$$= 1.04 \times 10^4\text{ m} = 10.4\text{ km}$$

This distance is approximately the same altitude at which muons are created. Hence, the detection of more muons than expected is a confirmation of time dilation.

Follow-Up Exercise. In this Example, what speed would enable the muons to travel 20.8 km relative to the Earth (that is, twice as far as the distance in the Example)? Would they have to travel twice as fast? Explain.

Problem-Solving Hint

In working time-dilation problems, the proper time interval Δt_o must be identified. To do this, first identify (1) the events that define the beginning and end of the interval and (2) a clock (real or imagined) that is present at both events. This "clock," and the observer at rest with respect to it, measures the proper time interval Δt_o. The "dilated" time interval can then be determined from $\Delta t = \gamma \Delta t_o$ for any inertial observer moving relative to this "proper" clock.

Length Contraction

To measure the length of a linear object not at rest in our reference frame, we must take care to mark both ends simultaneously. Consider again the two inertial reference frames used in the discussion of simultaneity, and imagine a measuring stick lying on the x-axis at rest in O (▸Fig. 26.7a). If the observer in O marks the ends simultaneously, the observer in O' will observe end A marked before B. Thus, for the observer in O' to make a correct length measurement, the observer in O must delay the marking of A relative to that of B (Fig. 26.7b). Imagine the observer in O setting off explosions that create burn marks in both reference frames (on both the x- and x'-axes). When the ends are marked so that the observer in O' agrees that they were done simultaneously, all that needs to be done is to subtract the two positions to get the length of the stick *as measured in O'*. Notice that it is going to be *less* than the length measured by O.

Again, it might be asked, "Which observer makes the correct measurement?" By now you know the answer: Both are correct. *Both have measured the length correctly*

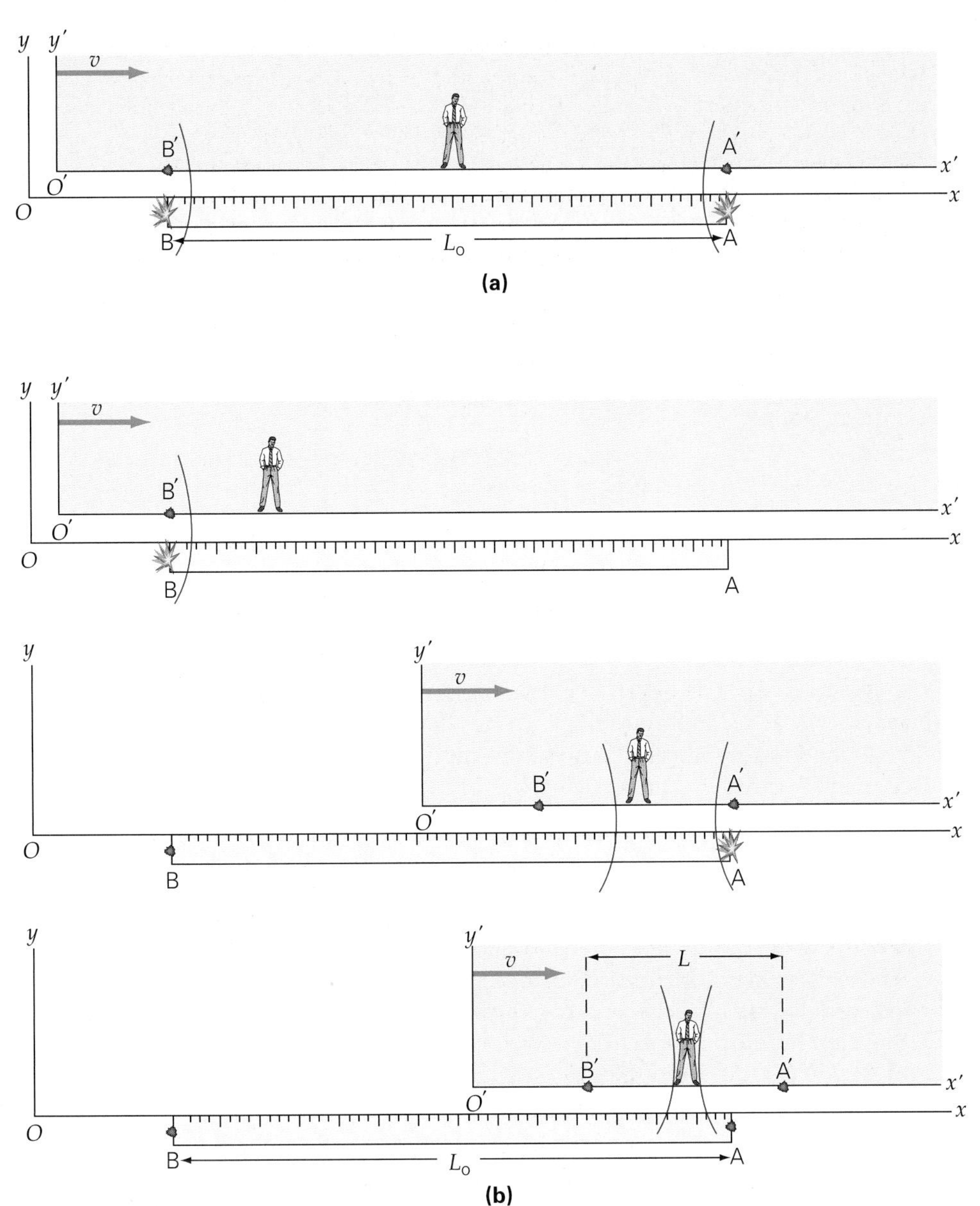

◀ **FIGURE 26.7 Measuring lengths correctly** To measure the length of a moving object correctly, the ends must be marked simultaneously. **(a)** When the observer in frame O marks the ends simultaneously, the observer in frame O' does not agree and observes A marked before B, resulting in too long a length from the viewpoint of O'. **(b)** When O delays the marking of A relative to that of B by just the correct amount (how do we know from the sketch?), the observer in O' measures the correct length from his point of view. The length measured by the observer in O' is less than the rest (or proper) length measured by the observer in O.

in their own frames. Neither thinks that the other has done things correctly, but each is satisfied with his or her own measurement. The observer in O' would have measured the same length as that in O *if he were willing to overlook the fact that the burn marks were not made simultaneously*, as in Fig. 26.7a. However, this is *not* the correct way to measure the length of the moving stick. The lack of agreement on simultaneity leads to the following qualitative statement about length contraction:

> An object's length is largest when measured by an observer at rest with respect to it (the "proper" observer). If the object is moving relative to an inertial observer, that observer measures a smaller length than the proper observer.*

As usual, this effect is entirely negligible at speeds that are slow compared with c.

The distance between two points as measured by the observer at rest with respect to them is designated by L_o and is called the **proper length**. The proper length (also known as the *rest length*) is the largest possible length. The term *proper*

Note: The proper length L_o is always greater than the contracted length L.

*Here, "length" refers to the dimension of the object that is in the direction of its relative velocity. That is, if a cylindrical stick is moving parallel to its long axis, then only its long-axis "length"—and not its diameter—would exhibit length contraction.

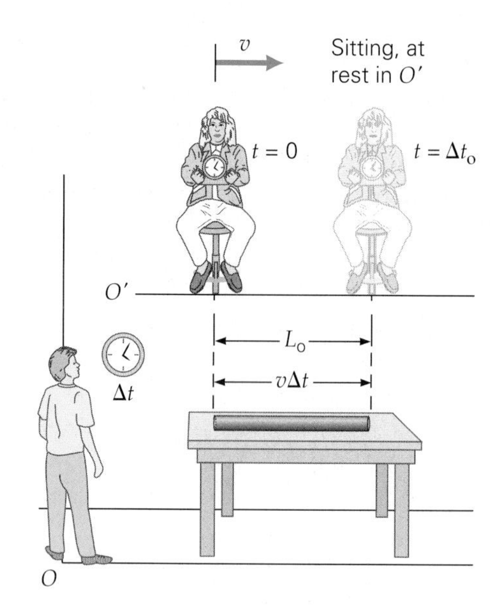

▶ **FIGURE 26.8 Derivation of length contraction** The observer in O measures the time it takes for the observer in O' to move past the ends of the rod. Similarly, the observer in O' measures the time it takes for the ends of the rod to pass her. The observer in O' is the *proper time measurer;* she measures the shortest possible time between these two events. The measured lengths of the rod are not the same. The observer in O is the *proper length measurer;* he measures the longest possible length.

has nothing to do with the correctness of the measurement, because each observer is measuring correctly from his or her point of view.

A *gedanken* experiment can help us develop an expression for length contraction. Consider a rod at rest in frame O. This means that the observer in frame O is the proper length measurer for this rod. Thus, the length of the rod that he measures is L_o. An observer in O', traveling at a constant speed v in the direction parallel to the stick, also measures the length of the rod (▲Fig. 26.8). She does this by using the clock she is holding to measure the time interval required for the two ends of the rod to pass her. Since she measures the proper time interval (how do we know this?), the time interval she measures is Δt_o. In her reference frame, the rod is moving to the left with speed v. With that information, she can determine the length L, since $L = v\Delta t_o$ (speed × time).

The observer in O could also measure the length of the rod by the same means. To him, the observer in O' is moving to the right past the rod. If he notes the times on his clock when O' passes the ends of the rod, he measures a time interval Δt. For him, the length of the stick is the proper length; therefore, $L_o = v\Delta t$. Then, dividing one length by the other, we have

$$\frac{L}{L_o} = \frac{v\Delta t_o}{v\Delta t} = \frac{\Delta t_o}{\Delta t} \tag{26.6}$$

But $\Delta t = \gamma \Delta t_o$ (Eq. 26.5), or $\Delta t_o/\Delta t = 1/\gamma$. Thus Equation 26.6 becomes

$$L = \frac{L_o}{\gamma} = L_o\sqrt{1 - (v/c)^2} \quad \textit{relativistic length contraction} \tag{26.7}$$

Since γ is always greater than 1, then $L < L_o$. This effect is called **relativistic length contraction.**

To see the effects of length contraction and time dilation, consider the following two high-speed Examples.

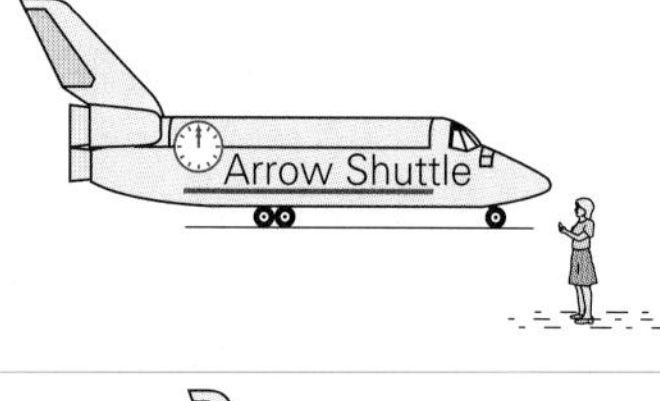

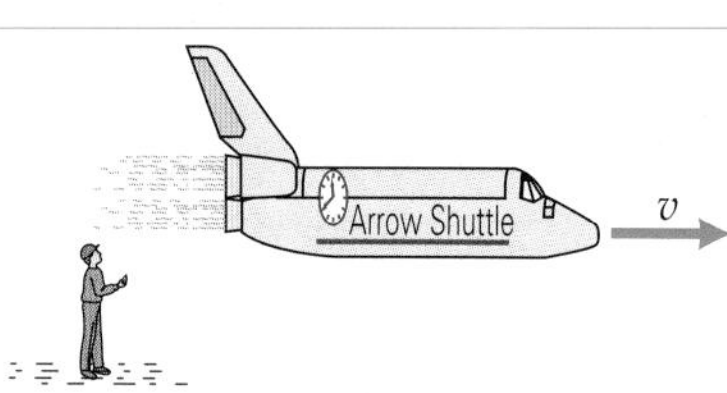

▲ **FIGURE 26.9 Length contraction and time dilation** As a result of length contraction, moving objects are observed to be shorter, or contracted, in the direction of motion, and moving clocks are observed to run more slowly because of time dilation. (See Example 26.3.)

Example 26.3 ■ Warp Speed? Length Contraction and Time Dilation

An observer sees a spaceship, measured to be 100 m long when at rest, pass completely by (that is, nose to tail) in uniform motion with a speed of 0.500c (◂Fig. 26.9). While this observer is watching the ship, a time of 2.00 s elapses on a clock onboard the ship. (a) What is the length of the ship as measured by the observer? (b) What time interval elapses on the observer's clock during the 2.00-s interval on the ship's clock?

Thinking It Through. One hundred meters is the proper length. (Why?) The time interval of 2.00 s is the proper time interval, because the same clock, in the same location, measures the beginning and end of the interval. In (a), Eq. 26.7 can be used to find the contracted length, and in (b), Eq. 26.5 will enable us to determine the dilated time interval.

Solution. Listing the given quantities:

Given: $L_o = 100$ m (proper length)
$v = 0.500c$
$\Delta t_o = 2.00$ s (proper time interval)

Find: (a) L (contracted length)
(b) Δt (dilated time interval)

(a) By calculation or from Table 26.1, $\gamma = 1.15$ for $v = 0.500c$, and the contracted length contraction is given by Eq. 26.7:

$$L = \frac{L_o}{\gamma} = \frac{100 \text{ m}}{1.15} = 87.0 \text{ m}$$

(b) The time interval Δt measured by the observer is longer than the proper time interval Δt_o and is given by Eq. 26.5:

$$\Delta t = \gamma \Delta t_o = 1.15(2.00 \text{ s}) = 2.30 \text{ s}$$

Follow-Up Exercise. In this Example, find the time it takes the spaceship to pass a given point in the observer's reference frame, as seen by (a) a person in the spaceship and (b) the observer watching the ship move by. Explain clearly why these time intervals are *not* the same.

Example 26.4 ■ Muon Decay Revisited: Alternative Explanations

Example 26.2 showed, using relativistic time dilation, how many more muons reach the Earth's surface than can be accounted for without relativistic considerations. Since a hypothetical observer *on the muon* could *not* use this argument (why not?), how would he or she explain the fact that the average muon does make it to the surface? Which explanation is "correct?"

Thinking It Through. A muon traveling at $v = 0.998c$ decays by its own clock (a proper time interval) in $\Delta t_o = 2.20 \times 10^{-6}$ s. In that time, we have seen that it would travel only 659 m, not nearly long enough to reach the Earth's surface. In Example 26.2, this apparent paradox was explained (at least for the Earth observer) by time dilation. According to the Earth observer, the muon's "clock" runs slowly, enabling it to travel farther than expected. But how is the "paradox" explained by a hypothetical observer on the muon? For such an observer, the muon clock is correct—and it registers a time interval that is not sufficient for the muon to reach the Earth's surface! How can this be reconciled with the experimental observation of the Earth observer that finds the muon making it to the surface? After all, two observers *cannot* disagree on this experimental result! To the observer on the muon, the distance to the surface moves by quickly, so the explanation from the muon reference frame must involve length contraction.

Solution.

Given: See Example 26.2

Find: the explanation for muons making it to the Earth's surface from the muon reference frame

The apparent paradox disappears when length contraction is taken into account. For the observer on the muon, its "clock" reads correctly ($\Delta t_o = 2.20 \times 10^{-6}$ s), but the travel distance is shorter because of length contraction. With $\gamma = 15.8$ for $v = 0.998c$, a length of 10.0 km in the Earth frame, which is the proper length L_o (why?), is measured by the observer on the muon to be considerably shorter because

$$L = \frac{L_o}{\gamma} = \frac{10.0 \text{ km}}{15.8} = 0.633 \text{ km} = 633 \text{ m}$$

To travel this distance would take a time (according to the muon observer) of

$$\Delta t = \frac{L}{v} = \frac{L}{0.998c} = \frac{633 \text{ m}}{0.998(3.00 \times 10^8 \text{ m/s})} = 2.11 \times 10^{-6} \text{ s}$$

This is approximately equal to the muon lifetime *in the muon's reference frame*. Thus, through relativistic considerations, both observers agree that many muons reach the Earth (the experimental result). The Earth observer explains this result by saying, "The muon clock is running slow" (time dilation). The observer on the muon says, "No, the clock is fine, but the distance we have to travel is considerably less than you claim" (length contraction). Who is correct? Both are. The reasoning is different for different observers, but the experimental result *must be* the same.

Follow-Up Exercise. Muons are actually created with a range of speeds. What is the speed of a muon if it decays 5.00 km from its creation point as measured by an Earth observer?

The Twin Paradox

Time dilation gives rise to a popular relativistic topic: the **twin paradox**, or **clock paradox**. According to special relativity, a clock moving relative to an observer runs more slowly than one in that observer's frame. Since heartbeat intervals and ages are proper time intervals, the question arises: Do you age more quickly than does a person moving relative to you?

One way to explore this question is through another *gedanken* experiment. Consider identical twins, one of whom goes on a high-speed journey into space. Will the space traveler come back younger than the Earth-bound twin? Or will the space twin see the Earth twin age more slowly? *Both can't be right*, and therein lies the apparent paradox.

The resolution of this apparent paradox lies in the fact that in leaving and returning to Earth, the space twin *must* experience accelerations and so is not always in an inertial reference frame. The stay-at-home twin does not feel the forces associated with speed and directional changes that the traveling twin experiences. Thus, the two twins are individually "marked," and their experiences are *not* symmetrical. However, if the acceleration periods (speedup at start, velocity reversal at turnaround, and slowdown at return) occupy only a negligible part of the total time of the trip, special relativity gives the correct result. Under these conditions, the result is that the traveling twin does indeed return younger than the Earth-bound twin, and both agree on that fact.

During the (constant-velocity) outward and return trips, the proper length of the trip is measured by the Earth twin with fixed beginning and end points. The traveling twin thus measures a length contraction, or a shorter distance for the trip. Traveling at the same relative speed, the space twin's heart thus beats for a shorter time than that of the Earth twin. The traveler *returns home younger* than the Earth-bound twin, *according to the traveling twin*. Now, according to the Earth twin, the traveling twin's heart beats more slowly (time dilation), so the traveling twin *returns home younger* than the Earth-bound twin, *according to the Earth twin*. There is no disagreement. The traveler returns having aged less than the nontraveler. At high speeds, it is possible for the traveler to return and find many generations of Earthlings long gone.

The twin "paradox" has been experimentally verified with extremely accurate atomic clocks flown around the world. Extreme accuracy is necessary because relativistic effects are very small at speeds much slower than c. The time recorded by the traveling clock was compared with that measured by an identical clock that stayed home. The difference in time intervals agreed with the theoretical predictions (with corrections for accelerations on landings and takeoffs). Thus time dilation in our everyday world, although extremely small, was experimentally verified.*

To see some of the details in the traveling twin situation, consider the next Example.

Example 26.5 ■ To the Stars: Relativity and Space Travel

Consider a high-speed round trip taken by one of a set of twins. Ignore accelerations at the start, end, and turnaround points, and assume that a negligible amount of time is spent at the turnaround. (a) Find the speed at which a space explorer would need to travel to make the round trip to a star 100 light-years away in only 20.0 years of traveler time. (*Note*: A light-year is defined as the distance light travels in a vacuum in 1 year.) (b) How much time elapses on the Earth during this trip?

Thinking It Through. Since the trip is symmetrical, we can calculate a one-way trip and double the result to get the answer for a round trip. The explorer measures a one-way *proper time* of 10.0 years, because he is present at the start and end of the trip. People on the Earth measure a one-way *proper length* of 100 light-years, since the beginning and end markers (the Earth and the star) of the trip are at rest with respect to the Earth. **(a)** To find the speed, either the Earth-bound explanation (time dilation) or that of the traveler (length contraction) can be used. **(b)** To determine the time from the Earth's viewpoint, we use the proper distance and the explorer's speed [from part (a)].

*J. Hafele and R. Keating, "Around-the-World Atomic Clocks: Relativistic Time Gains Observed," *Science*, 117 (July 14, 1972), 166–170.

Solution. Listing the given quantities:

Given: $L_o = 100$ light-years (proper length)
$\Delta t_o = 10.0$ years (one-way proper time)

Find: (a) v (speed of the traveler)
(b) Δt (time elapsed on Earth)

(a) For the traveler, the length is the contracted version of 100 light-years. According to the traveler, a one-way trip takes

$$\Delta t_o = \frac{\text{distance traveled}}{\text{speed}} = \frac{L}{v} = \frac{L_o\sqrt{1-\left(\frac{v}{c}\right)^2}}{v}$$

This equation can be solved for v:

$$v = \frac{c}{\sqrt{1+\left(\frac{c\Delta t_o}{L_o}\right)^2}} = \frac{c}{\sqrt{1+\left(\frac{10.0\text{ light-years}}{100\text{ light-years}}\right)^2}} = 0.995c$$

Thus, v is 99.5% the speed of light. Notice that we have used the fact that the distance light travels in 10.0 years ($c\Delta t_o$) is, by definition, 10.0 light-years. We did *not* have to convert to meters because both distances ($c\Delta t_o$ and L_o) were expressed in light-years.

(b) The people on the Earth observe the traveler covering a total of 200 light-years at $0.995c$, so the round-trip time is 200 light-years/$0.995c$, or about 201 years, compared with 20.0 years for the traveler. The traveler would come back to find that everyone who was alive on Earth when he left had long since died!

Follow-Up Exercise. Verify the same result as that obtained in part (b) of this Example, but from time-dilation considerations. Show, using that approach, that Earth observers find 201 of their years elapsed during the trip. [*Hint*: Carry intermediate results to five decimal places, and round to three significant figures at the end.]

26.4 Relativistic Kinetic Energy, Momentum, Total Energy, and Mass–Energy Equivalence

OBJECTIVES: To (a) understand the relativistically correct expressions for kinetic energy, momentum, and total energy, (b) understand the equivalence of mass and energy, and (c) use the relativistically correct expressions to calculate energy and momentum in elementary particle interactions.

The ramifications of special relativity are particularly important in particle physics, in which speeds routinely approach c. Many of the expressions from classical (low-speed) mechanics are incorrect at these high speeds. Kinetic energy, for example, is different from the familiar $K = \frac{1}{2}mv^2$ we routinely used in classical mechanics. Einstein showed that kinetic energy still increases with speed, but in a different way. He found that the **relativistic kinetic energy** of a particle of mass m moving with speed v is

$$K = \left[\frac{1}{\sqrt{1-(v/c)^2}} - 1\right]mc^2 = (\gamma - 1)mc^2 \quad \textit{relativistic kinetic energy} \quad (26.8)$$

It can be shown that this expression becomes the more familiar $K = \frac{1}{2}mv^2$ when $v \ll c$.

According to Eq. 26.8, as v approaches c, the kinetic energy of an object becomes infinite. In other words, to accelerate an object to $v = c$ would require an infinite amount of energy or work, which is not possible. Thus, no object can travel as fast as, or faster than, the speed of light.

Particle accelerators can accelerate charged particles to very high speeds. There is complete agreement between the experimentally measured kinetic energies of these charged particles and Eq. 26.8. A graphical comparison of the relativistically correct expression for kinetic energy and the classical (low-speed) expression is shown in ▸Fig. 26.10. As can be seen, the relativistic and classical expressions agree at low speeds.

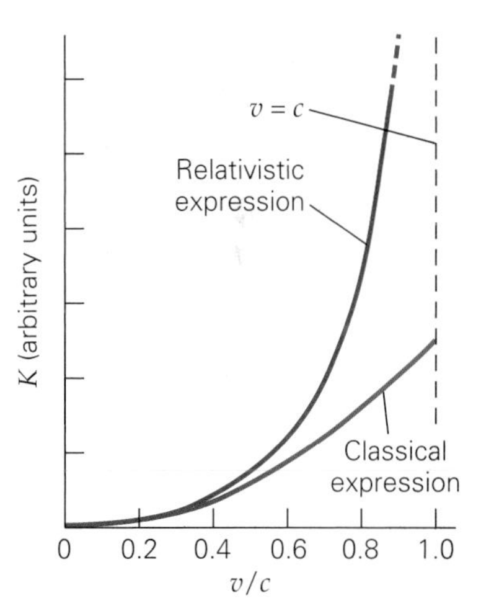

▲ **FIGURE 26.10 Relativistic versus classical kinetic energy** The variation in kinetic energy with particle speed, expressed as a fraction of c, is shown for the relativistically correct expression and for the classical expression. The classical expression becomes negligibly different from the relativistic one for speeds less than about $0.2c$. As objects approach the speed c, their kinetic energy becomes very large. Objects cannot have a speed of exactly c, because their kinetic energy would be infinite.

Relativistic Momentum

Just as for kinetic energy, the momentum of an object is different from the low-speed expression ($\vec{\mathbf{p}} = m\vec{\mathbf{v}}$). The expression for **relativistic momentum** is

$$\vec{\mathbf{p}} = \frac{m\vec{\mathbf{v}}}{\sqrt{1 - \left(\frac{v}{c}\right)^2}} = \gamma m\vec{\mathbf{v}} \quad \textit{relativistic momentum} \qquad (26.9)$$

Note that momentum is still a vector and total (vector) momentum is a conserved quantity under the proper conditions. (See Section 6.3.)

Relativistic Total Energy and Rest Energy: The Equivalence of Mass and Energy

In classical mechanics, the total mechanical energy of an object is the sum of its kinetic and potential energies ($E = K + U$). When there is no potential energy (a free object), this becomes $E = K$ since the total energy is all kinetic. However, Einstein was able to show that the **relativistic total energy** of such an object (that is, a free one) is instead given by

$$E = \frac{mc^2}{\sqrt{1 - \left(\frac{v}{c}\right)^2}} = \gamma mc^2 \quad \textit{relativistic total energy} \qquad (26.10)$$

Thus, according to relativity, when an object is at rest and has no kinetic energy ($K = 0, v = 0$) *it still has an energy of* mc^2, *not zero*. This minimum energy that an object always possesses is called its **rest energy** and is given by

$$E_o = mc^2 \quad \textit{rest energy} \qquad (26.11)$$

Unlike the classical result, when $v = 0$, the total energy of the particle is *not* zero, but E_o. Since $K = (\gamma - 1)mc^2$, the total energy of a particle can be expressed as the sum of its kinetic and rest energies. To see this, note that the relativistic kinetic energy expression (Eq. 26.8) can be rewritten as

$$K = (\gamma - 1)mc^2 = \gamma mc^2 - mc^2 = E - E_o$$

Thus, an alternative to Eq. 26.10 is

$$E = K + E_o = K + mc^2 \quad \begin{array}{l}\textit{relativistic total energy}\\ \textit{with no potential energy}\end{array} \qquad (26.12)$$

(In the more general case, when a particle also has potential energy U, its total energy is given by $E = K + U + mc^2$.)

A relationship between a particle's total energy and its rest energy can be obtained by replacing mc^2 with E_o in Eq. 26.10:

$$E = \gamma E_o \qquad (26.13)$$

As a double check, note that when $v = 0$ (that is, $K = 0$), $\gamma = 1$ and $E = E_o$ as expected.

Equation 26.11 expresses Einstein's famous **mass–energy equivalence**. *An object has energy even at rest*—its rest energy. Consequently, mass is a form of energy. In nuclear and particle physics, it is impossible for total energy to be conserved unless mass is treated as a form of energy.

Mass–energy equivalence does not mean that mass can be converted into useful energy at will. If so, our energy problems would be solved, since there is a lot of mass on the Earth. However, significant practical conversion of mass into other forms of energy, such as heat, does take place in nuclear reactors used to generate electric energy (Section 30.2).

The (rest) energy of an object depends on its mass. For example, the mass of an electron is 9.109×10^{-31} kg, and therefore its rest energy is

$$E_o = mc^2 = (9.109 \times 10^{-31}\ \text{kg})(2.998 \times 10^8\ \text{m/s})^2 = 8.187 \times 10^{-14}\ \text{J}$$

or, converted into electron-volts (eV) and megaelectron-volts (MeV),

$$E_o = (8.187 \times 10^{-14}\ \text{J})\left(\frac{1\ \text{eV}}{1.602 \times 10^{-19}\ \text{J}}\right) = 5.110 \times 10^5\ \text{eV} = 0.5110\ \text{MeV}$$

In high-energy, particle, and nuclear physics, it is common to express rest energies in terms of the electron-volt (eV) or multiples of it, such as the MeV. For example, the rest energy of an electron is said to be 511 keV or 0.511 MeV. This is very convenient because often the particles are accelerated by potential differences measured in volts.

How do you know whether you need to use the relativistic expressions or can "get away" with the classical ones? The rule of thumb is that if an object's speed is 10% of the speed of light or less, then the error in using the nonrelativistic kinetic energy expression is less than 1%. (The classical expression is lower than the relativistic one.) At $v < 0.1c$, the object's kinetic energy is less than 0.5% of its rest energy. Thus, a commonly accepted practice is to use this speed and kinetic-energy region as a dividing line:

> For speeds below 10% of the speed of light or kinetic energies less than 0.5% of an object's rest energy, the error in using the nonrelativistic formulas is less than 1%, and it is then usually acceptable to use the nonrelativistic expressions.

For example, an electron with a kinetic energy of 50 eV would qualify as a "nonrelativistic electron" because 50 eV is (0.01%) of its rest energy. An electron with a kinetic energy of 0.511 MeV would, however, be *highly* relativistic, because its kinetic energy is the same as its rest energy. Thus, what counts is the particle's kinetic energy relative to its rest energy, or its speed relative to that of light—even the dividing line is relative! The following Example shows how particle energies are calculated using relativistically correct expressions.

Example 26.6 ■ A Speedy Electron: Energy Required for Acceleration

(a) How much work is required to accelerate an electron from rest to a speed of $0.900c$? (b) How much error is made by using the nonrelativistic expression?

Thinking It Through. (a) By the work–energy theorem, the work needed is equal to the electron's gain in kinetic energy. The gain in kinetic energy is the same as the final kinetic energy, since the initial kinetic energy is zero. As we have seen, the rest energy of the electron is 0.511 MeV. From its speed, its kinetic energy can be determined with Eq. 26.8. (b) Here we will use the nonrelativistic kinetic energy, $K = \frac{1}{2}mv^2$, and compare it to the answer in part (a).

Solution. The data are as follows:

Given: $v = 0.900c$, $E_o = 0.5110$ MeV (from text)

Find: (a) W (work required), (b) K (nonrelativistic kinetic energy)

(a) The relativistic kinetic energy is given by Eq. 26.8:

$$K = (\gamma - 1)mc^2 = (\gamma - 1)E_o$$

With $v = 0.900c$, by calculation or from Table 26.1, $\gamma = 2.29$; therefore,

$$K = (\gamma - 1)E_o = (2.29 - 1)(0.511\ \text{MeV}) = 0.659\ \text{MeV}$$

Thus 0.659 MeV of work is required to accelerate an electron to a speed of $0.900c$.

(b) Many times, even with nonrelativistic expressions, a shortcut can be applied as follows: Essentially, you can work with mass expressed in energy units. The technique involves first multiplying and then dividing by c^2, which enables you to use E_o. Let's try it here:

$$K_{\text{nonrel}} = \tfrac{1}{2}mv^2 = \tfrac{1}{2}(mc^2)\left(\frac{v}{c}\right)^2 = \tfrac{1}{2}(0.5110\ \text{MeV})(0.900)^2 = 0.207\ \text{MeV}$$

In this case, the nonrelativistic expression for kinetic energy gives an answer that is low by a factor of more than 3.

Follow-Up Exercise. In this Example, (a) what is the relativistic total energy of the electron? (b) What would be the electron's total energy if it were treated nonrelativistically?

To see how relativistic momentum is applied, consider the next Example.

Integrated Example 26.7 ■ When 1 + 1 Doesn't Equal 2: Conservation of Relativistic Momentum and Energy

A particle of mass m, initially moving, collides with an identical particle initially at rest. The two stick together, forming a single particle of mass m'. (a) Do you expect (1) $m' > 2m$, (2) $m' < 2m$, or (3) $m' = 2m$? Explain. (b) If the incoming particle is initially moving at a speed $v = 0.800c$ to the right, what is m' in terms of m?

(a) Conceptual Reasoning. This is an example of an *inelastic* collision (Chapter 6). In such a collision, kinetic energy is not conserved. To conserve linear momentum, some, but not all, of the initial kinetic energy is lost. Since total energy is also conserved, any loss of kinetic energy is converted into mass. If this weren't true, then (3) would be the correct answer. However, the combined mass must include the mass equivalent of the "lost" kinetic energy, or

$$m' = 2m + \text{"some kinetic energy in the form of mass"}$$

Therefore, the correct answer is (1): $m' > 2m$.

(b) Quantitative Reasoning and Solution. Collisions are usually analyzed using conservation of momentum and total energy. After the collision, the combined particle must be moving to the right to conserve the direction of the total momentum. The magnitude of the moving particle's momentum before the collision must equal the magnitude of the single combined particle's momentum afterward. Total relativistic energy must also be conserved. From these considerations, we should be able to determine the combined particle's mass.

Given: $v = 0.800c$
m = mass of one particle

Find: m' (mass of combined particle)

For the incoming particle,

$$\gamma = 1/\sqrt{1 - (v/c)^2} = 1/\sqrt{1 - (0.800)^2} = 1.67$$

The total system momentum P is conserved, that is,

$$\vec{\mathbf{P}}_i = \vec{\mathbf{P}}_f$$

Using the expression for relativistic momentum (Eq. 26.9) and equating the magnitude of the momentum of the incoming particle to that of the combined particle, we have

$$\gamma m v = \gamma' m' v'$$

where γ' refers to the combined particle after the collision. Putting in the numbers,

$$1.67m(0.800c) = \gamma' m' v'$$

Next, we equate the total relativistic energy before the collision to that after the collision, remembering that the total energy of a particle is related to its rest energy by $E = \gamma E_o$. The initial total energy is the sum of the energy due to the moving particle and the rest energy of the "target," or $E_i = 1.67mc^2 + mc^2$. The final total energy of the combined particle is $E_f = \gamma' m' c^2$. Thus, energy conservation requires that

$$1.67mc^2 + mc^2 = \gamma' m' c^2 \quad \text{or} \quad 2.67m = \gamma' m'$$

(In the last step, the speed of light cancels out of both sides of the equation.) Dividing the last result into the momentum result, we obtain

$$\frac{1.67m(0.800c)}{2.67m} = \frac{\gamma' m' v'}{\gamma' m'} = v'$$

or

$$v' = 0.500c$$

Using this result, we find that $\gamma' = 1/\sqrt{1 - (0.500)^2} = 1.15$. Now the energy equation can be used to solve for the mass of the combined particle m':

$$2.67mc^2 = 1.15m'c^2 \quad \text{or} \quad m' = 2.32m$$

As expected, the mass of the combined particle is greater than $2m$ because some kinetic energy is converted into mass (energy).

Follow-Up Exercise. (a) In this Example, how much kinetic energy is lost? (b) What would be the mass of the combined particle if the two particles initially approached head-on each with a speed of $0.800c$ and stuck? (Your answers should be in terms of m and c.)

26.5 The General Theory of Relativity

OBJECTIVES: To (a) explain the principle of equivalence, and (b) examine some of the predictions of general relativity.

Special relativity applies to inertial systems, not to accelerating systems. Accelerating systems require a different approach, first described by Einstein about 1915. Called the **general theory of relativity**, it contains many implications about the theory of gravity.

The Principle of Equivalence

An important principle of general relativity was first envisioned by Einstein (after another *gedanken* experiment). He called it the **principle of equivalence**, which can be stated as follows:

> An inertial reference frame in a uniform gravitational field is physically equivalent to a reference frame that is not in a gravitational field, but that is in uniform linear acceleration.

What this means:

> No experiment performed in a closed system can distinguish between the effects of a gravitational field and the effects of an acceleration.

Thus, an observer in an accelerating system would find the effects of a gravitational field and those of their acceleration to be equivalent and indistinguishable. For simplicity, we will restrict consider only systems that are accelerating linearly (that is, those that have no rotational acceleration).

To understand this principle, consider the situations in ▸Fig. 26.11. Imagine yourself as an astronaut in a closed spaceship. Suppose when you drop a pencil, you observe that it accelerates to the floor. What does this mean? According to the principle of equivalence, it could mean (a) that you are in a gravitational field or (b) that you are in an accelerating system (Fig. 26.11a). Any experiment performed entirely inside the ship cannot determine one from the other. Whether the spaceship is in free space and accelerating with an acceleration $a = g$ or whether it is in a gravitational field with $g = -a$, the pencil has the same observed acceleration. In your closed system (remember that you cannot look outside), there is *no experiment* you could perform to distinguish whether the pencil's acceleration is a gravitational or a system's acceleration effect. According to the principle of equivalence, the two are physically indistinguishable.

As another example, suppose the pencil does not accelerate, but instead remains suspended next to you. This could mean that (a) you are in an inertial frame with no gravity or (b) you are in free fall in a gravitational field. Once again, in the closed spaceship (Fig. 26.11b), you have no way of distinguishing between these two possibilities. That is, they are *equivalent*.

From what you know about relativity at this point, you might conclude that it need be considered only at high speeds and in gravitational fields. Would you be surprised to find out that it is crucial to the development of many of our modern technological communications systems? Check out Insight 26.1 on Relativity in Everyday Living on page 838.

Light and Gravitation

The principle of equivalence leads to an important prediction: that a gravitational field bends light. To see how this prediction arises, let's use another *gedanken* experiment. Suppose a beam of light traverses a spaceship that is accelerating

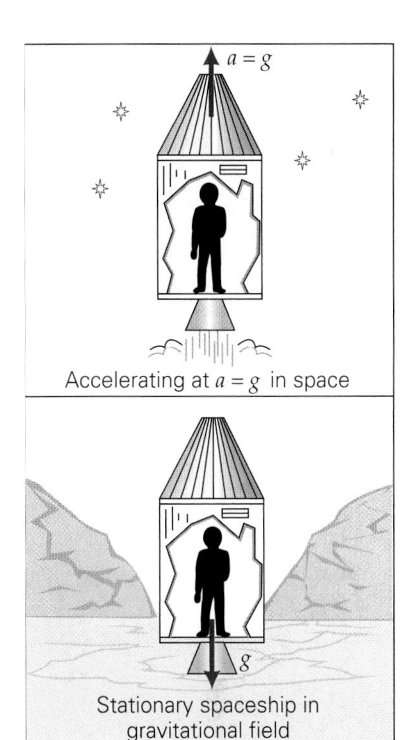

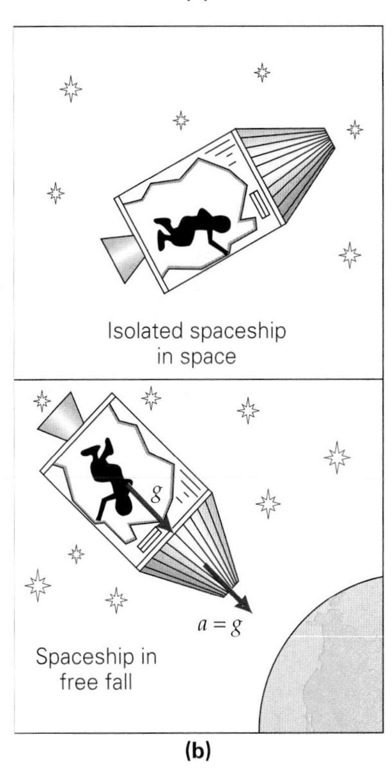

▲ **FIGURE 26.11 The principle of equivalence (a)** In a closed spaceship, the astronaut can perform no experiment that would determine whether he was in a gravitational field or an accelerating system. **(b)** Similarly, an inertial frame without gravity cannot be distinguished from free fall in a gravitational field.

INSIGHT 26.1 RELATIVITY IN EVERYDAY LIVING

You often hear that relativistic effects are observed only at speeds near the speed of light. This statement implies that these effects are irrelevant in everyday living. However, this is not true in our modern technological world. Relativistic effects, in fact, are crucial in global positioning systems (GPS), used for locating objects on the Earth to an accuracy of several meters. The system is important in many settings, such as modern airplane navigation and military operations. It is now included in many new cars for navigational purposes. The system works properly only because crucial corrections have been made for both special and general relativistic effects.

The GPS consists of an array of satellites, each with a very accurate atomic "clock" onboard. To determine the position of any object on the Earth's surface, that object must have a GPS receiver. The receiver detects light (radio) signals from several satellites. Knowing the speed of light and measuring the travel time, the receiver's computer can determine distances and directions to these satellites. Through triangulation, the GPS receiver's computer can rapidly calculate its location. As the location changes, the object's velocity can be calculated.

For GPS to work the clocks in orbit must be "in sync" with the corresponding clocks on the Earth. If they are not in sync, then the time of travel will be incorrect and the distances will be wrong. An error of just 100 ns (10^{-7} s) can lead to an error in location of several tens of meters—clearly unacceptable when you are trying to land a plane in bad weather, for example! Due to the satellites' orbital speeds (several kilometers per second) there are special relativistic time-dilation effects to account for. In addition, according to the theory of general relativity, the satellites will run at a faster rate because they are in a weaker gravitational field than the Earth-bound clocks. The net effect of that the orbiting clocks run at a faster rate. To keep them in synchronization with their counterparts on the Earth, these clocks are set to run at a slower rate before they are launched. When they reach the proper orbit, their rate increases, bringing them up to the same rate as that of the surface clocks. Relativity in everyday living—who would have thought it?

rapidly upward. If the spaceship were stationary, light entering the ship at point A would arrive at point B on the far wall; however, because the spaceship is accelerating, the light actually lands at point C (▼Fig. 26.12a).

From the point of view of the person onboard the ship (Fig. 26.12b), the light path is a downward-curving one. To this astronaut, it seems that the gravitational field she is in has bent the light path, much as a baseball's path would be. (Recall that according to the equivalence principle, the acceleration is indistinguishable from a gravitational field.) Although an outside observer "knows" that the rocket's acceleration produces this effect, we conclude by the principle of equivalence that a gravitational field should bend light paths. Notice that if the spaceship were moving up *without* accelerating, the light path in the ship would be a straight line, which is consistent with no gravitational effect. No gravity, no bending. Once again, the principle works.

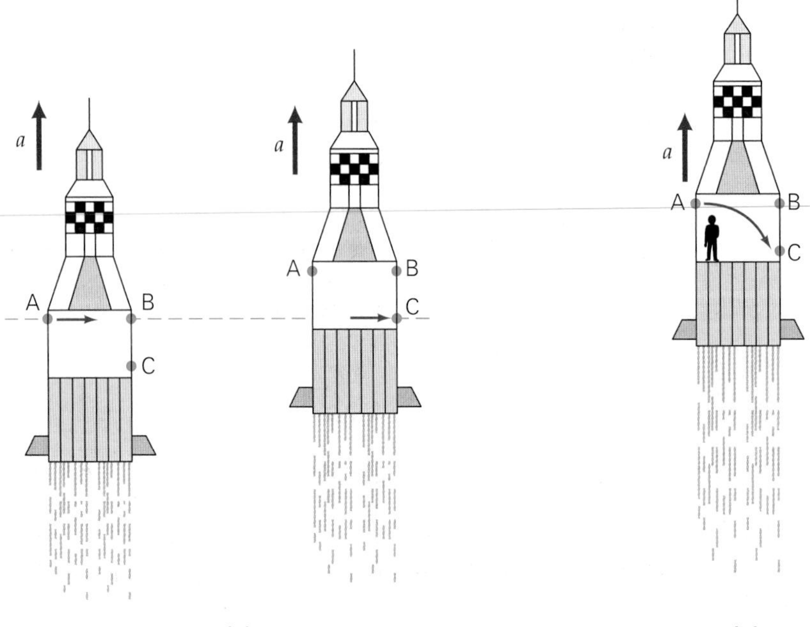

▶ **FIGURE 26.12** Light bending **(a)** Light traversing an accelerating, closed rocket from point A arrives at point C. **(b)** In the accelerating system, the light path would appear to be bent. Since an acceleration produces this effect, by the principle of equivalence, light should also be bent by a gravitational field.

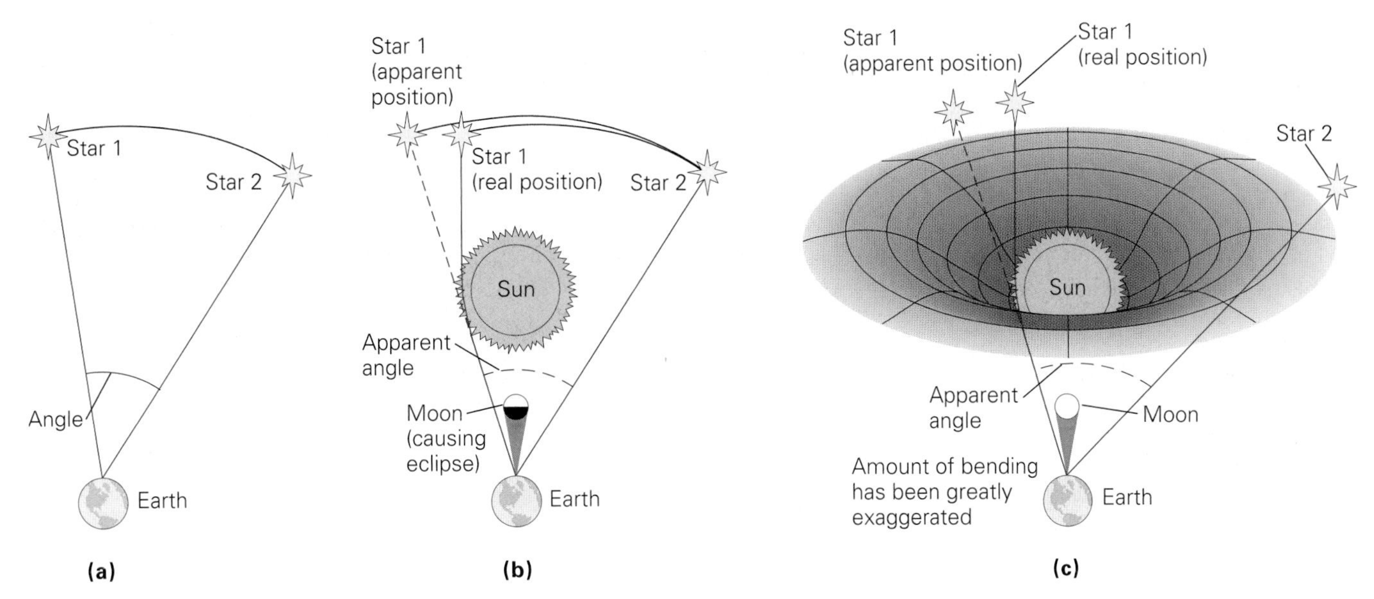

▲ **FIGURE 26.13 Gravitational attraction of light** **(a)** Normally, two distant stars are observed to have a certain angular separation. **(b)** During a solar eclipse, the star behind the Sun can still be seen because of the effect of solar gravity on starlight. The star has a larger measured angular separation than it has in the absence of such an eclipse. **(c)** General relativity views a gravitational field as a warping of space and time. A simplified analogy is the surface of a warped rubber sheet.

Under most conditions, this effect must be very small, since we don't observe everyday evidence of light bending in the Earth's gravitational field. However, this prediction was experimentally verified in 1919 during a solar eclipse. During times of the year when they aren't near the Sun and can be seen at night, distant stars have a constant angular separation between them (▲Fig. 26.13a). During other times of the year, light from some of these stars may pass near the Sun, which has a relatively strong gravitational field. However, any evidence of bending would not usually be observable on Earth because the starlight is almost always masked by the glare of the Sun.

However, stars may be seen during the day under the conditions of a *total solar eclipse*. When the Moon comes between the Earth and the Sun, an observer in the Moon's shadow (the umbra) can see stars not normally visible during the day (Fig. 26.13b). If the light from a star passing near the Sun is bent, then the star will have an *apparent* location that is different from its normal location. As a result, the angular distance between pairs of stars will be measured as slightly larger than normal. Einstein's theory predicted that the angular *difference* between the apparent positions of the stars during the eclipse of 1919 and their positions in the absence of solar gravity effects should be about 1.75 seconds of arc ($\approx 0.00005°$). The experimental difference was 1.61 ± 0.30 seconds of arc, in agreement within the uncertainty!

General relativity conceptually pictures a gravitational field as "warping" space and time, as illustrated in Fig. 26.13c. A light beam follows the curvature of space–time like a ball rolling on a curved surface. The bending of light by gravitational fields has been repeatedly verified during other solar eclipses and also by signals sent back to Earth by space probes passing near the Sun.

Gravitational Lensing

Another effect of gravity on light is called **gravitational lensing**. In the late 1970s, a double quasar was discovered. (A quasar is a powerful astronomical radio source.) The fact that it was a double quasar was not unusual, but everything about the two quasars seemed to be exactly the same, except that one was fainter than the other. It was suggested that perhaps there was only one quasar and that, somewhere between it and the Earth, a massive, but optically faint, object had

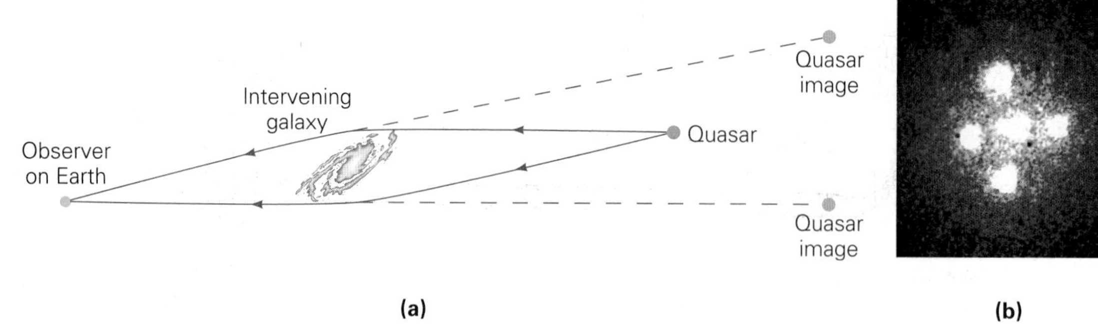

▲ **FIGURE 26.14** Gravitational lensing **(a)** The bending of light by a massive object such as a galaxy or a cluster of galaxies can give rise to multiple images of a more distant object. **(b)** The discovery of what appeared to be four images of the same quasar (the "Einstein Cross") suggested the possibility of gravitational lensing. On investigation, a faint intervening galaxy was found.

bent its light, producing multiple images. The subsequent detection of a faint galaxy between the two quasars confirmed this hypothesis, and other examples have since been discovered (▲Fig. 26.14).

The discovery of gravitational lenses gave general relativity a new role in modern astronomy. By examining multiple images of a distant galaxy or quasar and their relative brightness, astronomers can gain information about an intervening galaxy or cluster of galaxies whose gravitational field causes the bending of light.

Black Holes

The idea that gravity can affect light finds its most extreme application in the concept of a black hole. A **black hole** is thought to form from the gravitationally collapsed remnant of a massive star, typically many times the mass of our Sun. Such an object has a density so great and a gravitational field so intense that nothing can escape it. In terms of our space–time warp analogy, a black hole is graphically represented as a bottomless pit in the fabric of space–time. Even light can't escape the intense gravitational field of a black hole—hence the blackness.*

An estimate of the size of a black hole can be obtained by using the concept of escape speed. Recall from Section 7.6 that the escape speed from the surface of a spherical body of mass M and radius R is given by

$$v_{esc} = \sqrt{\frac{2GM}{R}} \qquad (26.14)$$

If light does not escape from a black hole, its escape speed must exceed the speed of light. The critical radius of a sphere around a black hole from which light cannot escape is obtained by substituting $v_{esc} = c$ into Eq. 26.14. Solving for R, we obtain

$$R = \frac{2GM}{c^2} \quad \textit{Schwarzschild radius} \qquad (26.15)$$

Note: A collapsing star becomes a black hole when it collapses to a radius less than its Schwarzschild radius.

The quantity R is called the **Schwarzschild radius**, after Karl Schwarzschild (1873–1916), a German astronomer who developed the concept. The boundary of a sphere of radius R defines the black hole's **event horizon**. Any event occurring within this horizon is invisible to an outside observer, since light cannot escape. The event horizon thus gives the limiting distance within which light cannot escape from a black hole. Information about what goes on inside the event horizon can never reach us, so questions such as, "What does it look like inside the event horizon?" are unanswerable (at least, given the current state of knowledge in physics†).

*It is speculated that black holes also may originate in other ways, such as from the collapse of entire star clusters in the center of a galaxy or at the beginning of the universe during the big bang. Some current theories (combining quantum mechanics and general relativity) hint that black holes might emit material particles and radiation and thus "evaporate," but none have been discovered to be evaporating. In this book, our discussion will be limited to stellar collapse.

†There is a discrepancy here that bothers physicists. According to quantum theory, information cannot be "lost." Theorists have proposed that black holes can actually radiate energy via a quantum phenomenon called "tunneling" (see Section 28.2). If true, this would bring relativity and quantum theory into better agreement, and is currently an active area of theoretical and experimental research.

Example 26.8 ■ If the Sun Were a Black Hole: Schwarzschild Radius

What would be the Schwarzschild radius if our Sun collapsed to a black hole? (The mass of the Sun is $M_S = 2.0 \times 10^{30}$ kg.)

Thinking It Through. This is a straightforward calculation using Eq. 26.15. We will need the gravitational constant and speed of light.

Solution.

Given: $M_S = 2.0 \times 10^{30}$ kg
$G = 6.67 \times 10^{-11}\ \mathrm{N \cdot m^2/kg^2}$
$c = 3.00 \times 10^8$ m/s

Find: R (Schwarzschild radius)

Thus we have

$$R = \frac{2GM_S}{c^2} = \frac{2(6.67 \times 10^{-11}\ \mathrm{N \cdot m^2/kg^2})(2.0 \times 10^{30}\ \mathrm{kg})}{(3.00 \times 10^8\ \mathrm{m/s})^2} = 3.0 \times 10^3\ \mathrm{m} = 3.0\ \mathrm{km}$$

This is less than 2 miles. The Sun's radius is about 7×10^5 km. Note however that the Sun will *not* become a black hole. This fate befalls only stars much more massive than the Sun.

Follow-Up Exercise. Once black holes form, they continuously draw in matter, increasing their Schwarzschild radius. How many times more massive would our Sun have to be for its Schwarzschild radius to extend to, and swallow up, Mercury, the innermost planet? (The average distance from the Sun to Mercury is 5.79×10^{10} m.)

If nothing, including radiation, escapes a black hole from inside its event horizon, how, then, might we observe or even merely locate a black hole? This question and some insight into current research into gravitational waves is addressed in Insight 26.2 entitled Black Holes, Gravitational Waves and LIGO, on page 842.

*26.6 Relativistic Velocity Addition

OBJECTIVES: To (a) understand the necessity for a relativistic velocity addition, and (b) investigate relative velocity addition through calculations.

As we have seen, the postulates of special relativity affect our everyday concepts of distance and time. Since velocity involves these quantities, relative velocities should also be affected, and that is indeed the case.

Recall that according to Newtonian relativity, there is a problem with vector addition when light is involved. In Fig. 26.1, the vector addition of velocities predicts a speed greater than c for the beam of light. Such a speed violates the second postulate of relativity.

The same problem occurs with objects moving at appreciable fractions of the speed of light. Consider the rocket separation shown in ▾Fig. 26.15. After separation, the jettisoned stage has a velocity $\vec{\mathbf{v}}$ with respect to the Earth, and the rocket has a velocity $\vec{\mathbf{u}}'$ with respect to the jettisoned stage. Then, from Newtonian relativity and vector addition, the velocity of the rocket payload with respect to the Earth is $\vec{\mathbf{u}} = \vec{\mathbf{v}} + \vec{\mathbf{u}}'$.

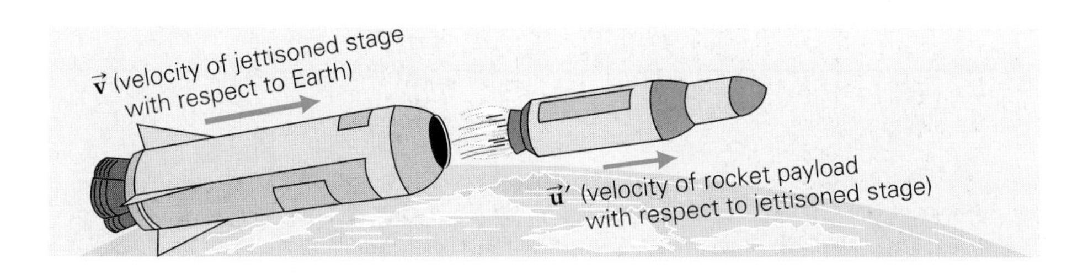

◂ **FIGURE 26.15** Relativistic velocity addition After jettisoning, the rocket payload has a velocity $\vec{\mathbf{u}}'$ with respect to the jettisoned stage, and the jettisoned stage has a velocity $\vec{\mathbf{v}}$ with respect to the Earth. For relativistic velocities, the velocity $\vec{\mathbf{u}}$ of the rocket with respect to the Earth is obtained from a relativistically correct version.

INSIGHT 26.2 BLACK HOLES, GRAVITATIONAL WAVES, AND LIGO

Try to imagine something so dense that nothing—not even light—can escape from it. According to stellar evolution theory, black holes could result from the collapse of stars having much greater mass than our Sun. But gathering experimental proof of the existence of a black hole is another matter.

The event horizon of a black hole is located at a certain distance (the Schwarzschild radius) from its center. The event horizon marks the location where gravity becomes sufficiently strong to keep even light from escaping. What form or size matter takes inside a black hole is not known and in principle can never be known. Even if a probe could be sent "inside" a black hole (that is, closer than its Schwarzschild radius), the probe could not send data back to us.

How, then, might a black hole be detected? Light passing near it would be bent, but it is unlikely that we would ever observe that, considering the vastness of space. The most likely possibility comes from observing a *binary star system*, which consists of two stars orbiting a common center of mass. Cygnus X-1, the first X-ray source discovered in the constellation Cygnus, provides the best evidence so far for a black hole in our galaxy.

Where Cygnus X-1 is located in the sky, we observe a giant star (visible light *cannot* detect two separate stars) whose spectrum shows periodic Doppler redshifts and blueshifts, indicating a periodic orbital motion away from us and toward us, respectively. This observation indicates a binary star system with an "invisible" companion to the giant star. Measuring the precise orbital data of both stars allow us to compute their masses. In Cygnus X-1, an apparent unseen companion to the giant star *seems* to have enough mass to qualify as a black hole.

Astronomers speculate that, in binary star systems such as Cygnus X-1, one member has evolved to become a black hole. Matter drawn from the other, more normal star would create an *accretion disk* of spiraling matter around the black hole (Fig. 1). The matter falling into the disk would be accelerated and heated. Collisions and deceleration would produce a characteristic spectrum of X-rays, and the black hole would appear to be the source of X-rays we observe. These X-rays are emitted by the hot matter before it gets closer than the Schwarzschild radius.

A similar mechanism may explain the enormous energy output of *active galaxies* and *quasars*. It has been proposed that the centers of these brilliant objects, which produce vast amounts of radiation, might contain black holes with masses millions or even billions of times that of our Sun. In fact, recent observations suggest that even many normal galaxies, including our own Milky Way galaxy, may harbor enormously massive black holes in their cores.

According to general relativity, a gravitationally violent event, such as black holes coalescing, should emit gravitational waves traveling at the speed of light. These waves should cause a displacement of the matter in the regions through which they travel. They are expected to be extremely weak, causing movements on the order of only 10^{-14} m (not much larger than the diameter of an atomic nucleus). As part of current research, U.S. scientists have collaborated in a project called LIGO (Laser Interferometer Gravitational-Wave Observatory). Each interferometer consists of two dangling weights with the distance between them monitored by laser. If a gravitational wave passes, the distance, and the laser interference pattern, should change.

When operational, there will be two identical installations, one in Washington and one in Louisiana (see Fig. 2). If a gravitational wave does pass through the Earth, both installations should detect the event simultaneously. Having two simultaneous signals reduces the likelihood that the event resulted from a local disturbance. In the future, a global network of these sensitive interferometers is proposed. If gravitational waves are finally detected, it will be a victory for relativity. Once understood, these waves could then be used as a cosmic probe—that is, to determine the details of the event that gave rise to them.

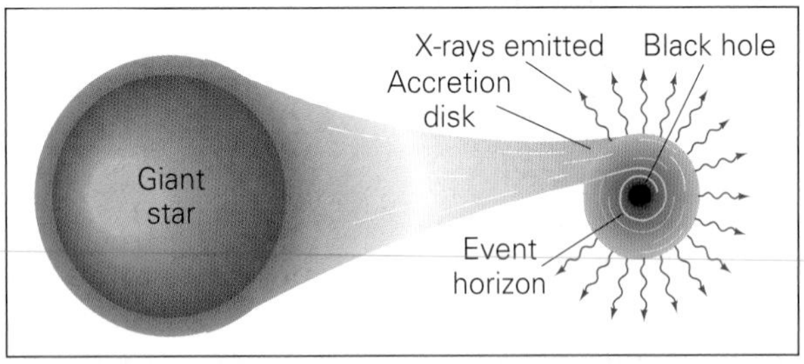

X-rays emitted
Black hole
Accretion disk
Giant star
Event horizon

FIGURE 1 X-rays and black holes Matter drawn from the "normal" member of a binary star system forms a spiraling accretion disk around the hole. Matter falling into the disk is accelerated, and collisions give rise to the emission of X-rays.

FIGURE 2 LIGO Sensitive laser interferometers accurately measure distances between weights at various locations in the United States, looking for changes in the distance due to passage of gravitational waves. This is the location in Louisiana.

However, at relativistic speeds, this addition law gives a result contradictory to special relativity. Suppose, for example, that the velocities are all in the same direction with magnitudes of $v = 0.50c$ and $u' = 0.60c$. Then $u = v + u' = 0.50c + 0.60c = 1.10c$. Thus, classical relativity predicts that an observer on the Earth would measure the rocket's speed to be greater than c, which, as we have seen, cannot happen.

Note: It is usually convenient to express speeds as fractions of c.

Einstein recognized that according to special relativity, lengths and times differ depending on the observer's reference frame. Thus classical velocity vector addition cannot be correct at relativistic speeds. He showed the correct equation (for motion in a straight line) to be

$$u = \frac{v + u'}{1 + \dfrac{vu'}{c^2}} \quad \textit{relativistic velocity addition (one dimension)} \tag{26.16}$$

where the velocities have the same meanings as in the preceding paragraph and sign notation is used to indicate velocity directions. Notice that the observed velocity u is *reduced* by a factor of $1/[1 + (vu'/c^2)]$ from that of the classical result. At very low speeds (that is, $v/c \ll 1$ and $u'/c \ll 1$), we have $vu'/c^2 = (v/c)(u'/c) \ll 1$. Thus, under low-speed conditions, the denominator in Eq. 26.16 is, for all practical purposes, equal to 1, and the Newtonian result, $u = v + u'$, is obtained.

In working out problems, it is important to identify the velocities clearly:

$\vec{\mathbf{v}}$ = velocity of object 1 with respect to an inertial observer

$\vec{\mathbf{u}}'$ = velocity of object 2 with respect to object 1

$\vec{\mathbf{u}}$ = velocity of object 2 with respect to an inertial observer

Since we will consider only one-dimensional-motion problems, plus/minus signs are used to indicate velocity directions. The next Example shows that Eq. 26.16 gives results that do *not* violate the second postulate.

Example 26.9 ■ Faster Than the Speed of Light? No! Thanks to Relativistic Velocity Addition

For the rocket separation in Fig. 26.15, let the speeds be $v = 0.50c$ and $u' = 0.60c$ (with directions shown in the figure). What is the velocity of the payload, as measured by an observer on Earth?

Thinking It Through. Clearly, Eq. 26.16 needs to be applied here because the speeds are relativistic, and the answer (u) must be less than c.

Solution. The velocities are taken to be in the positive direction.

Given: $v = +0.50c$ (speed of jettisoned stage with respect to Earth)
$u' = +0.60c$ (speed of rocket with respect to jettisoned stage)

Find: $\vec{\mathbf{u}}$ (velocity of rocket with respect to Earth)

Since both velocities are to the right, they are designated with plus signs. From Eq. 26.16, we have

$$u = \frac{v + u'}{1 + \dfrac{vu'}{c^2}} = \frac{+0.50c + 0.60c}{1 + \dfrac{(+0.50c)(+0.60c)}{c^2}} = \frac{+1.1c}{1.3} = +0.85c$$

As expected, u is less than c and is directed to the right, as indicated by the plus sign.

Follow-Up Exercise. Repeat this Example with both objects moving to the right at $0.40c$. According to Newtonian relativity, the velocity relative to the Earth is $0.80c$, which does *not* violate the second postulate. What is the correct result? [*Hint*: It should be lower than the Newtonian result of $0.80c$. Why?]

Chapter Review

- **Newtonian** or **classical relativity** supported the belief in the existence of an absolute **inertial reference frame** somewhere in the universe that was "at rest." In this reference frame was the material called the **ether,** the medium through which light could propagate.

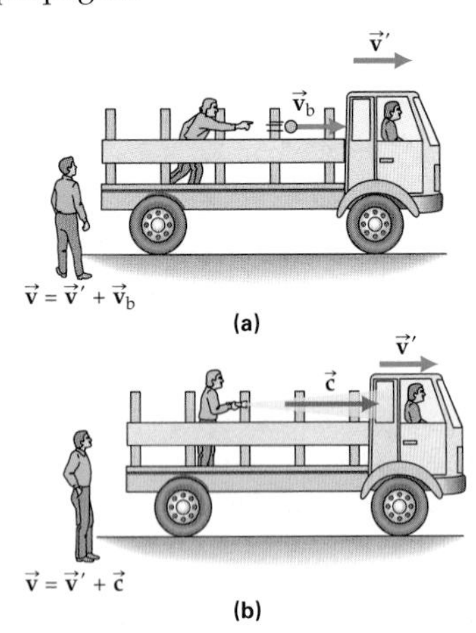

- **The Michelson–Morley experiment** attempted to measure the speed of the Earth relative to the ether frame. The results were always zero. Thus, physicists had to give up the idea of a reference frame that was absolutely at rest.
- **The special theory of relativity** involves inertial reference frames moving relative to one another and is based on two postulates:

Principle of relativity: All the laws of physics are the same in all inertial reference frames.

Principle of the constancy of the speed of light: The speed of light in a vacuum has the same value in all inertial systems.

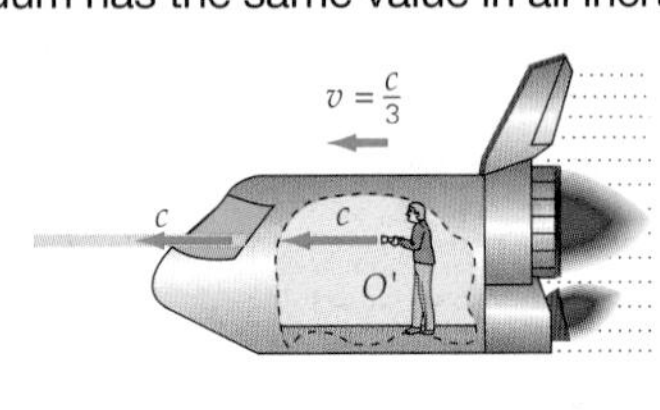

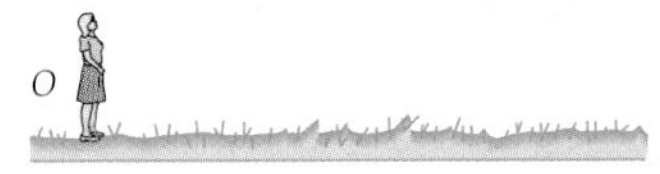

- A time interval measured by a clock that is present at both the starting and stopping events of the interval is a **proper time interval** Δt_o. The time interval Δt of an observer in any other inertial frame is larger than the proper time interval. The two intervals are related by

$$\Delta t = \frac{\Delta t_o}{\sqrt{1 - (v/c)^2}} \tag{26.3}$$

If γ (Greek letter "gamma") is defined as

$$\gamma \equiv \frac{1}{\sqrt{1 - (v/c)^2}} \tag{26.4}$$

Equation 26.3 can then be written more compactly as

$$\Delta t = \gamma \Delta t_o \tag{26.5}$$

This effect on time intervals is called **time dilation.**

- The length of an object, as measured by an observer at rest with respect to it, is called the object's **proper length** L_o. To an observer in any other inertial frame, the length L is smaller than the proper length by a factor of $1/\gamma$, or

$$L = \frac{L_o}{\gamma} = L_o\sqrt{1 - (v/c)^2} \tag{26.7}$$

This phenomenon is known as **length contraction.**

- At speeds near the speed of light, an object's **relativistic kinetic energy** is

$$K = \left[\frac{1}{\sqrt{1 - (v/c)^2}} - 1\right]mc^2 = (\gamma - 1)mc^2 \tag{26.8}$$

- At speeds near the speed of light, an object's **relativistic momentum** is

$$\vec{\mathbf{p}} = \frac{m\vec{\mathbf{v}}}{\sqrt{1 - \left(\frac{v}{c}\right)^2}} = \gamma m\vec{\mathbf{v}} \tag{26.9}$$

- At speeds near the speed of light, the **relativistic total energy** of an object is

$$E = \frac{mc^2}{\sqrt{1 - \left(\frac{v}{c}\right)^2}} = \gamma mc^2 \tag{26.10}$$

- Even when an object is at rest, it still has an energy of mc^2. This minimum energy called its **rest energy** and is

$$E_o = mc^2 \tag{26.11}$$

- An object's relativistic total energy can be written in terms of its rest energy as

$$E = K + E_o = K + mc^2 \tag{26.12}$$

- The **general theory of relativity** expresses how physical quantities are measured by observers in reference frames that are accelerating. The **principle of equivalence** states that

an inertial reference frame in a uniform gravitational field is physically equivalent to a reference frame that is not in a gravitational field, but that is in uniform linear acceleration.

this means that no experiment performed in a closed system can distinguish between the effects of a gravitational field and the effects of an acceleration.

- General relativity predicts many interesting phenomena that have been experimentally verified, such as stars called **black holes** and the bending of light by a gravitational field.

Exercises

MC = *Multiple Choice Question*, **CQ** = *Conceptual Question, and* **IE** = *Integrated Exercise. Throughout the text, many exercise sections will include "paired" exercises. These exercise pairs, identified with* **red numbers**, *are intended to assist you in problem solving and learning. In a pair, the first exercise (even numbered) is worked out in the Study Guide so that you can consult it should you need assistance in solving it. The second exercise (odd numbered) is similar in nature, and its answer is given at the back of the book.*

Note: Assume c to be exact at 3.00×10^8 m/s. *Consider speeds given in terms of c as having the same number of significant figures as the accompanying numerical coefficient. (For example, 0.85 c has two significant figures.) Gamma* (γ) *should be calculated explicitly.*

26.1 Classical Relativity and the Michelson–Morley Experiment

1. **MC** An object free of all forces exhibits a changing velocity in a certain reference frame. It follows that (a) the frame is inertial, (b) $\vec{\mathbf{F}} = m\vec{\mathbf{a}}$ applies in this frame, (c) the laws of mechanics are the same in this reference frame as in all inertial frames, (d) none of the preceding.

2. **MC** Car A is traveling eastward at 85 km/h Car B is traveling westward with a speed of 65 km/h. The velocity of car B as measured by the driver of car A is (a) 150 km/h eastward (b) 20 km/h westward (c) 150 km/h westward (d) 20 km/h eastward

3. **MC** A space probe is moving rapidly away from the Sun at 0.2*c*. As it passes the probe, the speed of the sunlight measured by the probe has what value? (a) 1.2*c* (b) exactly *c* (c) 0.8*c*

4. **CQ** We live on a rotating Earth and therefore are in an accelerating, noninertial system. How, then, can we apply Newton's laws of motion on the Earth?

5. **CQ** A car accelerating from rest is a noninertial system. Explain the "forces" that the driver uses to explain why things on the dashboard slide "backward." How does the inertial observer on the roadside explain this?

6. **CQ** A person rides with her package on the floor of an elevator that is moving upward at constant velocity. How does the free-body diagram of the package as drawn by the elevator person differ from that drawn by the observer remaining on the first floor? Explain.

7. • A person 1.20 km away from you fires a gun, and you observe the flash from the muzzle. A wind of 10.0 m/s is blowing. How long will the sound of gunfire take to reach you if the wind is (a) toward you and (b) toward the person who fired the gun? (Take the speed of sound to be 345 m/s.)

8. • A small airplane has an airspeed (speed with respect to air) of 200 km/h. Find the airplane's ground speed if there is (a) a headwind of 35 km/h and (b) a tailwind of 25 km/h.

9. • A speedboat can travel with a speed of 50 m/s in still water. If the boat is in a river that has a flow speed of 5.0 m/s, find the maximum and minimum values of the boat's speed relative to an observer on the riverbank.

10. **IE** •• A boat can make a round trip between two locations, A and B, on the same side of a river in a time *t* if there is no current in the river. (a) If there is a constant current in the river, the time the boat takes to make the same round trip will be (1) longer, (2) the same, (3) shorter. Why? (b) If the boat can travel with a speed of 20 m/s in still water, the speed of the river current is 5.0 m/s, and the distance between points A and B is 1.0 km, calculate the times when there is no current and when there is current.

11. ••• The apparatus used by the French scientist Armand Fizeau in 1849 for measuring the speed of light is illustrated in ▼Fig. 26.16. Teeth on a rotating wheel periodically interrupt a beam of light. The flashes of light travel to a plane mirror and are reflected back to an observer. Show that if the wheel is rotated at just the right frequency f, the light passing through one gap reaches the mirror and is reflected to the observer through the very next gap. When f is adjusted so this happens, show that the speed of light is given by $c = 2fNL$, where N is the number of gaps in the wheel, f is the frequency (in revolutions per second) made by the rotating wheel, and L is the distance between the wheel and the mirror.

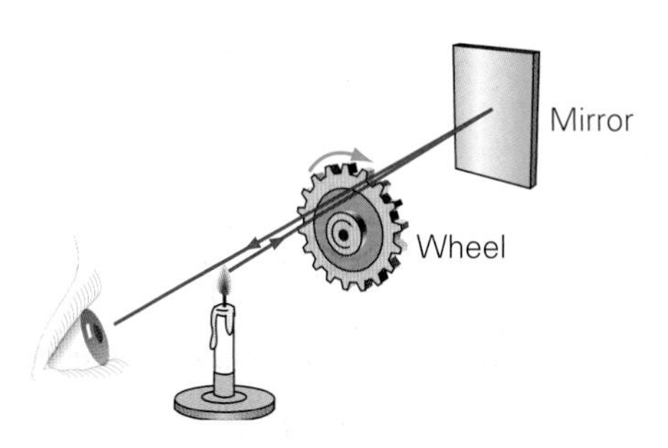

▲ **FIGURE 26.16 Fizeau's apparatus** See Exercise 11.

26.2 The Postulates of Special Relativity and the Relativity of Simultaneity

12. **MC** Events that are simultaneous in one inertial reference frame are (a) always simultaneous in other inertial reference frames, (b) never simultaneous in other inertial reference frames, (c) sometimes simultaneous in other inertial reference frames, (d) none of the preceding.

13. **MC** An object is at rest in an inertial reference frame. What will be the same about the object as measured by an inertial observer moving relative to the object: (a) its free-body diagram, (b) its velocity, (c) its kinetic energy, or (d) none of the preceding?

14. **MC** An object is moving in an inertial reference frame. What will be the same about the object as measured by an inertial observer moving relative to the object? (a) its velocity, (b) its speed, (c) its kinetic energy, (d) its acceleration, or (e) none of the preceding?

15. **CQ** In a space war, a warrior approaching his enemy head on at a speed of $0.85c$ escapes with a near miss as a high-energy laser beam barely misses him. What speed does the warrior observe for the light as it passes him?

16. **CQ** Is it possible for either observer on two approaching rocket ships to observe the other ship with a velocity greater than c? Why?

17. **CQ** In the *gedanken* experiment shown in ▼Fig. 26.17, two events in the same inertial reference frame O are related by cause and effect: (1) A gun at the origin fires a bullet along the x-axis with a speed of 300 m/s. (Assume that there are no gravitational or frictional forces.) (2) The bullet hits a target at $x = +300$ m. Show, using qualitative arguments, that the two events cannot be viewed simultaneously by any inertial observer. (*Note:* This shows that special relativity preserves the time *sequence* of two events if they are related as cause and effect. In this situation, this means that all observers agree that the gun fires before the bullet hits the target.)

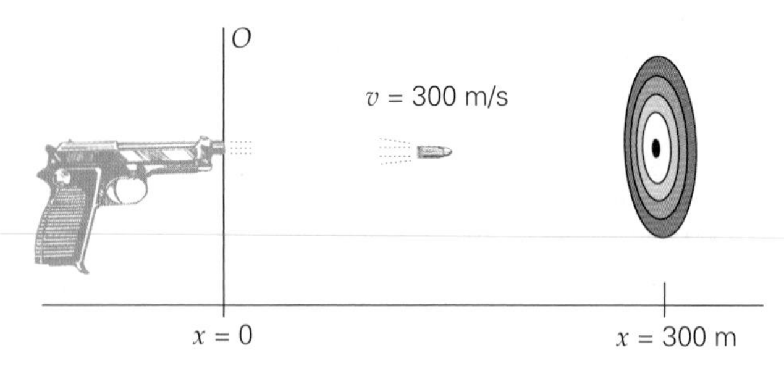

▲ **FIGURE 26.17 A thought experiment** See Exercise 17.

18. **CQ** In a *gedanken* experiment (▸Fig. 26.18), two events that cannot be related by cause and effect occur in the same inertial reference frame O: (1) Strobe light A, at the origin of the x-axis, flashes; (2) strobe light B, located at $x = +600$ m, flashes 1.00 μs later. B's flash cannot be caused by A's flash. (Light travels only 300 m in the time between the two events.) (a) Use qualitative arguments to show that there exists another inertial reference frame, traveling at less than c, in which these two events would be observed to occur simultaneously. (b) What is the direction of the velocity of the reference frame in part (a)? (*Note:* This question shows that relativity does *not* have to preserve the time order of events if they are not related by cause and effect. Since one event cannot cause the other, no physical principles are violated if they are seen in reverse order.)

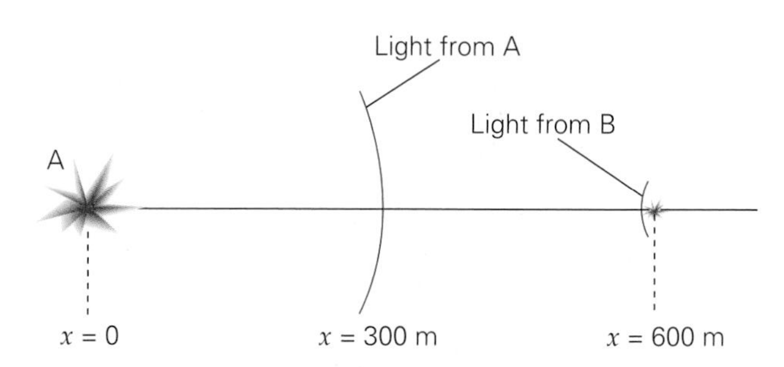

▲ **FIGURE 26.18 Another thought experiment** See Exercise 18.

26.3 The Relativity of Length and Time: Time Dilation and Length Contraction

19. **MC** An observer sees a friend passing her in a rocket ship that has a uniform velocity of $0.90c$. The observer knows her friend to be 1.45 m tall, and he is standing such that his length is perpendicular to their relative velocity. To the observer, her friend will appear (a) taller than 1.45 m, (b) shorter than 1.45 m, (c) exactly 1.45 m tall.

20. **MC** An observer sees a friend passing her in a rocket ship that has a uniform velocity of $0.90c$. Her friend claims that exactly 10 seconds have elapsed on his clock. How will the observer's identical clock measure the same time interval? The observer's clock will read (a) less than 10 seconds, (b) greater than 10 seconds, (c) exactly 10 seconds.

21. **MC** A speedboat completes one straight leg of a lap that the race organizers have carefully laid out to be exactly 2.55 km long. Which observer(s) measures the leg's proper length: (a) the organizers, (b) the driver of the speedboat, (c) they both do, or (d) neither does?

22. **CQ** You are standing on a road and observe a high-speed car as it passes you. Sketch the shape of the car when it is stationary and when it is moving at $0.50c$.

23. **CQ** A farm boy wants to store a 5-m-long pole in a shed that is only 4 m long (it does have both front and rear doors). He claims that if he runs through the shed sufficiently fast, according to an observer at rest, the pole will fit in the shed (both doors closed at least for an instant) as a result of length contraction. Can this be true? Show that from the boy's reference frame the pole could not possibly fit into the shed.

24. **CQ** You are standing on the Earth and observe a fast-moving spacecraft going by with your professor on board. (a) If both you and your professor are observing your wristwatch, who is measuring the proper time? (b) Who measures the proper length of the spacecraft?

25. ● A spacecraft moves past a student with a relative velocity of 0.90*c*. If the pilot of the spacecraft observes 10 min to elapse on his watch, how much time has elapsed according to the student's watch?

26. IE ● Your pulse rate is 80 beats/min, and your physics professor in a spacecraft is moving with a speed of $0.85c$ relative to you. (a) According to your professor, your pulse rate is (1) greater than 80 beats/min, (2) equal to 80 beats/min, (3) less than 80 beats/min. Why? (b) What is your pulse rate, according to your professor?

27. ● You fly your 15.0-m-long spaceship with a speed of $c/3$ relative to your friend. Your velocity is parallel to the ship's length. How long is your spaceship, as observed by your friend?

28. IE ● An astronaut in a spacecraft moves past a field 100 m long (according to a person standing on the field) and parallel to the field's length, with a speed of $0.75c$. (a) Will the length of the field, according to the astronaut, be (1) longer than 100 m, (2) equal to 100 m, or (3) shorter than 100 m? Why? (b) What is the length as measured by the astronaut? (c) Which length is the proper length?

29. ●● The proper lifetime of a muon is 2.20 μs. If the muon has a lifetime of 34.8 μs according to an observer on Earth, what is the muon's speed, as a fraction of c, relative to the observer?

30. ●● One of a pair of 25-year-old twins takes a round trip through space while the other twin remains on Earth. The traveling twin moves at a speed of $0.95c$ for 39 years, according to Earth time. Assuming that special relativity applies, what are the twins' ages when the traveling twin returns to Earth?

31. IE ●● Alpha Centauri, a binary star close to our solar system, is about 4.3 light-years away. Suppose a spaceship traveled this distance with a constant speed of $0.60c$ relative to Earth. (a) Compared with a clock on the spaceship, an Earth-based clock will measure (1) a longer time, (2) an equal time, (3) a shorter time. Why? (b) How much time would elapse on an Earth-based clock and on a clock on the spaceship?

32. ●● A cylindrical spaceship of length 35.0 m and diameter 8.35 m is traveling in the direction of its cylindrical axis (length). It passes by the Earth with a relative speed of 2.44×10^8 m/s. What are the dimensions of the ship, as measured by an Earth observer?

33. ●● A pole vaulter at the Relativistic Olympics sprints past you to do a vault with a speed of 0.65 c. When he is at rest, his pole is 7.0 m long. What length do you perceive the pole to be as he passes you, assuming his relative velocity is parallel to the length of the pole?

34. ●● A "flying wedge" spaceship is a right triangle in side view. When the ship is at rest, its base and altitude (which form the 90° angle) measure 40.0 m and 15.0 m, respectively. As the ship moves with a speed of 0.900 c past an observer on Earth, what does she (a) measure the area of the side of the ship to be? (b) calculate the angle of the triangular "nose" to be?

35. ●● How fast must a meterstick be moving, parallel to its length, relative to an observer so that he measures its length to be 50 cm?

36. ●● The distance to Planet X is 1.00 light-year. How long does it take a spaceship to reach X, according to the pilot of the spaceship, if the speed of the ship is $0.700c$ relative to X?

37. ●●● What is the length contraction (ΔL) of an automobile 5.00 m long when it is traveling at 100 km/h? [*Hint*: For $x \ll 1$, $\sqrt{1 - x^2} \approx 1 - (x^2/2)$.]

38. IE ●●● A student in a certain reference frame cannot understand why there is a problem in converting to the metric system, because she notes that the length of her professor's meterstick in another reference frame is the same length as her yardstick (parallel to the meterstick). (a) Which of the following is true: (1) The student is moving relative to the professor, (2) the professor is moving relative to the student, or (3) the only thing that matters is that they are moving relative to one another? Why? (b) What is their relative speed (assuming them to be moving in a direction parallel to their respective sticks)?

39. ●●● Sirius is about 9.0 light-years from Earth. To reach the star by spaceship in 12 years (ship time), how fast must you travel?

26.4 Relativistic Kinetic Energy, Momentum, Total Energy, and Mass–Energy Equivalence

40. **MC** How does an object's relativistically correct kinetic energy compare to its kinetic energy calculated from the Newtonian expression? (a) The relativistic result is always larger, (b) the Newtonian result is always larger, (c) they are the same, or (d) one can be larger or smaller than the other depending upon the object's speed.

41. **MC** The total energy E of a free-moving particle of mass m with speed v is given by which of the following: (a) mv^2, (b) γmc^2, (c) $1/2mc^2$, or (d) $K + \gamma mc^2$.

42. **MC** How does an object's relativistically correct linear momentum (magnitude) compare to its momentum calculated from the Newtonian expression? (a) The relativistic result is always less, (b) the Newtonian result is always less, (c) they are the same, or (d) one can be larger or smaller than the other, depending on the object's velocity.

43. CQ The special theory of relativity places an upper limit on the speed an object can have. Are there similar limits on energy and momentum? Explain.

44. CQ An object subject to a large constant force approaches the speed of light. Is its acceleration constant? Explain.

45. CQ If an electron has a kinetic energy of 2 keV, could the classical expression for kinetic energy be used to compute its speed accurately? What if its kinetic energy is 2 MeV? Explain.

46. ● An electron travels at a speed of $0.600c$. What is its total energy?

47. ● An electron is accelerated from rest through a potential difference of 2.50 MV. Find the electron's (a) speed, (b) kinetic energy, and (c) momentum.

48. ● How fast must an object travel for its total energy to be (a) 1% more than its rest energy and (b) 99% more than its rest energy?

49. ● An average home uses about 1.5×10^4 kWh of electricity per year. How much matter would have to be converted to energy (assuming 100% efficiency) to supply energy for one year to a city with 250000 such homes? (Are you surprised by the answer?)

50. ● The United States uses approximately 3.0 trillion kWh of electricity annually. If this electrical energy were supplied by nuclear generating plants, how much nuclear mass would have to be converted to energy, assuming a production efficiency of 25%?

51. ● To travel to a nearby star, a spaceship is accelerated to $0.99c$ in order to take advantage of time dilation. If the ship has a mass of 3.0×10^6 kg, how much work must be done to get it up to speed from rest? Compare this value with the annual electricity usage of the United States. (See Exercise 50.)

52. ●● An electron has a total energy of 2.8 MeV. What is its momentum?

53. ●● How much energy, in keV, is required to accelerate an electron from rest to $0.50c$?

54. ●● A proton moves with a speed of $0.35c$. What are its (a) total energy, (b) kinetic energy, and (c) momentum?

55. ●● A proton moving with a constant speed has a total energy 2.5 times its rest energy. What are the proton's (a) speed and (b) kinetic energy?

56. IE ●● The Sun's mass is 1.989×10^{30} kg and it radiates at a rate of 3.827×10^{23} kW. (a) Over time, must the mass of the Sun (1) increase, (2) remain the same, or (3) decrease? (b) Estimate the lifetime of the Sun from this data, assuming it converts all its mass into energy. (c) The actual lifetime of the Sun is predicted to be much less than the answer to part (b), even though its energy emission rate will remain constant. What does this tell you about the 100% conversion assumption?

57. ●● A nickel has a mass of 5.00 g. If this mass could be completely converted to electric energy, how long would it keep a 100-W lightbulb lit?

58. IE ●● Phase changes require energy in the form of latent heat (Chapter 11). (a) If 1 kg of ice at 0°C is converted to water at 0°C, will the water have (1) more, (2) the same, or (3) less mass compared to the ice? Why? (b) What is the difference in mass between the ice and the water? Do you think this difference would be detectable?

59. ●● A beam of electrons is accelerated from rest to a speed of $0.950c$ in a particle accelerator. In MeV, what are the (a) kinetic energy and (b) total energy of the electrons?

60. IE ●●● A particle of mass m, initially moving with speed v, collides head on elastically with an identical particle initially at rest. (a) Do you expect the total mass of the two particles after the collision to be (1) greater than $2m$, (2) equal to $2m$, or (3) less than $2m$? Why? (b) What are the total energy and momentum of the two particles after the collision, in terms of m, v, and c?

61. ●●● Using the relativistic expression for total energy E and the magnitude p of the momentum, show that E and p are related by $E^2 = p^2c^2 + (mc^2)^2$.

62. ●●● In a linear accelerator, protons are accelerated to an energy of 600 MeV. (a) How do we know that this quoted energy value must be kinetic, and not total, energy? (b) What is the speed of the protons? (c) What is their momentum?

26.5 The General Theory of Relativity

63. **MC** General relativity (a) provides a theoretical basis for explaining the gravitational force, (b) applies only to rotating systems, (c) applies only to inertial systems.

64. **MC** One of the predictions of general relativity is (a) the mass–energy equivalence, (b) time dilation, (c) the twin paradox, (d) the bending of light in a gravitational field.

65. **MC** Black hole A has three times the mass of black hole B. How do their Schwarzschild radii R compare? (a) $R_A = R_B$, (b) $R_B = 3R_A$, (c) $R_A = 3R_B$, (d) $R_A = 9R_B$.

66. CQ Suppose a meterstick is dropped toward a black hole. Describe what happens to its shape lengthwise as it gets close to the event horizon. [*Hint:* The force of gravity will be a lot different on one end of the stick than the other.]

67. CQ An apparatus like that in ▸Fig. 26.19 was given to Albert Einstein on his 76th birthday by Eric M. Rogers, a physics professor at Princeton University. The goal is to get the ball into the cup without touching the ball. (Jiggling the pole up and down will not do it.) Einstein solved the puzzle immediately and then confirmed his answer with an experiment. How did Einstein get the ball into the cup? [*Hint*: He used a fundamental concept of general relativity.*]

*This problem is adapted from R. T. Weidner, *Physics* (Boston: Allyn & Bacon, 1985).

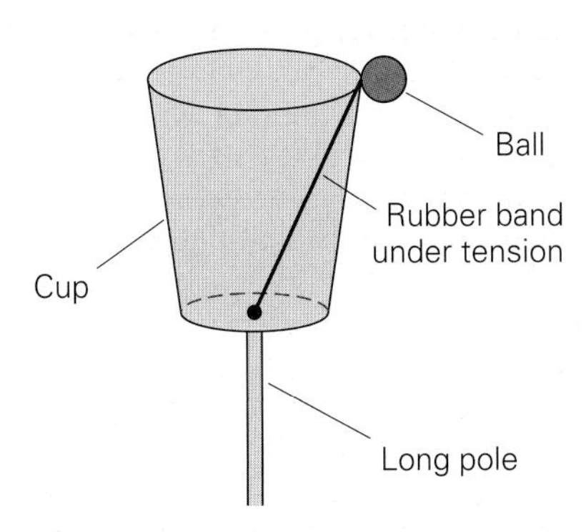

▲ **FIGURE 26.19 How to get the ball into the cup** See Exercise 67.

68. IE ●The mass of the planet Jupiter is about 318 times that of the Earth ($M_J = 318M_E$). (a) If these two planets could develop into black holes, would the event horizon for Jupiter be (1) larger than, (2) the same as, or (3) smaller than the Earth's? Why? (b) The mass of the Earth is $M_E = 6.0 \times 10^{24}$ kg; what are the radii of the event horizons for the Earth and Jupiter?

69. ●● If the Sun became a black hole, what would be its average density? (See Example 26.8.)

70. ●● A black hole has an event horizon of 5.00×10^3 m. (a) What is its mass? (b) Determine the lower limit on the density of the black hole.

*26.6 Relativistic Velocity Addition

71. ● After jettisoning a stage, a rocket has a velocity of $+0.20c$ relative to the jettisoned stage. An observer on Earth sees the jettisoned stage moving with a velocity of $+7.5 \times 10^7$ m/s, relative to her, in the same direction as the rocket. What is the velocity of the rocket relative to the Earth observer?

72. ● In moving away from Planet Z, a spacecraft fires a probe with a speed of $0.15c$, relative to the spacecraft, back toward Z. If the speed of the spacecraft is $0.40c$ relative to Z, what is the velocity of the probe as seen by an observer on Z?

73. ●● A rocket launched outward from Earth has a speed of $0.100c$ relative to Earth. The rocket is directed toward an incoming meteor that may hit the planet. If the meteor moves with a speed of $0.250c$ relative to the rocket and directly toward it, what is the velocity of the meteor as observed from Earth?

74. IE ●● Two spaceships, each with a speed of $0.60c$ relative to Earth, approach each other head on. (a) The speed of one ship relative to the other is (1) greater than c, (2) equal to c, or (3) less than c. Why? (b) What is the speed of one ship relative to the other?

75. ●●● In a colliding-beam apparatus, two beams of protons are aimed directly at each other. The first beam contains protons moving with a speed of $0.800c$ to the right, and the second beam's protons have a speed of $0.900c$ to the left. Both speeds are measured relative to the laboratory frame. What are (a) the velocity of the second beam's protons relative to that of the first, and (b) the velocity of the first beam's protons relative to that of the second?

Comprehensive Exercises

76. IE A spaceship containing an astronaut travels at a speed of $0.60c$ relative to a second inertial observer. (a) Which measures proper time intervals in the ship and the proper length of the ship: (1) the astronaut in the ship, (2) the second observer, (3) neither? (b) How much time does a clock onboard the spaceship appear to lose in a day, according to the second observer? (c) If the second observer measures a length of 110 m for the ship, what is its "proper" length"?

77. An electron is accelerated to a speed of 1.5×10^8 m/s. At that speed, compare the relativistic results to the classical results for (a) the electron's kinetic energy, (b) its total energy and (c) the magnitude of its momentum?

78. IE A relativistic rocket is measured to be 50 m long, 2.5 m high, and 2.0 m wide by its pilot. It is traveling at $0.65c$ (in the direction parallel to its length) relative to an inertial observer. (a) This observer will differ from the pilot in which dimension measurement: (1) length, width, and height, (2) only width and height, or (3) only length? (b) What are the dimensions of the rocket as measured by the inertial observer?

79. (a) If the mass of 1.0 kg of coal (or any substance) could be completely converted into energy, how many kilowatt-hours of energy would be produced? (b) Assume that the average U.S. family of four uses 600 kWh of electric energy each month and that modern power plants are 33% efficient in converting this released energy into electric energy. How long would this mass be able to supply the U.S. public with electric energy?

80. In proton–antiproton annihilation, a proton and an antiproton (which has the same mass as the proton, but carries a negative charge) interact, and both masses are completely converted to electromagnetic radiation. Assuming that both particles are at rest when they annihilate, what is the total energy emitted in the form of radiation?

81. At a typical nuclear power plant, refueling occurs about every 18 months. Assuming that a plant has operated continuously since the last refueling and produces 1.2 GW of electric power at an efficiency of 33%, how much less massive are the fuel rods at the end of the 18 months than at the start? (Assume 30-day months.)

82. IE Many radioactive sources emit neutrons. One way of detecting them is by measuring the light (energy) given off when they are captured by protons (for example, when they enter water, which contains many hydrogen atoms, each one of which has a proton for its nucleus). The released energy is 2.22 MeV and the combined neutron and proton is stable and called a *deuteron*. (a) A very slow neutron (neglect its kinetic energy) is captured by a proton. How should the resulting deuteron's mass compare to the sum of the neutron and proton masses? (1) $m_d = m_p + m_n$ (2) $m_d > m_p + m_n$ (3) $m_d < m_p + m_n$ (b) The proton and neutron masses are $m_p = 1.67262 \times 10^{-27}$ kg and $m_n = 1.67493 \times 10^{-27}$ kg. Determine the mass (in kilograms) of the deuteron (also to six significant figures).

83. A particle of mass m is initially traveling to the right at $0.900c$. It collides and sticks to a particle of mass $2m$ initially moving to the left at $0.750c$. Determine (a) the total mass (in terms of m) of the composite particle, (b) the amount of kinetic energy lost during the collision, and (c) the velocity (speed and direction) of the composite particle.

84. Three inertial observers are all moving in the $+x$-direction. Observer A has a speed of $0.50c$, observer B's speed is $0.90c$ relative to A, and observer C's speed is $0.50c$ relative to B. (a) If C's clock records a time interval of 1.00 hr, how much time has elapsed according to B? (b) In the same time interval, how much time has elapsed according to A? (c) A 10-m pole lies along the x-axis in reference frame A. (d) How long is it as measured by observers in B and C?

CHAPTER

27 QUANTUM PHYSICS

27.1 Quantization: Planck's Hypothesis 852

27.2 Quanta of Light: Photons and the Photoelectric Effect 854

27.3 Quantum "Particles": The Compton Effect 858

27.4 The Bohr Theory of the Hydrogen Atom 860

27.5 A Quantum Success: The Laser 866

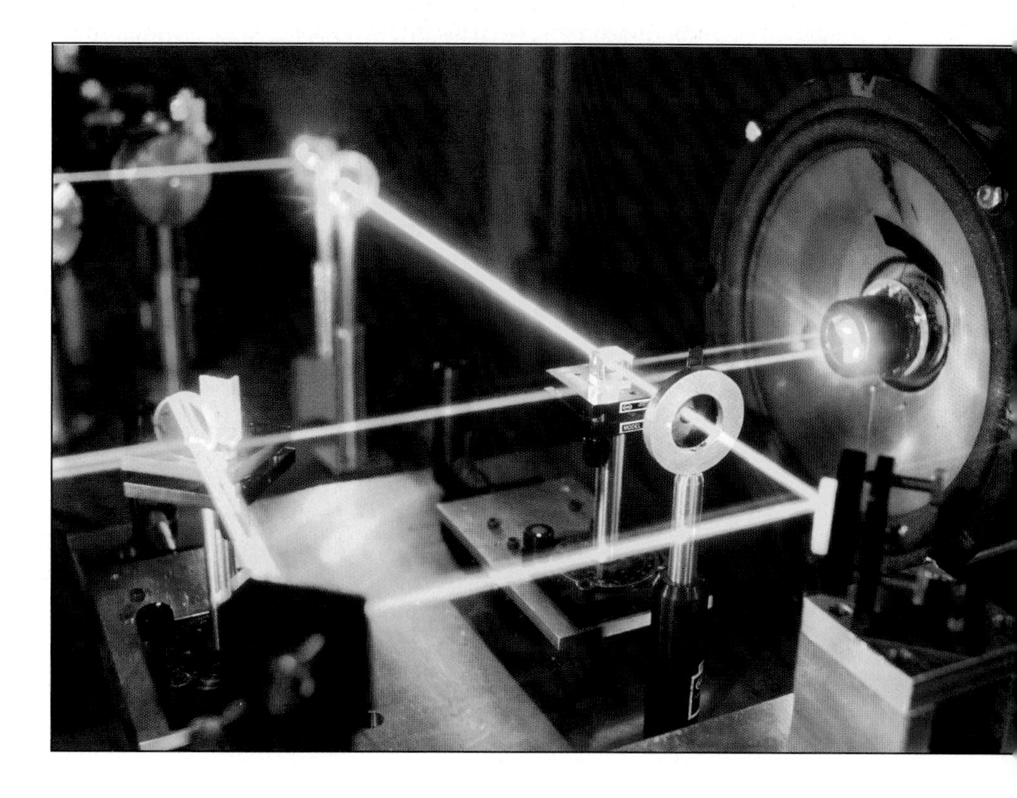

PHYSICS FACTS

- Theoretically, a hot sample of hydrogen gas can give off many different wavelengths. Of those, only four are in the visible range.
- It takes on the order of a million years for the high-energy gamma and X-ray photons released by nuclear fusion at the Sun's center to make their way to the surface of the Sun. This long trip is caused by the many random collisions that result in a reduction of photon energy down into the visible-light range. The trip to Earth from the solar surface takes only about 8 minutes.
- Helium was first discovered in the Sun's atmosphere using its characteristic photon absorption energies. Helium does not exist naturally in the Earth's atmosphere. We have significant amounts of it because it is trapped in deep wells with natural gas.
- Laser beams are so well defined spatially that bouncing them off a mirror located on the Moon (the Apollo astronauts placed the mirrors there in the early 1970s) enables us to determine the distance to the Moon to within a few meters.

Lasers are used in a variety of everyday applications—in bar code scanners at store checkout counters, in computer printers, in various types of surgery, and in laboratory situations, such as the one shown in the chapter-opening photo. You own a laser yourself if you have a CD or DVD player.

The laser is a practical application of principles that revolutionized physics. These principles were first developed in the early twentieth century, one of the most productive eras in the history of physics. For example, special relativity (Chapter 26) helped resolve problems faced by classical (Newtonian) relativity in describing objects moving at speeds near that of light. However, there were other troublesome areas in which classical theories did not agree with experimental results. To address these issues, scientists devised new hypotheses based on nontraditional approaches and ushered in a profound revolution in our understanding of the physical world. Chief among these new theories was the idea that light is *quantized* into discrete amounts of energy. This concept and others like it led to the formulation of a new set of principles and a new branch of physics called *quantum mechanics*.

Quantum theory demonstrated that particles often exhibit wave properties and that waves frequently behave as particles. Thus was born the *wave–particle duality* of matter. As a result of quantum theory, calculations in the realm of the very small—dimensions the sizes of atoms and smaller—must deal with *probabilities* rather than in the precisely determined values associated with classical theory.

A detailed treatment of quantum mechanics requires extremely complex mathematics. However, a general overview of the important results is essential to an understanding of physics as it is known today. Thus, the important developments of "quantum" physics are presented in this chapter, and an introduction to quantum mechanics is provided in Chapter 28.

27.1 Quantization: Planck's Hypothesis

OBJECTIVES: To (a) define blackbody radiation and use Wien's law, and (b) understand how Planck's hypothesis paved the way for quantum ideas.

One of the problems scientists faced at the end of the nineteenth century was how to explain the spectra of electromagnetic radiation emitted by hot objects—solids, liquids, and dense gases. This radiation is sometimes called **thermal radiation**. You learned in Chapter 11 that the total intensity of the emitted radiation from such objects is proportional to the fourth power of the absolute Kelvin temperature (T^4) of the object. Thus, all objects emit thermal radiation to some degree.

However, at everyday temperatures, this radiation is almost all in the infrared (IR) region and not visible to our eyes. However, at temperatures of about 1000 K, a solid object will begin to emit an appreciable amount of radiation in the long-wavelength end of the visible spectrum, observed as a reddish glow. A red-hot electric stove burner is a good example of this. Still-higher temperatures cause the radiation to shift to shorter wavelengths and the color to change to yellow-orange. Above a temperature of about 2000 K, an object glows yellowish-white, like the filament of a light bulb, and gives off appreciable amounts of all the visible colors, but with different percentages.

Although we observe a dominant color with our eyes, in actuality there is a *continuous spectrum*. A spectrum shows how the intensity of emitted energy depends on wavelength, as illustrated in ▾Fig. 27.1a for a hot object. Notice that practically all wavelengths are present, but there is a dominant color (wavelength region), which depends on the object's temperature.

Note: The concept of a blackbody was introduced in Section 11.4.

The curves shown in Fig. 27.1a are for an ideal **blackbody**. An ideal blackbody is an ideal object that absorbs all radiation that is incident on it. Although such a blackbody is not attainable, it can be experimentally approximated by a small hole that leads to a cavity inside a block of material (Fig. 27.1b). Radiation falling on the hole enters the cavity and is reflected back and forth off the cavity walls. If the hole is very small, only a small fraction of the incident radiation will make its way back out of the hole. Since nearly all the radiation incident on the hole is absorbed, as viewed from the outside the hole approximates a blackbody.

Two things happen to the spectrum as the temperature increases (Fig. 27.1a). As expected, more radiation is emitted at every wavelength, but also the wavelength

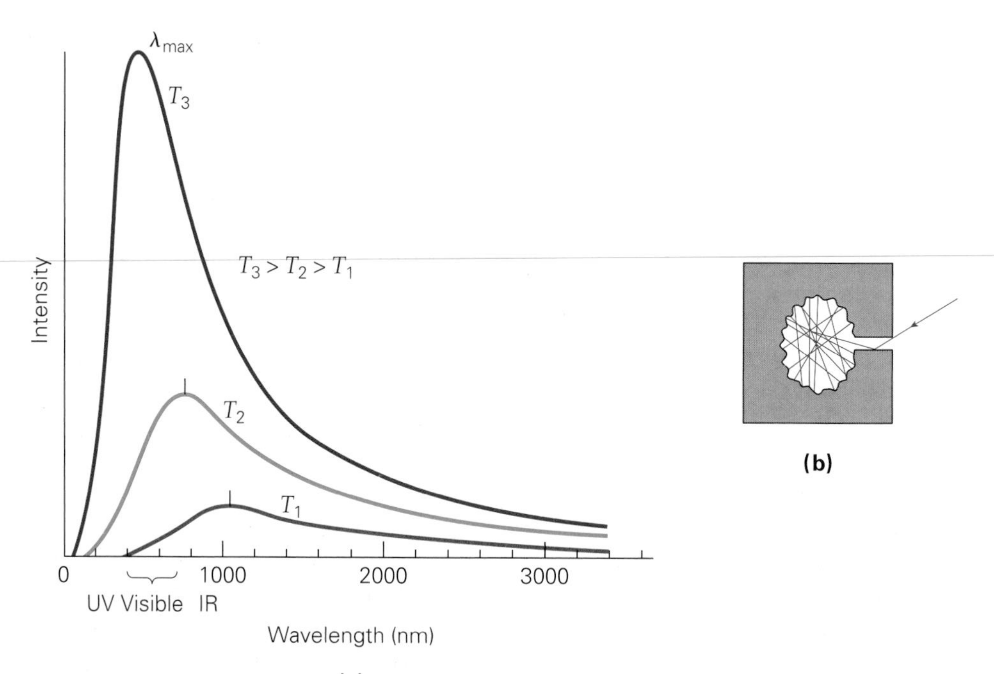

▶ **FIGURE 27.1 Thermal radiation** **(a)** Intensity-versus-wavelength curves for the thermal radiation from a blackbody at different temperatures. The wavelength associated with the maximum intensity (λ_{max}) becomes shorter with increasing temperature. **(b)** A blackbody can be approximated by a small hole leading to an interior cavity in a block of material.

of the maximum-intensity component (λ_{max}) becomes shorter. This wavelength shift is described experimentally by **Wien's displacement law,**

$$\lambda_{max}T = 2.90 \times 10^{-3}\ \text{m} \cdot \text{K} \tag{27.1}$$

where λ_{max} is the wavelength of the radiation (in meters) at which maximum intensity occurs and T is the temperature of the body (in kelvins).

Wien's law can be used to determine the wavelength of the maximum spectral component if the temperature of the emitter is known, or the temperature of the emitter if the wavelength of the strongest emission is known. Thus, it can be used to estimate the temperatures of stars (dense gases) from their radiation spectrum, as the following Example shows.

Example 27.1 ■ Solar Colors: Using Wien's Law

The visible surface of our Sun is the gaseous photosphere from which radiation escapes. At the top of the photosphere, the temperature is 4500 K; at a depth of about 260 km, the temperature is 6800 K. Assuming the Sun radiates energy as if it were a blackbody, (a) what are the wavelengths of the radiation of maximum intensity for these temperatures, and (b) to what colors do these wavelengths correspond?

Thinking It Through. Wien's displacement law (Eq. 27.1) enables us to determine the wavelengths.

Solution.

Given: $T_1 = 4500$ K, $T_2 = 6800$ K

Find: (a) λ_{max} (for the two different temperatures)
(b) Colors corresponding to these λ_{max} values

(a) At the top of the photosphere,

$$\lambda_{max} = \frac{2.90 \times 10^{-3}\ \text{m} \cdot \text{K}}{4500\ \text{K}} = (6.44 \times 10^{-7}\ \text{m})(10^9\ \text{nm/m}) = 644\ \text{nm}$$

and at the 260-km depth,

$$\lambda_{max} = \frac{2.90 \times 10^{-3}\ \text{m} \cdot \text{K}}{6800\ \text{K}} = (4.26 \times 10^{-7}\ \text{m})(10^9\ \text{nm/m}) = 426\ \text{nm}$$

(b) As the temperature increases with depth, the wavelength of the radiation of maximum intensity shifts from red toward the blue end of the spectrum. Thus the Sun's surface is orange-red and shifts towards the blue since temperature increases with depth. (For a discussion of the visible spectrum and color in relation to wavelength, see Section 20.4 and Fig. 20.23.) Combining all the depths and temperatures, we have a spectrum that shows all wavelengths (colors), but is dominated by yellow. Notice that some of the emitted radiation will be in the ultraviolet region, some of which is absorbed by the ozone layer in the Earth's atmosphere.

Follow-Up Exercise. What would be the dominant wavelengths (and corresponding colors) associated with the radiation emitted from the surface of the following stars: (a) Betelgeuse, with an average surface temperature of 3.00×10^3 K; (b) Rigel, with an average surface temperature of 1.00×10^4 K? *(Answers to all Follow-Up Exercises are at the back of the text.)*

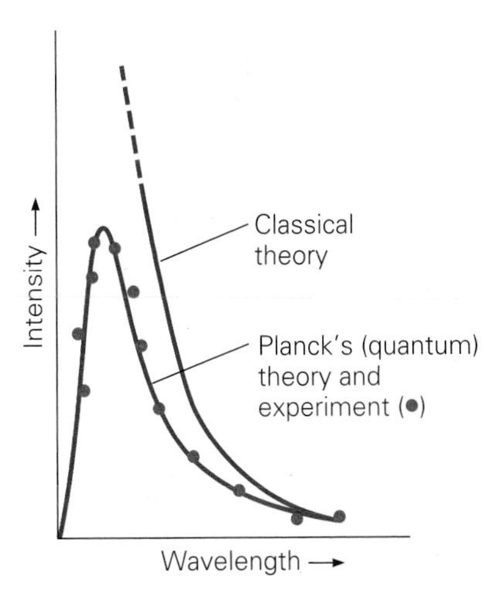

▲ **FIGURE 27.2 The ultraviolet catastrophe** Classical theory predicts that the intensity of thermal radiation emitted by a blackbody should be inversely related to the wavelength of the emitted radiation. If this were true, the intensity would become infinite as the wavelength approaches zero. In contrast, Planck's quantum theory agrees with the observed radiation distribution (solid dots).

The Ultraviolet Catastrophe and Planck's Hypothesis

Classically, thermal radiation results from the oscillations of electric charges associated with the atoms near the surface of an object. Since these charges oscillate at different frequencies, a continuous spectrum of emitted radiation is expected.

Classical calculations describing the radiation spectrum emitted by a blackbody predict an intensity that is inversely related to wavelength (actually $I \propto 1/\lambda^4$). At long wavelengths, the classical theory agrees fairly well with experimental data. However, at short wavelengths, the agreement disappears. Contrary to experimental observations, the classical theory predicts that the radiation intensity should increase without bound as the wavelength gets smaller. This is illustrated in ▸Fig. 27.2. The classical prediction is sometimes called the *ultraviolet catastrophe—ultraviolet* because the difficulty occurs for wavelengths shorter than that associated with violet, and *catastrophe* because it predicts that the emitted energy grows without limits at these wavelengths.

The failure of classical electromagnetic theory to explain the characteristics of thermal radiation led Max Planck (1858–1947), a German physicist, to re-examine the phenomenon. In 1900, Planck formulated a theory that correctly predicted the observed distribution of the blackbody radiation spectrum. (Compare the solid blue curve with the data points in Fig. 27.2.) However, his theory depended upon a radical idea. He had to assume that the thermal oscillators (the atoms emitting the radiation) have only *discrete*, or particular, amounts of energy rather than a continuous distribution of energies. Only with this assumption did his theory agree with experiment.

Planck found that these discrete amounts of energy were related to the frequency f of the atomic oscillations by

$$E_n = n(hf) \qquad \text{for } n = 1, 2, 3, \ldots \qquad \textit{Planck's quantization hypothesis} \qquad (27.2)$$

Note: As with other fundamental constants, we will take h to be *exact* at 6.63×10^{-34} J·s, for calculation purposes.

That is, the oscillator energy occurs only in integral multiples of hf. The symbol h is a constant called **Planck's constant**. Its experimental value (to three significant figures) is

$$h = 6.63 \times 10^{-34}\ \text{J}\cdot\text{s}$$

The idea expressed in Eq. 27.2 is called **Planck's hypothesis**. Rather than allowing the oscillator energy to have any value, Planck's hypothesis states that the energy is *quantized*; that is, it occurs only in discrete amounts. The smallest possible amount of oscillator energy, according to Eq. 27.2 with $n = 1$, is

$$E_1 = hf \qquad (27.3)$$

All other values of the energy are integral multiples of hf. The quantity hf is called a **quantum** of energy (from the Latin *quantus,* meaning "how much"). As a result, the energy of an atom can change only by the absorption or emission of energy in discrete, or *quantum,* amounts.

Although the theoretical predictions agreed with experiment, Planck himself was not convinced of the validity of his quantum hypothesis. However, the concept of quantization was extended to explain other phenomena that could not be explained classically. Despite Planck's hesitation, the quantum hypothesis earned him a Nobel Prize in 1918.

27.2 Quanta of Light: Photons and the Photoelectric Effect

OBJECTIVES: To (a) describe the photoelectric effect, (b) explain how it can be understood by assuming that light energy is carried by particles, and (c) summarize the properties of photons.

The concept of the quantization of light energy was introduced in 1905 by Albert Einstein in a paper concerning light absorption and emission, at about the same time he published his famous paper on special relativity. Einstein reasoned that energy quantization should be a fundamental property of electromagnetic waves (light). He suggested that if the energy of the thermal oscillators in a hot substance is quantized, then it necessarily followed that, to conserve energy, *the emitted radiation should also be quantized.* For example, suppose an atom initially had an energy of $3hf$ ($n = 3$ in Eq. 27.2) and ended up with a final (lower) energy of $2hf$ ($n = 2$ in Eq. 27.2). Einstein proposed that the atom *must* emit a specific amount (or *quantum*) of light energy—in this case, $3hf - 2hf = hf$. He named this quantum, or package, of light energy the **photon**. Each photon has a definite amount of energy E that depends on the frequency f of the light according to

$$E = hf \qquad \textit{photon energy and light frequency} \qquad (27.4)$$

This idea suggests that light can behave as discrete quanta (plural of "quantum"), or "particles," of energy rather than as a wave. One way to interpret Eq. 27.4 is as a mathematical "connection" between the *wave nature* of light (a wave of frequency f) and the *particle nature* of light (photons each with an energy E). Given light of a certain frequency or wavelength, Eq. 27.4 enables us to calculate the energy in each photon, or vice versa.

Einstein used the photon concept to explain the **photoelectric effect**, another phenomenon for which the classical description was inadequate. Certain metallic materials are *photosensitive;* that is, when light strikes their surface, electrons may be emitted. The radiant energy supplies the energy necessary to free the electrons from the material. A schematic representation of a photoelectric-effect experiment is shown in ▸Fig. 27.3a. A variable voltage (say by a variable battery or power supply) is maintained between the anode and the cathode. When light strikes the cathode (maintained at ground or zero voltage) which is photosensitive, electrons are emitted. Because they are released by absorption of light energy, these emitted electrons are called *photoelectrons.* They are collected at the anode, which is maintained initially at some positive voltage V to attract the photoelectrons. Thus, in the complete circuit a current is registered on the ammeter.

When a photocell is illuminated with monochromatic (single-wavelength) light of different intensities, characteristic curves are obtained as a function of the applied voltage (Fig. 27.3b). For positive voltages, the anode attracts the electrons. Under these conditions, the *photocurrent* I_p is the flow rate of the photoelectrons, which does *not* vary with voltage. This is because under a positive voltage the electrons are attracted to the anode and *all* the electrons reach it. As expected classically, I_p is proportional to the incident light intensity—the greater the intensity ($I_2 > I_1$ in Fig. 27.3b), the more energy is available to free electrons.

The kinetic energy of the photoelectrons can be measured by *reversing* the voltage across the electrodes, that is, reversing the battery terminals and making $V < 0$, and creating *retarding-voltage* conditions. Now the electrons are repelled from, instead of being attracted to, the anode. The electrons' initial kinetic energies are converted into electric potential energy as they approach the now negatively charged anode. As the retarding voltage is made more and more negative, the photocurrent decreases. This is because (by energy conservation) only electrons with initial kinetic energies greater than $e|V|$ can be collected at the negative anode and thus produce a photocurrent. At some value of retarding voltage, with a magnitude of V_o, called the **stopping potential**, the photocurrent drops to zero. No electrons are collected at that voltage or greater (more negative)—because even the fastest photoelectrons are turned around before reaching the anode. Hence, the maximum kinetic energy (K_{max}) of the photoelectrons is related to the magnitude of the stopping potential (V_o) by

$$K_{max} = eV_o \qquad (27.5)$$

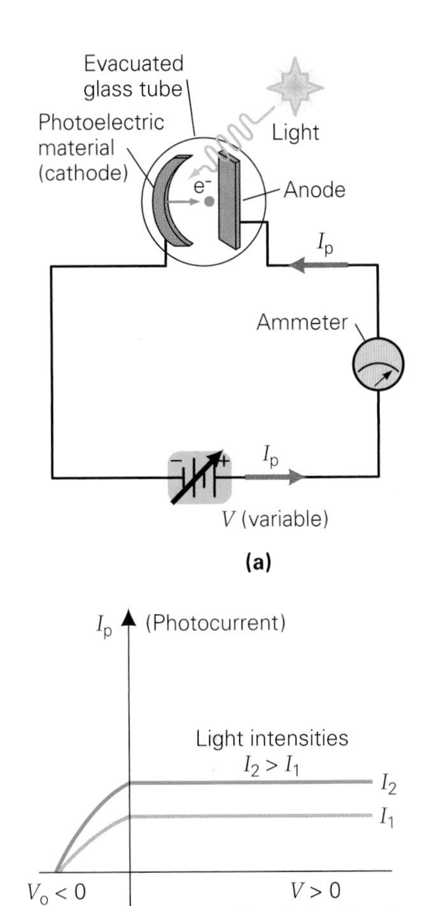

▲ **FIGURE 27.3 The photoelectric effect and characteristic curves** **(a)** Incident monochromatic light on the photoelectric material in a photocell (or phototube) causes the emission of electrons, which results in a current in the circuit. The applied voltage is variable. **(b)** As the plots of photocurrent versus voltage for two intensities of light show, the current stays constant as the voltage is increased. However, for negative voltages (using the battery with reversed polarity), the current goes to zero when the stopping potential has a magnitude of $|V_o|$, which, for a fixed frequency, depends only on the type of material, but is independent of intensity.

Note: Recall from Eq. 16.2 that $\Delta U_e = qV$.

Experimentally, when the frequency of the incident light is varied, this maximum kinetic energy increases linearly with the frequency (▾Fig. 27.4). No emission of electrons is observed for light with a frequency below a certain *cutoff frequency* f_o. Even if the light intensity is very low, the current begins essentially instantaneously with no observable time delay, as long as a material is being illuminated by light with a frequency $f > f_o$.

The important characteristics of the photoelectric effect are summarized in Table 27.1. Notice that only one of the characteristics is predicted correctly by classical wave theory, whereas Einstein's photon concept explains all of the results. When an electron absorbs a quantum of light energy, some of the photon's energy goes to free the electron, with the remainder showing up as kinetic energy. The work required to free the electron is designated by ϕ. So, when a photon of energy E is absorbed, conservation of energy requires that $E = K + \phi$, or, using Eq. 27.4 to replace E, we have

$$hf = K + \phi \qquad (27.6)$$

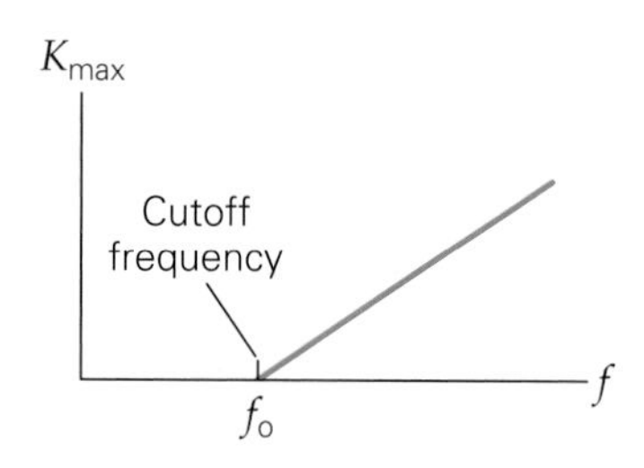

▲ **FIGURE 27.4 Maximum kinetic energy versus light frequency in the photoelectric effect** The maximum kinetic energy (K_{max}) of the photoelectrons is a linear function of the incident-light frequency. Below a certain cutoff frequency f_o, no photoemission occurs, regardless of the intensity of the light.

TABLE 27.1 Characteristics of the Photoelectric Effect

Characteristic	*Predicted by wave theory?*
1. The photocurrent is proportional to the intensity of the light.	Yes
2. The maximum kinetic energy of the emitted electrons depends on the frequency of the light, but not on its intensity.	No
3. No photoemission occurs for light with a frequency below a certain cutoff frequency f_o, regardless of the light intensity.	No
4. A photocurrent is observed immediately when the light frequency is greater than f_o, even if the light intensity is extremely low.	No

Since the energies are very small, the commonly used energy unit is the electron-volt (eV; see Chapter 16). Recall that 1.00 eV = 1.60×10^{-19} J.

The least tightly bound electron will have the maximum kinetic energy K_{max}. (Why?) The energy needed to free this electron is called the **work function (ϕ_o)** of the material. For this situation, Eq. 27.6 becomes

$$\underset{\text{incident photon energy}}{hf} = \underset{\text{maximum kinetic energy of freed electron}}{K_{max}} + \underset{\text{minimum work needed to free the electron}}{\phi_o} \quad (27.7)$$

Other electrons require more energy than the minimum to be freed, so their kinetic energy will be less than K_{max}. This concept is explored visually in the Learn by Drawing feature on this page. Some typical numerical values are shown in the next Example.

LEARN BY DRAWING

The Photoelectric Effect and Energy Conservation

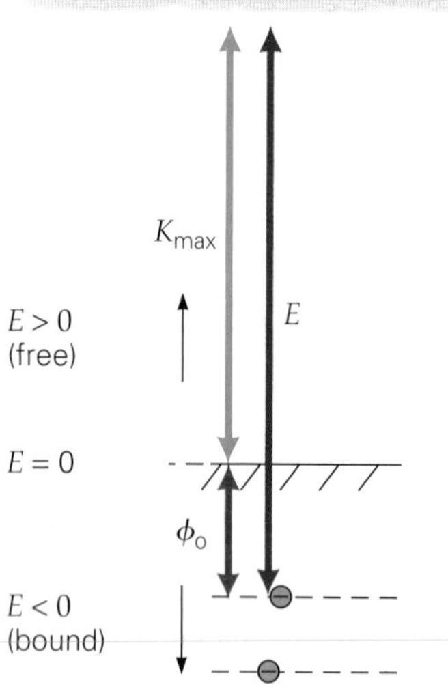

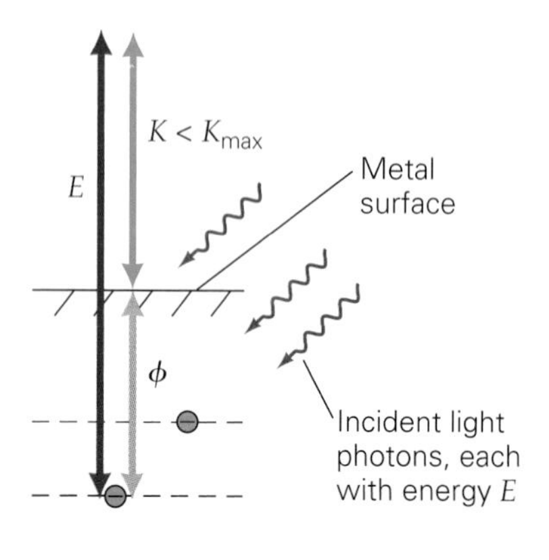

Example 27.2 ■ The Photoelectric Effect: Electron Speed and Stopping Potential

The work function of a particular metal is known to be 2.00 eV. If the metal is illuminated with light of wavelength 550 nm, what will be (a) the maximum kinetic energy of the emitted electrons and (b) their maximum speed? (c) What is the stopping potential?

Thinking It Through. (a) By energy conservation (Eq. 27.7), the maximum kinetic energy is the difference between the incoming photon energy and the work function. (b) Speed can be determined from kinetic energy, since the mass of an electron is known ($m = 9.11 \times 10^{-31}$ kg). (c) The stopping potential is found by requiring that all of the kinetic energy be converted to electric potential energy (Eq. 27.5).

Solution. First, converting the data into SI units.

Given: $\phi_o = (2.00 \text{ eV})(1.60 \times 10^{-19} \text{ J/eV}) = 3.20 \times 10^{-19}$ J
$\lambda = 550 \text{ nm} = 5.50 \times 10^{-7}$ m

Find: (a) K_{max} (maximum kinetic energy)
(b) v_{max} (maximum speed)
(c) V_o (stopping potential)

(a) Using $\lambda f = c$, we find that the photon energy of light with the given wavelength is

$$E = hf = \frac{hc}{\lambda} = \frac{(6.63 \times 10^{-34} \text{ J}\cdot\text{s})(3.00 \times 10^{8} \text{ m/s})}{5.50 \times 10^{-7} \text{ m}} = 3.62 \times 10^{-19} \text{ J}$$

Then

$$K_{max} = E - \phi_o = 3.62 \times 10^{-19} \text{ J} - 3.20 \times 10^{-19} \text{ J}$$
$$= (4.20 \times 10^{-20} \text{ J})\left(\frac{1 \text{ eV}}{1.60 \times 10^{-19} \text{ J}}\right) = 0.263 \text{ eV}$$

(b) v_{max} can be found from $K_{max} = \frac{1}{2}mv_{max}^2$:

$$v_{max} = \sqrt{\frac{2K_{max}}{m}} = \sqrt{\frac{2(4.20 \times 10^{-20} \text{ J})}{9.11 \times 10^{-31} \text{ kg}}} = 3.04 \times 10^{5} \text{ m/s}$$

(c) The stopping potential is related to K_{max} by $K_{max} = eV_o$; therefore,

$$V_o = \frac{K_{max}}{e} = \frac{0.420 \times 10^{-19}\ \text{J}}{1.60 \times 10^{-19}\ \text{C}} = 0.23\ \text{V}$$

Follow-Up Exercise. In this Example, suppose that a different wavelength of light is used and the new stopping voltage is found to be 0.50 V. What is the wavelength of this new light? Explain why this wavelength requires a larger stopping voltage.

Einstein's photon model of light is, in fact, consistent with *all* the experimental results of the photoelectric effect. In the photon model, an increase in light intensity means an increase in the number of photons and therefore in the number of photoelectrons (that is, the photocurrent). However, an increase in intensity would *not* mean a change in the energy of any one photon, since that energy depends only on the light frequency ($E = hf$). Therefore, K_{max} should be independent of intensity, but linearly dependent on the frequency of the incident light—as is observed experimentally.

Einstein's quantum theory of light also explains the existence of a cutoff frequency. In his interpretation, since photon energy depends on frequency, this means that below a certain (cutoff) frequency (f_o) the photons simply don't have enough energy to dislodge even the most loosely bound electrons. Therefore, no current is observed for those frequencies. Since, at the cutoff frequency, no electrons are emitted, the cutoff frequency can be found by setting $K_{max} = 0$ in Eq. 27.7:

$$hf_o = K_{max} + \phi_o = 0 + \phi_o$$

or

$$f_o = \frac{\phi_o}{h} \quad \textit{threshold, or cutoff, frequency} \qquad (27.8)$$

Note: A photon of energy hf_o will barely free an electron from a solid, but the electron will have essentially no kinetic energy.

The cutoff frequency f_o is sometimes called the **threshold frequency**. It represents the minimum frequency of light necessary to create photoelectrons. Below the threshold frequency, the binding energy of the least-bound electron exceeds the photon energy. Although the electron may absorb the photon energy, it will not have enough energy to be freed from the material and become a photoelectron. (How would you explain this, using a sketch such as the one in the accompanying Learn by Drawing?)

Example 27.3 ■ The Photoelectric Effect: Threshold Frequency and Wavelength

What are the threshold frequency and corresponding wavelength for the metal described in Example 27.2?

Thinking It Through. Example 27.2 gives the work function, so Eq. 27.8 allows us to determine the threshold frequency. The threshold wavelength can then be computed from $\lambda = c/f$.

Solution. Listing the data, we have

Given: $\phi_o = 2.00$ eV $= 3.20 \times 10^{-19}$ J (from Example 27.2)

Find: f_o (threshold frequency)
λ_o (wavelength at threshold frequency)

Solving for the threshold frequency f_o from $\phi_o = hf_o$ (Eq. 27.8), we get

$$f_o = \frac{\phi_o}{h} = \frac{3.20 \times 10^{-19}\ \text{J}}{6.63 \times 10^{-34}\ \text{J} \cdot \text{s}} = 4.83 \times 10^{14}\ \text{Hz}$$

The wavelength (the threshold wavelength) corresponding to this frequency is

$$\lambda_o = \frac{c}{f_o} = \frac{3.00 \times 10^{8}\ \text{m/s}}{4.83 \times 10^{14}\ \text{Hz}} = 6.21 \times 10^{-7}\ \text{m} = 621\ \text{nm}$$

Any frequency lower than 4.83×10^{14} Hz, or, alternatively, any wavelength longer than 621 nm, would not yield photoelectrons. Since this wavelength lies in the red-orange end of the electromagnetic spectrum, yellow light, for example, would dislodge electrons, but deep-red light would not.

Follow-Up Exercise. In this Example, what would be the stopping voltage if the frequency of the light were twice the cutoff frequency?

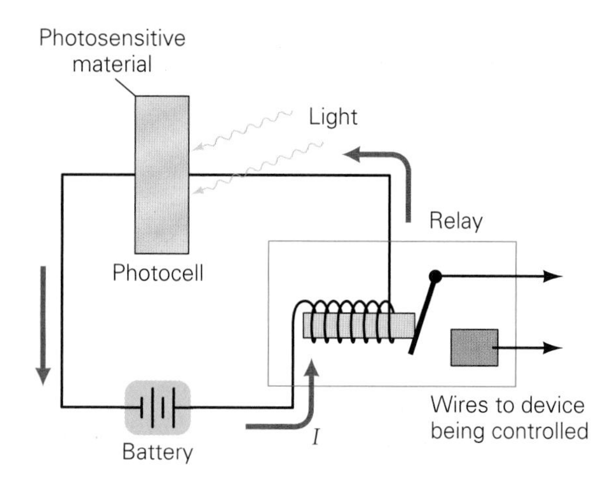

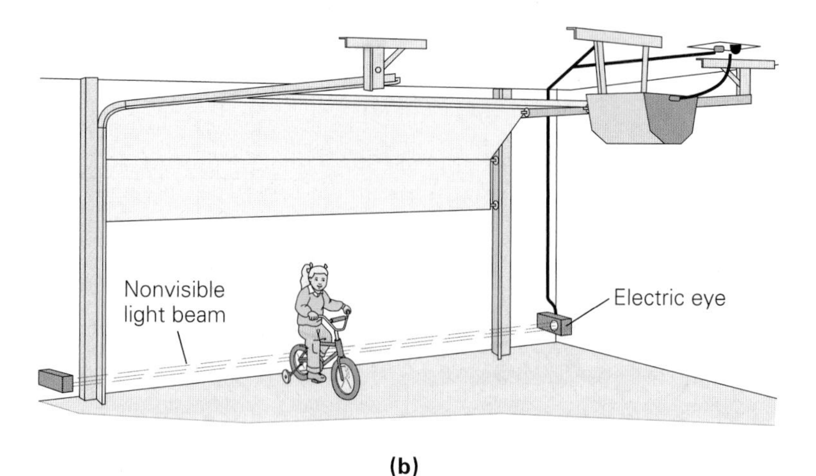

▲ **FIGURE 27.5 Photoelectric applications: The electric eye** **(a)** A diagram of an electric-eye circuit. When light strikes a photocell material (any photosensitive material), it frees electrons from their atoms (but not from the solid as a whole). In effect this lowers the material's resistance by enabling it to conduct current. The result is a current in the circuit. Interruption of the light beam opens the circuit in the relay (a magnetic switch) that controls the particular device. **(b)** Electric-eye circuits are used in automatic garage-door openers. When the door starts to move downward, any interruption of the electric-eye beam (usually IR light) causes the door to stop, protecting anything that may be under the descending door.

Problem-Solving Hint

In photon calculations, the wavelength of the light is often given rather than the frequency. Typically, what is needed is the photon *energy*. Instead of first calculating the frequency ($f = c/\lambda$), then the energy in joules ($E = hf$), and finally converting to electron-volts, this all can be done in one step. To do so, combine these two equations to form $E = hf = hc/\lambda$ and express the product hc in electron-volts times nanometers, or eV · nm. The value of this useful product is

$$hc = (6.63 \times 10^{-34}\,\text{J}\cdot\text{s})(3.00 \times 10^{8}\,\text{m/s}) = 1.99 \times 10^{-25}\,\text{J}\cdot\text{m}$$
$$= \frac{(1.99 \times 10^{-25}\,\text{J}\cdot\text{m})(10^{9}\,\text{nm/m})}{1.60 \times 10^{-19}\,\text{J/eV}}$$
$$= 1.24 \times 10^{3}\,\text{eV}\cdot\text{nm}$$

This shortcut can save time and effort in working problems and allows quick estimation of the photon energy associated with light of a given wavelength (or vice versa). Thus, if orange light ($\lambda = 600$ nm) is used, you need only divide in your head to realize that each photon carries approximately 2 eV of energy. A more exact result could be calculated if needed, as $E = \dfrac{hc}{\lambda} = \dfrac{1.24 \times 10^{3}\,\text{eV}\cdot\text{nm}}{600\,\text{nm}} = 2.07\,\text{eV}$.

There are many applications of the photoelectric effect. The fact that the current produced by photocells is proportional to the intensity of the light makes them ideal for use in photographers' light meters. Photocells are also used in solar-energy applications to convert sunlight to electricity.

Another common application of the photocell is the electric eye (▲Fig. 27.5a). As long as light strikes the photocell, there is current in the circuit. Blocking the light opens the circuit in the relay (magnetic switch), which in turn controls some device. A common application of the electric eye is to turn on streetlights automatically at night. A safety application of the electric eye is shown in Fig. 27.5b. Note that in many of these applications (including the garage-door safety mechanism) the IR light photons do not actually free the electrons from the *material*. All is required is that they free them from the *atoms* in the material—thus the electrons stay in the material but are free to move through the material. Once they are freed, the external voltage causes them to flow, resulting in an electrical current that can be detected and put to appropriate use depending on the application.

27.3 Quantum "Particles": The Compton Effect

OBJECTIVES: To (a) understand how the photon model of light explains scattering of light from electrons (the Compton effect), and (b) calculate the wavelength of the scattered light in the Compton effect.

In 1923, the American physicist Arthur H. Compton (1892–1962) explained the scattering of X-rays from a graphite (carbon) block by assuming the radiation to be composed of quanta. His explanation of the observed effect provided additional

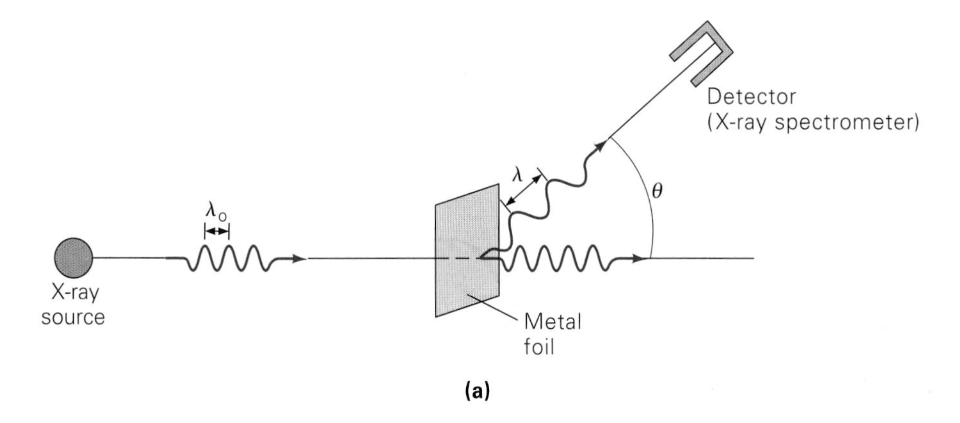

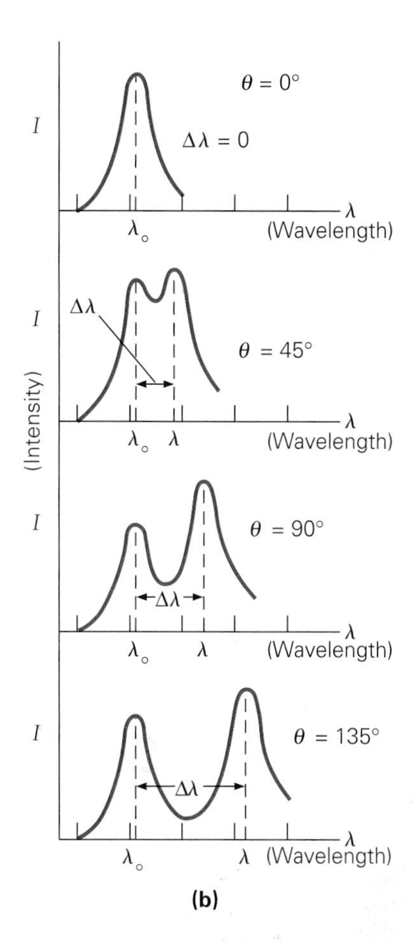

▲ **FIGURE 27.6 X-ray scattering** **(a)** When X-rays of a single wavelength are scattered by the electrons in metal foil, the scattered wavelength (λ) is longer than the incident wavelength (λ_o). Most of the incident X-rays pass through without interacting (and therefore undergo no change in wavelength). The scattered electrons are not shown, because they remain in the sample. **(b)** The change in wavelength increases with the scattering angle (θ). Note that an unshifted peak at λ_o remains at all angles. This corresponds to photons that scatter off electrons tightly bound to atoms. Thus, since the relatively massive atom (compared to an electron) recoils. In this situation, it takes a negligible energy, leaving the scattered photon with essentially the same energy as the incident one.

convincing evidence that, at least in certain types of experiments, light (electromagnetic) energy is carried by *photons*.

Compton had observed that when a beam of monochromatic (single-wavelength) X-rays was scattered by various materials, the wavelength of the scattered X-ray was longer than the wavelength of the incident X-ray. In addition, he noted that the change in the wavelength depended on the angle θ through which the X-rays were scattered, but *not* on the nature of the scattering material (▲Fig. 27.6). This phenomenon came to be known as the **Compton effect**.

According to the wave model, any scattered radiation should have the same frequency (and wavelength) as the incident radiation. In this model, the electrons in the atoms of the scattering material are accelerated by the oscillating electric field of the radiation and therefore oscillate at the same frequency as the incident wave. The scattered (or reradiated) radiation should thus have the same frequency, regardless of direction.

According to Einstein's photon picture, the energy of each photon is proportional to the frequency f of the associated light wave. Therefore a change in frequency or wavelength would indicate a change in photon energy. Because the wavelength increased (and the frequency decreased, because $f = c/\lambda$), the scattered photons had *less* energy than the incident ones. Moreover, the change in photon energy increased with the scattering angle—which reminded Compton of an elastic collision of two particles. Could the same principles apply in the scattering of these quantum "particles" called photons?

Pursuing this idea, Compton assumed that a photon behaves as a particle when it collides with electrons. He reasoned that if an incident photon collides with an electron initially at rest, the photon should transfer some energy and momentum to that electron. Thus we would expect that the energy of the scattered photon, as well as the frequency of the associated scattered light wave, should both decrease (since $E = hf$). Applying conservation of energy and linear momentum, Compton showed, using the photon model, that the shift in the wavelength of the light scattered at an angle θ from an electron is given by

$$\Delta\lambda = \lambda - \lambda_o = \lambda_C (1 - \cos\theta) \quad \textit{Compton scattering} \qquad (27.9)$$

where λ_o is the wavelength of the incident light and λ is that of the scattered light. The constant $\boldsymbol{\lambda_C}$ is called the **Compton wavelength** of the electron. It is inversely related to the mass of the electron by $\lambda_C = h/(m_e c)$. The Compton wavelength has a numerical value of $\lambda_C = 2.43 \times 10^{-12}$ m $= 2.43 \times 10^{-3}$ nm.* Equation 27.9 correctly predicts the observed wavelength shift. For his work, Compton was awarded a Nobel Prize in 1927.

Note that the *maximum* wavelength increase occurs when $\theta = 180°$ and has a value of $\Delta\lambda_{max} = 2\lambda_C = 4.86 \times 10^{-3}$ nm. (To see this, use Eq. 27.9 and note that for $\theta = 180°$, $\cos\theta = -1$ and $1 - \cos\theta = 2$.) The scattered wavelength is longest

Note: Elastic collisions are discussed in Section 6.4.

*This numerical value is the Compton wavelength of the *electron*. Compton scattering can occur from any particle; hence, there is a Compton wavelength of the proton, the neutron, and so on. These values are much smaller than the electron's Compton wavelength, because other particles are much more massive than electrons.

when the photon completely reverses direction, and the electron goes forward with the maximum amount of kinetic energy. Since this value is the maximum wavelength *change*, it is difficult to measure for incident wavelengths several thousand times this value or greater, such as for wavelengths larger than several nanometers (strong UV). In other words, the Compton effect is negligible for UV light and any other light with a longer wavelength, such as visible or IR light. It is significant only for X-ray and gamma-ray scattering.

Example 27.4 ■ X-Ray Scattering: The Compton Effect

A monochromatic beam of X-rays of wavelength 1.35×10^{-10} m is scattered by the electrons in a metal foil. By what percentage is the wavelength shifted if the scattered X-rays are observed at an angle of 90°?

Thinking It Through. The change, or shift, in the wavelength, $\Delta\lambda$, is given by Eq. 27.9 with $\theta = 90°$, and the fractional change is $\Delta\lambda/\lambda_o$. The change is positive, because the scattered light has a longer wavelength than that of the incident light.

Solution. Listing the data, we have

Given: $\lambda_o = 1.35 \times 10^{-10}$ m
$\theta = 90°$

Find: Percentage change in wavelength

Starting from Eq. 27.9, we can compute the fractional change directly, since $\cos 90° = 0$:

$$\frac{\Delta\lambda}{\lambda_o} = \frac{\lambda_C}{\lambda_o}(1 - \cos\theta) = \frac{2.43 \times 10^{-12}\ \text{m}}{1.35 \times 10^{-10}\ \text{m}}(1 - \cos 90°) = 1.80 \times 10^{-2}$$

So

$$\frac{\Delta\lambda}{\lambda_o} \times 100\% = 1.80\%$$

Follow-Up Exercise. In this Example, (a) what would be the maximum percentage change if gamma rays with a wavelength of 1.50×10^{-14} m were used instead? (b) Is this change larger or smaller than the maximum possible percentage change for the X-rays in the Example? Why?

Einstein's and Compton's success with photons left scientists with two apparently competing theories. Classically, light is a traveling wave, and this satisfactorily explains such phenomena as interference and diffraction. Conversely, quantum theory is necessary to explain the photoelectric and Compton effects. The two theories combined to give rise to a description called the **dual nature** (or **wave–particle duality**) **of light**, as follows:

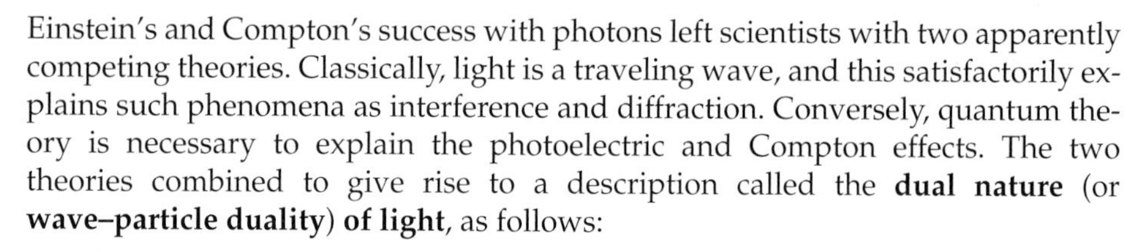

To explain all electromagnetic phenomena, light has to be considered sometimes as a wave and other times as a beam of photons. When it interacts with small (quantized) systems, such as atoms, nuclei, and molecules, the photon (quantized-energy) picture must be used. In everyday-sized systems—for example, slit systems causing diffraction and interference—the wave model is applicable.

(a)

(b)

▲ **FIGURE 27.7** Gas-discharge tubes **(a)** These luminous glass tubes are gas-discharge tubes, in which atoms of various gases emit light when electrically excited. Each gas radiates its own characteristic wavelengths. **(b)** Only some "neon lights" actually contain neon, which glows with a red hue; other gases produce other colors.

27.4 The Bohr Theory of the Hydrogen Atom

OBJECTIVES: To (a) understand how the Bohr model of the hydrogen atom explains that atom's emission and absorption spectra, (b) calculate the energies and wavelengths of emitted and absorbed photons for transitions in atomic hydrogen, and (c) understand how the generalized concept of atomic energy levels can explain other atomic phenomena.

In the 1800s, much experimental work was done with gas-discharge tubes—for example, those containing hydrogen, neon, and mercury vapor. Common neon "lights" are actually gas-discharge tubes (◂Fig. 27.7). Recall that light from an incandescent source, such as a lightbulb's hot filament, exhibits a *continuous spectrum* in which all wavelengths are present. However, when light emissions from gas-discharge tubes were

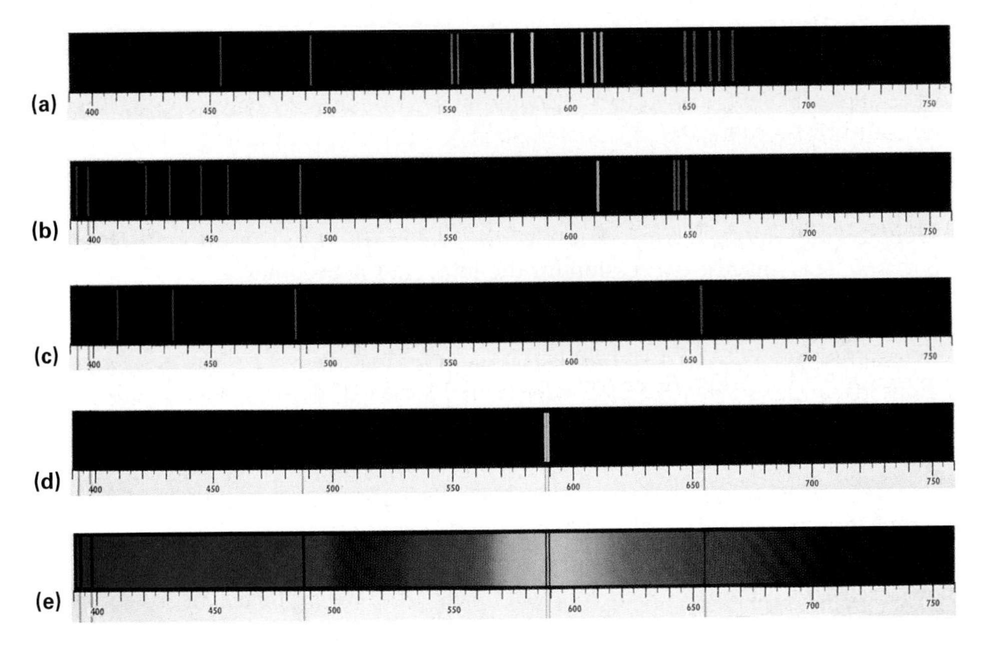

◀ **FIGURE 27.8** Emission and absorption spectra of gases When a gas is excited by heat or electricity, the light it emits can be separated into its various wavelengths by a prism or diffraction grating; the result is a bright-line, or emission, spectrum, such as the ones shown here from **(a)** barium, **(b)** calcium, **(c)** hydrogen, and **(d)** sodium, each with its own characteristic pattern. **(e)** When a continuous spectrum from a hot solid or dense gas is viewed after passing through a cool gas, a dark-line, or absorption, spectrum is observed. Each line represents a particular wavelength the gas has absorbed. The absorption spectrum of the Sun provided here shows several prominent absorption lines produced by the gases of the solar atmosphere before the sunlight makes it to the Earth. In fact, the inert gas helium was first discovered to exist on the Sun by this very method.

analyzed, discrete spectra with only certain wavelengths present were observed (▲Fig. 27.8). The spectrum of light coming from such a tube is called a *bright-line spectrum*, or **emission spectrum**. In general, the wavelengths present in an emission spectrum are characteristic of the individual atoms or molecules of the particular gas.

Atoms can absorb light as well as emit it. If white light is passed through a cool gas, the energy at certain frequencies or wavelengths is absorbed. The result is a *dark-line spectrum*, or **absorption spectrum**—a series of dark lines superimposed on a continuous spectrum (Fig. 27.8e). Just as in emission spectra, the missing wavelengths are uniquely related to the type of atom or molecule doing the absorbing. By determining the pattern of emitted and/or absorbed wavelengths, the type of atoms or molecules present in a sample can be identified. This method is called *spectroscopic analysis* and is widely used in physics, astrophysics, biology, and chemistry. For example, the element helium was first discovered on the Sun when scientists found that an absorption-line pattern in the sunlight did not match any known pattern on the Earth. That unknown pattern belonged to the helium atom.

Although the reason for line spectra was not understood in the 1800s, they provided an important clue to the electron structure of atoms. Hydrogen, with its relatively simple visible spectrum, received much of the attention. It is also the simplest atom, consisting of only one electron and one proton. In the late nineteenth century, the Swiss physicist J. J. Balmer found an empirical formula that gives the wavelengths of the four spectral lines of hydrogen in the visible region:

$$\frac{1}{\lambda} = R\left(\frac{1}{2^2} - \frac{1}{n^2}\right) \quad \text{for } n = 3, 4, 5, \text{and } 6 \quad \textit{visible spectrum of hydrogen} \qquad (27.10)$$

R is called the *Rydberg constant*, named after the Swedish physicist Johannes Rydberg (1854–1919), who also studied atomic emission lines. R has an experimental value of $1.097 \times 10^{-2}\ \text{nm}^{-1}$. The four spectral lines of hydrogen in the visible region (note four values for n in Eq. 27.10), are part of the **Balmer series**. There wavelengths fit the formula, but it was not understood why. Similar formulas were found to fit other spectral-line series that were completely in the ultraviolet and infrared regions.

An explanation of the spectral lines was given in a theory of the hydrogen atom put forth in 1913 by the Danish physicist Niels Bohr (1885–1962). Bohr assumed that the electron of the hydrogen atom orbits the proton in a circular orbit analogous to a planet orbiting the Sun. The attractive electrical force between the electron and proton supplies the necessary centripetal force for the circular motion. Recall that the centripetal force is given by $F_c = mv^2/r$, where v is the electron's orbital speed, m is its mass, and r is the radius of its orbit. The force between the proton and electron is given by Coulomb's law as $F_e = kq_1q_2/r^2 = ke^2/r^2$, where

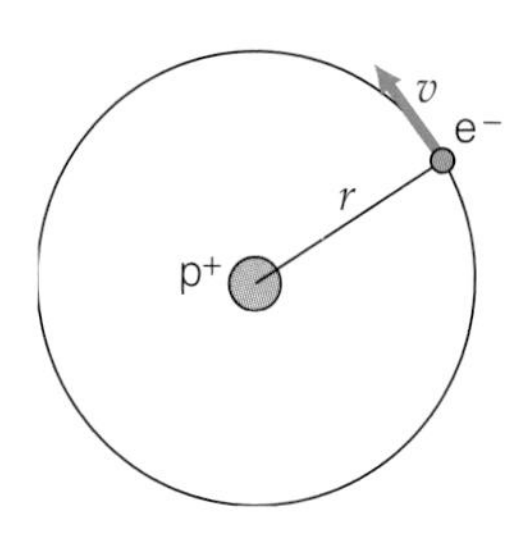

▲ **FIGURE 27.9 The Bohr model of the hydrogen atom** The electron is pictured as revolving around the much more massive proton in a circular orbit. The electric force of attraction provides the centripetal force.

e is the magnitude of the charge of the proton and the electron (◂Fig. 27.9). Equating these two forces, we have

$$\frac{mv^2}{r} = \frac{ke^2}{r^2} \tag{27.11}$$

The total energy of the atom is the sum of its kinetic and potential energies. Recall from Chapter 16 that the electric potential energy of two point charges is given by $U_e = kq_1q_2/r$. Since the electron and proton are oppositely charged, $U_e = -ke^2/r$. Thus, the expression for the total energy becomes

$$E = K + U_e = \tfrac{1}{2}mv^2 - \frac{ke^2}{r}$$

From Eq. 27.11, the kinetic energy can be written as $\frac{1}{2}mv^2 = ke^2/2r$. With this relationship, the total energy becomes

$$E = \frac{ke^2}{2r} - \frac{ke^2}{r} = -\frac{ke^2}{2r} \tag{27.12}$$

Note that E is negative, indicating that the system is bound. As the radius gets very large, E approaches zero. With $E = 0$, the electron would no longer be bound to the proton, and the atom, having lost its electron, would be *ionized*.

Up to this point, only classical principles had been applied. At this step in the theory, Bohr made a radical assumption—radical in the sense that he introduced a quantum concept to attempt to explain atomic line spectra:

> Bohr assumed that the angular momentum of the electron was quantized and could have only discrete values that were integral multiples of $h/2\pi$, where h is Planck's constant.

Recall that in a circular orbit of radius r, the angular momentum L of an object of mass m is given by mvr (Eq. 8.14). Therefore, Bohr's assumption translates into

$$mvr = n\left(\frac{h}{2\pi}\right) \quad \text{for } n = 1, 2, 3, 4, \ldots \tag{27.13}$$

The integer n is an example of a *quantum number*. Specifically, n is the atom's **principal quantum number**.* With this assumption, the orbital speed of the electron (v) can be found. Its (quantized) values are

$$v_n = \frac{nh}{2\pi mr} \quad \text{for } n = 1, 2, 3, 4, \ldots$$

Putting this expression for v into Eq. 27.11 and solving for r, we obtain

$$r_n = \left(\frac{h^2}{4\pi^2ke^2m}\right)n^2 \quad \text{for } n = 1, 2, 3, 4, \ldots \tag{27.14}$$

Here, the subscript n on r is used to indicate that only certain radii are possible—that is, the size of the orbit is also quantized. The energy for an orbit can be found by substituting this expression for r into Eq. 27.12, which gives

$$E_n = -\left(\frac{2\pi^2k^2e^4m}{h^2}\right)\frac{1}{n^2} \quad \text{for } n = 1, 2, 3, 4, \ldots \tag{27.15}$$

where the energy is also written with a subscript of n to show its dependence on n. The quantities in the parentheses on the right-hand sides of Eqs. 27.14 and 27.15 are constants and can be evaluated numerically. Because the radii are so small, they are typically expressed in nanometers (nm); similarly, energies are expressed in electron-volts (eV):

$$r_n = 0.0529n^2 \text{ nm} \quad \text{for } n = 1, 2, 3, 4, \ldots \tag{27.16}$$

orbital radii and energies for the hydrogen atom

$$E_n = \frac{-13.6}{n^2} \text{ eV} \quad \text{for } n = 1, 2, 3, 4, \ldots \tag{27.17}$$

The use of these expressions is shown in the next Example.

*The principal quantum number is only one of four quantum numbers necessary to completely describe each electron in an atom. See Chapter 28.

Example 27.5 ■ A Bohr Orbit: Radius and Energy

Find the orbital radius and energy of an electron in a hydrogen atom characterized by the principal quantum number $n = 2$.

Thinking It Through. Equations 27.16 and 27.17 are used with $n = 2$.

Solution. For $n = 2$,

$$r_2 = 0.0529n^2 \text{ nm} = 0.0529(2)^2 \text{ nm} = 0.212 \text{ nm}$$

and

$$E_2 = \frac{-13.6}{n^2} \text{ eV} = \frac{-13.6}{2^2} \text{ eV} = -3.40 \text{ eV}$$

Follow-Up Exercise. In this Example, what is (a) the speed and (b) the kinetic energy of the orbiting electron?

However, there was still a problem with Bohr's theory. Classically, any accelerating charge should radiate electromagnetic energy (light). For the Bohr circular orbits, the electron is accelerating centripetally. Thus, the orbiting electron should lose energy and spiral into the nucleus. Clearly, this doesn't happen in the hydrogen atom, so Bohr had to make another nonclassical assumption. He postulated that:

> The hydrogen electron does *not* radiate energy when it is in a bound, discrete orbit. It radiates energy only when it makes a *downward transition* to an orbit of lower energy. It makes an *upward transition* to an orbit of higher energy by absorbing energy.

Energy Levels

The "allowed" orbits of the electron in a hydrogen atom are commonly expressed in terms of their energy (▾Fig. 27.10). In this context, the electron is referred to as being in a particular "energy level" or state. The principal quantum number labels the particular energy level. The lowest energy level ($n = 1$) is the **ground state**. The energy levels above the ground state are called **excited states**. For example, $n = 2$ is the *first excited state* (see Example 27.5), and so on.

Note: Review the concept of potential-energy wells, presented in Sections 5.4 and 7.5. (See Figs. 5.14 and 7.18.)

The electron is normally in the ground state and must be given enough energy to raise it to an excited state. Since the energy levels have specific energies, it follows that the electron can be excited only by absorbing certain discrete amounts of energy. The energy levels can be thought of as the rungs of a ladder.* A person who goes up and down a ladder changes his or her gravitational potential energy by discrete amounts. Similarly, an electron goes up and down its own "energy ladder" in discrete steps. Notice, however, that the energy levels of the hydrogen atom are not evenly spaced, as they are on a ladder (Fig. 27.10).

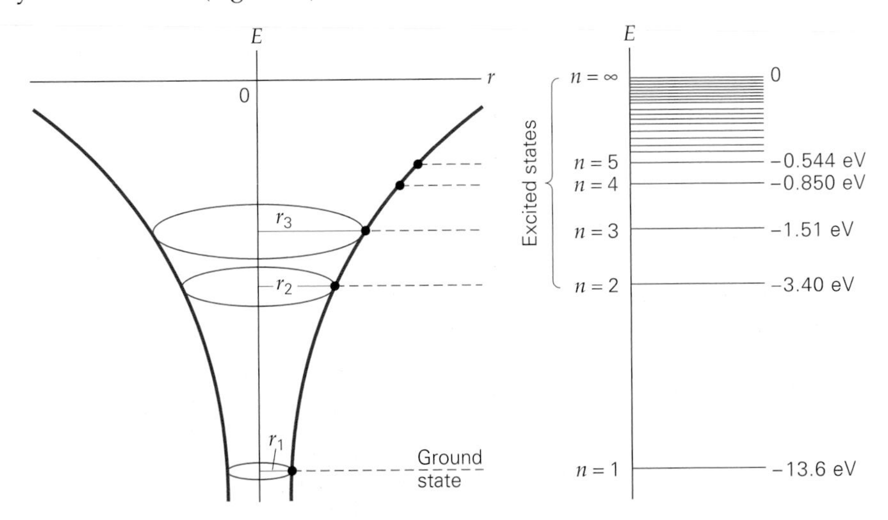

◂ **FIGURE 27.10 Orbits and energy levels of the hydrogen electron** The Bohr theory predicts that the hydrogen electron can occupy only certain orbits having discrete radii. Each allowed orbit has a corresponding total energy, conveniently displayed as an energy-level diagram. The lowest energy level ($n = 1$) is the ground state; the levels above it ($n > 1$) are excited states. The orbits are shown on the left, plotted in the $1/r$ electrical potential of the proton. The electron in the ground state is deepest in the potential-energy well, analogous to the gravitational potential-energy well of Fig. 7.18. (Neither r nor the energy levels are drawn to scale—can you tell why?)

*Care must be taken with this. As we step down or up a ladder, our gravitational potential energy takes on all values in between the end values. However, in an electron's "quantum" jump, it *never* has any intermediate energy value. That is, it has an energy to start and one to end, emitting or absorbing a quantum of energy in making the transition, but it never takes on an energy value in between the initial and final values.

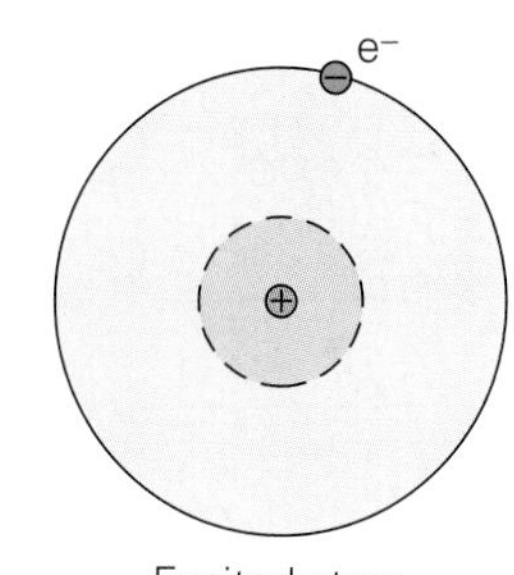

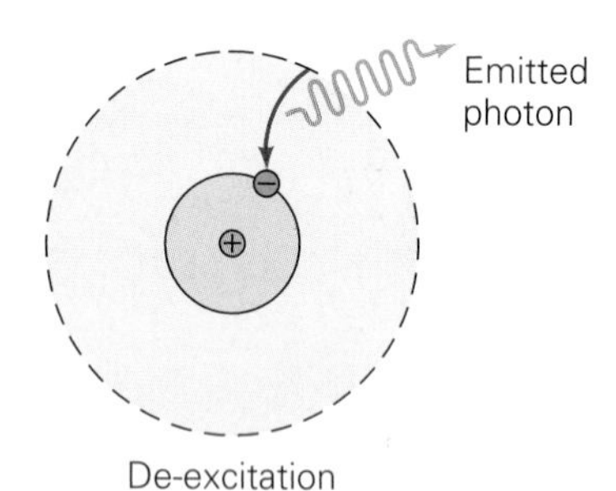

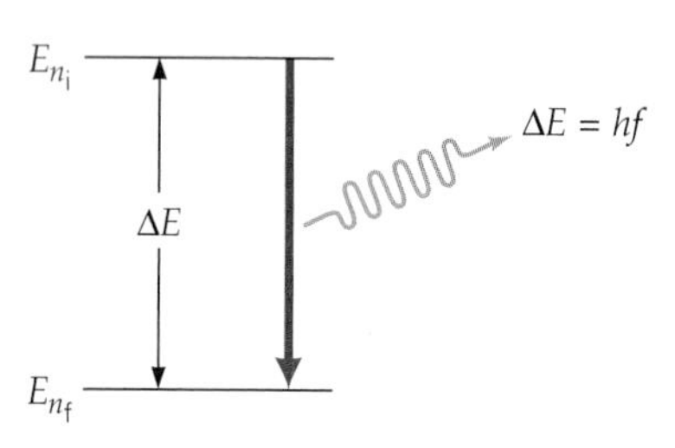

▲ **FIGURE 27.11 Electron transitions and photon emission** When a hydrogen atom emits light, its electron makes a downward transition to a lower orbit (with less energy), and a photon is emitted. The photon's energy is equal to the energy difference between the two levels.

If enough energy is absorbed, it is possible for the electron to no longer be bound to the atom; that is, it is possible for the atom to be *ionized*. For example, to ionize a hydrogen atom initially in its ground state requires a minimum of 13.6 eV of energy. This minimum process makes the final energy of the electron zero (since it is free), and it has a principal quantum number of $n = \infty$. However, if the electron is initially in an excited state, then less energy is needed to ionize the atom. Since the energy of the electron in any state is E_n, the energy needed to free it from the atom is $-E_n$. This energy is called the **binding energy** of the electron. Note that $-E_n$ is positive and represents the energy required to ionize the atom if the electron initially is in a state with a principal quantum number of n.

An electron generally does not remain in an excited state for long; it decays, or makes a downward transition to a lower energy level, in a very short time. The time an electron spends in an excited state is called the **lifetime** of the excited state. For many states, the lifetime is about 10^{-8} s. In making a transition to a lower state, the electron emits a quantum of light energy in the form of a photon (◂Fig. 27.11). The energy ΔE of the photon is equal in magnitude to the energy *difference* of the levels:

$$\Delta E = E_{n_i} - E_{n_f} = \left[\frac{-13.6}{n_i^2}\,\text{eV}\right] - \left[\frac{-13.6}{n_f^2}\,\text{eV}\right]$$

or

$$\Delta E = 13.6\left(\frac{1}{n_f^2} - \frac{1}{n_i^2}\right)\text{eV} \quad \textit{photon energy (in eV) emitted by an H atom} \tag{27.18}$$

Here, the subscripts i and f refer to the initial and final states, respectively. According to the Bohr theory, this energy difference is emitted as a photon with an energy E. Therefore, $E = \Delta E = hc/\lambda$, or $\lambda = hc/\Delta E$. Thus only particular wavelengths of light are emitted. These particular wavelengths (or alternatively, frequencies) correspond to the various transitions between energy levels and explain the existence of an emission spectrum.

The final principal quantum number n_f refers to the energy level in which the electron ends up during the emission process. The original *Balmer emission series* in the visible region corresponds to $n_f = 2$ and $n_i = 3, 4, 5$ and 6. There is only one emission series entirely in the ultraviolet range, called the *Lyman series*, in which all the transitions end in the $n_f = 1$ state (the ground state). There are many series entirely in the infrared region, most notably the *Paschen series*, which ends with the electron in the second excited state, $n_f = 3$. (These series take their names from their discoverers.)

Usually, the wavelength of the light is what is measured during the emission process. Since photon energy and light wavelength are related by Einstein's equation (Eq. 27.4), the wavelength of the emitted light, λ, can be obtained from

$$\lambda\ (\text{in nm}) = \frac{hc}{\Delta E} = \frac{1.24 \times 10^3\,\text{eV}\cdot\text{nm}}{\Delta E\ (\text{in eV})} \tag{27.19}$$

(See the Problem-Solving Hint on page 858.) Consider Example 27.6.

Example 27.6 ■ Investigating the Balmer Series: Visible Light from Hydrogen

What is the wavelength (and color) of the emitted light when an electron in a hydrogen atom undergoes a transition from the $n = 3$ energy level to the $n = 2$ energy level?

Thinking It Through. The emitted photon has an energy equal to the energy difference between the two energy levels (Eq. 27.18). The wavelength of the light can then be obtained by using Eq. 27.19.

Solution.

Given: $n_i = 3$, $n_f = 2$ *Find:* λ (wavelength of emitted light)

The energy of the emitted photon is equal to the magnitude of the atom's change in energy. Thus,

$$\Delta E = 13.6\left(\frac{1}{n_f^2} - \frac{1}{n_i^2}\right)\text{eV} = 13.6\left(\frac{1}{4} - \frac{1}{9}\right)\text{eV} = 1.89\ \text{eV}$$

Using Eq. 27.19 (and making sure that ΔE is expressed in electron-volts), we obtain

$$\lambda = \frac{1.24 \times 10^3 \text{ eV}\cdot\text{nm}}{\Delta E} = \frac{1.24 \times 10^3 \text{ eV}\cdot\text{nm}}{1.89 \text{ eV}} = 656 \text{ nm}$$

which is in the red portion of the visible spectrum. Refer to Fig. 27.8c and note the emission line right around 660 nm for hydrogen. The transition in this example is what gives rise to this red line.

Follow-Up Exercise. Light of what wavelength would be just sufficient to ionize a hydrogen atom if it started in its first excited state? Classify this type of light. Is it visible, UV, or IR?

In summary, since the Bohr model of hydrogen requires that the electron make transitions only between discrete energy levels, the atom emits photons of discrete energies (or light of discrete wavelengths), which results in emission spectra. This process is summarized in ▼Fig. 27.12.

Integrated Example 27.7 ■ The Balmer Series: Entirely Visible?

We know that four wavelengths of the Balmer series are in the visible range (Fig. 27.12). (a) There are more than just these four in this series. What type of light are they likely to be: (1) infrared, (2) visible, or (3) ultraviolet? (b) What is the longest wavelength of nonvisible light in the Balmer series?

(a) Conceptual Reasoning. The Balmer series of emission lines is given off when the electron ends in the first excited state, that is, $n_f = 2$ (Fig. 27.12). There are four distinct visible wavelengths, corresponding to $n_i = 3$, 4, 5, and 6. Any other lines in this series must start with $n_i = 7$ or higher and therefore represent a larger energy difference than those for the visible lines. This means that these photons carry more energy than visible-light photons, and the light will have a shorter wavelength than visible light. Wavelengths smaller than visible are UV. Thus the answer is (3); the other Balmer lines must be in the ultraviolet region.

(b) Thinking It Through. The longest nonvisible Balmer series wavelength corresponds to the smallest photon energy above the $n = 6 \rightarrow 2$ transition. (Why?). Hence, we are talking about $n_i = 7$ and $n_f = 2$. The energy difference can be computed from Eq. 27.18. Then λ can be calculated from Eq. 27.19.

Given: The Balmer emission series for hydrogen

Find: The longest nonvisible wavelength in the Balmer series

From Eq. 27.18,

$$\Delta E = 13.6\left(\frac{1}{n_f^2} - \frac{1}{n_i^2}\right)\text{eV} = 13.6\left(\frac{1}{4} - \frac{1}{49}\right)\text{eV} = 3.12 \text{ eV}$$

This energy difference corresponds to light of wavelength

$$\lambda = \frac{1.24 \times 10^3 \text{ eV}\cdot\text{nm}}{\Delta E} = \frac{1.24 \times 10^3 \text{ eV}\cdot\text{nm}}{3.12 \text{ eV}} = 397 \text{ nm}$$

This is just below the lower limit of the visible spectrum, which ends at 400 nm.

Follow-Up Exercise. In hydrogen, what is the longest wavelength of light emitted in the Lyman series? In what region of the spectrum is this light?

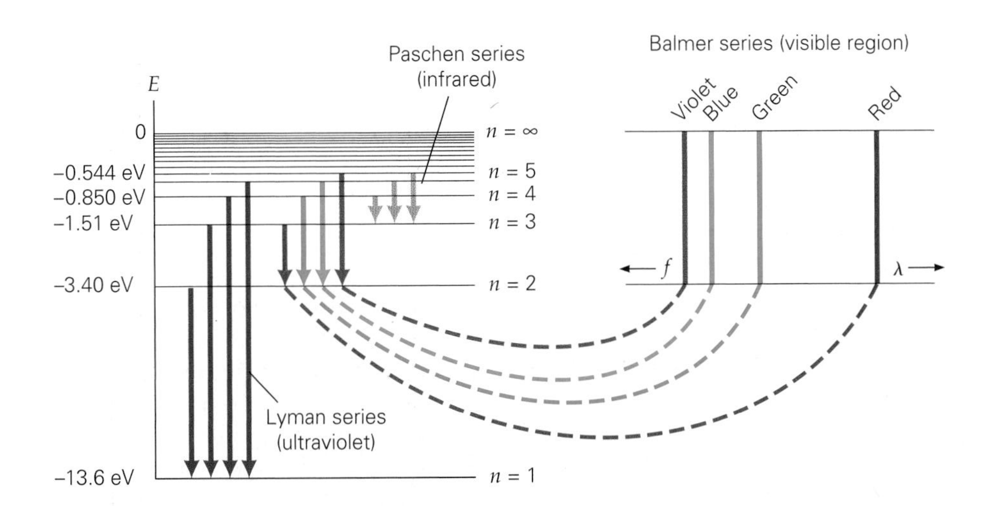

◀ **FIGURE 27.12 Hydrogen spectrum** Transitions may occur between two or more energy levels as the electron returns to the ground state. Transitions to the $n = 2$ state give spectral lines with wavelengths in the visible region (the Balmer series). Transitions to other levels give rise to other series (not in the visible region), as shown.

Conceptual Example 27.8 ■ Up and Down in the Hydrogen Atom: Absorbed and Emitted Photons

Assume that a hydrogen atom, initially in its ground state, absorbs a photon. In general, how many emitted photons would you expect to be associated with the de-excitation process back to the ground state? Can there be (a) more than one photon, or must there be (b) only one photon?

Reasoning and Answer. Since the difference between light emission and absorption is the direction of the transition (down for emission, up for absorption), you might be tempted to answer (b), because only one photon was required for the excitation process. However, if you look at the details of the two processes, you will realize that they are not necessarily symmetrical. In absorbing a photon's energy, a hydrogen atom will jump from its ground state to an excited state—let's say from $n = 1$ to $n = 4$. This process requires a single photon of a unique energy (that is, light of a unique wavelength).

However, in dropping back to the ground state, the atom may take any one of *several* possible routes. For example, the atom could go from the $n = 4$ state to the $n = 3$ state, followed by a transition to the ground state ($n = 1$). This process would involve the emission of two photons. Therefore, the answer is (a). In general, following the absorption of a single photon, several photons can be emitted.

Follow-Up Exercise. (a) How many different-energy photons may a hydrogen atom emit in de-exciting from the third excited state to the ground state? (b) Starting from the ground state, which excitation transition *must* result in only one emitted photon when the hydrogen atom de-excites? Explain your reasoning.

Bohr's theory gave excellent agreement with experiment for hydrogen gas as well as for other ions with just one electron, such as singly ionized helium. However, it could not successfully describe multielectron atoms. Bohr's theory was incomplete in the sense that it patched new quantum ideas into a basically classical framework. The theory contains some correct concepts, but a complete description of the atom did not come until the development of quantum mechanics (Chapter 28). Nevertheless, the idea of discrete energy levels in atoms enables us to qualitatively understand phenomena such as fluorescence.

In **fluorescence**, an electron in an excited state returns to the ground state in two or more steps, like a ball bouncing down a flight of stairs. At each step, a photon is emitted. Each such step represents a smaller energy transition than the original energy required for the upward transition. Therefore, each emitted photon must have a lower energy and a longer wavelength than the original exciting photon. For example, the atoms of many minerals can be excited by absorbing ultraviolet (UV) light and *fluoresce*, or glow, in the visible region when they de-excite (◂Fig. 27.13). A variety of living organisms, from corals to butterflies, manufacture fluorescent pigments that emit visible light.

(a) Visible illumination

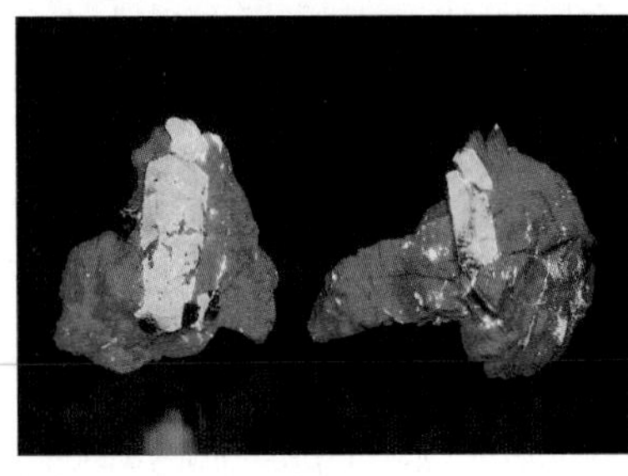

(b) UV illumination

▲ **FIGURE 27.13** Fluorescence Many minerals emit light of visible wavelengths when illuminated by invisible ultraviolet light (so-called black light). The visible light is produced when atoms excited by the UV light de-excite to lower energy levels in several smaller steps, yielding photons of less energy and longer (visible) wavelengths.

27.5 A Quantum Success: The Laser

OBJECTIVE: To understand some of the practical applications of the quantum hypothesis—in particular, the laser.

The development of the laser was a major technological success. Unlike the numerous inventions that have come about by trial and error or accident, including Roentgen's discovery of X-rays and Edison's electric lamp, the laser was developed on theoretical grounds. Through the use of quantum physics ideas, the **laser** was first predicted and then designed, built, and, finally, applied. (The word *laser* is an acronym that stands for *l*ight *a*mplification by *s*timulated *e*mission of *r*adiation. Stimulated emission will be discussed shortly.) The laser has found widespread applications, some of which are discussed at the end of this section.

The existence of atomic energy levels is of prime importance in understanding the laser's operation. Usually, an electron makes a transition to a lower energy level almost immediately, remaining in an excited state for only about 10^{-8} s. However, the lifetimes of some excited states are appreciably longer than this. An atomic state with a relatively long lifetime is called a **metastable state**. For example, **phosphorescent** materials are composed of atoms that have such metastable states. These materials are used on luminous watch dials, toys, and other items that "glow in the dark." When a phosphorescent material is exposed to light, the atoms are excited to higher energy levels. Many of the atoms return to their normal state very quickly. However, there are also metastable states in which the atoms may remain for seconds, minutes, or even more than an hour. Consequently, the material can emit light and glow for some time (▸Fig. 27.14).

▲ **FIGURE 27.14 Phosphorescence and metastable states** When atoms in a phosphorescent material are excited, some of them do not immediately return to the ground state, but remain in metastable states for longer than normal periods of time. In this exhibit at the San Francisco Exploratorium, the phosphorescent walls and floor continue to glow for about 30 seconds after being illuminated, retaining the shadows of children who were present when the phosphors were initially exposed to light.

A major consideration in laser operation is the emission process. As ▾Fig. 27.15 shows, absorption and spontaneous emission of radiation can occur between two energy levels. That is, a photon is absorbed and a photon is emitted almost immediately. However, when the higher energy state is metastable, there is another possible emission process, called **stimulated emission**. Einstein first proposed this process in 1919. If a photon with an energy equal to an allowed transition strikes an atom already in a metastable state, it may stimulate that atom to make a transition to a lower energy level. This transition yields a second photon that is identical to the first one. Thus, *two* photons with the same frequency and phase will go off in the same direction. Notice that stimulated emission is an amplification process—one photon in, two out. But this process is not a case of getting something for nothing, since the atom must be initially excited, and energy is required to do this.

Ordinarily, when light passes through a material, photons are more likely to be absorbed than to give rise to stimulated emission. This is because there normally are many more atoms in their ground state than in excited states. However, it is possible to prepare a material so that more of its atoms are in an excited metastable state than in the ground state. This condition is known as a **population inversion**. In this case, there may be more stimulated emission than absorption, and the net result is amplification. With the proper instrumentation, the result is a laser.

Today, there are many types of lasers capable of producing light of different wavelengths. The helium–neon (He–Ne) gas laser is probably the most familiar, since it is used for classroom demonstrations and laboratory experiments. The characteristic reddish-pink light produced by the He–Ne laser ($\lambda = 632.8$ nm) is also used in the optical scanning systems at supermarket checkouts. The gas mixture is about 85% helium and 15% neon. Essentially, the helium is used for energizing and the neon for amplification. The gas mixture is subjected to a high-voltage discharge of electrons produced by a radio-frequency power supply converted to direct current (dc). The helium atoms are excited by collision with these electrons (▾Fig. 27.16a). This process is referred to as *pumping*. Energy is pumped into the system, and the helium atoms are pumped into an excited state that is 20.61 eV above its ground state.

This excited state in helium has a relatively long lifetime of about 10^{-4} s and has almost the same energy as an excited state in the Ne atom at 20.66 eV. Because this lifetime is so long, there is a good chance that before an excited He atom can spontaneously emit a photon, it will collide with a Ne atom in its ground state. When such a collision occurs, energy can be transferred to the Ne atom. The lifetime of the 20.66-eV neon state is also relatively long. The delay in this metastable state of neon causes a population inversion in the neon atoms—a higher percentage of atoms in the 20.66-eV state than in ones below it. When the neon drops to the 18.70 eV state, it emits a photon with an energy equal to the difference in energy of the two levels, or about 2.0 eV, which corresponds to the red light with a wavelength of 633 nm that we observe.

▼ **FIGURE 27.15 Photon absorption and emission** **(a)** In the absorption of light, once a photon is absorbed, the atom is excited to a higher energy level. **(b)** After a short time, the atom spontaneously decays to a lower energy level with the emission of a photon. **(c)** If another photon with an energy equal to that of the *downward* transition strikes an excited atom, stimulated emission can occur. The result is *two* photons (the original plus the new one) with the same frequency. They travel in the same direction as that of the incident photon and are in phase with one another.

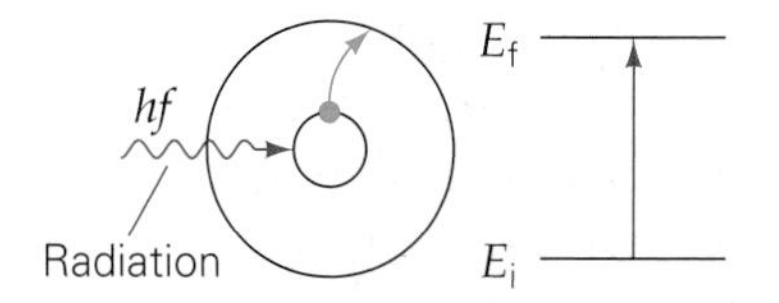

(a) Absorption

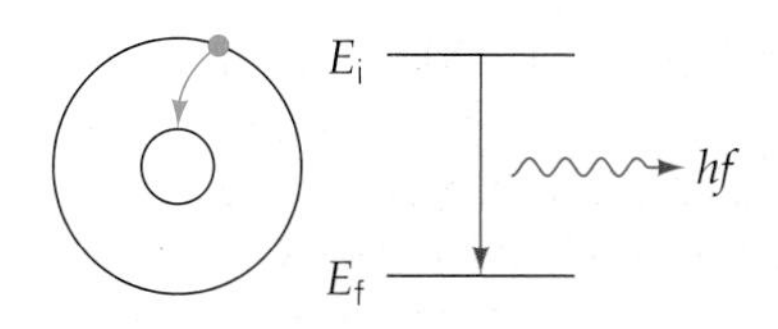

(b) Spontaneous emission

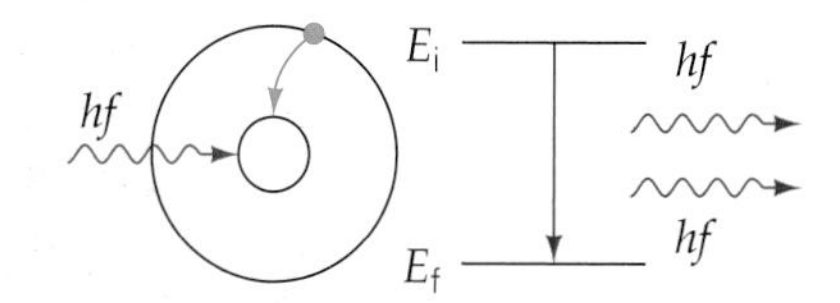

(c) Stimulated emission

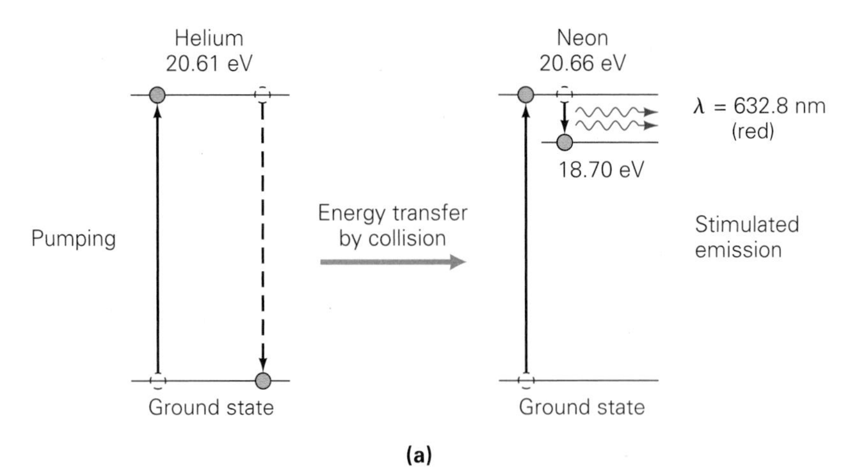

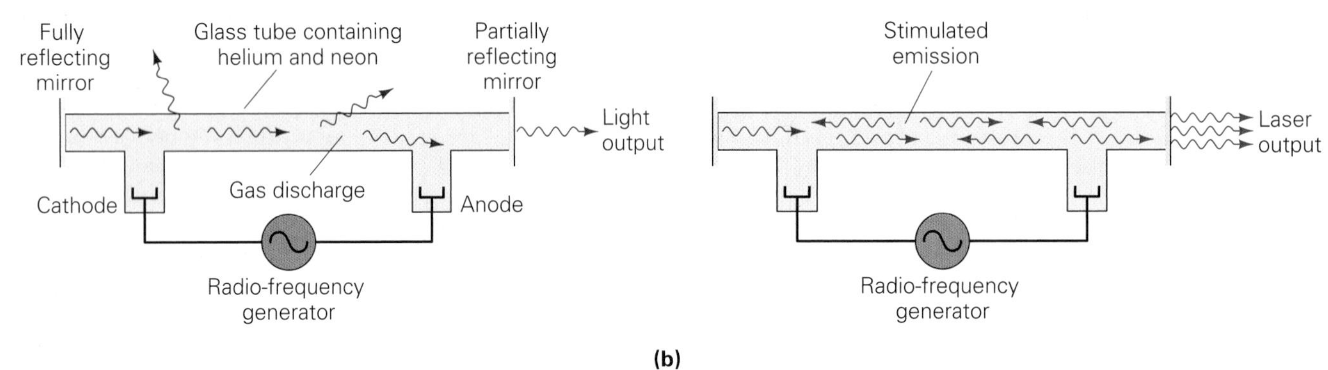

▶ **FIGURE 27.16 The helium–neon laser** **(a)** Helium atoms are first excited (or "pumped") by collision with electrons. This energy is then transferred from the helium atoms to the neon atoms in a metastable state. All that is needed is one photon from a downward transition to stimulate emission from the other excited neon atoms. When this process occurs, the result is the emission of an intense beam of red laser light. **(b)** End mirrors on the laser tube are used to enhance the light beam (from stimulated emission) in a direction along the tube's axis. One of the mirrors is only partially silvered, allowing for some of the light to come out of the tube and resulting in the beam of red light we observe.

The stimulated emission of the light emitted by these neon atoms is enhanced by reflections from mirrors placed at each end of the laser tube (Fig. 27.16b). Some excited Ne atoms spontaneously emit photons in all directions, and these photons, in turn, induce stimulated emissions. In stimulated emission, the two photons leave the atom in the same direction as that of the incident photon. Photons traveling in the direction of the tube axis are reflected back through the tube by the end mirrors. The photons, in reflecting back and forth, cause even more stimulated emissions. The result is an intense, highly directional, coherent (in phase), monochromatic (single-wavelength) beam of light traveling back and forth along the tube axis. Part of the beam emerges through one of the end mirrors, because it is only partially silvered.

The monochromatic, coherent, and directional properties of laser light are responsible for its unique properties (◂Fig. 27.17). Light from sources such as incandescent lamps is emitted from the atoms randomly and at different frequencies (a result of many different transitions). As a result, the light is out of phase, or incoherent. Such beams spread out and become less intense. The properties of laser light allow the formation of a very narrow beam, which with amplification can be very intense.

Some industrial laser applications are shown in ▾Fig. 27.18, and another is discussed in Insight 27.1 on CD and DVD Systems on page 869. For safety, remember that working with a laser beam can be quite hazardous. If the beam is focused on the retina in a very small area and if its intensity and viewing time are sufficient, the retina can be burned and either damaged or destroyed by the concentrated energy.

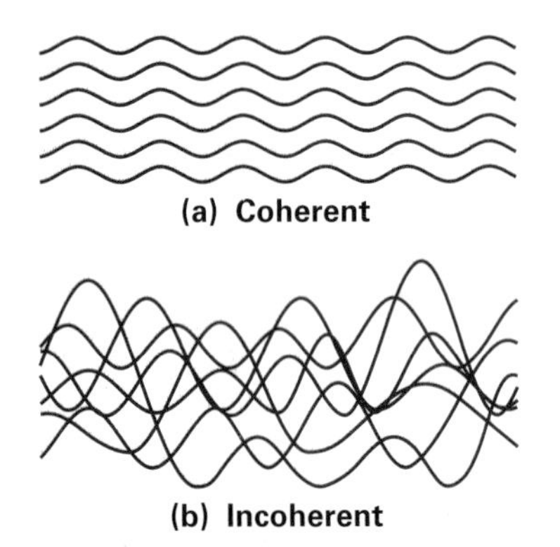

▲ **FIGURE 27.17 Coherent light** **(a)** Laser light is monochromatic (single-frequency and single-wavelength or color) and coherent, meaning that all the light waves are locked in phase. **(b)** Light waves from sources such as a lightbulb filament are emitted randomly. They consist of many different wavelengths (colors) and are incoherent or, on average, out of phase.

(a)

(b)

▶ **FIGURE 27.18 Some industrial laser applications** **(a)** Lasers are used for accurate car repairs. **(b)** An industrial laser cuts through a steel plate.

INSIGHT 27.1 CD AND DVD SYSTEMS

CD's (compact *d*iscs) and DVD (*d*igital *v*ideo *d*iscs) are methods of storing tremendous amounts of data, such as musical recordings and movies. The CD system was introduced around 1980, and the first DVD systems were sold in 1997. Since then, musical CDs and DVD movies have almost completely supplanted audiotapes and videotapes, not to mention vinyl records.

A CD is 12 cm in diameter and can store more than 6 billion bits of information. This capacity is the equivalent of more than 1000 floppy disks, or more than 275 000 pages of text! For audio use, a CD can store 74 min of music, and the sound reproduction is virtually unaffected by dust, scratches, or fingerprints on the disc.

Information on both types of disc is in the form of raised areas called *pits*, which are separated by flat areas called *land*. The surface of the disc is coated with a thin layer of aluminum to reflect the laser beam that "reads" the information (Fig. 1). The pits are arranged in a spiral track like the grooves you may or may not be used to in a phonograph record; however, CD tracks are about $\frac{1}{60}$ as far apart as are the grooves on a record. DVD tracks are even narrower and closer together than that, and can store up to *seven times* the information that is on a CD of the same size. (Software *video compression*, which eliminates the redundant and irrelevant information found on CD systems, also contributes to the increased storage available on a DVD.)

In the readout system which extracts the information, a laser beam from a small semiconductor (solid-state) laser is applied from below the disc and focused on the aluminum coating of the track. The disc rotates at about 3.5 to 8 revolutions per second as the laser beam follows the spiral track. The beam is reflected when it strikes a land area between two pits. When the beam spot overlaps a land area and a pit, the light reflected from the different areas interferes, causing fluctuations in the reflected beam. To make the fluctuations more distinct, the raised-pit thickness (t) is made to be one-fourth of the wavelength of the laser light. Then, reflected light from land areas travels an additional path length of half a wavelength. Destructive interference occurs when the two parts of the reflected beam combine. (See Section 24.1.) As a result, there is less reflected intensity when the beam passes over the edge of a pit than when it passes over a land area alone.

Then the reflected beam of varying intensity strikes a photodiode (a solid-state photocell), which reads the information. The fluctuations of the reflected light convey the coded information as a series of binary numbers (zeros and ones). A pit represents a binary 1, and a land area is read as a binary 0. The signals are then electronically converted into sound or video.

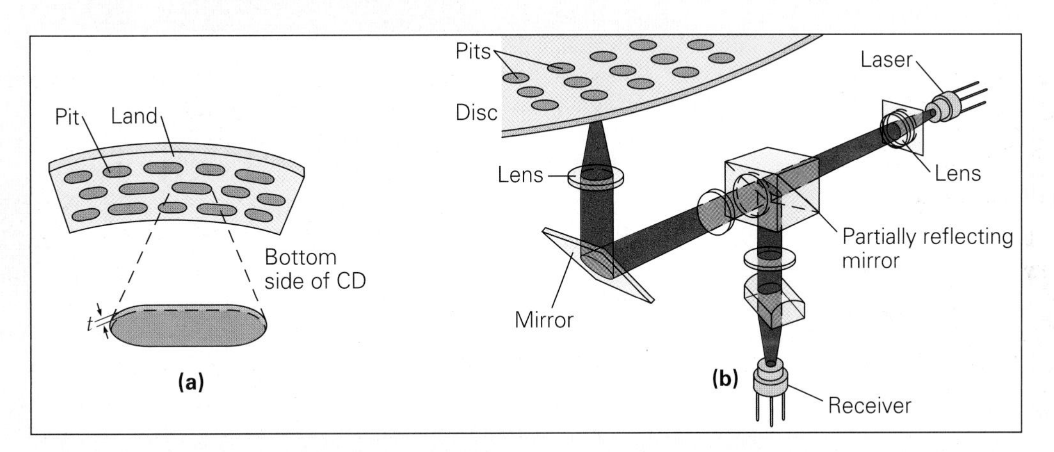

FIGURE 1 The CD and DVD (a) The information on the disc is recorded in the form of raised areas called *pits*, which are separated by flat areas called *land*. The pits are on the bottom of the disc. **(b)** The surface is coated with a thin layer of aluminum to reflect the laser beam, which "reads" the information. DVDs (for movies, for example) are physically similar to the audio CD, but the tracks are narrower and more closely spaced, allowing much more information to be stored.

However, the laser can also be used to promote health, as the applications in Insight 27.2 on Lasers in Modern Medicine on accompanying page.

Another interesting application of laser light is the production of three-dimensional images in a process called **holography**. The process does not use lenses, as ordinary image-forming processes do, yet it re-creates the original scene in three dimensions. The key to holography is the coherent property of laser light, which gives the light waves a definite spatial relationship to each other.

In the photographic process of making a *hologram*, an arrangement such as that illustrated in ▼Fig. 27.19 is used. Part of the light from the laser (the object beam) passes through a partially reflecting mirror to the object. The other part, or reference beam, is reflected to the film. The light incident on the object is also

INSIGHT 27.2 LASERS IN MODERN MEDICINE

The use of lasers has had a large impact on modern medicine, in areas ranging from elective cosmetic surgery to life-saving cancer surgery. You have already read about one of the most well-known medical laser applications in the area of vision correction. (See Chapter 25.) Another application for medical lasers is for *tattoo removal* without harming the surrounding cells. More and more lasers are being chosen over other methods, such as surgical excision, dermabrasion (sanding of the skin), chemical peels, and cryosurgery (freezing). Laser treatment is noninvasive and targets only the inks that make up the tattoo. This can be done by adjusting the wavelength (color) of the laser to match the color of the ink particle, enhancing absorption of the light energy. When the ink particles absorb the laser light, they are heated and fragment. These fragments are then absorbed through the bloodstream and eliminated from the body. This process generally takes a few weeks and may require multiple treatments if the ink particles are large (Figs. 1a and 1b).

Another common use of lasers is to treat painful *varicose veins*. A normal leg has many "one-way" valves in its veins that act to prevent blood from flowing backward (down), thus maintaining an uphill return flow back to the heart. With age (and sometimes pregnancy), these valves can become damaged to the point where they cannot perform. When this happens, the blood will pool in the lower leg, causing the large gnarly purple veins common in many patients (see Fig. 2a). Doppler ultrasound (see Chapter 14) is used to initially diagnose a valve problem by accurately determining the blood flow direction. If left untreated, this can become painful and debilitating. In the past, the only relief was surgical "stripping" of the veins (removal of the surface veins does not appreciably affect the return blood flow, as most of this flow is handled by the deep vein return system). Recently, lasers have been used to *ablate* (that is, thermally destroy) the afflicted vein using a procedure called EVLT (*e*ndo*v*enous *l*aser *t*herapy). EVLT is considerably less invasive, with little or no scarring, and has a lower complication rate than traditional surgical procedures. The vein usually responsible for the lack of upward blood flow is the *greater saphenous vein*, a large vein that extends from the knee to the hip area. In this procedure, a fiber-optic bundle is attached to a laser (typically an IR laser operating at 810-nm wavelength). After introduction of a local anesthetic, ultrasound images are used to guide the fiber as it is inserted through a small incision near the knee and pushed upward the full length of the saphenous vein. As the fiber is slowly pulled out, the laser is repeatedly fired, heating up the venous tissues and causing irreversible damage. Along with this procedure, it is common to remove the bulged smaller lower veins. Once the "downhill" blood flow is eliminated by destroying the saphenous vein, they should not reappear, and new ones are far less likely to form. After several weeks of continuous compression, the change in the leg can be dramatic, as shown in Figures 2b and 2c.

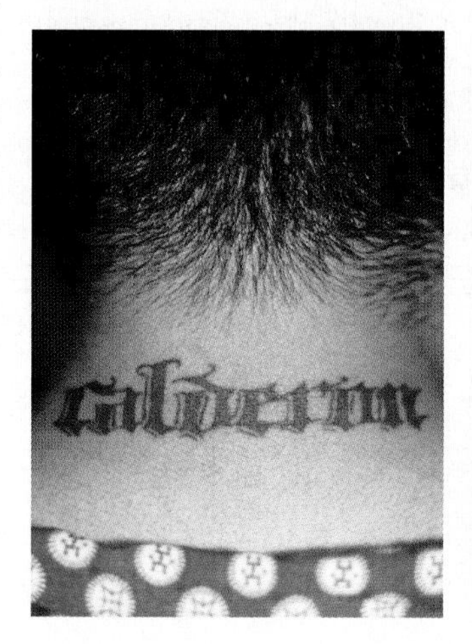
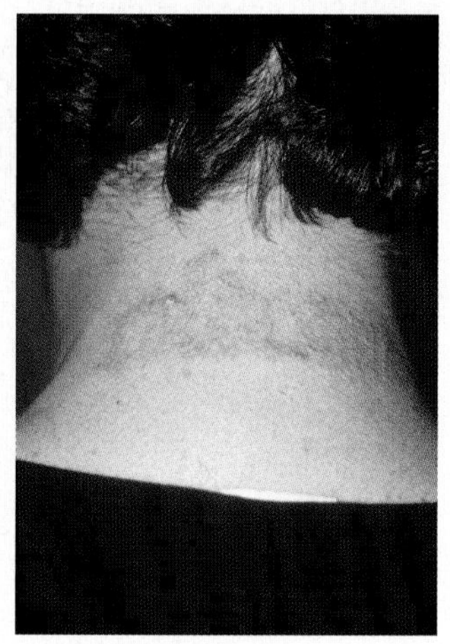

FIGURE 1 Laser and tattoo removal (a) After one laser treatment. **(b)** After three treatments.

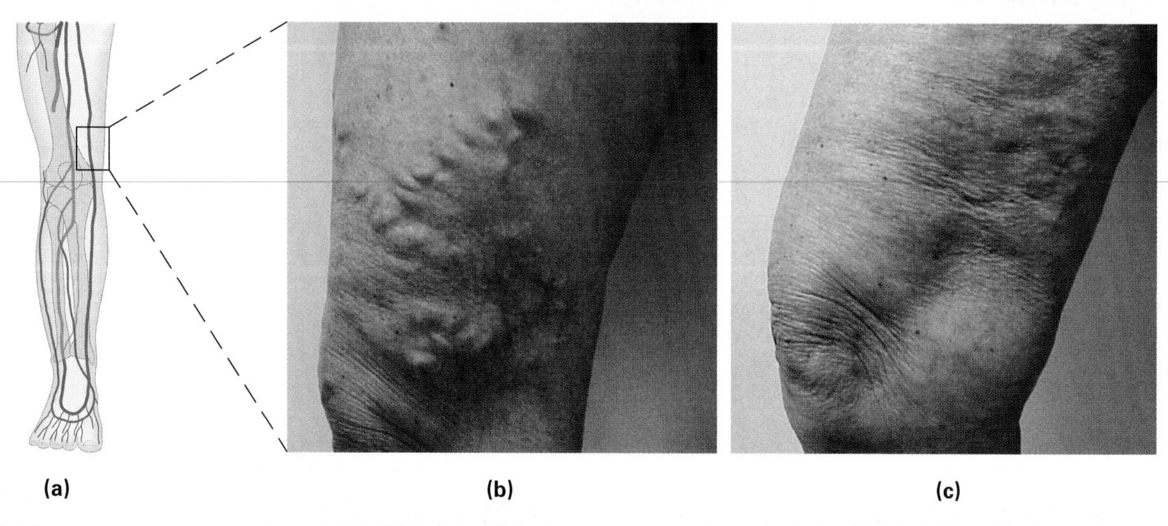

FIGURE 2 Laser and varicose vein treatment (a) A sketch of the venous system in the leg. **(b)** Serious and painful varicose veins before destruction by laser. **(c)** Marked improvement after the saphenous vein is destroyed by laser light.

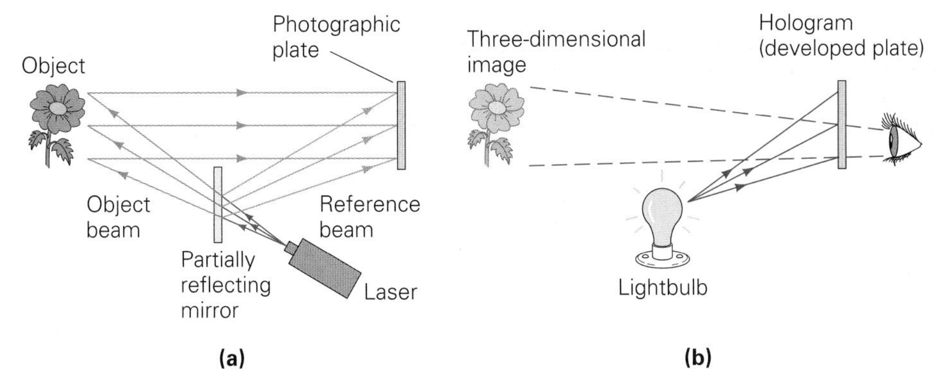

▲ **FIGURE 27.19 Holography** **(a)** The coherent light from a laser is split into reference and object beams. The interference pattern between these beams is recorded on a photographic plate. **(b)** When the developed plate, or hologram, is illuminated by normal light, the viewer sees a reconstructed three-dimensional image of the object.

▲ **FIGURE 27.20 Holographic images** A three-dimensional hologram can be viewed from any angle, just as if it were a real object.

reflected to the film, and it interferes with the reference beam. The film records the interference pattern of the two light beams, which essentially imprints on the film the information carried from the object by the light's wave fronts.

When the film is developed, the interference pattern bears no resemblance to the object and appears as a meaningless pattern of light and dark areas. However, when the wave-front information is reconstructed by passing light through the film, a three-dimensional image is seen (▸Fig. 27.20). If part of the three-dimensional image is hidden from view, you can see it by moving your head to one side, just as you would to see a hidden part of a real object.

Chapter Review

- All hot solids, liquids, and dense gases emit a **continuous spectrum** of electromagnetic (thermal) radiation. The maximum amount of energy is emitted at a wavelength (λ_{max}) determined by the material's absolute temperature T. **Wien's displacement law** gives the (inverse) relationship of wavelength and absolute temperature:

$$\lambda_{max} T = 2.90 \times 10^{-3}\ \text{m} \cdot \text{K} \qquad (27.1)$$

- The classical theory of thermal radiation predicted that an infinite amount of energy is emitted at the short wavelengths, a phenomenon called the **ultraviolet catastrophe,** which did not agree with experiment. To explain the observed spectrum from hot objects, **Planck's hypothesis** stated that the energy of the atoms in the material was quantized in multiples of their vibrational frequency, or

$$E_n = n(hf) \qquad \text{for } n = 1, 2, 3, \ldots \qquad (27.2)$$

where h, **Planck's constant,** has a numerical value of $6.63 \times 10^{-34}\ \text{J} \cdot \text{s}$.

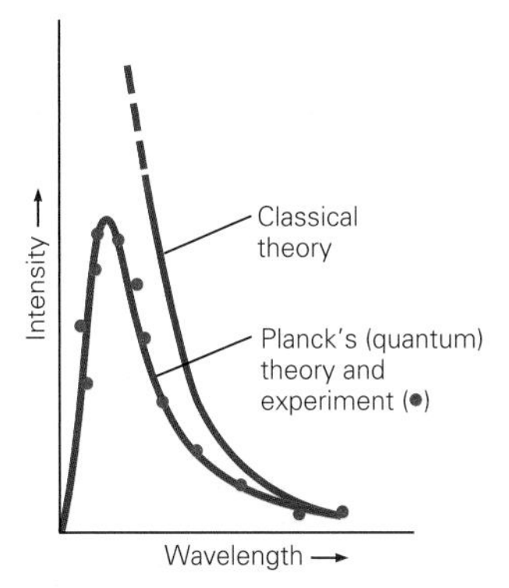

- In the **photoelectric effect,** light incident on a surface causes *photoelectrons* to come off that surface. To explain the experimental results, Einstein had to assume that light consisted of discrete quanta of energy, called **photons,** or particles of light. The energy of a photon associated with light of frequency f is

$$E = hf \qquad (27.4)$$

- Light can also interact with electrons via a scattering process called **Compton scattering**. To explain the details of such scattering, light (frequency f) must be treated as a photon carrying a quantum of energy, hf. When photons scatter off electrons, they impart kinetic energy to the electrons, and the scattered photons have less energy than the incident photons. In terms of waves, the scattered light has a longer wavelength λ than the wavelength of the incident light, λ_o. The relationship between the two wavelengths is given by the Compton scattering equation

$$\Delta\lambda = \lambda - \lambda_o = \lambda_C (1 - \cos\theta) \qquad (27.9)$$

where $\lambda_C = h/m_e c = 2.43 \times 10^{-12}$ m $= 2.43 \times 10^{-3}$ nm is the **Compton wavelength** of the electron.

- The **wave–particle duality of light** means that light must be thought of as having both particle and wave natures.
- The **Bohr theory** of hydrogen treats the electron as a classical particle held in circular orbit around the proton by the electrical force of attraction. In his theory, Bohr made two nonclassical, quantum postulates:

The angular momentum of the electron is quantized and can have only discrete values that are integral multiples of $h/2\pi$, where h is Planck's constant;

and

The electron does *not* radiate energy when it is in a bound, discrete orbit. It radiates energy only when it makes a downward transition to an orbit of lower energy. It makes an upward transition to an orbit of higher energy by absorbing energy.

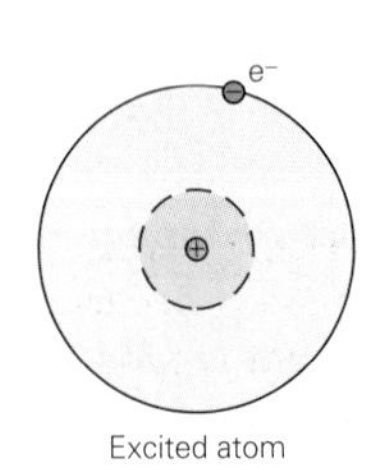

Excited atom

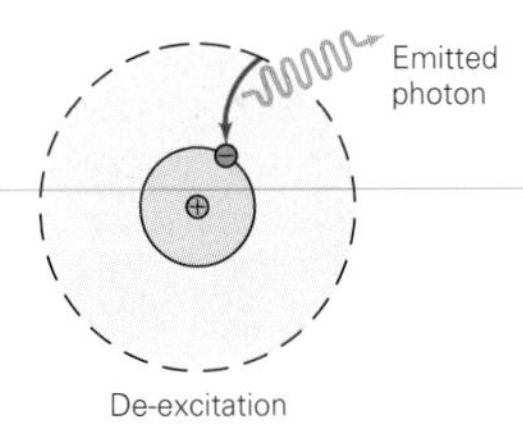

De-excitation

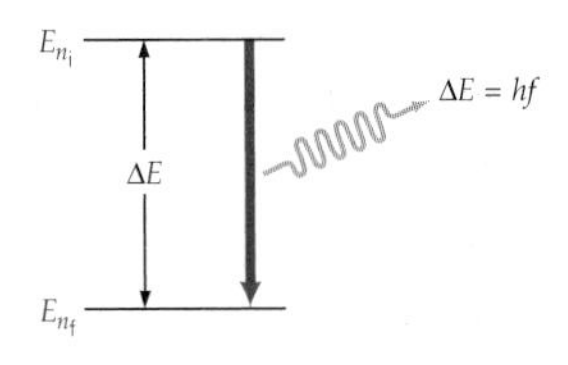

- The Bohr model of hydrogen led directly to the quantization of the radius and energy of the atom. Their values are

$$r_n = 0.0529n^2 \text{ nm} \qquad \text{for } n = 1, 2, 3, 4, \ldots \qquad (27.16)$$

and

$$E_n = \frac{-13.6}{n^2} \text{ eV} \qquad \text{for } n = 1, 2, 3, 4, \ldots \qquad (27.17)$$

where n is the **principal quantum number**. When the hydrogen atom is in the $n = 1$ state, it has its smallest size and lowest energy and is in its **ground state**. States of larger size and higher energy are **excited states**.

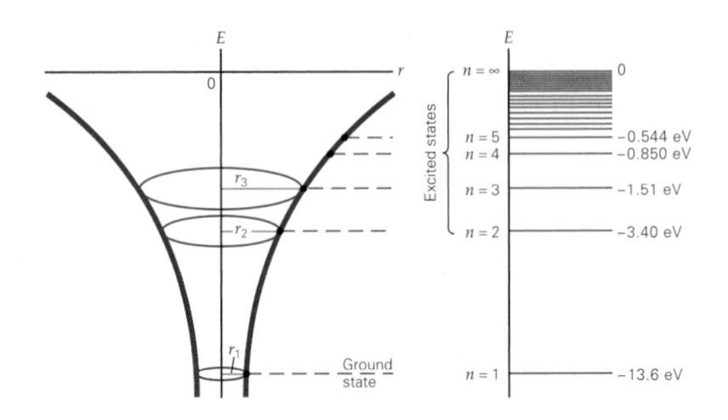

- Atoms emit light in an **emission spectrum,** which shows light emitted only at certain wavelengths characteristic of the atom. Since atomic energies are quantized, the emitted light must come off in quanta (photons) with fixed amounts of energy. Emitted photon energy (equal to the energy difference between the two atomic levels ΔE) is related to the wavelength of the emitted light wavelength by

$$\lambda \text{ (in nm)} = \frac{hc}{\Delta E} = \frac{1.24 \times 10^3 \text{ eV} \cdot \text{nm}}{\Delta E \text{ (in eV)}} \qquad (27.20)$$

- Atoms absorb light in an **absorption spectrum,** which shows light absorbed only at certain wavelengths characteristic of the atom. Since atomic energies are quantized, the absorbed light must be in the form of quanta (photons) with fixed amounts of energy. These amounts of energy correspond to absorbed light of only those certain wavelengths.
- A **metastable state** is a relatively long-lived excited atomic state.
- **Stimulated emission** can occur when a photon prematurely causes a downward atomic transition, yielding a photon identical to itself.

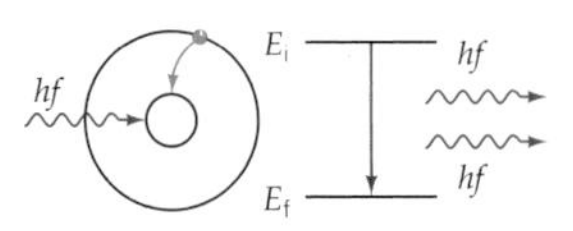

Stimulated emission

- A **laser** uses **population inversion**. This phenomenon is caused by metastable excited atomic states and results in light amplification through stimulated emission.

Exercises

MC = *Multiple Choice Question,* **CQ** = *Conceptual Question, and* **IE** = *Integrated Exercise. Throughout the text, many exercise sections will include "paired" exercises. These exercise pairs, identified with* **red numbers**, *are intended to assist you in problem solving and learning. In a pair, the first exercise (even numbered) is worked out in the Study Guide so that you can consult it should you need assistance in solving it. The second exercise (odd numbered) is similar in nature, and its answer is given at the back of the book.*

Note: Take h to have an exact value of 6.63×10^{-34} J·s *for significant-figure purposes, and use hc* $= 1.24 \times 10^3$ eV·nm *(three significant figures).*

27.1 Quantization: Planck's Hypothesis

1. **MC** Blackbody A is at a temperature of 2000 K and blackbody B is at 4000 K. What can you say about the total energy E they radiate: (a) $E_A = \frac{1}{2}E_B$, (b) $E_A = \frac{1}{4}E_B$, (c) $E_A = \frac{1}{8}E_B$, (d) $E_A = \frac{1}{16}E_B$, or (e) you can't tell from the information given?

2. **MC** Blackbody A is at a temperature of 3000 K and blackbody B is at 6000 K. What can you say about the wavelength at which they radiate the maximum energy: (a) $\lambda_{max,A} = \frac{1}{2}\lambda_{max,B}$, (b) $\lambda_{max,A} = 2\lambda_{max,B}$, (c) $\lambda_{max,A} = \lambda_{max,B}$, or (d) you can't tell from the information given?

3. **MC** The absolute temperature of a blackbody radiator is doubled. What can you say about the ratio of total energy$E$ emitted by this object: (a) $E_f/E_i = 2$, (b) $E_f/E_i = 4$, (c) $E_f/E_i = 8$, or (d) $E_f/E_i = 16$.

4. **CQ** Some stars appear reddish, and some others are blue. Which of these stars have the lower surface temperature?

5. **CQ** As a hot piece of iron is heated it begins to glow first red, then orange, then yellow, but then, instead of appearing green or blue as its temperature continues to rise, it becomes white to the eye. Explain.

6. **CQ** Make a graph showing how the wavelength of the most intense radiation component of blackbody radiation varies with the body's absolute temperature. By what ratio does λ_{max} change (final/initial) if the body's absolute temperature is tripled?

7. ● The walls of a blackbody cavity are at a temperature of 27°C. What is the frequency of the radiation of maximum intensity?

8. ● Find the approximate temperature of a red star that emits light with a wavelength of maximum emission of 700 nm (deep red).

9. ● What are the wavelength and frequency of the most intense radiation component from a blackbody with a temperature of 0°C?

10. **IE** ● (a) If you have a fever, will the wavelength of the radiation component of maximum intensity emitted by your body (1) increase, (2) remain the same, or (3) decrease as compared with its value when your temperature is normal? Why? (b) Assume that human skin has a temperature of 32°C. What is the wavelength of the radiation component of maximum intensity emitted by our bodies? In what region of the EM spectrum is this wavelength?

11. ●● What is the minimum energy of a thermal oscillator in a blackbody producing radiation at λ_{max} at a temperature of 212°F?

12. **IE** ●● The temperature of a blackbody increases from 200°C to 400°C. (a) Will the frequency of the most intense spectral component emitted by this blackbody (1) increase, but not double; (2) double; (3) be reduced in half; or (4) decrease, but not in half? Why? (b) What is the change in the frequency of the most intense spectral component of this blackbody?

13. ●● The temperature of a blackbody is 1000 K. If the intensity of the emitted radiation, 2.0 W/m^2, were due entirely to the most intense frequency component, how many quanta of radiation would be emitted per second per square meter?*

14. ●●● The wavelength at which the Sun emits its maximum energy light is about 550 nm. Assuming the Sun radiates as a blackbody, estimate (a) its surface temperature and (b) its total emitted power.

27.2 Quanta of Light: Photons and the Photoelectric Effect

15. **MC** In the photoelectric effect, classical theory predicts that (a) no photoemission occurs below a certain frequency; (b) the photocurrent is proportional to the light intensity; (c) the maximum kinetic energy of the emitted electrons depends on the light frequency, or (d) no matter how low the light intensity, photocurrent is observed immediately.

16. **MC** In the photoelectric effect, what happens to the stopping voltage when the light intensity is increased: (a) It increases, (b) it stays the same, or (c) it decreases.

17. **MC** In the photoelectric effect, what happens to the stopping voltage when the light frequency is increased: (a) It increases, (b) it stays the same, or (c) it decreases?

18. **MC** In the photoelectric effect, assuming photoemission continues, what happens to the stopping voltage when the emitting material is changed for one with a larger work function: (a) It increases, (b) it stays the same, or (c) it decreases?

19. **MC** In the SI system, the work function has what units? (a) joules (b) volts (c) coulombs (d) amperes

*Assume that each quantum has the minimum allowed energy.

20. CQ With respect to ionization (which can cause biological damage), is it more dangerous to stand in front of a beam of X-ray radiation with very low total energy or a beam of red light with much more total energy? How does the photon model of light explain this apparent paradox?

21. CQ Is it possible for a beam of IR radiation to contain more total energy than a beam of UV radiation? Explain.

22. CQ In the photoelectric effect, when the incident light is below the threshold frequency, a significant amount of its energy is still absorbed by the target material, but electrons are not emitted from the surface. Explain where this energy goes.

23. ● Each photon in a beam of light has an energy of 3.3×10^{-15} J. What is the light's wavelength? Use this to classify its type.

24. IE ● (a) Compared with a quantum of red light ($\lambda = 700$ nm), a quantum of violet light ($\lambda = 400$ nm) has (1) more, (2) the same amount of, (3) less energy. Why? (b) Determine the ratio of the photon energy associated with violet light to that related to red light.

25. ● A source of UV light has a wavelength of 150 nm. How much energy does one of its photons have expressed in (a) joules and (b) electron-volts?

26. IE● The work function of metal A is greater than that of metal B. (a) The threshold wavelength for metal A is (1) shorter than, (2) the same as, (3) longer than that of metal B. Why? (b) If the threshold wavelength for metal B is 620 nm and the work function of metal A is twice that of metal B, what is the threshold wavelength for metal A?

27. ● The photoelectrons ejected from a surface require a stopping voltage of 3.0 V. If the intensity of the light is tripled, what is the stopping voltage now?

28. ●● Assume that a 100-W lightbulb gives off 2.50% of its energy as visible light. How many photons of visible light are given off in 1.00 min? (Use an average visible wavelength of 550 nm.)

29. ●● ▼Figure 27.21 shows a graph of stopping potential versus frequency for a photoelectric material. Determine (a) Planck's constant and (b) the work function of the material from the data contained in the graph.

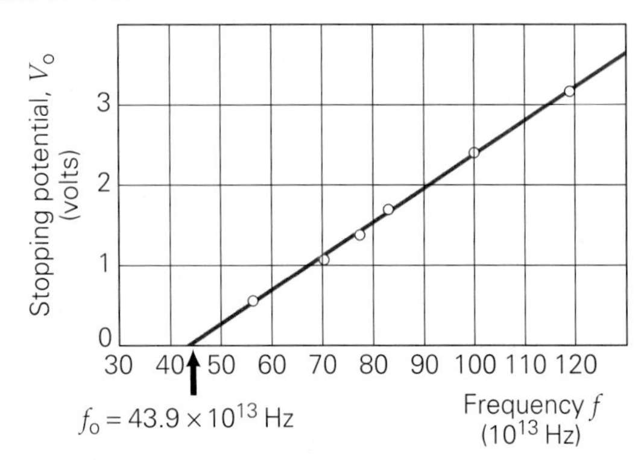

▲ **FIGURE 27.21 Stopping potential versus frequency** See Exercise 29.

30. ●● A metal with a work function of 2.40 eV is illuminated by a beam of monochromatic light. If the stopping potential is 2.50 V, what is the wavelength of the light?

31. ●● What is the lowest frequency of light that can cause the release of electrons from a metal that has a work function of 2.8 eV?

32. ●● The photoelectric effect threshold wavelength for a certain metal is 500 nm. Calculate the maximum speed of photoelectrons if we use light having a wavelength of (a) 400 nm, (b) 500 nm, and (c) 600 nm.

33. ●● In Exercise 32, what is the work function of the metal in eV?

34. ●● The work function of a material is 3.5 eV. If the material is illuminated with monochromatic light ($\lambda = 300$ nm), what are (a) the stopping potential and (b) the cutoff frequency?

35. ●● Blue light with a wavelength of 420 nm is incident on a certain material and causes the emission of photoelectrons with a maximum kinetic energy of 1.00×10^{-19} J. (a) What is the stopping voltage? (b) What is the material's work function? (c) What is the stopping voltage if red light ($\lambda = 700$ nm) is used instead? Explain.

36. ●●● When the surface of a particular material is illuminated with monochromatic light of various frequencies, the stopping potentials for the photoelectrons are determined to be the following:

Frequency (in Hz)			
9.9×10^{14}	7.6×10^{14}	6.2×10^{14}	5.0×10^{14}
Stopping potential (in V)			
2.6	1.6	1.0	0.60

Plot these data, and from the graph determine Planck's constant and the metal's work function.

37. ●●● When a certain photoelectric material is illuminated with red light ($\lambda = 700$ nm) and then blue light ($\lambda = 400$ nm), it is found that the maximum kinetic energy of the photoelectrons resulting from the blue light is twice that of those from red light. What is the work function of the material?

27.3 Quantum "Particles": The Compton Effect

38. MC The percentage change in wavelength in the Compton effect would be most easily observed using (a) visible light, (b) infrared radiation, (c) ultraviolet light, (d) X-rays.

39. MC The wavelength shift for Compton scattering is maximum when (a) the photon scattering angle is 90°, (b) the electron receives only a small fraction of its maximum possible recoil energy, (c) the electron is scattered directly forward.

40. MC At what photon-scattering angle will the electron receive the least recoil energy: (a) 20°, (b) 45°, (c) 60°, or (d) 80°?

41. CQ A photon can undergo Compton scattering from a neutron. How does the maximum wavelength shift for Compton scattering from a neutron compare with that from an electron? Explain.

42. CQ The Sun's energy production (near its the center) is initially X-rays and gamma rays. By the time it reaches the surface, it is mostly in the visible range. Use Compton scattering to explain how this happens.

43. CQ In Compton scattering, the recoiling electron must always be on the other side of the incident beam direction from the scattered photon. Why?

44. CQ In Compton scattering, how does the maximum wavelength shift for 0.100-nm X-ray photons compare to that of visible-light photons (500-nm wavelength)?

45. ● What is half the maximum wavelength shift for Compton scattering from a free electron?

46. ● What is the change in wavelength when monochromatic X-rays are scattered by electrons through an angle of 30°?

47. ● A monochromatic beam of X-rays with a wavelength of 0.280 nm is scattered by a metal foil. What is the wavelength of the scattered X-rays observed at an angle of 45° from the direction of the incident beam?

48. ● X-rays with a wavelength of 0.0045 nm are used in a Compton-scattering experiment. If the X-rays are scattered through an angle of 53°, what is the wavelength of the scattered radiation?

49. IE ●● A photon with an energy of 5.0 keV is scattered by a free electron. (a) The recoiling electron could have an energy of (1) zero, (2) less than 5.0 keV, but not zero, (3) equal to 5.0 keV. Why? (b) If the wavelength of the scattered photon is 0.25 nm, what is the recoiling electron's kinetic energy?

50. ●● X-rays scattered from a carbon atom show a wavelength shift of 0.000326 nm. What is their scattering angle?

51. ●● If the Compton shift for a photon scattered by an electron is 1.25×10^{-4} nm, what is the scattering angle?

52. IE ●●● The Compton effect can occur for scattering from any particle—for example, from a proton. (a) Compared with the Compton wavelength for an electron, the Compton wavelength for a proton is (1) longer, (2) the same size, (3) shorter. Why? (b) What is the value of the Compton wavelength for a proton? (c) Determine the ratio of the maximum Compton wavelength shift for scattering by an electron to that for scattering by a proton.

27.4 The Bohr Theory of the Hydrogen Atom

53. **MC** In his theory of the hydrogen atom, Bohr postulated the quantization of (a) energy, (b) centripetal acceleration, (c) light, (d) angular momentum.

54. **MC** An excited hydrogen atom emits light when its electron (a) makes a transition to a lower energy level, (b) is excited to a higher energy level, (c) is in the ground state.

55. **MC** A hydrogen atom in its first excited state absorbs a photon and makes a transition to a higher excited state. The longest-wavelength photon possible is absorbed. The quantum number of the final state is (a) 1, (b) 2, (c) 3, (d) 4.

56. CQ The Bohr theory is applicable only to the hydrogen atom and hydrogen-like atoms, such as singly ionized helium, and other one-electron systems. Why?

57. CQ Will it take more or less energy to ionize (remove the electron completely from) a hydrogen atom if the electron is in an excited state than if it is in the ground state? Explain.

58. CQ Very accurate measurements of the wavelengths emitted by a hydrogen atom indicate that they are all slightly longer than expected from the Bohr theory. Explain how conservation of linear momentum explains this. [*Hint*: Photons carry momentum and energy.]

59. ● Find the energy required to excite a hydrogen electron from (a) the ground state to the first excited state and (b) from the first excited state to the second excited state. (c) Classify the type of light needed to create each of the transitions.

60. ● Find the energy needed to ionize a hydrogen atom whose electron is in the (a) $n = 2$ state and (b) $n = 3$ state.

61. ● What is the frequency of light that would excite the electron of a hydrogen atom from (a) a state with a principal quantum number of $n = 2$ to that with a principal quantum number of $n = 5$? (b) What about from $n = 2$ to $n = \infty$?

62. ● Find the radius of the electron orbit in a hydrogen atom for states with the following principal quantum numbers: (a) $n = 2$, (b) $n = 4$, (c) $n = 5$.

63. ● Scientists are now beginning to study "large" atoms, that is, atoms with orbits that are almost large enough to be measured in our everyday units of measurement. For what excited state (give an approximate principal quantum number) of a hydrogen atom would the diameter of the orbit be a micron (10^{-6} m)—that is, close to the size of a dust particle?

64. ●● Find the binding energy of the hydrogen electron for states with the following principal quantum numbers: (a) $n = 3$, (b) $n = 5$, (c) $n = 10$.

65. IE ●● A hydrogen atom has an ionization energy of 13.6 eV. When it absorbs a photon with an energy greater than this energy, the electron will be emitted with some kinetic energy. (a) If the energy of such a photon is doubled, the kinetic energy of the emitted electron will (1) increase, but not necessarily double, (2) remain the same, (3) exactly double, (4) decrease. Why? (b) A photon associated with light of a frequency of 7.00×10^{15} Hz is absorbed by a hydrogen atom. What is the kinetic energy of the emitted electron?

66. ●● A hydrogen atom in its ground state is excited to the $n = 5$ level. It then makes a transition directly to the $n = 2$ level before returning to the ground state. (a) What are the wavelengths of the emitted photons? (b) Would any of the emitted light be in the visible region?

67. IE ●● (a) For which of the following transitions in a hydrogen atom is the photon of greatest energy emitted: (1) $n = 5$ to $n = 3$, (2) $n = 6$ to $n = 2$, or (3) $n = 2$ to $n = 1$? (b) Justify your answer mathematically.

68. ●● The hydrogen spectrum has a series of lines called the *Lyman series*, which results from transitions to the ground state. What is the longest wavelength in this series, and in what region of the EM spectrum does it lie?

69. ●● What is the binding energy for an electron in the ground state in the following hydrogen-like ions: (a) He^+, (b) Li^{2+}?

70. ●● A hydrogen atom absorbs light of wavelength 486 nm. (a) How much energy did the atom absorb? (b) What are the values of the principal quantum numbers of the initial and final states of this transition?

71. ●●● Show that the speeds of an electron in the Bohr orbits are given (to two significant figures) by $v_n = (2.2 \times 10^6 \text{ m/s})/n$.

72. ●●● By computing and adding the kinetic and electric potential energies, show that for an electron in the ground state of a hydrogen atom the total energy is −13.6 eV. [*Hint*: You will need to use the orbital radius.]

73. ●●● In Exercise 72, (a) how much of the energy is potential, and how much is kinetic? (b) How do the magnitudes of the two forms of energy compare?

27.5 A Quantum Success: The Laser

74. **MC** Which of the following is not essential for laser action: (a) population inversion, (b) phosphorescence, (c) pumping, or (d) stimulated emission?

75. **MC** Suppose a hypothetical atom had two metastable excited states, one 2.0 eV above the ground state and one 5.0 eV above it. If used in a laser with transitions only to the ground state, which laser would be in the visible range: (a) 5.0 eV, (b) 2.0 eV, (c) neither, or (d) both?

76. **MC** For the atom in Exercise 75, in order to create only the visible laser, the pumping light should not contain which type of light: (a) UV, (b) visible, or (c) IR?

77. **CQ** You have a choice of several lasers spanning the visible spectrum with which to remove a tattoo ink particle that is blue. What color laser would theoretically work the best? Explain. [*Hint*: See the Insight on Lasers in Modern Medicine, p. 870.]

78. **CQ** In what sense is a laser an "amplifier" of energy? Explain why this concept does not violate conservation of energy.

79. **CQ** Explain the difference between spontaneous emission and stimulated emission.

Comprehensive Exercises

80. Light of wavelength 340 nm is incident on a metal surface and ejects electrons that have a maximum speed of 3.5×10^5 m/s. (a) What is the work function of the metal? (b) What is its stopping voltage? (c) What is its threshold wavelength?

81. A 10.0-keV X-ray photon is successively scattered by two free electrons initially at rest. In the first case, it is scattered through an angle of 41°; the second scattering is through an angle of 72°. (a) What is the final photon energy? (b) How much kinetic energy does each electron receive?

82. **IE** (a) How many transitions in a hydrogen atom result in the absorption of red light: (1) one, (2) two, (3) three, or (4) four? (b) What are the principal quantum numbers of the initial and final states for this process? (c) What are the energy of the required photon and the wavelength of the light associated with it?

83. Under the right circumstances, if a photon contains above a minimum energy, that energy can be completely converted into creating an electron–positron pair (a positron is identical to an electron except that it has a positive charge). Recall from relativity that the energy equivalent of the electron mass is 0.511 MeV. Determine (a) the minimum-energy photon required to create such a pair, and (b) the wavelength of light associated with these photons. (c) If a photon of twice the minimum energy were used, what would be the total kinetic energy (electron + positron) of the two particles after their creation?

84. Consider an electron in its first excited state in a hydrogen atom. Determine its (a) speed, (b) angular speed, (c) linear momentum, and (d) angular momentum.

85. A gamma-ray photon scatters off a free proton (initially at rest) at an angle of 45°. The wavelength of the scattered light is measured to be 6.20×10^{-13} m. (a) What was the incoming photon energy? (b) How much kinetic energy did the proton receive? (You may need to carry an extra figure or two in your intermediate answers.)

86. In the nuclear version of the photoelectric effect (called the *photonuclear* effect), a high-energy photon is absorbed by an atomic nucleus and a proton is freed from that nucleus. If the minimum energy needed to free a proton from a particular nucleus is 5.00 MeV, (a) determine the maximum (threshold) wavelength of light that can cause this. (b) If a photon of half this wavelength is used instead, determine the kinetic energy of the ejected proton.

CHAPTER 28

QUANTUM MECHANICS AND ATOMIC PHYSICS

28.1 Matter Waves: The de Broglie Hypothesis 878

28.2 The Schrödinger Wave Equation 881

28.3 Atomic Quantum Numbers and the Periodic Table 885

28.4 The Heisenberg Uncertainty Principle 894

28.5 Particles and Antiparticles 896

PHYSICS FACTS

- Louis de Broglie's full name was Prince Louis-Victor Pierre Raymond de Broglie. Besides winning the Nobel Prize in physics in 1929 for his discovery of "the wave nature of the electron," he also excelled as a teacher/lecturer. In 1952 the first Kalinga Prize was awarded to him by UNESCO for his efforts to explain aspects of modern physics to laypeople.
- Hitler rose to power in Germany in 1933, the same year that Werner Heisenberg was awarded the Nobel Prize for his contributions to the creation of quantum mechanics. Heisenberg is associated with one of the most famous principles in modern physics: the Heisbenberg uncertainty principle. Protected by the Nobel Prize, Heisenberg became a spokesman for modern physics in Germany and remained there throughout World War II. He was not a Nazi, but he was a German citizen, and felt it his duty to preserve some of German science.
- All elementary particles have associated *antiparticles*. For the familiar electron, the antiparticle is the positive electron, or simply the positron. If electron and positron meet, they annihilate into gamma radiation, the result of converting their mass (energy) into electromagnetic (light) energy. Many of the properties of a particle and its antiparticle are opposite, such as electric charge. However, antiparticles have "regular" mass; that is, they are attracted downward to the Earth, not repelled.

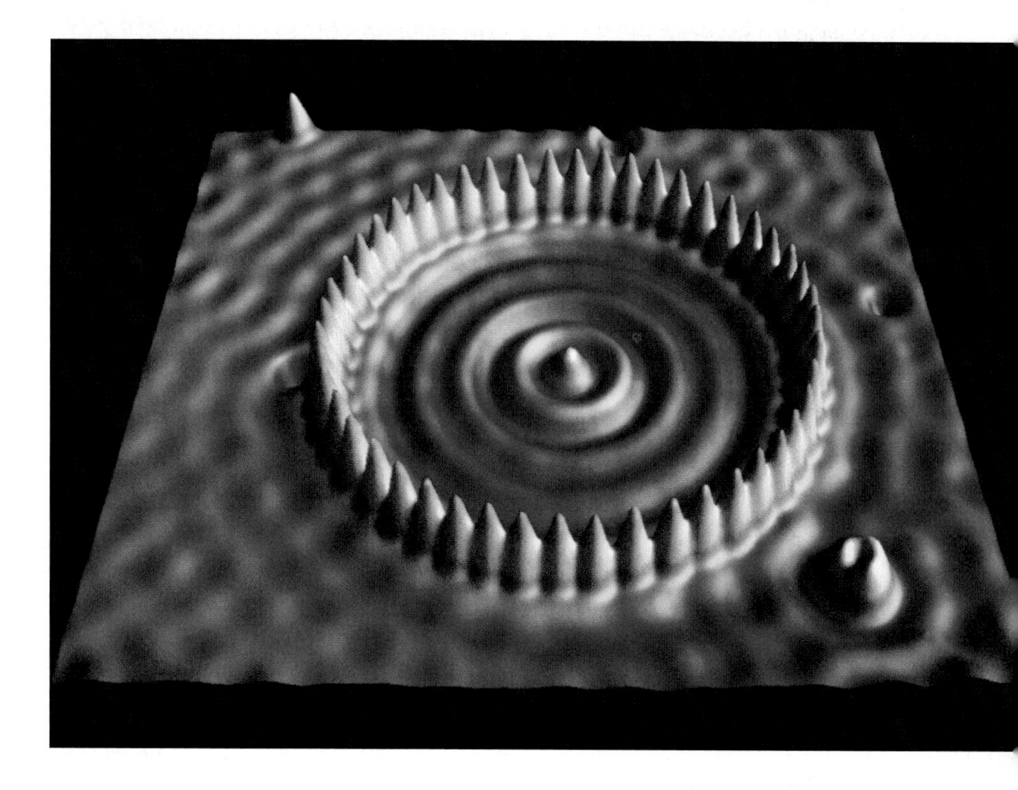

Just a few decades ago, if someone had claimed to have a photograph of an atom, people would have laughed. Today, a device called the scanning tunneling microscope (STM) routinely produces images such as shown in the photograph. The blue shapes represent iron atoms, neatly arranged on a copper surface.

The STM operates on a quantum-mechanical phenomenon called *tunneling*. Tunneling reflects some of the fundamental features of the subatomic realm: the probabilistic character of quantum processes and the wave nature of particles. These features explain how particles may turn up in places where, according to classical notions, they should not be.

In the 1920s, a new kind of physics, based on the synthesis of wave and quantum ideas, was introduced. This new theory, called **quantum mechanics**, combined the wave–particle duality of matter into a single consistent description. It revolutionized scientific thought and provides the basis of our understanding of phenomena that occur on the scale of molecular sizes and smaller.

In this chapter, some of the basic ideas of quantum mechanics are presented in order to show how they describe matter. Practical applications made possible by the quantum-mechanical view of nature, such as the electron microscope and magnetic resonance imaging (MRI), are also discussed.

28.1 Matter Waves: The de Broglie Hypothesis

OBJECTIVES: To (a) explain de Broglie's hypothesis, (b) calculate the wavelength of a matter wave, and (c) specify under what circumstances the wave nature of matter will be observable.

Note: These relationships are discussed in Section 26.4.

Since a photon travels at the speed of light, it must be treated relativistically as a particle with no mass. If this were not the case (that is, if it were to have mass), it would have infinite energy. This is because, at $v = c$, the relativistic factor γ becomes infinite, and its total energy $E = \gamma mc^2$ would be infinite unless $m = 0$. A photon's energy E and its momentum p are related by $p = E/c$. Recall from the Einstein relationship (Eq. 27.3) that the energy of a photon can be written in terms of associated wave frequency and wavelength as $E = hf = hc/\lambda$. Combining these two, we find that the magnitude of the momentum (p) of a photon is inversely related to the wavelength of the light by

$$p = \frac{E}{c} = \frac{hf}{c} = \frac{h}{\lambda} \quad \textit{(photon momentum)} \tag{28.1}$$

In the early part of the twentieth century, French physicist Louis de Broglie (1892–1987) suggested that there might be a symmetry between waves and particles. He conjectured that if light sometimes behaves like particles, then perhaps material particles such as electrons also might have wave properties. In 1924, de Broglie hypothesized that a moving particle has a wave associated with it. He proposed that the particle's wavelength is related to the magnitude of its momentum p (nonrelativistic) by an equation similar to that for a photon (Eq. 28.1), except that momentum is given by the expression for a particle with mass, $p = mv$ (see Chapter 6). The **de Broglie hypothesis** states the following:

> A (nonrelativistic) particle with momentum of magnitude p has a wave associated with it. This wave's wavelength is given by

$$\lambda = \frac{h}{p} = \frac{h}{mv} \quad \textit{(material particles only)} \tag{28.2}$$

The waves associated with moving particles were called **matter waves**, or, more commonly, **de Broglie waves**, and were thought to somehow influence or guide the particle's motion. The electromagnetic wave is the associated wave for a photon. However, de Broglie waves associated with particles such as electrons and protons are *not* electromagnetic waves. The de Broglie equation for the wavelength gives no clue as to the nature of the wave associated with a particle that has mass.

Needless to say, de Broglie's hypothesis met with great skepticism. The idea that the motions of photons were somehow governed by the wave properties of light seemed reasonable. But the extension of this idea to the motion of a particle *with mass* was difficult to accept. Moreover, there was no evidence at the time that particles exhibited any wave properties, such as interference and diffraction. (See Section 13.4 and Sections 24.1, 24.3, and 25.4.)

Note: The Bohr model is outlined in Section 27.4.

Note: Compare this standing wave with that in the discussion of standing waves in Section 13.5.

In support of his hypothesis, de Broglie showed how it could give an interpretation of the quantization of the angular momentum postulated by Bohr in his theory of the hydrogen atom. Recall that Bohr had to hypothesize that the angular momentum of the orbiting electron was quantized in integer multiples of $h/2\pi$ (Chapter 27). De Broglie argued that for a *free* particle, the associated wave would be a *traveling* wave. However, the *bound* electron of a hydrogen atom travels repeatedly in discrete circular orbits. The associated matter wave might therefore be expected to be a *standing* wave. For a standing wave to be produced, an integral number of wavelengths has to fit into the orbital circumference. The wave must reinforce itself constructively, much as a standing wave in a circular string (▶Fig. 28.1).

The circumference of a Bohr orbit of radius r_n is $2\pi r_n$, where n is the quantum number. De Broglie equated this circumference to an integer number of electron "wavelengths" in the following manner:

$$2\pi r_n = n\lambda \qquad \text{for } n = 1, 2, 3, \ldots$$

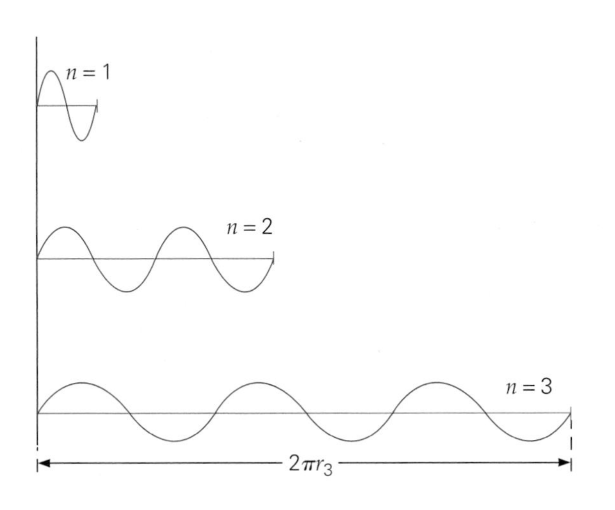

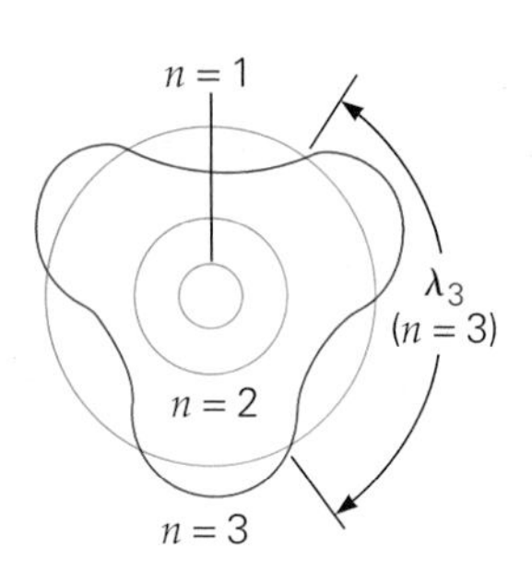

◀ **FIGURE 28.1 de Broglie waves and Bohr orbits** Similar to standing waves in a stretched string, de Broglie waves form circular standing waves on the circumferences of the Bohr orbits. The number of wavelengths in a particular orbit of radius r, shown here for $n = 3$, is equal to the principal quantum number of that orbit. (Not drawn to scale.)

Substituting for the wavelength λ from Eq. 28.2 yields

$$2\pi r_n = \frac{nh}{mv}$$

Now recall that, for a circular orbit, the angular momentum $L = mvr$; therefore we have

$$L_n = mvr_n = n\left(\frac{h}{2\pi}\right) \quad \text{for} \quad n = 1, 2, 3, \ldots$$

Note: See Eq. 8.14 for a reminder about angular momentum.

Thus, Bohr's angular-momentum quantization is equivalent to the de Broglie assumption that, in some way, the electron behaves like a wave.

For orbits other than those allowed by the Bohr theory, the de Broglie wave for the orbiting electron would not close on itself. This observation is consistent with the Bohr postulate of the electron being only in certain "allowed" orbits and implies that *the amplitude of the de Broglie wave might be related to the location of the electron.* This idea is actually a fundamental cornerstone of modern quantum mechanics, as will be seen.

If particles really have wavelike properties, why isn't their wave nature observed in everyday phenomena? It is because effects such as diffraction are significant only when the wavelength λ is on the order of the size of the object or the opening it meets. (See Section 24.3.) If λ is much smaller than these dimensions, then diffraction is negligible. The numbers in Examples 28.1 and 28.2 should convince you of the difference between the atomic world and our everyday world.

Example 28.1 ■ Should Ballplayers Worry about Diffraction? De Broglie Wavelength

A pitcher throws a fastball to the catcher at 40 m/s through a square strike-zone practice target. The square opening is cut in a sheet of canvas and is 50 cm on each side. If the ball's mass is 0.15 kg, (a) what is the wavelength of the de Broglie wave associated with the ball? (b) Should the catcher expect diffraction to occur as the ball passes through the opening to his mitt?

Thinking It Through. (a) The de Broglie relationship (Eq. 28.2) can be used to calculate the wavelength of the matter wave associated with the ball. (b) All that matters is whether the wavelength is much larger than, much smaller than, or approximately the same size as the opening.

Solution.

Given: $v = 40$ m/s, $d = 50$ cm $= 0.50$ m, $m = 0.15$ kg

Find: (a) λ (de Broglie wavelength) (b) whether diffraction is likely

(a) Equation 28.2 gives the wavelength of the baseball:

$$\lambda = \frac{h}{mv} = \frac{6.63 \times 10^{-34}\ \text{J}\cdot\text{s}}{(0.15\ \text{kg})(40\ \text{m/s})} = 1.1 \times 10^{-34}\ \text{m} \quad \textit{(extremely small)}$$

(b) For significant diffraction to occur when a wave passes through an opening, the wave must have a wavelength similar in size to that of the opening. Since $1.1 \times 10^{-34}\ \text{m} \ll 0.50$ m, the baseball travels straight into the catcher's mitt, with no noticeable wave diffraction.

(continues on next page)

Follow-Up Exercise. In this Example, (a) how fast would the ball have to be thrown for diffractive effects to become important? (b) At that speed, how long would the ball take to travel the 20 m to the plate? (Compare your answer with the age of the universe—about 15 billion years. Would someone watching this ball think that it is moving?) *(Answers to all Follow-Up Exercises are at the back of the text.)*

From Example 28.1, it is little wonder that the wave nature of matter isn't observed in our everyday lives. Notice, however, that a particle's de Broglie wavelength varies inversely with its mass and speed. So particles with very small masses traveling at low speeds might be another story, as the following Example shows.

Example 28.2 ■ A Whole Different Ball Game: De Broglie Wavelength of an Electron

Note: Single-slit diffraction is discussed in Section 24.3.

(a) What is the de Broglie wavelength of the wave associated with an electron that has been accelerated from rest through a potential of 50.0 V? (b) Compare your answer with the typical distance between atoms in a solid crystal, about 10^{-10} m. Would you expect diffraction to occur as these electrons pass between such atoms?

Thinking It Through. (a) We can use Eq. 28.2, but the electron's speed must first be calculated, using the given accelerating voltage. This computation involves consideration of energy conservation—the conversion of electric potential energy into kinetic energy. (b) For diffractive effects to be important, λ must be on the order of 10^{-10} m.

Solution. We list the data, as well as the mass of an electron, which can be found in Table 15.1.

Given: $V = 50.0$ V
$m = 9.11 \times 10^{-31}$ kg (from Table 15.1)

Find: (a) λ (de Broglie wavelength)
(b) whether diffraction is likely

(a) The magnitude of the potential energy lost by the electron is $|\Delta U_e| = eV$ and is equal to its gain in kinetic energy ($\Delta K = \frac{1}{2}mv^2$, because $K_o = 0$). Equating these two quantities enables us to calculate the speed:

$$\frac{1}{2}mv^2 = eV \quad \text{or} \quad v = \sqrt{\frac{2eV}{m}}$$

So,

$$v = \sqrt{\frac{2(1.60 \times 10^{-19}\,\text{C})(50.0\,\text{V})}{9.11 \times 10^{-31}\,\text{kg}}} = 4.19 \times 10^6\,\text{m/s}$$

Thus, the electron's de Broglie wavelength is

$$\lambda = \frac{h}{mv} = \frac{6.63 \times 10^{-34}\,\text{J}\cdot\text{s}}{(9.11 \times 10^{-31}\,\text{kg})(4.19 \times 10^6\,\text{m/s})} = 1.74 \times 10^{-10}\,\text{m}$$

(b) Since this result is the same order of magnitude as the opening between atoms, diffraction should be observed. Thus, passing electrons through a crystal lattice could prove de Broglie's hypothesis.

Follow-Up Exercise. In this Example, what would change if the particle were a proton? In other words, would the de Broglie wave of a proton be more or less likely to exhibit diffraction effects than would an electron under the same conditions? Explain, and give a numerical answer.

Problem-Solving Hint

In Example 28.2, if the accelerating voltage were changed, the calculation of v and λ would have to be repeated. This process would again involve the use of constants such as the electron's mass, its charge, and Planck's constant. In problems involving accelerating voltages, therefore, it is convenient to use the numerical values of m, e, and h to derive a nonrelativistic expression for the de Broglie wavelength of an electron when it is accelerated through a difference in potential, V. To begin, the kinetic energy is expressed in terms of momentum. Since $p = mv$, the kinetic energy, $\frac{1}{2}mv^2$, can also be written as $p^2/2m$. Then, by energy conservation,

$$\frac{p^2}{2m} = eV \quad \text{or} \quad p = \sqrt{2meV}$$

Thus, the de Broglie wavelength is

$$\lambda = \frac{h}{p} = \frac{h}{\sqrt{2meV}} = \sqrt{\frac{h^2}{2meV}}$$

Inserting the values of h, e, and m and rounding the result to three significant figures,

$$\lambda = \sqrt{\frac{1.50}{V\text{ (in volts)}}} \times 10^{-9}\text{ m} = \sqrt{\frac{1.50}{V\text{ (in volts)}}}\text{ nm} \quad \begin{array}{l}\textit{(nonrelativistic electron}\\ \textit{initially at rest; V in volts)}\end{array} \qquad (28.3)$$

If $V = 50.0$ V, then

$$\lambda = \sqrt{\frac{1.50}{50.0}} \times 10^{-9}\text{ m} = 0.173\text{ nm}$$

Note that this answer differs in the last digit from that found in Example 28.2, because there, the values for h, e, and m were rounded to three significant figures *before* the calculations were performed.

You may wish to derive a comparable formula for a proton, to use in solving similar problems. What quantities would have to be changed?

Note: This alternative de Broglie wavelength expression for electrons accelerated through a potential difference, in which V must be in volts, applies only to nonrelativistic electrons initially at rest.

In 1927, two physicists in the United States, C. J. Davisson and L. H. Germer, used a crystal to diffract a beam of electrons, thereby demonstrating a wavelike property of particles. A single crystal of nickel was cut to expose a spacing of $d = 0.215$ nm between the planes of atoms. When a beam of electrons was directed normally onto the crystal face, a maximum in the intensity of scattered electrons was observed at an angle of 50° relative to the surface normal (▸Fig. 28.2). The scattering was most intense for an accelerating potential of 54.0 V.

According to wave theory, constructive interference due to waves reflected from two lattice planes (a distance d apart) should occur at certain scattering angles θ. The theory predicts that the first-order maximum should be observed at an angle given by

$$\sin\theta = \frac{\lambda}{d}$$

This condition for the experimental setup in Fig. 28.2 requires a wavelength of

$$\lambda = d\sin\theta = (0.215\text{ nm})\sin 50^\circ = 0.165\text{ nm}$$

Using Eq. 28.3 to determine the de Broglie wavelength of the electrons after being accelerated through 54.0 V, we obtain

$$\lambda = \sqrt{\frac{1.50}{V}}\text{ nm} = \sqrt{\frac{1.50}{54.0}}\text{ nm} = 0.167\text{ nm}$$

The agreement was well within the experimental uncertainty. The Davisson–Germer experiment gave convincing proof of de Broglie's matter wave hypothesis. A practical application of the wavelike properties of electrons is discussed in Insight 28.1 on The Electron Microscope on page 883.

Note: Diffraction from crystal lattices is discussed in Section 24.3.

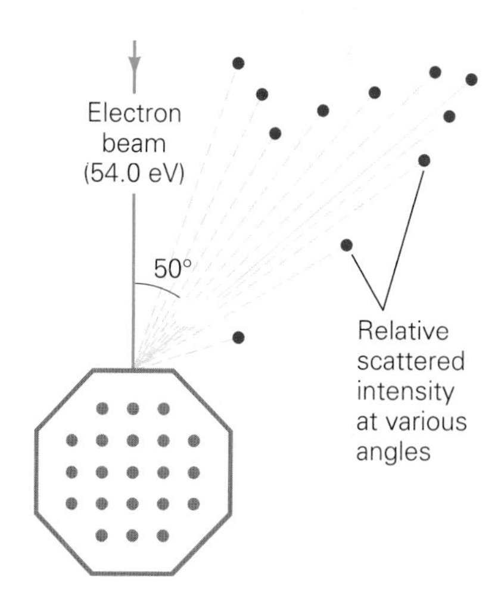

▲ **FIGURE 28.2 The Davisson–Germer experiment** When a beam of electrons of kinetic energy 54.0 eV is incident on the face of a nickel crystal, a maximum in the scattering intensity is observed at an angle of 50°.

28.2 The Schrödinger Wave Equation

OBJECTIVES: To understand qualitatively (a) the reasoning that underlies the Schrödinger wave equation, and (b) the equation's use in finding particle wave functions.

De Broglie's hypothesis predicts that moving particles have associated waves that somehow govern their behavior. However, it does not tell us the *form* of these waves, only their wavelengths. To have a useful theory, an equation that will give the mathematical form of these matter waves is needed. We also need to know how these waves govern particle motion. In 1926, Erwin Schrödinger, an Austrian physicist, presented a general equation that describes the de Broglie matter waves and their interpretation.

A traveling de Broglie wave varies with location and time in a similar way to everyday wave motion (Section 13.2). The de Broglie wave is denoted by ψ (the Greek letter psi, pronounced "sigh") and is called the **wave function ψ**, and is associated with the particle's kinetic, potential, and total energy. Recall that for a conservative mechanical system (Section 5.5), the total mechanical energy E, which is the sum of the kinetic and potential energies, is a constant; that is $K + U = E$. Schrödinger proposed a similar equation for the de Broglie matter waves, involving the wave function ψ. **Schrödinger's wave equation*** has the general form

$$(K + U)\psi = E\psi \tag{28.4}$$

Equation 28.4 can be solved for ψ, but doing so involves complex mathematical operations well beyond the scope of this text. For us, the more important question is related to the physical significance of ψ. During the early development of quantum mechanics, it was not at all clear how ψ should be interpreted. After much thought and investigation, Schrödinger and his colleagues hypothesized the following:

Note: More precisely, the square of the absolute value of the wave-function solution to Schrödinger's equation, $|\psi|^2$, gives the probability of finding a particle at a location.

> The square of a particle's wave function is proportional to the probability of finding that particle at a given location.

The interpretation of ψ^2 as a *probability* altered the idea that the electron in a hydrogen atom could be found only in orbits at discrete distances from the nucleus, as described in the Bohr theory. When the Schrödinger equation was solved for the hydrogen atom, there was a nonzero probability of finding the electron at almost any distance from the nucleus. The relative probability of finding an electron [in the ground state ($n = 1$)] at a given distance from the proton is shown in ▾Fig. 28.3a.

The *maximum* probability coincides with the Bohr radius of 0.0529 nm, but it is possible that, for instance, the electron could even be inside the nucleus. Notice that, although the ground-state wave function exists for distances well beyond 0.20 nm from the proton, there is little chance of finding an electron beyond this distance. The probability density distribution gives rise to the idea of an *electron cloud* around the nucleus (Fig. 28.3b). This cloud is actually a probability density cloud, meaning that the electron can be found in many different locations with varying probabilities.

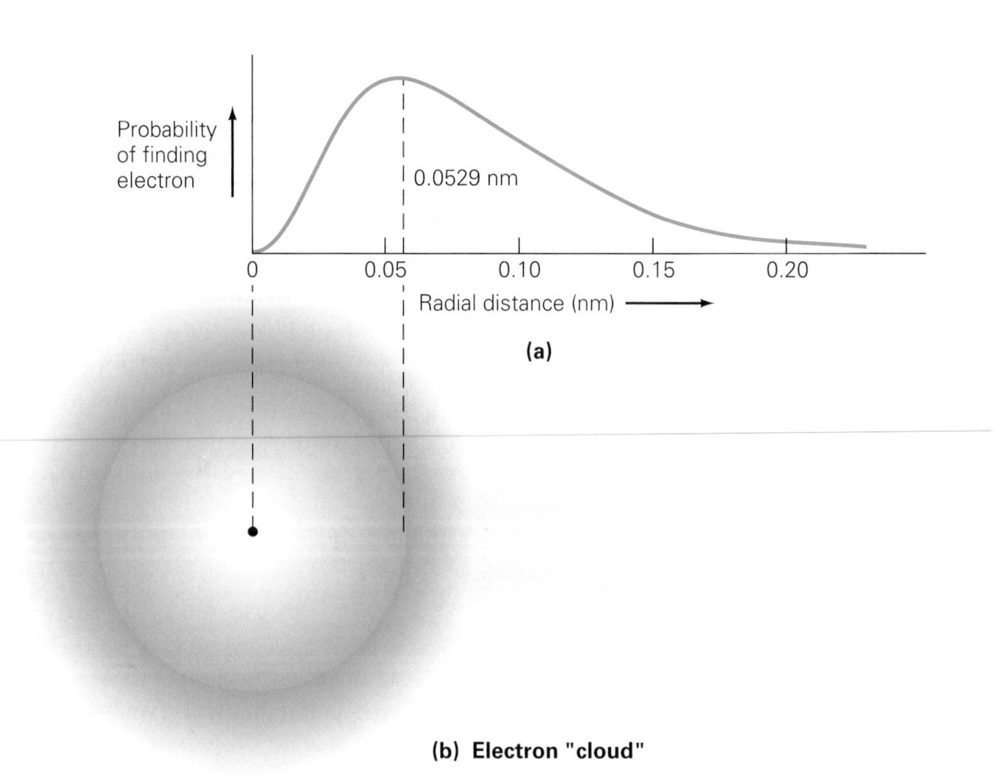

▶ **FIGURE 28.3 Electron probability for hydrogen-atom orbits (a)** The square of the wave function is the probability of finding the hydrogen electron at a particular location. Here, it is assumed to be in the ground state ($n = 1$), and its probability is plotted as a function of radial distance from the proton. The electron has the greatest probability of being at a distance of 0.0529 nm, which matches the radius of the first Bohr orbit. **(b)** The probability distribution gives rise to the idea of an electron probability cloud around the nucleus. The cloud's density reflects the probability density.

*Although Eq. 28.4 looks like a multiplication of $E = K + U$ by ψ, it is much more complex. For example, K is no longer $\frac{1}{2}mv^2$, but is instead replaced by a quantity (called an *operator*) that enables us to *extract* the kinetic energy from ψ.

INSIGHT 28.1 THE ELECTRON MICROSCOPE

The de Broglie hypothesis led to the development of an important practical application—the *electron microscope*. As the Davisson–Germer experiment demonstrated, under the right conditions, electrons experience diffraction, as do light waves. Does this mean that electrons might also be focused like light waves? The answer is yes. In fact, electron "waves" can be focused to form images, though the focusing mechanism is different than those used with light in a visible light microscope. As Example 28.2 shows, moving electrons have very short de Broglie wavelengths. With such short wavelengths, greater magnification and finer resolution can be obtained than with any light microscope. Recall the inverse relationship between resolving power, or the ability to see details, and wavelength—a smaller wavelength means greater resolving power (See Section 25.4.) In fact, the resolving power of standard electron microscopes is on the order of a few nanometers, which is only about 10 times larger than atomic sizes.

Knowledge of how to focus electron beams by using magnetic coils permitted the construction of the first electron microscope in Germany in 1931. In a *transmission electron microscope* (TEM), an electron beam is directed onto a very thin specimen. Different numbers of electrons pass through different parts of the specimen, depending on its structure. The transmitted beam is then brought into focus by a magnetic objective coil. The general components of electron and light microscopes are analogous (Fig. 1), but an electron microscope must be housed in a high-vacuum chamber to prevent the electrons from colliding with air molecules. As a result, an electron microscope looks nothing like a light microscope (Fig. 2). A normal light-microscope image (Fig. 3a) is limited to a magnification of about 2000×, while magnifications up to 100 000× can be achieved with an electron microscope (Fig. 3b).

Another difference between light microscopes and electron microscopes is that the final lens in an electron microscope, called the *projector coil*, has to project a real image onto a fluorescent screen or photographic film, since the eye cannot perceive an electron image directly. Because specimens for transmission electron microscopy must be very thin, special techniques for specimen preparation are required. These methods allow for specimen sections to be sliced as thin as 10 nm (only about 100 atoms thick).

The surfaces of thicker objects can be examined by the *reflection* of the electron beam from the surface. This process is accomplished with the *scanning electron microscope* (SEM). A beam spot is scanned across the specimen by means of deflecting coils, much as is done in a television tube. Surface irregularities cause directional variations in the intensity of the reflected electrons, which gives contrast to the image. Through such techniques, a scanning electron microscope gives pictures with a remarkable three-dimensional quality, such as those in Fig. 3c.

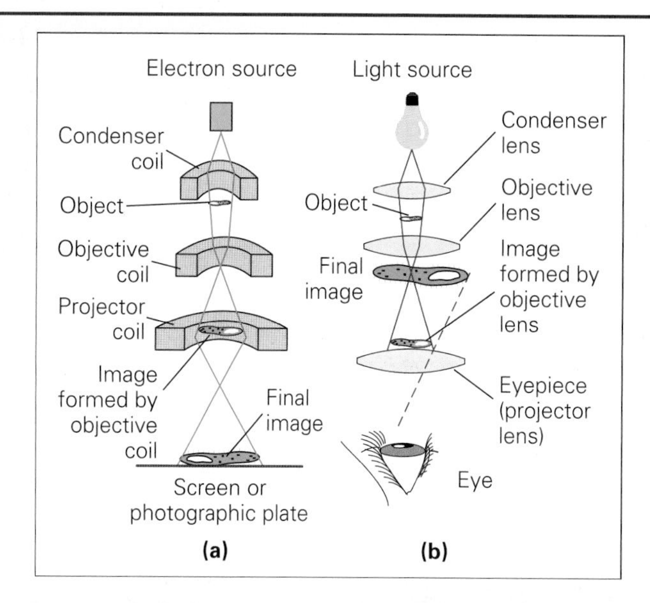

FIGURE 1 Electron and light microscopes A comparison of the elements of **(a)** an electron microscope and **(b)** a light microscope. The light microscope is drawn upside down for a better comparison.

FIGURE 2 An electron microscope The microscope is housed in a cylindrical vacuum chamber, to the left.

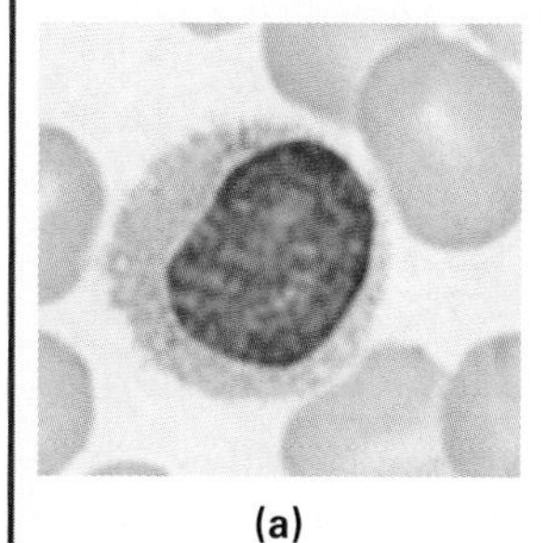
(a)

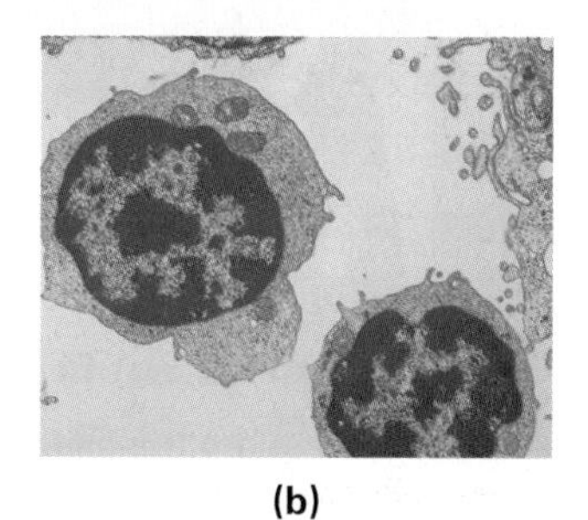
(b)

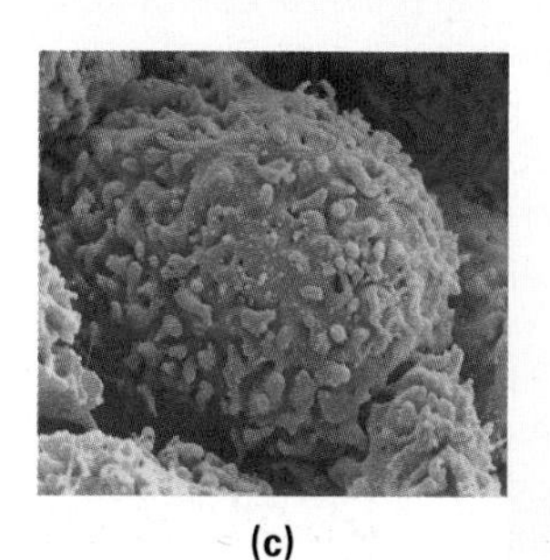
(c)

FIGURE 3 Lymphocytes (white blood cells) Images produced by **(a)** a light microscope, **(b)** a transmission electron microscope (TEM), and **(c)** a scanning electron microscope (SEM). Notice the enormous increase in magnification produced by the electron microscopes.

An interesting quantum-mechanical result that runs counter to our everyday experiences is *tunneling*. In classical physics, there are regions forbidden to particles by energy considerations. These regions are areas where a particle's potential energy would be greater than its total energy. Classically, the particle is not allowed in such regions because it would have a negative kinetic energy there ($E - U = K < 0$), which is impossible. In such situations, we say that the particle's location is limited by a *potential-energy barrier*.

In certain instances, however, quantum mechanics predicts a small, but finite, probability of the particle's wave function penetrating the barrier and thus of the particle being found on the other side of the barrier. Thus, there is a certain probability (which is practically zero for everyday objects) of the particle "tunneling" through the barrier, especially on the atomic level, where the wave nature of particles is exhibited. Such tunneling forms the basis of the scanning tunneling microscope (STM; see the accompanying Insight 28.2 on The Scanning Tunneling Microscope), which creates images with a resolution on the order of the size of a single atom. Barrier penetration also explains certain nuclear decay processes (Chapter 29).

INSIGHT 28.2 THE SCANNING TUNNELING MICROSCOPE (STM)

The *scanning tunneling microscope* (STM) was invented in the late 1970s and promptly revolutionized the field of surface physics. STMs use quantum-mechanical tunneling to produce stunning images of atoms, such as this chapter's opening photograph.

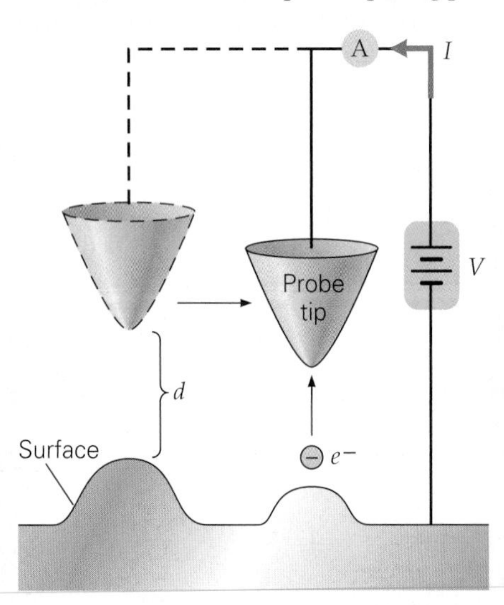

FIGURE 1 Schematic representation of the STM The tip of a probe is moved across the contours of a specimen's surface. A small voltage applied between the probe and the surface causes electrons to tunnel across the vacuum gap. The tunneling current is extremely sensitive to the separation distance between the probe tip and the surface. A feedback circuit (not shown) moves the probe up and down so as to keep constant the tunneling current as measured by the ammeter shown as (A). Thus, when the probe is scanned across the sample, surface features smaller than atoms (up to 0.001 nm vertically) can be detected. When the probe is passed over the surface in many successive and nearby parallel paths, the resulting data can be processed to produce three-dimensional images.

The STM produces atomic-sized images by positioning its sharp tip very close (about 1 nm) to a surface. A voltage is applied between the tip and the surface, causing electron tunneling through the vacuum gap (Fig. 1). This tunneling current is extremely sensitive to the separation distance. A feedback circuit monitors this current and moves the probe vertically to keep the current constant. The separation distance is digitized, recorded, and processed by computers for display. When the probe is passed over the surface of the specimen in successive nearby parallel movements, a three-dimensional image of the surface can be displayed. Typical vertical resolution for STMs is 0.001 nm, and lateral resolutions are about 0.1 nm. Because the diameters of individual atoms are on the order of several tenths of a nanometer, STMs allow detailed images of atoms to be created. Figure 2 shows the result of a surface scan on gallium arsenide, a semiconductor material.

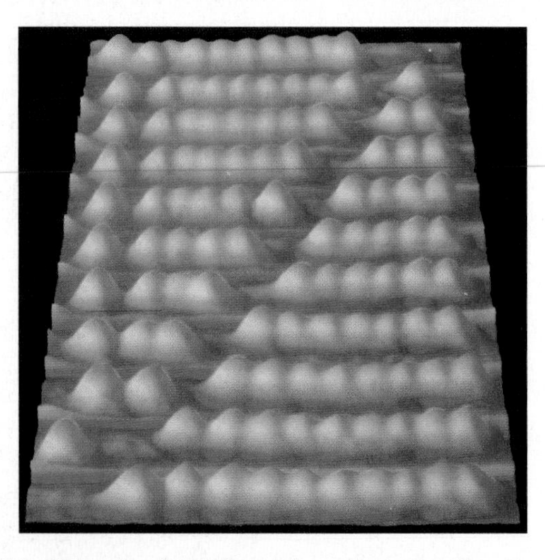

FIGURE 2 Scanning tunneling microscopy results Atoms of the semiconductor gallium arsenide.

28.3 Atomic Quantum Numbers and the Periodic Table

OBJECTIVE: To understand the structure of the periodic table in terms of quantum-mechanical electron orbits and the Pauli exclusion principle.

The Hydrogen Atom

When the Schrödinger equation was solved for the hydrogen atom, the results predicted the energy levels to be the same as those from the Bohr theory (Section 27.4). Recall that the Bohr-model energy values depended only on the **principal quantum number** n. However, in addition, the solution to the Schrödinger equation gave two other quantum numbers, designated as ℓ and m_ℓ. Three quantum numbers are needed because the electron can move in three dimensions.

The quantum number ℓ is called the **orbital quantum number**. It is associated with the orbital angular momentum of the electron. For each value of n, the ℓ quantum number has integer values from zero up to a maximum value of $n - 1$. For example, if $n = 3$, the three possible values of ℓ are 0, 1, and 2. ▸Figure 28.4 shows three orbits with different angular momenta, but the same energy. The number of different ℓ values for a given n value is equal to n. In the hydrogen atom, the energy of the electron depends only on n. Thus, orbits with the same n value, but different ℓ values, have the same energy and are said to be *degenerate*.

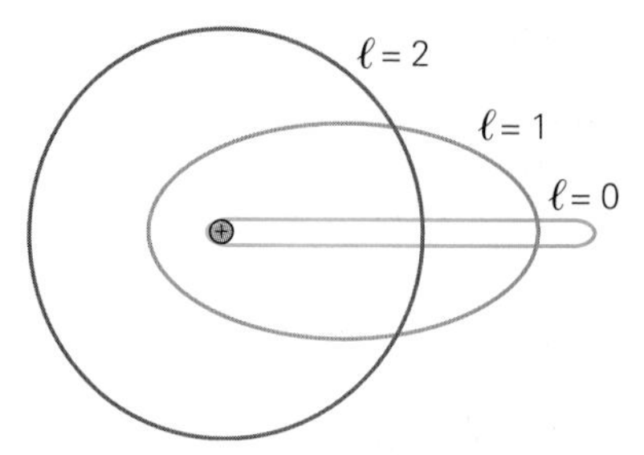

▲ **FIGURE 28.4 The orbital quantum number (ℓ)** The orbits of an electron are shown for the second excited state in hydrogen. For the principal quantum number $n = 3$, there are three different values of angular momentum (corresponding to the three differently shaped orbits and three different values of ℓ), but they all have the same total energy. The circular orbit has the maximum angular momentum; the narrowest orbit would actually pass through the proton classically and thus has zero angular momentum.

The quantum number m_ℓ is called the **magnetic quantum number**. The name originated from experiments in which an external *magnetic* field was applied to a sample. They showed that a particular energy level (one with given values of n and ℓ) of a hydrogen atom actually consists of several orbits that differ slightly in energy *only when in a magnetic field*. Thus, in the absence of a field, there was additional energy degeneracy. Clearly, there must be more to the description of the orbit than just n and ℓ. The quantum number m_ℓ was introduced to enumerate the number of levels that existed for a given orbital quantum number ℓ. Under zero-magnetic-field conditions, the energy of the atom does not depend on either of these quantum numbers.

The magnetic quantum number m_ℓ is associated with the orientation of the orbital angular-momentum vector $\vec{\mathbf{L}}$ in space (▸Fig. 28.5). If there is no external magnetic field, then all orientations of $\vec{\mathbf{L}}$ have the same energy. For each value of ℓ, m_ℓ is an integer that can range from zero to $\pm\ell$. That is, $m_\ell = 0, \pm1, \pm2, \ldots, \pm\ell$. For example, an orbit described by $n = 3$ and $\ell = 2$ can have m_ℓ values of $-2, -1, 0, +1$, and $+2$. In this case, the orbital angular-momentum vector $\vec{\mathbf{L}}$ has five possible orientations, all with the same energy if no magnetic field is present. In general, for a given value of ℓ, there are $2\ell + 1$ possible values of m_ℓ. For example, with $\ell = 2$, there are five values of m_ℓ, since $2\ell + 1 = (2 \times 2) + 1 = 5$.

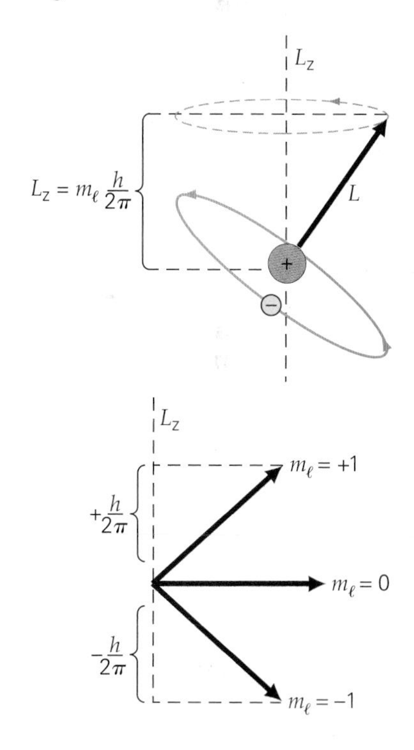

▲ **FIGURE 28.5 The magnetic quantum number m_ℓ** Here, $\vec{\mathbf{L}}$ is the vector angular momentum associated with the orbit of the electron. Because the energy of an orbit is independent of the orientation of its plane, orbits with the same ℓ that differ only in m_ℓ have the same energy. There are $2\ell + 1$ possible orientations for a given ℓ. The value of m_ℓ tells the *component* of the angular-momentum vector in a given direction, as shown (below) for $\ell = 1$.

However, this finding was not the end of the story. The use of high-resolution optical spectrometers showed that each emission line of hydrogen is, in fact, two very closely spaced lines. Thus, each emitted wavelength is actually two. This splitting is called *spectral fine structure*. Hence, a fourth quantum number was necessary in order to describe each atomic state completely. This number is called the **spin quantum number m_s** of the electron. It is associated with the *intrinsic* angular momentum of the electron. This property, called *electron spin*, is sometimes described by analogy to the angular momentum associated with a spinning object (▾Fig. 28.6). Because each energy level is split into only two levels, the electron's intrinsic angular momentum (or, more simply, its "spin") can possess only two orientations, called "spin up" and "spin down."

Thus, the fine structure of an atom's energy levels results from the electron spin's having two orientations with respect to the atom's *internal* magnetic field, produced by the electron's orbital motion. Analogous to a magnetic moment (such as a compass) in a magnetic field, the atom possesses slightly less energy when its electron's spin is "lined up" with the field than when the spin is aligned opposite to the field. However, keep in mind that spin is fundamentally a purely quantum-mechanical concept; it is not really analogous to a spinning top. This is because, as far as we know now, the electron possesses no size—that is, it is truly a point particle.

TABLE 28.1 Quantum Numbers for the Hydrogen Atom

Quantum Number	*Symbol*	*Allowed Values*	*Number of Allowed Values*
Principal	n	$1, 2, 3, \ldots$	no limit
Orbital angular momentum	ℓ	$0, 1, 2, 3, \ldots, (n-1)$	n (for each n)
Orbital magnetic	m_ℓ	$0, \pm 1, \pm 2, \pm 3, \ldots, \pm \ell$	$2\ell + 1$ (for each ℓ)
Spin	m_s	$\pm\frac{1}{2}$	2 (for each m_ℓ)

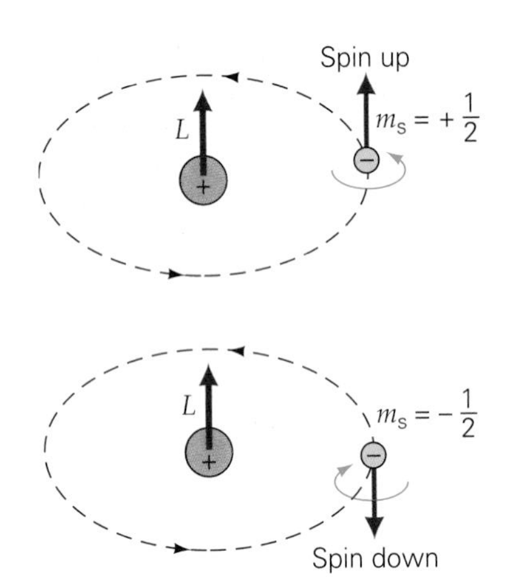

▲ **FIGURE 28.6 The electron-spin quantum number (m_s)** The electron spin can be either up or down. Electron spin is strictly a quantum mechanical property and should not be identified with the physical spin of a macroscopic body, except with respect to conceptual reasoning.

Thus, for the excited state of hydrogen with $n = 3$ and $\ell = 2$, each value of m_ℓ also has two possible spin orientations. For example, when $m_\ell = +1$, there are two possible sets of the four quantum numbers: $n = 3$, $\ell = 2$, and $m_\ell = +1$, with $m_s = +\frac{1}{2}$; and $n = 3$, $\ell = 2$, and $m_\ell = +1$, with $m_s = -\frac{1}{2}$. Both sets would have nearly the same energy. The orbit's energy is therefore almost independent of the electron's spin direction. This condition results in yet another (approximate) energy degeneracy. In summary, the energy of the various states of the hydrogen atom are, to a very high degree, determined solely by the primary quantum number n.

The four quantum numbers for the hydrogen atom are summarized in Table 28.1. Other particles, such a protons, neutrons, and composites of them called *atomic nuclei* also possess spin. A particularly useful property of the spin of a proton, and of atomic nuclei in general, is discussed in Insight 28.3 on Magnetic Resonance Imaging (MRI) on page 888.

Multielectron Atoms

The Schrödinger equation cannot be solved exactly for atoms with more than one electron (multielectron atoms). However, a solution can be found, to a workable approximation, in which each electron occupies a state characterized by a set of quantum numbers similar to those for hydrogen. Because of the repulsive forces between the electrons, the description of a multielectron atom is much more complicated. For one, the energy depends not only on the principal quantum number n, but also on the orbital quantum number ℓ. This condition gives rise to a subdivision (or "splitting") of the degeneracy seen in hydrogen atoms. In multielectron atoms, the energies of the atomic levels generally depend on all four quantum numbers.

Note: The n quantum number designates orbital shells; the ℓ quantum number designates orbital subshells.

It is common to refer to all the electron levels that share the same n value as making up a **shell** and to all the electron levels that share the same ℓ values within that shell as **subshells**. That is, electron levels with the same n value are said to be "in the same shell." Similarly, electrons with the same n and ℓ values are said to be "in the same subshell."

The ℓ subshells can be designated by integers; however, it is common to use letters instead. The letters $s, p, d, f, g, \ldots$ correspond to the values of $\ell = 0, 1, 2, 3, 4, \ldots$ respectively. After f, the letters go alphabetically (Table 28.2).

TABLE 28.2 Subshell Designations

Value of ℓ	*Designation*
$\ell = 0$	s
$\ell = 1$	p
$\ell = 2$	d
$\ell = 3$	f
$\ell = 4$	g
$\ell = 5$	h
...	...

Because in multielectron atoms an electron's energy depends on both n and ℓ, both quantum numbers are used to label atomic energy levels (a shell and subshell, respectively). The labeling convention is as follows: n is written as a number, followed by the letter that stands for the value of ℓ. For example, $1s$ denotes an energy level with $n = 1$ and $\ell = 0$; $2p$ is for $n = 2$ and $\ell = 1$; $3d$ is for $n = 3$ and $\ell = 2$; and so on. Also, it is common to refer to the m_ℓ values as representing *orbitals*. For example, a $2p$ energy level has three orbitals, corresponding to the m_ℓ values of -1, 0, and $+1$ (because $\ell = 1$).

The hydrogen atom energy levels are not evenly spaced, but do increase sequentially. In multielectron atoms, not only are the energy levels unevenly spaced, but their numerical sequence is also, in general, out of order. The shell–subshell (n–ℓ notation) energy-level sequence for a multielectron atom is shown in ▸Fig. 28.7a. Notice, for example, that the $4s$ level is, energywise, below the $3d$ level. Such variations result in part from electrical forces between the electrons. Furthermore, note

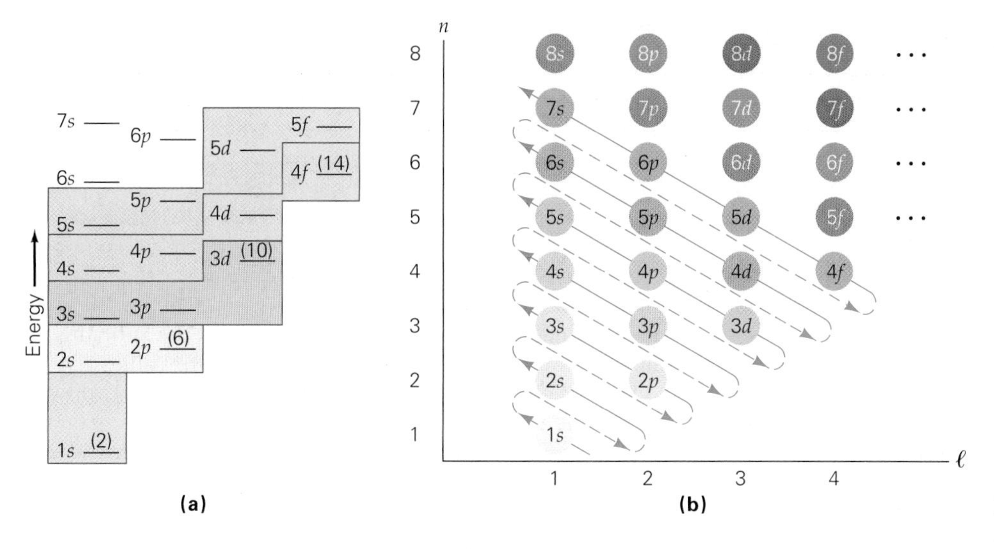

▲ **FIGURE 28.7 Energy levels of a multielectron atom** **(a)** The shell–subshell (n–ℓ) sequence shows that the energy levels are not evenly spaced and that the sequence of energy levels has numbers out of order. For example, the 4*s* level lies below the 3*d* level. The maximum number of electrons for a subshell, $2(2\ell + 1)$, is shown in parentheses on representative levels. (The vertical energy differences may not be drawn to scale.) **(b)** A convenient way to remember the energy-level order of a multielectron atom is to list the n-versus-ℓ values as shown here. The diagonal lines then give the energy levels in ascending order.

that the electrons in the outer orbits are "shielded" from the attractive force of the nucleus by the electrons that are closer to the nucleus. For example, consider the highly elliptical orbit (Fig. 28.4) of an electron in the 4*s* ($\ell = 0$) orbit. The electron clearly spends more time near the nucleus, and hence is more tightly bound, than if it were in the more circular 3*d* ($\ell = 2$) orbit. A convenient way to remember the order of the levels is given in Fig. 28.7b.

The ground state of a multielectron atom has some similarities to that of the hydrogen atom, with its one electron in the 1*s*, or lowest energy, level. In a multielectron atom, the ground state still has the lowest total energy, but is a combination of energy levels in this case. That is, the electrons are in the lowest possible energy levels. But to identify how the electrons fill these levels, we must know how many electrons can occupy a particular energy level. For example, the lithium (Li) atom has three electrons. Can they all be in the 1*s* level? As we shall see, the answer, given by the Pauli exclusion principle, is no.

The Pauli Exclusion Principle

Exactly how the electrons of a multielectron atom distribute themselves in the ground-state energy levels is governed by a principle set forth in 1928 by the Austrian physicist Wolfgang Pauli. The **Pauli exclusion principle** states the following:

> No two electrons in an atom can have the same set of quantum numbers (n, ℓ, m_ℓ, m_s). That is, no two electrons in an atom can be in the same quantum state.

This principle limits the number of electrons that can occupy a given energy level. For example, the 1*s* ($n = 1$ and $\ell = 0$) level can have only one m_ℓ value, $m_\ell = 0$, along with only two m_s values, $m_s = \pm\frac{1}{2}$. Thus, there are only two unique sets of quantum numbers (n, ℓ, m_ℓ, m_s) for the 1*s* level—$\left(1, 0, 0, +\frac{1}{2}\right)$ and $\left(1, 0, 0, -\frac{1}{2}\right)$—so only two electrons can occupy the 1*s* level. If this situation is indeed the case, then we say that such a shell is *full*; all other electrons are excluded from it by Pauli's principle. Thus, for a Li atom, with three electrons, the third electron must occupy the next higher level (2*s*) when the atom is in the ground state. This case is illustrated in ▸Fig. 28.8, along with the ground-state energy levels for some other atoms.

▼ **FIGURE 28.8 Filling subshells** The electron subshell distributions for several unexcited atoms (in their ground state) according to the Pauli exclusion principle. Because of spin, any *s* subshell can hold a maximum of two electrons, and any *p* subshell can hold a maximum of six electrons. What can you say about the *d* subshells?

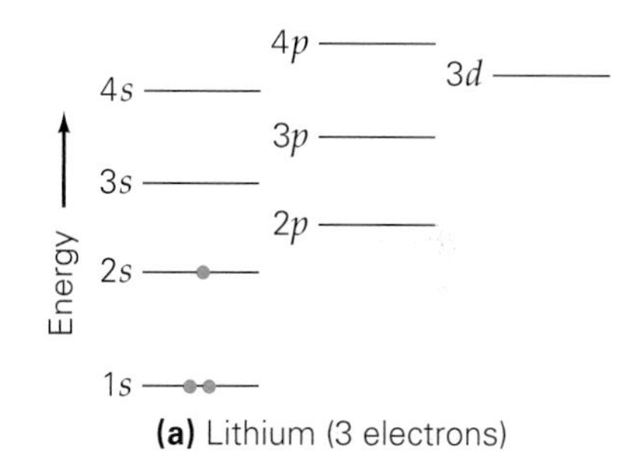

(a) Lithium (3 electrons)

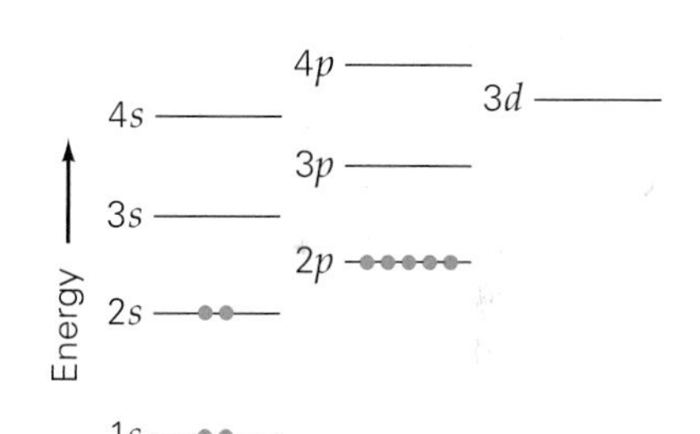

(b) Fluorine (9 electrons)

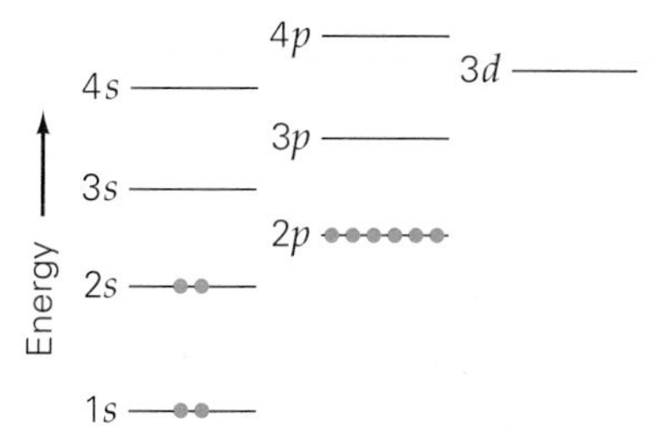

(c) Neon (10 electrons)

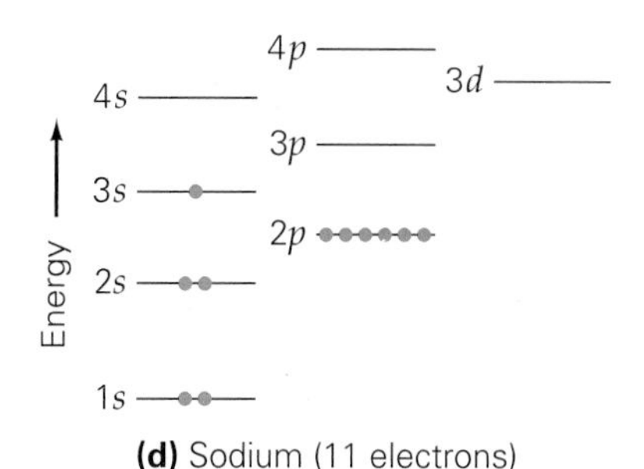

(d) Sodium (11 electrons)

INSIGHT 28.3 MAGNETIC RESONANCE IMAGING (MRI)

Magnetic resonance imaging, or MRI, has become a common and important medical technique for the noninvasive examination of the human body (Fig. 1). Originally known as NMR (for *nuclear magnetic resonance*), MRI is based on the quantum-mechanical concept of spin.

A current-carrying loop in a magnetic field experiences a torque that tends to orient the loop's magnetic moment parallel to the field, much like a compass (Chapter 20). Electrons possess an intrinsic angular momentum called *spin*. This condition gives rise to a *spin magnetic moment*, which aligns in a similar manner to the compass. When atoms are placed in a magnetic field, each of their energy levels is split into two levels (for the two possible spin orientations—parallel and antiparallel to the magnetic field), each with a slightly different energy, which results in the fine structure in the atoms' emission spectra.

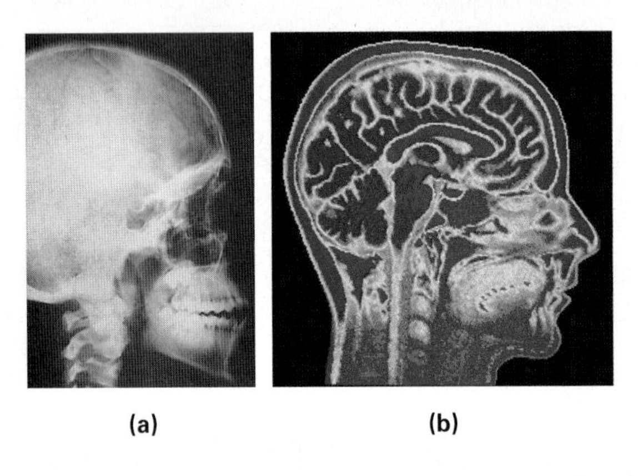

FIGURE 1 Diagnostic images (a) An X-ray of a human head. **(b)** A magnetic resonance image (MRI) of a human head. The amount of detail captured, especially in the soft tissues of the brain, makes such images very useful for medical diagnosis.

Nuclei exhibit similar spin effects when placed in magnetic fields. Because neutrons and protons possess spin, nuclei, which are composed of neutrons and protons, also possess magnetic moments. To understand the basics of MRI, let us concentrate on the simplest atom, hydrogen, with a nucleus composed of a single proton. The magnetic resonance of hydrogen is commonly measured in an MRI apparatus, since hydrogen is the most abundant element in the human body. The spin angular momentum of the proton can take on only two values, similar to that of electron spin, called "spin up" and "spin down." These values describe the orientation of the proton's magnetic moment relative to the direction of an external magnetic field (Fig. 2a). With spin up, the magnetic moment is parallel to the field, with spin down, it points opposite to the field. The spin-down orientation has a slightly higher energy, and the energy difference ΔE between the two levels is proportional to the magnitude of the magnetic field. The transition to the higher energy level can be made by allowing the proton to absorb a photon with energy equal to ΔE.

In a typical MRI apparatus, the sample (usually a region of the human body) is placed in a magnetic field $\vec{\mathbf{B}}$. The magnitude of the field determines the energy E (or light frequency f) of the photon needed to cause a transition. This is because the photon energy absorption requires that $E = hf = \Delta E$, which is proportional to the strength of the magnetic field, B. The photons that trigger this transition are supplied by a pulsed beam of radio-frequency (RF) radiation applied to the sample. If the frequency of the radiation is adjusted so that the photon energy equals the energy level difference, many nuclei will absorb the photon energy and be excited into the higher energy level. The frequency that creates such excitation is known as the *resonance frequency* ($f = \Delta E/h$), hence the name magnetic *resonance* imaging.

Figure 3 shows a typical MRI device. Large coils produce the magnetic field. (Notice the solenoid arrangement in Fig. 3a.) Other coils produce the RF signals ("RF photons") that cause the nuclei to

Integrated Example 28.3 ■ The Quantum Shell Game: How Many States?

(a) How does the number of possible electron states compare between the $2p$ and the $4p$ subshells: (1) The number of states in the $2p$ subshell is greater than that in the $4p$ subshell; (2) the number of states in the $2p$ subshell is less than that in the $4p$ subshell; or (3) the number of states in the $2p$ subshell is the same as that in the $4p$ subshell? (b) Compare the number of possible electron states in the $3p$ subshell to the number in the $4d$ subshell.

(a) Conceptual Reasoning. The term *subshell* refers to the states that share the same ℓ quantum number within a shell. That is, they all have the same principal quantum number n. All that matters is that their ℓ values are the same. Therefore, the number of states in the two subshells is the same, so the correct answer is (3).

(b) Thinking It Through. This part is a matter of following the quantum-mechanical counting rules. Also remember that each letter stands for a specific value of ℓ. The p level means that $\ell = 1$, and the d level means that $\ell = 2$. In each subshell, we replace the letter designation for ℓ by its number. Thus, the data given are as follows:

Given: $3p$ level means $n = 3$ and $\ell = 1$
$4d$ level means $n = 4$ and $\ell = 2$

Find: the number of quantum states in the $3p$ subshell as compared with the number in the $4d$ subshell

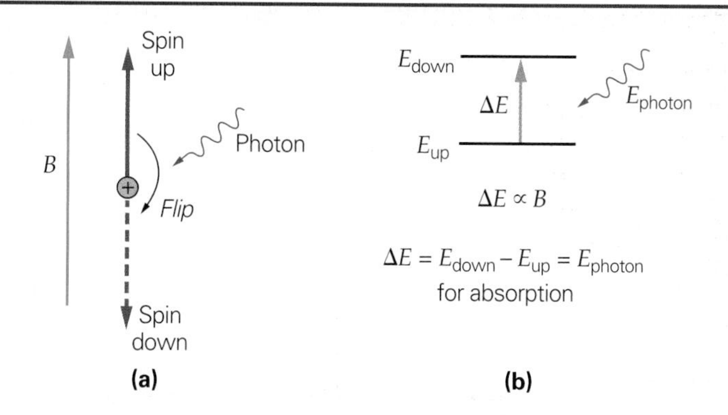

FIGURE 2 Nuclear spin (a) In a uniform magnetic field, the spin angular momentum, or spin magnetic moment, of a hydrogen nucleus (a proton) can have only two values—called "spin up" and "spin down," in reference to the direction of the external magnetic field. **(b)** This condition gives rise to two energy levels for the nucleus. Energy must be absorbed in order to "flip" the proton spin.

"flip their spin"—that is, to be excited from the lower to the upper energy level. The resulting absorption of energy is detected, as is emitted radiation coming from a return transition to the lower state. Regions that produce the greatest absorption (or re-emission) are those with the greatest concentration of the particular nucleus to which the apparatus is "tuned" by the choice of B and f. (Other nuclei besides hydrogen have their own characteristic intrinsic spins and can be imaged by tuning the frequency to match their resonance frequency.) Images are produced by means of computerized tomography, similar to that used in X-ray CT scans (see Section 20.4). The result is a two- or three-dimensional image (Fig. 1b) that can provide a great deal of diagnostic medical information.

Although a variety of atomic nuclei exhibit nuclear magnetic resonance, most MRI work is done with hydrogen, because body tissues vary in water content. For example, muscle tissue has more water than does fatty tissue, so there is a distinct contrast in the radiation intensity of the two materials. Similarly, fatty deposits in blood vessels are perceived distinctly from the tissue of the vessel walls. A tumor with a water content different than that of the surrounding tissue would also show up in an MRI image.

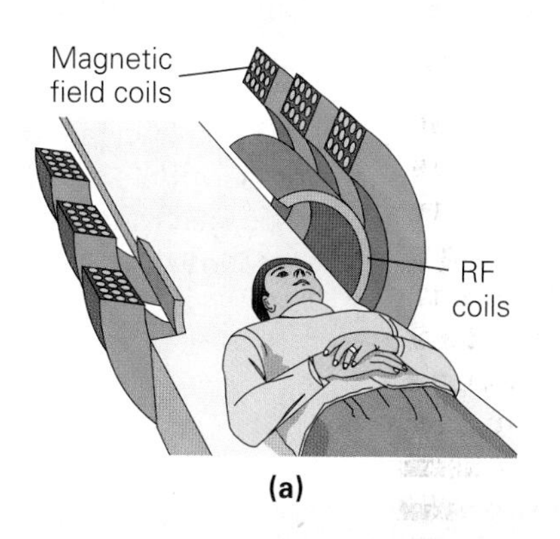

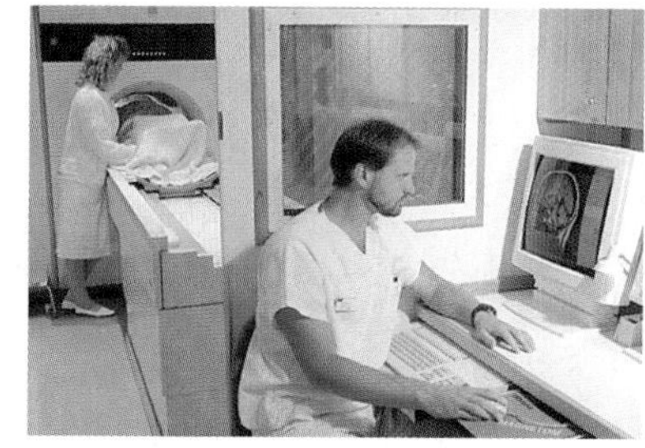

FIGURE 3 MRI (a) A diagram and **(b)** a photograph of the apparatus used for magnetic resonance imaging.

For a particular subshell, the ℓ value determines the number of states. Recall that there are $(2\ell + 1)$ possible m_ℓ values for a given ℓ. Thus, for $\ell = 1$, there are $[(2 \times 1) + 1] = 3$ values for m_ℓ. (They are $+1, 0,$ and -1.) Each of these values can have two m_s values $\left(\pm\frac{1}{2}\right)$, making six different combinations of (n, ℓ, m_ℓ, m_s), or six states.

In general, the number of possible states for a given value of ℓ is $2(2\ell + 1)$, taking into account the two possible "spin states" for each orbital state:

$$\text{Number of electron states for } \ell = 1 \text{ is } 2(2\ell + 1) = 2[(2 \times 1) + 1] = 6$$

$$\text{Number of electron states for } \ell = 2 \text{ is } 2(2\ell + 1) = 2[(2 \times 2) + 1] = 10$$

Comparison shows that the d subshell has more possible electron states than does the p subshell, regardless of which shell they are in (that is, their n value).

These results are summarized in Table 28.3. Notice that the total number of states in a given *shell* (designated by n) is $2n^2$. For example, for the $n = 2$ shell, the total number of states for its combined s and p subshells ($\ell = 0, 1$) is $2n^2 = 2(2)^2 = 8$. This means that up to eight electrons can be accommodated in the $n = 2$ shell: two in the $2s$ subshell and six in the $2p$ subshell.

Follow-Up Exercise. How many electrons could be accommodated in the $3d$ subshell if there were no spin quantum number?

TABLE 28.3 Possible Sets of Quantum Numbers and States

Electron Shell n	Subshell ℓ	Subshell Notation	Orbitals (m_ℓ)	Number of Orbitals (m_ℓ) in Subshell = $(2\ell + 1)$	Number of States m_s in Subshell = $2(2\ell + 1)$	Total Electron States for Shell = $2n^2$
1	0	1s	0	1	2	2
2	0	2s	0	1	2	8
	1	2p	1, 0, −1	3	6	
3	0	3s	0	1	2	18
	1	3p	1, 0, −1	3	6	
	2	3d	2, 1, 0, −1, −2	5	10	
4	0	4s	0	1	2	32
	1	4p	1, 0, −1	3	6	
	2	4d	2, 1, 0, −1, −2	5	10	
	3	4f	3, 2, 1, 0, −1, −2, −3	7	14	

Electron Configurations The electron structure of the ground state of atoms can be determined, so to speak, by putting an increasing number of electrons in the lower energy subshells [hydrogen (H), 1 electron; helium (He), 2 electrons; lithium (Li), 3 electrons; and so on], as was done for four elements in Fig. 28.8. However, rather than drawing diagrams, a shorthand notation called the **electron configuration** is widely used.

In this notation, the subshells are written in order of increasing energy, and the number of electrons in each subshell designated with a superscript. For example, $3p^5$ means that a $3p$ subshell is occupied by five electrons. The electron configurations for the atoms shown in Fig. 28.8 can thus be written as follows:

Li	(3 electrons)	$1s^2 2s^1$
F	(9 electrons)	$1s^2 2s^2 2p^5$
Ne	(10 electrons)	$1s^2 2s^2 2p^6$
Na	(11 electrons)	$1s^2 2s^2 2p^6 3s^1$

In writing an electron configuration, when one subshell is filled, you go on to the next higher one. The total of all the superscripts in any configuration must add up to the number of electrons in the atom.

The energy spacing between adjacent subshells is not uniform, as Figs. 28.7a and 28.8 show. In general, there are relatively large energy gaps between the s subshells and the subshells immediately below them. (Compare the 4s subshell with the 3p one in Fig. 28.7a.) The subshells just below the s subshells are usually p subshells, with the exception of the lowest subshell—the 1s subshell is below the 2s subshell. The gaps between other subshells—for example, between the 3s subshell and the 3p subshell above it, or between the 4d and 5p subshells—are considerably smaller.

This unevenness in energy differences gives rise to periodic large energy gaps, represented by vertical lines between certain subshells in the electron configuration:

$$1s^2 | 2s^2 2p^6 | 3s^2 3p^6 | 4s^2 3d^{10} 4p^6 | 5s^2 4d^{10} 5p^6 | 6s^2 4f^{14} 5d^{10} 6p^6 | \ldots$$

(number of states) (2) (8) (8) (18) (18) (32)

The subshells *between* the lines have only slightly different energies. The grouping of subshells (for example, $2s^2 2p^6$) that have about the same energy is referred to as an **electron period**.

Electron periods are the basis of the periodic table of elements. With your present knowledge of electron configurations, you are now in a position to understand the periodic table better than the person who originally developed it.

The Periodic Table of Elements

By 1860, more than sixty chemical elements had been discovered. Several attempts had been made to classify the elements into some orderly arrangement, but none were satisfactory. It had been noted in the early 1800s that the elements could be listed in such a way that similar chemical properties recurred periodically throughout the list. With this idea, in 1869, a Russian chemist, Dmitri Mendeleev (pronounced men-duh-*lay*-eff), created an arrangement of the elements, based on this periodic property. The modern version of his **periodic table of elements** is used today and can be seen on the walls of just about every science building (▾Fig. 28.9).

Mendeleev arranged the known elements in rows, called **periods**, in order of increasing atomic mass. When he came to an element that had chemical properties similar to those of one of the previous elements, he put this element below the previous similar one. In this manner, he formed both horizontal rows of elements and vertical columns called **groups**, or families of elements with similar chemical properties. The table was later rearranged in order of increasing atomic, or proton, number (the number of protons in the nucleus of an atom is the number at the top left of each of the element boxes in Fig. 28.9) in order to resolve some inconsistencies. Notice that if atomic masses were used, cobalt and nickel, atomic numbers 27 and 28, respectively, would fall in reversed columns.

With only 65 elements known at the time, there were vacant spaces in Mendeleev's table. The elements for these spaces were yet to be discovered. Because the missing elements were part of a sequence and had properties similar to those of other elements in a group, Mendeleev could predict their masses and chemical properties. Less than 20 years after Mendeleev devised his table, which showed chemists what to look for in order to find the undiscovered elements, three of the missing elements were, in fact, discovered.

The periodic table puts the elements into seven horizontal rows, or periods. The first period has only two elements. Periods 2 and 3 each have 8 elements, and periods 4 and 5 each have 18 elements. Recall that the s, p, d, and f subshells can contain a maximum of 2, 6, 10, and 14 electrons $[2(2\ell + 1)]$, respectively. You should begin to see a correlation between these numbers and the arrangements of elements in the periodic table.

The periodicity of the periodic table can be understood in terms of the electron configurations of the atoms. For $n = 1$, the electrons are in one of two s states (1s); for $n = 2$, electrons can fill the 2s and 2p states, which gives a total of 10 electrons; and so on. Thus, the period number for a given element is equal to the highest n shell containing electrons in the atom. Notice the electron configurations for the elements in Fig. 28.9. Also, compare the electron periods given earlier, as defined by energy gaps, and the periods in the periodic table (▾Fig. 28.10, p. 893). There is a one-to-one correlation, so the periodicity comes from energy-level considerations in atoms.

Chemists refer to *main group elements* as elements in which the last (least bound) electron enters an s or p subshell. In *transition elements,* the last electron enters a d subshell; and in *inner transition elements,* the last electron enters an f subshell. So that the periodic table is not unmanageably wide, the f subshell elements are usually placed in two rows at the bottom of the table. Each of the two rows is given a name—the *lanthanide series* and the *actinide series*—based on its position within the period.

Finally, we can also understand why elements in vertical columns, or groups, have similar chemical properties. The chemical properties of an atom, such as its ability to react and form compounds, depend almost entirely on the atom's outermost electrons—that is, those electrons in the outermost *unfilled* shell. It is these electrons, called *valence electrons,* that form chemical bonds with other atoms. Because of the way in which the elements are arranged in the table, the outermost electron configurations of all the atoms in any one group are similar. The atoms in such a group would thus be expected to have similar chemical properties, and they do. For example, notice the first two groups at the left of the table. They have one and two outermost electrons in an s subshell, respectively. These elements are all highly reactive metals that form compounds that have many similarities. The group at the far right,

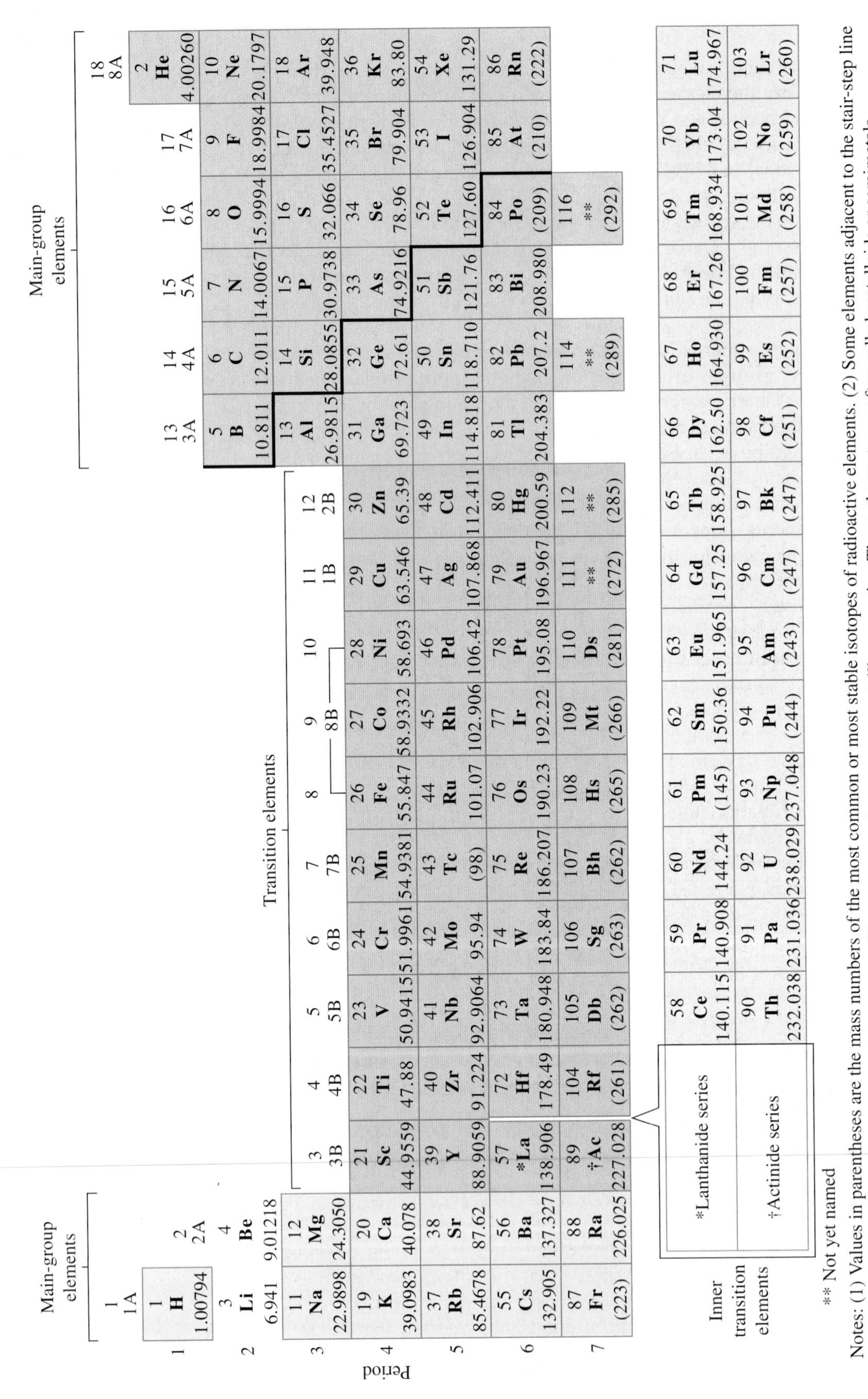

Period	1 1A	2 2A	3 3B	4 4B	5 5B	6 6B	7 7B	8	9 8B	10	11 1B	12 2B	13 3A	14 4A	15 5A	16 6A	17 7A	18 8A
1	1 **H** 1.00794																	2 **He** 4.00260
2	3 **Li** 6.941	4 **Be** 9.01218											5 **B** 10.811	6 **C** 12.011	7 **N** 14.0067	8 **O** 15.9994	9 **F** 18.9984	10 **Ne** 20.1797
3	11 **Na** 22.9898	12 **Mg** 24.3050											13 **Al** 26.9815	14 **Si** 28.0855	15 **P** 30.9738	16 **S** 32.066	17 **Cl** 35.4527	18 **Ar** 39.948
4	19 **K** 39.0983	20 **Ca** 40.078	21 **Sc** 44.9559	22 **Ti** 47.88	23 **V** 50.9415	24 **Cr** 51.9961	25 **Mn** 54.9381	26 **Fe** 55.847	27 **Co** 58.9332	28 **Ni** 58.693	29 **Cu** 63.546	30 **Zn** 65.39	31 **Ga** 69.723	32 **Ge** 72.61	33 **As** 74.9216	34 **Se** 78.96	35 **Br** 79.904	36 **Kr** 83.80
5	37 **Rb** 85.4678	38 **Sr** 87.62	39 **Y** 88.9059	40 **Zr** 91.224	41 **Nb** 92.9064	42 **Mo** 95.94	43 **Tc** (98)	44 **Ru** 101.07	45 **Rh** 102.906	46 **Pd** 106.42	47 **Ag** 107.868	48 **Cd** 112.411	49 **In** 114.818	50 **Sn** 118.710	51 **Sb** 121.76	52 **Te** 127.60	53 **I** 126.904	54 **Xe** 131.29
6	55 **Cs** 132.905	56 **Ba** 137.327	57 ***La** 138.906	72 **Hf** 178.49	73 **Ta** 180.948	74 **W** 183.84	75 **Re** 186.207	76 **Os** 190.23	77 **Ir** 192.22	78 **Pt** 195.08	79 **Au** 196.967	80 **Hg** 200.59	81 **Tl** 204.383	82 **Pb** 207.2	83 **Bi** 208.980	84 **Po** (209)	85 **At** (210)	86 **Rn** (222)
7	87 **Fr** (223)	88 **Ra** 226.025	89 **†Ac** 227.028	104 **Rf** (261)	105 **Db** (262)	106 **Sg** (263)	107 **Bh** (262)	108 **Hs** (265)	109 **Mt** (266)	110 **Ds** (281)	111 ** (272)	112 ** (285)		114 ** (289)		116 ** (292)		

Main-group elements: groups 1–2 and 13–18. Transition elements: groups 3–12.

Inner transition elements

*Lanthanide series	58 **Ce** 140.115	59 **Pr** 140.908	60 **Nd** 144.24	61 **Pm** (145)	62 **Sm** 150.36	63 **Eu** 151.965	64 **Gd** 157.25	65 **Tb** 158.925	66 **Dy** 162.50	67 **Ho** 164.930	68 **Er** 167.26	69 **Tm** 168.934	70 **Yb** 173.04	71 **Lu** 174.967
†Actinide series	90 **Th** 232.038	91 **Pa** 231.036	92 **U** 238.029	93 **Np** 237.048	94 **Pu** (244)	95 **Am** (243)	96 **Cm** (247)	97 **Bk** (247)	98 **Cf** (251)	99 **Es** (252)	100 **Fm** (257)	101 **Md** (258)	102 **No** (259)	103 **Lr** (260)

** Not yet named

Notes: (1) Values in parentheses are the mass numbers of the most common or most stable isotopes of radioactive elements. (2) Some elements adjacent to the stair-step line between the metals and nonmetals have a metallic appearance but some nonmetallic properties. These elements are often called metalloids or semimetals. There is no general agreement on just which elements are so designated. Almost every list includes Si, Ge, As, Sb, and Te. Some also include B, At, and/or Po.

▲ **FIGURE 28.9 The periodic table of elements** The elements are arranged in order of increasing atomic, or proton, number. Horizontal rows are called *periods*, and vertical columns are called *groups*. The elements in a group have similar chemical properties. Each atomic mass represents an average of that element's isotopes, weighted to reflect their relative abundance in our immediate environment. The masses have been rounded to two decimal places; more precise values are given in Appendices IV and V. (A value in parentheses represents the mass number of the best-known or longest-lived isotope of an unstable element. See Appendix IV for an alphabetical listing of elements.

Shell (last to be filled)	Subshells	Number of electrons in subshell, $2(2\ell + 1)$	Corresponding period in periodic table
$n = 7$	$7p$	6	Period 7 (32 elements)
	$6d$	10	
	$5f$	14	
	$7s$	2	
$n = 6$	$6p$	6	Period 6 (32 elements)
	$5d$	10	
	$4f$	14	
	$6s$	2	
$n = 5$	$5p$	6	Period 5 (18 elements)
	$4d$	10	
	$5s$	2	
$n = 4$	$4p$	6	Period 4 (18 elements)
	$3d$	10	
	$4s$	2	
$n = 3$	$3p$	6	Period 3 (8 elements)
	$3s$	2	
$n = 2$	$2p$	6	Period 2 (8 elements)
	$2s$	2	
$n = 1$	$1s$	2	Period 1 (2 elements)

Energy ⟶

◀ **FIGURE 28.10 Electron periods** The periods of the periodic table are related to electron configurations. The last n shell to be filled is equal to the period number. The electron periods and the corresponding periods of the table are defined by relatively large energy gaps between successive subshells (such as between $4s$ and $3p$) of the atoms.

the noble gases, includes elements with completely filled subshells (and thus a full shell). These elements are at the ends of electron periods, or just before a large energy gap. These gases are nonreactive and can form compounds (by chemical bonding) only under very special conditions.

Conceptual Example 28.4 ■ Combining Atoms: Performing Chemistry on the Periodic Table

Combinations of atoms, called *molecules*, can form if atoms come together and share outer electrons. This sharing process is called *covalent bonding*. In this "shared-custody" scheme, both atoms find it energetically beneficial (that is, they lower their combined total energies) to have the equivalent of a filled outer shell of electrons, if only on a part-time basis. Using your knowledge of electron shells and the periodic chart, determine which of the following atoms would most likely form a covalent arrangement with oxygen: (a) neon (Ne); (b) calcium (Ca); or (c) hydrogen (H).

Reasoning and Answer. Choice (a), neon, with a total of 10 electrons, can be eliminated immediately, because it has a full outer shell of 8 electrons and, as such, has nothing to be gained by losing or adding electrons. Looking at the periodic table, we see that oxygen, with 6 outer-shell electrons, is 2 electrons shy of having a full complement of 8 electrons. It *could* occasionally have those 2 electrons by sharing electrons with another atom or atoms. Choice (b), calcium, with its 20 electrons, is 2 electrons beyond the previous full shell of 10. You might think, therefore, that calcium is a possible covalent partner. However, you must remember that the covalent arrangement is a two-way street. In other words, the arrangement would also require calcium sometimes to have two *more* electrons than normal.

This situation would put the calcium atom in the awkward position of being 4 electrons beyond the full shell of 10 and 14 electrons away from the next complete shell. Hence, even though this attempt at covalent bonding might seem to work for oxygen, it certainly won't work well for calcium. The two species do, however, combine to form calcium oxide (CaO). The bonding that keeps calcium oxide together is based on the electrical attraction between the positive calcium ion, Ca^{+2}, and the negative oxygen ion, O^{-2}. In this type of bonding, the two atoms permanently exchange two electrons, making each a doubly charged ion of opposite signs. This bond is called an *ionic bond* and is *not* covalent. Thus, answer (b), calcium, is not correct.

The remaining candidate, hydrogen, has one electron fewer than a full shell of two. Thus, if a hydrogen atom could add one electron, it would attain an electron configuration like that of the lightest inert gas, helium. Since each hydrogen atom needs to share only one electron, two of them can accomplish this by sharing with a single oxygen atom. Part

(continues on next page)

of the time, the hydrogen atoms must share their electrons with the oxygen atom in order to create the latter's full outer shell of eight electrons. So, the correct answer is (c)—two hydrogen atoms covalently bound to a single oxygen atom. This combination has the molecular formula H_2O—water.

Follow-Up Exercise. In this Example, (a) what would be the electron configuration of the oxygen in the water molecule at some instant when it has "custody" of the electrons from both hydrogen atoms? (b) What would be the net charge on the oxygen in this case?

28.4 The Heisenberg Uncertainty Principle

OBJECTIVE: To understand the inherent quantum-mechanical limits on the accuracy of physical observations.

An important aspect of quantum mechanics has to do with measurement and accuracy. In classical mechanics, there is no limit to the accuracy of a measurement. Theoretically, by continual refinement of a measurement instrument and procedure, the accuracy could be improved to any degree so as to give *exact* values. This theoretical approach results in a *deterministic* view of nature. For example, if both the position and the velocity of an object are known *exactly* at a particular time, you can determine exactly where the object will be in the future and where it was in the past (assuming that you know all the forces that act on it).

However, quantum theory predicts otherwise and sets limits on the accuracy of measurements. This idea was introduced in 1927 by the German physicist Werner Heisenberg, who had developed another approach to quantum mechanics that complemented Schrödinger's wave theory. The **Heisenberg uncertainty principle** as applied to position and momentum (or velocity) can be stated as follows:

> It is impossible to know simultaneously an object's exact position and momentum.

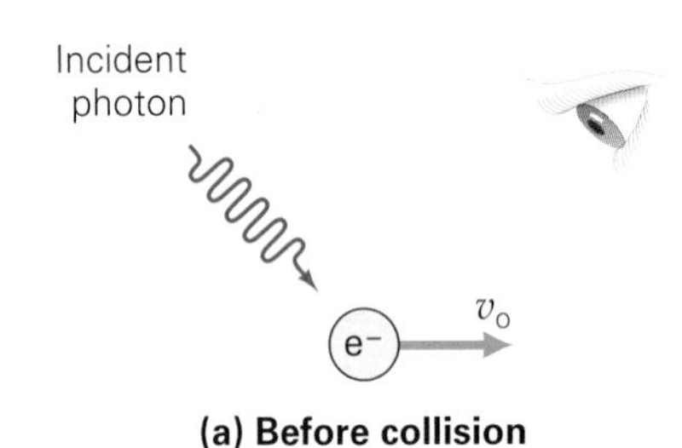

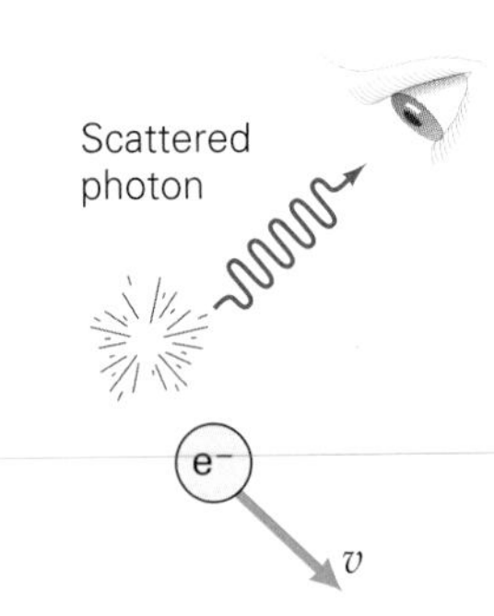

▲ **FIGURE 28.11 Measurement-induced uncertainty (a)** To measure the position and momentum (or velocity) of an electron, at least one photon must collide with the electron and be scattered toward the eye or detector. **(b)** In the collision process, energy and momentum are transferred to the electron, which induces uncertainty in its velocity.

This concept is often illustrated with a simple thought experiment. Suppose that you want to measure the position and momentum (actually, the velocity) of an electron. In order for you to "see," or locate, the electron, at least one photon must bounce off the electron and come to your eye (or detector), as illustrated in ◂Fig. 28.11. However, in the collision process, some of the photon's energy and momentum are transferred to the electron.

After the collision, the electron recoils. Thus, in the very process of locating the electron's position accurately, uncertainty is introduced into the electron's velocity (or momentum, because $\Delta\vec{\mathbf{p}} = m\Delta\vec{\mathbf{v}}$). This effect isn't noticed in our everyday macroscopic world, because the recoil produced by viewing an object with light is negligible. This is because the pressure exerted by the light cannot appreciably alter the motion or position of an object of everyday mass.

According to wave optics, the position of an electron can be measured *at best* to an uncertainty Δx of about the wavelength λ of the light used—that is, $\Delta x \approx \lambda$. The photon "particle" used for this location has a momentum of $p = h/\lambda$. Because the amount of momentum transferred during collision isn't determined, the final momentum of the electron would have an uncertainty on the order of the momentum of the photon, or $\Delta p \approx h/\lambda$.

Notice that the product of these two uncertainties is *at least* as large as h, because

$$(\Delta p)(\Delta x) \approx \left(\frac{h}{\lambda}\right)(\lambda) = h$$

This equation relates the *minimum* uncertainties, or maximum accuracies, of *simultaneous* measurements of the momentum and position. In actuality, the uncertainties could be worse, depending on the amount of light (number of photons) used, the apparatus, and the technique. Using more detailed considerations, Heisenberg found that the product of the two uncertainties would equal, *at a minimum*, $h/2\pi$. However, it could be higher. Hence, we can write the following:

$$(\Delta p)(\Delta x) \geq \frac{h}{2\pi} \quad (28.5)$$

That is, the product of the *minimum* uncertainties of simultaneous momentum and position measurements is on the order of Planck's constant divided by 2π (that is, about $10^{-34}\,\text{J}\cdot\text{s}$).

In order to locate the position of the particle accurately (that is, to make Δx as small as possible), a photon with a very short wavelength must be used. However, this type of photon carries a lot of momentum, which results in an increased uncertainty in the momentum. To take the extreme case, if the location of a particle could be measured exactly (that is, $\Delta x \to 0$), we would have no idea about its momentum ($\Delta p \to \infty$). Thus, it is the measurement process itself that limits the accuracy to which position and momentum can be measured simultaneously. In Heisenberg's words, "Since the measuring device has been constructed by the observer . . . we have to remember that what we observe is not nature in itself but nature exposed to our method of questioning."

To see how the Heisenberg uncertainty principle affects the microscopic and macroscopic worlds, consider the following Integrated Example.

Integrated Example 28.5 ■ An Electron versus a Bullet: The Uncertainty Principle

An electron and a bullet both have the same speed, measured to the same accuracy. (a) How would their minimum uncertainties in position compare: (1) The electron's location would be more uncertain than that of the bullet; (2) the bullet's location would be more uncertain than that of the electron; or (3) their location uncertainties would be the same? (b) If the bullet's mass is 20.0 g and both the bullet and the electron have a speed of 300 m/s, with an uncertainty of ±0.010%, determine the minimum uncertainty in the position of each.

(a) Conceptual Reasoning. The minimum uncertainty in location is related to the minimum uncertainty in momentum. Since both the bullet and the electron have the same uncertainty in velocity, the bullet's momentum is much more uncertain (momentum is proportional to mass, as is $\Delta\vec{\mathbf{p}}$, since $\Delta\vec{\mathbf{p}} = m\Delta\vec{\mathbf{v}}$). Therefore, the bullet's location will be much less uncertain than that of the electron, so the answer is (1).

(b) Thinking It Through. The uncertainty principle (Eq. 28.5) can be solved for Δx in each case, because $\Delta\vec{\mathbf{p}}$ can be determined from the uncertainty in velocity, $\Delta\vec{\mathbf{v}}$. The electron is affected much more, because of its very small mass, as seen in part (a). Listing the quantities, including the known electron mass,

Given: $m_e = 9.11 \times 10^{-31}$ kg
$m_b = 20\text{ g} = 0.020$ kg
$v_b = v_e = 300\text{ m/s} \pm 0.01\%$

Find: Δx_e and Δx_b (minimum uncertainties in position)

The uncertainty in speed is 0.01% (or 0.00010) for both the electron and the bullet. This uncertainty is $(300\text{ m/s})(0.00010) = 0.030$ m/s. Thus we have the same uncertainty for both:

$$v = 300\text{ m/s} \pm 0.030\text{ m/s}$$

The *total* uncertainty in speed is *twice* this amount, because the measurements can be off both above and below the measured values; hence, $\Delta v = 0.060$ m/s.

For the electron, the minimum uncertainty in position is

$$\Delta x_e = \frac{h}{2\pi\Delta p} = \frac{h}{2\pi m_e \Delta v} = \frac{6.63 \times 10^{-34}\,\text{J}\cdot\text{s}}{2\pi(9.11 \times 10^{-31}\,\text{kg})(0.060\text{ m/s})} = 0.0019\text{ m} = 1.9\text{ mm}$$

Similarly, the bullet's minimum uncertainty in position is

$$\Delta x_b = \frac{h}{2\pi m_b \Delta v} = \frac{6.63 \times 10^{-34}\,\text{J}\cdot\text{s}}{2\pi(0.020\text{ kg})(0.060\text{ m/s})} = 8.8 \times 10^{-32}\text{ m}$$

Notice that the uncertainty in the bullet's position is much smaller than the diameter of a nucleus. Its position uncertainty is also many orders of magnitude less than that of the electron. The lesson is that uncertainty in location for everyday objects traveling at ordinary speeds is negligible. However, for electrons, 1.9 mm is significant and measurable.

Follow-Up Exercise. In this Example, what would the minimum uncertainty in the electron's speed have to be for the minimum uncertainty in its position to be on the order of atomic dimensions, or 0.10 nm?

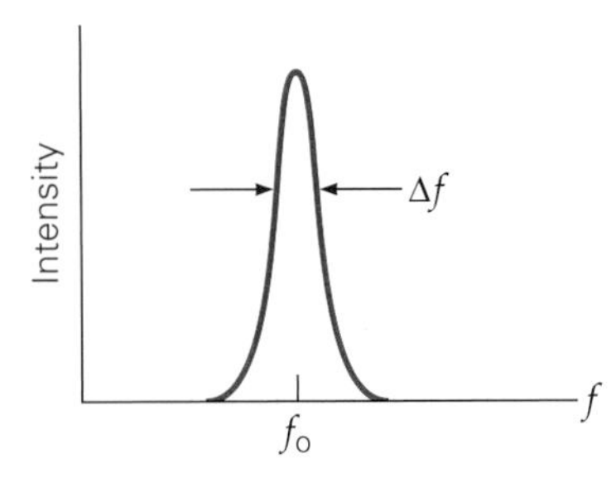

▲ **FIGURE 28.12 Natural line broadening** Because a measurement must be carried out in an amount of time comparable with the lifetime (Δt) of an excited atomic state, the energy of that state is uncertain by an amount $\Delta E = h\Delta f$. The observed emission line has a width of Δf, rather than being a line of single frequency f_o with zero width.

An equivalent form of the uncertainty principle relates uncertainties in energy and time. As with the position and momentum, a detailed analysis shows that, *at best*, the product of the uncertainties in energy and time is $h/2\pi$. Of course, it could be larger; hence, this form of the uncertainty principle is written as

$$(\Delta E)(\Delta t) \geq \frac{h}{2\pi} \tag{28.6}$$

This form of the Heisenberg uncertainty principle shows that the energy of an object may be uncertain by an amount ΔE, depending on the time taken to measure it, Δt. For longer times, the energy measurement becomes increasingly more accurate. Once again, these uncertainties are important only for very light objects. Such uncertainties are of particular importance in nuclear physics and elementary particle interactions (Chapter 30).

Notice that the energy of a particle cannot be measured exactly unless an infinite amount of time is taken to do so. If a measurement of energy is carried out in a time Δt, then the energy is uncertain by an amount ΔE. For example, the measurement of the frequency of light emitted by an atom is really the measurement of the energy of the photon associated with the transition from an excited state to the ground state. The measurement must be carried out in an amount of time comparable with the lifetime of the excited state. As a result, the observed emission line (Section 27.4) has a nonzero energy width, since $\Delta E = h\Delta f$ is *not* zero (◂Fig. 28.12). This *natural broadening* is generally small for atomic emissions and was therefore ignored in Chapter 27, in which spectral lines were considered to have exact frequencies.

28.5 Particles and Antiparticles

OBJECTIVES: To understand (a) the relationship between particles and antiparticles, and (b) the energy requirements for pair production.

When the British physicist Paul A. M. Dirac, in 1928, extended quantum mechanics to include relativistic considerations, something new and very different was predicted—a particle called the **positron**. The positron was predicted to have the same mass as the electron, but to carry a *positive* charge. The oppositely charged positron is the **antiparticle** of the electron. (Antiparticles will be discussed in more depth in Chapters 29 and 30.)

The positron was first observed experimentally in 1932 by the American physicist C. D. Anderson in cloud-chamber experiments with cosmic rays. The curvature of the particle tracks in a magnetic field showed two types of particles, with opposite charge and the same mass (◂Fig. 28.13). Anderson had, in fact, discovered the positron.

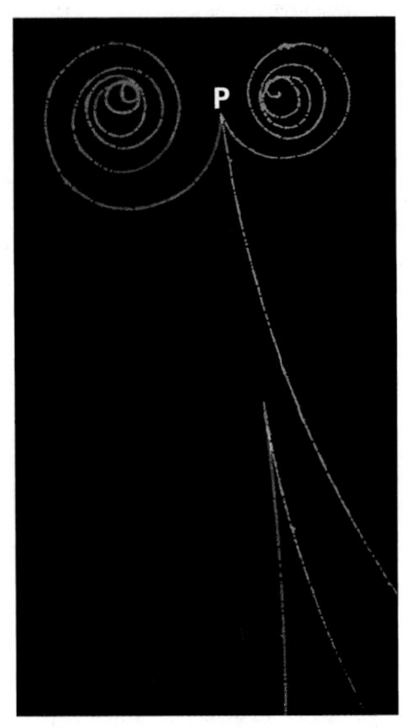

▲ **FIGURE 28.13 Cloud-chamber photograph of pair production** In this false-color cloud-chamber photograph, a gamma-ray photon (not visible, but it enters the region in a downward direction) interacts with a nearby atomic nucleus (point P) to produce an electron and a positron (green and red spiral tracks, respectively, at the top). In the process, the photon also dislodges an orbital electron (the nearly straight and vertical green track). An external magnetic field causes the electron and positron to be deflected in paths of opposite curvature. A similar event is recorded in the bottom half of the photo. (Why might the paths of the particles created in this case show less deflection?)

Because electric charge is conserved, a positron can be created only with the simultaneous creation of an electron (so that the net charge created is zero). This process is called **pair production**. In Anderson's experiment, positrons were observed to be emitted from a thin lead plate exposed to cosmic rays from outer space, which contain highly energetic X-rays. Pair production occurs when an X-ray photon comes near a nucleus—the nuclei of the lead atoms of the plate in the Anderson experiment. In this process, the photon goes out of existence, and an *electron–positron pair* (an electron and a positron) is created, as illustrated in ▸Fig. 28.14. This result represents a direct conversion of electromagnetic (photon) energy into mass. By the conservation of energy (neglecting the small recoil kinetic energy of the massive nearby nucleus),

$$hf = 2m_e c^2 + K_{e^-} + K_{e^+}$$

where hf is energy of the photon, $2m_e c^2$ is the total mass–energy equivalent of the electron–positron pair, and the Ks represent the kinetic energies of the produced particles.

The minimum energy to produce such a pair occurs when they are produced at rest—when K_{e^-} and K_{e^+} are zero. Thus, the *minimum* photon energy to produce an electron–positron pair is

$$E_{min} = hf = 2m_e c^2 = 1.022 \text{ MeV} \tag{28.7}$$

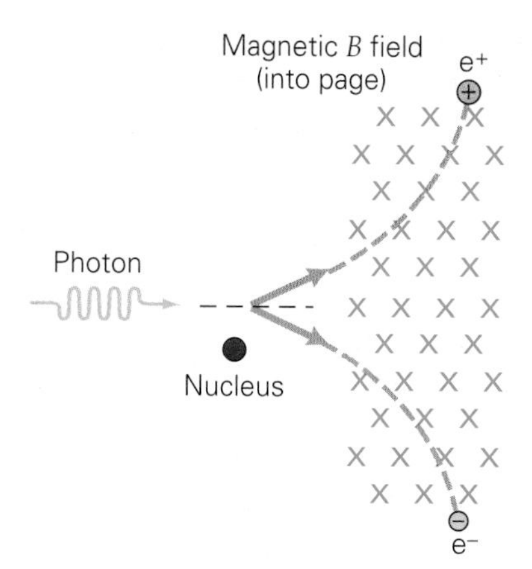

◀ **FIGURE 28.14 Pair production** An electron (e^-) and a positron (e^+) can be created when an energetic photon passes near a heavy nucleus.

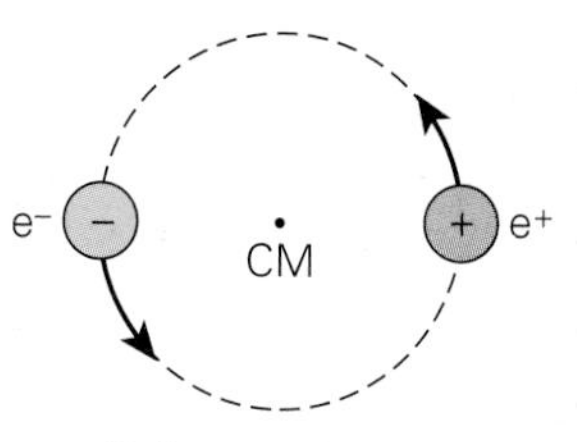

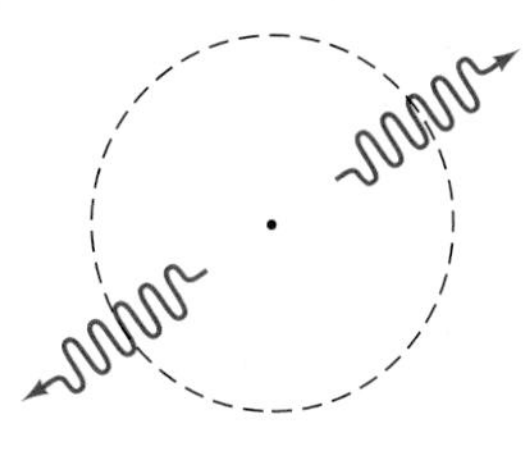

▲ **FIGURE 28.15 Pair annihilation** A slow positron and electron can form a system called a *positronium atom*. The disappearance of a positronium atom is signaled by the appearance of two photons, each with an energy of 0.511 MeV. Why would we not expect one photon with an energy of 1.022 MeV? (*CM* stands for *center of mass*.)

(Here, we have used the result from Section 26.5 that says that the mass of an electron is equivalent to an energy of $m_e c^2 = 0.511$ MeV.) This minimum photon energy is called the **threshold energy** for pair production.

But if they are created by cosmic rays, why aren't positrons commonly found in nature? This is because, almost immediately after their creation, positrons go out of existence by a process called **pair annihilation**. When an energetic positron appears, it loses kinetic energy by collision as it passes through matter. Finally, almost at rest, it combines with an electron and forms a hydrogen-like atom, called a *positronium atom*, in which a positron substitutes for a proton. The positronium atom is unstable and quickly decays ($\approx 10^{-10}$ s) into two photons, each with an energy of 0.511 MeV (▸Fig. 28.15). Pair annihilation is then a direct conversion of mass into electromagnetic energy—the inverse of pair production, so to speak. Pair annihilation is the basis for a medical diagnostic tool called a *positron emission tomography* (or PET) *scan*. This application will be discussed in more detail in Chapter 29 after you learn more about nuclear decay processes.

More generally, all particles have antiparticles. For example, there is an antiproton with the same mass as a proton, but with a negative charge. Even a neutral particle such as the neutron has an antiparticle—the antineutron. It is even conceivable that antiparticles predominate in some parts of the universe. If so, atoms made of the **antimatter** in these regions would consist of negatively charged nuclei composed of antiprotons and antineutrons, surrounded by orbiting positively charged positrons (antielectrons). It would be difficult to distinguish a region of antimatter visibly, since the physical behavior of antimatter atoms would presumably be the same as that of ordinary matter.

Chapter Review

- The magnitude of the momentum (p) of a photon carrying energy E is inversely related to the wavelength of the associated light wave by

$$p = \frac{E}{c} = \frac{hf}{c} = \frac{h}{\lambda} \tag{28.1}$$

- The **de Broglie hypothesis** assigns a wavelength to material particles, by analogy with the assignment of momentum to photons. The **de Broglie wavelength** of a particle is

$$\lambda = \frac{h}{p} = \frac{h}{mv} \tag{28.2}$$

- A quantum-mechanical **wave function** ψ is a "probability wave" associated with a particle. The probability density, the square of the wave function, gives the relative probability of finding a particle at a particular location.

- Electron-orbital energies are determined primarily by the **principal quantum number** n.

- The quantum number ℓ is called the **orbital quantum number** and is associated with the orbital angular momentum of the electron's orbit. For each value of n, the ℓ quantum number has one of n possible integer values from zero up to a maximum value of $n - 1$.

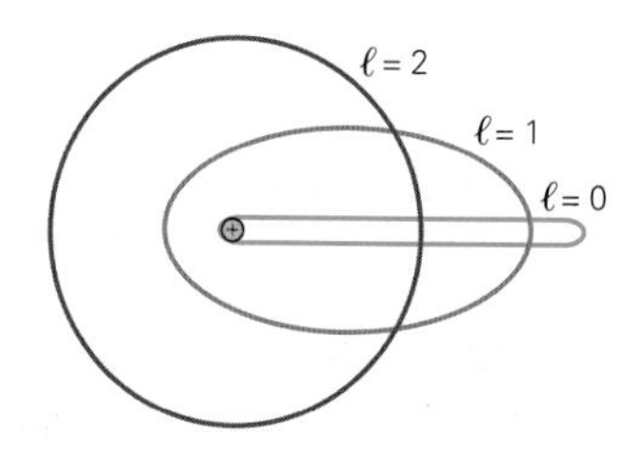

- The quantum number m_ℓ is called the **magnetic quantum number** and is associated with the z-component of the electron's orbital angular momentum. For a given ℓ, there are $2\ell + 1$ possible m_ℓ values (integers), given by $m_\ell = 0, \pm 1, \pm 2, \ldots, \pm \ell$.

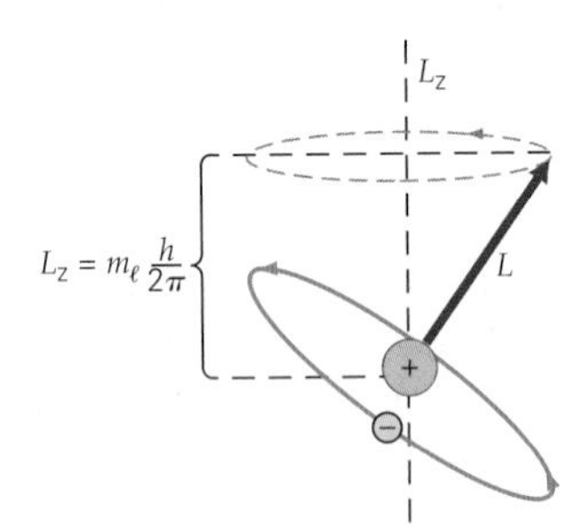

- The quantum number that describes an electron's intrinsic angular momentum is the **spin quantum number m_s**, which, for electrons, protons, and neutrons, can have only two values: $m_s = \pm\frac{1}{2}$. These values correspond to the two angular momentum directions of spin: up and down.

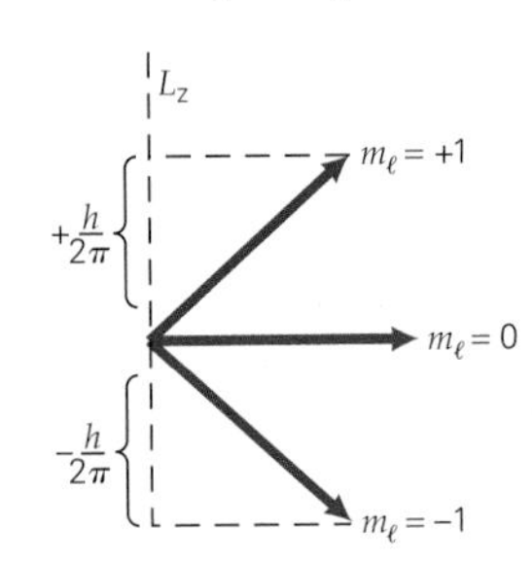

- Orbits with different quantum numbers that have the same energy are **degenerate**.

- Orbits that share a common principal quantum number n are in the same **shell**.

- Orbits that share a common orbital quantum number ℓ are in the same **subshell**.

- The **Pauli exclusion principle** states that in a given atom, no two electrons can have exactly the same set of quantum numbers.

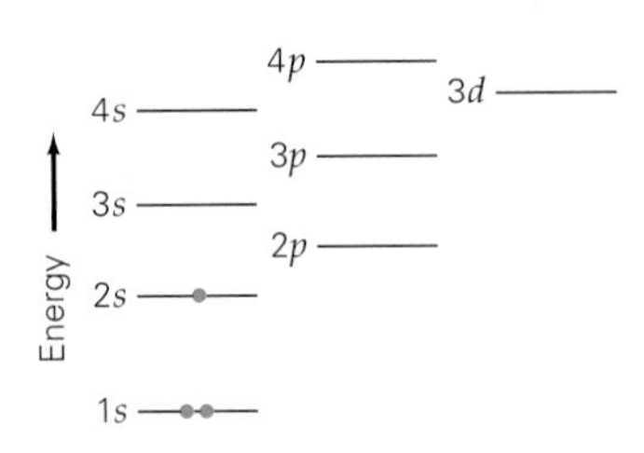

- The **Heisenberg uncertainty principle** states that you cannot simultaneously measure both the position and the momentum (or velocity) of a particle exactly. The same condition holds true for the particle's energy and the period of time during which that energy is measured. The uncertainties satisfy

$$(\Delta p)(\Delta x) \geq \frac{h}{2\pi} \tag{28.5}$$

and

$$(\Delta E)(\Delta t) \geq \frac{h}{2\pi} \tag{28.6}$$

- **Pair production** refers to the creation of a particle and its **antiparticle**. The reverse process is **pair annihilation**, in which a particle and its antiparticle annihilate and their energy is converted into two photons.

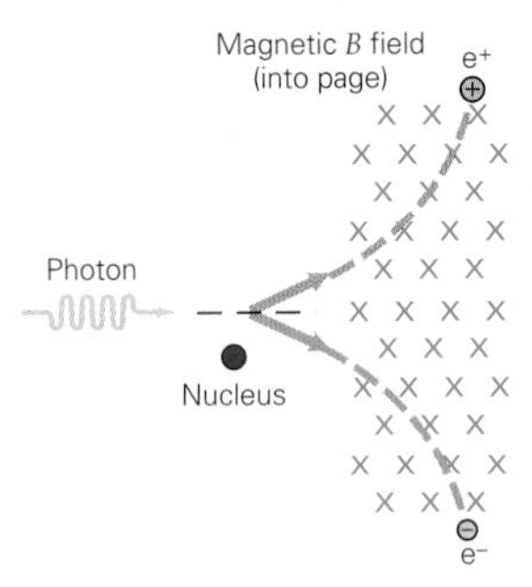

Exercises*

MC = *Multiple Choice Question,* **CQ** = *Conceptual Question, and* **IE** = *Integrated Exercise. Throughout the text, many exercise sections will include "paired" exercises. These exercise pairs, identified with* **red numbers**, *are intended to assist you in problem solving and learning. In a pair, the first exercise (even numbered) is worked out in the Study Guide so that you can consult it should you need assistance in solving it. The second exercise (odd numbered) is similar in nature, and its answer is given at the back of the book.*

28.1 Matter Waves: The de Broglie Hypothesis

1. **MC** Photons associated with which color light would have the largest momentum: (a) red, (b) green, or (c) violet?

2. **MC** If the following were traveling at the same speed which would have the shortest de Broglie wavelength: (a) an electron, (b) a proton, (c) a carbon atom, or (d) a hockey puck?

3. **MC** You have three neutrons traveling at different speeds. Which speed would be associated with the shortest neutron de Broglie wavelength: (a) 10^3 m/s, (b) 10^4 m/s, (c) 10^2 m/s, or (d) the wavelengths are the same at all speeds?

**Note:* Use Eq. 28.3 for λ where appropriate.

4. CQ The de Broglie hypothesis predicts that a wave is associated with any object that has momentum. Why don't we observe the wave associated with a moving car?

5. CQ If a baseball and a bowling ball were traveling at the same speed, which one would have a longer de Broglie wavelength? Why?

6. CQ An electron is accelerated from rest through an electric-potential difference. Will increasing the difference in potential result in a longer or shorter de Broglie wavelength? Why?

7. ● What is the de Broglie wavelength associated with a 1000-kg car moving at 25 m/s?

8. IE ● An electron and a proton are moving with the same speed. (a) Compared with the proton, will the electron have (1) a shorter, (2) an equal, or (3) a longer de Broglie wavelength? Why? (b) If the speed of the electron and proton is 100 m/s, what are their de Broglie wavelengths?

9. ●● A proton and an electron are accelerated from rest through the same difference in potential, V. What is the ratio of the de Broglie wavelength of an electron to that of a proton (to two significant figures)?

10. IE ●● Electrons are accelerated from rest through an electric-potential difference. (a) If this potential difference increases to nine times the original value, the new de Broglie wavelength will be (1) nine times, (2) three times, (3) one ninth, (4) one third that of the original. Why? (b) If the original potential is 250 kV and the new potential is 600 kV, what is the ratio of the new de Broglie wavelength to the original?

11. ●● An electron is accelerated from rest through a potential difference so that its de Broglie wavelength is 0.010 nm. What is the potential difference?

12. ●● A charged particle is accelerated through a potential difference V. If the voltage were doubled, what would be the ratio of the new de Broglie wavelength to the original value?

13. IE ●● A proton traveling at a speed of 4.5×10^4 m/s is accelerated through a potential difference of 37 V. (a) Will its de Broglie wavelength (1) increase, (2) remain the same, or (3) decrease, due to the potential difference? Why? (b) By what percentage does the de Broglie wavelength of the proton change?

14. ●● What is the energy of a beam of electrons that exhibits a first-order maximum at an angle of 25° when diffracted by a crystal grating with a lattice plane spacing of 0.15 nm?

15. ●● A scientist wants to use an electron microscope to observe details on the order of 0.25 nm. Through what potential difference must the electrons be accelerated from rest so that they have a de Broglie wavelength of this magnitude?

16. ●●● According to the Bohr theory of the hydrogen atom, the speed of the electron in the first Bohr orbit is 2.19×10^6 m/s. (a) What is the wavelength of the matter wave associated with the electron? (b) How does this wavelength compare with the circumference of the first Bohr orbit?

17. ●●● (a) What is the de Broglie wavelength of the Earth in its orbit about the Sun? (b) Treating the Earth as a de Broglie wave in a large "gravitational" atom, what would be the principal quantum number, n, of its orbit? (c) If the principal quantum number increased by 1, how would the radius of the orbit change? (Assume a circular orbit.)

28.2 The Schrödinger Wave Equation

18. MC The wave-function solution to the Schrödinger equation's description of a single particle (a) is the particle's de Broglie wavelength, (b) tells us the probability of finding a particle at a given location, (c) functionally describes the de Broglie wave of a particle, (d) none of the preceding.

19. MC The square of a particle's wave function (a) is the energy of the particle, (b) is the probability of locating the particle, (c) tells us the quantum number of its state, (d) provides the basis of the Pauli exclusion principle.

20. MC In the scanning tunneling microscope (STM), how does the tunneling current change if the tip gets nearer to the surface? (a) It increases, (b) it decreases, (c) it doesn't change, or (d) its change depends on the type of surface being explored.

21. CQ Explain how you would program the tip of an scanning tunnel microscope (STM) to move if it encounters a dip in the surface being explored and you want to keep the tunneling current constant.

22. CQ How would the maximum of the graph in Fig. 28.3 a change if the charge on the proton in the hydrogen atom were suddenly increased? Explain your reasoning

23. CQ According to modern quantum theory and the Schrödinger equation, there is a probability that if you ran into a wall you could end up on the other side. (Don't try this!) Explain the idea behind this and discuss why this has never been observed to happen (except in comic books and cartoons).

24. ●● A particle in box is constrained to move in one dimension, like a bead on a wire, as illustrated in ▼Fig. 28.16. Assume that no forces act on the particle in the interval $0 < x < L$ and that it hits a perfectly rigid wall. The particle will exist only in states of a certain kinetic energies that can be determined in analogy to a standing wave on a string (Chapter 13, Section 13.5). This means an integral number n of half-wavelengths "fit" into the box's length L, or $n\left(\frac{\lambda_n}{2}\right) = L$ where $n = 1, 2, 3, \ldots$. Using this relationship, show that the "allowed" kinetic energies K_n of the particle are given by $K_n = n^2\left[h^2/\left(8mL^2\right)\right]$ where $n = 1, 2, 3, \ldots$ and m is the particle's mass. [*Hint*: Recall that kinetic energy is related to momentum by $K = p^2/(2m)$ and that the de Broglie wavelength of the particle is also related to its momentum.]

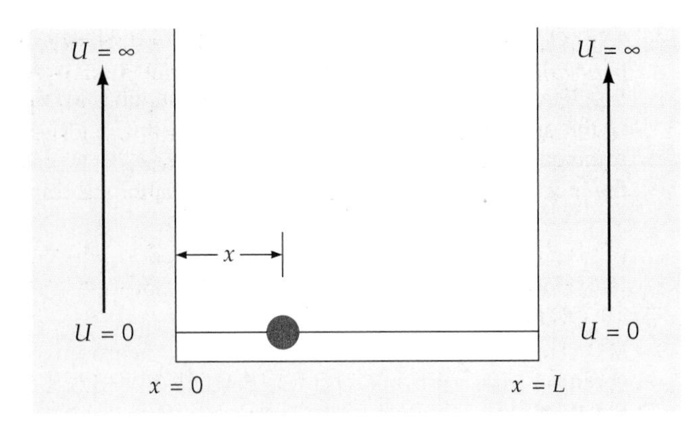

▲ **FIGURE 28.16 Particle in a box** See Exercises 24 and 25.

25. ●● Let's model a nucleus as a particle trapped in the one-dimensional box. Assume the particle is a proton and it is in a one dimensional nucleus of length of 7.11 fm (the approximate diameter of a Pb-208 nucleus). (a) Using the results of Exercise 24, determine the energies of the proton in the ground state and first two excited states. (b) The nucleus is to absorb a photon of just the right energy to enable the proton to make an upward transition from the ground state to the second excited state. How much energy would this be and what type of photon would it be classified as? (Neglect recoil of the absorbing nucleus after the photon energy is absorbed.)

28.3 Atomic Quantum Numbers and the Periodic Table

26. **MC** The quantum number ℓ (a) determines the total energy of a state, (b) is associated with the angular momentum of the electron in orbit, (c) is associated with the orientation of the angular-momentum vector, (d) is associated with electron spin.

27. **MC** The quantum number m_ℓ (a) determines the energy of the electron, (b) tells whether the electron is spinning up or down, (c) tells the orientation of the angular-momentum vector of the electron in orbit, (d) all of the preceding.

28. **MC** The quantum number m_s (a) can take on only two different values for an electron, (b) arises from the orbital motion of an electron, (c) is needed because the electron actually spins like a ball, (d) all of the preceding.

29. **CQ** What information does the quantum number n give for a hydrogen atom?

30. **CQ** Niels Bohr set forth a *correspondence principle*, which states that the results of quantum mechanics and classical physics approach agreement when quantum numbers become very large. Discuss this principle in terms of the hydrogen atom.

31. **CQ** What is the basis of the periodic table of elements in terms of quantum theory, and what do the elements in a particular group have in common?

32. ● (a) How many possible sets of quantum numbers are there for $n = 2$ and $n = 3$ shells? (b) Write the explicit values of all the quantum numbers (n, ℓ, m_ℓ, m_s) for these levels.

33. ● How many possible sets of quantum numbers are there for the subshells with (a) $\ell = 0$ and (b) $\ell = 3$?

34. **IE** ● (a) Which has more possible sets of quantum numbers associated with it, $n = 2$ or $\ell = 2$? (b) Prove your answer to part (a).

35. ● An electron in an atom is in an orbit that has a magnetic quantum number of $m_\ell = 2$. What are the minimum values that (a) ℓ and (b) n could be for that orbit?

36. ●● Draw the ground-state energy-level diagrams like those in Fig. 28.8 for (a) nitrogen (N) and (b) potassium (K).

37. ●● Draw schematic diagrams for the electrons in the subshells of (a) sodium (Na) and (b) argon (Ar) atoms in the ground state.

38. ●● Identify the atoms of each of the following ground-state electron configurations: (a) $1s^2 2s^2$; (b) $1s^2 2s^2 2p^3$; (c) $1s^2 2s^2 2p^6$; (d) $1s^2 2s^2 2p^6 3s^2 3p^4$.

39. ●● Write the ground-state electron configurations for each of the following atoms: (a) boron (B), (b) calcium (Ca), (c) zinc (Zn), and (d) tin (Sn).

40. **IE** ●●● (a) If there were no electron spin, the 1s state would contain a maximum of (1) zero, (2) one, (3) two electrons. Why? (b) What would be the first two inert or noble gases if there were no electron spin?

41. ●●● How would the electronic structure of lithium differ if electron spin were to have three possible orientations instead of just two?

28.4 The Heisenberg Uncertainty Principle

42. **MC** If the uncertainty in the position of a moving particle increases, (a) the particle may be located more exactly, (b) the uncertainty in its momentum decreases, (c) the uncertainty in its velocity increases, (d) none of the preceding.

43. **MC** According to the uncertainty principle, measurement of the exact energy of a particle requires (a) special equipment, (b) an infinite amount of time, (c) uncertainty in the momentum, (d) none of the preceding.

44. **CQ** Why is it impossible to simultaneously and accurately measure the position and velocity of a particle?

45. **CQ** A bowling ball has well-defined position and speed, whereas an electron does not. Why?

46. ● A 0.50-kg ball has a position of 5.0 m $\pm$ 0.01 m. To what minimum uncertainty can its momentum be measured?

47. **IE** ● An electron and a proton each have a momentum of 3.28470×10^{-30} kg·m/s $\pm$ 0.00025×10^{-30} kg·m/s. (a) The minimum uncertainty in the position of the electron compared with that of the proton will be (1) larger, (2) the same, (3) smaller. Why? (b) Calculate the minimum uncertainty in the position for each.

48. ● What is the minimum uncertainty in the velocity of an electron that is known to be somewhere between 0.050 nm and 0.10 nm from a proton?

49. ● What is the minimum uncertainty in the velocity of a 0.50-kg ball that is known to be at 1.0000 cm $\pm$ 0.0005 cm from the edge of a table?

50. ●● The energy of a 2.00-keV electron is known to within $\pm$3.00%. How accurately can its position be measured?

51. ●● If an excited state of an atom has a lifetime of 1.0×10^{-7} s, what is the minimum error associated with the measurement of the energy of this state?

52. ●● The energy of the first excited state of a hydrogen atom is -0.34 eV $\pm$ 0.0003 eV. What is the average lifetime for this state?

53. IE ●● (a) If the lifetime of excited state A is longer than that of state B, then the width of a spectral line due to natural broadening for a transition from state A to state B will be (1) smaller than, (2) the same as, (3) greater than that for a transition from state B. Why? (b) Calculate the ratio of the width of a spectral line due to natural broadening for a transition from an excited state with a lifetime of 10^{-12} s to that for a state with a lifetime of 10^{-8} s.

28.5 Particles and Antiparticles

54. **MC** Pair production involves (a) the production of two electrons, (b) the production of two positrons, (c) a positronium atom, (d) the production of a particle and its antiparticle.

55. **MC** Due to momentum considerations, pair annihilation cannot result in the emission of how many photons? (a) one (b) two (c) three

56. **MC** A new particle/antiparticle is found to require a minimum photon energy of 25 MeV to occur. How do their masses compare to that of an electron: (a) They are less massive than an electron, (b) they are more massive than an electron, (c) they would have the same mass as an electron, or (d) you can't tell from the data given?

57. CQ Why is the energy threshold for electron–positron pair production actually higher than the sum of their two masses (1.022 MeV in energy terms)? [*Hint*: Linear momentum must be conserved.]

58. CQ Can the production of two electrons and two positrons be accomplished using a high-energy photon? Explain. Why can't two electrons and one positron result?

59. IE ● A photon with an energy of 1.94 MeV comes near a heavy nucleus. (a) What will happen? (1) Pair production will not happen, (2) pair production can happen with the resulting particles at rest, or (3) pair production can happen but the resulting particles must be moving. (b) How much kinetic energy will each particle have after creation?

60. ● What is the energy of the photons produced in electron–positron pair annihilation, assuming that both particles are essentially at rest initially?

61. ● What is the threshold energy for the production of a proton–antiproton pair?

62. IE ●● A muon, or μ meson, has the same charge as an electron, but is 207 times as massive. (a) Compared with electron–positron pair production, the pair production of a muon and an antimuon requires a photon of (1) more, (2) the same amount of, (3) less energy. Why? (b) What would be the minimum energy for such a photon?

Comprehensive Exercises

63. An electron traveling at 3.00×10^4 m/s is further accelerated by a potential difference so as to reduce its de Broglie wavelength to one third of its original value. How much voltage is required to accomplish this?

64. There are negatively charged particles that have a spin quantum number, $m_s = 3/2$. (a) How many different spin orientiations would these have: (1) two, (2) three, or (3) four? (b) If you were building an atom with these particles orbiting the nucleus, how many would the "atom" have if it were to be the first "inert" atom in the periodic chart?

65. Suppose a starship had a mass of 1.25×10^9 kg and was initially at rest. If its "matter–antimatter engines" produced photons from electron–positron annihilation and focused them to travel backward out from the ship, how many photons would they have to emit to reach a speed of 3.00×10^5 m/s (0.100% the speed of light)? [*Hint*: Use conservation of linear momentum and remember that relativity is not needed here. (Why?)]

66. Using a typical nuclear diameter of 4.25×10^{-15} m as its location uncertainty, compute the uncertainty in momentum and kinetic energy associated with an electron if it were part of the nucleus. For energies greater than a few MeV, particles such as electrons would escape the nucleus. What does this tell you about the likelihood that an electron resides in the nucleus of an atom?

CHAPTER

29 THE NUCLEUS

29.1 Nuclear Structure and the Nuclear Force 903

29.2 Radioactivity 906

29.3 Decay Rate and Half-Life 911

29.4 Nuclear Stability and Binding Energy 917

29.5 Radiation Detection, Dosage, and Applications 922

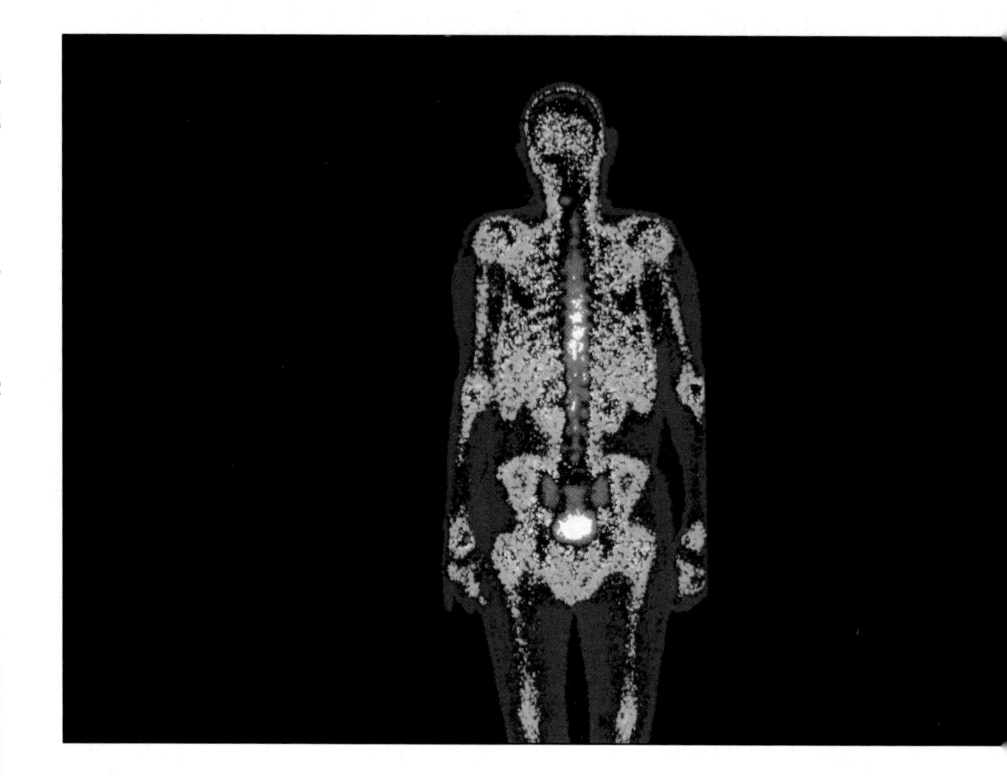

PHYSICS FACTS

- Spent nuclear fuel rods from nuclear reactors are laden with radioactive nuclei. Many of these are chemically separated and used in medical and industrial applications.
- Many fission fragments are potentially harmful to living things if ingested at high levels. For example, I-131, used as a diagnostic tool for thyroid cancer, can actually cause thyroid cancer at high levels of exposure.
- The radioactive nuclide americium-241, used in most smoke detectors, is actually created artificially. None exists naturally as its half-life is only about 400 years.
- A lengthy plane flight at high altitude can expose passengers to the amount of radiation energy dosage (from cosmic rays) comparable to that of a chest X-ray.
- Of the yearly "dose" of radiation, more than half is due to natural background radiation, the rest from sources such as medical X-rays.

The skeletal image (a bone scan) in the chapter-opening photograph was created by radiation from a radioactive source. *Radiation* and *radioactivity* are words that sometimes produce anxiety, but the beneficial uses of radiation are often overlooked. For instance, exposure to high-energy radiation can cause cancer—yet precisely the same sort of radiation, in relatively small doses, can be useful in the diagnosis and treatment of cancer.

The bone scan in the chapter-opening photo was created by radiation released when unstable nuclei spontaneously broke apart after being administered to and taken up by the body—a process we call *radioactive decay*. But what makes some nuclei stable, while others decay? What determines the rate at which they break down and the particles that they emit? These are some of the questions that will be explored in this chapter. We'll also learn how radiation is detected and measured, as well as more about its dangers and uses.

In addition, the study of radioactivity and nuclear stability helps us understand the nature of the nucleus, its structure, its energy, and how this energy can be released. Nuclear energy is one of our major energy sources, and will be considered in Chapter 30. In this chapter we concentrate on understanding the nucleus itself.

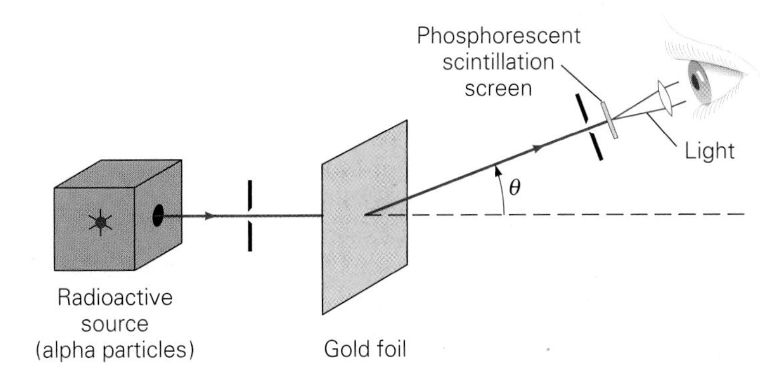

◀ **FIGURE 29.1 Rutherford's scattering experiment** A beam of alpha particles from a radioactive source was scattered by gold nuclei in a thin foil, and the scattering was observed as a function of the scattering angle θ. The observer detects the light (viewed through a lens) given off by a phosphorescent scintillation screen.

29.1 Nuclear Structure and the Nuclear Force

OBJECTIVES: To (a) distinguish between the Thomson and Rutherford–Bohr models of the atom, (b) specify some of the basic properties of the strong nuclear force, and (c) understand nuclear notation.

It is evident from the emission of electrons from heated filaments (called *thermionic emission*) and the photoelectric effect that atoms contain electrons. Since an atom is normally electrically neutral, it must contain a positive charge equal in magnitude to the total charge of its electrons. Since the electron's mass is small compared with the mass of even the lightest of atoms, most of an atom's mass appears to be associated with that positive charge.

Based on these observations, J. J. Thomson (1856–1940), a British physicist who had experimentally proven the existence of the electron in 1897, proposed a model of the atom. In his model, the electrons are uniformly distributed within a continuous sphere of positive charge. It was called a "plum pudding" model, because the electrons in the positive charge are analogous to raisins in a plum pudding. The region of positive charge was assumed to have a radius on the order of 10^{-10} m, or 0.1 nm, roughly the diameter of an atom.

As you probably know, our modern model of the atom is quite different. This model concentrates all the positive charge, and practically all of the mass, in a central *nucleus*, surrounded by orbiting electrons. The existence of such a nucleus was first proposed by British physicist Ernest Rutherford (1871–1937). Combining this idea with the Bohr theory of electron orbits (Section 27.4) led to the simplistic "solar system" model, or **Rutherford–Bohr model**, of the atom.

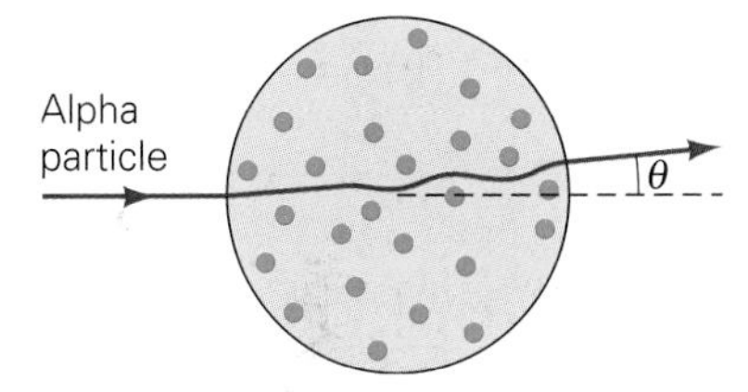

▲ **FIGURE 29.2 The plum pudding model** In Thomson's plum pudding model of the atom, massive alpha particles were expected to be only slightly deflected by collisions with the electrons (blue dots) in the atom. The experimental results were quite different.

Rutherford's insight came from the results of alpha-particle scattering experiments performed in his laboratory about 1911. An alpha (α) particle is a doubly positively charged particle ($q_\alpha = +2e$) that is naturally emitted from some radioactive materials. (See Section 29.2.) A beam of these particles was directed at a thin gold-foil "target," and the deflection angles and percentage of scattered particles were observed (▲Fig. 29.1).

An alpha particle is more than 7000 times as massive as an electron. Thus, the Thomson model predicts only tiny deflections—the result of collisions with the light electrons as an alpha particle passes through such a model of a gold atom (▶Fig. 29.2). Surprisingly, however, Rutherford observed alpha particles scattered at appreciable angles. In about 1 in every 8000 scatterings, the alpha particles were actually *backscattered;* that is, they were scattered through angles greater than 90° (▼Fig. 29.3).

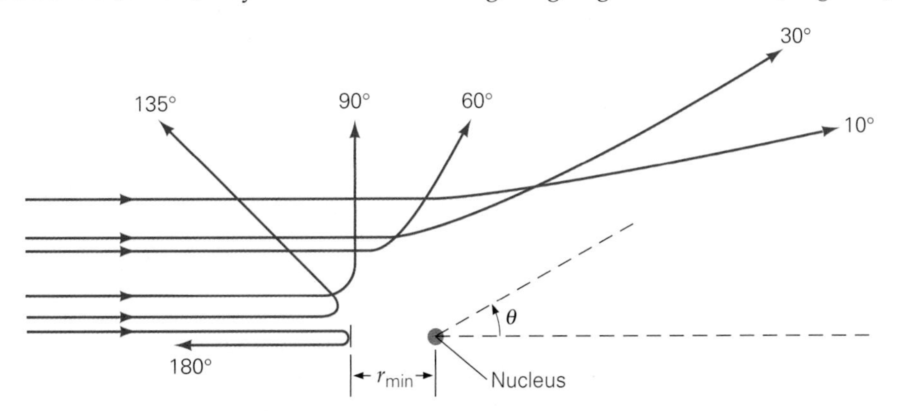

◀ **FIGURE 29.3 Rutherford scattering** A compact, dense atomic nucleus with a positive charge accounts for the observed scattering. An alpha particle in a head-on collision with the nucleus would be scattered directly backward ($\theta = 180°$) after coming within a distance r_{min} of the nucleus. At this scale, the electron orbits (about the nucleus) are too far away to be seen.

Calculations showed that the probability of backscattering in the Thomson model was minuscule—certainly much, much less than 1 in 8000. As Rutherford described the backscattering, "It was almost as incredible as if you had fired a 15-inch shell at a piece of tissue paper and it came back and hit you."

The experimental results led Rutherford to the concept of a nucleus: "On consideration, I realized that this scattering backward must be the result of a single collision, and when I made calculations I saw that it was impossible to get anything of that order of magnitude unless you took a system in which the greater part of the mass of the atom was concentrated in a minute nucleus. It was then that I had the idea of an atom with a minute massive center carrying a charge."

If all of the positive charge of a target atom were concentrated in a small region, then an alpha particle coming close to this region would experience a large deflecting (electrical repulsion) force. The mass of this positive "nucleus" would be larger than that of the alpha particle, and in this model backscattering is much more likely to occur than in the plum pudding model.

A simple estimate can give an idea of the approximate size of a nucleus. It is during a head-on collision that an alpha particle comes closest to the nucleus (a distance labeled as r_{min} in Fig. 29.3). That is, the alpha particle approaching the nucleus stops at r_{min} and is accelerated back along its original path. Assuming a spherical charge distribution, the electric potential energy of the alpha particle (α) and nucleus (n) when separated by a center-to-center distance r is $U = kq_\alpha q_n/r = k(2e)(Ze)/r$ (Eq. 16.5). Here, Z is the **atomic number**, or the number of protons in the nucleus. Therefore the charge of the nucleus is $q_n = +Ze$. By conservation of energy, the kinetic energy of the incoming alpha particle is completely converted into electric potential energy at the turnaround point, r_{min}. Using $q_\alpha = +2e$, we equate the two energies, which results in

$$\tfrac{1}{2}mv^2 = \frac{k(2e)Ze}{r_{min}}$$

or, solving for r_{min}, we obtain

$$r_{min} = \frac{4kZe^2}{mv^2} \qquad (29.1)$$

In Rutherford's experiment, the kinetic energy of the alpha particles from the particular source had been measured, and Z was known to be 79 for gold. Using these values, along with the constants in Eq. 29.1, Rutherford found r_{min} to be on the order of 10^{-14} m.

Although the nuclear model of the atom is useful, the nucleus is much more than a volume of positive charge. The nucleus is actually composed of two types of particles—protons and neutrons—collectively referred to as **nucleons**. The nucleus of the hydrogen atom is a single proton. Rutherford suggested that the hydrogen nucleus be named *proton* (from the Greek meaning "first") after he became convinced that no nucleus could be less massive than the hydrogen nucleus. A **neutron** is an electrically neutral particle with a mass slightly greater than that of a proton. The existence of the neutron was not experimentally verified until 1932.

The Nuclear Force

Of the forces in the nucleus, there is certainly the attractive gravitational force between nucleons. But in Chapter 15, this gravitational force was shown to be negligible compared with the repulsive electrical force between the positive protons. Taking only these repulsive forces into account, it would be predicted that the nucleus should fly apart. Yet the nuclei of many atoms are stable. Therefore, there must be an *attractive* force between nucleons that overcomes the electrical repulsion and thereby holds the nucleus together. This strong attractive force is called the **strong nuclear force**, or simply the *nuclear force*.

The exact expression for the nuclear force is extremely complex. However, some general features of it are as follows:

- The nuclear force is strongly attractive and much larger in magnitude than both the electrostatic force and the gravitational force between nucleons.

- The nuclear force is very short-ranged; that is, a nucleon interacts only with its nearest neighbors, over distances on the order of 10^{-15} m.
- The nuclear force is independent of electric charge; that is, it acts between *any* two nucleons—two protons, a proton and a neutron, or two neutrons.

Thus, nearby protons repel each other electrically, but attract each other (and nearby neutrons) by the strong force, with the latter winning the battle. Having no electric charge, neutrons only attract nearby protons and neutrons.

Nuclear Notation

To describe the nuclei of different atoms, it is convenient to use the notation illustrated in ▸Fig. 29.4a. The chemical symbol of the element is used with subscripts and a superscript. The subscript on the left is called the *atomic number* (Z), which indicates the number of protons in the nucleus. A more descriptive name for the symbol **Z** is **proton number**, which will be used in this book. For electrically neutral atoms, Z is equal to the number of orbital electrons. (Why?)

The number of protons in the nucleus of an atom determines the species of the atom—that is, the element to which the atom belongs. In Fig. 29.4b, the proton number $Z = 6$ indicates that the nucleus belongs to a carbon atom. The proton number thus defines which chemical symbol is used. Electrons can be removed from (or added to) an atom to form an ion, *but this does not change the atom's species.* For example, a nitrogen atom with an electron removed, N^+, is still nitrogen—a nitrogen *ion*. It is the proton number, rather than the electron number, that determines the species of atom.

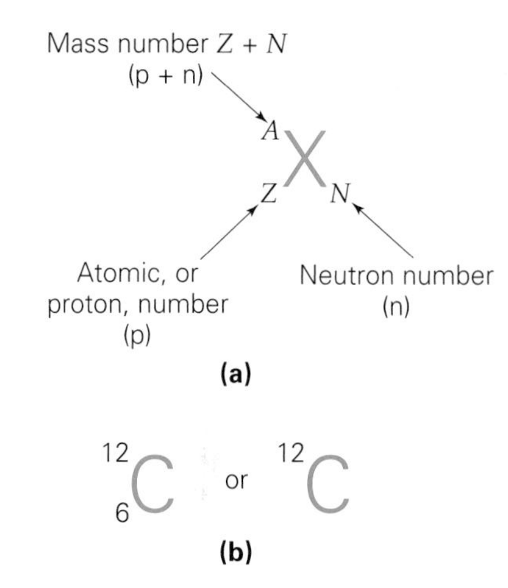

▲ **FIGURE 29.4 Nuclear notation**
(a) The composition of a nucleus is shown by the chemical symbol of the element with the mass number A (sum of protons and neutrons) as a left superscript and the proton (atomic) number Z as a left subscript. The neutron number N may be shown as a right subscript, but both Z and N are routinely omitted, because the letter symbol tells you Z, and $N = A - Z$.
(b) The two most common nuclear notations for a nucleus of one of the stable isotopes of carbon—carbon-12.

The superscript to the left of the chemical symbol is called the **mass number (*A*)**—the total number of protons and neutrons in the nucleus. Since protons and neutrons have roughly equal masses, the mass numbers of nuclei give a relative comparison of nuclear masses. For the carbon nucleus in Fig. 29.4b, the mass number is $A = 12$, because there are six protons and six neutrons. The number of neutrons, called the **neutron number (*N*)**, is sometimes indicated by a subscript on the right side of the chemical symbol. However, this subscript is usually omitted, because it can be calculated from A and Z; that is, $N = A - Z$. Similarly, the proton number is routinely omitted, because the chemical symbol uniquely specifies the value of Z.

Even though the atoms of an element all have the same number of protons in their nuclei, they may have different numbers of neutrons. For example, nuclei of different carbon atoms ($Z = 6$) may contain six, seven, or eight neutrons. In nuclear notation, these atoms would be written as ${}^{12}_{6}C_6$, ${}^{13}_{6}C_7$, and ${}^{14}_{6}C_8$, respectively. Atoms whose nuclei have the same number of protons, but different numbers of neutrons, are called **isotopes**. The three atoms we just listed are three isotopes of carbon.

Note: All the isotopes in one family have almost the same orbital-electron structure and thus very similar chemical properties.

Isotopes are like members of a family. They all have the same Z number and the same surname (element name), but they are distinguishable by the number of neutrons in their nuclei and, therefore by their mass. Isotopes are referred to by their mass numbers; for example, these isotopes of carbon are called *carbon-12*, *carbon-13*, and *carbon-14*, respectively. There are other isotopes of carbon that are unstable: ${}^{11}C$, ${}^{15}C$, and ${}^{16}C$. A particular nuclear species or isotope of any element is also called a **nuclide**. So far, we have mentioned six nuclides of carbon. Generally, only a few isotopes of a given species are stable. But this number can vary from none to several or more. In fact, in our carbon example, ${}^{14}C$ is unstable, although it is long lived. Only ${}^{12}C$ and ${}^{13}C$, in fact, are truly stable isotopes of carbon.

Another important family of isotopes is that of hydrogen, which has three isotopes: ${}^{1}H$, ${}^{2}H$, and ${}^{3}H$. These isotopes are given special names. ${}^{1}H$ is called *ordinary hydrogen*, or simply *hydrogen*; ${}^{2}H$ is called *deuterium*. Deuterium, which is stable, is sometimes known as *heavy hydrogen*. It can combine with oxygen to form heavy water (written D_2O). The third isotope of hydrogen, ${}^{3}H$, called *tritium*, is unstable.

29.2 Radioactivity

OBJECTIVES: To (a) define the term *radioactivity*; (b) distinguish among alpha, beta, and gamma decay; and (c) write nuclear-decay equations.

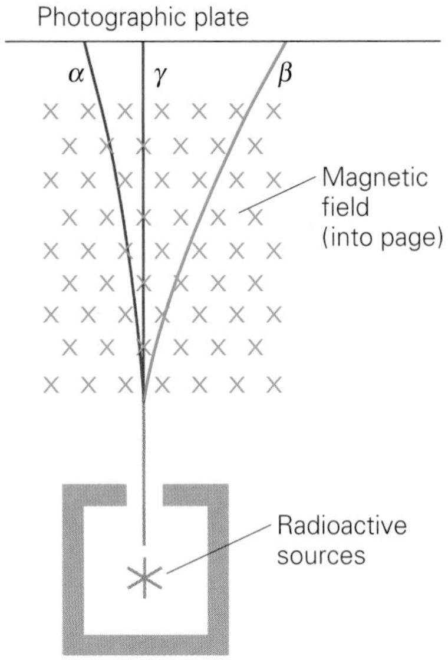

Photographic plate
α γ β
Magnetic field (into page)
Radioactive sources

▲ **FIGURE 29.5** Nuclear radiation Different types of radiation from radioactive sources can be distinguished by passing them through a magnetic field. Alpha and beta particles are deflected. From the right-hand magnetic-force rule, alpha particles are positively charged and beta particles are negatively charged. The radii of curvature (not drawn to scale) allow the particles to be distinguished by mass. Gamma rays are not deflected and thus are uncharged; they are quanta of electromagnetic energy.

Most elements have at least one stable isotope. It is atoms with stable nuclides with which we are most familiar in the environment. However, some nuclei are unstable and disintegrate spontaneously (or decay), emitting energetic particles and photons. Unstable isotopes are said to be *radioactive* or to exhibit **radioactivity**. For example, tritium (^3_1H) has a radioactive nucleus. Of all the unstable nuclides, only a small number occur naturally. Others can be produced artificially (Chapter 30).

Radioactivity is unaffected by normal physical or chemical processes, such as heat, pressure, and chemical reactions. Processes such as these simply do *not* affect the source of the radioactivity—the nucleus. Nor can nuclear instability be explained by a simple imbalance of attractive and repulsive forces within the nucleus. This is because, experimentally, nuclear disintegrations (of a given isotope) occur at a fixed rate. That is, the nuclei in a given sample do not all decay at the same time. According to classical theories, identical nuclei *should* decay at the same time. Therefore, radioactive decay suggests that the probability effects of quantum mechanics might be in play.

The discovery of radioactivity is credited to the French scientist Henri Becquerel. In 1896, while studying the fluorescence of a uranium compound, he discovered that a photographic plate near a sample had been darkened, even though the compound had not been activated by exposure to light and was not fluorescing. Apparently, this darkening was caused by some new type of radiation emitted from the compound itself. In 1898, Pierre and Marie Curie announced the discovery of two radioactive elements, radium and polonium, which they had isolated from uranium pitchblende ore.

Experiment shows that the radiation emitted by radioactive isotopes is of three different kinds. When a radioactive isotope is placed in a chamber so that the emitted radiation passes through a magnetic field to a photographic plate (◂Fig. 29.5), the various types of radiation expose the plate, producing characteristic spots by which the types of radiation may be identified. The positions of the spots show that some isotopes emit radiation that is deflected to the left; some emit radiation that is deflected to the right; and some emit radiation that is undeflected. These spots are characteristic of what came to be known as *alpha, beta,* and *gamma* radiations.

From the opposite deflections of two of the types of radiation in the magnetic field, it is evident that positively charged particles are associated with alpha decay and that negatively charged particles are emitted during beta decay. Because of their much smaller deflection, alpha particles must be considerably more massive than beta particles. The undeflected gamma radiation must be electrically neutral. (Why?)

Detailed investigations of the three different radiation types revealed the following:

- **Alpha particles** are actually doubly charged ($+2e$) particles that contain two protons and two neutrons. They are identical to the nucleus of the helium atom (^4_2He).
- **Beta particles** are electrons (positive electrons or *positrons* were discovered later).
- **Gamma rays** are high-energy quanta of electromagnetic energy (photons).

For a few radioactive elements, two spots are found on the film, indicating that the elements decay by two different modes. Let's now look at some details of each of these three modes of decay.

Alpha Decay

When an alpha particle is ejected from a radioactive nucleus, the nucleus loses two protons and two neutrons, so the mass number (A) is decreased by four ($\Delta A = -4$). The proton number (Z) is also decreased by two ($\Delta Z = -2$). Because the *parent nucleus* (the original nucleus) loses two protons, the *daughter nucleus* (the resulting nucleus) is the nucleus of a different element, defined by the new proton number. Thus, the **alpha-decay** process is one of nuclear *transmutation*, in which the nuclei of one element change into the nuclei of a lighter element.

An example of an isotope, or nuclide, that undergoes alpha decay is polonium-214. The decay process is represented as a nuclear equation (usually written without neutron numbers):

$$\underset{\textit{polonium}}{{}^{214}_{84}\mathrm{Po}} \rightarrow \underset{\textit{lead}}{{}^{210}_{82}\mathrm{Pb}} + \underset{\substack{\textit{alpha particle}\\ \textit{(helium nucleus)}}}{{}^{4}_{2}\mathrm{He}}$$

Notice that both the mass-number and proton-number totals are equal on each side of the equation: $(214 = 210 + 4)$ and $(84 = 82 + 2)$, respectively. This condition reflects the experimental facts that *two conservation laws apply to all nuclear processes*. The first is the **conservation of nucleons**:

The total number of nucleons (A) remains constant in any nuclear process.

The second is the familiar **conservation of charge**:

The total charge remains constant in any nuclear process.

These conservation laws allow us to predict the composition of the daughter nucleus, as the following Example illustrates.

Example 29.1 ■ Uranium's Daughter: Alpha Decay

A ${}^{238}_{92}\mathrm{U}$ nucleus undergoes alpha decay. What is the resulting daughter nucleus?

Thinking It Through. Nucleon conservation allows the prediction of the daughter's proton number. From that, the element's name can be determined from the periodic table.

Solution. Since $\Delta Z = -2$ for alpha decay, the parent uranium-238 (${}^{238}\mathrm{U}$) nucleus loses two protons, and the daughter nucleus has a proton number $Z = 92 - 2 = 90$, which is thorium's proton number (see the periodic table, Fig. 28.9). The equation for this decay can therefore be written as

$${}^{238}_{92}\mathrm{U} \rightarrow {}^{234}_{90}\mathrm{Th} + {}^{4}_{2}\mathrm{He} \quad \text{or} \quad {}^{238}_{92}\mathrm{U} \rightarrow {}^{234}_{90}\mathrm{Th} + {}^{4}_{2}\alpha$$

where the helium nucleus is finally written as ${}^{4}_{2}\alpha$ (sometimes just α).

Follow-Up Exercise. Using high-energy accelerators, it is possible to *add* an alpha particle to a nucleus—essentially the reverse of the reaction in this Example. Write the equation for this nuclear reaction, and predict the identity of the resulting nucleus if an alpha particle is added to a ${}^{12}\mathrm{C}$ nucleus. *(Answers to all Follow-Up Exercises are at the back of the text.)*

From experiments, it is found that the kinetic energies of alpha particles from radioactive sources are typically a few MeV. (See Section 16.2.) For example, the energy of the alpha particle from the decay of ${}^{214}\mathrm{Po}$ is about 7.7 MeV, and that from ${}^{238}\mathrm{U}$ decay is about 4.14 MeV. Alpha particles from such sources were used in the scattering experiments that led to the Rutherford nuclear model.

Outside the nucleus, the repulsive electric force increases as an alpha particle approaches the nucleus. Inside the nucleus, however, the strongly attractive nuclear force dominates. These conditions are depicted in ▸Fig. 29.6, which shows a graph of the potential energy U as a function of r, the distance from the center of the nucleus. Consider alpha particles (with kinetic energy of 7.7 MeV) from a ${}^{214}\mathrm{Po}$ source incident on ${}^{238}\mathrm{U}$ (Fig. 29.6). The alpha particles don't have enough kinetic energy to overcome the *electric potential-energy "barrier,"* whose height exceeds 7.7 MeV. Thus, Rutherford scattering occurs. On the other hand, we do know that the ${}^{238}\mathrm{U}$ nucleus does undergo alpha decay, emitting an alpha particle with an energy of 4.4 MeV, which is below the height of the barrier. How can these lower-energy alpha particles cross a barrier from the inside to the outside, when higher-energy alpha particles cannot cross from outside to the inside? According to classical theory, this is impossible, since it violates the conservation of energy. However, quantum mechanics offers an explanation.

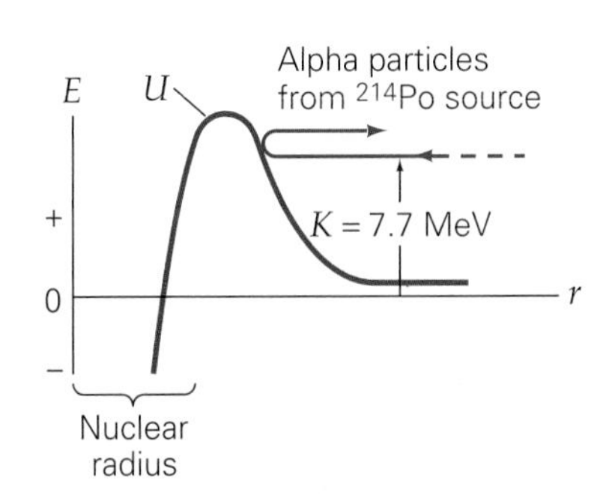

▲ **FIGURE 29.6 Potential-energy barrier for alpha particles** Alpha particles from radioactive polonium with energies of 7.7 MeV do not have enough energy to overcome the electrostatic potential-energy barrier of the ${}^{238}\mathrm{U}$ nucleus and are scattered.

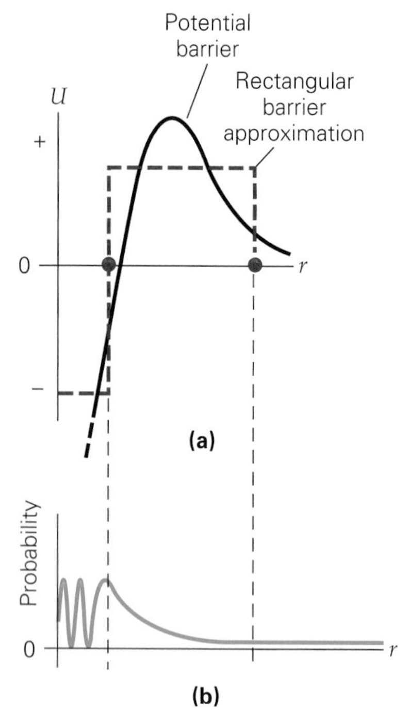

▲ **FIGURE 29.7 Tunneling or barrier penetration** **(a)** The potential-energy barrier presented by a nucleus to an alpha particle can be approximated by a rectangular barrier. **(b)** The probability of finding the alpha particle at a given location, according to quantum-mechanical calculations, is shown. If the particle is initially inside the nucleus, it has a likelihood of "tunneling" through the barrier and appearing outside the nucleus. Typically, this event has a very small, but nonzero, probability of occurring for elements above lead on the periodic table.

Quantum mechanics predicts a nonzero probability of an alpha particle, initially inside the nucleus, to be found *outside* the nucleus (◂Fig 29.7). This phenomenon is called **tunneling**, or **barrier penetration**, since the alpha particle has a certain probability of tunneling through the barrier. (As an example, recall that electrons do this in the scanning tunneling microscope [STM], in Chapter 28.)

Beta Decay

The emission of an electron (a beta particle) in a nuclear-decay process might seem contradictory to the proton–neutron model of the nucleus. Note, however, that the electron emitted in **beta decay** is *not* part of the original nucleus. *The electron is created during the decay.* There are several types of beta decay. When a negative electron is emitted, the process is called **β^- decay**. An example of this type of beta decay is that of ^{14}C:

$$\underset{\text{carbon}}{{}^{14}_{6}\text{C}} \rightarrow \underset{\text{nitrogen}}{{}^{14}_{7}\text{N}} + \underset{\substack{\text{beta particle}\\ \text{(electron)}}}{{}^{0}_{-1}\text{e}}$$

The parent nucleus (carbon-14) has six protons and eight neutrons, whereas the daughter nucleus (nitrogen) has seven protons and seven neutrons. Notice that the electron symbol has a nucleon number of zero (because the electron is not a nucleon) and a charge number of −1. Thus, both nucleon number (14) and electric charge (+6) are conserved.

In this type of beta decay, the neutron number of the parent nucleus decreases by one, and the proton number of the daughter nucleus increases by one. Thus, the nucleon number remains unchanged. In essence, it would appear that *a neutron within such an unstable nucleus decays into a proton and an electron (which is then emitted)*:

$$\underset{\text{neutron}}{{}^{1}_{0}\text{n}} \rightarrow \underset{\text{proton}}{{}^{1}_{1}\text{p}} + \underset{\text{electron}}{{}^{0}_{-1}\text{e}} \quad \textit{(basic } \beta^- \textit{ decay)}$$

Beta decay generally happens when a nucleus is unstable because of having too many neutrons compared to protons. (See Section 29.5, which discusses nuclear stability.) The most massive stable isotope of carbon is ^{13}C, with only seven neutrons. But ^{14}C has too many neutrons for a nucleus with six protons and is unstable. Since beta decay simultaneously decreases the neutron number *and* increases the proton number, the product is more stable. In this case, the product nucleus is ^{14}N, which is stable. For completeness, we note that another elementary particle, called a *neutrino*, is emitted in beta decay. For simplicity, it will not be shown in the nuclear-decay equations here. Its important role in beta decay will be discussed more fully in Chapter 30.

There are actually two modes of beta decay, β^- and β^+, as well as a third process called *electron capture*. Whereas β^- decay involves the emission of an electron, **β^+ decay**, or *positron decay*, involves the emission of a positron. The positron is a positive electron—the antiparticle of the electron (see Section 28.5). A positron is symbolized as ${}^{0}_{+1}\text{e}$. Nuclei that undergo β^+ decay have too many protons relative to the number of neutrons. The net effect of β^+ decay is to convert a proton into a neutron. As in β^- decay, this process serves to create a more stable daughter nucleus. An example of β^+ decay is the following:

$$\underset{\text{oxygen}}{{}^{15}_{8}\text{O}_7} \rightarrow \underset{\text{nitrogen}}{{}^{15}_{7}\text{N}_8} + \underset{\text{positron}}{{}^{0}_{+1}\text{e}}$$

Note: Remember that the positron or electron emitted during $\beta^{\pm}$ decay is *not* initially present in the neutron or proton that decays. Among other things, its presence before decay would violate conservation of energy, the uncertainty principle, and conservation of angular momentum. The electron or positron is *created at the time of the decay* and does not exist before that. See Chapter 30 for further details.

Positron emission is also accompanied by a neutrino (but a different type from that associated with β^- decay), which we will also discuss in Chapter 30.

A process that competes with β^+ decay is called **electron capture** (abbreviated as **EC**). This process involves the absorption of *orbital* electrons by a nucleus. The net result is the same daughter nucleus that would have been produced by positron decay—hence describes as *competing*. That is, there is usually a certain probability that *both* can happen. A specific example of electron capture is as follows:

$$\underset{\textit{orbital electron}}{{}^{0}_{-1}\text{e}} + \underset{\textit{beryllium}}{{}^{7}_{4}\text{Be}} \rightarrow \underset{\textit{lithium}}{{}^{7}_{3}\text{Li}}$$

As in β^+ decay, a proton changes into a neutron, but no beta particle is emitted in electron capture.

Gamma Decay

In **gamma decay**, the nucleus emits a gamma (γ) ray, a high-energy photon of electromagnetic energy. The emission of a gamma ray by a nucleus in an excited state is analogous to the emission of a photon by an excited atom. Most commonly, the nucleus emitting the gamma ray is a daughter nucleus left in an excited state after alpha decay, beta decay, or electron capture.

Nuclei possess energy levels analogous to those of atoms. However, nuclear energy levels are much farther apart and more complicated than those of an atom. The nuclear energy levels are typically separated by *kilo*electron-volts (keV) and *mega*electron-volts (MeV), rather than the few electron-volts (eV) that separate energy levels in atoms. As a result, gamma rays are very energetic, having energies larger than those of visible light and, thus, gamma rays have extremely short wavelengths. It is common to indicate a nucleus in an excited state with a superscript asterisk. For example, the decay of ^{61}Ni from an excited nuclear state (indicated by the asterisk) to one of lesser energy would be written as follows:

$$\underset{\substack{\textit{nickel}\\\textit{(excited)}}}{{}^{61}_{28}\text{Ni}^*} \rightarrow \underset{\textit{nickel}}{{}^{61}_{28}\text{Ni}} + \underset{\textit{gamma ray}}{\gamma}$$

Note that *in gamma decay, the mass and proton numbers do not change*. The daughter nucleus is simply the parent nucleus with less energy. As an example of gamma emission following beta decay, consider the following Integrated Example.

Integrated Example 29.2 ■ Two for One: Beta Decay and Gamma Decay

Naturally occurring cesium has only one stable isotope, ${}^{133}_{55}$Cs. However, the unstable isotope ${}^{137}_{55}$Cs is a common nucleus found in used nuclear fuel rods at power plants after their original uranium fuel has become depleted. (See Chapter 30.) When ${}^{137}_{55}$Cs decays, its daughter nucleus is sometimes left in an excited state. After the initial decay, the daughter emits a gamma ray to produce a final stable nucleus. (a) Does ${}^{137}_{55}$Cs first decay by (1) β^+ decay, (2) β^- decay, or (3) electron capture? Explain. (b) Find the final daughter product by writing the chain of decay equations. Show all the steps leading to the final stable nucleus.

(a) Conceptual Reasoning. The ${}^{137}_{55}$Cs isotope has too many neutrons to be stable, as ${}^{133}_{55}$Cs, with four fewer neutrons, is stable. Choices (1) and (3) both increase the number of neutrons relative to the number of protons. Lowering the number of neutrons calls for β^- decay, so the correct choice is (2), β^- decay.

(b) Thinking It Through. Since we know that ${}^{137}_{55}$Cs must decay by emitting a β^- particle, its daughter (in an excited state) can be determined from charge and nucleon conservation. The final state of the daughter will result after a gamma-ray photon is emitted.

The data is as follows:

Given: Initial nucleus of ${}^{137}_{55}$Cs

Find: The decay schemes that lead to the stable nucleus

During β^- decay, the proton number increases by one; thus, the daughter will be barium ($Z = 56$). (See the periodic table, Fig. 28.9.) The decay equation should indicate that barium is left in an excited state that is ready to decay via gamma emission. (As usual in this chapter, the neutrino is omitted.) Thus, the decay equation is

$$\underset{\textit{cesium}}{{}^{137}_{55}\text{Cs}} \rightarrow \underset{\substack{\textit{barium}\\\textit{(excited)}}}{{}^{137}_{56}\text{Ba}^*} + \underset{\textit{electron}}{{}^{0}_{-1}\text{e}}$$

(continues on next page)

This process is then quickly followed by the emission of a gamma ray from the excited barium nucleus:

$$\underset{\text{barium (excited)}}{{}^{137}_{55}\text{Ba}^*} \rightarrow \underset{\text{barium}}{{}^{137}_{55}\text{Ba}} + \underset{\text{gamma ray}}{\gamma}$$

Sometimes this process is written as a combined equation to show the sequential behavior:

$$\underset{\text{cesium}}{{}^{137}_{55}\text{Cs}} \rightarrow \underset{\text{barium (excited)}}{{}^{137}_{56}\text{Ba}^*} + \underset{\text{electron}}{{}^{0}_{-1}\text{e}}$$

$$\downarrow$$

$$\underset{\text{barium}}{{}^{137}_{56}\text{Ba}} + \underset{\text{gamma ray}}{\gamma}$$

Follow-Up Exercise. An unstable isotope of sodium, ^{22}Na, can be produced in nuclear reactors. The only stable isotope of sodium is ^{23}Na. ^{22}Na is known to decay by one type of beta decay. (a) Which type of beta decay is it? Explain. (b) Write down the beta-decay scheme, and predict the daughter nucleus.

Radiation Penetration

The absorption, or degree of penetration, of nuclear radiation is an important consideration in many modern applications. A familiar use of radiation is the radioisotope treatment of cancer. Radiation penetration is also important, for example, in determining the amount of nuclear shielding needed around a nuclear reactor. In our food industry, gamma radiation is now used to penetrate some foods in order to kill bacteria and thus sterilize the food. In industry, the absorption of radiation is used to monitor and control the thickness of metal and plastic sheets in fabrication processes.

The three types of radiation (alpha, beta, and gamma) are absorbed quite differently. As they move along their penetration paths, the electrically charged alpha and beta particles interact with the electrons of the atoms of a material and may ionize some of them. The charge and speed of the particle determine the rate at which it loses energy along its path (remember that ionizing an atom takes energy) and, thus, the degree of penetration. The degree of penetration also depends on properties of the material, such as its density. In general, what happens when the various particles enter a material is as follows:

- Alpha particles are doubly charged, have a relatively large mass, and move relatively slowly. Thus a few centimeters of air or a sheet of paper will usually completely stop them.
- Beta particles are much less massive and are singly charged. They can travel a few meters in air or a few millimeters in aluminum before being stopped.
- Gamma rays are uncharged and are therefore more penetrating than alpha and beta particles. A significant portion of a beam of high-energy gamma rays can penetrate a centimeter or more of a dense material, such as lead. Lead is commonly used as shielding against harmful X-rays and gamma rays. Photons can lose energy or be removed from a beam of gamma rays by a combination of Compton scattering, the photoelectric effect, and pair production (the latter occurs only for photon energies above about 1 MeV). (See Ch. 27.)

Radiation passing through matter can do considerable damage. Structural materials can become brittle and lose their strength when exposed to strong radiation, such as can happen in nuclear reactors (Chapter 30) and to space vehicles exposed to cosmic radiation. In biological tissue, the radiation damage is chiefly due to ionizations in living cells (Section 29.5). We are continually exposed to normal background radiation from radioisotopes in the environment and cosmic radiation from outer space. The energy we absorb and the damage inflicted to cells from exposure to everyday levels of such radiation is usually too low to be harmful. However, concern has been expressed about the radiation exposure of

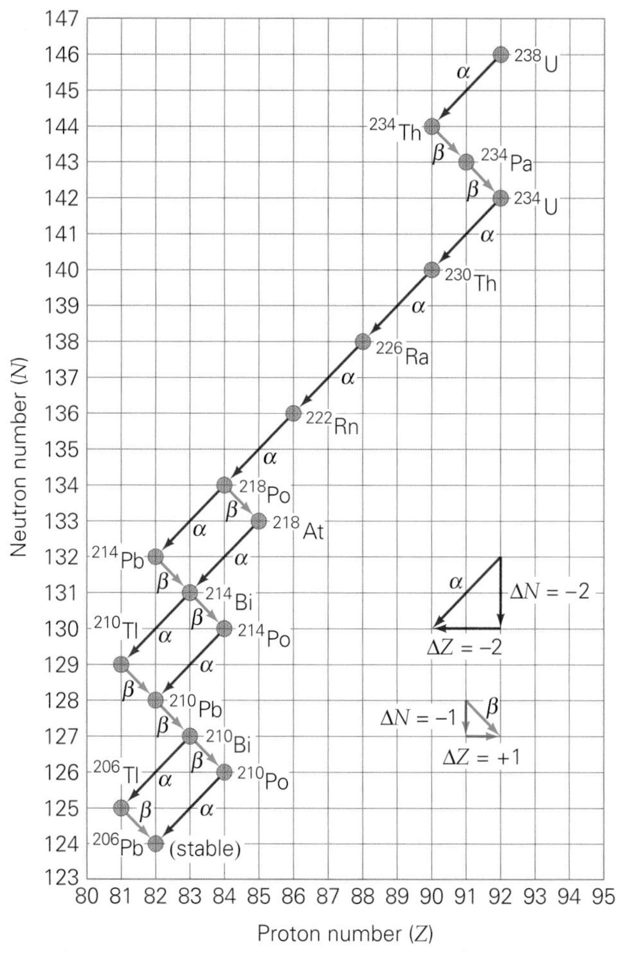

◀ **FIGURE 29.8 Decay series of uranium-238** On this plot of N versus Z, a diagonal transition from right to left is an alpha-decay process, and a diagonal transition from left to right is a β^--decay process. (How can you tell?) The decay series continues until the stable nucleus ^{206}Pb is reached.

people employed in jobs in which radiation levels may be considerably higher. For example, workers at nuclear power plants are constantly monitored for absorbed radiation and subject to strict rules that govern the amount of time for which they can work in a given period. Also, airplane flight crews who spend many hours aboard high-flying jet aircraft may receive significant exposure to radiation from cosmic rays. (Cosmic rays are discussed in more detail on page 915.)

Of the many unstable nuclides, only a small number occur naturally. Most of the radioactive nuclides found in nature are products of the decay series of heavy nuclei. There is continual radioactive decay progressing in a series into successively lighter elements. For example, the ^{238}U decay series (or "chain") is shown in ▲Fig. 29.8. It stops when the stable isotope of lead, ^{206}Pb, is reached. Note that some nuclides in the series decay by two modes and that radon (^{222}Rn) is part of this decay series. This radioactive gas has received a great deal of attention in the last few decades because it can accumulate in significant amounts in poorly ventilated buildings.

Note: See also Figs. 29.20 (^{237}Np) and 29.21 (^{239}Pu).

29.3 Decay Rate and Half-Life

OBJECTIVES: To (a) explain the concepts of activity, decay constant, and half-life of a radioactive sample; and (b) use radioactive decay to find the age of objects.

The nuclei in a sample of radioactive material do *not* decay all at once, but rather do so randomly at a rate characteristic of the particular nucleus and unaffected by external influences. It is impossible to tell exactly when a *particular* unstable nucleus will decay. What can be determined, however, is how many nuclei in a sample will decay during a given period of time.

The **activity (R)** of a sample of radioactive nuclide is defined as the number of nuclear disintegrations, or decays, per second. For a given amount of material, activity decreases with time, as fewer and fewer radioactive nuclei remain. Each nuclide has its own characteristic rate of decrease. The rate at which the number of parent nuclei (N) decreases is proportional to the number present, or $\Delta N/\Delta t \propto N$. This can be rewritten in equation form (using a constant of proportionality called λ) as follows:

$$\frac{\Delta N}{\Delta t} = -\lambda N$$

where the constant λ is called the **decay constant**. This quantity has SI units of s^{-1} (why?) and depends on the particular nuclide. The larger the decay constant λ, the greater is the rate of decay. The minus sign in the previous equation indicates that N is decreasing. The activity (R) of a radioactive sample is the magnitude of $\Delta N/\Delta t$, or the *decay rate*, expressed in decays per second, but without the minus sign (see the usage in Example 29.3):

Note: Activity = number of decays per second = $|\Delta N/\Delta t|$ and is a positive number.

$$R = \text{activity} = \left|\frac{\Delta N}{\Delta t}\right| = \lambda N \quad (29.2)$$

Using calculus, Eq. 29.2 (with the minus sign put back in) can be solved for the number of the remaining (or undecayed) parent nuclei (N) at any time t compared with the number N_o present at $t = 0$. The result is:

$$N = N_o e^{-\lambda t} \quad (29.3)$$

Thus the number of undecayed (parent) nuclei expressed as a fraction of the number initially present (N/N_o) decreases *exponentially* with time, as illustrated in ▾Fig. 29.9. This graph follows an exponentially decaying function $e^{-\lambda t}$. (Remember that $e \approx 2.718$ is the base of natural logarithms and should be available on your calculator.)

The decay rate of a nuclide is commonly expressed in terms of its *half-life* rather than the decay constant. The **half-life ($t_{1/2}$)** is defined as the time it takes for half of the radioactive nuclei in a sample to decay. This is the time corresponding to $N/N_o = \frac{1}{2}$ in Fig. 29.9. In the same amount of time, activity (decays per second) also is cut in half, since the activity is proportional to the number of undecayed nuclei present. Because of this proportionality, the decay rate is usually measured to determine half-life. In other words, what is usually measured is the rate at which the decay particles are emitted, and the time for that rate to drop in half.

For example, by plotting measured decay rates, ▸Fig. 29.10 illustrates that the half-life of strontium-90 (^{90}Sr) can be determined to be 28 years. An alternative way to view the concept of half-life is to consider the mass of parent material. Thus, if there were initially 100 micrograms (μg) of ^{90}Sr, only 50 μg of ^{90}Sr would remain after 28 years. The other 50 μg would have decayed by the following beta-decay process:

$$\underset{\textit{strontium}}{^{90}_{38}\text{Sr}} \rightarrow \underset{\textit{yttrium}}{^{90}_{39}\text{Y}} + \underset{\textit{electron}}{^{0}_{-1}\text{e}}$$

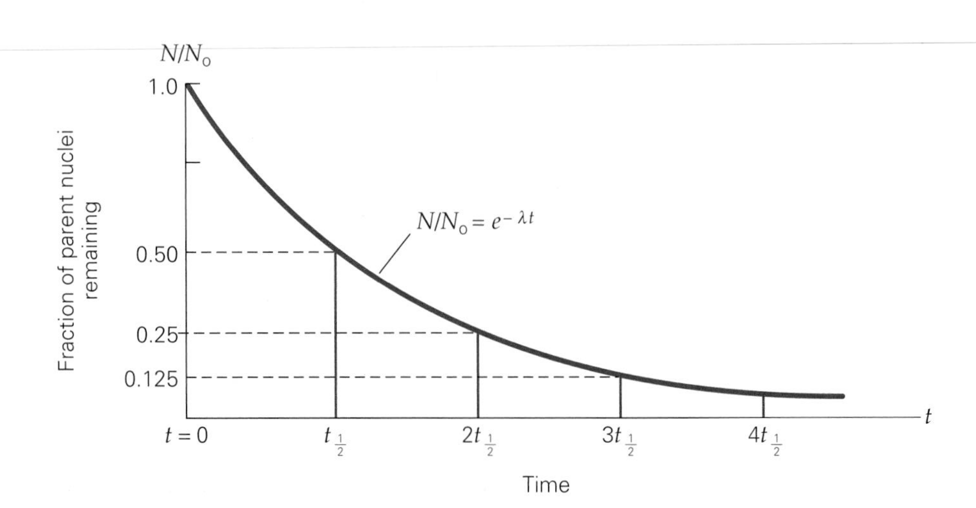

▶ **FIGURE 29.9** Radioactive decay The fraction of the remaining parent nuclei (N/N_o) in a radioactive sample plotted as a function of time follows an exponential-decay curve. The curve's shape and steepness depend on the decay constant λ or the half-life $t_{1/2}$.

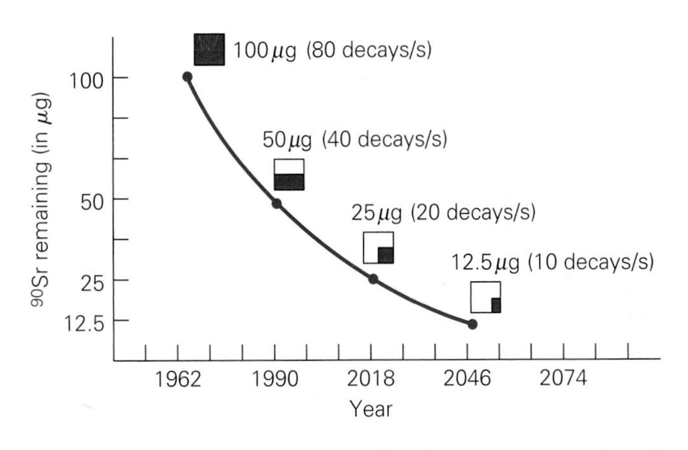

FIGURE 29.10 Radioactive decay and half-life As shown here for strontium-90, after each half-life ($t_{1/2} = 28$ y), only half of the amount of ^{90}Sr present at the start of that period of time remains, with the other half having decayed to ^{90}Y via beta decay. Similarly, the activity (decays per second) has also decreased by half after 28 years.

Thus, the sample would contain a mixture of both strontium and yttrium (and any decay products of yttrium). After another 28 years, half of these strontium nuclei would decay, leaving only 25 μg of ^{90}Sr, and so on.

The half-lives of radioactive nuclides vary greatly, as Table 29.1 shows. Nuclides with very short half-lives are generally created in nuclear reactions (Chapter 30). If these nuclides had existed when the Earth was formed (about 5 billion years ago), they would have long since decayed. In fact, this is the case for technetium (Tc) and promethium (Pm, not shown in Table 29.1). These elements do *not* exist naturally, as they have no stable configurations and their half-lives are short. However, they can be produced in laboratories. Conversely, the half-life of the naturally occurring ^{238}U isotope is about 4.5 billion years. This means that about half of the original ^{238}U present when the Earth was formed exists today. The longer the half-life of a nuclide, the more slowly it decays and the smaller is the decay constant λ. Thus the half-life and the decay constant have an inverse relationship, or $t_{1/2} \propto 1/\lambda$. To show the numerical relationship, consider Eq. 29.3. When $t = t_{1/2}$, then $N/N_o = \frac{1}{2}$. Therefore,

$$\frac{N}{N_o} = \frac{1}{2} = e^{-\lambda t_{1/2}}$$

TABLE 29.1 The Half-Lives of Some Radioactive Nuclides (in Order of Increasing Half-Life)

Nuclide	*Primary Decay Mode*	*Half-Life of Decay Mode*
Beryllium-8 ($^{8}_{4}$Be)	α	1×10^{-16} s
Polonium-213 ($^{213}_{84}$Po)	α	4×10^{-16} s
Oxygen-19 ($^{19}_{8}$O)	β^-	27 s
Fluorine-17 ($^{17}_{9}$F)	β^+, EC	66 s
Polonium-218 ($^{218}_{84}$Po)	α, β^-	3.05 min
Technetium-104 ($^{104}_{43}$Tc)	β^-	18 min
Iodine-123 ($^{123}_{53}$I)	EC	13.3 h
Krypton-76 ($^{76}_{36}$Kr)	EC	14.8 h
Magnesium-28 ($^{28}_{12}$Mg)	β^-	21 h
Radon-222 ($^{222}_{86}$Rn)	α	3.82 days
Iodine-131 ($^{131}_{53}$I)	β^-	8.0 days
Cobalt-60 ($^{60}_{27}$Co)	β^-	5.3 y
Strontium-90 ($^{90}_{38}$Sr)	β^-	28 y
Radium-226 ($^{226}_{88}$Ra)	α	1600 y
Carbon-14 ($^{14}_{6}$C)	β^-	5730 y
Plutonium-239 ($^{239}_{94}$Pu)	α	2.4×10^{4} y
Uranium-238 ($^{238}_{92}$U)	α	4.5×10^{9} y
Rubidium-87 ($^{87}_{37}$Rb)	β^-	4.7×10^{10} y

But because $e^{-0.693} \approx \frac{1}{2}$ (check this on your calculator), we can compare the exponents and determine (to three significant figures) the following result:

$$t_{1/2} = \frac{0.693}{\lambda} \qquad (29.4)$$

The concept of half-life is important in medical applications, as is shown in Example 29.3.

Example 29.3 ■ An "Active" Thyroid: Half-Life and Activity

The half-life of iodine-131 (^{131}I), used in thyroid treatments, is 8.0 days. At a certain time, about 4.0×10^{14} iodine-131 nuclei are in a hospital patient's thyroid gland. (a) What is the ^{131}I activity in the thyroid at that time? (b) How many ^{131}I nuclei remain after 1.0 day?

Thinking It Through. (a) Equation 29.4 enables us to determine the decay constant λ from the half-life, and then use Eq. 29.2 to find the initial activity. (b) To get N, Eq. 29.3 can be used in connection with the e^x button on a calculator.

Solution. Listing the data and converting the half-life into seconds,

Given: $t_{1/2} = 8.0 \text{ days} = 6.9 \times 10^5 \text{ s}$
$N_o = 4.0 \times 10^{14}$ nuclei (initially)
$t = 1.0$ day

Find: (a) R_o (activity at $t = 0$)
(b) N (number of undecayed nuclei after 1.0 day)

(a) The decay constant is determined from its relationship to the half-life (Eq. 29.4) as follows:

$$\lambda = \frac{0.693}{t_{1/2}} = \frac{0.693}{6.9 \times 10^5 \text{ s}} = 1.0 \times 10^{-6} \text{ s}^{-1}$$

Using the initial number of undecayed nuclei, N_o, we find that the initial activity R_o is

$$R_o = \left|\frac{\Delta N}{\Delta t}\right| = \lambda N_o = (1.0 \times 10^{-6} \text{ s}^{-1})(4.0 \times 10^{14}) = 4.0 \times 10^8 \text{ decays/s}$$

(b) With $t = 1.0$ day and $\lambda = 0.693/t_{1/2} = 0.693/8.0 \text{ days} = 0.087 \text{ day}^{-1}$,

$$N = N_o e^{-\lambda t} = (4.0 \times 10^{14} \text{ nuclei})e^{-(0.087 \text{ day}^{-1})(1.0 \text{ day})}$$
$$= (4.0 \times 10^{14} \text{ nuclei})e^{-0.087} = (4.0 \times 10^{14} \text{ nuclei})(0.917) = 3.7 \times 10^{14} \text{ nuclei}$$

The e^x-function calculator button is sometimes labeled as the inverse of the ln x function. Become familiar with it. Here we have $e^{-0.087} \approx 0.917$ to three significant figures.

Follow-Up Exercise. In this Example, suppose that the attending physician will not allow the patient to go home until the activity is $\frac{1}{64}$ of its original level. (a) How long would the patient have to remain in observation? (b) In practice, the amount of time is much shorter than your answer to part (a). Can you think of a possible biological reason(s) for this?

By the "strength" of a radioactive sample we really mean its activity R. A common unit of radioactivity is named in honor of Pierre and Marie Curie.* One **curie (Ci)** is defined as

$$1 \text{ Ci} \equiv 3.70 \times 10^{10} \text{ decays/s}$$

This definition is historical and is based on the known activity of 1.00 g of pure radium. However, the modern SI unit is the **becquerel (Bq)**, which is defined as

$$1 \text{ Bq} \equiv 1 \text{ decay/s}$$

Therefore,

$$1 \text{ Ci} = 3.70 \times 10^{10} \text{ Bq}$$

Even with the present-day emphasis on SI units, the "strengths" of radioactive sources are still commonly specified in curies. The curie is a relatively large unit, however, so the *millicurie* (mCi), the *microcurie* (μCi), and even smaller multiples such as the *nanocurie* (nCi) and *picocurie* (pCi) are used. Teaching laboratories, for

*Marie Sklodowska Curie (1867–1934) was born in Poland and studied in France, where she met and married physicist Pierre Curie (1859–1906). In 1903, Madame Curie (as she is commonly known) and Pierre shared the Nobel Prize in physics with Henri Becquerel (1852–1908) for their work on radioactivity. She was also awarded the Nobel Prize in chemistry in 1911 for the discovery of radium.

example, typically use samples with activities of one microcurie or less. The strength of a source is calculated in the following Example.

Example 29.4 ■ Declining Source Strength: Get a Half-Life!

A ^{90}Sr beta source has an initial activity of 10.0 mCi. How many decays per second will be taking place after 84.0 years?

Thinking It Through. Table 29.1 lists the half-life for the source. In this Example, we can use the fact that in each successive half-life, the activity decreases by half from what it was at the start of that interval. Thus, Eq. 29.3 need not be used, because the elapsed time is exactly three half-lives. (This approach is advisable only when the elapsed time is an integral multiple of the half-life, as it is here.)

Solution.

Given: Initial activity $= 10.0$ mCi
$t = 84.0$ y
$t_{1/2} = 28.0$ y (from Table 29.1)

Find: R (activity after 84.0 years)

Since 84 years is exactly three half-lives, the activity after that amount of time has elapsed will be one-eighth as great $\left(\frac{1}{2} \times \frac{1}{2} \times \frac{1}{2} = \frac{1}{8}\right)$, and the strength of the source will then be

$$R = \left|\frac{\Delta N}{\Delta t}\right| = 10.0 \text{ mCi} \times \frac{1}{8} = 1.25 \text{ mCi} = 1.25 \times 10^{-3} \text{ Ci}$$

In terms of decays per second, or becquerels, we have

$$R = \left|\frac{\Delta N}{\Delta t}\right| = (1.25 \times 10^{-3} \text{ Ci})\left(3.70 \times 10^{10} \frac{\text{decays/s}}{\text{Ci}}\right)$$
$$= 4.63 \times 10^{7} \text{ decays/s} = 4.63 \times 10^{7} \text{ Bq}$$

Follow-Up Exercise. For the material in this Example, suppose a radiation safety officer tells you that this sample can go into the low-level waste disposal only when its activity drops to one millionth of its initial activity. Estimate, to two significant figures, how long the sample must be kept before it can be disposed. [*Hint*: Your calculator may save some time: 2 raised to what power produces about a million?]

Radioactive Dating

Because their decay rates are constant, radioactive nuclides can be used as nuclear clocks. In the previous Example, the half-life of a radioactive nuclide was used to determine how much of the sample will exist in the future. Similarly, by using the half-life to calculate backward in time, scientists can determine the age of objects that contain known radioactive nuclides. As you might surmise, some idea of initial amount of the nuclide present must be known.

To illustrate the principle of radioactive dating, let's look at how it is done with ^{14}C, a very common method used in archeology. **Carbon-14 dating** is used on materials that were once part of living things and on the remnants of objects made from or containing such materials (such as wood, bone, leather, or parchment). The process depends on the fact that living things (including yourself) contain a known amount of radioactive ^{14}C. The concentration is very small—about one ^{14}C atom for every 7.2×10^{11} atoms of ordinary ^{12}C. Even so, the ^{14}C present in our bodies *cannot* be due to ^{14}C that was present when the Earth was formed. This is because the half-life of ^{14}C is $t_{1/2} = 5730$ yr, which is very short in comparison with the age of the Earth.

The ^{14}C nuclei that exist in living things are there because that isotope is continuously being produced in the atmosphere by cosmic rays. *Cosmic rays* are high-speed charged particles that reach us from various sources such as the Sun and nearby exploding stars called supernovae. These "rays" are actually primarily protons. When they enter our upper atmosphere they can cause reactions that produce neutrons (▸Fig. 29.11). These neutrons are then absorbed by the nuclei of the nitrogen atoms of the air, which, in turn, decay by emitting a proton (written as p or $^{1}_{1}$H) to produce ^{14}C by the reaction

$$^{14}_{7}\text{N} + {}^{1}_{0}\text{n} \rightarrow {}^{14}_{6}\text{C} + {}^{1}_{1}\text{H}$$

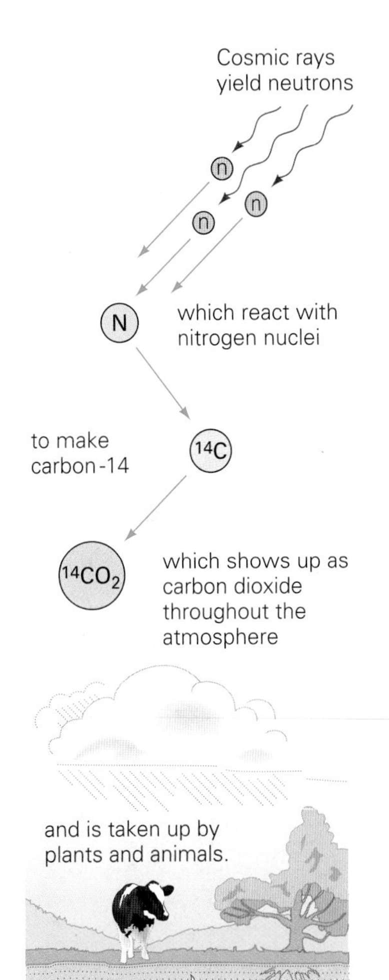
Cosmic rays yield neutrons
n
n
n
N
which react with nitrogen nuclei
to make carbon-14
14C
14CO2
which shows up as carbon dioxide throughout the atmosphere
and is taken up by plants and animals.
But when an organism dies, no fresh carbon-14 replaces the carbon-14 decaying in its tissues, and the carbon-14 radioactivity decreases by half every 5730 years.

▲ **FIGURE 29.11 Carbon-14 radioactive dating** The formation of carbon-14 in the atmosphere and its entry into the biosphere.

Note: The cosmic-ray production of ^{14}C is an example of a nuclear reaction that induces a nuclear transmutation. Such reactions will be studied in more detail in Chapter 30.

^{14}C eventually decays by β^- decay ($^{14}_{6}C \rightarrow {}^{14}_{7}N + {}^{0}_{-1}e$), because it is neutron rich. Although the intensity of incident cosmic rays may not be exactly constant over time, the concentration of ^{14}C in the atmosphere is relatively constant, because of atmospheric mixing and the fixed decay rate.

The ^{14}C atoms are oxidized into carbon dioxide (CO_2), so a small fraction of the CO_2 molecules in the air is radioactive. Plants take in this radioactive CO_2 by photosynthesis, and animals ingest the plant material. As a result, the concentration of ^{14}C in living organic matter is the same as the concentration in the atmosphere, one part in 7.2×10^{11}. However, once an organism dies, the ^{14}C in that organism is *no longer* replenished, and thus the ^{14}C concentration decreases. *Thus, the concentration of ^{14}C in dead matter relative to that in living things can be used to establish when the organism died.* Since radioactivity is generally measured in terms of activity, the ^{14}C activity in organisms now alive must somehow be found. The following Example shows how this is done.

Example 29.5 ■ Living Organisms: Natural Carbon-14 Activity

For ^{14}C, determine the average activity R in decays per minute per gram of natural carbon, found in living organisms, if the concentration of ^{14}C in the organisms is the same as that in the atmosphere.

Thinking It Through. From the previous discussion, we know the concentration of ^{14}C relative to that of ^{12}C. To calculate the ^{14}C activity, we need the decay constant (λ), which can be computed from the half-life of ^{14}C ($t_{1/2} = 5730$ years; see Table 29.1) and the number of ^{14}C atoms (N) per gram. Carbon has an atomic mass of 12.0, so N can be found from Avogadro's number (recall that $N_A = 6.02 \times 10^{23}$ atoms/mole) and the number of moles, $n = N/N_A$ (see Section 10.3).

Solution. Listing the known ratio and the half-life from Table 29.1 (and converting the half-life into minutes), we have

Given:

$$\frac{^{14}C}{^{12}C} = \frac{1}{7.2 \times 10^{11}} = 1.4 \times 10^{-12}$$

$$t_{1/2} = (5730\text{ years})(5.26 \times 10^5\text{ min/year}) = 3.01 \times 10^9\text{ min}$$

Carbon has 12.0 g per mole.

Find: Average activity R per gram

From the half-life, the decay constant is

$$\lambda = \frac{0.693}{t_{1/2}} = \frac{0.693}{3.01 \times 10^9\text{ min}} = 2.30 \times 10^{-10}\text{ min}^{-1}$$

For 1.0 g of carbon, the number of moles is $n = 1.0\text{ g}/(12\text{ g/mol}) = \frac{1}{12}\text{mol}$, so the number of atoms (N) is

$$N = nN_A = \left(\frac{1}{12}\text{mol}\right)(6.02 \times 10^{23}\text{ C nuclei/mol}) = 5.0 \times 10^{22}\text{ C nuclei (per gram)}$$

The number of ^{14}C nuclei per gram is given by the concentration factor

$$N\left(\frac{^{14}C}{^{12}C}\right) = (5.0 \times 10^{22}\text{ C nuclei/g})\left(1.4 \times 10^{-12}\frac{^{14}C\text{ nuclei}}{C\text{ nuclei}}\right) = 7.0 \times 10^{10}\ (^{14}C\text{ nuclei/g})$$

The activity in decays per gram of carbon per minute (to two significant figures) is

$$\left|\frac{\Delta N}{\Delta t}\right| = \lambda N = (2.30 \times 10^{-10}\text{ min}^{-1})(7.0 \times 10^{10}\ ^{14}C/g) = 16\frac{^{14}C\text{ decays}}{g \cdot \text{min}}$$

Thus, if an artifact such as a bone or a piece of cloth has a current activity of 8.0 decays per gram of carbon per minute, then the original living organism would have died about one half-life, or about 5700 years, ago. This would put the date of the artifact near 3700 B.C.

Follow-Up Exercise. Suppose that your instruments could measure ^{14}C beta emissions only down to 1.0 decays/min. How far back (to two significant figures) could you estimate the ages of dead organisms?

Now consider how the activity calculated in Example 29.5 can be used to date ancient organic finds.

Example 29.6 ■ Old Bones: Carbon-14 Dating

A bone is unearthed in an archeological dig. Laboratory analysis determines that there are 20 beta emissions per minute from 10 g of carbon in the bone. What is the approximate age of the bone?

Thinking It Through. Since the initial activity of a living sample is known (Example 29.5), we can work backward to determine the amount of time elapsed.

Solution. For comparison purposes, the activity *per gram* is the relevant number.

Given: Activity = 20 decays/min in 10 g of carbon = 2.0 decays/g · min

Find: Age of the bone

Assuming that the bone had the normal concentration of ^{14}C when the organism died, the ^{14}C activity at the time of death would have been 16 decays/g · min (Example 29.5). Afterward, the decay rate would decrease by half for each half-life:

$$16 \xrightarrow{t_{1/2}} 8 \xrightarrow{t_{1/2}} 4 \xrightarrow{t_{1/2}} 2 \text{ decay/g} \cdot \text{min}$$

So, with an observed activity of 2.0 decays/g · min, the ^{14}C in the bone has gone through approximately three half-lives. Thus, the bone is three half-lives old, or, to two significant figures,

$$\text{Age} \approx 3.0t_{1/2} = (3.0)(5730 \text{ y}) = 1.7 \times 10^4 \text{ y}$$
$$\approx 17\,000 \text{ y}$$

Follow-Up Exercise. Studies indicate that on Earth the stable isotope ^{39}K represents about 93.2% of all the potassium. A long-lived (but unstable) nuclide, ^{40}K, represents only about 0.010%. ^{40}K has a half-life of 1.28×10^9 y. (a) The remainder of the existing potassium (6.8%) is all one other isotope of potassium. What isotope is this most likely to be? (b) What would the percentage ^{40}K abundance have been when the Earth was first formed, assumed to be 4.7×10^9 years ago?

The limit of radioactive carbon dating depends on the ability to measure the very low activity in old samples. Current techniques give an age-dating limit of about 40 000–50 000 years, depending on the sample size. After about ten half-lives, the radioactivity is barely measurable (less than two decays per gram per *hour*).

Another radioactive dating process uses lead-206 (^{206}Pb) and uranium-238 (^{238}U). This dating method is used extensively in geology, because of the long half-life of ^{238}U. Lead-206 is the stable end isotope of the ^{238}U decay series. (See Fig. 29.8.) If a rock sample contains both of these isotopes, the lead is assumed to be a decay product of the uranium that was there when the rock first formed. Thus, the ratio of $^{206}Pb/^{238}U$ can be used to determine the age of the rock.

29.4 Nuclear Stability and Binding Energy

OBJECTIVES: To (a) state which proton- and neutron-number combinations result in stable nuclei, (b) explain the pairing effect and magic numbers in relation to nuclear stability, and (c) calculate nuclear binding energies.

Now that we have considered properties of unstable isotopes, let's turn to the stable ones. Stable isotopes exist naturally for all elements having proton numbers from 1 to 83, *except* those with $Z = 43$ (technetium) and $Z = 61$ (promethium). The nuclear interactions (forces) that determine nuclear stability are extremely complicated. However, by looking at some of the properties of stable nuclei, it is possible to obtain general criteria for nuclear stability.

Nucleon Populations

One of the first considerations is the relative number of protons and neutrons in stable nuclei. Nuclear stability depends on the dominance of the attractive nuclear

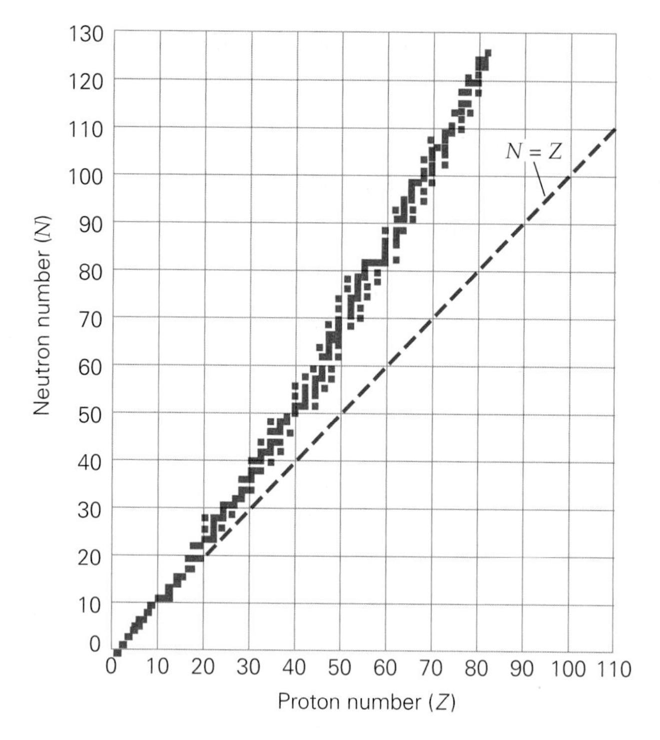

▶ **FIGURE 29.12 A plot of *N* versus *Z* for stable nuclei** For nuclei with mass numbers $A < 40$ ($Z < 20$ and $N < 20$), the number of protons and the number of neutrons are equal or nearly equal. For nuclei with $A > 40$, the number of neutrons exceeds the number of protons, so these nuclei lie above the $N = Z$ line.

force between nucleons over the repulsive Coulomb force between protons. This force dominance depends on the ratio of protons to neutrons.

For stable nuclei of low mass numbers (about $A < 40$), the ratio of neutrons to protons (N/Z) is approximately 1. That is, the number of protons and the number of neutrons are equal or nearly equal. As examples of this, we have $^{4}_{2}He$, $^{12}_{6}C$, $^{23}_{11}Na$, and $^{27}_{13}Al$. For stable nuclei of higher mass numbers ($A > 40$), the number of neutrons exceeds the number of protons ($N/Z > 1$). The heavier the nuclei, the higher is this ratio, that is, the more the neutrons outnumber the protons and the N/Z ratio increases with A.

This trend is illustrated in ▲Fig. 29.12, a plot of neutron number (N) versus proton number (Z) for stable nuclei. The heavier stable nuclei lie above the $N = Z$ line ($N > Z$). Examples of heavy stable nuclei include $^{62}_{28}Ni$, $^{114}_{50}Sn$, $^{208}_{82}Pb$, and $^{209}_{83}Bi$. In fact, bismuth-209 is the heaviest element that has a stable isotope.*

Radioactive decay "adjusts" the proton and neutron numbers of an unstable nuclide until a stable nuclide is produced—that is, until the product nucleus lands on the stability curve in Fig. 29.12. Since alpha decay decreases the numbers of protons and neutrons by equal amounts, alpha decay alone would give nuclei with neutron populations that are *larger* than those of the stable nuclides on the curve. However, β^- decay *following* alpha decay can lead to a stable combination, since the effect of β^- decay is to convert a neutron into a proton. Thus, very heavy unstable nuclei undergo a chain, or sequence, of alpha and beta decays until a stable nucleus is reached (recall Fig. 29.8 for ^{238}U).

Pairing

Many stable nuclei have even numbers of both protons and neutrons, and very few have odd numbers of *both* protons and neutrons. A survey of the stable isotopes (Table 29.2) shows that 168 stable nuclei have even–even combination, while 107 have even–odd or odd–even arrangements, and only 4 contain odd numbers of both protons and neutrons. These four are isotopes of the elements with the four lowest odd proton numbers: $^{2}_{1}H$, $^{6}_{3}Li$, $^{10}_{5}B$, and $^{14}_{7}N$.

The dominance of even–even combinations indicates that the protons and neutrons in nuclei tend to "pair up." That is, two protons pair up and, separately, two neutrons pair up. Aside from the four nuclei mentioned above, all odd–odd nuclei are unstable. Also, odd–even and even–odd nuclei tend to be less stable than the even–even variety.

TABLE 29.2

Pairing Effect of Stable Nuclei

Proton Number	*Neutron Number*	*Number of Stable Nuclei*
Even	Even	168
Even	Odd	107
Odd	Even	
Odd	Odd	4

*Bismuth-209 alpha decays, but with a half-life of 2×10^{18} years; for practical purposes, it is considered to be stable.

This **pairing effect** provides a qualitative criterion for stability. For example, you might expect the aluminum isotope $^{27}_{13}Al$ to be stable (even–odd), but not $^{26}_{13}Al$ (odd–odd). This is the case.

The *general criteria for nuclear stability* can be summarized as follows:

1. All isotopes with a proton number greater than 83 ($Z > 83$) are unstable.

2. (a) Most even–even nuclei are stable.
(b) Many odd–even and even–odd nuclei are stable.
(c) Only four odd–odd nuclei are stable (^2_1H, ^6_3Li, $^{10}_5B$, and $^{14}_7N$).

3. (a) Stable nuclei with mass numbers less than 40 ($A < 40$) have approximately the same number of protons and neutrons.
(b) Stable nuclei with mass numbers greater than 40 ($A > 40$) have more neutrons than protons.

Conceptual Example 29.7 ■ Running Down the Checklist: Nuclear Stability

Is the sulfur isotope $^{38}_{16}S$ likely to be stable?

Reasoning and Answer. We use the general criteria for nuclear stability to analyze this case:

1. *Satisfied*. Isotopes with $Z > 83$ are unstable. With $Z = 16$, this criterion is satisfied.
2. *Satisfied*. The isotope $^{38}_{16}S_{22}$ has an even–even nucleus, so it could be stable.
3. *Not satisfied*. Here, $A < 40$, but $Z = 16$ and $N = 22$ are not approximately equal.

Therefore, the ^{38}S isotope is likely to be unstable. (The nucleus is unstable and decays by β^- emission, since it is neutron rich.)

Follow-Up Exercise. (a) List likely isotopes of copper ($Z = 29$). (b) Apply the criteria for nuclear stability to see which of those isotopes should be stable. Use Appendix V to check your conclusions.

Binding Energy

An important quantitative aspect of nuclear stability is the *binding energy* of the nucleons. Binding energy can be calculated by considering the mass–energy equivalence along with known nuclear masses. Since nuclear masses are so small in relation to the kilogram, another standard, the **atomic mass unit (u)**, is used to measure them. The conversion factor (to six significant figures) between the atomic mass unit and the kilogram is

$$1\ \text{u} = 1.66054 \times 10^{-27}\ \text{kg}$$

Note: The concept of binding energy was introduced in Section 27.4.

The masses of the various particles are typically expressed in atomic mass units. (See Table 29.3.) The listed energy equivalents reflect Einstein's $E = mc^2$ mass–energy equivalence relationship from Eq. 26.11. Thus, a mass of 1 u has an energy equivalent to

$$mc^2 = (1.66054 \times 10^{-27}\ \text{kg})(2.9977 \times 10^8\ \text{m/s})^2 = 1.4922 \times 10^{-10}\ \text{J}$$

$$= \frac{1.4922 \times 10^{-10}\ \text{J}}{1.602 \times 10^{-13}\ \text{J/MeV}} = 931.5\ \text{MeV}$$

TABLE 29.3 The Atomic Mass Unit (u), Particle Masses, and Their Energy Equivalents

Particle	*u*	*Mass (kg)*	*Equivalent Energy (MeV)*
—	1 (exact)	1.66054×10^{-27}	931.5
Electron	5.48578×10^{-4}	9.10935×10^{-31}	0.511
Proton	1.007276	1.67262×10^{-27}	938.27
Hydrogen atom	1.007825	1.67356×10^{-27}	938.79
Neutron	1.008665	1.67500×10^{-27}	939.57

as given in the first entry of Table 29.3. We will use 931.5 MeV/u (to four significant figures) as a handy conversion factor (mass into its energy equivalent) to avoid having to multiply by c^2.

Note in Table 29.3, the proton and hydrogen atom (^1_1H) are listed separately, as they have slightly different masses. This is due to the mass of the atomic electron. We experimentally measure the masses of neutral *atoms* (nucleons plus Z electrons) rather than of their *nuclei*. Keep this factor in mind. Since nuclear-energy calculations usually involve very small differences in mass, the mass of the electron can be significant.

Nuclear stability can be looked at in terms of energy. For example, if the mass of a helium-4 nucleus is compared with the total mass of nucleons that compose it, a significant inequity emerges: A neutral helium atom (including its *two* electrons) has a mass of 4.002 603 u. (Atomic masses of various atoms are given in Appendix V.) The total mass of two hydrogen atoms (^1H) (including *two* electrons) and two neutrons is

$$\begin{aligned} 2m(^1H) &= 2.015\,650 \text{ u} \\ 2m_n &= \underline{2.017\,330 \text{ u}} \\ \text{Total} &= 4.032\,980 \text{ u} \end{aligned}$$

This total is greater than the mass of the helium atom (4.002 603 u). The helium nucleus is less massive than the sum of its parts by an amount

$$\begin{aligned} \Delta m &= [2m(^1H) + 2m_n] - m(^4He) \\ &= 4.032\,980 \text{ u} - 4.002\,603 \text{ u} = 0.030\,377 \text{ u} \end{aligned}$$

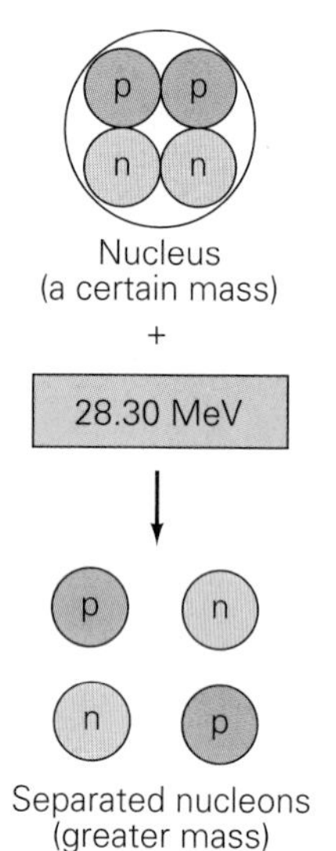

▲ **FIGURE 29.13** Binding energy 28.30 MeV is required to separate a helium nucleus into its constituent protons and neutrons. Conversely, if two protons and two neutrons combine to form a helium nucleus, 28.30 MeV of energy would be released.

(Note that the two electron masses of helium subtract out, since the mass of two hydrogen atoms also included two electrons.) This difference in mass, called the **mass defect**, has an energy equivalent of

$$(0.030\,377 \text{ u})(931.5 \text{ MeV/u}) = 28.30 \text{ MeV}$$

This energy is the *total binding energy* (E_b) of the ^4He nucleus.

In general, for any nucleus, the **total binding energy (E_b)** is related to the mass defect by

$$E_b = (\Delta m)c^2 \quad \textit{total binding energy} \tag{29.5}$$

where Δm is the mass defect. An alternative interpretation of binding energy is that it represents the energy required to separate the constituent nucleons completely into free particles. This concept is illustrated in ◂Fig. 29.13 for the helium nucleus, for which 28.30 MeV of energy is necessary to separate it into four nucleons.

An insight into the nature of the nuclear force can be gained by considering the *average binding energy per nucleon* for stable nuclei. This quantity is the total binding energy of a nucleus, divided by the total number of nucleons, or E_b/A, where A is the mass number. For the helium nucleus (^4He) in Fig. 29.13, the average binding energy per nucleon is

$$\frac{E_b}{A} = \frac{28.30 \text{ MeV}}{4} = 7.075 \text{ MeV/nucleon}$$

Compared with the binding energy of atomic electrons (13.6 eV for a hydrogen electron in the ground state), nuclear binding energies are millions of times larger, indicative of a very strong binding force.

Example 29.8 ■ The Stablest of the Stable: Binding Energy per Nucleon

Compute the average binding energy per nucleon of the iron-56 nucleus ($^{56}_{26}Fe$).

Thinking It Through. The atomic mass of iron-56 is found in Appendix V, and the other needed masses are in Table 29.3. We can then compute the mass defect Δm, the total binding energy E_b, and, lastly, the average binding energy per nucleon, E_b/A.

Solution.

Given: $^{56}_{26}Fe$ mass $= 55.934\,939$ u
^1_1H mass $= m(^1H) = 1.007\,825$ u
^1_0n mass $= m_n = 1.008\,665$ u

Find: E_b/A (average binding energy per nucleon)

(Notice the use of the masses of the iron *atom* and the hydrogen *atom* rather than the nuclear masses.)

The mass defect is the difference between the mass of the iron atom and the mass of its separated constituents. The total mass of the constituents (here, 26 hydrogen atoms and 30 neutrons) is found as follows:

$$26m(^1\text{H}) = 26(1.007\,825\ \text{u}) = 26.203\,450\ \text{u}$$
$$30m_n = 30(1.008\,665\ \text{u}) = 30.259\,950\ \text{u}$$
$$\text{Total} = 56.463\,400\ \text{u}$$

Thus, the mass defect is

$$\Delta m = [26m(^1\text{H}) + 30m_n] - m(^{56}\text{Fe}) = 56.463\,400\ \text{u} - 55.934\,939\ \text{u} = 0.528\,461\ \text{u}$$

The total binding energy is easily calculated using the energy equivalence of 1 u:

$$E_b = (\Delta m)c^2 = (0.528\,461\ \text{u})(931.5\ \text{MeV/u}) = 492.3\ \text{MeV}$$

This iron nuclide has 56 nucleons, so the average binding energy per nucleon is

$$\frac{E_b}{A} = \frac{492.3\ \text{MeV}}{56} = 8.791\ \text{MeV/nucleon}$$

Follow-Up Exercise. (a) To illustrate the pairing effect, compare the average binding energy per nucleon for ^{4}He (calculated previously to be 7.075 MeV) with that of ^{3}He. (Find the atomic masses in Appendix V.) (b) Which one is more tightly bound, on average, and how does your answer reflect pairing?

If E_b/A is calculated for various nuclei and plotted versus mass number, the values lie along the curve shown in ▾Fig. 29.14. The value of E_b/A rises rapidly with increasing A for light nuclei and starts to level off (around $A = 15$) at about 8.0 MeV/nucleon, with a maximum value of about 8.8 MeV/nucleon in the vicinity of iron, which has the most stable nucleus. (See Example 29.8.) For $A > 60$, the E_b/A values decrease slowly, indicating that the nucleons are, on average, less tightly bound.

The importance of the maximum in the curve cannot be understated. Consider what would happen on either side of it. If a massive nucleus split, or *fissioned*, into two lighter nuclei, the nucleons would be more tightly bound, and energy would be released. On the low-mass side of the maximum, if two nuclei could be fused, in a process called **fusion**, a more tightly bound nucleus would be created, and energy would be released. The details and application of these processes will be discussed in detail in Chapter 30.

The E_b/A curve shows that, except for very light nuclei, the binding energy per nucleon does not change a great deal and has a value of $E_b/A \approx 8$ MeV/nucleon. This means that, to a good approximation, we can write $E_b \propto A$. In other words, the *total* binding energy is (approximately) proportional to the total number of nucleons.

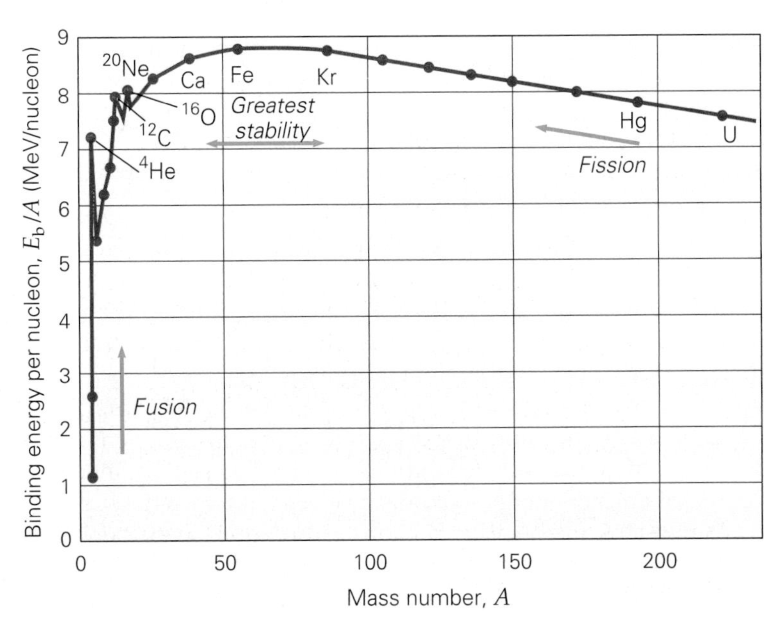

◀ **FIGURE 29.14 A plot of binding energy per nucleon versus mass number** If the binding energy per nucleon (E_b/A) is plotted versus mass number (A), the curve has a maximum near iron (Fe). This indicates that the nuclei in this region are, on average, the most tightly bound and have the greatest stability. Extremely heavy nuclei can release energy by splitting (fissioning). Extremely light nuclei can combine by fusion to release energy.

This proportionality indicates a characteristic of the nuclear force that is quite different from the electrical force. Suppose that the attractive nuclear force *did* act between all the pairs of nucleons in a nucleus. Each pair of nucleons would then contribute to the total binding energy. Considering all combinations, statistics tell us that in a nucleus containing A nucleons, there are $A(A - 1)/2$ pairs. Thus there would be $A(A - 1)/2$ contributions to the total binding energy. For nuclei with $A \gg 1$ (heavy nuclei), $A(A - 1) \approx A^2$, and we would expect the binding energy to be proportional to the *square* of A, or $E_b \propto A^2$ *if the nucleon–nucleon force were to act over a long range*. But as we have seen, in actuality, $E_b \propto A$. This relationship indicates that a given nucleon is *not* bound to all the other nucleons. This phenomenon, called *saturation*, implies that the nuclear forces act over a short range and that any particular nucleon interacts only with its nearest neighbors.

Magic Numbers

We are familiar with the concept of filled shells in atoms. There is an analogous effect in the nucleus. Although the concept of individual nucleon "orbits" inside the nucleus is hard to visualize, experimental evidence does indicate the existence of "closed nuclear shells" when the number of protons *or* neutrons is 2, 8, 20, 28, 50, 82, or 126. Important work on the nuclear-shell model was done by Nobel Prize winner Maria Goeppert-Mayer (1906–1972), a German-born physicist.

The number of stable isotopes of various elements provides solid evidence of the existence of such **magic numbers**. If an element has a magic number of protons, it has an unusually high number of stable isotopes. Elements whose proton number is far away numerically from a magic number may have only 1 or 2 (or even no) stable isotopes. Aluminum, for example, with 13 protons, has just 1 stable isotope, ^{27}Al. But tin, with $Z = 50$ (a magic number), has 10 stable isotopes, ranging from $N = 62$ to $N = 74$. Neighboring indium, with $Z = 49$, has only 2 stable isotopes, and antimony, with $Z = 51$, also has only 2.

Another piece of experimental evidence for magic numbers is related to binding energies. High-energy gamma-ray photons can be used to knock out single nucleons from a nucleus (a phenomenon called the *photonuclear effect*) in a manner analogous to the photoelectric effect in metals. Experimentally, a nuclide with a proton magic number, such as tin, requires about 2 MeV *more* photon energy to eject a proton from its nucleus than does a nuclide that does not have a magic number of protons. Thus, magic numbers are associated with extra large binding energies, another sign of higher-than-average stability.

29.5 Radiation Detection, Dosage, and Applications

OBJECTIVES: To (a) gain insight into the operating principles of various nuclear-radiation detectors, (b) investigate the medical and biological effects of radiation exposure, and (c) study some of the practical uses and applications of radiation.

Detecting Radiation

Since, in general, our senses cannot detect radioactive decay directly, detection must be accomplished through indirect means. For example, people who work with radioactive materials or in nuclear reactors usually wear film badges that indicate cumulative exposure to radiation by the degree of darkening of the film when developed. If more immediate ("real time") and quantitative methods are needed to detect radiation, a variety of instruments is available.

These instruments, as a group, are known as **radiation detectors**. Fundamentally, they are all based on the ionization or excitation of atoms, a phenomenon caused by the passage of energetic particles through matter. The electrically charged alpha and beta particles transfer energy to atoms by electrical interactions, removing electrons and creating ions. Gamma-ray photons can produce ionization by the photoelectric effect and Compton scattering (Section 27.3). They may also produce electrons and positrons by pair production (Section 28.5), if their energy is large enough. Regardless of the source, the particles produced by these interactions, not the actual radiated particles, are the objects "detected" by a radiation detector.

One of the most common radiation detectors is the *Geiger counter*, developed by Hans Geiger (1882–1945), a student and then colleague of Ernest Rutherford. The principle of the Geiger counter is illustrated in ▸Fig. 29.15. A voltage of about 1000 V is applied across the wire electrode and outer electrode (a metallic tube) of the Geiger tube that contains a gas (such as argon) at low pressure. When an ionizing particle enters the tube through a thin window, the particle ionizes some gas atoms. The freed electrons are accelerated toward the positive anode. On their way, they strike and ionize other atoms. This process snowballs, and the resulting "avalanche" produces a current pulse. The pulse is amplified and sent to an electronic counter that counts the pulses, or the number of particles detected. The pulses are sometimes used to drive a loudspeaker so that particle detection is heard as a click.

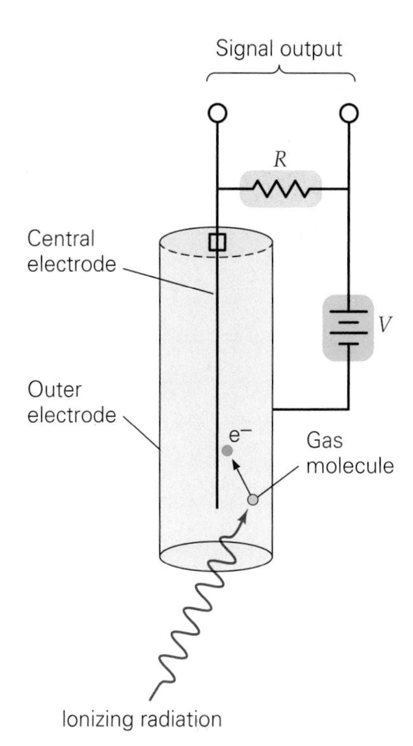

▲ **FIGURE 29.15 The Geiger counter** Incident radiation ionizes a gas atom, freeing an electron that is, in turn, accelerated toward the central (positive) electrode. On the way, this electron produces additional electrons through ionization, resulting in a current pulse that is detected as a voltage across the external resistor *R*.

Another method of detection is the *scintillation counter* (▾Fig. 29.16). Here, the atoms of a phosphor material [such as sodium iodide (NaI)] are excited by an incident particle. A visible-light pulse is emitted when the atoms return to their ground state. The light pulse is converted to an electrical pulse by a photoelectric material. The pulse is then amplified in a *photomultiplier tube*, which consists of a series of electrodes of successively higher potential. The photoelectrons are accelerated toward the first electrode and acquire sufficient energy to cause several secondary electrons from ionization to be emitted when they strike the electrode. This process continues, and relatively weak scintillations are converted into sizable electrical pulses, which are then counted electronically.

In a *solid-state*, or *semiconductor, detector*, charged particles passing through a semiconductor material produce electrons, because of ionization. When a voltage is applied across the material, the electrons are collected as an electric current, which can be amplified and counted.

The three previous detectors determine the number of particles that interact in their material. Other different methods allow the actual trajectory, or "tracks," of charged particles to be seen and/or recorded. Among this type of detector are the cloud chamber, the bubble chamber, and the spark chamber. In the first two, vapors and liquids are supercooled and superheated, respectively, by suddenly varying the volume and pressure.

The *cloud chamber* was developed early in the 1900s by C. T. R. Wilson, a British atmospheric physicist. In the chamber, supercooled vapor condenses into droplets on the sites of ionized molecules created along the path of an energetic particle. When the chamber is illuminated, the droplets scatter the light, making the path visible (▸Fig. 29.17).

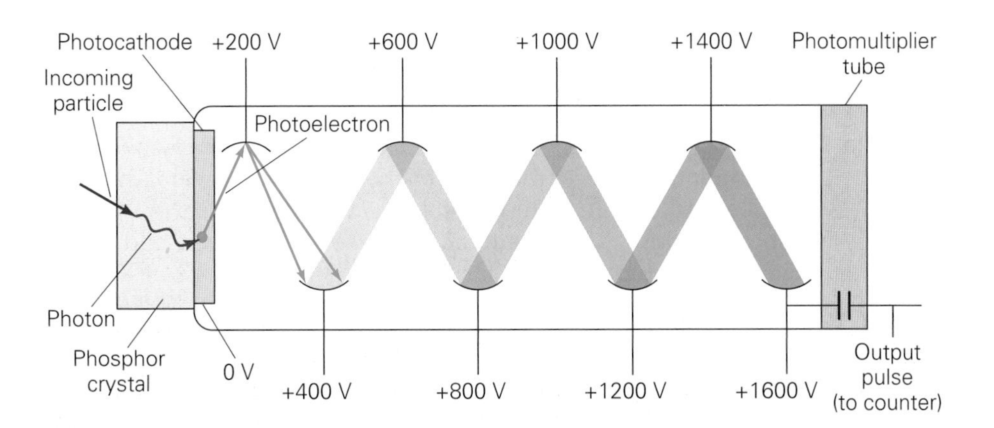

▲ **FIGURE 29.16 The scintillation counter** A photon emitted by a phosphor atom excited by an incoming particle causes the emission of a photoelectron from the photocathode. Accelerated through a difference in potential in a photomultiplier tube, the photoelectrons free secondary electrons when they collide with successive electrodes at higher potentials. After several steps, a relatively weak scintillation is converted into a measurable electric pulse.

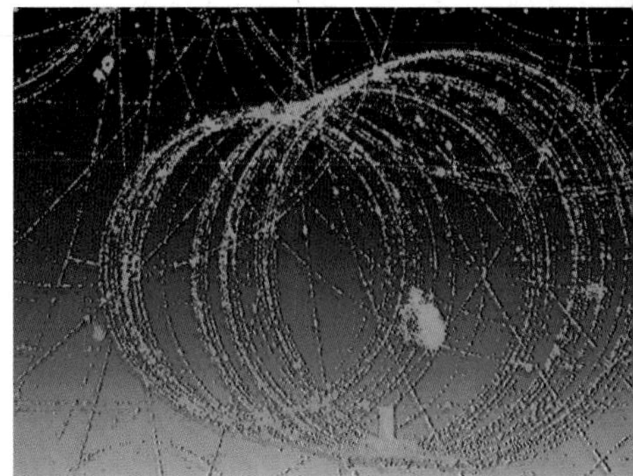

▲ **FIGURE 29.17 Cloud-chamber tracks** The circular track in this photograph of a cloud chamber was made by a positron in a strong magnetic field. (Can you explain the approximately circular path of the particle in terms of the orientation of the magnetic field relative to the positron's velocity?)

The *bubble chamber*, which was invented by the American physicist D. A. Glazer in 1952, uses a similar principle. A reduction in pressure causes a liquid to be superheated and able to boil. Ions produced along the path of an energetic particle become sites for bubble formation, and a trail of bubbles is created. Since the bubble chamber uses a liquid, commonly liquid hydrogen, the density of atoms in it is much greater than in the vapor of a cloud chamber. Thus, tracks are more readily observable in bubble chambers, and hence bubble chambers have largely replaced cloud chambers.

In a *spark chamber*, the path of a charged particle is registered by a series of sparks. The charged particle passes between a pair of electrodes that have a high difference in potential and are immersed in an inert (noble) gas. The charged particle causes the ionization of gas molecules, giving rise to a visible spark (flash of light) between the electrodes as the released electrons travel to the positive electrode. A spark chamber is merely an array of such electrodes in the form of parallel plates or wires. A series of sparks, which can be photographed, then marks the particle's path.

Once the particle's trajectory is displayed, the particle's energy can be determined. Typically, a magnetic field is applied across the chamber, any charged particles are deflected, and the energy of a particle can be calculated from the radius of curvature of its path. Gamma rays, of course, will not leave visible tracks in any of these detectors. However, their presence can be detected indirectly because they are able to produce electrons by such processes as the photoelectric effect and pair production. The gamma-ray energy can be determined from the measured energy of these electrons.

Biological Effects and Medical Applications of Radiation

In medicine, nuclear radiation can be used beneficially in the diagnosis and treatment of some diseases, but it also is potentially harmful if not handled and administered properly. Nuclear radiation and X-rays can penetrate human tissue without pain or any other sensation. However, early investigators quickly learned that large doses or repeated small doses can lead to reddened skin, lesions, and other conditions. It is now known that certain types of cancers can be caused by excessive exposure to radiation.

The chief hazard of radiation is damage to living cells, due primarily to ionization. Ions, particularly complex ions or radicals produced by radiation, may be highly reactive (for example, a hydroxyl ion [OH^-] produced from water). Such reactive ions interfere with the normal chemical operations of the cell. If enough cells are damaged or killed, cell reproduction might not be fast enough, and the irradiated tissue could eventually die. In other instances, genetic damage, or mutation, may occur in a chromosome in the cell nucleus. If the affected cells are sperm or egg cells (or their precursors), any children that they produce may have various birth defects. If the damaged cells are ordinary body cells, they may become cancerous, reproducing in a rapid and uncontrolled manner and eventually becoming a malignant tumor. The human cells most susceptible to radiation damage are those of the reproductive organs, bone marrow, and lymph nodes. To begin our discussion of radiation damage and applications, let us investigate how the radiation "dose" is quantified.

Radiation Dosage An important consideration in radiation therapy and radiation safety is the amount, or *dose,* of radiation energy absorbed. Several quantities are used to describe this amount in terms of *exposure, absorbed dose,* or *equivalent dose*. The earliest unit of dosage, the **roentgen (R)**, based on exposure and defined in terms of ionization produced in air. One roentgen is the quantity of X-rays or gamma rays required to produce an ionization charge of 2.58×10^{-4} C/kg in air.

The **rad** (**r**adiation **a**bsorbed **d**ose) is an *absorbed dose* unit. One rad is an absorbed dose of radiation energy of 10^{-2} J/kg of absorbing *material*. Note that the rad is based on energy absorbed from the radiation rather than simply ionization caused by the radiation in air (as the roentgen is). As such, it is more directly related to the biological damage caused by the radiation. Because of this characteristic, the rad has largely replaced the roentgen.

Note: The gray was named in honor of Louis Harold Gray, a British radiobiologist whose studies laid the foundation for measuring absorbed dose.

The rad is not an SI unit. The SI unit for absorbed dose is the **gray (Gy)**, defined as

$$1 \text{ Gy} = 1 \text{ J/kg} = 100 \text{ rad}$$

TABLE 29.4 Typical Relative Biological Effectiveness (RBE) Values of Various Types of Radiations

Type	*RBE (or QF)*
X-rays and gamma rays	1
Beta particles	1.2
Slow neutrons	4
Fast neutrons and protons	10
Alpha particles	20

However, the most meaningful assessment of the effects of radiation must involve measuring the *biological damage* produced, because it is well known that equal doses (in rads) of different types of radiation produce *different* effects. For example, a relatively massive alpha particle with a charge of $+2e$ moves through the tissue rather slowly, with a great deal of electrical interaction. The ionizing collisions thus occur close together along a short penetration path and are more localized. Therefore, potentially more dangerous damage is done by alpha particles than by electrons or gamma rays.

This *effective dose* is measured in terms of the **rem** (**r**ad **e**quivalent **m**an). The various degrees of effectiveness of different particles are characterized by a factor called **relative biological effectiveness (RBE)**, or *quality factor* (QF), which has been tabulated for various particles in Table 29.4. (Note in Table 29.4 that X-rays and gamma rays have, by definition, an RBE of 1.)

The effective dose is given by the product of the dose in rads and the appropriate RBE:

$$\text{effective dose (in rems)} = \text{dose (in rads)} \times \text{RBE} \tag{29.6}$$

Thus, 1 rem of *any* type of radiation does approximately the same amount of biological damage. For example, a 20-rem effective dose of alpha particles does the same amount of damage as a 20-rem effective dose of X-rays. However, note that to administer these doses, 20 rad of X-rays is needed, compared with only 1 rad of alpha particles.

Remember that the SI unit of *absorbed dose* is the gray. The SI unit of *effective dose* is the **sievert (Sv)**:

$$\text{effective dose (in sieverts)} = \text{dose (in grays)} \times \text{RBE} \tag{29.7}$$

Since 1 Gy = 100 rad, it follows that 1 Sv = 100 rem.

It is difficult to set a maximum permissible radiation dosage, but the general standard for humans is an average dose of 5 rem/yr after age 18, with no more than 3 rem in any three-month period. In the United States, the normal average annual dose per capita is about 200 mrem (millirem). About 125 mrem comes from the natural background of cosmic rays and naturally occurring radioactive isotopes in soil, building materials, and so on. The remainder is chiefly from diagnostic medical applications, mostly X-rays.

Medical Treatment Using Radiation Some radioactive isotopes can be used for medical treatment, typically for cancerous conditions. Since a radioactive isotope, or a *radioisotope* as it is sometimes called, behaves chemically like a stable isotope of the element, it can participate in chemical reactions associated with normal bodily functions. One such radioisotope, used to treat thyroid cancer, is ^{131}I. Under usual conditions, the thyroid gland absorbs normal iodine. However, if ^{131}I is absorbed in a large enough dose, it can kill cancer cells. To see how a dose of radiation to the thyroid from ^{131}I can be estimated, consider Example 29.9. For a discussion of further uses of radioisotopes, see Insight 29.1 on Biological and Medical Applications of Radiation (page 927).

Example 29.9 ■ Radiation Dosage: Iodine-131 and Thyroid Cancer

One method of treating a cancerous thyroid is to administer a hefty amount of the radioactive isotope ^{131}I. The thyroid absorbs this iodine, and the iodine's gamma rays kill cells in the thyroid. (For data on ^{131}I, see Example 29.3.) (a) Write down the decay scheme of ^{131}I, and predict the identity of the daughter nucleus, which, in this case, is stable after emitting a gamma ray. (b) The charged particle (part (a) tells the type) has an average kinetic energy of 200 keV. Assume that the patient was given 0.0500 mCi of ^{131}I and that the thyroid absorbs only 25% of this. Further assume that only 40% of that 25% actually decays in the thyroid. If all of the energy carried by the charged particles is deposited in the thyroid, estimate the dose received by the thyroid (50.0 g) due to the ionization created by the charged particle radiation only. (Do not include the effect of the gamma rays.)

Thinking It Through. (a) The decay will be β^- decay, because the initial nucleus contains too many neutrons (78 neutrons compared with the 74 for stable iodine). The daughter nucleus is determined by its proton number. The daughter is left in an excited state and emits a gamma ray in order to become stable. (b) The dose depends on the energy deposited per kilogram of thyroid. Hence, we need to know how many β^- particles are emitted (and therefore absorbed). This number is determined by the initial number of ^{131}I nuclei that are actually present in the thyroid. The effective dose will depend on the RBE for β^- particles, found in Table 29.4.

Solution.

Given: 0.0500 mCi of ^{131}I ingested
25% of the ^{131}I makes it to the thyroid
40% of the ^{131}I that makes it to the thyroid decays there
$\overline{K}_\beta = 200$ keV
$m = 50.0 \text{ g} = 0.0500 \text{ kg}$
RBE (see Table 29.4)

Find: (a) the decay scheme for ^{131}I
(b) the dose (in rem) from the emitted particles

(a) Looking up the element with $Z = 54$, we find that it is xenon (Xe). The decay scheme is therefore

$$\underset{\textit{iodine}}{{}^{131}_{53}\text{I}} \rightarrow \underset{\textit{xenon (excited)}}{{}^{131}_{54}\text{Xe}^*} + \underset{\textit{beta}}{\beta^-}$$

$$\downarrow$$

$$\underset{\textit{xenon}}{{}^{131}_{54}\text{Xe}} + \underset{\textit{gamma ray}}{\gamma}$$

(b) Of the 0.0500 mCi, only 0.0125 mCi makes it to the thyroid. Of that, only 40%, or 0.00500 mCi (5.00×10^{-6} Ci), actually decays in the thyroid. From this information and Eq. 29.2 the number of ^{131}I nuclei that decay in the thyroid can be found. From Example 29.3, the decay constant for ^{131}I is $\lambda = 1.0 \times 10^{-6}\ \text{s}^{-1}$. Thus, the number of ^{131}I nuclei, N, that decays in the thyroid is

$$N = \frac{R}{\lambda} = \frac{(5.00 \times 10^{-6}\ \text{Ci})\left(3.7 \times 10^{10}\,\dfrac{\text{nuclei/s}}{\text{Ci}}\right)}{1.0 \times 10^{-6}\ \text{s}^{-1}} = 1.85 \times 10^{11}\ {}^{131}\text{I nuclei}$$

Each ^{131}I nucleus releases one β^- particle, with an average kinetic energy of 200 keV. Remembering that there is 1.60×10^{-19} J/eV, or 1.60×10^{-16} J/keV, we find that the energy, E, deposited in the thyroid is

$$E = (1.85 \times 10^{11}\ {}^{131}\text{I nuclei})\left(200\,\frac{\text{keV}}{{}^{131}\text{I nuclei}}\right)\left(1.60 \times 10^{-16}\,\frac{\text{J}}{\text{keV}}\right)$$
$$= 5.92 \times 10^{-3}\ \text{J}$$

The absorbed dose is

$$\text{absorbed dose} = \frac{5.92 \times 10^{-3}\ \text{J}}{0.0500\ \text{kg}} = 0.118\,\frac{\text{J}}{\text{kg}}$$
$$= 0.118\ \text{Gy or } 11.8\ \text{rad}$$

The effective dose (in sieverts and rems) is this multiplied by the RBE for beta particles (1.2):

$$\text{effective dose} = (0.118\ \text{Gy})(1.2) = 0.142\ \text{Sv} = 14.2\ \text{rem}$$

Follow-Up Exercise. In this Example, determine the absorbed and effective dose from the gamma rays, assuming that 10% of the gamma rays are absorbed in the thyroid tissue and that their energy is 364 keV. Compare these results with the dose from the β^- radiation.

INSIGHT 29.1 BIOLOGICAL AND MEDICAL APPLICATIONS OF RADIATION

Radiation has always been a double-edged sword. We know of its potentially harmful side, yet sources of radiation can also provide solutions to problems. Exposure to high levels of gamma radiation is now an approved method of food sterilization in the United States. Chicken and beef are commonly sterilized this way, thus reducing the threat of *Salmonella* and *E. coli* contamination. In the aftermath of the terrorist attacks on the United States in the fall of 2001, some of this gamma-radiation technology is being retooled so that it can kill anthrax spores and other weapons based on organisms.

External radiation sources such as ^{60}Co are also used to treat cancer. ^{60}Co emits energetic gamma rays with energies of 1.17 and 1.33 MeV. Thus a sample of ^{60}Co, with its relatively long half-life, can provide an inexpensive and convenient source of penetrating radiation. Essentially, all you need is the ^{60}Co sample in a lead box with a hole to allow gamma rays to exit in one direction. One problem, however, with this "single-beam" method is that the gamma rays deposit energy in the healthy flesh both in front of and behind the targeted tumor.

An improved version of ^{60}Co treatment, called the *gamma knife*, is in use at several research hospitals (Fig. 1). Once the tumor is located, beams of gamma rays from ^{60}Co sources arranged in a ring are accurately aimed at it. Any one source is relatively weak and thus does not do too much damage outside the tumor itself. However, where the beams meet at the tumor site, a lot of energy is deposited. Thus, a large dose can be deposited at the tumor, with minimal damage to surrounding tissue. This instrument requires careful computer calculations and is still under development.

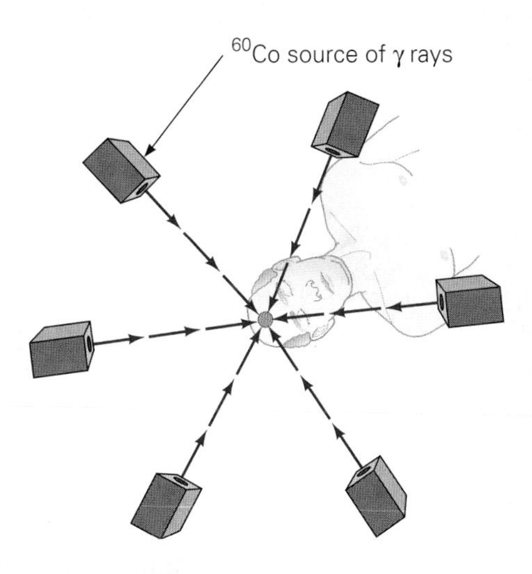

FIGURE 1 The gamma knife consists of many relatively weak beams of gamma rays (usually from ^{60}Co) that are calculated to converge on the location of an inoperable tumor. Thus, unlike in the traditional single-beam treatment, much more of the energy ends up at the tumor site, and less damage is done in front of or behind the tumor.

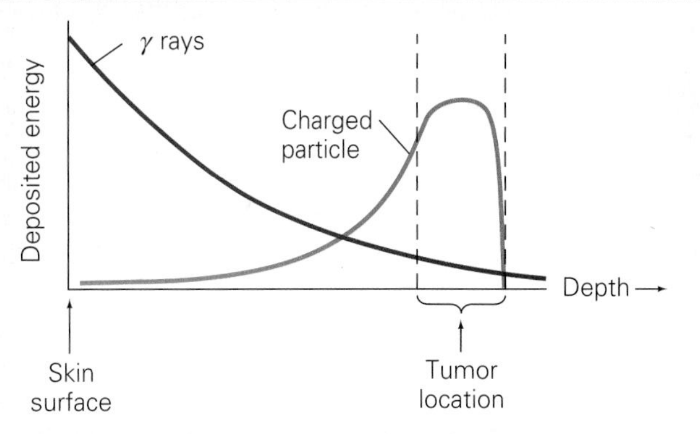

FIGURE 2 A comparison of the energy deposited for a gamma-ray beam with that of a charged particle, such as a negative pion, passing through tissue.

Even more exotic techniques are being tested in an effort to kill inoperable tumors. One such method is *pion therapy*. A pion is an unstable elementary particle (Chapter 30) that can be produced in accelerators by bombarding a target, such as carbon, with high-energy protons. Of medical interest are the negatively charged pions π^- (positive and neutral ones exist also). Pions of a specific kinetic energy can be selected and focused by magnetic fields onto a region of the body where a tumor exists. Unlike photons (gamma rays and X-rays), charged particles create most of their ionization "damage" at the end of their path, when they are moving slowly. By adjusting their kinetic energy, researchers can end the pion's path right at the tumor site (Fig. 2), thus causing maximum damage to the cancer cells. As a bonus, since the pions are unstable, they give off gamma rays that will do even more damage from inside the tumor (Fig. 3). Research on this technology-intensive technique is ongoing, with some success.

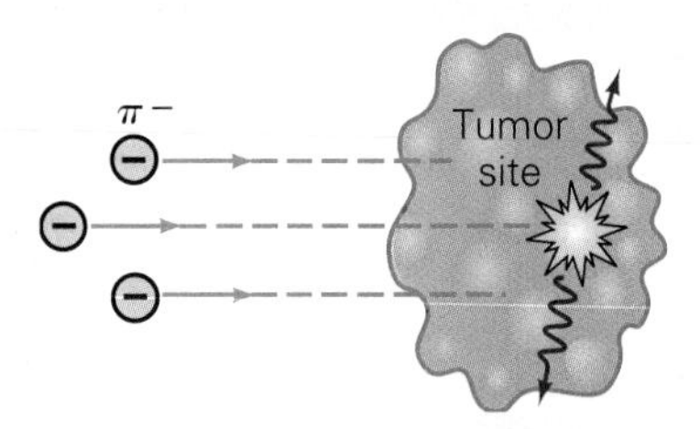

FIGURE 3 Pionic cancer therapy consists of focusing a beam of negative pions onto a tumor. Unlike energy from beams of gamma rays, most of the pion energy is deposited at the tumor site. In addition, when pions decay, gamma rays are released at the tumor, causing further destruction of cancer cells.

Medical Diagnosis Applications That Use Radiation Besides theraputic use, such as that previously described for iodine-131, radioactive isotopes can be used for diagnostic procedures. Since the radioisotope behaves chemically like a stable isotope, attaching radioisotopes to molecules enables the molecules to be used as tracers as they travel to different organs and regions of the body.

Many bodily functions can be studied by monitoring the location and activity of tracer molecules as they are absorbed during body processes. For example, the

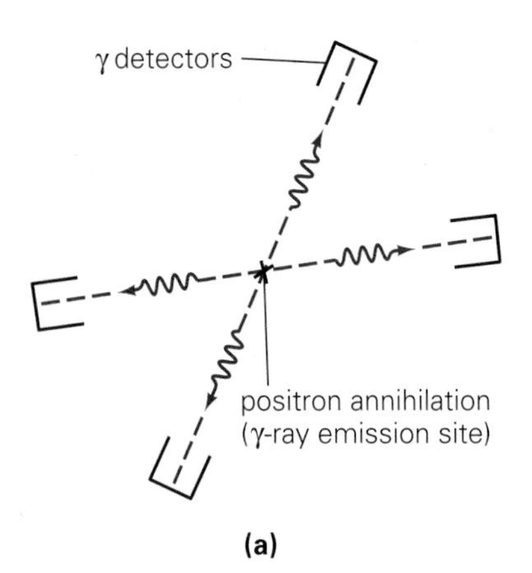

(a)

(b)

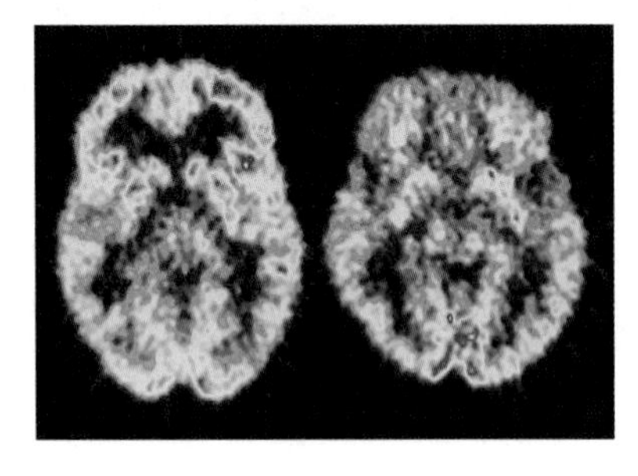

(c)

▶ **FIGURE 29.18** PET scan **(a)** and **(b)** A PET scanner can, for example, monitor brain activity after the administration of glucose-containing radioactive isotopes. Note the use of an array of detectors for the gamma radiation produced when a positron is annihilated. Any one detector pair pinpoints a line on which the source of the gamma emission was located. With many such pairs working together, the site can be determined to an accuracy of about a centimeter. **(c)** PET scans of a normal brain (left) and the brain of a schizophrenic patient (right).

activity of the thyroid gland can be determined by monitoring its iodine uptake with small amounts of radioactive iodine-123. This isotope emits gamma rays and has a half-life of 13.3 h. The uptake of radioactive iodine by a person's thyroid can be monitored by a gamma detector and compared with the function of a normal thyroid to check for abnormalities. Similarly, radioactive solutions of iodine and gold are quickly absorbed by the liver.

One of the most commonly used diagnostic tracers is technetium-99 (^{99}Tc). It has a convenient half-life of 6 h, emits gamma rays, and combines with a large variety of compounds. When injected into the bloodstream, ^{99}Tc will not be absorbed by the brain, because of the blood–brain barrier. However, tumors do not have this barrier, and brain tumors readily absorb the ^{99}Tc. These tumors then show up as gamma-ray emitting sites using detectors external to the body. Similarly, other areas of the body can be scanned and unusual activities noted and measured.

It is possible to image gamma-ray activity in a single plane, or "slice," through the body. A gamma detector is moved around the patient to measure the emission intensity from many angles. A complete image can then be constructed by using computer-assisted tomography, as in X-ray CT. This process is referred to as *single-photon emission tomography* (SPET). Another technique, *positron emission tomography* (PET), uses tracers that are positron emitters, such as ^{11}C and ^{15}O. When a positron is emitted, it is quickly annihilated, and two gamma rays are produced that travel in opposite directions. The gamma rays are recorded simultaneously by a ring of detectors surrounding the patient (◂Fig. 29.18).

In a common application, PET technology is used to detect fast-growing cancer cells. The positron emitter ^{18}F is chemically attached to glucose molecules and administered to the patient. Actively growing cells absorb glucose, but *very* active cancer cells absorb considerably more. By comparing the emissions coming from a given region on a potentially sick patient to that from a normal, healthy person, these "overactive" cancer cells can be detected. Such PET scans are now routinely done, for example, as a follow-up to chemotherapy treatment of lymphoma (cancer of the lymph system). A PET scan can detect even tiny leftover active tumors that can then be targeted for further treatment.

Domestic and Industrial Applications of Radiation

A common application of radioactivity in the home is the smoke detector. In this detector, a weak radioactive source ionizes air molecules. The freed electrons and the positive ions are collected using the voltage of a battery, thus setting up a small current in the detector circuit. If smoke enters the detector, the ions there become attached to the smoke particles, causing a reduction in the current. The drop in current is sensed electronically, which triggers an alarm (▸Fig. 29.19).

Industry also makes good use of radioactive isotopes. Radioactive tracers are used to determine flow rates in pipes, to detect leaks, and to study corrosion and wear. Also, it is possible to radioactivate certain compounds at a particular stage in a process by irradiating them with particles, generally neutrons. This technique is called **neutron activation analysis** and is an important method of identifying elements in a sample. Before the development of this procedure, the chief methods of identification were chemical and spectral analyses. In both of these methods, a fairly large amount of a sample has to be destroyed during the procedure. As a result, a sample may not be large enough for analysis, or small traces of elements in a sample may go undetected. Neutron activation analysis has the advantage over these methods on both scores. Only minute samples are needed, and the method can detect very minute trace amounts of an element.

A typical neutron activation process might start with californium-252, an unstable neutron emitter that can be produced artificially:

$$\underset{(source)}{{}^{252}_{98}\text{Cf}} \rightarrow {}^{251}_{98}\text{Cf} + {}^{1}_{0}\text{n}$$

These neutrons are used to bombard a sample and create characteristic gamma rays. A common target is nitrogen, consisting mostly of ^{14}N. When ^{14}N absorbs a neutron, the ^{15}N nucleus is usually created in an excited state. The excited nitrogen-15

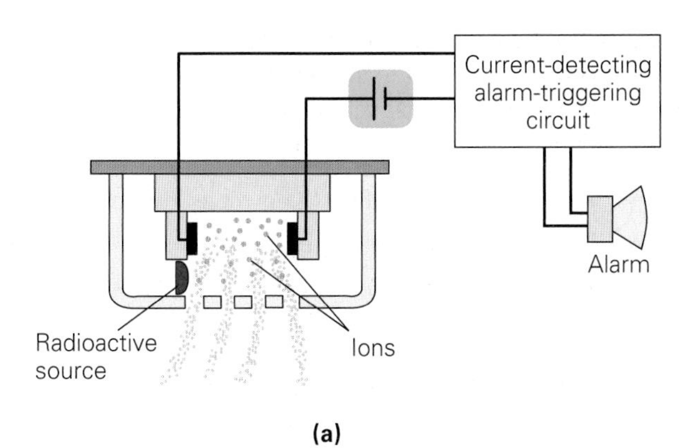

◀ **FIGURE 29.19** Smoke detector
(a) A weak radioactive source ionizes the air and sets up a small current. Smoke particles that enter the detector attach to some of the ionized electrons, thereby reducing the current, causing an alarm to sound. **(b)** Inside a real smoke detector, the ionization chamber is the aluminum "can" containing the americium-241. The slots allow airflow, and the can acts as one of the charged plates. Inside is a ceramic holder that contains the oppositely charged plate. Under that plate is the americium-241 source. Even though the activity of the source is small, caution should be taken never to touch the source if the detector is opened like this.

nucleus decays with the emission of a gamma ray with a distinctive energy. This reaction is shown as follows:

$$ {}^{1}_{0}\mathrm{n} + {}^{14}_{7}\mathrm{N} \rightarrow \underset{\textit{(excited nucleus)}}{{}^{15}_{7}\mathrm{N}^{*}} \rightarrow {}^{15}_{7}\mathrm{N} + \underset{\textit{(gamma ray)}}{\gamma} $$

When an energy-sensitive gamma-ray detector is placed to the side of the sample, the presence of nitrogen can be determined.

Nitrogen activation is commonly used as an important antiterrorist tool at airports. Virtually all explosives contain nitrogen. Thus, by using neutron activation and analyzing the energy of any gamma-ray emission coming from a suitcase, we can check for an explosive device in the suitcase. Other materials in the suitcase may contain nitrogen too, so manual checks are made to confirm any suspicious findings.

Recently, the U.S. government has given permission for the use of gamma radiation in the processing of poultry. The radiation kills bacteria, helps preserve the food, and in no way makes the food radioactive. There are, however, continuing concerns with this process from food health professionals. Even though the gamma-ray emission cannot make the meat radioactive, it *can* change some of the *chemical* bonding through ionization effects. This possibility has prompted enough concerns about whether this process can affect the chemical structure of the meat—making it unsafe to eat—to warrant further study.

Chapter Review

- The **nuclear force** is the short-range attractive force between nucleons that is responsible for holding the nucleus together.
- The nucleus of an atom contains protons and neutrons, collectively called **nucleons**.
- The nucleus is characterized by its **proton number** Z and its **neutron number** N. Its **mass number**, A, is the total number of nucleons, so $A = N + Z$.

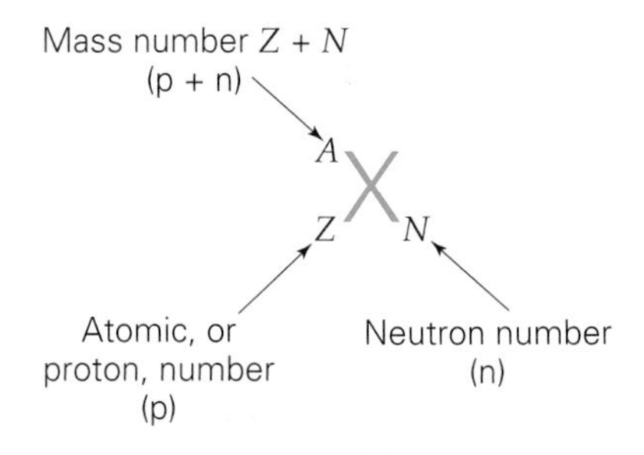

- **Isotopes** of a given element differ only in the number of neutrons in their nucleus.
- Nuclei may undergo **radioactive decay** by the emission of an **alpha particle** (a helium nucleus) **(α)**; a **beta particle**, which can be either an electron **(β^-)** or a positron **(β^+)**; or a **gamma ray (γ)**, a high-energy photon of electromagnetic radiation.

Some nuclei become more stable by capturing an orbital electron (electron capture, or **EC**).

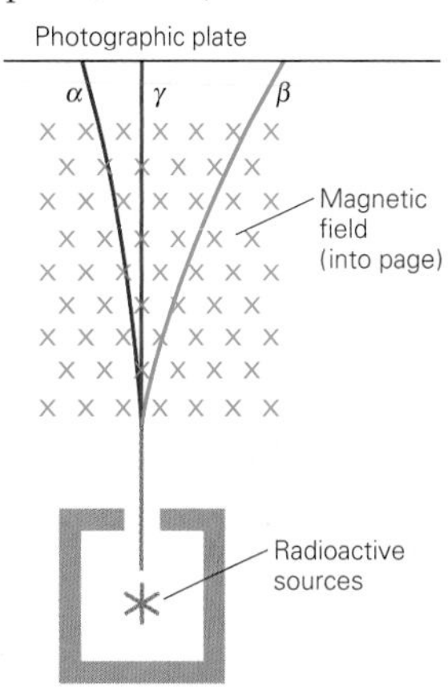

- In any nuclear process, two conservation rules pertain: **Conservation of nuclear number** and **conservation of charge**.
- The **half-life** of a nuclide is the time required for the number of undecayed nuclei in a sample to fall to half of its initial value. The number of undecayed nuclei remaining after a time t is given by the exponential-decay relationship

$$ N = N_0 e^{-\lambda t} \qquad (29.3) $$

where the **decay constant (λ)** is inversely related to the half-life by

$$t_{1/2} = \frac{0.693}{\lambda} \quad (29.4)$$

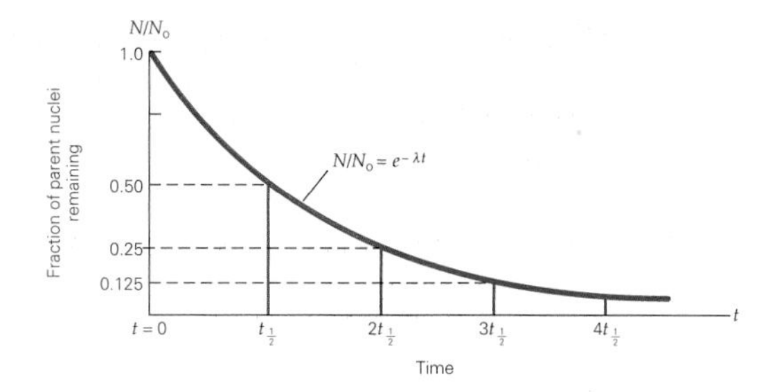

The **activity (R)** of a radioactive sample is the rate at which the nuclei in the sample decay. It is proportional to the number of undecayed nuclei and is given by

$$R = \text{activity} = \left|\frac{\Delta N}{\Delta t}\right| = \lambda N \quad (29.2)$$

Activity is measured in units of the **curie (Ci)** or the **becquerel (Bq)**. $1\text{ Ci} \equiv 3.70 \times 10^{10}$ decays/s, and $1\text{ Bq} \equiv 1$ decay/s; thus, $1\text{ Ci} = 3.70 \times 10^{10}$ Bq.

- Nuclear masses are usually measured in terms of the **atomic mass unit (u)**. The atomic mass unit is related to the kilogram by $1\text{ u} = 1.660\,54 \times 10^{-27}$ kg. In light of Einstein's mass–energy equivalence, the complete annihilation of 1 u of mass releases 931.5 MeV of energy.

- The **total binding energy (E_b)** of a nucleus is the minimum amount of energy needed to separate a nucleus into its constituent nucleons:

$$E_b = (\Delta m)c^2 \quad (29.5)$$

where Δm is the **mass defect**. The mass defect and is the difference between the sum of the masses of the constituent nucleons and the mass of the nucleus.

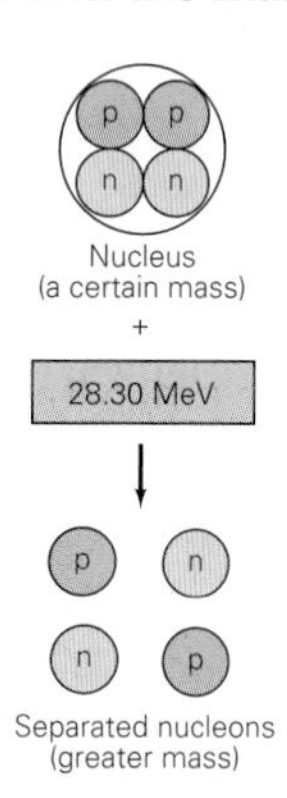

- The **effective dose** of radiation [in **rems** or **sieverts (Sv)**] is determined by the energy deposited per kilogram of material [in **rads** or **grays (Gy)**] and the type of particle depositing that energy (as expressed by the **relative biological effectiveness**, or **RBE**):

$$\text{effective dose (in rem)} = \text{dose (in rad)} \times \text{RBE} \quad (29.6)$$
$$\text{effective dose (in Sv)} = \text{dose (in Gy)} \times \text{RBE} \quad (29.7)$$

Exercises

MC = *Multiple Choice Question,* **CQ** = *Conceptual Question, and* **IE** = *Integrated Exercise. Throughout the text, many exercise sections will include "paired" exercises. These exercise pairs, identified with* **red numbers**, *are intended to assist you in problem solving and learning. In a pair, the first exercise (even numbered) is worked out in the Study Guide so that you can consult it should you need assistance in solving it. The second exercise (odd numbered) is similar in nature, and its answer is given at the back of the book.*

29.1 Nuclear Structure and the Nuclear Force

1. **MC** In the Rutherford scattering experiment, for which target nucleus would alpha particles of a given kinetic energy approach more closely: (a) carbon, (b) iron, or (c) uranium?

2. **MC** The nuclei of carbon-12 and carbon-13 (a) have the same number of nucleons, (b) have the same number of neutrons, (c) have the same number of protons, or (d) none of the preceding.

3. **MC** At the same close distance, between which pair of particles is the nuclear force the largest: (a) neutron–proton, (b) neutron–neutron, (c) proton–proton, or (d) the force is the same for all pairs?

4. **CQ** In the Rutherford scattering experiment, the minimum distance of approach for the alpha particle is given by Eq. 29.1. Explain why this does not necessarily represent the nuclear radius. Is it larger or smaller than the nuclear radius?

5. **CQ** Nuclei with the same number of neutrons are called *isotones*. What nitrogen nucleus is an isotone of carbon-13?

6. **CQ** Nuclei with the same number of nucleons are called *isobars*. What nuclide of nitrogen is an isobar of carbon-13?

7. • Determine the number of protons, neutrons, and electrons in a neutral atom with the following nuclei: (a) ^{43}Ca and (b) ^{206}Pb.

8. • Oxygen has three stable isotopes, with 8, 9, and 10 neutrons, respectively. Write these isotopes in nuclear notation.

9. • An isotope of potassium has the same number of neutrons as the nuclide argon-40. Write the nuclear notation for this potassium isotope.

10. • ^{24}Mg and ^{25}Mg are two isotopes of magnesium. What are the numbers of protons, neutrons, and electrons in each if (a) the atom is electrically neutral, (b) the ion has a −2 charge, and (c) the ion has a +1 charge?

11. • One isotope of uranium has a mass number of 235. What are the numbers of protons, neutrons, and electrons in a neutral atom of this isotope?

12. IE • (a) Isotopes of an element have the same (1) atomic number, (2) neutron number, (3) mass number. (b) Write two possible isotopes for ^{197}Au.

13. •• An approximate expression for the nuclear radius (R) is $R = R_o A^{1/3}$, where $R_o = 1.2 \times 10^{-15}$ m and A is the mass number of the nucleus. (a) Find the nuclear radii of atoms of the noble gases: He, Ne, Ar, Kr, Xe, and Rn. (b) Determine the mass density of each of these species. Does your answer surprise you?

14. IE ••• Assume Rutherford used alpha particles with a kinetic energy of 5.25 MeV. (a) To which of the following nuclei would the alpha particle come closest in a head-on collision: (1) aluminum, (2) iron, or (3) lead? (b) Determine the distance of closest approach for the three nuclei in part (a) and compare them to the nuclear radii given in Exercise 13. Are any of these distance comparable to the radius of the target nucleus? [*Hint*: In Equation 29.1, $mv^2 = 2K_\alpha$; why?]

29.2 Radioactivity

15. MC The conservation of nucleons and the conservation of charge apply to (a) only alpha decay, (b) only beta decay, (c) only gamma decay, or (d) all nuclear decay processes.

16. MC β^- decay can occur only in nuclei with what Z values: (a) $Z > 82$, (b) $Z \leq 82$, or (c) it can occur regardless of the Z value?

17. MC Aluminum has only one stable isotope, ^{27}Al. ^{26}Al would be expected to decay by which beta decay mode: (a) β^+, (b) β^-, or (c) beta decay would not be an option for ^{26}Al?

18. CQ The neutron number is not conserved in a beta decay. Is this a violation of the conservation of nucleons? Explain.

19. CQ ^{19}F is the only stable isotope of fluorine. What two possible decay modes would you expect ^{18}F to decay by? What would be the resulting nucleus in both cases?

20. CQ When an excited nucleus decays to a lower energy level by gamma-ray emission, the actual energy of the gamma-ray photon is a bit less than the difference in energy between the two levels involved in the transition. Explain why this is true.

21. IE • Tritium is radioactive. (a) Would you expect it to (1) β^+, (2) β^-, or (3) alpha decay? Why? (b) What is the identity of the daughter nucleus? Is it stable? [*Hint*: Write the nuclear equation for each.]

22. • Write the nuclear equations expressing (a) the beta decay of $^{60}_{27}Co$ and (b) the alpha decay of $^{226}_{88}Ra$.

23. • Write the nuclear equations for (a) the alpha decay of neptunium-237, (b) the β^- decay of phosphorus-32, (c) the β^+ decay of cobalt-56, (d) electron capture in cobalt-56, and (e) the γ decay of potassium-42.

24. IE • Polonium-214 can decay by alpha decay. (a) The product of its decay has how many fewer protons than polonium-214: (1) zero, (2) one, (3) two, or (4) four? Why? (b) Write the nuclear equation for this decay.

25. • A lead-209 nucleus results from both alpha–beta sequential decays and beta–alpha sequential decays. What was the grandparent nucleus? (Show this result for both decay routes by writing the nuclear equations for both decay processes.)

26. •• Complete the following nuclear-decay equations:
(a) $\_\_\_\_ \rightarrow {}^{4}_{2}He + {}^{4}_{2}He$
(b) $^{240}_{98}Pu \rightarrow \_\_\_\_ \rightarrow {}^{139}_{56}Pu + ({}^{1}_{0}n)$
(c) $^{47}_{21}Sc^* \rightarrow \_\_\_\_ + \gamma$
(d) $^{29}_{11}Na \rightarrow \_\_\_\_ + {}^{29}_{12}Mg$

27. •• Complete the following nuclear-decay equations:
(a) $^{238}_{92}U \rightarrow \_\_\_\_ + {}^{4}_{2}He$
(b) $^{40}_{19}K \rightarrow {}^{40}_{20}Ca + \_\_\_\_$
(c) $^{236}_{92}U \rightarrow {}^{131}_{53}I + \_\_({}^{1}_{0}n) + {}^{102}_{39}Y$
(d) $^{23}_{11}Na + \_\_ \rightarrow {}^{23}_{11}Na^*$
(e) $\_\_\_\_ + {}^{0}_{-1}e \rightarrow {}^{22}_{10}Ne$

28. •• Actinium-227 ($^{227}_{89}Ac$) decays by alpha decay or by beta decay and is part of a decay sequence like the one shown in Fig. 29.8. Each daughter nucleus then decays to radium-223 ($^{223}_{88}Ra$), which subsequently decays to polonium-215 ($^{215}_{84}Po$). Write the nuclear equations for the decay process in the decay series from ^{227}Ac to ^{215}Po.

29. ••• The decay series for neptunium-237 is shown in ▾ Fig. 29.20. Identify the decay modes and each of the nuclei in the sequence.

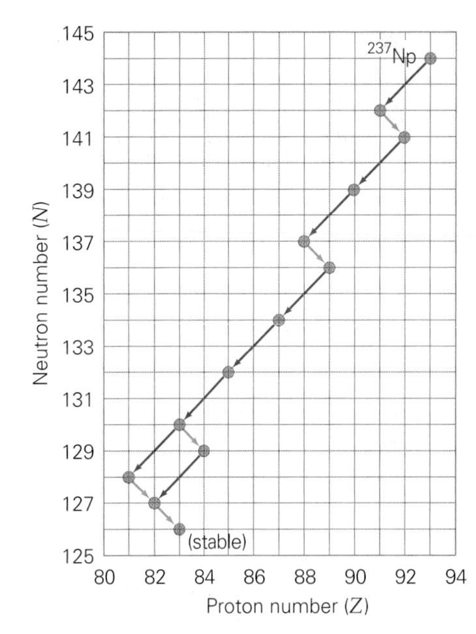

▲ **FIGURE 29.20 Neptunium-237 decay series** See Exercise 29.

29.3 Decay Rate and Half-Life

30. MC After one half-life, a sample of a particular radioactive material (a) is half as massive, (b) has its half-life reduced by half, (c) is no longer radioactive, (d) has its activity reduced by half.

31. **MC** In two half-lives, the activity of a radioactive sample will have decreased by what percent: (a) 25%, (b) 50%, (c) 75%, or (d) 87.5%?

32. **MC** An alpha emitter gives off alpha particles with an energy of 4.4 MeV. After three half-lives, what is the energy of the alpha particles being given off: (a) 2.2 MeV, (b) 1.1 MeV, (c) 8.8 MeV, or (d) 4.4 MeV?

33. **CQ** What physical or chemical properties affect the decay rate, or half-life, of a radioactive isotope?

34. **CQ** Nuclide A has a decay constant that is half that of nuclide B. If the two nuclides start with the same number of undecayed nuclei, will twice as many of nuclide A's nuclei as the number of nuclide B's nuclei decay in a given time? Explain.

35. **CQ** What are the (a) half-life and (b) decay constant for a stable isotope?

36. ● A particular radioactive sample undergoes 2.50×10^6 decays/s. What is the activity of the sample in (a) curies and (b) becquerels?

37. ● At present, a laboratory radioactive beta source has an activity of 20 mCi. (a) What is the present decay rate in decays per second? (b) Assuming that one beta particle is emitted per decay, how many are currently emitted per minute?

38. **IE** ● The half-life of a radioactive isotope is 1 h. (a) What fraction of a sample would be left after 3 h: (1) one third, (2) one eighth, or (3) one ninth? Why? (b) What fraction of a sample would be left after 1 day?

39. ● A 1.25-μCi alpha source gives off alpha particles with a kinetic energy of 2.78 MeV. At what rate (in watts) is kinetic energy being produced?

40. ●● A sample of technetium-104 has an activity of 10 mCi. Estimate the activity of the sample after one hour has elapsed ($t_{1/2} = 18$ min).

41. ●● What period of time is required for a sample of radioactive tritium (^{3}H) to lose 80.0% of its activity? Tritium has a half-life of 12.3 years.

42. ●● A certain amount of ^{131}I is introduced into a patient in a medical diagnostic procedure for her thyroid. What percentage of the sample remains after 1.00 day, assuming that all of the ^{131}I is retained in the patient's thyroid gland? (Answer to three significant figures.)

43. **IE** ●● Carbon-14 dating is used to determine the age of some buried bones. (a) If the activity of bone A is higher than that of bone B, then bone A is (1) older than, (2) younger than, (3) the same age as bone B. Why? (b) A sample of old bone is found to have 4.0 beta decays/min for each gram of carbon. Approximately how old is the bone?

44. ●● Show that the number N of radioactive nuclei remaining in a sample after n half-lives is given by

$$N = \frac{N_o}{2^n} = \left(\tfrac{1}{2}\right)^n N_o$$

where N_o is the initial number of nuclei.

45. ●● Some ancient writings on parchment are found sealed in a jar in a cave. If carbon-14 dating shows the parchment to be 28 650 years old, what percentage of the carbon-14 atoms still remains in the sample compared with the number that was present when the parchment was made?

46. ●● How long would it take (to the nearest whole year) for the activity of a sample of cobalt-60 to reduce to 20% of the original activity?

47. ●● A soil sample contains 40 μg of ^{90}Sr. Approximately how much ^{90}Sr will be in the sample 150 years from now?

48. ●● (a) What is the decay constant of fluorine-17 ($t_{1/2} = 66.0$ s)? (b) How long will it take for the activity of a sample of ^{17}F to decrease to 10% of its initial value?

49. ●● Francium-223 ($^{223}_{87}$Fr) has a half-life of 21.8 min. (a) How many nuclei are initially present in a 25.0-mg sample of this isotope? (b) How many nuclei will be present 1 h and 49 min later?

50. ●● A basement room containing radon gas ($t_{1/2} =$ 3.82 days) is sealed to be airtight. (a) If 7.50×10^{10} radon atoms are trapped in the room, estimate how many radon atoms remain in the room after one week. (b) Radon undergoes alpha decay. After 30 days, is the number of its daughter nuclei equal to the number of radon parents that have decayed? Explain. → Nope, gives off isotopes

on exam you can approx. or estimate.

51. ●● In 1898, Pierre and Marie Curie isolated about 10 mg of radium-226 from eight tons of uranium ore. If this sample had been placed in a museum, how much of the radium would remain in the year 2100?

52. ●● An ancient artifact is found to contain 250 g of carbon and has an activity of 475 decays per minute. What is the approximate age of the artifact, to the nearest thousand years?

53. ●●● The recoverable U.S. reserves of high-grade uranium-238 ore (high-grade ore contains about 10 kg of $^{238}U_3O_8$ per ton) are estimated to be about 500 000 tons. Neglecting any geological changes, what mass of ^{238}U existed in this high-grade ore when the Earth was formed, about 4.6 billion years ago? [*Hint*: See Appendix V.]

54. **IE** ●●● Nitrogen-13, with a half-life of 10 min, decays by positron emission. (a) The end product is (1) ^{13}N, (2) ^{13}C, or (3) ^{13}O. (b) If a sample of pure ^{13}N has a mass of 1.5 g at a certain time, what is the activity 35 min later? (c) What percentage of the sample is ^{13}N at this time?

29.4 Nuclear Stability and Binding Energy

55. **MC** For nuclei with a mass number greater than 40, which of the following statements is correct? (a) The number of protons is approximately equal to the number of neutrons; (b) the number of protons exceeds the number of neutrons; (c) all such nuclei are stable up to $Z = 92$; (d) none of the preceding.

56. **MC** The average binding energy per nucleon of the daughter nucleus in a decay process is (a) greater than, (b) less than, or (c) equal to that of the parent nucleus.

57. **MC** From which nucleus is it easier to remove a neutron: (a) ^{25}Mg, (b) ^{24}Mg, or (c) both require the same amount of energy?

58. **CQ** Why does aluminum have only one stable isotope (^{27}Al)?

59. **CQ** Explain why, of the two main uranium isotopes, ^{238}U is more abundant than ^{235}U. [*Hint:* Although they both are unstable, ^{238}U is closer to stability; why?]

60. **CQ** Compared to ^{3}He, the probability of absorbing a neutron is much less likely for ^{4}He. Explain this using what you know about odd/even proton and neutron numbers.

61. ● From which of the following pairs of nuclei would you expect it to be easier to remove a neutron: (a) $^{16}_{8}$O or $^{17}_{8}$O; (b) $^{40}_{20}$Ca or $^{42}_{20}$Ca; (c) $^{10}_{5}$B or $^{11}_{5}$B; (d) $^{208}_{82}$Pb or $^{209}_{83}$Bi? State your reasoning for your choice in each case.

62. ● Only two isotopes of Sb (antimony, $Z = 51$) are stable. Pick the two stable isotopes from the following list: (a) ^{120}Sb; (b) ^{121}Sb; (c) ^{122}Sb; (d) ^{123}Sb; (e) ^{124}Sb.

63. ● The total binding energy of $^{2}_{1}$H is 2.224 MeV. Use this information to compute the mass of a ^{2}H nucleus from the known mass of a proton and a neutron.

64. ●● Use Avogadro's number (see Section 10.3), $N_A = 6.02 \times 10^{23}$ atoms/mole, to show that $1\ u = 1.66 \times 10^{-27}$ kg. (Recall that a ^{12}C atom has a mass of exactly 12 u.)

65. ●● The mass of $^{12}_{6}$C is exactly 12 u. (a) What is the total binding energy of this nucleus? (b) What is the average binding energy per nucleon?

66. ●● The mass of $^{16}_{8}$O is 15.994 915 u. What is the average binding energy per nucleon (E_b/A) for this nucleus?

67. ●● Which isotope of hydrogen has the lower average binding energy per nucleon, deuterium or tritium? Justify your answer mathematically.

68. ●● Near high-neutron areas, such as a nuclear reactor, neutrons will be absorbed by protons (the hydrogen nucleus in water molecules) and will give off a gamma ray of a characteristic energy in the process. What is the energy of the gamma ray (to three significant figures)?

69. ●● How much energy (to four significant figures) would be required to completely separate all the nucleons of a nitrogen-14 nucleus, the atom of which has a mass of 14.003 074 u?

70. ●● Calculate the binding energy of the last neutron in the $^{40}_{19}$K nucleus. [*Hint*: Compare the mass of $^{40}_{19}$K with the mass of $^{39}_{19}$K plus the mass of a neutron.]

71. ●● If an alpha particle could be removed intact from an aluminum-27 nucleus ($m = 26.981\,541$ u), a sodium-23 nucleus ($m = 22.989\,770$ u) would remain. How much energy would be required to do this operation?

72. ●● On average, are nucleons more tightly bound in an ^{27}Al nucleus or in a ^{23}Na nucleus?

73. ●● The atomic mass of $^{235}_{92}$U is 235.043 925 u. Find the average binding energy per nucleon for this isotope.

74. ●●● The mass of $^{8}_{4}$Be is 8.005 305 u. (a) Which is less, the total mass of two alpha particles or the mass of the ^{8}Be nucleus? (b) Which is greater, the total binding energy of the ^{8}Be nucleus or the total binding energy of two alpha particles? (c) On the basis of your answers to parts (a) and (b) alone, do you expect the ^{8}Be nucleus to decay spontaneously into two alpha particles?

29.5 Radiation Detection, Dosage, and Applications

75. **MC** Which type of detector records the trajectory of charged particles: (a) Geiger counter, (b) scintillation counter, (c) solid-state detector, or (d) spark chamber?

76. **MC** A bubble chamber (in a magnetic field) event shows two tracks of equal curvature but in opposite directions, emanating from a point in space with no apparent incoming particle. More than likely this event is (a) alpha decay, (b) beta decay, or (c) pair production.

77. **MC** The same effective radiation dose (in rems) is given by a source of slow neutrons and an X-ray machine. What is the ratio of the dose (in rads) from the neutrons to that from the X-rays (also in rads): (a) 1:2 (b) 1:4, (c) 2:1, or (d) 4:1?

78. **CQ** A basic assumption of radiocarbon dating is that the cosmic-ray intensity has been generally constant for the last 40 000 years or so. Suppose it were found that the intensity was much less 100 000 years ago than it is today. How would this finding affect the results of carbon-14 dating?

79. **CQ** If X-ray and alpha particles give the same dose (in grays), how will their effective doses (in sieverts) compare?

80. **CQ** PET scans require extremely fast computers coupled with gamma-ray detectors capable of accurate energy measurements. Explain why both energy accuracy and comparison of arrival times are crucial to the success of a PET scan. [*Hint*: In a PET scan, two opposing detectors pick up pair-annihilation gamma-ray photons. See page 928.]

81. ● In a diagnostic procedure, a patient in a hospital ingests 80 mCi of gold-198 ($t_{1/2} = 2.7$ days). What is the activity at the end of one month if none of the gold is eliminated from the body by biological functions?

82. ● A technician working at a nuclear reactor facility is exposed to a slow neutron radiation and receives a dose of 1.25 rad. (a) How much energy is absorbed by 200 g of the worker's tissue? (b) Was the maximum permissible radiation dosage exceeded?

83. ● A person working with nuclear isotopes for a two-month period receives a 0.5-rad dose from a gamma source, a 0.3-rad dose from a slow-neutron source, and a 0.1-rad dose from an alpha source. Was the maximum permissible radiation dosage exceeded?

84. ●● Neutron activation analysis was performed on small pieces of hair that had been taken from the exiled Napoleon after he died on the island of St. Helena in 1821. The samples were found to contain abnormally high levels of arsenic, which supported a theory that his death was not due to natural causes. If this evidence was derived by studying beta emissions coming from arsenic-76 nuclei, what was the arsenic isotope present in the hair, and what was the final nucleus after the beta decay?

85. ●●● A cancer treatment called the gamma knife (see Insight 29.1 on page 927) uses a large number of ^{60}Co sources to treat tumors. ^{60}Co emits two gamma rays of energy 1.33 MeV and 1.17 MeV in quick succession. Assume that 50.0% of the total gamma-ray energy is absorbed by a tumor. The total activity of the ^{60}Co sources is 1.00 mCi, the tumor's mass is 0.100 kg, and the patient is exposed for an hour. Calculate the effective radiation dose received by the tumor. (Since the ^{60}Co half-life is 5.3 years, changes in its activity during treatment can be neglected.)

Comprehensive Exercises

86. The radioactive source in most smoke detectors is ^{241}Am, which has a half-life of 432 years. In a typical detector, only about 10^{-4} g of this material is needed. (a) Write down its alpha decay equation and predict the product nucleus. (b) What is the source activity? (c) What would be thesource's activity after 20 years in operation?

87. A sample of ^{215}Bi, which beta-decays ($t_{1/2}$ = 2.4 min), contains Avogadro's number of nuclei. (a) Write down the decay equation and predict the product nucleus. (b) How many bismuth nuclei are present after 10 min? (c) After 1.0 h? (d) What are the activities, in curies and becquerels, at these times?

88. IE High-energy photons (gamma rays) can remove nucleons from nuclei in a process called the *photonuclear effect*. (a) If you wanted to remove a single neutron from a nucleus, which nucleus would most likely require the higher energy photon: (1) ^{12}C; (2) ^{13}C; or (3) the energies would be the same? (b) Calculate the minimum energy of each photon required to eject a neutron from the two isotopes in part (a). (c) Determine the wavelengths of the light associated with the photons in part (b).

89. A very slow neutron (neglect its kinetic energy) is captured by an ^{10}B nucleus, which ends up in its ground state. Calculate the (a) energy and (b) wavelength of the photon given off. (Neglect the recoil kinetic energy of the daughter nucleus.)

90. The approximate radius of a nucleus is given by $R = R_o A^{1/3}$, where $R_o = 1.2 \times 10^{-15}$ m and A is the mass number of the nucleus. Assuming that nuclei are spherical (they are approximately so in many cases), (a) show that the average nucleon density in a nucleus is 1.4×10^{44} nucleons/m^3 and (b) estimate the nuclear density in kilograms per cubic meter.

91. ▼Figure 29.21 shows the decay series for plutonium-239. Use the information in the figure to (a) determine the two different decay schemes by which the isotope of radon associated with this chain can be formed, and (b) determine the subsequent decay scheme for that isotope of radon.

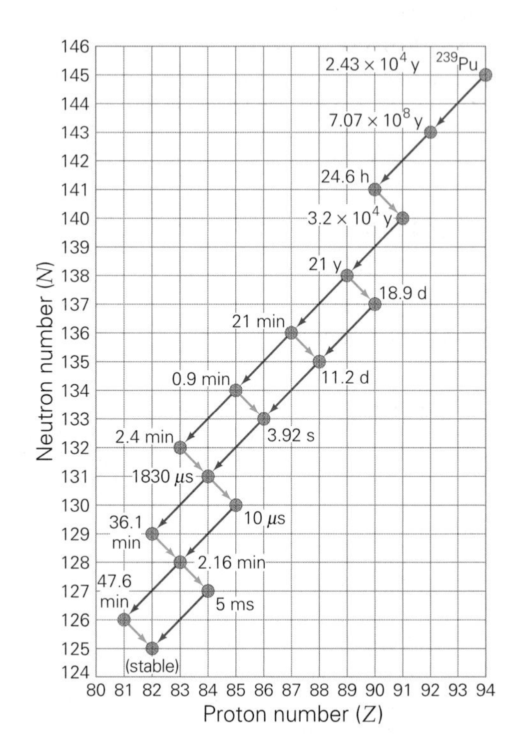

▲ **FIGURE 29.21 Plutonium-239 decay series** See Exercise 91.

92. Determine which of the following isotopes are likely to be stable and explain clearly and fully how you came to your conclusion, based on the "rules of stability." (a) ^{15}O; (b) ^{8}Li; (c) ^{222}Rn; (d) ^{27}Mg; (e) ^{41}Ca.

93. $^{3}_{1}$H (tritium) can be produced in water surrounding a strong source of neutrons, such as nuclear reactors. One of the ways is by neutron capture onto deuterium. (a) Write down the equation for this capture reaction. (b) Tritium has a half-life of 12.33 years. What percentage of a sample containing $^{3}_{1}$H will remain after 6.00 years? (c) Determine the gamma ray energy emitted during the capture (assuming the tritium ends up in its ground state and the incoming neutron kinetic energy is negligible). (d) Write down the reaction for the subsequent beta decay of the tritium and determine the stable daughter identity. (e) If all the energy released in the beta decay went into the beta particle, determine its energy.

CHAPTER

30 NUCLEAR REACTIONS AND ELEMENTARY PARTICLES

30.1 Nuclear Reactions 936
30.2 Nuclear Fission 939
30.3 Nuclear Fusion 944
30.4 Beta Decay and the Neutrino 946
30.5 Fundamental Forces and Exchange Particles 948
30.6 Elementary Particles 951
30.7 The Quark Model 953
30.8 Force Unification Theories, the Standard Model, and the Early Universe 954

PHYSICS FACTS

- Although nuclear power plants produce radioactive waste that presents long-term storage problems, they have advantages over fossil fuel plants. For example, they emit no greenhouse gases (which cause global warming) or oxides of sulfur and nitrogen (which cause acid rain). Nor do they require obtaining fossil fuels in environmentally sensitive areas. In fact, coal-fired plants emit much more radioactive material than nuclear plants due to the uranium in the coal.
- Photons are now known to be the only massless elementary particle. In the 1980s, the family of neutrinos, originally thought to all be massless, were determined to have mass. Recent experiments indicate that their mass is on the order of one millionth of the electron's mass. This has vast implications about the mass of the universe, and thus its (Big Bang) evolution.
- The smallest amount (quantum) of electric charge is not that of the electron (which has charge $-e$). Quarks have fractional charges of $\pm\frac{1}{3}e$ and $\pm\frac{2}{3}e$.
- Protons and neutrons are not elementary particles, but are each composed of three quarks. Because of the type of binding between quarks, current theories predict that free quarks may never be seen experimentally.

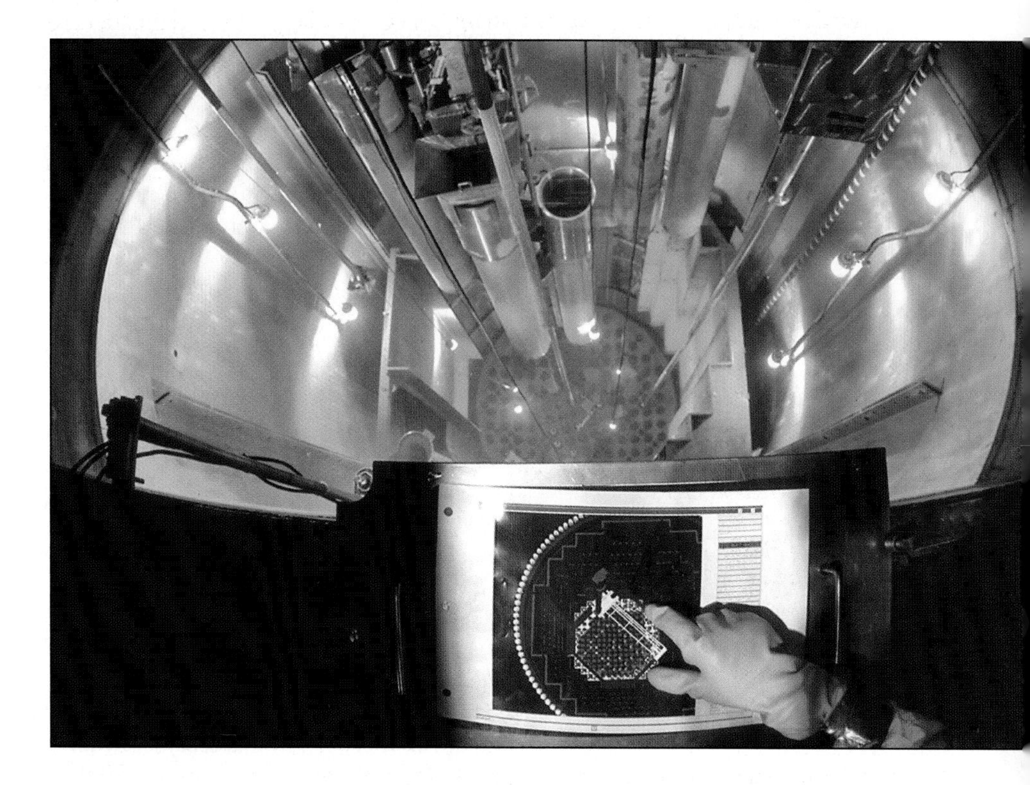

Today, more than half a century after the first nuclear reactor was built, these facilities remain a center of controversy. Some people think they are essential to solving the world's energy problems. To others, they embody all that is wrong and dangerous about modern technology. The reactor shown in the photo is a symbol of this ongoing debate. It is part of the Three Mile Island (TMI) plant in Pennsylvania, the site of one of the largest and most publicized nuclear accidents in the world, second perhaps only to Chernobyl (more on that later). Here, you are looking into the heart of one of the two remaining TMI reactors, both of which are still running safely.

In this chapter the nuclear fission reactions that power such reactors will be studied. Also we will study how scientists are trying to harness fusion reactions, which occur within the Sun and other stars and might one day provide a safe and virtually inexhaustible supply of clean energy for our planet.

Also considered are the elementary particles that make up our universe and whose interactions govern the structure and evolution of that universe. Investigations have shown that the proton and neutron are not elementary particles, but are composites of elementary particles. The discovery of families of elementary particles (and their properties and interactions) has given us broad insight into the structure of the nucleus and provided a clearer understanding of the nature, origin, and evolution of our universe.

30.1 Nuclear Reactions

OBJECTIVES: To (a) use charge and nucleon conservation to write nuclear reaction equations, and (b) understand and use the concepts of Q value and threshold energy to analyze nuclear reactions.

Ordinary chemical reactions between atoms and molecules involve only orbital electrons. The nuclei of these atoms do not participate in the process, so the atoms retain their identity. Conversely, in **nuclear reactions**, the original nuclei are converted into the nuclei of other elements. Scientists first became aware of this type of reaction during experimental studies that involved bombarding nuclei with energetic particles.

The first artificially induced nuclear reaction was produced by Ernest Rutherford in 1919. Nitrogen was bombarded with alpha particles from a natural source (^{214}Bi). The particles produced by the reactions were identified as protons. Rutherford reasoned that an alpha particle colliding with a nitrogen *nucleus* must sometimes be able to induce a reaction that produces a proton. We say that the nitrogen nucleus is *artificially transmuted* into an oxygen nucleus by the following reaction:

$^{14}_{7}\text{N}$	+	$^{4}_{2}\text{He}$	→	$^{17}_{8}\text{O}$	+	$^{1}_{1}\text{H}$
nitrogen		*alpha particle*		*oxygen*		*proton*
(14.003 074 u)		*(4.002 603 u)*		*(16.991 33 u)*		*(1.007 825 u)*

(The atomic masses are given for later use.)

This reaction and many others like it actually form a short-lived (intermediate or temporary) *compound nucleus* in an excited state. For example, the preceding reaction can be written more correctly as

$$^{14}_{7}\text{N} + {}^{4}_{2}\text{He} \rightarrow ({}^{18}_{9}\text{F}^*) \rightarrow {}^{17}_{8}\text{O} + {}^{1}_{1}\text{H}$$

The intermediate nucleus is a fluorine nucleus, $^{18}_{9}\text{F}^*$, formed in an excited state indicated by the asterisk. A compound nucleus typically loses excess energy by ejecting a particle (or particles)— in this case, a proton. Since a compound nucleus lasts only a very short time, it is commonly omitted from the nuclear reaction equation.

What Rutherford discovered was a way to change one element into another. This was the age-old dream of alchemists, although their main goal was to change common metals, such as mercury and lead, into gold. This seemingly profitable metamorphosis and many other transmutations can be initiated today with *particle accelerators*, machines that accelerate charged particles to very high speeds. When these particles strike target nuclei, they can initiate nuclear reactions. One reaction that can occur when a proton strikes a nucleus of mercury is

$^{200}_{80}\text{Hg}$	+	$^{1}_{1}\text{H}$	→	$^{197}_{79}\text{Au}$	+	$^{4}_{2}\text{He}$
mercury		*proton*		*gold*		*alpha particle*
(199.968 321 u)		*(1.007 825 u)*		*(196.966 56 u)*		*(4.002 603 u)*

In this reaction, mercury is converted into gold, so it would seem that modern physics has fulfilled the alchemists' dream. However, making such tiny amounts of gold in an accelerator costs far more than the gold is worth.

Reactions such as the foregoing ones have the general form

$$\text{A} + \text{a} \rightarrow \text{B} + \text{b}$$

where the uppercase letters represent the nuclei and the lowercase letters represent the particles. Such reactions are often written in a shorthand notation:

$$\text{A(a, b)B}$$

For example, in this form, the two previous reactions can be rewritten more compactly as

$$^{14}\text{N}(\alpha, \text{p})^{17}\text{O} \quad \text{and} \quad ^{200}\text{Hg}(\text{p}, \alpha)^{197}\text{Au}$$

The periodic table (see Fig. 28.9 and back cover of this text) lists more than 100 elements, but only 90 stable elements occur naturally on Earth. There are elements with unstable nuclei that exist from $Z = 83$ (bismuth) to $Z = 92$ (uranium) due to the decay chains that continually create them from heavier elements. Those with proton numbers greater than uranium ($Z = 92$), such as plutonium-239, as well as technetium ($Z = 43$) and promethium ($Z = 61$), are created artificially by nuclear reactions. (If technetium and promethium had been present when the Earth was formed, they would have long since decayed away.) The name *technetium* comes from the Greek word *technetos*, meaning "artificial"; technetium was the first unknown element to be created by artificial means. Elements with Z values up to about $Z = 114$ have been created artificially.*

Conservation of Mass–Energy and the Q Value

In every nuclear reaction, total (relativistic) energy ($E = K + mc^2$) must be conserved. (See Chapter 26.) Consider the reaction in which nitrogen is converted into oxygen: $^{14}N(\alpha, p)^{17}O$. By the conservation of total relativistic energy,

$$(K_N + m_N c^2) + (K_\alpha + m_\alpha c^2) = (K_O + m_O c^2) + (K_p + m_p c^2)$$

where the subscripts refer to the particular particle or nucleus. Rearranging the equation, we have

$$K_O + K_p - (K_N + K_\alpha) = (m_N + m_\alpha - m_O - m_p)c^2$$

The **Q value** of the reaction is defined as the change in kinetic energy, and

$$Q = \Delta K = (K_O + K_p) - (K_N + K_\alpha) \tag{30.1}$$

Q can be positive or negative, depending on whether the total kinetic energy of the system increases or decreases. Thus, the Q value is a measure of the kinetic energy released or absorbed in a reaction. Equation 30.1 can alternatively be expressed in terms of the masses:

$$Q = (m_N + m_\alpha - m_O - m_p)c^2 \tag{30.2}$$

In terms of a general reaction of the form $A + a \rightarrow B + b$,

$$Q = \Delta K = (m_A + m_a - m_B - m_b)c^2 = (\Delta m)c^2 \tag{30.3}$$

Note: Here $\Delta m = m_i - m_f$.

An alternative interpretation of Q is that it is the difference in the mass-equivalent energies of the reactants (initial) and the products (final) of a reaction. This reflects the fact that mass can be converted into kinetic energy and vice versa. Note that the mass difference Δm can be positive or negative. Notice also that $\Delta m = m_i - m_f$, the opposite of our usual convention. This is to guarantee that the value of Q is correctly related to ΔK. Thus, if the total mass of the system increases during the reaction and the kinetic energy therefore decreases, Q must be negative. Similarly, if the total mass decreases and the kinetic energy thus increases, then Q is positive.

If Q is negative, the reaction requires a minimum amount of kinetic energy before the reaction can happen. To see this, let us look at Rutherford's original reaction in some detail. Using the masses given under the $^{14}N(\alpha, p)^{17}O$ reaction equation on page 936, we have

$$\begin{aligned} Q &= (m_N + m_\alpha - m_O - m_p)c^2 \\ &= [(14.003\,074\ u + 4.002\,603\ u) - (16.999\,133\ u + 1.007\,825\ u)]c^2 \\ &= (-0.001\,281\ u)c^2 \end{aligned}$$

or, using the mass–energy equivalence factor from Section 29.4, we obtain

$$Q = (-0.001\,281\ u)(931.5\ \text{MeV/u}) = -1.193\ \text{MeV}$$

A reaction with a negative Q value is said to be **endoergic** (or *endothermic*). In endoergic reactions, the kinetic energy of the reacting particles is partially converted into mass.

When the Q value of a reaction is positive, energy is released, and the reaction is said to be **exoergic** (or *exothermic*). That is, energy is produced (*exo*) by the reaction. In this case, some mass is converted into energy in the form of increased kinetic energy of the reaction products.

*Beyond $Z = 112$, there are gaps. Extremely short-lived nuclei with $Z = 114$ (and possibly 116) have tentatively been discovered, but are yet unnamed. Can you explain why the even values of atomic numbers have been discovered and not the odd ones in between?

Example 30.1 ■ A Possible Energy Source: Q Value of a Reaction

Determine whether the following reaction is endoergic or exoergic, and calculate its Q value.

${}^{2}_{1}H$	+	${}^{2}_{1}H$	→	${}^{3}_{2}He$	+	${}^{1}_{0}n$
deuteron		*deuteron*		*helium*		*neutron*
(2.014 102 u)		*(2.014 102 u)*		*(3.016 029 u)*		*(1.008 665 u)*

Thinking It Through. The reaction is endoergic if $Q < 0$, and exoergic if $Q > 0$. We need the mass difference (Δm) to determine Q from Eq. 30.3.

Solution. Δm is calculated by subtracting the final masses from the initial masses. Therefore,

$$\begin{aligned}\Delta m &= 2m_D - m_{He} - m_n \\ &= 2(2.014\,102\text{ u}) - 3.016\,029\text{ u} - 1.008\,665\text{ u} = +0.003\,51\text{ u}\end{aligned}$$

Thus, mass has been lost, the total kinetic energy has increased, and the reaction is exoergic. The Q value is

$$Q = (0.003\,51\text{ u})(931.5\text{ MeV/u}) = +3.27\text{ MeV}$$

Follow-Up Exercise. Determine whether the following reaction is endoergic or exoergic, and calculate its Q value:

${}^{12}_{6}C$	+	${}^{4}_{2}He$	→	${}^{13}_{6}C$	+	${}^{3}_{2}He$
carbon		*helium*		*carbon*		*helium*
(12.000 000 u)		*(4.002 603 u)*		*(13.003 355 u)*		*(3.016 029 u)*

(Answers to all Follow-Up Exercises are at the back of the text.)

Problem-Solving Hint

Note in Example 30.1 that Q values are computed from the mass difference, expressed in atomic mass units, by using the mass–energy conversion factor derived in Chapter 29. (See Table 29.3.) This method eliminates the need to use c^2 and gives Q directly in MeV.

TABLE 30.1

Interpretation of Q Values

Q Value	*Effect*
Positive ($Q > 0$)	Exoergic, some mass converted into energy (mass of reactants greater than mass of products)
Negative ($Q < 0$)	Endoergic, some kinetic energy converted into mass (mass of products greater than mass of reactants)

Radioactive decay (Chapter 29) is a special type of nuclear reaction with one reactant nucleus and two (or more) products. The Q value of radioactive decay is always positive, because there is a gain in kinetic energy. For decay reactions, Q is called the *disintegration energy*. The interpretation of the sign of Q is summarized in Table 30.1.

When a reaction's Q value is negative, you might think that the reaction could occur if the incident particle had a kinetic energy at least equal to Q—that is, if it were to have $K_{min} = |Q|$.* However, if all the kinetic energy were converted to mass, the particles would be at rest after the reaction, which violates the conservation of linear momentum (Chapter 6).

Hence, in an endoergic reaction, to conserve linear momentum, the kinetic energy of the incident particle must be *greater* than $|Q|$. The minimum kinetic energy that a particle needs to initiate an endoergic reaction is called the **threshold energy** (K_{min}). For nonrelativistic energies, the threshold energy is

$$K_{min} = \left(1 + \frac{m_a}{M_A}\right)|Q| \quad \textit{(stationary target only)} \tag{30.4}$$

where m_a and M_A are the masses of the incident particle and the stationary target nucleus, respectively. In Eq. 30.4, the factor by which $|Q|$ is multiplied is greater than 1 (why?), so, as expected, $K_{min} > |Q|$. The calculation of a threshold energy is shown in the following Example.

*Kinetic energy is written in terms of $|Q|$, the absolute value of Q, because kinetic energy cannot be negative. The sign of Q arises from the mass difference and indicates the gain or loss of mass during the reaction. If Q is negative, we want K_{min} to be positive—hence, the use of absolute value.

Example 30.2 ■ Nitrogen into Oxygen: Threshold Energy

What is the threshold energy for the reaction $^{14}N(\alpha, p)^{17}O$?

Thinking It Through. The Q value for this reaction was calculated previously in the text. To get the threshold energy, we use Eq. 30.4.

Solution. The following data are taken from the text:

Given: $m_a = m_\alpha = 4.002\,603$ u
$M_A = m_N = 14.003\,074$ u
$Q = -1.193$ MeV

Find: K_{min} (threshold energy)

From Eq. 30.4,

$$K_{min} = \left(1 + \frac{m_\alpha}{M_N}\right)|Q| = \left(1 + \frac{4.002\,603\text{ u}}{14.003\,074\text{ u}}\right)|-1.193\text{ MeV}| = 1.534\text{ MeV}$$

Follow-Up Exercise. In this Example, how much of the threshold kinetic energy goes into increasing the mass of the system, and how much shows up as kinetic energy in the final state? Explain your reasoning.

Reaction Cross-Sections

In an endoergic reaction, when the incident particle has more than the threshold energies of several reactions, any of the reactions may occur, usually with differing probabilities according to the rules of quantum mechanics. A measure of the probability that a particular reaction will occur is called the **cross-section** for that reaction. The probability for a particular reaction depends on many factors. Usually, it depends on the kinetic energy of the initiating particle, sometimes very dramatically. For positively charged incident particles, the presence of the (repulsive) Coulomb barrier means that the probability of a given reaction occurring generally increases with the kinetic energy of the incident particle.

Being electrically neutral, neutrons are unaffected by the Coulomb barrier. As a result, the cross-section for a given reaction involving neutrons can be quite large, even for low-energy neutrons. Reactions involving neutrons such as $^{27}Al(n, \gamma)^{28}Al$ are called *neutron-capture reactions*. As the energy of the neutron increases, the cross-section can vary a great deal, as ▸Fig. 30.1 shows. The peaks in the curve, called *resonances*, are associated with nuclear energy levels in the nucleus being formed. If the neutron's energy is "just right" to create the final nucleus in one of its energy levels, there is a relatively high probability that neutron absorption will occur.

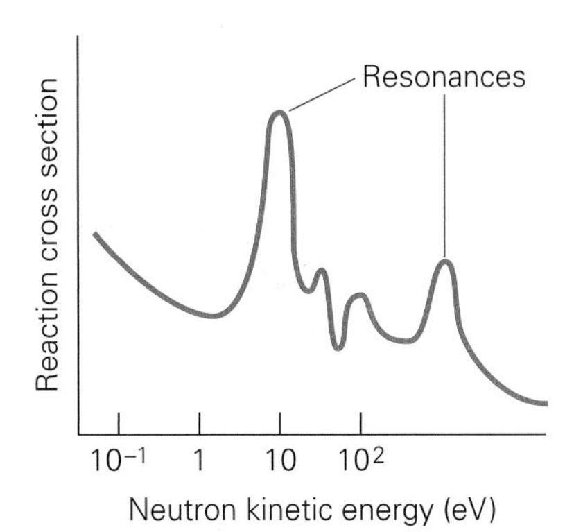

▲ **FIGURE 30.1 Reaction cross-section** A typical graph of a neutron reaction's cross-section versus energy. The peaks where the probabilities of reactions are greatest are called *resonances*. They correspond to energy levels in the compound nucleus formed when the neutron is temporarily captured.

30.2 Nuclear Fission

OBJECTIVES: To understand (a) the process of nuclear fission, (b) the nature and cause of a nuclear chain reaction, and (c) the basic principles involved in the operation of nuclear reactors.

In early attempts to make heavier elements artificially, uranium, the heaviest element known at the time, was bombarded with neutrons. An unexpected result was that the uranium nuclei sometimes split into fragments. These fragments were identified as the nuclei of lighter elements. The process was dubbed *nuclear fission*, after the biological fission process of cell division.

In a **fission reaction**, a heavy nucleus divides into two lighter nuclei with the emission of neutrons. Some of the initial mass is converted into kinetic energy of the neutron and fragments. Some heavy nuclei undergo *spontaneous fission*, but at very slow rates. However, fission can be *induced*, and this is the important process in practical energy production. For example, when a ^{235}U nucleus absorbs a neutron, it can fission into xenon and strontium by the reaction

$$^{235}_{92}U + ^{1}_{0}n \rightarrow (^{236}_{92}U^*) \rightarrow ^{140}_{54}Xe + ^{94}_{38}Sr + 2(^{1}_{0}n)$$

According to the *liquid-drop model*, due to the absorbed energy, this intermediate nucleus (^{236}U) undergoes oscillations and becomes distorted like a liquid drop (▾Fig. 30.2).

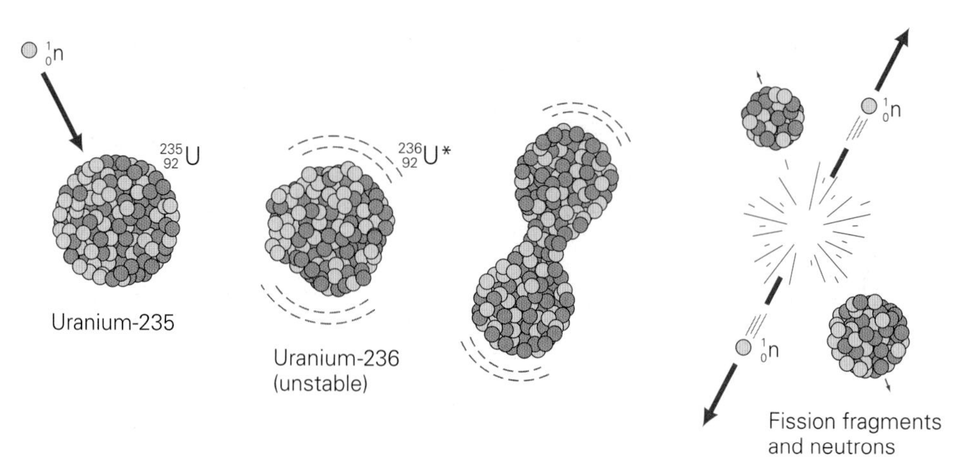

▶ **FIGURE 30.2 Liquid-drop model of fission** When an incident neutron is absorbed by a fissionable nucleus, such as ^{235}U, the unstable compound nucleus (^{236}U) undergoes violent oscillations and breaks apart like a liquid drop, typically emitting two or more neutrons and yielding two radioactive fragments.

The separation of the nucleons into different parts of the "drop" weakens the nuclear force, and the repulsive electrical force between the two parts of the "nuclear drop" causes it to split, or fission.

Note that the preceding reaction involving ^{235}U is *not* unique. There are many other possible outcomes, including the following (the compound nuclei are omitted):

$$^1_0\text{n} + ^{235}_{92}\text{U} \rightarrow ^{141}_{56}\text{Ba} + ^{92}_{36}\text{Kr} + 3(^1_0\text{n})$$

and

$$^1_0\text{n} + ^{235}_{92}\text{U} \rightarrow ^{150}_{60}\text{Nd} + ^{81}_{32}\text{Ge} + 5(^1_0\text{n})$$

Only certain nuclei undergo fission. For them, the probability of fissioning depends on the energy of the incident neutrons. For example, the largest probabilities for fission of ^{235}U and ^{239}Pu occur for "slow" neutrons, that is, neutrons with kinetic energies less than about 1 eV. However, for ^{232}Th, "fast" neutrons with energies of 1 MeV or greater are more likely to trigger a fission reaction.

An estimate of the energy released in a fission reaction can be obtained by considering the E_b/A curve for stable nuclei (Fig. 29.14). When a nucleus with a high mass number (A), such as uranium, splits into two nuclei, it is, in effect, moving inward and upward along the sloping tail of this curve toward more stable nuclei. As a result, the average binding energy per nucleon increases from about 7.8 MeV to approximately 8.8 MeV. Thus energy liberated is on the order of 1 MeV per nucleon in the fission products. In the reaction at the top of this page, $140 + 94 = 234$ nucleons are bound in the products. Thus, the energy release is approximately (1 MeV/nucleon) $\times$ 234 nucleons $\approx$ 234 MeV.

At first glance, this amount might not seem like much energy. 234 MeV is only about 3.7×10^{-11} J, which pales in comparison with everyday energies. In fact, 234 MeV is only about 0.1 percent of the energy equivalent of the mass of the ^{235}U nucleus, which is approximately (235 nucleons)(939 MeV/nucleon) $= 2.2 \times 10^5$ MeV. Nevertheless, on a percentage basis, it is many times larger than the amount of energy released in ordinary chemical reactions, such as in the burning of oil or coal.

Practical amounts of energy from fission can, however, be obtained when huge numbers of these fissions occur per second. One way of accomplishing this is by a **chain reaction**. For example, suppose a ^{235}U nucleus fissions (on its own or triggered by an external neutron) with the release of two neutrons (▸Fig. 30.3). Ideally, the released neutrons can then initiate two more fission reactions, a process that, in turn, releases four neutrons. These neutrons may initiate more reactions, and so on. Thus, the process can multiply, with the number of neutrons doubling with each generation. When this occurs, the neutron production rate (and, hence, the energy released from the sample) grows exponentially.

To maintain a sustained chain reaction, there must be an adequate quantity of fissionable material. The minimum mass required to produce a sustained chain reaction is called the **critical mass**. When the critical mass is attained, there is enough fissionable material such that at least one neutron from each fission event, on average, goes on to fission another nucleus.

Note: A chain reaction requires a critical mass.

Several factors determine critical mass. Most evident is the amount of fissionable material. If the quantity of material is small, many neutrons will escape from

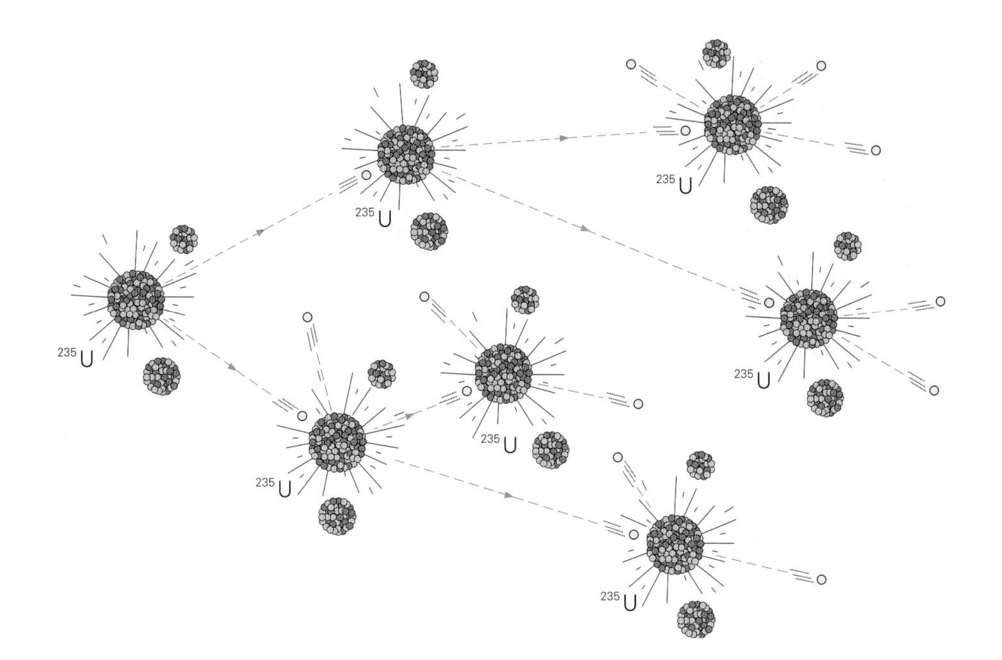

◀ **FIGURE 30.3 Fission chain reaction** The neutrons that result from one fission event can initiate other fission reactions, which, in turn, initiate further fission reactions, and so on. When enough fissionable material is present, the sequence of reactions can be adjusted to be self-sustaining (a chain reaction).

the sample (through the surface) before inducing a fission, and the chain reaction will die out. Also, nuclides other than ^{235}U in the sample may absorb neutrons, thereby limiting the chain reaction. As a result, the purity of the fissionable isotope affects the critical mass.

Natural uranium consists of the isotopes ^{238}U and ^{235}U. The natural concentration of ^{235}U is only about 0.7%. The remaining 99.3% is ^{238}U, which can absorb neutrons without fissioning, thereby inhibiting the chain reaction fission of ^{235}U. To have more fissionable ^{235}U nuclei in a sample and thus reduce the critical mass, the ^{235}U can be concentrated. This enrichment varies from 3% to 5% ^{235}U for nuclear reactor–grade material to more than 99% for weapons-grade material. This difference is important, because it is highly desirable that a nuclear reactor *not* explode like an atomic bomb nor use fuel capable of being made into a bomb.

Chain reactions take place almost instantly. If such a reaction proceeds uncontrolled, the quick and enormous release of energy can cause an explosion. This is the principle of the so-called atomic bomb. (A more descriptive name is the *fission or nuclear bomb*.) In such a bomb, several subcritical pieces of fuel are suddenly imploded to form a critical mass. The resulting chain reaction is then out of control, releasing an enormous amount of energy in a short period of time. For the steady production of energy from the fission process, the chain-reaction process must be controlled. We shall now see how this task is accomplished in nuclear reactors.

Nuclear Reactors

The Power Reactor Currently, the only practical design for generating electrical power from nuclear energy is based on the fission chain reaction. A typical design for a nuclear reactor is shown in ▼Fig. 30.4. There are five key elements to a reactor: fuel rods, core, coolant, control rods, and moderator.

Tubes packed with pellets of uranium oxide form the **fuel rods**, located in the central portion of the reactor called the **core**. A typical commercial reactor contains fuel rods bundled into assemblies of approximately 200 rods each. Coolant flows around the rods to remove the heat energy generated during the chain reaction. Reactors used in the United States are light-water reactors, which means that ordinary water is used as a **coolant** to remove heat. However, the hydrogen nuclei of ordinary water can capture neutrons to form deuterium, thus removing neutrons from the chain reaction. Hence, enriched uranium with 3–5% ^{235}U must be used.*

*Some Canadian reactors use heavy water, D_2O, as a coolant in place of light water. The advantage is that the deuterium (D) does not readily absorb neutrons, so the fuel can be of lower uranium enrichment. However, D_2O must first be separated from normal water, H_2O, an operation that takes energy.

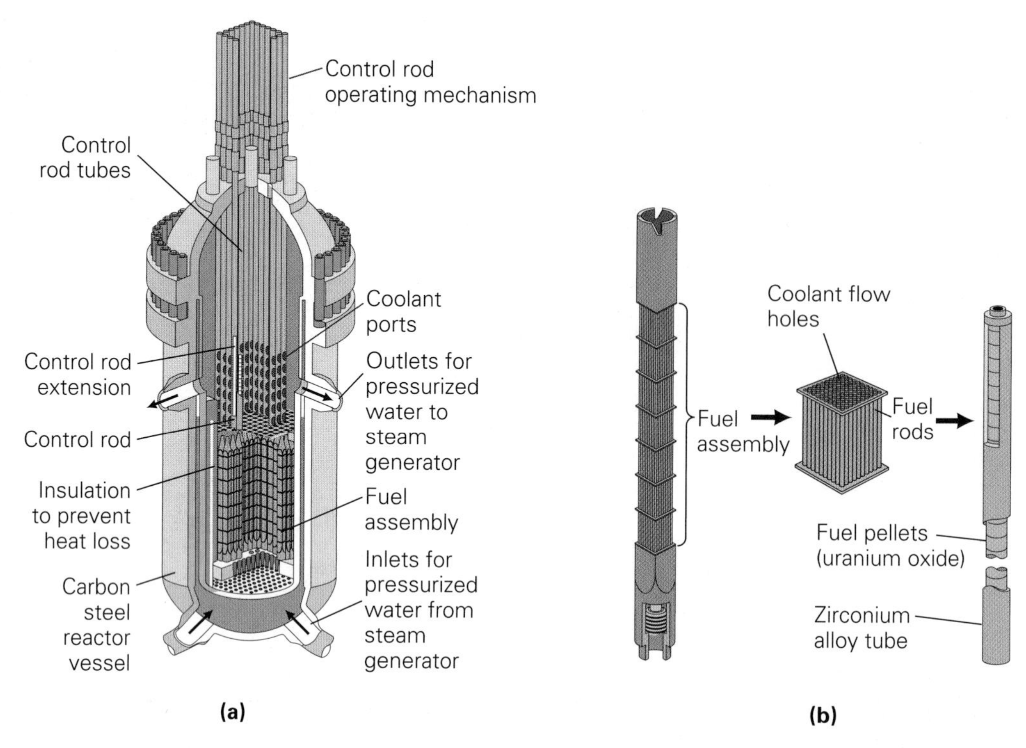

▶ **FIGURE 30.4 Nuclear reactor** **(a)** A schematic diagram of a reactor vessel. **(b)** A fuel rod and its assembly.

The chain-reaction rate and, therefore, the energy output of a reactor are controlled by boron or cadmium **control rods**, which can be inserted into or withdrawn from the reactor core. Cadmium and boron have a very high probability (cross-section) for absorbing neutrons. When these rods are inserted between the fuel-rod assemblies, neutrons are removed from the chain reaction. The control rods are adjusted so the chain reaction proceeds at a steady rate. The idea is to create a sustainable fission chain reaction in which the average fission produces only one more fission. For refueling, or in an emergency, the control rods can be fully inserted, and enough neutrons are removed to curtail the chain reaction and shut down the reactor. However, even with the chain reaction shut down, water must continue to circulate to prevent heat buildup due to the continuing decay of radioactive fission products in the fuel rods. If not, damage to the fuel rods can result. Melting and cracking of fuel rods due to inadequate cooling was the cause of the accident at Three Mile Island in 1979. The core of that reactor is still highly radioactive, because fission fragments were released from the broken rods.

The water flowing through the fuel-rod assemblies acts not only as a coolant, but also as a **moderator**. The fission cross-section for ^{235}U is largest for slow neutrons (sometimes called *thermal* neutrons with kinetic energies less than 1 eV). However, neutrons emitted from a fission are actually fast neutrons (with kinetic energies of about 2 MeV). Their speed is reduced, or moderated, by collisions with the water molecules. It takes about 20 collisions to moderate fast neutrons down to energies of about 1 eV.

The Breeder Reactor In a commercial power reactor, the ^{238}U goes along for the ride, so to speak. However, while it is unlikely to fission by absorbing a slow neutron, ^{238}U can be involved in reactions caused by *fast* neutrons. If a neutron's energy has not been completely moderated, reactions with ^{238}U can occur. For example, a conversion of ^{238}U to ^{239}Pu via successive beta decays after a fast neutron absorption can happen as follows:

$$\underset{(fast)}{{}^{1}_{0}n} + {}^{238}_{92}U \rightarrow ({}^{239}_{92}U^{*}) \rightarrow {}^{239}_{93}Np + \beta^{-}$$
$$\downarrow$$
$${}^{239}_{94}Pu + \beta^{-}$$

^{239}Pu, with a half-life of 24 000 years, *is* fissionable. Since it is possible to actively promote the conversion of ^{238}U to ^{239}Pu in a reactor by reducing the degree of moderation, the same amount of fissionable fuel (or more) can be produced (^{239}Pu) as is consumed (^{235}U). This is the principle behind the **breeder reactor.**

Note: A breeder reactor doesn't produce something from nothing. It uses the kinetic energy of the fast neutrons to create fissionable ^{239}Pu from the nonfissionable ^{238}U component of uranium.

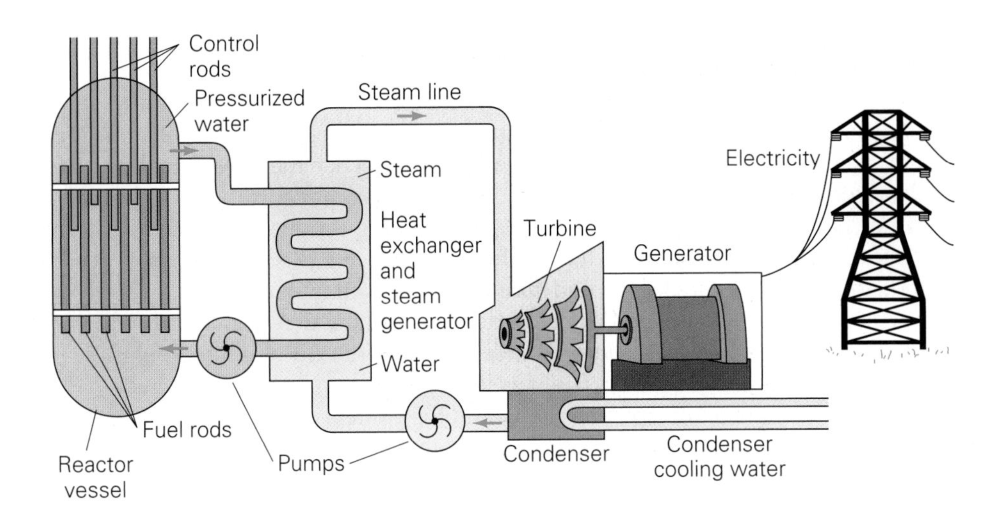

◀ **FIGURE 30.5 Pressurized water reactor** The components of a pressurized water reactor. The heat energy from the reactor core is carried away by the circulating water. The water in the reactor is pressurized so that it can be heated to high temperatures for more efficient heat removal. The energy is used to generate steam, which drives the turbine that turns the generator to produce electrical energy.

Notice that this isn't a case of getting something for nothing. Rather, the reactor is converting the unfissionable ^{238}U part of the fuel to the fissionable ^{239}Pu, while continuing to produce energy by ^{235}U fission.

Developmental work on the breeder reactor in the United States was essentially stopped in the 1970s. However, France went on to develop operational breeder reactors that provide nuclear fuel (^{239}Pu) for power reactors. Countries such as France that do not have huge natural resources of fossil fuels are highly dependent on nuclear energy and breeder reactors for their electrical energy needs.

Electricity Generation The components of a typical pressurized water reactor used in the United States are shown in ▲Fig. 30.5. The heat generated by the controlled chain reaction is carried away by the water passing through the rods in the fuel assembly. The water is pressurized to several hundred atmospheres so that it can reach temperatures over 300°C for more efficient heat removal. The hot water is then pumped to a heat exchanger, where the heat energy is transferred to the water of a steam generator. Notice that the reactor coolant and the exchanger water are in two separate and distinct closed systems. (Why?)

Note: Recall from Section 11.3 that the boiling point of water increases with increasing pressure.

Next, high-pressure steam turns a turbine that operates an electrical-energy generator, as is the case for any nonnuclear power plant. The steam is then cooled and condensed after turning the turbine. This final loop of coolant water typically carries the heat to a nearby ocean, lake, or river.

Nuclear Reactor Safety

Nuclear energy is used to generate a substantial amount of the electricity in the world. More than 25 countries now produce electricity by this method. Several hundred nuclear reactor units are in operation throughout the world, with more than 100 units in the United States. With the increasing number of nuclear facilities comes the fear of nuclear accidents and the subsequent release of radioactive materials into the environment.

If the coolant of a light-water reactor is lost, the chain reaction stops, because the coolant is also the moderator. However, the decay of the fission fragments, some with half-lives of hundreds of years, continues. In such a **LOCA** (an acronym for *loss-of-coolant accident*), the fuel rods might become hot enough (several thousand degrees Celsius) for the cladding (outside covering) to melt and fracture. Once this occurs, the hot, fissioning mass could fall into the water on the floor of the containment vessel and cause a steam explosion, a hydrogen explosion, or both. This explosion could rupture the walls of the containment vessel and allow radioactive fragments into the environment. Even if the walls were not breached, the hot "melt" of the fuel rods could burn through the floor of the building, eventually reaching ground level and the atmosphere (a situation called *China syndrome* because of the "melt" heading downward "towards China"). A partial **meltdown** did occur at the TMI generating plant in 1979. This meltdown was a LOCA, and a small amount of radioactive steam

▲ **FIGURE 30.6** Chernobyl, April 1986 An aerial photo showing damage to the reactor at Chernobyl. This accident released large amounts of radioactive materials into the environment, with dire consequences.

was vented to the atmosphere. Inside one reactor vessel, now sealed, electronic robots have discovered heavy damage to the fuel rods.

The April 1986 nuclear accident at Chernobyl was a meltdown following a LOCA caused by human error and magnified by an inherent instability resulting from the use of carbon as a moderator instead of water. When the flow of cooling water was inadvertently removed, the chain reaction went out of control—something that could not happen in a light-water reactor—producing a huge rise in temperature. The resulting explosions blew the top off the building (◂Fig. 30.6). When the fuel rods melted, the graphite blocks burned like a massive charcoal barbecue, spewing radioactive smoke into the air. Winds carried this radioactive smoke over much of Europe and over the North Pole into Canada and the United States, where significant amounts of core fission fragments (such as ^{131}I, ^{90}Sr, and ^{137}Cs) were detected.

Even if nuclear reactors operate safely (and their safety record, particularly in the United States, is a very good one), and although they emit virtually no pollutants such as greenhouse gases into the environment, there remains the problem of radioactive waste. Fission fragments are radioactive and have long half-lives. As a result, the safe handling of nuclear waste will be a problem for centuries to come. The United States is just beginning to come to grips with this. The consensus is that the only viable storage plan is one that buries the waste deep underground in geologically stable formations that keep it isolated from the atmosphere and ground water.

With its first shipment of low-level radioactive waste from the defense industry (*not* commercial power plants) in 1999, the United States opened its nuclear-waste site, in the New Mexico desert. Named the Water Isolation Pilot Plant, this plant stores such radioactive debris as plutonium-contaminated clothing, tools, and sludge. This material is housed several thousand feet below ground level in a hollowed-out salt formation. In 1987, Congress designated Yucca Mountain, about 90 miles northwest of Las Vegas, Nevada, to be the primary potential nuclear waste disposal site for the United States. A vertical shaft and excavated waste repository have been constructed there and the Department of Energy is proceeding with its application for a storage license. At the earliest, however, nuclear waste from commercial plants will not be received there until 2010. In fact there are proposals to locate waste storage in Utah rather than Nevada. Clearly, where and how to seal, safely transport, bury, and guard nuclear waste will be important decisions for generations to come.

30.3 Nuclear Fusion

OBJECTIVES: To (a) explain the fundamental difference between fusion and fission, (b) calculate energy releases in fusion reactions, and (c) understand how fusion might eventually provide a source of electric energy.

Another type of nuclear reaction that can release energy is *fusion*. In a **fusion reaction**, light nuclei fuse to form a more massive nucleus, releasing energy in the process ($Q > 0$). A simple fusion reaction—the fusion of two deuterium nuclei ($^{2}_{1}H$), sometimes called a D–D reaction—was examined in Example 30.1. There, it was shown that this reaction releases 3.27 MeV of energy per fusion.

Another example is the fusion of deuterium and tritium (a D–T reaction):

$$\underset{(2.014\,102\ u)}{^{2}_{1}H} + \underset{(3.016\,049\ u)}{^{3}_{1}H} \rightarrow \underset{(4.002\,603\ u)}{^{4}_{2}He} + \underset{(1.008\,665\ u)}{^{1}_{0}n}$$

Using the given masses, you should be able to show that this reaction involves a release of 17.6 MeV per fusion.

A fusion reaction releases much less energy in comparison with the more than 200 MeV released from a typical single fission. However, equal-mass samples of hydrogen and uranium have many, many more hydrogen nuclei than uranium nuclei. As a result, *per kilogram*, the fusion of hydrogen gives almost three times the energy released from uranium fission.

In a sense, our lives depend crucially on nuclear fusion, because it is the source of energy for most stars, including our Sun. One sequence of fusion reactions that is believed to be responsible for the Sun's energy output is as follows. First, there is proton–proton fusion,

$$^{1}_{1}H + ^{1}_{1}H \rightarrow ^{2}_{1}H + \beta^{+} + \nu$$

(where ν represents a neutrino, a particle to be discussed in the next section). Then another proton fuses with the deuteron:

$$^{1}_{1}H + ^{2}_{1}H \rightarrow ^{3}_{2}He + \gamma$$

Finally, two of the ^{3}He nuclei fuse:

$$^{3}_{2}He + ^{3}_{2}He \rightarrow ^{4}_{2}He + ^{1}_{1}H + ^{1}_{1}H$$

The net effect of this sequence, called the *proton–proton cycle*, is that four protons ($^{1}_{1}H$) combine to form one helium nucleus ($^{4}_{2}He$) plus two positrons (β^{+}), two gamma rays (γ), and two neutrinos (ν) with a release of about 25 MeV of energy:

$$4(^{1}_{1}H) \rightarrow ^{4}_{2}He + 2\beta^{+} + 2\gamma + 2\nu + Q \quad (Q = +24.7 \text{ MeV})$$

In a star such as our Sun, gamma-ray photons scatter off nuclei on their way to the surface. Each scattering results in a reduction in energy, until each photon has only a few electron-volts of energy. On reaching the surface, the photons are mostly visible-light photons. In our Sun, fusion involves only the central 10% of the Sun's mass. It has been going on for about 5 billion years and should continue, approximately as is, for another 5 billion years. As the next Example shows, an enormous number of fusion reactions per second is required to power the Sun.

Example 30.3 ■ Still Going: The Fusion Power of the Sun

Incoming sunlight energy falls on the Earth at the rate of 1.40×10^{3} W/m^{2}. Assuming that the Sun's energy is produced by the proton–proton cycle, calculate the mass lost by the Sun per second.

Thinking It Through. To find the mass-loss rate, we need to know the Sun's total power output. Imagine the power flow through a sphere centered on the Sun, with a radius equal to the distance between the Earth and Sun ($R_{E\text{-}S}$), and then calculate that sphere's area. The total power can then be determined from the power per square meter and the total area in square meters. The Sun's mass-loss rate can be calculated from the total power, based on the mass–energy equivalence (Table 29.3).

Solution. The data are as follows (using some solar system data given in Appendix III):

Given: $R_{E\text{-}S} = 1.50 \times 10^{8}$ km $= 1.50 \times 10^{11}$ m
$M_{S} = 2.00 \times 10^{30}$ kg
$P_{S}/A = 1.40 \times 10^{3}$ W/m^{2}

Find: $\Delta m/\Delta t$ (overall mass-loss rate)

The surface area of the imaginary sphere that intercepts all the Sun's energy is

$$A = 4\pi R_{E\text{-}S}^{2} = 4\pi(1.50 \times 10^{11} \text{ m})^{2} = 2.83 \times 10^{23} \text{ m}^{2}$$

Thus, the total power output of the Sun is

$$P_{S} = (1.40 \times 10^{3} \text{ W/m}^{2})(2.83 \times 10^{23} \text{ m}^{2}) = 3.96 \times 10^{26} \text{ W} = 3.96 \times 10^{26} \text{ J/s}$$

To find the equivalent mass-loss rate, we convert this power to MeV per second:

$$\frac{3.96 \times 10^{26} \text{ J/s}}{1.60 \times 10^{-13} \text{ J/MeV}} = 2.48 \times 10^{39} \text{ MeV/s}$$

Then, using the mass–energy equivalence (931.5 MeV/u), we find that the mass-loss rate is

$$\frac{\Delta m}{\Delta t} = \frac{(2.48 \times 10^{39} \text{ MeV/s})(1.66 \times 10^{-27} \text{ kg/u})}{931.5 \text{ MeV/u}} = 4.42 \times 10^{9} \text{ kg/s}$$

Follow-Up Exercise. In this Example, how many proton–proton cycles happen per second? [*Hint*: Each cycle releases 24.7 MeV.]

Fusion as a Source of Energy Produced on the Earth

In several ways, fusion appears to be an ideal energy source for the future. Enough deuterium exists in the oceans, in the form of heavy water (D_2O), to supply our needs for centuries. In addition, fusion does not depend on a chain reaction, so there is less danger of the release of radioactive material. Also, fusion products tend to have relatively short half-lives. For example, tritium has a half-life of only 12.3 years, compared with hundreds or thousands of years for some fission products.

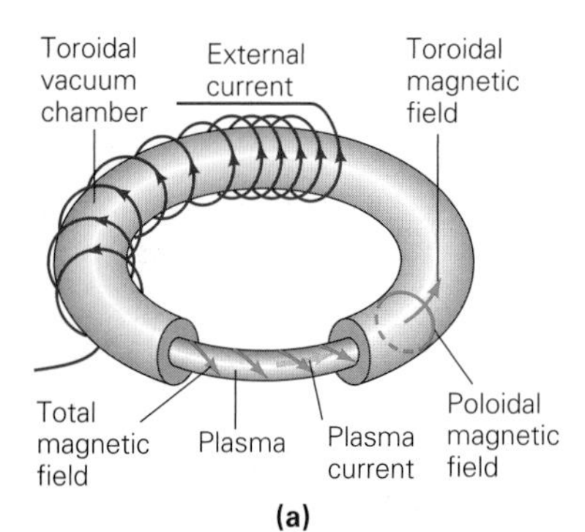

(a)

(b)

▲ **FIGURE 30.7 Magnetic confinement** Magnetic confinement is one method by which controlled nuclear fusion might be achieved. **(a)** Tokamak configuration, showing the *B* field generated by external currents. The magnetic field confines the plasma in the ring. **(b)** The Princeton Tokamak Fusion Test Reactor (TFTR).

However, many unresolved technical problems must be overcome before controlled fusion can be used commercially to produce electric energy. A primary problem is that very high temperatures are needed to *initiate* fusion reactions, because of the electrical repulsion between the nuclei. Temperatures on the order of millions of degrees are needed to initiate these *thermonuclear fusion reactions*. The problem is in confining sufficient energy in a reaction region to maintain these high temperatures. Because of this difficulty, practical fusion reactors have not yet been achieved. However, uncontrolled fusion has been demonstrated in the form of the *hydrogen (H) bomb*. In this case, the fusion reaction is initiated by an implosion created by a small atomic (fission) bomb. This implosion provides the necessary density and temperatures to begin the fusion process.

At the high temperatures required for fusion, electrons are stripped from their nuclei, and a gas is created that consists of positively charged ions and free, negatively charged electrons. Such a "gas" of charged particles is called a **plasma**.* Plasmas have a number of special physical properties and are sometimes referred to as the *fourth phase of matter*.

The technological problems involved in confining a plasma are being approached in at least two different ways: magnetic confinement and inertial confinement. Since a plasma is a gas of charged particles, it can be controlled and manipulated by using electric and magnetic fields. In **magnetic confinement**, magnetic fields are used to hold the plasma in a confined space, a so-called magnetic bottle. (See Fig. 19.33b.)

Once a plasma is confined, electric fields can be used to produce electric currents in it. The currents, in turn, raise the plasma's temperature. Temperatures of 100 million kelvins have been achieved in a design called a *tokomak* (◂Fig. 30.7). This design uses a magnetic field arranged in a donut shape to trap the charged particles.

In addition to high temperatures, the initiation of fusion has minimum requirements on plasma density and confinement time. The trick is to meet all these requirements at the same time. The generation of several megawatts of power for less than a second in a magnetically confined plasma is typical of the best results so far. Clearly, for commercial applications, much higher power levels must be attained at a continuous level.

Inertial confinement depends on implosion techniques. Hydrogen fuel pellets are either dropped or positioned in a reactor chamber. Pulses of laser, electron, or ion beams are then used to implode the pellet, producing compression and high densities and temperatures. Fusion can occur if the pellet stays together for a sufficient time, which depends on its inertia (hence the name *inertial confinement*). At present, laser and particle beams are not powerful enough to produce sustainable fusion by inertial confinement. In summary, practical energy production from fusion is not expected to be accomplished until well into this century, if at all.

30.4 Beta Decay and the Neutrino

OBJECTIVES: To (a) explain why the neutrino is necessary in order to account for observed beta-decay data, (b) specify some of the physical properties of neutrinos, and (c) write complete beta-decay equations.

At first glance, beta decay *appears* to be a two-body decay process in which unstable nuclei emit an electron ($^{0}_{-1}\text{e}$) or a positron ($^{0}_{+1}\text{e}$). Examples of both types of decay are

$$\underset{(14.003\,242\ \text{u})}{^{14}_{6}\text{C}} \rightarrow \underset{(14.003\,074\ \text{u})}{^{14}_{7}\text{N}} + {}^{0}_{-1}\text{e} \qquad \beta^{-}\ \textit{decay}$$

$$\underset{(13.005\,739\ \text{u})}{^{13}_{7}\text{N}} \rightarrow \underset{(13.003\,355\ \text{u})}{^{13}_{6}\text{C}} + {}^{0}_{+1}\text{e} \qquad \beta^{+}\ \textit{decay}$$

However, when analyzed in detail, these equations appear to violate the conservation of energy and linear momentum, as well as other conservation laws.

*Plasmas exist in our everyday world, such as in fluorescent lamps and lightning strokes.

The energy released (or disintegration energy) in the foregoing β^- process, as calculated from the mass defect, using the given masses*, is $Q = 0.156$ MeV. (You should show this.) Therefore, if the decay involves only two particles (electron and daughter), the electron, being much lighter than the daughter, should always have a kinetic energy of just slightly less than 0.156 MeV. However, this is *not* what happens. When the electron's kinetic energies are measured, a *continuous spectrum* of energies is observed up to $K_{max} \approx Q$, ▸Fig. 30.8. (See Conceptual Example 30.4.) That is, not all of the released energy is accounted for by the electron's kinetic energy. Nor is this the only difficulty. The emitted β^- and the daughter nucleus hardly ever leave the disintegration site in opposite directions. Thus, linear-momentum conservation appears to be violated as well.†

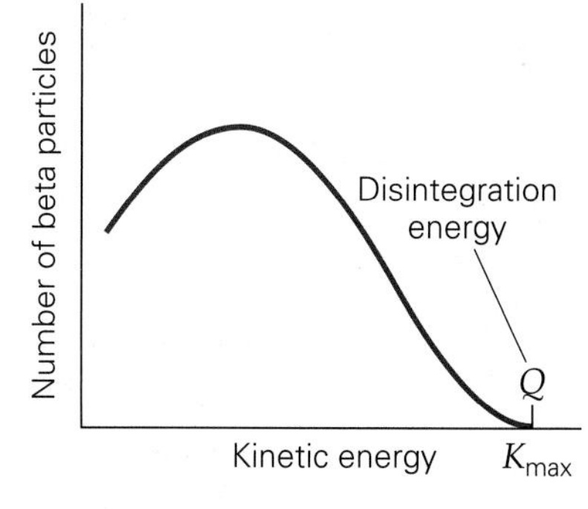

▲ **FIGURE 30.8 Beta ray spectrum** For a typical beta-decay process, all beta particles are emitted with $K < Q$, leaving unaccounted-for energy.

What, then, is the problem? It would be hoped that the conservation laws are not invalid. An alternative explanation is that these apparent violations are telling us something about nature that we do not yet recognize. All of the apparent difficulties can be resolved if it is assumed that an unobserved particle is also emitted during the decay. This explanation and the existence of such a particle were first proposed in 1930 by Wolfgang Pauli. Fermi christened this particle the **neutrino** (meaning "little neutral one" in Italian). For charge to be conserved, the neutrino has to be electrically neutral. Because the neutrino had been virtually impossible to observe, it must interact very weakly with matter. In fact, scientists eventually discovered that the neutrino interacts with matter through a second nuclear force, much weaker than the strong force, called the *weak interaction*, or the *weak nuclear force*. (See Section 30.5.)

Details of the initial experimental observations of beta decay suggested that the neutrino had zero mass and therefore traveled with the speed of light. It also had linear momentum p related to its total energy E by $E = pc$. Furthermore, it had a spin quantum number of $\frac{1}{2}$. In 1956, a particle with these properties was finally detected, and the neutrino's existence was firmly established.

The previous beta-decay equations can now be written correctly and completely as

$$^{14}_{6}\text{C} \rightarrow {}^{14}_{7}\text{N} + \beta^- + \bar{\nu}_e$$

and

$$^{13}_{7}\text{N} \rightarrow {}^{13}_{6}\text{C} + \beta^+ + \nu_e$$

where ν (the Greek letter nu) symbolizes the neutrino. The symbol with a bar over it represents an *antineutrino*. The overbar notation is a common way of indicating an antiparticle. Two *different* neutrinos are associated with beta decay. A neutrino is emitted in β^+ decay (ν_e), and an antineutrino is emitted in β^- decay ($\bar{\nu}_e$). The subscript e identifies the neutrinos as associated with electron–positron beta decay. As we will see in Section 30.5, additional types of neutrinos are associated with other decays triggered by the weak interaction.

Conceptual Example 30.4 ■ Having It All? Maximum Kinetic Energy in Beta Decay

Consider the decay of ^{14}C initially at rest:

$$^{14}_{6}\text{C} \rightarrow {}^{14}_{7}\text{N} + \beta^- + \bar{\nu}_e$$

What can be said about the maximum possible kinetic energy (K_{max}) of the beta particle: (a) $K_{max} > Q$, (b) $K_{max} = Q$, or (c) $K_{max} < Q$? Explain your reasoning clearly.

Reasoning and Answer. Answer (a) certainly cannot be correct, as it violates energy conservation. No one particle can have more than the total amount of energy available. So, can answer (b) be correct? Suppose that the beta particle did get all the released energy.

(continues on next page)

*Use of the atomic mass of the daughter ^{14}N (with seven electrons) is necessary in order to take into account the emitted electron, since the ^{14}N resulting from beta decay would have only the six electrons that orbit the parent ^{14}C nucleus.

†This process also violates the conservation of angular momentum. Careful analysis of the nuclear spin before the decay shows that it does *not* match the total spin (angular momentum) of the daughter plus the emitted electron.

That would mean that neither the neutrino nor the daughter nucleus had any energy, and therefore, neither would have any momentum. But to conserve momentum, at least one of these two particles must move off in the direction opposite to that of the beta particle. Hence, at least the neutrino or the daughter must have some energy. Therefore, the beta particle can't have it all, and answer (c) must be the correct one: $K_{max} < Q$.

Follow-Up Exercise. A typical energy spectrum of emitted beta particles is shown in Fig. 30.8. This spectrum indicates that there is a small probability of the beta particle having almost no kinetic energy. What would the decay products' trajectories look like after the decay if this were indeed the case?

30.5 Fundamental Forces and Exchange Particles

OBJECTIVES: To (a) understand the quantum-mechanical description of forces, and (b) classify the various forces according to their strengths, properties, ranges, and virtual particles.

The forces involved in everyday activities are complicated, because of the large numbers of atoms that make up ordinary objects. Contact forces between two hard objects, for example, are due to the repulsive electromagnetic forces between the atomic electrons. Looking at the *fundamental* interactions between particles makes things simpler. On this level, there are only four known **fundamental forces**: the *gravitational force*, the *electromagnetic force*, the *strong nuclear force*, and the *weak nuclear force*.

The most familiar of these forces are the gravitational and electromagnetic forces. Gravity acts between all particles, while the electromagnetic force is restricted to charged particles.* Both forces decrease with increasing particle separation distance and have a very large (essentially infinite) range.

To describe these forces classically, the concept of a field was employed. Modern physics, however, provides an alternative, more fundamental, description of how forces are transmitted. The force transmittal process is viewed as an exchange of particles. For example, a repulsive force would be analogous to you and another person interacting by tossing a ball back and forth. As you throw the ball and the other person catches it, for example, each of you feels a backward force. An observer who couldn't see the ball might conclude that there was a repulsive force between you and the other person.

The creation of such force-carrying particles would seem to violate energy conservation. However, because of the uncertainty principle (Section 28.4), a particle can be created for a *short time* with no outside energy input without violating energy conservation. Thus, over long time intervals, energy is conserved. For *extremely* short time intervals, the uncertainty principle *permits* a large *uncertainty* in energy ($\Delta E \propto 1/\Delta t$); creation of a particle is allowed; and energy conservation can be briefly violated. However, the created particle is absorbed before it is ever detected, so we never *observe* energy nonconservation. A particle created and absorbed in such a manner is called a **virtual particle**. In this sense, *virtual* means "undetected."

Thus, in the modern view, the fundamental forces are carried, or transmitted, by virtual **exchange particles**. The exchange particles for the four forces must differ in mass. Recall that the greater the mass of the particle, the greater is the energy ΔE required to create it, and thus the shorter is the time Δt for which it exists. Since a massive particle can exist for only a short time, the distance it can travel, and hence the range of the associated force, must be small. That is, *the range of a force associated with an exchange particle is inversely proportional to the mass of that exchange particle.*

The Electromagnetic Force and the Photon

The exchange particle of the electromagnetic force is a **virtual photon**. Here "virtual" refers to a photon that is exchanged but never detected directly. As a "massless" particle, it has an infinite range, as is required for the electromagnetic force.

*Both the electric and magnetic forces are actually components of a single force—the electromagnetic force.

A particle exchange can be visualized graphically by using a *Feynman diagram*, as in ▸Fig. 30.9. This graph shows the specific example of how the exchange idea can explain the electrical repulsion between two electrons. Such *space–time diagrams* are named after American scientist and Nobel Prize winner Richard Feynman (1918–1988), who used them to analyze electromagnetic interactions in his *quantum electrodynamics* theory. The important points are the vertices (intersections) of the diagram. One electron creates a virtual photon at point A, and the other electron absorbs it at point B. Each of the electrons undergoes a change in energy and in momentum (including direction) by virtue of the photon exchange and resulting force.

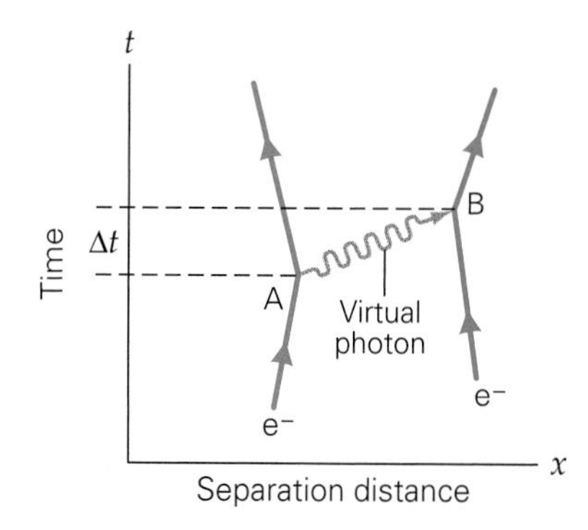

▲ **FIGURE 30.9 Feynman diagram of an electron–electron interaction** The interacting electrons undergo a change in energy and momentum, due to the exchange of a virtual photon, which is created at A and absorbed at B in an amount of time (Δt) that is consistent with the uncertainty principle.

The Strong Nuclear Force and Mesons

Japanese physicist Hideki Yukawa (1907–1981) proposed in 1935 that the short-range strong nuclear force between two nucleons is associated with an exchange particle called the **meson**. An estimate of the meson's mass can be made from the uncertainty principle. If a nucleon were to create a meson, conservation of energy would be (undetectably) violated by an amount of energy at least equal to the energy equivalent of the meson's mass, or

$$\Delta E = (\Delta m)c^2 = m_{\rm m}c^2$$

where $m_{\rm m}$ is the mass of the meson.

By the uncertainty principle, the meson would have to be absorbed in the exchange process in an amount of time on the order of

$$\Delta t \approx \frac{h}{2\pi \Delta E} = \frac{h}{2\pi m_{\rm m} c^2}$$

In this amount of time, the meson could travel a distance R (which stands for *range*) that must be less than that traveled by light in the same time; thus,

$$R < c\Delta t = \frac{h}{2\pi m_{\rm m} c} \tag{30.5}$$

Taking this distance to be the experimentally known range of the nuclear force ($R \approx 1.4 \times 10^{-15}$ m) and solving for $m_{\rm m}$ gives $m_{\rm m} \approx 270m_{\rm e}$, where $m_{\rm e}$ is the electron's mass. If Yukawa's meson exists, it should have a mass on the order of 270 times that of an electron.

Virtual mesons in an exchange process cannot be directly observed, because they are emitted and reabsorbed during the nucleon–nucleon interaction. Physicists reasoned, however, that if sufficient energy were involved in the *collision* of nucleons, real mesons might be created from the energy available in the collision. These mesons could then be detectable. At the time of Yukawa's prediction, there were no known particles with masses between that of the electron ($m_{\rm e}$) and the proton ($m_{\rm p} = 1836m_{\rm e}$).

In 1936, Yukawa's prediction seemed to come true when a new particle with a mass of about $200m_{\rm e}$ was discovered in cosmic rays. Originally called the μ (Greek letter mu) meson, and now just the **muon**, it was shown to have two charge varieties, $\pm e$, with a mass of $m_{\mu^\pm} = 207m_{\rm e}$. However, further investigations showed that the muon did *not* behave like the strongly interacting particle of Yukawa's theory.

This situation was a source of controversy and confusion for years. But, in 1947, more particles in this mass range were discovered in cosmic radiation. These particles (one positive, one negative, and one with no charge) were called π (pi) mesons (for *primary mesons*) and now are more commonly called **pions**. Measurement showed the masses of the pions to be $m_{\pi^\pm} = 273m_{\rm e}$ and $m_{\pi^{\rm o}} = 264m_{\rm e}$. Moreover, pions were found to interact strongly with matter. The pion fulfilled the requirements of Yukawa's theory. This meson was generally accepted as the particle primarily responsible for the transmission of the strong nuclear force. The Feynman diagram for a neutron and proton interacting via the exchange of a negative pion (the strong nuclear force) is shown in ▸Fig. 30.10.

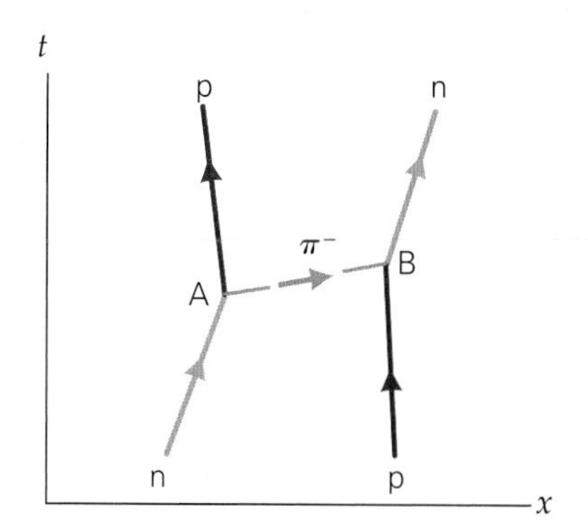

▲ **FIGURE 30.10 A Feynman diagram for nucleon–nucleon interactions via pion exchange** Nucleon–nucleon interactions via the strong nuclear force occur through the exchange of virtual pions, here a negative one. This diagram shows one of many possible n–p interactions. It is called a *charge exchange* reaction; can you tell why? What other possible diagrams for neutron–proton interactions are there?

Note: Muon decay was treated in Examples 26.2 and 26.4 as evidence of time dilation.

Free pions and muons are unstable. For example, the π^+ particle decays in about 10^{-8} s into a muon and another type of neutrino:

$$\pi^+ \rightarrow \mu^+ + \nu_\mu$$

The ν_μ is called a *muon neutrino*, and it differs from the electron neutrino (ν_e) produced in beta decay. The muons can also decay into positrons and electrons with the emission of both types of neutrinos. For example, the positive-muon decay scheme is

$$\mu^+ \rightarrow \beta^+ + \nu_e + \bar{\nu}_\mu$$

The Weak Nuclear Force and the W Particle

The discrepancies in beta decay discussed in the preceding section led to another discovery. Electrons and neutrinos are emitted from unstable nuclei, but they do *not* exist inside the nucleus before the decay takes place. Enrico Fermi proposed that these particles are actually created at the time the nucleus decays. For β^- decay, this means that a neutron is in some way changed, or *transmuted*, into three particles:

$$\mathrm{n} \rightarrow \mathrm{p}^+ + \beta^- + \bar{\nu}_e$$

Experiments confirmed that free neutrons disintegrate by this decay scheme, with a half-life of about 10.4 min. But which force could cause a neutron to disintegrate in this manner? Since the neutron is outside the nucleus (free) when it decays, none of the known forces, including the strong nuclear force, seemed applicable. Thus, some other fundamental force must be acting in beta decay. Decay-rate measurements indicated that the force was extraordinarily weak—weaker than the electromagnetic force, but still much stronger than the gravitational force. This force was dubbed the **weak nuclear force**.

Originally, it was thought that this *weak interaction* was extremely localized, without any measurable range. We now know that the weak force has a range of about 10^{-17} m. While this range is much smaller than that of the strong force, it isn't zero. This means that the exchange particles associated with the weak force (the virtual force carriers) must be much more massive than the pions of the strong force. The virtual exchange particles associated with the weak force were named **W particles**.* W (*weak*) particles have masses about 100 times that of a proton, a fact that correlates with the extremely short range of the weak force. The existence of the W particle was confirmed in the 1980s when accelerators were built with enough energy to create the first real (nonvirtual) W particles.

The weak force is the only force that acts on neutrinos, which explains why these particles are so difficult to detect. Research has shown that the weak force is involved in the transmutation of other subatomic particles as well. In general, the weak force is limited to transmuting the identities of particles within the nucleus. The only way it manifests its existence in the outside world is through the emitted neutrinos. For example, the Sun's fusion reactions create neutrinos that continuously pass through the Earth.

One highly noticeable, but infrequent, announcement of the weak force at work occurs during the explosion of a stellar *supernova*. In a supernova, the collapse of the core of an aging star gives rise to a huge energy release, accompanied by a great number of neutrinos. In a relatively "nearby" supernova (a mere 1.5×10^{18} km away) observed in 1987, a burst of neutrinos was detected.

Whereas the original thought was that neutrinos were massless, recent experiments have determined that they do, in fact, have some very small amount of mass, on the order of a millionth of the electron's mass. This fact could have vast implications to our understanding of the Big Bang theory and the unexplained predominance of regular matter over antimatter in our universe. At this point, neutrino research is a growing field that will play an important role in physics for years to come.

*The weak force is actually carried by *three* exchange particles: W^+, W^-, and Z^o (neutral).

TABLE 30.2 Fundamental Forces

Force	*Relative Strength*	*Action Distance*	*Exchange Particle*	*Particles That Experience the Interaction*
Strong nuclear	1	Short range ($\approx 10^{-15}$ m)	Pion (π meson)	Hadrons*
Electromagnetic	10^{-3}	Inverse square (infinite)	Photon	Electrically charged
Weak nuclear	10^{-8}	Extremely short range ($\approx 10^{-17}$ m)	W particle†	All
Gravitational	10^{-45}	Inverse square (infinite)	Graviton	All

*Hadrons are discussed in Section 30.6.
†Three particles are involved, as described in Section 30.8.

The Gravitational Force and Gravitons

The exchange particles associated with the gravitational force are called **gravitons**. There is still no firm evidence of the existence of this massless particle. (Why must gravitons be massless?) Ongoing experiments to detect the graviton have as yet proven unsuccessful, due to the relative weakness of its interaction.

A comparison of the relative strengths of the four fundamental forces is given in Table 30.2.

30.6 Elementary Particles

OBJECTIVES: To (a) classify the elementary particles into families and, (b) understand the different properties of the various families of elementary particles.

The fundamental building blocks of matter are referred to as **elementary particles**. Simplicity reigned when it was thought that an atom was an indivisible particle and therefore was *the* elementary particle. Early in the twentieth century, the proton, neutron, and electron were discovered to be constituents of atoms. It was hoped that these three particles were nature's elementary particles. However, scientists now know of a huge variety of subatomic particles and are working to simplify and reduce this list to a smaller set of truly elementary particles—building blocks for all the other "composite" particles—if this task is indeed possible.

Several systems classify elementary particles on the basis of their various properties. One classification uses the distinction of nuclear-force interactions. Particles that interact via the weak nuclear force, but not the strong force, are called leptons ("light ones"). The lepton family includes the electron, the muon, and their neutrinos. Other particles, called **hadrons**, are the only particles to interact by the strong nuclear force. These particles include the proton, neutron, and pion. Let's look briefly at the lepton and hadron families.

Leptons

The most familiar **lepton** is the electron. It is the only lepton that exists naturally in atoms. There is no evidence that it has any internal structure, at least down to 10^{-17} m. Thus, at present, the electron is considered to be a point particle.

Muons were first observed in cosmic rays. They appear not to have any internal structure and are therefore sometimes referred to as *heavy electrons*. Muons are unstable and decay in about 2×10^{-6} s, according to the following scheme:

$$\mu^- \rightarrow \beta^- + \bar{\nu}_e + \nu_\mu$$

A third charged lepton is known as a tau (τ^-) particle, or **tauon**. It has a mass about twice that of a proton. The electron, muon, and tauon are all negatively charged, have no apparent internal structure, and have positively charged antiparticles.

The remaining leptons are neutrinos, which are present in cosmic rays, emitted by the Sun, and appear in some radioactive decays. Neutrinos are now known to have a very small mass and travel close to, but slightly less than, the speed of light. They feel neither the electromagnetic force nor the strong force and pass through matter with very little interaction.

There are three types of neutrinos, each associated with a different charged lepton ($e^{\pm}, \mu^{\pm}, \tau^{\pm}$). They are named, not surprisingly, the *electron neutrino* (ν_e), the *muon neutrino* (ν_μ), and the *tau neutrino* (ν_τ). There is also an antineutrino for each of them, for a total of six different neutrinos. This completes the list of leptons. With a total of 6 leptons plus their antiparticles, there are 12 different leptons in all. Current theories predict that there should be no others.

Hadrons

Another family of elementary particles is the **hadrons**. All hadrons interact by the strong force, the weak force, and gravity. The electrically charged members can also interact by the electromagnetic force.

The hadrons are subdivided into *baryons* and *mesons*. **Baryons** include the familiar nucleons—the proton and neutron. They are distinguished from mesons in that they possess half-integer intrinsic spin values $\left(\frac{1}{2}, \frac{3}{2}, \ldots\right)$. Except for the stable proton, baryons decay into products that eventually include a proton. For example, recall from Section 29.2 the beta decay of a neutron into a proton.

Mesons, which include pions, have integer spin values (0, 1, 2, . . .) and eventually decay into leptons and photons. For example, the neutral pion decays into two gamma rays. (Why must there be at least two?)

The large number of hadrons suggests that they may be composites of other truly elementary particles. Some help in sorting out the hadron "zoo" came in 1963 when Murray Gell-Mann and George Zweig of the California Institute of Technology put forth the *quark theory*, which is discussed in the next section.

Leptons and hadrons and their properties are summarized in Table 30.3.

TABLE 30.3 Some Elementary Particles and Their Properties

Family Name	*Particle Type*	*Particle Symbol*	*Antiparticle Symbol*	*Rest Energy (MeV)*	*Lifetime* (s)*
Lepton					
	Electron	e^-	e^+	0.511	stable
	Muon	μ^-	μ^+	105.7	2.2×10^{-6}
	Tauon	τ^-	τ^+	1784	$\approx 3 \times 10^{-13}$
	Electron neutrino	ν_e	$\bar{\nu}_e$	$0^\dagger$	stable
	Muon neutrino	ν_μ	$\bar{\nu}_\mu$	$0^\dagger$	stable
	Tauon neutrino	ν_τ	$\bar{\nu}_\tau$	$0^\dagger$	stable
Hadron					
Mesons	Pion	π^+	π^-	139.6	2.6×10^{-8}
		π^o	same	135.0	8.4×10^{-17}
	Kaon	K^+	K^-	493.7	1.2×10^{-8}
		K^o	$\overline{K^o}$	497.7	8.9×10^{-11}
Baryons	Proton	p	$\bar{p}$	938.3	stable (?)§
	Neutron	n	$\bar{n}$	939.6	0.9×10^{2}
	Lambda	Λ^o	$\overline{\Lambda^o}$	1116	2.6×10^{-10}
	Sigma	Σ^+	$\overline{\Sigma^-}$	1189	8.0×10^{-10}
		Σ^o	$\overline{\Sigma^o}$	1192	0.6×10^{-20}
		Σ^-	$\overline{\Sigma^+}$	1197	1.5×10^{-10}
	Xi	Ξ^o	$\overline{\Xi^o}$	1315	2.9×10^{-10}
		Ξ^-	$\overline{\Xi^+}$	1321	1.6×10^{-10}
	Omega	Ω^-	Ω^+	1672	8.2×10^{-11}

*Lifetimes are expressed to two significant figures or fewer.

†Neutrinos are now known to possess a very small amount. Experiments yield upper limits and the mass of the electron neutrino is known to be less than 7×10^{-6} MeV.

§Electroweak theory predicts that the proton is unstable, with a half-life of 1000 trillion times the age of the universe.

30.7 The Quark Model

OBJECTIVES: To (a) become familiar with the quark model and properties of quarks, and (b) understand how the quark model accounts for the properties of baryons and mesons.

Gell-Mann and Zweig proposed that, in fact, hadrons are *not* elementary particles and therefore are not fundamental building blocks. Hadrons, they theorized, are composite particles composed of truly elementary (fundamental) particles. They named these particles **quarks** (taken from James Joyce's novel *Finnegan's Wake**). However, Gell-Mann and Zweig noted that because leptons and photons are essentially point particles, they *are* likely to be truly elementary particles with no internal structure.

Noting that some hadrons are electrically charged, Gell-Mann and Zweig reasoned that quarks must also possess charge. Their initial **quark model** consisted of three different quarks (with fractional charges) and their antiparticles (called *antiquarks*). Table 30.4 shows that, to account for hadrons that were undiscovered at the time, the list of quarks eventually had to be expanded to include six types. The original quark idea required only three quarks (plus three antiquarks) to construct the hadrons known at that time. These quarks were named the *up* quark (u), the *down* quark (d), and the *strange* quark (s). By using various combinations of these three types of quarks, the relatively heavy hadrons, the baryons (whose name means "heavy ones") could be built. In addition, quark–antiquark pairs could account for the lighter hadrons, the mesons. Thus, Gell-Mann and Zweig had proposed a radical new idea:

> Quarks are the fundamental particles of the hadron family.

Since several quarks have to combine to give the charge on the hadron, the quarks must have fractions of an electron charge e. The theory proposed that u, d, and s quarks have charges of $+\frac{2}{3}e$, $\pm\frac{1}{3}e$, and $-\frac{1}{3}e$, respectively. The antiquarks, designated by overbars, such as $\bar{u}$, have opposite charges. Thus, three quark combinations could produce any baryon. For instance, the quark composition of the proton and neutron would be uud and udd, respectively. Mesons could be constructed from various pairs of a quark and an antiquark, such as $u\bar{d}$ for the positive pion, π^+ (▸Fig. 30.11).

TABLE 30.4 Types of Quarks*

Name	*Symbol*	*Charge*
Up	u	$+\frac{2}{3}e$
Down	d	$-\frac{1}{3}e$
Strange	s	$-\frac{1}{3}e$
Charm	c	$+\frac{2}{3}e$
Top (Truth)	t	$+\frac{2}{3}e$
Bottom (Beauty)	b	$-\frac{1}{3}e$

*Antiquarks are designated by an overbar and have opposite charges compared with those of the corresponding quarks.

Note: Quark names (such as *up*, *down*, and *strange*) are arbitrary and therefore should not be taken literally. They serve simply to identify the quarks.

Note: Three quarks in proper combination account for all known baryons. Quark–antiquark pairs in proper combination account for all known mesons.

Conceptual Example 30.5 ■ Building Mesons: Quark Engineering, Inc.

Using the data in Tables 30.3 and 30.4, explain why it is not possible to build a meson from two quarks (in other words, without using any antiquarks).

Reasoning and Answer. It can be seen from the list of mesons in Table 30.3 that mesons are either neutral or if they are charged, it is an integral multiple of e. In Table 30.4, note that all the positively charged quarks have a charge of $+\frac{2}{3}e$. Thus, any two of them would add up to a meson charge of $+\frac{4}{3}e$, which is contrary to observation. Similar reasoning holds if we use two negatively charged quarks: Since they all have the same charge $\left(-\frac{1}{3}e\right)$, any two of them add up to a total charge of $-\frac{2}{3}e$, again in disagreement with experiment. Finally, consider combining one positively charged quark and one negatively charged quark. The net charge of this combination is $+\frac{1}{3}e$, again not in agreement with observation. Thus, no combination of two quarks (or two antiquarks) can be used to produce a meson. The combinations must include at least one antiquark and one quark.

Follow-Up Exercise. The antiparticle of the positive pion (π^+) is the negative pion (π^-). What is the quark structure of the negative pion? Show that it is composed of the antiquarks of the quarks that make up the positive pion.

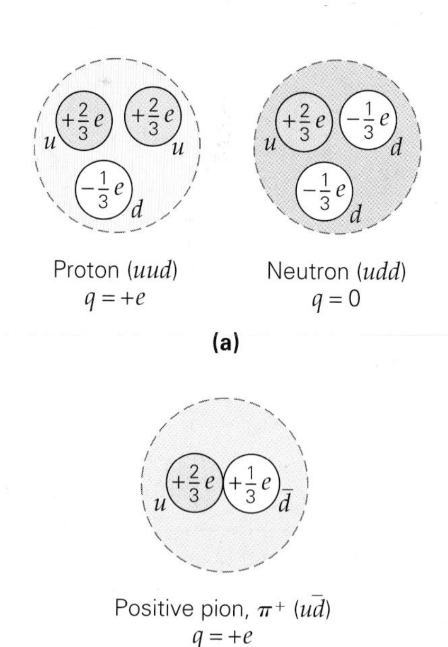

▲ **FIGURE 30.11 Hadronic quark structure (a)** Three combinations of quarks can be used to construct all the baryons, such as the proton and neutron. **(b)** Quark–antiquark combinations can be used to construct all the mesons, such as the positive pion.

The discovery of new subatomic particles in the 1970s led to the addition of the last three quark types: *charm* (c); *top*, or truth (t); and *bottom*, or beauty (b). Today, there is firm experimental evidence of the existence of all six quarks and their six antiquarks. How many more such particles will be needed to keep up

*A line in the novel exclaims, "Three quarks for Muster Mark!" The "three quarks" denote the three children of a character in the novel, Mister (Muster) Mark, who is also known as Mr. Finn.

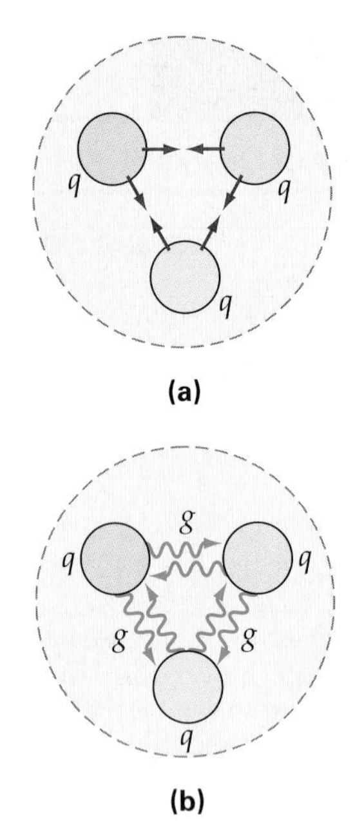

▲ **FIGURE 30.12 Quarks, color charge, confinement, and gluons** **(a)** Quarks of different color charge attract each other via the color force, which keeps them confined (here, inside a baryon—how do you know this is a baryon?). **(b)** Gluons (wiggly arrows) are exchanged between quarks of different color charge, creating the color force—analogous to virtual photons and the electric force.

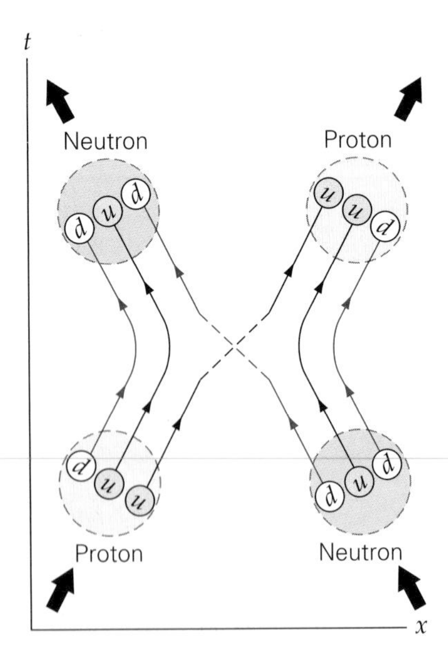

▲ **FIGURE 30.13 Quark depiction of the strong nuclear force** Instead of envisioning the n–p interaction as an exchange of a virtual π^- particle, we can use a model of quark exchange. In this exchange, a pair of quarks (equivalent to a π^- particle) is transferred. A proton becomes a neutron, and vice versa.

with our growing "zoo" of elementary particles? The hope is, of course, that this list will not need to be expanded. In summary, the present picture includes the following truly elementary particles: leptons (and antileptons), quarks (and antiquarks), and exchange particles such as the photon.

Quark Confinement, Color Charge, and Gluons

Thus far in our discussion on quarks, one thing has been missing—direct experimental observation of a quark. Unfortunately, even in the most energetic particle collisions, *a free quark has never been observed*. Physicists now believe that quarks are permanently confined within their particles by a springlike force. That is, a force exists between quarks that grows as they separate from one another. This force grows very rapidly with distance and prevents the ejection of a quark from its particle. This phenomenon is called **quark confinement**.

In order to explain the force between quarks and to clear up some problems with apparent violation of the Pauli exclusion principle (see the discussion of atomic structure in Section 28.3), quarks were endowed with another characteristic called **color charge**, or simply *color*. There are three types of color charge: red, green, and blue. (These names have nothing to do with visual color.) In analogy to electric charge, the quark confinement force exists because different color charges attract each other. Recall from Section 30.5 that the electromagnetic force is due to virtual photon exchanges between charged particles. Similarly, the force between quarks of different color is due to exchanges of virtual particles called **gluons** (◂Fig. 30.12). This force is sometimes called the **color force**. This theory is named *quantum chromodynamics (QCD); chromo* for "color", in analogy to Feynman's quantum *electrodynamics (QED)*, which successfully explains the electromagnetic force.

The concept of the force between quarks can be extended to explain the force between hadrons—the strong nuclear force. Consider our previous explanation of the strong force between a neutron and proton as shown in Fig. 30.10a: the exchange of a virtual negative pion. More fundamentally, we can qualitatively depict this force in terms of an exchange of quarks between the hadrons (◂Fig. 30.13).

30.8 Force Unification Theories, the Standard Model, and the Early Universe

OBJECTIVES: To (a) become familiar with current attempts to unify the four fundamental forces, and (b) understand why elementary-particle interactions might hold the key to understanding the very early evolution of the universe.

Unification Theories

Early in the twentieth century, Einstein was one of the first theorists to conjecture that it might be possible to unify the fundamental forces of nature—that these four, apparently very different forces, might really be just different manifestations of one force. Each manifestation would appear under a different set of physical conditions. For example, an electrically neutral particle would not exhibit the electromagnetic portion. Since then, it has been the dream of physicists to understand the "fundamental interactions" in the universe, using just one force. Attempts to unify the various forces are called **unification theories**.

Actually, the first step toward unification was taken in the nineteenth century by Maxwell when he combined the electric and magnetic forces into a single electromagnetic force. Einstein later showed that the two are connected by relativity. The next major step occurred in the 1960s when Sheldon Glashow, Abdus Salam, and Steven Weinberg successfully combined the electromagnetic force with the weak force into a single **electroweak force**. For their efforts, they were awarded a Nobel Prize in 1979.

How can such apparently very different forces be unified? After all, their exchange particles are so different—recall the massless photon (γ) for the electromagnetic force and the massive W particle for the weak force. However, it was reasoned that, at extremely high energies, the mass–energy of the W particle is negligible compared with its total energy, in effect making it massless, like the photon. To understand this reasoning, recall from Section 26.5 that the total energy of a particle is the sum of its kinetic and rest energies ($E = K + mc^2$). When $K \gg mc^2$ (that is, at high energies), the particle's rest energy is negligible compared with its kinetic energy—perhaps, Glashow, Salam, and Weinberg reasoned, making the W particle more "photonlike" than was previously thought.

In the electroweak theory, weakly interacting particles such as electrons and neutrinos carry a *weak charge*. This weak charge is responsible for the exchange of particles, thus creating the combined electroweak force. The electroweak unification theory predicted the existence of three electroweak exchange particles: W^+ and W^- when there is charge exchange, and the neutral Z^o when there is no charge transfer (▸Fig. 30.14). The eventual discovery of these particles led to a Nobel Prize for Carlo Rubbia and Simon van der Meer in 1984.

Scientists are now attempting to unify the electroweak force with the strong nuclear force. If successful, this **grand unified theory (GUT)** would reduce the number of fundamental forces to two. Most GUT attempts require several dozen exchange particles. In addition, leptons and quarks are combined into one family and can change into each other through the exchange of these particles. Scientists are mildly optimistic that experimental verification of one of the GUT candidates might occur in this century. For example, most of the current GUT theories predict that the proton should be unstable and decay with a half-life of about 10^{32} years, which is 10^{22} times the age of the universe. Experiments looking for this decay are currently underway, so far with no success. The experiments require huge amounts of water (protons) in order to make up for the expected very slow decay rate.

The ultimate unification would be to fold the gravitational force into the GUT, creating a single **superforce**. How this would be done, or even *if* it can ever be done, is not clear at this point. One problem is that, although the three components of the GUT can be represented as force fields in space and time, in our current view, gravity *is* space and time!

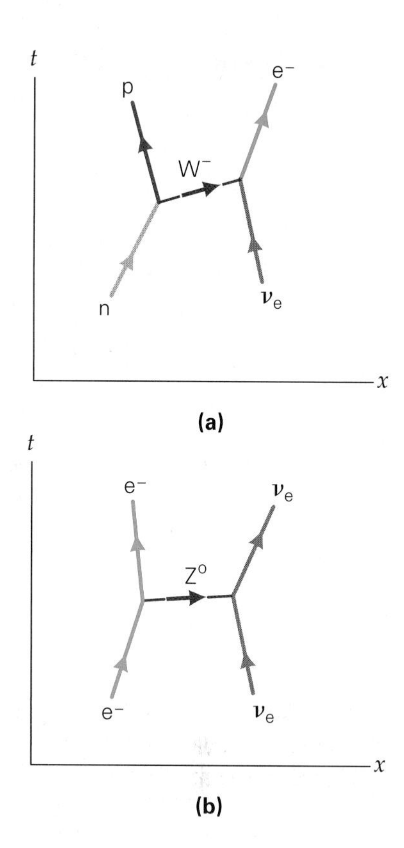

▲ **FIGURE 30.14 Weak-force interactions** The Feynman diagrams for **(a)** the neutrino-induced conversion of a neutron into a proton and an electron through the exchange of a W^- particle and **(b)** the scattering of a neutrino by an electron through the exchange of a neutral Z^o particle.

The Standard Model

Taken together, the electroweak theory and the QCD model for the strong interaction are referred to as the **standard model**. In this model, the gluons carry the strong force. This force keeps quarks together to form composite particles such as protons and pions. Leptons do not experience the strong force and participate only in the gravitational and electroweak interactions. Presumably, the gravitational interaction is carried by the graviton. The electromagnetic part of the electroweak interaction is carried by the photon, and the W and Z bosons carry the weak portion of the electroweak force.

Evolution of the Universe and the Superforce

An interesting connection now exists between elementary particle physicists and astrophysicists interested in the evolution of the universe—in particular, its very early evolution. This connection occurs because, according to present ideas, the universe began 10 billion to 15 billion years ago with the *Big Bang*. It is theorized that temperatures during the first 10^{-45} second of the universe were on the order of 10^{32} K. This corresponds to particle kinetic energies of about 10^{22} MeV, high enough that the rest energy of even the most massive elementary particles would be negligible. In effect, the particles would be massless, perhaps placing the four forces on an equal footing.

As the universe expanded and cooled, these early elementary particles condensed into what we see today. First, protons, neutrons, and electrons formed; in turn, they combined into atoms and, eventually, molecules. Over the billions of years, the average temperature of the universe has cooled off to its present value of about 3 K. In the process, scientists think that the superforce symmetry (the equal footing of all four forces) has been lost, leaving us with four very different-looking forces that are really components of a superforce.

It is hoped that future experiments might tell us more about the early moments of the universe and thus about the superforce. The ultimate goal of physics might even be within reach—to understand the basic interactions that govern the universe. While great strides are being made, it is likely to take well into this century, if not further, to achieve this goal.

Chapter Review

- A **nuclear reaction** (including decay) involves the interactions of nuclei and particles, usually resulting in different product nuclei or particles.

- The **Q value** of a reaction (or decay) is the energy released or absorbed in the process. For a two-body reaction of the form $A + a \rightarrow B + b$,

$$Q = (m_A + m_a - m_B - m_b)c^2 = (\Delta m)c^2 \qquad (30.3)$$

If $Q > 0$, the reaction is **exoergic** and energy is released.

If $Q < 0$, the reaction is **endoergic** and energy is absorbed.

- In **fission**, an unstable heavy nucleus decays by splitting into two fragments and several neutrons, which together have less total mass than that of the original nucleus; kinetic energy is thus released.

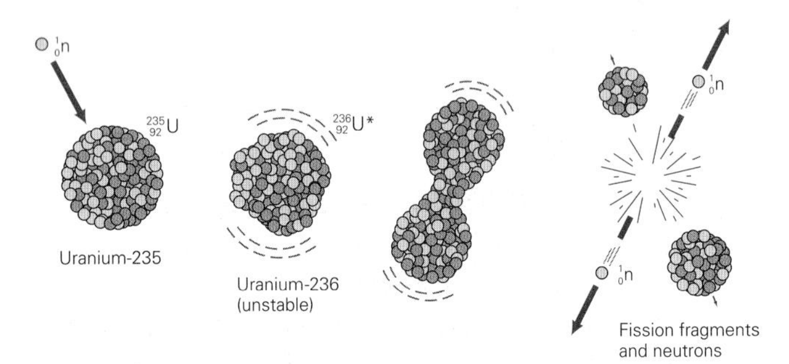

- A **chain reaction** occurs when the neutrons released from one fission trigger other fissions, which trigger further fissions, and so on.

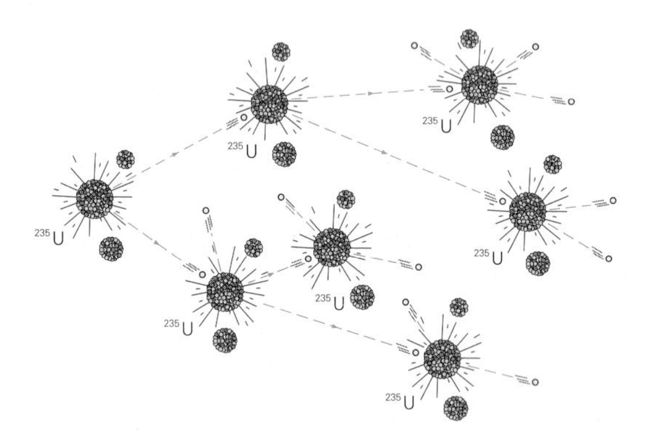

- In a **nuclear power reactor**, the fission chain reaction is kept from going out of control by a set of control rods. The heat energy released is usually used to create steam, which eventually turns generator turbine blades to create electricity.

- In **fusion**, two light nuclei fuse, producing a nucleus with less total mass than that of the original nuclei; energy is thus released. Controlled fusion reactions are, at present, not commercially feasible, because no one has achieved confinement of the gas of ionized atoms and free electrons, called a **plasma**, at the proper density and temperature.

- The **beta decay** of an unstable nucleus produces a daughter nucleus, an electron or positron, and an antineutrino or neutrino.

- **Exchange particles** are virtual particles associated with various forces. The **pion** (a meson) is the exchange particle primarily responsible for the strong nuclear force.

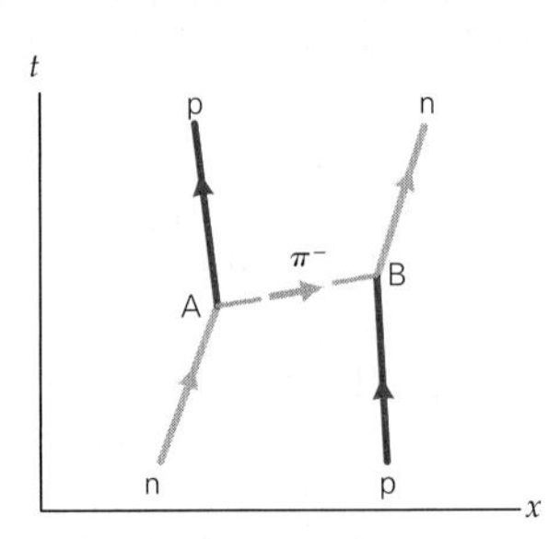

- The **weak nuclear force**, transmitted by the **W particle**, is primarily responsible for beta decay and the instability of the neutron.

- **Leptons** are the family of elementary particles that interact through the weak force, electromagnetism, and gravity, but not the strong force. Electrons, muons, tauons, and neutrinos make up the lepton family.

- **Hadrons** are the family of elementary particles that interact by the strong force, the weak force, and gravity. If electrically charged, hadrons also interact by the electromagnetic force. The hadrons are subdivided into baryons and mesons. **Baryons** include the familiar nucleons—the proton and neutron. Except for the stable proton, baryons decay into products that eventually include a proton. **Mesons**, which include pions, eventually decay into leptons and photons.

- **Quarks** are elementary, fractionally charged particles that make up hadrons.

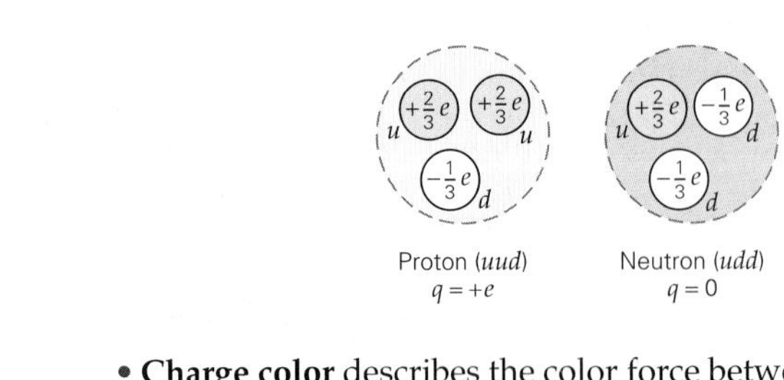

- **Charge color** describes the color force between quarks. Color force explains why a free quark is never likely to be seen, a phenomenon called **quark confinement**. Quark exchange is the fundamental explanation for the strong nuclear force.

- The **electroweak force** is the name given to the unified electromagnetic and weak forces.

- The **grand unified theory (GUT)** is an attempt to unify the electroweak force with the strong nuclear force.

- The **superforce** is the single force that will result if the long-pursued unification of the "fundamental" forces ever materializes.

- In the **standard model**, gluons carry the strong force, which keeps quarks together in composite particles such as protons. Leptons participate only in the gravitational and electroweak interactions. The former is carried by the graviton. The electroweak interaction is carried by the photon and the W and Z bosons.

Exercises

MC = *Multiple Choice Question,* **CQ** = *Conceptual Question, and* **IE** = *Integrated Exercise. Throughout the text, many exercise sections will include "paired" exercises. These exercise pairs, identified with* **red numbers**, *are intended to assist you in problem solving and learning. In a pair, the first exercise (even numbered) is worked out in the Study Guide so that you can consult it should you need assistance in solving it. The second exercise (odd numbered) is similar in nature, and its answer is given at the back of the book.*

30.1 Nuclear Reactions

1. **MC** In a particular nuclear reaction, the masses of the products add up to be less than the masses of all the initial nuclei involved. The total kinetic energy has (a) increased, (b) decreased, (c) stayed the same.

2. **MC** To initiate an endoergic reaction, a particle incident on a stationary nucleus must have (a) a kinetic energy equal to the Q value, (b) a kinetic energy less than a certain threshold energy, (c) a kinetic energy larger than the Q value.

3. **MC** The absorption of a slow neutron by a ^{238}U temporarily results in a compound nucleus which would be (a) ^{238}U, (b) $^{238}U^*$, (c) $^{239}U^*$, (d) $^{239}Np^*$.

4. **CQ** Using conservation of momentum, explain how a colliding beam reaction (in which both particles move toward a head-on collision) requires less incident kinetic energy than if one of them is at rest. [*Hint*: Linear momentum must be conserved.]

5. **CQ** For a given Q-value and incident particle (called a), how does the threshold energy required to initiate a specific nuclear reaction vary as the mass of the target (called A) changes? Sketch a graph of threshold energy versus target mass ranging from target masses on the order of the incident mass to target masses much more than the incident mass. What is the threshold energy if $M_A \gg m_a$? Explain what this means physically.

6. **CQ** In the capture of a slow neutron, which nucleus would probably have a larger capture cross-section: (a) ^{24}Mg, (b) ^{25}Mg, or (c) they would be about the same?

7. **CQ** What is the threshold energy for an alpha particle elastically scattering off a Ca-40 nucleus? Explain.

8. ● Complete the following nuclear reactions:
   (a) $^{1}_{0}n + ^{40}_{18}Ar \rightarrow$ ____ $+ \alpha$
   (b) $^{1}_{0}n + ^{235}_{92}U \rightarrow ^{98}_{40}Zr +$ ____ $+ 3(^{1}_{0}n)$
   (c) $^{1}_{0}n + ^{235}_{92}U \rightarrow ^{133}_{51}Sb + ^{99}_{41}Nb +$ ____$(^{1}_{0}n)$
   (d) $^{14}_{7}N(\alpha,$ __$)^{17}_{8}O$
   (e) ____$(n, p)^{137}_{55}Cs$

9. ● Complete the following nuclear reactions:
   (a) $^{13}_{6}C + ^{1}_{1}H \rightarrow$ __ $+ ^{14}_{7}N$
   (b) $^{10}_{5}B + ^{4}_{2}He \rightarrow ^{10}_{6}C +$ __
   (c) $^{27}_{13}Al(\alpha,$ __$)^{30}_{15}P$
   (d) $^{14}_{7}N(\alpha, p)$__
   (e) $^{13}_{6}C($__$, \alpha)^{10}_{5}B$

10. **IE** ● (a) Consider the reaction $^{13}_{6}C + ^{1}_{1}H \rightarrow ^{4}_{2}He + ^{10}_{5}B$. Use the masses of the nuclides involved to determine whether it is endoergic or exoergic. (b) If it is exoergic, find the amount of energy released; if it is endoergic, find the threshold energy.

11. ● What would the daughter nuclei in the following decay equations be? Can these decays actually occur spontaneously? Explain your reasoning in each case.
    (a) $^{22}_{10}Ne \rightarrow$ ____ $+ ^{0}_{-1}e$
    (b) $^{226}_{88}Ra \rightarrow$ ____ $+ ^{4}_{2}He$
    (c) $^{16}_{8}O \rightarrow$ ____ $+ ^{4}_{2}He$

12. ● Show that the Q value for the reaction $^{1}_{1}H + ^{2}_{1}H \rightarrow ^{3}_{2}He + \gamma$ is 5.49 MeV.

13. **IE** ● Uranium-238 undergoes alpha decay as follows:

$$^{238}_{92}U \rightarrow ^{234}_{90}Th + ^{4}_{2}He$$

(*238.050 786 u*) (*234.043 583 u*) (*4.002 603 u*)

(a) Would you expect the Q value to be (1) positive, (2) negative, or (3) zero? Why? (b) Find the Q value.

14. ●● Find the threshold energy for the following reaction:

$$^{16}_{8}O + ^{1}_{0}n \rightarrow ^{13}_{6}C + ^{4}_{2}He$$

(*15.994 915 u*) (*1.008 665 u*) (*13.003 355 u*) (*4.002 603 u*)

15. ●● Find the threshold energy for the following reaction:

$$^{3}_{2}He + ^{1}_{0}n \rightarrow ^{2}_{1}H + ^{2}_{1}H$$

(*3.016 029 u*) (*1.008 665 u*) (*2.014 102 u*) (*2.014 102 u*)

16. ●● Is the given reaction endoergic or exoergic? Prove your answer.

$$^{7}_{3}Li + ^{1}_{1}H \rightarrow ^{4}_{2}He + ^{4}_{2}He$$

(*7.016 005 u*) (*1.007 825 u*) (*4.002 603 u*) (*4.002 603 u*)

17. ●● Is the reaction $^{200}Hg(p, \alpha)^{197}Au$ endoergic or exoergic? Prove your answer. (The reaction on page 936 gives the mass values.)

18. ●● Determine the Q value of the following reaction:

$$^{9}_{4}Be + ^{4}_{2}He \rightarrow ^{12}_{6}C + ^{1}_{0}n$$

(*9.012 183 u*) (*4.002 603 u*) (*12.000 000 u*) (*1.008 665 u*)

19. ●● What is the minimum kinetic energy a proton must have in order to initiate the reaction $^{3}_{1}H(p, d)^{2}_{1}H$? ("d" stands for a deuterium nucleus and is called the *deuteron*.)

20. ●● ^{226}Ra decays and emits a 4.706-MeV alpha particle. Find the kinetic energy of the recoiling daughter nucleus from the decay of a stationary radium-226 nucleus.

21. IE ●●● The same type of incident particle is used for two endoergic reactions. In one reaction, the mass of the target nucleus is 15 times that of the incident particle, and in the other reaction, it is 20 times the mass. The Q value of the first reaction is known to be three times that of the second. (a) Compared with the second reaction, the first reaction has (1) greater, (2) the same amount of, or (3) less minimum threshold energy. (b) Prove your answer to part (a) by calculating the ratio of the minimum threshold energy for the first reaction to that for the second reaction.

22. ●●● Consider n (where n is an integer ≥ 1) ceramic pie plates of radius R randomly fixed (but none overlapping) on a rectangular wall with dimensions L and W. If you were to throw a very small (point) object at the wall, what would be the percentage probability of hitting a pie plate, in terms of the given parameters? Note that this is a crude analogy to the concept of a nuclear cross-section in determining, for example, the likelihood that a neutron might be absorbed by a given target. However, the nuclear cross-section is not necessarily related to the area of the target nucleus as it is for the pie plate. [*Hint*: Think in terms of the total area of plates compared to the total wall area.]

30.2 Nuclear Fission *and* 30.3 Nuclear Fusion

23. **MC** Nuclear fission (a) is endoergic, (b) occurs only for uranium-235, (c) releases about 500 MeV per fission, (d) requires a critical mass for a sustained reaction.

24. **MC** A nuclear reactor (a) can operate on natural (unenriched) uranium, (b) has its chain reaction controlled by neutron-absorbing materials, (c) can be partially controlled by the amount of moderator, (d) all of the preceding.

25. **MC** A nuclear fusion reaction (a) has a positive Q value, (b) may occur spontaneously, (c) is an example of "splitting" the atom.

26. CQ Explain clearly why the cooling water in a U.S. nuclear reactor helps keep the chain reaction from multiplying in addition to serving as a neutron moderator.

27. CQ Spent (used) fuel rods from most U.S. reactors are currently stored on site in pools of circulating cold water that contains a salt of boron (it is borated water). Why is the boron needed?

28. ● Find the approximate energy released in the following fission reactions: (a) $^{235}_{92}\text{U} + {}^{1}_{0}\text{n} \rightarrow$ fission products plus the release of five neutrons, (b) $^{235}_{94}\text{Pu} + {}^{1}_{0}\text{n} \rightarrow$ fission products plus the release of three neutrons.

29. ● Calculate the amounts of energy released in the following fusion reactions: (a) ${}^{1}_{1}\text{H} + {}^{1}_{0}\text{n} \rightarrow {}^{2}_{1}\text{H} + \gamma$, (b) ${}^{3}_{2}\text{He} + {}^{3}_{2}\text{He} \rightarrow {}^{4}_{2}\text{He} + 2\,{}^{1}_{1}\text{H}$.

30. ● Calculate the amounts of energy released in the following fusion reactions: (a) ${}^{2}_{1}\text{H} + {}^{2}_{1}\text{H} \rightarrow {}^{3}_{2}\text{He} + {}^{1}_{0}\text{n}$, (b) ${}^{2}_{1}\text{H} + {}^{3}_{1}\text{H} \rightarrow {}^{4}_{2}\text{He} + {}^{1}_{0}\text{n}$.

31. ●● In power reactors, using water as a moderator works well, because the proton and neutron have nearly the same mass. For a head-on elastic collision, we might expect a neutron to lose all of its kinetic energy in one collision, whereas for an "almost miss," we might expect it to lose essentially none. Assume that, on average, the neutron loses 60% of its kinetic energy during each collision. Estimate how many collisions are needed to reduce a 2.0-MeV neutron to a neutron with a kinetic energy of only 0.02 eV (approximately "thermal").

30.4 Beta Decay and the Neutrino

32. **MC** In the absence of a neutrino, which of the following quantities would not be conserved in beta decay: (a) total energy, (b) total linear momentum, (c) total angular momentum, or (d) all of the preceding?

33. **MC** A neutrino interacts with matter by (a) the electromagnetic interaction, (b) the strong interaction, (c) the weak interaction, (d) both b and c.

34. **MC** In β^- decay, if the daughter nucleus is left stationary, how do the momenta (magnitude) of the β^- particle and the antineutrino $\bar{\nu}_e$ compare: (a) the β^- has more momentum, (b) the $\bar{\nu}_e$ has more momentum, or (c) they have the same momentum?

35. CQ A $\bar{\nu}_e$ can interact with a proton in a target of water (although this is very rare) and the reaction is essentially the inverse of β^- decay; that is, the proton is converted into a neutron. Neutrino experimenters can detect that this reaction happened by using gamma-ray detectors. The neutrino "signature" occurs when the detector simultaneously records a gamma ray of 2.22 MeV and two others each with an energy of 0.511 MeV. Write the reaction and explain where the three gamma rays come from. [*Hint*: 2.22 MeV is the binding energy of the deuteron and 0.511 MeV is the rest energy of an electron.]

36. CQ Recently it has been determined that neutrinos actually possess a small amount of mass (which depends on the kind of neutrino). If they were massless (as previously assumed), neutrinos of all energies would travel at the same speed, c. Since they have mass, as we have seen from relativity, their speed is a bit less than c. Is it still true that all neutrinos of the same type would travel at the same speed regardless of their energy? Explain.

37. ● A neutrino created in a beta-decay process has an energy of 2.65 MeV. Assume that it is massless, and thus has the same relationship between its energy and momentum as a photon (that is, $E = pc$). Determine the de Broglie wavelength of the neutrino.

38. IE ●● In Exercise 37 with the same neutrino energy, what would be the maximum possible energy of the beta particle if the disintegration energy were 5.35 MeV? (a) (1) zero, (2) less than 5.35 MeV but not zero, (3) 5.35 MeV, or (4) greater than 5.35 MeV. (b) Under these conditions, determine the total and kinetic energy of the beta particle. (c) What is the direction, relative to the neutrino direction, of the momentum of the beta particle after the decay? Explain your reasoning.

39. ●● Show that the disintegration energy for β^- decay is $Q = (m_P - m_D - m_e)c^2 = (M_P - M_D)c^2$, where the m's represent the masses of the parent and daughter *nuclei* and the M's represent the masses of the neutral *atoms*.

40. ●● What is the maximum kinetic energy of the electron emitted when a ^{12}B nucleus beta decays into a ^{12}C nucleus? (See Exercise 39.)

41. ●● The kinetic energy of an electron emitted from a ^{32}P nucleus that beta decays into a ^{32}S nucleus is observed to be 1.00 MeV. What is the energy of the accompanying neutrino of the decay process? Neglect the recoil energy of the daughter nucleus. (See Exercise 39.)

42. ●●● Show that the disintegration energy for β^+ decay is $Q = (m_P - m_D - m_e)c^2 = (M_P - M_D - 2m_e)c^2$, where the m's represent the masses of the parent and daughter *nuclei* and the M's represent the masses of the neutral *atoms*.

43. ●●● The kinetic energy of a positron emitted from the β^+ decay of a ^{13}N nucleus into a ^{13}C nucleus is measured to be 1.190 MeV. What is the energy of the accompanying neutrino of the decay process? Neglect the recoil energy of the daughter nucleus. (See Exercise 42.)

44. IE ●●● The expressions for the Q values associated with both β^- and β^+ decay are given in Exercises 39 and 42. Assume that the daughter's atomic mass M_D is the same in both processes. (a) If both expressions are to be energetically possible, the mass of the parent atom, M_P, for β^- decay must be (1) greater than, (2) equal to, (3) less than that for β^+ decay. Why? (b) Write the mass requirements of the parent atoms for β^- and β^+ processes, respectively, in terms of the atomic masses of the daughter, parent, and electron.

30.5 Fundamental Forces and Exchange Particles

45. MC Virtual particles (a) form virtual images, (b) exist only in an amount of time as permitted by the uncertainty principle, (c) make up positrons, (d) can be observed in exchange processes.

46. MC The exchange particle for the strong nuclear force is the (a) pion, (b) W particle, (c) muon, (d) positron.

47. CQ If virtual exchange particles are unobservable by themselves, how is their existence verified?

48. CQ When a proton interacts with another proton, which of the four fundamental forces would be involved? How about when an electron interacts with another electron?

49. ● Assuming the range of the nuclear force to be on the order of 10^{-15} m, predict the mass (in kilograms) of the exchange particle that is related to this force.

50. ●● In a certain type of reaction, a high-speed proton collides with a nucleus and travels, on average, a distance of 5.0×10^{-16} m within the nucleus before the reaction takes place. What type of fundamental interaction is this most likely to be, and how much time elapses before the interaction takes place?

51. IE ●● (a) In an interaction using the virtual-particle model, the range of the interaction (1) increases, (2) remains the same, (3) decreases as energy of the exchange particle increases. Why? (b) A W particle in a weak interaction is found to have rest energy of 1.00 GeV. What is the approximate range for the weak interaction with this exchange particle?

52. ●● By what minimum amount of energy is the conservation of energy "violated" during a neutral pi-meson exchange process?

53. ●● How long is the conservation of energy "violated" in a neutral pi-meson exchange process?

30.6 Elementary Particles *and* 30.7 The Quark Model

54. MC Particles that interact by the strong nuclear force are called (a) muons, (b) hadrons, (c) W particles, (d) leptons.

55. MC Quarks are thought to make up which of the following particles: (a) hadrons, (b) muons, (c) Z particles, or (d) all of the preceding?

56. CQ What is meant by quark flavor and color? Can these attributes be changed? Explain.

57. CQ With so many types of hadrons, why aren't fractional electronic charges observed?

58. What are some of the important differences and distinctions between baryons and mesons?

59. ● Out of each of the following pairs, which particle has more mass: (a) π^+ and π^o; (b) K^+ and K^o; (c) Σ^+ and Σ^o; (d) Ξ^o and Ξ^-? [*Hint*: See Table 30.3.]

60. IE ● (a) The quark combination for a proton is (1) *uuu*, (2) *uud*, (3) *udd*, (4) *ddd*. (b) Prove your answer to part (a) given the correct electric charge.

61. IE ● (a) The quark combination for a neutron is (1) *uuu*, (2) *uud*, (3) *udd*, (4) *ddd*. (b) Prove your answer to part (a) given the correct electric charge.

30.8 Force Unification Theories, the Standard Model, and the Early Universe

62. MC The *grand unified theory* would reduce the number of fundamental forces to (a) one, (b) two, (c) three, (d) four.

63. MC The magnetic force is part of the (a) electroweak force, (b) weak force, (c) strong force, (d) superforce.

Comprehensive Exercises

64. Complete the following nuclear reactions:
(a) $^{6}_{3}\text{Li} + \_\_\_ \rightarrow ^{3}_{2}\text{He} + ^{4}_{2}\text{He}$
(b) $^{58}_{28}\text{Ni} + ^{2}_{1}\text{H} \rightarrow \_\_\_ + ^{1}_{1}\text{H}$
(c) $^{235}_{92}\text{U} + ^{1}_{0}\text{n} \rightarrow ^{138}_{54}\text{Xe} + 5^{1}_{0}\text{n} + \_\_\_$
(d) $^{9}_{4}\text{Be}(\alpha, \_\_)^{12}_{6}\text{C}$
(e) $\_\_\_(\text{n, p})^{16}_{7}\text{N}$

65. Compute the Q value (to three significant figures) of the following fusion reaction: ${}^2_1\mathrm{H} + {}^2_1\mathrm{H} \rightarrow {}^3_1\mathrm{H} + {}^1_1\mathrm{H}$.

66. (a) Write down the decay equation for the decay of ${}^{14}\mathrm{C}$. (b) What kind of neutrino is released in this decay? (c) Find the maximum kinetic energy of that neutrino.

67. Show that the Q value for electron capture is given by $Q = (m_\mathrm{P} + m_\mathrm{e} - m_\mathrm{D})c^2 = (M_\mathrm{P} - M_\mathrm{D})c^2$, where the m's represent the masses of the parent and daughter *nuclei* and the M's represent the masses of the neutral *atoms*.

68. **IE** Theoretically a ${}^7\mathrm{Be}$ nucleus can be converted into a ${}^7\mathrm{Li}$ nucleus by capturing an electron (see Exercise 67) or through β^+ decay (see Exercise 42). (a) From an energy point of view, which is more likely to occur: (1) electron capture, (2) β^+ decay, or (3) both are equally likely to occur? Why? (b) What is the energy of the emitted neutrino (neglect daughter-nucleus recoil) in electron capture?

69. Determine the threshold energy of the following reaction:

$$\underset{(15.994\,915\ u)}{{}^{16}_{8}\mathrm{O}} \; + \; \underset{(1.008\,665\ u)}{{}^{1}_{0}\mathrm{n}} \; \rightarrow \; \underset{(13.003\,355\ u)}{{}^{13}_{6}\mathrm{C}} \; + \; \underset{(4.002\,603\ u)}{{}^{4}_{2}\mathrm{He}}$$

70. Assume that the average kinetic energy of ions in a plasma is the same as the average kinetic energy of the atoms in an ideal gas $\left(\frac{1}{2}mv^2 = \frac{3}{2}k_\mathrm{B}T\right)$. If fusion can occur when the ions approach to within a distance of about 10^{-14} m, estimate the temperature required for fusion of two deuterium ions.

APPENDICES

APPENDIX I A Symbols, Arithmetic Operations, Exponents, and Scientific Notation
B Algebra and Common Algebraic Relationships
C Geometric Relationships
D Trigonometric Relationships
E Logarithms

APPENDIX II Kinetic Theory of Gases

APPENDIX III Planetary Data

APPENDIX IV Algebraic Listing of the Chemical Elements

APPENDIX V Properties of Selected Isotopes

ANSWERS TO FOLLOW-UP EXERCISES

ANSWERS TO ODD-NUMBERED EXERCISES

APPENDIX I* Mathematical Review (with Examples) for College Physics

A Symbols, Arithmetic Operations, Exponents, and Scientific Notation

Commonly Used Symbols in Relationships

$=$ means two quantities are equal, such as $2x = y$

$\equiv$ means "defined as," such as the definition of pi:

$$\pi \equiv \frac{\text{circumference of a circle}}{\text{the diameter of that circle}}.$$

$\approx$ means approximately equal, as in $30\frac{\text{m}}{\text{s}} \approx 60\frac{\text{mi}}{\text{h}}$.

$\neq$ means inequality, such as $\pi \neq \frac{22}{7}$.

$\geq$ means that one quantity is greater than or equal to another. For example, the age of the universe $\geq$ 5 billion years.

$\leq$ means that a quantity is less than or equal to another. For example, if a lecture room holds $\leq$ 45 students, the maximum is 45 students.

$>$ means that one quantity is greater than another, such as 14 eggs $>$ 1 dozen eggs.

$\gg$ means that one quantity is *much* greater than another. For example, number of people on Earth $\gg$ 1 million.

$<$ means that one quantity is less than another, such as $3 \times 10^{22} <$ Avogadro's number.

$\ll$ one quantity is *much* less than another, such as $10 \ll$ Avogadro's number.

$\propto$ means linearly proportional to. That is, if $y = 2x$ then $y \propto x$. This means that if x is increased by a certain multiplicative factor, y is also increased the same way. For example, if $y = 3x$, then if x is changed by a factor of n (that is, if x becomes nx), then so is y, because $y' = 3x' = 3(nx) = n(3x) = ny$.

ΔQ means "change in the quantity Q." This means "final minus initial." Hence if the value of an investor's stock portfolio in the morning is $V_i = \$10\,100$ and at the close of trading it is $V_f = \$10\,050$, then $\Delta V = \$10\,050 - \$10\,100 = -\$50$

The Greek letter capital sigma (Σ) indicates the sum of a series of values for the quantity Q_i where $i = 1, 2, 3, \ldots, N$, that is,

$$\sum_{i=1}^{N} Q_i = Q_1 + Q_2 + Q_3 + \cdots Q_N.$$

$|Q|$ denotes the absolute value of a quantity Q without a sign. If Q is positive then $|Q| = Q$; if Q is negative then $|Q| = -Q$. Thus $|-3| = 3$.

Appendix I-A Exercises

1. What values of x satisfy $3 \leq |x| \leq 8$?
2. What integer y comes closest to $y \approx |\sqrt{10}|$?
3. If at the end of the weekend you count your widgets and find $\Delta w = -10$ and the number of widgets on Friday was 500, how many do you have on Monday morning?
4. Give a reasonable number z that satisfies $1 < z \ll 100$.
5. If $y \propto x^2$ and the value of x doubles, what happens to the value of y?
6. What is $\dfrac{\sum_{i=1}^{3} 3^i}{10}$?

Arithmetic Operations and Their Order of Usage

Basic arithmetic operations are addition (+), subtraction (−), multiplication (× or ·), and division (/ or ÷). Another common operation, exponentiation (x^n), involves raising a quantity (x) to a given power (n). If several of these operations are included in one equation, they are performed in this order: (a) parentheses, (b) exponentiation, (c) division, (d) multiplication, (e) addition and subtraction.

A handy mnemonic used to remember this order is: "**P**lease **E**xcuse **M**y **D**ear **A**unt **S**ally," where the capital letters stand for the various operations: **P**arentheses, **E**xponents,

*This appendix does not include a discussion of significant figures, since a thorough discussion is presented in Chapter 1, Section 1.6.

Multiplication/Division, Addition/Subtraction. Note that operations within parentheses are always first, so to be on the safe side, appropriate use of parentheses is encouraged. For example, $24^2/8 \cdot 4 + 12$ could be evaluated several ways. However, according to the agreed-on order, it has a unique value: $24^2/8 \cdot 4 + 12 = 576/8 \cdot 4 + 12 = 576/32 + 12 = 18 + 12 = 30$. To avoid possible confusion, the quantity could be written using two sets of parentheses as follows: $(24^2/(8 \cdot 4)) + 12 = (576/(32)) + 12 = 18 + 12 = 30$.

Appendix I-B Exercises

1. Insert parentheses so $3^2 + 4^2 \cdot 1^3 - \sqrt{4} + 7$ yields 30 without any questions.
2. Evaluate $2 \cdot 3/4 + 5/2 \times 4 - 1$.
3. Evaluate $2 \times 4 + 7 - 6/3 \times 2$.
4. How would you use parentheses to write $3^2 + 4^2 \cdot 1^3 - \sqrt{4} + 7$ to guarantee that anyone evaluating the expression would obtain zero even if he or she didn't know the ordering rules?

Exponents and Exponential Notation

Exponents and exponential notation are very important when employing scientific notation (see the next section). You should be familiar with power and exponential notation (both positive and negative, fractional and integral) such as the following:

$$x^0 = 1$$

$$x^1 = x \qquad x^{-1} = \frac{1}{x}$$

$$x^2 = x \cdot x \qquad x^{-2} = \frac{1}{x^2} \qquad x^{\frac{1}{2}} = \sqrt{x}$$

$$x^3 = x \cdot x \cdot x \qquad x^{-3} = \frac{1}{x^3} \qquad x^{\frac{1}{3}} = \sqrt[3]{x} \qquad \text{etc.}$$

Exponents combine according to the following rules:

$$x^a \cdot x^b = x^{(a+b)} \qquad x^a/x^b = x^{(a-b)} \qquad (x^a)^b = x^{ab}$$

Appendix I-C Exercises

1. What is the value of $\frac{2^3}{2^4}$?
2. Evaluate $3^3 \times |9^{-1/2}|$.
3. Find the value(s) of $3^4 \times \sqrt{4^6}$.
4. What is $(\sqrt{10})^4$?

Scientific Notation (Also Known as Powers-of-10 Notation)

In physics, many quantities have values that are very large or very small. To express them, **scientific notation** is frequently used. This notation is sometimes referred to as powers-of-10 notation for obvious reasons. (See the previous section for a discussion of exponents.) When the number 10 is squared or cubed, we have $10^2 = 10 \times 10 = 100$ or $10^3 = 10 \times 10 \times 10 = 1000$. You can see that the number of zeros is equal to the power of 10. Thus 10^{23} is a compact way of expressing the number 1 followed by 23 zeros.

A number can be represented in many different ways—all of which are correct. For example, the distance from the Earth to the Sun is 93 million miles. This value can be written as 93 000 000 miles. Expressed in a more compact scientific notation, there are many correct forms, such as 93×10^6 miles, 9.3×10^7 miles, or 0.93×10^8 miles. Any of these is correct, although 9.3×10^7 is preferred, because in expressing powers-of-10 notation, it is customary to leave only one digit to the left of the decimal point, in this case 9. (This is called customary or standard form.) And so the exponent, or power of 10, changes when the decimal point of the prefix number is shifted.

Negative powers of 10 also can be used. For example, $10^{-2} = \frac{1}{10^2} = \frac{1}{100} = 0.01$. So, if a power of 10 has a negative exponent, the decimal point may be shifted to the left once for each power of 10. For example, 5.0×10^{-2} is equal to 0.050 (two shifts to the left).

The decimal point of a quantity expressed in powers-of-10 notation may be shifted to the right or left irrespective of whether the power of 10 is positive or negative. General rules for shifting the decimal point are as follows:

1. The exponent, or power of 10, is *increased* by 1 for every place the decimal point is shifted to the *left*.
2. The exponent, or power of 10, is *decreased* by 1 for every place the decimal point is shifted to the *right*.

This is simply a way of saying that as the coefficient (prefix number) gets smaller, the exponent gets correspondingly larger, and vice versa. Overall, the number is the same.

Appendix I-D Exercises

1. Express your weight (in pounds) in scientific notation.
2. The circumference of the Earth is about 40 000 km. Express this in scientific notation.
3. Evaluate and express the answer in scientific notation: $\frac{12.1}{1.10 \times 10^{-1}}$.
4. Find the value of $(1.44 \times 10^2)^{1/2}$ in scientific notation.
5. What is $(3.0 \times 10^8)^2$ in scientific notation?

B Algebra and Common Algebraic Relationships

General

The basic rule of algebra, used for solving equations, is that if you perform any legitimate operation on both sides of an equation, it remains an equation, or equality. (An example of an illegal operation is dividing by zero; why?) Thus adding a number to both sides, taking the square root of both sides, cubing both sides, and dividing both sides by the same number all maintain the equality.

For example, suppose you want to solve $\frac{x^2 + 6}{2} = 11$ for x. To do this, first multiply both sides by 2, giving $\left(\frac{x^2 + 6}{\cancel{2}}\right) \times \cancel{2} = 11 \times 2 = 22$ or $x^2 + 6 = 22$. Then subtract 6 from both sides to obtain $x^2 + 6 - 6 = 22 - 6 = 16$ or $x^2 = 16$. Finally taking the square root of both sides, the solution(s) are $x = \pm 4$ (two roots were expected; why?)

Some Useful Results

Many times *the square of the sum and/or difference of two numbers* is required. For any numbers a and b:

$$(a \pm b)^2 = a^2 \pm 2ab + b^2$$

Similarly *the difference of two squares* can be factored:

$$(a^2 - b^2) = (a + b)(a - b)$$

A quadratic equation is one that can be expressed in the form $ax^2 + bx + c = 0$. In this form it can always be solved (usually for two different roots) using the *quadratic formula*: $x = \frac{-b \pm \sqrt{b^2 - 4ac}}{2a}$. In kinematics this result can be especially useful as it is common to have equations of this form to solve: $4.9t^2 - 10t - 20 = 0$. Just insert the coefficients (making sure to include the sign) and solve for t (here t represents the time for a ball to reach the ground when thrown upward from a cliff; see Chapter 2). The result is

$$x = \frac{10 \pm \sqrt{10^2 - 4(4.9)(-20)}}{2(4.9)} = \frac{10 \pm 22.2}{9.8}$$
$$= +3.3\ \text{s} \quad \text{or} \quad -1.2\ \text{s}$$

In all such problems, time is "stopwatch" time and starts at zero; hence the negative answer can be ignored as physically unreasonable although it is a solution to the equation.

Solving Simultaneous Equations

Occasionally solving a problem might require solving two or more equations simultaneously. In general if you have N unknowns in a problem, you will need exactly N independent equations. If you have less than N equations, there are not enough for a complete solutions. If you have more than N equations, then some are redundant, and a solution is usually still possible, although more complicated. In general in this textbook, we will be concerned with two simultaneous equations, and both will be linear. Linear equations are of the form $y = mx + b$. Recall that when plotted on an x–y Cartesian coordinate system, the result is a straight line with a slope of m ($\Delta y/\Delta x$) and a y-intercept of b, as shown for the red line here.

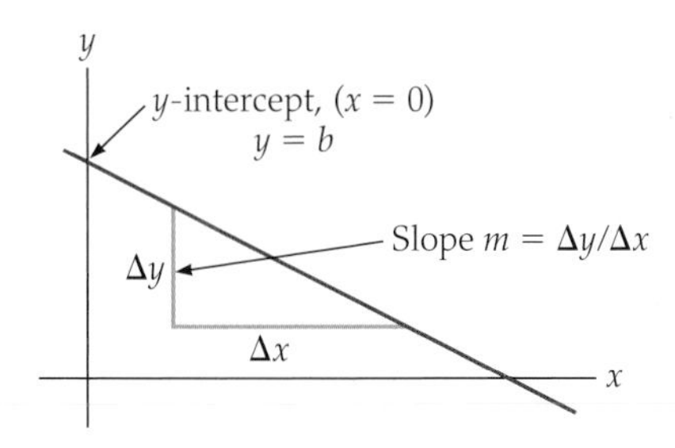

To solve two linear equations simultaneously *graphically*, simply plot them on the axes and evaluate the coordinates at their intersection point. While this can always be done in principle, it is only an approximate answer and usually takes quite a bit of time.

The most common (and exact) method of solving simultaneous equations involves the use of algebra. Essentially you solve one equation for an unknown and substitute the result into the other equation, ending up with one equation and one unknown. Suppose you have two equations and two unknown quantities (x and y), but in general, any two unknown quantities):

$$3y + 4x = 4 \quad \text{and} \quad 2x - y = 2$$

Solving the second equation for y, we have $y = 2x - 2$. Substituting this value for y into the first equation, we have $3(2x - 2) + 4x = 4$. Thus, $10x = 10$ and $x = 1$. Putting this value into the second of the original two equations, we have $2(1) - y = 2$ and therefore $y = 0$. (Of course, at this point a good double check is to substitute the answers and see if they solve both equations.)

■ Appendix I-E Exercises

1. Expand $(y - 2x)^2$.
2. Express $x^2 - 4x + 4$ as a product of two factors.
3. Solve the following equation for t: $4.9t^2 - 30t + 10 = 0$. How many physically reasonable roots are there?
4. Show that a quadratic equation has real roots only if $b^2 \geq 4ac$. Under what conditions (for a, b, and c) are the two roots identical?
5. Solve these equations simultaneously using algebra: $2x - 3y = 2$ and $3y + 5x = 7$.
6. Solve the two equations in Exercise 5 approximately using graphing methods.

C Geometric Relationships

In physics and many other areas of science, it is important to know how to find circumferences, areas, and volumes of some common shapes. Here are some equations for such shapes.

Circumference (c), Area (A), and Volume (V)

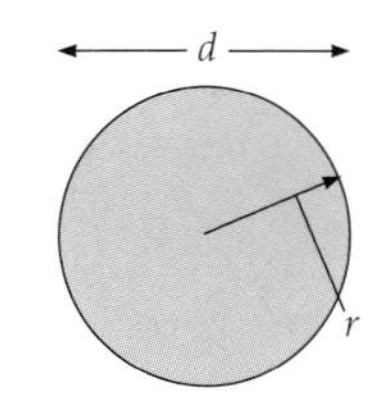

Circle: $c = 2\pi r = \pi d$

$A = \pi r^2 = \frac{\pi d^2}{4}$

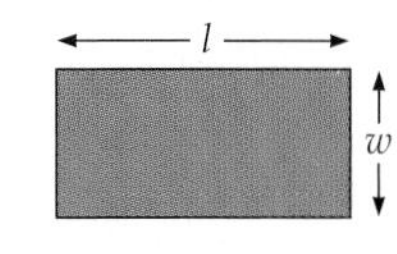

Rectangle: $c = 2l + 2w$

$A = l \times w$

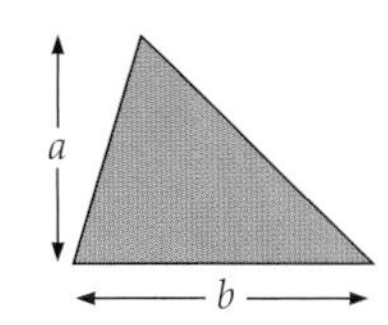

Triangle: $A = \frac{1}{2}ab$

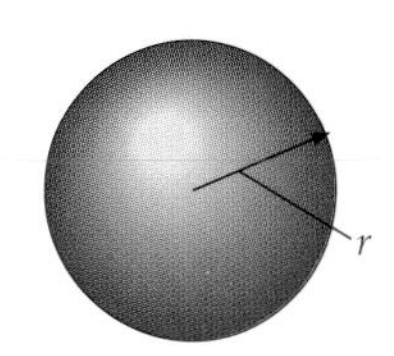

Sphere: $A = 4\pi r^2$

$V = \frac{4}{3}\pi r^3$

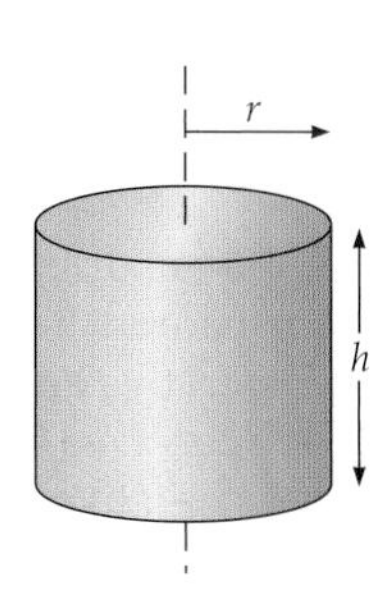

Cylinder: $A = \pi r^2$ (end)

$A = 2\pi rh$ (body)

$V = \pi r^2 h$

For practice, try the following exercises.

■ Appendix I-F Exercises

1. Estimate the volume of a bowling ball in cubic centimeters and cubic inches.

2. A square hole has a side measuring 5.0 cm. What is the area of the end of a cylindrical rod that will barely fit into this hole?
3. A glass of water has an interior diameter of 4.5 cm and contains a column of water 4.0 in. high. What volume of water does it contain in liters?
4. What is the total surface area of a pancake that is 16 cm in diameter and 8.0 mm thick?
5. Compute the volume of the pancake in Exercise 4 in cubic centimeters.

D Trigonometric Relationships

Understanding elementary trigonometry is crucial in physics, especially as many of the quantities are vectors. Here is a brief summary of common definitions, the first few of which you should commit to memory.

Definitions of Trigonometric Functions

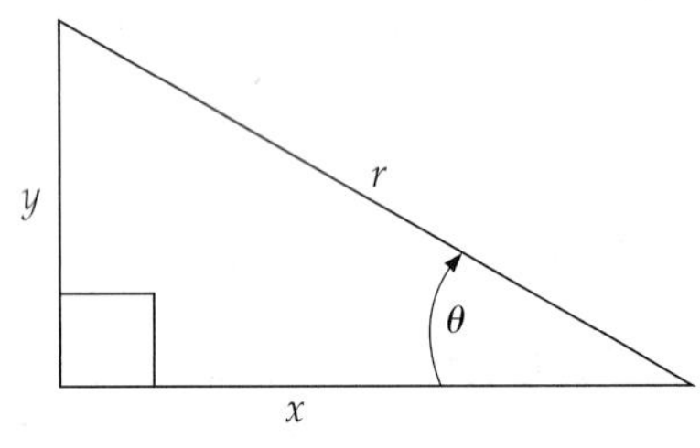

$$\sin\theta = \frac{y}{r} \qquad \cos\theta = \frac{x}{r} \qquad \tan\theta = \frac{\sin\theta}{\cos\theta} = \frac{y}{x}$$

$\theta°$ (rad)	$\sin\theta$	$\cos\theta$	$\tan\theta$
$0°(0)$	0	1	0
$30°\ (\pi/6)$	0.500	0.866	0.577
$45°\ (\pi/4)$	0.707	0.707	1.00
$60°\ (\pi/3)$	0.866	0.500	1.73
$90°\ (\pi/2)$	1	0	$\rightarrow\infty$

For very small angles,

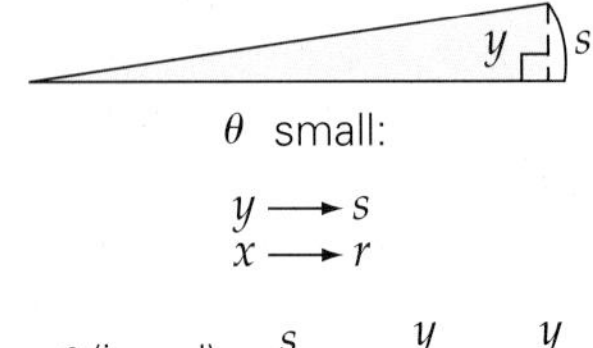

$$\theta \text{ (in rad)} = \frac{s}{r} \approx \frac{y}{r} \approx \frac{y}{x}$$

$$\theta \text{ (in rad)} \approx \sin\theta \approx \tan\theta$$

$$\cos\theta \approx 1 \qquad \sin\theta \approx \theta \text{ (radians)}$$

$$\tan\theta = \frac{\sin\theta}{\cos\theta} \approx \theta \text{ (radians)}$$

The sign of a trigonometric function depends on the quadrant, or the signs of x and y. For example, in the second quadrant x is negative and y is positive, therefore $\cos\theta = x/r$ is negative and $\sin\theta = y/r$ is positive. (Note that r is always taken as positive.) In the figure, the red lines are positive and the blue lines negative.

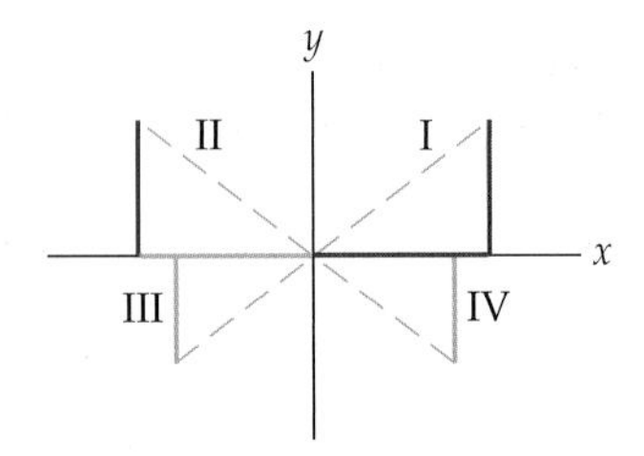

Some Useful Trigonometric Identities

$$\sin^2\theta + \cos^2\theta = 1$$
$$\sin 2\theta = 2\sin\theta\cos\theta$$
$$\cos 2\theta = \cos^2\theta - \sin^2\theta = 2\cos^2\theta - 1 = 1 - 2\sin^2\theta$$
$$\sin^2\theta = \frac{1}{2}(1 - \cos 2\theta)$$
$$\cos^2\theta = \frac{1}{2}(1 + \cos 2\theta)$$

For half-angle ($\theta/2$) identities, simply replace θ with $\theta/2$; for example,

$$\sin^2\theta/2 = \frac{1}{2}(1 - \cos\theta)$$
$$\cos^2\theta/2 = \frac{1}{2}(1 + \cos\theta)$$

Trigonometric values of sums and differences of angles are sometimes of interest. Here are several basic relationships.

$$\sin(\alpha \pm \beta) = \sin\alpha\cos\beta \pm \cos\alpha\sin\beta$$
$$\cos(\alpha \pm \beta) = \cos\alpha\cos\beta \mp \sin\alpha\sin\beta$$
$$\tan(\alpha \pm \beta) = \frac{\tan\alpha \pm \tan\beta}{1 \mp \tan\alpha\tan\beta}$$

Law of Cosines

For a triangle with angles A, B, and C with opposite sides a, b, and c, respectively:

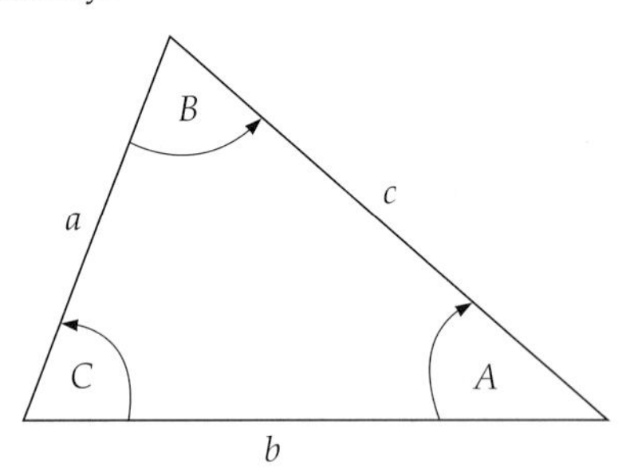

$$a^2 = b^2 + c^2 - 2bc\cos A \qquad \textit{(with similar results for } b^2 = \cdots \textit{ and } c^2 = \cdots\textit{).}$$

If $A = 90°$, this equation reduces to the Pythagorean theorem as it should:

$$a^2 = b^2 + c^2 \quad \textit{(of the form } r^2 = x^2 + y^2\textit{)}$$

Law of Sines

For a triangle with angles A, B, and C with opposite sides a, b, and c, respectively:

$$\frac{a}{\sin A} = \frac{b}{\sin B} = \frac{c}{\sin C}$$

Appendix I-G Exercises

1. From ground level you find you must look at an upward angle of 60 degrees to see the very top of a building that is 50 m away from you. How high is the building? How far is the top of the building from you?
2. On an x–y Cartesian set of axes, a point is at $x = -2.5$ and $y = -4.2$. What quadrant is it in? What is the angle of the line drawn between it and the origin? (Express the answer in degrees and radians.)

3. Use the sine equation for the sum of two angles, one angle being 30°, the other 60°, to show that the sine of a 90° angle comes out to be 1.00.
4. Assume that the radius of the Earth's orbit about the Sun is a circle with a radius of 150 million km. Calculate the arc-length distance traveled by the Earth about the Sun in four months. Using the laws of sines and cosines, determine the straight-line distance between the beginning and end points of this arc.
5. A right triangle has a hypotenuse of length 11 cm and one angle of 25°. Determine the two sides of the triangle, its area, and its perimeter.

E Logarithms

The following are fundamental definitions and relationships for logarithms. The logarithm is common in physics; you should know what they are and how to use them. Logarithms are very useful because they allow you to multiply and divide very large and very small numbers by adding and subtracting exponents (which we call the logarithms of the numbers).

General Definition of Logarithms

If a number x is written as another number a to some power n, as $x = a^n$, then n is defined to be the *logarithm of the number x to the base a*. This is written compactly as

$$n \equiv \log_a x.$$

Common Logarithms

If the base a is 10, the logarithms are called *common logarithms*. When the abbreviation *log* is used, without a base specified, base 10 is assumed. If another base is being used, it will be specifically shown. For example, $1000 = 10^3$; therefore $3 = \log_{10} 1000$, or simply $3 = \log 1000$. This is read "3 is the log of 1000."

Identities for Common Logarithms

For any two numbers x and y:

$$\log(10^x) = x$$
$$\log(xy) = \log x + \log y$$
$$\log\left(\frac{x}{y}\right) = \log x - \log y$$
$$\log(x^y) = y \log x$$

Natural Logarithms

The natural logarithm uses as its base the irrational number e. To six significant figures, its value is $e \approx 2.71828\ldots$. Fortunately most calculators have this number (along with other irrational numbers, such as pi) in their memories. (You should be able to find both e and π on yours.) The natural logarithm received its name because it occurs naturally when describing a quantity that grows or decays at a constant percentage (rate). The natural logarithm is abbreviated *ln* to distinguish it from the common logarithm, *log*. That is, $\log_e x \equiv \ln x$, and if $n = \ln x$, then $x = e^n$. Similarly to the common logarithm, we have the following relationships for any two numbers x and y:

$$\ln(e^x) = x$$
$$\ln(xy) = \ln x + \ln y$$
$$\ln\left(\frac{x}{y}\right) = \ln x - \ln y$$
$$\ln(x^y) = y \ln x$$

Occasionally you must convert between the two types of logarithms. For that, the following relationships can be handy:

$$\log x = 0.43429 \ln x$$
$$\ln x = 2.3026 \log x$$

For practice with logarithms of both types, try the following exercises:

■ Appendix I-H Exercises:

1. Use your calculator to find the following: log 20, log 50, log 2500, and log 3.
2. Explain why numbers less than 10 have a negative logarithm. Does it make sense to talk about $\log(-100)$? Explain.
3. Use your calculator to find the following: ln 20, log 2, ln 100, and log 3.
4. Double check your answers for ln 2 and log 2 from Exercises 1 and 3 using the relationships $\log x = 0.43429 \ln x$ and $\ln x = 2.3026 \log x$.
5. Show that the rules for combining logarithms work for the following by evaluating each side and showing an equivalence: $\log 1500 = \log(15 \times 100)$,

   $\log 6400 = \log\left(\frac{64}{0.01}\right)$, and $\log 8 = \log(2^3)$.
6. Show that the rules for combining logarithms work for the following by evaluating each side and showing an equivalence: $\ln 4 = \ln(2 \times 2)$, $\ln 20 = 2.3026 \log(2 \times 10)$, and $\log 49 = 0.43429 \ln(7^2)$.
7. In describing the growth of a bacteria colony, the number of bacteria N at any given time t (from the start of observation) can be written in terms of the number at the start, N_o, as follows: $N = N_o e^{0.020t}$, where t is in minutes. How many minutes does it take the colony to double in size?
8. In describing the decay of a radioactive sample of atomic nuclei, the number of undecayed nuclei N at any given time t (from the start of observation) can be written in terms of the number at the start, N_o, as follows: $N = N_o e^{-0.050t}$, where t is in years. How many years does it take until one-tenth the original number of nuclei remain?

APPENDIX II Kinetic Theory of Gases

The basic assumptions are as follows:

1. All the molecules of a pure gas have the same mass (m) and are in continuous and completely random motion. (The mass of each molecule is so small that the effect of gravity on it is negligible.)
2. The gas molecules are separated by large distances and occupy a volume that is negligible compared with these distances.
3. The molecules exert no forces on each other except when they collide.

4. Collisions of the molecules with one another and with the walls of the container are perfectly elastic.

The magnitude of the force exerted on the wall of the container by a gas molecule colliding with it is $F = \Delta p/\Delta t$. Assuming that the direction of the velocity (v_x) is normal to the wall, the magnitude of the average force is

$$F = \frac{\Delta(mv)}{\Delta t} = \frac{mv_x - (-mv_x)}{\Delta t} = \frac{2mv_x}{\Delta t} \quad (1)$$

After striking one wall of the container, which, for convenience, is assumed to be a cube with sides of dimensions L, the molecule recoils in a straight line. Suppose that the molecule reaches the opposite wall without colliding with any other molecules along the way. The molecule then travels the distance L in a time equal to L/v_x. After the collision with that wall, again assuming no collisions on the return trip, the round trip will take $\Delta t = 2L/v_x$. Thus, the number of collisions per unit time a molecule makes with a particular wall is $v_x/2L$, and the average force of the wall from successive collisions is

$$F = \frac{2mv_x}{\Delta t} = \frac{2mv_x}{2L/v_x} = \frac{mv_x^2}{L} \quad (2)$$

The random motions of the many molecules produce a relatively constant force on the walls, and the pressure (p) is the total force on a wall divided by the wall's area:

$$p = \frac{\Sigma F_i}{L^2} = \frac{m(v_{x_1}^2 + v_{x_2}^2 + v_{x_3}^2 + \cdots)}{L^3} \quad (3)$$

The subscripts refer to individual molecules.

The average of the squares of the speeds is given by

$$\overline{v_x^2} = \frac{v_{x_1}^2 + v_{x_2}^2 + v_{x_3}^2 + \cdots}{N}$$

where N is the number of molecules in the container. In terms of this average, Eq. 3 can be written as

$$p = \frac{Nm\overline{v_x^2}}{L^3} \quad (4)$$

However, the molecules' motions occur with equal frequency along any one of the three axes, so $\overline{v_x^2} = \overline{v_y^2} = \overline{v_z^2}$ and $\overline{v^2} = \overline{v_x^2} + \overline{v_y^2} + \overline{v_z^2} = 3\overline{v_x^2}$. Then

$$\sqrt{\overline{v^2}} = v_{\text{rms}}$$

where v_{rms} is called the root-mean-square (rms) speed. Substituting this result into Eq. 4 and replacing L^3 with V (since L^3 is the volume of the cubical container) gives

$$pV = \tfrac{1}{3}Nmv_{\text{rms}}^2 \quad (5)$$

This result is correct even though collisions between molecules were ignored. Statistically, these collisions average out, so the number of collisions with each wall is as described. This result is also independent of the shape of the container. A cube merely simplifies the derivation.

We now combine this result with the empirical perfect gas law:

$$pV = Nk_BT = \tfrac{1}{3}Nmv_{\text{rms}}^2$$

The average kinetic energy per gas molecule is thus proportional to the absolute temperature of the gas:

$$\overline{K} = \tfrac{1}{2}mv_{\text{rms}}^2 = \tfrac{3}{2}k_BT \quad (6)$$

The collision time is negligible compared with the time between collisions. Some kinetic energy will be momentarily converted to potential energy during a collision; however, this potential energy can be ignored, because each molecule spends a negligible amount of time in collisions. Therefore, by this approximation, the total kinetic energy is the internal energy of the gas, and the internal energy of a perfect gas is directly proportional to its absolute temperature.

APPENDIX III Planetary Data

Name	Equatorial Radius (km)	Mass (Compared with Earth's)*	Mean Density ($\times 10^3$ kg/m^3)	Surface Gravity (Compared with Earth's)	Semimajor Axis $\times 10^6$ km	Semimajor Axis AU†	Orbital Period Years	Orbital Period Days	Eccentricity	Inclination to Ecliptic
Mercury	2439	0.0553	5.43	0.378	57.9	0.3871	0.24084	87.96	0.2056	7°00′26″
Venus	6052	0.8150	5.24	0.894	108.2	0.7233	0.61515	224.68	0.0068	3°23′40″
Earth	6378.140	1	5.515	1	149.6	1	1.00004	365.25	0.0167	0°00′14″
Mars	3397.2	0.1074	3.93	0.379	227.9	1.5237	1.8808	686.95	0.0934	1°51′09″
Jupiter	71398	317.89	1.36	2.54	778.3	5.2028	11.862	4337	0.0483	1°18′29″
Saturn	60000	95.17	0.71	1.07	1427.0	9.5388	29.456	10760	0.0560	2°29′17″
Uranus	26145	14.56	1.30	0.8	2871.0	19.1914	84.07	30700	0.0461	0°48′26″
Neptune	24300	17.24	1.8	1.2	4497.1	30.0611	164.81	60200	0.0100	1°46′27″
Pluto	1500–1800	0.02	0.5–0.8	~0.03	5913.5	39.5294	248.53	90780	0.2484	17°09′03″

*Planet's mass/Earth's mass, where $M_E = 6.0 \times 10^{24}$ kg.

†Astronomical unit: 1 AU = 1.5×10^8 km, the average distance between the Earth and the Sun.

APPENDIX IV Alphabetical Listing of the Chemical Elements (The periodic table is provided inside the back cover.)

Element	Symbol	Atomic Number (Proton Number)	Atomic Mass
Actinium	Ac	89	227.0278
Aluminum	Al	13	26.981 54
Americium	Am	95	(243)
Antimony	Sb	51	121.757
Argon	Ar	18	39.948
Arsenic	As	33	74.9216
Astatine	At	85	(210)
Barium	Ba	56	137.33
Berkelium	Bk	97	(247)
Beryllium	Be	4	9.012 18
Bismuth	Bi	83	208.9804
Bohrium	Bh	107	(264)
Boron	B	5	10.81
Bromine	Br	35	79.904
Cadmium	Cd	48	112.41
Calcium	Ca	20	40.078
Californium	Cf	98	(251)
Carbon	C	6	12.011
Cerium	Ce	58	140.12
Cesium	Cs	55	132.9054
Chlorine	Cl	17	35.453
Chromium	Cr	24	51.996
Cobalt	Co	27	58.9332
Copper	Cu	29	63.546
Curium	Cm	96	(247)
Dubnium	Db	105	(262)
Dysprosium	Dy	66	162.50
Einsteinium	Es	99	(252)
Erbium	Er	68	167.26
Europium	Eu	63	151.96
Fermium	Fm	100	(257)
Fluorine	F	9	18.998 403
Francium	Fr	87	(223)
Gadolinium	Gd	64	157.25
Gallium	Ga	31	69.72
Germanium	Ge	32	72.561
Gold	Au	79	196.9665
Hafnium	Hf	72	178.49
Hahnium	Ha	105	(262)
Hassium	Hs	108	(265)
Helium	He	2	4.002 60
Holmium	Ho	67	164.9304
Hydrogen	H	1	1.007 94
Indium	In	49	114.82
Iodine	I	53	126.9045
Iridium	Ir	77	192.22
Iron	Fe	26	55.847
Krypton	Kr	36	83.80
Lanthanum	La	57	138.9055
Lawrencium	Lr	103	(260)
Lead	Pb	82	207.2
Lithium	Li	3	6.941
Lutetium	Lu	71	174.967
Magnesium	Mg	12	24.305
Manganese	Mn	25	54.9380
Meitnerium	Mt	109	(268)
Mendelevium	Md	101	(258)
Mercury	Hg	80	200.59
Molybdenum	Mo	42	95.94
Neodymium	Nd	60	144.24
Neon	Ne	10	20.1797
Neptunium	Np	93	237.048
Nickel	Ni	28	58.69
Niobium	Nb	41	92.9064
Nitrogen	N	7	14.0067
Nobelium	No	102	(259)
Osmium	Os	76	190.2
Oxygen	O	8	15.9994
Palladium	Pd	46	106.42
Phosphorus	P	15	30.973 76
Platinum	Pt	78	195.08
Plutonium	Pu	94	(244)
Polonium	Po	84	(209)
Potassium	K	19	39.0983
Praseodymium	Pr	159	140.9077
Promethium	Pm	61	(145)
Protactinium	Pa	91	231.0359
Radium	Ra	88	226.0254
Radon	Rn	86	(222)
Rhenium	Re	75	186.207
Rhodium	Rh	45	102.9055
Rubidium	Rb	37	85.4678
Ruthenium	Ru	44	101.07
Rutherfordium	Rf	104	(261)
Samarium	Sm	62	150.36
Scandium	Sc	21	44.9559
Seaborgium	Sg	106	(263)
Selenium	Se	34	78.96
Silicon	Si	14	28.0855
Silver	Ag	47	107.8682
Sodium	Na	11	22.989 77
Strontium	Sr	38	87.62
Sulfur	S	16	32.066
Tantalum	Ta	73	180.9479
Technetium	Tc	43	(98)
Tellurium	Te	52	127.60
Terbium	Tb	65	158.9254
Thallium	Tl	81	204.383
Thorium	Th	90	232.0381
Thulium	Tm	69	168.9342
Tin	Sn	50	118.710
Titanium	Ti	22	47.88
Tungsten	W	74	183.85
Uranium	U	92	238.0289
Vanadium	V	23	50.9415
Xenon	Xe	54	131.29
Ytterbium	Yb	70	173.04
Yttrium	Y	39	88.9059
Zinc	Zn	30	65.39
Zirconium	Zr	40	91.22

APPENDIX V Properties of Selected Isotopes

Atomic Number (Z)	Element	Symbol	Mass Number (A)	Atomic Mass*	Abundance (%) or Decay Mode† (if Radioactive)	Half-Life (if Radioactive)
0	(Neutron)	n	1	1.008 665	β^-	10.6 min
1	Hydrogen	H	1	1.007 825	99.985	
	Deuterium	D	2	2.014 102	0.015	
	Tritium	T	3	3.016 049	β^-	12.33 y
2	Helium	He	3	3.016 029	0.00014	
			4	4.002 603	≈100	
3	Lithium	Li	6	6.015 123	7.5	
			7	7.016 005	92.5	

Atomic Number (*Z*)	Element	Symbol	Mass Number (*A*)	Atomic Mass*	Abundance (%) or Decay Mode† (if Radioactive)	Half-Life (if Radioactive)
4	Beryllium	Be	7	7.016930	EC, γ	53.3 d
			8	8.005305	2α	6.7×10^{-17} s
			9	9.012183	100	
5	Boron	B	10	10.012938	19.8	
			11	11.009305	80.2	
			12	12.014353	β^-	20.4 ms
6	Carbon	C	11	11.011433	β^+, EC	20.4 ms
			12	12.000000	98.89	
			13	13.003355	1.11	
			14	14.003242	β^-	5730 y
7	Nitrogen	N	13	13.005739	β^-	9.96 min
			14	14.003074	99.63	
			15	15.000109	0.37	
8	Oxygen	O	15	15.003065	β^+, EC	122 s
			16	15.994915	99.76	
			18	17.999159	0.204	
9	Fluorine	F	19	18.998403	100	
10	Neon	Ne	20	19.992439	90.51	
			22	21.991384	9.22	
11	Sodium	Na	22	21.994435	β^+, EC, γ	2.602 y
			23	22.989770	100	
			24	23.990964	β^-, γ	15.0 h
12	Magnesium	Mg	24	23.985045	78.99	
13	Aluminum	Al	27	26.981541	100	
14	Silicon	Si	28	27.976928	92.23	
			31	30.975364	β^-, γ	2.62 h
15	Phosphorus	P	31	30.973763	100	
			32	31.973908	β^-	14.28 d
16	Sulfur	S	32	31.972072	95.0	
			35	34.969033	β^-	87.4 d
17	Chlorine	Cl	35	34.968853	75.77	
			37	36.965903	24.23	
18	Argon	Ar	40	39.962383	99.60	
19	Potassium	K	39	38.963708	93.26	
			40	39.964000	β^-, EC, γ, β^+	1.28×10^9 y
20	Calcium	Ca	30	39.962591	96.94	
24	Chromium	Cr	52	51.940510	83.79	
25	Manganese	Mn	55	54.938046	100	
26	Iron	Fe	56	55.934939	91.8	
27	Cobalt	Co	59	58.933198	100	
			60	59.933820	β^-, γ	5.271 y
28	Nickel	Ni	58	57.935347	68.3	
			60	59.930789	26.1	
			64	63.927968	0.91	
29	Copper	Cu	63	62.929599	69.2	
			64	63.929766	β^-, β^+	12.7 h
			65	64.927792	30.8	
30	Zinc	Zn	64	63.929145	48.6	
			66	65.926035	27.9	
33	Arsenic	As	75	74.921596	100	
35	Bromine	Br	79	78.918336	50.69	
36	Krypton	Kr	84	83.911506	57.0	
			89	88.917563	β^-	3.2 min
38	Strontium	Sr	86	85.909273	9.8	
			88	87.905625	82.6	
			90	89.907746	β^-	28.8 y
39	Yttrium	Y	89	89.905856	100	
43	Technetium	Tc	98	97.907210	β^-, γ	4.2×10^6 y

Atomic Number (Z)	Element	Symbol	Mass Number (A)	Atomic Mass*	Abundance (%) or Decay Mode† (if Radioactive)	Half-Life (if Radioactive)
47	Silver	Ag	107	106.905 095	51.83	
			109	108.904 754	48.17	
48	Cadmium	Cd	114	113.903 361	28.7	
49	Indium	In	115	114.903 88	95.7; β^-	5.1×10^{14} y
50	Tin	Sn	120	119.902 199	32.4	
53	Iodine	I	127	126.904 477	100	
			131	130.906 118	β^-, γ	8.04 d
54	Xenon	Xe	132	131.904 15	26.9	
			136	135.907 22	8.9	
55	Cesium	Cs	133	132.905 43	100	
56	Barium	Ba	137	136.905 82	11.2	
			138	137.905 24	71.7	
			144	143.922 73	β^-	11.9 s
61	Promethium	Pm	145	144.912 75	EC, α, γ	17.7 y
74	Tungsten	W	184	183.950 95	30.7	
76	Osmium	Os	191	190.960 94	β^-, γ	15.4 d
			192	191.961 49	41.0	
78	Platinum	Pt	195	194.964 79	33.8	
79	Gold	Au	197	196.966 56	100	
80	Mercury	Hg	202	201.970 63	29.8	
81	Thallium	Tl	205	204.974 41	70.5	
			210	209.990 069	β^-	1.3 min
82	Lead	Pb	204	203.973 044	β^-, 1.48	1.4×10^{17} y
			206	205.974 46	24.1	
			207	206.975 89	22.1	
			208	207.976 64	52.3	
			210	209.984 18	α, β^-, γ	22.3 y
			211	210.988 74	β^-, γ	36.1 min
			212	211.991 88	β^-, γ	10.64 h
			214	213.999 80	β^-, γ	26.8 min
83	Bismuth	Bi	209	208.980 39	100	
			211	210.987 26	α, β^-, γ	2.15 min
84	Polonium	Po	210	209.982 86	α, γ	138.38 d
			214	213.995 19	α, γ	164 μs
86	Radon	Rn	222	222.017 574	α, β	3.8235 d
87	Francium	Fr	223	223.019 734	α, β^-, γ	21.8 min
88	Radium	Ra	226	226.025 406	α, γ	1.60×10^3 y
			228	228.031 069	β^-	5.76 y
89	Actinium	Ac	227	227.027 751	α, β^-, γ	21.773 y
90	Thorium	Th	228	228.028 73	α, γ	1.9131 y
			232	232.038 054	100; α, γ	1.41×10^{10} y
92	Uranium	U	232	232.037 14	α, γ	72 y
			233	233.039 629	α, γ	1.592×10^5 y
			235	235.043 925	0.72; α, γ	7.038×10^8 y
			236	236.045 563	α, γ	2.342×10^7 y
			238	238.050 786	99.275; α, γ	4.468×10^9 y
			239	239.054 291	β^-, γ	23.5 min
93	Neptunium	Np	239	239.052 932	β^-, γ	2.35 d
94	Plutonium	Pu	239	239.052 158	α, γ	2.41×10^4 y
95	Americium	Am	243	243.061 374	α, γ	7.37×10^3 y
96	Curium	Cm	245	245.065 487	α, γ	8.5×10^3 y
97	Berkelium	Bk	247	247.070 03	α, γ	1.4×10^3 y
98	Californium	Cf	249	249.074 849	α, γ	351 y
99	Einsteinium	Es	254	254.088 02	α, γ, β^-	276 d
100	Fermium	Fm	253	253.085 18	EC, α, γ	3.0 d

*The masses given throughout this table are those for the neutral atom, including the Z electrons.

†"EC" stands for electron capture.

ANSWERS TO FOLLOW-UP EXERCISES

Chapter 1

1.1 L = 10 m.
1.2 Yes, [L] = [L], or m = m.
1.3 (a) 50 mi/h [(0.447 m/s)/(mi/h)] = 22 m/s.
(b) (1 mi/h)(1609 km/mi)(1 h/3600 s) = 0.477 m/s.
1.4 13.3 times.
1.5 $1\ m^3 = 10^6\ cm^3$.
1.6 European. 10 mi/gal ≈ 16 km/4 L = 4 km/L, as compared with 10 km/L.
1.7 (a) $7.0 \times 10^5\ kg^2$ (b) 3.02×10^2 (no units).
1.8 (a) 23.70. (b) 22.09.
1.9 $V = \pi r^2 h = \pi(0.490\ m)^2(1.28\ m) = 0.965\ m^3$
1.10 11.6 m.
1.11 A little steeper, $\theta = 31.3°$.
1.12 $750\ cm^3 = 7.50 \times 10^{-4}\ m^3 \approx 10^{-3}\ m^3$,
$m = \rho V \approx (10^3\ kg/m^3)(10^{-3}\ m^3) = 1\ kg$.
(By direct calculation, $m = 0.79$ kg.)
1.13 $V \approx 10^{-2}\ m^3$, cells/vol ≈ 10^4 cells/mm^3 (10^9 mm^3/m^3) = 10^{13} cells/m^3, and (cells/vol)(vol) ≈ 10^{11} white cells.

Chapter 2

2.1 $\Delta t = (8 \times 5.0\ s) + (7 \times 10\ s) = 110\ s$.
2.2 $s_1 = 2.00$ m/s; $s_2 = 1.52$ m/s; $s_3 = 1.72\ m/s \neq 0$, although the velocity is zero.
2.3 No. If the velocity is also in the negative direction, the object will speed up.
2.4 9.0 m/s in the direction of the original motion.
2.5 Yes, 96 m. (A lot quicker, isn't it?)
2.6 No, always more than one unknown variable.
2.7 No, changes x_o positions, but separation distance is the same.
2.8 $x = v^2/2a$, $x_B = 48.6$ m, and $x_C = 39.6$ m; the Blazer should not tailgate within at least 9.0 m.
2.9 1.16 s longer.
2.10 Time for bill to fall its length = 0.179 s. This time is less than the average reaction time (0.192 s) computed in the Example, so most people cannot catch the bill.
2.11 $y_u = y_d = 5.12$ m, as measured from reference $y = 0$ at the release point.
2.12 Eq. 2.8′, $t = 4.6$ s; Eq. 2.10′, $t = 4.6$ s.

Chapter 3

3.1 $v_x = -0.40$ m/s, $v_y = +0.30$ m/s; the distance is unchanged.
3.2 $x = 9.00$ m, $y = 12.6$ m (same).
3.3 $\vec{v} = (0)\,\hat{x} + (3.7\ m/s)\,\hat{y}$.
3.4 $\vec{C} = (-7.7\ m)\,\hat{x} + (-4.3\ m)\,\hat{y}$.
3.5 (a) $y_o = +25$ m and $y = 0$; the equation is the same. (b) $\vec{v} = (8.25\ m/s)\,\hat{x} + (-22.1\ m/s)\,\hat{y}$.
3.6 Both increase six fold.
3.7 (a) If not, the stone would hit to the side of the block. (b) Eq. 3.11 does not apply; the initial and final heights are not the same. $R = 15$ m, which is way off the 27-m answer.
3.8 The ball thrown at 45°. It would have a greater initial velocity.
3.9 At the top of the parabolic arc, the player's vertical motion is zero and is very small on either side of this maximum height. Here, the player's horizontal velocity component dominates, and he moves horizontally, with little motion in the vertical direction. This gives the illusion of "hanging" in the air.
3.10 4.15 m from the net.
3.11 $v_{bs}t = (2.33\ m/s)(225\ m) = 524\ m$
3.12 14.5° W of N.

Chapter 4

4.1 6.0 m/s in the direction of the net force.
4.2 (a) 11 lb. (b) Weight in pounds ≈ 2.2 lb/kg.
4.3 8.3 N
4.4 (a) 50° above the $+x$-axis. (b) x- and y-components reversed: $\vec{v} = (9.8\ m/s)\,\hat{x} + (4.5\ m/s)\,\hat{y}$.
4.5 Yes, mutual gravitational attractions between the briefcase and the Earth.
4.6 (a) $m_2 > 1.7$ kg. (b) $\theta < 17.5°$.
4.7 (a) 7.35 N. (b) Neglecting air resistance, 7.35 N, downward.
4.8 Increase. $\tan\theta = \dfrac{T}{mg} = \dfrac{55\ N}{(5.0\ kg)(9.8\ m/s^2)} = 1.1$, $\theta = 48°$
4.9 (a) $F_1 = 3.5w$. Even greater than F_2. (b) $\Sigma F_y = ma$, and F_1 and F_2 would both increase.
4.10 $\mu_s = 1.41\mu_k$ (for three cases in Table 4.1).
4.11 No. F varies with angle, with the angle for minimum applied force being around 33° in this case. (Greater forces are required for 20° and 50°.) In general, the optimum angle depends on the coefficient of friction.
4.12 Friction is kinetic, and f_k is in the $+x$ direction. Acceleration in the $-x$ direction.
4.13 Air resistance depends not only on speed, but also on size and shape. If the heavier ball were larger, it would have more exposed area to collide with air molecules, and the retarding force would increase faster. Depending on the size difference, the heavier ball might reach terminal velocity first, and the lighter ball would strike the ground first. Alternatively, the balls might reach terminal velocity together.

Chapter 5

5.1 −2.0 J
5.2 $d = \dfrac{W}{F\cos\theta} = \dfrac{3.80 \times 10^4\ J}{(189\ N)(0.866)} = 232\ m$
5.3 No, speed would decrease and it would stop moving.
5.4 $W_{x_1} = 0.034$ J, $W_x = 0.64$ J (measured from x_o)
5.5 No, $W_2/W_1 = 4$, or 4 times as much
5.6 Here we have $m_s = m_g/2$ as before. However, $v_s/\vec{v}_g = (6.0\ m/s)/(4.0\ m/s) = \frac{3}{2}$. Using a ratio, $K_s/K_g = \frac{9}{8}$, and the safety still has more kinetic energy than the guard. (Answer could also be obtained from direct calculations of kinetic energies, but for a relative comparison, a ratio is usually quicker.)
5.7 $W_3/W_2 = 1.4$, or 40% larger. More work, but a smaller percentage increase.
5.8 $\Delta U = mgh = (60\ kg)(9.8\ m/s^2)(1000\ m)\sin 10° = 10.2 \times 10^4$ J, yes doubled.
5.9 $\Delta K_{total} = 0$, $\Delta U_{total} = 0$
5.10 Without friction, the liquid would oscillate back and forth between the containers.
5.11 9.9 m/s
5.12 No. $E_o = E$ or $\frac{1}{2}mv_o^2 + mgh = \frac{1}{2}mv^2$. The mass cancels and the speed is independent of mass. (Recall that in free fall, all objects or projectiles fall with the same vertical acceleration g—see Section 2.5.)

5.13 0.025 m
5.14 (a) 59% (b) $E_{loss}/t = mg(y/t) = mgv = (60\ mg)$ J/s
5.15 Block would stop in rough area.
5.16 52%
5.17 (a) Same work in twice the time. (b) Same work in half the time.
5.18 (a) No. (b) Creation of energy.

Chapter 6

6.1 5.0 m/s. Yes, this is 18 km/h or 11 mi/h, a speed at which humans can run.
6.2 (1) Ship the greatest KE. (2) Bullet the least KE.
6.3 $(-3.0\ \text{kg}\cdot\text{m/s})\,\hat{\mathbf{x}} + (4.0\ \text{kg}\cdot\text{m/s})\,\hat{\mathbf{y}}$
6.4 It would increase to 60 m/s: greater speed, longer drive, ideally. (There is also a directional consideration.)
6.5 $F_{avg} = \dfrac{\Delta p}{\Delta t} = \dfrac{-310\ \text{kg}\cdot\text{m/s}}{0.600\ \text{s}} = -517\ \text{N}$
6.6 (a) No, for the m_1/m_2 system, external force on block. Yes, for the m_1/m_2 Earth system. But with m_2 attached to the Earth, the mass of this part of the system would be vastly greater than that of m_2, so its change in velocity would be negligible. (b) Assuming the ball is tossed in the +direction: for the tosser, $v_t = -0.50$ m/s; for the catcher, $v_c = 0.48$ m/s. For the ball: $p = 0$, $+25$ kg·m/s, $+1.2$ kg·m/s.
6.7 No. Energy went into work of breaking the brick, and some lost as heat and sound.
6.8 No.
6.9 No; all of the kinetic energy cannot be lost to make the dent. The momentum after the collision cannot be zero, since it was not zero initially. Thus, the balls must be moving and have kinetic energy. This can also be seen from Eq. 6.11: $K_f/K_i = m_1/(m_1 + m_2)$, and K_f cannot be zero (unless m_1 is zero, which is not possible).
6.10 $x_1 = v_1 t = (-0.80\ \text{m/s})(2.5\ \text{s}) = -2.0\ \text{m}$, $x_2 = v_2 t = (1.2\ \text{m/s})(2.5\ \text{s}) = 3.0\ \text{m}$
$\Delta x = x_2 - x_1 = 3.0\ \text{m} - (-2.0\ \text{m}) = 5.0$ m. Objects 5.0 m apart.
6.11 (a) $\Delta p_1 = p_{1_f} - p_{1_o} = 32\ \text{kg}\cdot\text{m/s} - 40\ \text{kg}\cdot\text{m/s} = -8.0\ \text{kg}\cdot\text{m/s}$
$\Delta p_2 = p_{2_f} - p_{2_o} = 13\ \text{kg}\cdot\text{m/s} - 5.0\ \text{kg}\cdot\text{m/s} = +8.0\ \text{kg}\cdot\text{m/s}$
(b) $\Delta p_1 = p_{1_f} - p_{1_o} = (-20\ \text{kg}\cdot\text{m/s}) - (12\ \text{kg}\cdot\text{m/s}) = -32\ \text{kg}\cdot\text{m/s}$
$\Delta p_2 = p_{2_f} - p_{2_o} = (8.0\ \text{kg}\cdot\text{m/s}) - (-24\ \text{kg}\cdot\text{m/s}) = +32\ \text{kg}\cdot\text{m/s}$
6.12 $p_{1_o} = mv_{1_o}$, $p_{2_o} = -mv_{2_o}$ and $p_1 = mv_1 = -mv_{2_o}$,
$p_2 = mv_2 = mv_{1_o}$, so conserved $K_i = \dfrac{m}{2}\left(v_{1_o}^2 + v_{2_o}^2\right)$ and
$K_f = \dfrac{m}{2}(v_1^2 + v_2^2) = \dfrac{m}{2}\left[(-v_{2_o})^2 + (v_{1_o})^2\right]$, so conserved
6.13 All of the balls swing out, but to different degrees. With $m_1 > m_2$, the stationary ball (m_2) moves off with a greater speed after collision than the incoming, heavier ball (m_1), and the heavier ball's speed is reduced after collision, in accordance with Eq. 6.16 (see Fig. 6.14b). Hence, a "shot" of momentum is passed along the row of balls with equal mass (see Fig. 6.14a), and the end ball swings out with the same speed as was imparted to m_2. Then, the process is repeated: m_1, *now moving more slowly*, collides again with the initial ball in the row (m_2), and another, but smaller, shot of momentum is passed down the row. The new end ball in the row receives less kinetic energy than the one that swung out just a moment previously, and so doesn't swing as high. This process repeats itself instantaneously for each ball, with the observed result that all of the balls swing out to different degrees.
6.14 $X_{CM} = \dfrac{(\text{same as in example}) + (8.0\ \text{kg})x_4}{(\text{same as in example}) + (8.0\ \text{kg})} = \dfrac{0 + (8.0\ \text{kg})x_4}{19\ \text{kg}} = +1.0\ \text{m}$

$$x_4 = \left(\frac{19}{8}\right)\text{m} = 2.4\ \text{m}$$

6.15 $(X_{CM}, Y_{CM}) = (0.47\ \text{m}, 0.10\ \text{m})$; same location as in Example, two-thirds of the length of the bar from m_1. Note: The location of the CM does not depend on the frame of reference.
6.16 Yes, the CM does not move.

Chapter 7

7.1 1.61×10^3 m = 1.61 km (about a mile)
7.2 (a) 0.35% for 10° (b) 1.2% for 20°
7.3 (a) 4.7 rad/s, 0.38 m/s; 4.7 rad/s, 0.24 m/s (b) To equalize the running distances, because the curved sections of the track have different radii and thus different lengths.
7.4 120 rpm
7.5 (a) 106 rpm (b) $a = \sqrt{2}g = 13.9\ \text{m/s}^2$, at 45° below plane of centrifuge.
7.6 The string cannot be exactly horizontal; it must make some small angle to the horizontal so that there will be an upward component of the tension force to balance the ball's weight.
7.7 No; it depends on mass: $F_c = \mu_s mg$.
7.8 No. Both masses have the same angular frequency or speed ω, and $a_c = r\omega^2$, so actually $a_c \propto r$. Remember, $v = 2\pi r/T$, and note that $v_2 > v_1$, with $a_c = v^2/r$.
7.9 $T = 5.2$ N
7.10 (a) The directions of ω and α would be downward, perpendicular to the plane of the CD. (b) Negative α, which means it is the opposite direction of ω.
7.11 $-0.031\ \text{rad/s}^2$
7.12 $2.8 \times 10^{-3}\ \text{m/s}^2$ (a large force, but a small acceleration)
7.13 $T^2 = \left(\dfrac{4\pi^2}{GM_E}\right)r^3 = \left(\dfrac{4\pi^2}{GM_E}\right)(R_E + h)^3 \approx \left(\dfrac{4\pi^2}{g}\right)R_E \approx 4R_E$

$T = 2\sqrt{R_E} = 2(6.4 \times 10^6\ \text{m})^{\frac{1}{2}} = 5.1 \times 10^3$ s (Why are the units not consistent?)
7.14 No, they do not vary linearly; $\Delta U = 2.4 \times 10^9$ J, only a 9.1% increase.
7.15 This is the amount of *negative* work done by an external force or agent when the masses are brought together. To separate the masses by infinite distances, an equal amount of positive work (against gravity) would have to be done.
7.16 $T^2 = \left(\dfrac{4\pi^2}{GM_S}\right)r^3$ and $M_S = \dfrac{4\pi^2 r^3}{GT^2} =$

$$\frac{4\pi^2(1.50 \times 10^{11}\ \text{m})^3}{(6.67 \times 10^{-11}\ \text{N}\cdot\text{m}^2/\text{kg}^2)(3.16 \times 10^7\ \text{s})^2} = 2.00 \times 10^{30}\ \text{kg}$$

Chapter 8

8.1 $s = r\omega = 5(0.12\ \text{m})(1.7) = 0.20\ \text{m}$;
$s = v_{CM}t = (0.10\ \text{m/s})(2.00\ \text{s}) = 0.20\ \text{m}$
8.2 The weights of the balls and the forearm produce torques that tend to cause rotation in the direction opposite that of the applied torque.
8.3 More strain.
8.4 $T \propto 1/\sin\theta$, and as θ gets smaller, so does $\sin\theta$ and T increases. In the limit, $\sin\theta \to 0$ and $T \to$ infinity (unrealistic).

8.5 $\Sigma\tau$: $Nx - m_1gx_1 - m_2gx_2 - m_3gx_3 = (200\text{ g})g(50\text{ cm}) - (25\text{ g})g(0\text{ cm}) - (75\text{ g})g(20\text{ cm}) - (100\text{ g})g(85\text{ cm}) = 0$, where $N = Mg$.
8.6 No. With f_{s_1}, the reaction force N would not generally be the same (f_{s_2} and N are perpendicular components of the force exerted on the ladder by the wall). In this case, we still have $N = f_{s_1}$, but $Ny - (m_1g)x_1 - (m_mg)x_m - f_{s_2}$, and $x_3 = 0$.
8.7 Hanging vertically.
8.8 Male: lighter upper torso. Female: heavier lower torso.
8.9 5 bricks
8.10 (d) no (equal masses) (e) Yes; with larger mass farther from axis of rotation, $I = 360\text{ kg}\cdot\text{m}^2$.
8.11 The long pole (or your extended arms) increases the moment of inertia by placing more mass farther from the axis of rotation (the tightrope or rail). When the walker leans to the side, a gravitational torque tends to produce a rotation about the axis of rotation, causing a fall. However, with a greater rotational inertia (greater I), the walker has time to shift his or her body so that the center of gravity is again over the rope or rail and thus again in (unstable) equilibrium. With very flexible poles, the CG may be below the wire, thus ensuring stability.
8.12 $t = 0.63\text{ s}$
8.13 $\alpha = \dfrac{2\,mg - (2\tau_f R)}{(2m + M)R}$; $\dfrac{\text{N}}{\text{kg}\cdot\text{m}}$; and $\dfrac{\text{N}}{\text{kg}\cdot\text{m}} = \dfrac{\text{kg}\cdot\text{m/s}^2}{\text{kg}\cdot\text{m}} = \dfrac{1}{\text{s}^2}$
8.14 The yo-yo would roll back and forth, oscillating about the critical angle.
8.15 (a) 0.24 m (b) The force of *static* friction, f_s, acts at the point of contact, which is always instantaneously at rest and so does no work. Some frictional work may be done due to rolling friction, but this is considered negligible for hard objects and surfaces.
8.16 $v_{CM} = 2.2\text{ m/s}$; using a ratio, 1.4 times greater; no rotational energy.
8.17 You already know the answer: 5.6 m/s. (It doesn't depend on the mass of the ball.)
8.18 $M_a = 75\text{ kg}(0.75) = 56\text{ kg}$. Then $L_1 = 13\text{ kg}\cdot\text{m}^2\text{/s}$ and $L_2 = (1.3\text{ kg}\cdot\text{m}^2)\omega$ [math not shown]. $L_2 = L_1$ or $(1.3\text{ kg}\cdot\text{m}^2)\omega = 13\text{ kg}\cdot\text{m}^2\text{/s}$ and $\omega = 10\text{ rad/s}$

Chapter 9

9.1 (a) +0.10% (b) 39 kg
9.2 2.3×10^{-4} L, or $2.3 \times 10^{-7}\text{ m}^3$
9.3 (1) Having enough nails, and (2) having them all of equal height and not so sharp a point. This could be achieved by filing off the tips of the nails so as to have a "uniform" surface. Also, this would increase the effective area.
9.4 3.03×10^4 N (or 6.82×10^3 lb—about 3.4 tons!) This is roughly the force on your back right now. Our bodies don't collapse under atmospheric pressure because cells are filled with incompressible fluids (mostly water), bone, and muscle, which react with an equal outward pressure (equal and opposite forces). As with forces, it is a pressure *difference* that gives rise to dynamic effects.
9.5 $d_o = \sqrt{\dfrac{F_o}{F_i}}d_i = \sqrt{\dfrac{1}{10}}(8.0\text{ cm}) = 2.5\text{ cm}$
9.6 Pressure in veins is lower than that in arteries (120/80).
9.7 As the balloon rises, the buoyant force decreases as a result of the temperature decrease (less helium pressure, less volume) and the less dense air ($F_b = m_fg = \rho_f gV_f$). When the net force is zero, the velocity is constant. The cooling effect continues with altitude and the balloon will start to sink when the net force is negative.
9.8 $r \approx 1.0$ m. $F_b = \rho gV = \rho g\left(\dfrac{4}{3}\pi r^3\right) = (0.18\text{ kg/m}^3)\left(\dfrac{4g\pi}{3}\right)(1.0\text{ m})^3 = 7.4\text{ N}$, much more.
9.9 (a) The object would sink, so the buoyant force is less than the object's weight. Hence, the scale would have a reading greater than 40 N. Note that with a greater density, the object would not be as large and less water would be displaced. (b) 41.8 N.
9.10 11%
9.11 −18%
9.12 $r = 9.00 \times 10^{-3}\text{ m}$, $v = \dfrac{\text{constant}}{A} = \dfrac{8.33 \times 10^{-5}\text{ m}^3\text{/s}}{\pi(9.00 \times 10^{-3}\text{ m})^2} = 0.327\text{ m/s}$; 23%
9.13 69%
9.14 As the water falls, speed (v) increases and area (A) must decrease to have Av = a constant.
9.15 0.38 m

Chapter 10

10.1 (a) 40° C (b) You should immediately know the answer—this is the temperature at which the Fahrenheit and Celsius temperatures are numerically equal.
10.2 (a) $T_R = T_F + 460$ (b) $T_R = \frac{9}{5}T_C + 492$ (c) $T_R = \frac{9}{5}T_K$
10.3 96°C
10.4 273° C; no, not on Earth
10.5 50 C°
10.6 It depends on the metal of the bar. If the thermal expansion coefficient (α) of the bar is less than that of iron, it will not expand as much and not be as long as the diameter of the circular ring after heating. However, if the bar's α is greater than that of iron, the bar will expand more than the ring and the ring will be distorted.
10.7 Basically, the situations would be reversed. Faster cooling would be achieved by submerging the ice in Example 10.7—the cooler water would be less dense and would rise, promoting mixing. For a lake with cooling at the surface, cooler, less-dense water would remain at the surface until minimum density was achieved. With further cooling, the denser water would sink and freezing would occur from the bottom up.
10.8 v_{rms}, 1.69%; K, 3.41%
10.9 The rotational kinetic energy for oxygen is the difference between the total energies, 2.44×10^3 J. The oxygen is less massive and has the higher v_{rms}.

Chapter 11

11.1 2.84×10^3 m
11.2 12.5 kg
11.3 (a) The ratio will be smaller because the specific heat of aluminum is greater than that of copper. (b) $Q_w/Q_{pot} = 15.2$
11.4 The final temperature T_f is expected to be higher because the water was at a higher initial temperature. $T_f = 34.4$°C
11.5 -1.09×10^5 J (negative because heat is lost)
11.6 (a) 2.64×10^{-2} kg or 26.4 g of ice melts. (b) The final temperature is still 0°C because the liver cannot lose enough heat to melt all the ice, even if the ice started at 0°C. The final result is an ice/water/liver system at 0°C, but with more water than in the Example.
11.7 1.1×10^5 J/s (difference due to rounding)
11.8 No, since the air spaces provide good insulation because air is a poor thermal conductor. The many small pockets of air between the body and the outer garment form an insulating layer that minimizes conduction and so retards the loss of body heat. (There is little convection in the small spaces.)

11.9 (a) -1.5×10^2 J/s or -1.5×10^2 W (b) The huge ear flaps have large surface area so more heat can be radiated out.
11.10 Drapes reduce heat loss by limiting radiation through the window and by keeping convection current away from the glass.

Chapter 12

12.1 0.20 kg
12.2 In both cases, heat flow is into the gas. During the isothermal expansion, $Q = W = +3.14 \times 10^3$ J. During the isobaric expansion, $W = +4.53 \times 10^3$ J and $\Delta U = +6.80 \times 10^3$ J, therefore $Q = \Delta U + W = +1.13 \times 10^4$ J.
12.3 753° C
12.4 As the air comes to lower elevations and higher pressures, it quickly compresses. This is approximately an adiabatic process, resulting in a temperature rise of the air.
12.5 (a) 142 K or -131° C. (b) For monatomic gas, $\Delta U = (3/2)nR\Delta T = -3.76 \times 10^3$ J. This should be the same as $-W$ since, for an adiabatic process, $Q = 0 = \Delta U + W$; therefore, $\Delta U = -W$. The slight difference is due to rounding.
12.6 -1.22×10^3 J/K
12.7 Overall zero entropy change requires $|\Delta S_w| = |\Delta S_m|$ or $|Q_w/T_w| = |Q_m/T_m|$. Because the system is isolated, the magnitudes of the two heat flows *must* be the same $|Q_w| = |Q_m|$. Thus no overall entropy change requires the water and the metal to have the same average temperature $\bar{T}_w = \bar{T}_m$. This is not possible, unless they are initially at the *same* temperature. Thus, this can only happen if there is no net heat flow.
12.8 If the basics of the cycles are kept, that is, the triangular shape and the volume doubling, then a way to increase the net work (the area inside the cycle) is to drop the pressure even further at the end of the isometric segment. If you allow the volume to more than double during the isobaric expansion, that would achieve the same end. Anything that increases the net area (work) will do.
12.9 (a) 150 J/cycle (b) 850 J/cycle
12.10 $Q_{34} = 610$ J and $Q_{23} = 730$ J, therefore $Q_c = Q_{23} + Q_{34} = 1.34 \times 10^3$ J. This agrees with $Q_c = Q_h - W_{net} = 59 \times 10^3\text{ J} - 245\text{ J} = 1.35 \times 10^3$ J (to within rounding errors).
12.11 (a) The new values are: $COP_{ref} = 3.3$ and $COP_{hp} = 4.3$. (b) The COP of the air conditioner has the largest percentage increase.
12.12 It would show an increase of 7.5%.

Chapter 13

13.1 No, its maximum speed is $(\sqrt{k/m})A = 4.0$ m/s. Thus it is traveling at 75% of the maximum speed.
13.2 0.49 J
13.3 (1) $y = -0.0881$ m, up. $n = 0.90$. (2) $y = 0$, going up. $n = 1.5$.
13.4 9.76 m/s^2; no. Since this is less than the accepted value at sea level, the park is probably at an altitude above sea level.
13.5 (a) 0.50 m (b) 0.10 Hz
13.6 440 Hz
13.7 increase the tension (by 44% as can be calculated).

Chapter 14

14.1 (a) 2.3 (b) 10.2
14.2 $v = (331 + 0.6T_C)$ m/s $= [331 + 0.6(38°)] = 354$ m/s Increase.
14.3 It would be greatest in He, because it has the smallest molecular mass. (It would be lowest in oxygen, which has the largest molecular mass.)
14.4 (a) The dB scale is logarithmic, not linear. (b) 3.16×10^{-6} W/m^2
14.5 No, $I_2 = (316)I_1$
14.6 65 dB
14.7 Destructive interference: $\Delta L = 2.5\lambda = 5(\lambda/2)$, and $m = 5$. No sound would be heard if the waves from the speakers had equal amplitudes. Of course, during a concert the sound would not be single-frequency tones but would have a variety of frequencies and amplitudes. Listeners at certain locations might not hear certain parts of the audible spectrum, but this probably wouldn't be noticed.
14.8 Toward, 431 Hz; past, 369 Hz
14.9 With the source and the observer traveling in the same direction at the same speed, their relative velocity would be zero. That is, the observer would consider the source to be stationary. Since the speed of the source and observer is subsonic, the sound from the source would overtake the observer without a shift in frequency. Generally, for motions involved in a Doppler shift, the word *toward* is associated with an *increase* in frequency and *away* with a *decrease* in frequency. Here, the source and observer remain a constant distance apart. (What would be the case if the speeds were supersonic?)
14.10 768 Hz; yes
14.11 $f_1 = \dfrac{v}{4L} = \dfrac{353\text{ m/s}}{4(0.0130\text{ m})} = 6790$ Hz

Chapter 15

15.1 1.52×10^{-20} %
15.2 No, if the comb were positive, it would polarize the paper in the reverse way and still attract it.
15.3 $\vec{F}_1$ has a magnitude of 3.8×10^{-7} N at an angle of 57° above the positive x axis. In unit vector notation: $\vec{F}_1 = (-0.22\ \mu\text{N})\ \hat{x} + (0.32\ \mu\text{N})\ \hat{y}$.
15.4 0.12 m or 12 cm.
15.5 $\dfrac{F_e}{F_g} = \dfrac{ke^2}{Gm_e^2} = 4.2 \times 10^{42}$ or $F_e = 4.2 \times 10^{42} F_g$. The magnitude of the electrical force is the same as that between a proton and electron (in the Example) because they have the same (magnitude) charge on them. However the gravitational force is reduced because the attracting masses are two electrons rather than an electron and a much more massive proton.
15.6 The field is zero to the left of q_1 at $x = -0.60$ m.
15.7 $\vec{E} = (-797\text{ N/C})\ \hat{x} + (359\text{ N/C})\ \hat{y}$ or $E = 874$ N/C at an angle of 24.2° above the negative x-axis.
15.8 In all three locations there are two fields to consider and add vectorially, one from the positive end and one from the negative end of the dipole. (a) Here the larger of the two fields is from the closer positive end and points upward. The smaller field due to the negative end points downward, thus the field direction is upward away from the positive end. (b) Here the larger of the two fields is from the closer negative end and points upward. The smaller field due to the positive end points downward, thus the field direction is upward towards the negative end. (c) Here both fields point downward, thus the net field is downward, away from the positive end and towards the negative end.
15.9 (a) The electric field is upward from ground to cloud. (b) 2.3×10^3 C.
15.10 Positive charge would reside completely on the outside surface, thus only the electroscope attached to the outside surface would show deflection.
15.11 Their sign is negative since the electric field lines end at negative charges, they are all inward relative to the Gaussian surface.

Chapter 16

16.1 (a) ΔU_e would double to $+7.20 \times 10^{-18}$ J because the particle's charge is doubled. (b) ΔV is unchanged because it is not related to the particle. (c) $v = 4.65 \times 10^4$ m/s.
16.2 6.63×10^7 m/s
16.3 (a) It has moved further from a positive charge (the proton) and thus has moved to a region of lower electric potential. (b) $\Delta U_e = +3.27 \times 10^{-18}$ J.
16.4 $U_{CO} = -3.27 \times 10^{-19}$ J. It is less stable, because it would take less work to break it apart than a water molecule.
16.5 (a) 2.22 m. (b) The one closest to the Earth's surface is at a higher potential. (c) No, you can only tell the separation distance between the two surfaces, not their absolute location.
16.6 (a) Surface 1 is at a higher electric potential than surface 2 because it is closer to the positively charged surface. (b) At large distances, the charged object would "look like" a point charge, thus the equipotential surfaces gradually becomes spherical as the distance from the object gets larger.
16.7 $d = 8.9 \times 10^{-16}$ m which is much smaller than the size of an atom (or a nucleus for that matter) and thus this design is entirely unfeasible.
16.8 7.90×10^3 V
16.9 The capacitance decreases as the spacing d increases. Since the voltage across the capacitor remains constant, this means the charge on the capacitor would have to decrease, thus charge would flow off of the capacitor. $\Delta Q = -3.30 \times 10^{-12}$ C.
16.10 $U_{parallel} = 1.20 \times 10^{-4}$ J and $U_{series} = 5.40 \times 10^{-4}$ J so the parallel arrangement stores more energy.
16.11 (a) $Q_1 = 8.0 \times 10^{-7}$ C; $Q_2 = 1.6 \times 10^{-6}$ C; $Q_3 = 2.4 \times 10^{-6}$ C. (b) $U_1 = 3.2 \times 10^{-6}$ J; $U_2 = 6.4 \times 10^{-6}$ J; $U_3 = 4.8 \times 10^{-6}$ J

Chapter 17

17.1 The result is the same, $V_{AB} = V$.
17.2 About 32 years.
17.3 100 V.
17.4 Our assumption is that $R = \dfrac{\rho L}{A}$. Thus, if resistivity is doubled and length halved, the numerator stays the same. If the diameter is halved, the area decreases by a factor of 4. The net result of these changes is that the resistance increases by a factor of 4, up to 3.0×10^3 Ω. Thus $I = \dfrac{V}{R} = \dfrac{400\ \text{V}}{3.0 \times 10^3\ \Omega} =$ 0.133 A.
17.5 $R = 0.67$ Ω. The material with the largest temperature coefficient of resistivity makes a more sensitive thermometer because it produces a larger (and therefore more accurate to measure) change in resistance for a given temperature change.
17.6 The heat needed is $Q = mc\Delta T = 1.67 \times 10^5$ J. Thus the power output of the heater needs to be $P = \dfrac{Q}{t} = \dfrac{1.67 \times 10^5\ \text{J}}{180\ \text{s}} = 930$ W. Since this is supplied by the joule heating, we have $R = \dfrac{V^2}{P} = \dfrac{(120\ \text{V})^2}{1.67 \times 10^5\ \text{J}} = 15.5\ \Omega$.
17.7 (a) $R_1 = \dfrac{V^2}{P_1} = \dfrac{(115\ \text{V})^2}{1200\ \text{W}} = 11.0\ \Omega$ and $R_2 = 0.900R_1 = 9.92\ \Omega$. (b) $I_1 = \dfrac{V}{R_1} = \dfrac{115\ \text{V}}{11.0\ \Omega} = 10.5$ A and $I_2 = 1.11I_1 = 11.6$ A.
17.8 8.3 hours.
17.9 At best, power plants produce electric energy with efficiencies of 35% (ignoring transmission losses). Thus in terms of primary fuels, the maximum efficiency of any electrical appliance is 35%. However, natural gas is delivered at essentially no energy cost. At the point of delivery, it is burned and can deliver, at least theoretically, up to 100% of its heat content to the task at hand. For example, a well-insulated water heater will be able to absorb about 95% of the energy heat delivered to it. Thus the overall electrical efficiency would be 0.95 (35%) or about 34%. For the gas version, it would be 95% efficient.

Chapter 18

18.1 (a) Series: $P_1 = 4.0$ W, $P_2 = 8.0$ W, $P_3 = 12$ W. Parallel: $P_1 = 1.4 \times 10^2$ W, $P_2 = 72$ W, $P_3 = 48$ W. (b) In series, the most power is dissipated in the largest resistance. In parallel, the most power is dissipated in the least resistance. (c) Series: total resistor power is 24 W, and $P_b = I_b V_b = (2.0\ \text{A})(12\ \text{V}) = 24$ W, so yes, as required by energy conservation. Parallel: total resistor power is $P_{tot} = 2.6 \times 10^2$ W, and $P_b = I_b V_b = (22\ \text{A})(12\ \text{V}) = 2.6 \times 10^2$ W (to two significant figures), so yes, as required by energy conservation
18.2 (a) The voltage across the open socket will be 120 V. (b) The voltage across the remaining bulbs will be zero.
18.3 $P_1 = I_1^2 R_1 = 54.0$ W, $P_2 = I_2^2 R_2 = 9.0$ W, $P_3 = I_3^2 R_3 = 0.87$ W, $P_4 = I_4^2 R_4 = 2.55$ W, and $P_5 = I_5^2 R_5 = 5.63$ W. Their sum is 72.1 W to three significant figures. We have agreement with the power output of the battery (difference due to rounding), that is, $P_b = I_b V_b = (3.00\ \text{A})(24.0\ \text{V}) = 72.0$ W.
18.4 (a) If R_2 is increased then the equivalent parallel resistance of R_2 and R_1 will increase. Thus the total circuit resistance should increase, resulting in a reduction in the total circuit current. Since the current in R_3 is the same as the total current, I_3 will decrease. From this, V_3 should decrease. Therefore V_1 and V_2 should increase since they are equal $V = V_2 + V_3 =$ a constant. Since R_1 has not changed, due to the voltage increase, I_1 should increase. Since I_3 decreases and I_1 increases, it must be (since $I_3 = I_1 + I_2$) that I_2 must decrease. (b) Recalculation confirms these predictions: $I_1 = 0.51$ A (increase), $I_2 = 0.38$ A (decrease), and $I_3 = 0.89$ A (decrease).
18.5 At the junction, we still have $I_1 = I_2 + I_3$ (Eq. 1). Using the loop theorem around loop 3 in the clockwise direction (all numbers are volts, deleted for convenience): $6 - 6I_1 - 9I_2 = 0$ (Eq. 2). For loop 1, the result is $6 - 6I_1 - 12 - 2I_3 = 0$ (Eq. 3). Solve Eq. 1 for I_2 and substitute into Eq. 2. Then solve Eq. 2 and Eq. 3 simultaneously for I_1 and I_3. All answers are the same as in the Example as they should be.
18.6 (a) The maximum energy storage at 9.00 V is 4.05 J. At 7.20 V, the capacitor stores only 2.59 J or 64% of the maximum. This is because the energy storage varies as the *square* of the voltage across the capacitor and $0.8^2 = 0.64$. (b) 8.64 V, because the voltage does not rise linearly, but levels off in an exponential fashion.
18.7 10 A.
18.8 0.20 mA.

Chapter 19

19.1 East, since reversing both the velocity direction and the sign of the charge leaves the direction the same.
19.2 (a) Using the force right-hand rule, the proton would initially deflect in the negative x direction. (b) 0.10 T

19.3 0.500 V
19.4 (a) At the poles the magnetic field is perpendicular to the ground. Since the current is parallel to the ground, according to the force right-hand rule the force on the wire would be in a plane parallel to the ground. Thus it would not be able to cancel the downward force of gravity. (b) The wire's mass is 0.041 g, which is unrealistically low.
19.5 (a) At 45°, the torque is 0.269 m · N or 70.7% of the maximum torque. (b) 30°.
19.6 (a) South (b) 38 A.
19.7 1500 turns
19.8 (a) The force becomes repulsive. You should be able to show this by using the right-hand source and force rules. (b) 0.027 m or 27 mm.
19.9 The permeability would only have to be 40% of the value in the Example, or $\mu \geq 480\mu_o = 6.0 \times 10^{-4}$ T · m/A.

Chapter 20

20.1 (a) Clockwise. (b) 0.335 mA.
20.2 Any way that will increase the flux, such as increasing the loop area or the number of loops. Changing to a lower resistance would also help.
20.3 7.36×10^{-4} T
20.4 1.5 m/s
20.5 0.28 m
20.6 (a) 6.1×10^3 J (b) 5.0×10^3 J so about 12 times more energy is used during startup.
20.7 (a) She would use a step-up transformer because European appliances are designed to work at 240 V, which is twice the U.S. voltage of 120 V. (b) The output current would be 1500 W/240 V or 6.25 A. Thus the input current would be 12.5 A. (Voltage would be stepped up by a factor of two, and thus the input current is twice as large as the output current.)
20.8 (a) Higher voltages allow for lower current usage. This in turn reduces joule heat losses in the delivery wires and in the motor windings, making more energy available for doing mechanical work and therefore a higher efficiency. (b) Since the voltage is doubled, the current is halved. The heat loss in the wire is proportional to the *square* of the current. Thus, losses will be cut by a factor of 4 or be reduced to 25% of their value at 120 V.
20.9 0.38 cm/s.
20.10 (a) With increasing distance, the Sun's light intensity (energy per second per unit area) drops. Therefore so would the force due to the light pressure on the sail. In turn, the ship's acceleration would be reduced. (b) You would need to somehow enlarge the sail area to catch more light.

Chapter 21

21.1 (a) 0.25A (b) 0.35 A (c) $9.6 \times 10^2\ \Omega$, larger than the 240 Ω required for a bulb of the same power in the United States. The voltage in Great Britain is larger than that in the United States. Thus to keep the power constant, the current must be reduced by using a larger resistance.
21.2 If the resistance of the appliance is constant, the power will quadruple since $P \propto V^2$. Even if the resistance increased, the power would probably be much more than the appliance was designed for and it would likely burn out, or at least blow a fuse.
21.3 (a) $\sqrt{2}(120\text{ V}) = 170$ V (b) 120 Hz
21.4 (a) $\sqrt{2}(2.55\text{ A}) = 3.61$ A (b) 180 Hz
21.5 (a) The current would increase to 0.896 A. (b) The capacitor is responsible; with a frequency increase, X_C decreases. Since resistance is independent of frequency, it remains constant and overall Z decreases.
21.6 (a) In an RLC circuit, the phase angle ϕ depends on the difference $X_L - X_C$. If you increase the frequency, X_L would increase and X_C would decrease, thus their difference would increase and so would ϕ. (b) $\phi = 84.0°$, an increase as expected.
21.7 6.98 W
21.8 (a) If you have a receiver tuned to a frequency *between* the two station frequencies, you would not receive the maximum strength signal from either station but there might be enough power from each to hear them simultaneously. (b) 651 kHz

Chapter 22

22.1 Light travels in straight lines and is reversible. If you can see someone in a mirror, that person can see you. Conversely, if you can't see the trucker's mirror, then he or she can't see your image in that mirror and won't know that your car is behind the truck.
22.2 $n = 1.25$ and $\lambda_m = 400$ nm
22.3 By Snell's law, $n_2 = 1.24$ so $v = c/n_2 = 2.42 \times 10^8$ m/s.
22.4 With a greater n, θ_2 is smaller so the refracted light inside the glass is toward the lower-left. Therefore the lateral displacement is larger. 0.72 cm.
22.5 (a) The frequency of the light is unchanged in the different media, so the emerging light has the same frequency as that of the source. (b) The wavelength in air is independent of the water and glass media, as can be shown by adding another step (medium) to the Example solution. By reverse analysis, $\lambda_{air} = n_{water}\lambda_{water} = (c/v_{water})\lambda_{water} = c/f$. Thus, the wavelength in air is c/f.
22.6 Because of total internal reflections, the diver could not see anything above water. Instead, he would see the reflection of something on the sides and/or bottom of the pool. (Use reverse ray tracing.)
22.7 $n = 1.4574$. Green light will be refracted more than red light as green has a shorter wavelength, thus greater n than red light. By Snell's law, green will have a smaller angle of refraction so it is refracted more.

Chapter 23

23.1 No effect. Note that the solution to the Example does not include the distance. The geometry of the situation is the same regardless of the distance from the mirror.
23.2 $d_i \approx 60$ cm; real, inverted, and magnified.
23.3 $d_i = d_o$ and $M = -1$; real, inverted, and same size
23.4 The image is also always upright and smaller than the object.
23.5 $d_i = -20$ cm (in front of the lens); virtual, upright, and magnified.
23.6 $d_o = 2f = 24$ cm
23.7 Blocking off half of the lens would result in half the *amount* of light focused at the image plane, so the resulting image would be less bright but still full size.
23.8 The image is also always upright and smaller than the object
23.9 3 cm behind L_2; real, inverted, and smaller than the object ($M_{total} = -0.75$)
23.10 If the lens is immersed in water, Eq. 23.8 should be modified to $\frac{1}{f} = (n/n_m - 1)\left(\frac{1}{R_1} + \frac{1}{R_2}\right)$, where $n_m = 1.33$ (water). Since $n = 1.52 > n_m = 1.33$, the lens is still converging. $P = \frac{1}{f} = (1.52/1.33 - 1)\left(\frac{1}{0.15\text{ m}} + \frac{1}{-0.20\text{ m}}\right) = 0.238\text{ 1/m} = 0.238$ D. $f = \frac{1}{0.238\text{ 1/m}} = 4.20$ m.

Chapter 24

24.1 $\Delta y = y_r - y_b = 1.2 \times 10^{-2} = 1.2$ cm
24.2 twice as thick, $t = 199$ nm
24.3 In brass instruments, the sound comes from a relatively large, flared opening. Thus there is little diffraction, so most of the energy is radiated in the forward direction. In wood-wind instruments, much of the sound comes from tone holes along the column of the instrument. These holes are small compared to the wavelength of the sound, so there is appreciable diffraction. As a result, the sound is radiated in nearly all directions, even backward.
24.4 The width would increase by a factor of $700/550 = 1.27$
24.5 $\Delta\theta_2 = \theta_2(700\text{ nm}) - \theta_2(400\text{ nm}) = 44.4° - 23.6° = 20.8°$.
24.6 45°
24.7 $\theta_2 = 41.2°$
24.8 589 nm; yellow

Chapter 25

25.1 It wouldn't work; a real image would form on person's side of lens. ($d_i = +0.75$ m).
25.2 For an object at $d_o = 25$ cm, the image for eye 1 would be formed at 1.0 m; this is beyond the near point for that eye, so the object could be seen clearly. The image for eye 2 would be formed at 0.77 m; this is inside the near point for that eye, so the object would not be seen clearly.
25.3 glass for near-point viewing, 2.0 cm longer
25.4 length doubles
25.5 $f_i = 8.0$ cm
25.6 The erecting lens (of focal length f_e) should go between the objective and the eyepiece, positioned a distance of $2f_e$ from the image formed by the objective, which acts as an object. The erecting lens then produces an inverted image of the same size at $2f_e$ on the opposite side of the lens, which acts as an object for the eyepiece. The use of the erecting lens lengthens the telescope by $4f_e$.
25.7 3.4×10^{-7} rad, an order of magnitude better than the typical 10^{-6} rad
25.8 2.9 cm

Chapter 26

26.1 Light waves from two simultaneous events on the y-axis meet at some midpoint receptor on the y-axis. Since there is no relative motion along that axis, a simultaneous recording of the two events will also be recorded along the y'-axis. Hence the two observers agree on simultaneity for this situation.
26.2 $v = 0.9995c$; no, not twice as fast. Travel is limited to less than c, so this is only about a 0.15% increase.
26.3 (a) 0.667 μs (b) 0.580 μs The observer watching the ship measures the proper time interval. To the person on the ship, that time interval is dilated.
26.4 $v = 0.991c$
26.5 The traveler measures the proper time interval of 20.0 y, but Earth inhabitants measure the dilated version of this (why?). The gamma factor is based on a recalculated value for traveler speed $v = 0.90504c$. Keeping five places after the decimal and then rounding to three significant figures we get $\gamma = 1/\sqrt{1 - (0.99504c/c)^2} = 10.04988$ and $\Delta t = \gamma\Delta t_o = (10.04988)(20.0\text{ y}) = 200.99\text{ y} \approx 201$ y.
26.6 (a) 1.17 MeV (b) 0.207 MeV
26.7 (a) $0.319mc^2$ (b) $3.33mc^2$
26.8 1.95×10^7 more massive
26.9 $u = +0.69c$ which, as expected, is lower in magnitude than the nonrelativistic (and wrong) result of $0.80c$.

Chapter 27

27.1 (a) The wavelength is 967 nm, which is infrared. With some emissions in the red region, Betelguese would appear reddish. (b) The wavelength is 290 nm, which is ultraviolet. With significant visible emissions, Rigel would appear bluish-white.
27.2 496 nm, which is shorter than the Example, indicating a higher photon energy. Thus the maximum kinetic energy of the photoelectrons is higher, requiring an increased stopping voltage.
27.3 2.00 V
27.4 (a) The ratio is $\Delta\lambda/\lambda_o = 324$ meaning a wavelength increase of 3.24×10^4%. (b) Percentagewise this is a much larger increase than in the Example because the wavelength of the incoming light (λ_o) is much smaller for gamma rays.
27.5 (a) 1.09×10^6 m/s. (b) 5.41×10^{-19} J = 3.40 eV.
27.6 365 nm (UV).
27.7 The least energetic photon in the Lyman series results in a transition from $n = 2$ (first excited state) to the ground state $n = 1$. This results in a photon of energy 10.2 eV. The wavelength of this light is 122 nm, which is UV.
27.8 (a) There are six possible transitions from the $n = 4$ state to the ground state, and thus the emitted light has six different possible wavelengths. (b) If the atom is excited from the ground state to the first excited state ($n = 1$ to $n = 2$), then it has no choice but to emit a single photon during de-excitation state ($n = 2$ to $n = 1$) because there are no intermediate states.

Chapter 28

28.1 (a) 8.8×10^{-33} m/s (b) 2.3×10^{33} s, or 7.2×10^{25} y. This is about 4.8×10^{15} times longer than the age of the universe. This movement would definitely not be noticeable.
28.2 The proton's de Broglie wavelength is 4.1×10^{-12} m, or about twenty times smaller than atomic spacing distances. With a wavelength much smaller than the atomic spacing, these protons would *not* be expected to exhibit significant diffraction effects.
28.3 Only five electrons could be accommodated in the 3d subshell if there were no spin (with spin there can be ten).
28.4 (a) $1s^2 2s^2 2p^6$ (b) $-2e$ or -3.2×10^{-19} C
28.5 1.2×10^6 m/s

Chapter 29

29.1 ${}^{12}_{6}\text{C} + {}^{4}_{2}\text{He} \longrightarrow {}^{16}_{8}\text{O}$, thus the resulting nucleus is oxygen-16.
29.2 (a) Since ${}^{23}_{11}\text{Na}$ is the stable isotope with 11 protons and 12 neutrons, ${}^{22}_{11}\text{Na}$ is one neutron shy of being stable. In other words, it is proton-rich or neutron-poor. Thus, the expected decay mode is β^+ or positron decay. (b) Neglecting the emitted neutrino, the decay is ${}^{22}_{11}\text{Na} \longrightarrow {}^{22}_{10}\text{Ne} + {}^{0}_{+1}\text{e}$. The daughter nucleus is neon-22.
29.3 (a) 48 d because reducing the activity by a factor of 64 requires six half lives; $1/2^6 = 1/64$. (b) The process of excretion from the body can also remove ${}^{131}\text{I}$.
29.4 The closest integer is 20, since $2^{20} \approx 1.05 \times 10^6$. Thus it takes 20 half-lives, or about 560 y.
29.5 The measurement can be made to four ${}^{14}\text{C}$ half-lives, or 2.3×10^4 y.
29.6 (a) ${}^{40}\text{K}$ is an odd-odd nucleus (19 protons, 21 neutrons) and thus is unstable. ${}^{41}\text{K}$, an odd-even potassium isotope (19 protons, 22 neutrons) is the likely candidate for the remainder of the stable potassium. ${}^{43}\text{K}$ would have too many neutrons

(24) compared to protons (19) for this region of the periodic chart. (b) Using $N = N_o e^{-\lambda t}$, it follows that $\frac{N_o}{N} = e^{\lambda t} = 12.8$. Thus there would have been about 13 times more ^{40}K (than exists now) at the time of the formation of the Earth.

29.7 (a) Starting with 29 protons and 29 neutrons (why?), we have the following candidates: $^{58}_{29}Cu$, $^{59}_{29}Cu$, $^{60}_{29}Cu$, $^{61}_{29}Cu$, $^{62}_{29}Cu$, $^{63}_{29}Cu$, $^{64}_{29}Cu$, etc. Now delete the odd–odd isotopes (why?) to get the most likely (stable) isotopes: $^{59}_{29}Cu$, $^{61}_{29}Cu$, $^{63}_{29}Cu$, $^{65}_{29}Cu$, etc. (b) Further trimming of the list can be done by deleting those with $N \approx Z$ (why?) and those with N significantly larger than Z (why?). Since Z should be just a bit smaller than N in this mass region, we expect neutron numbers in the mid-30s. Hence a good guess would be just $^{63}_{29}Cu$ and $^{65}_{29}Cu$. According to Appendix V, these are, in fact, the only two stable isotopes of copper.

29.8 (a) The result for ^3He is 2.573 MeV/nucleon, which is considerably smaller than the 7.075 MeV/nucleon for ^4He. (b) Thus ^4He is the more tightly bound of the two. Unlike ^3He, all protons and neutrons in ^4He *are* paired, resulting in a more tightly bound nucleus.

29.9 The absorbed dose is 0.0215 Gy or 2.15 rad. Since the RBE for gamma rays is 1, the effective dose is 0.0215 Sv or 2.15 rem. Both of these are about one-seventh of the dose from the beta radiation.

Chapter 30

30.1 $Q = -15.63$ MeV, so it is endoergic (takes energy to happen).

30.2 The increase in mass has an energy equivalent of 1.193 MeV. The rest of the incident kinetic energy (1.534 MeV − 1.193 MeV, or 0.341 MeV) must be distributed between the kinetic energies of the proton and of the oxygen-17.

30.3 There are 1.00×10^{38} proton cycles per second.

30.4 Because the beta particle has little energy and therefore little momentum, the neutrino and the daughter nucleus would have to recoil in opposite directions in order to conserve linear momentum. This assumes the original nucleus had zero linear momentum.

30.5 Since the negative pion (π^-) has a charge of $-e$, its quark structure must be $\bar{u}d$. Using similar reasoning, the quark structure of the positive pion (π^+) is $u\bar{d}$. Thus the antiparticle of the π^+ would have the composition of $\overline{u\bar{d}}$. However, the antiquark of an antiquark is just the original quark—for example $(\bar{\bar{u}}) = u$. Hence the antiparticle of the π^+ has the quark structure $\overline{u\bar{d}} = \bar{u}d$, which is the π^- quark structure, as expected.

ANSWERS TO ODD-NUMBERED EXERCISES

Chapter 1

1. **(c)**
3. **(b)**
5. Because there are no more fundamental quantities and all other quantities can be derived from the fundamental ones.
7. Mean solar day replaced the original second definition. No, atomic clocks are now used.
9. **(b)**
11. **(a)**
13. no, yes
15. Metric ton is defined as the mass of 1 m^3 of water. 1 m^3 = 1000 L and 1 L of water has a mass of 1 kg. So one metric ton is equivalent to 1000 kg.
17. **(a)** Different ounces are used for volume and weight measurements. 16 oz = 1 pt is a volume measure and 16 oz = 1 lb is a weight measure. **(b)** Two different pound units are used. Avoirdupois lb = 16 oz, troy lb = 12 oz.
19. **(d)**
21. **(a)**
23. No, unit analysis can only tell if it is dimensionally correct.
25. (Length) = (Length) + (Length)/(Time)× (Time) = (Length) + (Length)
27. $m^2 = (m)^2 = m^2$
29. no, $V = \pi d^3/6$
31. a: 1/m; b: dimensionless; c: m
33. kg/m^3
35. The first student, because $m/s = \sqrt{(m/s^2)(m)} = \sqrt{m^2/s^2} = m/s$.
37. **(a)** $kg \cdot m^2/s$ **(b)** The unit of $L^2/(2mr^2)$ is $(kg \cdot m^2/s)^2/(kg \cdot m^2) = kg \cdot m^2/s^2$, which is the unit of kinetic energy, K. **(c)** $kg \cdot m^2$
39. **(c)**
41. **(c)**
43. Yes, whether you multiply or divide should be consistent with units.
45. 39.6 m
47. 37 000 000 times
49. **(a)** 91.5 m by 48.8 m **(b)** 27.9 cm to 28.6 cm
51. 0.78 mi
53. **(a)** (1) 1 m/s **(b)** 33.6 mi/h
55. **(a)** 77.3 kg **(b)** 0.0773 m^3 or about 77.3 L
57. 6.5×10^3 L/day
59. **(a)** 59.1 mL **(b)** 3.53 oz
61. 6.1 cm
63. **(a)** 1.5×10^5 m^3 **(b)** 1.5×10^8 kg **(c)** 3.3×10^8 lb
65. **(a)**
67. **(b)**
69. No, there is always one doubtful digit, the last digit.
71. 5.05 cm; 5.05×10^{-1} dm; 5.05×10^{-2} m.
73. **(a)** 4 **(b)** 3 **(c)** 5 **(d)** 2
75. **(b)** and **(d)**. **(a)** has four and **(c)** has six.
77. 32 ft^3
79. **(a)** (2) three, since the height has only three significant figures. **(b)** 469 cm^2
81. **(a)** (1) zero, since 38 m has zero decimal place. **(b)** 15 m.
83. **(a)**
85. **(d)**
87. all six steps as listed in the chapter
89. The accuracy of the answer is expected to be within an order of 10.
91. 100 kg
93. **(a)** 8.72×10^{-3} cm **(b)** about 10^{-2} cm
95. 0.87 m
97. same area for both, 1.3 cm^2
99. 3.5×10^{10} white cells, 1.3×10^{12} platelets
101. 4.7×10^2 lb
103. about 10^{12} m^3
105. 17 m
107. **(a)** 283 mi **(b)** 45° north of east
109. 3.0×10^8 km
111. **(a)** (3) less than the 190 mi/h because more time spent at lower speeds, so affect average speed to be below average of all speeds. **(b)** 187 mi/h

Chapter 2

1. **(a)**
3. **(c)**
5. Yes, for a round-trip. No; distance is always greater than or equal to the magnitude of displacement.
7. The distance traveled is greater than or equal to 300 m. The object could travel a variety of ways as long as it ends up at 300 m north. If the object travels straight north, then the minimum distance is 300 m.
9. Yes, this is possible. The jogger can jog in the opposite direction during the jog (negative instantaneous velocity) as long as the overall jog is in the forward direction (positive average velocity).
11. 1.65 m down
13. **(a)** 0.50 m/s **(b)** 8.3 min
15. 0.17 m/s
17. **(a)** (2) greater than R but less than $2R$ **(b)** 71 m
19. **(a)** (3) between 40 m and 60 m **(b)** 45 m at 27° west of north
21. **(a)** 2.7 cm/s **(b)** 1.9 cm/s
23. **(a)** 90.6 ft at 6.3° above horizontal **(b)** 36.2 ft/s at 6.3° **(c)** Average speed depends on the total path length, which is not given. The ball might take a curved path.
25. **(a)** $\bar{s}_{0-2.0\,s} = 1.0$ m/s; $\bar{s}_{2.0\,s-3.0\,s} = 0$; $\bar{s}_{3.0\,s-4.5\,s} = 1.3$ m/s; $\bar{s}_{4.5\,s-6.5\,s} = 2.8$ m/s; $\bar{s}_{6.5\,s-7.5\,s} = 0$; $\bar{s}_{7.5\,s-9.0\,s} = 1.0$ m/s **(b)** $\bar{v}_{0-2.0\,s} = 1.0$ m/s; $\bar{v}_{2.0\,s-3.0\,s} = 0$; $\bar{v}_{3.0\,s-4.5\,s} = 1.3$ m/s; $\bar{v}_{4.5\,s-6.5\,s} = -2.8$ m/s; $\bar{v}_{6.5\,s-7.5\,s} = 0$; $\bar{v}_{7.5\,s-9.0\,s} = 1.0$ m/s **(c)** $v_{1.0\,s} = \bar{s}_{0-2.0\,s} = 1.0$ m/s; $v_{2.5\,s} = \bar{s}_{2.0\,s-3.0\,s} = 0$; $v_{4.5\,s} = 0$; $v_{6.0\,s} = \bar{s}_{4.5\,s-6.5\,s} = -2.8$ m/s **(d)** $v_{4.5\,s-9.0\,s} = -0.89$ m/s
27. 1 month
29. **(a)** 500 km at 37° east of north **(b)** 400 km/h at 37° east of north **(c)** 560 km/h **(d)** Since speed involves total distance, which is greater than the magnitude of the displacement, the average speed does not equal the magnitude of the average velocity.
31. **(d)**
33. **(c)**
35. Yes, although the speed of the car is constant, its velocity is not, because of the change in direction. A change in velocity signifies an acceleration.
37. Not necessarily. A negative acceleration can also speed up objects if the velocity is also negative (that is, in the same direction as the acceleration).
39. v_o. Since an equal amount of time is spent on acceleration and deceleration of the same magnitude.
41. 6.9 m/s^2
43. **(a)** (2) opposite to velocity as the object slows down **(b)** −2.2 m/s each second, opposite direction of velocity
45. −2.0 m/s^2
47. −70.0 km/h or −19.4 m/s, +2.78 m/s^2 (deceleration because the velocity is negative)
49. 4.8×10^2 m/s^2, this is a very large acceleration due to the change in direction of the velocity and the short contact time.
51. **(a)** $\bar{a}_{0-1.0\,s} = 0$; $\bar{a}_{1.0-s\text{-}3.0\,s} = 4.0$ m/s^2; $\bar{a}_{3.0\,s-8.0\,s} = -4.0$ m/s^2; $\bar{a}_{8.0\,s-9.0\,s} = 8.0$ m/s^2; $\bar{a}_{9.0\,s-13.0\,s} = 0$ **(b)** Constant velocity of −4.0 m/s
53. 150 s
55. **(d)**
57. It is zero because the velocity is a constant.
59. Consider the displacement $(x - x_o)$ as one quantity; there are four quantities involved in each kinematic equation (Eqs. 2.8, 2.10, 2.11, and 2.12). So three must be known before one can solve for any unknown. Or equivalently all but one have to be known.
61. no, acceleration must be 9.9 m/s^2
63. **(a)** 1.8 m/s^2 **(b)** 6.3 s
65. **(a)** 81.4 km/h **(b)** 0.794 s
67. 3.09 s and 13.7 s. The 13.7 s answer is physically possible but not likely in reality. After 3.09 s, it is 175 m from where the reverse thrust was applied, but the rocket keeps traveling forward while slowing down. Finally it stops. However, if the reverse thrust is continuously applied (which is possible, but not likely), it will reverse its direction and be back to 175 m from the point where the initial reverse thrust was applied; a process that would take 13.7 s.
69. no, $a = 3.33$ $m/s^2 < 4.90$ m/s^2
71. 2.2×10^5 m/s^2
73. no, 13.3 m > 13 m
75. **(b)** 96 m
77. **(a)** (3) $v_1 > \frac{1}{2}v_2$ **(b)** 9.22 m/s, 13.0 m/s
79. **(a)** −12 m/s; −4.0 m/s **(b)** −18 m **(c)** 50 m
81. **(a)** 12.2 m/s, 16.4 m/s **(b)** 24.8 m **(c)** 4.07 s
83. **(d)**
85. **(c)**
87. **(c)**
89. The ball moves with a constant velocity because there is no gravitational acceleration in deep space. If the gravitational acceleration is zero, $g = 0$, then v = constant.
91. First of all, the gravitational acceleration on the Moon is only 1/6 of that on the Earth.

Or $g_M = g_E/6$. Secondly, there is no air resistance on the Moon.

93. **(a)** (3) four times, height is proportional to the time squared. **(b)** 15.9 m, 3.97 m

95. no, not a good deal (0.18 s $<$ 0.20 s)

97. 67 m

99. $\Delta t = 0.096$ s

101. **(a)** (1) less than 95%, as the height depends on the initial velocity squared **(b)** 3.61 m

103. **(a)** 1.64 m/s^2 **(b)** 2.07 m/s

105. **(a)** 5.00 s **(b)** 36.5 m/s

107. 1.49 m above the top of the window

109. **(a)** 155 m/s **(b)** 2.22×10^3 m **(c)** 28.7 s

111. **(a)** 8.45 s **(b)** $x_M = 157$ m; $x_C = 132$ m **(c)** 13 m

113. **(a)** 38.7 m/s **(b)** 15.5 s **(c)** 19.2 m/s

115. **(a)** 119 m **(b)** 4.92 s **(c)** Lois: 48.2 m/s; Superman: 73.8 m/s

117. **(a)** −297 m/s **(b)** 3.66 m/s^2 **(c)** 108 s

Chapter 3

1. **(a)**

3. **(c)**

5. Yes, this is possible. For example, if an object is in circular motion, the velocity (along a tangent) is perpendicular to the acceleration (toward the center of the circle).

7. **(a)** (1) greater then because for $\theta < 45°$, $\cos\theta > \sin\theta$ and $v_x = v\cos\theta$ and $v_y = v\sin\theta$. **(b)** 28 m/s, 21 m/s

9. ± 6.3 m/s, there are two possible answers because the vector could be either in the first or the fourth quadrant.

11. **(a)** (2) north of east **(b)** 1.1×10^2 m, 27° north of east

13. $x = 1.75$ m, $y = -1.75$ m

15. **(a)** 75.2 m **(b)** 99.8 m

17. **(a)** $\theta = 56.3°$ below horizontal **(b)** 18.0 m/s

19. **(a)** 1.2 m/s **(b)** 49 m

21. **(c)**

23. **(d)**

25. Yes, when the vector is in the y-direction, it has a zero x-component.

27. Yes, if equal and opposite

29.

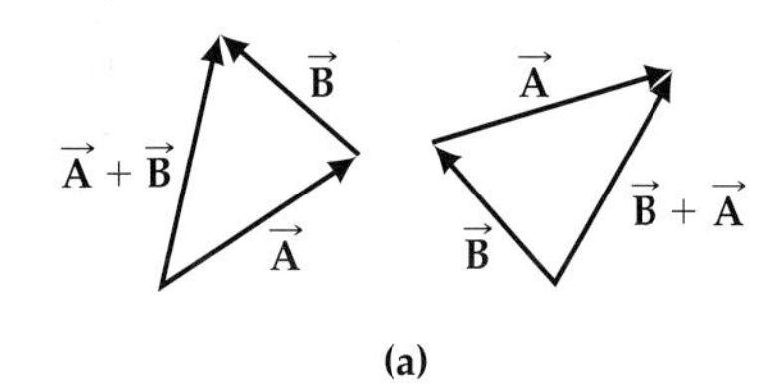

(a)

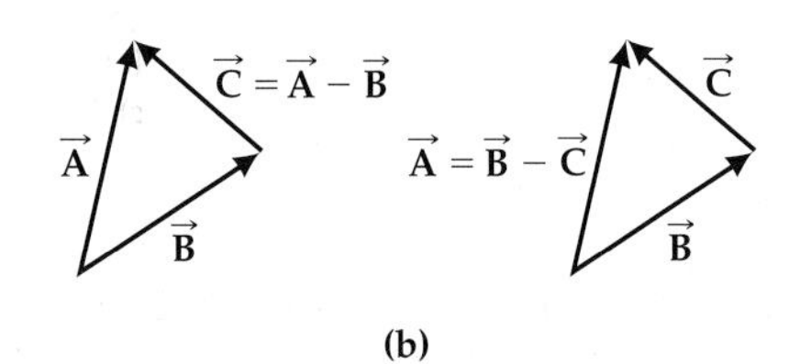

(b)

31. 4.9 m, 59° above $-x$-axis

33. 113 mi/h

35. **(a)** (−3.4 cm) $\hat{x}$ + (−2.9 cm) $\hat{y}$ **(b)** 4.5 cm, 63° above $-x$-axis **(c)** (4.0 cm) $\hat{x}$ + (−6.9 cm) $\hat{y}$

37. **(a)** (14.4 N) $\hat{y}$ **(b)** 12.7 N at 85.0° above $+x$-axis

39. **(a)** $\vec{v}_2 = (-4.0 \text{ m/s})\,\hat{x} + (8.0 \text{ m/s})\,\hat{y}$ **(b)** 8.9 m/s

41. 21 m/s at 51° below the $+x$-axis

43. **(a)** (−9.0 cm) $\hat{x}$ + (6.0 cm) $\hat{y}$ **(b)** 33.7°, relative to $-x$ axis

45. 8.5 N at 21° below $-x$-axis

47. parallel 30 N; perpendicular 40 N

49. The forces act on different objects (one on horse, one on cart) and therefore cannot cancel.

51. **(a)** (2) north of west **(b)** 102 mi/h at 61.1° north of west

53. **(a)** 42.8° south of west **(b)** 0.91 m **(c)** The reason is due to the fact that the ball might travel in a curve.

55. **(b)**

57. **(b)**

59. The horizontal motion does not affect the vertical motion. The vertical motion of the ball projected horizontally is identical to that of the ball dropped.

61. **(a)** 0.64 s **(b)** 0.64 m

63. 6.4 m

65. 40 m

67. **(a)** (2) Ball B collides with ball A because they have the same horizontal velocity. **(b)** 0.11 m, 0.11 m

69. **(a)** 0.77 m **(b)** ball would not fall back in.

71. 35° or 55°

73. 3.65 m/s^2

75. **(a)** 26 m **(b)** 23 m/s at 68° below horizontal

77. 1.4°

79. yes, at $x = 15$ m, $y = 0.87$ m $<$ 1.2 m

81. 8.7 m/s

83. The pass is short.

85. **(a)**

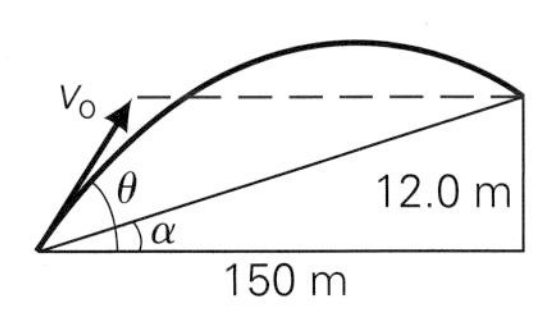

(b) 66.0 m/s **(c)** The shot is too long for the hole.

87. **(d)**

89. No, the Earth undergoes several movements such as orbiting the Sun and rotating about itself.

91. Since the rain is coming down at an angle relative to you, you should hold the umbrella so it is tilted forward.

93. You throw the ball straight up. This way, both you and the ball have the same horizontal velocity relative to the ground or you have zero horizontal velocity relative to each other, so the object returns to your hand.

95. 6.7 s

97. **(a)** +85 km/h **(b)** −5 km/h

99. 146 s = 2.43 min

101. **(a)** Same both ways, 4.25° upstream **(b)** 44.6 s

103.

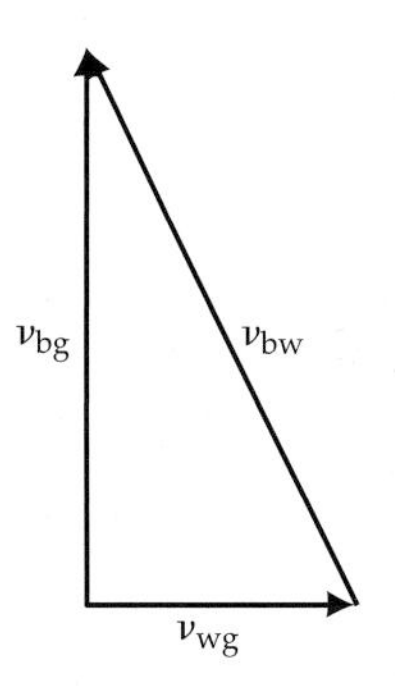

Use the following subscripts: b = boat, w = water, and g = ground. For the boat to make the trip straight across, v_{bw} must be the hypotenuse of the right-angle triangle. So it must be greater in magnitude than v_{wg}. So if the reverse is true, that is, if $v_{wg} > v_{bw}$, the boat cannot make the trip directly across the river.

105. 1.21 m/s

107. **(a)** 24° east of south **(b)** 1.5 h

111. **(a)** 47.8 s **(b)** 5.32×10^3 m **(c)** 310 m/s

Chapter 4

1. **(d)**

3. **(c)**

5. **(d)**

7. According to Newton's first law, your tendency is to remain at rest or move with constant velocity. However, the plane is accelerating to a velocity faster than yours so you are "behind" and feel "pushed" into the seat. The seat actually supplies a forward force to accelerate you to the same velocity as the plane.

9. **(a)** The bubble moves forward in the direction of velocity or acceleration, because the inertia of the liquid will resist the forward acceleration. So the bubble of negligible mass or inertia moves forward relative to the liquid. Then it moves backward opposite the velocity (or in the direction of acceleration) for the same reason. **(b)** The principle is based on the liquid inertia.

11. According to Newton's first law, or the law of inertia, the dishes at rest tend to remain at rest. The quick pull of the tablecloth required a force that exceeds the maximum static friction (discussed in Section 4.6) so the cloth can move relative to the dishes.

13. 0.40 kg

15. 0.64 m/s^2

17. **(a)** (3) the upward force is the same in both situations. In both situations, there is no acceleration in the vertical direction so the net force in the vertical direction is zero or the normal force is equal to the weight. **(b)** 0.50 lb

19. **(a)** (3) either (1) or (2) is possible, because "at rest" or "constant velocity" both have zero acceleration. **(b)** yes, 2.5 N at 36° above the $+x$ axis

21. **(a)** no **(b)** $F_5 = 4.1$ N at 13° above the $-x$ axis

23. **(b)**

25. **(d)**

27. There will be extra acceleration. A pickup truck in snow (mass increases) will have less acceleration due to the extra mass and launched rocket (mass decreases) will have greater acceleration.

29. "Soft hands" here result in longer contact time between the ball and the hands. The increase in contact time decreases the magnitude of acceleration. From Newton's second law, this, in turn, decreases the force required to stop the ball and its reaction force, the force on the hands.

31. 1.7 kg

33. **(a)** (3) 6.0 kg, because mass is a measure of inertia, and it does not change. **(b)** 9.8 N

35. **(a)** (1) on the Earth. 1 lb is equivalent to 454 g, or 454 g has a weight of 1 lb. **(b)** 5.4 kg (2.0 lb)

37. **(a)** (4) one-fourth as great. **(b)** 4.0 m/s^2

39. 2.40 m/s^2

41. **(a)** 30 N **(b)** −4.60 m/s^2

43. 8.9×10^4 N

45. Lois is saved.

47. **(c)**

49. The forces act on different objects (one on horse, one on cart) and therefore cannot cancel.

51. **(a)** (2) two forces acting on the book, the gravitational force (weight, w) and the normal force by the surface, N. **(b)** The reaction of w is an upward force on the Earth by the book, and the reaction force of N is a downward force on the horizontal surface by the book.

53. **(a)** (3) the force the blocks exerted forward on him. **(b)** 3.08 m/s^2

55. **(a)** (4) the pull of the rope on the girl. The pull of the rope on the girl is the reaction of the pull of the girl on the rope. **(b)** 264 N

57. **(d)**

59.

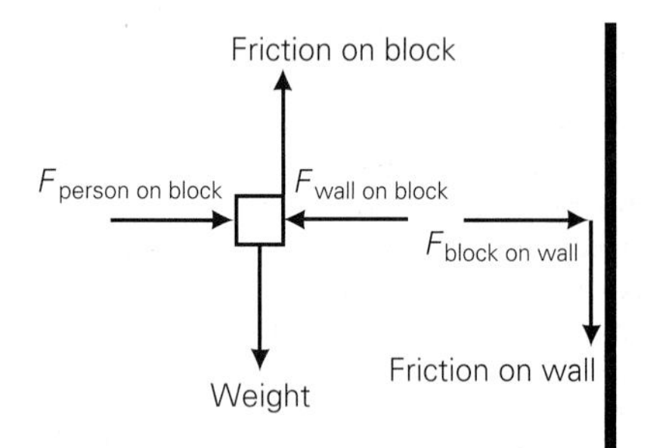

The force on the block by the wall $F_{\text{wall on block}}$ and the force on the wall by the block $F_{\text{block on wall}}$ are an action-reaction pair. The friction on the block and on the wall are also an action-reaction pair.

61. **(a)** (1) less than the weight of the object, $N = w \cos\theta < w$ (for any $\theta \neq 0$). **(b)** 98 N and 85 N

63. 585 N

65. **(a)** 1.7 m/s^2 at 19° north of east **(b)** 1.2 m/s^2 at 30° south of east

67. **(a)** 0.96 m/s^2 **(b)** 2.6×10^2 N

69. 123 N up the incline

71. 64 m

73. **(a)** (3) both the tree separation and sag. **(b)** 6.1×10^2 N

75. 2.63 m/s^2

77. 6.25×10^5 N

79. 2.0 m/s^2

81. 1.1 m/s^2 up

83. **(a)** (1) $T > w_2$ and $T < F$ **(b)** 1.70×10^3 N **(c)** 1.13×10^3 N

85. **(a)** 1.2 m/s^2, m_1 up and m_2 down **(b)** 21 N

87. **(c)**

89. **(b)**

91. This is because kinetic friction (sliding) is less than static friction (rolling). A greater friction force can decrease the stopping distance.

93. **(a)** No, there is no inconsistency. Here the friction force OPPOSES slipping. **(b)** Wind can increase or decrease air friction depending on wind directions. If wind is in the direction of motion, friction decreases, and vice versa.

95. **(a)** (3) increase, but more than double. There is constant friction involved in this exercise so the net force more than doubles. So the acceleration is more than double. **(b)** 7.0 m/s^2

97. 2.7×10^2 N

99. **(a)** $\theta_{min} = \tan^{-1} 0.65 = 33° > 20°$, so it will not move.

101. 0.064

103. **(a)** (2) pulling at the same angle. **(b)** 296 N; 748 N

105. **(a)** 30° **(b)** 22°

107. **(a)** 6.0 kg **(b)** 1.2 m/s^2

109. **(a)**

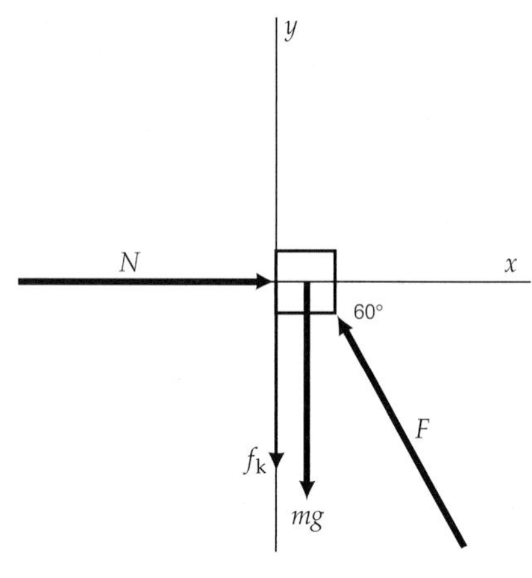

(b) 30 N **(c)** 23 N

111. **(a)** 0.179 kg **(b)** 0.862 m/s^2

113. **(a)** (4) the force of static friction on A due to the top surface of B. **(b)** 5.00 N, 17.5 N

115. **(a)** 5.5 m/s^2 **(b)** 173 N

117. **(a)** 10.3 N **(b)** 0.954 kg

Chapter 5

1. **(d)**

3. **(b)**

5. **(a)** No, the weight is not moving, so there is no displacement and therefore, no work. **(b)** Yes, positive work is done by the force exerted by the weightlifter. **(c)** No, as in (a), no work is done. **(d)** Yes, but the positive work is done by gravity, not the weightlifter.

7. Positive on the way down and negative on the way up. No, it is not constant.

9. −98 J

11. 8.31 m

13. 3.7 J

15. 2.3×10^3 J

17. **(a)** (2) one, the only force that does nonzero work is the kinetic friction force. **(b)** −62.5 J

19. **(a)** 1.48×10^5 J **(b)** -1.23×10^5 J **(c)** 2.50×10^4 J.

21. **(a)** Pulling requires the student to do (1) less work. Compared with pushing, pulling decreases the normal force on the crate. This, in turn, decreases the kinetic friction force. **(b)** pulling: 1.3×10^3 J; pushing: 1.7×10^3 J

23. No, it takes more work. This is because the force increases as the spring stretches, according to Hooke's law. Also the displacement is greater.

25. 80 N/m

27. 1.25×10^5 N/m

29. **(a)** (1) $\sqrt{2}$, because when W doubles, x becomes $\sqrt{2}$ as much. Therefore, it will stretch by a factor of $\sqrt{2}$. **(b)** 900 J

31. **(a)** (1) more than, because the force required is greater (while the displacement is the same) to stretch from 10 cm to 20 cm, according to Hooke's law. **(b)** 0–10 cm: 0.25 J; 10–20 cm: 0.75 J

33. **(a)** 4.5 J **(b)** 3.5 J

35. 6.0 J

37. **(d)**

39. **(c)**

41. Reducing speed by half will reduce kinetic energy by $\frac{3}{4}$, whereas reducing the mass by half will only reduce kinetic energy by half.

43. $\sqrt{2}\,v$

45. -1.3×10^3 N

47. **(a)** 45 J **(b)** 21 m/s

49. 200 m

51. 2.0×10^3 m

53. **(d)**

55. **(d)**

57. They will have the same potential energy at the top because they have the same height.

59. **(a)** (4) only the difference between the two heights, because the change in potential energy depends only on the height difference, not positions. **(b)** position is lowered 0.51 m

61. **(a)** (4) all the same, because the change in potential energy is independent of the reference level. **(b)** $U_b = -44$ J, $U_a = 66$ J **(c)** -1.1×10^2 J

63. 1/25

65. **(a)** 0.154 J **(b)** −0.309 J

67. **(d)**

69. The initial potential energy is equal to the final potential energy so the final height is equal to the initial height.

71. Yes. When the thrown-up ball is at its maximum height, its velocity is zero so it has the same energy as the dropped ball. Since the total energy of each ball is conserved, both balls will have the same mechanical energy at half the height of the window. As a matter of fact, both balls will have the same kinetic and potential energy at the half height position.

73. 5.10 m

75. **(a)** (3) at the bottom of the swing, because the lower the potential energy (at the bottom is the lowest), the higher the kinetic energy, therefore speed. **(b)** 5.42 m/s
77. 0.176 m
79. **(a)** 1.03 m **(b)** 0.841 m **(c)** 2.32 m/s
81. **(a)** 11 m/s **(b)** no **(c)** 7.7 m/s
83. **(a)** 2.7 m/s **(b)** 0.38 m **(c)** 29°
85. -7.4×10^2 J
87. 12 m/s
89. **(b)**
91. No, paying for energy because kWh is the unit of power × time = energy. 9.0×10^6 J
93. They are doing the same amount of work (same mass, same height). So the one that arrives first will have expended more power due to shorter time interval.
95. 97 W
97. 5.7×10^{-5} W
99. 6.0×10^2 J
101. **(a)** 5.5×10^2 W **(b)** 0.74 hp
103. 48.7%
105. 5.0×10^2 W
107. 0.536
109. 10 m

Chapter 6

1. **(b)**
3. **(d)**
5. Not necessarily. Even if the momentum is the same, different mass can still have different kinetic energies.
7. **(a)** 1.5×10^3 kg · m/s **(b)** zero
9. **(a)** 85 kg · m/s **(b)** 3.0×10^4 kg · m/s
11. 31 m/s
13. 4.05 kg · m/s in the direction opposite v_o
15. **(a)** 13 kg · m/s **(b)** 43 kg · m/s
17. $\Delta\vec{\mathbf{p}} = (-3.0 \text{ kg}\cdot\text{m/s})\,\hat{\mathbf{y}}$
19. 16 N
21. 68 N
23. **(a)** 10.6 kg · m/s opposite v_o **(b)** 2.26×10^3 N
25. **(a)** 2.09 kg · m/s **(b)** 24.0 N upward
27. **(c)**
29. By stopping, the contact time is short. From the impulse momentum theorem ($F_{avg}\Delta t = \Delta p = mv - mv_o$), a shorter contact time will result in a greater force if all other factors (m, v_o, v) remain the same.
31. In (a), (b), and (c), it is to have greater contact time—less average force. This is because ($F_{avg}\Delta t = \Delta p = mv - mv_o$ so the greater the Δt, the less the $\overline{F}$. It can also decrease the pressure on the body because the force is spread over a larger area.
33. 6.0×10^3 N
35. **(a)** 1.2×10^3 N **(b)** 1.2×10^4 N
37. **(a)** (2) the driver putting on the brakes, because it is to reduce the speed. **(b)** 28.7 m/s
39. **(a)** Hitting it back requires a greater force. Force is proportional to the change in momentum. When a ball changes its direction, the change in momentum is greater. **(b)** 1.2×10^2 N in direction opposite v_o
41. 1.1×10^3 N, 4.7×10^2 N
43. 15 N upward
45. **(a)** 36 km/h, 56° north of east **(b)** 49% lost
47. 8.06 kg · m/s, 29.7° above the $-x$ axis
49. **(d)**
51. Air moves backward and the boat moves forward according to momentum conservation. If a sail were installed behind the fan on the boat, the boat would not go forward because the forces between the fan and the sail are internal forces of the system.
53. No, it is impossible. Before the hit, there is some initial momentum of the two-object system due to the one moving. According to momentum conservation, the system should also have momentum after the hit. Therefore it is not possible for both to be at rest (zero total system momentum).
55. moves 0.083 m/s in opposite direction
57. 1.08×10^3 s = 18.1 min
59. 7.6 m/s, 12° above $+x$-axis
61. **(a)** 45 km/h **(b)** 15 km/h **(c)** 105 km/h
63. 0.33 m/s
65. 0.78 m/s
67. **(a)** 4.0 m/s **(b)** 4.0 m/s
69. 82.8 m
71. **(a)** 9.7° **(b)** 0.10 m/s
73. **(c)**
75. **(a)**
77. This is due to the fact that momentum is a vector and kinetic energy is a scalar. For example, two objects of equal mass traveling with the same speed in opposite directions have positive total kinetic energy but zero total momentum. After they collide inelastically, both stop, resulting in zero total kinetic energy and zero total momentum. Therefore, kinetic energy is lost and momentum is conserved.
79. Yes, it can. For example, when two objects of equal masses are approaching each other with equal speeds, the total initial momentum is zero. After they collide, if they both are stationary, the total momentum is still zero. However, after the collision, the two-object system has no kinetic energy left.
81. $v_p = -1.8 \times 10^6$ m/s; $v_a = 1.2 \times 10^6$ m/s
83. $v_1 = -0.48$ m/s; $v_2 = +0.020$ m/s
85. $v_p = 38.2$ m/s, $v_c = 40.2$ m/s
87. 0.94 m
89. 1.1×10^2 J
91. **(a)** (1) south of east, according to momentum conservation. The initial momentum of the minivan is to the south, and the initial momentum of the car is to the east, so the two-vehicle system has a total momentum to the southeast after the collision. **(b)** 13.9 m/s, 53.1° south of east
93. no, 1.1 kg · m/s, 249°
95. **(a)** 28% **(b)** 1.3×10^7 m/s
97. $\theta_1 = 9.94°$, $\theta_2 = 19.8°$
99. **(d)**
101. The flamingo's center of mass is directly above the foot on the ground for it to be in equilibrium.
103. **(a)** (0, −0.45 m) **(b)** no, only that they are equidistant from CM due to the equal masses of the particles.
105. **(a)** 4.6×10^6 m from the center of the Earth **(b)** 1.8×10^6 m below the surface of the Earth
107. 82.8 m
109. The CM of both the square sheet and the circle are at the center of the square. So from symmetry, the CM of the remaining portion is still at center of sheet.
111. 0.175 m
113. **(a)** the 65 kg travels 3.3 m and the 45 kg travels 4.7 m **(b)** same distances as in part (a)
115. massive end: 5.59 m; less massive end: 6.19 m
117. **(a)** 0.222° **(b)** 199 m/s
119. **(a)** 4.65 m/s, $\theta = 2.13°$ **(b)** $K < K_o$, inelastic

Chapter 7

1. **(c)**
3. 2π rad = 360°, so 1 rad = 57.3°
5. (2.5 m, 53°)
7. 1.4×10^9 m
9. **(a)** 30° **(b)** 75° **(c)** 135° **(d)** 180°
11. **(a)** 4.00 rad **(b)** 229°
13. 10.7 rad
15. **(a)** 129° **(b)** about 1.4×10^4 km
17. 28.3 m
19. **(a)** 3.0×10^2 rad **(b)** 91 m
21. **(b)**
23. **(d)**
25. Viewing from opposite sides would give different circular senses, i.e., make clockwise counterclockwise and vice versa.
27. 21 rad/s to 47 rad/s
29. 1.8 s
31. particle B
33. **(a)** 0.84 rad/s **(b)** 3.4 m/s, 4.2 m/s
35. **(a)** The rotating angular speed is greater because the time is less (angular displacement is the same). **(b)** 7.27×10^{-5} rad/s for rotation, 1.99×10^{-7} rad/s for revolution
37. **(a)** 60 rad **(b)** 3.6×10^3 m
39. **(d)**
41. **(d)**
43. The floats of the little mass will move in direction of acceleration, inward. It works the same way as the accelerometer in Fig. 4.24. No, it does not make a difference since the centripetal acceleration is always inward.
45. Centripetal force is required for a car to maintain its circular path. When a car is on a banked turn, the horizontal component of the normal force on the car is pointing toward the center of the circular path. This component will enable the car to negotiate the turn even when there is no friction.
47. 1.3 m/s
49. 2.69×10^{-3} m/s²
51. 11.3°
53. **(a)** weight is supplying the centripetal force **(b)** 3.1 m/s
55. 29.5 N, string will work
57. **(a)** $v = \sqrt{rg}$ **(b)** $h = (5/2)r$
61. **(d)**
63. Yes, a car in circular motion always has centripetal acceleration. Yes, it also has angular acceleration as its speed is increasing.
65. The answer is no. When the tangential acceleration increases, the tangential speed

increases, resulting in an increase in centripetal acceleration.

67. 1.1×10^{-3} rad/s^2

69. **(a)** (3) both angular and centripetal accelerations. There is always centripetal acceleration for any car in circular motion. When the car increases its speed on a circular track, there is also angular acceleration. **(b)** 53 s **(c)** $\vec{a} = -(8.5 \text{ m/s}^2)\,\hat{r} + (1.4 \text{ m/s}^2)\,\hat{t}$

71. 6.69 rad/s^2

73. **(a)** 2.45 rad/s^2 **(b)** 17.0 m/s^2 **(c)** 38.2 N

75. **(d)**

77. No, these terms are not correct. Gravity acts on the astronauts and the spacecraft, providing the necessary centripetal force for the orbit, so g is not zero and there is weight by definition ($w = mg$). The "floating" occurs because the spacecraft and astronauts are "falling" ("accelerating" toward Earth at the same rate).

79. Yes, if you also know the radius of the Earth. The acceleration due to gravity near the surface of the Earth can be written as $a_g = GM_E/R_E^2$. By simply measuring a_g, you can determine $M_E = a_g R_E^2/G$.

81. 2.0×10^{20} N

83. 8.0×10^{-10} N, toward opposite corner

85. 3.4×10^5 m

87. 1.5 m/s^2

89. **(a)** -2.5×10^{-10} J **(b)** 0

91. **(c)**

93. **(d)**

95. **(a)** Rockets are launched eastward to get more velocity relative to space because the Earth rotates toward the east. **(b)** The tangential speed of the Earth is higher in Florida because Florida is closer to the equator than California hence a greater distance from the axis of rotation. Also, the launch is over the ocean for safety.

97. **(a)** 3.7×10^3 m/s **(b)** 34%

99. 4.4×10^{11} m

101. 1.53×10^9 m

103. **(a)** 8.91 m/s^2, toward center **(b)** 1.40×10^4 N, toward car **(c)** 2.27 m/s^2, opposite velocity **(d)** 9.19 m/s^2, $\theta = 75.7°$

105. 31 s **(b)** 19 rev

107. 2.97×10^{30} kg

Chapter 8

1. **(a)**

3. **(b)**

5. **(b)**

7. If v is less than $R\omega$, the object is slipping. Yes, it is possible for v to be greater than $R\omega$ when the object is sliding.

9. At the nine-o'clock position, the velocity is straight upward. So it is a "free-fall" with an initial upward velocity. It will rise, reach a maximum height, and then fall back down.

11. 0.10 m

13. 1.7 rad/s

15. **(a)** 0.0331 rad/s^2 **(b)** 1.99×10^{-3} m/s^2

17. **(b)**

19. **(a)**

21. This is to lower the center of gravity so the weight will have a shorter lever arm therefore smaller torque.

23. Frictional torque causes the motorcycle to rotate upward until balanced by the torque of weight.

25. 5.6×10^2 N

27. 3.3×10^2 N

29. **(a)** Yes, the seesaw can be balanced if the lever arms are appropriate for the weights of the children because torque is equal to force times the lever arm. **(b)** 2.3 m

33. 1.6×10^2 N

35. **(a)** clockwise. **(b)** 4.66 m · N

37. 0, 9.80 m · N, 17.0 m · N, 19.6 m · N

39. **(a)** (2) toward the scale at the person's head, because the mass distribution of the human body is more toward the upper body than the lower body. **(b)** 0.87 m from the feet

41. Yes. The center of gravity of every stick is at or to the left of the edge of the table.

43. 1.2 m from left end of board

45. $T_1 = 21$ N, $T_2 = 15$ N

47. **(a)** (2) the board with the clown on it. At first glance, there seem to be two factors resulting in a greater tension in the rope. One is the extra weight of the crown. The other is the shortened lever arm of the tension in the rope when it is at an angle. However, the weight of the board and crown will also have shortened the lever arms by the same factor, so the effects cancel. **(b)** 172 N, 539 N

49. **(d)**

51. **(a)**

53. **(a)** Yes. Moment of inertia has a minimum value calculated about the center of mass. **(b)** No, mass would have to be negative.

55. The hard-boiled egg is a rigid body, while the raw egg is not.

57. This is to increase the moment of inertia. If walker starts to rotate (fall), the angular acceleration will be smaller, thus giving more time to recover.

59. 0.64 m · N

61. **(a)** 2.4 kg · m^2 **(b)** 0.27 kg · m^2 **(c)** 2.4 kg · m^2 (same)

63. 1.1 rad

65. 1.2 m/s^2

67. **(a)** 2.93 m · N **(b)** 58.7 N

69. **(a)** $1.5g$ **(b)** 67-cm position

71. 6.5 m/s^2

73. **(a)** (3) $7\mu_s/2$ **(b)** 63.8°

75. **(c)**

77. Yes. The rotational kinetic energy depends on the moment of inertia, which depends on both the mass and the mass distribution. Translational kinetic energy depends only on the mass.

79. According to the work-energy theorem, rotational work is required to produce a change in rotational kinetic energy. Rotational work (W) is done by a torque (τ) acting through an angular displacement (θ).

81. **(a)** 28 J **(b)** 14 W

83. 0.47 m · N

85. 0.16 m

87. 78.5 N

89. cylinder goes higher by 7.1%

91. **(a)** 1.31×10^8 J **(b)** 1.46×10^6 W

93. **(a)** 29% **(b)** 40% **(c)** 50%

95. **(a)** $v = \sqrt{gR}$ **(b)** $h = 2.7R$ **(c)** weightlessness

97. **(d)**

99. Walking toward the center decreases the moment of inertia and so increases the rotational speed.

101. In each case, the change in the wheel's angular momentum vector is compensated by the rotation of the person to conserve the total angular momentum, so the vertical angular momentum remains constant.

103. **(a)** zero **(b)** Linear kinetic energy is converted into rotational kinetic energy

105. 1.4 rad/s

107. $L_{rot} = 2.4 \times 10^{29}$ kg · m^2/s; $L_{rev} = 2.8 \times 10^{34}$ kg · m^2/s

109. 1.18 rad/s

111. **(a)** 4.3 rad/s **(b)** $K = 1.1K_o$ **(c)** work done by skater

113. $d = b(v_o/v)$

115. **(a)** (2) rotate in the direction opposite that in which the cat is walking, because from angular momentum conservation, the lazy Susan will rotate in the opposite direction. **(b)** 0.56 rad/s **(c)** no, 2.1 rad

117. **(a)** 33.1 rad/s **(b)** 5.29 m/s **(c)** 28.6 rad/s, 4.58 m/s

119. 0.104 m/s

Chapter 9

1. **(c)**

3. **(d)**

5. Steel wire has a greater Young's modulus. Young's modulus is a measure of the ratio of stress over strain. For a given stress, a greater Young's modulus will have a smaller strain. Steel will have smaller strain here.

7. Through capillary action, the wooden peg absorbs water and it swells and splits the rock.

9. 3.1×10^4 N/m^2

11. **(a)** 9.4×10^4 N/m^2 **(b)** 1.2×10^5 N/m^2

13. 47 N

15. **(a)** (1) a cold day, because tracks expand when temperature increases. **(b)** 1.9×10^5 N

17. **(a)** bends toward brass, because the stresses are the same for both, and brass has a smaller Young's modulus. Brass will have a greater strain $\Delta L/L_o$, so it will be compressed more. Therefore, the brass will be shorter than the copper. **(b)** brass: $\Delta L/L_o = 2.8 \times 10^{-3}$; copper: $\Delta L/L_o = 2.3 \times 10^{-3}$

19. 4.2×10^{-7} m

21. **(a)** Ethyl alcohol has the greatest compressibility, because it has the smallest bulk modulus B. The smaller the B, the greater the compressibility. **(b)** $\Delta p_w/\Delta p_{ea} = 2.2$

23. 7.28×10^5 N/m^2

25. **(d)**

27. **(a)**

29. Bicycle tires have a much smaller contact area with the ground so they need a higher pressure to balance the weight of the bicycle and the rider.

31. **(a)** The pressure is determined by only depth so the effect is none—same depth, same pressure. **(b)** The dams are usually thicker at the bottom because of the greater pressure at greater depth.

33. **(a)** The pressure inside the can is equal to the atmospheric pressure outside. When the liquid is poured from an unvented can, a partial vacuum develops inside, and the pressure difference causes the pouring to be difficult. By opening the vent, you are allowing air to go into the can and the pressures are equalized so the liquid can be easily poured.
(b) When you squeeze a medicine dropper before inserting it into a liquid, you are forcing the air out and reducing the pressure inside the dropper. When you release the top with the dropper in a liquid, the liquid rises in the dropper due to atmospheric pressure.
(c) To inhale, the lungs physically expand, the internal pressure decreases, and air flows into the lungs. To exhale, the lungs contract, the internal pressure increases, and air is forced out.
35. **(a)** (1) a higher height than the mercury barometer due to its low density. **(b)** 10 m
37. **(a)** (1) a higher column height because it has a lower density. **(b)** 22 cm
39. 6.39×10^{-4} m^2
41. **(a)** (3) A lower pressure inside the can as steam condenses. Since there is a partial vacuum inside the can, the net force exerted by the atmospheric pressure on the outside crushes the can. **(b)** 1.6×10^4 N = 3600 lb
43. $p = -1.1 \times 10^4$ Pa. Obviously, pressure cannot be negative, so this calculation shows that the air density is not a constant, but air density decreases rapidly with altitude.
45. 0.51 N
47. 2.2×10^5 N (about 50 000 lb)
49. 1.9×10^2 m/s
51. **(a)** 1.1×10^8 Pa **(b)** 1.9×10^6 N
53. 549 N, 1.37×10^6 Pa
55. 0.173 N
57. **(c)**
59. **(a)**
61. Ice has a larger volume than water of equal mass. The level does not change. As the ice melts, the volume of the newly converted water decreases; however, the ice, which was initially above the water surface, is now under the water. This compensates for the decrease in volume. It does not matter whether the ice is hollow or not.
63. Same buoyant force, because buoyant force depends only on the volume of the fluids displaced and is independent of the mass of the object.
65. **(a)** (3) stay at any height in the fluid, because the weight is exactly balanced by the buoyant force. Wherever the object is placed, it stays at that place. **(b)** sink, $W > F_b$
67. 2.6×10^3 kg
69. no, $14.5 \times 10^3 \text{ kg/m}^3 < \rho_g = 19.3 \times 10^3 \text{ kg/m}^3$
71. **(a)** 0.09 m **(b)** 8.1 kg
73. 1.00×10^3 kg/m^3 (probably H_2O)
75. 17.7 m
77. 8.1×10^2 N
79. **(a)**
81. **(d)**
83. There are many capillaries and only a few arteries. The total area of the capillaries is greater than that of the arteries. So if A increases, v decreases.
85. **(a)** The concave bottom makes the air travel faster under the car. This increase in speed will reduce the pressure under the car. The pressure difference forces the car on to the ground more to provide a greater normal force and friction for traction. **(b)** The spoiler is to produce a downward force for better traction or more friction.
87. 0.98 m/s
89. **(a)** 3.5 cm^3/s **(b)** 0.031% **(c)** It is a physiological need. The slow speed is needed to give time for the exchange of substances such as oxygen between the blood and the tissues.
91. 53.6 Pa
93. **(a)** 0.13 m^3/s **(b)** 1.8 m/s
95. 2.2 Pa
99. **(c)**
101. The 10 and 40 measure viscosity and the "W" means winter.
103. 3.5×10^2 Pa
105. 13.5 s
107. 8.0×10^2 kg/m^3
109. **(a)** 3.83 m/s **(b)** 399

Chapter 10

1. **(b)**
3. **(a)**
5. incandescent lamp filament, up to 3000°C
7. Celsius
9. **(a)** 302°F **(b)** 90°F **(c)** −13°F **(d)** −459°F
11. **(a)** 245°F **(b)** 375°F
13. 56.7°C and −62°C
15. **(a)** (3) $T_F = T_C$, because we want to find the one temperature at which the Celsius and Fahrenheit scales have the same reading. **(b)** −40°C = −40°F
17. **(a)** −101 F° **(b)** 558 F°
19. **(a)** (2) $T_C = 0$, because $T_F = 9/5\, T_C + 32$, compared to $y = ax + b$. To find the y-intercept, we set x (T_C) = 0. **(b)** 32°F **(c)** 5/9; −18°C
21. **(a)**
23. The volume of the gas is held constant. So if the temperature increases, so does the pressure and vice versa, according to the ideal gas law. Therefore, temperature can be determined by measuring pressure.
25. Absolute zero implies zero pressure or volume. Negative absolute temperature implies negative pressure or volume.
27. The same, because mole is defined in terms of the number of molecules.
29. **(a)** −273°C **(b)** −23°C **(c)** 0°C **(d)** 52°C
31. **(a)** 53 541°F; 29 727°C **(b)** 0.910%
33. **(a)** (2) decrease, because $p_1V_1/T_1 = p_2V_2/T_2$. With $V_1 = V_2$, $T_2/T_1 = p_2/p_1$ or temperature is proportional to pressure. **(b)** 167°C
35. 1.7×10^{23}
37. 0.0370 m^3
39. 0.16 L
41. 33.4 lb/in.^{2}
43. 2.31 atm
45. **(a)** (1) increase, because with $p = p_o$, $p_oV_o/T_o = pV/T$ becomes $V/V_o = Tp_o/(T_op) = T/T_o$ or volume is proportional to temperature. **(b)** 10.6%
47. 5.1 cm^3
49. **(c)**
51. **(a)** ice moves upward **(b)** ice moves downward **(c)** copper
53. When the ball alone is heated, it expands and cannot go through the ring. When the ring is heated, it expands and the hole gets larger so the ball can go through again.
55. Metal has a higher coefficient of thermal expansion than glass. The lid expands more than glass so it is easier to loosen the lid.
57. **(a)** (1) high, because the tape shrinks. One division on the tape (it is now less than one division due to shrinkage) still reads one division. **(b)** 0.060%
59. 0.0027 cm
61. **(a)** 60.1 cm **(b)** 3.91×10^{-3} cm^2, yes
63. **(a)** (1) the ring, so it expands, then the ball can go through. **(b)** 353°C
65. 5.52×10^{-4}/C°
67. **(a)** larger, because it expands. **(b)** 5.5×10^{-6} m^3
69. **(a)** 116°C **(b)** no
71. yes, 79°C
73. **(a)**
75. The gases diffuse through the porous membrane, but the helium gas diffuses faster because its atoms have a smaller mass. Eventually, there will be equal concentrations of gases on both sides of the container.
77. **(a)** 6.07×10^{-21} J **(b)** 7.72×10^{-21} J
79. **(a)** 6.21×10^{-21} J **(b)** 1.37×10^3 m/s
81. increases by a factor of $\sqrt{2}$
83. **(a)** 1.82×10^7 J **(b)** 1.55×10^3 m
85. 273°C
87. 899°C
89. **(a)** (1) $^{235}UF_6$, because at the same temperature, the smaller the mass, the greater the average rms speed. **(b)** 1.00429
91. **(b)**
93. nRT
95. 6.1×10^3 J
97. **(a)** 1.21×10^7 J **(b)** 3.03×10^7 J
99. 0.272%
101. **(a)** (3) helium, because at the same temperature, helium has the smallest mass so the highest rms speed. **(b)** 425 m/s < 1100 m/s

Chapter 11

1. **(d)**
3. 1 Cal = 1000 cal
5. 6.279×10^6 J
7. 4
9. **(a)**
11. **(b)**
13. Since $Q = cm\Delta T = cm(T_f - T_i)$, specific heat and mass can cause the final temperature of the two objects to be different, if Q and T_i are the same.
15. **(a)** (1) more heat, because copper has a higher specific heat. **(b)** copper requires 2.1×10^4 J more
17. 1.7×10^6 J
19. **(a)** (3) less than, because aluminum has a higher specific heat than copper. From $Q = mc\Delta T$, if Q and ΔT are the same, a higher specific heat results in a lower mass. **(b)** 1.27 kg

21. 84°C
23. 0.13 kg
25. **(a)** (1) more heat than the iron, because aluminum has a higher specific heat, when all other factors (m and ΔT) are the same. **(b)** Al by 1.8×10^4 J more
27. **(a)** (1) higher, because if some water splashed out, there will be less water to absorb the heat. The final temperature will be higher and the measured specific heat value will be in error and appear to be higher than the value calculated for the case in which the water does not splash out. **(b)** 3.1×10^2 J/(kg·C°)
29. **(a)** 7.0×10^2 W **(b)** 9.4×10^2 W
31. 20.0°C
33. **(d)**
35. **(c)**
37. This is due to the high value of the latent heat of vaporization. When steam condenses, it releases 2.26×10^6 J/kg of heat. When 100°C water drops its temperature by 1 C°, it releases only 4186 J/kg.
39. 1.13×10^6 J
41. 2.5×10^6 J and 2.5×10^5 J; yes
43. 1.2×10^6 J
45. **(a)** (2) only latent heat, because the boiling point of mercury is 357°C = 630 K so it is already at the boiling temperature. **(b)** 4.1×10^3 J
47. 11°C
49. 1.8×10^{-2} kg
51. 2.1×10^5 J
53. **(a)** (2) some of the ice will melt. Answer (3) is eliminated because water at 10°C will give off enough heat to raise the temperature of the ice, and melt some of the ice because ice has a smaller specific heat than water. Answer (1) is also eliminated because of the high value of the latent heat of fusion for ice (3.3×10^5 J/kg). The 10 C° drop in temperature of the water does not release enough heat to melt the ice completely. **(b)** 0.5032 kg
55. 0.17 L
57. **(c)**
59. Metal has a higher heat conductivity, so metal conducts heat away from your hand more quickly.
61. Air is a poor heat conductor so the hollow hair minimizes heat loss.
63. 4.54×10^6 J
65. 13 J
67. **(a)** (1) longer, because copper has a higher thermal conductivity **(b)** 1.63
69. **(a)** 5.5×10^5 J/s **(b)** 73 kg; yes, see ISM
71. 411°C
73. **(a)** (3) thinner than, because glass wool has a lower thermal conductivity. **(b)** 4.9 in.
75. 1.0×10^8 J
77. **(a)** 1.5×10^3 J/s **(b)** 46 J/s
79. 2.3 cm
81. 23°C
83. 7.8 h
85. 4.0×10^2 m/s
87. 0.49 kg

Chapter 12

1. **(a)**
3. **(d)**
5. No. All it means is that the intermediate steps are non-equilibrium states so you cannot retrace the process exactly.
7. **(c)**
9. 1: isothermal expansion; 2: isobaric compression; 3: isometric pressure increase

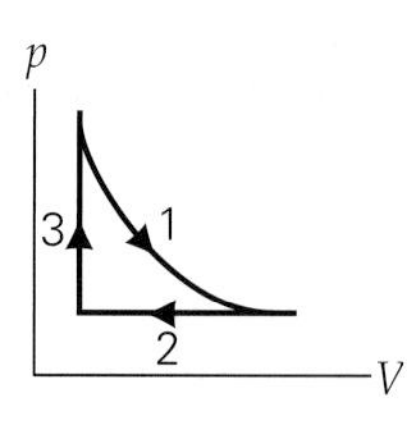

11. When you play a game of basketball, you lost heat, did work, and decreased your internal energy.
13. Work: 1, 2, 3. Work is equal to the area under the curve in p-V diagram. The area under 1 is the greatest and the area under 3 is the smallest. Final temperature: 1, 2, 3. According to the ideal gas law, the temperature of a gas is proportional to the product of pressure and volume, $pV = nRT$. Since the final volume is the same for all three processes, the higher the pressure, the higher the final temperature.
15. **(a)** (2) the same as, because $\Delta U = 0$ for a cyclic process. **(b)** added, 400 J
17. **(a)** (3) decreases, as $Q = 0$ and W is positive **(b)** −500 J
19. **(a)** 3.3×10^3 J **(b)** yes, 5.1×10^3 J
21. **(a)** (2) zero, because $\Delta T = 0$. **(b)** $-p_1V_1$ (on the gas) **(c)** $-p_1V_1$ (out of the gas)
23. 3.6×10^4 J
25. **(a)** (2) isobaric, because the pressure is maintained at 1.00 atm. **(b)** 146 J
27. **(a)** (2) process 2, because more heat is added and the change in internal energy is the same (the initial and final temperatures are the same). **(b)** $\Delta U = 2.5 \times 10^3$ J for both; $W_1 = 0$, $W_2 = 5.0 \times 10^2$ J
29. **(a)** Path AB, -1.66×10^3 J; path BC, 0; path CD, 3.31×10^3 J; path DA, 0 **(b)** $\Delta U = 0$, $Q = W = 1.65 \times 10^3$ J **(c)** 800 K
31. **(c)**
33. **(a)** increases since heat is added. **(b)** decreases since heat is removed. **(c)** increases since heat is added. **(d)** decreases since heat is removed.
35. No, this is not a valid challenge, because ice or water itself is not an isolated system. When water freezes into ice, it gives off heat and that causes the entropy of the surroundings to increase. This increase actually is more than the decrease that occurred in the water-ice phase change. So the net change in entropy of the system (ice plus water) still increases.
37. **(a)** (1) positive, because heat is added in the process (positive heat). **(b)** $+1.2 \times 10^3$ J/K
39. -2.1×10^2 J/K
41. 126°C
43. **(a)** (1) increase, because $Q > 0$. **(b)** +11 J/K
45. **(a)** (1) positive, according to the second law of thermodynamics. **(b)** +1.33 J/K
47. **(a)** (2) zero, $\Delta S = 0$ **(b)** 2.73×10^4 J
49. **(a)** 61.0 J/K **(b)** −57.8 J/K **(c)** 3.2 J/K
51. **(b)**
53. It is unchanged, because it returns to its original value for a cyclic process. This is true for many quantities such as temperature, pressure, and volume.
55. This is important because heat can be completely converted to work for a single process (not a cycle), such as an isothermal expansion process of an ideal gas.
57. No, as the warm air rises to the higher altitude, both gravity and buoyancy forces do work. Since it is a natural process with work input, the entropy increases and the second law is not violated.
59. 25%
61. 1.47×10^5 J
63. **(a)** 6.6×10^8 J **(b)** 27%
65. **(a)** (1) increases, because $\varepsilon = 1 - Q_c/Q_h$. When ε increases, the ratio of Q_c/Q_h decreases or the ratio Q_h/Q_c increases. **(b)** +0.024
67. **(a)** 6.1×10^5 J **(b)** 1.9×10^6 J
69. 3.0 kW
71. 6.0 h
73. **(a)** 1800 **(b)** 3.4×10^7 J **(c)** 2.7×10^7 J
75. **(a)**
77. The more efficient one is water cooled, because efficiency of cooling depends on ΔT, and water can maintain a large ΔT. Also water has high specific heat so it can absorb more heat.
79. Diesel engines run hotter because diesel fuel has a higher spontaneous combustion temperature. According to Carnot efficiency, the higher the hot reservoir temperature, the higher the efficiency, for a fixed low temperature reservoir.
81. 0°C
83. **(a)** 6.7% **(b)** Probably not at the moment, due to low efficiency and still relatively cheap fossil fuels.
85. 9.1×10^3 J
87. **(a)** (3) higher than 327°C. From $\varepsilon_C = 1 - T_c/T_h$, we can see that T_h must be higher for ε_C to increase while T_c is kept constant. **(b)** 427°C
89. **(a)** no, $\varepsilon_C = 57\%$ while $\varepsilon = 67\%$ **(b)** 17.5 kW
91. **(a)** 42% **(b)** 39 kW
93. 53%
95. **(a)** (2) zero, because many quantities such as temperature, pressure, volume, internal energy, and entropy return to original value after each cycle. **(b)** 3750 J
97. **(a)** 64% **(b)** ε_C is the upper limit of efficiency. In reality, a lot more energy is lost than in the ideal situation.
99. **(a)** 13 **(b)** no, $COP_C = 11$
101. 20 mi/gal
103. 2.7 kg
105. 0.157

Chapter 13

1. **(b)**
3. **(b)**
5. **(a)** four times as large **(b)** twice as large
7. $T/4$, $T/2$

9. $4A$
11. 0.025 s
13. 41 N/m
15. **(a)** 10^{-12} s **(b)** 63 m/s
17. **(a)** (1) $x = 0$, because at $x = 0$ there is no elastic potential energy, so all the energy of the system is kinetic, thus maximum speed. **(b)** 2.0 m/s
19. **(a)** 0.77 m/s **(b)** 1.2 N
21. 1.08 m/s
23. **(a)** 2.5 m/s **(b)** 2.5 m/s **(c)** 2.7 m/s, equilibrium position
25. **(a)** 17.6 N/m **(b)** 1.04 m/s
27. **(d)**
29. **(b)**
31. This could be done by tracing out the path of the object on a scrolling horizontal paper.
33. In an upward accelerating elevator, the effective gravitational acceleration increases. According to $T = 2\pi\sqrt{L/g}$, the period would be decreased.
35. 10 kg
37. **(a)** 1.7 s **(b)** 0.57 Hz
39. **(a)** $x = A \sin \omega t$ **(b)** $x = A \cos \omega t$
41. **(a)** 5.0 cm **(b)** 10 Hz **(c)** 0.10 s
43. **(a)** (3) less, because $E \propto 1/T^2$ and System A has a longer period. **(b)** 1.8×10^2
45. **(a)** 0.188 m **(b)** 3.00 m/s^2
47. **(a)** (3) $1/\sqrt{3}$, because $T = 2\pi\sqrt{m/k}$ so $T_2/T_1 = \sqrt{k_1/k_2} = \sqrt{1/3}$ **(b)** 2.8 s
49. **(a)** using the period of vibration **(b)** 76 kg
51. 2.70 J
53. 0.279 m/s, 0.897 m/s^2 = $(0.0915)g$
55. **(a)** (1) increase, because $T = 2\pi\sqrt{L/g}$, a smaller g will have a longer T. **(b)** 4.9 s
57. **(a)** $y = (-0.10 \text{ m})\sin(10\pi/3)t$ **(b)** $k = 27$ N/m
59. **(a)** 1.21 m **(b)** 0.301 m/s **(c)** 0.248 rad/s
61. **(d)**
63. **(c)**
65. The one on the top is transverse and the one on the bottom is longitudinal.
67. 0.340 m
69. 0.47 m/s
71. 1.7 cm to 17 m
73. No, it is not in vacuum. $v = \lambda f = (500 \times 10^{-9}\text{ m})(4.00 \times 10^{14}\text{ Hz}) = 2.00 \times 10^8 \text{ m/s} < 3.00 \times 10^8 \text{ m/s}$.
75. 6.00 km
77. **(a)** 3.8×10^2 s **(b)** yes, 1.9×10^3 km > 30 km **(c)** 1.6×10^3 s
79. **(a)** 0.20 s **(b)** 0.40 s
81. **(d)**
83. **(d)**
85. Reflection (this is called echolocation), because the sound is reflected by the prey.
87. **(d)**
89. **(c)**
91. **(a)** This is caused by the glass vibrating in a resonance mode. **(b)** The frequency will increase because the wavelength will decrease due to the shortened air column. This results in an increase in frequency as $v = \lambda f$ and v is a constant.
93. A thinner string will sound higher frequency. Since $v = \sqrt{F_T/\mu}$, and a thinner string has a smaller μ, the speed is higher so is frequency ($v = \lambda f$).
95. 150 Hz
97. **(a)** (1) increases by $\sqrt{2}$, because $v = \sqrt{F_T/\mu}$, so $v_2/v_1 = \sqrt{F_{T2}/F_{T1}} = \sqrt{2/1} = \sqrt{2}$. **(b)** 8.49 m/s **(c)** $f_n = (0.425)n$ Hz; $n = 1, 2, 3, \ldots$
99. $n = 5$
101. 16.5 N
103. 1/4
105. 0.016 kg
107. **(a)** released form rest, downward **(b)** 1.05 s **(c)** $y = (0.100 \text{ m}) \cos(6t)$ **(d)** 3.6 m/s^2
109. **(a)** 5.00 N **(b)** 12.5 Hz **(c)** 0.40 m from one end
111. 3.0 s

Chapter 14

1. **(b)**
3. **(a)**
5. Some insects produce sounds with frequencies that are not all in our audible range.
7. They arrive at the same time because sound is not dispersive, i.e., speed does not depend on frequency.
9. **(a)** 1.0 km **(b)** 0.60 mi
11. 32°C
13. The unit of v in a liquid is

$$\sqrt{\frac{\text{N/m}^2}{\text{kg/m}^3}} = \sqrt{\frac{\text{N}\cdot\text{m}}{\text{kg}}} = \sqrt{\frac{\text{kg}\cdot\text{m}^2/\text{s}^2}{\text{kg}}} = \sqrt{\frac{\text{m}^2}{\text{s}^2}} = \text{m/s}.$$

Y has the same unit as B, so the unit of v in a solid is also m/s.
15. **(a)** (1) increases), because the speed of sound increases with temperature and $v = \lambda f$. So if v increases and f remains the same, λ increases. **(b)** +0.047 m
17. **(a)** 7.5×10^{-5} m **(b)** 1.5×10^{-2} m
19. **(a)** 1.08 s **(b)** 1.04 s
21. 90 m
23. **(a)** 0.107 m **(b)** 1.43×10^{-4} s **(c)** 4.29×10^{-4} m
25. **(a)** (1) less than double, because the total time is the sum of the time it takes for the stone to hit the ground (free fall motion) and the time it takes sound to travel back that distance. While the time for sound is directly proportional to the distance, the time for free fall is not. Since $d = \frac{1}{2}gt^2$ or $t = \sqrt{2d/g}$ (see Chapter 2), doubling the distance d will only increase the time of fall by a factor of $\sqrt{2}$. **(b)** 1.0×10^2 m **(c)** 8.7 s
27. 4.5%
29. **(b)**
31. (1) by a factor of 2
33. Yes. Since $\beta = 10 \log I/I_o$ and $\log x < 0$ for $x < 1$, if $I < I_o$. So for an intensity below the intensity of threshold of hearing, β is negative.
35. **(a)** (4) 1/9, because I is inversely proportional to the square of R. Tripling R will reduce I to $1/3^2 = 1/9$. **(b)** 1.4 times
37. **(a)** 3.0 Hz **(b)** not enough information given
39. **(a)** 100 dB **(b)** 60 dB **(c)** −30 dB
41. **(a)** 1.1×10^{-5} W/m^2
43. **(a)** 3.72×10^{-4} W/m^2; 1.00×10^{-1} W/m^2 **(b)** 9.55×10^{-3} W/m^2; 6.03×10^{-2} W/m^2
45. **(a)** 2.82 m **(b)** 2.82×10^3 m, unreasonable
47. five
49. 10 bands
51. $I_B = 0.563I_A$, $I_C = 0.250I_A$, $I_D = 0.173I_A$
53. **(a)** 2.5×10^2 m **(b)** 2.5×10^5 m away from the charge
55. 10^5 bees
57. **(a)**
59. No. The beat of music has to do with tempo. Beats are physical phenomena related to the frequency difference between two tones.
61. The varying sound intensity is caused by the interference effect. At certain locations there is constructive interference and at other locations, there is destructive interference.
63. 0.172 m
65. **(a)** (4) both (1) and (3), because the beat frequency measures only the frequency difference between the two, and it does not specify which frequency is higher. So the frequency of the violin can be either higher or lower than that of the instrument. **(b)** 267 Hz or 261 Hz
67. **(a)** (1) moving toward the siren, because the frequency heard is higher than the siren's. **(b)** 14 m/s
69. 3.3 Hz
71. **(a)** (2) less than 300 Hz, because the observer and the source are moving toward each other and the frequency heard is greater than the source frequency. **(b)** 251 Hz
73. 30°, yes
75. **(a)** 2.0 **(b)** 638 m/s
77. **(a)** 103 Hz approaching, 97.0 Hz receding **(b)** 6 Hz
79. **(a)** 36.3 kHz **(b)** 37.6 kHz **(c)** yes
81. **(b)**
83. **(d)**
85. **(a)** The snow absorbs sound so there is little reflection. **(b)** In an empty room, there is less absorption. So the reflections die out more slowly; and therefore, the sound seems hollow and echoing. **(c)** Sound is reflected by the shower walls, and standing waves are set up, giving rise to more harmonics and therefore richer sound quality.
87. For an open pipe, $f_n = nv/(2L)$ for $n = 1, 2, 3, \ldots$. For a closed pipe, $f_m = mv/(4L)$ for $m = 1, 3, 5, \ldots$. So $f_n/f_m = (n/2)/(m/4) = 2n/m$. For $f_n = f_m$, $2n = m$. No, this is not possible because $2n$ is an even integer and m is an odd integer.
89. For a pipe closed at one end, the closed end must be a node. So only odd harmonics are possible.
91. 510 Hz
93. **(a)** f_2 does not exist, only odd harmonics **(b)** 0.30 m
95. **(a)** 0.635 m **(b)** 265 Hz
97. 0°C
99. $f_{air} = 552$ Hz, $f_{He} = 1610$ Hz
101. 0.249 m and 0.251 m
103. 5.55×10^4 Hz
105. **(a)** (1) yes, because the observer will hear beats between the source and the reflection. **(b)** 1.03×10^3 Hz **(c)** 12 Hz

Chapter 15

1. **(c)**
3. **(c)**
5. No. Charges are simply moved from the object to another object.

7. Bring the charged object near the electroscope and observe how the leaves move. If the repulsion between the leaves increases, the charge on the object has the same sign as the one on the electroscope; if the repulsion between the leaves decreases, then the charge on the object has opposite sign as the one on the electroscope.
9. -1.6×10^{-13} C
11. $+6.40 \times 10^{-19}$ C
13. **(a)** (1) positive, because of the conservation of charge. When one object becomes negatively charged, it gains electrons. These same electrons must be lost by another object, and therefore it is positively charged.
(b) $+4.8 \times 10^{-9}$ C, 2.7×10^{-20} kg
(c) 2.7×10^{-20} kg
15. **(d)**
17. This is to remove excess charge due to friction of rubber on road. If the excess charge is not removed, this could result in a spark, causing a gasoline explosion.
19. If you bring a negatively charged object near the electroscope, the induction process will charge the electroscope with positive charges. You can prove the charges are positive by bringing the negatively charged object near the leaves and see if the leaves are attracted by the negatively charged object.
21. **(a)**
23. Although the electric force is fundamentally much stronger than the gravitational force, both the Earth, our bodies, and other objects are electrically neutral so there are no noticeable electric forces.
25. 9
27. **(a)** 1 **(b)** 1/4 **(c)** 1/2
29. **(a)** 5.8×10^{-11} N **(b)** zero
31. 2.24 m
33. **(a)** 50 cm **(b)** 50 cm
35. **(a)** $x = 0.25$ m **(b)** nowhere
(c) $x = -0.94$ m for $\pm q_3$
37. **(a)** 8.2×10^{-8} N **(b)** 2.2×10^{6} m/s
(c) 9.2×10^{21} g
39. **(a)** 96 N, 39° below positive x-axis
(b) 61 N, 84° above negative x-axis
41. **(c)**
43. **(a)**
45. It is determined by the relative density or spacing of the field lines. The closer the lines, the greater the magnitude.
47. If a positive charge is at center of the spherical shell, the electric field is *not* zero inside. The field lines run radially outward to the inside surface of the shell where they stop at the induced negative charges on this surface. The field lines reappear on the outside shell surface (positively charged) and continue radially outward as if emanating from the point charge at the center. If the charge were negative, the field lines would reverse their directions.
49. **(a)** Yes, this is possible. For example, when the electric fields created by the two charges are equal in magnitude and opposite in direction at some locations. At the midway point in between and along a line joining two charges of the same type and magnitude, the electric field is zero. **(b)** No, this is not possible.
51. 2.0×10^{5} N/C
53. 1.2×10^{-7} m away from the charge
55. 1.0×10^{-7} N/C upward, 5.6×10^{-11} N/C downward
57. $\vec{E} = (2.2 \times 10^5 \text{ N/C})\hat{x} + (-4.1 \times 10^5 \text{ N/C})\hat{y}$
59. 5.4×10^{6} N/C toward the charge of $-4.0\ \mu$C
61. 3.8×10^{7} N/C in the $+y$-direction
63. 15 μC/m^2
65. $\vec{E} = (-4.4 \times 10^6 \text{ N/C})\hat{x} + (7.3 \times 10^7 \text{ N/C})\hat{y}$
67. **(b)**
69. **(b)**
71. The surface must be spherical.
73. **(a)** (1) negative due to induction **(b)** zero **(c)** $+Q$ **(d)** $-Q$ **(e)** $+Q$
75. **(a)** zero **(b)** kQ/r^2 **(c)** zero **(d)** kQ/r^2
77.

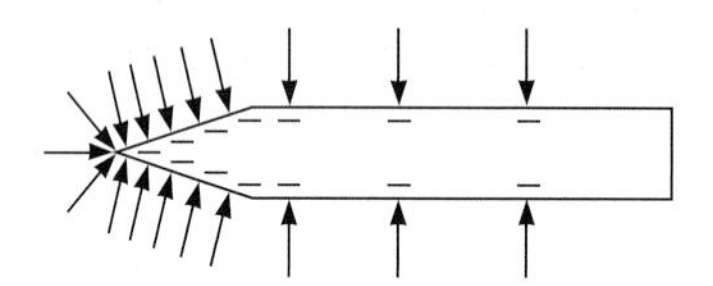

79. **(c)**
81. Since the number of lines is proportional to charge, the net charges are equal and opposite in sign.
83. -6 lines, or net of 6 lines entering it
85. 10 field lines entering (negative)
87. **(a)** bottom **(b)** 4.90×10^{-4} kg
89. **(a)** positive on right plate and negative on left plate **(b)** from right to left
(c) 1.13×10^{-13} C
91. 5.5×10^{3} N/C at 66° below positive x-axis
93. **(a)** 2.02×10^{4} N/C **(b)** 1.79×10^{-7} C/m^2

Chapter 16

1. **(d)**
3. **(b)**
5. **(a)** The kinetic energy of the approaching proton decreases as its electric potential energy increases since its total energy is constant. **(b)** The electric potential energy of the system increases because the distance between the charges decreases. **(c)** The total energy of the system remains the same because of the conservation of energy.
7. It will move to the right or toward the higher potential region because the electron has negative charge. The higher the potential region, for the electron, the lower the potential energy.
9. It takes zero work. Since $W = q_o \Delta V$, if $\Delta V = 0$, $W = 0$.
11. **(a)** 2.7 μC **(b)** negative to positive
13. 1.6×10^{-15} J
15. **(a)** 5.9×10^{5} m/s, down **(b)** lose potential energy
17. **(a)** (2) 3, because electric potential is inversely proportional to the distance.
(b) 0.90 m **(c)** -6.7 kV
19. **(a)** gains 6.2×10^{-19} J **(b)** loses 6.2×10^{-19} J **(c)** gains 4.8×10^{-19} J
21. 1.1 J
23. **(a)** $+0.27$ J **(b)** no
25. -0.72 J
27. **(a)** 3.1×10^{5} V **(b)** 2.1×10^{5} V
29. **(a)** (3) a lower, because electrons have a negative charge. They move toward higher potential regions where they have lower potential energy. **(b)** 4.2×10^{7} m/s
(c) 6.0×10^{-9} s
31. **(b)**
33. The ball would accelerate in the direction from the beach to the ocean (from higher potential energy to lower potential energy).

Higher gravitational potential energy
Beach
Ball accelerating
Ocean
Lower gravitational potential energy

35. It takes zero work, because there is no change in kinetic or potential energy. The net work is zero.
37. **(a)** cylindrical **(b)** near the outer surface **(c)** near the inner surface
39. **(a)** 1.60×10^{-13} J **(b)** it would double
41. 12.6 m
43. 70 cm
45. 1.7 mm away from the positive plate, toward the negative plate
47. **(a)** (1) concentric spheres, because the electric potential depends only on the distance from the charge. **(b)** $+298$ eV
49. **(a)** 2.0×10^{7} eV **(b)** 2.0×10^{4} keV
(c) 20 MeV **(d)** 2.0×10^{-2} GeV **(e)** 3.2×10^{-12} J
51. 6.2×10^{7} m/s (proton), 4.4×10^{7} m/s (alpha)
53. **(a)** 3.5 V, 1.1×10^{6} m/s **(b)** 4.1 kV, 3.8×10^{7} m/s **(c)** 5.0 kV, 4.2×10^{7} m/s
55. **(a)** $+0.40$ V **(b)** -0.40 V **(c)** zero
57. **(c)**
59. **(a)**
61. **(b)**
63. **(a)** Since $Q = CV$, it doubles. **(b)** Since $U_C = \frac{1}{2}CV^2$, it quadruples.
65. 2.4×10^{-5} C
67. 0.71 mm
69. **(a)** 4.2×10^{-9} C **(b)** 2.5×10^{-8} J
71. 2.2 V
73. **(a)** 2.2×10^{4} V/m **(b)** 1.1×10^{-5} C
(c) 5.7×10^{-4} J **(d)** $E = 6.7 \times 10^{4}$ V/m, $\Delta Q = 0$, $\Delta U_C = -1.7 \times 10^{-3}$ J
75. **(d)**
77. We cannot maintain a nonzero voltage on a conductor; charges will move from the positive to the negative immediately so they cannot be stored.
79. When the power supply is not connected, the charge remains the same but the capacitance increases once the dielectric material is inserted. Therefore the potential difference decreases ($V = Q/C$) and so does the electric field ($E = V/d$). When the power supply remains connected, the potential difference remains constant, so does the electric field.

81. 3.1×10^{-9} C; 3.7×10^{-8} J
83. (a) $\kappa = 2.4$ **(b)** decreased **(c)** -6.3×10^{-5} J
85. (b)
87. (b)
89. They have the same charge when they have equal capacitance.
91. (a) C/N **(b)** NC **(c)** $4C/N$
93. (a) (1) more, because the equivalent capacitance is higher and the energy stored (drawn) is proportional to capacitance. **(b)** 6.0 μF
95. (a) (3) $Q/3$, because $Q_{total} = Q_1 + Q_2 + Q_3$. Also $Q_1 = Q_2 = Q_3$ because the capacitors have the same capacitance. Therefore each capacitor has only 1/3 of the total charge. **(b)** 3.0 μC **(c)** 9.0 μC
97. max. 6.5 μF; min. 0.67 μF
99. C_1: 2.4 μC, 6.0 V; C_2: 2.4 μC, 6.0 V; C_3: 1.2 μC, 6.0 V; C_4: 3.6 μC, 6.0 V
101. (a) $K_o = 29.2$ eV, total $\Delta U = 75$ eV, so it can't **(b)** 30.6 cm from bottom
103. (a) -1.7×10^{-17} J **(b)** 6.9×10^{23} m/s^2 **(c)** -8.5×10^{-18} J **(d)** 8.5×10^{-18} J
105. (a) (1) a higher potential, because the electron has negative charge. It will experience an upward force if the potential is higher at the top. **(b)** 8.37×10^{-13} V **(c)** any location
107. (a) 2.9 pF **(b)** 0.20 pC

Chapter 17

1. (b)
3. (b)
5. (c)
7. No. Any battery has internal resistance, and there will be a voltage across the internal resistance when the battery is in use. The terminal voltage is lower than the emf of the battery when it is in use.
9. (a) 4.5 V **(b)** 1.5 V
11. (a) 24 V **(b)** two 6.0-V in series, together in parallel with the 12-V
13. (a) (2) the same, because the total voltage of identical batteries in parallel is the same as the voltage of each individual battery, and the total voltage of the batteries in series is the sum of the voltages of each individual battery. Each arrangement has one parallel and one series so they have the same total voltage. **(b)** 3.0 V, 3.0 V
15. (a)
17. (a) upward **(b)** downward **(c)** upward
19. 0.25 A
21. (a) 0.30 C **(b)** 0.90 J
23. 56 s
25. (a) (2) to the left, because the current due to the protons will be to the left, and the current due to the electrons will also be to the left because electrons have negative charge. **(b)** 3.3 A
27. (a)
29. (a)
31. From $V = (R)I$ ($y = mx$ is the equation for a straight line where m is the slope) we conclude that the one with the shallower slope is less resistive.
33. (a) same **(b)** one-quarter the current
35. (a) 11.4 V **(b)** 0.32 Ω
37. (a) (1) a greater diameter, because aluminum has a higher resistivity. Its area (diameter) must be greater, if the length of the wire is the same, to have the same resistance as copper according to $R = \rho L/A$. **(b)** 1.29
39. 1.0 V
41. 1.3×10^{-2} Ω
43. (a) 4 **(b)** 4
45. (a) 0.13 Ω **(b)** 0.038 Ω
47. (a) 4.6 mΩ **(b)** 8.5 mA
49. 5.4×10^{-2} m
51. (a) (1) greater than, because after the stretch, the length L increases and the cross-sectional area A decreases, so R increases according to $R = \rho L/A$. **(b)** 1.6
53. (a) 7.8 Ω **(b)** 0.77 A **(c)** 16.4°C
55. (d)
57. (d)
59. Since $P = V^2/R$, the bulb of higher power has smaller resistance or thicker wire. So the wire in the 60-W bulb would be thicker.
61. 144 Ω
63. 2.0×10^3 W
65. 1.2 Ω
67. (a) (4) 1/4, because if the voltage is halved, the current is also halved. Power is equal to voltage times current, so power becomes 1/4 of its original value. **(b)** 10 W
69. (a) 4.3×10^3 W **(b)** 13 Ω
71. (a) 58 Ω **(b)** 86 Ω
73. (a) 0.60 kWh **(b)** \$0.09
75. (a) 0.15 A **(b)** 1.4×10^{-4} Ω·m **(c)** 2.3 W
77. (a) 1.1×10^2 J **(b)** 6.8 J
79. 21 Ω
81. $R_{120}/R_{60} = 4/3$
83. \$152
85. 117°C for copper or −72.6°C for aluminum
87. yes, R is a constant ($R + r = 6.00$ Ω)
89. about one-half (1 power plant delivers about 1000 MW)
91. 6.6×10^{-6} m/s
93. 1.6×10^3 Ω
95. (a) 400 A **(b)** 4.5×10^{-3} Ω **(c)** 1.8 V **(d)** 250 kV

Chapter 18

1. (b)
3. (b)
5. (a)
7. No, not generally. However if all resistors are equal, the voltages across them are the same.
9. If they are in series, the effective resistance will be closer in value to that of the large resistance because $R_s = R_1 + R_2$. If $R_1 \gg R_2$, then $R_s \approx R_1$. If they are in parallel, the effective resistance will be closer in value to that of the small resistance because $R_p = R_1R_2/(R_1 + R_2)$. If $R_1 \gg R_2$, then $R_p \approx R_1R_2/R_1 = R_2$.
11. (a) The third resistor has the largest current, because the total current through the two other resistors is equal to the current through the third resistor. **(b)** The third resistor also has the largest voltage, because the current through it is the largest and all the resistors have the same resistance value ($V = IR$). **(c)** The third resistor also has the largest power output, because it has the largest voltage and current, and power is equal to the product of current and voltage.
13. (a) in series, 60 Ω **(b)** in parallel, 5.5 Ω
15. 30 Ω
17. (a) 30 Ω **(b)** 0.30 A **(c)** 1.4 W
19. (a) 0.57 Ω **(b)** 6.0 V **(c)** 9.0 W
21. (a) (1) $R/4$. Each shortened segment has a resistance of $R/2$ because resistance is proportional to length (Chapter 17). Then two $R/2$ resistors in parallel gives $R/4$. **(b)** 3.0 $\mu\Omega$
23. (a) 1.0 A **(b)** 1.0 A **(c)** 2.0 W, 4.0 W, 6.0 W **(d)** $P_{sum} = P_{total} = 12$ W
25. 1.0 A (for all); $V_{8.0} = 8.0$ V; $V_{4.0} = 4.0$ V
27. 2.7 Ω
29. (a)

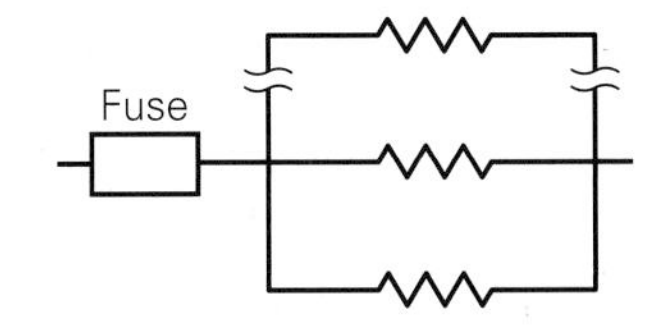

(b) 31
31. (a) 1.0 A; 0.50 A; 0.50 A **(b)** 20 V; 10 V; 10 V **(c)** 30 W
33. no, since $I = 14.6$ A < 15 A
35. 100 s = 1.7 min
37. (a) 0.085 A **(b)** 7.0 W, 2.6 W, 0.24 W, 0.41 W
39. (a) 0.67 A, 0.67 A, 1.0 A, 0.40 A, 0.40 A **(b)** 6.7 V, 3.3 V, 10 V, 2.0 V, 8.0 V
41. 8.1 Ω
43. (a)
45. (d)
47. No, it does not have to be. An example is charging a battery. When a battery is connected to a charger (with a higher electromotive force), the current is forced through the battery.
49. The 60 W bulb has a higher resistance than the 100 W bulb. When these are in series, they have the same current. Therefore, the 60 W bulb will have a higher voltage. Thus, the 60 W bulb has more power because $P = IV$.
51. Around loop 1 (reverse), $-V_1 + I_3R_3 + V_2 + I_1R_1 = 0$. If we multiply by -1 on both sides, it is the same as the equation for loop 1 (forward). Around loop 2 (reverse), $I_2R_2 - V_2 - I_3R_3 = 0$. Again, if we multiply by -1 on both sides, it is the same as the equation for loop 2 (forward).
53. $I_1 = 1.0$ A; $I_2 = I_3 = 0.50$ A
55. $I_1 = 0.33$ A (left); $I_2 = 0.33$ A (right)
57. $I_1 = 3.75$ A (up); $I_2 = 1.25$ A (left); $I_3 = 1.25$ A (right)
59. $I_1 = 0.664$ A (left); $I_2 = 0.786$ A (right); $I_3 = 1.450$ A (up); $I_4 = 0.770$ A (down); $I_5 = 0.016$ A (down); $I_6 = 0.664$ A (right)
61. (c)
63. (b)
65. It takes shorter than one time constant because the time constant is defined as the time it takes to charge the capacitor charged to 63% of its maximum charge.
67. (a) $V_C = 0$; $V_R = V_o$ **(b)** $V_C = 0.86V_o$; $V_R = 0.14V_o$ **(c)** $V_C = V_o$; $V_R = 0$

69. **(a)** (1) increase the capacitance, because $\tau = RC$ **(b)** 2.0 MΩ

71. **(a)** 1.50 MΩ **(b)** 11.4 V

73. **(a)** 9.4×10^{-4} C **(b)** $V_C = 24$ V; $V_R = 0$

75. **(a)** 2.0×10^{-3} A at $t = 0$ **(b)** 0.080% **(c)** 1.7×10^{-6} C at a very long time after connection **(d)** 99.9%

77. **(b)**

79. **(a)** An ammeter has very low resistance, so if it were connected in parallel in a circuit, the circuit current would be very high and the galvanometer could burn out.
(b) A voltmeter has very high resistance, so if it were connected in series in a circuit, it would read the voltage of the source because it has the highest resistance (most probably) and therefore the most voltage drop among the circuit elements.

81. An ammeter is used to measure current when it is connected in series to a circuit element. If it has very small resistance, there will be very little voltage across it, so it will not affect the voltage across the circuit element, nor its current.

83. **(a)** (3) a multiplier resistor, because a galvanometer cannot have a large voltage across it, the large voltage has to be across a series resistor (multiplier). **(b)** 7.4 kΩ

85. 50 kΩ

87. 0.20 mA

89. **(a)** (1) zero, because an ammeter is connected in series with a circuit element. If its resistance is zero, it will not affect the current through the circuit element.
(b) The current reading I is the current through R, and the voltage reading is the total voltage across R and R_a so V/I gives the resistance of the series combination.
(c) The voltage reading is $V = I(R + R_a)$, so $R = V/I - R_a$.
(d) An ideal ammeter has R_a approaching 0, then $R = V/I$, i.e., the measurement is "perfect."

91. **(c)**

93. No, a high voltage can produce high harmful current, even if resistance is high because current is caused by voltage (potential difference).

95. It is safer to jump. If you step off the car one foot at a time, there will be a high voltage between your feet. If you jump, the voltage between your feet is zero because your feet will be at the same potential all the time.

97. $I_1 = 2.6$ A (right); $I_2 = 1.7$ A (left); $I_3 = 0.86$ A (down)

99. **(a)** $I_1 = 1.0$ A; $I_2 = 0.40$ A; $I_3 = 0.20$ A; $I_4 = 0.40$ A **(b)** $P_1 = 100$ W; $P_2 = 4.0$ W; $P_3 = 2.0$ W; $P_4 = 4.0$ W

101. 6.0 Ω

103. 10 mΩ, 2.0 mΩ, and 1.0 mΩ

105. two in parallel with each other and in series with the other resistor

107. **(a)** 12.1 ms **(b)** 1.21 kΩ **(c)** 13.0 ms

Chapter 19

1. **(a)**

3. **(c)**

5. Near the north pole of a permanent bar magnet, the north pole of a compass will point away from the bar magnet so the field lines leave the north pole. Near the south pole of a permanent bar magnet, the south pole of a compass will point toward the bar magnet so the field lines enter the south pole.

7. **(a)**

9. **(d)**

11. Not necessarily, because there still could be a magnetic field. If the magnetic field and the velocity of the charged particle make an angle of either 0° or 180°, there is no magnetic force because $F = qvB \sin \theta$.

13. **(a)** The bottom half would have a magnetic field directed into the page and the top half would have a magnetic field directed out of the page. **(b)** They are the same since centripetal force does not change the speed of the particle.

15. 3.5×10^3 m/s

17. 2.0×10^{-14} T, left, looking in the direction of the velocity

19. **(a)** 3.8×10^{-18} N **(b)** 2.7×10^{-18} N **(c)** zero **(d)** zero

21. **(a)** 8.6×10^{12} m/s^2, horizontal and south **(b)** 8.6×10^{12} m/s^2, horizontal and north **(c)** same magnitude but the direction is horizontal and north

23. **(b)**

25. The magnetic force on the electron beam, which "prints" pictures, causes the deflection of the electrons.

27. The electric force is $F_e = qE$ and the magnetic force is $F_B = qvB$. The purpose of the velocity selector is for the electric force to equal the magnetic force. Since $qE = qvB$, $v = E/B$, independent of the charge.

29. **(a)** 1.8×10^3 V **(b)** same voltage, independent of charge

31. 5.3×10^{-4} T

33. **(a)** 4.8×10^{-26} kg **(b)** 2.4×10^{-18} J **(c)** no, work equals zero

35. **(d)**

37. **(b)**

39. It shortens because the coils of the spring attract each other due to the magnetic fields created in the coils. (Parallel wires with current in same direction will attract each other.)

41. Pushing the button in both cases completes the circuit. The current in the wires activates the electromagnet, causing the clapper to be attracted and ring the bell. However, this breaks the armature contact and opens the circuit. Holding the button causes this to repeat, and the bell rings continuously. For the chimes, when the circuit is completed, the electromagnet attracts the core and compresses the spring. Inertia causes it to hit one tone bar, and the spring force then sends the core in the opposite direction to strike the other bar.

43. 1.2 N perpendicular to the plane of $\vec{B}$ and I.

45. 5.0×10^{-3} T north to south

47. **(a)** zero **(b)** 4.0 N/m in $+z$ **(c)** 4.0 N/m in $-y$ **(d)** 4.0 N/m in $-z$ **(e)** 4.0 N/m in $+y$

49. 0.40 N/m; $+z$

51. **(a)** (1) attractive **(b)** 6.7×10^{-6} N/m

53. 0.53 N northward at an angle of 45° above the horizontal

55. 2.7×10^{-5} N/m toward wire 1

57. 7.5 N upward in the plane of the paper

59. zero, yes

61. **(a)**

63. **(b)**

65. Because $B = \mu_o I/(2\pi d)$, you need to double the current and reverse the direction.

67. There are two wires carrying the current into and out of the appliances. These two currents are in opposite directions. When the two wires are very close to each other, the magnetic fields created by the two opposite currents essentially cancel.

69. 3.8 A

71. 0.25 m

73. **(a)** 2.0×10^{-5} T **(b)** 9.6 cm from wire 1

75. both 2.9×10^{-6} T

77. 3.3×10^{-5} T

79. 1.0×10^{-4} T, away from the observer

81. 4.0 A

83. **(a)** 8.8×10^{-2} T **(b)** to the right

85. $\sqrt{2}\,\mu_o I/(\pi a)$ at 45° toward the lower left wire

87. **(b)**

89. The direction of the magnetic field is away from you, according to the right-hand source rule (electron has negative charge).

91. You can destroy or reduce the magnetic field of a permanent magnet by hitting or heating it.

93. 12 T

95. **(b)**

97. **(a)**

99. It will be the north magnetic pole. Right now, the pole near the Earth's geographical North pole is the south magnetic pole.

101. 0.44 T

103. **(a)** 5.9×10^{-21} kg · m/s **(b)** 1.0×10^{-14} J

105. 0.682 V

107. **(a)** (2) into the page **(b)** 0.030 m from left wire

109. **(a)** 3.74×10^{-3} T · m/A **(b)** 3.0×10^3

Chapter 20

1. **(d)**

3. **(d)**

5. **(d)**

7. The direction is counterclockwise (in head-on view).

9. No, it does not depend on the magnetic flux. It depends on the rate of the flux change with time.

11. Sound waves cause the resistance of the button to change as described. This results in a change in the current, so the sound waves produce electrical pulses. These pulses travel through the phone lines and to a receiver. The receiver has a coil wrapped around a magnet, and the pulses create a varying magnetic field as they pass through the coil causing the diaphragm to vibrate and thus produce sound waves as the diaphragm vibrates in the air.

13. 42° or 138°

15. 3.3×10^{-2} T·m^2

17. 1.3×10^{-6} T·m^2

19. 1.6 V

21. 0.30 s

23. **(a)** (1) counterclockwise **(b)** 0.35 V

25. **(a)** (1) at the equator, because the velocity of the metal rod is parallel to the magnetic field at the equator. **(b)** 0.50 mV at the pole, zero at the equator

27. **(a)** 0.60 V **(b)** 0 A

29. 4.0 V

31. **(a)** 0.037 T·m^2 (lower incline); 0.034 T·m^2 (upper incline) **(b)** −0.071 T·m^2 **(c)** zero **(d)** zero; this means that the net flux is equal to zero or there are as much flux leaving the box as entering.

33. **(c)**

35. The magnet moving through the coil produces a current. As the magnet moves up and down in the coil it will induce a current in the coil that will light the bulb. However, the magnet produces the current (Faraday's law of induction) at the expense of its kinetic energy and potential energy. The magnet's motion will therefore damp out.

37. $\mathscr{E} = \mathscr{E}_o = \sin \omega t$, where $\mathscr{E}_o = NBA\omega$. ($N$ is the number of turns, B is the magnetic field strength, and ω is the angular speed) You could increase N, B, or ω.

39. **(a)** 0.057 V **(b)** 0.57 V

41. **(a)** (2) two, because the initial voltage direction was not specified, thus, there are two possible directions. **(b)** ±104 V

43. **(a)** 100 V **(b)** 0 V

45. 16 Hz

47. **(a)** (3) lower than 44 A (110 V/2.50 Ω = 44 A), because of the back emf induced when the motor turns. The back emf lowers the effective voltage of the motor, thus, the current is lower than 44 A. **(b)** 4.00 A

49. **(a)** 216 V **(b)** 160 A **(c)** 8.1 Ω

51. **(b)**

53. Yes, a step-up transformer can be used as a step-down transformer. You just need to reverse the roles primary and secondary coils, so there are more turns on the high voltage side.

55. **(a)** 16 **(b)** 5.0×10^2 A

57. 24:1

59. **(a)** 17.5 A **(b)** 15.7 V

61. **(a)** (2) non-ideal, because $P_s < P_p$ (the power in the secondary is lower than that in the primary). **(b)** 45%

63. **(a)** N_s/N_p is 1:20 **(b)** 2.5×10^{-2} A

65. **(a)** 128 kWh **(b)** $1840

67. **(a)** 1:2, 1:14, 1:30 **(b)** 2.0, 14, 30 **(c)** 833

69. **(a)** 53 W **(b)** $N_p/N_s = 200$

71. **(d)**

73. **(d)**

75. UV radiation causes sunburn and it can still go through the clouds. You feel cool because infrared (heat) radiation is absorbed by clouds (water molecules).

77. According to $c = \lambda f$, wavelength and frequency are inversely proportional to each other. Thus, radar frequencies are much higher, because wavelengths are much shorter, speeds are the same.

79. 326 m and 234 m

81. 2.6 s

83. AM: 67 m; FM: 0.77 m

85. **(a)** (1) up, according to Lenz's law. **(b)** 25 mA

87. **(a)** no, input power is greater than output power **(b)** 90.9%

89. **(a)** (1) clockwise **(b)** 5.00×10^{-3} V **(c)** 0.0879 s

91. 3.79 m, no

93. 0.159 Ω

Chapter 21

1. **(a)**

3. **(a)**

5. That means the voltage and current both reach maximum or minimum at the same time.

7. No, the circuit element cannot be a resistor. Voltage and current should be in phase for a resistor. Yes, the frequency is 60 Hz, because $\omega = 2\pi f = 120\pi$.

9. 7.1 A

11. 1.2 A

13. **(a)** 10.0 A **(b)** 14.1 A **(c)** 12.0 Ω

15. **(a)** 4.47 A, 6.32 A **(b)** 112 V, 158 V

17. $V = (170\text{ V})\sin(119\pi t)$

19. 0.33 A; 0.47 A

21. **(a)** 20 Hz, 0.050 s **(b)** 2.4×10^2 W

23. **(a)** 60 Hz **(b)** 1.4 A **(c)** 1.2×10^2 W **(d)** $V = (120\text{ V})\sin 380t$ **(e)** $P = (240\text{ W})\sin^2 380t$ **(f)** $P = (240\text{ W})[1 - \cos 2(380t)]/2 = 120\text{ W} - (120\text{ W})\cos 2(380t)$. The average of a sine or cosine function is zero. So $\overline{P} = 120$ W, the same as in (c).

25. **(b)**

27. For a capacitor, the *lower the frequency*, the longer the charging time in each cycle. If the frequency is very low (dc), then the charging time is very long, so it acts as an ac open circuit. For an inductor, the *lower the frequency*, the more slowly the current changes in the inductor The more slowly the current changes, the less back emf is induced in the inductor, resulting in less impedance to current.

29. At $t = 0$, $I = 120$ A, or at maximum. The voltage is then zero, because current leads voltage by 90° in a capacitor. When current is maximum, voltage is 1/4 period behind, or at zero. They are out of phase.

31. 1.3×10^3 Ω

33. **(a)** 19 Ω **(b)** 6.4 A **(c)** Voltage leads current by 90°

35. an increase of 60%

37. 255 Hz

39. **(a)** 90 V **(b)** voltage leads current by 90°

41. 4.4 μF

43. **(d)**

45. **(d)**

47. No, there is no power delivered to capacitors or inductors in an ac circuit. For either a pure capacitive or inductive circuit, the phase angle $\phi = 90°$, and so the power factor is $\cos\phi = 0$.

49. **(a)** 1.7×10^2 Ω **(b)** 2.0×10^2 Ω

51. **(a)** 38 Ω; 1.1×10^2 Ω **(b)** 1.1 A

53. **(a)** (3) negative, because this is a capacitive circuit. **(b)** −27°

55. **(a)** (3) in resonance, because $X_L = X_C$, so $Z = R$. **(b)** 72 Ω

57. 50 W

59. 5.3×10^{-11} F

61. ab: 1.3 A; ac: 1.2 A; bc: 4.0 A; cd: 1.8 A; bd: 1.6 A; ad: 2.9 A

63. 13 A

65. $(V_{rms})_R = 12$ V; $(V_{rms})_L = 2.7 \times 10^2$ V; $(V_{rms})_C = 2.7 \times 10^2$ V

67. **(a)** (2) equal to 25 Ω. At resonance, $X_L = X_C$, so $Z = R$. **(b)** 362 Ω

69. 30%

71. **(a)** 38 Ω **(b)** 63 Ω **(c)** 1.8 A **(d)** zero **(e)** 37°

73. **(a)** (2) zero, as $X_L = X_C$, according to $\phi = \tan^{-1}[(X_L - X_C)/R]$. **(b)** 9.4 μF

Chapter 22

1. **(c)**

3. **(d)**

5. This is irregular or diffuse reflection, because the paper is microscopically rough.

7. 70°

9. **(a)** (2) $90° - \alpha$, because $\theta_i = \theta_r$ and $\alpha + \theta_i = 90°$, $\theta_r = \theta_i = 90° - \alpha$ **(b)** 47°

11. **(a)** (3) $\tan^{-1}(w/d)$ **(b)** 27°

13. When the mirror rotates through a small angle of θ, the normal will rotate through an angle of θ and the angle of incidence is $35° + \theta$. The angle of reflection is also $35° + \theta$. Since the original angle of reflection is 35°, the reflected ray will rotate through an angle of 2θ. If the mirror rotates in the opposite direction, the angle of reflection will be $35° - \theta$. However, the normal will again rotate the through an angle of θ but also in the opposite direction. Thus, the reflected ray still rotates through an angle of 2θ.

15. 90°, any θ_{i_1}

17. **(d)**

19. It is because light speed depends on the medium. For example, light speed is different in air than in water. Because of the speed difference, light changes direction when entering a different medium at an angle of incidence that is not zero.

21. This severed look is because the angle of refraction is different for air-glass interface than for water-glass interface. The top portion refracts from air to glass, and the bottom portion refracts from water to glass. This is different from what's on Fig. 22.13b. In that figure, we see the top portion in air directly and the bottom portion in water through refraction from water to air.

The angle of refraction made the pencil appear to be bent.

23. The laser beam has a better chance to hit the fish. The fish appears to the hunter at a location different from its true location due to refraction. The laser beam obeys the same law of refraction and retraces the light the hunter sees from the fish. The arrow goes into the water in a near-straight line path and thus passes above the fish.

25. **(a)** (1) greater than, because its index of refraction is smaller. **(b)** 1.26

27. **(a)** (1) greater than, because water has a lower index of refraction. **(b)** 17°

29. **(a)** (2) a diamond to air, because diamond has a higher index of refraction. **(b)** 24.4°

31. 47°

33. 6.5×10^{14} Hz, 2.8×10^{-7} m

35. **(a)** (3) less than, because its index of refraction is higher. **(b)** 15/16

37. **(a)** This is caused by refraction of light in the water-air interface. The angle of refraction in air is greater than the angle of incidence in water so the object immersed in water appears closer to the surface.

39. 75.2%

41. **(a)** (3) less than, because it is equal to $90° - \theta_1$. $\theta_1 > 45° = \theta_2$ and $n_1 < n_2$. **(b)** 20°

43. seen for 40° but not for 50°, $\theta_c = 49°$

45. **(a)** yes, $\theta_c = 32° < 45°$ **(b)** no, $\theta_c = 46° > 45°$

47. We can measure the angles of incidence and refraction from the photography and calculate the index of refraction of the fluid from the law of refraction. It is about 1.3.

49. 1.64

51. **(a)** (3) total internal reflection **(b)** $\theta_c = 39° < 45°$, no **(c)** $\theta_c = 56° < 71°$, still not transmitted

53. 2.0 m

55. **(a)** 12.5° **(b)** 26.2°

57. **(b)**

59. In a prism, there are two refractions and two dispersions because both refractions cause the refracted light to bend downward, therefore doubling the effect or dispersion.

61. To see a rainbow, the light has to be behind you. Actually, you won't see a primary rainbow if the Sun's angle above the horizon is greater than 42°. Therefore you cannot look up to find a rainbow, thus you cannot walk under a rainbow.

63. **(a)** Usually $\theta \approx 0°$, so there is no dispersion because the angle of refraction for all colors is also zero. **(b)** as explained in (a). With $\theta \approx 0$, light of any wavelength will experience no refraction. (No, the speeds are actually different.)

65. 1.498

67. **(a)** 21.7° **(b)** 0.22° **(c)** 0.37°

69. **(a)** 49° **(b)** 1.5 **(c)** 1 **(d)** 42°

71. **(a)** (1) more than, because red light will have a smaller index of refraction and thus a higher speed of light than blue light. **(b)** 1.3 mm

73. 1.41 to 2.00

Chapter 23

1. **(b)**

3. **(c)**

5. During the day, the reflection is mainly from the silvered back surface. During the night, when the switch is flipped, the reflection comes from the front side. And so there is a reduction of intensity and glare because the front side reflects only about 5% of the light, which is more than enough to see due to the dark background.

7. When viewed by a driver through a rear view mirror, the right-left reversal property of the image formed by a plane mirror will make it read "AMBULANCE."

9. **(a)** 4.0 m **(b)** upright, virtual, and same size

11. 5.0 m

13. **(a)** 1.5 m behind the mirror **(b)** 1.0 m/s

15. **(a)** You see multiple images caused by reflections off two mirrors. **(b)** 3.0 m behind the *north* mirror, 11 m behind the *south* mirror, 5.0 m behind the *south* mirror, 13 m behind the *north* mirror

17. The two triangles (with d_o and d_i as base, respectively) are similar to each other because all three angles of one triangle are the same as those of the other triangle due to the law of reflection. Furthermore, the two triangles share the same height, the common vertical side. Therefore the two triangles are identical. Hence $d_o = d_i$.

19. **(d)**

21. **(a)**

23. **(a)** A spoon can behave as either a concave or a convex mirror depending on which side you use for reflection. If you use the concave side, you normally see an inverted image. If you use the convex side, you always see an upright image. **(b)** In theory, the answer is yes. If you are very close (inside the focal point) to the spoon on the concave side, an upright image exists. However, it might be difficult for you to see the image in practice, because your eyes might be too close to the image. Eyes cannot see things that are closer than the near point (Chapter 25).

25. The image of a distant object (at infinity) is formed on a screen in the focal plane. The distance from the vertex of the mirror to the plane is the focal length. The same cannot be done for a convex mirror because the image is a virtual one and it cannot be formed on a screen.

27. **(a)** It is seen from the ray diagram that the image is virtual, upright, and reduced.

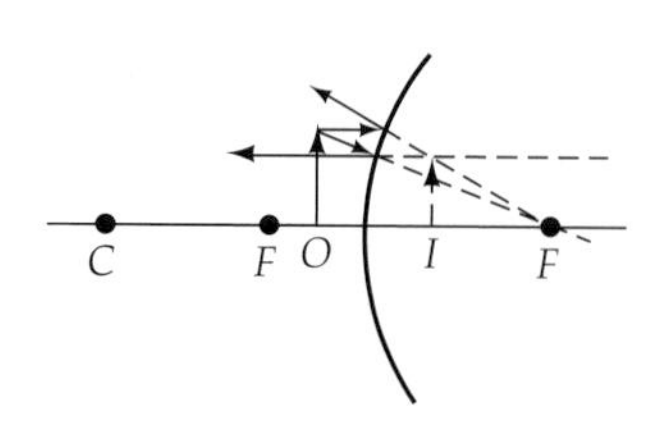

(b) $d_i = -20$ cm; $h_i = +0.67h_o$

29. $d_i = -30$ cm; $h_i = 9.0$ cm; image is virtual, upright, and magnified

31. From the mirror equation, $1/(2f) + 1/d_i = 1/f$. So $1/d_i = 1/f - 1/(2f) = 1/(2f)$, or $d_i = 2f$. $M = -d_i/d_o = -(2f)/(2f) = -1$. Therefore, the image is inverted ($M < 0$), and the same size as the object ($|M| = 1$).

33. **(a)** convex, because a concave mirror can only form magnified virtual images. **(b)** 14 cm

35. **(a)** concave, because only concave mirror can form magnified images. **(b)** 13.3 cm

37. f is negative for a convex mirror. So $d_i = d_of/(d_o - f) = d_o(-|f|)/(d_o + |f|) < 0$. Also $M = -d_i/d_o = -d_o(-|f|)/[d_o(d_o + |f|)] = |f|/(d_o + |f|) < +1$. Therefore the image is virtual (negative d_i), upright (positive M, and reduced ($|M| < 1$).

39. **(a)** concave, because only concave mirror can form real images (formed on a screen). **(b)** 24 cm

41. **(a)** virtual and upright **(b)** 1.5 m

43. 2.3 cm

45. **(a)**

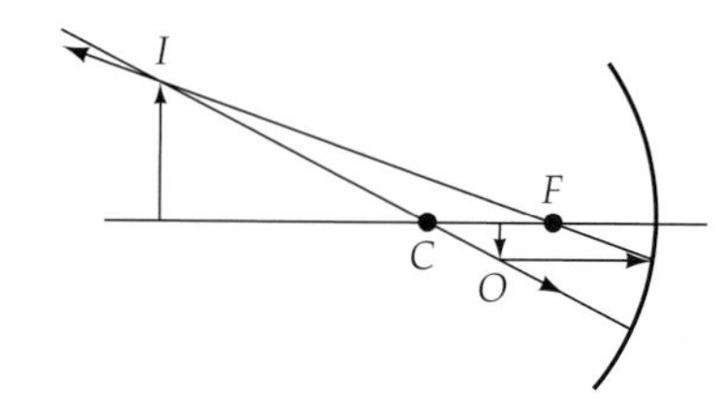

(b) $d_i = 60$ cm, $M = -3.0$, real and inverted

47. 10 m

49. **(a)** Front surface: 60 cm, real, inverted, and magnified; back surface: 46.7 cm, real, inverted, and magnified **(b)** no, the image of the cube is no longer a cube because different parts of the cube have different magnifications.

51. **(a)** two, one for a real image and another for a virtual image. **(b)** 5.0 cm, 15 cm

53. Yes, it is possible. One is a real image and the other is a virtual image. 13 cm; 27 cm

55. **(d)**

57. When the fish is inside the focal point, the image is upright, virtual, and magnified.

59. You can locate the image of a distant object. The distance from the converging lens to the image is the focal length. No, the same method won't work for a diverging lens because a diverging lens does not form real images of real objects.

61. $d_i = 12.5$ cm; $M = -0.250$

63. 22 cm

65. **(a)** $f = 5.9$ cm **(b)** 67 cm, inverted

67. **(a)** 18 cm **(b)** 6.0 cm

69. 14 cm

71. 0.55 mm

73. **(a)** 20 cm **(b)** $M = -1.0$

75. **(a)** $d = 4f$ **(b)** approaches 0

77. **(a)** −18 cm **(b)** −63 cm

79. 4.2 cm

81. 18 cm to the left of the eyepiece; virtual image

83. $M_1 = -h_{i1}/h_{o1}$, $M_2 = -h_{i2}/h_{o2}$, and $M = h_{i2}/h_{o1}$. Since $h_{o2} = h_{i1}$ (the image formed by the first lens is the object for the second lens), $M_1 M_2 = (h_{i1}/h_{o1})(h_{i2}/h_{o2}) = h_{i2}/h_{o1} = M_{total}$.

85. **(b)**

87. **(b)**

89. Our eyes are "designed" or used to seeing things clearly when our surroundings are air. When you are underwater, the index of refraction of the surroundings (water now) changes. From Exercise 23.80(a), the focal length of the eyes changes so everything is blurry. When you wear goggles, the surroundings of the eyes are again air, so you can see things clearly.

91. **(a)** yes, increases **(b)** diverging lens becomes converging lens and vice versa

93. +4.0 D

95. −0.70 D

97. 85 cm

99. The image formed by the converging lens is at the mirror. This image is the object for the diverging lens. If the mirror is at the focal point of the diverging lens, the rays refracted after the diverging lens will be parallel to the axis. These rays will be reflected back parallel to the axis by the mirror and will form another image at the mirror. This second image is now the object for the converging lens. By reversing the rays, a sharp image is formed on the screen located where the original object is. Therefore the distance from the diverging lens to the mirror is the focal length of the diverging lens.

101. 20 cm on object side of first lens, inverted, $M_{total} = -1.0$

103. 60 cm to right of second lens; real and upright, $M_{total} = 4.0$

Chapter 24

1. **(b)**

3. **(b)**

5. Since $\Delta y = L\lambda/d \propto \lambda$, the spacing between the maxima would increase if wavelength increases.

7. 3.4%

9. 0.37°

11. 489 nm

13. **(a)** 440 nm **(b)** 4.40 cm

15. **(a)** (1) increase, because $\Delta y = L\lambda/d \propto 1/d$, the distance between the maxima would decrease if the distance between the slits were increased. **(b)** 1.9 mm **(c)** 2.4 mm

17. **(a)** $\lambda = 402$ nm, violet **(b)** 3.45 cm

19. 450 nm

21. **(a)**

23. The wavelengths that are not visible in the reflected light are all wavelengths except bluish purple.

25. It is always dark because of destructive interference due to the 180° phase shift. If there had not been the 180° phase shift, zero thickness would have corresponded to constructive interference.

27. **(a)** 30λ **(b)** destructively

29. 54.3 nm

31. **(a)** (2) 600 nm, because $t_{min} = \lambda/(4n_1)$ or $t \propto \lambda$. **(b)** 160 nm; 200 nm

33. **(a)** 158.2 nm **(b)** 316.4 nm

35. 1.51×10^{-6} m

37. **(a)**

39. no; yes (barely, at $\theta = 90°$)

41. According to $d \sin\theta = n\lambda$, the advantage is a wider diffraction pattern, as d is smaller.

43. **(a)** 5.4 cm **(b)** 2.7 cm

45. **(a)** 4.3 mm **(b)** microwave

47. 1.24×10^3 lines/cm

49. blue, 18.4°, 39.2°; red: 30.7°, not possible

51. **(a)** 2.44×10^3 lines/cm **(b)** 11 (maximum n is 5)

53. **(a)** (1) blue, because it has a shorter wavelength. From $d \sin\theta = n\lambda$, we can see that the shorter the wavelength, the smaller the $\sin\theta$ or θ. **(b)** blue: 18.7°; red: 34.1°

55. From $d \sin\theta = n\lambda$, $\theta = \sin^{-1} n\lambda/d$. For violet, $\theta_{3v} = \sin^{-1}(3)(400\text{ nm})/d = \sin^{-1}(1200\text{ nm})/d$. For yellow-orange, $\theta_{2y} = \sin^{-1}(2)(600\text{ nm})/d = \sin^{-1}(1200\text{ nm})/d$. So $\theta_{3v} = \theta_{2y}$, that is, they overlap.

57. **(d)**

59. **(c)**

61. **(a)** twice **(b)** four times **(c)** none **(d)** six times

63. The numbers appear and disappear as the sunglasses are rotated because the light from the numbers on a calculator is polarized.

65. **(a)** (1) also increase, because $\tan\theta_p = n_2/n_1 = n_2\,(n_1 = 1)$. If n_2 increases, so does θ_p. **(b)** 58°, 61°

67. **(a)** (2) decrease, because the transmitted light intensity depends on $\cos^2\theta$. As θ increases from 0° to 90°, $\cos\theta$ decreases. **(b)** $0.500I_o$, $0.375I_o$ **(c)** $0.500I_o$, $0.125I_o$

69. 55°

71. 57.2°

73. In water, $\theta_p = \tan^{-1}(n_2/n_1)$. The angle of incidence at the water–glass interface must be $\theta_p = \tan^{-1}(1.52/1.33) = 48.8°$. For the air–water interface, $n_1 \sin\theta_1 = n_2 \sin\theta_2$, so $\sin\theta_1 = n_2 \sin\theta_2/n_1 = (1.33)\sin 48.8° > 1$. Since the maximum of $\sin\theta_1$ is 1, the answer is no.

75. **(c)**

77. **(a)** This is caused by the variable air molecule density. **(b)** There is no air on the surface of the Moon, and so an astronaut would see a black sky.

79. no, because $n_2 = 1$ (air), $\tan\theta_p = n_2/n_1 = 1/n_1$. For total internal reflection, $\sin\theta_c = n_2/n_1 = 1/n_1$. That means $\tan\theta = \sin\theta$. This is not possible for any angle that is not equal to zero. **(b)** 29.8°

81. $\Delta y = 0.25\Delta y_o$

83. $n = 1$ for red; $n = 2$ for violet

Chapter 25

1. **(b)**

3. **(a)**

5. The pre-flash occurs before the aperture is open and the film exposed. The bright light causes the iris to reduce down (giving a small pupil) so that when the second flash comes momentarily, you don't have a wide opening through which you get the red-eye reflection from the retina.

7. **(a)** The eye is nearsighted because the far point is not at infinity. **(b)** The eye is farsighted because the near point is not 25 cm. **(c)** (a), diverging; for (b), converging

9. **(a)** +5.0 D **(b)** −2.0 D

11. **(a)** (1) converging, because the person is farsighted. **(b)** +2.0 D

13. diverging, −0.500 D

15. **(a)** take them out **(b)** +3.0 D

17. **(a)** (1) converging, because he is farsighted. **(b)** +3.3 D

19. **(a)** −0.505 D **(b)** −0.500 D

21. 6.7 m

23. **(a)** −0.67 D **(b)** yes, 21 cm **(c)** 30–40 years old

25. right: +1.42 D, −0.46 D; left: +2.16 D, −0.46 D

27. **(d)**

29. The object should be inside the focal length. When the object is inside the focal length, the image is virtual, upright, and magnified.

31. **(a)** 2.3× **(b)** 2.5×

33. 2.5×

35. **(a)** (1) high powered, because a high powered lens has short focal length and the magnification is $1 + (25\text{ cm})/f$. **(b)** 1.9× and 1.6×

37. −375×

39. **(a)** (2) The one with the shorter focal length, because the total magnification is inversely proportional to the focal length of the objective. **(b)** −280× and −360×

41. **(a)** −340× **(b)** 3900%

43. 25×

45. **(a)** greatest: 1.6 mm/10×; least: 16 mm/5× **(b)** $M_{max} = -930×$; $M_{min} = -42×$;

47. **(b)**

49. No, the whole star can still be seen. The obstruction will reduce the intensity or brightness of the image.

51. The one with the shorter focal length should be used as the eyepiece for a telescope. The magnification of the telescope is inversely proportional to the focal length of the eyepiece ($m = -f_o/f_e$).

53. **(a)** −4.0× **(b)** 75 cm

55. 1.00 m and 2.0 cm

57. 5.0 cm

59. **(a)** 60.0 cm and 80.0 cm; 40.0 cm and 90.0 cm **(b)** −75×; −44×

61. **(a)**

63. Smaller minimum angle of resolution corresponds to higher resolution because smaller angle of resolution means more details can be resolved.

65. From a resolution point of view, the smaller camera (lens) has lower resolution. The smaller the lens, the greater the minimum angle of resolution, the lower the resolving power.

67. 550 nm

69. 1.32×10^{-7} rad; θ_{min} by Hale is 1.6 times as great

71. **(a)** (3) blue, because the minimum angle of resolution is proportional to wavelength and blue has the shorter wavelength. **(b)** 9.6×10^{-5} rad and 1.1×10^{-4} rad

73. 17 km

75. 4.1×10^{16} km
77. (a) 5.55×10^{-5} rad **(b)** blue **(c)** 33.3%
79. (d)
81. With red light, red and white appear red; blue appears black. With green light, only white appears green; both red and blue appear black. With blue light, red appears black; white and blue appear blue.
83. The liquid is dark or colored because it absorbs all light except that color. The amount of light absorbed by an object always depends on how much material is absorbing the light. Foam has very low material density and it can only absorb very little light or almost all light is reflected; therefore, the foam is generally white.
85. From thin lens equation:
$d_o = d_i f/(d_i - f)$, we have
$d_i/d_o = (d_i - f)/f = [-(D - d) - f]/f$. By small angle approximation:
$m = \theta_i/\theta_o = (y_i/D)/[y_o/(25 \text{ cm})]$
$= (y_i/y_o) \times [(25 \text{ cm})/D]$. By similar triangles: $y_i/y_o = -d_i/d_o$, the negative sign is introduced because d_i is negative (virtual image). So
$m = \{[(D - d) + f]/f\} \times [(25 \text{ cm})/D]$
$= (25/f) \times (1 - d/D) + 25/D$.
87. (a) (1) B, (2) A **(b)** $-110\times$, 8.95×10^{-7} rad
89. (a) 6.3 and 0.25 **(b)** 1/120 s

Chapter 26

1. (d)
3. (b)
5. The driver invokes a "fictitious" force in the backward direction so the objects appeared to accelerate backward and Newton's law is still valid. Observer on street simply observes the car accelerates out from under the objects which stay at rest since the net force on them is zero.
7. (a) 3.38 s **(b)** 3.58 s
9. 55 m/s and 45 m/s
13. (a)
15. The speed is c, since the speed of light is the same regardless of motion of the source or observer.
17. In frame O, the bullet takes 1 s to hit target. Light takes 10^{-6} s to get to the target (the time for light of speed 3.00×10^8 m/s to travel 300 m). Frame O' would have to travel to the right. The light flashes from the gun reaches the target in 10^{-6} s. The observer in frame O' would have to cover the 300 m in less than 10^{-6} s to intercept the signals at the same time, which means $v > c$. Since $v > c$ is not possible, all observers agree that the gun fires before the bullet hits the target.
19. (c)
21. (a)
23. No, this is not possible. From the boy's view, the barn is moving at the same speed, so it would appear to contract and be even shorter than 4.0 m.
25. 23 min
27. 14.1 m
29. $0.998c$
31. (a) (1) a longer time due to time dilation **(b)** 7.2 years and 5.7 years
33. 5.3 m
35. $0.87c$
37. 2.14×10^{-14} m
39. $0.60c$
41. (b)
43. No, there are no such limits on momentum and energy as $p = \gamma m v$ and $E = \gamma m c^2$. γ and m can be anything from 0 to ∞.
45. Yes, because the kinetic energy is much less than the rest-energy of the electron (2 keV $\ll$ 511 keV). No, because the kinetic energy is considerably greater than the rest-energy of the electron (2 MeV $\gg$ 0.511 MeV).
47. (a) $0.985c$ **(b)** 2.50 MeV **(c)** 1.56×10^{-21} kg·m/s
49. 0.15 kg
51. 1.6×10^{24} J, or 150000 times more!
53. 79 keV
55. (a) $0.92c$ **(b)** 1.4×10^3 MeV
57. 1.43×10^5 years
59. (a) 1.13 MeV **(b)** 1.64 MeV
63. (a)
65. (c)
67. Drop the cup with the pole vertical. By the principle of equivalence, the weight of the ball appears to be zero in the falling reference frame. The ball is then subject only to the tension force of the stretched rubber band and is pulled inside the cup.
69. 1.8×10^{19} kg/m^3
71. $0.43c$
73. $-0.154c$ (toward Earth)
75. (a) $0.988c$ to the left **(b)** $0.988c$ to the right
77. (a) 79 keV **(b)** 590 keV **(c)** 1.6×10^{-22} kg·m/s
79. (a) 2.5×10^{10} kWh **(b)** 6 days
81. 1.9 kg
83. (a) $5.31m$ **(b)** $2.31mc^2$ **(c)** $0.0382c$ left

Chapter 27

1. (e)
3. (d)
5. When it is really hot, it actually radiates in all wavelengths (although not in equal amounts). The brain interprets the combination of all wavelengths as "white."
7. 3.10×10^{13} Hz
9. 1.06×10^{-5} m, 2.83×10^{13} Hz
11. 2.56×10^{-20} J
13. 2.9×10^{19} quanta/(s·m^2)
15. (b)
17. (a)
19. (a)
21. Yes, it is possible. Energy depends on two things: the number of photons and frequency. A beam of lower frequency can have more energy if it has a lot more photons.
23. 6.0×10^{-11} m; X-ray
25. (a) 1.32×10^{-18} J **(b)** 8.27 eV
27. 3.0 V
29. (a) 6.7×10^{-34} J·s **(b)** 2.9×10^{-19} J
31. 6.8×10^{14} Hz
33. 2.48 eV
35. (a) 0.625 V **(b)** 2.33 eV **(c)** zero
37. 0.44 eV
39. (c)
41. The maximum wavelength shift for Compton scattering from a neutron is smaller compared with that from a free electron because the Compton wavelength is inversely proportional to the mass of the scattering particle and $\Delta\lambda_{max} = 2\lambda_C = 2h/(mc) \propto 1/m$. Since the mass of a neutron is about 1836 times that of electron, it is smaller by a factor of 1836.
43. to conserve total linear momentum
45. 2.43×10^{-3} nm
47. 0.281 nm
49. (a) (2) less than 5.0 keV but not zero, according to the conservation of momentum and energy, the electron must recoil so it has some kinetic energy at the expense of the photon energy. The photon has some energy but less than its initial amount. **(b)** 20 eV
51. 18.5°
53. (d)
55. (c)
57. It takes less energy to ionize the electron that is in an excited state than in the ground state. The excited state already has more energy.
59. (a) 10.2 eV **(b)** 1.89 eV **(c)** The first is UV; the second visible (red)
61. (a) 6.89×10^{14} Hz **(b)** 8.21×10^{14} Hz
63. $n \approx 100$
65. (a) (1) increase, but not necessarily double, because part of the energy is used to overcome the ionization energy which does not double. Actually the energy of the emitted electron will more than double because the portion of energy left for the electron (after ionization energy) will be more than double. **(b)** 15.4 eV
67. (a) $n = 2$ to $n = 1$ **(b)** $\Delta E_{53} = 0.967$ eV; $\Delta E_{62} = 2.97$ eV; $\Delta E_{21} = 10.2$ eV
69. (a) 54.4 eV **(b)** 122 eV
73. (a) potential is -27.2 eV and kinetic is $+13.6$ eV **(b)** $|U| = 2K$, potential energy is twice as large in magnitude.
75. (b)
77. Since tattoo is blue it reflects blue, red end of spectrum would work the best to maximize absorption.
79. In a spontaneous emission, electrons jump from a higher-energy state to a lower-energy state without any external stimulation, and a photon is released in the process.
Stimulated emission is an induced emission. The electron in the higher-energy orbit can jump to a lower-energy orbit when a photon of energy equals the difference of the energy between the two orbits is introduced. Once atoms are prepared with enough electrons in the higher-energy state, stimulating photons trigger them to jump down to the lower-energy state. The emitted photons trigger the rest of the electrons and eventually all the electrons will be in the lower energy state.
81. (a) 9.82 keV **(b)** first: 48 eV, second: 134 eV
83. (a) 1.02 MeV **(b)** 1.21×10^{-3} nm **(c)** 1.02 MeV
85. (a) 2.00 MeV **(b)** about 1 keV

Chapter 28

1. (c)
3. (b)
5. The baseball, as its momentum is smaller due to its smaller mass. The de Broglie wavelength is inversely proportional to the mass, $\lambda = h/mv$.

7. 2.7×10^{-38} m
9. $\lambda_e/\lambda_p = 43$
11. 1.5×10^4 V
13. (a) (3) decrease due to the potential difference. The proton gains speed from the potential difference and the de Broglie wavelength is inversely proportional to speed, $\lambda = h/mv$. **(b)** -53% (a decrease)
15. 24 V
17. (a) 3.71×10^{-63} m **(b)** 1.18×10^{72} **(c)** There would be an increase but undetectable because 1 compared to 1.18×10^{72} is negligible.
19. (b)
21. When the distance increases due to the dip, the current decreases. To keep the current constant, the tip should move toward the surface to make up for decreasing current.
23. According to modern quantum theory and the Schrödinger equation, there is a probability that if you ran into a wall you could end up on the other side. (Don't try this!)
25. (a) 4.07 MeV, 16.3 MeV, 36.6 MeV **(b)** 32.5 MeV, gamma ray
27. (c)
29. The principle quantum number, n, gives information about the total energy and orbit radius of an orbit in a hydrogen atom.
31. The periodic table groups elements according to the values of quantum numbers n and l. Within a group, the elements have the same or very similar electronic configurations for the outmost electrons.
33. (a) 2 **(b)** 14
35. (a) $\ell = 2$ **(b)** $n = 3$
37.

Na has 11 electrons

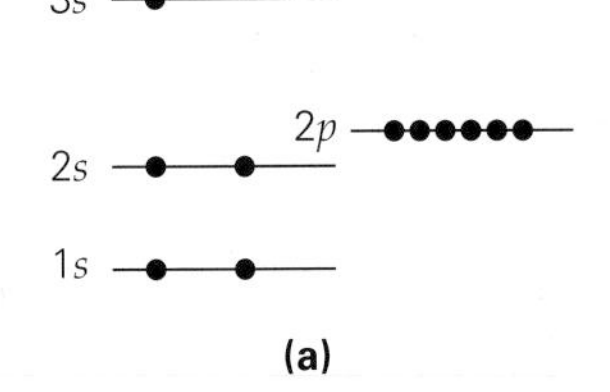

(a)

Ar has 18 electrons

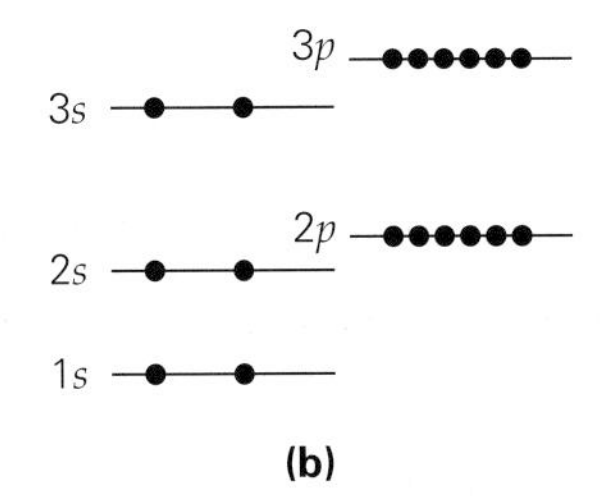

(b)

39. (a) $1s^2 2s^2 2p^1$ **(b)** $1s^2 2s^2 2p^6 3s^2 3p^6 4s^2$ **(c)** $1s^2 2s^2 2p^6 3s^2 3p^6 3d^{10} 4s^2$ **(d)** $1s^2 2s^2 2p^6 3s^2 3p^6 3d^{10} 4s^2 4p^6 4d^{10} 5s^2 5p^2$
41. It would be a $1s^3$ and would be the first closed shell inert gas. With three spin orientations, s states can contain three electrons without violating the Pauli exclusion principle.
43. (b)
45. According to the uncertainty principle, the product of uncertainty in position and the uncertainty in momentum is in the order of Plank's constant. A bowling ball's large diameter and momentum (mostly due to its large mass) make the uncertainty in them (determined by the extremely small value of Planck's constant) negligible. However for an electron, with its very small size and momentum (mostly due to its very small mass), the uncertainties cannot be ignored.
47. (a) (2) the same, because both have the same uncertainty in momentum. **(b)** both 0.21 m
49. 2.1×10^{-29} m/s
51. 1.1×10^{-27} J
53. (a) (1) smaller than, according to the uncertainty principle, $\Delta E \Delta t \geq h/(2\pi)$. The greater the lifetime, Δt, the smaller the energy uncertainty, therefore the smaller the width of spectral line. **(b)** 10^4
55. (a)
57. Linear momentum needs to be conserved so the particles will be moving afterwards because the initial photon has momentum. Therefore the electron-positron pair carry kinetic energy. The supplied energy needs to create the masses plus the kinetic energy.
59. (a) (3) Pair production can happen but the resulting particles must be moving to conserve momentum. **(b)** 0.46 MeV
61. 1.9 GeV
63. 27.6 V
65. 1.37×10^{36} photons

Chapter 29

1. (a)
3. (d)
5. Carbon-13 has 7 neutrons. Since nitrogen has 7 protons, the nitrogen isotone of carbon-13 should have 7 neutrons so it is nitrogen-14.
7. (a) 20p, 23n, 20e **(b)** 82p, 124n, 82e
9. ${}^{41}_{19}\mathrm{K}$
11. 92p, 143n, 92e
13. (a) $R_{He} = 1.9 \times 10^{-15}$ m; $R_{Ne} = 3.3 \times 10^{-15}$ m; $R_{Ar} = 4.1 \times 10^{-15}$ m; $R_{Kr} = 5.3 \times 10^{-15}$ m; $R_{Xe} = 6.1 \times 10^{-15}$ m; $R_{Rn} = 7.3 \times 10^{-15}$ m **(b)** $M_{He} = 6.68 \times 10^{-27}$ kg; $M_{Ne} = 3.34 \times 10^{-26}$ kg; $M_{Ar} = 6.68 \times 10^{-26}$ kg; $M_{Kr} = 1.40 \times 10^{-25}$ kg; $M_{Xe} = 2.20 \times 10^{-25}$ kg; 3.71×10^{-25} kg **(c)** 2.3×10^{17} kg/m^3; yes, the answer surprises because the density is huge.
15. (d)
17. (a)
19. β^+ decay and/or electron capture; in β^+ decay, ${}^{18}_{9}\mathrm{F} \rightarrow {}^{18}_{8}\mathrm{O} + {}^{0}_{1}\mathrm{e}$, so the resulting nucleus is ${}^{18}\mathrm{O}$; in an electron capture, ${}^{18}_{9}\mathrm{F} + {}^{0}_{-1}\mathrm{e} \rightarrow {}^{18}_{8}\mathrm{O}$, so the resulting nucleus is ${}^{18}\mathrm{O}$.
21. (a) Tritium, ${}^{3}_{1}\mathrm{H}$, has one extra neutron compared to the stable deuterium (${}^{2}_{1}\mathrm{H}$). We expect (2) β^- decay. **(b)** helium-3; yes, it's stable
23. (a) ${}^{237}_{93}\mathrm{Np} \rightarrow {}^{233}_{91}\mathrm{Pa} + {}^{4}_{2}\mathrm{He}$ **(b)** ${}^{32}_{15}\mathrm{P} \rightarrow {}^{32}_{16}\mathrm{S} + {}^{0}_{-1}\mathrm{e}$ **(c)** ${}^{56}_{27}\mathrm{Co} \rightarrow {}^{56}_{26}\mathrm{Fe} + {}^{0}_{+1}\mathrm{e}$ **(d)** ${}^{56}_{27}\mathrm{Co} + {}^{0}_{-1}\mathrm{e} \rightarrow {}^{56}_{26}\mathrm{Fe}$ **(e)** ${}^{42}_{19}\mathrm{K}^* \rightarrow {}^{42}_{19}\mathrm{K} + \gamma$
25. (a) α–β: ${}^{209}_{82}\mathrm{Pb} + {}^{0}_{1}\mathrm{e} \rightarrow {}^{209}_{81}\mathrm{Tl}$; ${}^{209}_{81}\mathrm{Tl} + {}^{4}_{2}\mathrm{He} \rightarrow {}^{213}_{83}\mathrm{Bi}$. **(b)** β–α: ${}^{209}_{82}\mathrm{Pb} + {}^{4}_{2}\mathrm{He} \rightarrow {}^{213}_{84}\mathrm{Po}$; ${}^{213}_{84}\mathrm{Po} + {}^{0}_{-1}\mathrm{e} \rightarrow {}^{213}_{83}\mathrm{Bi}$.
27. (a) ${}^{234}_{90}\mathrm{Th}$ **(b)** ${}^{0}_{-1}\mathrm{e}$ **(c)** 3 **(d)** γ **(e)** ${}^{22}_{11}\mathrm{Na}$
29. α to ${}^{233}\mathrm{Pa}$; β^- to ${}^{233}\mathrm{U}$; α to ${}^{229}\mathrm{Th}$; α to ${}^{225}\mathrm{Ra}$; β^- to ${}^{225}\mathrm{Ac}$; α to ${}^{221}\mathrm{Fr}$; α to ${}^{217}\mathrm{At}$; α to ${}^{213}\mathrm{Bi}$; α to ${}^{209}\mathrm{Tl}$; β^- to ${}^{209}\mathrm{Pb}$; β^- to ${}^{209}\mathrm{Bi}$.
Or
α to ${}^{233}\mathrm{Pa}$; β^- to ${}^{233}\mathrm{U}$; α to ${}^{229}\mathrm{Th}$; α to ${}^{225}\mathrm{Ra}$; β^- to ${}^{225}\mathrm{Ac}$; α to ${}^{221}\mathrm{Fr}$; α to ${}^{217}\mathrm{At}$; α to ${}^{213}\mathrm{Bi}$; β^- to ${}^{213}\mathrm{Po}$; α to ${}^{209}\mathrm{Pb}$; β^- to ${}^{209}\mathrm{Bi}$.
31. (c)
33. None, it is totally independent of temperature, environment, and chemistry.
35. (a) infinite **(b)** zero
37. (a) 7.4×10^8 decays/s **(b)** 4.4×10^{10} betas/min
39. 2.06×10^{-8} W
41. 28.6 years
43. (a) (2) younger than, because the activity decreases as time passes so a higher activity indicates a short time, that is, younger. **(b)** 1.1×10^4 years
45. 3.1%
47. 0.98 μg
49. (a) 6.75×10^{19} nuclei **(b)** 2.11×10^{18} nuclei
51. 9.2 mg
53. 7.6×10^6 kg
55. (d)
57. (a)
59. ${}^{238}_{92}\mathrm{U}$ is even-even so tends to be more stable than ${}^{235}_{92}\mathrm{U}$ which is even-odd. Even so, both are unstable because $Z > 82$, but ${}^{238}_{92}\mathrm{U}$ has a longer half-life, that is, it is closer to stability.
61. (a) ${}^{17}_{8}\mathrm{O}$ **(b)** ${}^{42}_{20}\mathrm{Ca}$ **(c)** ${}^{10}_{5}\mathrm{B}$ **(d)** approximately the same
63. 2.013 553 u
65. (a) 92.2 MeV **(b)** 7.68 MeV/nucleon
67. deuterium; H-2: 1.11 MeV/nucleon; H-3: 2.83 MeV/nucleon
69. 104.7 MeV
71. 10.1 MeV
73. 7.59 MeV/nucleon
75. (d)
77. (b)
79. RBE for X-rays is 1 and 20 for alphas from Table 29.4. Equal absorbed doses mean 20 times more effective dose for alphas, thus the effective dose of an alpha particle would be 20 times that of an X-ray.
81. 36 μCi
83. yes
85. 0.266 Sv or 26.6 rem
87. (a) ${}^{215}_{83}\mathrm{Bi} \rightarrow {}^{215}_{84}\mathrm{Po} + {}^{0}_{-1}\mathrm{e} + \bar{\nu}_e$ **(b)** 3.4×10^{22} nuclei **(c)** 1.8×10^{16} nuclei **(d)** 10 min: 1.6×10^{20} Bq $= 4.4 \times 10^9$ Ci; 1 h: 8.6×10^{13} Bq $= 2.3 \times 10^3$ Ci
89. (a) 11.5 MeV **(b)** 1.08×10^{-13} m
91. (a) ${}^{223}_{88}\mathrm{Ra} \rightarrow {}^{219}_{86}\mathrm{Rn} + {}^{4}_{2}\mathrm{He}$ and ${}^{219}_{85}\mathrm{At} \rightarrow {}^{219}_{86}\mathrm{Rn} + {}^{0}_{-1}\mathrm{e} + \bar{\nu}_e$ **(b)** ${}^{219}_{86}\mathrm{Rn} \rightarrow {}^{215}_{84}\mathrm{Po} + {}^{4}_{2}\mathrm{He}$.
93. (a) ${}^{2}_{1}\mathrm{H} + {}^{1}_{0}\mathrm{n} \rightarrow {}^{3}_{1}\mathrm{H} + \gamma$ **(b)** 71.4% **(c)** 6.26 MeV **(d)** ${}^{3}_{1}\mathrm{H} \rightarrow {}^{3}_{2}\mathrm{He} + {}^{0}_{-1}\mathrm{e} + \bar{\nu}_e$ **(e)** 18.6 keV

Chapter 30

1. (a)
3. (c)
5.

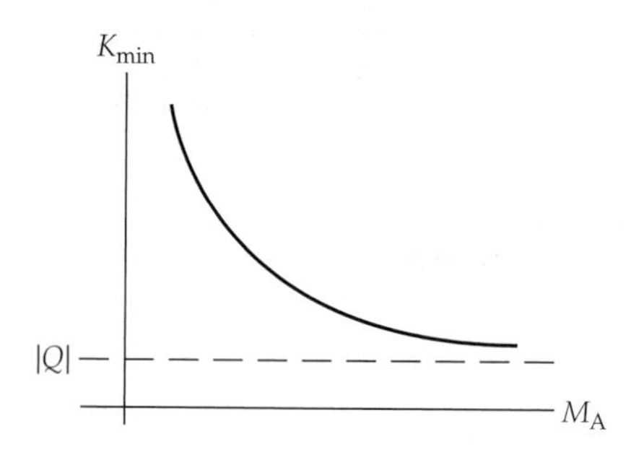

As the target mass M_A becomes very large ($M_A \gg m_a$), $K_{min} \to |Q|$. Physically, the target would remain at rest so no energy is "lost" to recoil.

7. Elastic scattering occurs at all energies so $K_{th} = 0$. Another way to look at this is that there is no mass change in the scattering.

9. **(a)** γ **(b)** ${}^2_1$H **(c)** n or ${}^1_0$n **(d)** ${}^{17}_{8}$O **(e)** p or ${}^1_1$H

11. For a reaction to occur spontaneously, the Q value must be greater than zero. **(a)** ${}^{22}_{11}$Na, no **(b)** ${}^{222}_{86}$Rn, yes **(c)** ${}^{12}_{6}$C, no

13. **(a)** (1) positive, because this is a decay (spontaneous) process. **(b)** +4.28 MeV

15. 4.36 MeV

17. exoergic, $Q = +6.50$ MeV

19. 5.38 MeV

21. **(a)** (1) greater, because $K_{min} = (1 + m_a/M_A)|Q|$. The difference in Q value (3 times) is greater than the difference caused by the target mass (15 and 20). **(b)** 3.05

23. (d)

25. (a)

27. This is because boron has a high probability of absorbing neutrons and it absorbs many neutrons that might create fission in other fuel rods.

29. **(a)** 2.22 MeV **(b)** 12.9 MeV

31. 36 collisions

33. (c)

35. $p + \bar{\nu}_e \to n + \beta^+$ and $p + n + \to {}^2_1H + \gamma$. The 2.22 MeV gamma ray is from the formation of deuterium. The two 0.511 MeV gamma rays are the result of the pair annihilation when the positron meets an electron.

37. 4.69×10^{-13} m

41. 0.71 MeV

43. 9 keV

45. (b)

47. The forces created by the virtual exchange particles can be predicted and these predictions agree with experiment (scientific method).

49. 3.5×10^{-28} kg

51. **(a)** (3) decreases, because if energy increases, the mass increases, and the range is inversely proportional to the mass, $R = h/(2\pi m_m c)$. **(b)** 1.98×10^{-16} m

53. 4.71×10^{-24} s

55. (a)

57. All hadrons contain quarks and/or antiquarks. Quarks are not believed to exist freely outside the nucleus.

59. **(a)** π^+ **(b)** K^o **(c)** Σ^o **(d)** Ξ^-

61. **(a)** (3) *udd* **(b)** the neutron has no charge, $\frac{2}{3}e - \frac{1}{3}e - \frac{1}{3}e = 0$

63. (a)

65. 4.03 MeV

69. 2.35 MeV

PHOTO CREDITS

FM—AU1 Jerry Wilson **AU2** Anthony Buffa **AU3** Bo Lou

Chapter 1—CO.1 Donald Miralle/Donald Miralle/Allsport Concepts/Getty Images **Fig. 1.2b** International Bureau of Weights and Measures, Sevres, France **Fig. 1.3b** National Institute of Standards and Technology (NIST) **Fig. 1.4** IBM Research, Almaden Research Center. **Fig. 1.6** Frank Labua/Pearson Education/PH College **Fig. 1.8a** Jerry Wilson/Lander University **Fig. 1.8b** Jerry Wilson/Jerry Wilson **Fig. 1.9** T. Kuwabara/Don W. Fawcett/Visuals Unlimited **Fig. 1.9.1** JPL/NASA **Fig. 1.10** John Smith/Jerry Wilson **Fig. 1.15** Peter Brock/Getty Images, Inc./Liaison **Fig. 1.17** Frank Labua/Pearson Education/PH College **Fig. 1.18** Dennis Kunkel/©Dennis Kunkel Microscopy Inc.

Chapter 2—CO.2 David A. Northcott/Corbis/Bettmann **Fig. 2.2** John Smith/Jerry Wilson **Fig. 2.3** JPL/NASA Headquarters **Fig. 2.5** Royalty-Free/© Royalty-Free/CORBIS **Fig. 2.14b** James Sugar/Stockphoto.com/Black Star **Fig. 2.14.1** North Wind Picture Archives **Fig. 2.14.2** AP Wide World Photos **Fig. 2.15a** Frank Labua/Pearson Education/PH College **Fig. 2.15b** Frank Labua/Pearson Education/PH College **Fig. 2.27** AP Wide World Photos

Chapter 3—CO.3 C. E. Nagele/Edmund Nagele F. R. P. S. **Fig. 3.10b** Richard Megna/Educational Development Center/© Richard Megna Fundamental Photographs, NYC **Fig. 3.12** Paul Thompson; Ecoscene/© Paul Thompson; Ecoscene/CORBIS **Fig. 3.16b** John Garrett/John Garrett © Dorling Kindersley **Fig. 3.17a** JAVIER SORIANO/AFP/Javier Soriano/Agence France Presse/Getty Images **Fig. 3.17b** Andrew D. Bernstein/Photo by Andrew D. Bernstein/NBAE/Allsport Concepts/Getty Images. **Fig. 3.19c** Photri-Microstock, Inc.

Chapter 4—CO.4 AP Wide World Photos **Fig. 4.1** The Granger Collection **Fig. 4.4** John Smith/Jerry Wilson **Fig. 4.6.1** Science Photo Library/Photo Researchers, Inc. **Fig. 4.12.2a** AP Wide World Photos **Fig. 4.12.2b** Francois Moussis/Sygma/Corbis/Sygma **Fig. 4.14a** Ronald Brown/Arnold & Brown Photography **Fig. 4.14b** Ronald Brown/Arnold & Brown Photography **Fig. 4.18a** Vandystadt/Photo Researchers, Inc. **Fig. 4.18b** Charles Krebs/Corbis/Stock Market **Fig. 4.23** M. Ferguson/PhotoEdit **Fig. 4.25** Jump Run Productions/Getty Images Inc./Image Bank **Fig. 4.27a** Prentice Hall School Division **Fig. 4.27b** Prentice Hall School Division **Fig. 4.29** Tom Hauck/Allsport Concepts/Getty Images **Fig. 4.30** Tony Buffa **Fig. 4.38** Michael Dunn/Corbis/Stock Market **Fig. 4.39L** The Goodyear Tire & Rubber Company **Fig. 4.39R** The Goodyear Tire & Rubber Company **Fig. 4.40** Agence Zoom/Agence Zoom/Allsport Concepts/Getty Images

Chapter 5—CO.5 Harold E. Edgerton/© Harold & Esther Edgerton Foundation, 2002, courtesy of Palm Press, Inc. **Fig. 5.10** Rae Cooper/Construction Photography.com **Fig. 5.12a** Reuters/Corbis/Reuters America LLC **Fig. 5.12b** Vince Streano/The Image Works **Fig. 5.20** The Image Works **Fig. 5.23** SuperStock, Inc. **Fig. 5.24a** Bob Daemmrich/Stock Boston **Fig. 5.24b** Bob Daemmrich/Stock Boston

Chapter 6—CO.6 Jed Jacobsohn/Allsport Concepts/Getty Images **Fig. 6.1a** Gary S. Settles/Photo Researchers, Inc. **Fig. 6.1b** David G. Curran/Rainbow **Fig. 6.1c** Tom & Susan Bean, Inc. **Fig. 6.5a** Omikron/Science Scource/Photo Researchers, Inc. **Fig. 6.8a** Vandystadt/Photo Researchers, Inc. **Fig. 6.8b** Globus Brothers/Corbis/Stock Market **Fig. 6.8.1** Llewellyn/Pictor/ImageState/International Stock Photography Ltd. **Fig. 6.8.2a** NASA Headquarters **Fig. 6.8.2b** Dan Maas/Animation by Dan Maas, Maas Digital LLC ©2002 Cornell University. All rights reserved. This work was performed for the Jet Propulsion Laboratory, California Institute of Technology, sponsored by the United States Government under Prime Contract # NAS7-1 **Fig. 6.8.2c** Dan Maas/Animation by Dan Maas, Maas Digital LLC ©2002 Cornell University. All rights reserved. This work was performed for the Jet Propulsion Laboratory, California Institute of Technology, sponsored by the United States Government under Prime Contract # NAS7-1 **Fig. 6.11a** Carl Purcell/Photo Researchers, Inc. **Fig. 6.11b** H.P. Merten/Corbis/Stock Market **Fig. 6.15a** Richard Megna/Fundamental Photographs, NYC **Fig. 6.15b** Richard Megna/Fundamental Photographs, NYC **Fig. 6.16** Richard Megna/Fundamental Photographs, NYC **Fig. 6.20L** Paul Silverman/Fundamental Photographs, NYC **Fig. 6.20C** Paul Silverman/Fundamental Photographs, NYC **Fig. 6.20R** Paul Silverman/Fundamental Photographs, NYC **Fig. 6.22** John McDermott Photography **Fig. 6.23** Jeff Rotman/Nature Picture Library **Fig. 6.25b** NASA Headquarters **Fig. 6.25c** NASA Headquarters **Fig. 6.28** Jonathan Watts/Science Photo Library/Photo Researchers, Inc. **Fig. 6.30** Runk/Schoenberger/Grant Heilman Photography, Inc. **Fig. 6.34** Fritz Polking/© Fritz Polking/Peter Arnold, Inc. **Fig. 6.37** Charles Krebs/Getty Images Inc./Stone Allstock

Chapter 7—CO.7 Tony Savino/The Image Works **Fig. 7.6b** Tom Tracy/Corbis/Stock Market **Fig. 7.10** Chris Priest/Science Photo Library/Photo Researchers, Inc. **Fig. 7.21** NASA Headquarters **Fig. 7.23a** NASA/Science Source/Photo Researchers, Inc. **Fig. 7.24T** Photo Researchers, Inc. **Fig. 7.33** NASA Headquarters **Fig. 7.34** Frank LaBua/Pearson Education/PH College

Chapter 8—CO.8 AP Wide World Photos **Fig. 8.5a** STEVE SMITH/Getty Images, Inc./Taxi **Fig. 8.10a** David Madison/Getty Images Inc./Stone Allstock **Fig. 8.13a** Jean-Marc Loubat/Agence Vandystadt/Photo Researchers, Inc. **Fig. 8.13b** Richard Hutchings/Photo Researchers Inc. **Fig. 8.14a** Holly Drummond/Holly Drummond **Fig. 8.14b** Holly Drummond/Holly Drummond **Fig. 8.14.1** Albert Einstein(TM) Represented by The Roger Richman Agency, Inc. Beverly Hills, CA 90212. www.hollywoodlegends.com. Photo courtesy of the Archives, California Institute of Technology. **Fig. 8.15a** Gavin Hellier/Nature Picture Library **Fig. 8.15b** Owen Franken/CORBIS/NY **Fig. 8.16b** Beaura Kathy Ringrose/Michael Tobin **Fig. 8.28.a1** Frank LaBua/ Pearson Education/PH College **Fig. 8.28.a2** Frank LaBua/Pearson Education/PH College **Fig. 8.28b1** Brian Bahr/Brian Bahr/Allsport Concepts/Getty Images **Fig. 8.28b2** Brian Bahr/Brian Bahr/Allsport Concepts/Getty Images **Fig. 8.28c** AFP Photo/NOAA/Getty Images, Inc./Agence France Presse **Fig. 8.32.a** Vince Streano/Corbis/Stock Market **Fig. 8.32.b** NASA Headquarters **8.33L** Jerry Wilson **Fig. 8.33R** Courtesy of Arbor Scientific **Fig. 8.49b** Tony Savino Photography/The Image Works **Fig. 8.51** Gerard Lacz/NHPA Limited

Chapter 9—CO.9 Masterfile Corporation **Fig. 9.6.1** ESRF-CREATIS/Photo Researchers, Inc. **Fig. 9.6.3** Yoav Levy/Phototake NYC **Fig. 9.11** Science Photo Library/Photo Researchers, Inc. **Fig. 9.11.2** Blair Seitz/Photo Researchers, Inc. **Fig. 9.12** REUTERS/Alexander Demianchuk/Getty Images, Inc./Liaison **Fig. 9.14** Compliments of Clearly Canadian Beverage Corporation **Fig. 9.15** Ralph A. Clevenger/© Ralph A. Clevenger/CORBIS **Fig. 9.16b** Tom Pantages **Fig. 9.22c** Hermann Eisenbeiss/Photo Researchers, Inc. **Fig. 9.23a** David Spears/Science Photo Library/Photo Researchers, Inc. **Fig. 9.23b** Richard Steedman/Corbis/Stock Market **Fig. 9.25** Patrick Watson/Stockphoto.com/Medichrome/The Stock Shop, Inc. **Fig. 9.27** PORNCHAI KITTIWONGSAKUL/AFP/PORNCHAI KITTIWONGSAKUL/Agence France Presse/Getty Images **Fig. 9.28** Raymond Gehman/CORBIS-NY **Fig. 9.29** Frank LaBua/Pearson Education/PH College **Fig. 9.30L** Charles D. Winters/Photo Researchers, Inc. **Fig. 9.30R** Charles D. Winters/Photo Researchers, Inc. **Fig. 9.37a** Stephen T. Thornton **Fig. 9.37R** Stephen T. Thornton **Fig. 9.38** Michael J. Howell/Stock Boston **Fig. 9.39a** Stephen T. Thornton **Fig. 9.39b** John Smith/Jerry Wilson **Fig. 9.41** Stephen T. Thornton

Chapter 10—CO.10 Jim Corwin/Photo Researchers, Inc. **Fig. 10.2b** Frank LaBua/Pearson Education/PH College **Fig. 10.3a** Leonard Lessin/© Leonard Lessin/Peter Arnold Inc. **Fig. 10.3b** Richard Megna/Fundamental Photographs, NYC **Fig. 10.5.1** Maximilian Stock Ltd./Science Photo Library/Photo Researchers, Inc. **Fig. 10.5.2** Spitzer Science Center, California Institute of Technology (NASA Infrared Processing and Analysis Center). **Fig. 10.6a** Sinclair Stammers/Science Photo Library/Photo Researchers, Inc. **Fig. 10.6b** Sinclair Stammers/Science Photo Library/Photo Researchers, Inc. **Fig. 10.11a** Richard Choy/© Richard Choy/Peter Arnold Inc. **Fig. 10.11b** Joe Sohm/The Image Works **Fig. 10.15a** Paul Silverman/Fundamental Photographs, NYC **Fig. 10.15b** Paul Silverman/Fundamental Photographs, NYC **Fig. 10.15c** Paul Silverman/Fundamental Photographs, NYC **Fig. 10.20a,b,c** Stephen T. Thornton

Chapter 11—CO.11 AP Wide World Photos **Fig. 11.2** Jerry Wilson **Fig. 11.4** John Smith/Jerry Wilson **Fig. 11.8** Frank LaBua/Pearson Education/PH College **Fig. 11.9c** Richard Lowenberg/Science Source/Photo Researchers, Inc. **Fig. 11.9.1** McClain Finlon Advertising Inc. **Fig. 11.13** Jerry Wilson **Fig. 11.14** AFP/STR/Agence France Presse/Getty Images **Fig. 11.15T** Collection CNRI/Phototake NYC **Fig. 11.15B** Collection CNRI/Phototake NYC **Fig. 11.16** John Smith/Jerry Wilson **Fig. 11.18** Carl Glassman/The Image Works **Fig. 11.19b** Bo Lou/Bo Lou **Fig. 11.21** John Smith/Jerry Wilson

Chapter 12—CO.12 Helen Marcus/Photo Researchers, Inc. **Fig. 12.5** Michael Marzelli/Index Stock Imagery, Inc. Royaly Free

Fig. 12.13.1 Francois Gohier/Photo Researchers, Inc. **Fig. 12.16.1** Samuel Ashfield/Photo Researchers, Inc.

Chapter 13—CO.13 CHARLES BENES/Getty Images, Inc./Taxi **Fig. 13.7b** Robert Reiff/Robert Reiff **Fig. 13.8** John Matchett/ACE Photo Agency/©2003 John Matchett/ACE Photo Agency/Picture Perfect **Fig. 13.12.1** Peter Menzel/Peter Menzel Photography **Fig. 13.12.2** H. Yamaguchi/Corbis/Sygma **Fig. 13.13a** Education Development Center, Inc. **Fig. 13.15a** Mark Wagner/Getty Images Inc./Stone Allstock **Fig. 13.18a** Fundamental Photographs, NYC **Fig. 13.18b** Fundamental Photographs, NYC **Fig. 13.19b** Richard Megna/Fundamental Photographs, NYC **Fig. 13.20R1** Richard Megna/Fundamental Photographs, NYC **Fig. 13.20R2** Richard Megna/Fundamental Photographs, NYC **Fig. 13.20R3** Richard Megna/Fundamental Photographs, NYC **Fig. 13.21** SuperStock, Inc. **Fig. 13.21.1** University of Washington Libraries, Special Collections, UW21422 **Fig. 13.22** Lawrence Migdale/Stock Boston **Fig. 13.26T** Courtesy of Cenco **Fig. 13.26B** Courtesy of Cenco **Fig. 13.28** Courtesy of Cenco

Chapter 14—CO.14 Susan McCartney/Photo Researchers, Inc. **Fig. 14.1b** Leonard Lessin/© Leonard Lessin/Peter Arnold, Inc. **Fig. 14.3b** Merlin D. Tuttle/Photo Researchers, Inc. **Fig. 14.3.1** Howard Sochurek/Woodfin Camp & Associates **Fig. 14.3.2** Matt Meadows/Science Photo Library/Photo Researchers, Inc. **Fig. 14.6** Reed Kaestner/© Reed Kaestner/CORBIS **Fig. 14.12b** Philippe Plailly/Science Photo Library/Photo Researchers, Inc. **Fig. 14.12.1a** JAMES KING-HOLMES/Photo Researchers, Inc. **Fig. 14.12.1b** Yoav Levy/Phototake NYC **Fig. 14.12.2a** David R. Frazier/Photo Researchers, Inc. **Fig. 14.12.2b** David P. Wojtowicz, Jr./Department of Atmospheric Sciences, University of Illinois at Urbana-Champaign **Fig. 14.14** Jeff Greenberg/Visuals Unlimited **Fig. 14.15c** Jerry Wilson **Fig. 14.16a** George De Sota/Photo by George De Sota/Getty Images **Fig. 14.18b** Richard Megna/Fundamental Photographs, NYC

Chapter 15—CO.15 Zefa/London/Corbis/Stock Market **Fig. 15.7a** Jerry Wilson/Jerry Wilson **Fig. 15.7c** Charles D. Winters/Photo Researchers Inc. **Fig. 15.16.1b** Keith Kent/© Keith Kent/Peter Arnold Inc. **Fig. 15.16.1c** Philippe Wojazer/Reuters/Corbis/Bettmann **Fig. 15.16.2a** Taser International via Getty Images/Photo Courtesy of Taser International via Getty Images. **Fig. 15.16.2b** David Hoffman/Alamy Images **Fig. 15.16.3a** Pat Morris/Ardea London Limited

Chapter 16—CO.16 Richard Megna/Fundamental Photographs, NYC **Fig. 16.15** Spencer Grant/Photo Researchers, Inc. **Fig. 16.18b** Paul Silverman/©1989 Paul Silverman, Fundamental Photographs

Chapter 17—CO.17 Adam Hart Davis/Photo Researchers, Inc./Walter Dickenman/Sandia National Laboratories **Fig. 17.9** Tom Pantages **Fig. 17.13a** Ronald Brown/Arnold & Brown Photography **Fig. 17.13b** Frank Labua/Pearson Education/PH College **Fig. 17.13c** Frank Labua/Pearson Education/PH College **Fig. 17.14** NASA/Mark Marten/Photo Researchers, Inc. **Fig. 17.15** Frank Labua/Pearson Education/PH College **Fig. 17.19** Tom O'Neil/Photo courtesy of Thomas O'Neil.

Chapter 18—CO.18 Richard Megna/Fundamental Photographs, NYC **Fig. 18.15a** Courtesy of Cenco **Fig. 18.16.2a** Bruce Ayres/Getty Images Inc./Stone Allstock **Fig. 18.18c** Tom Pantages/Tom Pantages **Fig. 18.20b** Frank Labua/Pearson Education/PH College **Fig. 18.21b** Frank LaBua/Pearson Education/PH College **Fig. 18.24a** Frank Labua/Pearson Education/PH College **Fig. 18.24b** Frank Labua/ Pearson Education/PH College **Fig. 18.41** M. Antman/The Image Works

Chapter 19—CO.19a Peter Menzel/Stock Boston/Courtesy: Transrapid International/USA, Inc. **CO.19b** Fritz Stoiber Productions/Photo courtesy of Transrapid International GmbH & Co. KG. **Fig. 19.1** Courtesy of Cenco **Fig. 19.3a** Richard Megna/Fundamental Photographs, NYC **Fig. 19.3b** Richard Megna/Fundamental Photographs, NYC **Fig. 19.3c** Richard Megna/Fundamental Photographs, NYC **Fig. 19.7** Leonard Lessin/©Leonard Lessin/Peter Arnold, Inc. **Fig. 19.10** Hewlett Packard/Fundamental Photographs, NYC **Fig. 19.20a** Richard Megna/Fundamental Photographs, NYC **Fig. 19.22a** Richard Megna/Fundamental Photographs, NYC **Fig. 19.24a** Richard Megna/Fundamental Photographs, NYC **Fig. 19.25a** Richard Megna/Fundamental Photographs, NYC **Fig. 19.28c** Grant Heilman Photography, Inc. **Fig. 19.29.1** Richard Blakemore/Richard Blakemore, University of New Hampshire **Fig. 19.29.2** Richard Blakemore, University of New Hampshire **Fig. 19.32** Pekka Parviainen/Science Photo Library/Photo Researchers, Inc.

Chapter 20—CO.20 Tim Barnwell/Stock Boston **Fig. 20.12.1a** Richard Megna/Fundamental Photographs, NYC **Fig. 20.12.2** Jack Hollingsworth/Getty Images, Inc./Photodisc. **Fig. 20.13.1** Neal & Molly Jansen/AGE Fotostock America, Inc. **Fig. 20.13.2a** DaimlerChrysler AG **Fig. 20.14a** Tim Matsui/Photo by Tim Matsui/Liaison **Fig. 20.14b** U.S. Department of the Interior **Fig. 20.25a** Fotopic,/Index Stock Imagery, Inc. **Fig. 20.25b** Scott Camazine/Photo Researchers, Inc.

Chapter 21—CO.21 Christopher J. Morris/© Christopher J. Morris/CORBIS All Rights Reserved **Fig. 21.5** Frank Labua/Pearson Education/PH College **Fig. 21.14** Frank LaBua/Pearson Education/PH College **Fig. 21.14.1** Rip Griffith/Photo Researchers, Inc.

Chapter 22—CO.22 Anya Levinson/Anya Levinson **Fig. 22.5b** Peter M. Fisher/Corbis/Stock Market **Fig. 22.7.2b** E. R. Degginger/Color-Pic, Inc. **Fig. 22.8** Richard Megna/Fundamental Photographs, NYC **Fig. 22.9a** Richard Megna/Fundamental Photographs, NYC **Fig. 22.12a** Kent Wood/Photo Researchers, Inc. **Fig. 22.12.1** Minas Tanielian/Boeing Phantom Works **Fig. 22.13b** Bo Lou **Fig. 22.14a** ©Edward Pascuzzi **Fig. 22.15b** Ken Kay/Fundamental Photographs, NYC **Fig. 22.17** Stuart Westmorland/© Stuart Westmorland/CORBIS **Fig. 22.18a** ©Gemological Institute of America. Reprinted by permission. **Fig. 22.19a** Courtesy of Cenco **Fig. 22.20a** Nick Koudis/Getty Images, Inc./Photodisc. **Fig. 22.20b** Hank Morgan/Photo Researchers, Inc. **Fig. 22.20.1a** Southern Illinois University/Photo Researchers, Inc. **Fig. 22.20.1b** CNRI/SCIENCE PHOTO LIBRARY/Photo Researchers, Inc. **Fig. 22.21a** David Parker/SPL/Photo Researchers, Inc. **Fig. 22.21.1** Doug Johnson/ SPL/Photo Researchers, Inc. **Fig. 22.22** Photri/Corbis/Stock Market **Fig. 22.24** SIU/Photo Researchers, Inc. **Fig. 22.25L** Frank LaBua/Pearson Education/PH College **Fig. 22.25R** Frank LaBua/Pearson Education/PH College

Chapter 23—CO.23 Craig Tuttle/Corbis/Stock Market **Fig. 23.6** Paul Silverman/Fundamental Photographs, NYC **Fig. 23.5.1** Reproduced by permission from Modern Magic by Professor Hoffmann, American edition. Copyright ©1865 by George Routledge & Sons. Image courtesy of Taylor and Francis LTD. **Fig. 23.5.2** Courtesy of the Library of Congress **Fig. 23.12b** Richard Hutchings/Photo Researchers, Inc. **Fig. 23.15T** John Smith/Jerry Wilson **Fig. 23.15B** Richard Megna/Fundamental Photographs, NYC **Fig. 23.18.1b** John Smith/Jerry Wilson **Fig. 23.18.1c** Bohdan Hrynewych/Stock Boston **Fig. 23.23** Frank Labua/Pearson Education/PH College **Fig. 23.24b** Michael Freeman/Michael Freeman **Fig. 23.25** Tom Tracy/Corbis/Stock Market **Fig. 23.26** Michael W. Davidson, National High Magnetic Field Laboratory, The Florida State University. **Fig. 23.27** John Smith/Jerry Wilson

Chapter 24—CO.24 Lawrence Migdale/Lawrence Migdale/Pix **Fig. 24.1** Richard Megna/Fundamental Photographs, NYC **Fig. 24.2b** From the "Atlas of Optical Phenomena," Michel Cagnet, Maurice Francon, Jean Claude Thrierr. © by Springer-Verlag OHG, Berlin, 1962. **Fig. 24.6c** Peter Aprahamian/SPL/Photo Researchers, Inc. **Fig. 24.7a** David Parker/ Science Photo Library/Photo Researchers, Inc. **Fig. 24.7b** Gregory G. Dimijian, MD/Photo Researchers, Inc. **Fig. 24.10b** Ken Kay/Fundamental Photographs, NYC **Fig. 24.10.1** Kristen Brochmann/ Fundamental Photographs, NYC **Fig. 24.11** Sabina Zigman, Junior at Cornell University. **Fig. 24.12a** Fundamental Photographs, NYC **Fig. 24.12b** Education Development Center, Inc./ PSSC Physics ©1965, Education Development Center, Inc.; D. C. Heath & Company. **Fig. 24.13T** Ken Kay/ Fundamental Photographs, NYC **Fig. 24.13B** Ken Kay/Fundamental Photographs, NYC **Fig. 24.14** Springer Verlag GmbH & Co. **Fig. 24.15** Springer Verlag GmbH & Co. **Fig. 24.17** ACE STOCK LIMITED/ Alamy Images **Fig. 24.19b** A.J. Stosick **Fig. 24.19c** Courtesy of Dr. M.F. Perutz/Medical Research Council **Fig. 24.22c** Diane Schiumo/ Fundamental Photographs, NYC **Fig. 24.24b1** Nina Barnett Photography **Fig. 24.24b2** Nina Barnett Photography **Fig. 24.25b** Ed Degginger/Color-Pic, Inc. **Fig. 24.26b** Peter Aprahamian/Sharples Stress Engineers Ltd./SPL/Peter Aprahamian/Sharples Stress Engineers Ltd./Science Photo Library/Photo Researchers, Inc. **Fig. 24.26.2L** Leonard Lessin/© Leonard Lessin/Peter Arnold Inc. **Fig. 24.26.2R** Leonard Lessin/© Leonard Lessin/Peter Arnold Inc. **Fig. 24.28** Roger Ressmeyer/© Roger Ressmeyer/CORBIS **Fig. 24.28.1** Sercomi/Photo Researchers, Inc.

Chapter 25—CO.25 Joe McNally Photography **Fig. 25.3.1** Robert Dyer/Robert Dyer **Fig. 25.4b** Frank Labua/Pearson Education/PH College **Fig. 25.8b** Rocher/Jerrican/Photo Researchers, Inc. **Fig. 25.13a** SAGEM/European Southern Observatory **Fig. 25.13b** SAGEM/European Southern Observatory **Fig. 25.14** NASA Headquarters **Fig. 25.15.1** Tony Craddock/Science Photo Library/Photo Researchers, Inc. **Fig. 25.16a** Reproduced by permission from Michel Cagnet, Maurice

Franzon, and Jean Claude Thierr, Atlas of Optical Phenomena. New York: Springer-Verlag, 1962. © 1962 by Springer-Verlag GmbH & Co. **Fig. 25.16b** Reproduced by permission from Michel Cagnet, Maurice Franzon, and Jean Claude Thierr, Atlas of Optical Phenomena. New York: Springer-Verlag, 1962. © 1962 by Springer-Verlag GmbH & Co. **Fig. 25.17a** The Image Finders 2003 **Fig. 25.17b** The Image Finders 2003 **Fig. 25.17c** The Image Finders 2003 **Fig. 25.18** Bill Bachmann/Photo Researchers, Inc. **Fig. 25.20a** Fritz Goro/Fritz Goro, Life Magazine ©TimePix

Chapter 26—CO.26 W. Couch/W. Couch, University of New South Wales, Australia/Space Telescope Science Institute OPO/NASA **Fig. 26.2** Science Photo Library/Photo Researchers, Inc. **Fig. 26.14b** Space Telescope Science Institute/Photo Researchers, Inc. **Fig. 26.15.2** NASA/Ames Research Center/Ligo Project

Chapter 27—CO.27 Hank Morgan/Photo Researchers, Inc. **Fig. 27.7a** Larry Albright/Larry Albright Etc **Fig. 27.7b** Ann Purcell/Photo Researchers, Inc. **Fig. 27.8a** Wabash Instrument Corp./Fundamental Photographs, NYC **Fig. 27.8b** Wabash Instrument Corp./Fundamental Photographs, NYC **Fig. 27.8c** Wabash Instrument Corp./Fundamental Photographs, NYC **Fig. 27.8d** Wabash Instrument Corp./Fundamental Photographs, NYC **Fig. 27.8e** Wabash Instrument Corp./Fundamental Photographs, NYC **Fig. 27.13a** Mark A. Schneider/©Mark A. Schneider/Photo Researchers, Inc. **Fig. 27.13b** Mark A. Schneider/©Mark A. Schneider/Photo Researchers, Inc. **Fig. 27.14** Dan McCoy/Rainbow **Fig. 27.18a** AP Wide World Photos **Fig. 27.18b** Rosenfeld Images Ltd/Science Photo Library/Photo Researchers, Inc. **Fig. 27.18.2L** JPD/Custom Medical Stock Photo/Custom Medical Stock Photo, Inc. **Fig. 27.18.2R** JPD/Custom Medical Stock Photo/Custom Medical Stock Photo, Inc. **Fig. 27.18.3b** Steven E. Zimmet/Photos courtesy of Steven E. Zimmet, MD FACPh. **Fig. 27.18.3c** Steven E. Zimmet/Photos courtesy of Steven E. Zimmet, MD FACPh. **Fig. 27.20** Philippe Plailly/SPL/Photo Researchers, Inc.

Chapter 28—CO.28 Almaden Research Center/Research Division/NASA/Media Services **Fig. 28.3.3** Bob Thomason/Stone/Getty Images Inc./Stone Allstock **Fig. 28.3.4a** Manfred Kage/© Manfred Kage/Peter Arnold Inc. **Fig. 28.3.4b** Dr. R. Kessel/© Dr. R. Kessel/Peter Arnold Inc. **Fig. 28.3.4c** David Scharf/© David Scharf/Peter Arnold Inc. **Fig. 28.3.5** Courtesy of International Business Machines Corporation. Unauthorized use not permitted. **Fig. 28.8.1a** Omikron/Science Source/Photo Researchers, Inc. **Fig. 28.8.1b** Mehau Kulyk/Science Photo Library/Photo Researchers, Inc. **Fig. 28.8.3c** Will & Deni McIntyre/Photo Researchers, Inc. **Fig. 28.13** Lawrence Berkeley National Laboratory/Science Photo Library/Photo Researchers, Inc.

Chapter 29—CO.29 Roger Tully/Getty Images Inc./Stone Allstock **Fig. 29.17** Lawrence Berkeley National Laboratory/Photo Researchers, Inc. **Fig. 29.18b** Dan McCoy/Rainbow **Fig. 29.18c** Monte S. Buchsbaum, M.D., Mt. Sinai School of Medicine, New York, NY. **Fig. 29.19b** Jeff J. Daly/Fundamental Photographs, NYC

Chapter 30—CO.30 Alexander Tsiaras/Stock Boston **Fig. 30.6** Shone/Getty Images, Inc./Liaison **Fig. 30.7b** Plasma Physics Laboratory, Princeton University

INDEX

Note: Entries having page numbers with n, f, or t refer to material in a footnote, figure, or table.

A

Abera, Gezahgne, 367
Aberration(s)
astigmatism, 753, 797, 798
chromatic, 752
lens, 752–53
in reflecting telescope, 806
spherical mirror, 733, 740, 752
Absolute pressure, 310
Absolute reference frame, 821, 822
Absolute temperature, 343–49, 355
Absolute value, 78
Absolute zero, 338, 346–49, 355, 425
Absorbed dose unit, 924
Absorption
emitter and, 387
photon, 867
resonance, 385
spectrum, 861, 866
ac (alternating current), 663, 687–88. *See also* Circuits, ac
Accelerated rolling without slipping, 258
Acceleration, 40–44
angular, 228–31
apparent, 245
average, 40–42, 44
centripetal, 223–25, 230
of charge, 539
constant (*See* Constant acceleration)
due to gravity, 49–56, 233–34
of electron, 540, 835
force and, 104
horizontal projections, 81–82
instantaneous, 41, 116
Newton's second law on, 106–11
signs of, 42
in simple harmonic motion, 444–45
speed and kinetic energy, 152
tangential, 229–30
ac circuits. *See* Circuits, ac
Accommodation, 794
Accretion disk, 842
Acetylcholine, 576
ac generators, 663–65, 667, 668
Achilles-tendon force, 120
Achromatic doublet, 752
ac power, 688
ac power distribution system, 671–72
Acrobats, and center of gravity, 268
Actinide series, 891, 892f
Action-at-a-distance forces, 105
Action-reaction forces, 113–14, 506
Active noise cancellation, 452
Activity, of radioactive isotope, 911–12, 914
Acuity, visual, 798–99
ac voltage, 688–89
Adapters (wall socket), 689
Addition, with significant figures, 19
Addition of vectors. *See* Vector addition
Additive method of color production, 811
Additive primary colors, 811
Adhesion, local, 122
Adiabat, 406, 423
Adiabatic process, 406–7, 408, 413, 423
Aerobraking, of spacecraft, 129
Aerosol cans, disposing of, 406
Air. *See also* atmosphere (atm)
composition of, 358
speed of sound in, 471–74
Air bags, 186
Air Canada, 16
Air conditioners, 421, 422
Air foils, 128
Airplanes
airspeed of, 94
flight dynamics, 94, 322–23
shock wave and, 488–90
wind shear and, 491
Airport screening, 664
Air purifiers, and electric force, 505, 512
Air resistance, 127–29
free fall and, 50, 51, 163
range of projectile and, 87–88
Alchemists, 936
Alcohol, in liquid-in-glass thermometers, 340–41
Algebraic relationships, A-2–A-3
All-optical atomic clock, 5n
Alpha decay, 906–8, 918
Alpha particles, 903–4, 906–8, 910, 922
Alternating current (ac), 663, 687–88. *See also* Circuits, ac
Alternator emf, 665
Alternators, 663–65, 667, 668
Altitude, boiling point and, 378
Aluminum
calorimetry and, 372–73
resistivity and temperature coefficient, 577t
stable nuclei, 918, 919, 922
AM (amplitude-modulated) radio band, 676, 700, 760, 770–71
American electrical system, 689
American Journal of Physics, 326n
Americium-241, 902
Ammeters, 591, 607–9, 611f, 635
amp, 572
Ampére, Andre, 568, 572
ampere (A), 7, 568
defined, 572, 623, 641
for household appliances, 582
Amplification, stimulated emission, 867
Amplitude, 434–35, 441, 447, 483
Analytical component method (vectors), 73, 75–80
Analyzers, polarized light, 777
Anderson, C. D., 896
Aneroid barometer, 311
Angle(s)
critical, 717–18
of incidence, 707, 711, 717
phase, 694–96
polarizing (Brewster), 779
projectile motion at, 82–89
of reflection, 707
of refraction, 709, 711, 712, 717
of resolution, minimum, 808–9
shear, 301
Angular acceleration, 228–31
Angular displacement, 217, 277
Angular distance, 218
Angular frequency of oscillating object, 440
Angular magnification of microscopes, 799–801, 802
Angular measure, 217–19
Angular momentum, 280–86
conservation of, 281–83, 947n
of electron, 885
quantization, 878–79
real-life, 283–86
of rigid body, 281
Angular separation, 807, 839
Angular speed, 217, 219–21
Angular velocity, 217, 219–21
Anisotropic optical property, 781
Anodes, 569, 629, 678
Anthrax spores, 927
Antilock brakes, 256, 281
Antimatter, 897, 950
Antineutrino, 947, 952
Antinodes, 454, 468, 492
Antiparticles, 877, 896–97
Antiquarks, 953
Antiterrorism, 664
Aphelion, 236, 239
Apparent acceleration, 245
Apparent location of stars, 839
Apparent weight, 245
Apparent weightlessness, 243–44
Appliances, household
efficiency limits, 583–84
electrical safety, 611–14
power and requirements, 581, 582
power ratings of, 611–12
refrigerators, 421, 583
Applied force *vs.* force of friction, 123
Approximations, 23–24
Archimedes, 297, 314
Archimedes' principle, 314–16
Architectural design, passive solar design, 390
Arc length, 217–19
Arcs, parabolic, 83–84
Arctangent notation, 75f
Area equations, A-3
Area expansion (thermal), 350, 351, 352–53
Area(s)
converting units of, 15
of cylindrical container, 21
Kepler's law of, 239–40, 282–83
of rectangle, finding, 22
Area vector, 658
Aristotle, 32, 50, 51, 105
Arithmetic operations, A-1–A-2
Armature, 636, 663–65, 667–68
Arteriography, 490
Arthroscope, 720
Artificial gravity, 245–46
Artificial transmutation of nucleus, 936. *See also* Transmutation, nuclear
Asperities, 123
Astigmatism, 753, 797, 798
Astronauts, and gravity, 108, 245
Astronomical telescopes, 803–5
Astronomy, and general theory of relativity, 840–41
Atmosphere, standard, 310
atmosphere (atm), 306
Atmospheric lapse rate, 362
Atmospheric pressure, 306–7, 309–13, 311
Atmospheric refraction, 716–17

Atmospheric scattering of light, 782, 784
Atomic bomb, 941
Atomic clocks, 5, 6f, 819, 832, 838
Atomic energy levels, 863–66, 866
atomic mass unit (u), 345, 919
Atomic nuclei, 886
Atomic number, 904, 905
Atomic orbitals, 886
Atomic physics. *See* Quantum physics
Atomic quantum numbers, 885–94
hydrogen atom, 885–86
multielectron atoms, 886–87
Pauli exclusion principle, 887–90
Atom(s). *See also* specific elements
Bohr theory of, 860–66, 878–79, 882, 885, 903
combining into molecules, 893–94
ionized, 862, 864
multielectron, 886–87
nuclear-shell model, 922
plum pudding model of, 903–4
Rutherford-Bohr model of, 903–4, 907
solar system model of, 506, 903
Attenuation of sound in air, 474
Attractive forces, 506, 542
Atwood machine, 136, 292
Audible region, 468
Aurora australis and borealis, 647
Automobiles. *See* Vehicles
Average acceleration, 40–42, 44
Average angular acceleration, 228–29
Average angular speed, 219
Average angular velocity, 220
Average binding energy per nucleon, 920, 940
Average flow rate, 327
Average power, 687–88, 696
Average speed, 33–34
average velocity *vs.*, 37
Average velocity, 36–37, 39, 44
Avogadro's number, 345–46
Axes, Cartesian coordinate, 35
Axes of symmetry, 273–74
Axis, transmission, 776–78
Axis of rotation
of Earth, 285
instantaneous, 257, 258
parallel axis theorem, 273–74, 278
rotational directions, 263
rotational inertia, 270–71, 286

B

Back emf, 667–68
inductors and, 691
oscillator circuits and, 699
in transformers, 669
Back-scattering, 903–4
Bacteria, magnetotactic, 645
Balance, electronic, 636–37
Balanced forces, 104, 261. *See also* Net force
Balanced torques, 261
Balance point, 199
Ballistic pendulum, 212
Balloons
hot-air, 338
weather, 314–15
Balls
angular momentum, at end of string, 282
collisions and, 196, 197–98
completely inelastic collision, 194
energy exchanges, 161
kinetic energy and, 154
moment of inertia, hitting baseballs, 273
speed and conservation of energy, 160
Balmer, J. J., 861
Balmer series, 861, 864–65
Band, radio frequencies, 700
Banked roads, 226–27, 250
bar, 311n
Barium, 861, 909–10
Barometer(s), 309f, 310, 311
Barrier, potential-energy, 884
Barrier penetration (tunneling), 908
Baryons, 952, 953
Base units, 3, 6–7
Basketball. *See* Sports
Bats, echolocation in, 469–70
Batteries, 569–71
action of, 569
capacitance and, 549
dielectrics and, 553–55
direct current and, 569–71
emf and terminal voltage, 570–71
in parallel, 570–71
in series, 570
terminal voltage of, 545, 570–71, 572f
Beat frequency, 483–84
Beats, 483–84
Becquerel, Henri, 906, 914n
becquerel (Bq), 914
bel (B), 476–79
Bell, Alexander Graham, 477
Bernoulli, Daniel, 322
Bernoulli's equation, 319, 322–23
Beta decay, 908–9, 909–10, 946–48
β^+ decay, 908–9, 946
β^- decay, 908, 916, 918, 946
Beta particles, 906, 908–9, 910, 922
Biconcave lenses, 740, 741, 747
corrective eyeglasses, 795–96
Biconvex lenses, 740, 741, 744, 745–46, 748
corrective eyeglasses, 795, 797–96
Bifocal lenses, 796
Big Bang, 950, 955
"Big Bertha" gun, 67
Billion, 8
Bimetallic coils, 340
Binary star system, 842
Binding energy, 864, 919–22, 940
E_b/A curve, 921, 940
Binoculars, prism, 804
Bioelectrical impedance analysis (BIA), 578
Bio-generation of high voltage, 575–76
Biological heat engine, 420
Biological radiation effects, 924–28
Biomedical scattering, 785
Biopsy, optical, 785
Bird-leg syndrome, 245
Birefringence, 780–82
Blackbody, 387, 852, 853
Blackbody radiation, 852–54
Black holes, 819, 840–41, 842
Blacking out, 108
Black light (ultraviolet light), 866
Blakemore, R. P., 645n
Blinker circuits, 606–7
Blood. *See also* Heart, human
calculating density, 23–24
capillary system, 14–15, 357
diffusion, 357
flow rate, 321, 490
pooling in lower body, 108
pressure, 312–13
separating components, 224–25
transfusions, 312–13, 328
weightlessness effects on, 245
Blue end of visible spectrum, 782
Blue shift, Doppler, 488
Body heat, 367, 380, 385, 387
Body waves, 450
Bohr, Niels, 861–63
Bohr orbit, 878–79, 882
Bohr theory of hydrogen atom, 860–66, 878–79, 882, 885, 903
Boiling point, 375, 378
Boltzmann's constant, 345
Bombs, atomic and hydrogen, 941, 946
Bones
bone mineral density (BMD), 304–5
bone tissue density, 304
carbon-14 dating, 917
stress on, 300–301
Boundaries, and waves, 452–53
Boundary conditions, 454
Bow waves, 488, 489f
Boyle, Robert, 343
Boyle's law, 343
Bragg, W. L., 774
Bragg's law, 774
Brahe, Tycho, 239
Braking car, with antilock brakes, 256, 281
Braking (reverse) thrust, 206
Branch (in electrical circuit), 599
Branch currents, 603–4
Brass instruments, 493
Breakdown voltage, 607
Breeder reactors, 942–43
Brewster, David, 779
Brewster angle (polarizing angle), 779
Bridges, resonant vibration in, 458
Bright-line spectrum (emission spectrum), 861, 864
Brilliance of diamonds, 718, 721
British electrical system, 689
British thermal unit (Btu), 368–69
British units, 3, 5, 7, 9, 10. *See also* Measurement
grain, 179
of power, 165
of pressure, 303
of speed, 33
of work, 142
Broadcast frequencies, 698–99, 700
Broadening, natural (line), 895
Bubble chamber, 923, 924
Bulk modulus, 301–2, 471
Buoyancy, 313–18
in air, 315
Archimedes' principle, 314–16
buoyant force, 313–16
and density, 316–18

C

Calcite, 781
Caloric, 338
calorie, gram (cal), 368
Calorie, kilogram (Cal), 368
Calorimeter, 372
Calorimetry, 372–73
Camera
lens, 705, 729, 768
operation of, 793
RC circuits in, 606
resolution of, 792, 808
Cancer
radiation therapy for, 927
thyroid, iodine-131 and, 914, 925–26
UV radiation danger, 677
candela (cd), 7

Capacitance, 549–52
defined, 550, 690
equivalent parallel, 558–59
equivalent series, 558
Capacitive ac circuits, 695
completely, 696
purely, 689–90, 692
Capacitive reactance, 689–91, 693, 697
Capacitor(s), 522, 549
in cardiac defibrillators, 522, 551
charging through resistor, 604–5
compared in ac and dc circuits, 686, 689
in computer keyboards, 556
dielectric material in, 552–56
discharging through resistor, 605–7
energy storage in, 549–51
oscillator circuits and, 699
in parallel, 557, 558–60
parallel-plate, 549–51
power loss and, 696
in RLC circuits, 693–94
in series, 557, 558, 559
in series-parallel combination, 560
variable air, 698
Carbon
isotopes, 905, 908
resistivity and temperature coefficient, 577t
as semiconductor, 579
stable nuclei, 918
Carbon-14 dating, 915–17
Carbon dioxide (CO_2), 356, 357, 388, 916
Cardiac defibrillators, 522, 551
Cardioscope, 720
Carnot, Sadi, 423
Carnot cycle, 422–24
Carnot efficiency, 423–25
Cartesian coordinates, 35, 36
Cassini, G. D., 238
***Cassini-Huygens* spacecraft,** 238
Cathode ray tube (CRT), 629
Cathodes, 569, 629, 678
Cats, 469
Cavendish, Henry, 232, 233, 513
CDs (compact disc), 222, 229, 772–73, 869
Celestial mechanics, 32
Cellular phones, 676
Celluloid, transparent, 776
Celsius, Anders, 338
Celsius temperature scale, 340–42
conversion to Fahrenheit, 341–42
conversion to Kelvin, 347
Center of curvature, 732, 741
Center of gravity (CG)
center of mass and, 203–4
in human body, 204, 261, 268
locating, 272
stable equilibrium and, 266–69
Center of mass (CM), 198–204
center of gravitational force and, 232n
center of gravity and, 203–4
summation process of finding, 200
centipoise (cP), 326
Central maximum
double slit, 762–63
in optical instruments, 807
single slit, 769–71, 807–8
Central ray (chief ray), 742, 743
Centrifuge, 224–25
Centripetal acceleration, 223–25, 230
Centripetal force, 216, 225–28, 232
spectral lines of hydrogen atom, 861
Cesium atoms, 5, 909
CFCs (chlorofluorocarbons), 677
cgs system, 7
cgs units
specific gravity and density, 318
of viscosity, 326
Chain reaction, 940–41, 942
Change in length, 299–301
Characteristic frequencies, 454
Charge, electric, 505–8
conservation of, 507–8, 907
electric force per unit, 538
net charge, 507, 572
point, 540–42
quantized charge, 507, 508
storm clouds and, 523, 524–25
test charge, 517
Charge configurations, electric potential energy of, 542–43
Charged particles
applications in magnetic fields, 629–32
auroras in Van Allen belts, 647
motion in magnetic fields, 646–47
moving, and right-hand force rule, 627–28
Charge-force law (law of charges), 506
Charging, electrostatic, 508–12
by contact or conduction, 510–11
by friction, 510
by induction, 511
by polarization, 511–12
Charging time, 605, 608
Charles, Jacques, 344
Charles' law, 344, 345f
Chernobyl reactor, 944
Chief ray
central, 742, 743
radial, 734
Chimneys, 322
China
Great Wall, view from space, 810
passive solar house design, 390
China syndrome, 943
Chlorofluorocarbons (CFCs), 677
Cholesterol, 321
Christmas tree lights, 596–97
Chromatic aberration, 752
Circuit breakers, 612–13
Circuit diagrams
ac circuits, 687, 689, 690, 692
ammeters and voltmeters, 607, 610, 611
back emf, 667
basic circuit, 572, 574
basics of drawing, 571
electromagnetic induction, 657, 658
magnetic field and, 637, 643
multiloop circuits, 599, 601, 603
oscillating LC circuit, 699
RC circuits, 604, 606, 693, 694, 695
resistors, 592, 593, 596, 597
Circuit impedance, 693–97
Circuit reduction, 560f, 597–98
Circuits, ac, 686–704
analogy to spring-mass system, 686
capacitive reactance, 689–91
impedance and, 693–97
inductive reactance, 691–92
resistance in, 687–89
resonance in, 697–700
RLC circuits (*See* RLC circuits)
Circuits, basic, 591–622. *See also* Kirchhoff's rules; Resistor(s)
ammeters and voltmeters, 591, 607–10, 611f
blinker, 606–7
complete, 569n, 571
dc compared to ac, 686
household, 611–14
multiloop, 599–604
open, 593, 596, 612
RC, 604–7, 608–9, 693–94
resistances, 592–99
safety and, 611–14
Wheatstone bridge, 618
Circular aperture resolution, 808–9
Circular motion, 216–55. *See also* Gravity; Kepler's laws of planetary motion
angular acceleration, 228–31
angular measure, 217–19
angular speed, 217, 219–22
angular velocity, 217, 219–22
centripetal acceleration, 223–25, 230
centripetal force, 225–28
geosynchronous satellites, 234, 236
relationship of radius, speed, and energy, 243
uniform, 223–28, 439, 441
Circumference equations, A-3
Classical electromagnetic theory, 853–54
Classical relativity (Newtonian relativity), 820–21, 822
Classical wave theory, 855
Clausius, Rudolf, 411
Clock, light pulse, 825
Clock paradox (twin paradox), 832
Clocks, atomic, 5, 6f, 819, 832, 838
Closed (isolated) systems, 156, 187, 192, 413
Closed-tube manometer, 309f, 310
Cloud chamber, 896, 923
Clouds, 129, 523, 524–25
Coatings on lenses, nonreflective, 766–67, 768
Cobalt, as ferromagnetic material, 641
Coefficient(s)
of friction, 122–27, 124f, 227
of kinetic friction, 123, 124f
of performance (COP), 421–22
of static friction, 122–23, 124f
temperature, of resistivity, 577t, 578–79
thermal, 350–51
of viscosity, 326
Coherent sources, 761, 868
Coil inductance, 691–92
Cold-blooded creatures, 344
Collagen, 301
Collision(s)
conservation of momentum and, 196
defined, 177, 185
elastic, 191–92, 195–98
inelastic, 191, 192–94, 836
Color, 810–13
and temperature, 852–53
Color blindness, 811
Color charge, quark characteristic, 954
Color force, 954
Colors, solar, 853
Color vision, 810–13
Comets, 675
Common logarithms, 404, 476–79, A-5
Communications, global, 676–77
Commutator, split-ring, 636
Compact disc (CD), 229, 772–73, 869
Compass, 625, 637
Complementary colors, 811
Complete circuits, 569n, 571
Completely capacitive circuits, 696
Completely inductive circuits, 696
Completely inelastic collisions, 193
Completely resistive circuits, 696

Component form of unit-vector, 76
Component method of vector addition, 75–80
Components of force, 111–12, 118
Components of motion, 68–73
kinematic equations for, 70–73
Newton's second law on, 111–12
relative velocity and, 92–93
Compound lens systems, 748–50
Compound microscopes, 801–3
Compound nucleus, 936
Compressibility, 302
Compressional stress, 298
Compressional waves (longitudinal waves), 448–49, 450, 475
Compton, Arthur H., 858–60
Compton effect (scattering), 858–60, 922
Compton wavelength, 859
Computational method. *See* Analytical component method (vectors)
Computer
keyboards, 556
microchips, 509
monitors, 629
screens, 783
Computerized tomography (CT), 678, 928
Concave (converging) mirrors, 732, 735–36, 738
mirror ray diagram, 734
reflecting telescope, 806
Concurrent forces, 117, 261, 262f
Condensation point, 374
Condensation(s) of sound waves, 468, 482
Condenser, 802–3
Condition for
constructive interference, 762, 766, 767, 774
destructive interference, 763, 766, 767
interference maxima, 763, 773
rolling without slipping, 258
rotational equilibrium, 261–62
stable equilibrium, 267
translational equilibrium, 119, 261
Conduction
electrostatic charging by, 510–11
of heat on atomic scale, 400
heat transfer by, 379–83, 384, 386
of light, 718–19
Conductivity, thermal, 380–81, 382f
Conductors, 508–9
charged, equipotential surfaces outside, 548
electric fields and, 526–27
Gauss's law and, 529
induced currents and emf, 662
resistivities and temperature coefficients, 577t
thermal, 367, 379
Cones, of eye, 793f, 794, 811
Configuration, energy of, 152
Configuration(s)
of charges, 517, 518, 526
electric potential energy of various, 542–43
of masses, 237
Confinement
inertial, 946
magnetic, 647, 946
quark, 954
Conical shock wave, 490
Conservation
of angular momentum, 281–83, 947n
of charge (electric), 507–8, 600, 907
of energy, 141, 155–64, 319, 400, 660, 946–47
of linear momentum, 185–91, 946–47
of mass, and fluid flow, 319
of mass-energy, 937–38
of momentum and kinetic energy, elastic collisions, 196
of nucleons, 907
of relativistic momentum and energy, 836
of total energy, law of, 156
of total mechanical energy, 158–61
Conservative forces, 157–58
mechanical energy of spring, 160–61
Conservative gravitational forces, 157–58
Conservative systems, 158
Constancy of speed of light, 822, 823f
Constant acceleration, 42–44
free fall, 49–56, 108–9
kinematic equations, 45–49, 52, 70–72
Newton's second law and, 116
Constant forces, 116, 141–45
Constant-pressure process, 404
Constant(s)
Boltzmann's, 345
decay, 912
dielectric, 552, 553, 554
Planck's, 854, 862, 895
Rydberg, 861
spring, 146–47, 434, 437–38, 441
Stefan-Boltzmann, 387
thermal conductivity, 380–81
time, for RC circuits, 605, 608
universal gas, 345
universal gravitational, 231–32
Constant-temperature process, 403, 408
Constant velocity, 38
Constant-volume gas thermometer, 346
Constant-volume process, 405
Construction industry, energy conservation and, 384
Constructive interference, 451
condition for, 762, 766, 767, 774
double-slit, 761–63
scattering of electrons, 881
sound waves, 482, 488
thin-film, 765–67
Contact, electrostatic charging by, 510–11
Contact forces, 105
Contact lenses, 796
Continuity, equation of, 320–21
Continuous spectrum, 852, 860, 861f, 947
Contraction, thermal, 340
Control rods, 942
Convection, 383–85, 386
Conventional current, 571–72, 632
Converging (biconvex) lenses, 740, 741, 744, 745–46, 748
compound microscope, 801, 802
corrective eyeglasses, 795, 797–96
Converging (concave) mirrors, 732, 735–36, 738
mirror ray diagram, 734
Conversion factors, 13–15
Conversions
Celsius-to-Fahrenheit, 341–42
Celsius-to-Kelvin, 347
Fahrenheit-to-Celsius, 341–42
unit, 12–16
Converters, 689
Convex (diverging) mirrors, 732, 739–40
Cooking, 378, 381
Coolants in nuclear reactors, 941–43
LOCA and, 943–44
Coordinates
Cartesian, 35, 36
(*p*, *V*, *T*) in ideal gas law, 398–99
polar, 217
Core, of nuclear reactor, 941
Cornea of eye, 793f, 794, 797
Correspondence principle, 900
Cosines, law of, A-4
Cosmic rays, 915–16, 949
Cost of electric power, 583
Coulomb, Charles Augustin de, 505, 506, 512
Coulomb barrier, 939
coulomb (C), 506–7
Coulomb force, 861, 918
Coulomb's law, 513–14, 517
Counter emf, 667–68
Countertorque, 668
Couple (pair of equal and opposite forces), 261
Covalent bonding, 893
Crane hoist, 165
Critical angle, 717–18
Critical mass, 940–41
Crossed polarizers, 777f, 781f, 782, 783
Crossover point, 736
Cross section, of reactions, 939
Crum, Lawrence A., 471n
Cryogenic experiments, 425
Crystals
anisotropic and birefringence, 781
diffraction of electrons, 881
polarization by selective absorption, 776
X-ray diffraction, 774–75
CT (computerized tomography), 678, 928
cubic meter (m), 8
cubic meters per second (m^3/s), 327
Curie, Marie, 906, 914
Curie, Pierre, 623, 644, 906, 914
curie (Ci), 914
Curie temperature, 623, 644, 646
Currency, U.S., 8
Current. *See* Electric current
Current-carrying loops
magnetic field at center of, 638–39
torque on, 634–35
Current-carrying solenoid, 639–41, 643
Current-carrying wires
electrical resistance of, 574–75, 577, 579
ground (neutral), 611f, 612–13
high-potential (hot wires), 611–14
magnetic field, and applications of, 635–37
magnetic field near, 637–38
magnetic forces between two parallel, 640–41
magnetic forces on, 632–35
solenoid *vs.*, 640
Curvature
center of, 732, 741
radius of, 732, 733, 736, 741
Curvilinear motion, 67, 71–72, 223
Cutoff frequency, 855, 857
Cycle, four-stroke, 417–18
Cycle of object in circular motion, 221–22
cycle per second (cycle/s), 434–35
Cyclic heat engine, 417, 422, 424
Cygnus X-1, 842
Cylinder surface area, 21

D

d'Alibard, Thomas François, 523
Damped harmonic motion, 445–46

Damping, 308
Dark-line spectrum (absorption spectrum), 861, 866
Dating, radioactive, 915–17
Daughter nucleus, 906–7, 908
Davisson, C. J., 881
Davisson-Germer experiment, 881, 883
Day-night atmospheric convection cycles, 384
dc circuits. *See* Circuits, basic
dc (direct current), 569–71, 572, 604–7
dc motors, 636, 667–68
D-D reaction, 944
de Broglie, Louis, 877, 878
de Broglie equation, 878
de Broglie hypothesis, 878–81
de Broglie wavelength, 879–81, 883
de Broglie waves, 878–81
Decay, radioactive, 902
  alpha, 906–8, 918
  beta, 908–10, 918, 946–48
  gamma, 909–10
  muon, 827–28, 831
  positron, 908
  Q value of, 938
Decay constant, 912
Decay rate, 911–17
Decay series of uranium-238, 911
Deceleration, 42
decibel (dB), 476–80
decibel level, 477
Decimal system, 7–8
Declination, magnetic, 646
Dedicated grounding, 611f, 612–13
Deformation, 105
Degenerate orbits, 885, 886
Degree of freedom, 358
Degrees, and radians, 217–19
Delta (Δ), 33, 35–36
Density
  of blood, calculating, 23–24
  buoyancy and, 316–18
  of common substances, 304
  defined, 303
  determination of, 12
  linear turn, 639
  optical, 711
  probability, 882
  temperature and, 354
Deposition, 374
Depth and pressure, 305–7
Derived units, 3
Descartes, Rene, 32
Destructive interference, 451–52
  in CDs and DVDs, 869
  condition for, 763, 766, 767
  double-slit, 761–63
  single-slit diffraction, 769
  sound waves, 482
  thin-film, 765–67
Deterministic view of nature, 894
Deuterium, 905, 941, 944, 945
Diagnostic tracers, 927–28
Diamonds, 718, 721
Diastolic pressure, 312–13
Diatomic gases, 354, 357–59
Dichroic crystal, 776
Dichroism, 776–78
Dielectric constant, 552, 553, 554
Dielectric permittivity, 555
Dielectric(s), 552–56
Diesel-powered vehicles, 397
Diffraction, 453, 768–75
  de Broglie wavelength and, 879–80
  defined, 768
  in optical instruments, 807–10
  radio reception and, 770–71
  single-slit, 769–71, 808–9
  of sound, 481, 769
  of visible light, 769
  X-ray, 774–75
Diffraction gratings, 772–74
Diffraction limit, 715
Diffuse reflection (irregular reflection), 707–8, 709
Diffusion, 355–57
Dilated time interval, 827
Dimensions, 10
Diopters, 751, 794
Dipoles
  electric, 511, 521, 522f
  magnetic, 624, 645
Dirac, Paul A. M., 896
Direct current (dc), 569–71, 572, 604–7
Direction, 35–36
  acceleration and, 40
  of vector quantities, 52–55
Dish (parabolic collector), 808
Disintegration energy, 938, 947
Disorder, measure of, 411, 413–14
Dispersion, 453, 721–23
Displacement, 35–36, 47
  angular, 217, 277
  resolving into components of motion, 68–69, 70–72
  waves and, 434–36, 443–45, 449, 451, 454
  work and, 141–42
Displacement reference, 148
Distance, 33
  angular, 218
  image, 730, 736, 745
  object, 730, 736, 745
Diverging (biconcave) lenses, 740, 741, 747
  corrective eyeglasses, 795–96
  in telescopes, 804
Diverging (convex) mirrors, 732, 739–40
Division, and significant figures, 18, 19
Domains, magnetic, 641, 642f, 643, 644
Doppler, Christian, 484
Doppler effect, 484–88
  applications of, 490–91, 677
  for light waves, 488
  for ultrasound, 470
Doppler radar, 491
Doppler shifts (red and blue), 488
Double refraction, 780–82
Double-slit experiment, Young's, 761–64
Drift velocity, 572–73
Dry ice (solid carbon dioxide), 374
D-T reaction, 944
Dual energy X-ray absorptiometry (DXA), 305
Duality, wave-particle, 851, 855, 860
Dual nature of light, 860
DVDs (digital video discs), 772–73, 869
Dynamics. *See also* Thermodynamics
  fluid, 319–23
  rotational (*See* Rotational dynamics)
  study of, 32, 103
Dynamometer, 420
Dynamos, ac, 656

E

Ear, human. *See also* Hearing
  anatomy of, 475
  audible region of sound, 468
  damage exposure times, 480
  earaches and atmospheric pressure, 311
  standing waves and, 495
Earth. *See also* Equator
  electric field and equipotential surfaces, 546
  gravitational attraction between Moon and, 105, 232
  magnetic field of, 624, 644–47
  plate tectonic motion, 644
  rotational motion of, 285–86
Earthquake(s)
  effect on the Earth's rotation, 256
  infrasound and, 467, 468, 469
  seismic waves, 446, 450
Earth's satellites, 241–46
E_b/A curve, 921, 940
Echograms of fetus, 470
Echolocation, 469–70
Eclipse, solar, 839
Eddy currents, 670–71
Edison-base fuse, 612
Effective current (rms current), 688
Effective dose, 925
Effective voltage (rms voltage), 688
Efficiency, 166–67
  Carnot, 423–25
  electrical, 583–84
  of human body, 397, 420
  mechanical, 166–67
  thermal, 416–22, 424
Effusion, 356
Einstein, Albert, 271, 819, 822–23, 825, 833, 837, 843, 854–55, 857, 859, 860, 867, 954
***e* (irrational number, natural logarithm),** 605
Elastic collisions, 191–92, 195–98
Elastic limit, 146, 297, 299
Elastic moduli, 298–302
Electrical efficiency, 583–84
Electrical heating, 422
Electrical plugs, 613–14
Electrical resistance, 571, 573–79, 592–99. *See also* Resistor(s)
  in an ac circuit, 687–89
  bioelectrical impedance analysis (BIA), 578
  defined, 573
  factors influencing, 574–75
  in human body, 574, 577
  in parallel, 593–97
  resistivity, 576–79
  in series, 592–93
  temperature and, 574, 577–79
  of a wire, 574–75, 577, 579
Electrical safety, 611–14
Electrical shock, 591, 613, 614
Electrical signals in optical fibers, 720
Electrical systems, British *vs.* American, 689
Electrical thermometer, 579
Electric charge, 505–8. *See also* Charge, electric
Electric circuits. *See* Circuits, ac; Circuits, basic
Electric current, 568–90. *See also* Current-carrying loops/solenoid/wires
  alternating (ac), 663, 665, 687–88
  batteries and, 569–71
  branch, 603–4
  charge and, 572
  conventional, 571–72, 632
  defined, 569, 570, 572
  direct (dc), 569–71, 572, 604–7
  drift velocity and, 571–73
  eddy, 670–71
  in generators, 663
  induced, 657–58, 660, 661–62
  magnetic fields and, 637
  magnetic forces and, 632–35
  photo, 855
  power, electric, 580–84
  resistance and Ohm's law, 573–79
  rms (effective), 688
Electric dipoles, 511, 521, 522f

Electric eels, 505, 524, 536, 568, 575–76, 577–78
Electric energy generation, mechanical work into electrical current, 662–63
Electric energy transmission, 572–73
Electric eye, 858
Electric field(s), 517–26
conductors and, 526–27
drift velocity and, 573
electric lines of force, 520–23
equipotential surfaces and, 543–48
Gauss's law on, 528–29
line pattern, 520, 521, 547
in one dimension, 519
in storm clouds, 523, 524–25
superposition principle for, 518–20
time-varying, 661, 673
in two dimensions, 519–20
Electric fish, 505, 524–25, 536. *See also* Electric eels
Electric force, 512–16
gravitational force *vs.*, 506, 516
per unit charge, 538
Electric generators, 663–65, 667, 668
Electricity, 505–35, 536–67. *See also* Charge, electric; Circuits, ac; Circuits, basic; Electrical resistance; Electric current; Electric field(s); Electric force
capacitance, 549–52
dielectrics, 552–56
electric potential difference, 538–40
electric potential energy, 537, 542–43
electrostatic charging, 508–12
energy harvesting from human body, 156
equipotential surfaces, 543–48
generation by nuclear reactor, 943
safety and, 611–14
Electric lines of force (electric field lines), 520–22
Electric motor, 636
Electric potential, 539, 552–53
Electric potential difference, 538–40. *See also* Voltage
batteries and direct current, 569
common voltages, 546
due to point charge, 540–42
Electric potential energy, 537
barrier, 907
of charge configurations, 542–43
Electric power, 580–84
Electric power distribution system, 671–72
Electric voltage. *See* Electric potential difference; Voltage
Electrocardiogram, 490
Electrocommunication, 524
Electrodes, 569
Electrolocation, 524–25, 536
Electrolytes, 569
Electromagnetic field, and Maxwell's equations, 673
Electromagnetic force, 505
force unification theories and, 954
hadrons and, 952
photon and, 948–49
Electromagnetic induction, 656–85. *See also* Electromagnetic waves
applications of, 664, 666
back emf, 667–68, 669, 691
charging by, 511
defined, 657
Faraday's law on, 659–63, 669, 691
generators and, 663–65, 667, 668
hazard to electrical equipment, 661–62
induced emf, 657–63, 659, 664–65, 669–71
Lenz's law on, 659–63, 670, 671
power transmission and, 671–72
transformers and, 662–72
Electromagnetic inertia, 659, 660
Electromagnetic radiation, 433, 672
Electromagnetic spectrum, 676
Electromagnetic theory, and special relativity, 822
Electromagnetic waves, 672–78
classification of, and frequencies and wavelengths, 676
gamma rays (*See* Gamma rays)
infrared radiation, 385–86, 677
microwaves, 677
oscillator circuits and, 699
photons, 878
power waves, 675–76
radiation pressure, 674–75
radio waves, 676–77
source of, 673f
speed in vacuum, 674
TV waves, 676–77
types of, 433, 675–78
ultraviolet radiation, 677–78
visible light, 677, 864–65, 866
X-rays, 678
Electromagnetism, 623, 626
Ampere's law, 568
source of magnetic fields, 637–41
Electromagnets, 642–44
Electromotive force (emf), 570–71
alternator, 665
back, 667–68, 669
induced, 657–63, 659, 664–65, 669–71
motional, 662
self-induced, 691
Electron capture, 908
Electron clouds, 882, 896
Electron configuration, 890, 891
Electron-electron interaction, Feynman diagram, 949f
Electron flow, 572–73
Electronic balance, 636–37
Electron microscope, 883
Electron neutrino, 950, 952
Electron-positron pair, 896
Electron(s)
Compton wavelength of, 859–60
current and charge, 572
de Broglie wavelength of, 879–81, 883
electric charge of, 506, 507
energy level of, 863–66
Heisenberg uncertainty principle and, 894–95
in lepton family, 951
pair production and, 896–97
photoelectrons, 855
scattering by nickel crystal, 881
speed of, 549, 856–57
valence, 508, 891, 892f
X-rays and, 678
Electron spin, 641, 885–86
electron-volt (eV), 548–49, 835, 856
Electroplates, on eels, 575–76
Electroplating, 591
Electroscope, 509
Electrostatic charging, 508–12
by contact or conduction, 510–11
by friction, 510
by induction, 511
by polarization, 511–12
Electrostatic force, 509, 512, 513, 515–16, 904
Electrostatics, 506, 526
Electroweak force, 954–55
Electroweak unification theory, 955
Elementary particles in nuclear reactions, 951–52
Elements
alphabetical listing of, A-7
periodic table of, 891–94
Elevator, and apparent weight, 244–45
Elliptical orbits, 239
emf. *See* Electromotive force (emf)
Emission
fluorescence, 629
of light (auroras), 647
line of hydrogen, 885
photoelectric, 854–56
stimulated, 867–68
thermionic, 903
Emission spectrum, 861, 864–65
Emissivity, 387
Emphysema, 325
***Endeavor* space shuttle,** 807f
Endoergic (endothermic) reaction, 937–38, 939
Endoscope, 720
Energy, 140–41
binding, 864, 919–22, 940
of configuration, 237
conservation of, 141, 155–64, 319, 400, 660, 946–47
disintegration, 938, 947
in elastic collisions, 195–98
electric, 568, 572–73, 583
fusion as a source of, 945–46
gravitational potential, 152–54, 235–37
in inelastic collisions, 195–95
internal, 339
internal, in thermodynamics, 399–402
kinetic (*See* Kinetic energy)
law of conservation of total, 156
measurement of, and uncertainty principle, 896
mechanical, 158–61, 162–63
of position (or configuration), 152
potential, 152–55 (*See also* Potential energy)
quantum of, 854
radiant, 385
relationship to radius, speed, and circular motion, 243
relativistic kinetic, 833
relativistic total, 834
rest, 834–35
of a spring-mass system in simple harmonic motion, 435–38
storage in capacitor, 549–51
thermal insulation for, 382–83, 384
threshold, 897, 938–39
total, 155–56, 162–64
total mechanical, 158–61
transfer, propagation of disturbance, 446
transfer of heat and, 368
transported by waves, 447
in work-energy theorem, 148–52, 277–79
zero-point, 355
Energy guide for consumers, 584
Energy harvesting from human body, 156
Energy levels, 863–66, 885–87
Engineering system of measurement, 7
Engines
biological heat, 420
diesel, 397

Engines *(cont.)*
gasoline, 417
heat, 414–22
ideal heat, 422–24
internal combustion, 415, 418
steam, 397, 415
Entropy, 411–14
Envelopes (wave outlines), 483–84
Epicenter, of earthquake, 450
Equal and opposite force, 113
Equal-loudness contours, 494–95
Equation(s)
Bernoulli's, 319, 322–23
of continuity, 320–21
de Broglie, 878
flow rate, 320–21
kinematic, 45–49, 52
kinematic, components of motion, 70–73
lens maker's, 750–51
Maxwell's, 656, 673, 821
of motion, 439–46
Schrödinger's wave, 882, 885
sound intensity level, 477
spherical mirror, 736, 737–38
of state, 398
thermodynamic processes, 409
thin-lens, 745, 749
translational and rotational, 278
Equator, and magnetic field/forces, 633–34, 646
Equilibrium, 256, 261–66
mechanical, 262
rotational, 261–62
rotational static, 264–65
stable, 266–69
static, 262–66
thermal, 340, 374
translational, 119–21, 256, 261
translational static, 119–21, 262–63
unstable, 266–69
Equilibrium position, 147
Equipartition theorem, 358
Equipotential surfaces, 543–48
Earth's electric field and, 546
electric field lines and, 547
gravitational, 544–45
outside charged conductor, 548
Equivalence principle, 837
Equivalence statement, 13
Equivalent length, 13
Equivalent parallel capacitance, 558–59
Equivalent parallel resistance, 594–95
Equivalent series capacitance, 558
Equivalent series resistance, 592–93
Erecting lens (inverting lens), 804
Erythrocite sedimentation rate (ESR), 225
Escape speed, 241, 840
Ether, 821, 822
Evaporation, 374, 379, 380, 411
Event horizon, 840–41, 842
Exact numbers, 17
Exchange particles, 948–51
Excited state, 863–66
emission of gamma ray, 909
of hydrogen, 886
lifetime of, 864
Exercising, and heat and work, 367, 369, 401
Exoergic (exothermic) reaction, 937–38
Expansion, thermal, 338, 340, 350–53
coefficients for some materials, 351
Expansion gaps, 351–52
Exponents, A-2
External forces, 199
Extraordinary ray, 781
Eye, human, 792, 793–99. *See also* Eyeglasses
color vision, 810–11
focusing adjustment, 794
polarized *vs.* unpolarized light, 777
refraction and wavelength, 713–14
resolution and Rayleigh criterion, 809
structure of, 793–94
ultraviolet protection, 677–78
vision defects, 795–99
Eyeglasses
corrective, 795–98
lens power, 751
photogray sunglasses, 677–78
polarized, 777, 779
Eyepiece (ocular), 801, 803

F

Fahrenheit, Daniel Gabriel, 338
Fahrenheit temperature scale, 340–42
absolute zero and, 348
converstion to Celsius, 341–42
Faraday, Michael, 536, 658
faraday (F), 536
Faraday's ice pail experiment, 527
Faraday's law of induction, 659–63, 669, 691
farad (F), 536, 550
Far point, 794, 795–96
Farsightedness (hyperopia), 795, 797–98
feet per second (ft/s), 33
feet per second squared (ft/s^2), 40
Fermi, Enrico, 947, 950
Ferromagnetic materials, 641–44
Feynman, Richard, 949
Feynman diagram, 949
Fiber optics, 705, 718–20, 870
Fields. *See also* Electric field(s); Magnetic field(s)
electric, force, and vector, 517
electromagnetic, 673
First law of motion, Newton's, 105–6, 107n. *See also* Newton's laws of motion
conservation of linear momentum, 186
inertial reference frame and, 820
translational equilibrium, 261
First law of thermodynamics, 399–402
adiabatic process, 406–7, 408
heat engine cycle and, 417
heat pumps and, 421
isobaric process, 404
isometric process, 405–6
isothermal process, 403, 408, 409
summary of processes, 409
Fish, and buoyancy, 317
Fission reaction, 921, 939–44
Flashlights, 664
Fletcher-Munson curves, 494–95
Floating (apparent weightlessness), 243–44
Flow of fluids, 319–21
blood, 321, 490
incompressible, 319
irrotational, 319
laminar, 326
nonviscous, 319
steady, 319
turbulent, 326
Flow rate, 320f
average, 327
pressure and, 322f
from tank, 323
Flow rate equation, 320–21
Fluid(s), 297–98, 302–37. *See also* Flow of fluids; Gas(es); Liquid(s)
Archimedes' principle, 314–16
Bernoulli's equation, 319, 322–23
buoyancy and, 313–18
diffusion, 355–57
dynamics, 319–23
ideal, 319
Pascal's principle, 307–9
Poiseuille's law, 327–28
pressure and, 302–7
pressure measurement, 309–13
surface tension and, 324–25
viscosity of, 319, 325–27
Fluid volume expansion, 353
Fluorescence, 584, 629, 866
Fluorine, 936
Flux, magnetic. *See* Magnetic flux
FM (frequency-modulated) radio band, 676, 700, 760, 770
Foam insulation, 381, 385
Focal length, 733, 736, 741
Focal point, 732, 735–36, 741
Focal rays, 734, 742, 743
Focus, earthquake, 450
Foghorns, 474
Following through, 184–85
Food sterilization, gamma radiation and, 927, 929
Football, 151, 284–85
foot-pound (ft-lb), 142
foot-pound per second (ft • lb/s) 165
Force and motion, 103–39. *See also* Force(s)
concept of, 104–5
defined, 104
free-body diagrams, 116–18
inertia and, 105–6
Newton's first law, 105–6
Newton's second law, 106–12
Newton's third law, 112–15
translational equilibrium, 119–21
Force component, and work, 141
Force constant (spring constant, *k*), 146–47, 434, 437–38, 441
Forced-air heating systems, 385
Forced convection, 385
Force field, 517
Force multiplied at expense of distance, 309
Force of friction, 121, 122–27, 124f
Force of kinetic (or sliding) friction, 122, 124f
Force of static friction, 122, 124f
Force pairs, 113–14
Force per unit area (pressure), 302–3
Force per unit mass, 108
Force right-hand rule. *See* Right-hand force rule
Force(s). *See also* Force and motion
action-at-a-distance, 105
action-reaction, 113–14
attractive, 506, 542
balanced, 104, 261
buoyant, 313–16
centripetal, 216, 225–28, 232
color, 954
components of, 111–12, 118
concurrent, 117, 261, 262f
conservative, 157–58, 160–61
constant, 116, 141–45
contact, 104
Coulomb, 861, 918
electric, 512–16

electromagnetic (*See* Electromagnetic force)
electroweak, 954–55
external, 199
as a function of position, 146
fundamental, 948–51, 954
gravitational (*See* Gravitational force)
g's of, 108
instantaneous, 116
magnetic (*See* Magnetic force)
momentum and, 182
net, 104 (*See also* Net force)
nonconservative, 157–58, 162–64
normal, 113–14, 122, 126–27
nuclear, 904–5, 917–18
pressure and, 303, 306–7
radiation pressure and, 674
repulsive, 506, 515–16, 542
strong nuclear (*See* Strong nuclear force)
unbalanced, 104, 106, 261
van der Waals, 324
variable, and work done by, 145–47
weak nuclear (*See* Weak nuclear force)
Force transmittal process, 948
Force unification theories, 954–55
Forensics, 630
Formula mass, 345
Fosbury flop, 204
Four-stroke cycle, 417–18
Fourth phase of matter, 946
fps system, 7
Fractional change in length, 299, 350
Frame of reference, 90, 820–25. *See also* Reference frame
Frankel, R. B., 645n
Franklin, Benjamin, 523, 796
Free-body diagrams, 116–18
Free fall, 49–56, 108–9
Free fall distance *vs.* time, 105
Free space, magnetic permeability of, 638
Free space, permittivity of, 550
Freezing point, 374, 375, 378
Freon, 677
Frequency(ies), 221–22
angular, of mass oscillating on spring, 441
beat, 483–84
broadcast, 698–99, 700
color of light and, 811–12
cutoff (threshold), 855, 857
of electromagnetic wave classifications, 676
fundamental, 455, 456, 493–96
for musical instruments, 494–96
of pendulum, 443
resonance, 697, 888
resonant (natural), 454–59, 493
wave, 434–35, 447
Fresnel, Augustin, 748
Fresnel lenses, 748
Friction, 121–29
air resistance, 127–29
centripetal force and, 225–27
coefficients of, 122–23, 124f, 227
electrostatic charging by, 510
forces of, 122–27
kinetic (sliding), 122, 125
as nonconservative force, 157–58
rolling, 122
snow and car tire example, 103
static, 122, 125, 127
total energy and, 162
and walking, 121
Frost, 374
Fuel rods, 941
Fundamental forces, 948–51, 954. *See also* Electromagnetic force; Gravitational force; Strong nuclear force; Weak nuclear force
Fundamental frequency, 455, 456, 493–96
Fuses, 612–13
Fusion, latent heat of, 375, 376
Fusion reaction, 921, 944–46

G

g (acceleration due to gravity), 49
Gadolinium, as ferromagnetic material, 641
Galaxies, active, 842
Galilean telescope, 804
Galileo Galilei, 32, 50, 51, 105, 310, 792
Galileo's experiment with rolling balls, 105
Gallium arsenide, 884
Galvanometer, 607–10, 635, 636f, 657
Gamma decay, 909–10
Gamma knife, 927
Gamma radiation, and poultry, 927, 929
Gamma ray detector, 928–29
Gamma rays
defined, 678
radiation detection, 922, 924
radioactivity decay, 906, 909–10
from Sun to Earth, 851
telescopic observation of, 808
Gas-discharge tubes, 860–61
Gaseous diffusion, 356–57
Gaseous (vapor) phase, 374
Gas(es), 297. *See also* Fluid(s)
compressibility for, 302
diatomic, 354, 357–59
in first law of thermodynamics, 400
greenhouse, 367, 388
kinetic theory of, 354–57, A-5–A-6
monatomic, 354, 355, 357, 359, 419
noble (inert), 357
specific heat of, 373–74
speed of sound in, 471–74
as thermal conductors, 380
Gas laws, 343–49
Gas leaks, 365
Gasoline, density of, 305
Gasoline engines, 417
Gas-solid phase change, 374
Gas thermometer, 342
Gauge, pressure, 310
Gauss, Karl Friedrich, 528
gauss (G), 626
Gaussian surfaces, 528
Gauss's law, 528–29
Gedanken (thought) experiments, 823–25, 830, 832, 837–38
Geiger, Hans, 923
Geiger counter, 923
Gell-Mann, Murray, 952, 953
General theory of relativity, 837–41
black holes, 819, 840–41, 842
equivalence principle, 837
gravitational lensing, 839–40
light and gravity in, 837–39
Generators, electric, 663–65, 667, 668
Geographic *vs.* magnetic poles, 646
Geomagnetic field, 644
Geomagnetism, 644–47
Geometrical optics, 705–6. *See also* Lens(es); Mirrors; Reflection; Refraction
Geometric methods of vector addition, 73–74
Geometric relationships, A-3–A-4
Geosynchronous satellite orbit, 234, 236
Germer, L. H., 881
gigaelectron-volt (GeV), 548
giga- prefix, 8
Gilbert, William, 644
"Gimli Glider," 16
Glacier, momentum of, 179–80
Glare reduction, 779, 780f
Glashow, Sheldon, 954–55
Glass
index of refraction, 710, 712
optically active, 782
resistivity and temperature coefficient, 577t
viscosity of, 326n
Glasses for eyes. *See* Eyeglasses
Glazer, D. A., 924
Global communications, 676–77
Global Positioning System (GPS), 646, 819, 838
Global warming, 388
Gluons, 954, 955
Goeppert-Mayer, Maria, 922
Golden Gate Bridge, 338, 352
Golfing. *See* Sports
Goose bumps, 343
GPS (Global Positioning System), 646, 819, 838
grad (angular unit), 219
Grain (weight), 179
Grand unified theory (GUT), 955
Graphical analysis
of kinematic equations, 49
of motion and velocity, 38–40
of motion with constant acceleration, 43–44
Graphs
position-*versus*-time, 39
velocity-*versus*-time, 43, 49
Gratings, diffraction, 772–74
Gravitational analogies to electricity
electric equipotential surface, 544–45
parallel-plate electric field, 537
potential energy *vs.* potential, 540
resistors in series and in parallel, 593, 594
Gravitational constant, universal, 231–37
Gravitational force, 948
in attraction of the Earth and Moon, 105, 206, 232
compared to nuclear force, 904
conservative, 158
electric force *vs.*, 506, 516
grand unified theory and, 955
gravitons and, 951, 955
hadrons and, 952
pendulums and, 442, 443
working against, 144, 154
Gravitational lensing, 839–40
Gravitational potential energy, 152–54, 235–37, 569
Gravitons, 951, 955
Gravity. *See also* Center of gravity (CG); Circular motion
acceleration due to, 49–56
artificial, 246–46
in general theory of relativity, 837–41, 842
horizontal projections, 81–82
Newton's law of, 231–37, 242
specific, 318
zero, 243, 245
Gravity assists, 238, 312–13, 328
gray (Gy), 924–25
Great Wall of China, view from space, 810
Greenhouse effect, 386, 388, 677
Greenhouse gases, 367, 388
Grounded plugs, 613–14
Ground (electric charge), 511, 523
Ground (neutral) wires, 611f, 612–13

Ground state of atoms
electron configurations, 890
energy levels and, 863–66
of multielectron atom, 887
probability density, 882
Group, of elements, 891, 892f
***g*'s of force,** 108
***g*-suits,** 108
Gymnastics, iron cross position, 266
Gyroscope, 284–85

H

Hadrons, 951, 952, 952t, 953
Hafele, J., 832n
Hale Observatory, 806
Half-life, 912–15
radioactive dating, 915–17
of radioactive nuclides, 913
Hall effect, 649
Halley, Edmond, 103
Halley's Comet, 103
Hard iron, 644
Harmonic motion. *See* Simple harmonic motion (SHM)
Harmonic series, 455–56, 493, 496
Hearing. *See also* Ear, human
anatomy of ear, 475
audible region of sound, 468
protecting one's, 480
threshold of, 476, 494–95
Heart, human. *See also* Blood
defibrillators, 522, 551
pacemaker cells and electrical signals, 591, 608–9
as pump, 312–13
Heat, 367–96
body, 367, 380, 385, 387
calorimetry, 372–73
defined, 368
joule, 580–83 (*See also* Joule heat)
latent, 374–78
mechanical equivalent of, 369
phase changes and, 374–79
specific, 370–74
temperature distinguished from, 338, 339–40
units of, 368–69
Heat capacity, specific, 370
Heat engines, 414–22
biological, 420
cyclic, 417, 422, 424
ideal, 422–24
thermal efficiency, 416–22
Heat exchanger, 385
Heat of fusion, 375, 376
Heat of sublimation, 375
Heat of vaporization, latent, 375, 376, 411
Heat pumps, 420–22
Heat radiation, 386
Heat rays, 677
Heat reservoirs, 398
Heat stroke, 380
Heat transfer, 379–90
by conduction, 379–83, 384, 386
by convection, 383–85, 386
by radiation, 385–90
Heavy hydrogen, 905
Heavy water (deuterium), 905, 941, 944, 945
Heisenberg, Werner, 877, 894
Heisenberg uncertainty principle, 894–96
Helium atom
discovery of, 851, 861
expansion, adiabatic *vs.* isothermal, 408
mass compared to hydrogen, 920
molecular speed, 355
proton-proton cycle, 945
stable nuclei, 918
Helium-neon (He-Ne) gas laser, 867–68
Henry, Joseph, 658, 691
Henry I, and length of yard, 1
henry (H), 691
Herapath, W., 776
Herschel Space Observatory, 729
Hertz, Heinrich, 222, 434
hertz (Hz), 222, 434
Hibernation, and body temperature, 344
High-intensity focused ultrasound (HIFU), 471
High-potential (hot) wires, 611–14
***Hindenburg* zeppelin,** 297, 334
Hologram, 869, 871
Holography, 869, 871
Homogeneous spheres, 232–33
Hooke, Robert, 146, 299
Hooke's law, 146, 299, 434, 441
Horizontal projectile motion, 81–82
horsepower (hp), 165
Hot-air balloons, 338
Household circuits, 611–14
Hubble Space Telescope, 807, 808
Human body. *See also* Medical applications
air pressure and earaches, 311
bioelectrical impedance analysis (BIA) of, 578
blood (*See* Blood)
body heat, 367, 380, 385, 387
bones, 300–301, 304–5
brain waves, 433
center of gravity, 204, 261, 268
diffusion in life processes, 357
ears (*See* Ear, human)
effects of electric current on, 591, 614
efficiency of, 397, 420
electrical resistance of, 574, 577
energy harvesting, 156
energy needs, 140
exercising, 367, 369, 401
eyes (*See* Eye, human)
g's of force and effects on, 108
heart (*See* Heart, human)
impulse force and injury, 184
lungs, 325, 357, 407
muscles and torque, 260–61
nerve signal transmission, 552–53
temperature, 343, 380
thermodynamics and, 420
weightlessness effects, 245
Humidity, 474
Hurricanes, 283
Huygens, Christian, 238
Hybrid cars, 164, 666
Hydraulic lift, 307–8
Hydroelectric power, 665, 667, 672
Hydrogen atom
atomic quantum numbers and, 885–86
Bohr theory of, 860–66, 878–79, 882
electric potential difference and, 541–42
energy levels, 863–66, 885–86
isotopes, 905
magnetic resonance of, 888
mass compared to proton, 920
probability density cloud, 882
solar system model of, 506
Hydrogen (H) bomb, 946
Hydrogen spectrum, 851, 864–65
Hyperopia (farsightedness), 795, 797–98
Hyperthermic therapy, 642
Hypothermia, 343, 380

I

Ice
anisotropic and birefringence, 781
as thermal conductor, 389
Iceberg, 318
Iceland, hot-spring heat, 397
Ice pail experiment, 527
Ice point, 341
Ideal Carnot efficiency, 423
Ideal fluid, 319
Ideal gas law, 344–45, 398
absolute temperature and, 347, 348–49
adiabatic process for, 406–7
isobaric process for, 404–5
isometric process for, 405–6
isothermal process for, 403, 408, 409
macroscopic form of, 345–46
Ideal heat engines, 422–24
Ideal spring force, 146–48, 434
Image distance, 730, 736, 745
Image(s)
holographic, 871
real, 730, 734, 736, 744
virtual, 730, 734, 736, 744
Image side of lenses, 741
Impedance
bioelectrical analysis (BIA), 578
in RLC circuits, 693–97
Impulse, 182–85
Impulse-momentum theorem, 182–85
Incandescent lamps, 868
Incidence
angle of, 707, 711, 717
plane of, 707
Inclined plane, 117
Incoherent sources, 761
Incompressible flow, 319
Independent of path, 155, 157
Index of refraction, 710–11, 714–15, 723
negative, 715
thin-film interference, 764–66
Induced battery, 667
Induced-current right-hand rule, 659
Induced currents, 657–58, 660, 661–62
Induced emf, 657–63, 659, 664–65, 669–71
Induced fission, 939
Inductance, 691
Induction. *See* Electromagnetic induction
Inductive ac circuits, 695
completely, 696
purely, 692
Inductive reactance, 691–92, 697
Inductor(s), 691, 694, 696
Industrial Revolution, 397
Inelastic collisions, 191, 192–94, 836
Inertia
electromagnetic, 659, 660
moment of, 270–73, 273–74, 283–84
in Newton's first law of motion, 106
pulley, 275
rotational, 270–71, 286
Inertial confinement, 946
Inertial reference frame, 820
principle of equivalence, 837
relativity of simultaneity and, 823–25
special relativity and, 822
time intervals and, 825
Inertial reference system, 107n
Inert (noble) gases, 357
Infrared radiation, 385–86, 388, 677, 808
Infrared region, 852, 864–65
Infrasonic region, 468

Infrasound, 468–69
Initial condition (equation of motion), 440, 443–44
Inner transition elements, 891, 892f
Input coil (primary coil), 669
Instantaneous
acceleration, 41, 116
angular speed, 219
angular velocity, 220
axis of rotation, 257, 258
force, 116
power, 687
speed, 34
velocity, 37–38, 40
Insulation
R-values (thermal resistance values), 384, 395
and thermal conductivity, 382–83
Insulators
dielectric, 552
of electric charge, 508–9
foam, 367, 381, 385
resistivities and temperature coefficients, 577t
thermal, 367, 379, 382
thermal conductivities of substances, 381
Intensity, of sound, 474–80, 494–95
Intensity levels, 476–79
Interference, 449–52
constructive, 451 (*See also* Constructive interference)
destructive, 451–52 (*See also* Destructive interference)
sound, 482–84
total constructive, 451, 482
total destructive, 451–52, 482
Young's double-slit experiment, 761–63
Interference maxima, 762–63, 773
Interferometers, 821n, 842
Internal combustion engines, 415, 418
Internal energy, 339
of diatomic gases, 358–59
of monatomic gases, 355
in thermodynamics, 399–402
Internal motion, 202
Internal reflection, total, 717–18, 719
Internal resistance, 570
International Bureau of Weights and Measures, 4
International Committee on Weights and Measures, 348
International System of Units (SI), 3–10
Intravenous (IV) blood transfusions, 312–13, 328
Intrinsic angular momentum of electron, 885
Inverse-square law, 231, 242
Inversion, population, 867
Inverted image, 794
Inverting lens (erecting lens), 804
Ionic layers in atmosphere, 676
Ionization, 524
Ionized atoms, 862, 864
Ions, mass spectrometer and, 630–31
I^2R losses, 580, 612, 643, 668, 670, 671, 687
Iron
core, and electromagnets, 642–44
core, and transformers, 669, 670
as ferromagnetic material, 641, 646
filings, 624–25
resistivity and temperature coefficient, 577t
Iron cross gymnastic position, 266
Irregular reflection (diffuse reflection), 707–8, 709
Irreversible process, 399
Irrotational flow, 319
Isobaric expansion, 405
Isobaric process, 404
Isobars, 404–5, 415
Isochoric process, 405
Isogonic lines, 646
Isolated systems (closed systems), 156, 187, 192, 413
Isomet, 406, 415
Isometric process, 405–6
Isothermal process, 403, 408, 409, 423
Isotherm(s), 403, 405, 408, 409, 423
Isotope(s), 905, 908, A-7–A-9. *See also* Half-life
Isotropic expansion, 350
Isotropic optical property, 780
Isovolumetric process, 405
IV (intravenous) injection, 312–13, 328

J

Jansky, Carl, 808
Jefferson, Thomas, and length standard, 1
Jet fighter pilots and g's of force, 108
Jet propulsion, 114, 204–6
Jet streams, 384–85
Joule, James, 142, 369
Joule heat, 580–83
ac current and, 687
electromagnets and, 643
in household circuits, 612
power transmission losses, 668, 670, 671
joule (J), 142
conversion to electron volt, 548
food values, 368–69
joule per kelvin (J/K), 411
joule per second (J/s), 580
Junctions (nodes), electrical, 594, 599
Junction theorem, 600, 608
Jupiter, 240–41, 646

K

Keating, R., 832n
Keck telescopes, 807
Kelvin, Lord, 347, 417
kelvin (K), 7, 347
Kelvin temperature scale, 343, 346–49
conversion to Celsius, 347
Kepler, Johannes, 216, 238
Kepler's laws of planetary motion, 232, 238–46
first law (law of orbits), 239
second law (law of areas), 239–40, 282–83
third law (law of periods), 240
Keyboards, computer, 556
kilocalorie (kcal), 368
kiloelectron-volt (keV), 548
kilogram (kg), 4–5, 7, 8–9
kilogram-meters squared ($kg \cdot m^2$), 270
kilogram-meters squared per second ($kg \cdot m^2/s$), 280
kilogram per cubic meter (kg/m^3), 303
kilohms ($k\Omega$), 573
kilometers per hour (km/h), 33
kilowatt-hour (kWh), 583
kilowatt (kW), 165
Kinematic equations, 45–49, 52. *See also* Constant acceleration
for components of motion, 70–73
Kinematics, 32–66. *See also* Displacement; Distance; Motion; Speed; Velocity
acceleration, 40–44 (*See also* Acceleration)
constant acceleration, 45–49
free fall, 49–56, 108–9
rotational, 230
scalar quantities: distance and speed, 33–34
vector quantities: displacement and velocity, 35–40
Kinetic energy
of alpha particles, 904
in beta decay, 947–48
defined, 149
elastic collisions and, 196
electron-volt, 548
gravitational potential energy and, 154
momentum *vs.*, 185
of orbiting satellite, 243
photoelectrons and, 855–57
Q value and, 937–38
relativistic, 833
relativistic total energy and, 833–35
rotational work and, 277–80
translational, 278–79, 339, 354, 356, 358, 359
work-energy theorem and, 148–52, 277–80
Kinetic friction (sliding friction), 122, 123, 124f
Kinetic theory of gases, 354–59, A-5–A-6. *See also* Temperature
absolute temperature in, 355
diffusion in, 355–57
equipartition theorem in, 358
evaporation and, 379
internal energy of diatomic gas, 358–59
internal energy of monatomic gases, 355
Kirchhoff, Gustav, 599n
Kirchhoff's rules, 599–604, 608–9, 610
application of, 603
first rule (junction theorem), 600, 608
second rule (loop theorem), 600–602, 604, 693–94
Klystrons, 677

L

Laminar flow, 326
Land, Edwin H., 776, 811
Lanthanide series, 891, 892f
Laplace's law, 325
Lapse rate, atmospheric, 362
Laser beam, 761, 773
Laser Interferometer Gravitational-Wave Observatory (LIGO), 842
Laser(s)
applications and operations of, 851, 866–71
for correcting nearsightedness, 797
in eye surgery, 792, 797
helium-neon, 867–68
in holography, 869, 871
in modern medicine, 870
Latent heat, 374–78
of fusion, 375, 376
of sublimation, 375
of vaporization, 375, 376, 411
Lateral magnification factor (M), 730, 731f, 737
Law
of areas (Kepler's second law), 239–40, 282–83
of charges (charge-force law), 506
Charles', 344, 345f
of conservation of linear momentum, 186
of conservation of mechanical energy, 158
of conservation of total energy, 156
of cosines, A-4

Law *(cont.)*
of gravitation (Newton), 231–37, 242
ideal gas (*See* Ideal gas law)
of inertia, 106 (*See also* First law of motion, Newton's)
Laplace's, 325
Lenz's (*See* Lenz's law)
of mechanics, and inertial reference frames, 820
of motion (*See* Newton's laws of motion)
of nature, and reference frame, 820
of orbits (Kepler's first law), 239
of periods (Kepler's third law), 240
of poles (pole-force law), 624
of reflection, 707, 708, 734
of sines, A-4
Stefan's, 386–87
thermodynamics (*See* Thermodynamics)
Wien's displacement, 853
Law enforcement, and electric fields, 524
LC circuit, oscillating, 699
LCDs (liquid crystal displays), 374, 783
Leaning Tower of Pisa, 50, 51, 269
LEDs (light-emitting diodes), 783
Leibniz, Gottfried, 103
Length
arc, 217–19
British unit, 10
change in, and Young's modulus, 299–301
equivalent, 13
focal, 733, 736, 741
fractional change in, 299, 350
proper, 829–30
SI unit, 3–4, 10
Length contraction, 828–31
Lens(es), 729, 740–53
aberrations in, 752–53
combinations of, 748–50
converging (biconvex), 740, 741 (*See also* Converging (biconvex) lenses)
convex, simple microscope, 799
crystalline, 793f, 794
diopters, 751, 794
diverging (biconcave), 740, 741 (*See also* Diverging (biconcave) lenses)
erecting (inverting), 804
Fresnel, 748
half an image, 746
historical, 792
image side, 741
meniscus, 741, 751
nonreflecting, 766–67, 768
objective, 801, 803, 808
object side, 741
oil immersion, 810
perfect, 715
polarized, 777, 779
ray diagrams for, 741, 743, 744, 747
thin-film interference and symmetry, 767–68
thin lens sign conventions, 745
Lens maker's equation, 750–51
Lens power, 751
Lenz, Heinrich, 659
Lenz's law
ac circuits and, 691, 692, 699
Faraday's law of induction and, 659–63
self-induced emf, 670, 671
Leptons, 951–52, 952t, 955
Lever arm (moment arm), 259
Life, and second law of thermodynamics, 414
Lifetime, of excited state, 864
Lift
of airplane, 322–23
hydraulic, 307–8
Light. *See also* Physical optics (wave optics); Reflection; Refraction
atmospheric scattering of, 782, 784
color determination, 811–13
dispersion, 453, 721–23
dual nature of, 851, 855, 860
electromagnetic wave origin, 673
in general theory of relativity, 837–40
monochromatic, 721, 855, 859
quantization of, 851, 854–58
relativistic velocity addition for, 841, 843
speed of (*See* Speed of light)
ultraviolet (black light), 866
visible, 677, 705, 769, 864–65, 866
wavelength of, 763–64, 773–74
white, 721, 723, 812
Lightbulbs
in complete circuit, 569, 573
efficiency of, 581, 584
filaments and resistance, 574
rms and peak values, 688–89
Light-emitting diode (LED), 664, 783
Lighthouse lenses, 748
Light microscopes, 883
Lightning, 505, 523, 524
Lightning rods, 523, 527
Light pipes, 719
Light polarization. *See* Polarization
Light pulse clock, 825
Light rays, 706, 708, 711. *See also* Ray(s)
Light waves. *See* Physical optics (wave optics)
light-year (ly), 832
LIGO (Laser Interferometer Gravitational-Wave Observatory), 842
Linear expansion (thermal), 350
Linearly polarized light (plane polarized light), 776
Linear momentum, 178–82. *See also* Momentum
conservation of, 185–91, 946–47
Linear motion, 38–40, 46
Linear relative velocity, 91–92
Linear turn density, 639
Linear wave front, 706
Liquid crystal displays (LCDs), 374, 629n, 783
Liquid crystal(s), 374
Liquid-drop model, 939–40
Liquid-gas phase change, 375
Liquid-in-glass thermometers, 340–42
Liquid phase, 374
Liquid(s), 297. *See also* Fluid(s); Thermal expansion
diffusion of, 355–57
specific heat of, 370–72
speed of sound in, 471–72
as thermal conductors, 380
volume stress and bulk modulus and, 302
liter (L), 8–9
Lithotripsy, 470
Load (normal force), 122
LOCA (loss-of-coolant accident), 943–44
Lodestone, 623, 644
Logarithms, 404, A-5
common, 404, 476–79
natural, 404, 605
and sound intensities, 476–79
Longitudinal waves (compressional waves), 448–49, 450, 475
Long jump event, 88–89
Looming effect, 716
Loops, current-carrying
magnetic field at center of, 638–39
torque on, 634–35
Loop theorem (Kirchhoff), 600–602, 604, 693–94
Loss-of-coolant accident (LOCA), 943–44
Loudness, 476, 494–95
Lubrication, 123
Luminiferous ether, 821, 822
Lungs, 325, 357, 407
Lyman series, 864–64
Lymphocytes (white blood cells), 883

M

Mach, Ernst, 490
Mach number, 490–91
Macroscopic form of ideal gas law, 345–46
Macroscopic heat transfer, 400
Magic numbers, 922
Magnetically levitated trains, 623
Magnetic confinement, 647, 946
Magnetic declination, 646
Magnetic dipoles, 624, 645
Magnetic domains, 641, 642f, 643, 644
Magnetic echo, 664
Magnetic field(s), 624–25, 657. *See also* Electromagnetism
charged particles in, 629–32
direction, 624–25
of the Earth, 624, 644–47
electromagnetism and current, 637–41
inductors and, 691
line, 625
relationship to electric fields, 673
strength of, 626–28
time-varying, 661, 673
Magnetic flux, 658–59
in electric generators, 664–65
inductors and, 691, 692
leakage of, 669–71
in Lenz's law, 660–61
Magnetic force, 626–28
applications of, on charged particles, 629–32
on current-carrying wires, 632–35
magnetic field strength and, 626–28
in medicine, 642
between two parallel wires, 640–41
Magnetic materials, 641–44
Magnetic moment, 634, 635
Magnetic moment, spin, 888
Magnetic moment vector, 634
Magnetic monopole, 624
Magnetic north pole, 646
Magnetic permeability
of core material, 643
of free space, 638
Magnetic poles, 624, 646
Magnetic quantum number, 885
Magnetic resonance imaging (MRI), 642, 888–89
Magnetic south pole, 646
Magnetic tape/disk, 656
Magnetism, 623–55. *See also* Electromagnetism
geomagnetism, 644–47
magnetic fields (*See* Magnetic field(s))
magnetic force (*See* Magnetic force)
magnetic materials, 641–44
magnetic poles, 624, 646
pole-force law (law of poles), 624
Magnetite, 642, 645
Magnetohydrodynamics, 631–32
Magnetotactic bacteria, 645
Magnetron, 677
Magnets

bar, 624
electromagnets, 642–44
horseshoe, 626
permanent, 624, 625, 626, 641, 644
superconducting, 579, 626, 631, 643–44

Magnification
angular, 799–801, 802
factor, 737, 745, 749
factor, lateral, 730, 731f, 737
of refracting telescope, 803

Magnifying glass, 799–801
Magnifying power, 799–801, 802
Magnitude, 36
Magnitude-angle form (for vectors), 75
Malus, E. L., 777
Malus's law, 777
Manometers, 309–10
Maps, topographic, 544–45
Mariana Trench, 297
Mars
distance and *Viking* probes, 674
free fall and Mars Polar Lander, 56
red sky of, 760, 784

Mars Climate Orbiter, unit conversions and failure of, 16
Mars Exploration Rovers, 34, 186–87
Mass
British unit, 9
center of, 198–204
conservation of, 937–38
critical, 940–41
formula, 345
as fundamental property, 108
inertia and, 106
molecular, 346, 631
momentum and, 178
Newton's second law on, 106–10
SI unit, 4–5, 10

Mass defect, 920
Mass distribution, and rotational inertia, 270–71
Mass-energy equivalence, 834–36, 919–20
Masses and measures, as phrase, 5
Mass number, 905
Mass reduction and jet propulsion, 205
Mass spectrometer, 629–31
Mass spectrum, 631
Mass-spring system, 435–38
Materials
densities of, 304
dielectric constants for, 553
elastic moduli for, 300
latent heats and phase change temperatures of, 375
magnetic, 641–44
radioactive nuclide half-lives, 913
resistivities of, 577
specific heat of, 370
speed of sound, 472
temperature coefficients of resistivities, 577
thermal conductivity of, 381
thermal expansion coefficients of, 351
viscosities of, 327
voltages, common, 546

Mathematical relationships, A-1–A-5
Matter, and antimatter, 896–97
Matter waves (de Broglie), 878–81
Maxima (interference)
diffraction gratings, 773
double slit, 762–63
single slit, 769–71, 807–8

Maximum range (projectile), 87–88
Maxwell, James Clerk, 656, 673, 821, 954
Maxwell's equations, 656, 673, 821
Measured numbers, 17
Measurement, 1–20. *See also* British units; SI unit(s)
angular, 217–19
metric system, 3, 7–10
of pressure, 309–13
significant figures, 17–20
SI units, 3–7
standard units, 3
and uncertainty principle, 894
unit analysis, 10–12, 13
unit conversions, 12–16

Measure of disorder, 411, 413
Mechanical efficiency, 166–67
Mechanical energy, conservation of, 158–61, 162–63
Mechanical equilibrium, 262
Mechanical equivalent of heat, 369
Mechanical load, motors and back emf, 667
Mechanical resonance, 458–59
Mechanical work, 141, 142–43
into electrical current, 662–63, 663–65, 667
in thermal cycle, 410

Mechanics. *See also* Dynamics; Kinematics
study of motion, 32

Medical applications. *See also* Human body; Surgery
bioelectrical impedance analysis (BIA), 578
body temperature and surgery, 343
centrifuge, 224–25
Doppler and blood flow, 490
EEG (electroencelphalograph) and brain waves, 433
fiber optics, 720
infrared thermometers, detecting SARS, 386
laser surgery, 792, 797, 870
magnetic force and, 642
magnetic resonance imaging (MRI), 888–89
microscopic images of blood cells, 883
optical biopsy, 785
organ transplants, 378
pneumatic massage, 108
of radiation, 914, 924–28
RC circuits and cardiac pacemakers, 608–9
ultrasound, 305, 470–71, 490
weightlessness effects, 245
X-rays, 304–5, 540, 775

megaelectron-volt (MeV), 548, 835
megohms (MΩ), 573
Meltdown, 943–44
Melting point, 374
Mendeleev, Dmitri, 891
Meniscus lenses, 741, 751
Mercury
barometer, 309f, 310, 311
converted into gold, 936
density of, 305
in liquid-in-glass thermometers, 340–41
magnetic fields, 646
resistivity and temperature coefficient, 577t

Merry-go-rounds, 221
Meson(s), 949–50, 952, 953
Metal detectors, 664, 666
Metals, as thermal conductors, 379
Metastable state, 867
meter (m), 3–4, 7
meter-newton (m • N), 259
Meter of the Archives, 4
Meters. *See* Ammeters; Galvanometer; Voltmeters
meters per second (m/s), 33
meters per second squared (m/s²), 40
Methane, 388, 536
Method of mixtures, 372–73
Metric system, 3, 7–10. *See also* SI unit(s)
multiples and prefixes for metric units, 7

metric ton, 9
mho, 568
Michelson, Albert A., 821
Michelson interferometer, 821n
Michelson-Morley experiment, 821, 822, 823
microampere (μA), 572
Microchips, computer, 509
microcoulomb (μC), 507
microcurie (μCi), 914–15
microfarad (μF), 550
Microgravity, 244n
Microscopes, 799–803
compound, 801–3
compound-lens system, 748
electron, 883
light, 883
magnifying glass, 799–801
resolving power of, 808–9
scanning electron, 883
scanning tunneling, 877, 884, 908
transmission electron, 883

Microscopic form of ideal gas law, 345
Microscopic heat transfer, 400
microteslas (μT), 626
Microwave ovens, 230
Microwaves, 677
miles per hour (mi/h), 33
Milky Way, and Doppler shifts of light, 488
milliampere (mA), 572
millibar (mb), 311n
millicurie (mCi), 914
millihenry (mH), 691
milliliter (mL), 9
milliteslas (mT), 626
Minima (interference)
double slit, 762
single slit, 769–71

Minimum angle of resolution, 808–9
Mirage, 714–16
Mirror images, 730
Mirror length, 731
Mirrors, 729–40
parabolic, 740, 806
plane, 730–32
spherical (*See* Spherical mirrors)

***Mir* space station,** 243
Mixed units, 12
Mixtures, method of, 372–73
mks system, 7
Moderator, 942
***Modern Magic*,** 733
Moduli
bulk, 301–2, 471
elastic, 298–302
shear, 301
Young's, 299–301, 471

Molecular mass, 346, 631
Molecular speed, 355, 356
Molecules
combining atoms, 893
mass of, and mass spectrometer, 631

mole (mol), 7, 345
Moment arm (lever arm), 259
Moment of inertia, 270–73
ice skaters and, 283–84
parallel axis and, 273–74
for symmetric diatomic molecule, 358
of uniform-density objects, 273

Momentum, 177–215. *See also* Angular momentum; Collision(s)
center of mass and, 198–99
change in, 181
conservation of, and kinetic energy, 196
conservation of linear, 185–91, 946–47
defined, 178
in elastic collisions, 195–98
force and, 182
impulse and, 182–85
in inelastic collisions, 192–95
jet propulsion and, 114, 204–6
kinetic energy *vs.*, 185
linear, 178–82
magnitude of, 185
measurement of, and uncertainty principle, 894–95
of photon, and de Broglie hypothesis, 878
relativistic, 834, 836
total, 178, 180–81
total linear, 178
Monatomic gases, 354, 355, 357, 359, 419
Monochromatic light, 721, 855, 859
Monopoles, magnetic, 624
Moon
angular momentum of, 286
experiment on acceleration due to gravity, 50
gravitational attraction between the Earth and, 105, 206, 232
laser beams and distance to, 851
mass on, 5
weight on, 108
Morley, Edward W., 821
Motion. *See also* Circular motion; Components of motion; Force and motion; Friction; Kinematics; Newton's laws of motion; Projectile motion; Rotational motion; Simple harmonic motion (SHM)
damped harmonic, 445–46
defined, 33
equations of, 439–46
internal, 202
nonuniform linear, 39–40
periodic, 434
translational motion, 257–59, 278
uniform circular, 223–28, 439, 441
uniform linear, 38–39
Motional emf, 662
Motion detectors, 556
Motion in two dimensions, 67–102
components of, 68–73
curvilinear, 67, 71–72, 223
projectile, 81–89
relative velocity, 90–94
in two dimensions, 69
vector addition and subtraction, 73–80
Motor oils, and viscosity, 327
Motors, 656
back emf of, 667–68
dc, 636
Moving frame of reference, 90
MRI (magnetic resonance imaging), 888–89
Mufflers, and destructive interference, 452
Multielectron atoms, 886–87
Multiloop circuits, 599–604
Multimeters, 610, 611f
Multiplication, and significant figures, 18, 19
Multiplier resistors, 610
Multirange meters, 610, 611f
Muon decay, 827–28, 831
Muon neutrino, 950, 952
Muon(s), 949–50, 951
Muscles, and torque, 260–61
Musical instruments, 455–58, 484, 491–96
Mutual induction, 657n, 658f
Myopia (nearsightedness), 795–96, 797

N

Na/K ATPase molecular pump, 553
nanoampere (nA), 572
nanocoulombs (nC), 507
nanocurie (nCi), 914
nanofarad (nF), 550
nanometer (nm), 8, 710n
nano- prefix, 8
Nanotechnology, 8, 156
National Institute of Standards and Technology (NIST), 5f, 6f, 641
Natural convection, 383
Natural frequencies (resonant frequencies), 454–59, 493
Natural line broadening, 896
Natural logarithms, 404, 605, A-5
Natural resources, and electrical efficiency, 583–84
Near point, 794, 795–96, 798, 799, 800
Nearsightedness (myopia), 795–96, 797
Negative charges, 540
Negative index of refraction, 715
Negative lens, 745
Neodymium, as ferromagnetic material, 641
Neon, in gas laser, 867–68
Neon lights, 860–61
Neon-tube relaxation oscillator (blinker circuit), 606–7
Nerve signal transmission, 552–53
Net charge, 507
Net external force of system, 199
Net force
change in momentum, 182
defined, 104
equilibrium and, 261
in second law of motion, 106–9
Net internal force of closed system, 187
Net inward force, 225
Net rotational work, 277
Net torque, 270, 280, 281
Net work per cycle, 415–16
Net work (total work), 144–45, 149
Neurons, 552–53
Neutral wires (ground), 611f, 612–13
Neutrino(s), 946–48
antineutrino, 947, 952
beta decay and, 908, 947–48
electron, 950, 952
leptons, 951–52, 952t, 955
muon, 950, 952
in nuclear fusion, 945
tau, 952
weak force and, 950
Neutron activation analysis, 928
Neutron-capture reactions, 939
Neutron number, 905
Neutrons
in beta decay, 908
electric charge of, 506, 507
as hadrons, 951, 952
nuclear stability and, 917–19
in nuclear structure, 904, 905
in radioactive dating, 915
thermal, 942
Newton, Isaac, 32, 103, 104, 178, 231, 232, 721, 767
Newtonian focus, 806
Newtonian mechanics theory, and special relativity, 822
Newtonian relativity (classical relativity), 820–21, 822
newton-meter (N · m), 142. *See also* joule (J)
newton (N), 107
newton per ampere-meter [N/(A · m)], 626
newton per meter (N/m), 146
newton per square meter (N/m²), 298, 299, 301, 302–3
newton-second (N · s), 182–83
Newton's law of gravitation, 231–37, 242
Newton's laws of motion, 103, 105–14
in component form, 111–12
first, 105–6 (*See also* First law of motion, Newton's)
free-body diagrams and, 116–18
second, 106–12 (*See also* Second law of motion, Newton's)
third, 112–15 (*See also* Third law of motion, Newton's)
weight in, 107–11
Newton's rings, 767–68
Nickel
in Davisson-Germer experiment, 881
as ferromagnetic material, 641, 646
resistivity and temperature coefficient, 577t
stable nuclei, 918
Nitrogen
molecules in air, 358, 359
transmuted into oxygen, 936, 939
Nitrogen activation, 928–29
Noble gases (inert), 357, 892f, 893
Nodes
electrical (junctions), 599
in standing waves, 454
Noise, 467
Noise reduction, 452, 482–83
Nonconservative forces, 157–58, 162–64
Nondispersive waves, 453
Noninertial reference frame, 820
Nonmetals, as thermal conductors, 379
Nonreflective coatings on lenses, 766–67, 768
Nonuniform linear motion, 39–40
Nonviscous flow, 319
Nonvisual radiation, and telescopes, 808
Normal force, 113–14, 122, 126–27
Normal modes of vibration (resonant modes), 454
North Star, 256, 285
Nuclear chain reactions, 940–41, 942
Nuclear explosions, 469, 941
Nuclear force, 904–5, 917–18
Nuclear fuel rods, 902
Nuclear magnetic resonance (NMR), 888

Nuclear notation, 905
Nuclear reactions, 935–60
- defined, 936
- elementary particles in, 951–52
- fission, 921, 939–44
- force unification theories, 954–55
- fundamental forces and exchange particles in, 948–51
- fusion, 921, 944–46
- quark model, 953–54

Nuclear reactors, 935, 941–44
- breeder reactor, 942–43
- electricity generation, 943
- power reactor, 941–42
- safety measures, 943–44

Nuclear-shell model, 922
Nuclear stability, 917–22
Nuclear structure, 903–5
Nuclear Test Ban Treaty, 469
Nuclear transmutation, 907, 936, 950
Nuclear waste, 944
Nucleon-nucleon interaction, 949
Nucleon populations, 917–18
Nucleons, 904
- baryons, 952
- binding energy per, 920–21
- conservation of, 907

Nucleus, 902–34. *See also* Decay, radioactive; Half-life
- atomic, 886
- compound, 936
- decay rate, 911–17
- nuclear stability, 917–22
- radioactivity, 906–11 (*See also* Radioactivity)
- stability, 917–22
- structure, 903–5

Nuclide, 905, 911, 913t

O

Object, virtual, 736n, 748
Object distance (d), 730, 736, 745
Objective lens, 801, 803, 808
Object side, of lenses, 741
Ocular (eyepiece), 801, 803
Oersted, Hans Christian, 637
Ohm, Georg, 568, 573
Ohmic resistance, 574
ohm (Ω), 568, 573, 690, 692, 693
Ohm's law, 574, 687, 693
Oil firm, 765
Oil immersion lens, 810
Oil-water interface, 765
One-dimensional relative velocity, 90–92
Onnes, Heike Kamerlingh, 579
Open circuit, 593, 596, 612
- compared in ac and dc circuits, 686

Open-circuit terminal voltage, 572f
Open-tube manometer, 309–10
Operating voltage of battery, 570
Operations, arithmetic, A-1–A-2
Optical activity, 781–82
Optical biopsy, 785
Optical density, 711
Optical fibers, 705, 718–20, 870
Optical flats, 767
Optical instruments, 792–818
- color, 810–13
- diffraction and resolution in, 807–10
- human eye, 793–99 (*See also* Eye, human)
- microscopes, 799–803 (*See also* Microscopes)
- telescopes, 803–7 (*See also* Telescopes)

Optical lens, 740
Optical stress analysis, 782
Optic axis, 732, 781
Optics, 705. *See also* Lens(es); Light; Mirrors; Physical optics (wave optics); Vision
- geometrical optics, 705–6

Orbital motion, 234, 236, 243. *See also* Circular motion
Orbital quantum number, 885, 886
Orbitals, atomic, 886
Orbits
- Kepler's law of, 239
- satellites, 241–43

Order, and entropy of system, 411, 413
Order-of-interference maximum, 773
Order-of-magnitude calculations, 23–24, 179
Ordinary ray, 781
Organ, pipe, 492, 494
Organ transplants, preserving organs for, 378–79
Orthokeratology (Ortho-K), 797
Oscillation(s), 433
- compared with waves, 447
- decaying/damping out, 445
- described, 434
- of electrons and electromagnetic waves, 673
- and energy, 435–36
- equation of motion and, 439–42
- in and out of phase, 443–44
- in parabolic potential well, 437
- phase shifts, 443–44

Oscillator circuits, 699
Oscillator energy, 854
Oscilloscope screens, 629
Osmosis, 357
Osteoporosis, 304–5
Otto, Nickolaus, 417
Otto cycle, 417–18
Output coil (secondary coil), 669
Overtones, 496
Oxygen
- diatomic molecules in air, 358, 359
- diffusion of, 356, 357
- transmuting nitrogen into, 936, 939

Ozone layer depletion, 677

P

Pain, threshold of, 476, 494
Paint, and pigment mixing, 812–13
Pair annihilation, 897
Paired forces, 113–14
Pairing effect of stable nuclei, 918–19
Pair production, 896–97
Parabola, 83
Parabolic arcs, 83–84
Parabolic collector (dish), 808
Parabolic mirrors, 740, 806
Parabolic potential well, 437
Parallel axis theorem, 273–74, 278
Parallel connections
- batteries, 570–71
- capacitors, 557, 558–60
- resistors, 593–97
- resistors, and series combinations, 595–96, 597–99

Parallel current-carrying wires, and magnetic force, 640–41
Parallel LC circuit, oscillating, 699
Parallel plates
- as capacitor, 549–51
- charge on storm clouds, 524–25
- electric field between, 522
- electric potential energy, 537
- electric potential energy *vs.* potential, 539
- equipotential surfaces and, 544

Parallel rays, 734, 741, 743
Parent nucleus, 906, 908
Partially polarized light, 776
Particle accelerators, 833, 935
Particle nature of light, 855
Particle(s)
- alpha, 903–4, 906–8, 910, 922
- antiparticles and, 877, 896–97
- beta, 906, 908–9, 910, 922
- elementary, 951–52
- exchange, 948–51
- quantum, 858–60
- scattering of light and, 782, 784, 785
- system of, 198–200
- virtual, 948
- W, 950, 951t, 954

Pascal, Blaise, 303
pascal (Pa), 303
pascal-second (Pa • s), 326
Pascal's principle, 307–9
Paschen series, 864–65
Passive solar house design, 390
Path independence, 155, 157
Path-length difference, 482, 762–63, 774, 775f
Pauli, Wolfgang, 887, 947
Pauli exclusion principle, 887–90, 955
Peak current, 687
Peak voltage, 687
Pendulum, 441–42, 443, 459
Penetration, of nuclear radiations, 910–11
Perception of sound, 467, 494–95
Perfect gas law, 345. *See also* Ideal gas law
Performance, coefficient of, 421–22
Perihelion, 236, 239
Periodic motion, 434
Periodic table of elements, 891–94
Periodic waves, 447
Period(s), 221–22
- electron, 890, 893f
- of elements, 891, 892f, 893f
- Kepler's law of, 240
- of object or mass oscillating on spring, 441
- of pendulum, 441, 443
- of periodic wave, 448
- in simple harmonic motion, 434–35

Permanent magnets, 624, 625, 626, 641, 644
Permeability, magnetic, 638
Permittivity of free space, 550
Perpendicular (normal) force, 113
Perpetual motion machines, 410, 414
Perspiration, evaporation of, 380, 411
PET (positron emission tomography) scan, 897, 928
Phase, in ac circuits
- in current and voltage, 687, 690, 697
- phase angle, 694–96
- phase diagrams, 693–95

Phase, in wave motion
- initial conditions, 443–44
- oscillations in and out of, 444–45
- phase differences, 443, 482

Phase, in wave optics
double-slit interference, 761
thin-film interference, 764–65
Phase changes, 374–79
evaporation, 374, 379, 380, 411
latent heat, 374–78
Phases of matter, 374
Phasors, 693
Phosphorescent materials, 867
Photocells, 855, 858, 869
Photocurrents, 855
Photodiode, 869
Photoelectric effect, 855–58
characteristics of, 856
detecting radiation, 922
electron speed and, 856–57
threshold frequency and wavelength, 857
Photoelectrons, 855
Photography. *See* Camera
Photomultiplier tube, 923
Photon model of light, 855–57
Photon(s), 854–58
absorption and emission, 867
Compton effect and, 858–60
de Broglie waves and, 878
defined, 854
electromagnetic force and, 948–49
lasers and, 867
mass of, 935
in standard model, 955
Photonuclear effect, 922
Photosensitive materials, 855
Photosynthesis, 357
Physical optics (wave optics), 760–91
atmospheric scattering of light, 782, 784
diffraction, 768–75 (*See also* Diffraction)
Doppler effect, 488
Newton's rings, 767–68
optical flats, 767
polarization, 775–82 (*See also* Polarization)
thin-film interference, 764–68
wave nature of light, 760, 855
Young's double-slit experiment, 761–64
Physics, reason to study, 2
Physiological diffusion, 357
Physiological regulation of body temperature, 380
picocoulomb (pC), 507
picocurie (pCi), 914
picofarad (pF), 550
Piezoelectric devices, 156, 470
Pigments, 812–13
Pi mesons, 949
Pion(s), 927, 949–50, 952, 955
Pion therapy, 927
Pipe organ, 492, 494
Piston, cylindrical, 401
Pitch, 484, 494
Planck, Max, 854
Planck's constant, 854, 862, 895
Planck's hypothesis, 854
Plane, inclined, 117
Plane mirrors, 730–32
Plane of incidence, 707
Plane polarized light (linearly polarized light), 776
Planetary data, A-6
Planetary fly-by, 238
Planetary motion, Kepler's laws of, 232, 238–46
Plane wave front, 706
Plane waves, 673
Plaque, and cholesterol, 321
Plasma, 569, 946
Plasma phase of matter, 374
Plastics, optically active, 782
Plimsoll mark, 334
Plugs, electrical, 613–14
Plum pudding model of atom, 903–4
Plutonium, 936
Point charges, 540–42
Point source, intensity of, 475–76
poise (P), 326
Poiseuille, Jean, 326, 327
poiseuille (Pl), 326
Poiseuille's law, 327–28
Polar coordinates, 217
Polaris (North Star), 256, 285
Polarity, reverse
in ac circuits, 690, 692, 697, 699
ac generators, 663
dc motors, 636, 667
of electric eels, 576
Polarity reversal of the Earth's poles, 644, 646
Polarization, electrostatic charge separation, 511–12
Polarization direction, 776–78
Polarization of light, 775–82
by double refraction, 780–82
LCDs and, 783
by reflection, 778–80
by selective absorption (dichroism), 776–78
Polarized plugs, 613f, 614
Polarizers, crossed, 777f, 781f, 782, 783
Polarizing angle (Brewster angle), 779
Polarizing sheets, 777
Polaroid, 775, 776, 811
Pole-force law (law of poles), 624
Poles, electric, 521
Poles, magnetic, 624, 646
Polygon method, 74
Polymers, 385, 698, 776
Population inversion, 867
Position
energy of, 152
measurement of, and uncertainty principle, 894–95
Position-*versus*-time graphs, 39
Positive charges, 540
Positive ions, 507
Positive lens, 745
Positive-muon decay, 950
Positive test charge, 517, 538
Positron decay, 908
Positron emission tomography (PET), 897, 928
Positronium atom, 897
Positron(s), 896–97
Postulates of special relativity, 822–23
Potential difference, electric, 538–40, 570–71
Potential energy, 152–55
barrier, 884
barrier for alpha particles, 907
chemical, in battery, 569
conservative forces and, 157
electric, 537, 542–43
gravitational, 152–54, 235–37
of a spring, 152
of spring-mass system in SHM, 435–38
well, 155
zero reference point and, 155
Potentiometer, 622
pound per square inch (lb/in^2), 303
Power, 164–67
ac current and, 687–88
defined, 164
electric, 580–84
horse, 165
household appliance ratings, 611–12
of lens (P), 794
resolving, of microscope, 808–10
rotational, 277
Power distribution system, 671–72
Power factor, 696–97
Power lines, electric, 661
Power plants, 584, 665, 667, 686. *See also* Nuclear reactors
Power reactors, 941–42
Powers-of-10
in metric system, 7–8
notation (scientific notation), 23, A-2
Power transmission, 671–72
losses, 668, 672
Power waves, 675–76
Precession, 285
Precision gratings, 772–73
Prefixes for metric units, 7f
Presbyopia, 795
Pressure, 302–13
absolute, 310
atmospheric, 306–7, 309–13
blood, 312–13
defined, 302
depth and, 305–7
flow rate and, 322f
force and, 303, 306–7
gauge, 310
measurement of, 309–13
Pascal's principle and, 307–9
radiation, 674–75
temperature *vs.*, 346, 378
Pressure cookers, 378
Primary coil (input coil), 669
Primary colors, 811–12
Primary mesons, 949
Princeton Tokamak Fusion Test Reactor (TFTR), 946
Principal quantum number, 862, 886
Principle
of classical (Newtonian) relativity, 820–21
of equivalence, 837
Heisenberg uncertainty, 894–96
of relativity, 822
of superposition, electric fields, 518–20
of superposition, wave interference, 449, 451f, 454
Prism, 760
diffraction grating *vs.*, 773–74
dispersion by, 721, 723
in fiber optics, 720
internal reflection in, 717
reflection and refraction, 709
Prism binoculars, 804
Probability density, 882
Problem solving, 20–24
Problem-solving hints and strategies
on Celsius-Fahrenheit conversions, 342
on charge force law in vector addition, 515
on components of motion, 70
on "correct" answer, 19–20
on coupled rotational and translational motions, 276
on current in junctions, 600
on de Broglie wavelength, 880–81
on Doppler effect, 487
on equations for simple harmonic motion, 442
on free-body diagrams, 118
on Kelvin temperatures and ideal gas law, 348
on kinematic equations, 46, 73
on natural logarithms, 404
on Newton's second law, 118
on phase changes, 376, 377, 379
on photon calculations, 858
on projectile motion, 87
on Q values, 938
on ratios of acceleration due to gravity, 234
on relative velocity, 92
on resistors in parallel, 595
on simplifying equation through cancellation, 145
on spherical mirror equation, 739
on spring force, 147

on static equilibrium, 266
on temperature fourth powers, 388
on time dilation, 828
on trigonometric functions, 219
on vertical projectile motion, 55
on voltage sign convention, 600
on work-energy theorem, 151
Process curve, 402
Processes, thermodynamic, 399
adiabatic process, 406–7, 408
direction of, 410–11
for ideal gas, 403–9
irreversible, 399
isobaric process, 404
isometric process, 405–6
isothermal process, 403, 408, 409, 423
reversible, 399, 423
Projectile motion, 81–89
at arbitrary angles, 82–89
horizontal, 81–82
and momentum change, 182
vertical, 52–55
Projectile range, 83–85, 87–88
Projectiles, 67
Propagation, of waves, 433, 446, 449, 472
Propagation of errors, 17
Proper length, 829–30
Proper time interval, 827
Proportional limit, 299
Propulsion
jet, 114, 204–6
via magnetohydrodynamics, 631–32
Proton number, 905
Proton-proton cycle, 944–45
Proton(s), 904–5
electric charge of, 506, 507
electric potential, 539–40, 541–42
as hadrons, 952
mass compared to hydrogen atom, 920
pairing effect, 918–19
unstability of, 955
***p-T* diagram,** 399
Puffy-face syndrome, 245
Pulleys, and inertia, 275
Pulmonary disease, 325
Pulmonary embolism, 108
Pulse, of human body, 313
Pulse, wave, 433, 447
Pulsed induction (PI), 664
Pumping, laser, 867
Pump(s)
heat, 420–22
human heart as, 312–13
***p-V* diagram,** 398f, 399, 402
ideal heat cycle, 423
isobaric process on, 404
P waves (primary), 450. *See also* Compressional waves
Pythagorean theorem, 22

Q

Quadratic formula, A-3
Quality factor (QF), 925
Quality of tone, 495–96
Quantative ultrasound (QUS), 305
Quantities, units of, 12
Quantization
of angular momentum, 878–79
of light, 851, 854–58
Planck's hypothesis and, 854
thermal radiation, 852–54
Quantized charge, 507, 508
Quantum chromodynamics (QCD), 955
Quantum electrodynamics (QED), 949, 955
Quantum mechanics, 851, 866, 877–901
atomic quantum numbers, 885–94 (*See also* Atomic quantum numbers)
de Broglie hypothesis, 878–81
Heisenberg uncertainty principle, 894–96
particles and antiparticles, 877, 896–97
Schrödinger's wave equation, 881–84, 885
Quantum numbers, 862, 885–87, 890
Quantum of energy, 854
Quantum particles, 858–60
Quantum physics, 851–76
Bohr theory, 860–66, 878–79, 882, 885
Compton effect, 858–60
lasers, 866–71
Planck's hypothesis, 854
quantization, 852–54
Quantum theory, 840n, 851, 857
Quark confinement, 954
Quark model, 953–54
Quarks, 506n, 935, 953–54, 955
Quasars, 839, 842
Quinine sulfide periodide (herapathite), 776
Q value, 937–38

R

rad (unit), 924
Radar, 469, 488. *See also* Doppler effect
Radial acceleration (centripetal), 223–25, 230
Radial ray (chief ray), 734
Radian (rad), 217–19
Radiant energy, 385
Radiation, 902
biological effects and medical application, 924–28
blackbody, 387, 852–54
damage, 910–11
detection of, 922–24
domestic and industrial applications, 928–29
dosage, 924–25
electromagnetic, 672–78 (*See also* Electromagnetic waves)
heat transfer by, 385–90
nonvisible, for telescopes, 808
thermal, 852–54
Radiation detectors, 922–24
Radiation dosage, 924–25
Radiation penetration, 910–11
Radiation pressure, 674–75
Radio
and oscillator circuits, 699
reception, 760, 770–71
resonance frequency and, 698–99, 700
waves, 676–77, 771
Radioactive dating, 915–17
Radioactive decay. *See* Decay, radioactive
Radioactive half-lives. *See* Half-life
Radioactive tracers, 928–29
Radioactive waste, 944
Radioactivity, 902, 906–11
alpha decay, 906–8, 918
beta decay, 908–9, 916, 918, 946–48
decay rate, 911–17
defined, 906
gamma decay, 909–10
radiation penetration, 910–11
Radio-frequency (RF) radiation, 888
Radioisotopes, 925
Radio telescopes, 808
Radium, 906
Radius, Schwarzschild, 840–41
Radius of curvature, 732, 733, 736, 741
Radon (Ra), 359, 911
Rainbows, 722
Range, projectile, 83–85, 87–88
Rankine scale, 348
Rarefactions of sound waves, 448–49, 468, 482
Ray diagrams
for lenses, 741, 743, 744, 747
for mirrors, 734–40
Rayleigh, Lord, 782, 807
Rayleigh criterion, 807, 809
Rayleigh scattering, 782, 784
Ray(s)
chief, 734
extraordinary, 781
focal, 734, 742, 743
gamma, 678, 906, 909–10, 922, 924
light, 706, 708, 711
ordinary, 781
parallel, 734, 741, 743
RC circuits, 604–7, 608–9, 693–94
Reactance
capacitive, 689–91, 693, 697
inductive, 691–92, 697
Reaction cross-sections, 939
Reaction forces, 113–14
Reactions. *See* Nuclear reactions
Reaction time, 53
Real image, 730, 734, 736, 744, 804
Rectangle, finding area of, 22
Rectangular vector components, 75
Rectifier circuit, 664
Red end of visible spectrum, 784
Red shift, Doppler, 488
Reference, displacement, 148
Reference circle, 439–41
Reference circle object, 440
Reference frame, 90
absolute, 821, 822
inertial, 820, 822–25, 837
noninertial, 820
Reference points, 155
Reflecting telescopes, 805–7
Reflection, 707–8
angle of, 707
diffuse (irregular), 707–8, 709
law of, 707, 708, 734
phase shifts and, 764
polarization by, 778–80
of sound, 481
specular (regular), 707, 709
thin-film interference, 764–65
total internal, 717–18, 719
wave, 452–53
Reflection gratings, 772–73
Refracting telescopes, 803–5
Refraction, 708–17
angle of, 709, 711, 712, 717
atmospheric, 716–17
defined, 708
index of, 710–11, 714–15, 723
negative, 715
polarization by double, 780–82
of sound, 481
wave, 453
and wavelength in human eye, 713–14
Refrigerators, 421, 583
Regular reflection (specular reflection), 707, 709
Relative biological effectiveness (RBE), 925
Relative permeability, 643
Relative velocity, 90–94, 820, 821, 841
in one dimension, 90–92
in two dimensions, 92–94
Relativistic kinetic energy, 833
Relativistic length contraction, 830
Relativistic momentum, 834, 836
Relativistic time dilation, 825–28
Relativistic total energy, 834

Relativistic velocity addition, 841, 843
Relativity, 819–50
classical (Newtonian), 820–21, 822
general theory of, 837–41
length contraction, 828–31
mass-energy equivalence, 834–36
Michelson-Morley experiment, 821, 822, 823
relativistic kinetic energy, 833
relativistic momentum, 834, 836
relativistic total energy, 834
relativistic velocity addition, 841, 843
rest energy, 834–35
of simultaneity, 823–25
space travel and, 832–33
special theory of, 822–23, 824
time dilation, 825–28, 830–31, 832
twin paradox, 832
rem, 925
Replica gratings, 773
Representative elements, 891
Repulsive forces, 506, 515–16, 542
Resistance. *See also* Air resistance; Electrical resistance
in ac circuits, 687–89
thermal, 384
Resistive ac circuits
completely, 696
purely, 687f
Resistivity, 576–79
Resistor(s), 592–604. *See also* Electrical resistance
charging capacitor through, 604–5
defined, 573
discharging capacitor through, 605–7
multiplier, 610
in parallel, 593–97
in RLC circuits, 693–94, 696
in series, 592–93
in series-parallel combinations, 595–96, 597–99
shunt, 607–9
Resolution in optical instruments, 792, 807–10
Resolving curved motion, 67
Resolving power
electron microscopes and, 883
of microscope, 809–10
scanning tunneling microscope, 884
Resolving vector into rectangular components, 75–76
Resonance
of ac circuits, 697–700
of nuclear reaction, 939
of waves, 458–59
Resonance absorption, 385
Resonance frequency, 697, 888
Resonant frequencies (natural frequencies), 454–59, 493
Resonant modes of vibration (normal modes), 454
Rest energy, 834–35
Rest length (proper length), 829–30
Restoring force, 433, 434
propagation of sound wave, 472
torque on loop, 634
wave motion, 446, 448
Retarding voltage, 855
Retina, 793, 794, 795, 811
Reverse (braking) thrust, 206
"Reverse" electric field, 554
Reverse osmosis, 357
Reverse polarity. *See* Polarity, reverse
Reverse thrust, 243
Reversible processes, 399, 423
Revolution, 216, 221–22, 257
revolutions per minute (rpm), 219–20
Right-hand force rule
for current-carrying wire, 632–33
for moving charges, 627–28
radiation pressure and, 674
Right-hand rule, 220, 260
induced-current, 659
Right-hand source rule, 638, 639
Right-left reversal, 731
Rigid body
angular momentum of, 281
defined, 257
in equilibrium, 262, 266, 268
moment of inertia and, 270
motions of, 257–59
rotational kinetic energy, 277–80
RLC circuits, 693–97
power factor for, 696–97
resonance in, 697, 698f
series R, 693–94
series RL, 694
series RLC, 694–96
RL circuits, 694
rms current (effective current), 688
rms speed of gas molecules, 354, 356
rms voltage (effective voltage), 688
Rockets, 204–6
Roentgen, Wilhelm, 678
roentgen (R), 924
Rolling, and components of motion, 70
Rolling friction, 122
Rolling motion, 257–59, 276, 279–80, 281
Root-mean-square (rms)
current, 688
speed of gas molecules, 354, 356
voltage, 688
Rotation, 216, 257, 258. *See also* Axis of rotation
Rotational acceleration, 260
Rotational dynamics, 270–76
applications of, 274–76
moment of inertia and, 270–73
parallel axis theorem and, 273–74
Rotational equilibrium, 261–62
Rotational inertia, 270–71, 286
Rotational kinematics, 230
Rotational kinetic energy, 277–79
Rotational moment of inertia for symmetric diatomic molecule, 358
Rotational motion, 256–96. *See also* Rotational dynamics
angular momentum and, 280–86
defined, 257
equations for, 278
equilibrium, 261–64
rigid bodies, translations, and rotations, 257–59
stability and center of gravity, 266–69
torque, 259–61
translational motion *vs.*, 257–58
Rotational power, 277
Rotational static equilibrium, 264–65
Rotational work, 277–80
Rounding, 18–20
ROY G BIV, 721
rpm (revolutions per minute), 219–20
Rubbia, Carlo, 955
Rumford, Count (Benjamin Thompson), 369
Russell traction, 289
Rutherford, Ernest, 903–4, 923, 936, 937
Rutherford-Bohr model of the atom, 903–4, 907
R-values, 384, 395
Rydberg, Johannes, 861
Rydberg constant, 861

S

Safety
airplane, and Doppler radar, 491
driving on rainy night, 709
electrical, 611–14
nuclear reactors, 943–44
seatbelts and air bags, 112–13, 186
Sailing, and force components, 115
Salam, Abdus, 954–55
San Andreas fault, 450
SARS (Severe Acute Respiratory Syndrome), 386
Satellites. *See also* Spacecraft
Earth's, 241–46
geosynchronous orbit, 234, 236
kinetic energy of orbiting, 243
observatories of nonvisible radiation, 808
sailing the solar system, 675
Saturation, 922
Scalar quantities, 33–34. *See also* Distance; Length; Speed
energy, 149
work, 141–42, 149
Scanning electron microscope (SEM), 883
Scanning tunneling microscope (STM), 877, 884, 908
Scattering
of atmospheric light, 782, 784
back-, 903–4
biomedical, 785
Compton, 858–60, 922
electrons, 881
Rayleigh, 782, 784
X-ray, 858–60
Schrödinger, Erwin, 881
Schrödinger's wave equation, 881–84, 885
Schwarzschild, Karl, 840
Schwarzschild radius, 840–41
Scientific method, 51
Scientific notation (powers-of-10 notation), 23, A-2
Scintillation counter, 923
Seatbelt safety, 112–13, 186
Secondary coil (output coil), 669
Second law of motion, Newton's, 106–12, 118, 149. *See also* Newton's laws of motion
airplane lift, 323
centripetal force and, 227–28
law of gravitation and, 233
momentum in, 182, 185–86
noninertial reference frame and, 820
rotational form of, 270, 274
Second law of thermodynamics, 410–14, 417, 420, 422
second (s), 5, 6, 7
Sedimentation, 225
Seismic waves, 450
Seismograph, 450, 556f
Seismology, 450
Selective absorption
color filters and, 812
dichroism, polarization by, 776–78
Selector, velocity, 630
Self-induction, 670, 691
Semiconductor detector (solid state detector), 923

Semiconductors, 508–9, 577t, 579
Series
batteries in, 570
capacitor connection, 557, 558, 559, 560
resistors in, 592–93
Series circuits
impedance and phase angles of, 695
power factor for RLC, 696–97
RLC impedance, 693–97
Series-parallel capacitor combination, 560
Series-parallel resistor combinations, 595–96, 597–99
Series RC circuits, 694–96
Series RLC circuits, 693–94, 695, 696–97
Series RL circuits, 694
Severe Acute Respiratory Syndrome (SARS), 386
Shadow zones, 450
Shear angle, 301
Shear modulus, 301
Shear strain, 301
Shear stress, 301
Shear waves (transverse waves), 448–49, 450, 673
Shell(s), electron, 886–87, 889, 890, 893–94, 922
Shifts of light. *See* Doppler effect
SHM. *See* Simple harmonic motion (SHM)
Shock absorbers, 307, 446
Shock waves, 488–90
Shortwave bands, 676
Shunt, 596
Shunt resistor, 607–9
sievert (Sv), 925
Significant figures, 17–20
Simple harmonic motion (SHM), 434–38
damped, 445–46
energy and speed of spring-mass system in, 435–38
equations of motion for, 439–46
initial conditions and phase, 443–44
velocity and acceleration in, 444–45
Simple microscope (magnifying glass), 799–801
Simultaneity, relativity of, 823–25
Simultaneous equations, solving, A-3
Simultaneous measurement, of momentum and position, 894–95
Sines, law of, A-4
Single-photon emission tomography (SPET), 928
Single-slit diffraction, 769–71, 807–8
Sinusoidal functions, 440, 447, 449
SI unit(s), 3–10. *See also* Measurement
of acceleration, 40
of angular momentum, 280
of angular speed, 219
of atmospheric pressure, 311n
base units, 3, 6–7
British units compared with, 10
of bulk modulus, 302
of capacitance, 550
of capacitive reactance, 690
of charge, 506–7
of current, 568, 572, 623
of decay constant, 912
of density, 303
derived units, 3
of effective dose, 925
of elastic modulus, 299
of electrical resistance, 568, 573
of electric field, 517
of electric potential difference, 538
of electric power, 580
of energy, 152, 153
of entropy, 411
of flow rate, 327
of force, 107
of frequency, 434
of heat, 368
of impulse and momentum, 182
of inductive reactance, 692
of intensity, 474
of length, 3–4
of magnetic field, 623, 626
of magnetic flux, 658
of magnetic moment, 634
of mass, 4–5, 10
of moment of inertia, 270
of power, 165
of pressure, 302–3
of radiation dosage, 924–25
of radioactivity, 914
of resistivity, 576
of shear modulus, 301
of speed, 33
of spring constant, 146
of stress, 298
of time, 5–6
of torque, 259
of velocity, 36
of viscosity, 326
of volume, 8–10
of work, 142
of Young's modulus, 299
Sky, color of, 782, 784
Skylight, polarization of, 760
Sliding friction (kinetic friction), 122, 123
Sliding motion, 279–80, 281
Slingshot effect, 238
Slipping, 258–59
Slope of line, 39
slug, 7
Small-angle approximation, 219f
Smoke detector, 928, 929f
Smokestacks, 322
Smudge pots, and heat transfer, 389
Snell, Willebord, 709
Snell's law, 709, 711, 712, 717
Soap bubble/film, 765
Soft iron, 643
Solar cells, 768, 858
Solar clock, 5
Solar collectors, 414
Solar colors, 853
Solar eclipse, 839
Solar flares, 647
Solar house design, 390
Solar panels, 389
Solar system model of atom, 506, 903. *See also* Bohr theory of hydrogen atom
Solenoids, 639–41, 643, 664
Solid-gas phase change, 375
Solid-liquid phase change, 375
Solid phase, 374
Solids, 297–302
elastic moduli and, 298–302
specific heat of, 370–71
speed of sound in, 471–72
as thermal conductors, 380
thermal expansion of, 340, 350–52
Solid-state detector (semiconductor), 923
Solid-state devices, in radios, 698
Sonar, 469, 488
Sonic booms, 488–91
Sound, 467–504. *See also* Doppler effect
diffraction of, 481, 769
interference in, 482–84
of musical instruments, 491–96
phenomena, 481–84
speed of, 471–74
upper limit, 471
waves, 448, 468–71
Sound frequency spectrum, 468
Sound intensity, 474–80
Sound intensity level, 476–79
equation for, 477
hearing damage, 480
Sound waves, 448, 468–71
Space colonies, 245–46
Spacecraft. *See also* Satellites
Cassini-Huygens, 238
Endeavor space shuttle, 807f
Mars Climate Orbiter, 16
Mars Exploration Rovers, 34, 186–87
Mars Polar Lander, 56
Odyssey, 129
space shuttle, 177
Viking probes, 674
Space diagrams, 116
Space exploration, and gravity assists, 238
Space-time diagrams (Feynman diagrams), 949
Space travel, 832–33
Spark chamber, 923, 924
Special theory of relativity, 822–23, 824
Specific gravity, 318
Specific heat, 370–74
of gases, 373–74
of solids and liquids, 370–72
of various substances, 370
Specific heat capacity, 370
Spectral fine structure, 885
Spectrometer, 774
mass, 629–31
Spectroscope, 599n
Spectroscopic analysis, 861
Spectroscopy, 773–74
Spectrum
absorption, 861, 866
of colors, 721, 763
color vision, 811–13
continuous, 852, 860, 861f
electromagnetic, 676
emission, 861, 864
hydrogen, 851, 864–65
of light in diffraction gratings, 773–74
mass, 631
sound frequency, 468
Specular reflection (regular reflection), 707, 709
Speed, 33–34
acceleration and, 40
angular, 217, 219–21
average, 33–34
conservation of energy and, 160
of electromagnetic waves in vacuum, 673–74
of electron, accelerated, 549
of electrons, 856–57
escape, 241
of fluid flow, 319–21
ground, 94
instantaneous, 34
kinetic energy and, 150–52
of light (*See* Speed of light)
mass *vs.*, 151
molecular, 355, 356
relationship to radius, energy, and circular motion, 243
root-mean-square (rms), 354, 356
of sound, 471–74
of spring-mass system in SHM, 435–38
tangential, 220–21, 241–43, 258
of various animals, 32
wave, 448
Speed of light, 4
constancy of, 822
and electric and magnetic fields, 673
escape, 840

Speed of light *(cont.)*
ether and, 821
and index of refraction, 710, 721
relativistic velocity addition *vs.*, 843
in special relativity, 822
warp, 830–31
SPET (single-photon emission tomography), 928
Spherical aberration, 733, 740, 752
Spherical mirror equation, 736, 737–38
Spherical mirrors, 730, 732–40
aberrations, 740, 752
concave (converging), 732, 734, 735–36, 738
convex (diverging), 732, 739–40
ray diagrams and, 734–40
sign conventions for, 736
Sphygmomanometer, 313
Spin magnetic moment, 888
Spin quantum number, 885–86
Split-ring commutator, 636
Spontaneous fission, 939
Sports
baseball, 177
basketball jumping, 88
bicycle stability, 271
boating, 44, 92–93, 115
contact times, 183
discus throwing, 67
downhill skiing, 162–63
drag racing, 42–43, 122
exercising, 367, 369, 401
golfing, 83–84, 183, 185
high jump, 204
hockey, 89
hot-air balloons, 338
ice skaters, 190–91, 256, 283–84
javelin throw, 88
race car stability, 268
sailing, 115
scuba diving, 305–7
shuffleboard, 150
skiing, downhill, 162–63
skydiving, 128–29, 163
Spring
mechanical energy of, 160–61
potential energy of, 152, 160
Spring constant (force constant), 146–47, 434, 437–38, 441
Spring force, 146–48, 434
Spring-mass system, 435–38
analogy to ac circuit, 686
Stability. *See also* Equilibrium
and center of gravity, 266–69
riding a bicycle, 271
Stable elements, 936
Stable equilibrium, 266–69
Stable isotopes, 917, 918
Standard acceleration (g), 108
Standard atmosphere, 310
Standard model, 955
Standard temperature and pressure (STP), 345
Standard units, 3
Standing waves, 454–59, 492, 878
Stars, twinkling of, 717
State, equations of, 398
State of a system, 398–99
State variables, 398
Static charges, 515
Static cling, 512
Static equilibrium, 262–66
Static friction, 122–27
Static translational equilibrium, 119–21, 262–63
Stationary frame of reference, 90
Stationary interference patterns, 761
Steady flow, 319
Steam engines, 397, 415
Steam point of water, 341
Stefan-Boltzmann constant, 387
Stefan's law, 386–87
Stem cells, 797
Step-down transformers, 669–70, 672
Step-up transformers, 669–70
Stimulated emission, 867–68
Stopping distance of a vehicle, 48–49, 127
Stopping potential, 855, 856–57
STP (standard temperature and pressure), 345
Strain, 298
shear, 301
tensile, 298
volume, 302
Streamlines, 319
Streamlining automobiles, 128
Stress, 298
compressional, 298
on femur, 300
shear, 301
tensile, 298
thermal expansion and, 352
volume, 301, 302
vs. strain, 299f
String, pulse in, reflection and phase shifts, 764f
String, stretched
natural frequencies and, 454–55, 459
wave speed, 456, 471
Stringed musical instruments, 455–58, 484, 491–92
Strong nuclear force, 904, 948
hadrons and, 951, 952, 954
mesons and, 949–50, 951t
Stun gun, 505, 524
Subatomic particles, and electric charge, 507
Sublimation, 374
latent heat of, 375
Submarines, and silent propulsion, 631–32
Subshells, 886–87, 888–90, 893
Subtraction
with significant figures, 19
vector, 74
Subtractive method of color production, 812–13
Subtractive primary pigments, 812–13
Subway railcars, and braking, 671
Sun
as a blackbody, 853
colors of, 853
fusion power of, 944–45
perihelion and aphelion of Earth, 236, 239
photon energy, 851
Sunlight, 780, 782, 784
Sunrises/sunsets, 782, 784
Supercold ice, 376–77
Superconducting magnets, 579, 626, 631, 643–44
Superconducting phase of matter, 374
Superconductivity, 579
Superforce, 955
Superheated steam, 375, 376–77
Supernova, 950
Superposition principle
for electric fields, 518–20
in wave interference, 449, 451f, 454
Surface tension, 324–25
Surface waves, 450
Surgery
eye, 792, 797
hip replacement, 108
laser, 792, 797, 870
lowering body temperature, 343
S waves (secondary), 450. *See also* Shear waves
Swing-by (gravity assist), 238
Swirling currents (eddy currents), 670
Switches, electrical, 612, 614
Symbols, A-1
Synchronous satellite orbit, 234, 236
System(s)
center of mass and, 199–200
conservative, 158
defined, 156
entropy of isolated, 413
isolated (closed), 156, 187, 192, 413
nonconservative, 158
of particles, 198–200
thermally isolated, 398
thermodynamic, 398
of units, 3
Systolic pressure, 312–13

T

Tacoma Narrows Bridge, 458
Tangential acceleration, 229–30
Tangential speed, 220–21, 241–43, 258
Tangential velocity, 220
Taser stun gun, 505, 524
Tattoo removal, 870
Tau neutrino, 952
Tauon(s), 951
Telescopes, 803–7
astronomical, 803–5
compound-lens system, 748
constructing, 805
Galilean, 804
Hubble Space, 807, 808
Keck, 807
nonvisible radiation used for, 808
radio, 808
reflecting, 805–7
refracting, 803–5
resolution and, 807–10
terrestrial, 804–5
VLT (Very Large Telescope), 807
Television
oscillator circuits and, 699
picture tubes, 629, 678
waves, 676–77
Temperature, 338–66. *See also* Kinetic theory of gases
absolute, 343–49, 355
in Big Bang, 955
Celsius scale, 340–42
Curie, 623, 644, 646
defined, 339
density and, 354
difference and heat transfer, 410–11
doubling, 349
electrical resistance and, 574, 577–79
Fahrenheit scale, 340–42, 348
gas laws, 343–49
heat distinguished from, 339–40
of human body, 343, 380
Kelvin scale, 343, 346–49
phase change, 374–78
pressure *vs.*, 346, 378
specific heat and, 370–71
speed of sound and, 472, 474
thermal expansion, 340, 350–53
thermal radiation, 852–54
Temperature coefficient of resistivity, 578–79
Tensile strain, 298
Tensile stress, 298
Tension
of string, 110–11
surface, 324–25
tera- prefix, 8
Terminal velocity, 128–29
Terminal voltage of batteries, 545, 570–71, 572f
Terrestrial telescopes, 804–5
Tesla, Nikola, 623, 626, 656
Tesla coil, 656
tesla (T), 626

Test charge, 517
Theory of relativity
general, 837–41
special, 822–23, 824
Thermal coefficients
of area expansion, 351
of linear expansion, 350
of volume expansion, 351
Thermal conductivity, 380–81, 382f
Thermal conductors, 367, 379
Thermal contact, 340
Thermal contraction, 340
Thermal convection cycle, 384
Thermal cycles, 410, 415
Thermal efficiency of heat engine, 416–22, 424
Thermal equilibrium, 340, 374
Thermal expansion, 338, 340, 350–53
coefficients for some materials, 351
Thermal insulators. *See* Insulation; Insulators
Thermally isolated systems, 398
Thermal neutrons, 942
Thermal oscillators, 853–54
Thermal pumps, 414, 421
Thermal radiation, 852–54
Thermal resistance, 384
Thermionic emission, 903
Thermodynamics, 397–432
Carnot cycle, 422–24
entropy, 411–14
first law of, 399–402, 403–9, 417, 421
of heat engines, 414–22, 422–24 (*See also* Heat engines)
processes, 399
processes for ideal gas, 403–9
second law of, 410–14, 417, 420, 422
state of a system, 398–99
systems, 398
thermal efficiency of heat engine, 416–22
third law of, 425
Thermoelectric materials, 156
Thermograms, 382, 386, 387f
Thermography, 386, 387f
Thermometer(s), 340
constant-volume gas, 346
electrical, 579
gas, 342
history of, 338
liquid-in-glass, 340–42
Thermonuclear fusion reactions, 946
Thermos bottle, 389
Thermostats, 340
Thin-film interference, 764–68
Newton's rings, 767–68
nonreflective coatings, 766, 768
optical flats, 767
Thin-lens equation, 745, 749
Thin lenses, sign conventions, 745
Third law of motion, Newton's, 112–15. *See also* Newton's laws of motion
airplane lift, 323
electric charge, 506
jet propulsion, 204–5
momentum and, 187
silent propulsion, 632
Third law of thermodynamics, 425
Thompson, Benjamin (Count Rumford), 369
Thomson, J. J., 903
Thomson, William (Lord Kelvin), 347n
Thought (gedanken) experiments, 823–25, 830, 832, 837–38, 894
3D image, 871
3D movies, 777
Three Mile Island (TMI) reactor, 935, 942, 943
Three-wire system, 611–12
Threshold energy, 938–39
for pair production, 897
Threshold frequency, 857
Threshold of hearing, 476, 494–95
Threshold of pain, 476, 494
Thyroid treatments, and iodine-131, 914, 925–26, 928
Tidal waves, 433
Tightrope walker, 256
Time
and components of motion, 83–85
measurement of, and uncertainty principle, 896
reaction, 53
SI unit, 5–6
work and, 166
Time constant, for RC circuits, 605, 608
Time dilation, 825–28, 830–31, 832, 838
Time interval, proper, 827
Time rate
of change of angular momentum, 280
of doing work (power), 164
of heat flow, 380
Time-varying fields, electric and magnetic, 661, 673
Tinnitus, 467, 475
Tip-to-tail method (triangle method), 73–74, 77
Tire gauges, 309
TMI (Three Mile Island) reactor, 935, 942, 943
Tobin, Thomas William, 733
Tokamak configuration, 946
ton, metric, 9
Tone (sound), 468, 495–96
Topographic maps, 544–45
Tornadoes, 468, 469, 491
Torque, 259–61
angular momentum and, 281–83, 285, 286
countertorque, 668
in dc motor, 636
defined, 259
door opening and, 274
magnetic, on current-carrying loop, 634–35
moment of inertia and, 270–73
net, 270, 280, 281
rotational work and, 277
yo-yo roll and, 276
torr, 311
Torricelli, Evangelista, 310, 311
Total binding energy, 920–21
Total constructive interference, 451, 482
Total destructive interference, 451–52, 482
Total energy, 162–64
law of conservation of, 156
Total flux, 659
Total gravitational potential energy, 237
Total internal reflection, 717–18, 719
Total kinetic energy on rotating rigid body, 278
Total linear momentum, 178
Total magnification, 801–2
Total mechanical energy, conservation of, 158–61
Total momentum, 178, 180–81
Total work (net work), 144–45, 149
net work per cycle, 415–16
Tourmaline, 776
Tracers, radioactive, 928–29
Traction, 119
Transducers, 470
Transfer of energy, and heat, 368
Transfer of heat, 397, 399–400
Transformers, 656, 668–72
Transition elements, 891, 892f
Translational equations, 278
Translational equilibrium, 119–21, 256, 261–63
condition for, 119, 261
Translational kinetic energy, 278–79, 339, 354, 356, 358, 359
Translational motion, 257–59, 278
Translational static equilibrium, 119–21, 262–63
Transmission axis (polarization direction), 776–78
Transmission electron microscope (TEM), 883
Transmission gratings, 772
Transmutation, nuclear, 907, 936, 950
Transverse waves (shear waves), 448–49, 450, 673
Traveling wave (de Broglie), 878, 881
Triangle method (tip-to-tail method), 73–74, 77
Trifocal lenses, 796
Trigonometric functions
definitions of, A-4
problem solving using, 22
radian computation, 219
Trip elements, circuit breakers, 612f
Triple point of water, 348
Tritium, 905, 906, 944
***T-S* diagram, ideal heat cycle,** 423
Tsunami, 256, 433
Tumors, 927–28
Tuning forks, 468, 484
Tunneling, 840n, 877, 884, 908
Turbines, 414, 667
Turbulent flow, 326
***T-V* diagram,** 399
20/20 vision, 798–99
Twin paradox (clock paradox), 832
Twisted-nematic display, 783
Two-dimensional relative velocity, 92–94
Tyndall, John, 718
Type-S fuse, 612

U

UHF (ultrahigh frequency) band, 676
Ultrasonic generators (transducers), 470
Ultrasonic region, 469
Ultrasonic scalpel, 471
Ultrasound, 467, 469–71
applications of, 470–71
medical, 470, 490
quantative ultrasound (QUS), 305
Ultraviolet catastrophe, 853–54
Ultraviolet light (black light), and fluorescence, 866
Ultraviolet radiation, 677–78
telescopic observation of, 808
Ultraviolet range, emission series, 864–65
Unbalanced force, 104, 106, 261. *See also* Net force
Unbalanced torque, 261
Uncertainty principle, Heisenberg, 894–96, 948, 949
Unification theories, 954–55
Uniform circular motion, 223–28
simple harmonic motion and, 439, 441
Uniform motion, 38–39
Unit analysis, 10–12, 13
Unit conversions, 12–16
Unit(s). *See also* British units; Measurement; SI unit(s)
of heat, 368–69
of quantities, 10, 12
system of, 3

Unit-vector component form, 76
Unit vectors, 75–76
Universal gas constant (*R*), 345
Universal gravitational constant (*G*), 231–32
Universal law of gravitation, 231–37, 242
Universe
entropy of, 413, 414
evolution of, 955
Unpolarized light, 775
Unresolved images, 807
Unstable equilibrium, 266–69
Uranium, 906, 907, 911, 917, 936, 939–41

V

Vacuum, speed of electromagnetic waves in, 674
Vacuum tubes, 629
Valence electrons, 508, 891, 892f
Van Allen belts, 647
Van der Meer, Simon, 955
Van der Waals forces, 324
Vaporization, latent heat of, 375, 376, 411
Vapor phase, 374
Variable air capacitor, 698
Variable forces, work done by, 145–47
Variables, state, 398
Varicose veins, laser treatments, 870
Vector, area, 658
Vector addition, 73–80
analytical component method, 75–80
Coulomb's law on, 514–15
geometric methods, 73–74
linear momentum, 180–81, 187
in three dimensions, 79
of velocities, 841, 843
Vector components
curving path, 72
rectangular, 75
unit vectors, 75–76
vector drawing, 80
Vector field, 517, 625
Vector quantities, 35–40. *See also* Displacement; Velocity
directions of, 52–55
force and acceleration, 104
Vector subtraction, 74
Vector sums. *See* Vector addition
Vehicles
acceleration, 42–43, 47, 152
air bags, 186
antilock brakes, 256, 281
batteries, 570–71
braking, 112–13
diesel-powered, 397
efficiency of automobile, 167
friction and centripetal force, 226–27
gasoline engines, 417
heat exchange in engine, 385
hybrid cars, 164, 666
internal combustion engines, 415, 418
mufflers, and destructive interference, 452
race car stability, 268
relative velocity, 90–91
shock absorbers, 446
stopping distance, 48–49, 127
thermodynamic efficiency, 397
Velocity, 36–38
angular, 217, 219–21
average, 36–37, 39, 44
constant, 38
drift, 572–73
graphical analysis of, 38–40
horizontal projections, 81–82
instantaneous, 37–38, 40
momentum and, 178
of object in simple harmonic motion, 436
of projectile motion, 86–87
relative, 90–94, 820, 821, 841
signs of, 42
in simple harmonic motion, 444–45
tangential, 220
terminal, 128–29
Velocity addition, relativistic, 841, 843
Velocity selector, 630
Velocity vector, resolving into components of motion, 68–69, 70–72
Velocity-*versus*-time graphs, 43, 49
Venturi tunnel, 335
Vertex of spherical mirror, 732
Vertical projectile motion, 52–55
Vertical velocity and acceleration in SHM, 444–45
Very Large Array (VLA), 808
Very Large Telescope (VLT), 807
Vibrations, 433, 468. *See also* Oscillation(s); Wave(s)
Videotape, 656
Virtual image, 730, 734, 736, 744
Virtual object, 736n, 748, 804
Virtual particle, 948
Viscosity, 319, 325–27
coefficient of, 326
of various fluids, 327
Visible light, 677, 705, 769, 864–65, 866
Visible spectrum, 723
Vision, 792. *See also* Eye, human
color, 810–13
defects of, 795–99
Visual acuity, 798–99
Viviani, Vincenzo, 51
VLA (Very Large Array), 808
VLT (Very Large Telescope), 807
Volta, Allesandro, 536, 538, 569
Voltage, 538, 545–46, 569. *See also* Electric potential difference
ac, 688–89
across capacitor, 690
across inductor, 692
across resistor, 696
breakdown, 607
drop, 558
emfs in transformer design, 669
in household wiring, 611–12
on liquid crystal, 783
maintaining, 607
open-circuit terminal, 572f
peak, 687
resistors and, 592, 598–99
retarding, 855
reverse, 691
rms (effective), 688
sign convention in circuit loop, 600
terminal, 570–71
Voltmeters, 591, 607, 610, 611f, 635
volts per meter (V/m), 545
volt (V), 538
electron-, 548
Volume equations, A-3
Volume expansion, 350, 351, 353
Volume rate of flow, 320
Volume(s)
British unit, 10
bulk modulus and, 301–2
SI unit, 8–10
Volume strain, 302
Volume stress, 301
von Laue, Max, 774
Vortices, 458n

W

Warm-blooded creatures, 344
Warping space and time, 839
Waste, radioactive, 944
Water. *See also* Ice
density of, 305
electric polarization, 512f
electrostatic potential energy of molecule, 542–43
evaporation, 374, 379, 380, 411
flow of stream, 320, 323, 324
Gauss' law analogy, 528
ice point, 341
mass and weight, 1, 9
phase-change temperatures and pressure on, 378
reverse osmosis and purification of, 357
specific heat of, 371–72
steam point, 341
thermal equilibrium, 374–75
thermal expansion of, 353
triple point, 348
waves, 761, 768
Water waves, 448, 449
Watt, James, 165
watts per square meter (W/m^2), 474
watt (W), 165, 580, 582
Wave fronts, 706
Wave function, 882, 884
Wave interference. *See* Interference
Wavelength
color of light and, 811–12
Compton scattering, 858–60
de Broglie, 879–81
defined, 447
Doppler effect and, 485
of electromagnetic wave classifications, 676
of light, double-slit experiment, 763–64
of light, measurement of, 773–74
photoelectric effect and, 856–57
refraction in human eye, 713–14
and resolving power of optical instruments, 810
thermal radiation, 852–54
ultraviolet catastrophe and, 853–54
Wave motion, 433, 446–49
Wave nature of light, 760, 855. *See also* Physical optics (wave optics)
Wave optics. *See* Physical optics (wave optics)
Wave-particle duality
of light, 855, 860
of matter, 851, 877
Wave pulse, 433, 447
Wave(s), 433–66
body, 450
characteristics of, 447–48
compared with oscillations, 447
defined, 447
diffraction (*See* Diffraction)
electromagnetic (*See* Electromagnetic waves)
equations of motion, 439–46
infrasonic, 468–69
light, 488
longitudinal (compressional), 448–49, 450, 475
motion, 446–49
nature of matter (de Broglie), 878–81, 883
nondispersive, 453
periodic, 447
P (primary), 450
properties of, 449–53
Schrödinger's equation, 881–84, 885

seismic, 450
shock (bow), 488, 489
simple harmonic motion (*See* Simple harmonic motion (SHM))
sound, 448, 468–71
S (secondary), 450
standing, 454–59, 492
standing (de Broglie), 878
surface, 450
transverse (shear), 448–49, 450
traveling (de Broglie), 878, 881
types of, 448–49
ultrasonic, 469–71
water, 448, 449, 761, 768

Wave speed, 448, 456
Weak charge, 955
Weak interaction, 947, 950
Weak nuclear force, 947, 948
hadrons and, 952
leptons and, 951
W particle and, 950, 951t
Weather
balloons, 314–15
and Doppler radar, 491
"red sky" saying, 784
Weber, Wilhelm Eduard, 658n
weber (Wb), 658
Weight
apparent, 245
artificial gravity and, 245–46
buoyant force and, 316
mass *vs.*, 5, 107–8
in Newton's second law of motion, 107–11
Weightlessness
apparent, 243–44
effects on human body, 245
Weight loss, first law of thermodynamics applied to, 401
Weights and measures, as phrase, 5
Weinberg, Steven, 954–55
Well
parabolic potential, 437
potential energy, 155, 235
Wheatstone, Sir Charles, 618
Wheatstone bridge, 618
White blood cells (lymphocytes), 883
White light, 721, 723, 812
Wien's displacement law, 853
Wilson, C. T. R., 923
Wind instruments, 492–93
Wind shears, 491
Wire loop, induced emf, 657–61, 663
Wires. *See* Current-carrying wires
Work, 140–52
by a constant force, 141–45
electric power and, 583
as measure of transfer of kinetic energy, 149
mechanical, 141, 142–43
mechanical efficiency and, 166–67
mechanical into electrical current, 662–63
net, 415, 416
power and, 164–67
rotational, 277–80
in thermodynamics, 399–402
time and, 166
total (net), 144–45
by a variable force, 145–47
work-energy theorem, 148–52

Work-energy theorem, 148–52, 185, 277–79, 835
Work function, 856
World's Columbian Exposition at Chicago, 656
W particle, and weak nuclear force, 950, 951t, 954

X

X-rays
accelerating electrons, 540
biological effects of, 924
CT, 928
diffraction, 774–75
dual energy X-ray absorptiometry (DXA), 305
radiation from, 902
scattering, 858–60
source in constellation Cygnus, 842
in Sun, 851
telescopic observation of, 808
as type of electromagnetic wave, 678

X-ray tube, 678

Y

Yerkes Observatory, 806
Young, Thomas, 299n, 760, 761, 763
Young's double-slit experiment, 761–64
Young's modulus, 299–301, 471
Yucca Mountain, 944
Yukawa, Hideki, 949

Z

Zanotto, E. D., 326n
Zeppelins, 297, 334
Zero, absolute, 338, 346–49, 355, 425
Zero gravity, 243, 245
Zero-point energy, 355
Zero reference point, 155
Zweig, George, 952, 953

Wolfgang Christian and Mario Belloni
Physlet Physics: Interactive Illustrations, Explorations, and Problems for Introductory Physics
CD-ROM
0-13-140334-6

Pearson Prentice Hall
Pearson Education, Inc.
Upper Saddle River, NJ 07458

YOU SHOULD CAREFULLY READ THE TERMS AND CONDITIONS BEFORE USING THE CD-ROM PACKAGE. USING THIS CD-ROM PACKAGE INDICATES YOUR ACCEPTANCE OF THESE TERMS AND CONDITIONS.

Pearson Education, Inc. provides this program and licenses its use. You assume responsibility for the selection of the program to achieve your intended results, and for the installation, use, and results obtained from the program. This license extends only to use of the program in the United States or countries in which the program is marketed by authorized distributors.

LICENSE GRANT

You hereby accept a nonexclusive, nontransferable, permanent license to install and use the program ON A SINGLE COMPUTER at any given time. You may copy the program solely for backup or archival purposes in support of your use of the program on the single computer. You may not modify, translate, disassemble, decompile, or reverse engineer the program, in whole or in part.

TERM

The License is effective until terminated. Pearson Education, Inc. reserves the right to terminate this License automatically if any provision of the License is violated. You may terminate the License at any time. To terminate this License, you must return the program, including documentation, along with a written warranty stating that all copies in your possession have been returned or destroyed.

LIMITED WARRANTY

THE PROGRAM IS PROVIDED "AS IS" WITHOUT WARRANTY OF ANY KIND, EITHER EXPRESSED OR IMPLIED, INCLUDING, BUT NOT LIMITED TO, THE IMPLIED WARRANTIES OR MERCHANTABILITY AND FITNESS FOR A PARTICULAR PURPOSE. THE ENTIRE RISK AS TO THE QUALITY AND PERFORMANCE OF THE PROGRAM IS WITH YOU. SHOULD THE PROGRAM PROVE DEFECTIVE, YOU (AND NOT PEARSON EDUCATION, INC. OR ANY AUTHORIZED DEALER) ASSUME THE ENTIRE COST OF ALL NECESSARY SERVICING, REPAIR, OR CORRECTION. NO ORAL OR WRITTEN INFORMATION OR ADVICE GIVEN BY PEARSON EDUCATION, INC., ITS DEALERS, DISTRIBUTORS, OR AGENTS SHALL CREATE A WARRANTY OR INCREASE THE SCOPE OF THIS WARRANTY.

SOME STATES DO NOT ALLOW THE EXCLUSION OF IMPLIED WARRANTIES, SO THE ABOVE EXCLUSION MAY NOT APPLY TO YOU. THIS WARRANTY GIVES YOU SPECIFIC LEGAL RIGHTS AND YOU MAY ALSO HAVE OTHER LEGAL RIGHTS THAT VARY FROM STATE TO STATE.

Pearson Education, Inc. does not warrant that the functions contained in the program will meet your requirements or that the operation of the program will be uninterrupted or error-free. However, Pearson Education, Inc. warrants the CD-ROM(s) on which the program is furnished to be free from defects in material and workmanship under normal use for a period of ninety (90) days from the date of delivery to you as evidenced by a copy of your receipt. The program should not be relied on as the sole basis to solve a problem whose incorrect solution could result in injury to person or property. If the program is employed in such a manner, it is at the user's own risk and Pearson Education, Inc. explicitly disclaims all liability for such misuse.

LIMITATION OF REMEDIES

Pearson Education, Inc.'s entire liability and your exclusive remedy shall be:

1. the replacement of any CD-ROM not meeting Pearson Education, Inc.'s "LIMITED WARRANTY" and that is returned to Pearson Education, or
2. if Pearson Education is unable to deliver a replacement CD-ROM that is free of defects in materials or workmanship, you may terminate this agreement by returning the program.

IN NO EVENT WILL PEARSON EDUCATION, INC. BE LIABLE TO YOU FOR ANY DAMAGES, INCLUDING ANY LOST PROFITS, LOST SAVINGS, OR OTHER INCIDENTAL OR CONSEQUENTIAL DAMAGES ARISING OUT OF THE USE OR INABILITY TO USE SUCH PROGRAM EVEN IF PEARSON EDUCATION, INC. OR AN AUTHORIZED DISTRIBUTOR HAS BEEN ADVISED OF THE POSSIBILITY OF SUCH DAMAGES, OR FOR ANY CLAIM BY ANY OTHER PARTY.

SOME STATES DO NOT ALLOW FOR THE LIMITATION OR EXCLUSION OF LIABILITY FOR INCIDENTAL OR CONSEQUENTIAL DAMAGES, SO THE ABOVE LIMITATION OR EXCLUSION MAY NOT APPLY TO YOU.

GENERAL

You may not sublicense, assign, or transfer the license of the program.

Any attempt to sublicense, assign or transfer any of the rights, duties, or obligations hereunder is void.

This Agreement will be governed by the laws of the State of New York.

Should you have any questions concerning this Agreement, you may contact Pearson Education, Inc. by writing to:

ESM Media Development
Higher Education Division
Pearson Education, Inc.
1 Lake Street
Upper Saddle River, NJ 07458

Should you have any questions concerning technical support, you may write to:

New Media Production
Higher Education Division
Pearson Education, Inc.
1 Lake Street
Upper Saddle River, NJ 07458

YOU ACKNOWLEDGE THAT YOU HAVE READ THIS AGREEMENT, UNDERSTAND IT, AND AGREE TO BE BOUND BY ITS TERMS AND CONDITIONS. YOU FURTHER AGREE THAT IT IS THE COMPLETE AND EXCLUSIVE STATEMENT OF THE AGREEMENT BETWEEN US THAT SUPERSEDES ANY PROPOSAL OR PRIOR AGREEMENT, ORAL OR WRITTEN, AND ANY OTHER COMMUNICATIONS BETWEEN US RELATING TO THE SUBJECT MATTER OF THIS AGREEMENT.

Program Instructions

–Windows:

To access the Physlet Physics curricular material content, follow the following steps:

–Insert the "Physlet Physics" CD into your CD-ROM drive.
–Browse to the "Physlet_Physics" CD-ROM using your Windows explorer window.
–Double-click the "start.html" file.

To access the Physlet Physics Exploration Worksheets, double-click the "exploration_worksheets" folder, choose a chapter folder to view the selection of associated Exploration PDF worksheets, and double-click a PDF file to view.

Minimum System Requirements

–Windows:
400MHz Intel Pentium processor
Windows 2000/XP
32 MB or more of available RAM
800×600 monitor resolution set to 16 bit color
Mouse or other pointing device (?)
4× CD-ROM Drive
Active Internet connection (optional)

Requires a browser supporting the Java 1.4 Virtual Machine and JavaScript to Java communication. Internet Explorer 5.5 or higher with the Sun Java plugin 1.4 or higher, or Mozilla 1.3 or higher with the Sun Java plugin 1.4 or higher are recommended.

–Macintosh
PowerPC G3, G4, or G5 processor
Mac OS X.3/X.4
Safari 1.2
Macintosh Java plugin 1.4.2 or higher

Adobe Acrobat Reader 5.0 © 2002 or above is required to view and print the pdfs of the Exploration Worksheets.

TECHNICAL SUPPORT

If you are having problems with this software, call (800) 677-6337 between 8:00 a.m. and 8:00 p.m. Monday through Friday, and 5:00 p.m. to 12:00 a.m. on Sunday (all times listed are Eastern). You can also get support by filling out the web form located at:
http://247.prenhall.com/mediaform

Our technical staff will need to know certain things about your system in order to help us solve your problems more quickly and efficiently. If possible, please be at your computer when you call for support. You should have the following information ready:

- Textbook ISBN
- CD-ROM ISBN
- corresponding product and title
- computer make and model
- Operating System (Windows or Macintosh) and Version
- RAM available
- hard disk space available
- Sound card? Yes or No
- printer make and model
- network connection
- detailed description of the problem, including the exact wording of any error messages

NOTE: Pearson does not support and/or assist with the following:

- third-party software (i.e. Microsoft including Microsoft Office suite, Apple, Borland, etc.)
- homework assistance
- Textbooks and CD-ROMs purchased used are not supported and are non-replaceable. To purchase a new CD-ROM, contact Pearson Individual Order Copies at 1-800-282-0693

The Periodic Table of Elements

(See Chapter 28.3)

- Metals
- Nonmetals
- Noble gases

Main-group elements (groups 1–2, 13–18); Transition elements (groups 3–12)

Period	1 1A	2 2A	3 3B	4 4B	5 5B	6 6B	7 7B	8 8B	9 8B	10 8B	11 1B	12 2B	13 3A	14 4A	15 5A	16 6A	17 7A	18 8A
1	1 **H** 1.00794																	2 **He** 4.00260
2	3 **Li** 6.941	4 **Be** 9.01218											5 **B** 10.811	6 **C** 12.011	7 **N** 14.0067	8 **O** 15.9994	9 **F** 18.9984	10 **Ne** 20.1797
3	11 **Na** 22.9898	12 **Mg** 24.3050											13 **Al** 26.9815	14 **Si** 28.0855	15 **P** 30.9738	16 **S** 32.066	17 **Cl** 35.4527	18 **Ar** 39.948
4	19 **K** 39.0983	20 **Ca** 40.078	21 **Sc** 44.9559	22 **Ti** 47.88	23 **V** 50.9415	24 **Cr** 51.9961	25 **Mn** 54.9381	26 **Fe** 55.847	27 **Co** 58.9332	28 **Ni** 58.693	29 **Cu** 63.546	30 **Zn** 65.39	31 **Ga** 69.723	32 **Ge** 72.61	33 **As** 74.9216	34 **Se** 78.96	35 **Br** 79.904	36 **Kr** 83.80
5	37 **Rb** 85.4678	38 **Sr** 87.62	39 **Y** 88.9059	40 **Zr** 91.224	41 **Nb** 92.9064	42 **Mo** 95.94	43 **Tc** (98)	44 **Ru** 101.07	45 **Rh** 102.906	46 **Pd** 106.42	47 **Ag** 107.868	48 **Cd** 112.411	49 **In** 114.818	50 **Sn** 118.710	51 **Sb** 121.76	52 **Te** 127.60	53 **I** 126.904	54 **Xe** 131.29
6	55 **Cs** 132.905	56 **Ba** 137.327	57 ***La** 138.906	72 **Hf** 178.49	73 **Ta** 180.948	74 **W** 183.84	75 **Re** 186.207	76 **Os** 190.23	77 **Ir** 192.22	78 **Pt** 195.08	79 **Au** 196.967	80 **Hg** 200.59	81 **Tl** 204.383	82 **Pb** 207.2	83 **Bi** 208.980	84 **Po** (209)	85 **At** (210)	86 **Rn** (222)
7	87 **Fr** (223)	88 **Ra** 226.025	89 **†Ac** 227.028	104 **Rf** (261)	105 **Db** (262)	106 **Sg** (263)	107 **Bh** (262)	108 **Hs** (265)	109 **Mt** (266)	110 **Ds** (281)	111 ** (272)	112 ** (285)		114 ** (289)		116 ** (292)		

Series														
*Lanthanide series	58 **Ce** 140.115	59 **Pr** 140.908	60 **Nd** 144.24	61 **Pm** (145)	62 **Sm** 150.36	63 **Eu** 151.965	64 **Gd** 157.25	65 **Tb** 158.925	66 **Dy** 162.50	67 **Ho** 164.930	68 **Er** 167.26	69 **Tm** 168.934	70 **Yb** 173.04	71 **Lu** 174.967
†Actinide series	90 **Th** 232.038	91 **Pa** 231.036	92 **U** 238.029	93 **Np** 237.048	94 **Pu** (244)	95 **Am** (243)	96 **Cm** (247)	97 **Bk** (247)	98 **Cf** (251)	99 **Es** (252)	100 **Fm** (257)	101 **Md** (258)	102 **No** (259)	103 **Lr** (260)

** Not yet named

Notes: (1) Values in parentheses are the mass numbers of the most common or most stable isotopes of radioactive elements. (2) Some elements adjacent to the stair-step line between the metals and nonmetals have a metallic appearance but some nonmetallic properties. These elements are often called metalloids or semimetals. There is no general agreement on just which elements are so designated. Almost every list includes Si, Ge, As, Sb, and Te. Some also include B, At, and/or Po.

Physical Data*

Quantity	*Symbol*	*Approximate Value*
Universal gravitational constant	G	$6.67 \times 10^{-11}\ N\cdot m^2/kg^2$
Acceleration due to gravity (generally accepted value on surface of Earth)	g	$9.80\ m/s^2 = 980\ cm/s^2 = 32.2\ ft/s^2$
Speed of light	c	$3.00 \times 10^8\ m/s = 3.00 \times 10^{10}\ cm/s = 1.86 \times 10^5\ mi/s$
Boltzmann's constant	k_B	$1.38 \times 10^{-23}\ J/K$
Avogadro's number	N_A	$6.02 \times 10^{23}\ mol^{-1}$
Gas constant	$R = N_A k_B$	$8.31\ J/(mol\cdot K) = 1.99\ cal/(mol\cdot K)$
Coulomb's law constant	$k = 1/4\pi\epsilon_o$	$9.00 \times 10^9\ N\cdot m^2/C^2$
Electron charge	e	$1.60 \times 10^{-19}\ C$
Permittivity of free space	ϵ_o	$8.85 \times 10^{-12}\ C^2/(N\cdot m^2)$
Permeability of free space	μ_o	$4\pi \times 10^{-7}\ T\cdot m/A = 1.26 \times 10^{-6}\ T\cdot m/A$
Atomic mass unit	u	$1.66 \times 10^{-27}\ kg \leftrightarrow 931\ MeV$
Planck's constant	h	$6.63 \times 10^{-34}\ J\cdot s$
	$\hbar = h/2\pi$	$1.05 \times 10^{-34}\ J\cdot s$
Electron mass	m_e	$9.11 \times 10^{-31}\ kg = 5.49 \times 10^{-4}\ u \leftrightarrow 0.511\ MeV$
Proton mass	m_p	$1.672\,62 \times 10^{-27}\ kg = 1.007\,276\ u \leftrightarrow 938.27\ MeV$
Neutron mass	m_n	$1.674\,93 \times 10^{-27}\ kg \times 1.008\,665\ u \leftrightarrow 939.57\ MeV$
Bohr radius of hydrogen atom	r_1	0.053 nm

*Values from NIST Reference on Constants, Units, and Uncertainty.

Solar System Data*

Equatorial radius of the Earth	$6.378 \times 10^3\ km = 3963\ mi$
Polar radius of the Earth	$6.357 \times 10^3\ km = 3950\ mi$
	Average: 6.4×10^3 km (for general calculations)
Mass of the Earth	$5.98 \times 10^{24}\ kg$
Diameter of Moon	$3500\ km \approx 2160\ mi$
Mass of Moon	$7.4 \times 10^{22}\ kg \approx \frac{1}{81}$ mass of Earth
Average distance of Moon from the Earth	$3.8 \times 10^5\ km = 2.4 \times 10^5\ mi$
Diameter of Sun	$1.4 \times 10^6\ km \approx 864\,000\ mi$
Mass of Sun	$2.0 \times 10^{30}\ kg$
Average distance of the Earth from Sun	$1.5 \times 10^8\ km = 93 \times 10^6\ mi$

*See Appendix III for additional planetary data.

Mathematical Symbols

$=$	is equal to
$\neq$	is not equal to
$\approx$	is approximately equal to
$\sim$	about
$\propto$	is proportional to
$>$	is greater than
$\geq$	is greater than or equal to
$\gg$	is much greater than
$<$	is less than
$\leq$	is less than or equal to
$\ll$	is much less than
$\pm$	plus or minus
$\mp$	minus or plus
$\bar{x}$	average value of x
Δx	change in x
$\lvert x \rvert$	absolute value of x
Σ	sum of
∞	infinity

The Greek Alphabet

Alpha	A	α	Nu	N	ν
Beta	B	β	Xi	Ξ	ξ
Gamma	Γ	γ	Omicron	O	o
Delta	Δ	δ	Pi	Π	π
Epsilon	E	ε	Rho	P	ρ
Zeta	Z	ζ	Sigma	Σ	σ
Eta	H	η	Tau	T	τ
Theta	Θ	θ	Upsilon	Υ	υ
Iota	I	ι	Phi	Φ	ϕ
Kappa	K	κ	Chi	X	χ
Lambda	Λ	λ	Psi	Ψ	ψ
Mu	M	μ	Omega	Ω	ω

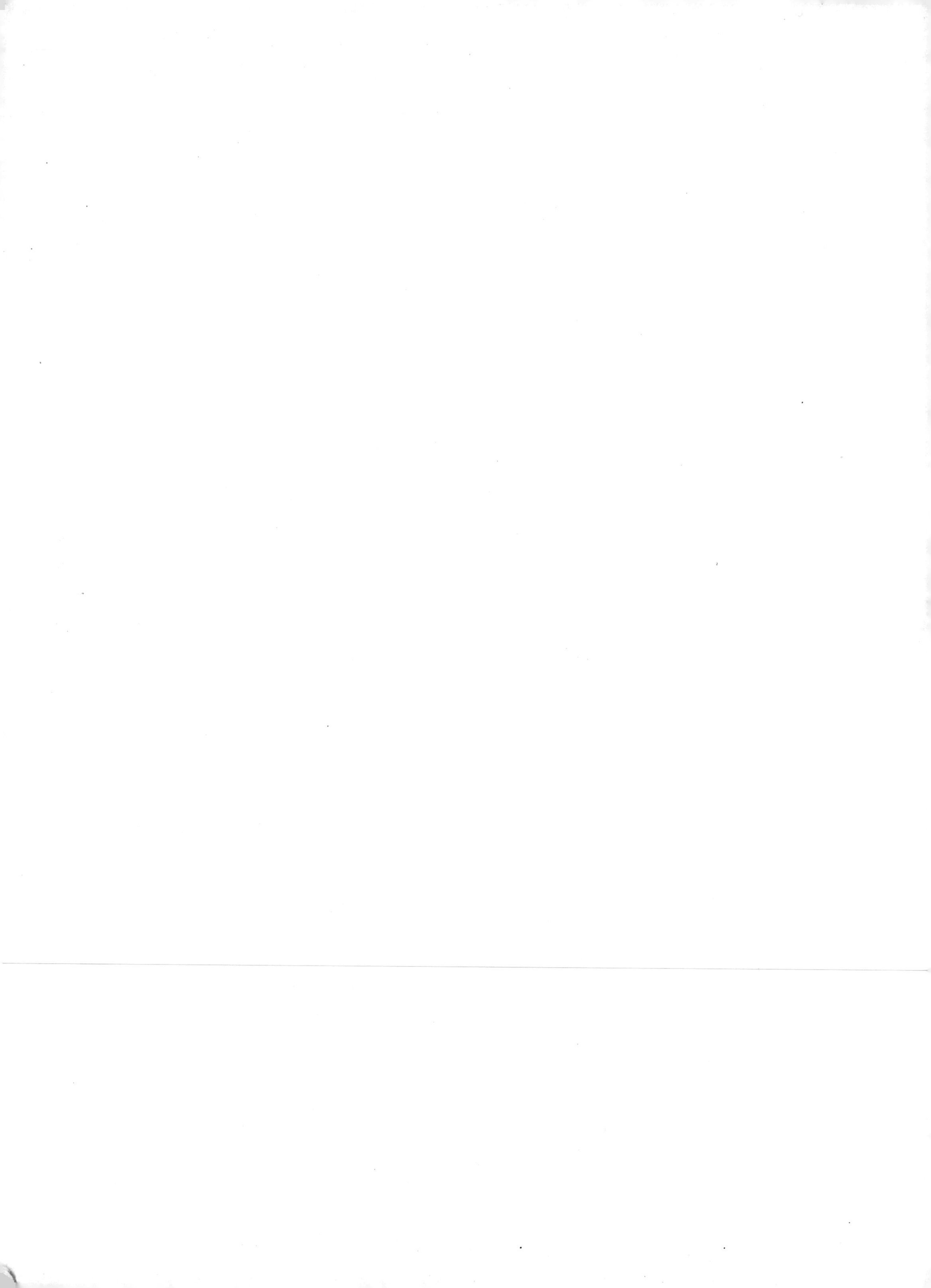